Data Network Design Strategies

McGRAW-HILL/DATA COMMUNICATIONS BOOK SERIES   VOLUME **4**

Data Network Design Strategies

BY THE EDITORS OF DATA COMMUNICATIONS MAGAZINE

McGraw-Hill Information Services Company
1221 Avenue of the Americas
New York, NY 10020

Cover Photograph by Walter Wick

Copyright © 1990 by McGraw-Hill Information Services Co. All rights reserved. Printed in the United States of America. Except as permitted under the United States Copyright Act of 1976, no part of this publication may be reproduced or distributed in any form or by any means, or stored in a data base or retrieval system, without the prior written permission of the publisher.

Library of Congress Cataloging-in-Publication Data
Data network design strategies / edited by the staff of Data
  communications magazine.
     p. cm. — (Data communications book series)
   ISBN 0-07-607023-9 : $39.95
   1. Computer networks   2. Data transmission systems.   I. Data
communications.   II. Series: McGraw-Hill data communications book
series.
TK51055.D3378 1990
650'.0285'46—dc20   90-5753
                                                                                                CIP

McGraw-Hill Data Communications BOOK SERIES

McGraw-Hill Information Services Company
1221 Avenue of the Americas
New York, NY 10020

- **Basic Guide to Data Communications, Edition 2**
- **Cases in Network Design**
- **Connectivity and Standards, Volume 3**
- **Data Communications: A Comprehensive Approach, Edition II**
- **Data Communications: Beyond the Basics**
- **Data Network Design Strategies, Volume 4**
- **Inside X.25: A Manager's Guide**
- **Integrating Voice and Data, Volume 3**
- **Linking Microcomputers**
- **The Local Area Network Handbook, Volume 3**
- **Network Management and Maintenance, Volume 4**
- **Networking Software**
- **Telecommunications and Data Communications Factbook**

Table of Contents

ix **Preface**

1 **SECTION 1**
ARCHITECTURE
2 Time-staged delivery networks save time, enhance productivity, *Jeri Edwards* (February 1986)
6 A blueprint for business architectures, *Rudolf Strobl and Bill Stackhouse* (March 1986)
16 The how and where of switched 56-kbit/s service, *S. Helga Eriksen* (August 1986)
23 A new transport architecture for sluggish networks, *Morris Neuman* (August 1986)
30 Understanding IBM's electronic mail architectures, *Donald H. Czubek* (November 1986)
37 What parallel processing can offer information networking, *Paulina M. Borsook* (January 1987)
48 Distributed SNA: A network architecture gets on track, *Thomas J. Routt* (February 1987)
62 Open networks free users from their mainframe architectures, *John T. Becker and Curtis G. Gray* (February 1987)
70 BTAM, VTAM, X.25: Uneasy alliance, *David G. Matusow* (March 1987)
75 Developing a user-oriented networking architecture, *Shel Blauman* (May 1987)
82 Networking for greater metropolitan areas, *James F. Mollenauer* (February 1988)
90 New proposal extends the reach of metro area nets, *John L. Hullet and Peter Evans* (February 1988)
95 How to grow a world-class X.25 network, *Cormac O'Reilly* (May 1988)
105 From TOP (3.0) to bottom: Architectural close-up, *Thomas J. Routt* (May 1988)
118 The formula for network immortality, *Peter M. Haverlock* (August 1988)
125 A network architecture for retail bank networks, *James A. Craig and William L. Gerwitz* (October 1988)
134 Knocking on users' doors: Signaling System 7, *Walter Roehr* (February 1989)
142 Mastering SS7 takes a special vocabulary, *Daniel R. Seligman* (February 1989)

145 **SECTION 2**
DESIGN AND OPERATION
146 Computer-aided engineering enhances network design, *James Broughton* (February 1986)
150 The planning and managing of packet-switching networks, *Robert Cameron* (March 1986)
154 Microcomputer programs can be adapted for data network design, *James H. Green* (April 1986)
162 Computers design networks by imitating the experts, *Larry Cynar, Don Mueller and Andrew Paroczai* (April 1986)
167 Choosing a key management style that suits the application, *C. R. Abbruscato* (April 1986)
177 Internetworking in an OSI environment, *David M. Piscitello, Alan J. Weissberger, Scott A. Stein and A. Lyman Chapin* (May 1986)
190 A user implements a T1 network, *Robert W. Junke* (June 1986)
195 The choices in designing a fiber-optic network, *Marc A. Wernli* (June 1986)
200 Can users really absorb data at today's rates? Tomorrow's?, *Hal B. Becker* (July 1986)
209 Ways to avoid the stumbling blocks that foil T1 network success, *Anthony Badalato and Paul Drimmer* (September 1986)
214 Combining data and voice network management, *Nicholas Papadopoulos* (December 1986)
219 Network designers face shifting industry givens, *John T. Becker and Curtis G. Gray* (December 1986)
224 Metropolitan area networking: Past, present, and future, *William E. Bracker and Benn R. Konsynski III and Timothy W. Smith* (January 1987)
229 ADPCM offers practical method for doubling T1 capacity, *Nathan J. Muller* (February 1987)
223 Interconnectivity: Risks may be substantial, but the benefits are great, *David W. Campt* (March 1987)
243 A dozen ways to beef up your network with a port selector, *Gilbert Held* (June 1987)
250 Zap data where it really counts — Direct-to-host connections, *Michael W. Cerruti and Maurice Voce* (July 1987)
255 Orchestrating people and computers in their networks, *Lee Mantelman* (September 1987)
267 Protecting networks from power transients, *Kevin R. Sharp* (September 1987)
275 An all-purpose model to aid in all phases of network design, *Larry L. Duitsman and Madeline M. Pinelli* (September 1987)
284 ISDN when? What your firm can do in the interim, *James G. Herman and Mary A. Johnston* (October 1987)
291 Another use for the versatile protocol analyzer: Capacity planning, *Gilbert Held* (October 1987)
295 Data networks' endangered and protected species, *Charlie Bass* (October 1987)
303 Implementing automatic route selection in T1 networks, *Robert S. Alford and Carl Holmberg* (November 1987)
310 Making the most of your existing telephone wiring for data traffic, *James F. Kearney* (November 1987)
315 Packet-satellite networks: Updating and expanding the hybrid concept, *Alan B. Taffel* (November 1987)
324 T1 to ISDN: How to get there from here, *Nick Berberi* (February 1988)
329 Don't just guess: How to figure delays in private packet networks, *Joseph Fernandez and David E. Liddy* (March 1988)
339 A user's guide to network design tools, *Harrell J. Van Norman* (April 1988)
354 Designing network control centers for greater productivity, *Kerry Kosak* (April 1988)
361 Building a private wide-area, fiber backbone network, *Warren Cohen and Nicholas John Lippis III* (April 1988)
366 Is your packet network a 'model' of response-time efficiency?, *Joseph I. Fernandez and David E. Liddy* (May 1988)
375 Five years after deregulation, FM subcarrier comes of age, *Alvaro Andraca* (May 1988)
379 Wide area network: Sewing a patchwork of pieces together, *Dennis Rose* (May 1988)
386 Coming soon: A cabling standard for buildings, *Bryan F. Gearing* (August 1988)
391 T1 network design and planning made easier, *Ranjana Sharma* (September 1988)
396 Designing gateways into metropolitan area networks, *Terence D. Todd* (October 1988)
403 Digital signaling: Which techniques are best — and why it matters to you, *William Stallings* (November 1988)
409 Moving to new quarters? Don't trip over the cabling issue, *Fred Yarusso* (November 1988)
418 The shots seen round the world, *Robert Rosenberg* (December 1988)
428 High-flying Hughes earns its wide area networking wings, *Jack Covert, Tom Nakamura and Nilo A. Niccolai* (February 1989)
437 OSI-based net management: Is it too early, or too late?, *Jill Ann Huntington* (March 1989)
448 Remote-control software: PCs apart, but not alone, *Sue J. Lowe* (April 1989)
456 How to make the PBX-to-ISDN connection, *Roger L. Koenig* (May 1989)
468 Taking your SNA network abroad, *Lee Mantelman* (June 1989)
475 NetView shows highs and lows of net mangement (July 1989)
478 How to design and build a 16-Mbit/s Token Ring, *Robert D. Love and Thomas Toher* (July 1989)
488 Making the most of ISDN now, *Dean Wolf and Steven S. King* (August 1989)
494 Get ready for T3 networking, *Stephen Fleming* (September 1989)
502 Dialing data via mobile telephones, *Ira Brodsky* (October 1989)
510 Routers as building blocks for robust internetworks, *John M. McQuillan* (September 1989)
516 Router roundup: Tools for network segmentation come of age, *Edward R. Teja* (September 1989)
522 Understanding XNS: The prototypical internetwork protocol, *Dale Neibaur* (September 1989)

vii

529 Is ISDN an obsolete data network?, *Gilbert Held* (November 1989)
534 New signs of life for packet switching, *Edwin E. Mier* (December 1989)
545 Managing networks of the '90s, *Kumar Shah* (December 1989)

553 **SECTION 3**
ECONOMICS

554 Rapid deregulation spawns conflicts, realignments, *Paul R. Strauss* (January 1986)
561 Comparing the long-distance carriers, *Edwin E. Mier* (August 1986)
585 Voice mail: Key tool or costly toy?, *David M. Rappaport* (October 1986)
594 Here is one way to get a close estimate of a data link's efficiency, *William Stallings* (October 1986)
602 Making the most of post-divestiture tariffs for data network design, *Peter Hansen* (January 1987)
610 Squeezing line costs via data compression, *Darlane Hoffman* (August 1987)
616 1987: The year when networking became part of the bottom line, *Joseph Braue* (January 1988)
621 Looking ahead: Network planning for the 1990s, *Donald F. Blumberg* (February 1988)
626 Turning your network into a profit center: Users as vendors (July 1988)
633 Is a private T1 network the right *business* decision?, *Timothy G. Zerbiec and Rosemary M. Cochran* (July 1988)
647 Ratings reveal carriers are worlds apart (Aug. 1988)
652 Making your network make money, *James Randall* (June 1989)
660 Coping with the cost realities of multipoint networks, *Nicholas John Lippis III* (July 1989)
667 The hidden costs of using an SNA backbone for X.25 traffic, *Cheryl L. Sommer* (November 1989)

Preface

The information-technology explosion has brought to network design a wealth of options that can be downright bewildering not only to the general manager but also the communications professional. Indeed, by the late 1980s, computer-aided network design had become a reality, if not a necessity.

Technological progress wasn't the only source of concern for network designers of the era, however. Deregulation was afoot; disaggregation and globalization were watchwords for the coming decade.

It is that decade — the 1990s — that was uppermost on the minds of the contributors to this collection of articles. Presented here as edited by the staff of McGraw-Hill's Data Communications Magazine from 1986 through 1989, these articles show the wide variety of network design alternatives that will come into play over the next 10 years. They also provide rich insights into the thinking of designers whose work will shape data networks well into the next century.

Section 1
Architecture

Jeri Edwards, Tandem Computers Inc., Cupertino, Calif.

Time-staged delivery networks save time, enhance productivity

When resources, facilities, and recipients are unavailable, a new type of network allows data to reach its appointed rounds.

Time-staged delivery allows a sender to input information to a network and then go on to other tasks while the information is being sent. Unlike interactive communications, time-staged delivery does not require that the information receiver be present at the time data is transmitted. A delivery mechanism transfers the information to a location in the network designated as the receiver's reception "depot" and stores it there. Message receivers can then pick up messages at their convenience. Time-staged delivery is often termed "asynchronous," or without regard to time, because the sender and receiver do not have to be in lockstep for information to pass between them. Several requests may be sent at once. The responses are returned later, without regard to time or sequence.

Before time-staged delivery, all information exchange was interactive. That is, senders and receivers exchanged information in real time (Fig. 1). This meant that the sender, receiver, and network all had to be active at the same time. Such setups are generally called synchronous communications because the sender and receiver match, or synchronize, responses with requests. A response is the information returned directly after a request, and it must be received before another request can be made.

Interactive and time-staged delivery parallel the development of the modern telephone network and electronic-mail implementations, respectively. That is, people use the telephone when they must talk directly to another person. The sender, receiver, and communications resource must all be active for the message to get through. But often it is more productive to use an electronic-mail facility, especially if time is not critical. This type of implementation accepts messages (from the sender) and guarantees that they will reach the intended receiver, thereby eliminating the need for direct contact between sender and receiver.

This does not mean that all electronic-mail implementations are time-staged delivery networks. Conventional electronic-mail networks incorporate time-staged delivery but only for a particular terminal type and software application. General time-staged delivery networks, on the other hand, are multipurpose implementations that can be used by any application.

Occasionally, time-staged delivery networks are confused with store-and-forward implementations, such as Arpanet. Packet-switching networks have been labeled store-and-forward networks because they take packets off incoming communications lines, put them on an outgoing queue, and then transmit them (see "X.400 compared"). Time-staged delivery networks often operate on top of such packet-switching networks, providing an added layer of service. The delivery network, however, manages the entire transmission (which may consist of multiple packets) through the packet-switching network on behalf of the sender.

Transfer vs. SNADS

Tandem Computers' time-staged delivery network is called Transfer. Tandem customers write time-staged applications that use Transfer for the distribution and storage of information. Tandem's electronic-mail and facsimile transport products use Transfer for office communications.

IBM also has a time-staged delivery capability known as Systems Network Architecture Delivery Systems (SNADS). Like IBM's Systems Network Architecture (SNA), SNADS is also a network architecture. To date, it has been implemented in IBM's Distributed Office Support System (DISOSS) version 3.2, System/36, and the 5520 word processing system. These products

1. Interactive vs. time-staged. *With interactive communications, sender and receiver exchange data in real time. In time-staged, the network is independent.*

INTERACTIVE COMMUNICATIONS

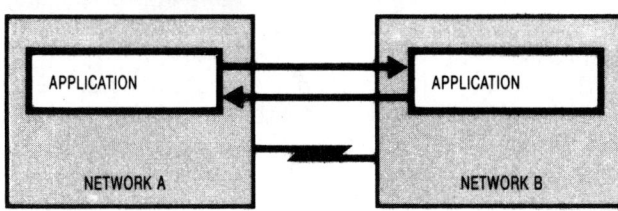

TIME-STAGED DELIVERY

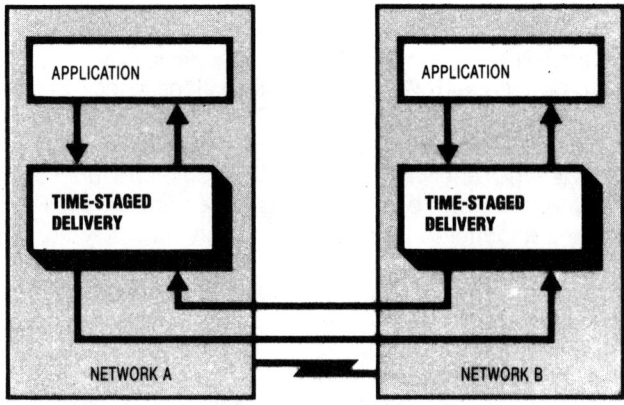

implement SNADS for document distribution. The SNADS implemention in such instances is closed: Applications cannot use the SNADS facility. The recent SNADS implementation in IBM System/38, however, supports a method for applications to use its SNADS capabilities.

Any applications that interact with each other could probably be enhanced by using a time-staged delivery network, and there are four categories of applications that can be immediately identified:

X.400 compared

The International Telegraph and Telephone Consultative Committee (CCITT) approved the X.400 message handling standard in October 1984 (DATA COMMUNICATIONS, May 1984, p. 159). Since then, several vendors have announced support for the standard. Electronic-mail vendors are examining X.400 for use in linking private electronic-mail systems to public services.

X.400, Systems Network Architecture Delivery System (SNADS) coupled with Distributed Office Support System, and Transfer have certain elements in common. Conceptually, all three involve a delivery system with a postman, envelopes, and document contents. It appears there may have to be gateway products to link X.400 with SNADS and other products. Transfer is similar to X.400, but it is not currently compatible at all X.400 levels.

- Job networking.
- Distributed database maintenance.
- Document distribution.
- Intelligent networking.

Job networking. This first category is also known by the terms "extended transactions" and "functional distribution." There are instances where processes need to be tied together for full automation into a sequence of steps. Information then flows between the steps and triggers the execution of dependent processes. An example of job networking might be a chain store where orders taken at branch offices are then relayed to regional offices. At the regional office daily sales records are taken and supply orders are filled out and sent to a regional warehouse. When the stock arrives at the store's receiving dock, the receiving clerk records its arrival (Fig. 2).

The Japanese are developing a new twist to job networking, a time-staged method of manufacturing they call "Kanban"—also known as just-in-time manufacturing in the United States. With Kanban, factory parts are worked on in various locations and ultimately arrive at predetermined locations just in time to be incorporated with other arriving parts. Accompanying the parts is documentation regarding identification, history, and the next station destination for the parts. Once the parts reach the next station, the information regarding them is updated.

Robots and cards

Although the Kanban method promised to save manufacturing companies millions of dollars, it has not done so yet. As it turns out, the task of automating parts flow is easier than automating information flow. The Japanese used cards to record parts information (*Kanban* means card), which worked well when people manned stations, but not when robots manned stations.

Time-staged delivery solves this problem by automating the flow of information between processes. After information is delivered to a process-receiving depot, an application can get the information and display it for a human operator, or a robot can access the same information. After the part is processed, the information can be updated and handed back to the time-staged delivery operation for transport to the next process.

Distributed database maintenance. Managing a network containing a distributed database can be a difficult job. The entire database must be available and accessible when updates occur, otherwise it can become inconsistent. For example, a worker jackhammers through all eight lines in a communications path, or computers are taken off an application after business hours in London to work on another job while updates are being sent from California during business hours. Thus, vital updates during such downtime can be lost.

A time-staged delivery network can alleviate this maintenance headache by delivering information when resources are available to handle it. The network can hold the update until the links damaged by the jackhammer are repaired, deliver the update to the destina-

2. Chain store. *Time-staged delivery networks are useful for chain stores where orders must be coordinated with merchandise shipments and delivery dates.*

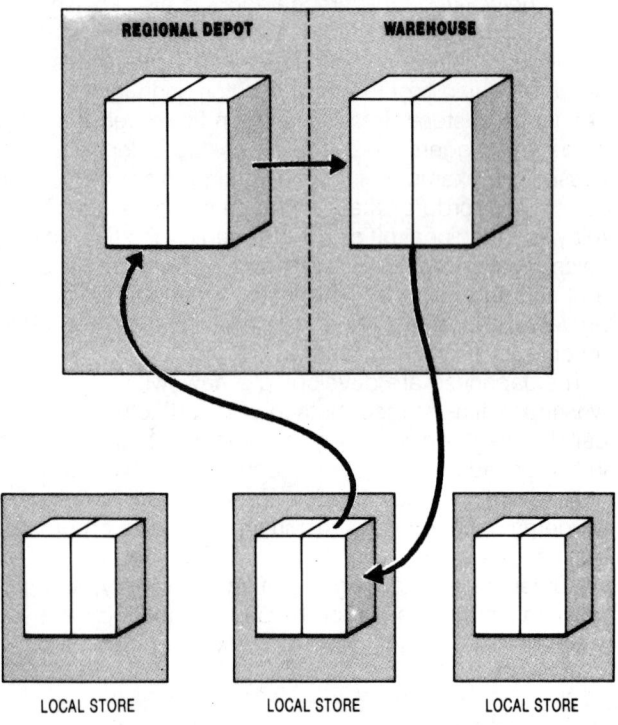

tion node, and hold it there until the application has returned.

Document distribution. This application covers information transportation between individuals. These applications are the most obvious for timed-staged delivery because people are usually too busy to wait for information. A time-staged delivery system would take information from the sender and hold it until the receiver is ready to pick it up.

Electronic mail is the most common document distribution application. File transfer and forms routing are also document distribution applications. File transfer applications take documents created outside the system (for example, a microcomputer or word processing equipment) and transport them to receivers (one person or a group of people). Forms-routing applications typically automate the flow of business forms through a corporation.

Tools that automate business activity

Some document distribution applications can be similar to job networking in office environments. For example, a person fills out an expense report that then gets routed by several individuals in management for approval before it reaches the payroll department. Unlike job networking or distributed database maintenance (which is transaction based), document distribution applications are productivity tools that automate business activity.

Intelligent networking. These applications connect diverse computers and devices through a central network of processors. Applications on diverse computers usually are created to work in standalone environnts. Connecting them through a central network facilitates communications, while placing the burden of connection on the network. This arrangement also tends to be least disruptive to the diverse computer applications.

Requiring synchronized communications for passing information on an intelligent network would often require programming modifications. In many cases, a time-staged delivery network eliminates most of the need for programming modification. The network handles the communications for the applications. Because the network holds the information at the destination for the recipient, it allows the receiving station to respond to the information when it is convenient to do so, rather than interrupting the recipient's local work flow.

Transfer and SNADS represent solutions to the requirements for time-staged delivery. The goals for the two architectures are similar. Both move data between the sender and receiver asynchronously; both add functionality to a network built for interactive communications; both are meant to be used in mixed interactive/time-staged applications, as well as "pure" time-staged environments; and both have goals of application independence, ease of use, manageability, efficiency, and extendability.

As with any network, there is a trade-off between functionality and resource use. A full-function time-staged delivery network offers a variety of features. A basic network (distributed spooler) offers a skeleton of the functionality of the full system, but it consumes fewer resources. This is not a bargain, however, if you end up adding the missing features yourself.

Automatic process invocation

Automatic process invocation is an important feature for time-staged delivery systems. Three of the four types of applications for which time-staged delivery is appropriate (job networking, distributed database maintenance, and intelligent networking) require processes to be triggered because of the arrival of certain types of messages.

Even document distribution applications can be enchanced by having such a facility. A document distribution application could use this feature to invoke a filing process that would automatically file each incoming document in an appropriate folder.

Figure 3 compares the architectures of SNADS and Transfer at a high level. With the Transfer architecture, a user can be a person, device, or application. An application program, called a "client," interfaces to Transfer for the user. When the user is an application, it can be its own "client" and interface to Transfer directly. Clients interface to Transfer through a well-defined, high-level language called "Units of Work." There are more than 40 Units of Work available for clients to use (although most clients will not use all of them).

After Transfer has accepted the message from the sender's client, it sends that message through the network. At the destination node, another Transfer

3. Comparison. *The operational architecture of IBM's SNADS and Tandem Computers' Transfer are similar. Both are time-staged delivery networks.*

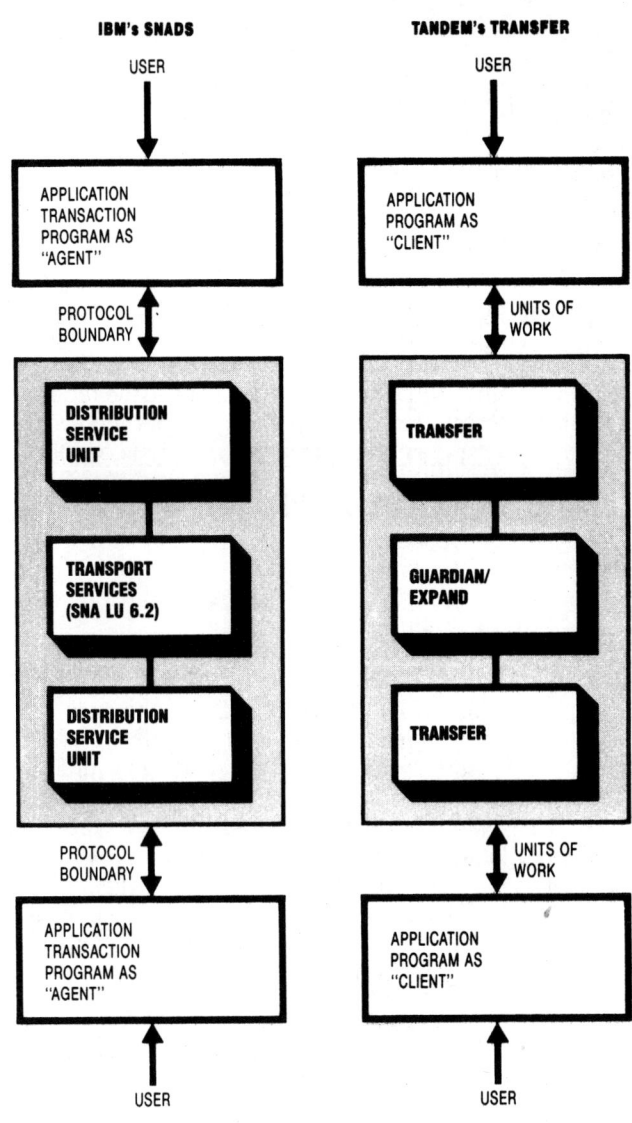

subsystem holds the messages until they are picked up by the recipient user's client.

SNADS performs document delivery between DISOSS applications; that is, when DISOSS needs to distribute a document outside its node, it uses SNADS. SNADS requires several other IBM architectures to distribute a document for DISOSS. It uses SNA (Logical Unit 6.2 sessions) to move the documents. The documents themselves use DIA/DCA (Document Interchange Architecture/Document Content Architecture) formats to define tabs, pages, paragraphs, centering commands, and so forth and to indicate what the recipient should do with the documents. ∎

Jeri Edwards holds a master's degree in cybernetic systems from San Jose State University. Before working for Tandem Computers Inc., Edwards was PBX specialist for Rolm Inc. of Santa Clara, Calif.

Rudolf Strobl and Bill Stackhouse, Software Research Corp., Natick, Mass.

A blueprint for business architectures

IBM's SNA Distribution Services represent one path to information network architectures, which promise to free users from the constraints of time and location.

Today, service-industry applications consist of heterogeneous environments that arose because buyers favored "specialized" computer vendors. Engineering groups tended to purchase from Hewlett-Packard and Digital Equipment Corp., office processing groups from Wang, and accounting-oriented groups from IBM. Accordingly, each vendor developed protocols that worked only with its gear.

To overcome such isolation and allow users or programs from any one environment or on any piece of equipment to communicate with users in any other environment, information network architectures (INAs) will have to be established in five to 10 years. Certain users, the "Citibanks" of the world, have already begun to develop their own. IBM has tipped its INA hand with strategic directions in its office product line, based on its Systems Network Architecture (SNA). Other vendors are likely to follow the path being blazed by international standards bodies.

The purpose of an INA is to separate business application programs from the communications support structure and to provide these programs with a common interface to obtain end-to-end services. This necessitates dividing what is called the application layer in many existing network architectures into three distinct layers. Properly executed, an INA could allow users to share files and information, based not on which minicomputer or mainframe host (or file server) the users are connected to, but on a common area of interest or activity.

To understand the need for an INA, consider the predicament of many users. Business application programs were written for individual departments having different hosts (minicomputers or mainframes), each of which had its own unique communications software components. Application programs (the "A" programs in Figure 1) assumed such communications functions as routing, protocol and device management, and polling. In some cases, the A program itself would perform the communications functions. In others, the communications subprogram of the A program, the "C" portion, would do so. Boundaries between A and C programs were not uniform across terminal types.

Whether located in the A or the C portion, one element in the host application controls and acts on behalf of the terminal. In order to access the application, the terminal must communicate with this element. Note that the logical connection is established in the host, where both the terminal's element and the application processing reside, although the physical connection is out in the real world. Therefore, there is a logical star "network" in which all logical connections terminate in the host.

With this type of logical network, each application program controlled its own terminal resources, making it impossible to share communications programs or information across different applications or across different corporate departments. Even if a connection could be made, information content could not necessarily be exchanged intact. Each application had its own telephonelike network (see "Evolution of a fractured mish-mash"). For one terminal type, all the polling and message decomposition occurred within the A portion. For another, the C function might have done the polling, but the A itself did the message decomposition. Another problem could have been the use of different versions of the same data communications protocol.

The type of communications interface that evolved, that is, each of the legs of the logical network, was point-to-point. This approach corresponded with the corporate structure of independent business units

1. Discombobulation. *In today's environment, the communications portions of application programs are dedicated and inconsistent from one to the other.*

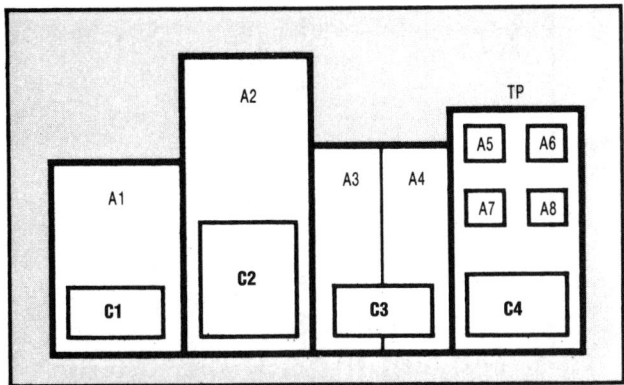

HOST (MINICOMPUTER OR MAINFRAME)

A = APPLICATION PROGRAM TP = TELEPROCESSING MONITOR
C = COMMUNICATIONS PROGRAM

responsible for their own profit and loss as well as for computer resources. For example, large banks had different departments, such as credit cards, personal loans, and checking and saving accounts, operating completely independently from each other. In support of such applications and of data processing at large, companies developed point-to-point schemes.

In addition, two other areas fostered this development: office computers and microcomputers. Office computers and their associated word processing terminals were introduced into corporations primarily as flexible and highly productive "typewriters." They were not implemented with the expectation that they would later actively participate in corporate networks. However, the view is becoming more widely held that office computers can be used for correspondence among professionals.

Although this perception is creating a need for electronic mail and document interchange, broader acceptance of these services must await computers that function as post offices—distributing mail to other post office computers, which is then accessed by users through their terminal devices. Any company with two or more offices and computers in each will need to share information between the machines. A document entered at a terminal will likely be sent to another user, with the computers acting as intermediary.

The explosive proliferation of microcomputers has brought the problems associated with point-to-point arrangements into sharp focus. Previously, users depended totally on mainframes or minicomputers for business applications, even for word processing. With the advent of the microcomputer, they could access friendly, sophisticated, powerful applications of all types right at their desks.

Most users have became aware that, to realize the full potential of the microcomputer, they also need access to corporate databases. In the first year of using the microcomputer, for example, users might wish to

Evolution of a fractured mish-mash

The earliest application programs supporting terminal networks integrated the logic for business-oriented task processing (for which the application was initially designed) with the logic used for communications. There was no separate communications program. The communications program was embedded in the application program and ran as a subroutine, as depicted in the figure (part a). Lines and terminals could not be shared with other programs.

For example, BTAM (Basic Telecommunications Access Method), IBM's oldest communications program, required that lines and terminals be dedicated to the application served by the access method. Therefore, each application had to have its own copy of BTAM. Perhaps more importantly, users had to have a separate terminal on their desks for each application.

To alleviate this problem, communications software, such as teleprocessing (TP) monitors that permitted the sharing of communications resources, was developed. TP monitors were large programs containing smaller ones (see figure, part b). Customer Information Control System (CICS) is one example of a TP monitor.

This intermediate implementation provided considerable relief because the smaller programs could share the communications resources that were now owned on their behalf by the TP monitor. However, while programs running under a common TP monitor could now share lines and terminals, no sharing was possible between different TP monitors.

The next step was to have a separate program for communications with business programs connected to it (see figure, part c). An early IBM example is the Telecommunications Access Method (TCAM). Whenever the application program needed to communicate with terminals, it asked the communications program to provide the appropriate resources.

For user organizations with monolithic, self-contained needs, this approach sufficed for a while. However, even such organizations ran into limitations as they developed a mix of TP monitors and other types of programs. The need to share terminals between multiple applications in the same computer or with applications in another computer stimulated intense development activities by manufacturers. These efforts resulted in a complete overhaul of the means by which information is exchanged between terminals and application programs.

Communications architectures are an outgrowth of this era. They were developed to segment communications functions and to standardize interfaces. These functions were broken down into layers, of which the lower ones were responsible for the actual movement of data and the upper ones for the establishment of connections, or sessions, between terminals and programs. Each layer has a correspondent, or peer, layer in the other node, with which it believes it is in direct dialog. The actual flow, however, goes

however goes through the layers and back again.

Since each vendor developed its own architecture, different implementations that were incompatible with one another became available. To ensure compatibility, vendors realized that they would have to keep up with the enormous resources of the dominant vendor (IBM) and with continuous enhancements to IBM's Systems Network Architecture (SNA). Such knowledge is both rare and expensive. Given the exorbitant costs faced by non-IBM vendors, most have begun to adopt the architecture sponsored by the International Organization for Standardization (ISO), namely, the Open Systems Interconnection (OSI) reference model. It is likely that, by the end of the decade, implementations of IBM's SNA and ISO's OSI will have come to dominate the marketplace.

The most important consideration for users is that both architectures, or at least interfaces into them, will be running on their computers. SNA will be supplied by the dominant vendor; OSI implementations will represent the rest of the world. But such implementations may be available even from IBM in the future. IBM recently announced a new product called "Open Systems Transport and Session Support," which complies with OSI layers four and five. The company is currently evaluating the top two layers of the OSI model: six (presentation) and seven (application). As the specifications of these layers are completed, IBM may find it advantageous to develop products supporting those layers.

A major dividend of the development of formal architectures was the interconnection of multiple hosts. It thus became necessary for the programs responsible for terminals on one host to access application programs of other hosts. Routing and resource scheduling between computers became a major issue. IBM initially accomplished interconnection within a single network using Multiple Systems Networking Facility (MSNF) software and later, across independent networks, using the SNA Interconnect (SNI) package.

Since its announcement, IBM's SNA (today the most successful network architecture in the marketplace) took 10 years to become workable, reliable, and acceptable. Yet, with the notable exception of advanced program-to-program communications (APPC), SNA has only solved the problem of communications that are sent from a terminal to programs in one or more host computers.

With APPC, IBM is finally beginning to respond to the proliferation of intelligent nodes in the network. (Such nodes include microcomputers, distributed office computers like the System/36 and System/38, and even IBM local area networks, which can be considered distributed computers.) APPC is a protocol under SNA that uses logical unit (LU) 6.2 sessions to enable a program to exchange information with another program rather than with a terminal.

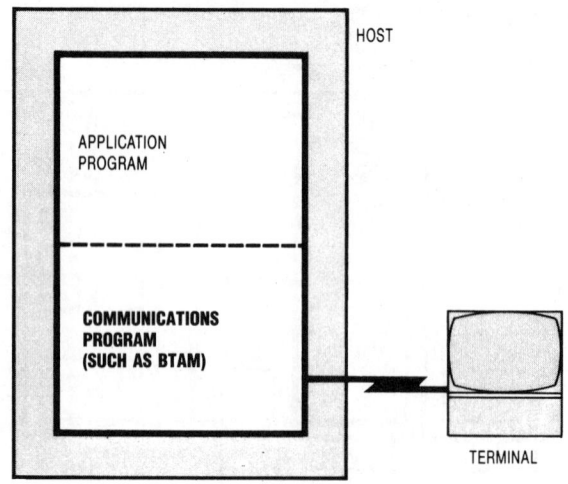

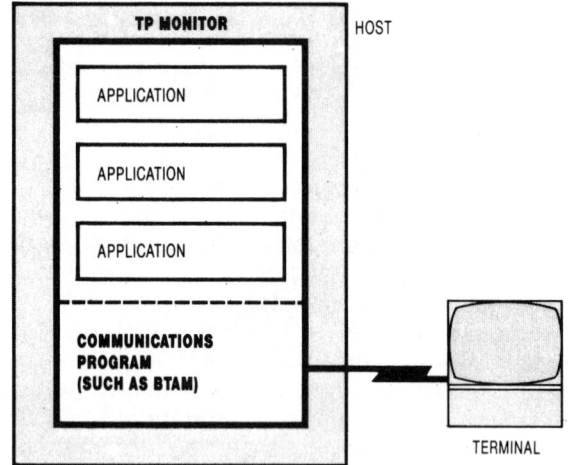

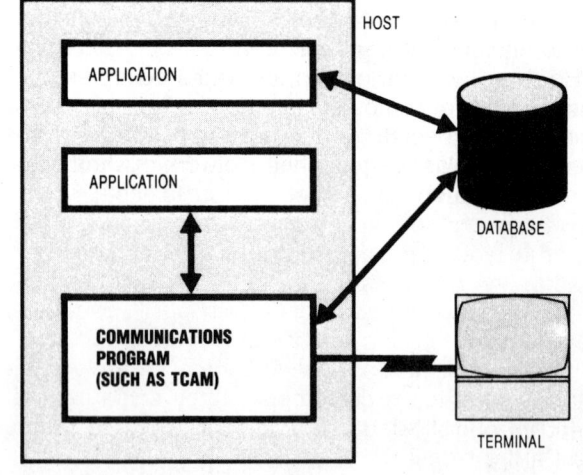

BTAM = BASIC TELECOMMUNICATIONS ACCESS METHOD
TCAM = TELECOMMUNICATIONS ACCESS METHOD

2. Architectural solution. *Dividing the application layer into three parts gives independence from lower layers and better interprogram communications. The amount of code in the lowest of the three new layers, which rest on top of existing architectures, varies with the maturity of the underlying layer.*

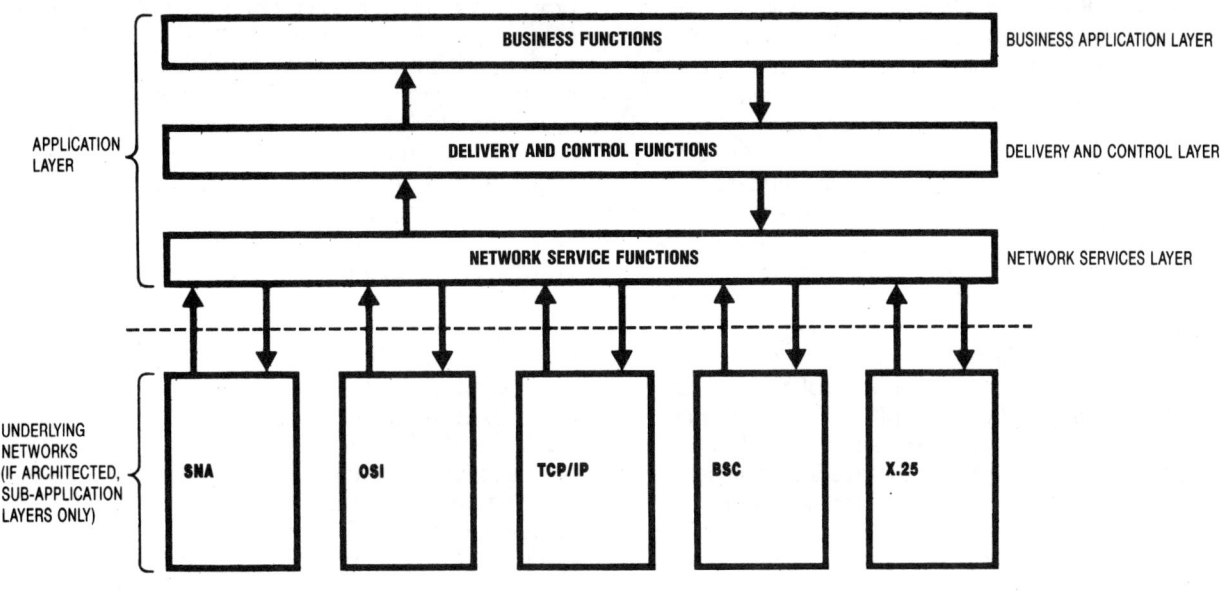

BSC = BINARY SYNCHRONOUS COMMUNICATIONS
OSI = OPEN SYSTEMS INTERCONNECTION
SNA = SYSTEMS NETWORK ARCHITECTURE
TCP/IP = TRANSMISSION CONTROL PROTOCOL/INTERNET PROTOCOL

massage mainframe-resident data with Lotus 1-2-3. But in the second year, they look for an alternative to hand delivering the massaged data. Only then do users begin to demand the ability to exchange information with each other and with computers other than the attached mainframe.

Almost all of the smaller machines had some communications capability but only with the IBM host and mostly by emulating nonintelligent 3270 terminals. Manufacturers and software houses rushed to provide quick, patched, short-sighted, and yet relevant solutions to some of the information-exchange problems faced by the user community. As a result, many point-to-point implementations linking a certain microcomputer to a certain mainframe or one communicating word processor to another became available.

However, with the growth and rapid acceptance of various types of computers and applications as well as rising user expectations, the point-to-point products created the classical spaghetti network of short, dead-end roads. Such roads were both physical, linking two machines, and logical, as between host software components. No access was provided to the electronic "interstate highways" of X.25 and T1, since these backbones were not well integrated with the point-to-point spaghetti.

Networks were highly specialized, separate entities. Terminals clustered around a mainframe or minicomputer that acted as the network host. Alternately, self-contained, standalone local area networks (LANs) connected different hardware via a common medium, such as fiber, twisted pair, or coaxial cable, with no information sharing across networks. Communications in these LANs, including physical interface, data link control, and physical connection services, have been essentially standardized. However, the problems of information sharing, including the structure of content (such as common file and document formats) and internetwork delivery, are not yet resolved. Moreover, each network is managed separately.

The way out

To help corporations operate their businesses in the future, networks will have to allow for the free transfer of information, regardless of manufacturer and product type. From the user's perspective, the next step in computer evolution that needs to occur should be the development of a common, networkwide function known as "information delivery." This refers to the interconnection of applications, regardless of their location or internal format or the type of computer on which they reside.

Figure 2 shows the layers of INA that users will need to secure for themselves in the coming decade. (IBM's SNA Distribution Services, or SNADS, will become a de facto INA standard. Users too small to develop their own INAs should request them from their vendors.) The separation of the application layer into three layers each having a discrete function can be seen as an extension of current architectural specifications. It would provide a common "pipeline" supporting services for storing, routing, protecting, and accounting for information between application programs, roughly comparable to the services of the U. S. Post Office for delivering mail.

Typically, standard architectural reference models depict the application layer as one single entity. This view resulted from the fact that, in the past, each application program had a unique communications interface—as discussed above.

IBM has begun to build a mechanism for information delivery based on SNADS (see "SNADS: The beginning of IBM's INA?"). While such delivery today includes only text documents, there is little doubt that it will be extended to, for example, voice, graphics, and imagery. An important question is whether or not other computer companies will also adopt the same technique in a compatible manner to solve the same problem. As IBM is evolving its delivery infrastructure, other standards—such as X.400 from the International Telegraph and Telephone Consultative Committee (CCITT)—are being accepted by some companies. This may only add to the confusion surrounding SNA versus Open Systems Interconnection (OSI).

User organizations will find it extremely difficult to begin converting millions of dollars worth of software to utilize SNADS with LU (Logical Unit) 6.2 or X.400 with OSI today. However, to wait and see which will be the predominant way to implement a delivery mechanism will only increase the risk and cost. Choosing one now makes better sense. Users could either develop their own delivery infrastructure, if they have the resources (Citibank, for example, has already begun to do this), or locate one developed and supported by an independent third party who could support the multiple vendors (or equipment types) in the user's installation.

Details of the architecture

INA breaks down the functions of the application layer and redefines their relationship with presentation layer functions for the purpose of information sharing. Today, the relationship between the application and presentation layers is not uniform because the boundaries and functions between the two layers have not been clearly defined. Different vendors and standards bodies are working together, through the International Organization for Standardization (ISO), to define the upper layers of the OSI architecture.

Since OSI is a reference model, each vendor implementation of it is likely to be different. Even when the reference model is fully defined, implementations may be different or inconsistent because programmatic interfaces between applications and the presentation layer are different. And even if these interfaces were the same, both within and across vendors, there is yet another level of the application functions to consider: the decomposition and mapping of transactions into atomic business functions.

For example, airline reservations consist of several atomic transactions. An incoming ticket request is decomposed into three transactions: one to update the seat inventory on the flight, one for the financial exchange, and one for issuing the ticket. Each of these atomic transactions is individually processed by its respective business program. Each atomic business program, in turn, responds to the code responsible for orchestrating the process. The orchestrator initiates each atomic business program as it is required. When all atomic transactions have been processed, the orchestrator responds to an airline terminal with the result of the request (for example, no seats, credit denial, printer broken, request completed). However, this division of labor among discrete routines might exist only in a limited number of application programs today. Where it does exist, no two applications use the same structure.

Users within both single- and multivendor environments must protect themselves from this lack of uniformity. One way of doing so is by breaking down and carefully defining the application layer into three layers known as business application, delivery and control (D&C), and network services.

What's at the layers

The top layer is responsible for the business functions of application programs. For example, one business function would be the reading and writing of files (credit inquiries, account balances) to and from disk. It should not have to be concerned about where the files or the target disk reside in the network. On behalf of business functions, the D&C layer requires access to a directory that knows the name and location of the appropriate computer, whether local or remote. Application programs can address local and remote resources automatically.

Business functions use the services provided by the D&C functions. One D&C function might be to package a file and ship it to its destination. Services provided by the delivery layer will allow business functions to communicate on a peer-to-peer basis with other business functions. For example, a transfer of funds in one business function may trigger a debit in another in the same computer or on a distant computer. This process requires proper routing decisions by the delivery layer. Moreover, different business functions may require different classes of services, which corporations should be able to assign.

The last layer, network services, is responsible for functions typically thought of as communications. It must provide interfaces to architected protocols, such as Systems Network Architecture (SNA), Open Systems Interconnection (OSI), and Transport Control Protocol/Internet Protocol (TCP/IP), as well as nonarchitected protocols, such as binary synchronous communications (BSC). The network services layer packages messages into the appropriate format and attaches whatever routing information (headers and so forth) is required.

The primary purpose of this layer is to insulate investments in code written for business functions and D&C functions from the computer hardware environment as well as from public and proprietary data communications protocols. This allows business functions to be redesigned or rewritten without affecting the underlying transport mechanism. Moreover, it lets corporations integrate business functions throughout multiple interconnected networks, regardless of the protocol used by each network.

The amount of code in the network services layer is related to the lack of uniformity of the presentation-layer implementations available today. As presentation layers become more uniform, the need for the network services

SNADS: The beginning of IBM's INA?

To date, Systems Network Architecture (SNA) has mostly addressed problems particular to terminals sharing application programs in one or more mainframes. However, IBM has begun to respond to the market's demand for new ways to use its gear and software for information networking. As part of this strategy, IBM has created and is starting to implement enhancements corresponding to delivery and control functions under the name of SNA Distribution Services (SNADS). As such, SNADS may represent an IBM bid to supply an information network architecture (INA) to businesses.

IBM announced SNADS for store-and-forward document distribution between multiple mainframes running the Distributed Office Support System (DISOSS), IBM's host-based archive. SNADS, currently bundled with DISOSS, allows documents to be scheduled for distribution to another DISOSS host even when the path to that host is not active. In addition to its archival or library functions, DISOSS also serves as an intermediate storage location.

SNADS uses IBM's Document Interchange Architecture (DIA) and Document Content Architecture (DCA) for message handling. DIA is a specialized communications protocol for mailing objects. It specifies the rules for delivery and defines the envelope carrying the attributes both of the object and of its recipients. DCA specifies and defines the structure of documents (with such parameters as layout, indentation, and margin justification). Currently, this structure is quite simple. DIA and DCA can only deliver documents consisting of text. However, in the future, incoming data will contain various types of information objects, such as text, voice, graphics, and imagery, forming a compound document.

DISOSS can be considered a local post office or an intermediate mailroom. When one DISOSS mainframe talks to another, it uses SNADS as an interpost-office protocol when the transaction is store-and-forward. (For immediate relaying of messages, standard SNA sessions are established.) Links between DISOSS machines through SNADS are set up using the Logical Unit (LU) 6.2. All user messages are packaged in DIA and submitted to the node that implements the local post office functions of DISOSS (a subset of DISOSS known as the internal messaging application). The messages may then either be distributed locally by that DISOSS host or they can be handed over to SNADS for delivery to another.

The figure shows that every IBM application with its associated devices presently requires its own unique communications program and SNA logical connection (represented by different LUs). For example, remote job entry (RJE) or network job entry (NJE) provide access for remote printers and workstations accessing the Job Entry Subsystem (JES), a batch-oriented application. At the transport level, JES and remote devices depend on LU type 1. Other applications and devices similarly rely on highly product-dependent LU implementations. Another disadvantage is that the LUs require logical point-to-point connections. These connections are an outgrowth of multiple delivery mechanisms, each of which is both application- and device-dependent.

For IBM to paint itself out of this corner, it needed to move toward another type of LU. This LU would let any program or user talk to any other without requiring a direct logical connection. Instead, it would

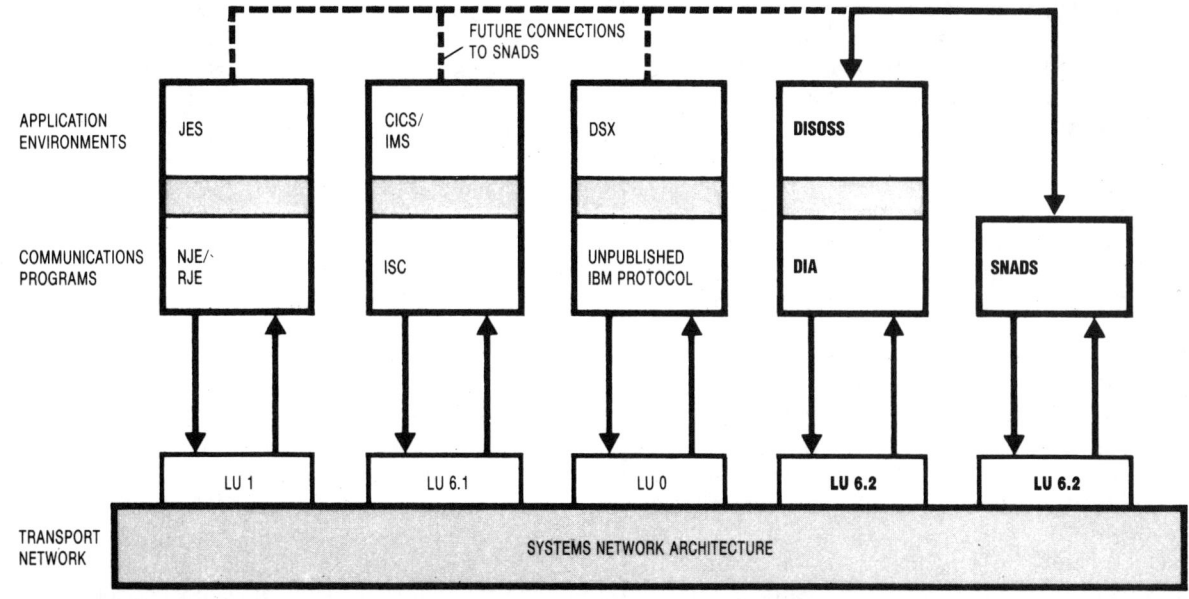

CICS = CUSTOMER INFORMATION CONTROL SYSTEM
DIA = DOCUMENT INTERCHANGE ARCHITECTURE
DISOSS = DISTRIBUTED OFFICE SUPPORT SYSTEM
DSX = DISTRIBUTED SYSTEMS EXECUTIVE
IMS = INFORMATION MANAGEMENT SYSTEM
ISC = INTER-SYSTEM COMMUNICATIONS
JES = JOB ENTRY SUBSYSTEM
LU = LOGICAL UNIT
NJE = NETWORK JOB ENTRY
RJE = REMOTE JOB ENTRY
SNADS = SNA DISTRIBUTION SERVICES
(SNA = SYSTEMS NETWORK ARCHITECTURE)

let a user address a common piece of code in each computer. LU 6.2 is a step in this direction.

However, the way LU 6.2 is implemented today still requires that every node in the network implement a logical session manager. For LU 6.2 to work, this implies that a virtual point-to-point connection must still be set up between two nodes. Currently, advanced program-to-program communication (APPC) does not address the higher-level functions needed to form a real information network, such as alternate routing, time-independent delivery, and the mapping of canonical forms. Thus program-to-program communications and LU 6.2 do not as yet solve the problem of information delivery. In order to reap the true benefits of LU 6.2, IBM will have to modify the logical unit so that it is decoupled from its current environment and transformed into a common resource that can be addressed by entities throughout the network. This resource will provide a set of network services in support of information delivery and control.

SNADS, which is evolving into a delivery and control layer, will utilize this networkwide code in order to locate programs, establish and control the connections, and deliver information. SNADS also needs to be decoupled (from DISOSS) because each node does not need a full version of DISOSS but merely its routing portion.

By means of SNADS, a user at a microcomputer that is not configured as an RJE device will eventually be able to submit a request to JES, even if the microcomputer is not directly connected to the mainframe running JES. A batch user could send print requests to any printer in the network regardless of whether that printer supports a JES device (such as 2770, 3770, 2780, or 3780) and regardless of whether the necessary resources are active at the time. SNADS will also permit network operators to dynamically configure new printers without having to reconfigure JES, as is required with RJE and LU 1.

SNADS is one of the first industry implementations of delivery and control functions. Since IBM has opened up SNADS to the non-IBM world (that is, published the SNADS formats and protocols so that non-IBM vendors can implement compatible functions), SNADS has a good chance to become a de facto standard just like SNA. But, typically, objects must be moved between more than just an IBM and a non-IBM environment. The real question is, who will deliver objects from DEC to Wang?

The authors believe that non-IBM vendors that are promoting their processors at the departmental level will need to implement DISOSS-like archival and SNADS-like distribution services. It is unlikely that these companies will develop their own proprietary versions of such a delivery mechanism that do not interface with the evolving IBM delivery and control layer. Such software is currently being developed by the non-IBM vendors, often in conjunction with third-party software houses.

layer will be reduced. The extent of the layer is also related to the sophistication of the underlying protocol. Today, for example, the amount of code within the network services layer for BSC would be greater than that for SNA to compensate for BSC's lack of a presentation layer.

The business application layer

Eventually, there will be no need at all for a network services layer per se, as standards are set (or de facto ones accepted) and as communications architectures become more prevalent. To understand the role of the other two new layers, it is helpful to consider examples of the types of entities and concerns at each.

"Business functions" refers to a set of common facilities for issuing calls to send and receive messages. These calls can be characterized as a request for information and the information in response, each of which is an atomic transaction. Real-time exchanges, such as inventory queries and updates or automated teller machine (ATM) transactions, require the business functions that are driving the initial need for INAs. These business functions include deposits to an account, balance inquiries, updates of spreadsheets, or sending someone a message or a document. They should therefore be able to support a wide variety of sources, such as terminals, programs, microcomputers, and workstations.

Today a bank customer can go to an ATM and say "give me this" or "send this from here to there." It is not similarly possible to simply send a file from any computer to any other, since this calls for a particular logical connection between the two machines. Each such connection has its own protocol, and the number of them grows in proportion to the number of device types to be joined. With an INA, however, the business function "requesting a file" has a local dialog with the D&C layer. Since this layer uses the network services layer, which knows all the required protocols, files can travel from any point in the network to any other. Without that, each business function has to have a separate connection.

Some components of the business application layer are listed below:

■ Atomic transactions. A business function may consist of several atomic transactions. For example, a person drawing from a checking account with overdraft protection may cause the business function of "debiting" to occur on one computer. The business functions of "overdraft protection" and "crediting" may involve a funds transfer from a reserve credit line on another computer to the processor on which the debiting function initially occurred. The process appears as one transaction to the user. This example can be applied to reservations in the airline industry and to inventory control in manufacturing, retail, and wholesale. As changes to inventories on one computer are made, they affect accounting records on another.

■ Error management. Failures of business application code must be detected, contained (the atomic transactions insulated from each other), and recovered from by business functions. These functions should take responsibility for the successful performance of all related atomic transactions. Stand-in business functions must be provided for critical applications.

- File transfer services. Such services, in support of the business function, are already among the bread-and-butter services required by data-center applications. Today, the distribution of batch files is a manual operation and may include sending a tape in a cab across town. Instead, an INA should provide automatic distribution. Once a batch job is completed, the batch program should send a transaction to a business function that would cause the results of the batch job to be sent to the other data center. This should occur without operator intervention. While this example concerns large files, microcomputer and workstation users should also be able to submit files for delivery across the network without having to establish a connection to the destination.
- Software distribution. Computers will need some means to guarantee that all common software is operating at the same version level. Version maintenance and the distribution and inventory of software can make up one or more business functions.
- Universal electronic mailboxes. These locations for users to receive all electronic correspondence will allow users in an IBM DISOSS environment to create a document and send it to another user without having to know whether the addressee resides in the same, or another, DISOSS environment, Wang Office, or DEC All-In-One environment.
- Library services. These could contribute to the maturation of electronic mail. Libraries are electronic filing cabinets where users store different types of information objects, such as files, documents, messages, software, and extracts of data. Each object is stored with a set of attributes that may include owner, date, type (such as report or manual), and a descriptive summary. Users can search, display, and select any library item as needed. The location and number of libraries can be organized depending on organizational and business needs. IBM's DISOSS is the best-known library product available. Although DISOSS is capable of storing any type of object, IBM has restricted its use to accept only text documents when accessed from IBM office environments.

Library services should enable businesses to eliminate the need to continuously distribute thousands of pieces of printed literature and to store thousands of individual copies of information, such as updates of policy and procedure manuals.

The delivery and control layer
Services at this layer deal with the following D&C functions: directories, store-and-forward processing, routing, security, statistics, notifications, and logging. The layer determines which business functions are available and can reach them through application interfaces. For example, some business functions may be available only during certain times of day. The D&C function must therefore know where they reside and when they are available. When the business functions are not available, the D&C layer should be able to route transactions to stand-in business functions on the same, or another, node.

Below are some examples of entities and activities of the D&C layer:

- Levels of service. Businesses and users should be able to specify the time of day when delivery should occur and how quickly. They should be able to execute or transmit, for example, microcomputer files (or other objects) or host batch jobs at any given time, without requiring human intervention. D&C functions must provide these capabilities. They should also enable businesses to select the most appropriate level of service to most closely match business operating requirements (for example, for business functions requiring interactive, as opposed to batch, processing).

Alternate network paths must be available to accommodate the different levels of service. The cost of each service may vary relative to the capabilities that a single path can provide, with value-added services being more costly. For example, the cost of a service will vary depending on if a request requires expedited processing and, therefore, a direct link, or if it can be stored at intermediate nodes and forwarded throughout the network until resources become available.

Moreover, the structure of telecommunications tariffs makes it desirable for businesses to be able to schedule information transfers at the time of day when links are cheapest. Intermediate store-and-forward processing is transparent to the user who initiated the information transfer. Of course, businesses should be able to charge back users (divisions, departments, or individuals) for their use of the network. Such a mechanism should also enable corporations to differentiate classes of users and to assign costs to them based on how closely their requested network-service priority levels correspond with business needs.

- Global network directories. The information network must maintain a corporate directory of all nodes and their addresses independent of the directory that the underlying communications network uses to establish connections in the first place. Global node and address directories are roughly comparable to post office directories that contain all the zip code information needed to locate a recipient. The global directory must also include the names and addresses of nodes within subnetworks, such as IBM's SNA networks, Wangnets, Decnets, Tandem networks, LANs, and so on.
- Translation and transformations. It is necessary that the D&C functions know the internal formats of the objects, such as documents, that may be passed to them by network services. Document structures must be converted into a canonical form prior to their submission to the delivery layer. In this way the business application layer can always work with the same object, regardless of the originating computer. Only business functions dealing with word processing documents need to understand these canonical forms. Another business function, like a checking account inquiry, might need to understand a different set of canonical forms.
- Notification procedures. What is commonly known to users of the postal service as return-receipt-

3. Layers deployed. *Some current host environments are sophisticated enough to separate applications from the supporting communications (A). Even so, one of these two functions would have to be modified to handle pass-through "postal" messages. Alternatively, separate layers would let this host handle such traffic (B).*

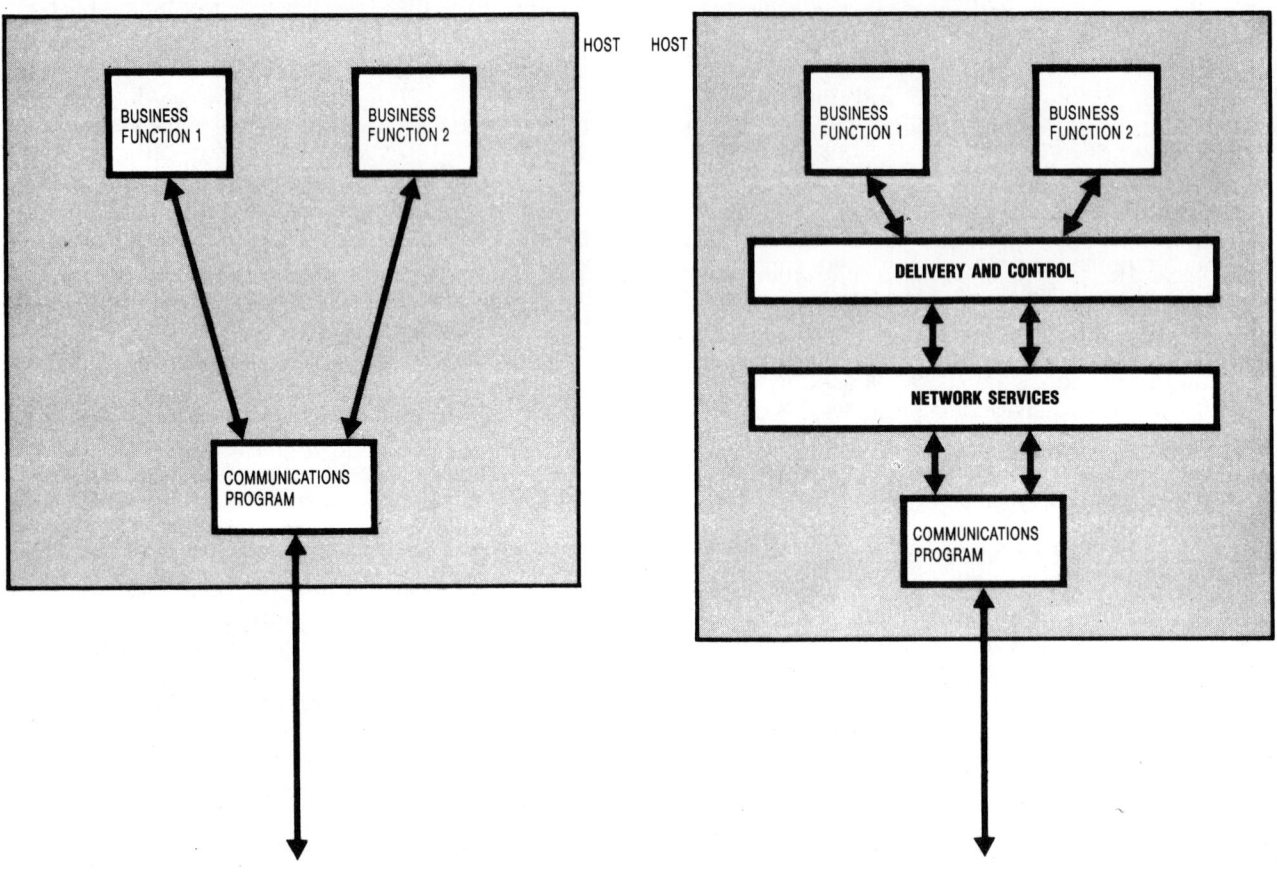

requested (for example, with certified mail) should be automatically available in a network environment. Both senders and receivers should be automatically notified on completion of such business functions as file or document transfers. Besides automatically notifying the actual recipient, a user should be able to request that additional parties be notified (as in "send this report to Charlie and also notify his boss"). Also, failures that can occur during deliveries and prevent their completion may require notification of different individuals so that the problem can be resolved by the most knowledgeable person. For example, supervisors might need to be informed if certain vital communications of their underlings went awry.

■ Security. It is assumed that the required physical security (access to computers and their peripherals) is provided as part of overall premises security. Each individual network within the larger corporate network must be permitted to choose its own local security strategy, such as Resource Access Control Facility (RACF), Access Control Facility (ACF-2), or Top Secret on IBM mainframes. Each network can expect the larger network to respond to these requirements. Security functions must, therefore, validate such things as user identification, data encryption, as well as access rights with the local security environment.

■ Recovery management. This corresponds to recovery management for internal business-code failures. The D&C layer must provide error and recovery management for failures that affect delivery, such as the routing of transactions, files, documents, or messages. Software at the D&C layer must be robust enough to be able to detect and recover from failures without affecting other components (or business functions). In the case of a file or document transfer, users should be allowed to resume their transfers at a checkpoint rather than having to re-establish an entire file or document transfer. The same applies to programs that may handle transactions without operator intervention.

Formatted messages flow from the underlying transport network into the network services layer. This layer transforms these messages into a canonical form that can be processed by the delivery and control layer. Another set of canonical forms should be used between the D&C layer and the business application layer. The network services layer will eventually be rendered unnecessary, so the previous discussion of this layer should suffice.

Recognizing that the next evolutionary step in computer communications is to interconnect or integrate architectures is one thing. Implementing it is another. However, information network administrators should begin to consider the process, as many of them are reaching an opportune juncture. Most existing code is terminal-to-application oriented and will eventually have to be modified to take advantage of distributed processing, placing an enormous burden on developers. Besides, user organizations will have to rewrite applications both to keep up with evolving user requirements and because some applications will have reached the end of their life cycle.

Depending on the magnitude of the effort needed, user organizations could take advantage of the opportunity to implement the business information architecture discussed in this article. A major benefit of the architecture is that it insulates business and D&C functions from the network services and their underlying communications protocols. This fact provides a migration path to gradually modify existing software.

By moving to this architecture, information network administrators will be able to better insulate their business functions from future changes. The goal of this insulation is to develop new business functions that let users group and regroup dynamically into "circles of interest" at the discretion of the enterprise without having to be in the same place or attached to the same computer.

Moreover, one application program will be able to initiate the execution of another according to the business need. With code at the new layers in each computer, every application program will be able to talk to every other such program (Fig. 3). The only requirement would be an adequate transport network beneath them.

Implementations of the information delivery concept will continue to evolve over the next five to 10 years and will become an important method for integrating diverse applications and diverse computer equipment. Large user organizations will be able to use these implementations to isolate their applications from the underlying network mechanics. This will allow them to choose the most appropriate hardware and software for future applications. ■

Rudolf Strobl, Ph. D., is a senior consultant at Software Research Corp. (SRC). Strobl is involved in the planning and project management of information delivery products for SRC and Fortune 500 *companies. Previously, Strobl was with the Yankee Group, a Boston-based market research and consulting company. Bill Stackhouse is SRC's director of product planning. Prior to joining SRC, Stackhouse was a vice president at the Bank of America. In that position, he was responsible for branch delivery systems.*

S. Helga Eriksen, Brookside Communications, Leesburg, Va.

The how and where of switched 56-kbit/s service

This cents-per-minute offering is available today from three vendors, with a fourth in the wings. Here's how to make connections and a cost comparison.

The availability of low-cost broadband capacity will be a major issue in the next wave of new office technology. Support for this statement comes, for example, from the experience of Chicago-based AM International (AMI). One reason the company dissolved its Electronic Character Recognition manufacturing division, after developing a "Cadillac" Document Delivery System for MCI/SBS (Satellite Business Systems) in the early 1980s, may have been the prohibitive price tag: more than $100,000. But close behind in contributing to the demise was the lack of affordable (cents-per-minute) broadband in the 56-kbit/s range.

Today we find that data speeds have inched up, but this movement is due more to technology improvements in error-correction techniques than to the availability of cheaper high-speed (56 kbit/s and above) facilities. In other words, we are pumping more bits over the same old "pipes." However, it is the global availability and worldwide compatibility of those old pipes that has made the microcomputer such a continuing success. But today's microcomputer typically deals in character-bit representations that are simpler—as compared to image representations—and much more bandwidth-sparing than those required to represent high-resolution images (defined as those that contain at least 32 shades of gray).

The transfer of high-resolution images opens up the next frontier in office automation—to the extent that it can be used as easily and cheaply as the microcomputer. Exponential improvements in compression techniques have made possible the transmission of television-quality images over 56-kbit/s facilities, thereby avoiding the choppy-motion effect of the traditional freeze-frame schemes. This videoconferencing-compression breakthrough is typified in a recent offering from Peabody, Mass.-based Pictel Corp., which, together with other companies, promotes the use of switched 56-kbit/s service with its product.

What these advanced videoconferencing and document-delivery applications could use is something equivalent to a global dial-up 56-kbit/s service approaching the same price range as typical microcomputer-usage costs—that is, switched 56-kbit/s service that is close to the cents-per-minute cost experienced by the telephone user.

Gaining access

Switched 56-kbit/s service is available from two basic sources in the United States: the long-distance carriers and the regional Bell operating companies (RBOCs). Availability of, and compatibility with, an RBOC implementation is strategic to end-to-end delivery. With RBOC availability absent in a given region, switched 56-kbit/s service is limited to less than a two-mile radius surrounding the long-distance carrier's "point of presence" (POP) in a major city. This radius can be extended by the purchase of Dataphone digital service (DDS)—subject, of course, to the availability of DDS. It's available today from about 50 percent of telephone company central offices (Table 1) but is too costly ($400 to $500 per month) for most customer premises equipment (CPE) customers.

There are essentially three vendors who currently offer switched 56-kbit/s service: AT&T, McLean, Va.-based MCI/SBS, and Argo, based in New Rochelle, N. Y. The American Satellite Co. of Rockville, Md., is in an early developmental stage.

Table 2 compares the features, availability, and compatibility of the long-distance vendors' 56-kbit/s offerings. A long-distance implementation of a circuit between major metropolitan areas involves testing for compatibility between services. This testing involves AT&T, MCI/SBS, Argo, and American Satellite (a test participant) at an RBOC's switched 56-kbit/s service POP. The testing is also done to ensure equal access to interexchange carriers between Local Access and Transport Areas (LATAs) nationwide.

Table 1: Dataphone digital service areas

STATE	CITIES OR AREAS	STATE	CITIES OR AREAS
ALABAMA	BIRMINGHAM, HUNTSVILLE, MONTGOMERY, MOBILE	MONTANA	GREAT FALLS, BILLINGS
ARIZONA	PHOENIX, TUCSON, NAVAJO INDIAN RESERVATION	NORTH CAROLINA	ASHEVILLE, CHARLOTTE, GREENSBORO, RALEIGH, WILMINGTON, FAYETTEVILLE, ROCKY MOUNT
ARKANSAS	FORT SMITH, LITTLE ROCK, PINE BLUFF	NEBRASKA	OMAHA, GRAND ISLAND, LINCOLN
CALIFORNIA	SAN FRANCISCO, CHICO, SACRAMENTO, FRESNO, LOS ANGELES, SAN DIEGO, BAKERSFIELD, MONTEREY, STOCKTON, SAN LUIS OBISPO, PALM SPRINGS	NEVADA	RENO, PAHRUMP
		NEW HAMPSHIRE	(ENTIRE STATE)
COLORADO	DENVER, COLORADO SPRINGS	NEW JERSEY	ATLANTIC COAST, DELAWARE VALLEY, NORTHERN AREA
CONNECTICUT	(ENTIRE STATE)		
DISTRICT OF COLUMBIA	WASHINGTON	NEW MEXICO	(ENTIRE STATE)
DELAWARE	(PART OF PHILADELPHIA AREA)	NEW YORK	METROPOLITAN NEW YORK CITY, POUGHKEEPSIE, ALBANY, SYRACUSE, BINGHAMTON, BUFFALO, FISHERS ISLAND, ROCHESTER
FLORIDA	PENSACOLA, PANAMA CITY, JACKSONVILLE, GAINESVILLE, DAYTONA BEACH, ORLANDO, SOUTHEAST AREA, FORT MYERS, GULF COAST, TALLAHASSEE	OHIO	CLEVELAND, YOUNGSTOWN, COLUMBUS, AKRON, TOLEDO, DAYTON, CINCINNATI (BELL), MANSFIELD
GEORGIA	ATLANTA, SAVANNAH, AUGUSTA, ALBANY, MACON	OKLAHOMA	OKLAHOMA CITY, TULSA
IDAHO	(ENTIRE STATE)	OREGON	EUGENE, PORTLAND
ILLINOIS	CHICAGO, ROCKFORD, CAIRO, STERLING, FORREST, PEORIA, CHAMPAIGN, SPRINGFIELD, QUINCY, MATTOON, GALESBURG, OLNEY	PENNSYLVANIA	HARRISBURG AREA, PHILADELPHIA, ALTOONA, NORTHEAST AREA, PITTSBURGH, ERIE
		RHODE ISLAND	(ENTIRE STATE)
INDIANA	EVANSVILLE, SOUTH BEND, AUBURN/HUNTINGTON, INDIANAPOLIS, BLOOMINGTON, RICHMOND, TERRE HAUTE	SOUTH CAROLINA	GREENVILLE, FLORENCE, COLUMBIA, CHARLESTON
IOWA	SIOUX CITY, DES MOINES, DAVENPORT, CEDAR RAPIDS	TENNESSEE	MEMPHIS, NASHVILLE, CHATTANOOGA, KNOXVILLE, BRISTOL
KANSAS	WICHITA, TOPEKA	TEXAS	EL PASO, MIDLAND, LUBBOCK, AMARILLO, WICHITA FALLS, ABILENE, DALLAS, LONGVIEW, WACO, AUSTIN, HOUSTON, BEAUMONT, CORPUS CHRISTI, SAN ANTONIO, BROWNSVILLE, HEARNE, SAN ANGELO
KENTUCKY	LOUISVILLE, OWENSBORO, WINCHESTER		
LOUISIANA	SHREVEPORT, LAFAYETTE, NEW ORLEANS, BATON ROUGE	UTAH	(ENTIRE STATE)
MAINE	(ENTIRE STATE)	VIRGINIA	ROANOKE, CULPEPER, RICHMOND, LYNCHBURG, NORFOLK, HARRISONBURG, CHARLOTTESVILLE, EDINBURG
MASSACHUSETTS	(WESTERN AND EASTERN AREAS)		
MARYLAND	BALTIMORE, HAGERSTOWN, SALISBURY	VERMONT	(ENTIRE STATE)
MICHIGAN	DETROIT, UPPER PENINSULA, SAGINAW, LANSING, GRAND RAPIDS	WASHINGTON	SEATTLE, SPOKANE
		WISCONSIN	NORTHEAST, NORTHWEST, SOUTHWEST, AND SOUTHEAST AREAS
MINNESOTA	ROCHESTER, DULUTH, ST. CLOUD, MINNEAPOLIS		
MISSISSIPPI	JACKSON, BILOXI	WEST VIRGINIA	CHARLESTON, CLARKSBURG, BLUEFIELD
MISSOURI	ST. LOUIS, WESTPHALIA, SPRINGFIELD, KANSAS CITY	WYOMING	(ENTIRE STATE)

Table 2: Comparing the offerings

	AT&T	MCI/SBS	AMERICAN SATELLITE	ARGO	COMSAT	TYMSTAR	DAMANET
SWITCHED 56-KBIT/S SERVICE AVAILABILITY	NOW	NOW	1986 (?)	NOW	DEDICATED ONLY	DEDICATED ONLY	DEDICATED ONLY
SWITCH	#4ESS	SATELLITE COMMUNICATIONS CONTROLLER AND SELECTOR "500"	EARTH STATION	DMS250 AND CHANNEL UNIT			
USER INTERFACE	V.35	RS-449, DS-1, V.35, OR OTHER		V.35			
SERVICE NAME AND TYPE	ACCUNET 56	SBS DATA SERVICE		ARGONET SWITCHED 56-KBIT/S SERVICE			
NUMBER OF LOCATIONS	24	21	8	20			
SERVICE EXTENSION	RBOC SWITCHED 56-KBIT/S OR DDS LINKS	RBOC SWITCHED 56-KBIT/S OR DDS LINKS		RBOC SWITCHED 56-KBIT/S OR DDS LINKS			
RBOC SWITCHED 56-KBIT/S COMPATIBLE	TESTING WITH RBOCs	TESTING WITH RBOCs	TESTING WITH NYNEX	TESTING WITH NYNEX AND PACIFIC BELL			
OCC COMPATIBLE	NO DIRECT INTERFACE	NO DIRECT INTERFACE	NO DIRECT INTERFACE	NO DIRECT INTERFACE			
TYPE OF SWITCHING PROTOCOL	POINT-TO-POINT MEASURED TIME	TDMA WITH IN-BAND SIGNALING	TDMA	TDMA 4 WIRE, PLUS VOICE CONNECTION			

DDS = DATAPHONE DIGITAL SERVICE
OCC = OTHER COMMON CARRIER
RBOC = REGIONAL BELL OPERATING COMPANY
TDMA = TIME-DIVISION MULTIPLE ACCESS

The testing is scheduled for completion this year by Ameritech, Pacific Telesis, and Nynex. Testing by U. S. West, Southwestern Bell, and BellSouth is under way.

The service's pricing by AT&T, MCI/SBS, and Argo is based on a proposed traffic model. The monthly usage charges range from 30 to 70 cents per minute on the average. This is the first time a 56-kbit/s data service has been priced in cents per minute. (Before switched 56 kbit/s became available, users desiring that data rate had to lease lines at thousands of dollars per month.) Pricing by the three carriers in 1985 is compared in Table 3.

The RBOCs are in various stages of implementing switched 56-kbit/s services. The RBOCs refer to the service as Public Switched Digital Service (PSDS) or as Circuit-Switched Digital Capability (CSDC) in those regions where the implementation was accomplished mainly through an upgrade to the Western Electric (W. E.) 1AESS (Electronic Switching System) switch. In those regions where the Northern Telecom DMS-100 switch was used, the service is referred to as Datapath. Nynex is an exception: It uses the DMS-100 switch but calls the service Switchway. Switchway is one of a family of new services called Pathways, all based on Northern Telecom technology in which Nynex recently invested $300 million. Other regions are considering different names for the service.

Almost all the RBOCs anticipate having some implementation of switched 56-kbit/s service this year—the one exception is Bell Atlantic. This RBOC has tested the service in Murray Hill, N. J., but declined to disclose any plans for 1986 at the time of this writing. Some of the Bell operating

Table 3: The switched 56-kbit/s offerings (1985)

	MCI/SBS	AT&T	ARGO
A. MONTHLY PORT CHARGES	$1,066*	$550	$650
B. USAGE CHARGE (PER MINUTE)	0.45	0.70	0.30
C. USAGE SCENARIO:			
100,000 MINUTES/MONTH			
100 PORTS—COST/MONTH	$151,600	$125,000	$95,000
COST/PAGE	0.38	0.31	0.24
400 PORTS—COST/MONTH	$471,400	$290,000	$290,000
COST/PAGE	1.18	0.73	0.73
400,000 MINUTES/MONTH			
100 PORTS—COST/MONTH	$286,600	$335,000	$185,000
COST/PAGE	0.18	0.21	0.12
400 PORTS—COST/MONTH	$606,400	$500,000	$380,000
COST/PAGE	0.38	0.31	0.24

*INCLUDES CHARGES FOR SELECTOR 500 AND 56-KBIT/S DATAPHONE DIGITAL SERVICE ACCESS LINE

companies in the Bell Atlantic region (such as Bell of Pennsylvania and New Jersey Bell) do not see a market for the service. Although the Chesapeake & Potomac Telephone company, based in the Washington, D. C., area, is interested, Bell Atlantic is concerned about compatibility issues between PSDS and Datapath. These issues have been taken up with Bell Communications Research Corp. (Bellcore, the RBOCs' jointly owned advisory organization) and are discussed later. Table 4 summarizes the RBOCs' switched 56-kbit/s services availability and their tentative plans for this year.

Implementations

The user interface to switched 56 kbit/s can be obtained both from Sunnyvale-based California Microwave and AT&T Information Systems (ATTIS) as the PSDS implementation and from Northern Telecom as the Datapath implementation (Fig. 1). There are three interfaces presented to the user: V.35, RS-449, and RS-232-C. The Datapath implementation was initially limited in distance from the main switch because of the wire gauge it used. But the use of channel banks extends the distance by 50 miles. And the use of remote DMS-100 switches can extend this to 150 miles within the public switched telephone network.

The PSDS implementation on the 1AESS switch involves the Network Channel Terminal Equipment (NCTE) and the Terminal Interface Equipment (TIE). The NCTE is a W. E. 554 channel service unit (CSU) that does loopback testing, detects "on-hook" and "off-hook" conditions, converts DTMF (dual-tone multifrequency) signaling, and derives timing from the 1AESS.

The TIE is a W. E. 504 data service unit that does self-test, conversion of four-wire to two-wire interfaces, and near-end (telephone company central office) loopback testing. A button disconnects voice mode and starts time compression multiplexing (TCM). The buffer holds 72 bytes. A switch enables 9.6-kbit/s transmission.

Call setup is accomplished via analog telephone, using DTMF signaling. The telephone set is controlled by the use of off-hook, dialing, and on-hook. (The start/stop control signals are initiated by off-hook/on-hook, respectively.) With no provision for digitizing voice, the telephone cannot be used for voice in this application.

California Microwave produces a PSDS-compatible unit called the Flextie 56, which essentially combines the functions of the NCTE and the TIE. The range of pricing is typically $3,500 to $5,000, depending on such options as autodial and extended diagnostics.

Datapath's data unit is situated between the customer terminal and the telephone local loop. Under a TCM scheme, the data unit transmits signals on two separate logical channels: control signals at 8 kbit/s and data signals at a user rate of 56 kbit/s. When the subscriber loop consists of 26-gauge wire, the data unit can be 2.5 miles from the data line card; 22-gauge wire, 3.4 miles away. This distance can be extended up to 150 miles using the remote DMS-100 switch.

Call setup is accomplished via the keypad on the data unit: The calling user dials the number of the intended data recipient. The user receives an indication that the connection is completed. The user then starts to transmit.

The DMS-100 converts the TCM-formatted data to 64-kbit/s pulse code modulation (PCM) form, including an 8-kbit/s supervisory-message channel. The data line card permits loopback and bit-error-rate testing, takes up two physical slots in a family of new DMS-100 peripherals, and can be mixed with voice line cards.

In addition to the data units, Datapath offerings are complemented by the following:
- 3270 terminal emulation/concentration.
- Coaxial-cable elimination with 3270 switching.
- IBM/Systems Network Architecture connectivity for bisynchronous or synchronous data-link-control data transmission.
- EIA RS-422 interfacing for microcomputers.

PSDS basically employs TCM, which is really an analog scheme modified to be unipolar rather than bipolar. (A bipolar signal is a logical data input represented by a voltage polarity opposite that representing noise or a nondata signal. If both data and nondata signals are represented by the same polarity, they are defined as unipolar.) TCM—also known as ping-pong transmission—uses a full-duplex channel. TCM is created by alternating one-way bursts in each direction at a clock rate of around 2.5 times the user's information-transfer rate.

Datapath employs TCM in the same way, except that PSDS uses a 144-kbit/s data rate; Datapath uses 160 kbit/s. Both use 8-kbit/s control signaling outside the customer bandwidth of 56 kbit/s.

Long-distance carriers and the RBOCs are conducting interexchange-carrier tests in several cities (Table 4). The incompatibility on the network level involves changing from the previously required four-state signaling to two-state signaling (see "Four- to two-state signaling"). This is facilitated by the CDU (CSDC data unit), which is plugged in on the trunk side of the 1AESS switch. The CDU does the analog-to-digital conversion to digitize the TCM stream to DS1 (1.544 Mbit/s) digital format (Fig. 2).

Accunet (AT&T's data-oriented digital services) is implemented on a 4ESS switch with two-state signaling. The Datapath design does not need a CDU because the

Table 4: The RBOCs and switched 56 kbit/s

	AMERITECH	NYNEX	PACIFIC TELESIS GROUP	BELLSOUTH	US WEST	BELL ATLANTICOM	SOUTHWEST BELL
SWITCH	1AESS	NTI DMS100 DATAPATH	1AESS	1AESS AND SOME DMS100	1AESS AND SOME DMS100	1AESS AND SOME DMS100	1AESS (20%) DMS100 (80%)
USER INTERFACE	V.35; RS-449; RS-232-C USING FLEXTIE 56 FROM CALIF. MICRO	V.35; RS-232-C; RS-449 USING DATA LINE CARD AND DATA UNIT	V.35; RS-449; RS-232-C WITH TIE (AT&T)	V.35; RS-449; RS-232-C USING TIE OR FLEXTIE 56	V.35; RS-449; RS-232-C WITH TIE OR FLEXTIE 56	NO PLANS FOR 1986	PROBABLY THE SAME AS OTHER REGIONS
SERVICE NAME AND TARIFF	PSDS TARIFFED	SWITCHWAY (NOT TARIFFED)	PSDS TARIFFED	TARIFF FILED FOR MIAMI	PSDS TARIFF FILED FOR MINNEAPOLIS	PSDS NO PLANS	UNKNOWN
LOCATIONS (INTRA-LATA)	CHICAGO CHAMPAIGN FRANKLIN CLEVELAND COLUMBUS DETROIT (EST) INDIANAPOLIS (EST) MILWAUKEE SPRINGFIELD	NYC AND BOSTON TESTS	SAN FRANCISCO AND LOS ANGELES	ATLANTA* KNOXVILLE CHATTANOOGA MEMPHIS* NASHVILLE* MIAMI (ALL IN TEST)	MINNEAPOLIS DES MOINES* (EST) OMAHA* (EST)	TEST COMPLETED IN MURRAY HILL, NJ (NEAR BELL LABS), NEW BRUNSWICK, NEWARK	WICHITA (EST) TOPEKA (EST)
INTER-LATA SERVICE	TESTED WITH PACIFIC TELESIS AND NYNEX	TEST IN-HOUSE	YES	NO FILINGS	YES TARIFF 52	NO PLANS	UNKNOWN, NO NATIONAL OR INTERNATIONAL STANDARD
ATT ACCUNET COMPATIBLE	YES; WILL TEST		YES	PLANS FOR MIAMI	YES; LOOKING FOR ATT AS A SUBSCRIBER	NO PLANS	YES, WILL BE
OCC COMPATIBLE	YES, WILL BE	TESTING WITH AMSAT AND MCI/SBS	YES, BUT NOT TESTED	EQUAL ACCESS FOR ALL	WORKING WITH MCI/SBS AND AT&T	NO PLANS	YES, WILL BE
TYPE OF SWITCHING PROTOCOL	CSDC BELLCORE 2-WIRE SERVICE	NTI DATAPATH 2-WIRE SERVICE	CSDC BELLCORE 2-WIRE SERVICE	CSDC AND DATAPATH 2-WIRE SERVICE	CSDC BELLCORE 2-WIRE SERVICE	NO PLANS	CSDC OR NTI DATAPATH 2-WIRE SERVICE

CSDC = CIRCUIT-SWITCHED DIGITAL CAPABILITY
EST = ESTIMATED
LATA = LOCAL ACCESS AND TRANSPORT AREA
NTI = NORTHERN TELECOM INC.
OCC = OTHER COMMON CARRIER
PSDS = PUBLIC SWITCHED DIGITAL SERVICE
RBOC = REGIONAL BELL OPERATING COMPANY
TIE = TERMINAL INTERFACE EQUIPMENT
*DMS100 SITES

DMS-100 switch is already digital in operation and uses two-state signaling for Datapath interoffice trunking.

Both types of implementation—CSDC (PSDS) and Datapath—are, of course, composed of compatible communications elements from end to end, provided they are configured with their own equipment. Should the pathway contain equipment from the other type of implementation, there would be a problem in that Datapath uses its own link-level message protocol, T-Link, while CSDC lacks a specific end-to-end message protocol.

AT&T and the RBOCs are considering adding T-Link, or an equivalent minimum-complexity message-handshaking protocol, to their switched 56-kbit/s offerings, and they have discussed this with Bellcore. Due to the implementation schedule involved, this requirement was waived for interexchange testing, although vendors generally agree that such a protocol is desirable, considering such user benefits as speed calling and ring again.

1. The choices. There are three 56-kbit/s interfaces presented for the user—V.35, RS-449, and RS-232-C—and two services—Public Switched Digital Service and Datapath. The user interface is smaller than a traditional standalone modem. Its suppliers are California Microwave, AT&T, and Northern Telecom.

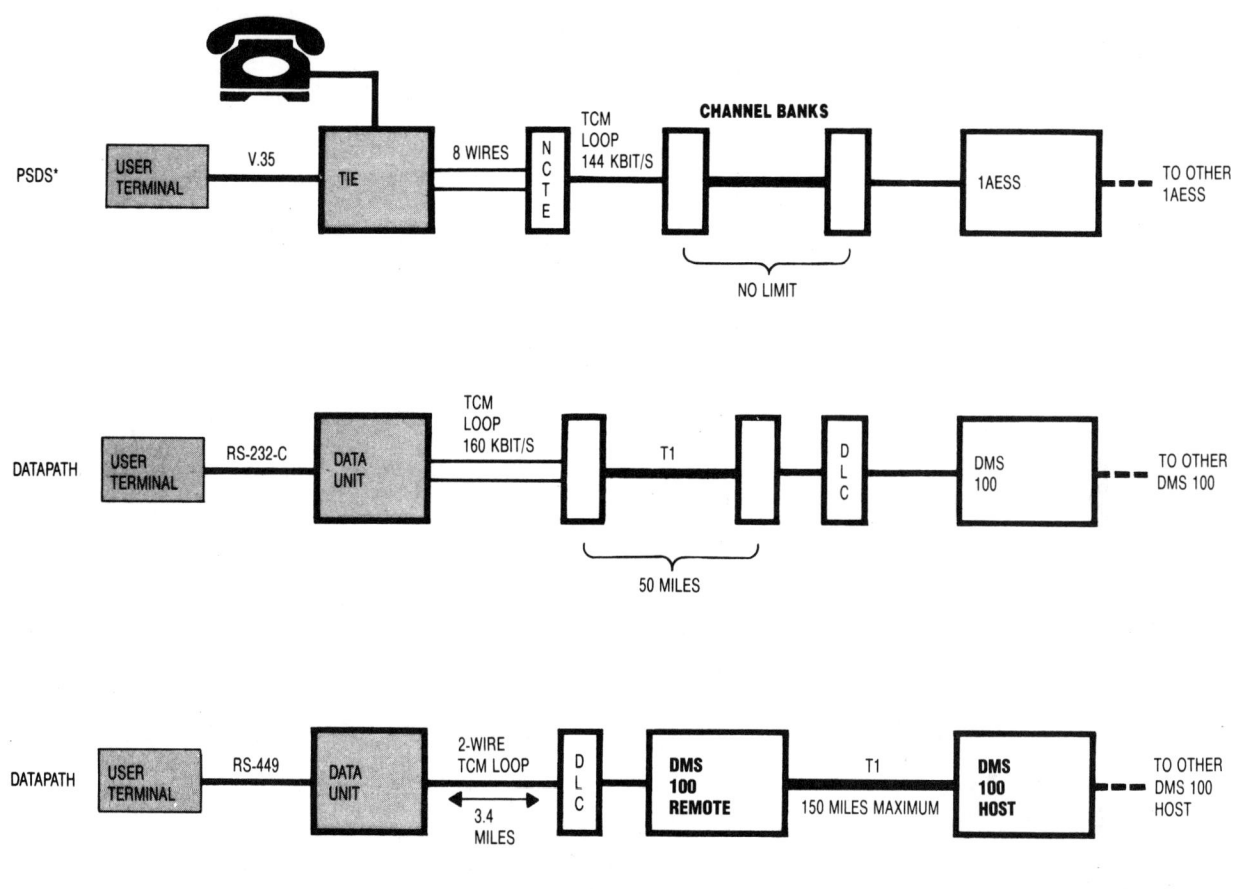

Datapath's T-Link is a byte-oriented end-to-end protocol designed to transfer either synchronous or asynchronous data over a 64-kbit/s digital circuit at user rates up to 56 kbit/s. There is a continuous flow of out-of-band control messages over the circuit. Therefore, once a network connection has been established, it must be dedicated for the duration of the end-to-end communications. T-Link treats the 64-kbit/s channel as if it consists of 8,000 eight-bit bytes per second.

Before the actual start of transmission, either end can turn off the clear-to-send signal at the remote end to prevent transmission to it. There is no explicit end-of-transmission indication in T-Link. Data will continue to flow end-to-end until the connection is broken or re-initialized or the user equipment is turned off. T-Link is a transparent protocol and does not modify any higher levels of protocol that may be in use.

Datapath's transparent FDHP (full-duplex handshaking protocol) is a byte-oriented envelope or wraparound protocol that provides error detection, error correction, and flow control. (Each byte contains both data and handshaking-control information. The three types of information in each byte are the transmit state, the receive state, and the data "nibble," which is four bits of data.) FDHP is a higher-level protocol that wraps around T-Link.

To provide a circuit-switching capability with a minimum number of changes to the existing switches in the public network—the 1AESS switch is an analog switch—CSDC call setup is done in the analog mode using DTMF signals. (Datapath is entirely digital. The CSDC data scheme is a hybrid of analog and digital.) End-to-end compatibility between CSDC and Datapath is possible if the T-Link protocol is deleted, thus making the call setup analog. This means that basic circuit switching can be accomplished, but user benefits such as speed calling and ring again are not then available.

Four- to two-state signaling

CSDC (Circuit-Switched Digital Capability) loop transmission between the user and the network is accomplished through four possible network states: idle, voice, data, and conflict. The idle state occurs when the channel is not in use. The voice state provides a 4-kHz voice channel between the communicating parties. The data state provides a full-duplex digital circuit—using time compression multiplexing, or TCM—between parties. The conflict state occurs when the two users of a connection are in dissimilar states.

■ **Idle state.** In the idle state, no call is in progress and the user's instrument is "on-hook." If the user is being called while the network interface is in the idle mode, the network applies a ringing voltage. This ringing voltage is the same as that used in the public switched network.

■ **Voice state.** Immediately following the successful completion of a call setup—via DTMF (dual-tone multifrequency) signals—the end-to-end network is in the voice state. This end-to-end voice connection must be established to ensure a successful entrance into the data state.

■ **Data state.** When both network interfaces are in the data mode, the user has a full-duplex digital circuit (using TCM) established, and the network can carry user data and user control bytes. The network remains in this state until either user changes mode.

■ **Conflict state.** When users are in dissimilar modes, the network is said to be in the conflict state. In this case, the network alerts each user to the mode of the other end. This state persists until either or both users change modes.

The CDU (CSDC data unit), on the network side of the Western Union 1AESS (Electronic Switching System), resolves this scheme of four states into one of two states that are acceptable to the DS1 format of T1 facilities. In the DS1 format, the A-bit (high-order bit) is either on or off, denoting an idle or active state, respectively. The idle state has the same meaning in both schemes. All other states are translated into two-state signaling as an active state. They are differentiated via the specific bit settings of the standard DS1 format for T1 trunk facilities.

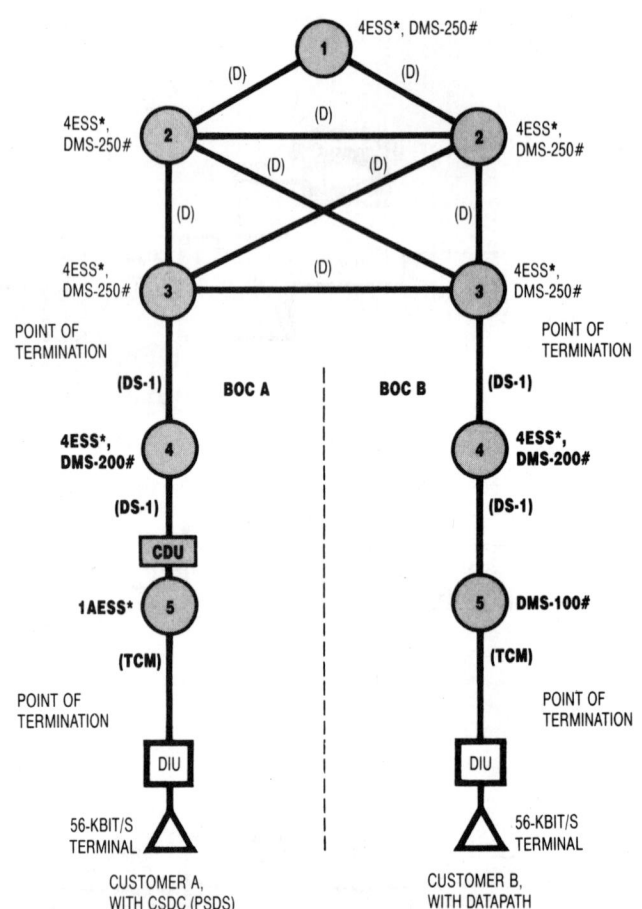

2. Two paths. For full local capabilities, a mix of Datapath and Public Switched Digital Service should be used only in the network's long-distance sections.

* = AT&T SWITCH
# = NORTHERN TELECOM SWITCH
BOC = (FORMER) BELL OPERATING COMPANY
CDU = CSDC DATA UNIT
CSDC = CIRCUIT-SWITCHED DIGITAL CAPABILITY
(D) = DIGITAL 64-KBIT/S CHANNEL
(DS-1) = T1 CHANNEL
DIU = DATA INTERFACE UNIT
PSDS = PUBLIC SWITCHED DIGITAL SERVICE
TCM = TIME COMPRESSION MULTIPLEXING

Figure 2 illustrates a public network's switched digital service using a mixture of Datapath and PSDS technologies. The mix works in the long-distance section (between points of termination). If full local capability is to be maintained, there can be no technology mix in the local areas—those leading to and from the long-distance sections. The upper section of the figure, which includes the switches cross-connected by 64-kbit/s channels, represents the lowest level of the connection hierarchy at a telephone company's central office.

Fitting 56-kbit/s digital technology into an analog telephone network has brought about certain problems and constraints on the local-loop level. The long-distance portion of switched 56-kbit/s service is well on its way to solving these problems with three or four strong contenders for the market and good coverage nationwide. The pricing is very attractive (Table 3) on both the local and long distance level.

Bellcore is doing a good job demonstrating user applications that require switched 56 kbit/s and continues to address the compatibility issues as well. Its role is consultative only, however; implementation is up to the RBOCs. Each RBOC has to struggle with its considerable investment in analog technology and select only those implementations that appear to have a guaranteed profit return. ■

Helga Eriksen, with over 18 years in data communications, operates her own consulting firm. She recently completed an M. S. degree in Special Studies of Telecommunications Policy at George Washington University.

Morris Neuman, Special to DATA COMMUNICATIONS

A new transport architecture for sluggish networks

One bank had to decide between upgrading existing hardware and replacing it with newer models. The planning group saw an alternative: Extended Slotted Ring Architecture.

The symptoms were unmistakable: The teleprocessing network could grow no further on existing hardware. Response time was very poor at critical peak periods; it was impossible to connect new advanced-protocol devices; control of the network was poor; and the software was a nightmare to modify. While it still functioned as intended, the network being used at Irving Trust Co. had become obsolete. As the bank's senior telecommunications analyst of Advanced Communication Planning in the Telecommunications Division, it was my job to present senior management with all possible alternatives, taking technology, function, and cost into account.

In the end, we turned to an innovative but relatively unknown transport architecture called Extended Slotted Ring Architecture (ESRA). But before ESRA emerged as the clear-cut answer, the planning group I headed engaged in some tough detective work to pinpoint not only the electronic options available, but also the bank's current and future networking needs.

The first decision was easy: upgrade or replace? Since investing in obsolete hardware that would again reach its limits in the near future made no sense, replacement was the only logical choice. For the operations group, replacement meant going to IBM Corp. to check out its latest and greatest gadgetry. For the planning group, it meant investigating the existing alternatives.

To begin, we needed to know the communications needs of the organization on two levels. The first was a simple question of data flow—how many bits of information were flowing from every user site in a network that served all of New York State? (Its major node centers were at Rochester, Syracuse, Albany, and Scarsdale.) To find out, we tabulated by the day and by the week a random week's worth of hourly traffic going in and out of every link at every node (Fig. 1). This helped identify traffic density from all geographical locations. It was important to know byte, rather than block or packet, counts since the actual size of the latter is not constant across all environments.

The second level of traffic analysis concerned the business-information flow, and this data was much more difficult to obtain. To gather it meant getting information from the many diverse business sectors of the bank to determine what types of data were being communicated and their requisite traffic paths within the organization. But the effort told us whether the existing data flow was meeting the needs of the users and helped us estimate future data-flow growth areas by means of projected business-growth figures.

Next we documented the current environment's hardware, software, and functionality per application. The old network supported remote 327X terminal access to the IMS database, which ran on hosts located in New York City, and digitized voice-response queries for account status and balances, which used the same IMS database.

The third step involved documenting as realistically as possible the future desired position of the bank, detailing expected growth, future applications, and network capabilities. Finally, we compiled a list of specific technical issues to be addressed, including the capacity and performance of networking products, the protocols they supported, expandability, line requirements, control and maintenance, software support, SNA/SDLC (Systems Network Architecture/synchronous data link control) support, maintenance throughout New York State, pricing, and availability. We combined all this material in a detailed request for information package, which we submitted to the top 20 telecommunications vendors (see table).

The vendors now knew exactly what we were looking for. To support future business growth and help the bank remain competitive, our networks would have to be able to:
 1. Grow in a modular fashion.
 2. Interconnect with existing standard public networks.

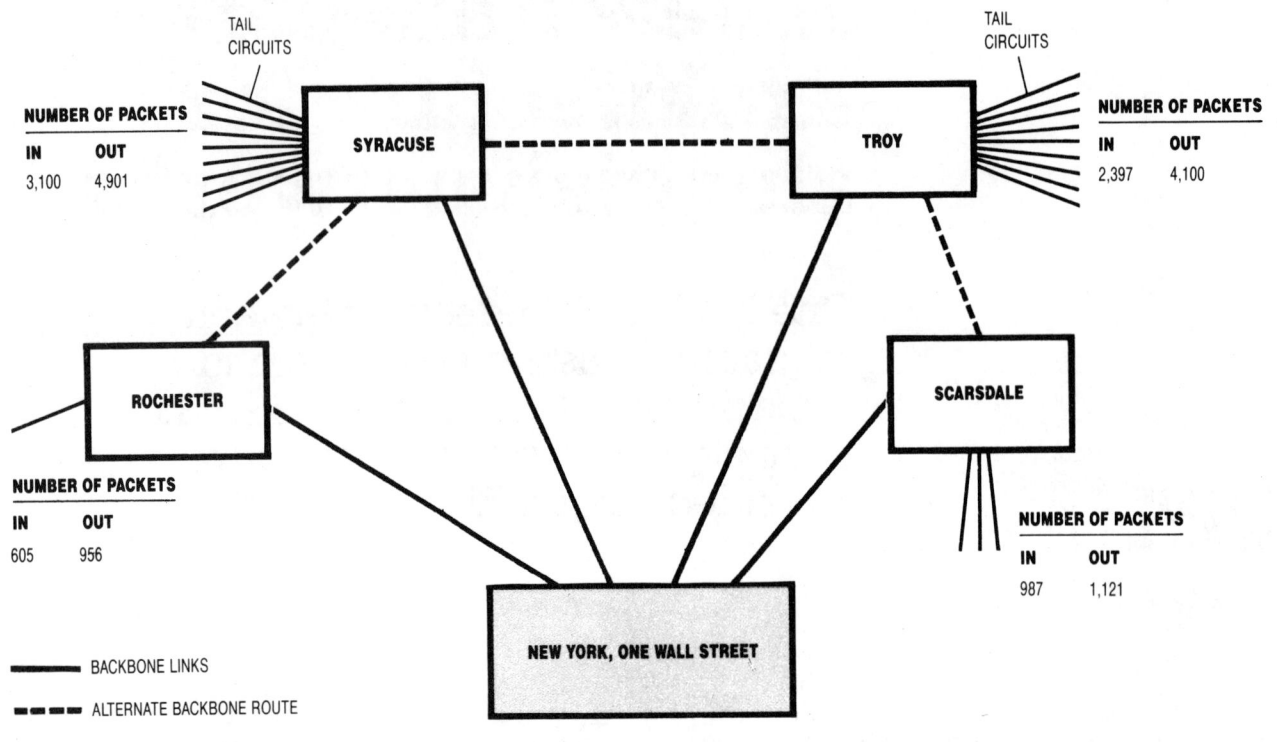

1. A day's traffic volume. *A topological map of the bank's statewide packet network indicates the flow of data traffic for a single day. Traffic from tail circuits was concentrated at nodes. It then took the shortest route to New York City. The alternate backbone route was used only in the event of a link failure.*

3. Support multiple protocols, present and future.
4. Support multiple vendor hardware.
5. Exploit technological advances as they occurred.
6. Be easy to maintain and control.
7. Minimize communications costs.

In sum, we needed the ideal network—a data-transport mechanism that was independent of what was connected to it and what kind of data was sent on it, one that could grow without effort and always maintain complete control. A simple request from a very progressive industry.

The possibilities were vast: There were SNA networks and X.25 networks, packet networks and concentration networks, proprietary networks and public networks. Of the 20 vendors we queried, 10 responded. Of those vendors who responded, five presented acceptable networking strategies for this situation: BBN Communications, Codex, Doelz, Infotron, and Paradyne. One of them—Doelz Networks Inc., Irvine, Calif., a company heretofore unknown to me—proposed a truly innovative solution.

The extended slotted ring

New networking architectural schemes arise infrequently. When they do, they must offer new capabilities or satisfy a pending need in order to deserve attention. Doelz's Extended Slotted Ring Architecture (ESRA), which the company calls a fourth-generation networking architecture, does both. Here are ESRA's capabilities:

1. Mixed independent protocol support.
2. Media-independent.
3. Universal architecture for local area, wide area, and distributed networks.
4. Optimal network performance under all traffic conditions.
5. Best cost performance.

ESRA is one of the newest solutions addressing the need for a universal network architecture for future ISDNs (Integrated Services Digital Networks) and providing the capability for mixed independent protocol support. By decentralizing contention management of its primary resource—the high-speed backbone links that connect any network nodes—it's adaptable for local area, wide area, and dedicated distribution networks. Most important, it boasts optimal utilization of the primary data-transport mechanism, allowing for greater throughput and minimal time delay.

Most networking architectures are tailored to satisfy the operational requirements of one of the network types but not the others. For example, in local area networks the backbone link is shared by all user devices in a centrally managed manner, as in CSMA/CD (carrier-sense multiple access with collision detection) and token passing. This satisfies the need to dynamically configure and expand in a simple and very flexible manner when, for example, people move from desk to desk or during floor-plan revisions. The trade-off for this flexibility is an underutilized primary resource. Wide area networks have dedicated backbone links with high utilization but are not suitable for dynamic growth and reconfiguration. Distribution networks serve a many-to-one communications need in a data-concentration fashion. But they're highly inflexible and do not support mixed independent protocols.

If networks of the future are to act as ISDNs—that is,

Acceptable networking strategies

BOLT BERANEK AND NEWMAN, BBN COMMUNICATIONS

TRANSPORT ARCHITECTURE	**X.25 AND PROPRIETARY HDLC FEATURING FULLY DISTRIBUTED AUTOMATIC DYNAMIC ADAPTIVE ROUTING AND LOAD SHARING.**
HARDWARE IMPLEMENTATION	FULLY DISTRIBUTED MICROCOMPUTER NODES WITH PROPRIETARY HARDWARE USING ECL/TTL GATE ARRAY TECHNOLOGY (BBN COMM C/30 C/300).
NETWORK CONTROL SCHEME	DISTRIBUTED NETWORK OPERATION IN THE COMPONENTS AND A PASSIVE NETWORK OPERATIONS SYSTEM (ISOLATION OR FAILURE OF SYSTEM DOES NOT IMPACT NETWORK PERFORMANCE) BASED ON BBN COMM C/70 NETWORK OPERATIONS CENTER, RUNNING PROPRIETARY NETWORK UTILITY SOFTWARE UNDER UNIX.
NETWORK CONTROL OPTION	$200,000 PER NETWORK, INCLUDING COMPUTER EQUIPMENT.
MIXED PROTOCOL SUPPORT	ASCII, 3270 BSC, 3780/2780 BSC, SNA/SDLC, X.25, X.75.
DYNAMIC GROWTH AND RECONFIGURATION	NETWORK AUTOMATICALLY DETECTS AND "FOLDS" IN NEW COMPONENTS, LINES, AND TOPOLOGIES WITHOUT AFFECTING OPERATIONAL COMPONENTS. CONFIGURATION CAN BE PERFORMED AUTOMATICALLY OR UNDER OPERATOR CONTROL.
COST PER PORT	ACCESS $400 PER PORT, SWITCHING $2,000 PER PORT.

CODEX

TRANSPORT ARCHITECTURE	**CDLSC, PROPRIETARY SDLC PROTOCOL.**
HARDWARE IMPLEMENTATION	MOTOROLA 6800 BASED. 1 PROCESSOR PER NETWORK PORT. 6809 PROCESSOR PER TERMINAL PORT CARD.
NETWORK CONTROL SCHEME	INTEGRATED INTO NETWORK HARDWARE ACCESSED BY DUMB TERMINAL, CONTROLS OVER 100 NODES.
NETWORK CONTROL OPTION	$1,200 FOR DUMB TERMINAL.
MIXED PROTOCOL SUPPORT	ASYNCHRONOUS, BISYNCHRONOUS, HDLC, X.25 LEVEL 3, BATCH AND INTERACTIVE.
DYNAMIC GROWTH AND RECONFIGURATION	YES, MENU DRIVEN FROM CONTROL TERMINAL.
COST PER PORT* (100-PORT, 6-NODE NETWORK)	$1,400 PER TERMINAL.

DOELZ

TRANSPORT ARCHITECTURE	**EXTENDED SLOTTED RING.**
HARDWARE IMPLEMENTATION	HIGHLY DISTRIBUTED MOTOROLA 6800, PER NODE, PER LINE, PER BACKBONE LINK.
NETWORK CONTROL SCHEME	INTEGRATED IN NETWORK FIRMWARE, OPTIONAL EXTERNAL UNIX-BASED WITH COLOR GRAPHICS, HARD DISK, TAPE, AND PRINTER.
NETWORK CONTROL OPTION	$1,000 FOR DUMB TERMINAL OR $75,000 FOR ADDITIONAL HARDWARE FOR LARGE NETWORK.
MIXED PROTOCOL SUPPORT	ASYNCHRONOUS, BISYNCHRONOUS, BITSYNCHRONOUS, BATCH, INTERACTIVE.
DYNAMIC GROWTH AND RECONFIGURATION	YES. ONLINE AND TABLE DRIVEN.
COST PER PORT	VARIABLE BASED ON NETWORK CONFIGURATION.

INFOTRON

TRANSPORT ARCHITECTURE	**X.25 WITH PROPRIETARY HDLC WITH DYNAMIC ADAPTIVE ROUTING AND LOAD SHARING.**
HARDWARE IMPLEMENTATION	LOOSELY COUPLED BUS ARCHITECTURE. MICROPROCESSOR-BASED MULTINODE DESIGN.
NETWORK CONTROL SCHEME	MULTILEVEL NODAL AND CENTRAL SITE CONTROL, FROM CONSOLE OR OPTIONAL GRAPHICS SOFTWARE ON AN IBM PC AT.
NETWORK CONTROL OPTION	$6,000 FOR OPTIONAL IBM PC AT.
MIXED PROTOCOL SUPPORT	ALL ASYNCHRONOUS, BISYNCHRONOUS, BITSYNCHRONOUS, INTERACTIVE, AND BATCH. 99 PERCENT OF ALL EXISTING PROTOCOLS.
DYNAMIC GROWTH AND RECONFIGURATION	YES. ONLINE CONSOLE-SUPPORTED RECONFIGURATION, UP TO 32 NODES.
COST PER PORT (100-PORT, 6-NODE NETWORK)	$500-$1,500 PER PORT.

PARADYNE

TRANSPORT ARCHITECTURE	**X.25 WITH ENHANCED LINK-LEVEL HDLC. SUPPORT FOR MULTIPLE PHYSICAL LINES PER TRUNK WITH LOAD BALANCING.**
HARDWARE IMPLEMENTATION	MULTIPLE DISTRIBUTED MODULAR MICROPROCESSOR ARCHITECTURE. UP TO 40 MODULES PER SWITCH WITH UP TO 15 MICROCOMPUTERS PER MODULE.
NETWORK CONTROL SCHEME	DISTRIBUTED MANAGEMENT SCHEME WITH MINICOMPUTERS PERFORMING DATA COLLECTION, ANALYSIS, CONTROL, AND MONITORING.
NETWORK CONTROL OPTION	$100,000, INCLUDING COMPUTER EQUIPMENT.
MIXED PROTOCOL SUPPORT	ASYNCHRONOUS, BISYNCHRONOUS 3270 2780 3780, SDLC/SNA, BATCH, INTERACTIVE, X.32, X.25, VIDEOTEX.
DYNAMIC GROWTH AND RECONFIGURATION	YES. WITHOUT AFFECTING ANY NETWORK OPERATIONS.
COST PER PORT (100-PORT, 6-NODE NETWORK)	$900-$1,600 PER PORT.

*AMOUNTS SHOWN ARE NOT FOR COMPARABLE UNITS, AS EACH COMPANY CALCULATES COST BASED ON DIFFERENT PARAMETERS.

Contention management schemes

The primary resource of any data network is the transmission bandwidth of its backbone links. Backbone links are usually high-speed transmission media with associated hardware that interconnect a network's nodes. In a local area network, user devices are connected to the backbone through network interface devices, which usually transmit or modulate the data into a format suitable for high-speed media (coaxial cable, fiber optics, or twisted copper). One or more users may be served by a network interface device, depending on the manufacturer's design. In wide area networks, backbone links connect the various nodes, and user devices connected to the nodes perform routing and flow control for the network.

But in all types of networks, devices contend for bandwidth, and which ones will get it when is settled by the transport layer's contention resolution management. As with any management scheme, many different approaches have evolved. In local area networks, CSMA/CD (carrier-sense multiple access with collision detection) and token passing have been widely accepted and are part of the IEEE 802 standard series. In a wide area packet-switching network, contention is resolved with priority assignments and fairness algorithms—first come, first served on a packet basis.

In the slotted TDMA (time-division multiple access) approach, total transmission time over a broadcast medium is divided into smaller periods when individual users have permission to transmit data. The periods are called slots and are usually of fixed and equal lengths. The number of slots in a cycle depends directly on the speed of the network and the number of potential devices. The slots are then numbered serially, wrapping around when the end of a sequence is reached to start the cycle over again.

Devices can transmit only during a time slot that has been assigned to them, which means messages may have to be broken up for transmission as devices wait for their slot to come around again. Priority devices may be assigned more than one slot per cycle, if the user wishes. And a new architectural scheme, "Unilink," by Applitek, can assign more than one user to a slot, letting them contend in a CSMA/CD fashion to transmit. This hybridization of CSMA and TDMA allows a network to be tuned to specific traffic conditions.

The origin of the slots is a scheme called "Slotted Aloha," which was originally used on the islands of Hawaii to manage access to a radio network. Aloha is a pure random multiple-access contention scheme, designed to avoid the collisions that occurred as users vied for radio time. Originally, anyone who wanted to broadcast did so, but if someone else started transmitting at the same time, both were forced to stop. This required users to listen to existing broadcasts and wait until they were finished. Finally, dividing the access time into slots and allowing transmission to start only at slot boundaries was found to further improve collision avoidance.

transport binary data from any source to any destination in the network—in an unrestricted and transparent fashion, they will have to support independent protocols of different vendor hardware and grow in any shape and size while performing at an optimal level of throughput and time delay. ESRA satisfies these needs by means of its decentralized backbone-contention management, topology-independent ring-path configuration, very small packet size, and single user per packet features.

Developed by Doelz and the basis of the company's networking products, ESRA is described in a paper titled "Extended Slotted Ring Architecture for a Fully Shared and Integrated Communication Network," written by M. L. Doelz & R. L. Sharma. But the name "Extended Slotted Ring Architecture" is something of a misnomer, since ESRA doesn't use the centralized control of typical slotted-ring networks.

The term "slotted ring" refers to a local area network with the topology of a ring or loop, which employs a slotted time-division multiple access (TDMA) backbone contention management scheme (see "Contention management schemes"). In such an arrangement, the total transmission time over a broadcast medium is divided into smaller periods, during which individual users of the network have permission to transmit data. Synchronization is accomplished by a master control device, which transmits very short timing messages, and by requiring each network device to time its slot lengths precisely. The allocation of users to slots is also managed by the central control device, usually a master node in the network.

It is the idea of centralized control that is uprooted in the ESRA scheme, which allows for a truly distributed or decentralized network management. This level of distributed control has been achieved in the public switched (telephone) network, which has evolved over the last 90 years. A call being processed in one central office has no interdependence with a call to any other central office in the network. Yet any two users of the network can connect to each other. If the more than 80 million telephones in the United States had to be managed by one central computer or network, the calls could not be processed in a timely fashion.

The "ring" in slotted ring implies a topology in which devices are connected to each other in a daisy-chain with the first one connected to the last node, forming a physical loop or ring. The topology of a local area network configured in a loop imposes restrictions on the transmission of data. In a serial transmission loop, nodes receive data from the previous node and transmit it to the next node in the loop; the communication travels one-way around the loop. Data must travel through every node on the loop for a node to receive data from the next node or to transmit to the previous node in the network. Transmission can be in either a baseband or a broadband mode.

Token passing is an example of a contention management scheme in a serial loop network. In a broadcast transmission loop, users simply contend for the bandwidth, and it may be managed with either a CSMA/CD, TDMA (time-division multiple access), or FDMA (frequency-division multiple access) scheme. The loop configuration has no restrictions on the data path; however, only broadband transmission can be used.

ESRA is a packet-oriented transport mechanism employing a priority TDMA access scheme. It addresses network issues at the transport layer, offering end-to-end transfer services to the session and presentation layers of user devices. It is designed to be independent of topologies and transmission media. The name ESRA is derived from the idea of regular slotted rings. But ESRA extends the slotted TDMA contention management scheme in a distributed, or decentralized, fashion to a packet-switching environment: hence, the name Extended Slotted Ring Architecture.

With ESRA each user device is connected to a network interface device or node. Also entering each node is a link from the previous and next nodes. Each node must then manage the flow of incoming traffic, from both locally connected devices and "on-network traffic" from the previous node. The slotted TDMA scheme is applied in a decentralized or local government fashion at each node. Each node is required to assign time slots only on the out-bound link, to all the traffic coming into the node (from all sources), and for the purpose of getting it out. There is no need for synchronization throughout the network since each node is responsible for synchronizing only the data streams local to it. Queuing is implemented at both the local and network levels to smooth out data flows, allowing nodes to act independently. In this way, ESRA achieves a totaly decentralized contention management scheme allowing independent control of each high-speed backbone link in optimal manner.

The basic topology of ESRA is a dual-directional serial loop. The extended feature of having the serial loop operate in both directions eliminates the problem that serial loops pose while adopting their positive aspect of being independent of media and transmission mode (broadband or baseband). A daisy-chain approach is still used, but the last node does not have to connect to the first one. Instead, it loops back to the previous node (Fig. 2). Closing the loop creates a redundant data path that can be deployed in the event of a single link failure. In the hardware implementation of the architecture, Doelz allows for the loop to close with a dial-up connection. This offers the network contingency capabilities without wasting transmission resources. The daisy chain is simply made up of high-speed full-duplex links of any media type. In conjunction with a bridging node (which forms a sort of "T" in the loop), the user can therefore build networks of any topology. Being media- and topology-independent is a highly desirable feature for any network scheme.

ESRA packet switching

On the data link control level, small fixed-length packets are the basic transmission component. All packets are fixed to be either 12 or 24 bytes long. The decision to make packets of a fixed length and a small size was fundamental to the ESRA concept. This is one of the parameters that was fixed at the design stage to be constant. It simplifies the architecture and allows for a smooth flow of data through the network. When packets are small compared with the link transmission speed, there is a lower probability of packet error since there are fewer bits in the packet.

The probability of error in a packet is simply the bit error rate of the transmission media times the number of bits in a packet. Small packet size also results in short service

2. One case of ESRA. *This configuration of ESRA shows a tree topology made by using a bridge node, which allows ESRA to deviate from a pure ring topology.*

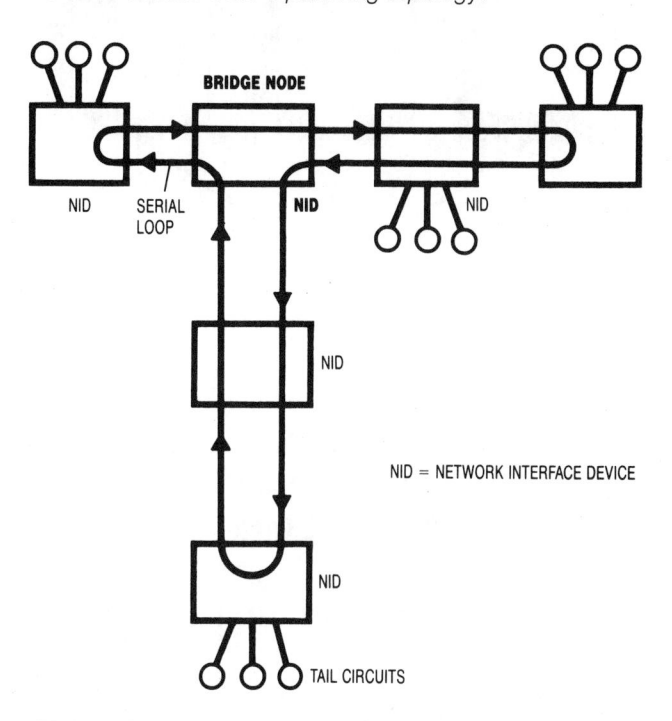

time at each node, allowing for good queuing performance. The only drawback of small packet size is that more packets have to be processed. ESRA overcomes this drawback by minimizing packet overhead to two or three bytes for most packets and up to five for others and a single user per packet. Instead of unpacking and processing a packet at each node the packet visits, the first few bytes of each packet determine if it is destined for this node or not. Transient packets are simply left untouched and are assigned a slot to be transmitted on the next node. Packets that are destined for this node are taken off the network and delivered to the locally attached device.

The other fundamental aspect of the packets in ESRA is that they contain data from only one source or user device. This is known as "single user per packet." Many other packet-switching schemes will perform blocking to achieve a multiplexing affect (Fig. 3). Blocking or grouping of many sources of data into larger blocks of data for transmission in a large packet has a negative consequence in that the packet must be dissected at each node traversed. Dissection is necessary to determine if any data is destined for the node's user devices, in which case it would be stripped off and delivered to the desired destination. This packet processing overhead reduces packet throughput and increases propogation delay through the network. Single user per packet eliminates much of the overhead of packet processing in ESRA.

Finally, error control at the data link level is eliminated and delegated to the higher-level transport layer in ESRA. In regular packet-switching schemes, error checking also introduces throughput delays at the data link level, since a packet is checked for transmission errors every time it traverses a link in the network. If an error is detected, the

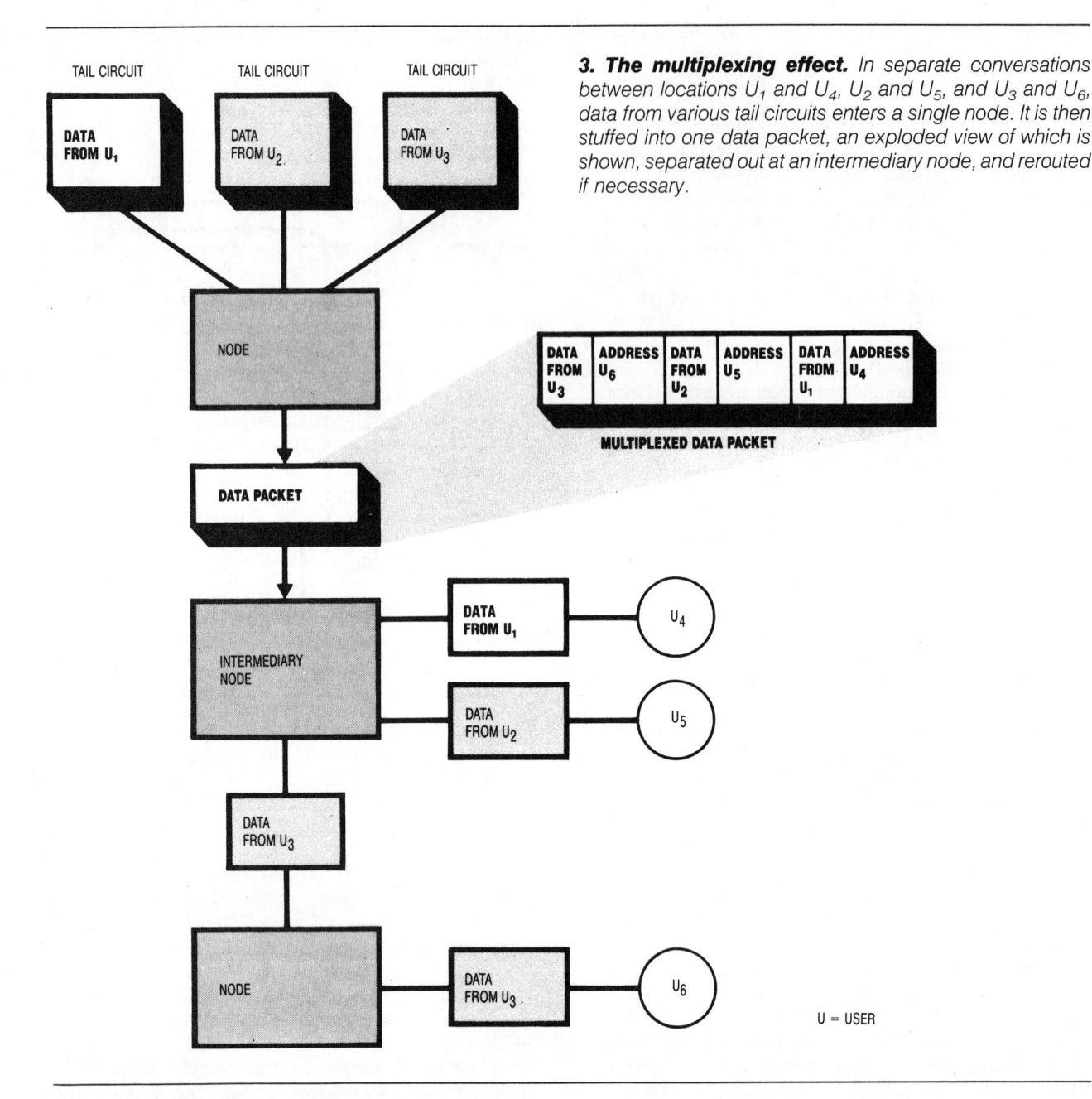

3. The multiplexing effect. In separate conversations between locations U_1 and U_4, U_2 and U_5, and U_3 and U_6, data from various tail circuits enters a single node. It is then stuffed into one data packet, an exploded view of which is shown, separated out at an intermediary node, and rerouted if necessary.

U = USER

packet is retransmitted from the last node that received it correctly. For a packet to travel through the network without error, it is checked at each intermediary node and ACKed or NACKed, as the protocol calls for. In ESRA, error checking is done on an end-to-end basis. Only when the final destination node receives the data is it checked for errors. If an error exists, the packet must be retransmitted from the source node. In a very noisy transmission environment, data link error control will enable packets to eventually get through the network. However, in the typical transmission environment, with bit error rates in the range of 1 in 10^6 or better for conditioned telephone circuits, ESRA will perform better since the packets are small in size (reducing the probability of error). Only about one in 10,000 packets will be in error. Error control processing can, therefore, be eliminated at all intermediary nodes, increasing throughput. ESRA also allows for end-to-end error checking to be disabled entirely, letting the application program be responsible for error detection if this is desirable in the situation. Applications like digital voice can thus ride on the network. Errors in voice packets are tolerable since they affect the resulting speech in a minor way.

Network control

ESRA supports switched virtual circuits, as do most packet-switching schemes. However, path routing is established at call-setup time and is not dynamic during a session as it is in some other packet-switching architectures. Dynamic routing is the ability for a node to select a different path for packets in the same session. The paths are usually based on traffic loading and are chosen by virtue of the load level of the traffic over the different links to a destination. Static routing sets up a fixed path, which all packets of the same session take to get to their destination. Static routing during sessions further eliminates packet delay at each node since no processing time is wasted in figuring out which route each packet that is just passing through an intermediary node should take. A simple and fast table look-up is performed, thus minimizing packet delay. A secondary alternate route is supported in the event of a primary link failure, enhancing end-to-end connectivity. Both permanent and switched virtual circuits are supported. Switched virtual circuits are established by a user device requesting a connection to another device by simply dialing the destination address. Permanent circuits are logical paths that are established once and can be used at any time without dialing in an address. This is analogous to switched telephone connections and private leased lines.

Because ESRA is media-independent, any transmission medium can be employed—copper, fiber, coaxial cable, microwave, or satellite. This independence is achieved by eliminating the use of a broadcast environment, which is usually employed in local-area slotted rings, and instead building the network up from simple point-to-point connections (actually full-duplex daisy chains). Real media independence is actually a function of the hardware that implements the ESRA. Manufacturers must decide to support various media interfaces for their products. ESRA, itself, imposes no logical media restrictions, as CSMA/CD does by favoring a broadcast media environment, which inhibits the use of fiber optics.

At the highest and most visible layer of ESRA is the transport layer, with which user devices must interface. It must manage address mapping, end-to-end network flow control, and transmission contention resolution.

Addressing in ESRA conforms to the CCITT X.121 International Data Number standard. An address is a 14-digit number made up of a DNIC (Data Network ID Code) and a network terminal number, which, in turn, is composed of an area code and a port address. This allows addressing down to the port level in a data network. Addresses are assigned to user devices by the user at network configuration time. This facilitates speedy call setup times along with static routing at the network level.

Flow control in ESRA is accomplished at the network, and at the user device level incorporating the transmission priority scheme. On the network side, when buffer conditions warrant flow restraint, user data having the lowest priority is affected first. This allows higher-priority data to flow unrestricted. Regular user device flow control, EIA, and in-band X-on/X-off is supported.

Transmission contention resolution is probably the most important function of the transport layer. In ESRA, contention can arise in two ways. First, there is contention among the user devices that are connected to the same node processor to get out on the backbone. In this scenario, user devices with the highest priority go first. If user devices with the same priority are contending, then they are handled first come, first served on a packet basis. The second area of contention is from packets that are just passing through a node, contending with local packets to be transmitted to the next node in the link. Priority again wins out with first come, first served treatment resolving contention among equal priority packets.

More important than the specifics of an architecture are the benefits ESRA offers to the user. Its most important feature is openness, which in this case is the ability to adapt to most types of user-device protocols. Due to the design of the data-link network and transport layers, ESRA can support both synchronous and asynchronous protocols. The concept of a single user per packet also helps, since it ensures that protocols with different timing and synchronization requirements are not bundled together. At the same time, overall network performance, as measured by throughput and packet-time delay, improves because of several factors. Among them are:
- Small fixed packet size.
- End-to-end versus per-link basis error control.
- Static routing.
- Priority contention management.

Finally, media and topology independence allow ESRA to be implemented for almost any type of network. By fixing various network parameters (that is, packet size, static routing, end-to-end error control, single user per packet, and slotted TDMA), ESRA allows a network to go about its business of transmitting data over the associated links and nodes. ESRA does not get involved with any managerial functions that can be better handled at a higher level in the protocol. It will increase network performance while offering increased functionality and openness.

While most networks are designed to operate under specific traffic loads, ESRA will run optimally under all traffic-load conditions. By way of comparison, CSMA/CD is very good under light to moderate traffic loading, where there is a utilization of up to 40 percent of the transmission bandwidth; but under heavy loads above 40 percent, a CSMA/CD network will cease to operate. Token-passing networks work just fine under heavy loads—40 percent to 80 percent utilization—but when traffic is light, they introduce unnecessary delays.

In an ESRA, user devices see a continuous level of service under all traffic conditions. Statistical time-division multiplexing cannot support the mix of protocols, and FDMA has media restrictions. The priority aspect of ESRA serves many functions, including allowing for the mix of both batch and interactive data transmission without degradation to the interactive traffic.

ESRA is not a revolutionary concept but an evolutionary one. Like other fourth-generation transport architectures, it's basically an enhancement of third- and second-generation schemes. Some such systems are hybrids of two or more schemes; others offer new ideas and new approaches. It is the user's responsibility to select an appropriate scheme to serve his needs until a standard or universal architecture is adopted or accepted by default. Unless users make a commitment to investigate their options, they will be dictated to by large vendors with dominant market shares. ■

Morris Neuman is an independent computer speech specialist. Formerly a research analyst in telecommunications applications with Chase Manhattan Bank and Irving Trust Co., he has an M. S. E. E. in data communications and a B. S. E. E. in digital electronics from Columbia University, and a B. A. in computer science from Queens College. He lectures on and teaches data/voice communications at Columbia University's Division of Special Programs.

Donald H. Czubek, Communications Solutions Inc., San Jose, Calif.

Understanding IBM's electronic mail architectures

DIA and SNADS both perform application-level electronic mail functions. This guide for the perplexed differentiates between the two methodologies.

IBM markets a wide range of office automation products capable of supporting electronic mail users. These products are implemented on three different levels (Fig. 1). On the mainframe is the software cornerstone of IBM's office strategy, the Distributed Office Support System (DISOSS). At the departmental processor level are the Personal Services products for System/36 and System/38, as well as the older 5520 Administrative System and the 8100's DOSF (Distributed Office Support Facility) software. At the single-user workstation level are the Personal Services/PC package for the IBM PC, the 6580 Displaywriter, and the 8815 Scanmaster.

In an effort to ensure that all of its office electronic mail products are compatible, IBM has defined standard architectures at both the data transport and application levels. The structured data transport protocols are delineated by Systems Network Architecture (SNA). Most of IBM's office products support SNA's new Advanced Program-to-Program Communications, but some use the older Logical Unit Type 2 support defined for 3270s and asynchronous ASCII communications.

At the applications level, all IBM office products support Document Interchange Architecture (DIA) or SNA Distribution Services (SNADS), or both. There is considerable confusion about why two different application-level architectures are used to perform a single function and about which network components use SNADS and which use DIA. This article will attempt to show how these architectures work together to perform electronic mail functions.

The role of Document Interchange Architecture

The IBM office network is made up of products that are either requesters or providers of services. These services include document distribution (electronic mail), document libraries, application processing, and file transfers. Some products, particularly departmental processors, are hybrid devices that act as both requesters and servers. The requesters act on behalf of network users to request such services as electronic mail. The requested functions are performed by the service providers (see table).

DIA defines the interaction between a user of an office network and the components of the network that actually supply the network services. In DIA terms, the user makes requests for office services such as electronic mail and library access through a workstation called a Source/Recipient Node (SRN).

The components of a DIA network that actually perform the requested office functions are called Office System Nodes (OSNs). The most important example of an OSN is IBM's mainframe-based DISOSS package. Departmental processors, such as System/36, System/38, 5520, and 8100, have OSN capability either available or announced by IBM. The role of DIA is to standardize the way SRNs interact with OSNs to request office services.

While Figure 1 shows the architecture of document distribution, Figure 2 provides a simple example of document distribution showing the actual products in an office network. User A wants to distribute a document from a Displaywriter to User B, who is also a Displaywriter user. Both Displaywriters are connected to the same DISOSS.

User A will first initiate a DIA SIGN-ON REQUEST, which will cause a DIA session, or logical connection, to be established between the Displaywriter (a DIA SRN) and DISOSS (a DIA OSN). Note that the DIA session is just between User A's Displaywriter and DISOSS, and is not an end-to-end connection between the two Displaywriters.

At this point, User A has made a request for distribution, and the document to be distributed resides in DISOSS. DISOSS will respond to the request for distribution by putting the document into an output queue that is associated with User B.

User B will issue a DIA SIGN-ON REQUEST to begin a

1. Office architectures. *Architectural relationships of IBM electronic mail products, including mainframes, minicomputers, and microcomputers, are shown.*

IBM MAINFRAME

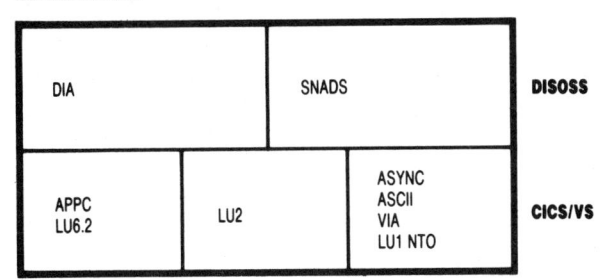

DIA	SNADS		**DISOSS**
APPC LU6.2	LU2	ASYNC ASCII VIA LU1 NTO	**CICS/VS**

SYSTEM/36

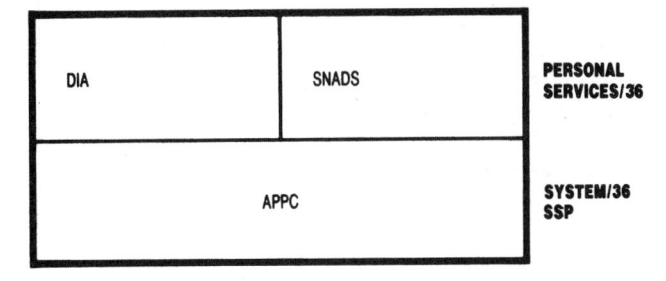

DIA	SNADS	**PERSONAL SERVICES/36**
APPC		**SYSTEM/36 SSP**

IBM PERSONAL COMPUTER

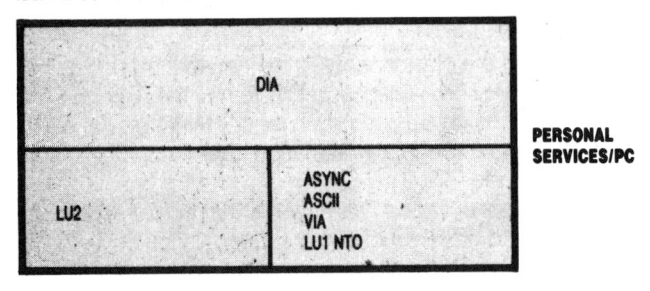

DIA		**PERSONAL SERVICES/PC**
LU2	ASYNC ASCII VIA LU1 NTO	

APPC = ADVANCED PROGRAM-TO-PROGRAM COMMUNICATIONS
CICS/VS = CUSTOMER INFORMATION CONTROL SYSTEM/VIRTUAL STORAGE
DIA = DOCUMENT INTERCHANGE ARCHITECTURE
DISOSS = DISTRIBUTED OFFICE SUPPORT SYSTEM
LU = LOGICAL UNIT
NTO = NETWORK TERMINAL OPTION
SNADS = SYSTEM NETWORK ARCHITECTURE DISTRIBUTION SERVICES
SSP = SYSTEM SUPPORT PROGRAM

2. Office network. *When only one Office System Node is used, DIA is used to handle document transactions. When two or more are involved, SNADS must come into play.*

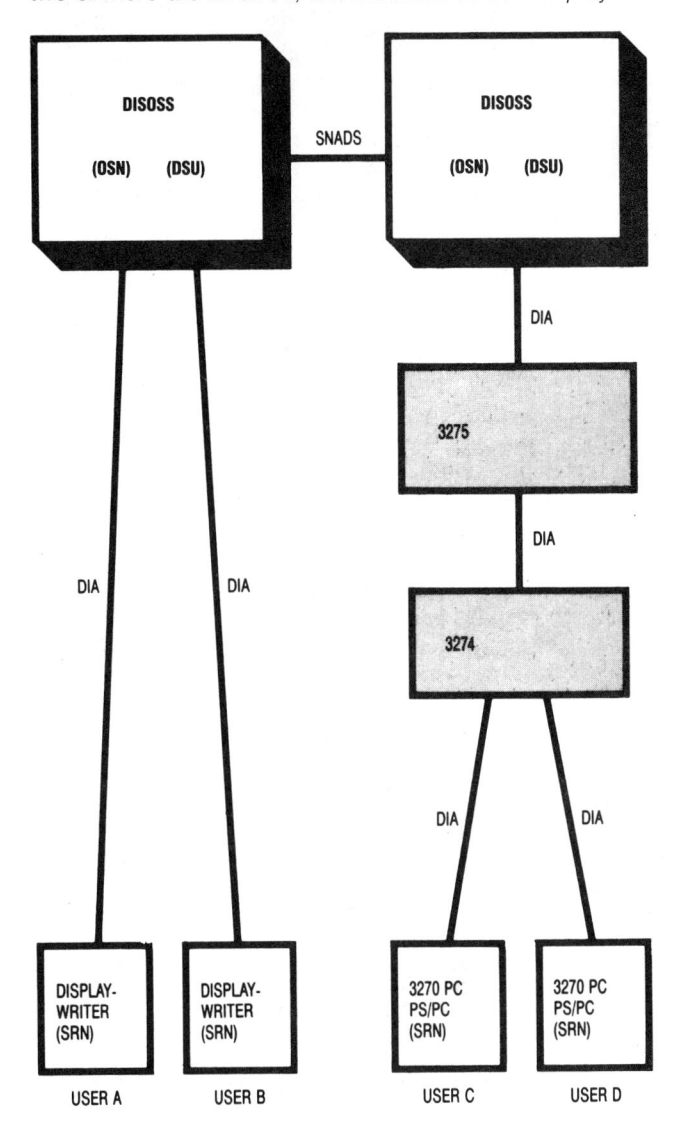

DIA = DOCUMENT INTERCHANGE ARCHITECTURE
DISOSS = DISTRIBUTED OFFICE SUPPORT SYSTEM
DSU = DISTRIBUTION SERVICE UNITS
OSN = OFFICE SYSTEM NODES
PS/PC = PERSONAL SERVICES/PERSONAL COMPUTER
SNADS = SYSTEM NETWORK ARCHITECTURE DISTRIBUTION SERVICES
SRN = SOURCE/RECIPIENT NODE

DIA session with DISOSS. User B can request a list of documents in the DISOSS output queue, which is, in effect, B's mail box. User B can then retrieve the document that was previously distributed by User A.

DIA, by itself, is capable only of performing document distributions that involve only a single OSN and the SRNs directly attached to it. To perform document distributions in more complex office networks with more than one OSN, SNADS must also be used.

SNADS provides generalized delayed delivery. It defines a store-and-forward network made up of components called Distribution Service Units (DSUs). As the data to be delivered moves through the SNADS network, it may be

Requests and products

REQUESTORS	REQUESTOR/SERVERS	SERVERS
PERSONAL SERVICES/PC	PERSONAL SERVICES/36	DISOSS
DISPLAYWRITER	PERSONAL SERVICES/38	
SCANMASTER	5520	

DISOSS = DISTRIBUTED OFFICE SUPPORT SYSTEM

Glossary of terms

APPC Advanced Program-to-Program Communications. A subset of SNA (Systems Network Architecture) functions and protocols used specifically to allow programs on one site to communicate with programs at another using LU 6.2. APPC is designed to make SNA more usable in a distributed application.

Conversation Verb Interface (CIV) IBM's structured application program interface for distributed LU 6.2 application programs.

DIA Document Interchange Architecture. An application-level architecture that defines protocols and data structures for the consistent exchange of documents and files amoung distributed office applications, such as DISOSS and Personal Services products. DIA supports such functions as distribution, application services, and document library.

DISOSS Distributed Office Support System. An application subsystem that provides a variety of office automation functions, including document distribution services and library services. Allows users to send, receive, distribute, and file documents. Utilizes various IBM protocols, such as DIA and SNADS.

DIU Document Interchange Unit. The data structures used to carry out operatons in SNADS and DIA networks.

DOSF Distributed Office Support Facility. DOSF manages text and data used on IBM 8100 equipment.

DSL Distribution Service Level. Used to select a route in a SNADS network based on priority, capacity, and safe delivery of messages.

DSU Distribution Service Unit. The components of a SNADS network that are responsible for data delivery, routing, and queuing.

EDD Electronic Document Distribution. EDD is a set of software services for the Displaywriter allowing documents to be filed and distributed to other Displaywriters and hosts running DISOSS.

LU 6.2 Logical Unit 6.2. Part of IBM's Advanced Program-to-Program Communications that provides a standard communications protocol for distributed application programs.

OSU Office System Node. The OSN performs document distribution, library services, application services, and file transfers on behalf of network users.

Personal Services A series of IBM software products (PS/PC, PS/36, and PS/370) that run on the PC, System/36, and 370-type mainframes, respectively, and provide a connection to DISOSS. The Personal Services products allow users to request distribution and library services.

SNADS SNA Distribution Services. An application-level architecture that provides asynchronous (delayed delivery) distribution of documents and files between users. SNADS implements a store-and-forward mechanism that allows information to be stored until the path to the receiver is available.

SRN Source/Recipient Node. The SRN requests through DIA that the OSN perform certain functions on behalf of network users. These functions include document distribution, library services, application services, and file transfers.

---

queued indefinitely in intermediate DSUs. It is important to remember that SNADS is separate from DIA and is a broadly applicable architecture not limited to document delivery functions.

The role of SNA Distribution Services

The delayed delivery capability of SNADS means users do not need a complete end-to-end connection across a SNADS network before they can begin to send data. Each of the nodes that makes up a SNADS network contains queues that can hold data indefinitely. DSUs are the SNADS network nodes, and as data moves through the SNADS network it is forwarded from one DSU to the next. If a communications path between two DSUs is not currently active or available, the sending DSU will queue the data until the communications path becomes available. The data may be queued several times for indefinite periods as it moves through a SNADS network. This is why SNADS is said to provide delayed delivery.

Looking at Figure 2 from the SNADS point of view, the DISOSS units in our sample network are SNADS DSUs, while the Displaywriters are simply users of the SNADS network. The Displaywriters and their users are beyond the scope of the SNADS networks; they are simply the sources and destinations of documents being distributed.

SNADS comes into play when document distributions are performed across multiple DIA OSNs that are also SNADS DSUs. This is the case when we distribute a document from User A to User C in Figure 2. The distribution operation starts the same way as in the first example—User A initiates a DIA session with the adjacent DISOSS setup and used DIA protocols to request that DISOSS perform the distribution. The document to be distributed now resides in a DISOSS queue and User A ends the DIA session. Note that DIA is still used to request that the distribution be initiated.

When DISOSS begins the process of distributing the document, it finds that the intended recipient is on the other DISOSS. The document must first be sent to the other DISOSS before User C can gain access to it. SNADS is used to send documents between DISOSS units.

If the communications facilities connecting the two

3. SNADS point of view. *The same office network that appears in Figure 2, expanded to show how SNADS operates with DIA. (A) and (B) show the sign-on and sign-off mechanisms operating in DISOSS; (C) shows the operating structures in Personal Services. Receiving and sending Distribution Service Units orchestrate the process.*

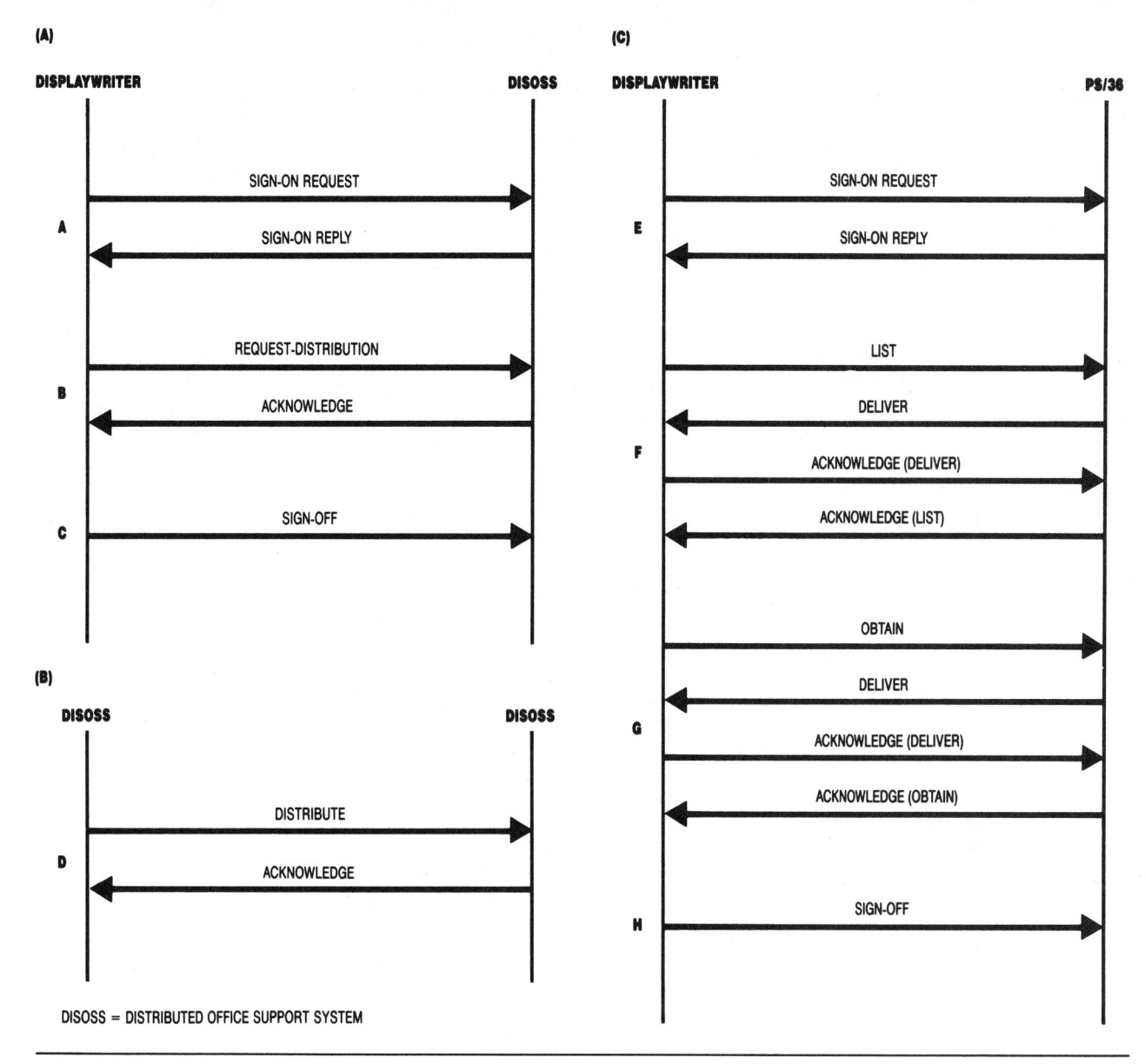

(A)

DISPLAYWRITER → DISOSS

A: SIGN-ON REQUEST →
   ← SIGN-ON REPLY

B: REQUEST-DISTRIBUTION →
   ← ACKNOWLEDGE

C: SIGN-OFF →

(B)

DISOSS → DISOSS

D: DISTRIBUTE →
   ← ACKNOWLEDGE

DISOSS = DISTRIBUTED OFFICE SUPPORT SYSTEM

(C)

DISPLAYWRITER → PS/36

E: SIGN-ON REQUEST →
   ← SIGN-ON REPLY

F: LIST →
   ← DELIVER
   ACKNOWLEDGE (DELIVER) →
   ← ACKNOWLEDGE (LIST)

G: OBTAIN →
   ← DELIVER
   ACKNOWLEDGE (DELIVER) →
   ← ACKNOWLEDGE (OBTAIN)

H: SIGN-OFF →

DISOSS facilities are not available, the DISOSS on the right will queue the document for transmission and it will be sent whenever a communications path becomes available. The document may remain queued for milliseconds or days. If there were multiple DISOSS DSUs between the sender of the document and the recipient, delays could be introduced at each intermediate DSU.

In Figure 2, DISOSS is an implementation of the SNADS architecture. SNADS is already widely implemented, not only in DISOSS, but in IBM's departmental processors, including System/36, System/38, 5520 Administrative System, and 8100 Information System. All of these products have, or will have, SNADS support.

Let's look at how all these technologies work together. User A at the Displaywriter on the left will be sending a document to User D, who is using an IBM Personal Computer running the Personal Services/PC software. This distribution operation will involve the use of SNA's LU 6.2 (Advanced Program-to-Program Communications, or APPC) and LU 2 at the data transport level, and DIA and SNADS at the application level.

A typical electronic mail operation

DIA initiates the distribution. The first protocol exchange is between User A's Displaywriter and its adjacent DISOSS. The purpose of this interchange is to request initiation of distribution to User D. The Displaywriter, since it is a Source/Recipient, can request that distribution be initiated, but *it* is performed by the OSN, DISOSS.

The data transport protocol used by the Displaywriter's Electronic Document Distribution (EDD) software is LU 6.2. Each of the DIA protocol exchanges between the Display-

4. APPC/DIA Interaction. *APPC, SNADS, and DIA work together to implement an electronic mail network. Both DIA and SNADS use APPC as the transport mechanism.*

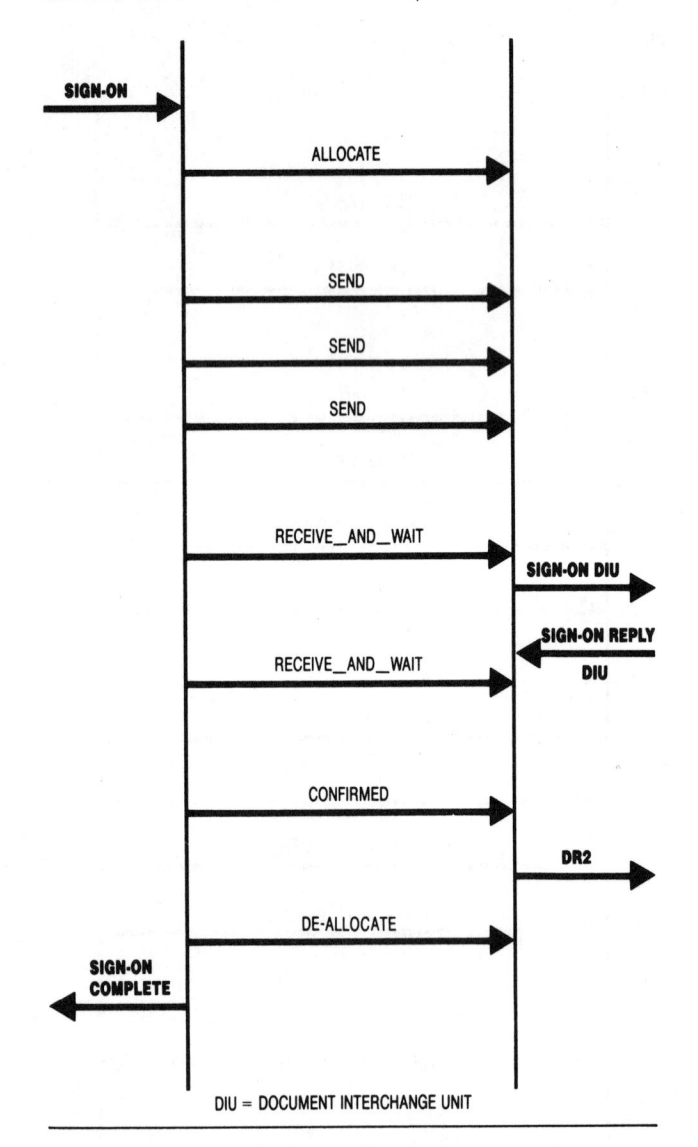

```
SIGN-ON ──►
           ├──── ALLOCATE ────────►
           │
           ├──── SEND ────────────►
           │
           ├──── SEND ────────────►
           │
           ├──── SEND ────────────►
           │
           ├──── RECEIVE_AND_WAIT ►
           │                        ├──── SIGN-ON DIU ────►
           │                        │
           │                        ◄──── SIGN-ON REPLY DIU
           ├──── RECEIVE_AND_WAIT ►
           │
           ◄──── CONFIRMED ────────
           │                        ├──── DR2 ────────────►
           │
           ◄──── DE-ALLOCATE ──────
SIGN-ON
COMPLETE ◄──
```

DIU = DOCUMENT INTERCHANGE UNIT

writer and DISOSS will be carried on a separate LU 6.2 conversation. LU 6.2 uses conversations to manage and control access to SNA sessions.

First, a logical connection must be established between User A and DISOSS. This connection, called a DIA session, is initiated by a DIA SIGN-ON REQUEST (Fig. 3). The SIGN-ON REQUEST contains the user's I. D. and password as well as a list of the DIA function sets to be used on the DIA session. The SIGN-ON REPLY indicates whether the sign-on operation was successfully completed or an error condition occurred. The document distribution operation is initiated by a REQUEST-DISTRIBUTION command. This DIA command carries the list of recipients as well as indicators that request confirmation of delivery and priority. The REQUEST-DISTRIBUTION command also carries the document to be distributed. The ACKNOWLEDGE command indicates whether the distribution operation was successfully initiated. It is important to understand that this is not an acknowledgment of delivery, but that the distribution was initiated by DISOSS. Confirmation of delivery will, optionally, occur later, when the document is actually received by the recipient. The DIA session is terminated by a SIGN-OFF.

At this point the document distribution process, requested by the Displaywriter, is ready to begin. The request for distribution was made using DIA protocols; the distribution process itself will involve the use of SNADS. DISOSS and SNADS actually perform the distribution.

The actual document distribution process now begins in DISOSS. DISOSS checks the destination address to see whether the recipient is local (attached to the same DISOSS equipment) or remote (attached to another DISOSS node or some other OSN/DSU). If the distribution is local, the document is simply put into the output queue that is associated with the local user.

If the recipient is attached to another OSN/DSU, SNADS will be used to deliver the document to the target OSN. To determine the target OSN/DSU, DISOSS performs the SNADS directing function, determining the destination DSU name, based on a user I. D. In our example, the destination DSU will be DISOSS User B. After the directing function is completed, the DSU will perform the SNADS routing function. This involves looking up the destination DSU name in a routing table to find the path to the destination. Multiple routes may be defined between DSUs. The most appropriate route is determined by matching the requested Distribution Service Level (DSL). The DSL specifies the priority, level of data protection, and the required queuing capacity of intermediate DSUs.

In Figure 3, DISOSS uses the SNADS DISTRIBUTE command to send the document to its destination DSU. In this example, the document is sent directly to the destination DSU. The DISTRIBUTE command may, in some networks, travel through several intermediate nodes before arriving at the destination DSU. The DISTRIBUTE command is used to move the document through these intermediate store-and-forward nodes. When the document is successfully delivered to a DSU, it replies to the sending DSU with an ACKNOWLEDGE command.

In solicited delivery of mail, the document will remain in the destination DSU until the user solicits delivery to the workstation. The remainder of this example shows how a user at a DIA SRN can check the output queue (mailbox) and retrieve selected documents from the OSN/DSU.

In Figure 3, User B must first logically connect with the adjacent DISOSS. This is again accomplished by sending the DIA SIGN-ON command and waiting for its reply. Next the user will check the contents of the output queue in the adjacent OSN. This is the user's mailbox. The DIA LIST command solicits a list of documents that have arrived for this user. This list is built by the OSN/DSU and sent to the user's SRN in a DELIVER command. Receipt of the list is acknowledged by the SRN; then the OSN/DSU acknowledges completion of the entire LIST operation. The user uses this list to select mail to be retrieved from the OSN/DSU.

The documents the user wants to read are retrieved by sending an OBTAIN command from PS/PC to DISOSS. The document is sent to the SRN in a DELIVER command. Receipt of the document is acknowledged by the SRN, then the OSN/DSU acknowledges completion of the entire obtain operation. Multiple documents can be retrieved in a

APPC, DIA, and SNADS

Most of IBM's mainstream office products use Systems Network Architecture's (SNA's) Advanced Program-to-Program Communications (APPC) to support DIA and SNADS operations. APPC is the underlying communications protocol used by DISOSS, System/36-Personal Services/36, System/38-Personal Services/38, Displaywriter, Scanmaster, 5520 Administrative System, and 8100.

When a DIA SRN that supports APPC signs on to its adjacent OSN, the DIA command processor uses the APPC Conversational Verb Interface. The application software in the SRN creates a SIGN-ON Document Interchange Unit (DIU) and issues an APPC ALLOCATE verb which will allocate an LU 6.2 conversation on an existing SNA session. The ALLOCATE verb also carries the name of the transaction software program to be started by the remote logical unit. This transaction software program will be run inside the mainframe Customer Information Control System (CICS) program that DISOSS will use to parse and execute the SIGN-ON command that follows.

The transaction program name is carried in a Type 5 Function Management Header, a data structure sent to initiate a remote program. This header is built by the logical unit and placed in the LU 6.2 output buffer. Note that it is not actually sent until either the output buffer is full or a verb is issued that forces transmission of the output buffer.

The SIGN-ON DIU itself is written by issuing one or more APPC SEND verbs. Again, the data is not actually transmitted until either the output buffer is full or a verb that forces transmission is issued. After the SIGN-ON DIU is written to DISOSS, receipt of a SIGN-ON reply is anticipated, so an APPC RECEIVE-AND-WAIT verb is issued.

The reply is read by issuing an appropriate number of RECEIVE-AND-WAIT verbs. The last RECEIVE-AND-WAIT verb will also receive an indicator that requests an APPC confirmation. The CONFIRMED verb satisfies this request by sending an SNA DR2, a definite response indicator, which serves as an acknowledgment to DISOSS. The APPC conversation is then de-allocated, and the DISOSS sign-on operation is complete for the network user.

---

single operation. Each document is sent in a separate DELIVER command. The DIA session is then terminated by a SIGN-OFF command from the SRN (Fig. 4).

It is expected that most vendors who compete with IBM in the workstation marketplace will provide similar support within two years. Two vendors, Digital Equipment Corp. and Data General, are already supplying Source/Recipient Node capability through their implementations of APPC and DIA. Wang and Honeywell have also announced products supporting these architectures. ∎

Donald H. Czubek is a founder and vice president of Communications Solutions Inc. Czubek has 15 years of data communications product design experience, most recently with Four-Phase Systems.

Paulina M. Borsook, DATA COMMUNICATIONS

What parallel processing can offer information networking

As an esoteric technology gains mainstream acceptance, it holds the promise of surprising benefits for communications users.

Imagine a communications processor that can handle thousands of network transactions orders of magnitude faster than anything currently on the market. This is the promise parallel processing holds for data communications. Today these machines are largely found in the realm of scientific and engineering computing, but commercial development is well on its way, limited only by time and marketing. With parallel processing, far more bits can be computed than can be communicated, but as such advanced technologies as fiber and T3 take hold, the computers and communications see-saw should come into balance.

Parallel processing is already being used for communications processors and in machines linked on networks. Applications range from simulations of battlefield performance using a parallel machine as a server on a Digital Equipment Corp. (DEC) network to central sites in hospitals that use barcodes to collect information on patients. Network management databases are also being run on parallel processors. Artificial intelligence applications, which require exponentially more power than conventional programs, can be run cost-effectively on supermicrocomputers with the efficiencies of parallel processing, making AI-based network management equipment an economic reality. The technology already spans the range from Cray competitors to engineering workstations.

Serial processing, the predominant computer architecture of the last 30 years, is based on the Von Neumann model, where one processor shuttles back and forth between data and memory, and communications between processors is forbidden. VLSI (very large scale integration) has made serial processors cheaper, cooler, and smaller, but has also begun to show the limits to increased performance arising from conventional computing techniques (Fig. 1). As computer memories have gotten larger, however, the lost processing time involved in the link between memory and processor has begun to significantly affect processing performance. Parallel computing is a way around the Von Neumann bottleneck, the passageway between a CPU (central processing unit) and memory. Having multiple processors accessing memory at one time in communication with each other is parallel processing's radical solution to the decreasing gains in performance. The technology is the latest development in the path computing has followed for 30 years, becoming exponentially faster, cheaper, and more powerful.

Benefits present and future
In fact, Electronic Trend Publications, a market research group based in Saratoga, Calif., predicts that parallel processing machines will take 48 percent of the market in high-performance computers by 1990, in part because uniprocessor architectures are approaching theoretical limits in processing speeds. Slow input/output linkages inherent in nonparallel processing are a hindrance to array processors and networks; and users will increasingly demand higher performance than conventional minicomputers and mainframes can provide. There are kinds of problems and solutions that can't be attempted without parallel processing. Calculations whose output are complex graphic images instead of text or digital signals might overwhelm serial processors. A full-scale simulation of an entire airplane in a simulated wind tunnel cannot be performed without parallel processing. An analysis in real time of seismic waves generated by detonations involved in an oil exploration cannot be managed without individual sensors on the ocean floor transmitting data back to the ship's on-board host parallel processor. With a kind of second-generation time-sharing, parallel machines offer very fast, powerful, and cheap computing available to many users at the same time.

The significant differences in the numbers of MIPS

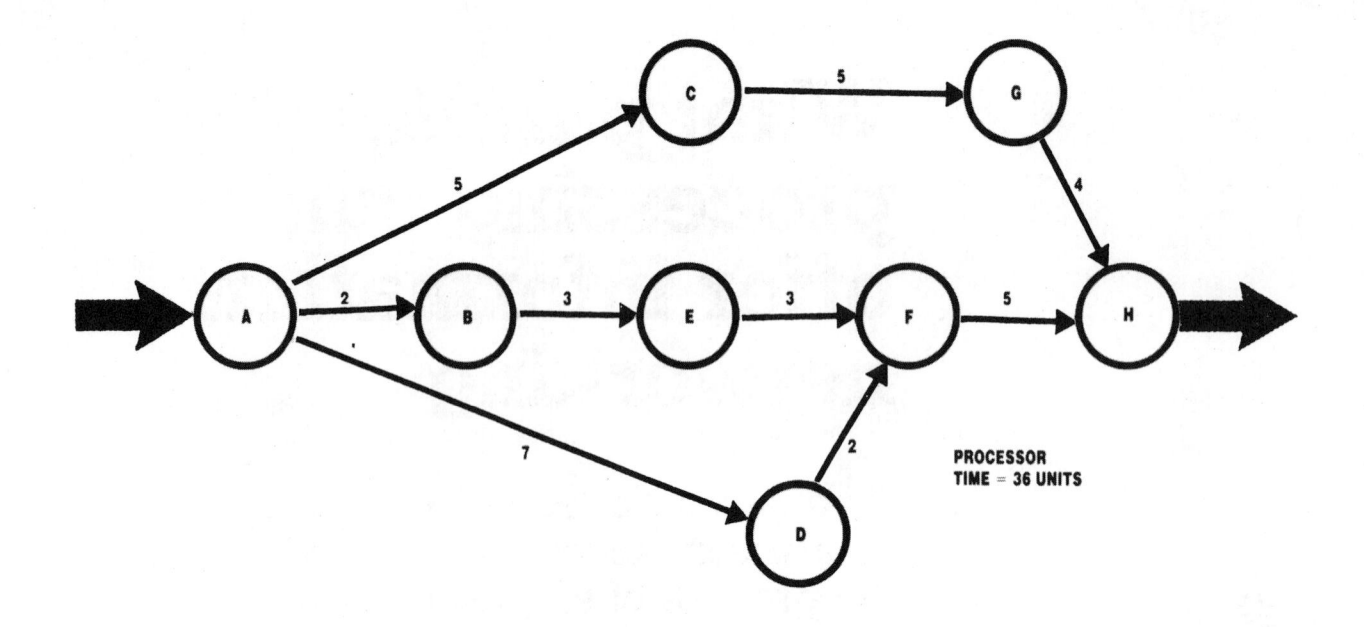

PROCESSOR TIME = 36 UNITS

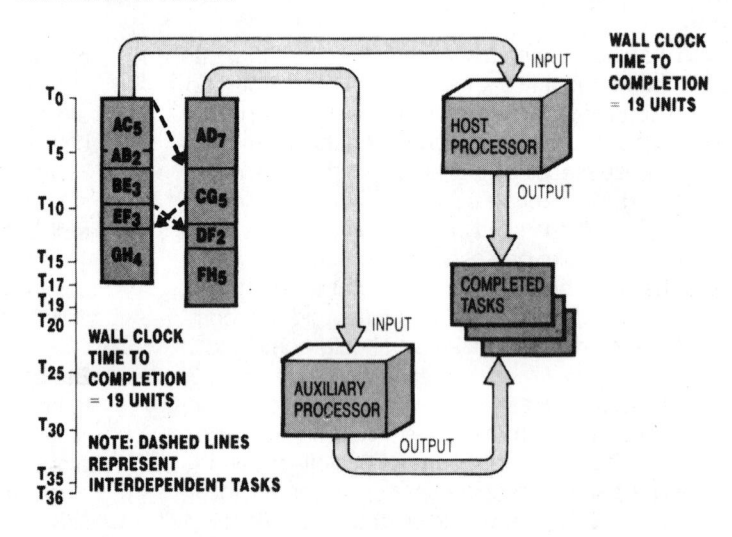

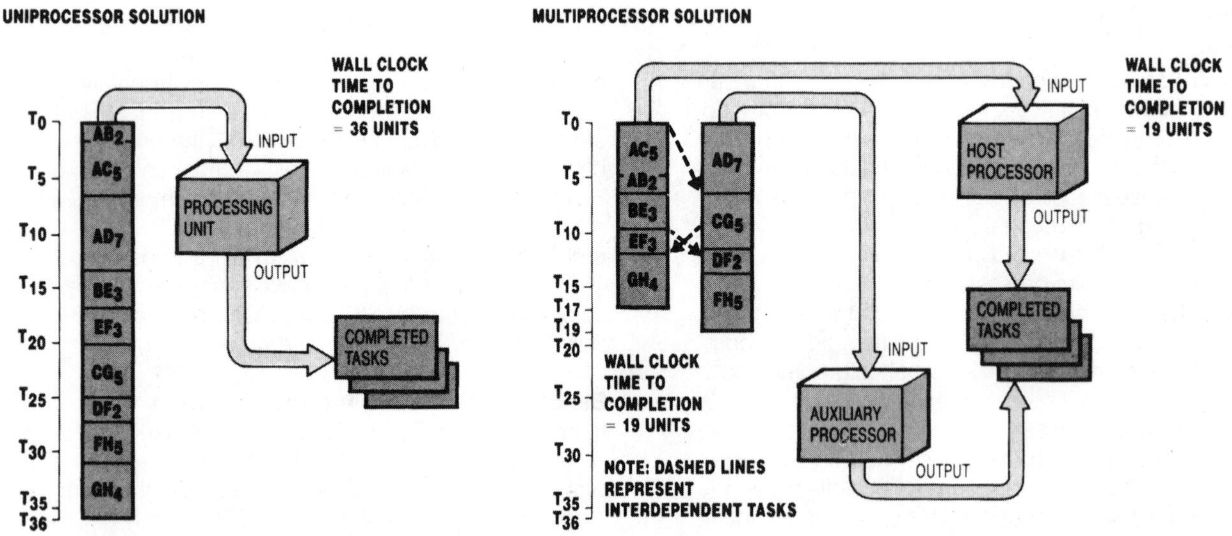

UNIPROCESSOR SOLUTION

WALL CLOCK TIME TO COMPLETION = 36 UNITS

WALL CLOCK TIME UNITS	= 36
AVAILABLE PROCESSOR TIME UNITS	= 36
PROCESSOR TIME UNITS USED BY APPLICATION	= 36
PROCESSOR TIME UNITS NOT USED	= 0

MULTIPROCESSOR SOLUTION

WALL CLOCK TIME TO COMPLETION = 19 UNITS

NOTE: DASHED LINES REPRESENT INTERDEPENDENT TASKS

WALL CLOCK TIME UNITS	= 36
AVAILABLE PROCESSOR TIME UNITS	= 2 × 36 = 72
PROCESSOR TIME UNITS USED BY APPLICATION	= 36
PROCESSOR TIME UNITS NOT USED	= 36

(million instructions per second) per dollar that serial and parallel processing can deliver are already apparent. While comparisons of computing performance are always tricky, it may be instructive to use that mainstay of departmental computing, the DEC VAX, as a rough measure. The high-end VAX, the 8800, delivers about 12 MIPS for about $650,000; the most powerful Alliant FX, manufacutered in Acton, Mass., comes in about $1 million for 34.5 MIPS. Price/performance analyses can be particularly misleading with the two different kinds of machines, for the parallel processor generally performs spectacularly better on applications tailored for parallelism, but will lose its commanding lead when asked, for example, to perform office automation tasks that require multitasking. Nonetheless, parallel machines can act as more-familiar multiprocessing engines performing tasks in foreground and in background, some in parallel and some in serial.

Simulations run on Ethernet

Tom Dunnigan, a data communications expert at Oak Ridge National Laboratory in Oak Ridge, Tenn., is convinced that the parallel processing research being done on interprocessor routing and flow-control information will come to benefit the networking community. "X.25 boxes

1. Uniprocessor, multiprocessor, parallel processor.
Each circle node is an element of solution data. Task numbers represent units of execution time. Execution can proceed only in the direction of arrows, and all tasks entering a node must be completely processed before an exit path becomes available. Each path, such as AB or AC, is a task, so that task AB uses data element A as input to produce solution data B as output. Paths in each node define data dependencies. In the multiprocessor solution, execution time is cut by 50 percent and processor power is doubled; in the parallel solution, execution time is reduced from 36 units to 14 units while increasing total work capability to 72 units.

PARALLEL SOLUTION

WALL CLOCK TIME UNITS	= 36
AVAILABLE PROCESSOR TIME UNITS	= 3 × 36 = 108
PROCESSOR TIME UNITS BY APPLICATION	= 36
PROCESSOR TIME UNITS NOT USED	= 72

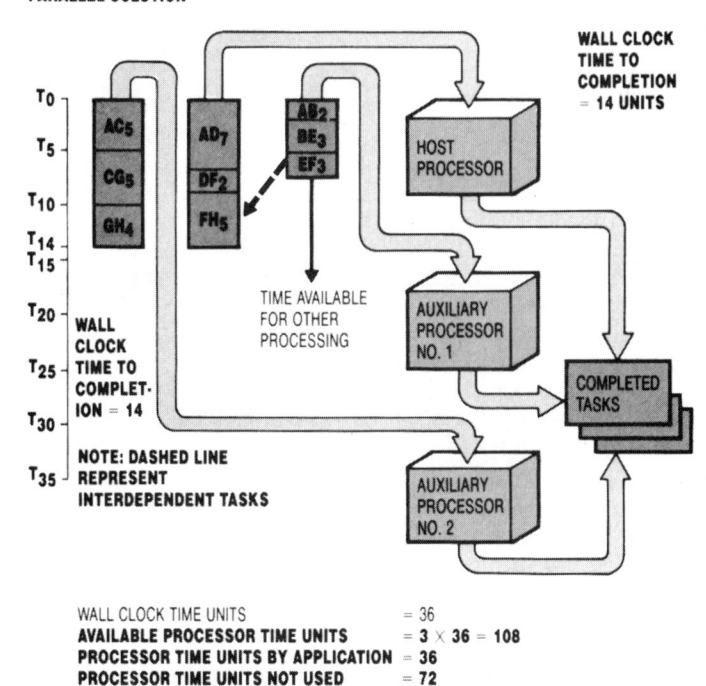

with multiple circuits having dedicated 68000 processors are already a sort of parallel machine," Dunnigan says. "And the economics are so good, there's no reason not to have a CPU and memory on each line."

Parallel-processing subroutines are like Fortran send and receives, so that writing software for parallel processing is much like writing Decnet or TCP/IP (transmission control protocol/internet procotol) task-to-task message-passing services. That messages are being sent in the same box is irrelevant, so that simulations of parallel-processing architectures are customarily run on Ethernet. Dunnigan believes, as does Jeffrey Canin, an analyst at Hambrecht & Quist in San Francisco, that "machines like the Sequent or Encore could be used like BBN's Butterfly communications processor to drive any fast, real-time application, with lots of processors working on the same task."

There is a basic connection between data communications and parallel processing. Charles Seitz, professor of computer science at Caltech in Pasadena, Calif., and father of one of the currently popular parallel architectures, the hypercube, is optimistic about the place of parallel processing in data communications. The original Caltech hypercube was connected to Arpanet and other Caltech Ethernets from the start, with a gateway called Zaphod, named after a five-headed monster in a science fiction novel. Users across the country have always employed the network, and the same programs run inside the cube as on networked VAXs or Suns. Message-passing and synchronization issues are the same, whether on a network or inside a computer.

Many of the conceptual underpinnings of the hypercube were taken from Leonard Kleinrock's series of books on queueing theory, *Queueing Systems*. Kleinrock contributed to the algorithms that supported the development of Arpanet. The UCLA professor addressed questions of how speeds of information fed into flow performance on store-and-forward networks. Latency as a function of applied load on a network, throughput as a function of applied load, and the basic methods of packet routing, all bore relevance to research on the hypercube.

Seitz predicts that, as there are higher- and higher-performance gateways, message passing will be adapted to share protocol functions. The megabits per second of satellite transmission and bursty communications will have gateways powered by parallel processing that serial CPUs can't handle. Commercial development of the communications-oriented hypercube has progressed very quickly, from a research project in 1983 to the point where most licensees are working on second-generation projects. Seitz mentions a current commercial product under development using an Intel hypercube implementation, where the cube channels are bisected so that some remain within the computer and some are sent to outside networks.

Current implementations

Parallelism isn't new. Some of the first experimental computing machines of the 1940s were distributed machines; the dominance of serial processing had more to do with the physical limitations (expense, unreliability, size, and heat generation) of multiple processors operating simultaneously than with their computational deficiencies.

Parallel processing is rife with competing schemes, of course, and a subvernacular is being created to describe the growing field (see "Parallel-computing glossary"). Because the field is rapidly evolving, definitions have not been set on what strictly falls into the category of parallel processing, so the table, "a sampling of parallel processing developments," reflects the diversity in the marketplace but does not include fault-tolerant machines. All the processors have communications capabilities, starting with Ethernet and ranging from X.25 to RS-232-C. "It's ridiculous to have a computer that you can't log into remotely," says a spokesman at Ncube Corp., a parallel-processing manufacturer in Beaverton, Ore.

Parallel-computing glossary

Asymmetry. Architecture where one processor acts as a host, performing functions for other processors, which in turn may be dedicated to specific communications tasks or to other functions.

Binary n-cube. A generalized term for terms in the geometric series *point, line, square, hypercube* where binary means that two processors share the same edge (see *hypercube*, below).

Closely coupled. Also, *tightly coupled*. Multiple processors sharing memory and the load of arithmetic computation rather than managing input/output activities. The main CPU is not differentiated from the rest of the processors so that, for example, time-sharing operating systems can maintain a list of processes waiting for service from the CPU. To overcome technological bounds on the number of ports possible on multiported memory, closely coupled architectures connect processors through a shared bus or switching network. Compare with *loosely coupled*.

Coarse-grain parallelism. Partitioning an application or function into logically parallel units that can be processed concurrently so that jobs can be mapped onto a computer. Entire programs or large routines execute concurrently on different processors. May be used to describe handling of unrelated tasks on MIMD machines.

Concurrency. Refers to the use of many processors working in tandem on a single problem. Concurrent processors each act simultaneously on a private set of instructions and data.

Control-level parallelism. Instructions in a control sequence can operate independently, executing in parallel on multiple processors.

Dataflow. Architecture with multiple execution units of various types and a stored program control processor.

Data-level parallelism. Operations of independent data elements can be carried out simultaneously in parallel by multiple processors.

Distributed memory. Each processor has its own dedicated local memory, working on a small portion of an overall computation problem and thus distributing the load. The computer's entire memory is nothing more than the sum of individual memories. When a processor attempts to reach a memory location, the address of the desired location determines the individual memory module where the looked-for memory location is situated. Communications arrangements therefore become important for programming and performance. Also referred to as local memory, or private memory.

Fine-grain parallelism. Requires decomposition of an application at the instruction level or below, so that computing is efficient at the level of operations or groups of operations. The various processing elements or pipeline stages must be able to synchronize their actions in submicroseconds.

Floating-point arithmetic. Arithmetic that allows the decimal point to float as needed to allow numbers to range from the very large to the very small.

Floating-point operand. A real fractional number represented in scientific notation, for example 2.45×10^5.

Full crossbar. An interconnection network that directly connects everything to everything else.

Gather-scatter. Refers to the need to collect or disperse data so that a vector computer can always deal with quantities that are organized in lists.

Global memory. Separate real physical memory that can be read from and written to by a number of processors. Data need only be loaded once from memory so that processors can interact with it and use memory to communicate with each other. However, processors using this scheme generally have to spend much of their time absorbing the initial data and then passing intermediate results to each other. The number of processors sharing common memory effectively is a function of the connectivity between the processors.

GFLOPS. Giga (1 billion) floating-point operations per second.

Harness. A construct describing how multiple programs fit onto multiple processors, detailing how programs and data reside on each processor.

Hypercube. A higher-dimension analog of a cube, developed as a parallel computer architecture at Caltech.

Interprocessor communications. Resemble conventional intraprocessor data communications, but the correspondence is not absolute. Resolving the communications problems between processing elements is most technically diffcult component of parallel processing.

Kernel MFLOPS. Computer performance on a simple concentrated task, such as summing a list of numbers.

Local memory. See *distributed memory*.

Loosely coupled. Parallel processors using local area network technology. Intramachine communications cannot be allowed to cause a bottleneck, so that algorithms nned to be coded to avoid excessive nonlocal communications. Contrast with *closely coupled*.

Medium-grain parallelism. Involves concurrent execution of sequential operations (large-scale interactive processing). Concurrent execution of sequential DO-Loop iterations are an example. Underlying machines must have synchronization primitives that can be set and tested in microseconds.

MFLOPS. Millions of floating-point operations per second.

MIMD. Multiple instruction stream, multiple data stream. The architecture of most multiprocessor and parallel computers. Several instruction cycles may be at work at any moment, each independently fetching instructions and operands into multiple processing units and operating on them concurrently.

MISD. Multiple instruction stream, single data stream. A theoretical construct where several processors perform various operations on the same data.

Multistage. An architecture that avoids connecting each processor to each memory module by inserting two or more intermediate switching nodes.

NVON. Non-Von Neumann architectures.

Partitioning. Dividing up a problem into smaller pieces so that many processors may work on it simultaneously.

Peak MFLOPS. A speed rating of a computer based on

all function units operating under ideal conditions.
Percentage parallel. The fraction of a job that can be rendered faster by user parallel processors.
Scalar processing. Computing with one number at a time.
Sequential processing. Processing where operations must be performed serially. See *Von Neumann machine*.
Shared memory. Logically addressed and directly accessed by all processors. May cause a bottleneck because of simultaneous demands from many processors. Requires high-performance memory components, and ultimately limits computer scalability.
SIMD. Single instruction stream, multiple data stream. Processing so that a machine with multiple processing units is governed by a single control unit with limited memory-to-processsor communications. These machines offer simpler synchronization by keeping processors in lockstep by broadcast instructions. Array, vector, and pipeline processors fall into this category.
SISD. Single instruction stream, single data stream. The architecture of conventional serial computers that execute instructions sequentially but may overlap or pipeline execution.
Symmetry. Architecture where all processors have identical capabilities and functional equality.
Sustained MFLOPS. Also called Application MFLOPS or JOB MFLOPS, it is the speed a computer achieves in actual uses. Contrast with *Peak MFLOPS*.
Systolic array. A set of homogeneous processing elements, each capable of performing simple arithmetic operations, typically interconnected in regular geometric patterns.
Tightly coupled. High degree of communications through shared memory and processor interaction. Contrasted with *loosely coupled*.
Vector processing. Using the same operation for a list of numbers. May be multidimensional as in the form of arrays.
Vector registers. Special storage locations close to a CPU (central processing unit) which permit entire lists of numbers to be used in a computer's scratchpad calculations as if they were single entities.
Vectorize. Reorganizing a program to reveal repetitious operations.
Von Neumann bottleneck. The critical path in communications between an instruction processor and its random-access storage.
Von Neumann machine. A computation model devised by John Von Neumann and his associates 40 years ago. Its essential features are a processor that performs instructions; a memory that stores both the instructions and data of a program in cells having unique addresses; and a control scheme that fetches one instruction after another from memory for execution by the processor, and that shuttles data between memory and processor one word at a time. Several accesses to data are required to complete a single operation. Interprocessor communications and synchronization are outlawed in the scheme.

Maturing parallel-processing technology already has some of the hallmarks of a mainstream industry. It has had its first spectacular bankruptcy and is undergoing scrutiny by the Japanese. Bankrupt Denelcor of Boulder, Colo., manufacturer of the HEP machine, a Cray workalike, was one of the first major players in the middle 1970s. The much-publicized Japanese Fifth Generation computer project is devoting considerable resources to research in parallel processing. The Japanese are attempting to provide machines with the power to handle the natural parallelism found in the problems handled by human beings, and attempting to achieve the high-speed parallel processing necessary to support what they term "intelligent human activities."

Commercial parallel processing is still young enough that there is no tidy way to characterize its different architectures. While there are many competing approaches (see "Different Approaches"), there is as yet no universally accepted ordering scheme. Consequently, there are many hybrid and proprietary processing techniques that cannot be easily characterized and that do not fit into ready-made definitions.

In assessing the various parallel computer architectures, Jack Dongarra, scientific director of the Advanced Computer Research Facility at the University of Chicago's Argonne National Laboratory, places the Cray, a general-purpose machine using four processors at most, at one extreme of the range. "It's a dumb use of parallelism," Dongarra says. "Its strength comes from the speed of the processors, not the number." At the other end of the spectrum is the Connection Machine, equipped with 65,536 processors, which do not even implement floating-point operations, mathematical constructs used in technical computing, in hardware.

Pros and cons

Parallel processing is not without its detractors. Critics base their objections on, among other things, the difficulties surrounding the development of applications software. Supporters, however, think that adapting parallelism for everyday use presents no more problems than any other technology.

Lou Piazza, director of advanced system development at BBN Communications in Cambridge, Mass., and a promoter of parallelism, believes that one of the advantages of parallel processing over serial processing is that it can extend the limits of machine design within the limits of any given technology. While there is some loss of efficiency due to the synchronization overhead involved with the linking together of multiple processors, users can normally expect about 75 percent to 80 percent of the function per processor that would be expected if the processors were not linked in parallel. So with 128 processors working in parallel all within the same computer, users can get the power of about 100 isolated serial processors.

Parallel processing also extends the range of performance. "You have to invest a higher cost at the low end if you want to have capacity at the high end. If you want the difference between the minimum and maximum configuration to be a factor of two or three, then a uniprocessor or CPU with slaves will do just fine. But if you need orders of magnitude of difference, with performance varying by tens

Different approaches

Homogeneous architecture is a computer design that has communications, processors, amd memory evenly distributed throughout, with any number of identical nodes. Typically, each node is a complete computer with its own arithmetic unit and memory. Contrast with *heterogeneous architecture* where a multiprocessor consistsof any number of different processors.

Crossbar switches, as exemplified in the University of Illinois Cedar pilot parallel processing project, create direct connections between each processor and each memory. These links can be circuit- or packet-switched.

A *grid* is the architecture associated with the National Aeronautics and Space Administration-Goddard's Goodyear Massively Parallel Processor, the prototype for the celebrity Connection Machine created by Thinking Machines Corp. of Cambridge, Mass. Processors and their associated memories are laid out in a grid: Each processor is connected to its north, south, east, and west neighbors, with the exception of the edges of the grid.

Omega networks, as used in Bolt, Beranek, and Newman's Butterfly, consists of a column of processors and a column of memory modules. Each unit of memory contains addresses in a particular range of addresses; moving a message from a processor to a memory requires a number of steps, each bringing the message closer. The address of a memory location can be used to determine the memory module by inspecting the address one bit at a time. If a different piece of hardware is used at each stage, several memory requests can move through the network concurrently, gaining performance benefits similar to pipelining.

The *dataflow* model clusters tasks into code blocks, which can be executed in parallel on the same or a different set of processors. Each instruction in a code block is allocated to a processor at compile time according to its iteration and statement number. All output can be connected to any input through a switchable interconnecting network. Any instruction becomes enabled, or ready to execute, as soon as all input values, or tokens, become available at the outputs of some execution unit.

Much as three-dimensional cubes can be represented on a two-dimensional sheet of paper, a higher-dimensional cube can be conceptualized in three dimensions, the theory behind the *hypercube*. Point-to-point connections are avoided by using the processors themselves as switching nodes. Each processor is attached to only some of its immediate neighbors, so that if the need arises to communicate with a more remote processor, messages are sent through one or more intermediate nodes as in a store-and-forward network. Routing in a hypercube is simple. If a relative node address of a message is all zeros, the message is at the correct node. Each step of a transfer of a message zeros one component of a node address.

One offbeat approach being developed by AT&T, Texas Instruments, and TRW, among others, treats parallel processing as an analog to the the *neural networks* in the brain. Each chip acts as a neuron, acting in associative rather than linear ways. Connections are patterned after nerve cell interconnections, so that individual processors have multiple, three-dimensional connections, existing both as autonomous units and as components in a larger machine.

(A) LOOSELY COUPLED ARCHITECTURE SHARES I/O WITH LOCAL MEMORY AND INDEPENDENT PROCESSING

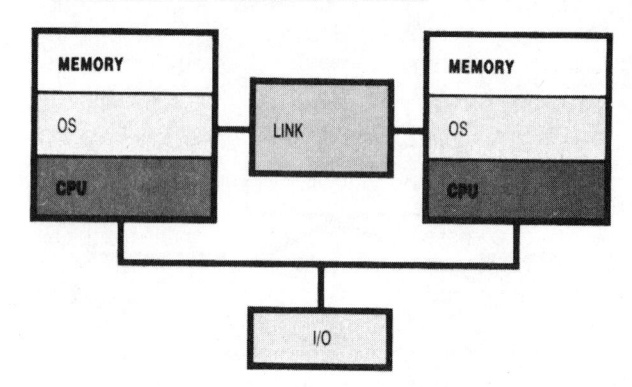

(B) PROCESSING NODES IN TIGHTLY COUPLED, HOMOGENEOUS ARCHITECTURES ARE THEORETICALLY EQUAL AND SHARE GLOBAL MEMORY

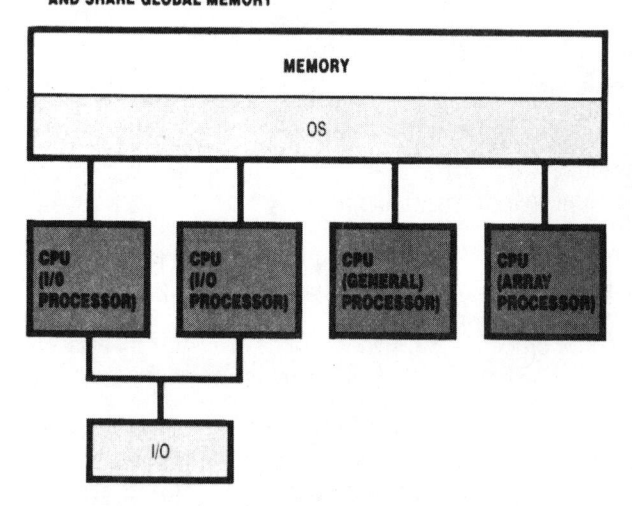

(C) TIGHTLY COUPLED, HETEROGENEOUS ARCHITECTURE

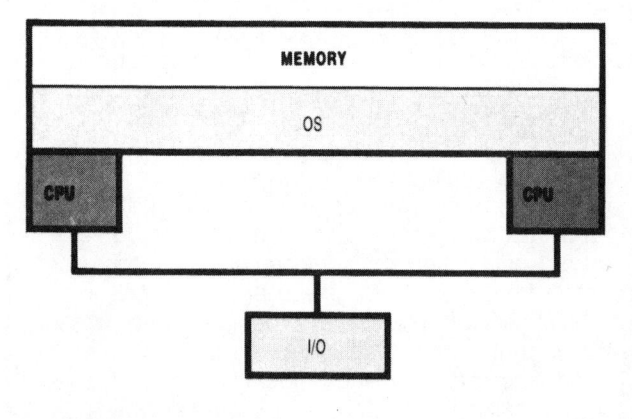

CPU = CENTRAL PROCESSING UNIT
I/O = INPUT/OUTPUT
OS = OPERATING SYSTEM

2. Alcoa. *Graffic and Photon software will be migrated to the Convex, striving to achieve as much vectorization as possible. Several Tektronix programs are supported.*

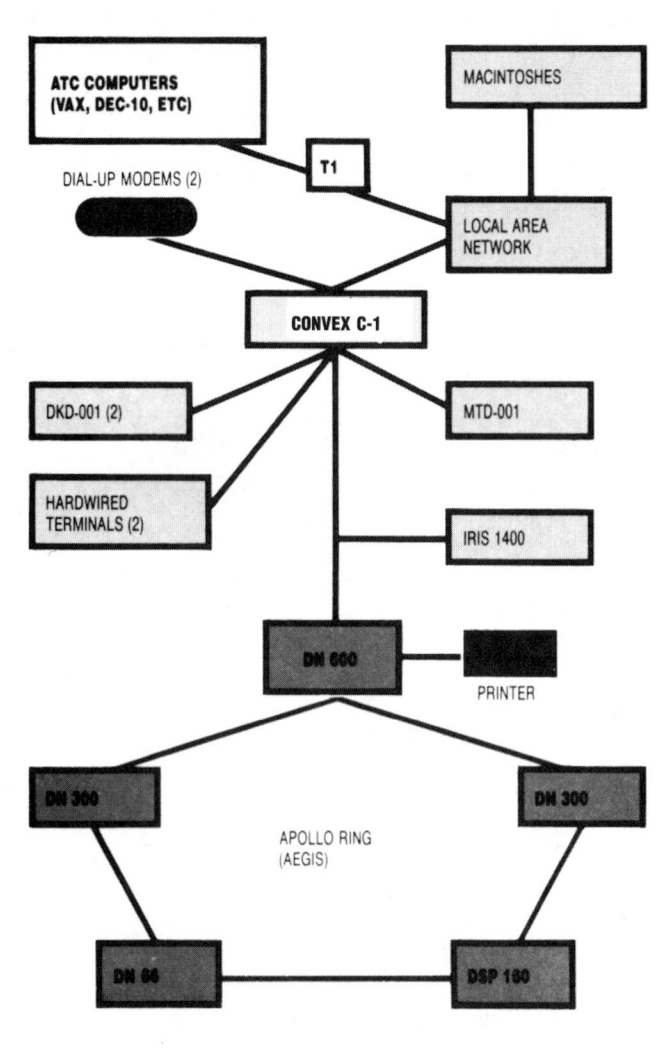

ATC = ALCOA TECHNICAL CENTER
DKD-001 = FUJII EAGLE 400-MBYTE DISK DRIVE
DN = DOMAIN NODE
DSP = DOMAIN SPECIAL PROCESSOR
MTD-001 = STC MAG TAPE DRIVE

or hundreds, you need parallel processing. It increases the range of span."

Jim Clark, senior product manager at Digital Communications Associates in Roswell, Ga., is more cautious. In his view, parallel processing would be useful only in the most exotic communications applications. Theoretically it might be used to improve the throughput of very large packet switches that go above the 2,000- to 10,000-packets-per-second threshold, to manage networks with horrendous peaks, or oversee two Hyperchannels on the same site. Its best use might be wideband packet switching, particularly if voice is involved. In this case, parallel processing would be helpful in processing packets to keep the delay down. "But software is a problem. And hardware is generally less expensive, and we'd rather go with a larger, faster microprocessor than invest in reprogramming costs."

Piazza disagrees, believing that marketing, and not technology, is the problem. "Lots of vendors are stuck with uniprocessor architecture, and they need to protect their investment. Any implementation has its difficulties. Any kind of porting is difficult to think about."

Programmers familiar with interprocess communications, such as Unix sockets and streams, or with message handling and protocols, will probably have few problems working with the processor-to-processor communications that parallel processing implies. "There's a natural fit with those experienced with communications and distributed programming," says Charles Seitz.

Communications problems lend themselves very well to parallel processing because they involve data switching for a number of lines and a number of users, each user running relatively independently. For example, a processor running a relational network management database performs a variety of functions. A single CPU or set of CPUs is dedicated to retrieving information in real time. Second, the database stores data and reads data into several files several times, causing multiple reads and partitioning of the disk if the data is to be read quickly. At the same time, a piece of the machine needs to be dedicated to graphics and display in real time, choosing colors, mapping points, and creating lines. Furthermore, data needs to be efficiently transferred from the collection processors to the display processors. Finally, the report function on a single CPU machine is a drain, much like batch processing. If each of these processes can be run on a dedicated CPU, there can be vastly improved performance.

To manage networks eventually
Parallel processing applications that use data communications range from service in process control industries to management of data from satellites. Generally, communicating parallel machines act as hosts on networks, but with time they will be used more and more frequently to manage the networks themselves:

■ *Aluminum Co. America.* Alcoa's Chemical Systems Division in New Kensington, Pa., uses a Convex C-1 for flow modeling of the Hall electrolytic aluminum manufacturing process. Because of magnetic stirring, there is lots of fluid flow in the liquid aluminum at the bottom of Hall cells (Fig. 2). The machine is connected through an Ethernet to an Apollo Domain local area network (LAN) and through a Sytek LAN to a Silicon Graphics Iris 2400 machine. A T1 link consisting of 10 outgoing ports and six incoming ports connects the departmental network to another TRW LAN 12 miles away at the Alcoa Technical Center, a major research center where there are more than 1,000 computers strung together on a network.

Alcoa staff scientist Walter Wahnsiedler explains that communications between the Convex machine and the Apollo equipment is not transparent; users typically execute compute-intensive procedures on the larger machine, downloading the results to the Apollo, which is "better for graphics, editing, and interaction."Graphics and finite element analysis programs used with the Convex also run on the Apollo network, as well as FXP file transfer software providing connections to DEC VAX computers, among others. The four serial ports on the Convex plug directly into the Sytek LAN. The machine's dial-up ports are used infrequently, in part by an employee who dials in at night to

A sampling of parallel processing developments

MANUFACTURERS	ARCHITECTURE	PERFORMANCE	LANGUAGES
ALLIANT COMPUTER SYSTEMS LITTLETON, MASS.	VECTOR AND PARALLEL	11.8-14.4 MFLOPS	C, FORTRAN, PASCAL
AMETEK COMPUTER RESEARCH DIVISION ARCADIA, CALIF.	HYPERCUBE	12-200 MIPS 0.8-12 MFLOPS	C, FORTRAN
BBN ADVANCED COMPUTERS INC. CAMBRIDGE, MASS.	BUTTERFLY (MULTISTAGE)	1-256 MIPS	C, FORTRAN, LISP
CONCURRENT COMPUTER CORP. TINTON FALLS, N.J.	TIGHTLY COUPLED, LARGE GRAIN	6 MIPS-34 MIPS	ASSEMBLY, COBOL, C, BASIC II, FORTRAN, SIBOL, RPG II
CONVEX COMPUTER CORP. RICHARDSON, TEX.	CXS (7-BOARD ASYMMETRICAL)	60 MIPS	ADA, C, FORTRAN
CULLER SCIENTIFIC SYSTEMS CORP. SANTA BARBARA, CALIF.	VECTOR AND MULTIPROCESSOR PARALLELISM EXTRACTED IN REPETITIVE AND NONREPETITIVE CODE	18-36 MIPS 3.4-22 MFLOPS	FORTRAN (SUN MICROSYSTEMS COMPATIBLE)
ELXSI SAN JOSE, CALIF.	GIGABUS	10-100 WHETSTONE MIPS	C, FORTRAN
ENCORE COMPUTER CORP. MARLBOROUGH, MASS.	NANOBUS	UP TO 20 MIPS	C, COBOL, FORTRAN, PASCAL
ETA SYSTEMS INC. ST. PAUL, MINN.	VECTOR/SCALAR/PARALLEL	10 GFLOPS	FORTRAN
FLEXIBLE COMPUTER CORP. DALLAS, TEX.	MIMD/MULTIPLE I/O STREAM	3-60 MIPS	ADA, CONCURRENT C, CONCURRENT FORTRAN
FLOATING POINT SYSTEMS BEAVERTON, ORE.	HYPERCUBE	6 MFLOPS-262 GFLOPS	C, FORTRAN
INTEL SCIENTIFIC COMPUTERS BEAVERTON, ORE.	HYPERCUBE	2.5-1,280 MFLOPS	C, FORTRAN, LISP
INTERNATIONAL PARALLEL MACHINE NEW BEDFORD, MASS.	MULTI-ACCESS MEMORY	4-60 MIPS 100-500 MFLOPS	C, FORTRAN
MASSCOMP WESTFORD, MASS.	TRIPLEBUS	.7-10 MIPS	C, FORTRAN, FRANZ LISP, ASSEMBLER
NCUBE BEAVERTON, ORE.	HYPERCUBE	8-2,000 MIPS 2-500 MFLOPS	C, FORTRAN
SCIENTIFIC COMPUTER SYSTEMS CORP. SAN DIEGO, CALIF.	CRAY XMP	UP TO 18 MIPS UP TO 44 MFLOPS	FORTRAN
SEQUENT COMPUTER SYSTEMS INC. PORTLAND, ORE.	TIGHTLY COUPLED	1.4-21 MIPS	ADA, C, ASSEMBLER, PASCAL, FORTRAN
THINKING MACHINES CORP. CAMBRIDGE, MASS.	DATA PARALLEL	10,000 MIPS	C*, *LISP
VITESSE ELECTRONICS CORP. CAMARILLO, CALIF.	DATA-DRIVEN SUPPORTING AUTOMATIC RUN-TIME OPTIMIZATION; DATA FLOW-LIKE	25-150 MFLOPS	C, FORTRAN, PASCAL

MFLOPS = MILLION FLOATING POINT OPERATIONS PER SECOND
MIMD = MULTIPLE INSTRUCTION STREAM, MULTIPLE DATA STREAM
MIPS = MILLION INSTRUCTIONS PER SECOND
UUCP = UNIX-TO-UNIX COPY

COMMUNICATIONS	OPERATING SYSTEM	PRICE RANGE
TCP/IP, ETHERNET	CONCENTRIX	$300,000-$1 MILLION
EIGHT 3-MBIT/S CHANNELS, RS-232-C DIAGNOSTIC PORT	CONCURRENT OPERATING AND MESSAGING	$75,000-$890,000
WIDEBAND ATLANTIC SATELLITE NET, ARPANET, PROTEON RING, RS-232-C, HDLC, RS-422, X.25, ETHERNET, TCP/IP	CHRYSALIS, UNIX 4.2 BSD	$40,000-$375,000
SDLC, HDLC, ADCCP, BISYNCHRONOUS, SNA, X.25, X.29, ETHERNET	OS/32	$265,000-$1 MILLION
VT-100, RS-232-C, TCP/IP	UNIX 4.2 BSD	$98,000-$1 MILLION
ETHERNET, TCP/IP, NFS, ARPANET, UUCP, RS-232-C	UNIX 4.2, BSD	$98,000-$1 MILLION
UUCP, ASYNCHRONOUS, ETHERNET, TCP/IP, DR11W, HYPERCHANNEL, XNS, HASP, 3770, 2780/3780	UNIX 4.3 BSD, EMBOX, UNIX V V.2	$399,000-$3 MILLION
ETHERNET, TCP/IP	UMAX 4.2, UMAX V	$144,000-$300,000
ETHERNET	VOS, UNIX, VSOS	$8 MILLION-$20 MILLION
ETHERNET	UNIX V, MSOS	$24,000-$600,000
ETHERNET, TCP/IP, NFS	SJE	$100,000-$1 MILLION
ETHERNET, TCP/IP	XENIX 3.0	$150,000-$850,000
ETHERNET	RUNIX	$83,000
ETHERNET, TCP/IP, X.25	RTU	$15,000-$250,000
ETHERNET, DR11W	XENIX, MS-DOS, AXIS	$10,000-$200,000
HYPERCHANNEL	CTSS, VAX	FROM $595,000
DDN, APPLETALK, RS-232-C, ETHERNET, TCP/IP, NFS, X.25, SNA	DYNIX	$60,000-$500,000
ETHERNET	VAX OR SYMBOLICS FRONT-END	$1 MILLION-$3 MILLION
ETHERNET, TCP/IP	UNIX V.3	$300,000

practice his Fortran and Unix skills, and in part by Convex customer engineers performing routine diagnostic checks.

As Wahnsiedler's staff has become more proficient in Unix, they have been taking greater advantage of the Convex's hard-wired terminals; when they want to "move things around, they have a choice of either typing Telnet on their terminals, or sliding their chairs over" and going to work directly on the Convex. The department intends to implement the Apollo virtual streams interface, allowing transparent access to Convex files from Apollo across the Ethernet.

■ *Argonne National Laboratory.* At the Department of Energy's Argonne National Laboratory, Jack Dongarra has become known as something of a Wise Man on the subject of computer performance and parallel processing. The Linpack benchmark created under his direction has become one of the most widely used measures of computing performance, gauging the time it takes computers ranging from the Cray X-MP to the IBM PC to solve a dense set of linear equations in full- and half-precision Fortran. The laboratory's Advanced Computing Research Facility (Fig. 3), a testbed for parallel computing architectures, was established in 1984 and is operated by the laboratory's Mathematics and Computer Science Division.

The ACRF network links up five commercially available parallel-processing machines. It consists of an Alliant FX/8 with eight vector processors, a Sequent Balance 2100 with 24 processors, an Encore Multimax with 20 processors, an Intel iPSC with 32 nodes, and an Intel iPSC with 16 nodes plus vectorization. The computers are all on Ethernet, use TCP/IP, and run on some flavor of Unix. They are linked through a dual-processor VAX 11/780, which acts as a host machine, performing day-to-day, nonparallel applications. The DEC computer was at ACRF before the parallel machine arrived, and acts as a gateway to Arpanet/Milnet and Tymnet. Another VAX 11/750 serves as a gateway to the Department of Energy's Magnetic Fusion Energy Network, or MFENET (see "New Cray refines research

3. Argonne. *This network supports classes on parallel computing. ACRF also runs performance evaluations on supercomputers.*

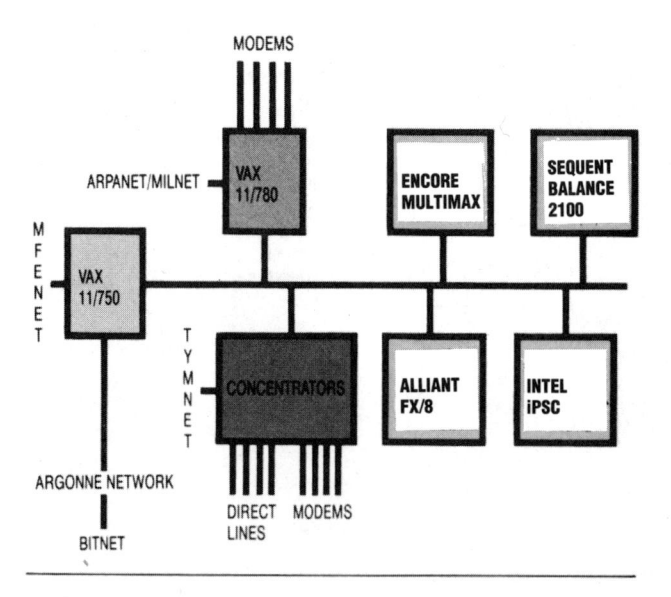

4. Wideband network. The BBN wideband Arpanet network (A) uses Butterflies as gateways. (B) shows a close-up of a network site. The network routinely carries data, packetized voice, and video, and supports real-time teleconferencing. The prototype was the 64-kbit/s Atlantic Packet Satellite Network, SATNET.

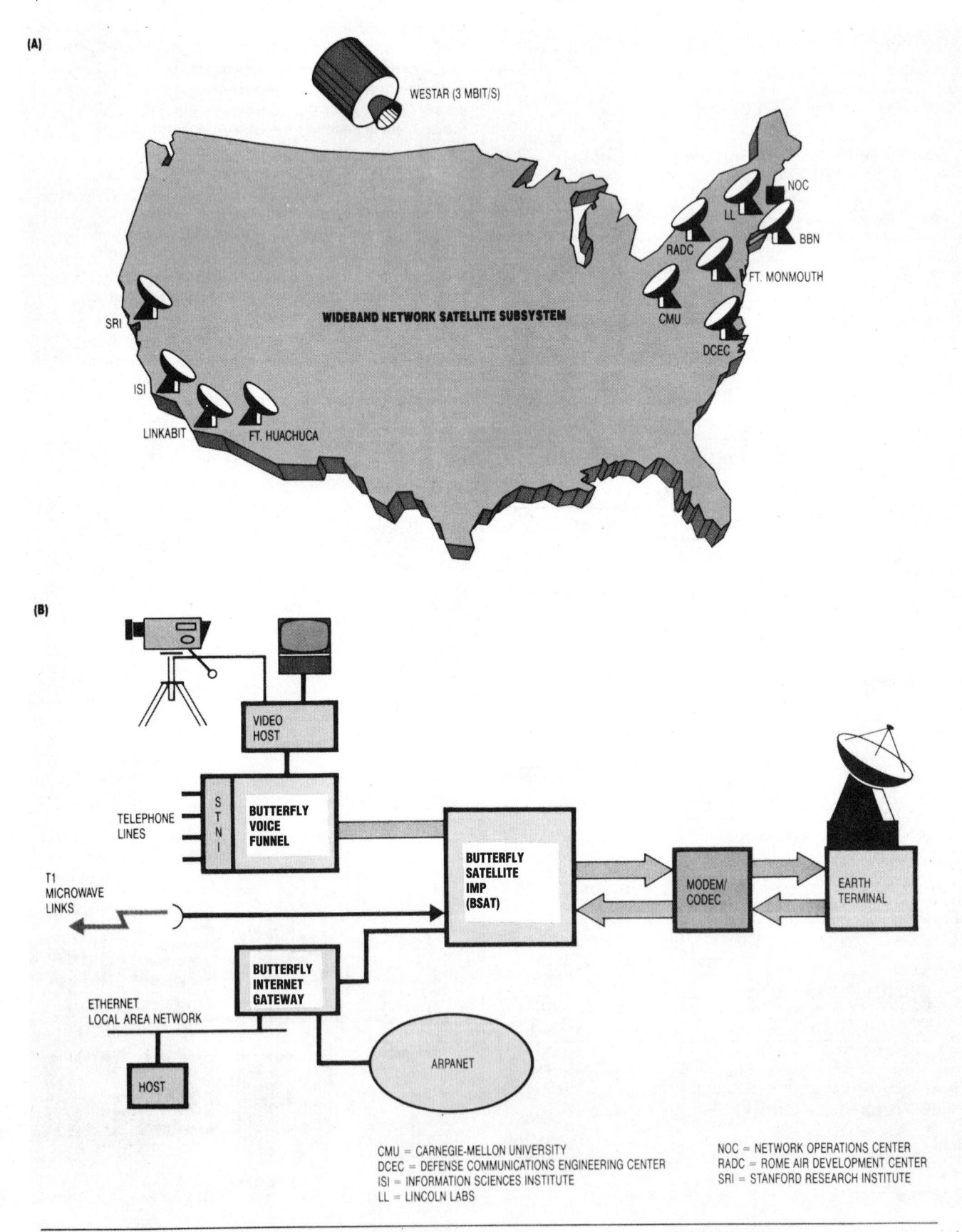

CMU = CARNEGIE-MELLON UNIVERSITY
DCEC = DEFENSE COMMUNICATIONS ENGINEERING CENTER
ISI = INFORMATION SCIENCES INSTITUTE
LL = LINCOLN LABS
NOC = NETWORK OPERATIONS CENTER
RADC = ROME AIR DEVELOPMENT CENTER
SRI = STANFORD RESEARCH INSTITUTE

network for fusion energy scientists," DATA COMMUNICATIONS, July 1985, p. 62).

■ *Arpanet.* Among the first and most famous uses of parallel processing on networks is the Arpanet (Fig. 4), the 100-plus-node network developed for the Department of Defense by BBN. Early in the 1970s, the Pluribis was used as a six-to-eight processor packet-switching node, its parallel architecture chosen because it could handle high traffic volumes, large numbers of hosts, and high-speed trunks.

The Pluribus IMP (Interface Message Processor) has been replaced in the last year or so in gateway applications by the BBN Butterfly, which connects the Arpanet to different packet-switched networks such as the Defense Data Network, LANs, wideband networks, and such satellite networks as the Atlantic Satellite Network. The Butterfly was originally developed in the late 1970s as a voice and video multiplexer in an application called the voice funnel. These machines were connected to the Pluribus Satellite IMP (PSAT). Each Butterfly handles up to four or five networks, and can provide routing and packet switching for Ethernet and Token Ring LANs.

There are more than 150 networks registered on Arpanet, although not all of them are active, according to Dennis Perry, program manager of networking C-3 technology at DARPA-ISTO (Defense Advanced Research Projects Agency-Information Science and Technology Office) in Washington, D. C. The number of networks was limited by the number of slots on the core gateway setup; this is soon to be upgraded to 300.

■ *Oak Ridge National Laboratory.* The Computer Sciences Research Group of the Mathematical Sciences Section at the Department of Energy's Oak Ridge National Laboratory in Tennessee has a parallel-processing network with extensive connections to the outside world. An Intel 310 is the host, equipped with an Ethernet board and Xenix from Excelan. A DEC VAX 780 running Berkeley 4.2 BSD Unix on a Unibus is included on the network, as is a Sequent Balance 8000 and three Sun 3 machines. Four or five IBM PCs plus a Compaq crop up on the network; "it varies from day to day," says group leader Mike Heath. The PCs use Excelan and 3Com hardware and Massachusetts Institute of Technology-created free software (PC/IP) operating under MS-DOS, and the PC ATs run Xenix and use Excelan boards. The research group obtains Unix 4.2 and 4.3 support from Mt. Xinu in Berkeley, Calif.

Tom Dunnigan, the communications expert on the research staff, explains that Applitek bridges provide connections to Ethernet, and that the group had to do a little TCP/IP programming. Its Sun graphics applications talk transparently to the hypercube nodes, creating a true distributed parallel application.

Heath says that at this point in the network's evolution, each machine functions separately—there is no file server as yet. Each user has a home machine, but can log in remotely to other machines using the Unix remote shell (rsh) utility. There are no peripherals hanging off of the Intel hypercube computer, so that these users log in remotely to the VAX and give the DEC machine, for example, commands to send printouts to its laser printer.

The network enlarges its communications universe in four ways. First, it is connected to Oak Ridge's backbone broadband network, which covers the laboratory's campus, spread out over eight miles. Second, a 9.6-kbit/s twisted-pair network connects "everybody and his dog" to 55 different networks throughout the laboratory, serving the 7,000 to 8,000 members of the Oak Ridge community. This second network provides terminal access but not file access and is also connected to the University of Tennessee's network 20 miles away in Knoxville.

The Computer Research Group's VAX 780 is a gateway to Arpanet, the Class C gateway for all of Oak Ridge. Finally, the network can log on to the network at Argonne National Laboratory. The group also has individual remote users who reach the network through Arpanet or through dial-up (as from the University of Virginia). They dial through the twisted-pair network, which has the appropriate modem connections and directs them to select System 24. That system then logs them on to the VAX, which connects them to the parallel machines. "It sounds like a kluge, but it works just fine," says Heath. A second hypercube machine manufactured by Ncube is "a bit marooned." So far, it is connected only to the 9.6-kbit/s LAN, but as soon as Ethernet connections become available for it, it will be connected to the broadband backbone.

■ *Southwestern Research Institute.* The Southwestern Research Institute is a nonprofit San Antonio, Tex., organization employing 2,200 people with an annual budget of about $135 million, working particularly on satellite technology for the National Aeronautics and Space Administration and the Air Force. David Winningham's responsibility as institute scientist for the division of space sciences and instrumentation is to build instruments and turn them into products.

The space sciences and instrumentation's network consists of three Unix PCs (which people take home occasionally), three Masscomp machines (two of which have dual processors), and one Convex C-1. Ethernet TCP/IP strings them all together, and NFS (Network File System) is used on the Masscomp and Convex machines. The network handles most flavors of Unix, including Xenix, System V, and Berkeley 4.2 BSD. Winningham says that the division is in the process of migrating from TCP/IP to NFS, while keeping Ethernet, for NFS will make SWRI's communications transparent.

One parallel application at SWRI involved the European Giotto space project, the mission that recently went past Halley's comet. As real-time information poured into the network, it was written to disk, where, through shared memory, a second processor fed the data out to the appropriate applications and to real-time graphic presentations. A second application of parallel processing involved a loosely coupled multiprocessor loaner machine sent to Germany, which broke up Fortran DO loops while working on the task of modeling instruments before they are built.

Artificial intelligence, graphics workstations, and computer accelerators are already taking advantage of parallel processing. In his book *Computer Futures*, computer scientist and writer Jonathan Post defines a supercomputer as "any machine that is only one generation behind the computing requirements of leading-edge efforts in science and engineering." If this is true, parallel processing can be the way for data communications managers to put supercomputing power to work on their networks. ■

Thomas J. Routt, Network Systems Consulting, Seattle, Wash.

Distributed SNA: A network architecture gets on track

Here is the definitive look at the evolution of IBM's Systems Network Architecture, including APPC, SNADS, and LEN.

Users are increasingly presented with distributed transaction processing requirements. These requirements derive primarily from (1) generalized decentralization trends within the organization, (2) technological migration of data processing and communications capabilities to desktop workstations, and (3) the resultant need to provide peer-to-peer connectivity and resource sharing among distributed, hostless processing environments.

Host-controlled, centralized network architectures traditionally lend themselves well to batch processing and generally maximize computational marginal cost economies of scale. However, interactive user information throughput and response times degrade rapidly as central host utilizations increase.

Rapid hardware price decelerations and workstation computational power improvements have resulted in the proliferation of distributed, desktop, and departmental small-system databases across all conceivable corporate, governmental, and university landscapes. User requirements are therefore to place computational, analytic, storage/retrieval, and distribution capabilities at the location of the decision maker, and to do so while presenting the user with a transparent, single-system image into target applications, irrespective of target machine vendor type or geographic location.

IBM has responded to these distributed transaction processing requirements by formalizing extensions to its Systems Network Architecture (SNA) specifications. These distributed extensions are based on Advanced Program-to-Program Communications (APPC) and include SNA Distribution Services (SNADS), Document Interchange Architecture (DIA), Document Content Architecture (DCA), Distributed Data Management (DDM), and Low Entry Networking (LEN). This article describes the evolution of SNA to APPC, defines the APPC-based distributed SNA architectures, and presents implementation cases of each.

SNA, announced by IBM in 1974, is IBM's strategic communications architectural blueprint from which to define, design, and implement interconnection and resource sharing among communications network products. SNA is a set of formalized architectural specifications that define the ways in which data communications products can be interconnected. These specifications provide the set of rules, logical structures, procedures, formats, and protocols that are implemented in various hardware and software products. The SNA architectural rule sets specify:
■ SNA end users. End users are application programs, Input/Output (I/O) devices, and terminal operators. They are not part of the SNA architecture, but they use SNA for interconnection and resource sharing.
■ The interface and relationship between SNA and logical network components.
■ The logical components required to interconnect and share resources among end users.
■ The operational sequences for controlling the configuration and operation of a network.

The SNA architecture is implemented in a variety of hardware and software products. The distinction between the architecture and its product expressions is that, while SNA in and of itself is a set of formal specifications, products represent implementations of these rules.

The SNA architecture defines Network Addressable Units (NAUs) as the logical network components that provide communications ports through which end users can gain access to each other. There are three types of NAUs: Logical Unit (LU), Physical Unit (PU), and System Services Control Point (SSCP). All NAUs reside within SNA "nodes," which are end points that connect links and perform SNA-defined functions. NAUs are logical locations within an SNA network that have associated with them unique network names and addresses. Figure 1 indicates

1. Porting in. *Network addressable units (NAUs) reside within SNA nodes and provide end users access to SNA networks. End users port into Logical Units.*

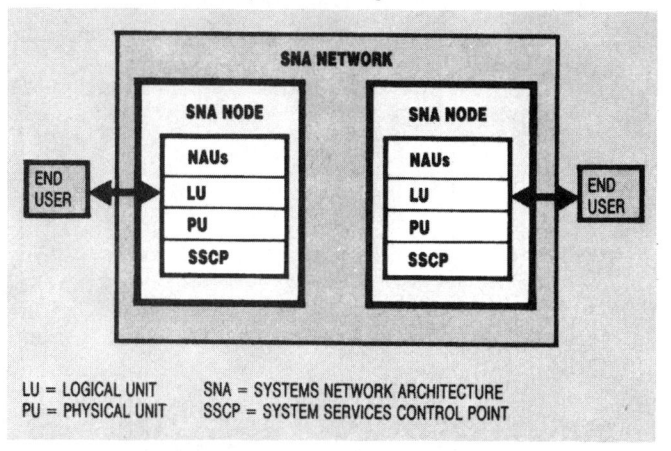

```
┌─────────────────── SNA NETWORK ───────────────────┐
│   ┌─── SNA NODE ───┐      ┌─── SNA NODE ───┐      │
│   │     NAUs       │      │     NAUs       │      │
┌───┤├──│     LU         │      │     LU         │──┤├───┐
│END││  │     PU         │◄────►│     PU         │  │END │
│USER│  │     SSCP       │      │     SSCP       │  │USER│
└────┘  └────────────────┘      └────────────────┘  └────┘
└───────────────────────────────────────────────────┘

LU = LOGICAL UNIT      SNA = SYSTEMS NETWORK ARCHITECTURE
PU = PHYSICAL UNIT     SSCP = SYSTEM SERVICES CONTROL POINT
```

2. Breakdown. *Here are LU Categories and Session Types. Non-Program-to-Program Types assume a hierarchical session between host applications and I/O devices.*

CATEGORY	TYPE	FUNCTION	USER CASES
NOT SPECIFIED BY SNA	0	IMPLEMENTATION DEFINED SPECIALIZED TRANSACTIONS	IMS/VS TO 3600
TERMINALS	1	HOST APs ACCESS NON-DISPLAY I/O KEYBOARD PRINTERS USE SCS	CICS/VS TO 3767
	2	HOST APs ACCESS 3270 DISPLAY TERMINALS USE 3270 DSC	CICS/VS TO 3278
	3	HOST APs ACCESS PRINTERS USE 3270 DSC SUBSET	CICS/VS TO 3289
	4	HOST APs ACCESS DP TERMINALS USE SCS	IMS/VS t0 6670
	7	APs ACCESS DISPLAY TERMINALS USE SCS OR 5250 DS	S/34 TO 5251
PROGRAM-TO-PROGRAM	6.0	TWO CICS/VS APs IN SESSION SAME HOST	CICS/VS TO CICS/VS
	6.1	CICS/VS APs, IMS/VS APs, CICS/VS-IMS/VS APs, SAME OR REMOTE HOST	IMS/VS TO CICS/VS
	6.2 (APPC)	GENERAL-PURPOSE PROGRAM-TO-PROGRAM	CICS/VS TO CICS/VS, DISOSS TO PS/36, APPC/PC TO APPC/PC

APPC = ADVANCED PROGRAM-TO-PROGRAM COMMUNICATIONS
APs = APPLICATION PROGRAMS
CICS/VS = CUSTOMER INFORMATION CONTROL SYSTEM/VIRTUAL STORAGE
IMS/VS = INFORMATION MANAGEMENT SYSTEM/VIRTUAL STORAGE
I/O = INPUT/OUTPUT
LU = LOGICAL UNIT
SCS = SNA CHARACTER STRING
SNA = SYSTEMS NETWORK ARCHITECTURE

the relationship between end users and NAUs within an SNA network.

LUs are the logical ports through which end users gain access to SNA network resources. The primary LU function is allocation of resources to end users. These resources include machine processor cycles, real and virtual storage, Direct Access Storage Devices (DASDs), I/O devices, terminal keyboards and displays, queues, database records, and sessions.

Sessions in SNA are temporary, logical connections between two NAUs. Sessions that directly connect end-user application programs or devices are called LU-LU sessions. All other sessions are control sessions between the SSCP and other SSCPs, PUs, or LUs. Figure 2 illustrates the three LU categories and the LU session types within those categories. Note that all of the non-LU 6.n (Program-to-Program) LU session types assume a hierarchical session between a host-resident application program and a dumb I/O device (a terminal or printer).

PUs are software-based resource managers within SNA nodes. PUs provide a location for configuration services, request software downloads into nodes, and generate node diagnostic information, including storage dumps and activity trace information. PUs are classified into PU Types 5, 4, 2.1, 2.0, and 1, depending on the SNA node function provided.

A PU Type 5 node is a host node. Host nodes are any System/370-architecture host processor, such as the 3090, 3080, 3030, 4300, or 9370 series machines, which support SSCP software. A PU Type 4 node is a communications controller (COMC), such as the Model 3705, 3725/26, and 3720/21, which support Network Control Program (NCP) software. Both PU Type 5 and 4 nodes are also called subarea nodes because they support global (subarea) addressing capabilities and are cognizant of global networked environments. Also, subarea nodes are involved in end-point as well as intermediate network routing.

A PU Type 2.0 node is a cluster controller (CLC) node, such as the Model 3174, 3274, and 3276 CLCs, which support PU Control Point (PUCP) software. PU Type 2.0 nodes support a single link between themselves and their immediately upstream subarea (Boundary Function) node. This link can be either a Synchronous Data Link Control (SDLC) link for remote CLCs downstream from a PU Type 4 COMC running NCP, or a host data channel for local CLCs directly attached to a PU Type 5 host running SSCP.

A PU Type 2.1 node can be either a Network Node (NN) or a Peripheral Node (PN). PU Type 2.1 is also referred to as an Advanced CLC Node. This node is implemented by System/36, System/38, 5520, 8100, Series/1, and the IBM Personal Computer, which support Peripheral Node Control Point (PNCP) software. PNCP implements a major SSCP subset, and permits PU Type 2.1 nodes to control SNA sessions directly.

A PU Type 1 node is a "Pre-SNA Device," and is also referred to as a Terminal Node (TN). PU Type 1 TNs support non-SNA link-level protocols, such as Asynchronous and 3270 Binary Synchronous Communications (BSC), as well as SDLC. Implementation examples include the 3271 and 3272 devices. Figure 3 illustrates the SNA PU types.

The third NAU, SSCP, is a host-resident NAU, which is responsible for network management and is a subset of

Virtual Telecommunications Access Method (VTAM). SSCP is the controller of an SNA domain. In SNA, a "domain" is defined as one SSCP and all of the PUs, LUs, links, and link station addresses defined to it, initialized by it, and deactivated by it. The host node is the only SNA node that contains all three NAU types (SSCP, PU, LU).

SNA nodes and their constituent NAUs are logical network components implemented within physical devices that are physically and logically interconnected in various ways. Subarea nodes (PU Types 5 and 4) include hosts and COMCs, and may be interconnected in either channel-attached (host-to-host, host-to-COMC) or link fashion. Subarea nodes may be connected to other subarea nodes, or to peripheral nodes (PU Types 2.0, 2.1, 1) via dedicated or switched circuits over privately owned or public data networks.

Control within a pre-APPC SNA network is hierarchically cascaded from the SSCP to attached NCPs, and from each NCP to downstream CLCs running PUCP. Each NCP is a Boundary Function (BF) subarea for directly attached CLCs, TNs, and devices, as is each SSCP for channel-attached peripheral nodes. Each NCP is also an Intermediate Network Node (INN) to support the routing of traffic through its subarea on to other subarea nodes. Figure 4 depicts the arrangement of the various NAUs (SSCP, PU, LU) within a two-domain, four-subarea SNA network. Note also that the System/36 attached over an SDLC link to the right PU Type 4 372X appears to this BF subarea node as a PU Type 2.0, whereas the System/38 attached downstream of the System/36 recognizes the System/36 as a PU Type 2.1 node.

SNA is organized into functional layers, each of which provides specific services. The major advantage of layering is the creation of functional modularity, thus ensuring that changes to one layer do not adversely affect another. This functional modularity provides enhancement flexibility and allows for functions that are logically independent to be designed, implemented, and invoked independently.

IBM's SNA architectural reference document, "Systems Network Architecture Format and Protocol Reference Manual: Architectural Logic," most recently published in 1980, delineates six SNA layers. Other IBM documents published during 1985 indicate that there are seven SNA layers (see references).

Layer 7, Transaction Services, provides the end-user logical interface. During October 1986, IBM formalized the Transaction Services Architectures, which structure this interface. The concern at this layer is with providing end users with a single-system image into the network services, through common command-language formats.

Layer 6, Presentation Services, formats data for various presentation media, and, as such, is concerned with syntactic and semantic data representations. Layer 5, Data Flow Control, is concerned with the correlation of data exchange and flow synchronization during sessions.

Layer 4, Transmission Control, is responsible for end-to-end connectivity, session activation/deactivation, and session pacing. Layer 3, Path Control, provides network route selection, class of service, and message segmentation/blocking.

Layer 2, Data Link Control, initializes, transfers data over, and disconnects logical links between adjacent nodes.

3. PU Types. *This chart shows the Physical Unit Types, including the host node, Type 5, and the pre-SNA Type 1. PUs are software-based resource managers within SNA.*

PU TYPE	NODE TYPE	HARDWARE	SOFTWARE
5	HOST	S/370 309X, 308X, 303X, 43XX, 937X	ACF/VTAM ACF/VTAME SSCP
4	COMC	3725, 3726, 3720, 3721, 3705, 3706, 3704 (37XX)	ACF/NCP/VS
2.0	CLC	3174, 3274, 3276	PUCP
2.1	NN OR PN	S/36, S/38, PC, DISPLAYWRITER, 5520, SERIES/1, 8100	PNCP
1	TN	3271, 3272	PUCP

ACF/VTAM = ADVANCED COMMUNICATIONS FUNCTION/VIRTUAL TELECOMMUNICATIONS ACCESS METHOD
ACF/VTAME = ACF/VTAM ENTRY
CLC = CLUSTER CONTROLLER
COMC = COMMUNICATIONS CONTROLLER
NCP = NETWORK CONTROL PROGRAM
NN = NETWORK NODE
PN = PERIPHERAL NODE
PNCP = PERIPHERAL NODE CONTROL POINT
PU = PHYSICAL UNIT
PUCP = PHYSICAL UNIT CONTROL POINT
SSCP = SYSTEM SERVICES CONTROL POINT
TN = TERMINAL NODE

4. Relationships. *SNA network with two domains and four subareas shows the relationship between end-user devices and application programs.*

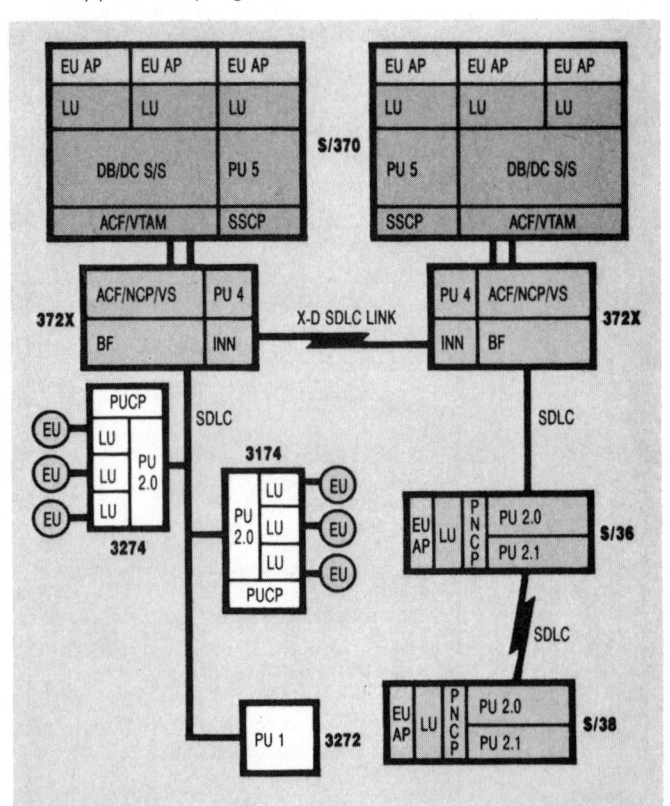

ACF/VTAM = ADVANCED COMMUNICATIONS FUNCTION/
    VIRTUAL TELECOMMUNICATIONS ACCESS METHOD
AP = APPLICATION PROGRAM
BF = BOUNDARY FUNCTION
EU = END USER
INN = INTERMEDIATE NETWORK NODE
LU = LOGICAL UNIT
NCP = NODE CONTROL POINT
DB/DC S/S = DATABASE/DATA
    COMMUNICATIONS SUBSYSTEM
PNCP = PERIPHERAL NODE CONTROL POINT
PU = PHYSICAL UNIT
PUCP = PHYSICAL UNIT CONTROL POINT
VS = VIRTUAL STORAGE

5. Layers and Services. *SNA is organized into seven functional layers, each providing specific services. This ensures that changes to one layer don't affect another.*

NUMBER	LAYER NAME	SERVICES
7	TRANSACTION SERVICES	END-USER LOGICAL INTERFACE SINGLE-SYSTEM INTERFACE
6	PRESENTATION SERVICES	DATA PRESENTATION SYNTACTIC, SEMANTIC TRANSFORMS
5	DATA FLOW CONTROL	SESSION DATE EXCHANGE CORRELATION HALF-SESSION DATA FLOW
4	TRANSMISSION CONTROL	END-TO-END CONNECTIVITY SESSION ACTIVATION/DEACTIVATION SESSION PACING
3	PATH CONTROL	NETWORK ROUTE SELECTION CLASS OF SERVICE MESSAGE SEGMENTATION/BLOCKING
2	DATA LINK CONTROL	LOGICAL LINK INITIALIZATION DATA TRANSFER LINK DISCONNECTION
1	PHYSICAL CONTROL	DTE/DCE PHYSICAL INTERFACE ELECTRICAL, MECHANICAL, FUNCTIONAL, PROCEDURAL

DTE/DCE = DATA TERMINAL EQUIPMENT/DATA CIRCUIT-TERMINATING EQUIPMENT

6. Construction and encapsulation. *The Request/Response Unit (RU) is constructed first by the highest SNA layers (7, 6, 5, 4) and is a variable-length field.*

| LH | TH | RH | FMH | RU | LT |

BIU spans RH, FMH, RU
PIU OR BTU spans TH, RH, FMH, RU
BLU OR SDLC FRAME spans all

BIU = BASIC INFORMATION UNIT
BLU = BASIC LINK UNIT
BTU = BASIC TRANSMISSION UNIT
FMH = FUNCTION MANAGEMENT HEADER
LH = LINK HEADER
LT = LINK TRAILER
PIU = PATH INFORMATION UNIT
RH = REQUEST/RESPONSE HEADER
RU = REQUEST/RESPONSE UNIT
TH = TRANSMISSION HEADER

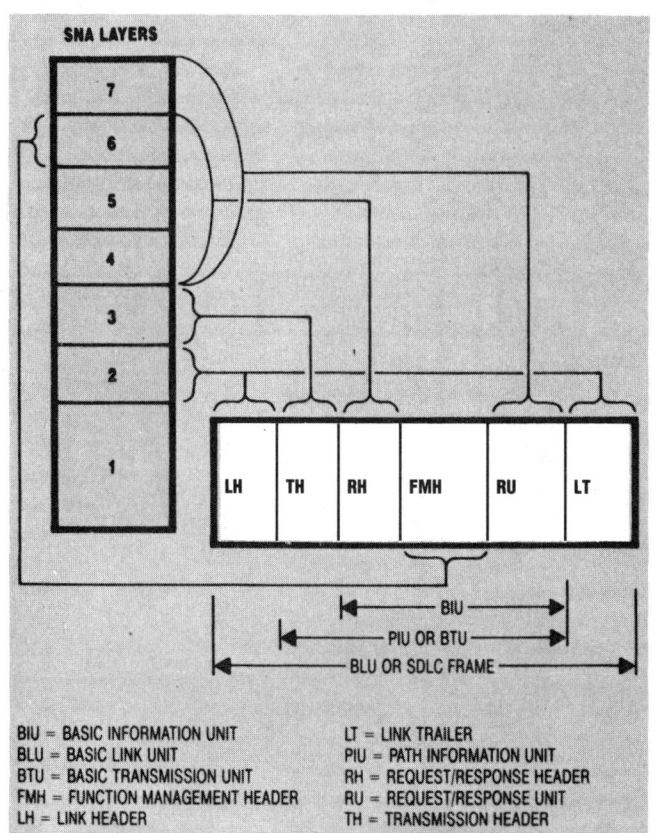

Layer 1, Physical Control, physically connects data terminal equipment (DTE) to data circuit-terminating equipment (DCE) devices.

Figure 5 depicts the seven SNA layers and the services they provide.

The construction of SNA message units

SNA message units are constructed by the various layers, from the highest operative layers, and successively encapsulated within headers provided by lower-layer protocols within SNA nodes. In this way, peer layers in adjacent or end-point nodes recognize, decapsulate, operate on, and encapsulate the message unit constructed by their architectural counterparts elsewhere within the network. Figure 6 illustrates the relationship among the SNA architectural layers, as well as the message units for which each is responsible. Note that the Request/response Unit (RU) is the fundamental SNA message unit, containing end user or network control information. The RU consists of field-formatted binary data assembled by any of Layers 4 through 7 within either a PU Type 5, 2.1, 2.0, or 1 node.

Layer 6 can also, for certain LU session types (such as LU Type 1, 6.0, 6.1, 6.2), construct a Function Management Header (FMH), which is a mechanism to pass control information within sessions that convey precise instructions (selection of an end-user destination device for a multi-device LU) or specific data operations, such as compression, compaction, or output instructions.

The Request/response Header (RH) is constructed by Layer 4, 5, or 6, and conveys and enforces session-unique profiles and protocols. Note in Figure 6 that the combination of RU, optional FMH, and RH creates a Basic Information Unit (BIU).

The Transmission Header (TH) is constructed by Layer 3 and is used by Path Control to perform node routing. It supports various Class of Service (COS) levels based on Virtual Route (VR) Transmission Priority (TP) definitions. The encapsulation of a BIU behind a TH creates a Path Information Unit (PIU).

Layer 2 encapsulates the PIU with a Link Header (LH) and Link Trailer (LT). The result is a Basic Link Unit (BLU). If the BLU traverses an SDLC link, it defines an SDLC frame. If, however, the Layer 2 link protocol is not SDLC, such as a Token-Ring Network, the PIU would be encapsulated within the appropriate Layer 2 protocol header and trailer (in the case of the Token-Ring Network, within a Physical Header and Physical Trailer defined by IEEE 802.5 Token-Passing Ring Medium Access Control, and, optionally, IEEE 802.2 Logical Link Control, frame).

Centralized SNA network evolution

SNA networks were originally characterized as simple, tree-oriented, single-host, single COMC environments. While these earliest versions supported only one host and COMC node, the major enhancements over pre-SNA included:

■ Definition of VTAM within the host, which consolidated the functionality of pre-SNA access methods and offloaded many network overhead functions from application programs and subsystems.

■ Definition of NCP within the COMC, which provided software programmability not found in pre-SNA environments.

- Elimination, to a great extent, of device and protocol dependencies that had previously required separate terminals and lines for end-user access to separate, host-resident applications.

Figure 7 depicts a single-host (single-domain) SNA network.

During 1976, IBM announced Advanced Communications Function (ACF) for VTAM and NCP. ACF/VTAM and ACF/NCP, together with host-resident Multisystems Networking Facility (MSNF), provide the capability to define multihost, multi-COMC networks (multidomain networks). Figure 8 depicts an early, multidomain SNA network.

By 1979, SNA network connectivity had been extended to support:
- Cross-domain links between remote as well as local NCPs.
- Parallel links between adjacent NCPs.
- Logical consolidation of adjacent-NCP links into Transmission Groups (TGs).
- Multiple and alternate routing between end-point subarea nodes over Explicit Routes (ERs), Reverse Explicit Routes (RERs) and VRs. Figure 9 depicts this extended connectivity and routing.

SNA network connectivity was further extended during 1983 when IBM introduced SNA Network Interconnection (SNI). SNI is a logical extension of cross-domain networking, which allows end-user devices and application programs to gain access to each other from distinct and separate SNA networks. Each participating, interconnected SNA network is able to preserve its unique naming and addressing characteristics, as well as unique internal network management procedures and controls. Figure 10 illustrates an SNI environment in the simplest SNI configuration. In it, the Gateway NCP, which physically interconnects the two SNA networks, contains two subarea addresses, one for each network. Also note that the Gateway VTAM host can reside within either network.

The strategic objective of SNI is to interconnect separate networks, each of which supports internally unique global naming and addressing conventions and provides cross-network connectivity and resource sharing transparent to these differences. A tactical objective of SNI was to remove addressing constraints from large SNA networks. This was made possible by breaking the network address (at that time, a 16-bit field divided into MAXSUBA, or maximum number of subareas per network, and element addresses per subarea) into smaller fields within subnetworks interconnected through SNI.

During 1984, IBM introduced Extended Network Addressing (ENA), which expands a single network address space to allow for up to 255 subareas and up to 32,000 elements per subarea, thereby supporting 8.35 million element addresses per single SNA network. ENA was provided to relieve the addressing constraints within larger SNA networks. Figure 11 depicts the ENA addressing approach.

Additional SNA extensions during this period included:
- Virtual Storage Constraint Relief, which allows Multiple Virtual Storage/Extended Architecture (MVS/XA) applications using 31-bit addresses (addressing 2 gigabytes of virtual memory) to communicate with ACF/VTAM through a standard Application Program Interface (API).
- Native Virtual Machine/Support Program (VM/SP) SNA support, which allows application programs running under the VM/370 operating system to communicate natively through SNA.
- Network Concentration and Protocol Conversion. During this period, the Model 3710 was announced as a network concentrator, supporting up to 31 downstream lines, and protocol converter (supporting Asynchronous, 3270 BSC, and CCITT X.25). Several other hardware protocol converters and software terminal emulators (e.g., network terminal option and non-SNA interconnect) have also been provided.

APPC SNA

Despite the evolution of subarea node connectivity, routing, and addressing extensions, SNA networks have been classically optimized to support connection of dumb terminals (end users) to host-resident application programs (end users). All global design parameters, including application subsystem, VTAM, and NCP buffer sizes and dispatch algorithmic Path Control Network routing over ERs, RERs, and VRs, and session-pacing techniques, are tuned to assume continuous connection of dumb devices to hosts.

These classical networks were never intended to provide for timely access of nonhost, standalone, data processing subnetworks. Today, however, end-user environments are increasingly characterized by PCs and/or other hostless (small-system) environments. These dumb-terminal replacements can behave within networks as dumb-terminal emulation devices, standalone data processing machines, and/or file transfer nodes. The latter two characteristics of small-system nodes fundamentally violate classical SNA network design assumptions. As a result, their integration into these classical networks has resulted in poor user information throughput and response times, even under lightly loaded network conditions; and highly over- or underused aggregate network load profiles, depending upon whether small systems users are engaged in file transfer or standalone processing, respectively.

IBM's strategic response to support distributed, small systems within classical, subarea networks was the evolution of APPC. APPC is a distributed extension to the SNA architecture and defines the rules of operation of LU 6.2. LU 6.2 is a member of the Program-to-Program category of SNA LUs, and is the centerpiece to IBM's continuing evolution of the SNA architecture. Whereas all previous, non-LU 6.n Types logically port dumb I/O devices and host-resident application program end users, LU 6.2 structures an API to port-distributed application transaction programs (ATPs) and enable program-to-program access and resource sharing. The APPC API defines a structured protocol boundary between ATPs and APPC.

Figure 12 depicts the LU 6.2 relationship to ported applications. Note that LU 6.2 (APPC) invokes the services of SNA Layers 6, 5, and 4, and that the ATP is external to the SNA architecture.

ATPs communicate with APPC over the API through a generic set of commands called verbs. Verbs are distributed processing programming statements that structure "conversations" over which distributed ATPs communicate. Conversations represent serialized, time-slice segments of LU 6.2 sessions, and are provided to ATPs as an LU 6.2

7. Eliminated incompatibilities. Single-domain SNA network circa 1974 supported single-host connection of application programs to end-user devices.

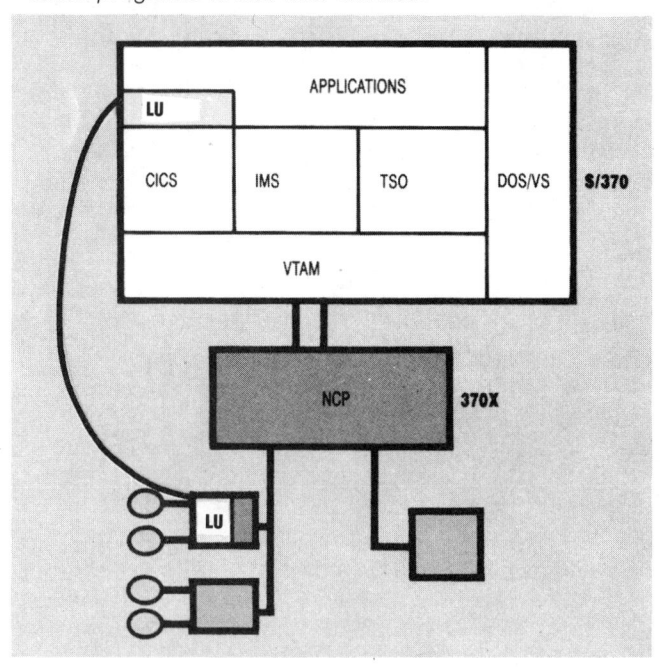

8. Multiple-host log-on. Early multidomain SNA network, circa 1976, supports end-user devices in logging-on to applications programs in multiple hosts.

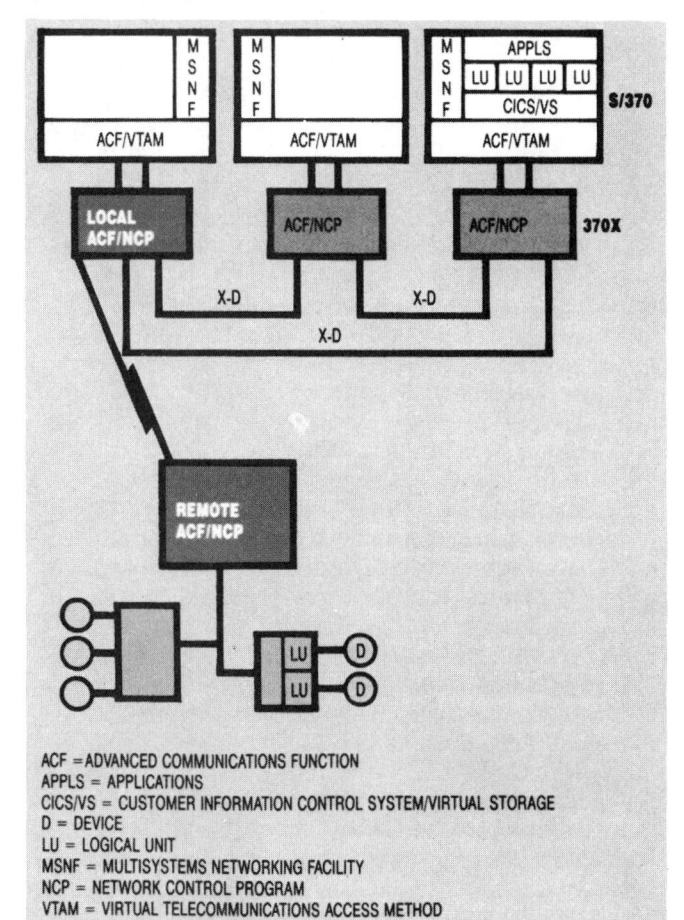

ACF = ADVANCED COMMUNICATIONS FUNCTION
APPLS = APPLICATIONS
CICS/VS = CUSTOMER INFORMATION CONTROL SYSTEM/VIRTUAL STORAGE
D = DEVICE
LU = LOGICAL UNIT
MSNF = MULTISYSTEMS NETWORKING FACILITY
NCP = NETWORK CONTROL PROGRAM
VTAM = VIRTUAL TELECOMMUNICATIONS ACCESS METHOD
X-D = CROSS-DOMAIN

resource to enable interprogram communications.

Figure 13 differentiates the logical relationships among ATP conversations, LU 6.2 sessions, and SNA Path Control Network connectivity. Conversations are recognized only by peer ATPs; sessions between adjacent APPC (LU 6.2) environments invoke SNA Layers 6, 5, and 4, and provide conversation resources to the ATPs; and the SNA Path Control Network provides the underlying SNA Layers 3, 2, and 1 connectivity. Note also that LU 6.2 is supported with PU Type 5 (Host) and PU Type 2.1 (Advanced CLC) nodes.

The API, defined as a structured protocol boundary between APPC and the ATPs, provides a standard transaction protocol boundary among distributed interprogram environments; permits interprogram communications to proceed transparent to the intricacies of the underlying APPC session and Path Control Network environments; and, in so doing, isolates application developers from network issues.

The verb statements passed across the APPC API comprise the following categories:
- Basic Conversation Verbs;
- Mapped Conversation Verbs;
- Type-Independent Conversation Verbs; and
- Control Operator Verbs.

Whereas all verbs are passed between programs and APPC, only conversation verbs are carried between LUs over an LU 6.2 session. Control Operator verbs are passed only over the structured protocol boundary within a single LU.

Figure 14 distinguishes the relationships between conversation and control operator environments. It is shown that conversation verbs passed between ATPs delineate conversations, and are serially time-sliced within LU-to-LU sessions.

Basic Conversations correspond to the most basic services provided by LU 6.2, and are intended for use by LU Service Transaction Program (STPs). LU STPs include, for example, SNA Distribution Services (SNADS), Document Interchange Architecture (DIA), Change Number of Sessions (CNOS) Model, and Resynchronization (Resync) Model. LU STPc can additionally provide protocol boundaries for end-user ATPs. For example, Basic Conversation Verbs can be issued during the processing of Mapped Conversation Verbs. The Basic Conversation Verbs and their respective functions are:

ALLOCATE. Begins, or allocates, a conversation that logically connects two transaction programs.

CONFIRM. Sends a confirmation request from a local to a remote transaction program. The local transaction program then waits for a reply to enable the two transaction programs to synchronize their execution.

CONFIRMED. Sends a confirmation reply to the transaction program, which issued the CONFIRM verb, and therein enables synchronization of local and remote transaction program execution.

DEALLOCATE. Ends, or deallocates, a conversation from the transaction program.

FLUSH. Sends all buffered data from the local LU.

GET-ATTRIBUTES. Returns conversation-specific information, such as local and remote LU names and conversation synchronization levels to the issuer.

POST-ON-RECEIPT. Requests the posting of a specified

conversation when such information as data, conversation status, or request for confirmation or sync point processing are available for the transaction program to receive.

PREPARE-TO-RECEIVE. Changes the conversation from send to receive states in preparation for the reference transaction program to receive data. The underlying LU session at Data Flow Control (SNA Layer 5) supports conversation with a Half-Duplex Flip-Flop protocol.

RECEIVE-AND-WAIT. Waits for information to arrive on the conversation, and subsequently receives the information.

RECEIVE-IMMEDIATE. Receives any information that is available on a specified conversation. If none is available, this verb does not allow waiting for information to arrive.

REQUEST-TO-SEND. The local transaction program notifies the remote transaction program that the local is requesting to enter a conversation send state.

SEND-DATA. Sends data from a local to a remote transaction program. The data format is defined in logical records under an LU 6.2 data stream called General Data Stream (GDS). A GDS logical record contains a length field and data field. The length field is 2 bytes long, and references the length of the data field, which can range from 0 to 32,765 bytes.

SEND-ERROR. Sends detected-error notification from a local to a remote transaction program.

TEST. Tests the conversation to ascertain whether it has been "posted" or whether a REQUEST-TO-SEND notification has been received by the local LU.

Mapped Conversations are distinguished from Basic Conversations in that Mapped Conversations are easier for communicating transaction programs to use, because they provide data formatting services that programs would have to perform for themselves during Basic Conversations. In other words, transaction programs that allocate Mapped Conversations do not need to ensure that data sent over the conversation conform to the format of the conversation GDS-defined data stream. Specifically, transaction programs sending over Mapped Conversations need not prefix data with GDS logical length fields nor complete the logical record that they are sending before leaving the SEND state.

Mapped Conversation Verbs are designed to support conversations between two ATPs, and are essentially the same set of verbs as the Basic Conversation Verbs. The syntax distinction is that Mapped Conversation Verbs begin with the prefix MC-. Whereas Basic Conversations are bounded by the verbs ALLOCATE and DEALLOCATE, Mapped Conversations are bounded by the verbs MC-ALLOCATE and MC-DEALLOCATE. These Mapped Verbs are particularly suitable for application programs written in higher-level languages because the verbs support data mapping and make the requirements for managing GDS and logical record formats transparent to the applications.

Type-Independent Verbs support both Basic and Mapped Conversations and include Syncpoint and Backout processing.

Control Operator Verbs define the structured protocol boundary for control operator transaction programs to:
- Define local LU and PU resources.
- Initiate the LU 6.2 session.
- Control the LU 6.2 session through modification sequences (such as CNOS). Figure 15 summarizes the major

9. Multiple support. Extended connectivity and routing are possible, including across domains and between remote and local NCPs and between adjacent NCPs.

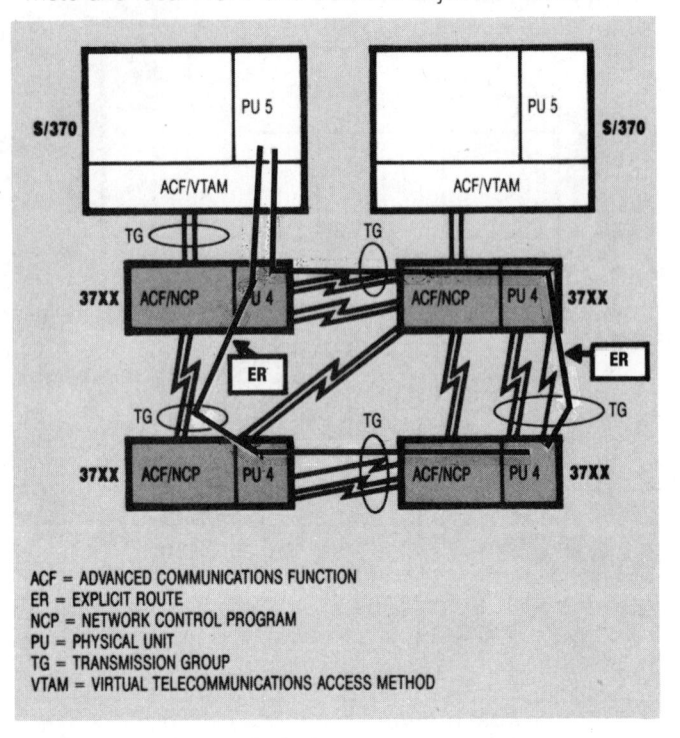

ACF = ADVANCED COMMUNICATIONS FUNCTION
ER = EXPLICIT ROUTE
NCP = NETWORK CONTROL PROGRAM
PU = PHYSICAL UNIT
TG = TRANSMISSION GROUP
VTAM = VIRTUAL TELECOMMUNICATIONS ACCESS METHOD

10. Session extension. SNA network interconnection (SNI) logically extends cross-domain session capabilities through multiple networks.

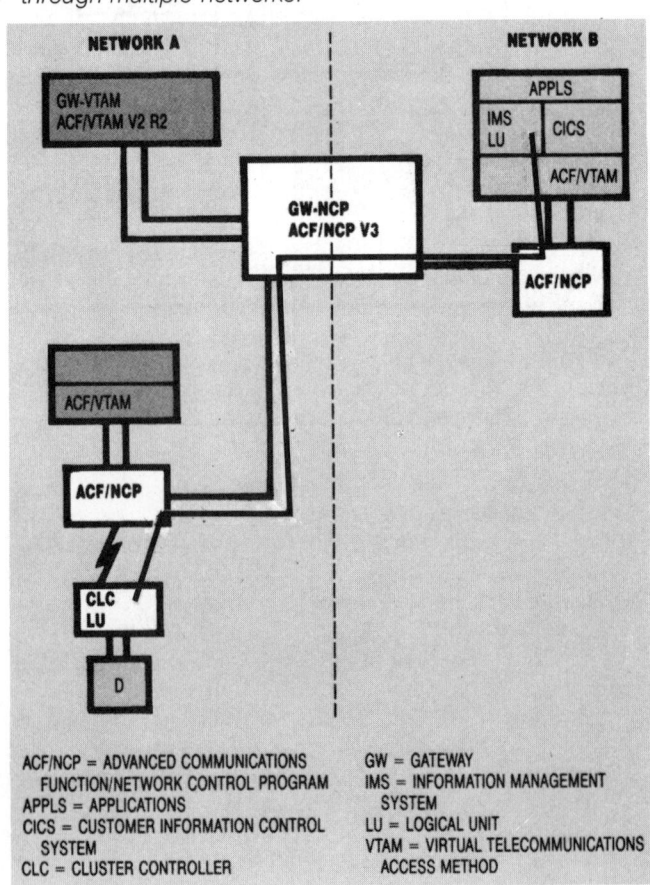

ACF/NCP = ADVANCED COMMUNICATIONS FUNCTION/NETWORK CONTROL PROGRAM
APPLS = APPLICATIONS
CICS = CUSTOMER INFORMATION CONTROL SYSTEM
CLC = CLUSTER CONTROLLER
GW = GATEWAY
IMS = INFORMATION MANAGEMENT SYSTEM
LU = LOGICAL UNIT
VTAM = VIRTUAL TELECOMMUNICATIONS ACCESS METHOD

verb categories that structure protocol boundaries from ATPs, STPs, and Control Operator Programs (COPs) into LU 6.2.

APPC supports a Base Set of nine conversation verbs, and each LU 6.2 must support the Base Set. LU 6.2 also defines several Option Sets that are product implementation-unique. APPC permits peer LU verb option-set communications only when the LUs in session support both the Base Set and identical Option Sets.

SNA Distribution Services
IBM has defined APPC as the distributed architectural baseline for the Transaction Services Architectures, which include SNADS, DIA, and Distributed Data Management (DDM). These architectures rely on APPC to provide a synchronous, logical point-to-point service for cooperating ATPs within distributed transaction processing networks.

SNADS is a formalized set of Distribution Transaction Programs that run within an APPC LU 6.2 environment to provide users with a store-and-forward distribution service. APPC provides for the synchronous transmission of SNADS distributions between adjacent nodes within a SNADS-defined network.

The SNADS store-and-forward distribution service is considered to be asynchronous in the sense that SNADS users do not have to be concurrently available to the network, and because SNADS is a network of queues that provides a connectionless, noninteractive service, allowing transaction programs that cooperate to perform distributions.

SNADS users can include individual workstations, small systems, ATPs, and databases. They are defined as addressable locations on whose behalf ATPs originate and accept delivery of distributions. Users are further characterized as either "originating users," who request a SNADS Distribution Function to be performed, and "destination users," who receive distributions from the SNADS network.

The unit of work performed by SNADS is a "distribution." SNADS delivers a distribution on behalf of SNADS users, in the form of a "distribution object," which includes user information and SNADS control information. SNADS distribution objects are contained within Interchange Units, which are characterized as either a Distribution Interchange Unit (DIU) or an Acknowledge Interchange Unit (AIU).

DIUs and AIUs are passed between SNADS Distribution Service Units (DSUs) within a SNADS network. A DSU is a collection of DTPs and data structures that provide distribution functionality. Figure 16 illustrates the architectural relationship between ATPs that issue SNADS verbs to perform distribution functions (these verbs are DISTRIBUTE-DATA, RECEIVE-DISTRIBUTION, and DISTRIBUTION-STATUS) across the APPC API, LU 6.2 APPC, and SNADS Presentation Services and Service Transaction Programs (which are defined within LU Presentation Services).

Figure 17 depicts a SNADS network. Note that SNADS DSUs can be implemented within Distributed Office Support System/370 (DISOSS/370) under Customer Information Control System/Virtual Storage (CICS/VS) on an MVS host, System/36, System/38, 5520, Series/1, or 8100. Also note that Originating and Destination ATPs are associated with SNADS users, which can be implemented with PCs, DisplayWriters, and Scanmasters.

Document Interchange Architecture (DIA) is an APPC-based Transaction Services Architecture that provides document interchange capabilities through distributed office systems. DIA defines the protocols and data structures that enable distributed office functions, such as electronic mail, document storage and retrieval, document distribution, file and bulk data transfer, and distributed application processing. These functions are provided for:
- Document Distribution Services (DDS), which distributes documents and messages to one or more distribued recipients.
- Document Library Services (DLS), which enables distributed users to file, retrieve search, and delete documents in a DIA document library.
- File Transfer Services (FTS), which supports DLS.
- Application Processing Services (APS), which provides for document processing, such as format transforms, descriptor modification, and user-supplied program execution.

These four DIA Services are enabled through DIA Function Sets, which define DIA Session command and response formats to allow users to perform desired functions with a DIA Network.

A DIA Network consists of Office System Nodes (OSNs) and Source/Recipient Nodes (SRNs). OSNs support all four DIA Services on behalf of end users located at SRNs. Figure 18 depicts a DIA/SNADS Network. Note that DIA SRNs (which can be implemented within PCs, Scanmasters, and DisplayWriters) do not communicate directly, but rather through OSNs, which can be implemented within DISOSS/370 in a CICS/VS MVS host, System/36, System/38, 5520, Series/1, or 8100.

It can also be seen that OSNs communicate with SNADS, so the distinction between DIA and SNADS is that SNADS provides remote document distribution between adjacent SNADS DSUs, including DIA OSNs, while DIA supports local Document Distribution, Document Library, File Transfer, and Application Processing Services between SRNs and OSNs.

DIA defines a data stream that is structured by a DIU. The DIA DIU is encapsulated within an SNA RU for exchange within an APPC LU 6.2 session. The logical relationship between DIA and LU 6.2 sessions is shown in Figure 19. Note that a "Requester," a DIA Process, can "ask" for DIA Services (DDS, DLS, FTS, or APS) and that a "Server," also a DIA Process, provides the requested service through either a local or a remote process.

Document Content Architecture
Document Content Architecture (DCA) is an APPC-based data stream that standardizes the content of document objects transmitted through distributed office networks. There are three forms of DCA: Revisable-Form Text (RFT) DCA, Final-Form Text (FFT) DCA, and Mixed-Form FFT. All DCA forms support distributed office functions, such as electronic mail, word processing, and document distribution.

RFT DCA supports the generation of revisable documents designed for editing. Revisable documents are seen by end users as documents that contain both text and the format information that directs textual presentation. RFT DCA textual format information includes general document characteristics, such as page width, page depth, page numbering scheme, and other page and line definitions that are represented by control codes embedded within the text. These control codes are based on the SNA Character String (SCS) EBCDIC control codes.

Figure 20 indicates the relationship between end user-intelligible revisable documents and the RFTDCA data stream that generates them. The figure further illustrates:
- Encapsulation of the RFTDCA data stream within a DIA Document Unit.
- Encapsulation of a DIA Document Unit within a DIA DIU.
- Encapsulation of a DIA DIU within an LU 6.2 APPC GDS logical record (assumes a Mapped Conversation).

4. Encapsulation of a GDS record within an LU 6.2 Basic Information Unit (BIU), which begins with a Request Header (RH) that signifies either Begin Bracket Indicator (BBI), Conditional End Bracket (CEB), or Implied Middle of Bracket. The Bracketing protocol is a Layer 5 SNA Data Flow Control protocol that delimits conversations within APPC sessions. Also, since APPC session conversations are logically half-duplex and send/receive half-sessions relinquish send state by flipping RH bits, there is an RH setting that signifies Half-Duplex Flip-Flop. The FMH is either FMH Type 5 (for normal data flows) or FMH Type 7 (for session error-condition reporting).
- Encapsulation of the BIU within an SNA Basic Link Unit (BLU) that includes a Transmission Header (TH), Link Header (LH), and Link Trailer (LT).

FFTDCA defines a data stream that generates final-form documents destined for final display or print. FFTDCA does not support further document modification, as does RFT DCA. Mixed-Form FFT supports integration of multiple information types within the same data stream. Examples include text plus facsimile, voice-annotated text, and imbedded graphics within text.

Distributed Data Management

Distributed Data Management (DDM) is a Transaction Services Architecture. Like SNADS and DIA, it can operate independently of APPC, wich provides data connectivity for distributed, record-oriented files and provides a data connectivity language that facilitates file interchange among various IBM systems, such as CICS/VS, System/36, and System/38. While DDM can be described as a distributed data management architecture, it does not qualify as a distributed database management system (DDBMS). A DDBMS permits:
- Distribution of a segmented file over multiple, dissimilar systems.
- Reassembly of a segmented file to any single system.
- Updates to a specific database to synchronously and transparently update other, geographically dispersed databases that reside in heterogenous processing environments.

DDM does provide a data-sharing language common to systems, and it supports file creating, deleting, renaming, loading and unloading, and locking and unlocking on interconnected systems. However, these distributed files must each be intact file entities.

The DDM Architecture distinguishes between Source DDM (SDDM) — which resides on the system that executes a program, translates program requests into DDM com-

11. Extended Network Addressing. *ENA, announced with ACT/VTAM V3, extends the global network address to 23 bits, supporting up to 255 subarea nodes.*

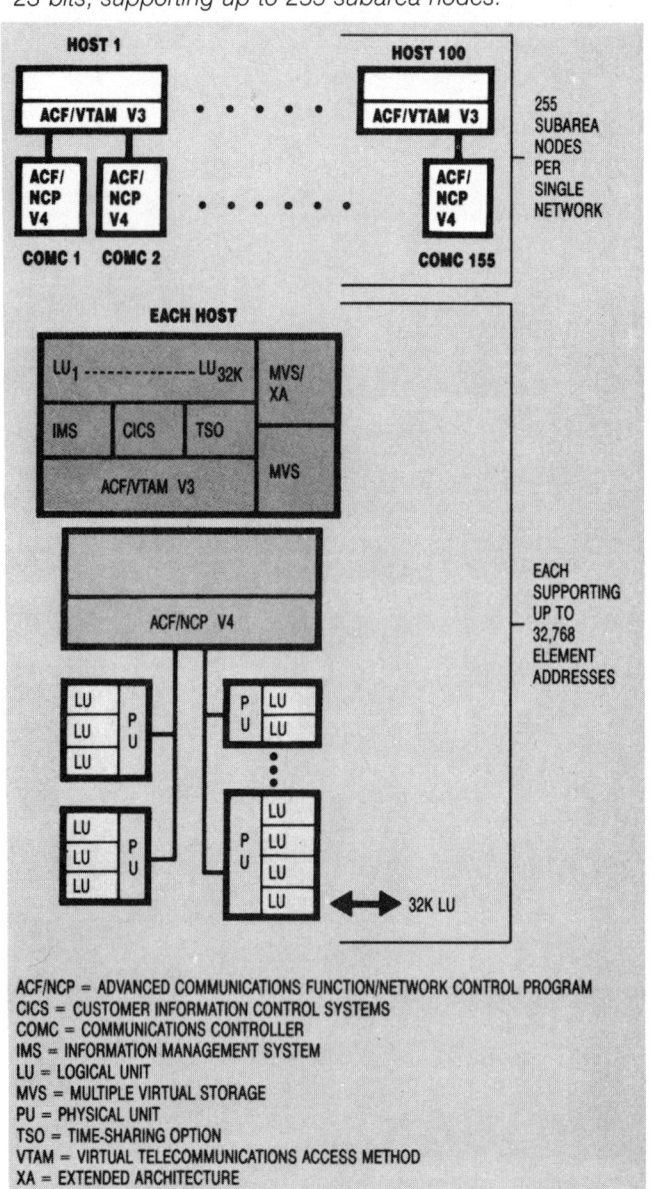

ACF/NCP = ADVANCED COMMUNICATIONS FUNCTION/NETWORK CONTROL PROGRAM
CICS = CUSTOMER INFORMATION CONTROL SYSTEMS
COMC = COMMUNICATIONS CONTROLLER
IMS = INFORMATION MANAGEMENT SYSTEM
LU = LOGICAL UNIT
MVS = MULTIPLE VIRTUAL STORAGE
PU = PHYSICAL UNIT
TSO = TIME-SHARING OPTION
VTAM = VIRTUAL TELECOMMUNICATIONS ACCESS METHOD
XA = EXTENDED ARCHITECTURE

12. Distributed ATPs. *ATPs communicate locally with LU 6.2 Presentation Services through the Advanced Program-To-Program Communications Application Program Interface.*

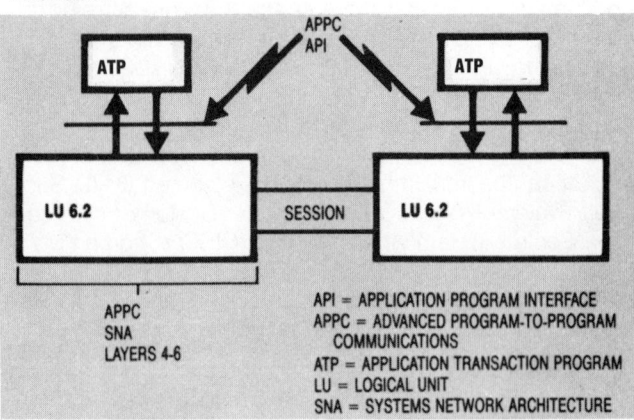

API = APPLICATION PROGRAM INTERFACE
APPC = ADVANCED PROGRAM-TO-PROGRAM COMMUNICATIONS
ATP = APPLICATION TRANSACTION PROGRAM
LU = LOGICAL UNIT
SNA = SYSTEMS NETWORK ARCHITECTURE

13. Conversations. *Conversations between distributed ATPs are serial time-slices of LU 6.2 sessions. The LU Type 6.2 NAUs are located with PU Type 5 or Type 2.1.*

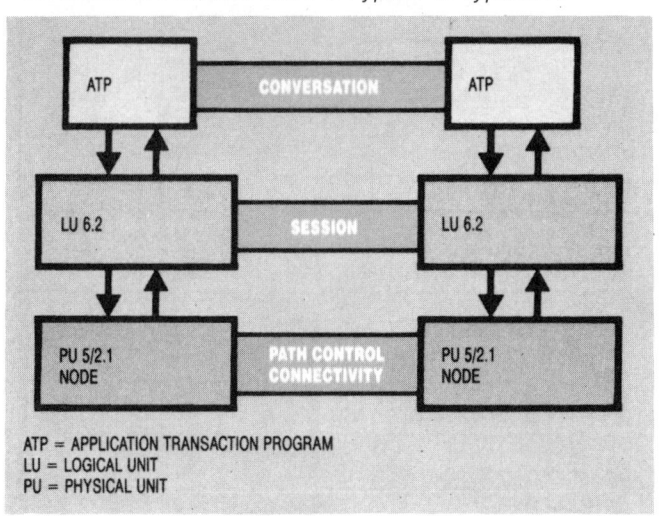

14. Verbs and statements. *ATPs communicate through conversations by issuing and receiving Conversation Verbs and other Program Statements over LU Type 6.2 sessions.*

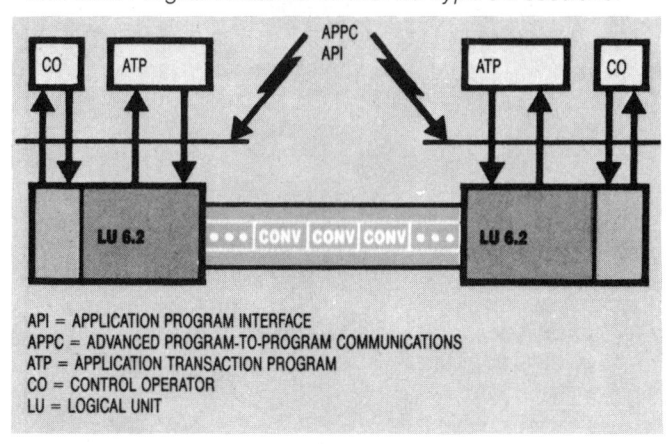

15. Verb forms. *There are three major classes of Verbs across the APPC API: Mapped Conversation Verbs, Basic Conversation Verbs, and Control Operator Verbs.*

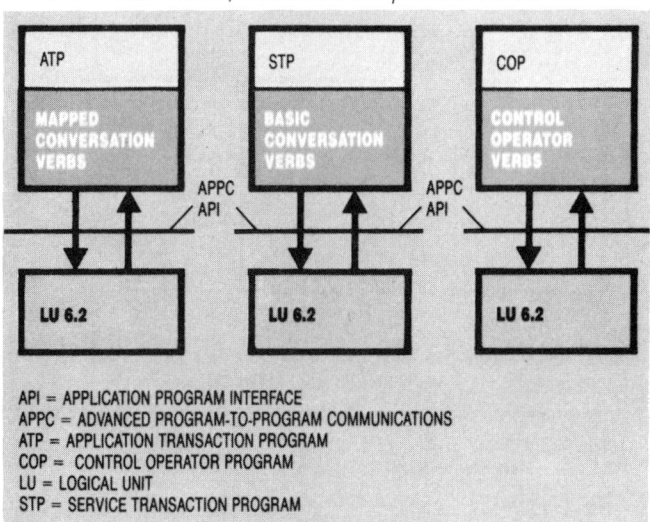

mands, and forwards these—and Target DDM (TDDM), which resides on the system that supports the targeted intact data file. TDDM intercepts SDDM commands and translates them into the local system data management interface for execution.

DDM defines four record-oriented file models: Sequential, Direct, Keyed, and Alternate Index. It also defines seven access methods to apprehend data within the file models: relative by record number, random by record number, combined relative and random by record number, relative by key, random by key, combined relative and random by key, and combined relative and random by record number and key.

Figure 21 depicts a DDM Model
- The Source System contains the ATP that requests access to a data file residing in another system.
- The Target System contains the requested data file.
- The Requesting ATP requests a specific record from the Target File, which is within the Target System, and sends this request to the Source System Local Data Management Interface (LDMI).
- When it is determined that the Target File does not reside on the Source System, the SDDM Server (DDM Process that provides for data usability and availability) translates the request into DDM commands.
- The SDDM Communications Manager passes these DDM Commands to the TDDM Communications Manager.
- The TDDM Server intercepts and interprets the DDM Commands, locates the Target File, translates the DDM Commands to the Target System LDMI, and requests execution of the Requesting ATP request.
- The Target System LDM retrieves the requested Target File records and passes them to the TDDM Server.
- The TDDM Process generates the Target files records as DDM records, and passes them to the Target System Communications Manager.
- The DDM records are passed to the Source System Communications Manager and to the SDDM Server.
- The SDDM Server translates the DDM records into the the format expected by the Source System LDM, and the LDM passes them to the Source System requesting the ATP.

Note that with DDM, each Source and Target System's internal ATPs need only communicate record formats with the local DDM Server, whereas prior to DDM within an IBM data management environment, it was necessary for each distributed instance of ATP to communicate and translate record formats to every other instance, therein requiring $N(N-1)/2$ translation instances for N Source and Target ATP occurrences. Therefore, while DDM does not qualify as a DDBMS, it does support relatively transparent distributed file record connectivity among cooperating networks.

Low Entry Networking

Low Entry Networking (LEN) is an architectural extension of the PU Type 2.1 Node, and builds upon the PU Type 2.1 and LU Type 6.2 base to define enhancements that allow small networks, traditionally regarded within an SNA subarea network as Peripheral Nodes (PNs), to function as full network nodes. In this sense, LEN defines a Network Node (NN) architecture as an extension to the PN architecture.

LEN derives its name from the fact that small systems

users often prefer to design and implement networks in bottom-up, as opposed to top-down, fashion, in that they tend to install and interconnect departmental processors and desktop workstations first and integrate host processors later. Hence, low entry.

LEN architecturally addresses these requirements while providing APPC SNA peer-to-peer functionality.

LEN enhances the PU Type 2.1 Node architecture in three fundamental ways:
- Each processor, under the control of a Peripheral Node Control Point (PNCP), provides autonomous, rather than centralized, control.
- Network resource fluctuations are dynamically and automatically defined quite independently of the manual system generations characteristic of classical SNA networks.
- LEN NNs perform Intermediate Network Node (INN) functions, a feature classically reserved for PU Type 4 COMC Nodes.

Figure 22 illustrates a LEN network and its relationship to an SNA subarea network. The LEN NN appears to the upstream ACF/NCP BF Node as a PU Type 2.0 CLC running PUCP, but that all LEN Nodes recognize themselves as peer PU Type 2.1 Nodes running PNCP. This distinction is recognized by a single-bit difference in the Layer 3 Path Control Transmission Header, which, for PU Type 2.1, supports the use of the Origin Address Field-Destination Address Field Assignment Indicator bit set high in the Format Identifier 2 Transmission Header.

The above describes a PU Type 2.1 Node environment. LEN adds five Network Node services to this baseline architecture: Connectivity Services (CS), Directory Services (DS), Route Selection Service (RSS), Session Services (SS), and Data Transport (DT). Note back in Figure 22 that the NN supports all five of these services, whereas the PN supports all but RSS. These five LEN NN services enable new nodes and applications to be added to and deleted from a LEN network without the accompanying classical SNA network requirement to notify all changes to centralized tables through manual system generations. All LEN network changes are reported on an as-needed basis to nodes that request access, through a stochastic, flooded broadcast algorithm (quite similar to Arpanet in this regard).

Each LEN NN maintains two directories under PNCP control—a Local Directory and a Cache Directory. The Local Directory lists all local NN resources and all resources contained within PNs attached to each NN. The Cache Directory is a list of most recently located remote LEN network resources. When a Network Node seeks a requested resource, it begins by searching its Local and Cache Directories. If neither directory contains the requested resource reference, the NN initiates a distributed search of all NNs. The requested resource is located stochastically, without reference to a centralized system directory (as is the case in an SSCP Domain Resource Table or a Cross-Domain Resource Table lookup).

Once the requested resource has been located, the NN RSS determines the preferred route to the resource through PNCP reference to a topology database, which calculates preferred routes for various classes of service (for example, batch or interactive). When the route has been selected, the NN SS activates an LU Type 6.2 session between the requesting and selected resources.

16. SNADS. *SNA Distribution Services is accessed by user Application Transaction Programs (ATPs) issuing SNADS Verbs across the APPC API.*

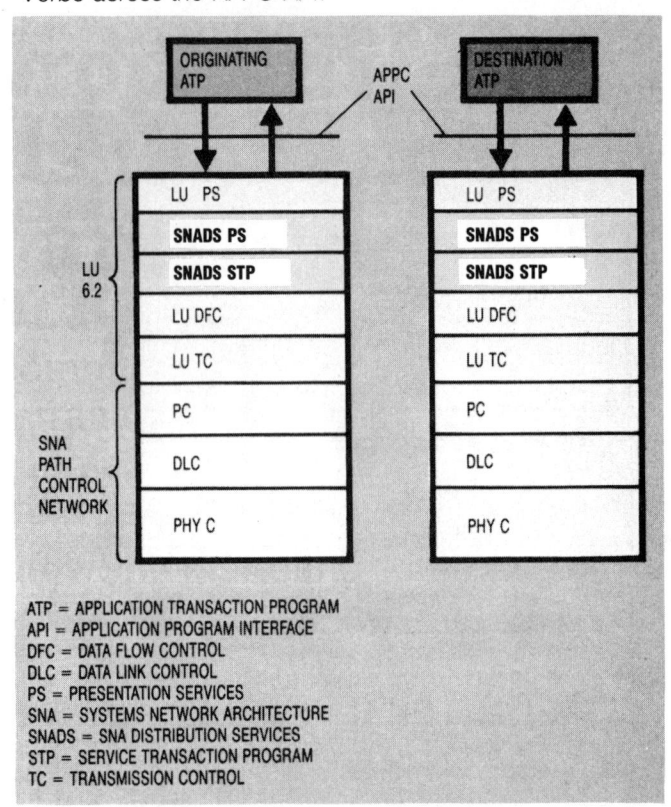

ATP = APPLICATION TRANSACTION PROGRAM
API = APPLICATION PROGRAM INTERFACE
DFC = DATA FLOW CONTROL
DLC = DATA LINK CONTROL
PS = PRESENTATION SERVICES
SNA = SYSTEMS NETWORK ARCHITECTURE
SNADS = SNA DISTRIBUTION SERVICES
STP = SERVICE TRANSACTION PROGRAM
TC = TRANSMISSION CONTROL

17. APPC connections. *The SNADS Distribution Network comprises Distribution Service Units (DSUs), which perform the SNADS asynchronous delayed delivery service.*

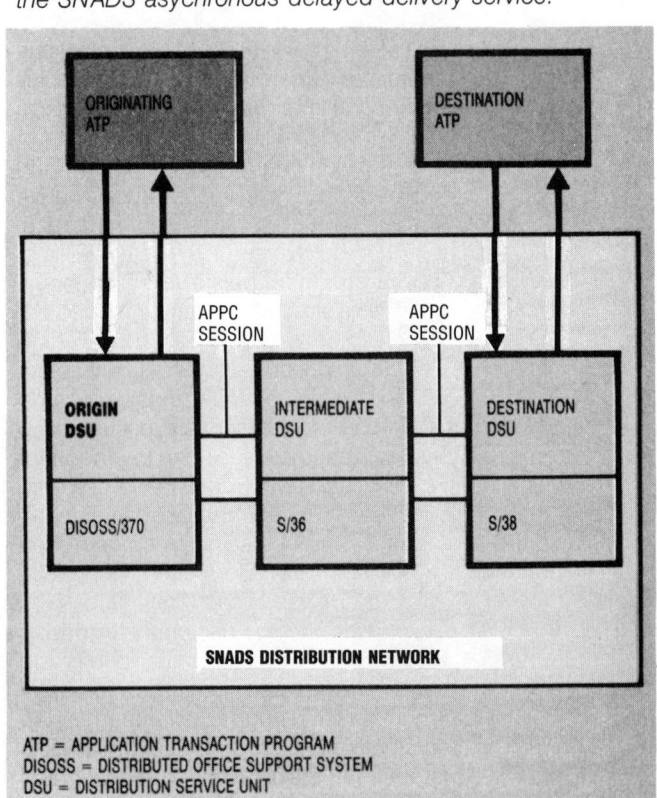

ATP = APPLICATION TRANSACTION PROGRAM
DISOSS = DISTRIBUTED OFFICE SUPPORT SYSTEM
DSU = DISTRIBUTION SERVICE UNIT

18. DIA sessions and nodes. *SNADS users can participate in DIA Networks by establishing DIA Sessions to DIA Office System Nodes.*

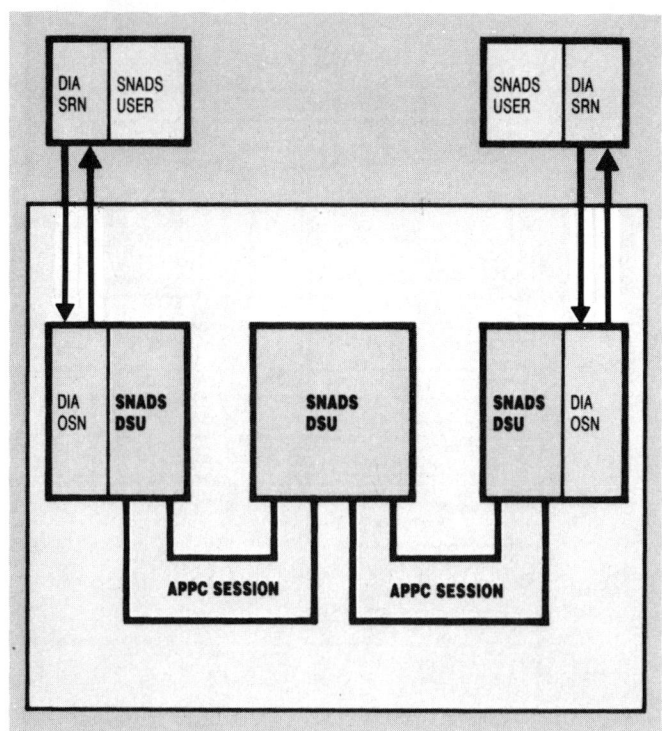

APPC = ADVANCED PROGRAM-TO-PROGRAM COMMUNICATIONS
DIA = DOCUMENT INTERCHANGE ARCHITECTURE
DSU = DISTRIBUTION SERVICE UNIT
OSN = OFFICE SYSTEM NODE
SNADS = SNA DISTRIBUTION SERVICES
SRN = SOURCE/RECIPIENT NODE

19. Using a process. *DIA Sessions can run within LU 6.2 Sessions. DIA Users, located at SRNs, utilize a DIA Process to request access into various services.*

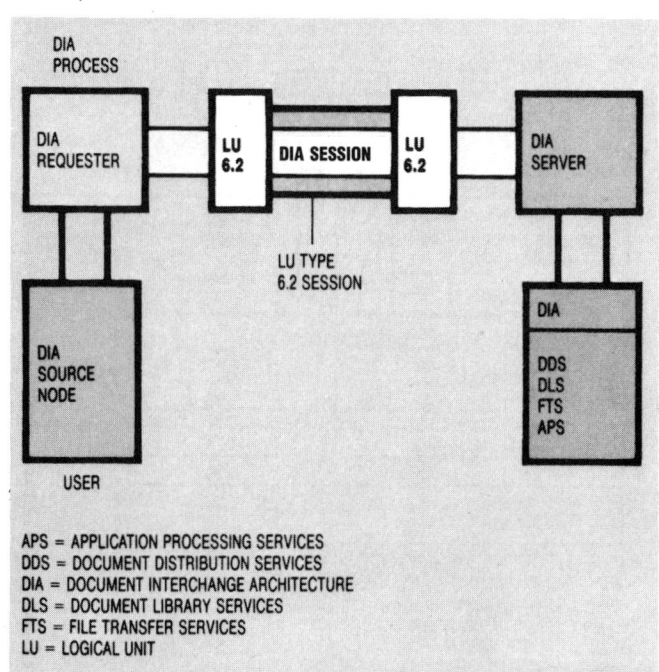

APS = APPLICATION PROCESSING SERVICES
DDS = DOCUMENT DISTRIBUTION SERVICES
DIA = DOCUMENT INTERCHANGE ARCHITECTURE
DLS = DOCUMENT LIBRARY SERVICES
FTS = FILE TRANSFER SERVICES
LU = LOGICAL UNIT

Session BIND Request/response Units (RUs) "hop" from NN to NN, with successive node references providing the route determination for the next hop. Before hopping to the next node, table entries (called session connectors) are made, which cause subsequent packets for the same session to traverse the same path. Once the resources are in session, the DT service activates data transport over the session.

LEN also supports intermediate node routing functions, similar to those provided through PU Type 4 Nodes. The first LEN implementation, Advancerd Peer-To-Peer Networking (APPN), supports this capability among interconnected System/36 and System/38 environments. Before this capability was available, it was necessary to establish distinct LU Type 6.2 sessions between each adjacent PU Type 2.1 Node. The LEN architecture provides for one end-to-end session between nonadjacent nodes.

Directions

LU Type 6.2 is the architectural centerpiece of distributed function SNA, and it provides the basis for the Transaction Service Architectures. LU Type 6.2 behaves as a distributed operating system, in that LUs can be connected by multiple sessions (parallel sessions) and in that LU Type 6.2 provides functionality beyond merely porting end users. These additional functions include interprogram communications, which enable local programs to view the LU as a local operating system, and the network as distributed LUs, each providing symmetric, interprogram communications.

Development emphasis within the APPC-based distributed architectures is on elaboration of peer-to-peer connectivity and resource sharing independent of host notification and control. The Low Entry Networking architecture promises to find its way into System/370 as well as desktop workstation environments. LEN represents a fundamental and significant departure from classical SNA, in that LEN network node and application changes do not need to be reflected in centralized tables through manual system generations, and LEN nodes provides intermediate routing node capabilities previously reserved for PU Type 4 nodes.

The introduction of the Token-Ring Network provides further insight into distributed SNA directions. PCs and other ring stations support APIs onto the ring:

■ Direct API, which conforms to the Institute of Electrical Engineers (IEEE) 802.5 Token-Passing Ring deterministic LAN access protocol at the Layer 2A Medium Access Control (MAC) sub-layer.
■ IEEE 802.2 Logical Link Control (LLC), which is a Layer 2B access protocol that defines connectionless as well as logically connection-oriented link connections over the ring.
■ Network Basic Input/Output System (Netbios), which provides ring stations with File, Print, Message, and Communications (Gateway) Server functions.
■ APPC/PC Program, which provides PCs with LU Type 6.2 and PU Type 2.1 Node functions. The Token-Ring Network appears to be positioned not simply as a standalone LAN, but rather as an integral-distributed transaction processing component, from which end users located at ring stations can gain access to distributed SNA target application processors regardless of subnetwork location.

Formalizing the Network Management Architecture and unveiling NetView (the first major Network Management

Architecture implementation) are also significant. NetView consolidates, within five modules, disparate SNA host-resident network management products:
- Network Communications Control Facility (NCCF), a VTAM application program that provides single-and multidomain physical management.
- Network Problem Determination Application (NPDA), a NCCF program that supports Problem-Determination data-base capabilities on DASD.
- Network Logical Data Manager (NLDM), an NCCF/VTAM application that provides logical session trace functions.
- VTAM Node Control Application (VNCA).
- Network Management Productivity Facility (NMPF).

An "open-API" has also been provided from NetView into LU Type 6.2 (NetView/PC for the Token-Ring Network) and non-IBM network management systems, enabling multi-vendor networks with functionally disparate network management systems to communicate with NetView.

SNA access into Open Systems Interconnection (OSI) environments has also been formalized at Layers 1 through 3 within Open Communications Architectures. The Open Communications Architectures are implemented at this time within a range of SNA/X.25 products. IBM has also committed to gateways from SNA to OSI through:
- The Manufacturing Automation Protocol (MAP) Communications Server (MCS), which is currently based on a Series/1 that interconnects SNA and distributed factory applications. MCS provides gateways between SNA Layers 7 through 1 into MAP Layers 7 through 1.
- Open Systems Transport and Session Support (OTSS) and Open Systems Network Support (OSNS) implement SNA to OSI protocol gateways through Layer 5, and have been successfully tested in Europe.
- SNA gateways to CCITT X.400 Message Handling System (MHS) have been successfully tested in Europe.

SNA development directions can be summarized by describing APPC as the strategic centerpiece of the Transaction Services Architectures. Further, LEN implementations are significant to distributed architectural developments, and will proceed into the mainframe and workstation environments, therein enabling cooperative processing among peer systems. Although SNA gateways to OSI environments will be further formalized, SNA is IBM's strategic proprietary network architecture, and, as such, will remain the centerpiece of communications systems development. DDM provides further development toward APPC-based DDBMS systems. The Token-Ring Network offers local communications to collocated PCs that have high file transfer requirements within a ring and distributed ATP access requirements through APPC throughout all interconnected environments. Network management standardization provides open-API network management and control. ■

Thomas J. Routt, president of Network Systems Consulting, specializes in SNA. Routt previously was manager of Boeing Network Architecture for Boeing Computer Services Co., where he managed global network planning, design, and implementation for the Boeing Co. He holds an M.B.A. in information systems from Southern Illinois University and a bachelor's degree in environmental science from Western Washington University.

20. Encapsulated documents. *End user Revisable Documents are encapsulated with the RFT DCA Data Stream, DIA Document Units, and DIA Document Interchange Units.*

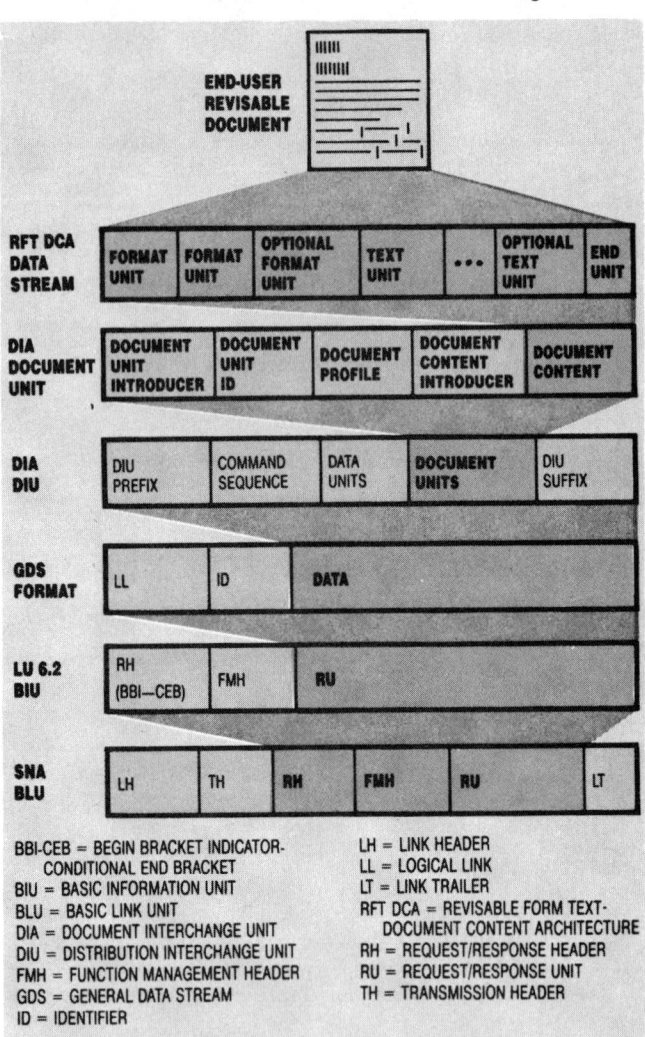

21. Passing files. *Distributed Data Management provides connectivity for record-oriented files and a language that facilitates file transfer.*

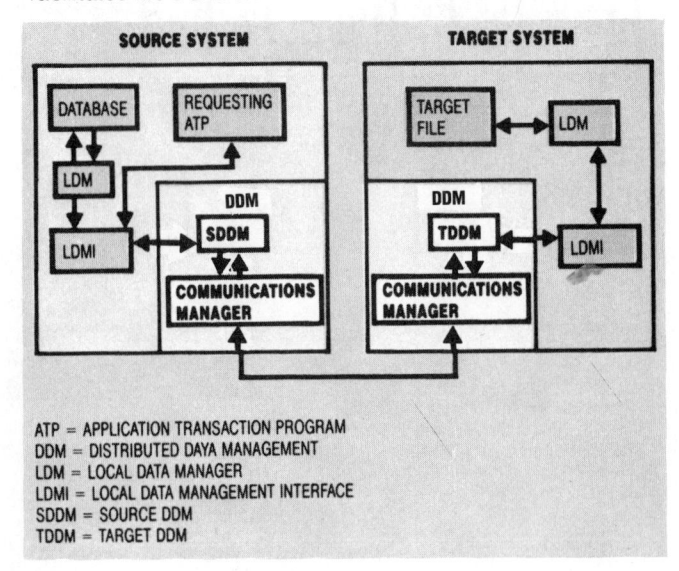

22. LEN's work. *Low Entry Network defines an extension to PU Type 2.1 and supports LU Type 6.2. LEN Networks can co-exist with SNA subarea networks.*

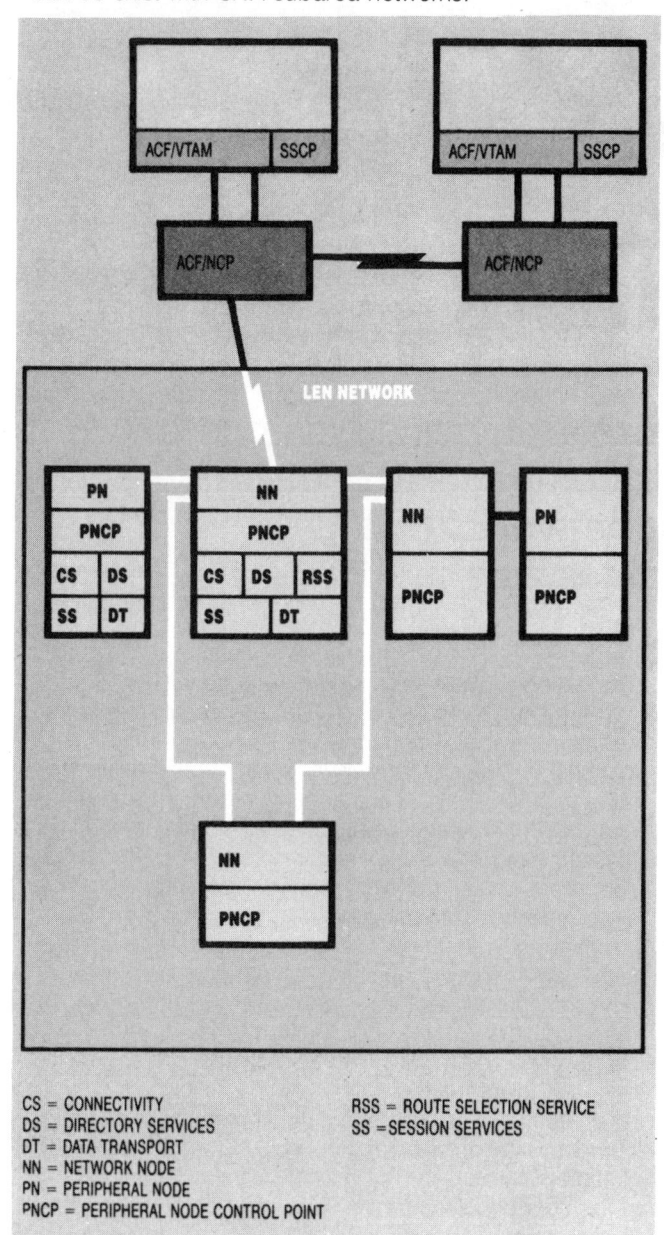

CS = CONNECTIVITY
DS = DIRECTORY SERVICES
DT = DATA TRANSPORT
NN = NETWORK NODE
PN = PERIPHERAL NODE
PNCP = PERIPHERAL NODE CONTROL POINT
RSS = ROUTE SELECTION SERVICE
SS = SESSION SERVICES

For further reading

Systems Network Architecture Format and Protocol Reference Manual: Architecture Logic, IBM, SC30-3112-1, Third Edition, November 1980.

Systems Network Architecture Technical Overview, IBM, GC30-3073-1, Second Edition, May 1985.

Systems Network Architecture Transaction Programmer's Reference Manual for LU Type 6.2, IBM, GC30-3084-2, Third Edition, November 1985.

Systems Network Architecture Format and Protocol Reference Manual: Architecture Logic for LU Type 6.2, IBM, SC30-3269-3, Fourth Edition, December 1985. ■

John T. Becker and Curtis G. Gray, Network Synergies Inc., West Lafayette, Ind.

Open networks free users from their mainframe architectures

Users can spread information access through their companies by building fault-tolerant networks that treat mainframes as a resource.

Many organizations have benefitted by acquiring data communications and data processing equipment from multiple vendors, since different offerings excel in different areas. People tend to bring in whatever computing tools they find most appropriate to their work, which helps the business's overall productivity. However, information on one computer is often inaccessible to users of another. In an ideal world the opposite would be true. Even specialized equipment needed only by one department, such as floating-point processors or a lab analyzer, should be capable of sending and receiving information to any other device with minimal disruption.

Both human and technical shortcomings may forestall information sharing in a multivendor environment. On the human side, many users tend to horde the information residing on their departmental processors. To combat such provincialism, management must decree that information is owned by the organization, not by individual departments.

On the technical side, most companies base their networks on the architectural scheme of their mainframe manufacturers (Fig. 1a). In such a network, the mainframe controls all interaction and acts as a central information repository—but often a bottleneck. Departmental data managers with links to the mainframe then have a solid technical justification for not sharing information with other departments: They can't.

Limiting the technology that employees are allowed to acquire might seem like the only way to ensure interoperability. However, a different approach may offer organizations a more palatable alternative. An open network architecture, in which all communication is handled by a specially designed network, provides a common medium by which any authorized user can access data stored on any computer (Fig. 1b). Such an approach can become the key to management's long-range strategy for making information available throughout the organization.

The centralized approach that many companies have adopted by default is inadequate as a long-term strategy for achieving the goal of multivendor support and information accessibility. With centralization, all terminals, printers, and other computers are attached to the mainframe in a master/slave relationship. The links may be governed by something similar to IBM's Systems Network Architecture as it existed prior to the availability of Logical Unit 6.2, which allows peer-to-peer communications. Although SNA is the most widely understood, other centralized network environments are available from various mainframe manufacturers.

The following are some of the advantages offered by the centralized approach:
■ All information is held at a single point, the mainframe computer. Giving everyone access to one master set of data furthers consistency.
■ The user can select a single vendor that will have developed many of the technical interfaces needed for joining network components (such as the links from host to controller or from controller to terminals).
■ Access is regulated by a closed architecture. A minimal number of interfaces coming into the mainframe means that the user need worry about only a few protocols.

Top-heavy
Centralization, however, also has disadvantages. Using such a scheme, devices not supported by the primary vendor's architecture attach directly to the mainframe communications controller or via relatively simple network interface computers (Fig. 2). Departmental computers look like terminals to the mainframe. Information on the local machines is usually updated daily in batch mode, since it is too difficult to update it in real time. Thus information on local processors may differ from that on the mainframe. For

1. Competing frameworks. *Users can build networks based on mainframes (a) or on the idea that, in accesssing information, all devices are created equal (b).*

(A) TRADITIONAL HIERARCHY

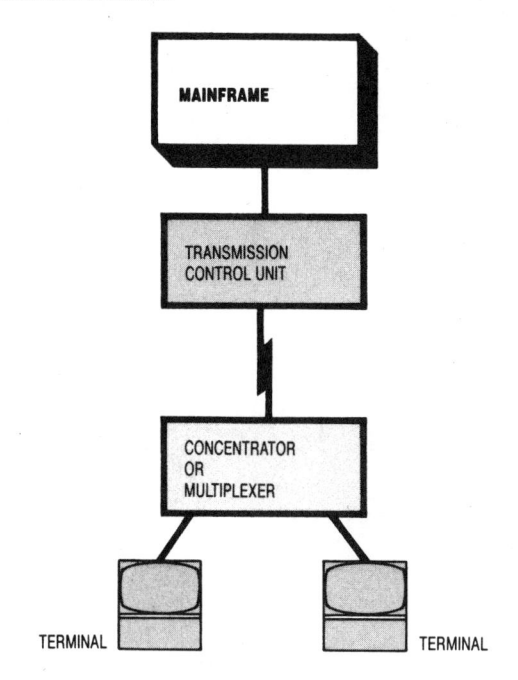

[Diagram: MAINFRAME → TRANSMISSION CONTROL UNIT → CONCENTRATOR OR MULTIPLEXER → TERMINAL, TERMINAL]

(B) OPEN NETWORK APPROACH

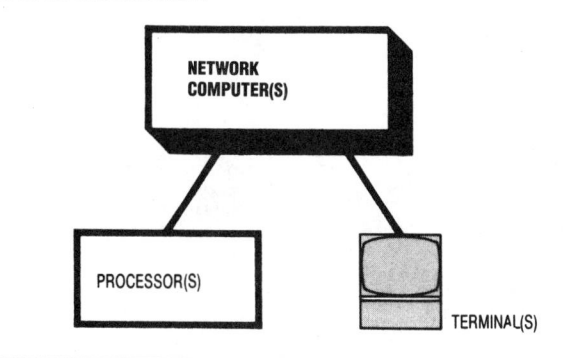

[Diagram: NETWORK COMPUTER(S) connected to PROCESSOR(S) and TERMINAL(S)]

example, if activity at automated teller machines (ATMs) is not reflected immediately on the bank's host, then someone could withdraw a large sum from the bank twice: once from an ATM on a departmental machine and again from a teller at a host terminal.

Mixing mainframes with departmental devices also complicates data access. All requests for access to the complete set of information in the centralized database are directed to the mainframe. Certain information, however, may reside on a foreign or departmental processor without being duplicated on the mainframe. To retrieve such information, a user might have to be at a specialized terminal attached directly to the foreign machine. Alternately, if the foreign computer supports the mainframe vendor's network architecture, terminals attached to the mainframe may establish a session with the departmental machine. These two types of links often use different protocols and data formats.

Centralization can also yield the following redundancies:
- *Storage.* If information (such as a file) must be on the central processor to be accessible to every user, it duplicates the copy on the departmental processor.
- *Input.* Without multivendor communications capability, users may have to retype data to transfer it.
- *Processing.* If they cannot accept summary records or information in a preprocessed format, computers at two or more locations may have to do the same work on the same raw data.
- *Applications development.* In a hodge-podge environment, application programmers have to worry about handling different device types. Thus a driver for a given type of terminal may be written again from scratch for each new application.

The open alternative

To ensure that their network is an enduring and sound fiscal investment, many organizations have adopted a philosophy of open networks. This approach takes advantage of developments in data processing and communications technology to provide an alternative to centralization. An "open network" is defined as a fault-tolerant computer or a private network of such computers to which all applications, terminals, departmental processors, and other communicating devices attach by means of standard communications interfaces (any type of interface besides a channel attachment). This approach is gaining acceptance for the following reasons:
- The business community increasingly depends on timely and correct information delivered electronically to arbitrary locations.
- Over the past several years, the cost of computers has dropped significantly. Minicomputers and microcomputers have made it possible to move processing power closer to the user. Departmental computers from a variety of vendors must be interconnected as various locations become increasingly interdependent. The alternative to developing numerous costly, separate point-to-point links is to establish a common ground.
- Communications is consuming an ever-larger percentage of the data processing budget. Therefore, a shared, multipurpose backbone network is becoming economically unavoidable. It would be too expensive to support a network for each application.
- Selecting all equipment from one vendor is impossible. Specialized devices are being developed rapidly, and users in many organizations are buying them as soon as they are released. A vehicle is needed to integrate all this equipment in as short a time as possible.
- Divestiture means that the average user no longer has a single contact for problem resolution. As a result, designers and users are more responsible for the proper functioning of the networks. The multivendor environment calls for a new source of centralized network management information.

This environment also raises the question of how to work with numerous common carriers. One network may have been set up to serve certain applications using AT&T, while another may have been built using MCI with local service from the Bell operating companies. Keeping track of vendor contacts for all these networks would be a network

2. Typical. *If they are connected at all, departmental processors in most firms are tied together through the mainframe, perhaps via simple interface devices. While an adequate processing host, the mainframe is less than perfect as a network server, since information flow between foreign computers is tortuous at best.*

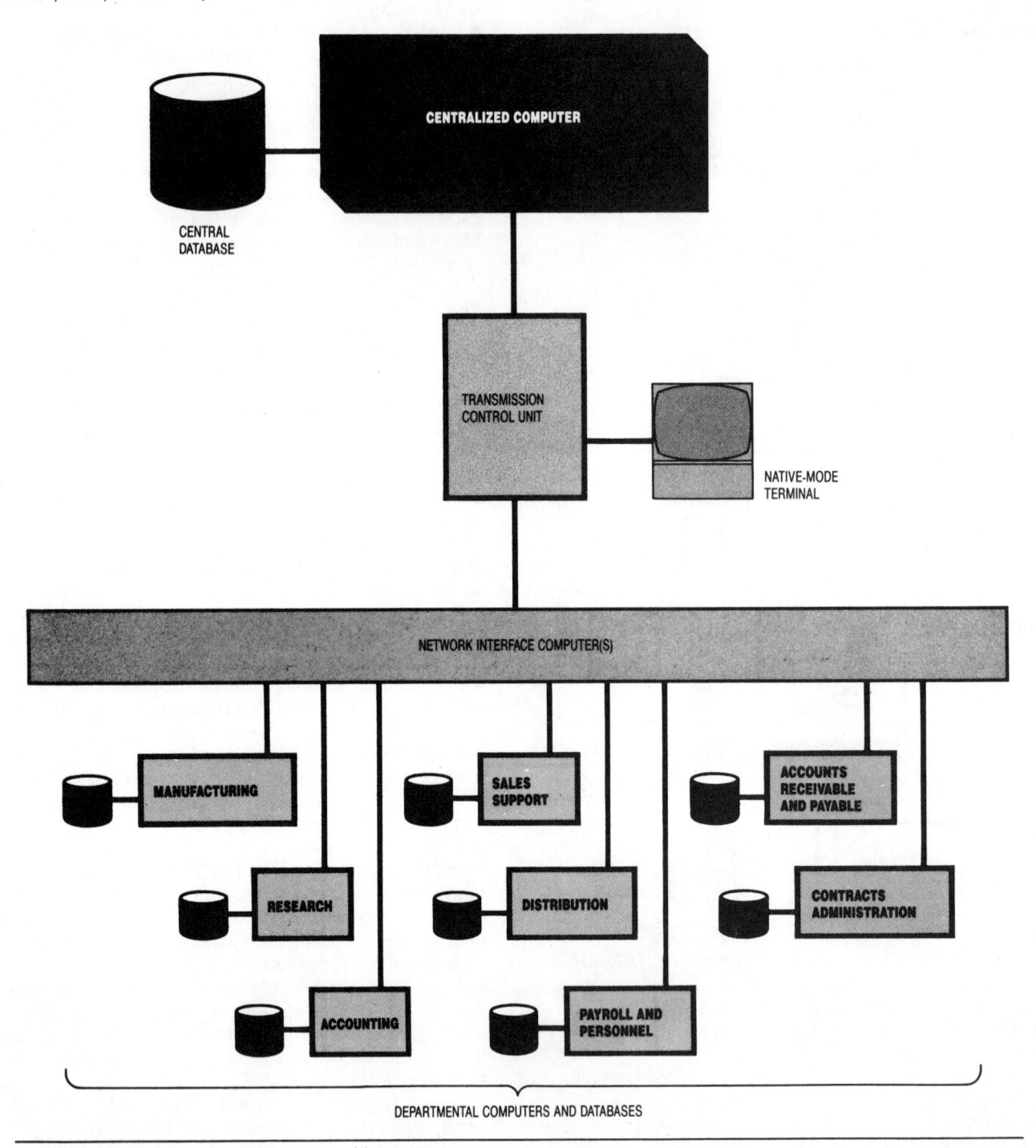

manager's nightmare. The alternative is a common network for all applications, which would allow networking managers to decide on vendors for each location based on cost. It would also allow them to better manage information about the common carriers.

Underlying the philosophy of open networks is the perspective from which information is viewed. With this approach information is treated as a precious organizational resource, making it far more important than just the input to or end-product of application software. Programs, terminals, and computers are reduced to the status of entities operating on that information, the flow of which is governed by the open network (Fig. 3).

Of course, since each company has particular problems and aspirations, it is no more possible to devise a network architecture that will solve all needs of all users than it is to build a computer that is ideal for all purposes. However, even though individual industries have specific concerns,

3. A linguistic specialist. *Displacing the mainframe from its role as bottleneck is done by one or more network computers acting as universal translators. Control of information flow rests in the network rather than in applications processors. Hosts and terminals are treated equally, as appendages to the network.*

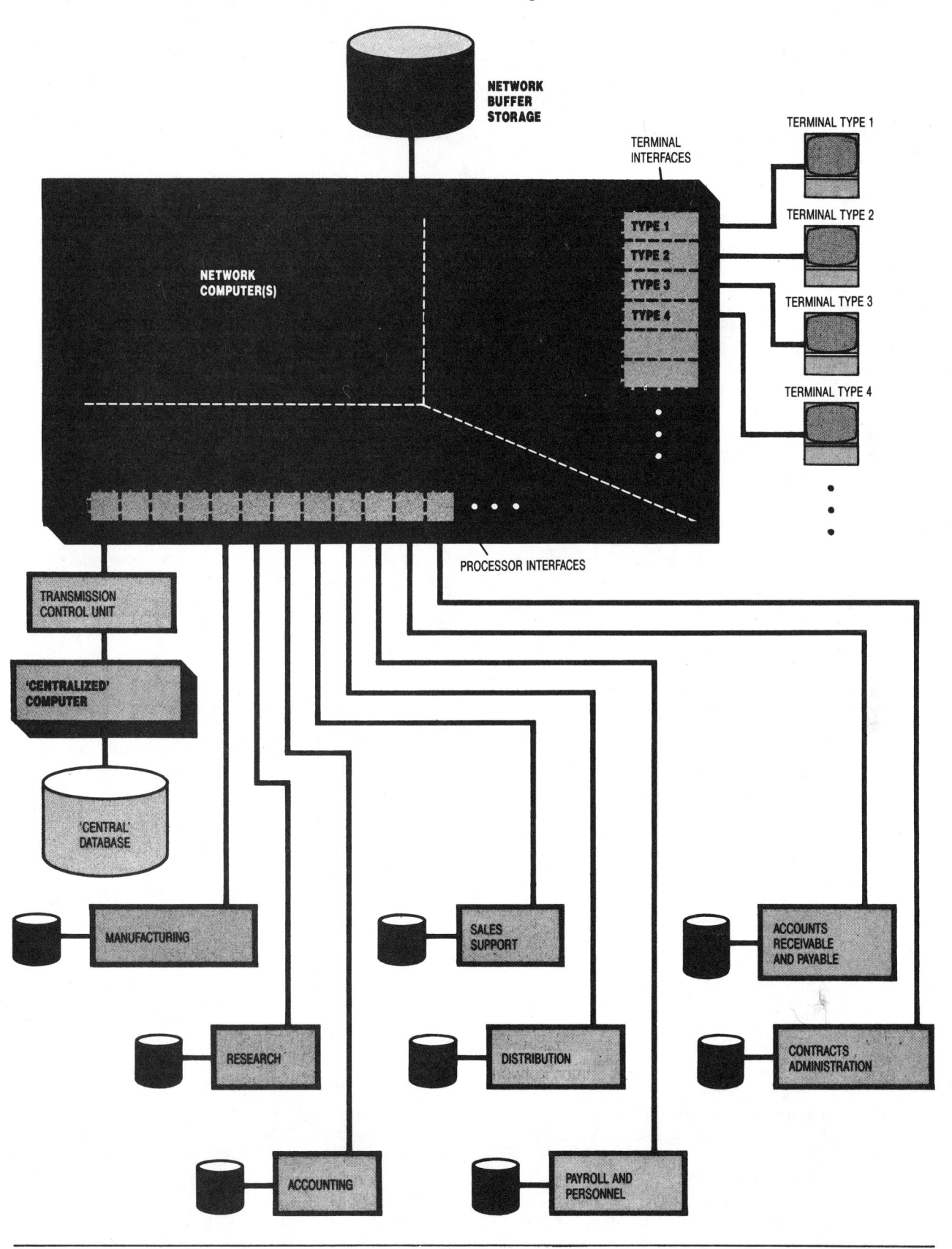

all users have a number of reasons for considering open network designs. An open network should do all of the following:

- Allow access to any information resident on any processor attached to the network (under the appropriate security guidelines, of course).
- Allow access from any given terminal, eliminating the need for distinct terminals to access data from distinct computers.
- Provide a limited set of interface definitions, on the basis of which users can select processors that can easily attach them to the network.
- Support a set of standard terminal interfaces to the network, thereby allowing multiple vendor sources for terminals. Availability of multiple vendors will tend to minimize the impact a single vendor can have on the environment as a whole.

An open network also offers the following benefits:

- Terminal users need not be familiar with the operation of all attached computers. For instance, they may access information without knowledge of log-on procedures, screen elements, or other details. All these are handled by the network.
- The network may perform some security screening functions for the individual processors. Security can be invoked on a user-, terminal-, or time-sensitive basis or on a combination of these criteria. For instance, the network's security function could prevent specific terminals from accessing certain data. Another restriction, based on time and user criteria, might prohibit students from logging on before five o'clock.
- The network, not the user, formats the transaction. Thus users need not remember a multitude of transaction formats for different computers within the organization. They must only know the general network transaction format.
- Each user entry may be split into several transactions by the network and presented to different computers. For instance, a single order may be divided into three transactions: one for the sales-support computer, another for a research machine, and the last for the financial computer—all without the user's intervention or in-depth knowledge of each computer.

A single log-on could allow a sales representative to consult various departmental processors from the field. With the information so obtained, the sales representative could say to the customer: "We've got the widgets you want in stock, so I've just placed your order with the sales control computer and notified inventory that they can go ahead and ship. Your credit's been approved and, by the way, the research and development people have the enhanced widget you inquired about scheduled for customer availability in the second quarter of 1987."

- If desired, the network or the user device could generate the application's background screens (which contain name fields, explanations, and so on). Since the network is intelligent, it can also strip out screen information coming from the host before it is transmitted. This reduction in the number of characters sent can often save enough on transmission charges to pay for the network processors. It may also improve response time by having less unnecessary traffic clogging up the network.
- The open network approach lends itself to serving locations remote from the base of installed terminals or host processors. For example, connecting a new sales office to the network simply means placing a terminal at the new location and installing a communications link to the closest network machine. Such a terminal has the same capability as any terminal at the primary processing location.

Technical foundations

At the heart of an open network is at least one network computer, which can be considered a unit called simply "the network." The network consists of a security module, switching logic, data-format translation capability, and interface modules for hosts and terminals (Fig. 4). It allows information on any departmental computer or mainframe to be presented, upon request and with adequate security clearance, to any connected terminal.

Access takes place on either a transaction or a passthrough basis. In a passthrough connection, the user is in control of establishing a direct virtual path to the remote computer. The path may run under asynchronous, bisynchronous, or some other protocol, but the user specifies the endpoints of the connection. In passthrough mode, the destination computer must perform security screening.

Transaction-based access means that the terminal user does not have to know about the individual computers attached to the network. The network interprets the transaction request and determines where (using its switch logic) and whether (using its security module) to route it. If the terminal sends transactions in a format that is not native to the destination host, the network reformats the messages.

The hardware needed to implement an open network is largely available in the form of reliable, redundant processors, which are becoming affordable due to declining chip prices. Much of the software also exists.

For example, fault-tolerant processor vendors are offering standard protocol-conversion software. Moreover, a number of specialized software houses are providing protocol interfaces and, in some cases, macro capability for text-format translation.

Useful traits

When implementing the open network concept, users should attempt to endow the network with the following six properties:

1. Device flexibility. Users should be free to acquire the specific departmental equipment they need with the knowledge that the equipment will be able to exchange data with any other device on the network. To this end, the network must be able to deliver information from different computers, files, or databases to a single workstation or terminal device. It must also support a growth path to handle new devices, services, and interfaces.

Therefore, the computer on which the network is based should support such widely used protocols as asynchronous teletypewriter (TTY), binary synchronous communications, synchronous data link control, high-level data link control, and X.25. In addition, it should be programmable on a bit-by-bit basis, since there are unusual 6-, 7-, 8-, 10-,

4. Layered functions. *The network can provide simple "passthrough"-style links where data reformatting may or may not be performed. Alternately, users may let the network handle transaction-style switching between terminal and destination host, in which case the network is also responsible for security screening when the call is set up.*

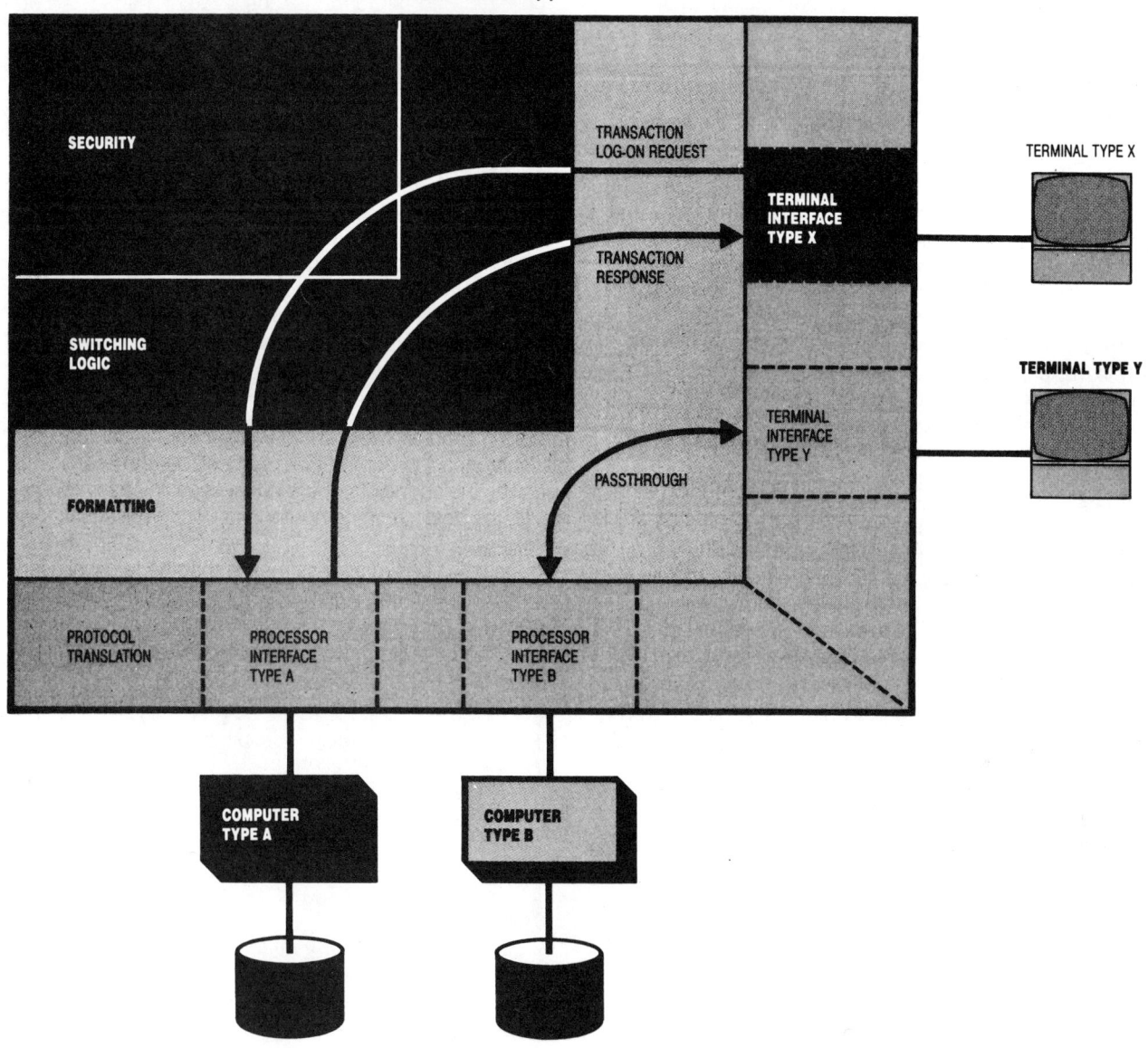

and 12-bit protocols in use. If the network computer can be adapted to any of these, management can decide whether the programming effort to support a new device is worthwhile.

2. Functional isolation. Users can separate specific processor technologies from their proprietary data communications techniques by taking the open network approach. Applications programs should serve every user, regardless of terminal type or location.

Splitting applications from communications has several benefits. Formal interface standards are established between the two areas (corresponding to the interface between layers six and seven of the Open Systems Interconnection, or OSI, model). Management can thus dedicate applications staff to application development and communications staff to network growth and operations. The availability of both types of professional should improve, since they will not need cross-training in each other's specialties.

Isolating the two areas will also make it easier to identify costs involved in bringing new applications to the network. These applications can be implemented more quickly and easily because applications programmers need not concern themselves with the current communications environment.

The open network approach also protects the installed base of applications software, since established interface specifications allow communications layers to change without impact to the programs that use them. Thus, as higher-capacity or more economical communications alternatives become available, they may be implemented less expensively and on a more timely basis.

3. Network buffering. The network must be able to act as a temporary storage site for information. Some applications, particularly batch tasks, can wait on the delivery of information for an hour, a day, or longer. If the network can buffer information that is not time-critical, it can give priority to information with more immediate value.

Buffering also simplifies restart conditions when terminals, processors, or other components fail. A device that has recovered from a failure can be more easily synchronized by an intelligent network than by a complex end-to-end protocol between the restored device and all the devices with which it had been in contact. Moreover, the device at one end of the circuit can be formally notified when the device at the other end expires, which is impossible with direct connection.

4. Modular growth. Adding terminals or processors to the network should be relatively easy. Equally important is the ability to make capacity changes without having to replace equipment or redevelop software. The network owner should have the choice of whether or not to retain hardware and software in order to grow.

5. Reliability. The network must provide an environment where a single component failure will not cause a catastrophic network outage. The more information that passes through the network, the greater the benefit to the organization that built the network and the greater the value of that information. Since devices attached to the network rely on network services, the network can be made fault-tolerant and present a high degree of availability even though the devices themselves may fail. For this reason, users should plan their network support operation carefully (see "Open network control").

In the monolithic approach, either the whole hierarchical, mainframe-centered structure must be made redundant (usually at high cost) or there is no fault tolerance. High availability can be achieved at lower cost in an open network environment, since attached devices may or may not be made fault-tolerant depending on cost and need.

6. Phased installation. Since the transition to the new network cannot be made in one step, the architecture must make it possible to migrate gradually toward the desired environment. Once the decision to implement an open network has been made, the means of integrating existing components into that network must be planned. Migration of the communications capability can be phased in on an application-by-application basis. Attempting a complete or one-step conversion will usually lead to immense problems.

Each application or function to be added to the open network should be viewed as an independent entity and judged on its own merits. Although the establishment of an open network will normally be justified on the basis of shared communications facilities, each application should be reviewed to ensure that its migration to the network occurs economically and at the appropriate time.

Making it happen

Planning, developing, and implementing an open network requires three basic elements:

1. Management involvement. Upper management must be aware of the commitment necessary when undertaking a major networking effort. Throughout implementation, many departments will be affected. Thus support by top management is critical to the success of the project. Information sharing usually crosses departmental boundaries and responsibilities. If the information to be shared is considered a corporate resource rather than an individual departmental resource, no conflicts will arise. In most cases, however, interdepartmental conflict on information sharing is real and must be resolved through corporate policy and upper management action.

Open network control

Effective control of an expanding organizational network is critically important, since user acceptance hinges on the availability and comprehensibility of the network. As with any office machine, portions of the network will occasionally fail. Users may find themselves cut off from the rest of the network. Even so, operations must not be disrupted or information lost. Network users must be presented with a network that is reliable but begins to fade into the background.

User confidence can result from several factors, including the following:

■ Fault-tolerant network design, one that will not allow a single component failure to put the network out of operation for any given user.

■ Completely documented operational procedures on how to interface with the environment.

■ Comprehensive user education, especially for applications people who have to be taught what the network provides. This keeps them from having to get involved in communications programming.

■ Quick fault isolation, due to the obvious need for responsive network control.

■ Fallback procedures—that is, what to do when failsafe components fail.

■ An effective preventive maintenance plan, including regular diagnostic checking, cleaning of mechanical devices on the network, and software testing with dummy transactions.

■ A good problem reporting, recording, and tracking mechanism, consisting of procedural steps to identify each problem as well as where it originated, how to resolve it, and where to record it.

There should be a single point of responsibility with authority to act on these factors. The network control organization should have the following features:

■ A trouble desk to screen incoming trouble calls and determine their nature.

■ A technical-control center capable of locating failures and initiating solutions.

■ Ongoing reliability meetings with constituent vendors to resolve component failures or performance problems.

■ A preventive maintenance strategy and plan to keep the network running in a planned rather than a reactive mode driven by user complaints and equipment failure. This entails fault prediction and a comprehensive program of preventive maintenance coupled with a philosophy of planned spare equipment.

2. A programming framework. In the future, the now-prevalent separation of programming staff responsibilities in terms of equipment will disappear. It will become logical to train communications programmers to work on multiple machines rather than training applications programmers in networking. The logical division of responsibilities will follow the traditional data processing segregation of application from system programming departments. Organizations will be better able to link different types of equipment if there is a single point of responsibility for all necessary network programming.

3. Interface definitions. While it is the goal of the open network to serve a wide variety of processors and terminals, the manner in which the network will accept information must be clearly defined. Initial design efforts should define specific protocols and data formats that devices must present to the network.

Even with the capability to handle virtually any communications device by custom programming, most objectives of the network can be met through a limited set of interface specifications. The interface definitions must address such areas as:

- Audit control (ensuring that the anticipated delivery of information takes place).
- Status reporting (registering the delivery of information and the physical status of devices).
- Protocol capabilities (informing each partner in a circuit about the nature of the other).
- Restart/recovery (specifying when and where the network should intercept the data that passes back and forth during a session to minimize the effect of a failure on either of the devices in the session).

In all likelihood, when a company adopts the open network approach, some technical responsibilities will be the vendor's, others will be the user's, and the rest will be shared. In terms of the OSI model, the vendor will be expected to provide the lowest three layers and the user the top two, with the middle two shared. This is the division of responsibilities that seems to arise in practice, although the user and the vendor usually want each other to accept responsibility for all seven layers.

Although any organization can implement an open network, each must still formulate its own business needs and decide how the architecture will fulfill those needs. In the end, though, most companies will find that the open approach outlined above has an undeniable appeal and, economically, is almost inevitable. ■

John T. Becker, president of Network Synergies Inc., has been implementing computer networks for 20 years. He has written a manual called Data Communications Survey and Procedures. *He worked for five years in data communications at IBM, founded a Midwest manufacturing software firm, and has had his own consulting corporation since 1972. Becker has a B. A. from the University of Iowa.*

Curtis G. Gray, executive vice president of Network Synergies, has a B. S. E. E. and an M. A. in computer science. He has worked as a voice switch designer for a major common carrier and has been involved in the implementation of data communications networks since 1976. Gray has spent a great deal of time developing the simulation models needed to ensure the success of client networks.

David G. Matusow, Computer Task Group Inc., Phoenix, Ariz.

BTAM, VTAM, X.25: Uneasy alliance

Linking different machines over an X.25 network requires patience and hard work. And though problems may never disappear completely, the results can be worth the hassles.

Be careful in interconnecting machines with different access methods, whether from the same or different vendors. A lot can go wrong. In particular, it is important to be knowledgeable about both environments. Even so, however, linking such machines, especially with a sophisticated network protocol, can be an arduous process.

One recent example of such a struggle occurred when a worldwide financial services company (call it "company A") had to draw on data owned by a Compuserve-like information provider ("company B"). The need to access the data regularly and readily pointed to a direct connection between the company's respective mainframes. However, although both utilized IBM mainframes, operating systems, and access methods, the connection was anything but straightforward.

Company A's host ran Systems Network Architecture (SNA) under the Virtual Telecommunications Access Method (VTAM), while company B's host used the Basic Telecommunications Access Method (BTAM), which is an older, non-SNA environment (Fig. 1). Since both of these machines already had X.25 interfaces to a public packet network for various other purposes, it was decided that the links (the existing leased lines from both company A and company B to the packet network) could be used to save both time and money in creating a new way of joining the two companies.

The X.25 link from A's host was IBM's NCP Packet Switching Interface (NPSI) running under the Network Control Program (NCP). It would be used to dial out to company B's host, which ran X.25 software from Comm-Pro Associates Inc. in Redondo Beach, Calif.

IBM has had little experience doing dial-out from NPSI. Thus, the support and documentation it provided were far from optimal. In fact, very little hands-on experience with such environments was to be found anywhere. This was compounded by the fact that NCP Packet Switching Interface is not widely used and has many operating problems.

Even after these problems were resolved, however, the connection was not totally stable, nor is it to this day. For example, if the host processor is loaded more heavily than usual, thereby causing a change in the operating speed of the teleprocessing monitor, the connection does not always work. Almost all of the problems occur during the initial handshaking that is required to start the connection. Once the connection gets beyond this point, problems usually disappear.

Plan ahead
Early investigation showed that data could be passed from SNA to non-SNA environments using one of the following three methods:
- *Full SNA emulation.* Emulations of full SNA subarea nodes were the most flexible but also the most complex and hard to implement of the methods. Few off-the-shelf products were available.
- *Passthrough applications.* These programs, software emulations of cluster controllers, were simpler and more readily available. Each half of the software communicated with one environment, such as ASCII and 3270. However, only single sessions or groups of sessions could be set up between machines. Increasing throughput between the machines would require meshing data from multiple sessions. The delicate balance between the two halves of the implementation would become increasingly difficult to support.
- *Hardware emulations.* Protocol converters are effective for low-speed (less than 9.6-kbit/s) data transfer and simple interactive traffic but offer limited speed and throughput for heavy traffic between computers.

An easier solution to the requirements for both data

1. The plan. *Company A needed information that company B had. A public data network made sense. In fact, more problems arose at the endpoints than in the middle. These were due to difficulties in getting two machines with different operating environments coordinated, even if one vendor supplies both the hardware and software.*

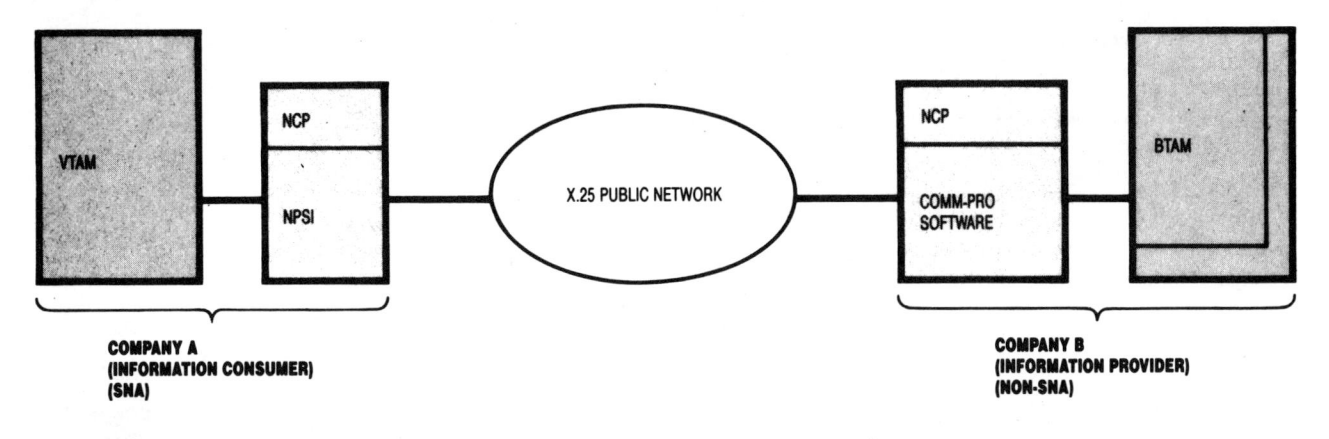

```
┌──────┐  ┌─────┐                            ┌──────────┐  ┌──────┐
│      │  │ NCP │                            │   NCP    │  │      │
│ VTAM │  ├─────┤──── X.25 PUBLIC NETWORK ───┤──────────│  │ BTAM │
│      │  │ NPSI│                            │ COMM-PRO │  │      │
│      │  │     │                            │ SOFTWARE │  │      │
└──────┘  └─────┘                            └──────────┘  └──────┘
  COMPANY A                                        COMPANY B
  (INFORMATION CONSUMER)                           (INFORMATION PROVIDER)
  (SNA)                                            (NON-SNA)
```

BTAM = BASIC TELECOMMUNICATIONS ACCESS METHOD
NCP = NETWORK CONTROL PROGRAM
NPSI = NCP PACKET-SWITCHING INTERFACE
SNA = SYSTEMS NETWORK ARCHITECTURE
VTAM = VIRTUAL TELECOMMUNICATIONS ACCESS METHOD

---

transfer and interactive traffic was needed. The answer was sought in the X.25 communications standard. It was felt that an X.25-based solution could avoid many low-level "bits-and-bytes" implementation details.

As with all intercomputer connections, even between machines in a single company, a great deal of detailed preplanning was necessary. The companies held many discussions to ensure that link- and packet-level parameters, such as packet-size and logical-channel numbers, were consistent between the parties. Even such careful planning might not have covered every parameter that could have caused a glitch, though luckily no problems were encountered at this stage.

In configuring the X.25 parameters for the connection, each company was constrained in some way by the software components it used. For example, company B's range of logical channel numbers was limited by the Comm-Pro software. Only a certain packet size offered by the network vendor could be used. Moreover, the equipment owned by each company used a different character code. The ASCII data favored by B's environment had to be translated into EBCDIC (Extended Binary Coded Decimal Interchange Code) for company A's SNA gear.

It was decided that the link would utilize switched virtual circuits (SVCs). This method of connection allows the creation of multiple, nonpermanent, virtual point-to-point links from fewer permanent physical circuits. Using SVCs allows multiple logical sessions to be multiplexed between the two endpoints, which reduces connection cost.

Life's hard for NPSI

Company A's NPSI software, residing in the 37X5 front-end processor, performs protocol conversion between SNA and the X.25 network. This conversion consists of two basic parts:

■ NPSI takes outbound messages, strips off the SNA headers, packetizes the remaining (raw) data, adds X.25 headers, and sends the packets into the X.25 network. The opposite is done for arriving data.

■ The software also simulates an SNA device for the host-based SNA software products. This includes implementing such SNA logical unit (LU) functions as session setup and problem notification.

The most difficult task facing NPSI is to keep each network pleased with the requests and responses it provides. For example, it must satisfy pacing requirements on both the SNA and X.25 sides. With X.25, the software must keep track of the number of outstanding packets, both on the link as a whole and on each virtual circuit.

The physical link is paced to a maximum of seven unacknowledged packets. Upon reaching this threshold, acknowledgment must be made before any more packets can be passed. In addition, two outstanding packets are permitted on each SVC. The need to satisfy both the physical and virtual links results in a very complex acknowledgment scheme.

On the SNA side, NPSI is simpler, but also more constrained. It appears to the host as an unsophisticated 3767 SNA printer, a physical unit (PU) type-1 device. PU 1 is the simplest physical-unit type in SNA, supporting few of the higher SNA enhancements. One of its limitations is that it allows only single request units (RUs) to be sent and received. (By contrast, a device of PU and LU types 2 can handle multiple outstanding RUs.) Thus, NPSI cannot keep track of the sequence numbers used to count outstanding packets. This restriction reduces the throughput of the session by requiring constant acknowledgments.

A buggy ride

With very few companies running NPSI, the amount of support from IBM for both procedural questions and technical difficulties varies from little to nonexistent. As a result, there are several bugs in the software. Some of these are easy to find and resolve, while others can be quite elusive. One major difficulty involved a departure of the "call out" methodology from the documented procedures.

One of the most troublesome tasks was to call out from

2. Startup. *One task calling another allows company A to establish a call request, via a logically defined terminal (A), through a packet network (B). Events that follow the generation of the requested task include a link to the access method, from there to the front-end processor, and then an X.25 call-request sequence.*

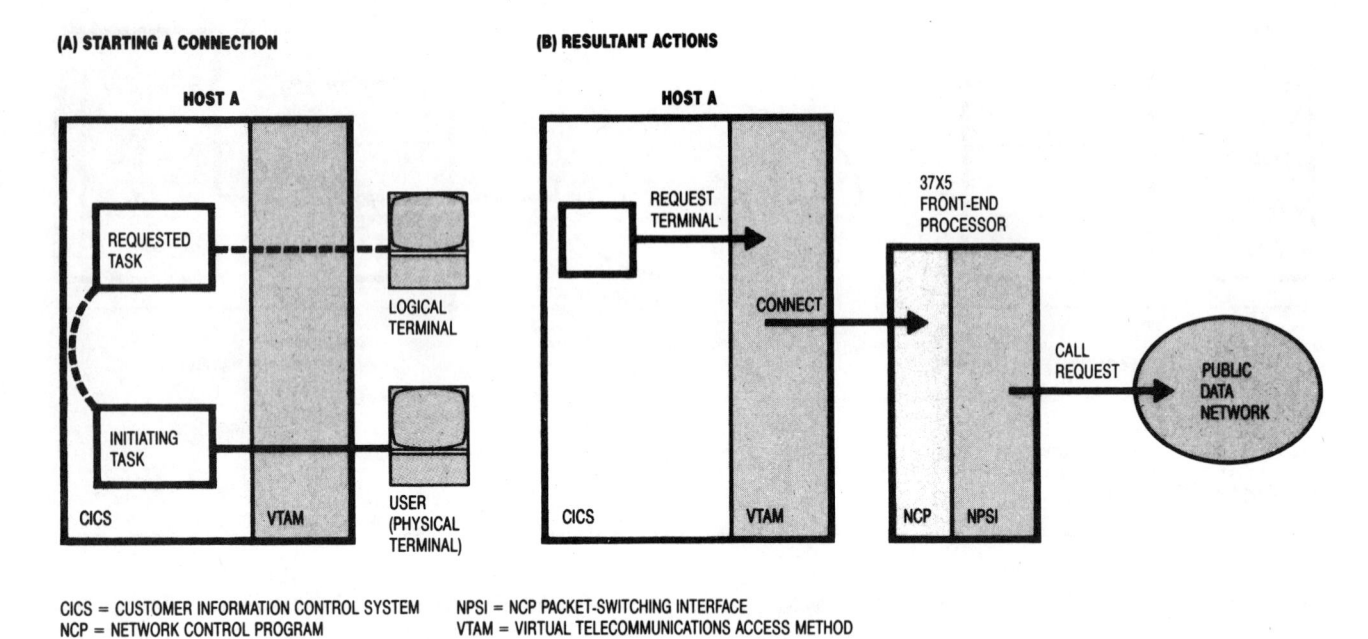

(A) STARTING A CONNECTION

(B) RESULTANT ACTIONS

CICS = CUSTOMER INFORMATION CONTROL SYSTEM
NCP = NETWORK CONTROL PROGRAM
NPSI = NCP PACKET-SWITCHING INTERFACE
VTAM = VIRTUAL TELECOMMUNICATIONS ACCESS METHOD

NPSI through the packet network to company B. Trying to implement this function led to many confusing problems that still defy resolution. A problem in the parameter definitions between NPSI/NCP and VTAM meant that the session would not complete, even when the documentation was followed explicitly.

The parameters in question were the DIALNUM parameter in the VTAM switch major node (VBUILD = TYPESWNET) and certain parameters in the NPSI generation. The NPSI manual was extremely confusing, making conflicting statements in different places. Here, it stated that the two numbers should be the same. There, it stated that VTAM would respond to NPSI by taking the IDNUM from NPSI and subtracting one from it.

By experimentation, it was discovered that VTAM did subtract the incoming IDNUM by one. However, the IDNUM had to be specified properly in the NPSI generation. The solution was to specify what was actually an improper IDNUM in the VTAM definition. Circumventing this problem was an exercise in illogic that allowed the session to be set up and completed.

Later, a method was found of specifying the same number in the NPSI generation and in VTAM. Though this took care of the original problem, it created another one: Half the SVCs that NPSI generated could not be used because the IDNUM would result in a hexidecimal number, something that NPSI would not tolerate.

The network connection between the two computers was

3. Trouble. *After a successful X.25 connection, company B's software sent two problematic messages, a log-on request and a command that should have stayed internal to company B's gear (the WRITE INITIAL command, meant for supporting terminals, was supposed to be intercepted by B's front-end processor).*

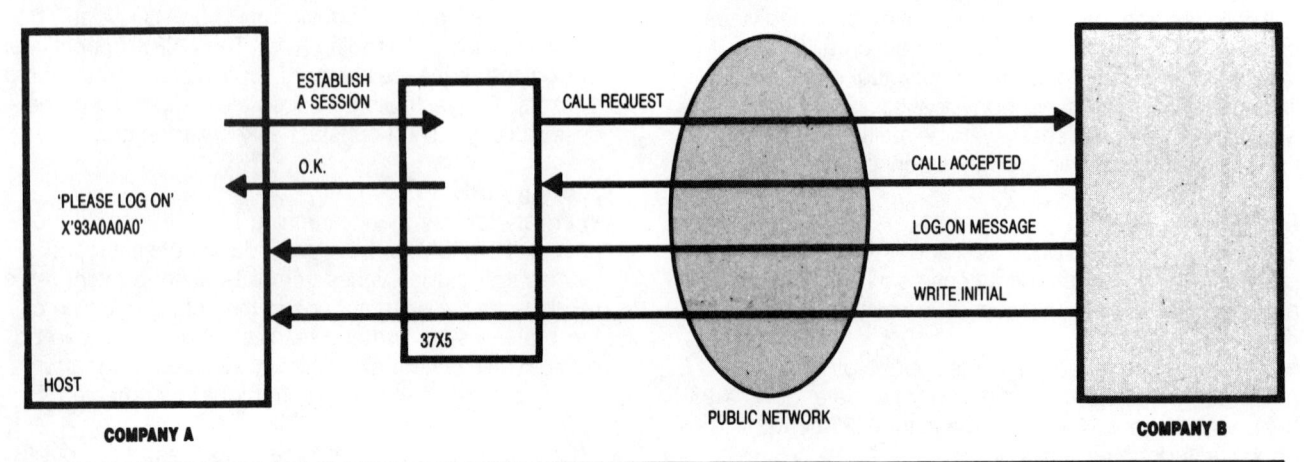

4. Mixup. The hosts think of each other as terminals, so Host A believes Host B's message includes a log-on destination, but it does not. What's worse, A can't figure out B's ASCII message. And all of this is happening while A's software is trying to set up an internal session. This problem was resolved by keeping A quiet for a while.

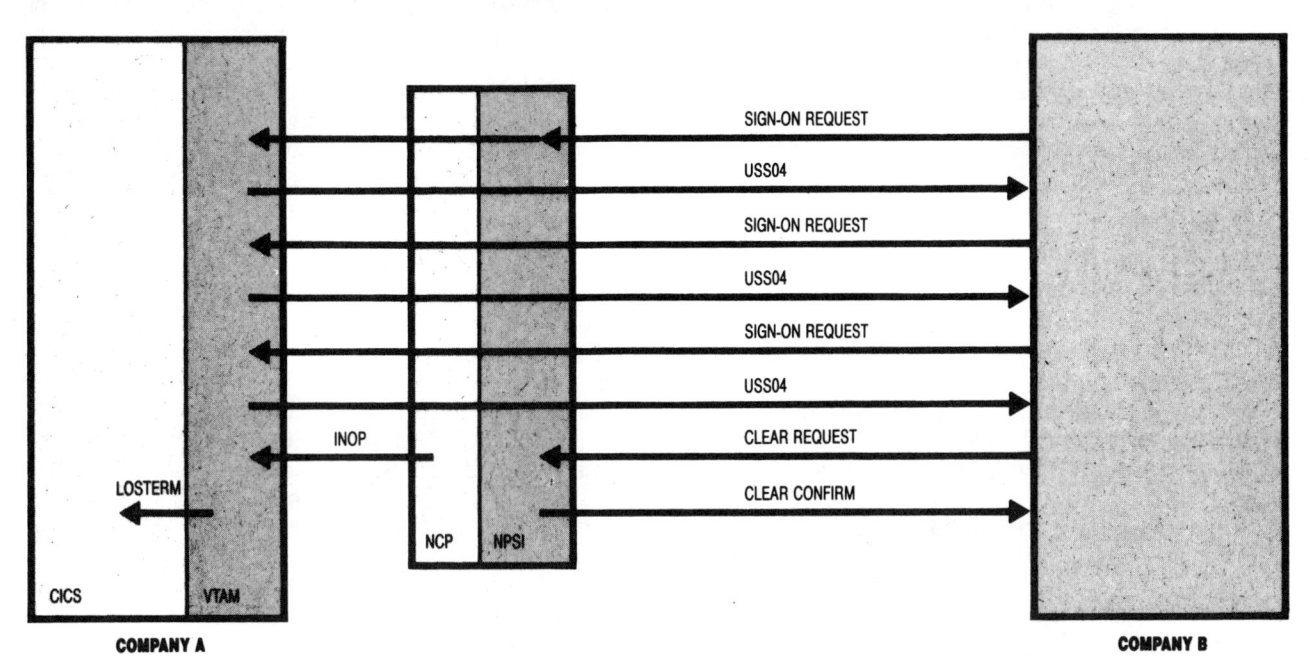

CICS = CUSTOMER INFORMATION CONTROL SYSTEM
INOP = INOPERABLE (MEANS B HAS DISCONNECTED)
LOSTERM = LOST TERMINAL
NCP = NETWORK CONTROL PROGRAM
NPSI = NCP PACKET-SWITCHING INTERFACE
USS = UNFORMATTED SYSTEM SERVICES
VTAM = VIRTUAL TELECOMMUNICATIONS ACCESS METHOD

5. Bottom line. If the hex message arrived when the application was ready, the connection could be made (A). Otherwise, the session ended in error (B). Even after all the network and access-method problems were solved, this millisecond timing problem remains, causing failure on about 1 percent of log-on attempts.

(A) CODE ARRIVES AFTER APPLICATION ATTACHED

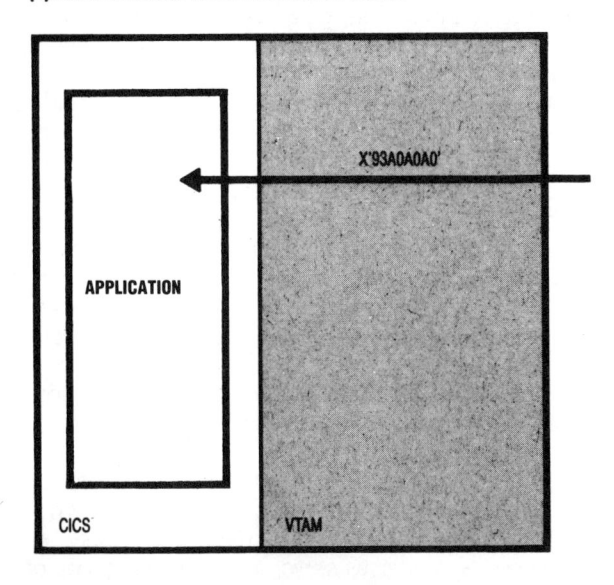

(B) CODE ARRIVES BEFORE APPLICATION ATTACHED

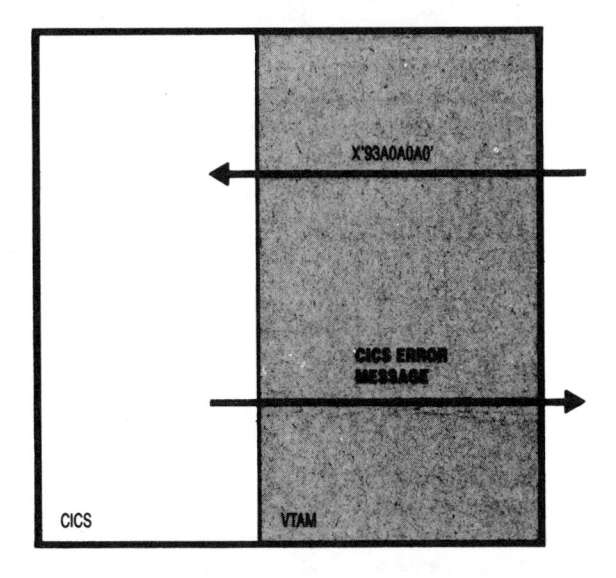

CICS = CUSTOMER INFORMATION CONTROL SYSTEM
VTAM = VIRTUAL TELECOMMUNICATIONS ACCESS METHOD
X'93A0A0A0' = THE ODD-PARITY ASCII EQUIVALENT OF WRITE INITIAL

complete at last. Unfortunately, the difficulties thus far encountered were only the first of a tortuous set of problems caused by the time-independent or unsynchronized operation of the large software components. Two such major headaches arose: one in establishing an SNA session between VTAM and NPSI, and another caused by an errant message packet sent from company B to company A.

■ Session setup. The first major synchronization problem, getting a connection through VTAM, was due to the fact that data activity on the X.25 link (which provides the logical connection between A and B) is not synchronized with the session setup between VTAM and NPSI, although the SNA session relies on the intermainframe link.

The configuration involved an X.25 call initiated from company A to company B. To accomplish this, the following sequence was to be carried out:

■ A Customer Information Control System (CICS) task requested the attachment of another, new task to one of a set of terminals logically defined for connecting to company B (Fig. 2A).

■ CICS recognized that it does not have an active session with the requested terminal and asked VTAM to make the connection.

■ VTAM, in turn, asked NPSI to send an X.25 CALL REQUEST command, which included the address of the destination computer, to the public network (Fig. 2B).

■ Company B responded with a CALL ACCEPTED packet (Fig. 3) and the X.25 switched virtual circuit was completed.

■ At this point, any data that the 37X5 receives from company B is held until NPSI responds ("O.K.") to VTAM's initial request for session establishment.

■ Once the SVC is active to VTAM, the buffered data continues through to CICS. VTAM normally expects this data to be log-on data.

Unfortunately, this is not the way it happened.

Company B's host was sending out a log-on request message as it normally would to terminals. It then expected log-on data to be sent back. Also, BTAM is idiosyncratic in that it must issue and process a WRITE INITIAL command to turn the line around before it can read incoming data. This command is supposed to be handled by B's NCP and not go any farther, but instead it was packetized and sent to company A. Thus, as Figure 3 shows, company A received an X.25 packet containing company B's log-on request and a second packet immediately thereafter containing the hexidecimal code X'93A0A0A0' (the odd-parity ASCII equivalent of WRITE INITIAL).

The first packet (the log-on request) was sent into VTAM. VTAM, expecting this data to be an SNA log-on request, tried to find the specified destination. However, VTAM had trouble with the incoming ASCII data. When it searched its tables to find the specified destination, VTAM was not able to match the input request. VTAM would then send back a message ("USS04") that the destination was unknown (Fig. 4). This would happen three times in a row, since B was set up to request a log-on three times. There was no synchronization between these exchanges and the rest of the session setup to CICS.

At the same time, company B would receive the response, a CICS message, from VTAM. Company B's computer was expecting a response to the log-on request of a user I.D. and password. Since this did not occur, company B's computer would break the session.

Many attempts were made to circumvent this sequence. The first effort was trying to get VTAM to set up the session more quickly. If it could be done before the three messages went out through VTAM, the messages could be isolated in CICS. This was unsuccessful.

However, a solution was finally found: to eliminate the error message VTAM was issuing. This was done by specifying in the unformatted system services (USS) table that no error message be issued. Since VTAM would not signal that it did not accept the data coming in, company B's host had no knowledge of a problem. This allowed enough time for the session to CICS to be completed.

■ Hex message. Now the CICS session was in place, but the second major synchronization problem appeared in trying to support the X'93A0A0A0' message, which CICS did not know how to interpret. It was at this point that very difficult timing problems started appearing. These problems, caused by related events occurring at unpredictable times with respect to each other, were highly sensitive to loading conditions of the processors, which affected timing. For example, sometimes CICS would receive the hexidecimal X'93A0A0A0' after the transaction program was attached to the terminal (Fig. 5A). In this case, the application program would request a log-on I.D. and password and the connection would proceed as planned.

At other times, however, CICS would receive the hexadecimal message before the application was attached (Fig. 5B). CICS, in this instance, would think the data included a transaction code in the first four bytes (CICS's default location for the transaction code). When this happened, CICS would transmit an error message, which company B's host would misinterpret as a command meaning "disconnect." To handle these cases, a dummy program (a sort of "null subroutine") was written and associated with a new CICS transaction code of X'93A0A0A0'. The dummy program would simply discard the hexidecimal message and return control back to CICS.

There was one other problem at an even higher level. Since the access method runs in a way uncoordinated with the application processes, the applications would have to interrogate the condition of the CICS application program interface (API) to know whether it should be receiving data from the line or was free to send data. However, between checking the API and actually issuing the appropriate command, the interface condition would sometimes change. This brought on a form of deadlock when the application would try to send but VTAM would not allow it to because new data had arrived on the line for processing.

To work around the uncertainty in the API, the application had to be modified to carefully interrogate some of the operating information kept by CICS in its internal control blocks. Normally, a user program would never have to look at these structures. However, this was the only way that the application would work at all with any regularity. ■

David G. Matusow, senior telecommunications consultant, works in the Phoenix, Ariz., office of the Computer Task Group Inc. in Buffalo, N.Y. Matusow holds a B.A. in political science and a B.S. in computer science, both from Purdue University in Lafayette, Ind.

Shel Blauman, Boeing Computer Services, Seattle, Wash.

Developing a user-oriented networking architecture

Tired of vendors' proprietary network architectures? Then develop your own. Large user Boeing explains why, and how.

Mention the words "network architecture" and most users think of a specific vendor's communications framework, one that ties together that vendor's products and, invariably, locks out those of competitors. In this context "architecture" may seem more intimidating than inviting. But the notion of a structured architecture can actually be quite useful. Indeed, for users with mixed-vendor networking environments, an architectural framework can be enormously helpful in shaping future computing and communications strategies.

Architectures come in various shapes and sizes. With its Systems Network Architecture (SNA), Document Interchange Architecture, and Document Content Architecture, IBM, of course, has made an art of creating architectures. And IBM is not alone: All major vendors have promulgated identifiable communications architectures.

The Boeing Company, a major manufacturer of military and commercial aircraft which is also involved in space and defense projects and electronics design and manufacture, has pioneered the notion of a user-driven network architecture. There has been considerable attention paid lately to Boeing's Technical and Office Protocol (TOP). What is not widely known, though, is that TOP was a direct result of the company's work toward such a user-oriented communications architecture.

A protocol 'stack'

The Boeing Network Architecture (BNA) began in 1979 as an attempt to bring a layered structure to Boeing's internal data communications activities. TOP, as well as General Motors' Manufacturing Automation Protocols (MAP), provide detailed specifications for implementing certain of the International Organization for Standardization's (ISO's) Open Systems Interconnection (OSI) standards. Anyone implementing TOP, MAP, or some other OSI protocol "stack" should seriously consider expanding on this effort and developing a complete layered architecture, which can make more efficient and effective use of existing multivendor networks.

BNA was a project of Boeing Computer Services, Boeing's technical consulting arm. The concept was to establish a family of standards that would provide interconnection among Boeing's diverse computing environments. Originally, IBM's SNA was the protocol suite of choice for BNA. Boeing now is migrating to OSI, but both SNA and OSI are recognized as members of the BNA family. The architecture is not based on any single set of proprietary protocols; in fact, protocols come and go within the context of BNA. Moreover, the communications environment encompassed by both SNA and OSI is just one aspect of the Boeing architecture.

Today the term "network" in network architecture has been eclipsed as the scope of BNA has progressed to include databases, operating systems, and graphics, as well as any other technology that could be involved in a distributed environment. The Boeing Network Architecture name remains, but it now encompasses all technologies dealing with distributed computing resources. The BNA architects have implemented the user-oriented networking structure across local area networks (LANs) and have also incorporated departmental computing into it. Perhaps the most significant achievement, however, has been the establishment of BNA as a working architecture within a production environment where computing and communications technologies are constantly changing.

Data highways

As an outgrowth of Boeing's architectural development program, TOP has received much more publicity than the overall architecture itself. But TOP alone is not enough. Network implementers are learning that the availability of such common communications protocols is a necessary

1. Three-tiered. *In its simplest and original form, the Boeing Network Architecture specifies three processing levels with two intervening layers of communications.*

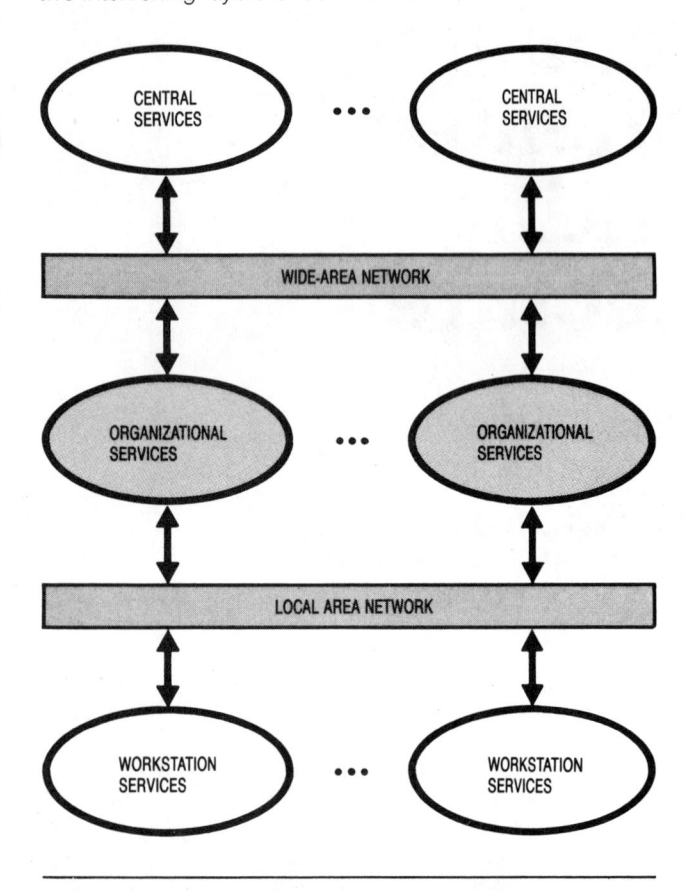

[Figure 1: Diagram showing CENTRAL SERVICES ovals connected via WIDE-AREA NETWORK to ORGANIZATIONAL SERVICES ovals, connected via LOCAL AREA NETWORK to WORKSTATION SERVICES ovals.]

ingredient for the achievement of an integrated network, but these protocols must be made a part of a more comprehensive plan.

The infrastructure of highways and roadways is a good example of an architecture that can be compared to computer networks. There are local streets, arterials, regional freeways, and interstate highways. The different categories exist mainly because the alternative, direct point-to-point paths between all end points, is either impractical or impossible.

A network architecture similarly can establish categories of networks, such as clusters, local area networks, campus-wide networks, and wide-area networks, with different access mechanisms and traffic volumes specified for each. Just as in the highway analogy, the goal is the efficient and effective utilization of resources, which is achieved by hierarchical organization.

Boeing's network architecture is based on a model that divides information-processing functions (services) into three hierarchical levels: central, organizational, and workstation. Mapped between these are wide-area and local communications (Fig. 1). The three processing tiers and two communications levels are sufficient to describe most networked applications.

The same three-tier model can also be mapped to generic applications and environments, such as the office or the factory. Although each has unique requirements, they are similar in that they use databases, communications, and operating systems, all of which must be compatible. The same model can also be applied to finance, in-house publishing, and engineering design and analysis.

The simplicity of the three-tier model can be deceptive. While there is nothing revolutionary about three levels of computing mainframes, minicomputers, and desktop microprocessors have been with us for years, for example, this three-tier model is logical, and not physical. This means that the software performing an office function does not have to actually reside on the served organization's minicomputer. An organizational processing function may be performed on a microcomputer that serves as an organizational server on a LAN, or on a mainframe that is accessed via remote communications. For example, IBM's Professional Office System, which runs on IBM hosts located in data centers, is considered within the Boeing architecture to be an organizational capability.

Organizational servers have been useful in promoting the network-architecture concept. Sharing expensive peripherals, multiplexing transactions over communications lines, and using the server as a workstation switch have been a great benefit in advancing BNA. The organizational server is a prime example of the change in thinking that must occur when making a transition from host-driven to peer-level processing environments. In the past, today's organizational servers were miniature hosts driving local terminals. And the transition from dumb terminals to microcomputer-based workstations has shifted the role of the mini-computer from one of local processor to local service center.

New wrinkles

BNA was developed initially for data-center interoperability. Local area networking was introduced as an intrinsic, rather than an incidental, component part of the three-tier model. Conveniently, LANs fit well between the organization level and the workstation level of the model (second and third tiers). At the same time, long-haul or wide-area networks (WANs), such as SNA, were categorized as residing between the organizational and the central tiers.

The LAN of choice for the office environment within the Boeing architecture was identified as Ethernet (later IEEE-802.3); the local network for the factory and campus environments was selected as EIA TR40.1 (later to become the basis for the broadband specification of the token-bus-based IEEE-802.4).

In the three-tier conceptual framework, how should microcomputer-to-host connections be viewed? Rigidly applied, the three-tier model would require an intelligent workstation to go through an organizational server to access a central application program. This scenario, however, fails to account for nonintelligent, directly connected terminals.

Obviously, Boeing's architecture could not be implemented overnight. It might be 10 years before all users have an intelligent workstation with OSI peer-to-peer communications. So, to be accepted, any architecture must provide a migration path to accommodate the organization's long-range plan. Trying to explain that path in the context of existing, as well as future, usage led Boeing's network architects to a model that accommodated all possible communications between the three tiers (Fig. 2).

The two new contributions of this Venn diagram (where

intersecting circles are used to explain exclusive and overlapping domains) are the inclusion of remote communications and common protocol standards. The three-tier model could not explain how digital private branch exchanges (PBXs) and other telecommunications gear related to LANs or where they fit in the architecture.

Although remote communications had traditionally been the most prevalent means of networking, it was not stressed in the design of the architecture. Such connections still play a part, to be sure, but these are handled via Boeing Telephone Services, the firm's private telephone network. Still, as the private analog switches in this network are converted to support the Integrated Services Digital Network (ISDN), they are expected to play a larger role in BNA, complementing local and wide-area networks. To access central services, workstation users eventually may choose to traverse either an ISDN or LAN path, with OSI protocols providing universal higher-level connectivity in either case.

The BNA developers recognized from the start that there could not be one communications solution for all applications. As a result, such issues as broadband versus baseband or digital switch versus LAN never predominated. The objective was a family of communications solutions that would offer compatible networking despite the particular technology employed.

Other network-categorization questions remained, of

2. No circus. *Communications tasks fall neatly into the rings' overlap. Remote links, now by phone, will soon be via ISDN. OSI is, or will be, common to all connections, and is being deployed as products become available. The wide-area and local-area distribution elements now in place follow the standards appropriate to such elements.*

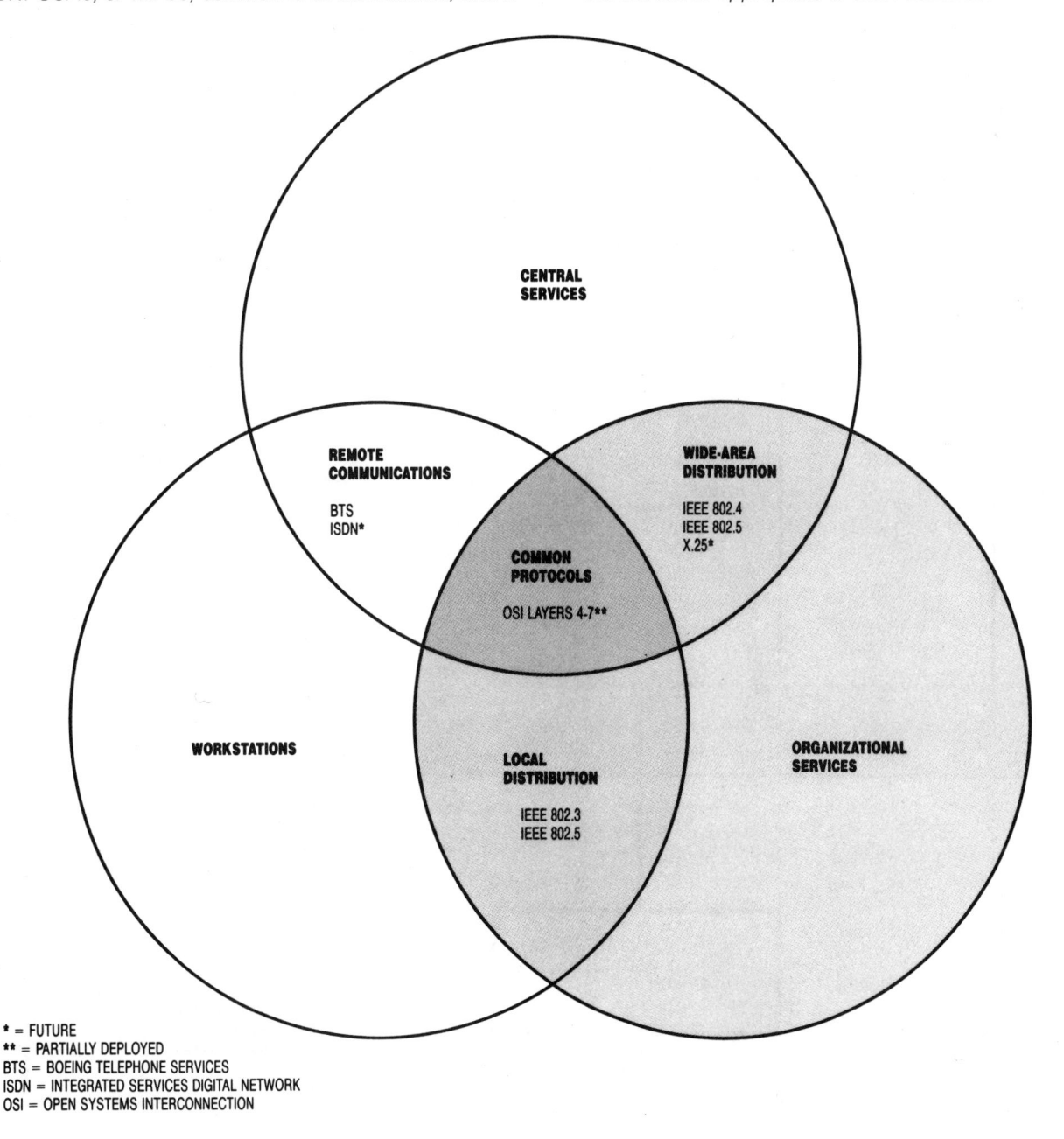

CENTRAL SERVICES

REMOTE COMMUNICATIONS
BTS
ISDN*

WIDE-AREA DISTRIBUTION
IEEE 802.4
IEEE 802.5
X.25*

COMMON PROTOCOLS
OSI LAYERS 4-7**

WORKSTATIONS

LOCAL DISTRIBUTION
IEEE 802.3
IEEE 802.5

ORGANIZATIONAL SERVICES

* = FUTURE
** = PARTIALLY DEPLOYED
BTS = BOEING TELEPHONE SERVICES
ISDN = INTEGRATED SERVICES DIGITAL NETWORK
OSI = OPEN SYSTEMS INTERCONNECTION

3. Stacks. *In a network node, local devices, applications, and utilities receive administrative support from the operating system. The node does most of its work, such as storing and retrieving data and processing communications, in the operating tasks column. The management stack oversees the processing and operations of the other stack.*

NETWORK NODE

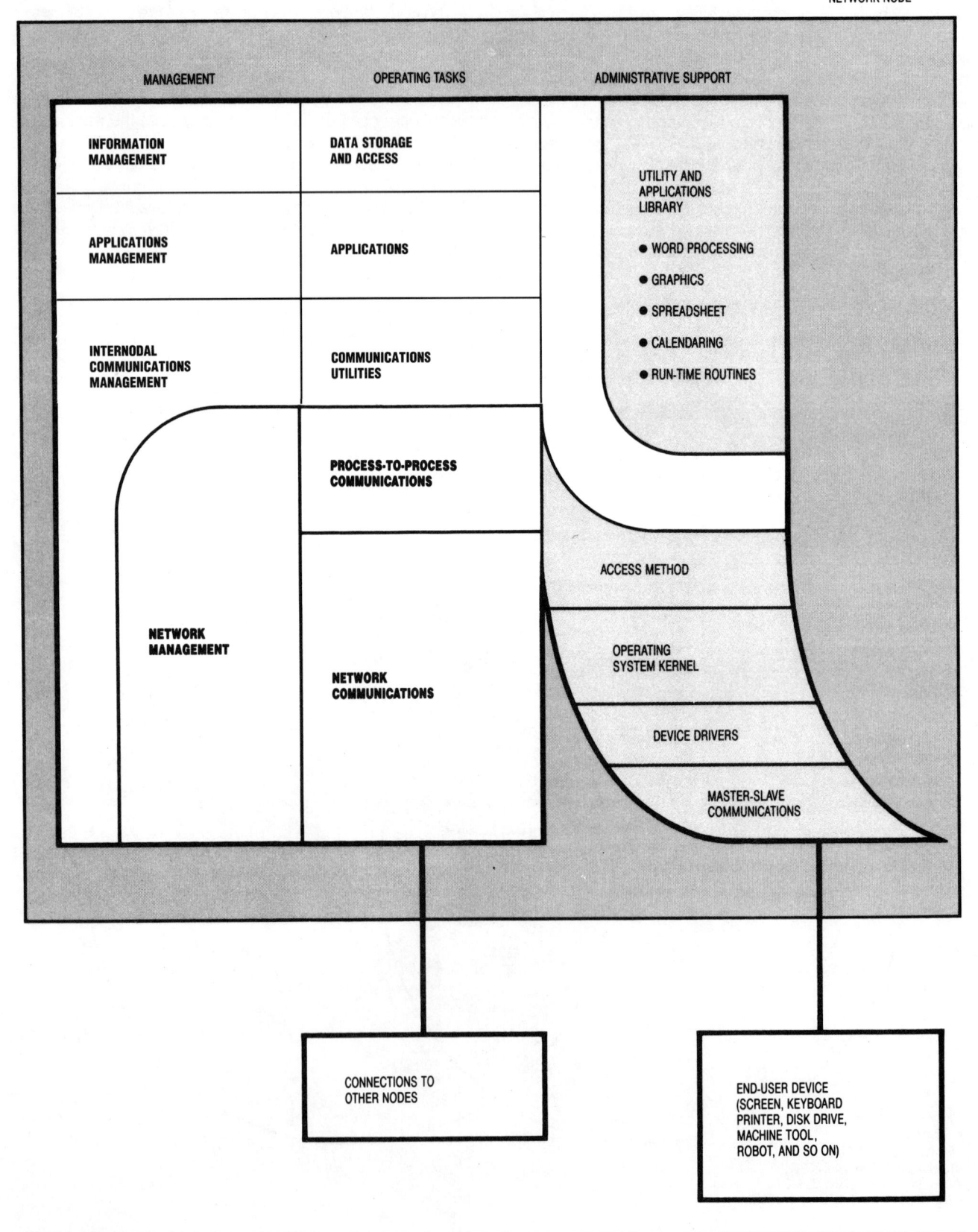

MANAGEMENT	OPERATING TASKS	ADMINISTRATIVE SUPPORT
INFORMATION MANAGEMENT	DATA STORAGE AND ACCESS	UTILITY AND APPLICATIONS LIBRARY • WORD PROCESSING • GRAPHICS • SPREADSHEET • CALENDARING • RUN-TIME ROUTINES
APPLICATIONS MANAGEMENT	APPLICATIONS	
INTERNODAL COMMUNICATIONS MANAGEMENT	COMMUNICATIONS UTILITIES	
NETWORK MANAGEMENT	PROCESS-TO-PROCESS COMMUNICATIONS	ACCESS METHOD
	NETWORK COMMUNICATIONS	OPERATING SYSTEM KERNEL
		DEVICE DRIVERS
		MASTER-SLAVE COMMUNICATIONS

CONNECTIONS TO OTHER NODES

END-USER DEVICE (SCREEN, KEYBOARD, PRINTER, DISK DRIVE, MACHINE TOOL, ROBOT, AND SO ON)

course. For example, where do 50-Mbit/s, computer-to-computer, data-center networks fit into the architecture? Are they LANs or WANs? As it turns out, they are neither. As they are LANs local to the data centers, they are within a tier, not between tiers. Because they are within a tier, they are considered "closed clusters," which are acknowledged by the architecture but not bound by it.

Network nodes

The model still was not complete. Almost all computing at Boeing had been host-based, centered around either a mainframe or minicomputer. LANs and the ISO's OSI development, however, were introducing the concept of peer-level networking. It became necessary therefore to account for differences in the communications paths between host-based and peer-to-peer, transaction-based services. This led to the development of the network-node model (Fig. 3).

Not as apparently simplistic as the other models, the network-node model explains the role of operating systems, libraries, management functions, and database managers. The technologies depicted in Figure 3 actually developed, broadly speaking, from bottom to top and from the right column to the left.

The model depicts just one nodal element out of a network of communicating elements. Administrative support refers to the operating-system services provided by the node, such as terminal drivers, small software routines that enable specific devices to access applications. The operating tasks perform the actual work of the node, including end-to-end communications and transaction processing. Modules in the management stack are responsible for knowing and updating the status of the node and of ther relevant nodes in the network.

In Boeing's architecture, as in most others, the management-stack functions are still largely under development. Typically, management is only needed to make exception decisions. Thus, the network node could operate without this stack, though not indefinitely. Information management refers to knowledge of the location of data and to the management of it as a resource. Applications management, similarly, implies knowledge of the status and location of applications and also the ability to initiate and terminate them. The internodal communications manager needs the ability to network both the resources local to the node as well as those located elsewhere in the network. Network management entities, associated with each network layer, are currently being defined under OSI.

The stack of operating tasks is the heart of the model and is where most of the node's networking activity actually takes place. The data storage and access module knows how to locate desired data, both internal and external to the node, and can request external data from a peer entity in another machine. And since it is the end point for the storage and retrieval of any data, this module is the last location of intelligence in the data path. It resides therefore at the top of the column, above applications, which merely process the data on its way to or from storage. The applications module consists of data processing programs.

The modules below applications, which move preprocessed or processed data, comprise the local OSI implementation. Communications utilities encompass OSI's Layer 7, the actual communications interface, and Layer 6, which presents data to Layer 7. In most implementations appearing today, these two layers are usually combined (or at least closely integrated).

The process-to-process module contains Layer 4 (transport), designed for setting up peer links, and Layer 5 (session), a refinement of transport which creates sessions between processes and holds onto them. Although Layer 4 transports a session's data over a peer link, it does not require a peer network beneath it to do its job. Network communications refers to Layers 1, 2, and 3, which may actually be implemented, for example, as either a LAN or a packet network.

Finally, the administrative-support stack performs the node's utility routines. It can best be described from the bottom up, looking from the terminal's point of view. Data from a terminal flows through a device driver, then the operating system kernel, and finally through the access method to get to the user interface, which, in principle, allows the user access to any of the applications or utilities in the node.

Interaction

If the user wanted, for example, to access local administrative resources (such as to write a letter or work on a spreadsheet), the data path would be set up and remain within the administrative support stack. If the user wanted to get at data but did not know where the data was, the session would be diverted to the data storage and access module, which might in turn go to information management's data dictionary to locate it. Depending on where the data was stored, the session might then be routed out over the network, or possibly back down the administrative stack to a locally attached storage device.

Applications and utilities stored in the local library can be accessed via the user interface. And since the three modules atop the first two stacks are applications, they should in principle be similarly accessible. For example, if the user needs the services of a distributed-processing application, the session would veer left from the user interface to the applications module. To access another node directly, the user could go to communications utilities—for example, an OSI Layer 7 protocol, such as File Transfer, Access, and Management (FTAM). The session would then proceed down the operating-tasks stack and out over the network to a corresponding FTAM process in another node.

This network-node model is, in fact, rich enough to be compared to a conventional business structure. Management in a node corresponds to corporate management; operating tasks are equivalent to line management in an organization, the locus of day-to-day operations. Administrative support maps to the secretarial staff.

Implementation issues

Once the definition of an architecture is underway (a never-ending process), there is the small remaining detail of implementation. Implementing an architecture requires acceptance by the user organization, as well as by the vendor, or vendors.

Management can rarely impose on users an architecture

Migration strategy

Boeing Computer Services currently provides internal Boeing users with a document interchange and library services capability between many of its approved office-automation vendors' equipment. This service is based on IBM's Distributed Office Support System (Disoss) product. The interfaces from other vendors' environments to Disoss are, in most cases, supplied by the vendors themselves. Two of the approved vendors were unable to provide such interfaces, however, and Boeing itself developed an interim adaptation, pending full X.400 implementation.

In cooperation with the vendors, Boeing developed document transformation and communications capabilities. The vendors' products communicate in an IBM 3270 mode with the Boeing software, which transforms the documents into IBM Document Content Architecture format and then passes them to Disoss.

BNA's migration strategy is to allow vendors to individually convert to Open Systems Interconnection (OSI) while continuing to support their own architectures. It is not practical to expect vendors to change overnight from proprietary protocols to international standards. Nor can a user dictate a specific date for full OSI compliance. An architectural gameplan must be able to bring OSI solutions into play as they become available; they will not be available from all vendors at the same time.

Non-IBM vendors can only exchange mail-type documents via the Disoss backbone. However, they may directly exchange files and records via OSI protocols such as File Transfer, Access, and Management. X.400 is not expected to displace Disoss for intervendor document exchange in the Boeing environment until sometime into the 1990s.

The Boeing-developed transform capability will eventually be eliminated as vendors develop their own DCA transform capabilities and later progress to an as-yet-undefined OSI document-content specification. Because of this, it will not be necessary to wait until the 1990s for an intervendor document exchange capability to become available. This capability is available now under BNA, in a form that will allow migration to OSI.

OSI implementation by vendors apparently is occurring from the bottom up, starting with local area networks (LANs) and progressing to the higher-level protocols. In fact, some OSI-to-LAN connections (IEEE 802.3) are available today: Digital Equipment Corp. is now offering the OSI Transport Class 4 protocol as an option.

Most vendors are expected to implement the higher-level OSI protocols as a capability within their products independently of their own architectures, waiting for a complete protocol suite before substituting OSI for their proprietary protocols. Users will then be able to write applications using any available OSI capabilities while the vendor can support existing applications.

This migratory approach has proven useful in responding to real-world requirements at Boeing. If Boeing had refused to purchase equipment pending the development and implementation of OSI standards, there would have been no new equipment purchased for the last three years. Granting preferential treatment to vendors who commit to Boeing architectural goals helps promote rapid OSI development without the need to halt computer purchases until the OSI products are available. Vendors are now being required to formally commit to OSI-implementation schedules, and users generally will be expecting those commitments before placing orders.

---

that they are loath to accept. The easiest approach is to ensure user acceptance first and then use their collective purchasing power to influence vendors. Users, however, have immediate requirements to fulfill and cannot wait for a long-range plan to become effective. Users' purchasing power can be tapped, therefore, only if the users know that the developed architecture is consistent with the vendor's strategic intentions.

Convincing vendors to abide by a set of architectural guidelines, when the vendors' only incentive to do so is the voluntary cooperation of users who must continue to purchase their products anyway, is truly a chicken-and-egg situation. The solution, for Boeing Computer Services, was to first identify a set of vendors recommended or endorsed by the parent Boeing Company. This vendor-approval process has been useful in being able to get through to the right person within the vendor organizations.

The architecture that Boeing outlined to office vendors was one that allowed them to proceed with their proprietary development efforts within their own computing environments but required them to adopt specific interfaces for communications outside those environments. The original list of approved office vendors included Digital Equipment Corp., Hewlett-Packard, IBM, Wang, and Xerox.

IBM's SNA was selected as the backbone communications architecture among vendors, though OSI was specified as the long-range migration plan. Vendors were required to commit to the OSI concept. SNA was selected because of its popularity, its layered structure, and the fact that most vendors were already developing SNA interfaces on their own.

The many advantages offered by the resulting Boeing architecture include that:
■ It allowed the continuing use of many host-based applications during the development of distributed computing.
■ It enabled the use of the IBM-based data centers as intermediate nodes between vendors and allowed a corporation-wide document-distribution scheme to be established (see "Migration strategy").
■ It allowed the use of vendor-proprietary internal protocols, at least for the short term.
■ It offered a palatable plan to the vendors, most of whom had already intended to develop SNA gateways and had seen the OSI writing on the wall.

The vendors would have cooperated even without the threat of lost sales. However, some pressure was necessary to drive home the fact that, from Boeing's point of view, the days of proprietary communications were over and that, eventually, every vendor would have to be able to communicate directly with every other vendor. Without that leverage,

Supporting the nonconformist

In the absence of Open System Interconnection (OSI) products, it has been necessary to purchase computing equipment that not only does not conform to OSI but that, in some cases, does not conform to the other objectives of Boeing's network architecture, such as being able to interface to Disoss. This is not done willingly, but it is done nevertheless when a product fits a particular business need.

Boeing's goal is to manufacture its products in the most cost-effective manner possible, and OSI is seen as the long-range means to cost-effective computing. However, implementing OSI at the expense of productivity is counterproductive, so long-range planning must be tempered by practicality.

As a result, some islands of nonconformity have come to exist within the company. Still, Boeing's architectural standards expressly prohibit the direct interconnection of these islands with each other by bypassing the backbone. Thus, turnkey packages, incompatible with BNA, are relegated to isolation unless interfaces necessary to connect to the network provided by Boeing Computer Services are developed. The cost of interconnecting nonstandard equipment is borne by the organization requesting it. As a result, many nonconforming purchases have been avoided when the full implications and costs were examined. Still, some of these islands do exist.

And BNA cannot ignore these islands. Ignoring a fact of life does not make it go away. Architectural anomalies have to be monitored and brought into the network whenever possible. The mechanism Boeing has developed for this is the cluster. Clusters are families of compatible equipment from a single vendor, which might be connected via a turnkey local area network. They, in all likelihood, use proprietary protocols and, in some cases, provide neither a Systems Network Architecture nor an OSI gateway. For the most part, clusters consist only of locally interconnected workstations, which would conceptually reside within the third tier of the Boeing three-tier architectural model. If LAN servers are included, these are, by definition, tier-three equipment as well.

The cluster concept, though developed to categorize nonconforming equipment, may also have a place in the Boeing architecture for conforming gear. A few workstations in an immediate work area may share a disk or a printer, perhaps allowing the use of diskless workstations. Workstations in this category are not fully standalone, but are not terminals either. The clusters may, however, in turn, be interconnected to the organization's network via a single interface.

Real-world requirements and the rule of supply and demand have led to the occasional purchase at Boeing of nonconforming equipment. Nonapproved vendors can sell to Boeing, but on a case-by-case, buyer-beware basis under the cluster concept. These users, however, are prohibited from linking their clusters via any method other than that prescribed under the Boeing architecture.

it would have have been difficult, if not impossible, to get vendors to deliver the required SNA gateways and to commit to OSI within the desired timeframe. These responses on the part of vendors have convinced most Boeing users (though not all; see "Supporting the nonconformist") of the advantages of a uniform user architecture.

Typically, nonconforming clusters are introduced because they provide needed functionality at the user level, not at the network level. Users may generally expect to see the same functionality on an established base from vendors that insist on retaining compatibility with former products within a year or two. In general, users who hop from one fad product to another are on the forefront of capability, but they are also continually purchasing new equipment and most probably spending a good deal of time reprogramming.

The relationship of a user networking architecture to OSI and to other standards, both de facto and de jure, is a sensitive one. While OSI is a family of communications protocols, an architecture (such as BNA) is a structure that can make use of such standards while still allowing vendor-proprietary protocols. OSI is not, it should be noted, concerned with the organization of the office or with office servers but only with the interconnection of computing devices and the exchange of information. Architectures, on the other hand, address technologies other than network communications standards. (BNA, for example, covers operating systems and database management.)

Since its inception, BNA has seen SNA become a de facto standard, LANs become a significant communications alternative, and OSI near the threshold of reality. BNA has had to address the growing importance of the departmental computer as a server for independent workstations, rather than as a local computing resource. In so doing, BNA has established itself within the Boeing community as a valid template for networked computing.

Interest in architectural definition is on the rise, especially since advances in technology have led some companies to purchase such new products as LANs, with little or no operational improvement. Implementors of automated offices have learned that there is much more to be done than just connecting boxes with a wire.

But just as protocols alone are not sufficient for network-node implementation, architecture alone does not ensure interoperability. Users who are about to implement a multivendor network or are planning to do so in the near future should have two objectives: They should determine which standards will provide intervendor interoperability, and they should endeavor to define an architecture that will supply the structure for a complex multivendor network. ∎

Shel Blauman is manager of networked applications architecture and integration technology at Boeing Computer Services in Seattle, Wash. He has worked on the Boeing Network Architecture for five years and in the computer industry for more than 25 years. In 1981, Blauman founded the Network Users' Association, focusing on making Open Systems Interconnection a reality, and was the group's first president. Among his accomplishments is the development of a 20-Mbit/s fiber-optic ring network in 1979. He holds a B. S. in mathematics from Arizona State University and an M. A. in cinema from the University of Southern California.

James F. Mollenauer, Computervision Corp., Bedford, Mass.

Networking for greater metropolitan areas

A new age of MAN. Recent developments foreshadow rapid emergence of city-size networks

High-speed metropolitan area networks (MANs) able to carry both voice and data are ready to challenge the traditional star topology of the voice-only telephone network. Based on rings or buses and tuned to a city-sized scale, these networks will provide high-speed services that have previously been available only on the smaller scale of a local area network (LAN).

Standards for MANs are being developed as part of the Institute of Electrical and Electronic Engineers (IEEE) 802 LAN project, and several promising designs that employ high-speed fiberoptic technology are now being evaluated.

The MAN working group of IEEE 802 began developing a standard in 1982, about two years after the inception of the 802 project. The intent was to provide a specification for a citywide network. Within a 50-kilometer diameter, the MAN must be able to carry digital voice and video, as well as interconnect to LANs and other data sources.

The MAN group — known as the IEEE 802.6 committee — is unique within the IEEE 802 organization. It is not tied to a specific protocol, in the way that 802.5, for example, is limited to token-passing rings. Another unique feature of the 802.6 work is that because MANs are designed to be high-speed digital backbones, the 20-Mbit/s data-rate limit imposed on other 802 LAN standards has been removed.

Conferences on MAN technology have been sponsored by the IEEE; the first of these was held last April, and a second was held last fall. All of this activity indicates that MAN standards development has reached critical mass.

Leading contender

The 802.6 standards committee is now focusing on a single technology for metropolitan networks. This is the dual bus proposal from Telecom Australia. For more than a year, this proposal — and a slotted ring proposal known as MST — were debated and refined in committee meetings. In last November's meeting in Fort Lauderdale, Fla., a consensus emerged in favor of the dual bus, and it received votes well in excess of the required three-quarters majority. The strength of the consensus is indicated by the participation of a majority of the regional Bell operating companies, AT&T, and Bellcore.

In addition, the committee accepted a proposal by AT&T for interconnecting dual bus systems. This interconnection scheme would bridge together multiple bus systems using intelligent switching fabrics that are presently being prototyped by the telephone industry.

MAN technology, which evolved from local area network designs, uses digital backbones to carry the traffic and provide links to wide area networks. In some proposals, subsidiary rings running at lower speeds enable connection of individual voice or data terminals. A highly schematic view of a metropolitan area network is shown in Figure 1.

MANs and LANs are, in fact, a result of applying computer technology to communications, though perhaps they were not an expected result. The roots of this new phenomenon are worth examining.

Computer technology has already heavily infiltrated the telephone industry. Though it has accelerated the pace of automation, it has not conceptually altered the nature of the telephone network. Even the growing data communications load on the telephone system has not resulted in any radical overhaul.

Data traffic has been accommodated only by making it look like voice traffic: first with modems that convert data into an analog signal that can pass through networks designed for speech and more recently with digital trunks that reverse the situation — by digitizing voice to give it the form of data.

Nevertheless, the telephone industry has not changed its network topology. It is still configured as a hierarchy of stars based on PBXs, local offices, toll offices, and so on. This has many good features: It operates with very simple and

1. Family of MAN. *MAN technology, which evolved from local area network designs, uses digital backbones to carry the traffic and provide links to wide area networks.*

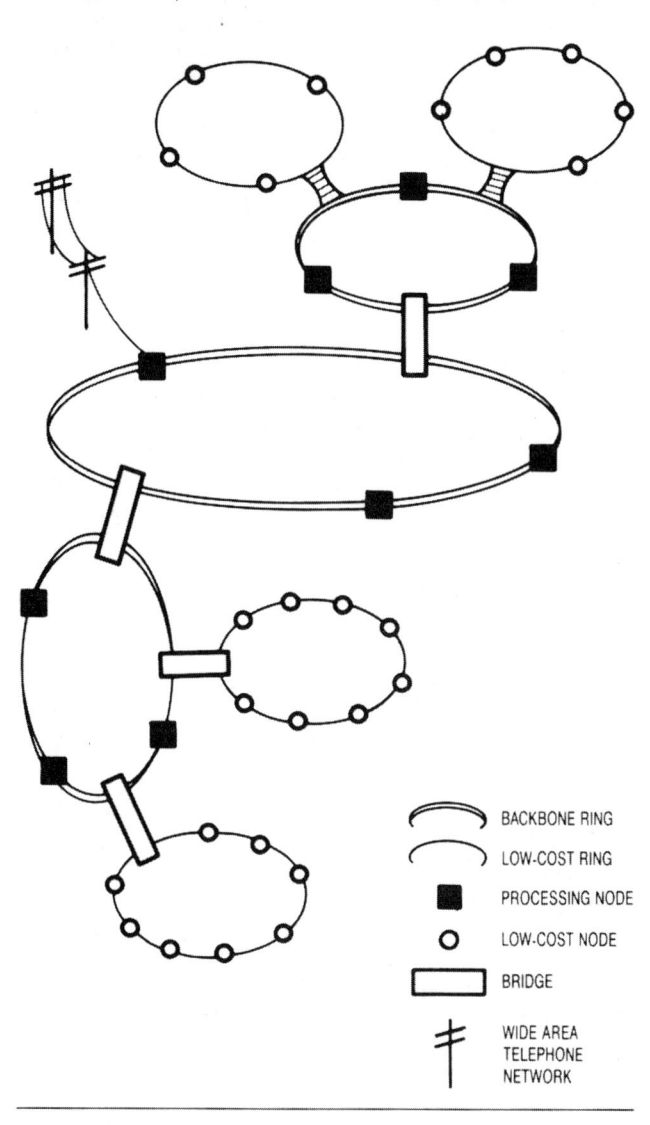

- ⌒ BACKBONE RING
- ⌒ LOW-COST RING
- ■ PROCESSING NODE
- ○ LOW-COST NODE
- ▭ BRIDGE
- ⌁ WIDE AREA TELEPHONE NETWORK

rugged end-user equipment, it is highly automated and reliable, and it distributes the required intelligence in a sensible way. So why change it?

The answer is that technology has moved too far for the original networking concept to remain valid as the best solution in all communications situations. Intelligence in the form of microprocessors has become too cheap to be confined to central offices, and the disparity in requirements between voice and data is accelerating.

The processing overhead inherent in complex data communications protocols is substantial. Computers cannot absorb or generate packets nearly as fast as the transmission medium can carry them. Therefore, it makes sense to share the communications medium and—very important—to obtain simultaneous distributed switching.

In this context, it is not surprising that the competition to the conventional star topology arose from the computer industry rather than from telecommunications. This competition is represented by the LAN and, more recently, the MAN. These networks stem from computer architecture and make heavy use of computer technology that would have been far too costly a few years ago.

A LAN is not a scaled-down telephone network; rather, it is a scaled-up and serialized computer bus. Almost every computer is designed with an internal bus structure where all data flows between the central processor, memory, and peripherals. Such buses must work on a time-shared basis carrying data now to the central processing unit (CPU), now to memory, now to an input/output (I/O) device. With or without priority features, it can adapt to bursty requirements while serving a large number of simultaneous data streams.

Making the bus

An extremely effective way for a set of computers to communicate would be to extend a common bus between them. This is not quite practical because bus-allocation schemes usually assume very short buses, and the parallel cabling would be unduly expensive and cumbersome. LANs do the next best thing: They serialize the data for transmission over a single line, and they use an allocation scheme suitable for building-size or larger buses. Finally, the allocation logic is distributed rather than centralized, eliminating single points of failure and lowering the initial installation expense.

Computers attached to LANs share a high-speed transmission facility that no single processor dominates. When a computer needs high performance, it can use a LAN for file transfer, bit-map graphics, and the like. This situation has clear advantages, so LANs are a popular means of communications between buildings or clusters of buildings.

Local area networks, however, have been optimized only for distances up to several kilometers and will not work with the same capacity over longer distances. This is where MANs come in. Long-distance connections are still the province of telephone-based technology or earth satellites, though MANs are ready to serve areas the size of a city and its suburbs. MANs represent local area network technology optimized for the longer distances. Some technical aspects of LAN technology must be changed, though much technology remains the same.

A MAN need not handle all the computer and voice traffic of a city. Rather, it is an overlay on existing communication facilities, one that can interconnect networks scattered throughout a city. In addition, it can serve to interconnect the large numbers of personal computers in businesses and homes. The MAN's packet capability is much better adapted to file-transfer or interactive applications than the fixed-bandwidth, circuit-switched connections of the telephone network.

The corporate MAN

Corporate MANs may come into existence in many areas before public MANs. A single corporation can fund the installation of a network from the beginning, while public networks will require widespread installation of facilities before revenue comes in. A typical corporate network might look like the one in Figure 2, with branch offices and manufacturing plants tied in to the corporate headquarters via telephone offices.

While the topology of the corporate network may be a ring, the wiring geometry approximates multiple stars. This

2. Corporate connection. *Corporate MANs will come into existence in many areas before public MANs. Telephone offices facilitate connections of corporate sites.*

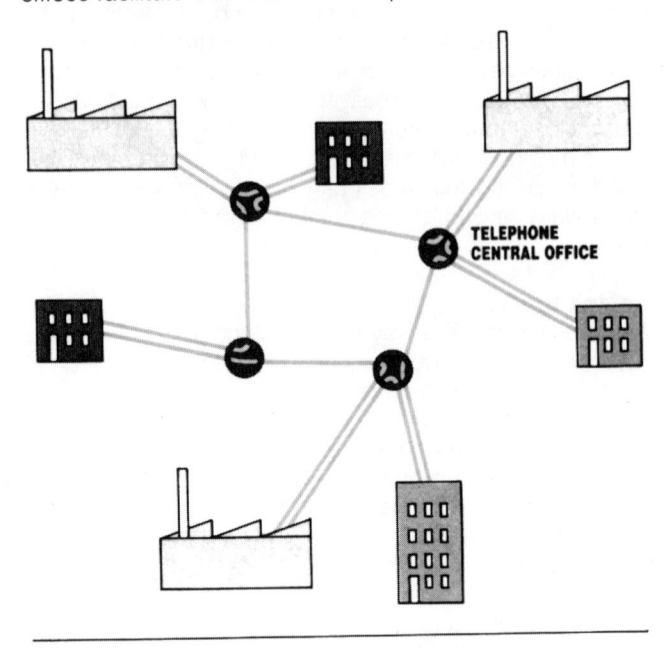

permits network maintenance and management to be handled from attended telephone offices.

Most of the information flow in a corporation is voice rather than computer-generated data. PBX interconnection between a corporation's buildings still represents more bits per second than data, usually by a margin of four to one or more. This will not always be the case—since computer communications will eventually take over a greater share of the information flow—but for the next few years it is a fact of life.

Thus, a priority during a MAN introduction will be to connect the internal voice networks at various sites. Voice has received much more attention in the MAN committee than it has in the LAN standards committees. Though voice has more stringent technical requirements than data, the 802.6 committee has attempted to give roughly equal weight to both data and voice requirements. MANs work well for both, and any inefficiencies are more than compensated for by the high bandwidth available from fiber optics.

Typical MAN traffic is expected to include:
- LAN interconnection.
- Graphics and digital images.
- Bulk data transfer.
- Digitized voice.
- Compressed video.
- Conventional terminal traffic.

The LAN traffic component is likely to consist of today's interactive file-transfer applications and high-resolution graphics. The graphics will originate on workstations attached to LANs. The loads they generate will be considerable. For example, a high-quality computer-aided design (CAD) display has a screen with a million pixels in color (roughly 1,000 by 1,000 points) and requires a megabyte of data to refresh it. Both the screen images and the databases from which they are generated require megabytes of transfer capacity between locations, such as between engineering and production departments of manufacturing corporations. Surveys have shown these types of transfer operations to be a major unfilled need.

Another type of load expected in the future is medical imaging. The medical arts could be improved by routing X rays, CAT scans, and other images between hospitals and physicians' offices and using multicast transmissions for consultations. Like CAD displays, they will require megabytes of data for image refreshing.

Video for conferencing and instructional purposes can be compressed to a few megabits per second and transmitted without serious loss of quality. While this accounts for only a small volume of traffic now, it is expected to grow considerably in the next decade or two.

Of all the traffic envisioned for MANs, only conventional data terminal traffic is well served by existing facilities. PBX-to-PBX digital voice traffic can be routed over leased T1 circuits. But T1 modularity is fixed at 24 voice calls, so flexibility is limited. Since digitized voice comprises the bulk of the bits sent between locations, any MAN must deal efficiently with it.

Optimizing a network protocol design to run well with such a variety of traffic is a challenge. As Figure 3 shows, the speed/burstiness matrix is fully populated. The network must offer fixed bandwidth as well as high bandwidth on demand.

Privacy and security

A significant distinction between a LAN and a MAN is in the area of privacy and security. A LAN is intended to carry only data belonging to the organization that owns it. Operating on shared media raises no concern that competitors will potentially have access to a company's data or that disaffected customers may damage the system.

A MAN in which a dedicated fiber connects only a single company's sites would be akin to a network of private lines and likewise would not have such security problems. However, one serving multiple customers would need to take additional steps to guarantee privacy and security.

The architectures considered in 802.6 assume that the full MAN will not traverse each customer's premises. The MAN is divided, in effect, into two networks: an access

3. Traffic patterns. *Of all the traffic envisioned for MANs, only conventional data terminal traffic is well served by existing data communications facilities.*

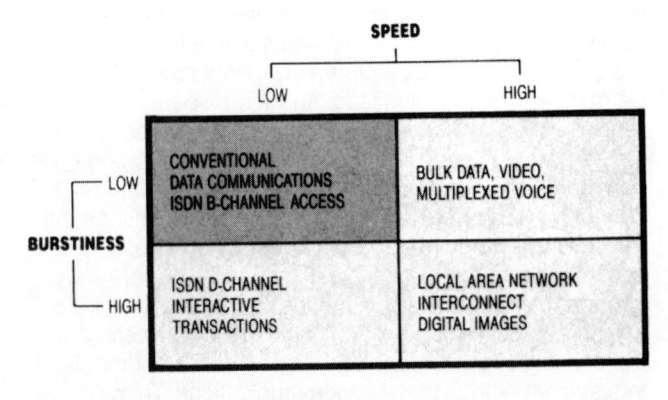

BURSTINESS	SPEED LOW	SPEED HIGH
LOW	CONVENTIONAL DATA COMMUNICATIONS ISDN B-CHANNEL ACCESS	BULK DATA, VIDEO, MULTIPLEXED VOICE
HIGH	ISDN D-CHANNEL INTERACTIVE TRANSACTIONS	LOCAL AREA NETWORK INTERCONNECT DIGITAL IMAGES

ISDN = INTEGRATED SERVICES DIGITAL NETWORK

4. Access and transport. *The need for privacy and network security within the MAN environment dictates that data from one customer's network not pass through another's. To meet this goal, metropolitan area network planners envision a demarcation between the MAN and access networks carrying a single customer's data.*

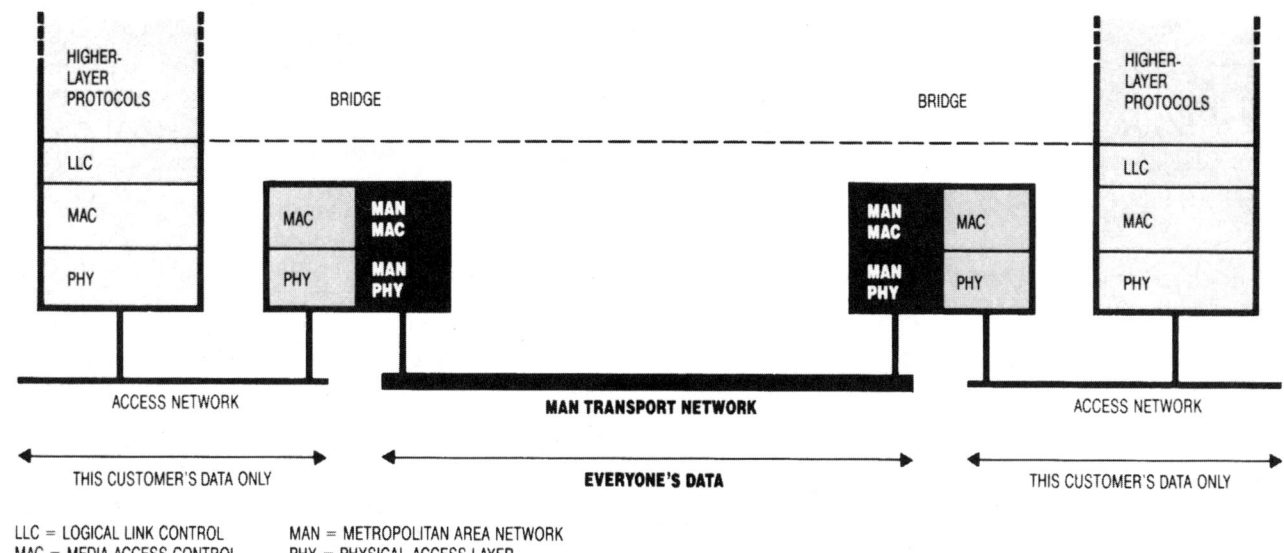

LLC = LOGICAL LINK CONTROL MAN = METROPOLITAN AREA NETWORK
MAC = MEDIA ACCESS CONTROL PHY = PHYSICAL ACCESS LAYER

network carrying only one customer's data and the transport network carrying many people's traffic. Bridging between the two takes place off-premises on telephone poles, in vaults, or even in the central offices (Fig. 4). The access networks can be 802.6 technology, or a LAN can extend out from the customer's building to the bridging point between the networks. The choice depends on traffic volume, the distance involved, and whether voice/video capability is needed.

Other MAN design requirements reflect the fact that the network will not be limited to one building or adjacent buildings. It may possibly provide public or shared user service, and it surely must cross public rights-of-way. It requires a central organization to install it, operate it, and bill for services. These requirements are met in most places by only two organizations, the cable television (CATV) operator and the telephone company. Both have widely installed cables that could be utilized, and both are familiar with providing services to a large number of customers.

However, the types of cabling are very different: broadband radio frequency for CATV and fiber optics for telephony. Twisted-pair wiring, the traditional medium of the telephone industry, does not have the capacity needed for MANs, though the newer fiber optic cables do. Therefore, the choice for standards was to aim for either broadband or fiber optic media.

The 802.6 committee spent considerable time working on broadband technology. However, interest on the part of most CATV operators has been slow to materialize. Despite the fact that they have extra cables installed (either due to contractual requirements or as spares), they have ignored the opportunity to carry data and improve the revenue from these cables.

The success Manhattan Cable had carrying data between Midtown and Wall Street in New York did not draw many imitators. Consumer trials were seldom economically successful, and, as a result, cable industry interest in MAN standards was low until recently. However, a resurgence of interest in CATV-based networks is on the horizon: Broadband-oriented presentations have been increasing in recent 802.6 meetings.

The other transmission choice, fiber optics, has proved to be more fruitful. This is currently where the committee is investing much energy.

The fiber optic proposals have brought many new players into the MAN standardization arena. In addition to Bell Communications Research, which has been represented right along, strong support has been given by most of the Bell operating companies, as well as AT&T Bell Laboratories. AMP Inc. and AT&T have been very active in the connector and cabling areas. Burroughs, now Unisys, has been the main player among mainframe builders; and IBM has participated since 802.6's inception. Internationally, a host of telephone suppliers and operating companies have been involved, with consistent participation on the part of Hasler, Plessey, Ericsson, NEC, and British Telecom; proposals have also been made by Telecom Australia and NTT.

Needs slotting

The design requirements for a MAN include the provision of fixed-bandwidth voice channels of 64 kbit/s with a round-trip delay of 1.5 milliseconds or less, as well as on-demand transmission of data packets. The 1.5-millisecond constraint is imposed because the echo-canceling circuits used in long-haul telephone lines cannot tolerate longer delays.

The timing constraint causes trouble in conventional LAN technology. In Ethernet, for example, there is no upper bound on delay as traffic load approaches saturation. In IEEE token bus and token ring LANs, the normal token holding times result in potential waiting time much greater than 1 or 2 milliseconds when several units attempt to transmit.

The solution is to impress a timing pattern, or slots, on the underlying transmission protocol. The monitor (master) station emits empty frames every 125 microseconds. This time corresponds to the voice digitizing rate of one 8-bit sample 8,000 times per second. Each active voice station on the ring is assigned a byte position within the frame; by counting from the start of the frame, it is able to determine which byte position, or channel, it is to use for sending and receiving. No explicit addressing is needed; once assigned, addressing is implicit in the byte position.

This scheme provides isochronous, or equal-time, transmission. In addition to the fixed bandwidth and low delay, it provides evenly spaced transmission of bytes just as a simple speech digitizer would generate them—eliminating the need for buffering mechanisms. A diagram of the frame structure used in one proposal is shown in Figure 5.

Data, on the other hand, does not require the low delay or fixed bandwidth needed for voice traffic. Data loads are usually very bursty, whether in LAN interconnection or in the transmissions required for file transfer or screen refreshing. For such "nonisochronous" traffic, a different kind of slot is needed. In this case, the network behaves more like a conventional token ring.

A schematic diagram of the functional layers in a typical MAN implementation is shown in Figure 6. The segmentation and reassembly (if required) are fully within the media access control (MAC) layer. MAC management is divided into two parts—one for the normal data functions equivalent to those used in local area networks and isochronous management needed for voice or video control. The isochronous data stream bypasses the logical link control (LLC) layer, which provides error control not needed for voice or video.

Until very recently, there were competing proposals before the 802.6 committee. These include the the dual bus, or Queued Packet and Synchronous Exchange

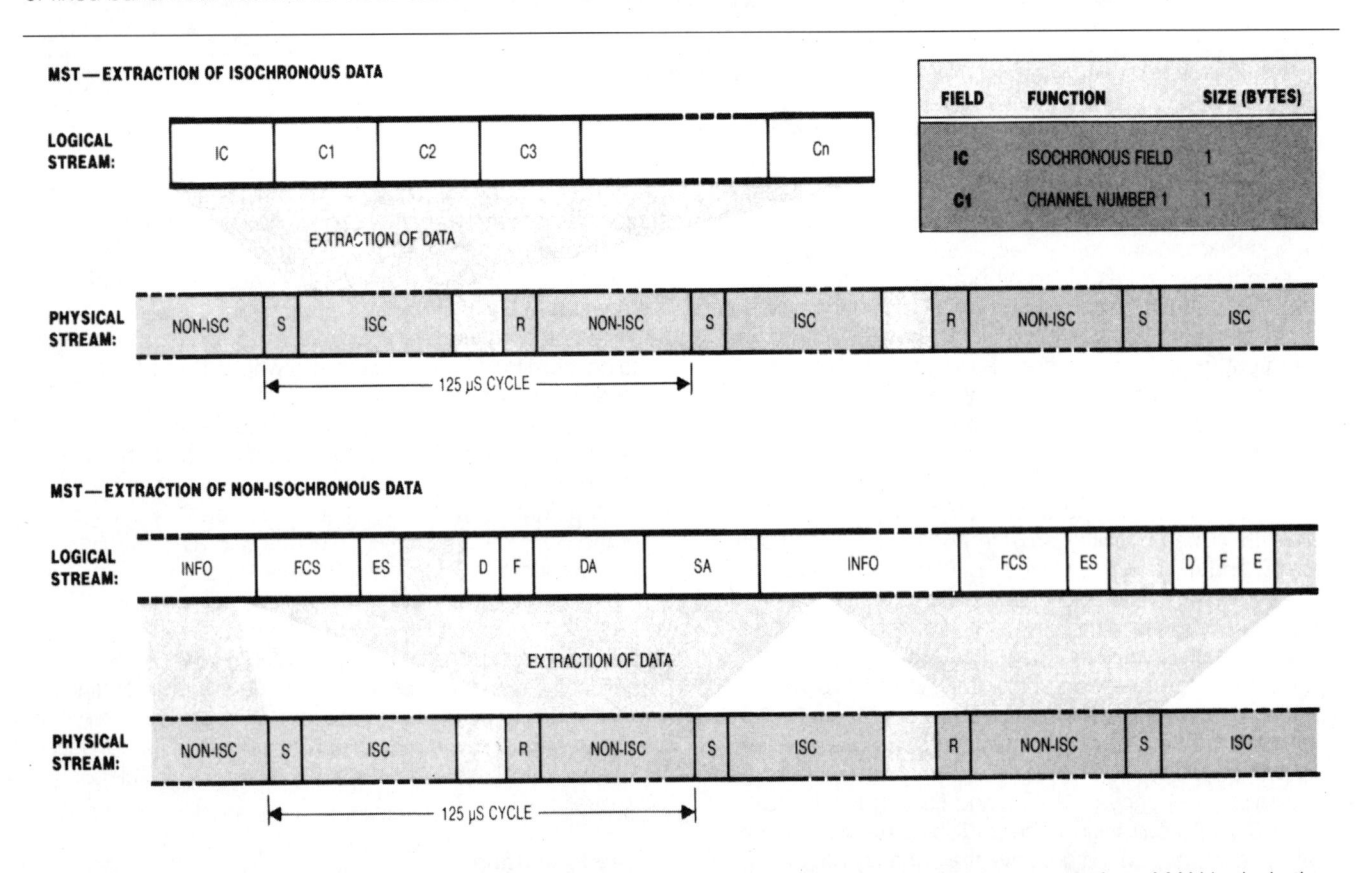

MST—EXTRACTION OF ISOCHRONOUS DATA

FIELD	FUNCTION	SIZE (BYTES)
IC	ISOCHRONOUS FIELD	1
C1	CHANNEL NUMBER 1	1

MST—EXTRACTION OF NON-ISOCHRONOUS DATA

FIELD	FUNCTION	SIZE (BYTES)
D	START DELIMITER	1
F	FRAME CONTROL	1
DA	DESTINATION ADDRESS	2 OR 6
SA	SOURCE ADDRESS	2 OR 6
INFO	INFORMATION	0-4,096
FCS	FRAME CHECK SEQUENCE	4
ES	END DELIMITER AND STATUS	2 OR MORE
E	END DELIMITER OF TOKEN	1

5. Timing. *The design requirements for a MAN include the provision of fixed-bandwidth voice channels of 64 kbit/s with a round-trip delay of 1.5 milliseconds or less, as well as on-demand transmission of data packets. The MST scheme provides isochronous, or equal-time, transmission, ensuring fixed bandwidth and low delay.*

6. Protocol layers. *Nonisochronous traffic on a MAN behaves more like a conventional token ring. Segmentation and reassembly (if required) are fully within the media access control (MAC) layer. MAC management is divided into two parts: one for the normal data functions and isochronous management needed for voice or video control.*

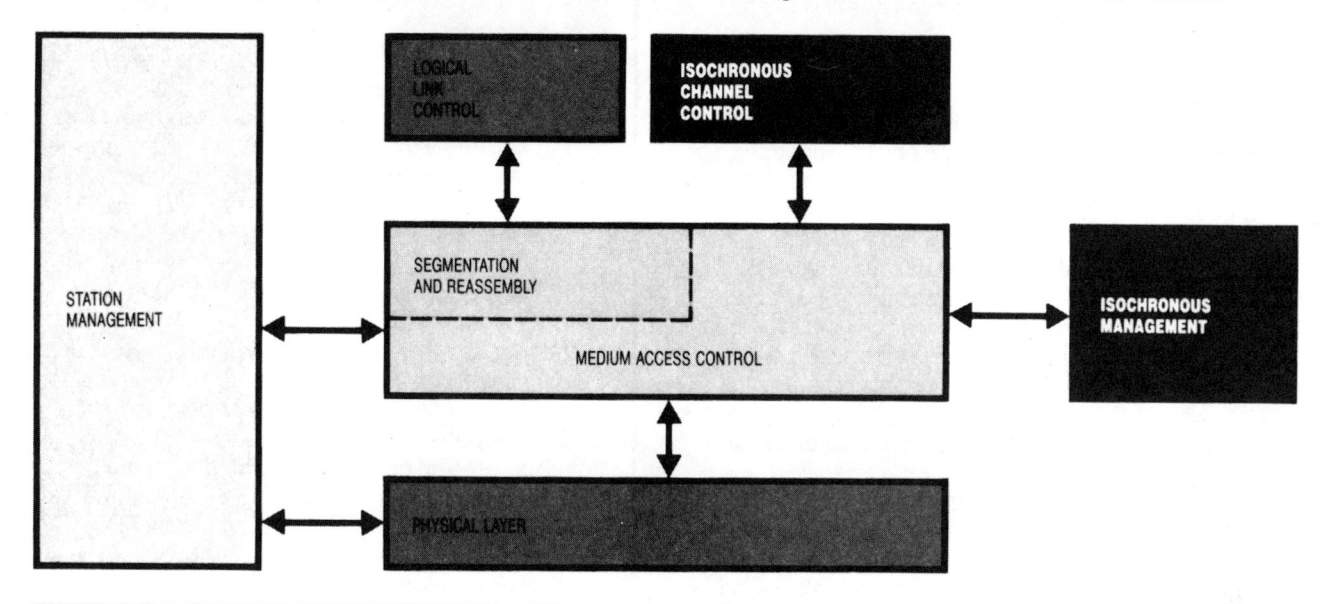

(QPSX), which is a bus bent into a ring; the multiplexed slot and token (MST), which is an enhancement of the 802.5 token ring; and other slotted ring proposals. Each proposal had its relative strengths and weaknesses. In the past year, the committee focused on the QPSX and MST proposals; work proceeded on both while the telephone industry had time to study them and consider how they would mesh with existing facilities.

QPSX proposal . . .

The QPSX design, which was recently adopted by 802.6, represents a new topology for IEEE 802 networks. It is a dual, nonbranching bus that may be looped into a ring. It is nominally fiber-based, but it can also use coaxial cable or even radio segments.

When laid out as a ring, one unit does not repeat the incoming data. It terminates the logical bus on both ends. This provides a unique, fail-safe property: If a node or cable segment fails, the logical opening is moved to the physical outage, and operation continues with no degradation of performance. This system is now going into field trials in Australia and is drawing substantial interest in the North American telephone industry. It is fully detailed in "New proposal extends the reach of metro area nets" (in this issue).

. . . versus MST

The MST proposal (see table) was developed by Integrated Networks Inc., of San Diego, Calif., in collaboration with Hasler AG of Switzerland. It overlays an isochronous capability on a token-based packet MAC.

Compared to fully slotted mechanisms, the MST has the advantage of avoiding segmentation of data into small slots and their later reassembly. The fixed timing required for the isochronous traffic is enforced by sending suspend and resume symbols around the ring at the appropriate time. When a suspend signal is received by a transmitting station, it stops and permits the isochronous traffic to move through. When the resume signal is received, it completes the packet transmission.

When an MST controller is expecting to receive isochronous data, it waits for the signal indicating that isochronous data is coming. When it receives the isochronous data, it routes the stream away from the normal MAC layer into the isochronous controller, where the channels are counted until the assigned one is found. No addressing or error-control overhead is imposed; the assumption is that the traffic is voice. If error control is desired, it can be overlaid by higher layers; this situation would apply where a data application requires a fixed-bandwidth allocation.

The packet layer for MST is a version of the token-passing ring protocol. This variation, known as early token release, is now being considered in 802.5. In the present standard, the token is not released until it circulates around the whole ring back to the station that transmitted it. For a very large ring, or for speeds well in excess of the present 4 Mbit/s, this is inefficient. If the token is released sooner, efficiency is much better under the conditions of MAN operation.

MST proposed rates

RATE	APPLICATION
1.544 MBIT/S (T1)	CAN RUN ON TWISTED PAIR
2.048 MBIT/S	EUROPEAN EQUIVALENT OF T1
34 MBIT/S	EUROPEAN FIBER OPTIC RATE
44.7 MBIT/S (DS3)	NORTH AMERICAN FIBER OPTIC RATE
150 MBIT/S (APPROXIMATELY)	PROPOSED JOINT EUROPEAN AND AMERICAN RATE (BROADBAND ISDN)
ISDN = INTEGRATED SERVICES DIGITAL NETWORK	

In MST, the packet protocols are overlaid with isochronous slots, optimized for voice at 64 kbit/s. Any number of isochronous channels can be assigned, so long as enough bandwidth is left to do call setup. The overlay is illustrated in Figure 6.

Slotted rings

The first fiber optic network reviewed in the 802.6 committee was a slotted ring proposal championed by Burroughs, Plessey, and National Semiconductor. While not actively pursued in 802.6, it has continued to be developed in Japan, and it indicates the range of options available in fiber-based networks (See "Other slotted ring proposals").

The proposal includes two tiers, a backbone ring constructed of dual fiber optic cables running at 43 Mbit/s, and a lower-cost ring running at one-quarter of the speed of the main ring. The speeds were chosen to match the existing telephone network DS-3 interface, which has a user data speed of 43 Mbit/s. Synchronization to the wide area digital network is done at one gateway.

The frame rate is set by the need to digitize voice once per 125 microseconds. These frames are subdivided into slots. In the case of the high-speed ring there are seven slots per frame, while the low-speed ring has two slots. One slot is reserved for nonisochronous use since it is required for call setup; the other slots can be either isochronous or nonisochronous.

The slot size is 77 bytes if 2-byte packet addresses are used and 85 bytes if 6-byte addresses are used with the same information field. (Both address options must be available according to Project 802 rules.) This slot size permits 76 isochronous channels if the slot is for isochronous use or 64 information bytes plus header and check bytes in the nonisochronous case. As with QPSX, segmentation within the MAC layer is required to accommodate the relatively small slot size.

A ring topology fits well with fiber optics. Unlike electrical currents, light signals cannot be tapped at will or split easily into branching trees. However, optical signals must be converted to an electrical form and regenerated at each location that intends to use the signal.

The ring is subject to complete failure if any node or fiber segment fails or is powered off. To provide recovery, a dual ring architecture is used. The secondary ring is normally idle in the slotted ring and MST proposals. When a node fails to receive any signal (data or idle) from its upstream neighbor, it assumes that either the link has failed or that the neighbor has failed. Either way, it reconfigures the ring, taking its input from the secondary fiber and sending via the primary fiber. This recovery method is different from that used by QPSX, which operates on both fibers at all times.

The failure can then be repaired, while the network continues to function. If two failures occur, reconfiguration results in two separate networks, both of which can communicate within themselves.

The second item accepted at the IEEE 802.6 meeting was AT&T's proposal to interconnect MAN dual buses with high-speed multiport bridges. These bridges will carry data from one bus to any of several others. This proposal will provide a growth path for MANs for several decades.

Since the combined data rate of the dual bus can reach

Other slotted ring proposals

The Cambridge Ring and FDDI-II standards also provide the fixed bandwidth and repetition period of a slotted protocol. Though these standards have some philosophical similarity to the 802.6 work, they aim at different application areas.

The Cambridge Ring was first developed at Cambridge University in England nearly 10 years ago. It has recently surfaced as a British and International Standards Organization standard (ISO 8802/7). Though the ISO standard bears a number similar to the ISO versions of the IEEE Project 802 family (8802/3, 8802/4, and so forth), it was developed independent of IEEE Project 802.

This ring runs at 10 Mbit/s on a wide variety of copper wire cable types, such as twisted pair or coax. The standard specifies pulse quality rather than cable length or type. The packets are quite small, ranging from 2 bytes to an optional 8-byte length.

The other major standard offering slotted data transmission is the FDDI-II proposal. The parent FDDI, or Fiber Distributed Data Interface, is a 100-Mbit/s fiber optic token ring standard developed by the American Standards Committee X3T9.5. In contrast to the Cambridge Ring, FDDI conforms to the IEEE 802 guidelines and is intended to run under 802's logical link protocol.

FDDI was originally proposed as a computer-room network connecting large computers and peripherals at 100 Mbit/s. Its technology permits much longer distances than the IEEE 802 LANs can accommodate. The cable length can be 200 kilometers or more, making it a potential vehicle for metropolitan- or campus-scale networks.

However, FDDI does not provide the isochronous property needed for voice and video. This has been remedied in a proposal known as FDDI-II. Here the monitor station holds the token indefinitely and emits slotted frames.

A major difference between FDDI-II and the 802.6 proposals is that the 100-Mbit rate is not compatible with existing or proposed telephone company installations. "Dark fibers" dedicated to FDDI would have to be provided, while 802.6 is intended to be used with DS3 or higher-speed "light fibers" that make use of normal telephone-system control and maintenance procedures. The 802.6 standard is also intended to be compatible with emerging telephone industry standards such as Sonet and broadband ISDN.

In addition, in the 802.6 proposals, there is provision for isochronous channels down to the individual voice circuit (64 kbit/s), while FDDI-II directly manages the channels only down to the 6-Mbit level, equivalent to 96 voice circuits. The channels can be sub-allocated with proprietary PBX protocols, or other protocols, such as CCITT's Q.931 standard, may be used.

However, given that both MST (multiplexed slot and token) and FDDI-II are based on the same FDDI technology, there is room for compromise between the two camps, with a stronger design emerging.

300 Mbit/s, a single loop might serve most customers initially. As usage expands, the loop can be subdivided, with the pieces connected by two-port bridges similar to those being developed in the LAN industry. Alternatively, a second-level or backbone loop could connect all the other ones. But as capacity requirements increase, it makes sense to use a very high capacity multiport bridge constructed from a newly developed family of switching components.

Such intelligent crosspoints are now being prototyped by the telephone industry. These devices read the headers on small packets (which may actually carry data or digitized voice) and route the packets accordingly, interpreting the header one bit per binary switching stage. Because of the shape of the paths traced through these arrays, they are sometimes refered to as banyan networks.

Banyan networks are very fast, operating at fiber optic speeds. They also have the advantage of being easily adapted to silicon, so they can be fabricated relatively inexpensively. While they have been developed as components in a telephone network based on a hierarchy of switches, they can be used to interconnect MAN loops equally well. The job of 802.6 will be to define packet and segment structures for the MAN that will work well with such components in the future. The committee does not expect to write standards for the internal operation of these bridges but, rather, will define the interfaces between them and the MAN bus.

Compatibility with ISDN

The MAN falls between local area networks and wide area networks in its scope, so moving from LAN to MAN is no more complex than moving data from one LAN to another.

The telephone industry's wave of the future is the Integrated Services Digital Network (ISDN). While it may not be optimal for data, it is good for digital voice, and the fixed-bandwidth 64-kbit/s B channels are well suited for PCs. There is no question that ISDN will be widely introduced over the next decade. Header compatibility between QPSX and broadband ISDN slots is now being worked out.

The proposed MANs are compatible with ISDN. The isochronous channels can be mapped to ISDN's B channels with no difficulty; both are 64-kbit/s, full-duplex digital channels. The nonisochronous (packet) slots can interface to ISDN's D channels, although there is a speed mismatch problem in moving from the multimegabit capability of the MAN to the 16-kbit/s D channels. A solution here would be to send the packets over groups of B channels, as in the primary-rate ISDN circuit with 23 B channels (31 in Europe) and one D channel.

As ISDN circuits become available on an intercity basis, they may become the interconnection of choice between MANs, handling the portion of corporate data and voice that requires wide area networking. The details of the interfacing between the two types of networks will require close work in the future between the committees drafting the standards. ∎

James F. Mollenauer, a technology manager for networks and communications at Computervision Corp., chairs the IEEE Project 802 Working Group on Metropolitan Area Networks. He received his B. A. from Amherst and a Ph. D. from the University of California, Berkeley.

John L. Hullett, QPSX Communications Pty. Ltd., Western Australia, and
Peter Evans, Telecom Australia Research Laboratories, Western Australia

New proposal extends the reach of metro area nets

IEEE 802.6 committee adopts Telecom Australia plan for 'dual bus' scheme; said to work faster and farther than ring architectures

The metropolitan area network, an evolutionary step beyond the local area network, promises high-speed communications at distances greater than any LAN can handle. But in another sense, the MAN is what standards groups will make it. Just what standards will define a MAN is being hammered out by the 802.6 committee of the Institute of Electrical and Electronic Engineers (IEEE).

A good standard is crucial to MAN development, since interoperability between computer and telecommunications networks is a prerequisite to a successful launch of the new technology. Several key requirements for MANs have already been fashioned by 802.6. These include employing a shared medium capable of operating over areas of at least 50 kilometers in diameter, providing high-speed packet data transmission and voice capability as well as other services requiring guaranteed bandwidth and constrained delay.

The large area envisioned for MANs, as well as the ability to handle voice, places the MAN squarely in the arena of such public network providers as the Postal, Telegraph, and Telephone agencies (PTTs) and the Bell operating companies (BOCs). They welcome that position and are taking an active role in its development.

In addition to its being a highly versatile customer-access network, telephone companies see the MAN as meeting the market for multimegabit data services. The types of services envisioned for such bit rates include computer-to-computer communications and image transfers. For example, graphics-intensive applications such as computer-aided design, computer-aided manufacturing, and medical imaging require data rates up to 150 Mbit/s or more.

The need to make the MAN compatible with the public network environment was brought into sharp focus at a recent meeting of the 802.6 committee. The committee adopted these goals for a MAN standard:

- It should accommodate fast and robust signaling schemes for MANs.
- It should guarantee security and privacy and permit establishment of virtual private networks within MANs.
- It must ensure high network reliability, availability, and maintainability.
- It should promote efficient performance for MANs regardless of their size.

The Queued Packet and Synchronous Switch (QPSX) dual bus MAN proposal sponsored by Telecom Australia was adopted in November 1987 by the 802.6 committee because it meets or exceeds those requirements. Figure 1 illustrates how LAN standards and the proposed dual bus MAN standard relate to the Open Systems Interconnection (OSI) model. The dual bus network can operate with any synchronous transmission interface including DS-3, CCITT G.703, and Sonet.

A dual bus MAN employing the QPSX access protocol is the ideal networking technology for a MAN operating in the public network. It performs more reliably and efficiently than other protocols that were proposed. This is because its signaling, multilevel priority structure, synchronization, and signaling are unaffected by congestion.

The dual bus architecture of QPSX is shown in Figure 2. It consists of two lines, or buses, which carry traffic in opposite directions. Each node, with connections to both buses, has read and unidirectional write capability. Every node can communicate to every other node sending information on one bus and receiving it on the other.

In contrast to ring architecture, data does not pass through each node. Nodes on the bus read the addresses of passing packets and copy data if there is an address match between the node and the destination address within the packet.

By way of comparison, ring architectures are being used in the Fiber Distributed Data Interface (FDDI), the multi-

1. OSI relationship. *The queued packet and synchronous switch dual bus proposal, sponsored by Telecom Australia, fits neatly into the Open Systems Interconnection model.*

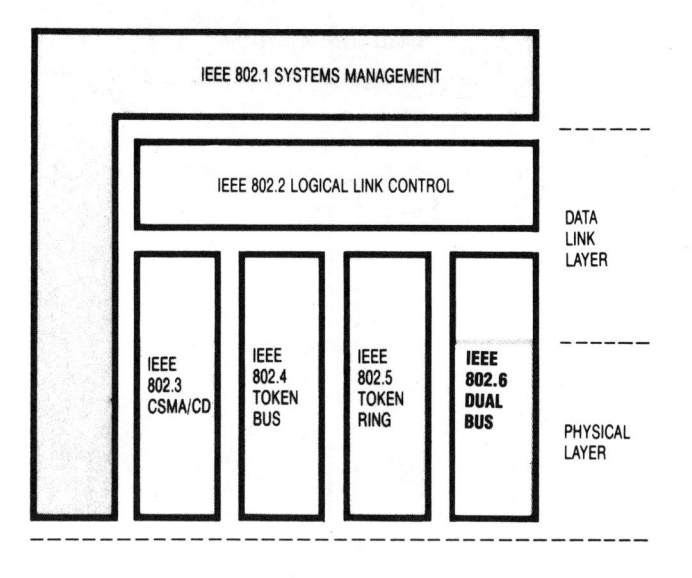

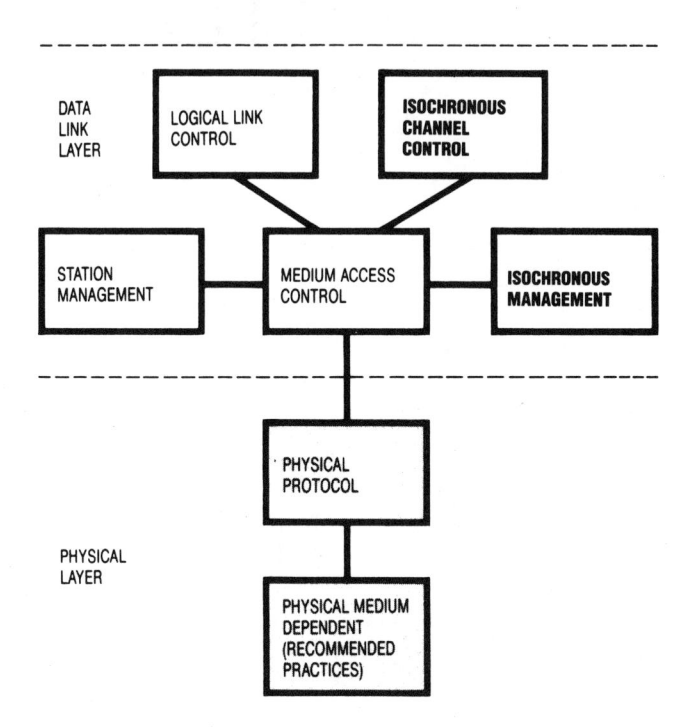

CSMA/CD = CARRIER-SENSE MULTIPLE ACCESS WITH COLLISION DETECTION

plexed slotted and token (MST) network, and the Cambridge Ring. Nodes in ring architectures are serially connected on a single line. Each node receives and retransmits or repeats the messages in turn. Messages are generally removed by the source node. The bus architecture offers several significant advantages over the ring in terms of robustness and reliability.

In the dual bus architecture, the unidirectional reading and writing to each bus is implemented electronically using simple OR gates and shift registers, as shown in Figure 3. Writing is accomplished by transmitting a logical OR onto a synchronized and formatted time structure generated by the frame generator at the head of each bus.

It is possible to loop the dual buses, a configuration shown in Figure 4. Its similarity to the ring architecture is in form only, since the looped bus can use a master clock located, for example, in the telephone network.

Ticka ticka timing

Time on each bus is divided into fixed-length slots with a fixed number of slots allocated to each 125-microsecond frame. The arrangement is depicted in Figure 5. The actual number of slots in each frame depends on the bit rate on the buses. The bit rate depends on the particular transport adopted for the physical medium-dependent sublayer.

In North America, the slot size would be 45 octets, although size can be varied once broadband ISDN standards emerge. The overhead in logically linking packet segments in QPSX is only about 7 percent.

Each node may engage in packet- or channel-switched (asynchronous or synchronous) communications. Slots not reserved for synchronous traffic use are available for packet switching. The allocation of bandwidth between synchronous and packet use is dynamically controlled according to demand.

Because the information flow on the buses is unidirectional, all nodes on the bus may be synchronized without regard to distance. Slots used for packet switching are accessed using the distributed queuing protocol. The distributed queuing protocol's efficiency is unaffected by distance. As a consequence, QPSX bus networks can be arbitrarily extended. The medium used for the buses is also arbitrary, but in most cases it would be optical fiber. Cable TV technology could operate as a variant of QPSX distributed queuing (see "Networking for greater metropolitan areas," in this issue).

Reliability built in

The reliability of QPSX stems both from its architecture and its distributed queuing medium-access control (MAC) protocol. There are two reasons for the inherent reliability of the bus architecture. First, the network nodes are logically adjacent to the bus, not serially and logically con-

2. Dual bus. *In the QPSX dual bus, every node can communicate to every other node by sending information on one bus and receiving it on the other.*

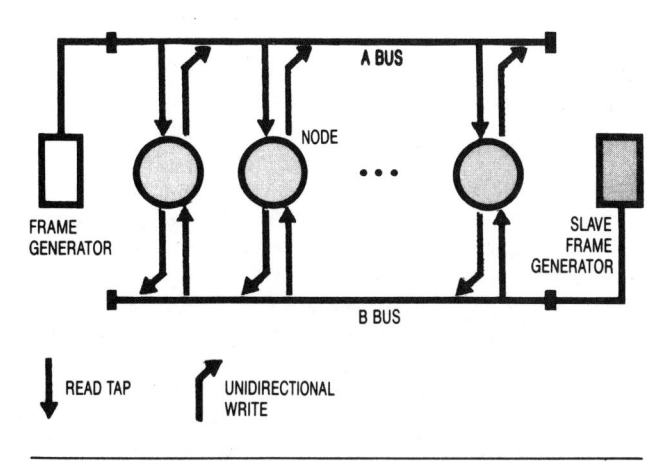

3. Write stuff. *Unidirectional reading and writing to each bus is implemented electronically, using OR gates and shift registers. Writing is accomplished by transmitting a logical OR onto a synchronized and formatted time structure generated by a frame generator located adjacent to the head end of each queued packet and synchronous switch bus.*

nected as in a ring. Thus, data does not pass through the nodes, and passive MAC failures have no effect. Second, because the node's writing function is by logical OR (that is, without overwrites), active MAC failures and bus failures are easily detected by input-to-output coincidence checking (Fig. 6). Any failures that occur are bypassed at the node. The operation can be totally hardware-based with immediate response, no loss of synchronization, and minimal network disruption. With QPSX, no node monitoring functions are required to check the MAC.

In ring-based networks, active- and passive-node MAC failures are catastrophic. This is because both types of failure must invoke healing mechanisms associated with their redundant rings. Healing involves considerable network disruption because it entails a reconfiguration, which causes a loss of synchronization. Any additional node failure can then produce isolated network islands.

QPSX architecture node failures do not invoke any network healing mechanism that requires reconfiguration of the network. Faulty nodes are easily bypassed. Failures in the physical transport, however, do involve reconfiguration. But once the healing is completed, the network captures the same synchronization, so disruption is minimal. Physical faults or line breaks are healed by repositioning the natural break in the loop to the position of the break.

There is no loss of network capacity involved in this type of network reconfiguration. This contrasts sharply with ring-type networks. If a ring network carries traffic on a

4. Looped bus. *Though the looped bus looks similar to the ring architecture, its fault-healing and synchronization is superior. Its master clock is located in the telephone network.*

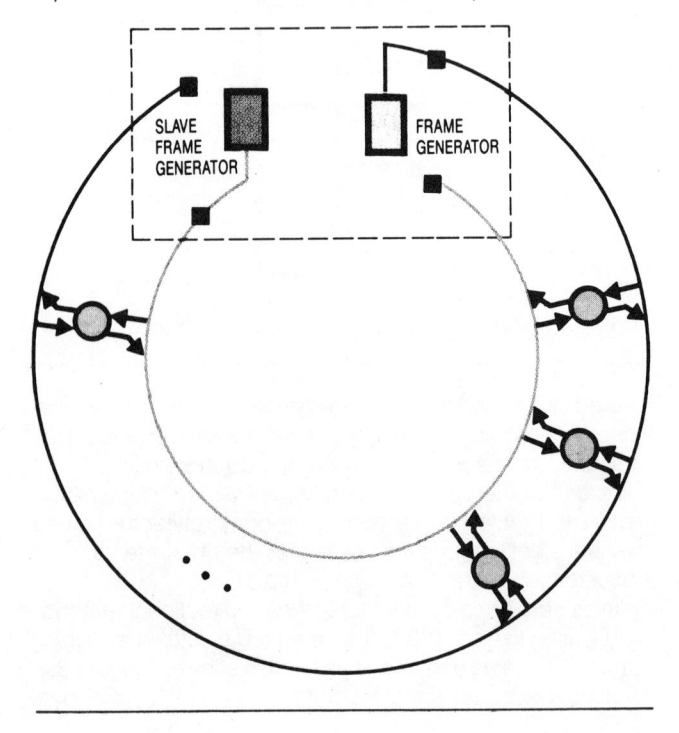

5. Timing. *Fixed-length slots with a fixed number of slots allocated to each frame are used to keep time across the bus. In North America, slot size is 45 octets.*

| FRAME HEADER | SLOT 0 | SLOT 1 | ... | SLOT 54 | STUFFING FIELD |

backup ring, then network reconfiguration due to either node or transport failure introduces a 50 percent reduction in packet data capacity. Alternatively, if the backup ring carries no data traffic, there is always a 50 percent inefficiency factor relative to the QPSX bus network.

QPSX invokes healing only after a real disaster, such as a backhoe through physical transport. Node failures do not invoke reconfiguration in QPSX, since faulty nodes are excised using simple low-cost hardware.

Speedy access

The distributed queuing data-access protocol of QPSX enables looped bus networks to operate at maximum efficiency at very high speeds and over unlimited distance. Here is how it works.

The operation is fundamentally different from all existing LAN packet-access protocols. In all random- or controlled-access protocols, each station has no record of network loading. Only when a station has a packet for transmission does it ascertain, either explicitly or implicitly, the information required for scheduling access.

By contrast, each distributed queuing node keeps a current-state record of the number of packet segments across the network awaiting access. When a station has a packet for transmission, it uses this count to determine its position in the distributed queue. If no packet segments are waiting, access is immediate. Otherwise, deference is given to those packets that are queued first.

With this arrangement, capacity is never wasted, and minimum access delay at all levels of loading is guaranteed right up to 100 percent utilization of the bus. This performance is achieved with negligible control overhead (less than 1 percent), and it is effectively independent of the bit rate and physical extent of the network. Access is uniform across the network, so there is no possibility of network hogging.

6. Reliability. *Because network nodes are logically adjacent to the bus, not serially and logically connected as in a ring, data will not be lost with a node failure.*

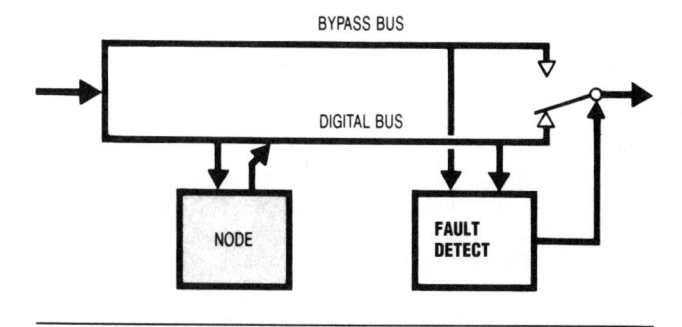

The distributed queuing access control is completely distributed. It requires only simple logic hardware at each station. The only shared function is the frame timing generation that must be at the head of the bus.

The queuing protocol uses just two bits of control overhead in the access control field of each packet. The Busy and Request Control bits order the access of packets to the bus. The Busy bit indicates that a packet is filled with data, so it is not available for access. The Request Control bit is used to indicate that a node has a packet queued for transmission.

7. Distributed queuing. *A position in the distributed queuing access control is signified by a sequence number. A station uses this count to determine its position in the queue.*

(A) QUEUE FORMATION ON BUS A

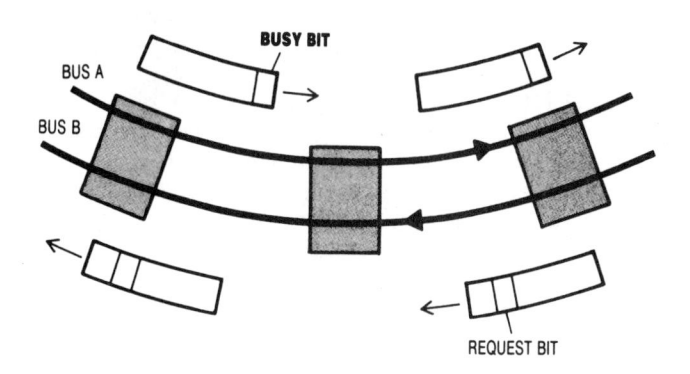

(B) NODE NOT QUEUED TO SEND

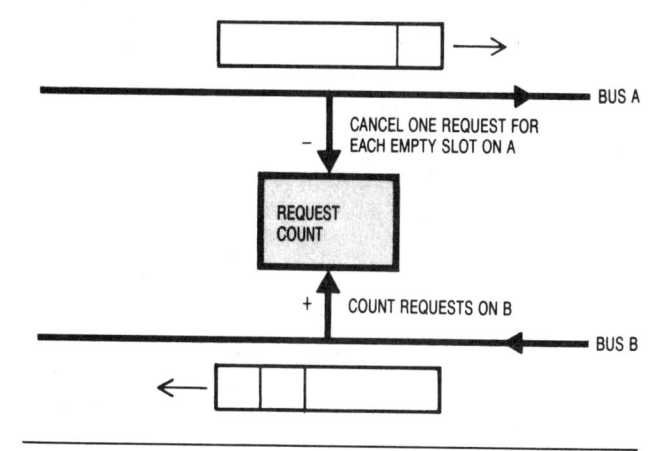

How does the distributed queue operate? In Figure 7A, a queue is formed in relation to nodes using bus A, which flows to the right. The position in the queue is signified by a sequence number. At a node on the network, the sequence number is held in a request counter. When that node has no packet to send on line A, the request counter tracks the Request Control bits sent on line B. Line B request signaling is used for line A access and vice versa. The counter is incremented for each Request Control bit received on line B and decremented for each empty packet slot passing on line A (Fig. 7B). Decrementing the count recognizes that the passing empty packet will serve a node queued downstream.

When a node has a packet ready for transmission, it obtains its position in the queue by reading its request counter. The request counter is reset, and the count is transferred to a countdown counter. In the countdown state, when a node is awaiting access, passing an empty packet slot decrements the countdown counter, while new requests increment the request counter. The situation is shown in Figure 8. When the countdown count is zero, the node may access the bus using the next non-busy packet. If the node has another packet for the distributed queuing protocol, the process is repeated using the current request count.

The QPSX distributed queuing protocol can be used to assign priority to packets. By operating separate distributed queues for each level of priority, queued packets with high priority gain access to the bus before lower-priority packets. In the MAN, the highest priority would be reserved for signaling, control, and fault configuration. Even when the network is highly loaded, with many nodes simultaneously transmitting large, 8-kbyte packets, high-priority access will be immediate

In this context it is worth noting that the packet segmentation used in QPSX offers a near-immediate response to sharing of network capacity or changes in loading. A similar situation in terms of processing capacity exists in time-shared computing. In the computing environment, queued jobs cyclically receive a quantum of service each. Token rings use nonsegmented access that is analogous to batch processing.

Performance

By using request counters, the distributed queuing protocol provides ideal ordered-access characteristics. If the queue size is zero, a packet will gain immediate access, but if the queue size is anything other than zero, the packet waits only while those queued ahead of it access the bus. Thus, in contrast to all existing media-access protocols, there

8. Countdown. *When a node has a packet ready for transmission, it obtains its position in the queue by reading its request counter.*

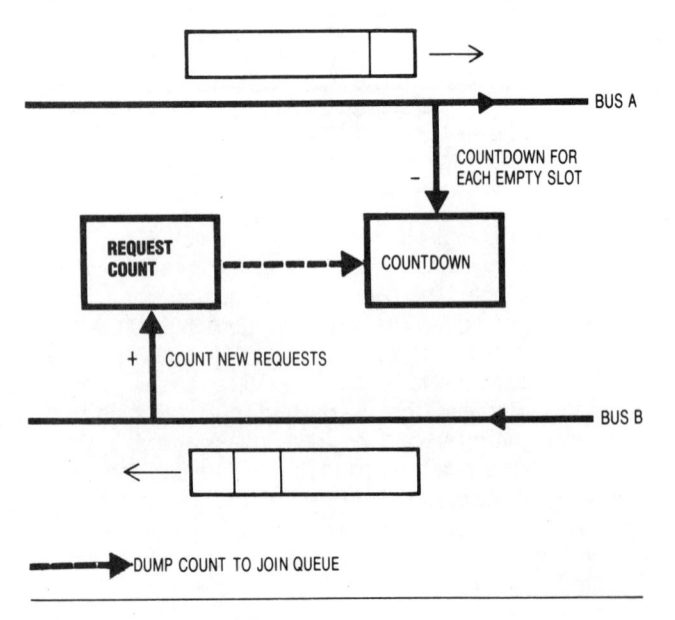

9. No token. *Distributed queuing has a shorter access delay than token ring access since there is no overhead waiting for a token to transfer.*

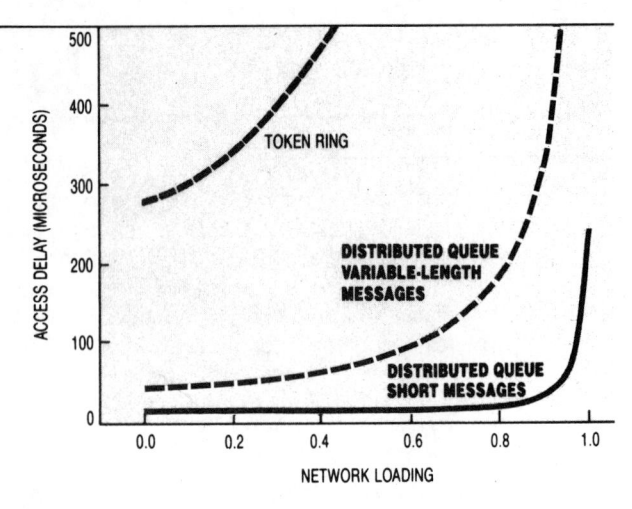

is never any wasted capacity with distributed queuing.

The ideal access characteristics of distributed queuing are illustrated by comparing its expected transmission delay with that of a token ring (Fig. 9). In the example, a network of 50 stations is shown operating at 100 Mbit/s over a loop length of 100 kilometers. The average packet size is 100 bytes. The figure also shows a graph of the delay for short packets (acknowledgment or control packets).

Distributed queuing provides two clear advantages over the token ring. First, it has a shorter access delay since there is no overhead waiting for a token to transfer. Access can be immediate if no other packets are queued. Second, fair treatment is given in the distributed queue to short messages, providing all the advantages of time-shared processors. The big problem with the token ring is the delay in token transfers: It takes an average access delay of at least half of the ring latency—even when the network is lightly loaded. This delay increases with the network size and eventually limits the physical size of the network.

Public network architecture

Since the QPSX can be used in either a dual bus or a looped configuration, many network architecture variations can be considered. It can be used in a looped-network hierarchy or a star-network hierarchy based on dual buses. Or a topology of loops and stars can be combined.

Considerable savings can be made in transmission and switch resources by running multiple dual bus connections in conjunction with a single looped bus return. By creating a number of virtual looped buses, reliability and global healing of the looped bus can be achieved at lower cost. This achitecture may suit many applications, particularly on intercity routes. ■

Peter Evans, a senior engineer with the local-access section of Telecom Australia, received a B. S. E. E. from the University of Adelaide, Southern Australia. John L. Hullett received his Ph.D. in electrical engineering from the University of Western Australia in 1970. He worked for Telecom Australia before joining QPSX as its technical director.

Cormac O'Reilly, Schlumberger Ltd., Houston, Tex.

How to grow a world-class X.25 network

It took three years to revamp a multiple-architecture network of more than 1,200 computers. Installing company-wide electronic mail was no small feat.

Successful corporate networks are flexible and economical; they have added value and are forward-looking. By definition, they embrace a wide range of network architectures, provide the necessary security, and exploit economies of scale. The information network at Schlumberger Ltd. (Houston, Tex.) belongs in this category. How it got there is a story of obstacles overcome and challenges met.

Built between 1984 and 1987, SINet (Schlumberger Information Network) is a wide area X.25 packet-switching network that links, directly or indirectly, more than 1,200 computers that operate in a variety of networking environments, as follows:
- TCP/IP (Transmission Control Protocol/Internet Protocol, which is largely used by the Unix community).
- Digital Equipment Corp.'s DCA (Digital Communications Architecture, which supports DEC equipment).
- IBM's SNA (Systems Networking Architecture).
- Workstations (more than 500 of these are from Sun Microsystems Inc., Mountain View, Calif.).
- More than 6,000 personal computers (PCs).

Moreover, SINet provides company-wide electronic mail, which is used by upwards of 11,000 people to replace telephone, Telex, facsimile, and courier communications. It is also a means of transmitting files to and from PCs.

SINet transports more than 250 million segments of data (one segment is 128 bytes) between connected computers each month, a figure that is growing more than 40 percent each year. This translates into a reliance on computer processing and data communications.

A *Fortune* 100-size company founded in 1920, Schlumberger employs nearly 50,000 people of 75 nationalities in major cities and field sites worldwide. Its operations include three overall business groups (Oilfield Services, Industries, and Technologies), which are further broken down into business units. The common denominator throughout is information—capturing data and refining it into information, which is processed, transmitted, graphically depicted, stored, and integrated.

Computer processing affects all of Schlumberger's business functions. The company uses a variety of computers, ranging from personal computers (IBM PCs and Apple Macintoshes) to departmental minicomputers (IBM System/38s and System/36s) to mainframes (IBM 30XX and 43XX), to support a substantial portion of the business data processing activity (see table). In addition, DEC VAX computers are in evidence by the hundreds. There are also three supercomputers from Cray Research Inc. (headquartered in Minneapolis, Minn.) plus a growing population of workstations from Sun Microsystems and other vendors.

Rapid transit

Moving data around the world and back is key to the company's success, and fast communications represents a competitive edge. Data communications can be divided into three major categories. The first represents the needs of individual business units and includes both IBM (SNA) and non-IBM (usually DEC) traffic. On the IBM side, this means networks that include mainframes plus departmental and personal computers.

The second category of data communications consists of inter-unit transmission within the same business organization. While units operate independently, they sometimes work together and have the potential to exchange data with one another, particularly the research and engineering division.

The third category involves exchange of information across organizational boundaries. An example of this is electronic mail. For several years, electronic mail has been a highly effective means of information exchange in geo-

95

Schlumberger's computer environment

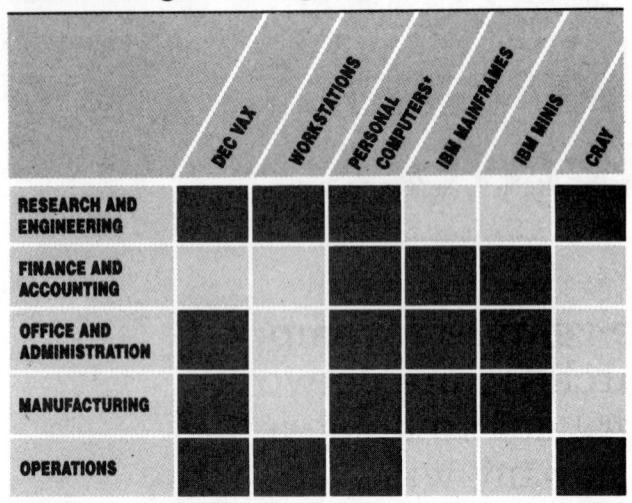

	DEC VAX	WORKSTATIONS	PERSONAL COMPUTERS*	IBM MAINFRAMES	IBM MINIS	CRAY
RESEARCH AND ENGINEERING	■	■	■			■
FINANCE AND ACCOUNTING			■	■	■	
OFFICE AND ADMINISTRATION	■		■	■	■	
MANUFACTURING	■		■	■	■	
OPERATIONS	■	■	■		■	

*PREDOMINANTLY IBM PCs AND CLONES BUT A GROWING PERCENTAGE OF MACINTOSH COMPUTERS

graphically dispersed Schlumberger engineering communities. Providing compatible electronic mail throughout Schlumberger represented a major opportunity to cut cost and improve productivity. It also presented the challenge of connecting different types of computers before electronic mail, documents, and files could be exchanged.

Schlumberger's data communications evolution was complicated by the wide variety of computers used throughout the company, the decentralized nature of its business functions (see "Oil and data *do* mix"), the absence of widely accepted networking conventions, and the absence of common industry standards.

These factors led to multiple networks (from the standpoint of circuits, protocols, and organizational locales) that, rather than serving communications needs, addressed specific network architectures (IBM and DEC). Amplifying the problem, engineering, operations, and administrative networks were often separated for security or control reasons.

The limited number of common standards and conventions found throughout the fragmented networks stemmed from the largely informal working arrangements in Schlumberger's decentralized company structure. The computer experts in the various parts of Schlumberger had loyalties and past experiences that influenced their judgment. The three camps (IBM with SNA, DEC with Decnet, and the Unix world) were all convinced that they alone were on the road to the future. Their individual approaches worked against a common goal.

Network strategy

In 1984, Schlumberger began studying company-wide networking needs and what could be done to implement electronic mail throughout the company. Three major objectives evolved:
■ Improve data communications both within and across business units.
■ Share long-haul communications facilities to increase utilization and reduce costs.
■ Provide common, cost-effective, electronic mail.

At the time, Schlumberger had limited activity in its international SNA networks (mainly remote job entry, or RJE, and occasional 3270 dial-in access), extensive DEC networks (several Decnets encompassing 200 to 300 computers), and an emerging presence of TCP/IP networks. Most data communications efforts went into local area rather than wide area networking; and Ethernet had become the de facto LAN standard within Schlumberger.

BBN Communications (Cambridge, Mass.) was commissioned to conduct a communications study for the company because it had previously successfully completed several research and engineering projects on behalf of Schlumberger. In the later stages of the project, Schlumberger also used the services of Logica Ltd. (a consulting company based in London, England), Wang Laboratories Inc. (Lowell, Mass.), and France Cable and Wireless PLC (Paris, France) to supplement BBN's efforts or to provide local services.

The study consisted of a survey of data communications uses, the associated costs, and functions plus a requirements analysis to determine the needs of each business unit. More than 200 sites were included in the survey. Among other information, the survey provided the company its first inventory showing the aggregate cost of its data communications equipment and services.

The value of the survey cannot be overstated. It provided a perspective from which to consider high-level, organized solutions. The opportunities for increasing resource efficiency, reducing costs, and improving operations emerged when networking was viewed in this way. When a pattern emerged, three problems became clear: overcapacity, unnecessarily parallel resources, and unfulfilled needs.

■ *Overcapacity.* The first problem involved sizing and duplication of telecommunications resources. For example, separate units were allocated their own circuits, each dedicated to a particular protocol or application. Each circuit was justified in terms of an individual business unit but not as part of an overall company network.

Geographically overlapping (even identically located) lines and networks were found that were sometimes underutilized. In one extreme case, a pair of cities with eight 9.6-kbit/s leased lines connecting them was identified. The lines were owned by five different units; no line was used more than 15 percent. Clearly, with the right approach and technology, one higher-bandwidth link could carry all this traffic, simplifying network management and control in the process. Simple consolidation would give immediate cost savings; a more sophisticated solution that increased redundancy of data communications paths would add reliability and flexibility.

■ *Unnecessarily parallel resources.* The second problem was that, despite overcapacity and duplication, there was substantial dial-up carrier service and use of the public network. In many cases, existing networks could have carried this traffic, but the appropriate connections, interfaces, and protocols were not in place. The right network

> ## Oil and data *do* mix
>
> Oil drilling provides a good illustration of the information-intensive nature of Schlumberger's business and the role of networks.
>
> Say an oil rig somewhere has drilled several thousand feet in search of oil. So far, nothing. The owner has to decide whether or not to keep the rig going. It is costing thousands of dollars a day to "keep making hole." The rig operators hire Schlumberger to analyze the situation.
>
> A large blue logging truck (a computer installation on wheels) that has a big reel of cable (known as a wireline) on the back arrives. Work on the well is stopped; the truck backs up to the hole and tools are lowered into the hole to probe for the presence of oil.
>
> A Schlumberger engineer then "logs" the well, which can take several hours. During logging, the truck-board computer processes the information captured by the tool sensors. This produces a vast amount of data that will be used to decide whether or not to keep on drilling. It must be processed and turned into a format that can be read at the well site or processed remotely at a field computer center.
>
> The key network task is moving data and results rapidly from the well site to the client. This data is often processed by either Schlumberger or the client for further interpretation. Information has to be transmitted to a decision maker, typically not at the well site and sometimes thousands of miles away.
>
> Data communications provides the link between data acquisition, processing, and reporting. In transmitting data from remote well sites in North America to Schlumberger's computer centers, Schlumberger uses LOGNet, a Schlumberger developed VSAT (very small aperture terminal) network. The computers analyze the data and proclaim the presence of oil; drilling continues; a gusher appears; and everyone goes home—hopefully, very rich and happy.

structure here would offer worthwhile savings plus a more efficient operation.
- *Unfulfilled needs*. The third problem was the "have-nots." These are best described as the folks who really need electronic mail or some kind of data communications but whose part of the company cannot justify the cost of dedicated networks. Here, potentially valuable interactions (data transfer, access to computers, and electronic mail) simply did not happen.

BBN's study proposed implementing a private, wide area X.25 packet-switching network to serve as the backbone for Schlumberger's data communications needs. This one network would replace many of the dedicated communications lines with a single, shared facility, supporting computer equipment from multiple vendors and environments, including SNA and Decnet. A single, shared facility meant economies of scale and cost savings.

The proposed solution provided the necessary cross-company, cross-vendor capabilities. It also provided connectivity, which is the foundation for company-wide electronic mail. In addition, by being able to interconnect with public packet-switching networks (Packet Switching Services, or PSS, in England, Transpac in France, and Tymnet in the United States), it provided a low-cost means (based on volume) of reaching many traffic locations that need only one or two hours of connection daily.

Figure 1A represents a population of networks, dedicated leased lines, and other communications methods similar to Schlumberger's original data-traffic environment. Despite the duplication of circuits between cities, the "have-not" companies could not take advantage of surplus capacity. The figure shows different organizations and computer groups having physically separate data communications circuits between the same cities.

Figure 1B shows, in contrast, a stylized representation of the improved Schlumberger network. It illustrates the concept of connecting many computers and terminal clusters to a packet-switched node (PSN) and then linking PSNs to one another using 56- to 64-kbit/s trunks. The organization of PSNs in the figure illustrates that one of SINet's goals in terms of topology is to have multiple paths connect the PSNs. Working in harmony, the PSNs take care of traffic balancing and accommodate trunk failures.

Will it work?
The BBN proposal was followed by a broader in-house feasibility study. As part of this exercise, alternative approaches and suppliers were considered, a financial model of the overall data communications costs for 107 Schlumberger sites was constructed, and other companies that had built similar networks and had extensive electronic mail facilities were interviewed.

The feasibility study recommendations concentrated on the business and implementation issues. They also identified that the cost of a new network would be offset by:
- Putting PSNs in cities where data traffic was heaviest, connecting those PSNs with high-speed (56- or 64-kbit/s) trunks, and eliminating underutilized circuits. IBM and DEC computers would connect to the PSNs using the internationally accepted X.25 standard; X.25 interface software (Packet Switching Interface, or PSI, from DEC; Network Packet Switching Interface, or NPSI, and others from IBM) would handle the interface. Terminals and older computers, for which an X.25 interface was not available, would be connected to PADs (packet assembler/disassemblers).
- Using high-speed trunks that are cheaper per unit of data than ordinary lines. In the clearest example that we found, a 56-kbit/s digital circuit between the United States and Europe cost only 40 percent more than a 9.6-kbit/s circuit and could transport five times the data.
- Lower telephone, Telex, and courier bills would result from company-wide electronic mail. Schlumberger's targets were intentionally unambitious: 10 percent savings for telephone, 50 percent for Telex, and 30 percent for courier.
- Centralizing the wide area network operation functions would conserve personnel efforts because traffic load balancing is automatic and failed circuits and equipment

1. Redesign. *Some sites cannot use the surplus capacity (A). In SINet, multiple paths link packet-switched nodes, which handle traffic balancing and trunk failures (B).*

(A) UNUSED SURPLUS CAPACITY

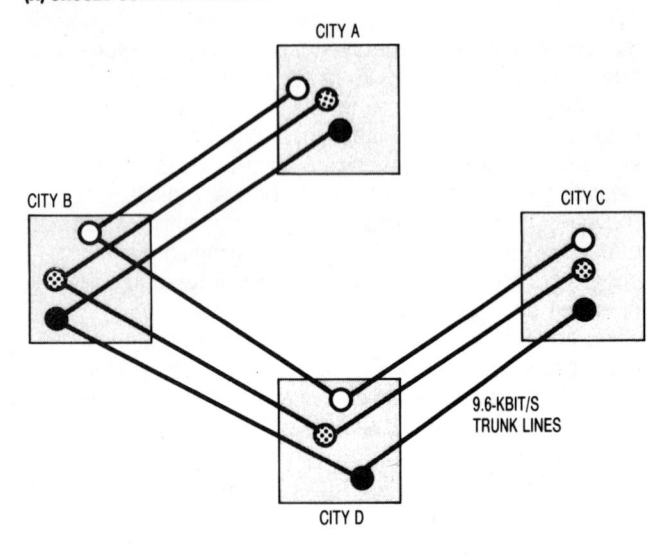

(B) PACKET-SWITCHED SOLUTION

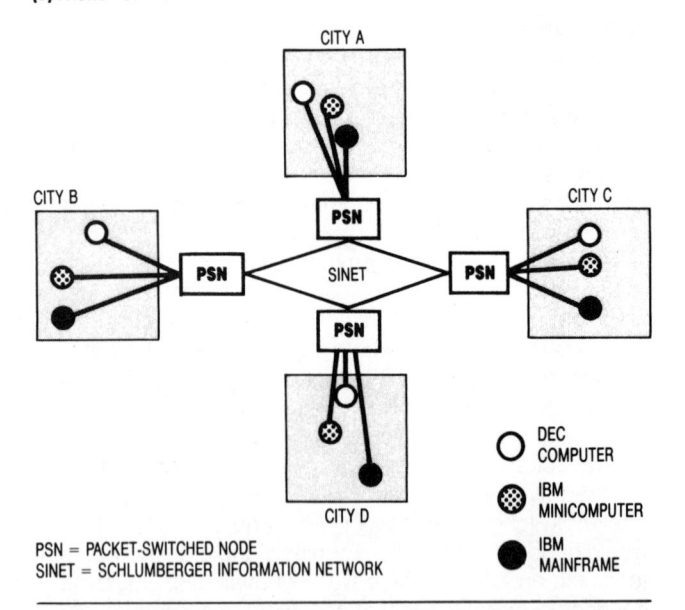

PSN = PACKET-SWITCHED NODE
SINET = SCHLUMBERGER INFORMATION NETWORK

are bypassed via the dynamic adaptive routing features of the PSNs.

Flexibility was another focal point of the feasibility study. For the future, with a consolidated network in place, new users, applications, computers, and sites could be easily integrated into Schlumberger's overall scheme of information technology. With the OSI architecture, of which X.25 is part, this integration would be simple, easy to plan, and comparatively economical. New requirements could be met by a combination of access lines, additional PSNs, additional trunk circuits, and increased capacity of existing trunk circuits.

Independence from specific computer vendor protocols for data transmission was another important advantage of the proposed scenario. OSI is clearly the road to the future, and committing to X.25 for data transmission and X.400 for electronic mail put Schlumberger on that road. By adopting a "universal" approach for all of Schlumberger's data transportation, performance and capacity management could be performed for the network as a whole. Information gathered during network operation could be processed by BBN's modeling tools (based on artificial intelligence) to give advanced warning of traffic needs.

Perhaps as important as the immediate benefits were the longer-term strategic implications. This new networking approach would provide a data communications platform for continued growth (estimated at 30 to 40 percent per year by the sites included in the financial model) and improved customer service. It would provide access to computer facilities, improve the speed of data processing and the delivery of results, and replace much of the paperwork with electronic transactions and mail exchange.

Schlumberger selected BBN as the vendor to provide the X.25 network on the strength of its ability, its reputation, and its security advantage. The latter is a particularly strong point, since BBN's equipment was developed for defense networks. In addition, BBN had built up a large knowledge base about Schlumberger during the study. Other vendors were considered but rejected on the same grounds. Although BBN's equipment was competitively priced, with a project of this scale, equipment cost is not the most important factor. Track record and reputation take precedence.

Pilot network

After the feasibility study, a pilot network was established both to prove the strategy in practical terms and to confirm the benefits. More important, the project team wanted to be convinced that a common network was a workable solution for Schlumberger. Pilot testing made it possible to address legitimate concerns and objections, such as performance, the transparency of the X.25 interface, and security.

Acceptance and test procedures (ATPs) were developed for each computer and terminal that would use the network. In addition, ATPs were developed for network operations, maintenance, support, and administration services. Each specified what Schlumberger required and defined a detailed test plan with specific measurement criteria. Responsibility for ATPs was shared by the sites involved in the pilot network. A consultant from Wang who had been involved in installing the Wang BBN network helped coordinate the activity and provided assistance to the pilot sites.

The pilot network configuration involved PSNs and PADs in Europe (one city in England and one in France) and the United States (two in the Northeast, one in the South, and one on the West Coast). The geographic spread tested BBN's ability to handle the international aspects of installation, maintenance, and support. At the same time, it included key Schlumberger sites already involved in networking. When proven, the pilot formed the nucleus of SINet. There were also gateways to PSS in England and to Transpac in France.

In addition to the usual network installation, integration, and start-up, BBN Communications provided continuing network management from its own network operations center in Cambridge, Mass. This arrangement continued after the pilot phase; over time, these administrative functions have been transferred to Schlumberger, which will take over the operation of SINet in 1988.

Infrastructure and cooperation

The decentralized, independent operations of the business units increased the need for a participatory approach, since people must be involved from the beginning as active members of the evaluation team in order to be strongly committed to it.

At an early stage, a coordinator's committee was established that included a cross-section of current and future network users. Its purpose was to oversee the project and represent the interests of each Schlumberger business group.

The committee was supplemented by two working groups. One addressed electronic mail; another dealt with developing and operating acceptance test procedures. Group members were drawn from the pilot test sites. The coordinators' committee and the working groups resolved several issues, such as defining how the different network communities were to be interconnected, delegating network security responsibilities across the organization, and reviewing cost recovery and budgeting.

Each monthly meeting of these groups was held at a different pilot site. A presentation of the site business, its use of computer technology, and the role of networking were part of each agenda, as was a site tour. Through this process, committee and working group members saw beyond their own needs and gained a better understanding of the company. One side benefit was to come up with the network's name, SINet.

Security

When users migrate from a dedicated network to a consolidated common-user network, the access and security challenges increase. Instead of addressing the issue of how to provide access, the concern becomes keeping people out: facilitating access by authorized users while ensuring that all others are excluded. The multiple-user communities should have at least the same security level as they had with the dedicated network.

Coordinators, reflecting the views within the part of Schlumberger that they represented, had different notions of how important security was and what degree of protection was necessary. Some tended to want "one big happy family" when it came to Schlumberger's internal computer activity, often not wanting segregation at all. On the other hand, all were very protective of customers, clients, and proprietary product information. All agreed on the need for cross-communications for such areas as engineering and manufacturing, which would want to share control procedures and integrate CAD/CAM (computer-aided design and manufacturing) resources. But they were guarded and concerned about making product databases and other key databases widely available.

To resolve the issue, Logica was brought in to independently study security needs and to make recommendations. The goals and expectations of the study were set. The study was overseen by the coordinators, and the working groups helped review the recommendations.

Everybody participated in the development of network security standards. In the end, the project team was able to prepare a booklet explaining the security framework and rules that are used throughout Schlumberger. During the pilot phase, it was proved that the network provided the necessary security features, including communities of interest, network user authentication, and access profiles.

Based on practical experience gained during the pilot phase, the project team determined specific approaches to certifying network equipment and computer/terminal equipment for connection to SINet; formalized installation procedures; developed operation guidelines; and implemented administrative procedures.

By introducing an X.25 packet-switching network, Schlumberger hoped to interface with and exploit the many emerging national and international, publicly available networks. In parallel with the pilot phase, Schlumberger commissioned France Cable and Wireless, a consulting division of France Telecomm, to study this aspect. France Cable and Wireless, because of its involvement with Transpac (the French public packet-switching service) and other networks, was able to consider the technical and legal aspects. Based on this review, Schlumberger confirmed that access was possible in most of the 49 countries in which it had a presence and identified the best access method. The France Cable and Wireless report from this study is, to this day, widely used at Schlumberger.

2. Network security. *With a feature called community of interest, computer communications is restricted. Each logical network is inaccessible and invisible to the other.*

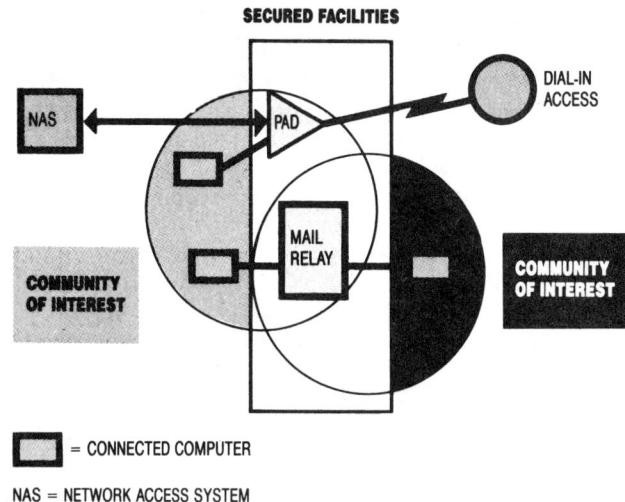

The single most difficult application problem was to devise an economical, company-wide electronic mail solution that would operate on existing DEC and IBM computers, could be used for office procedures, and would be easily accessible to the targeted 20,000 Schlumberger personnel who would benefit from electronic mail. Solving this problem took several months, a lot of cooperation, and trial and error. In the end, Schlumberger's DEC VAX-based solution used a combination of network features and available DEC software: most notably, the product called Message Router.

Electronic mail solution

Schlumberger's electronic mail solution uses existing terminals and computers and operates with software provided by the vendor; thus, it requires little or no investment and is inexpensive to use. The electronic mail implementation addresses 90 percent of Schlumberger's needs and embraces the IBM and Unix worlds as well.

The basic functions it provides are sending and receiving messages and computer files, replying to messages received, forwarding messages, and electronically filing messages. Operated around five MicroVAX computers, which are known as Mail Relays and are located in the United States and Europe, SINet provides users with the following electronic mail services:

■ *Simplified mail addressing.* To send mail, users need only know the mail host address (the author's, for example, is ASLVX6) and the user name. The Mail Relays deliver mail using the most efficient route and do not rely on Decnet.

■ *Store-and-forward facilities.* If a mail host computer is not available, Mail Relays retain mail and regularly retry delivery. After 24 hours, undelivered mail is returned to the sender with an appropriate message. Two other classes of delivery service can be selected by the user: immediate and confirmed delivery.

■ *Interface between different computers.* Over SINet, electronic mail can be exchanged with Unix machines. A number of IBM interfaces exist, and IBM's SNADS and X.400 interfaces will be implemented in the coming months. In addition, mail can be directly exchanged with CSNet. (CSNet is a nonprofit computer network dedicated to the support of computer-related research and development in universities, industry, and government. It is a project of the University Corporation for Atmospheric Research under contract to the National Science Foundation; the CSNet Coordination and Information Center is located at BBN Laboratories, Cambridge, Mass.)

An external message interface allows outgoing electronic mail messages to be delivered to Telex and facsimile machines; electronic mail can also be delivered as cablegrams. Telex traffic is refiled to obtain the lowest costs, and incoming Telex messages can be optionally directed to electronic mailboxes.

Electronic bulletin boards allow self-registration of electronic mail users and automatic posting/distribution of messages. SINet electronic bulletin boards are used to circulate information on everything from CAD/CAM to DEC product announcements to buying and selling equipment.

■ *Directory services.* A company-wide electronic mail directory is updated regularly and distributed electronically to mail host computers.

The biggest challenge in introducing electronic mail was making it simple to use. To this end, Schlumberger produced a training video, published a simple electronic mail tutorial in the form of a users' guide, and developed a menu-driven computer interface, all aimed at the new user. For experienced users, Schlumberger provided a small, plastic-laminated reference card that summarized key features and instructions.

Electronic mail's success at Schlumberger owes much to the dedication of a few clever computer engineers and the enthusiasm and support of all the members of the electronic mail working group.

Teamwork

When the pilot network became the live network and the real network-building process began, the network team grew within and outside of Schlumberger.

Schlumberger computer and data communications personnel came into contact with SINet through the integration process (adding terminals and computers to the network) and by implementing electronic mail. Outside Schlumberger, the SINet team extended to AT&T, Cable and Wireless, IBM, DEC, Sun, and many other organizations involved in the building process. At any one time during 1985, 1986, and 1987, from 65 to 100 people were directly involved with developing SINet, though most on a part-time basis. To keep people informed, a SINet users' guide was published that explained the basic principles, functions, and benefits of the network and electronic mail. A SINet newsletter also appeared every three to six months to keep the dialogue going. SINet presentations were also given to audiences that varied from computer technicians to lawyers, and from accountants to field engineers.

What follows are a few of the networking applications and problems the SINet team faced and an account of what was achieved with new network. The applications and problems involve linking large IBM mainframes used for manufacturing applications, separating two secure networks, linking IBM minicomputers (System/36 and System/38) together, and how Schlumberger coped with the extensive DEC-based engineering data communications.

IBM mainframes

Schlumberger often uses IBM mainframes (43XX and 30XX) for production planning and control. This was the case for two large manufacturing plants, one in Paris and one in Houston. Together, the plants created an overlapping range of equipment for the field locations. The major network applications were queries to the manufacturing inventory database and parts ordering.

Despite having comparable hardware (IBM 3083s) and software, both sites prior to SINet had updated their databases manually, and orders were sent by courier for re-entry and batch processing. Linking the two IBM 3083s did not appear to be justified because of the high cost of a dedicated transatlantic circuit ($12,000 per month at the

Secure access to SINet

When Schlumberger lets its customers access SINet so that they can interpret the data collected at, for example, an oil field, it faces the challenge of security. The larger SINet grows, the more people will need access to it.

Thus, Schlumberger created its Smart Card and Systems Division. It has developed a network security product for internal use to validate user computer access, encrypt sessions, and allow application programs to create, update, and access computer files held in the Smart Card. The card is a networking technology that operates in parallel with existing security features of SINet. It enables Schlumberger users to differentiate and protect computer data and processing.

The Smart Card and Systems Division implements a comprehensive identification process for computer users. This process, which overcomes the limitations associated with simple passwords, operates at two levels: the initial computer connection and the application.

User confidentiality is ensured via encryption of the interactive session between user and computer. This process is equivalent to encapsulating the interactive dialogue in an impenetrable casing that extends from user terminal to computer. The techniques involve advanced encryption and use of public/private keys.

Because of the Smart Card operation, the SINet user community can control the networking environment without endangering the confidentiality of the individual customers. A combination of infrastructure, computer software, and minimal additional hardware and audit trails contributes to the Smart Card approach.

In addition, the Schlumberger Smart Card security solution serves as the platform on which future computer applications can create, access, and update secure information residing on smart cards. This means that multiple data files can be held and used by individual users during computer access. In this context, it provides almost unlimited new computer application opportunities and ways of reducing paperwork.

---

time). On the other hand, the engineering departments, located at the same two sites, had a leased line connecting their VAX computers. An early attempt to multiplex the transatlantic circuit had proved unsuccessful because the equipment was limited and the analog circuit unreliable.

Based on a traffic analysis, both locations were designated to be SINet sites and PSNs were installed. The leased line previously connecting the VAX hosts was upgraded (from 9.6 to 56 kbit/s) and became a SINet trunk connecting the PSNs. Both VAX and IBM computers were connected to the PSNs; NPSI was implemented on the IBM computers and PSI on the DEC VAX machines. In addition to linking the IBM and DEC equipment, at the PSN sites, other IBM and DEC computers in Houston and Paris were connected via local leased lines to the PSNs.

The result was faster, more direct connections for the VAX users. For the IBM users, it was now possible to access remote databases directly and to place orders. The "have-nots" got their connectivity, and a substantial amount of data rekeying was eliminated.

As implementation of the network proceeded, other lines were combined in the same way, and a second link between the United States and Europe was added to increase reliability. Two digital 56-kbit/s transatlantic circuits replaced three 9.6-kbit/s analog circuits. This resulted in reduced line cost (the 56-kbit/s circuits cost only 40 percent more than the 9.6-kbit/s circuits) and more traffic capacity (112 kbit/s compared with 8.8 kbit/s). The elimination of data rekeying and associated clerical activity provided an unexpected bonus of $12,000 per month in cost savings.

Having provided connectivity, the network planners then had to control who could use it. Access security meant ensuring that different user communities were unaware of each other's existence from the standpoint of gaining access to unauthorized computer resources. This aspect of network security is essential to Schlumberger's operations and, therefore, a prerequisite of any effort to increase overall connectivity within the network.

Two facilities are available to solve this problem: one, a feature of X.25 known as closed user groups, implemented by the computers; the other, a feature of the network equipment called communities of interest (COIs), which restrict a computer to talking only to other designated computers (COIs are shown in Figure 2).

The COIs made it possible to define different logical networks within the overall network environment. These logical networks shared the same physical network facilities: packet switches, trunk lines, and so forth. But each was inaccessible and invisible to the other.

The Mail Relays (secured using the DEC product PSI Mail) appear in all COIs to allow mail exchange. Direct user connection to Mail Relays is not allowed by PSI Mail, and restrictions are implemented so that it is only possible to send and receive electronic mail.

Most users are at sites connected directly to a host. However, many engineers and managers who travel and rely on electronic mail need dial-in access. For this, PADs with dial-in modems are installed in key cities; dial-in PADs exist in all of the COIs. The manager or engineer, using a laptop computer, can dial to the PADs from a hotel room or client's office. Alternatively, access to the PAD can be gained through another computer in the same community of interest via an X.29 call, which effectively establishes the equivalent of dial-in terminal connection using packet-switching features.

Once a dial-in or X.29 connection is made, BBN's Network Access System (NAS) comes into play. The NAS consists of a Master Database (MDB) host and a loading-sharing Access Control Server (ACS) host (both based on DEC's MicroVAX), operating in cooperation with BBN's PAD and X.29 calls.

When a PAD or X.29 access request occurs, the NAS requires the user to provide a user identification code and a password, which are checked against the NAS database to verify authentication. Without the verification, access to the network itself is not possible.

Once authenticated, users are automatically connected to a specific computer as defined in their MDB. Other entries in the MDB may include restrictions of time, geography, and so forth. Computer center managers control new entries in the NAS database. The NAS also provides all the security reporting and management features used to police the security procedures.

In addition to these and other access restrictions that the connected computers may enforce, additional security is possible by using smart cards developed by the Smart Card and Systems Division of Schlumberger. This added security, involving intelligent user identification using secret keys, also incorporates economical link encryption (see "Secure access to SINet").

IBM minicomputers

Schlumberger's MIS environment closely matches its decentralized organization. IBM System/38s and 36s provide distributed computing horsepower, sometimes supplemented by a large mainframe.

MIS computers are typically configured with IBM mainframes at headquarters or manufacturing centers and IBM System/38s or System/36s at regional sites; these are used mainly for consolidated financial and business processing. At district and branch offices, there are System/36s with connected IBM 5251 and/or IBM Personal Computers for field data processing.

Historically, information was collected in the field and then transferred via divisional and regional offices to headquarters at month and quarter ends, using dial-up access or, where affordable, leased lines. IBM Personal Computers are now used to gather and initially process data in the field; they communicate as a terminal to the IBM System/36.

The System/36 consolidates and processes data and then communicates with an IBM System/38 as a "slave." System/38 and System/36 can, depending on the application and where it is based, communicate on a peer-to-peer basis. Both machines, emulating IBM 3270 or remote job entry terminals, communicate with the manufacturing centers as well as with mainframes at the headquarters locations.

One Schlumberger company consists of several independent business units, each having a separate sales-order processing application running on an IBM System/38. Each sales office sells all products. This requires IBM 5251 workstations (or IBM PCs emulating a 5251) at each office to access multiple IBM System/38s. As an interim solution, the company bought software (Comware International's Lode Star 3438) that enables an IBM Series/1 to let the IBM 5251 workstations do dynamic terminal switching. The IBM PCs were also used to access an external timesharing service for electronic mail (Fig. 3A).

The SINet challenges were: improving performance, adding capability, improving connectivity, and reducing cost. Figure 3B illustrates how SINet replaced a collection of lines and removed the need for the IBM Series/1.

Computers and terminal clusters supporting X.25 connect to the closest PSN; others are connected through PADs. The PSNs are connected to one another within SINet with multiple paths to improve reliability.

The SNA computers use SINet as a single, dedicated entity. Some VAX computers within the company have the SNA Bridge and they, too, figure in the SNA environment. Connections are made between DEC terminals and IBM mainframes (3270 emulation), between System/36 workstations and DEC VAX computers (Teletype, or TTY, terminal emulation) on SINet, and by a variety of dial-in microcomputers for access to the VAX machine for electronic mail.

The standard SINet X.25 interface allows this flexible computer-to-computer and terminal-to-computer connection, which, when combined with PADs for dial-in access and peer-to-peer features, allows the System/38 to access VAX computers through a System/36.

In the IBM environment, the VAX SNA connections over X.25 are usually used to allow sites with DEC equipment to access databases held in IBM mainframes. (Through protocol converters, Schlumberger has also managed to connect VAX users to System/36 and System/38 computers.)

The other concern was mainframe communications. Better access to the mainframes and improved access among the System/38s was the goal. By using an X.25 link to connect System/38s and System/36s, both to each other and to the IBM 43XX hosts, Sclumberger was able to give workstation users and the System/38s better access to the mainframes and to the other System/38s. One result was that electronic mail was transferred from the timesharing service to existing VAX computers that made use of the SINet electronic mail services. This alone saved $25,000 each month in timesharing costs.

SINet helped improve and simplify the computer center organization and support functions. Because the connections proved so reliable, the two computer centers were consolidated and more advantage was taken of the time differences between the United States and Europe to exploit processing capacity. By having the high level of connectivity, technical support of remote computers has become routinely possible, meaning that specialist computer support groups can be concentrated without incurring heavy travel overhead.

Engineering

A lot of computing horsepower in Schlumberger is dedicated to engineering applications. These typically include "farms" or clusters of DEC VAX computers, powerful workstations, and the occasional Cray.

VAX computers are usually linked via Decnet for wide area networking. In addition to file transmissions and interactive sessions, the DEC community has traditionally used electronic mail extensively for correspondence and electronic bulletin boards.

In the DEC environment, everybody implemented networking independently, on an as-needed basis. This resulted in many separate physical networks. Moreover, the engineering networks were often kept separate from the DEC operations networks for security purposes. In a typical Decnet configuration before SINet, if any of the computers were overburdened or not operating, the network was impaired. Because of the daisy-chain approach to connecting the DEC computers, a single low-speed circuit will affect the network performance and negate the benefit of higher-speed circuits.

SINet has replaced long-distance point-to-point lines with X.25 connections. Decnet has largely been retained as a logical architecture, and separate logical networks were established where necessary for security purposes using the COI feature.

SINet improved the efficiency of Schlumberger's Decnet environment; the X.25 approach reduced the Decnet daisy-chain to one hop in most cases.

When Schlumberger started to use the network statistics generated by BBN's network operations center to review traffic loading, a number of long-term, hidden Decnet configuration inefficiencies became obvious, mostly a result of poor coordination between different computing groups. These inefficiencies were caused by problems with Decnet areas, inconsistent profiles, incorrect circuit-costing information, and inconsistent routing tables.

All of this resulted in the unnecessary generation of control traffic. With help from DEC and a number of the talented DEC specialists within Schlumberger and BBN, these problems were overcome and redundant traffic was reduced. This fine-tuning involved trial-and-error testing and ongoing performance measurements.

In 1986, DEC, based on this work, certified the SINet PSI profile (the parameters that define the function of the X.25 interface). This means that DEC is able to provide standard support internationally.

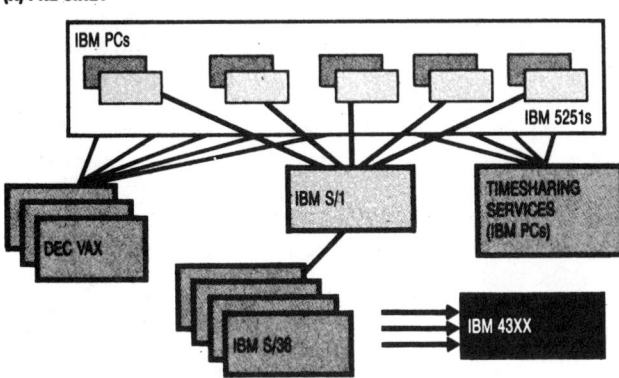

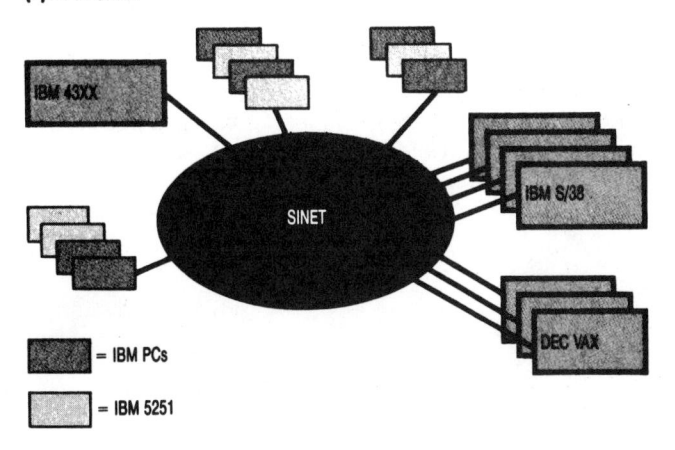

Overall results

Looking at the relative costs of data communications circuits in 1984 compared with those in 1987, SINet has achieved at least a 10 percent savings. If costs are measured on a unit-of-traffic basis (in Schlumberger's case, kilosegments), the 1987 circuit costs are 36 percent of those in 1984, reflecting the traffic growth and a 10 percent cost reduction.

However, these figures are misleading, since they reflect circuit costs alone and do not take into consideration other SINet costs (equipment, operation, support, management, and so forth). For example, SINet line costs represent 50 percent of the total 1987 budget.

Aside from cutting down the circuit costs, SINet has absorbed significant dial-up and timesharing traffic, saving an estimated $600,000 to $1,000,000 per year. Savings in support staff resulting from SINet's consolidated and more automatic network management features is estimated at one quarter of a person per connected site annually, and more than 100 sites are connected at present.

No assessment has been made of cost-saving impact in the following data communications-related areas:
- *Computer usage.* Where SINet handles data-traffic routing, it has eliminated the middleman function associated with processing the traffic between three or more computers.
- *Productivity.* SINet, with its alternative paths and management control, is more reliable than Schlumberger's 1984 networks. This reliability translates into increased productivity for computer users as well as the support and development staff.
- *Connectivity.* Improved connectivity has a positive impact on the way support staffs are organized, on how applications are developed, on computer utilization, and on new application opportunities.
- *Circuit-rate reductions.* The cost of 56-kbit/s and higher-speed trunks, particularly on international routes, continues to decrease at a faster rate than that of 9.6-kbit/s circuits. Between the United States and France, for example, 56-kbit/s circuit rates have decreased by 60 percent from 1985 to present. In contrast, 9.6-kbit/s circuit rates have declined about 25 percent.

As a result of SINet, the number of electronic mail users

3. Before and after. *PCs used a timesharing service, and Series/1 let the 5251s do dynamic switching (A). SINet replaced many lines and the need for the Series/1 (B).*

in Schlumberger grew from 3,000 to 11,000 during the past 18 months (to mid-1987), and is estimated to reach a total potential user base of 20,000 by 1989.

Based on an assessment in the United States, Schlumberger saves 10 percent on its telephone costs. This savings is expected to increase to more than 20 percent in the next year or two as more use is made of electronic mail. An electronic mail message costs approximately 1 percent of the equivalent long-distance telephone call and prevents the unproductive game of telephone tag. The savings in telephone costs alone each year more than pays for SINet.

A 50 percent saving of Telex costs has resulted from: (a) using electronic mail as an alternative for internal company correspondence; (b) the cheaper tariffs that were obtained by using a refiling agent for the remaining Telex traffic carried through electronic mail; (c) avoiding line and equipment costs as a result of being able to receive Telex messages through electronic mail; and (d) reduced clerical effort because a Telex operator is no longer involved in the sending and receiving process. These figures are based on a small sample of sites taking into account points a, b, and c; if d is taken into account, greater than 50 percent of the true cost of Telex is more likely.

A 30 percent saving in courier costs is possible, but because of the difficulty of collecting comparable data, this has not been validated.

Software and hardware used to connect computers to the network consists of off-the-shelf X.25 interfaces provided by IBM, DEC, or Sun. And the company-wide electronic mail solution makes use of standard DEC and IBM software. DEC VAX computers, because of the available software (VAX Mail) and the numbers deployed throughout Schlumberger, are used as mail hosts. IBM computer users are connected, using the features of the X.25 network, to the DEC mail hosts.

Standards

Electronic mail is fast becoming the standard means of internal communications at Schlumberger. Major savings in productivity are being made. Messages handled by the Mail Relays were approximately 100,000 per month in 1987 and this number is growing steadily.

SINet has become the standard wide area network for all company-wide data communications, except the highly specialized applications. And as part of the SINet project, a consistent set of electronic mail and security standards reflect company-wide needs and are consistently followed.

With the additional security layer provided by the Network Access Security feature of SINet, Schlumberger business units are starting to encourage key customers and suppliers to access Schlumberger's computers in a controlled and secure way. New network-based customer services are being planned and more use is being made of the network to improve existing services. (One example is faster transmission of processing results for international clients.) For Schlumberger personnel who travel, it is becoming common to dial a local PAD from a hotel or office and, using a laptop PC, access an electronic mail host computer to receive and send messages.

New applications

The project team had an intuitive feeling that the network would lead to greater productivity opportunities. Only now, however, have they become apparent. The problem is a classic chicken-and-egg dilemma: Until you build the network, you do not see all the uses it can have. Although SINet is still in its early stages, the Schlumberger business units have found new data communications applications for the network.

A good example is DEC. The X.25 electronic mail link to DEC allows direct communications with DEC's Schlumberger support team and all the local branch offices. DEC product announcements are made this way, as is information on hardware and software. (One interesting and unexpected application of the DEC link came in the form of maintenance scheduling. By directing the output of diagnostic tools based on artificial intelligence to the local DEC office, computers have started to order replacement parts and to schedule maintenance with DEC directly.)

In analyzing where SINet is today and looking back over the process that got it here, it appears that taking the long-term strategic perspective was the most important ingredient to the success of the project. SINet was a business investment first, a technical exercise second. The business investment focused on the strategic, revenue-generating opportunities as well as the real cost savings. Intangible benefits, such as productivity improvements, were largely in the background. The technical exercise took advantage of standards, such as X.25, and advanced technology.

The process proved that future success for data communications networks and the companies that have them is not about how to make the technology work but how to harness it in a meaningful and profitable way. ■

Cormac O'Reilly is a business manager with Schlumberger Industries, based at the headquarters in Houston, Tex. Before joining Schlumberger, O'Reilly held a number of technical, senior management, and consulting positions with British Oxygen International, Unilever, Abbey National, and Computer Service Center Ltd.

Thomas J. Routt, contributing editor, DATA COMMUNICATIONS

From TOP (3.0) to bottom: Architectural close-up

Presented here is a detailed look at the way international standards are shaping up as they continue to make inroads into U. S. user firms.

The business mission of computer networks is to serve as an enterprise-wide infrastructure that can deliver, format, present, and interchange appropriate data on a highly available, timely, and cost-effective basis to distributed decision makers in a sufficiently detailed or abstracted level, at whatever organizational, geographic, or temporal distance from one another and regardless of which computer interface is currently available.

Enabling networks to meet this challenge within a multi-vendor setting has been the goal of international standards, specifically the International Organization for Standardization's (ISO's) Open Systems Interconnection (OSI) efforts. In the United States, these efforts have been embodied principally in the work of the Technical and Office Protocols (TOP) and Manufacturing Automation Protocol (MAP).

However, the TOP specification Version 1.0, published by Boeing Computer Services in November 1985 to be compatible with the then-current MAP Versions 2.1 and 2.2, provided only a narrow range of services. While several vendors implemented TOP 1.0 and users began to employ TOP in limited settings, more widespread acceptance of TOP would await Version 3.0, which appeared in April 1987 (see "Where TOP comes from").

The major distinction between the two versions is that TOP 3.0 provides a relatively complete set of application services. This set includes standard graphics application interfaces, electronic mail, and remote file and terminal access as well as network directory and management services. Such services are considered essential to many user environments.

This article, the second in a series (see "Under the Big TOP at Enterprise '88: TOP 3.0's debut," DATA COMMUNICATIONS, April, p. 155), examines the TOP 3.0 architecture at all its levels and focuses on the application-level protocols and interfaces that will bring these much-called-for services to users' multivendor networks.

Rationale
Technical and administrative offices represent perhaps the most significant settings for distributed computing. These settings have witnessed a dramatic growth in productivity tools in recent years. The proliferation of distributed office tools has matched exponential technology and attendant price-performance improvements and has led to increased user dependence upon proprietary computer network architectures, protocols, and products.

Unfortunately, user networks often contain components from multiple vendors, which may have been engineered to be fundamentally incompatible at the application-interface and communications levels. Such incompatibility engenders technical, economic, and organizational inefficiencies that make it impossible for computer networks to fulfill their business mission.

The primary goal of TOP is to accelerate the availability of cost-effective, off-the-shelf, interoperable, computer-based products. These products should satisfy the requirements of technical and administrative office users of multivendor networks. The resulting TOP specification represents the collective requirements of the MAP/TOP Users Group, which in turn is understood by vendors to represent a significant market segment. This group includes the following:

■ Hughes Aircraft, Alcoa, Jaguar Cars, General Electric, Kaiser Aluminum, Rockwell International, Ford Motor Co., General Dynamics, Chrysler Motors, Lockheed, Kraft, Grumman, Scott Paper Co., Monsanto Co., Boeing, General Motors, Du Pont, Weyerhaeuser, McDonnell Douglas, U. S. Air Force, Deere & Co., Frito-Lay, Mobil, Eastman-

Kodak, Northrop, Shell Oil, Nabisco, TRW, Proctor and Gamble, Martin Marietta, and Union Carbide.

The current TOP specification, Version 3.0, defines a suite of protocols for the seven OSI layers and for the interfaces between applications and Layer 7.

Newer and richer

The major architectural concern of TOP 1.0 was to provide error-free file transfers between cooperating end systems (Layer-4-and-above protocol environment). TOP 1.0 certainly provided connectivity within multiple-vendor environments. Noticeably absent, however, was the extended interoperation that cooperating applications could achieve with common data formats. Furthermore, there was little agreement on application-layer functions during discussions of TOP 1.0 at the NBS Implementers' Workshops.

Figure 1 illustrates TOP Version 1.0, which essentially provides for file transfer services at Layer 7 over an Ethernet-like, 10-Mbit/s local area network (LAN) using an ISO carrier sense multiple access with collision detection (CSMA/CD) at Layers 1 and 2. Note that TOP 1.0 defines a single OSI Layer 7 protocol, File Transfer, Access, and Management (FTAM), and does not specify an OSI Layer 6 protocol.

Due to these limitations, several proprietary vendor architectures at the time provided a richer set of application services than TOP 1.0. It was widely recognized that a robust suite of application-layer protocols was necessary. That recognition, and resulting upper-layer protocol agreements, led to the endorsement and publication of TOP Version 3.0. With TOP 3.0, networks can come close to fulfilling their business mission.

APIs

Increased flexibility in meeting users' networking needs is made possible by application program interfaces (APIs), standard routines that provide network-based services to application programs. Technical office applications, such as computer-aided design (CAD) packages, simulation, and scientific data analysis, frequently use graphical services. Many business office applications are also increasingly dependent on graphics. These include desktop publishing, illustration, and drafting programs, as well as integrated spreadsheets that create business graphs for presentation.

TOP 3.0 regards computer graphics application programs as those that produce or manipulate computer graphics pictures and requiring computer graphics services. This includes graphics output or display, input from an operator, interactive graphics-based applications, storage and retrieval of items in graphics databases, and graphics metafiles, the generation, interchange, and interpretation of picture descriptions.

Figure 2 illustrates TOP 3.0. The Standard TOP Application Program Interface Library provides the data-interchange formats that allow heterogeneous graphics applications to interconnect and share resources in an application-independent and device-independent manner.

Note in the figure that both the Computer Graphics

Acronyms

ACSE Association Control Service Element
ADMD Administrative Management Domain
AE Application Entity
APDU Application Protocol Data Unit
ANS American National Standard
AP Application Process, Application Profile
API Application Program Interface
ASN.1 Abstract Syntax Notation One
CASE Common Application Service Element
CCR Commitment, Concurrency, and Recovery
CGI Computer Graphics Interface
CGM Computer Graphics Metafile
CLNP Connectionless-Mode Network Protocol
CLNS Connectionless-Mode Network Service
CMIP Common Management Information Protocol
CMIS Common Management Information Services
CONS Connection-Mode Network Service
DDGL Device-Dependent Graphics Layer
DIB Directory Information Base
DIGL Device-Independent Graphics Layer
DIGS Device-Independent Graphics Services
DIS Draft International Standard
DP Draft Proposal
DSA Directory Service Agent
DUA Directory User Agent
EDIF Electronic Design Interchange Format
FDDI Fiber Distributed Data Interface
FTAM File Transfer, Access, and Management
GKS Graphical Kernel System
IDU Interface Data Unit
IGES Initial Graphics Exchange Specification
IS International Standard
MHS Message Handling System
MOTIS Message Oriented Text Interchange System
MTA Message Transfer Agent
MTAE Message Transfer Agent Entity
NPDU Network Protocol Data Unit
NSAP Network Service Access Point
ODA Office Document Architecture
ODIF Office Document Interchange Format
PDES Product Data Exchange Standard
PDU Protocol Data Unit
PPDU Presentation Protocol Data Unit
PRMD Private Management Domain
ROS Remote Operation Service
RTS Reliable Transfer Service
SMAE System Management Application Entity
SPDU Session Protocol Data Unit
STEP Standard for the Exchange of Product Model Data
TPDU Transport Protocol Data Unit
TSAP Transport Service Access Point
UA User Agent
UAE User Agent Entity
VDI Virtual Device Interface
VT Virtual Terminal

1. TOP 1.0. *The initial TOP specification provided file transfer services at Layer 7 and little else over an Ethernet-like LAN using a contention scheme at Layers 1 and 2.*

OSI LAYER	TOP 1.0 PROTOCOLS
7 APPLICATION	ISO FTAM 8571
6 PRESENTATION	NULL (ASCII AND BINARY ENCODING)

5 SESSION	ISO SESSION 8327
	   BASIC COMBINED SUBSET AND SESSION KERNEL,
	   FULL DUPLEX

4 TRANSPORT	ISO TRANSPORT 8073 CLASS 4

3 NETWORK	ISO INTERNET 8473
	   CONNECTIONLESS AND FOR X.25
	   SUBNETWORK DEPENDENT CONVERGENCE PROTOCOL

2 DATA LINK	**ISO LOGICAL LINK CONTROL 8802/2** **(IEEE 802.2) TYPE 1, CLASS 1**

1 PHYSICAL	**ISO CARRIER SENSE MULTIPLE ACCESS/COLLISION**
	   DETECTION MEDIUM ACCESS CONTROL
	     (IEEE 802.3 MAC) 8802/3
	     10 BASE5 (10 MBIT/S, BASEBAND, 5 MHz)

Meta-file (CGM) and Graphics Kernel System (GKS) constitute TOP 3.0 APIs, which, in turn, provide the functional interface between graphics applications and OSI Layer 7.

CGM and GKS are part of the TOP 3.0 computer graphics standards that were selected from those developed by ISO Technical Committee 97/Subcommittee 21/Working Group 2 (the formal document references are listed at the end of this article). CGM provides an Application Profile (AP, a TOP 3.0-specific OSI protocol stack) and a set of Device-Independent Graphics Services (DIGS, which provide graphics-device transparencies) for 2-D graphics (3-D ISO DIGS standards are currently under development). GKS defines a language-independent API to provide interoperability between graphic applications and hardware.

The TOP 3.0 architecture internally structures an API as the interface between the graphics application process and computer graphics services. Figure 3 shows the TOP Graphics Reference Model. APIs are implemented to provide the services of the Device-Independent Graphics Layer (DIGL, part of the API, not part of Layer 7).

The DIGL provides services that are common to all graphics architectures and hardware, and its use makes device dependencies transparent to the user. GKS specifies the DIGL component of the TOP 3.0 API.

TOP 3.0 achieves this graphics-device independence within the API by segregating all of the device-dependent graphics functionality into a separate API component, architecturally subordinate to the DIGL, called the Device-Dependent Graphics Layer (DDGL). The service provided by DDGL enables the DIGS to support specific graphics devices and workstations. The DIGL-DDGL interface is called the device interface and is specified as CGM Interchange Format within TOP 3.0.

CGM provides a picture-description metafile that contains, in device- and installation-independent form, the pictorial description of information represented by the graphics functions invoked through the API. In essence, the CGM stores output primitives and attributes. These compose the picture and can be used for archiving and transferring picture-description information. CGM is graphics-device independent and is created by a CGM generator residing at the device-driver level.

Note in Figure 3 that GKS is interfaced to CGM-related device drivers. A CGM generator residing at the device-driver level is invoked by the application-callable layer (that is, the DIGL). Correspondingly, CGM may be invoked by an application-callable, device-independent graphics system, such as GKS.

TOP 3.0 selects the Initial Graphics Exchange Specification (IGES) Version 3.0 (as shown in Figure 2) to enable exchanges of product-definition data (a subset of engineering-drawing applications that align with the U. S. Government's Computer-Aided Acquisition Logistics Support program). In essence, IGES exchanges human-interpretable product data. This data is defined by IGES as entities such as point, line, circular arc, and so on.

Product Data Exchange Specification/Standard for the Exchange of Product Model Data (PDES/STEP) is a follow-on activity to the NBS IGES effort. The PDES/STEP effort is intended to develop a richer set of data descriptions than those provided by IGES for a range of application areas. These areas include mechanical, electrical, and electronic products, finite-element modeling, piping and plant design, architecture, engineering, and construction, as well as in the preparation of technical publications.

PDES/STEP Version 1.0 is anticipated to become an ISO standard. The current direction is to incorporate PDES/STEP into the TOP specification when it is sufficiently mature but also to retain IGES.

TOP 3.0 defines Office Document Architecture/Office Document Interchange Format (ODA/ODIF) for formatting and interchanging compound electronic documents. These documents may be composed of multiple content types, such as character-, geometric-, and raster-graphics data. TOP 3.0's selection and recommendation of ODA/ODIF is formally characterized as the TOP Document Architecture (DA) Application Profile (AP). Specifically, the TOP DA AP defines an implementation of ODA/ODIF that supports the interchange of revisable- or formatted-form compound documents. The graphics objects within these documents are specified according to the TOP Computer

2. TOP 3.0. The successor to TOP 1.0 provides a rich set of application services, including electronic mail, remote file access, remote terminal access, network directory, and network management. These application services are provided to assist in graphical application data interchange. TOP 3.0 also defines a standard application program interface.

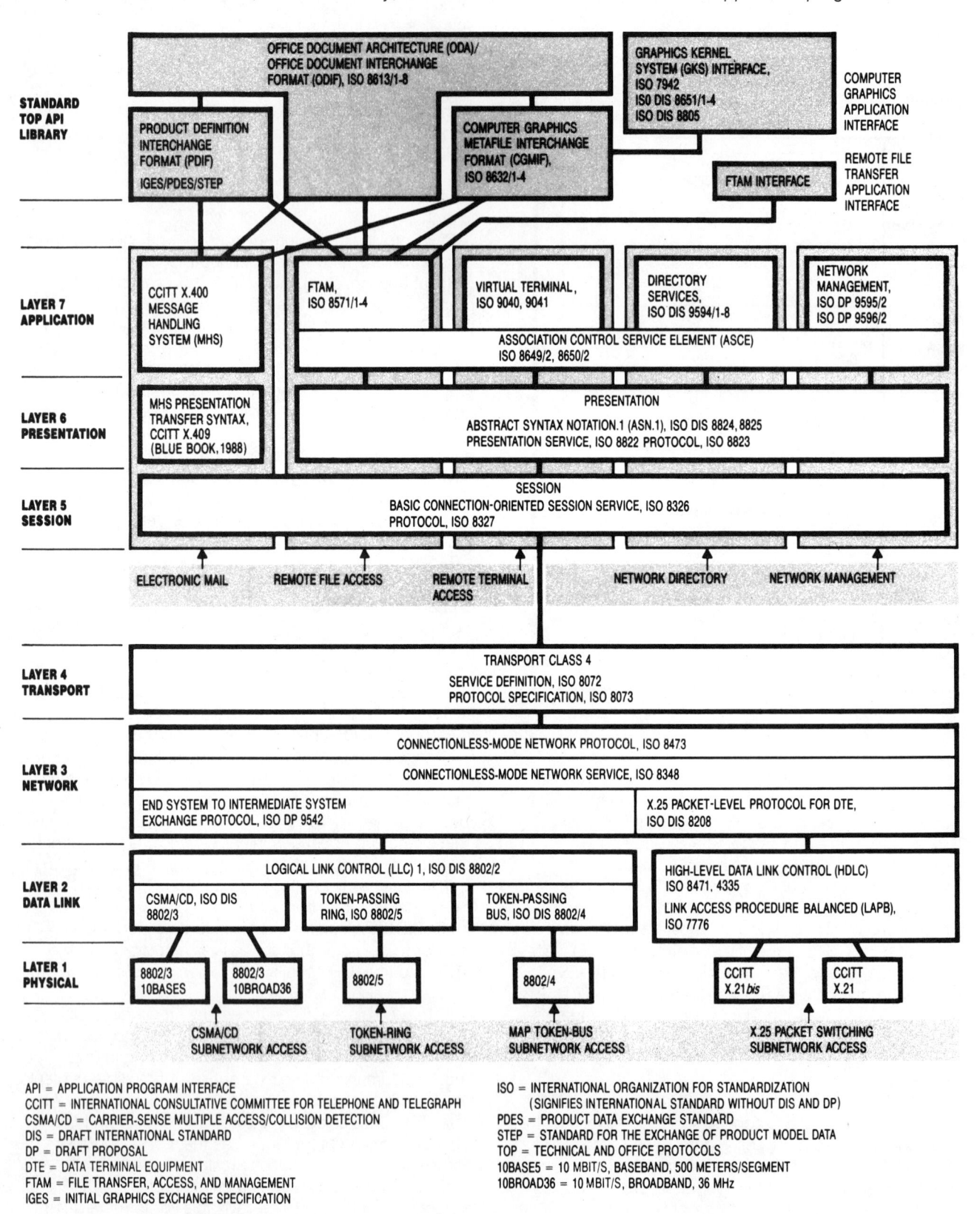

STANDARD TOP API LIBRARY	OFFICE DOCUMENT ARCHITECTURE (ODA)/OFFICE DOCUMENT INTERCHANGE FORMAT (ODIF), ISO 8613/1-8; PRODUCT DEFINITION INTERCHANGE FORMAT (PDIF) IGES/PDES/STEP; COMPUTER GRAPHICS METAFILE INTERCHANGE FORMAT (CGMIF), ISO 8632/1-4; GRAPHICS KERNEL SYSTEM (GKS) INTERFACE, ISO 7942, ISO DIS 8651/1-4, ISO DIS 8805 — COMPUTER GRAPHICS APPLICATION INTERFACE; FTAM INTERFACE — REMOTE FILE TRANSFER APPLICATION INTERFACE
LAYER 7 APPLICATION	CCITT X.400 MESSAGE HANDLING SYSTEM (MHS); FTAM, ISO 8571/1-4; VIRTUAL TERMINAL, ISO 9040, 9041; DIRECTORY SERVICES, ISO DIS 9594/1-8; NETWORK MANAGEMENT, ISO DP 9595/2, ISO DP 9596/2; ASSOCIATION CONTROL SERVICE ELEMENT (ASCE) ISO 8649/2, 8650/2
LAYER 6 PRESENTATION	MHS PRESENTATION TRANSFER SYNTAX, CCITT X.409 (BLUE BOOK, 1988); PRESENTATION — ABSTRACT SYNTAX NOTATION.1 (ASN.1), ISO DIS 8824, 8825; PRESENTATION SERVICE, ISO 8822 PROTOCOL, ISO 8823
LAYER 5 SESSION	SESSION — BASIC CONNECTION-ORIENTED SESSION SERVICE, ISO 8326; PROTOCOL, ISO 8327
	ELECTRONIC MAIL — REMOTE FILE ACCESS — REMOTE TERMINAL ACCESS — NETWORK DIRECTORY — NETWORK MANAGEMENT
LAYER 4 TRANSPORT	TRANSPORT CLASS 4; SERVICE DEFINITION, ISO 8072; PROTOCOL SPECIFICATION, ISO 8073
LAYER 3 NETWORK	CONNECTIONLESS-MODE NETWORK PROTOCOL, ISO 8473; CONNECTIONLESS-MODE NETWORK SERVICE, ISO 8348; END SYSTEM TO INTERMEDIATE SYSTEM EXCHANGE PROTOCOL, ISO DP 9542; X.25 PACKET-LEVEL PROTOCOL FOR DTE, ISO DIS 8208
LAYER 2 DATA LINK	LOGICAL LINK CONTROL (LLC) 1, ISO DIS 8802/2; CSMA/CD, ISO DIS 8802/3; TOKEN-PASSING RING, ISO 8802/5; TOKEN-PASSING BUS, ISO DIS 8802/4; HIGH-LEVEL DATA LINK CONTROL (HDLC) ISO 8471, 4335; LINK ACCESS PROCEDURE BALANCED (LAPB), ISO 7776
LATER 1 PHYSICAL	8802/3 10BASE5; 8802/3 10BROAD36; 8802/5; 8802/4; CCITT X.21bis; CCITT X.21
	CSMA/CD SUBNETWORK ACCESS — TOKEN-RING SUBNETWORK ACCESS — MAP TOKEN-BUS SUBNETWORK ACCESS — X.25 PACKET SWITCHING SUBNETWORK ACCESS

API = APPLICATION PROGRAM INTERFACE
CCITT = INTERNATIONAL CONSULTATIVE COMMITTEE FOR TELEPHONE AND TELEGRAPH
CSMA/CD = CARRIER-SENSE MULTIPLE ACCESS/COLLISION DETECTION
DIS = DRAFT INTERNATIONAL STANDARD
DP = DRAFT PROPOSAL
DTE = DATA TERMINAL EQUIPMENT
FTAM = FILE TRANSFER, ACCESS, AND MANAGEMENT
IGES = INITIAL GRAPHICS EXCHANGE SPECIFICATION
ISO = INTERNATIONAL ORGANIZATION FOR STANDARDIZATION (SIGNIFIES INTERNATIONAL STANDARD WITHOUT DIS AND DP)
PDES = PRODUCT DATA EXCHANGE STANDARD
STEP = STANDARD FOR THE EXCHANGE OF PRODUCT MODEL DATA
TOP = TECHNICAL AND OFFICE PROTOCOLS
10BASE5 = 10 MBIT/S, BASEBAND, 500 METERS/SEGMENT
10BROAD36 = 10 MBIT/S, BROADBAND, 36 MHz

3. Graphics model. *TOP 3.0 achieves graphics-device independence within the API by segregating device-dependent graphics functionality into a separate API component.*

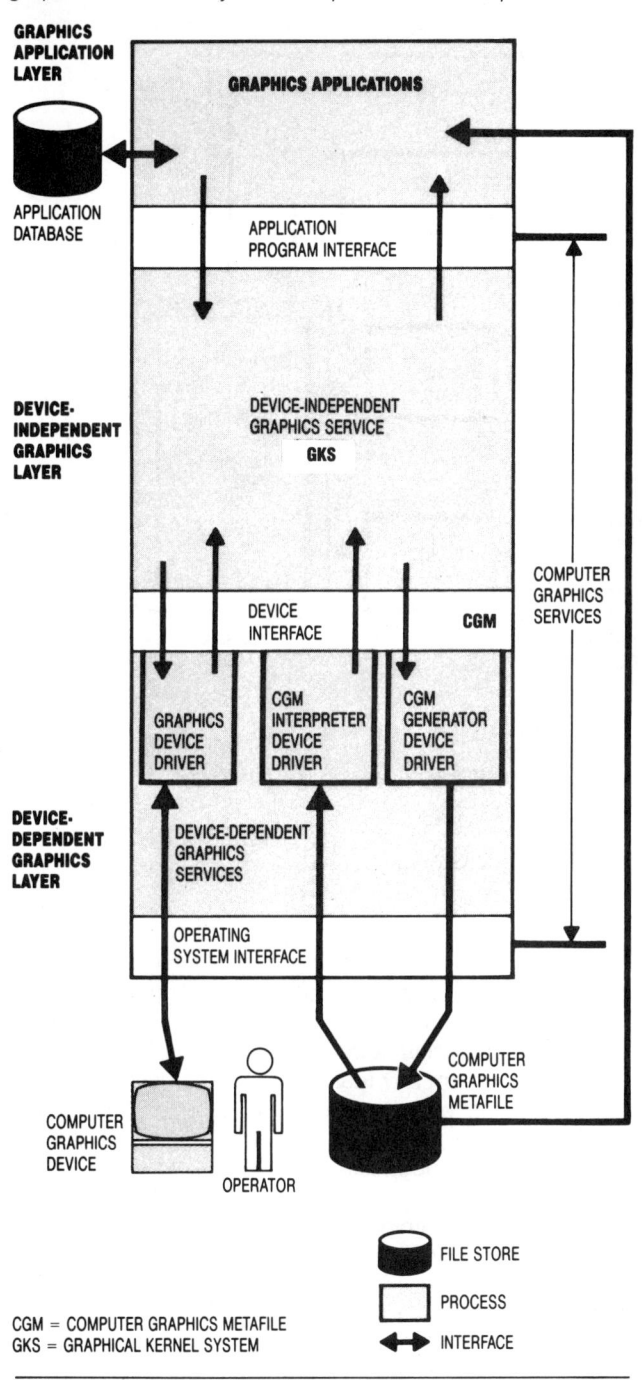

presentation styles. The TOP DA AP specifies a combination of conformance levels for document architecture, document profile, and interchange format.

Application layer

Note in Figure 2 that all members of the Standard TOP API Library (such as CGM, GKS, IGES, and ODA/ODIF) are architecturally independent of the particular electronic delivery service invoked. Therefore, within the TOP 3.0 architecture, files created by the API Library members can be transferred through Data Transfer Services (a TOP term for application-layer specifications). These services include Message Handling System (MHS), FTAM, Virtual Terminal, Directory Services, Network Management, and Association Control Service Element, as described below.

■ **MHS.** The TOP 3.0 MHS AP is based upon CCITT Recommendation X.400 (Red Book, 1984), the X.400 Implementers' Guide, and the NBS Implementation Agreements for OSI Protocols, which came out of the NBS Implementers' Workshops. The primary objectives of X.400-series MHS is to enable messages of any content to be transparently encapsulated within a standard envelope for subsequent delivery to an electronic mail destination. These messages, which can include text, graphics, facsimile, voice, and other data types, may be transferred within and among public and private networks.

TOP 3.0 endorses simple lines of ASCII text as the MHS envelope contents, in conformance with the NBS Implementation Agreements. TOP will likely accommodate future editions of the NBS implementation agreements, including binary, facsimile, and other document types. Also, TOP working groups have determined that there is a requirement to carry other data interchange formats, such as ODA, CGM, and IGES.

The TOP 3.0 MHS AP specifies the use of Private Management Domain (PRMD)-to-PRMD communications to connect proprietary mail systems. PRMDs are defined as privately owned and operated Message Handling Systems grouped together for purposes of common administration. Figure 5 shows that a given Message Transfer Agent (MTA) may support one or more User Agents (UAs) on behalf of users. Moreover, MTAs can communicate directly within the same PRMD and between different PRMDs.

Vendor implementations that support PRMD-to-PRMD services accessed by the TOP profile provide two X.400-defined protocols: P1 Message Transfer Protocol (X.411) for exchanges between MTA entities, and P2 Interpersonal Messaging Protocol (X.420) for exchanges between UA entities. While direct PRMD-to-PRMD connections support the exchange of messages without the need for an Administrative Management Domain (ADMD, a registered public provider of X.400-based Message Handling Services), the TOP MHS AP also specifies support for message interfaces from PRMD to ADMD and from ADMD to ADMD.

In general, a product that supports the TOP MHS AP can connect in an MHS messaging environment to any other vendor's product as long as they both define gateways conforming to the TOP MHS Application Profile. Although

Graphics Metaphile Application Profile (TOP CGM AP).

The TOP 3.0 architecture specifies a TOP Document Processing Reference Model (Fig. 4). In this model, a document progresses through three processing phases: editing, which includes both content and logical structure editing; layout, which defines a page-oriented, physical-layout organization; and imaging, which displays a specific layout structure with associated formatted content and

4. Document processing. The TOP Document Architecture Application Profile supports the interchange of revisable- or formatted-form compound documents using the model shown. In this model, a document progresses through editing, layout, and imaging stages on its way to final printout or display.

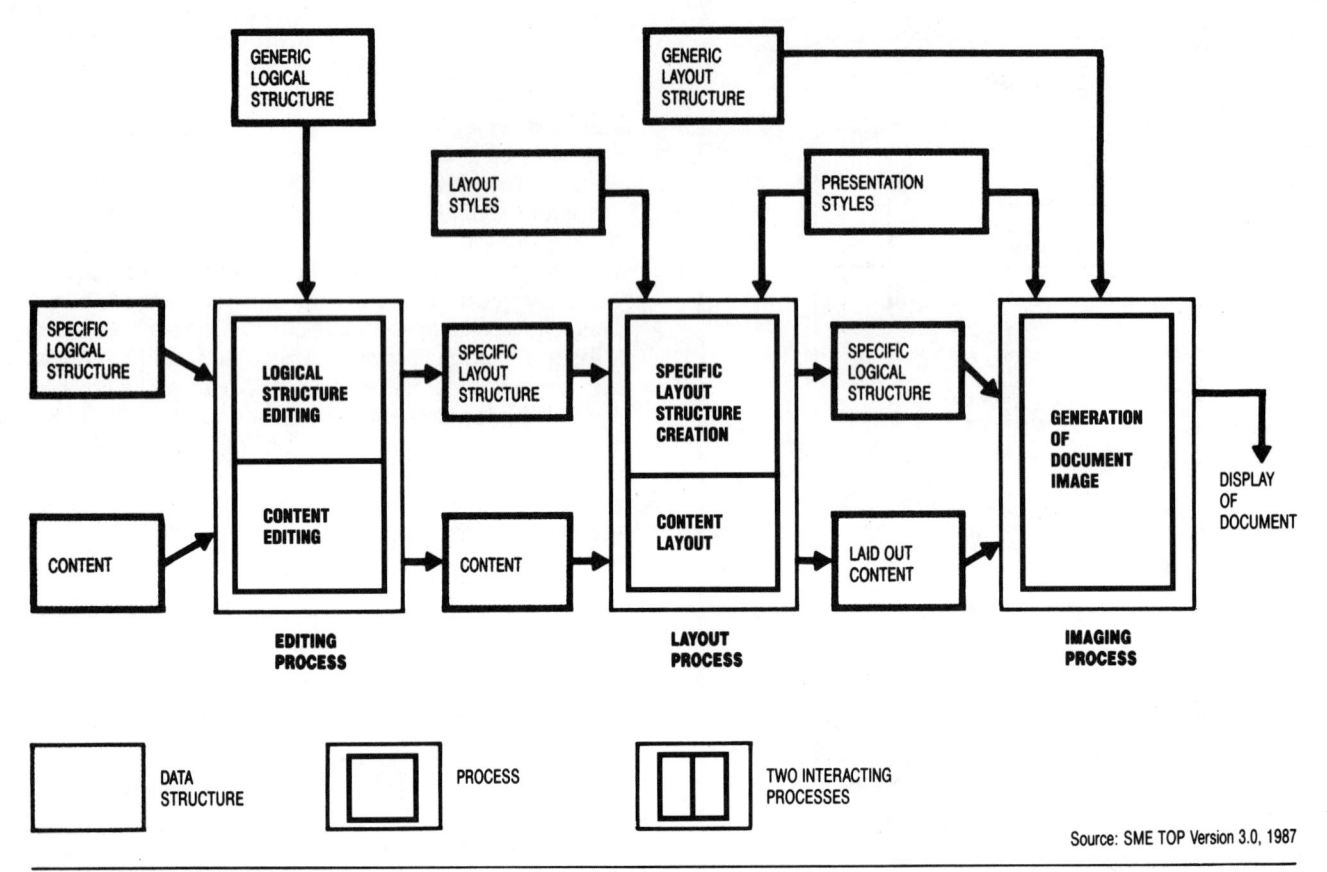

Source: SME TOP Version 3.0, 1987

the AP conforms to CCITT X.400-1984, that CCITT recommendation uses different presentation and session protocols than does the ISO endorsement, Message-Oriented Text Interchange Standard. X.400 1988 Blue Book will align with Abstract Syntax Notation One (ASN.1) and Session Protocol.

At present, TOP 3.0 has no choice but to follow the CCITT's 1984 agreements. Note that PRMD was not specified in the CCITT 1984 recommendations and, furthermore, that X.400 1984 specifies Transport Class 0. By contrast, Class 4 was selected by NBS and Government OSI Profile for PRMD to provide a highly reliable transport service over a connectionless internet protocol.

■ **FTAM.** The TOP FTAM AP specifies a set of services to transfer information between application processes and filestores (which consist of file-storage and file-service mechanisms). It also supports the ability to create, delete, and transfer entire binary or text files. Moreover, it enables a process to read or change file attributes, erase file contents, locate specific records, and read or write file records.

The ISO FTAM specifications are concerned with the manipulation of identifiable bodies of information that can be treated as files and stored within open systems or passed between OSI application processes. The FTAM File Service definitions and protocol specifications are expressed in the context of a virtual filestore (a file-mapping structure commonly referred to by open systems), as distinguished from any real, local filestore definitions that exist within real end systems. Each real end system that behaves as an open system must map the virtual filestore descriptions and operations into real, local file-management functions.

The ISO FTAM protocol is asymmetrical. It defines two types of peer entities: an initiator, which requests most activities as the master of an interaction, and a responder, which responds to initiator requests as slave. The protocol is connection- and transaction-oriented in that an initiator must establish a connection with a responder prior to request/response flows. Figure 6 illustrates the FTAM initiator/responder relationship between two TOP end systems.

The TOP FTAM AP includes nine document types. These identify file characteristics such as structure, syntax, content semantics, and permissible operations.

Whereas FTAM defines a virtual filestore used to describe the FTAM services and the mapping of these services onto the local filestore, neither the FTAM standard nor the NBS implementation agreements specify an interface to FTAM services. Note in Figure 2 that a TOP 3.0

FTAM interface is specified, shown above Layer 7. This programming interface is intended to provide application program portability.

■ The TOP Virtual Terminal (VT) Application Profile, based on the ISO VT-Basic Class, maps a specific device into a canonical form called a virtual terminal and then back into the device format expected by the application. It supports the interactive transfer and manipulation of graphic elements. TOP VT AP provides an object-based, remote, interactive, connection-based terminal service between heterogeneous TOP end systems. These are shown in Figure 7 as local terminal and remote host environments.

The TOP VT service uses an object-based model. In it, the virtual terminal is represented by an object that has a structure characterized as a 1-, 2-, or 3-dimensional array residing in the Application Entity (AE, such as instances of FTAM, VT, Network Management, or Directory Services).

Each array element may contain a single character. Attributes are defined for each element, such as color of screen foreground/background, emphasis, character repertoire, and font. The TOP VT service is interactive, and users can read or update the VT object by simulating the capabilities of a character-mode terminal.

■ TOP Directory Services support high-level references to network objects and provide information to users about the systems and services that can be invoked on a network. The relationship between directory references and directory objects is based on directory entries, which consist of sets of attributes.

The directory environment includes a collection of directory entry information in the form of a directory information base (DIB). The DIB can be accessed remotely through a directory service AP known as a Directory Service Agent (DSA). Each user application interacts with a DSA through

5. Message handling. *The TOP MHS Application Profile enables interconnection of proprietary mail systems that are privately owned and operated for purposes of common administration under private management domains. These domains support the exchange of messages without the need for an administrative management domain.*

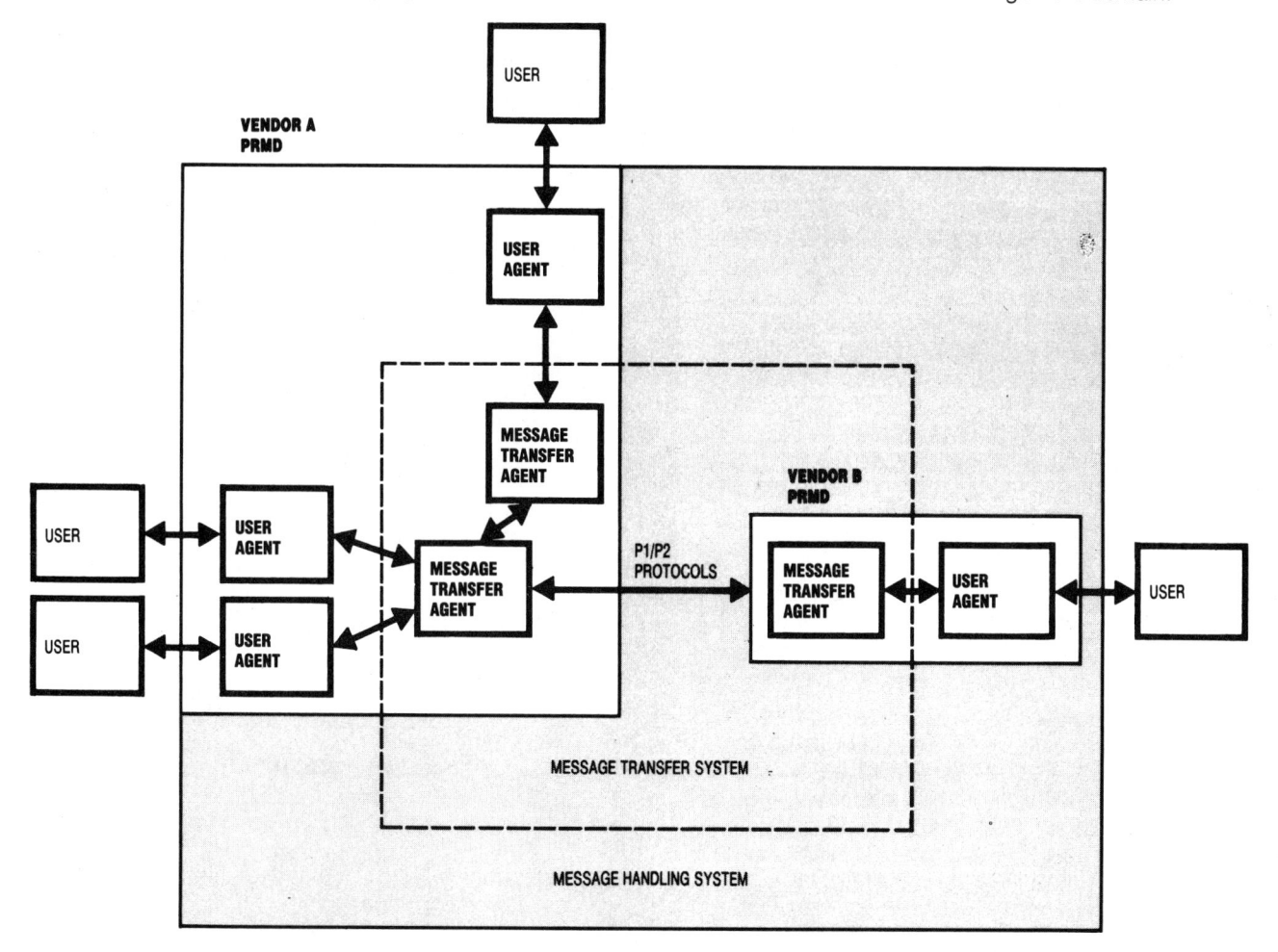

P1 = MESSAGE TRANSFER PROTOCOL (X.411)
P2 = INTERPERSONAL MESSAGING PROTOCOL (X.420)
PRMD = PRIVATE MANAGEMENT DOMAIN

Source: SME TOP Version 3.0, 1987

6. Initiator/responder. *TOP's File Transfer, Access, and Management Application Profile uses an asymmetrical protocol, which defines two types of peer entities. The FTAM initiator requests transactions and the responder reacts to these requests. This protocol is connection- and transaction-oriented.*

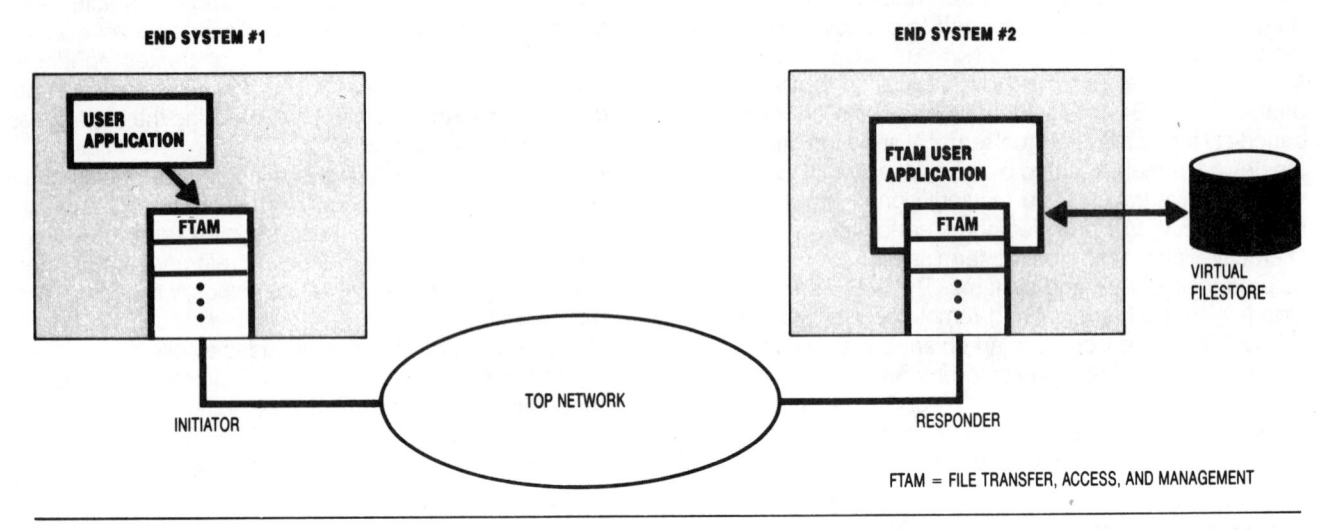

---

a Directory User Agent (DUA) local to the application.

The ISO directory standards specify two principal directory protocols. Directory Access Protocol defines the exchange of requests and responses between a DUA and a DSA. Directory Systems Protocol defines the exchange of requests and responses directly between two DSAs.

The DIB internally classifies directory objects according to a Directory Information Tree that exploits the hierarchical relationship commonly found among objects (for example, a person works for a department that is part of an enterprise that is headquartered in a specific country).

There are three basic directory services. A name-to-attribute binding, analogous to a "white pages" directory, binds a name to a related piece of information. A name-to-list-of-names binding, analogous to a distribution list, is used in applications such as electronic mail and routing tables. Finally, an attribute-to-set-of-names binding, analogous to a "yellow pages" directory, lists the names of objects having a given attribute.

■ TOP Network Management defines information services, Common Management Information Services (CMIS), and Common Management Information Protocol (CMIP). The overall framework supports three levels of TOP resource management: protocol management, which manages all resources associated within a specific TOP protocol layer; layer management, which provides insight into layer conditions (for example, such states as waiting or transmitting); and system management, which is supported by System Management Application Entities (SMAEs).

TOP CMIS services enable SMAEs to intercommunicate and provide TOP/OSI protocol-stack state status and management, including configuration, performance, and fault management. Configuration management involves addition, deletion, and modification of network resources. Performance management includes network measurement, analysis and adjustment procedures to determine congestion, message delay, service reliability, routing throughput, bridge throughput, and link utilization. Fault management includes fault detection, notification, isolation, and recovery.

The TOP CMIP provides the request/response service between an initiator SMAE in one TOP end system and a responder SMAE in another, which is executing management activities on behalf of the initiator SMAE. The protocol also supports an unsolicited event-reporting service between an event-intiator SMAE and a collector (responder) SMAE, the one responsible for event logging.

■ Association Control Service Element (ACSE) provides services to such application contexts as ISO FTAM and Virtual Terminal Services, TOP Network Management, ISO/CCITT Directory Services, and PRIVATE (a default value). Note in Figure 2 that ACSE is not required for MHS.

The TOP ACSE protocol provides standard services for establishing and terminating application associations. It enables TOP/OSI applications to intercommunicate common parameters, such as titles, addresses, and application context, during an application association.

Below the top of TOP

To summarize, the major building blocks provided by TOP's upper layers are as follows:
■ Electronic mail, which uses MHS services from GKS, CGM, IGES, or ODA/ODIF APIs.
■ Remote file access, which uses FTAM services from GKS, CGM, IGES, or ODA/ODIF APIs.
■ Remote terminal access, which makes use of the Virtual Terminal ACSE-ASN.1-Session Protocol stack.
■ Network directory, which makes use of the Directory Services ACSE-ASN.1-Session Protocol stack.
■ Network management, which is enabled through the Network Management ACSE-ASN.1-Session Protocol stack.

At the layers beneath the application layer, TOP 3.0 specifies OSI protocols, just as it does at the application layer. Generally, Layer 1 through Layer 6 specifications are at the International Standards stage, with the exception of some network and data-link protocols, which are still draft international standards. The layers are as follows:

- Presentation layer. The TOP ACSE protocol makes use of the Kernel functional unit (defined below) of the Presentation Service to pass information, in the form of ACSE Application Protocol Data Units (APDUs), between peer AEs. Note in Figure 2 that the TOP 3.0 architecture uses Abstract Syntax Notation One (ASN.1) as the presentation-layer facility to specify a notation for abstract syntax definition.

TOP 3.0 selects ASN.1 because the nature of user-data parameters within the presentation layer can change. That is, the application layer may require that the presentation layer carry the value of complex data types, including character strings from a variety of character sets. These values reside within multiple character sets.

ASN.1 is a generic character notation that negotiates which transfer syntaxes are to be used. Data types referenced with ASN.1 include, for example, Boolean, integer, bit string, octet string, null, sequence, tagged, any, and character string. Note also, in Figure 2, that the set of ASN.1 documents is technically aligned with the relevant parts of CCITT Recommendation X.409 (Blue Book, 1988), shown as the MHS Presentation Transfer Syntax for Layer 7 MHS.

- Session layer. The purpose of the session layer is to provide dialogue-management services to the presentation and application layers and, in so doing, to provide peer-session users with the means to exchange data in an organized and synchronized way. The TOP 3.0 session agreements support the use of the ISO Basic Connection Oriented Session Service and Session Protocol. These agreements define the session layer into three phases, the session-connection-establishment, data-transfer, and session-connection-release phases.

TOP Session Services for FTAM are enabled through the use of two types of functional units. Kernel functional units support the basic session services required to establish a session connection, transfer normal data, and release the connection. The other type, the duplex functional unit, supports two-way session transfer services. In addition, VT and MHS use the Basic Activity Subset and the Basic Synchronized Subset of the ISO Session Protocol.

- Transport layer. TOP 3.0 requires ISO Transport Class 4 to support the connection-oriented transport service, which provides flow control and the ability to multiplex user transmissions to the network. The ISO Transport Class 4 connection-oriented transport service is required within the TOP architecture because Connectionless Network Service is selected at Layer 3.

- Network layer. The network layer, in essence, makes transparent to the transport layer the way in which supporting communications resources are utilized to accomplish data transfer. A primary TOP 3.0 network-layer requirement is to specify protocols that support TOP end systems attached to LANs. The TOP 3.0 LANs, specified in Layers 1 and 2, are carrier-sense multiple access/collision detection (CSMA/CD), token-passing ring, and MAP token-passing bus. It is recognized, however, that not all TOP end systems that in an open technical and administrative office environment will be LAN-attached but rather may interface through wide area networks (WANs). To this end, TOP 3.0 additionally provides for the X.25 packet-level protocol at the network layer where TOP end or intermediate systems are connected to packet-switching networks.

TOP 3.0 specifies three network-routing levels: the enterprise portion, which uniquely identifies the highest level group or organization (for example, corporation, governmental agency, or university); the intermediate system (IS)-to-IS portion; and the end system (ES)-to-IS portion. When ES-to-IS routing is provided on LAN subnetworks, TOP 3.0 dictates the use of the ISO ES-IS Exchange Protocol.

- Data link layer. The OSI data link layer (Layer 2) performs frame formatting, error checking, addressing, and link management to ensure accurate data transmission over links between nodes. The TOP data link layer distinguishes between LAN access, which provides Logical Link Control (LLC) over CSMA/CD, token-passing ring, and token-passing bus LANs, and WAN access, which provides for High-Level Data Link Control (HDLC) and Link Access Procedure-Balanced (LAPB) if the network layer is defined by the X.25 Packet Level Protocol.

The data link layer for LAN use is subdivided into Link Layer Control and Media Access Control. Figure 8 presents the TOP 3.0 model for the data link and physical layers and indicates the Link Layer Control and Media Access Control (MAC) divisions of the data link layer for LANs. The figure also indicates that the LAN Link Layer Control sublayer is specified by IEEE 802.2.

The primary TOP 3.0 LAN is specified by IEEE 802.3 as

Where TOP comes from

TOP 3.0 consists of numerous technical specifications and agreements that derive from the collective work of various TOP Users Group technical subcommittees. Although these subcommittees have specialized in various functional areas, each of them collects, elaborates, and studies technical and administrative office requirements and specifies protocol solutions that satisfy those requirements. In addition, each committee works closely with standards organizations, such as ISO, CCITT, IEEE, and the National Bureau of Standards OSI Implementers' Workshops. All solutions proposed by TOP technical subcommittees are subject to public review and approval before being incorporated into the TOP specification.

One major positive result of that specification is to promote the design and testing of TOP products needed by the significant market segment represented by the TOP Users Group. TOP 3.0 also functions as a procurement template through which users can effectively measure vendors' responses to requests for proposals, quotations, and information. Not least, TOP serves to educate users.

7. Virtual Terminal. *The VT Application Profile maps a specific device into a virtual terminal and then back into the format expected by the application.*

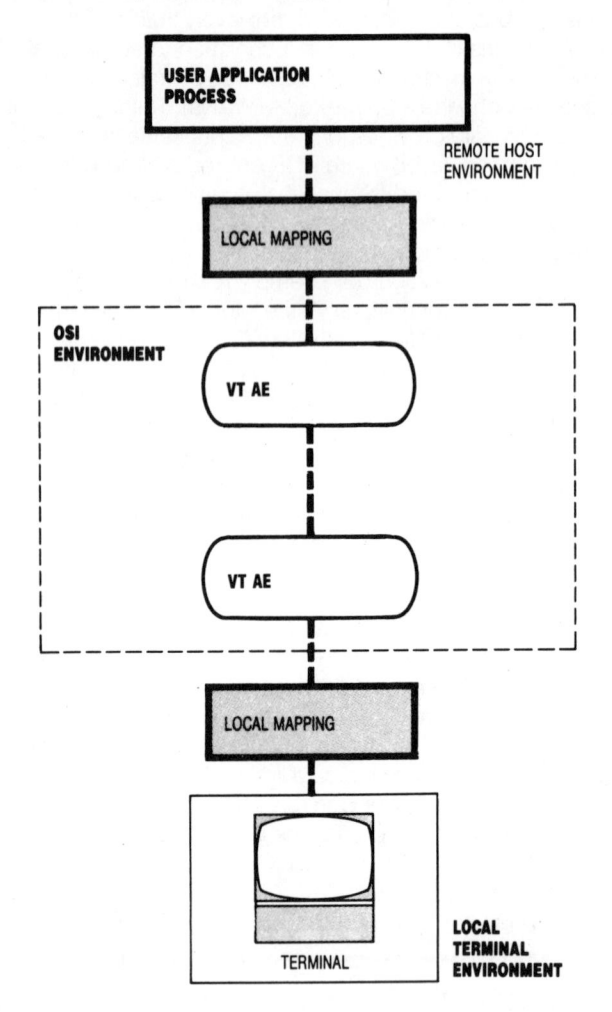

AE = APPLICATION ENTITY
OSI = OPEN SYSTEMS INTERCONNECTION
VT = VIRTUAL TERMINAL

CSMA/CD, which uses a LAN contention access method. CCITT X.25 LAPB (1984 Red Book) specifies access to public switched data networks (PSDNs). However, the IEEE 802.5 ring and IEEE 802.4 bus LANs, both of which use token passing as a deterministic access method, are also named as alternate TOP 3.0 LANs.

The reasoning behind TOP's selection of 802.3 CSMA/CD as the initial MAC sublayer protocol is the ease of migration from Ethernet Version 2.0 LANs to IEEE 802.3 LANs and the fact that a wide base of components already exists for Ethernet LANs. Additionally, CSMA/CD has been shown to be quite effective in handling a broad variety of distributed office applications, including document and graphics interchange. TOP 3.0 also supports the token-passing IEEE 802.5 and 802.4 protocols because the token-passing subnetwork access method provides a reasonably predictable, collision-free allocation of network capacity with stable throughput and response times under varying network loads and utilizations.

■ Physical layer. The primary TOP 3.0 physical-layer protocol is 802.3 baseband CSMA/CD, called 10Base5 (for 10 Mbit/s, baseband, 500 meters per segment), which utilizes the CSMA/CD media access protocol over a shielded coaxial cable. TOP 3.0 selected 10Base5 as its primary physical layer because of the large installed base of 10Base5 cabling from widespread use of Ethernet LANs, because of the graceful migration possible from Ethernet Version 2.0 to 802.3 CSMA/CD (to the extent that the two can coexist on the same cable), and because 10Base5 cable is a proven technology that provides reliable, high-speed data transfer.

One alternate TOP 3.0 physical-layer selection (refer to Figure 8) is 802.3 broadband CSMA/CD (10Broad36), which can link back to 10Base5 and provide multiple-vendor-attachment support within a broadband cable technology using CSMA/CD as the MAC protocol. Another is 802.5 token-passing ring on shielded twisted-pair cable, which links to 10Base5 subnetworks and supports multiple-vendor attachments with a performance-oriented office automation protocol under high sustained network loading conditions. A third is MAP 802.4 token-passing bus on coaxial cable (not recommended as a TOP subnetwork), which links to 10Base5 subnetworks and supports multiple-vendor attachments over both carrier band and broadband options under the MAP Version 3.0 specification. Yet another option is the PSDN permanent-circuit interface specified by CCITT X.21*bis*, which incorporates EIA-232-D, CCITT Recommendation V.35, and EIA-449.

Directions

Graphical Kernel System and Computer Graphics Metafile were the two graphics data interchange standards selected for TOP 3.0. Several graphics standards are under consideration for incorporation into TOP, as follows:

■ Computer Graphics Interface defines an interactive virtual-device interface between the DIGL and the DDGL. Recall that TOP 3.0 specifies the use of the CGM standard to provide 2-D picture interchange between end systems and the use of the GKS standard to provide an API and DIGS for 2-D graphics. The elaboration of CGI within the TOP specification will serve to standardize the API mechanism that enables end-to-end resource and data sharing between heterogeneous graphics applications.

■ Three-dimensional graphical functionality will be addressed by GKS-3D.

■ Whereas IGES 3.0, as described earlier, provides a subset of a full product data-exchange capability, future versions of TOP will likely incorporate the capabilities of PDES/STEP, to be used in conjunction with IGES 3.0. Since not all CAD/CAM setups will generate complete data, and since received product model data will continue to be interpreted by a person either as a display or as a generated plot, it is likely that IGES files will continue to be transferred to maintain data-exchange compatibility between current and anticipated environments.

■ Initial Graphics Exchange Specification 3.0 will eventually

include binary and compressed ASCII file formats.
- Electronic Design Interchange Format (EDIF) will likely be included to provide for transfer of integrated circuit information.

Another major TOP data interchange direction is referred to as Transaction Processing. The TOP Distributed Transaction Processing Subcommittee is defining a transaction-processing model service and protocol consistent with the efforts of ISO and ANSI. TOP requirements for future incorporation of a transaction-processing protocol are that the protocol must be consistent with the ISO OSI application-layer architecture, must support the processing of subtransactions (transactions that are initiated by other transactions), must use the services of ACSE and Commitment, Concurrency, and Recovery (CCR) primitives in communicating with lower layers, and must support the multiple use and reuse of an application association.

At the data link layer, high-speed protocol standards that use optical fiber backbones, such as the ANSI Fiber Distributed Data Interface (FDDI) and IEEE 802.6 metropolitan area networks, may find their way into the TOP specification. While the FDDI and 802.6 fiber protocols are competing efforts, it is likely either that they will converge or that TOP will select one.

Future physical-layer standards under consideration for incorporation into TOP include IEEE 802.3 1Base5 (1 Mbit/s, baseband, 500 meters per segment), support for token-passing ring (IEEE 802.5) on unshielded twisted-pair wire, protocols that may result from the efforts of the CCITT on Integrated Services Digital Networks, and ANSI FDDI fiber physical interface protocols.

In all these cases, as is true of the existing TOP 3.0 physical-layer protocols such as 802.3, 802.5, and 802.4, IEEE 802.3 CSMA/CD 10Base5 remains the primary TOP subnetwork access protocol. Therefore, all other physical-layer protocols are required to provide internetworking connections back to IEEE 802.3 10Base5.

A near-future release of TOP is expected to consolidate the Transaction Processing service and protocol agreements and to incorporate stabilized text on Directory Services and Network Management. To summarize, the TOP Version 3.0 architecture provides a comprehensive set of application services to enable standardized data interchange and communications over a wide range of LAN and WAN subnetwork environments. It thereby provides a funtionally cost-effective alternative to the proliferation of proprietary networking environments. For these reasons, TOP 3.0 represents a significant step toward fulfilling the accomplishment of the user technical and administrative distributed office user business case. ∎

Thomas J. Routt was recently named a contributing editor to Data Communications *and is president of Network Systems Consulting, a firm that provides worldwide network architecture consulting to Fortune 1,000 corporations concerned with migration to OSI and SNA. Previously, he was manager of Boeing Network Architecture for Boeing Computer Services Company. In this capacity, he managed global network planning, design, and implementation for the Boeing Company. Routt holds an M. B. A. in information systems from Southern Illinois University and a B. S. in environmental science from Western Washington University.*

The author wishes to acknowledge the input of Laurie Bride, Boeing's manager of network architecture .nd program manager of the Boeing TOP program, a .d of her staff. Their organizational assistance, interface coordination to TOP users and vendors, and review of the manuscript proved invaluable to this effort.

Selected documents that form the basis of and are related to TOP 3.0:

General
TOP: Technical and Office Protocols Specification Version 3.0, Society of Manufacturing Engineers, July 1987.
MAP: Manufacturing Automation Protocol Specification 3.0, Society of Manufacturing Engineers, July 1987.
OSI: ISO 7498, CCITT X.200, Information Processing Systems-Open Systems Interconnection-Basic Reference Model.

Application Program Interfaces
GKS: ISO 7942, Information Processing Systems-Computer Graphics-Graphical Kernel System Functional Description.
  —ISO DIS 8651, Information Processing Systems-Computer Graphics-GKS Language Bindings-Part 1: FORTRAN, Part 2: Pascal, Part 3: Ada, Part 4: C.
  —ISO DIS 8805, Information Processing Systems-Computer Graphics-GKS for Three Dimensions (GKS-3D) Functional Description.
CGM: ISO 8632, Information Processing Systems-Computer Graphics-Metafile for the Storage and Transfer of Picture Description Information-Part 1: Functional Specification, Part 2: Character Encoding, Part 3: Binary Encoding, Part 4: Clear Text Encoding.
ODA/ODIF: ISO 8613, Information Processing-Text and Office Systems-Office Document Architecture and Interchange Format-Part 1: General Information, Part 2: Document Structures, Part 3: Document Processing Reference Model, Part 4: Document Profile, Part 5: Office Document Interchange Format, Part 6: Computer Graphics Content Architectures, Part 7: Raster Graphics Content Architectures, Part 8: Geometric Graphics Content Architectures.
IGES: ANS Y14.26 M-1987 (Identical to NBSIR 86-3359, April 1986), Digital Representation for the Communication of Product Definition Data (V3.0).

Application-layer specifications
MHS (Message Handling Systems, Red Book, 1984): CCITT X.400 : System Model-Service Elements.
  —CCITT X.401: Basic Service Elements and Optional User Facilities.
  —CCITT X.408: Encoded Information Type Conversion Rules.
  —CCITT X.409: Presentation Transfer Syntax

8. Lower layers. The primary TOP 3.0 subnetwork is IEEE 802.3 Carrier Sense Multiple Access/Collision Detection 10Base5. However, three other LAN subnetworks may be used (IEEE 802.5 token-passing ring, 802.4 token-passing bus, and 802.3 CSMA/CD 10Broad36), as can a wide area network access method (X.25/X.21/X.21 bis).

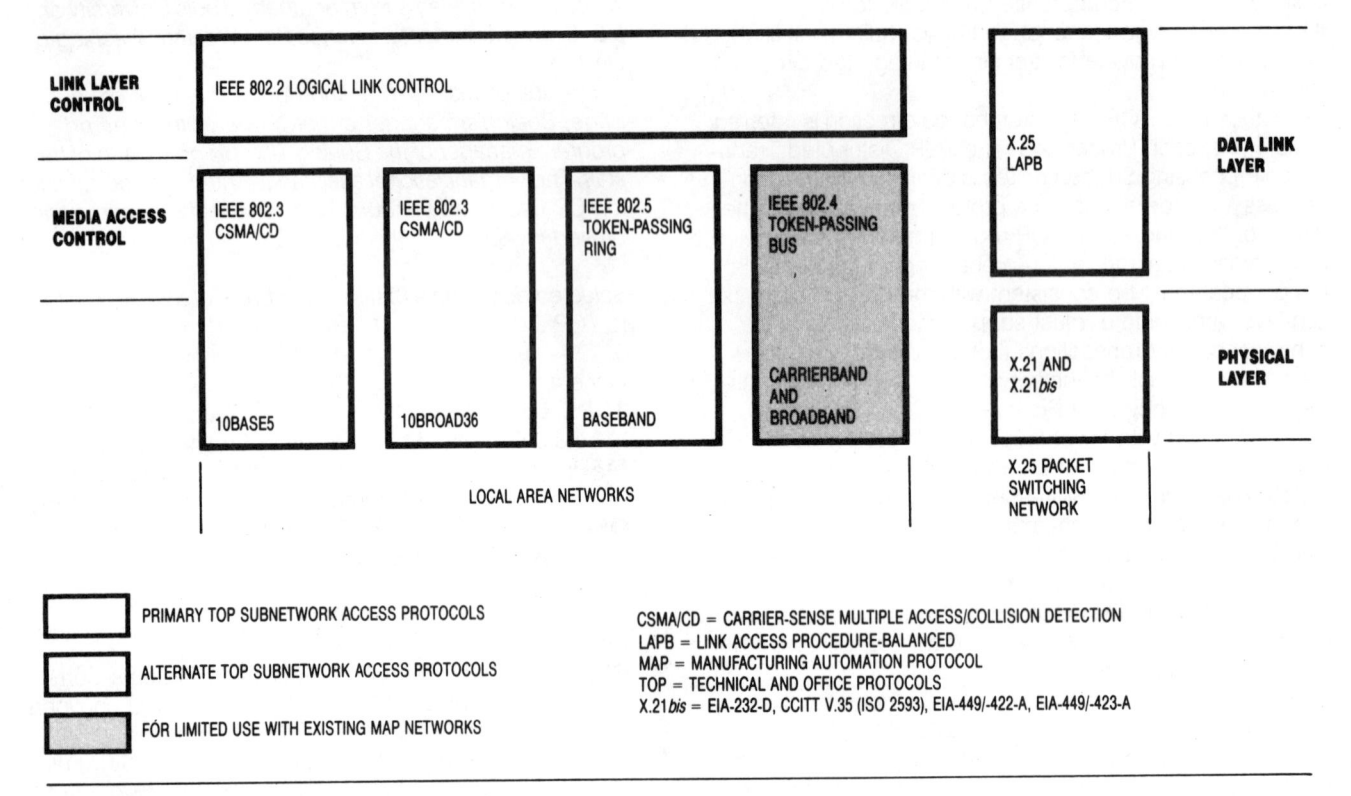

LINK LAYER CONTROL	IEEE 802.2 LOGICAL LINK CONTROL	
MEDIA ACCESS CONTROL	IEEE 802.3 CSMA/CD — 10BASE5 · IEEE 802.3 CSMA/CD — 10BROAD36 · IEEE 802.5 TOKEN-PASSING RING — BASEBAND · IEEE 802.4 TOKEN-PASSING BUS — CARRIERBAND AND BROADBAND	X.25 LAPB · X.21 AND X.21 bis · X.25 PACKET SWITCHING NETWORK
	LOCAL AREA NETWORKS	

Legend:
- □ PRIMARY TOP SUBNETWORK ACCESS PROTOCOLS
- □ ALTERNATE TOP SUBNETWORK ACCESS PROTOCOLS
- ▨ FOR LIMITED USE WITH EXISTING MAP NETWORKS

CSMA/CD = CARRIER-SENSE MULTIPLE ACCESS/COLLISION DETECTION
LAPB = LINK ACCESS PROCEDURE-BALANCED
MAP = MANUFACTURING AUTOMATION PROTOCOL
TOP = TECHNICAL AND OFFICE PROTOCOLS
X.21 bis = EIA-232-D, CCITT V.35 (ISO 2593), EIA-449/-422-A, EIA-449/-423-A

Notation.
—CCITT X.410: Remote Operations and Reliable Transfer Server.
—CCITT X.411: Message Transfer Layer.
—CCITT X.420: Interpersonal Messaging User Agent Layer.

FTAM: ISO 8571, FTAM-Part 1: General Introduction, Part 2: The Virtual Filestore, Part 3: The File Service Definition, Part 4: The File Protocol Specification.

VT: ISO 9040, Information Processing Systems-Open Systems Interconnection-Virtual Terminal Service-Basic Class.
—ISO 9041, Information Processing Systems-Open Systems Interconnection-Virtual Terminal Protocol-Basic Class.

Directory: ISO DIS 9594, Information Processing Systems-Open Systems Interconnection-The Directory, Parts 1-8; the eight ISO Directory Services sections include: (1) Concepts and services, (2) Information framework, (3) Access and system services definition, (4) Procedures for distributed operation, (5) Access and system protocols, (6) Selected attribute types, (7) Selected object classes, and (8) Authentication framework (related to security).

Network Management: ISO DP 9595, Management Information Service Definition, Part 1: Overview, Part 2: Common Management Information Service (CMIS) Definition.
—ISO DP 9596, Management Information Protocol Specification, Part 1: Overview, Part 2: Common Management Information Protocol (CMIP).

ASCE: ISO 8649/2, Information Processing Systems-Open Systems Interconnection-Service Definition for Common Application Service Elements-Part 2: Association Control.
—ISO 8650/2, Information Processing Systems-Open Systems Interconnection-Protocol Specification for Common Application Service Elements-Part 2: Association Control.

Sub-application layers

ASN.1: ISO DIS 8824, Information Processing Systems-Open Systems Interconnection-Specification of Abstract Syntax Notation One.
—ISO DIS 8825, Information Processing Systems-Open Systems Interconnection-Specification of Basic Encoding Rules for Abstract Syntax Notation One.

Presentation: ISO 8822, Information Processing Systems-Open Systems Interconnection-Connection Oriented Presentation Service Definition.
—ISO 8823, Information Processing Systems-Open Systems Interconnection-Connection Oriented Presentation Protocol Definition.

Session: ISO 8326, Information Processing Systems-Open Systems Interconnection-Basic Connection Oriented Session Service Definition.

—ISO 8327, Information Processing Systems-Open Systems Interconnection-Basic Connection Oriented Session Protocol Definition.

Transport: ISO 8072, Information Processing Systems-Open Systems Interconnection-Transport Service Definition.

—ISO 8073, Information Processing Systems-Open Systems Interconnection-Connection Oriented Transport Protocol Specification.

Network: ISO 8473 (Final Text), Information Processing Systems-Data Communications-Protocol for Providing the Connectionless-Mode Network Service.

—ISO DP 9542, Information Processing Systems-Data Communications-End System to Intermediate System Exchange Protocol for Use in Conjunction with ISO 8473.

—ISO 8348 DAD2, Information Processing Systems-Data Communications-Network Service Definition Addendum 2-Covering Network Layer Addressing.

—ISO DIS 8208, Information Processing Systems-X.25 Packet Level Protocol for Data Terminal Equipment.

—CCITT Recommendation X.25 (Red Book, 1984), Interface between Data Terminal Equipment and Data Circuit-Terminating Equipment for Terminals Operating in the Packet Mode and Connected to Public Data Networks by Dedicated Circuit.

Data Link: ISO 8802/2, IEEE 802.2, Local Area Networks-Logical Link Control.

—ISO 8802/3, IEEE 802.3, Local Area Networks-Carrier Sense Multiple Access with Collision Detection (CSMA/CD) Access Method and Physical Layer Specifications.

—ISO 8802/4, IEEE 802.4, Local Area Networks-Token-Passing Bus Access Method and Physical Layer Specification. Options and parameters as per MAP V3.0 Specification, 1987.

—ISO 8802/5, IEEE 802.5, Local Area Networks-Token-Passing Ring Access Method and Physical Layer Specification.

Physical: CCITT Recommendation X.21 (Red Book, 1984), Interface between Data Terminal Equipment and Data Circuit-Terminating Equipment for Synchronous Operation on Public Data Networks.

—CCITT Recommendation X.21*bis* (Red Book, 1984), Use on Public Data Network of Data Terminal Equipment that is Designed for Interfacing to Synchronous V-Series Modems.

—CCITT Recommendation V.35 (Red Book, 1984), Data Transmission at 48 kbit/s Using 60-108 KHz Group Band Circuits.

—EIA-449 (1977), General Purpose 37-Position and 9-Position Interface for Data Terminal Equipment and Data Circuit-Terminating Equipment Employing Serial Binary Data Interchange.

—EIA-232-D (1986), Interface between Data Terminal Equipment and Data Communication Equipment Employing Serial Binary Data Interchange (Supercedes EIA-232-C, 1969).

Peter M. Haverlock, IBM Corp., Thornwood, N.Y.

The formula for network immortality

Availability is the sine qua non of networking, but at what cost? Increasing network reliability and serviceability is a better, and more affordable, path to high availability.

Availability, performance, and cost are three of the most important attributes of a well-run network, according to user surveys. In many businesses, availability is often *the* most important system requirement. Indeed, the cost associated with maintaining high availability is deemed unimportant if availability itself is threatened. But communications professionals can strike a balance between availability and cost without sacrificing performance—provided they understand the relationship between each of the attributes.

Too often, availability and reliability are used interchangeably. Reliability is an element of availability, as is serviceability. Network availability, in its broadest sense, is that characteristic of a communications link that provides connectivity to a user. Availability can be expressed as a value between one and zero, with one corresponding to 100 percent availability and zero to no availability. Availability is composed of two factors: reliability, which is expressed as the mean time between failures (MTBF), and serviceability, which is expressed as the mean time to repair (MTTR).

$$\text{AVAILABILITY} = \frac{\text{MTBF}}{\text{MTBF} + \text{MTTR}}$$

RELIABILITY—MEAN TIME BETWEEN FAILURES (MTBF)
SERVICEABILITY—MEAN TIME TO REPAIR (MTTR)

Availability is presented as the total operative and inoperative days and hours, which are noted as availability and unavailability in the table. The numbers cited in the table assume 24-hour-per-day operation over one year of 365 days.

In some organizations, availability is calculated on the basis of the hours that a network is actually in operation. In such cases, outages that take place outside operational hours are not included in the reported availability figures. Calculating availability in this fashion increases overall availability numbers, but it may not present a totally accurate picture.

Equation 1 shows that availability can be increased by either increasing reliability, expressed as the mean time between failures, or decreasing the time for serviceability, expressed as the mean time to repair. By lengthening the MTBF, the network will become more available if the MTTR is not increased. It also follows that if the MTBF is not affected and the MTTR is reduced, the availability of the network will be increased. Many networks make use of both techniques to achieve increased availability; redundancy is built into the network by deploying parallel trunks and alternate routes as well as automating failure-recovery systems.

Network communications links are frequently composed of complex serial components. Serial components usually add functions or reduce the cost of the network link, but they can also decrease availability. For example, a line multiplexer allows the concentration of multiple lines into one line; however, if the multiplexer fails, all the attached lines may become unavailable. Building parallelism into network designs is often the best way to offset any decrease in availability produced by serial component failures.

The use of automated service restoration also increases availability by decreasing the service time required to establish an alternate connection after a component or link failure. For example, a short power failure at a remote location may require the network operator at the host location to go through a set of procedures to reactivate the modem and the remote control units. The procedure can be accomplished through an automatic programmed interface that invokes the commands required to bring the remote location on line; the human operator is notified only if the procedure is not successful.

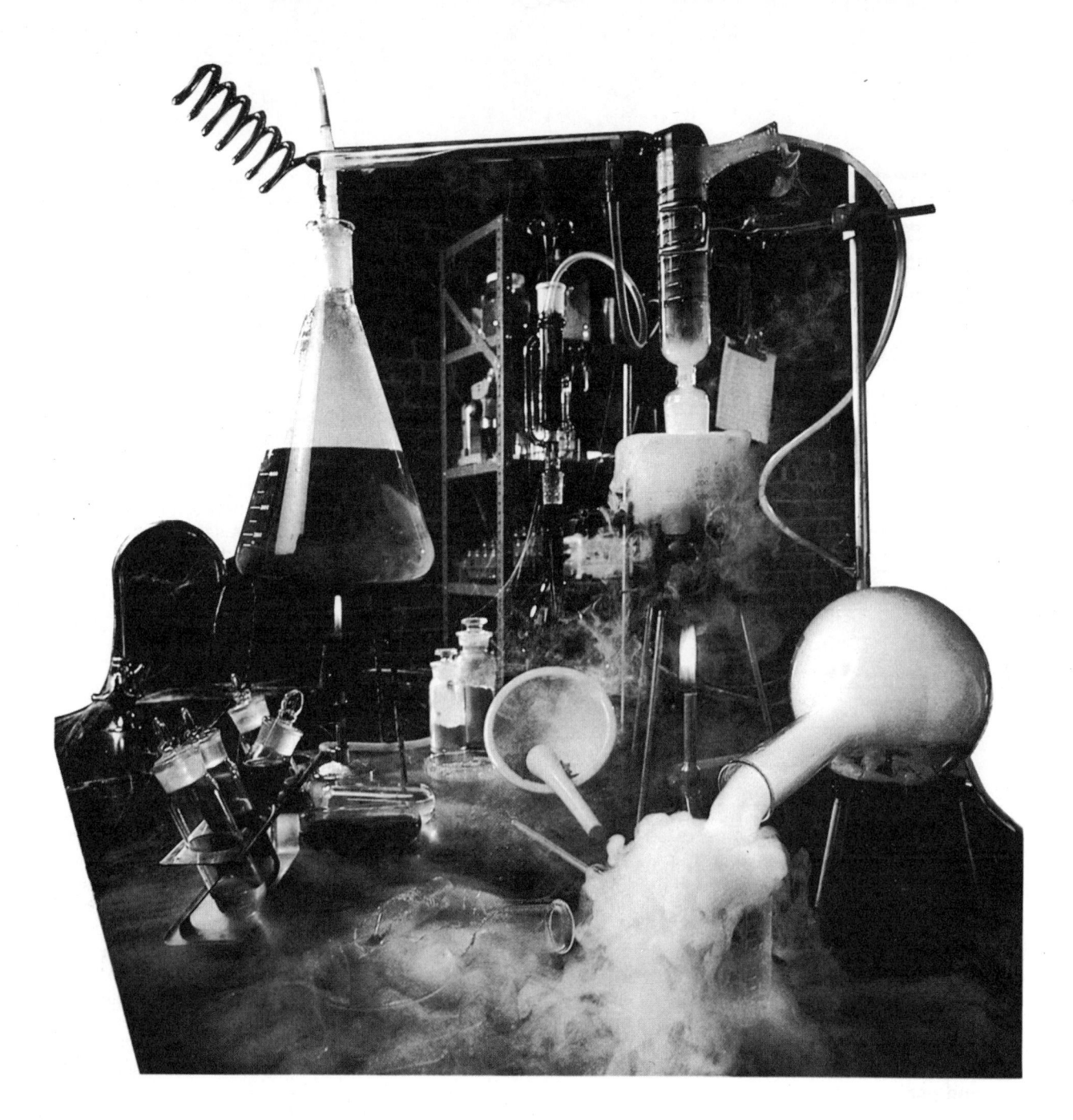

As a network grows, more serial components are added. Adding serial components only decreases availability since failure of a lone component can cause total system failure. System availability is the product of the individual availabilities shown in Equation 2.

$$\text{AVAILABILITY OF A AND B} = \left(\text{AVAILABILITY OF COMPONENT A}\right) \times \left(\text{AVAILABILITY OF COMPONENT B}\right)$$

In a network that is organized in parallel fashion, all components must fail to cause total system failure. System availability and failure (system unavailability) are complements of each other, as shown in Equation 3.

$$\text{AVAILABILITY} = 1 - (\text{UNAVAILABILITY})$$

The availability of a system with two parallel components is shown in Equation 4. According to the equation, the probability of system availability is 1 minus the product of the individual failure probabilities.

$$\text{AVAILABILITY OF A AND B} = 1 - \left(\text{UNAVAILABILITY OF COMPONENT A}\right) \times \left(\text{UNAVAILABILITY OF COMPONENT B}\right)$$

Figure 1 presents some examples of serial and parallel component availability. Components linked in serial fashion in the upper portion of the figure compare unfavorably to availability of components linked in parallel.

The figure uses Equations 2 and 4 to show that highly available systems can be created from unreliable components used in parallel. Failures are reduced because system failure will require the simultaneous failure of both parallel components (redundancy). This is highlighted in the parallel example with 0.8 availability, which produces a system availability of 0.96. Referencing Table 1, an availability of 0.8 has an unavailability of 73 days. Using the

Network availability by the numbers

AVAILABILITY	NON-OPERATING TIME DAYS AND HOURS		OPERATING TIME DAYS AND HOURS	
.64	131	10	233	14
.80	73	0	292	0
.95	18	6	346	18
.96	14	15	350	9
.97	10	23	354	1
.98	7	8	357	16
.9801	7	7	357	17
.99	3	16	361	8
.995	1	20	363	4
.997	1	3	363	21
.999	0	9	364	15
.9999	0	1	364	23
1.000	0	0	365	0

simple parallel design shown in Figure 1 decreases this outage to 14 days and 15 hours, with an availability increased to 0.96. The other example in Figure 1 shows the use of highly available components, 0.99 availability, in a parallel design producing an availability of 0.9999. If parallel components replace the serial components, the system availability will be the product of the improved parallel components, that is, 0.96 x 0.96 or 0.9999 x 0.9999. Thus, by using highly available components and parallel design, a very high degree of availability can be achieved.

Topology and performance

A well-designed network topology can increase network availability. The topology of a network is the routing of its physical or logical links. A logical circuit can be viewed as the ruler distance between two points. The physical circuit is the actual connecting path between the two points. The same network can have different logical and physical topologies. The topology of a network can affect network availability, performance, and cost because the pricing of a circuit is not always related to the actual physical routing.

Users can employ their knowledge of physical-line routing to improve network availability through redundancy, by using components with higher availability, and by decreasing the time required for network-problem determination.

Communications services can be divided into two major types: private or leased lines and the public network of switched lines.

The private line is a dedicated communications link between two or more locations; it offers volume-insensitive pricing for voice and data traffic in the United States. When first developed, Wide Area Telephone Service (WATS) was priced at a flat rate; however, within a few years, its pricing became traffic-sensitive. Because of the flat-rate pricing of private lines, other options are compared against their fixed monthly cost.

Figure 2 shows the monthly cost of a private line compared with direct dialing and two other discount offerings. Common carriers provide different types of bulk-discount offerings; for simplicity, they are referred to here as Discount A and Discount B. The slopes and Y-axis intercept may change for different types of communications services and varying physical or logical topologies since there is never a guarantee that one offering will be priced lower under all conditions.

Heavy communications traffic between two company locations will invariably lead to the selection of private-line configurations because of the fixed line costs. The chart only presents the tariff link cost of a connection: The private network's hardware and other support costs would need to be added to get a complete picture. This could affect the position and slope of the lines in the chart but not the general picture.

The basic routing of a private interstate circuit is shown in Figure 3 by the solid lines. The dial-access lines are shown as dashed lines between customer location A and the Local Exchange Carrier (LEC) central office in Local Access and Transport Area (LATA) 1. The routing of the circuit passes through the facilities of three carriers: the LEC at location A, the Interexchange Carrier (IXC) connecting the two central offices' exchanges near location A and B, and the LEC at location B. The circuit can be leased from the IXC, which in turn leases both ends from the LEC, or through separate arrangements for each service.

The LATA, which is sometimes described as a calling zone, is the geographic area that is serviced by the local common carrier. The LEC provides services between points within one LATA. An IXC provides services between LATAs as well as lines that carry interstate traffic.

The lines are available in either analog or digital formats.

1. Robust designs. By changing the configuration from a serial to parallel design, poorly performing serial components (top) operate more reliably (bottom).

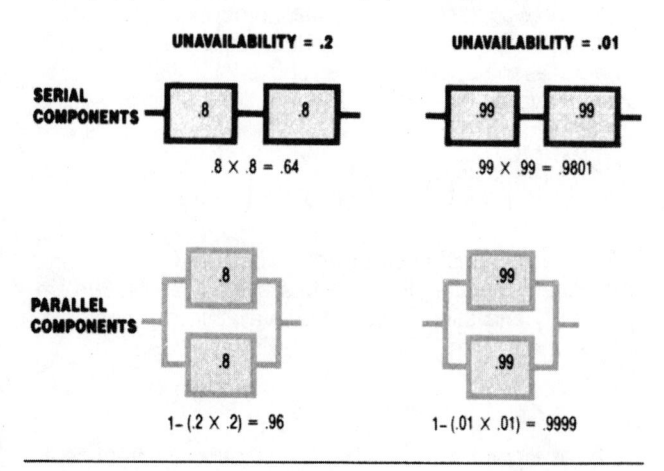

2. Private versus public. *Physical-line routing can be used to improve network availability. When heavy traffic is expected, private lines with fixed costs are best.*

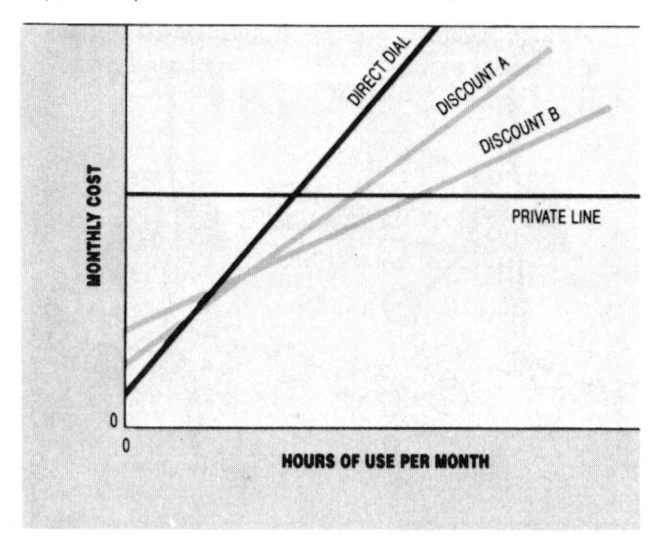

The termination interface at the customer's premises determines the format. An analog line may be transmitted in digital format over part of its path, but a digital line is digital over its entire length.

The digital line tends to be more accurate because it uses repeaters to precisely replicate the transmitted signal. Analog, on the other hand, is subject to greater errors because the analog wave form may be amplified or attenuated slightly out of phase or frequency as it travels across a link. Distortions also occur because the transmission properties of the line differ over a range of frequencies.

The goal of AT&T Dataphone digital service (DDS) digital lines is 99.9 percent availability. Such high availability is achieved through common carrier network redundancy and unique maintenance—despite the fact that DDS sometimes uses T1 circuits for transport. T1 circuits are designed for lower availability goals than DDS circuits.

T1's availability goal is 99.7 percent or greater. Nevertheless, a single failure can knock out the multiple communications links being multiplexed over the 1.544 Mbit/s line. AT&T's Technical Reference 62411 of October 1985 states T1's quality goals for error-free seconds of operation are 99.6 percent for less than 250 miles, 96.0 percent for 250 to 1,000 miles, and 95 percent for over 1,000 miles.

The public switched network, also known as direct dial, connects users to their destinations via a dialing sequence that is composed of an area code, exchange, and subscriber number. The public switched network may offer a different interoffice line each time a call is placed. Users, therefore, can never be certain of the quality of their connection.

For dial-in users, the direct-dial connection offers the convenience of dependable connectivity with reasonable availability. The actual degree of availability is dependent on line quality and modem technology. The public switched network can be used to back up private line speeds varying from 1.2 kbit/s to as high as 19.2 kbit/s for dial calls that are routed over high-quality circuits. The use of higher speed may require two calls; one sending, the other receiving.

When components are connected serially, they can adversely affect availability. Availability can be increased by configuring the components in parallel. In fact, unreliable components configured in parallel can produce a reliable system. Parallel components, configured for redundancy, provide the basis for high network availability. Network managers can use one of several options offered by common carriers to increase the availability of a network path. Those options include:

■ *Diversity.* Diversity is used to prevent two or more circuits from being routed through the same distribution cables and central offices. This option is not always available because of topology constraints, distribution cable routing, central office functions, and central office trunk routing. Sometimes diversity is implied because of the routing of different services (analog and digital) by the common carrier. The diversity option has a fixed monthly charge and an installation charge.

■ *Avoidance.* The user can specify that a line not be routed through a specific geographic area, such as new construction zones or areas subject to flooding. Again, because of topology constraints, this is not always practical.

■ *Multiple-carrier circuit routing.* This technique generally provides diverse routing for the interexchange portion of the circuit. The local channel portion of a point-to-point circuit is routed through the LEC's central office without diversification and then to the IXC central office. At the ICX, multiple-carrier routing is provided. However, the service is not always available because all locations are not serviced by multiple communications carriers. It is important to remember that some ICX carriers resell their services, so

3. Point-to-point? *The routing of a private interstate circuit can be circuitous: from dial-access lines to the LEC, then to the IXC and, finally, back to an LEC.*

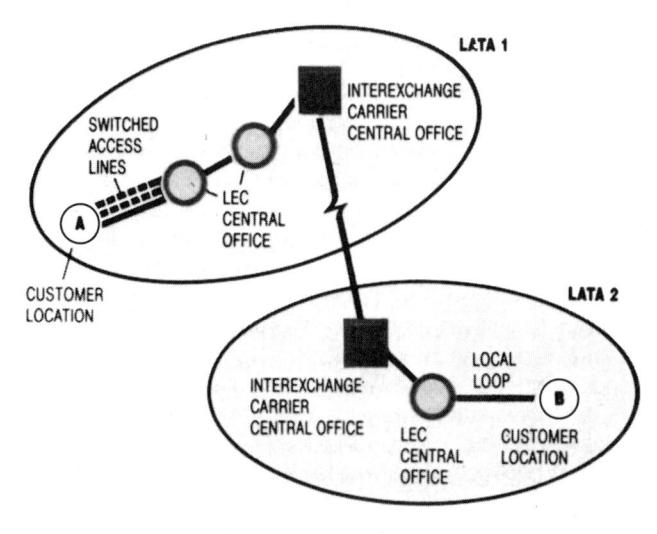

LATA = LOCAL ACCESS AND TRANSPORT AREA
LEC = LOCAL EXCHANGE CARRIER

verify that the multiple-carrier option is not a repackaged form of single-carrier routing.

■ *Parallel circuits.* This option is almost always available. To control costs users can mix high-speed 56-kbit/s circuits with slower, less-expensive, 9.6-kbit/s circuits. However, if diversity is not specified, the telephone company may route the circuits through the same cables and distribution frames from the customer's premises to the LEC. T1 circuits cannot ensure true parallelism because their multiple circuits use the same media and the same physical topology. Without route diversification, parallel circuits cannot offer protection against a cable-seeking backhoe.

■ *Switched network backup.* While not strictly an option, if a private line fails, public switched access lines can be used to provide low-speed transmission between two points. With a properly configured modem, a user can make the public switched network a backup for a failed private circuit.

A modem's performance depends on many factors, including the modem's technology, physical topology of the network, and the distance between nodes. Throughput and performance may suffer in a switched line environment, however, since switched lines tend to produce more errors than private lines. Just as with private lines, there is no guarantee that the switched circuits will be route-diversified. Also, multiple call attempts may be required in order to acquire a high-quality circuit, since all interoffice trunks are not created equal.

■ *Digital circuits.* Digital circuits provide better availability and fewer line errors than most analog lines. Digital data circuits have a design goal of 99.9 percent availability with 99.5 percent error-free seconds of operation. This specification is for end-to-end service.

A study of 6,500 point-to-point private analog line customers reported .56 failure incidents per month for each analog line. It is estimated that about 75 percent of the failures on an analog line are caused by the analog termination equipment. The cost for digital service for the ICX portion is about equal to that of analog service. The digital portion of the local circuit usually costs more than a comparable analog circuit.

■ *Line conditioning.* Higher modem operating speeds may require the control of amplitude, envelope delay distortion, harmonic distortion, and noise. This is accomplished by the installation of line conditioning or by building the electronics into the modem to provide similar line quality controls.

Common carriers provide two types of line conditioning. Type C is designed to minimize the effects of amplitude and envelope delay distortion; Type D conditioning controls the amount of harmonic distortion and also controls noise to tighter limits than Type C conditioning. There are various levels of Type C and D conditioning. Not all levels of conditioning are provided at each LEC central office. Many modems have automatic equalization; this provides some of the functions of line conditioning.

■ *Bridging options of multipoint lines.* Bridging connects two or more private lines at a telephone central office to provide a multipoint circuit. In a voice network, it is the basis for a conference call.

4. Better bridge. *A multipoint interstate private line is usually bridged at the Interexchange Carrier's central office, but using the Local Exchange Carrier's central office costs less.*

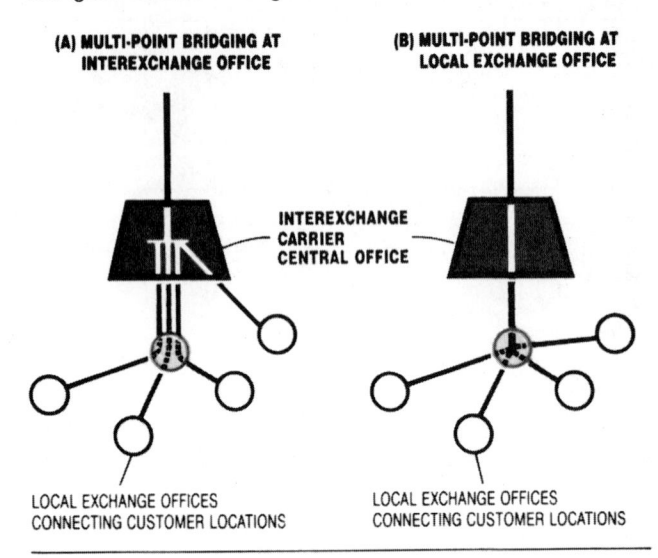

A multipoint interstate private line that has multiple locations in the same LATA will usually be configured as a mini-star network with the interconnect points being bridged at the IXC's central office (Fig. 4A). However, it can also be configured using one of the LEC's central offices as the bridging hub configuration (Fig. 4B). This option usually produces a lower tariff link cost.

The question of which option provides greater availability cannot be answered from Figure 4 if more stations are to be added to the circuit. Assume that the circuit is complete except for a leg to a network processor. If the ICX is used as the contact point—the contact point is the carrier that will be called at the first sign of line trouble—then the MTTR should be shorter in the configuration shown in Figure 4A. If the LEC is the contact point, the configuration shown in Figure 4B should have a shorter MTTR. The point is that the repair time for the circuit may be lengthened when one carrier has to call the other carrier to do problem determination.

The assumptions made about repairs are based on failures caused by one of the end points on the star topology. The repair procedure would isolate the failed point and make the other points on the circuit operational. The failed component then would be repaired and the circuit reconfigured to its original topology. Again, an understanding of the topology and repair procedures used by the designated carrier can help in the availability design of the circuit. The carrier must also be able to isolate the trouble leg. The network processor leg is the most important; if that link fails, other stations on the circuit may not be available.

Adding it up

Communications cost is the aggregate of the amortized cost for capital equipment and communications services or tariff costs. The proportion of total communications costs also varies with the type of service and whether voice or data is being transmitted.

In voice networks, approximately 75 percent of the cost is allocated to tariff charges. After all, standard telephone instruments have little bearing on the total system communications cost.

Generally, data networks have a reversed relationship, with approximately 25 percent being spent on link tariff charges and the rest being spent on front-end processors, communications controllers, modems, or channel service units, remote control units, multiplexers, host and remote communications software, remote terminal hardware, hardware and software maintenance, and, of course, help-desk support for the remote network. Data networks also use such hardware as concentrators, multiplexers, modem technologies, and multipoint line bridging to lower network costs.

The percentage of communications costs allocated to links should start to shift as voice and data networks are integrated. In a digitally integrated network, the differentiation of voice and data line charges will blur. The Integrated Services Digital Network (ISDN) will also blur voice and data networks by combining voice and data connections over one physical line. The availability optimization processes for voice and data networks have similar goals.

Even though a line may be point-to-point, there are many private-line tariffs. Before AT&T's divestiture in 1984, there were about 200 special tariffs for pricing local circuits. Today there are more than 800 special private line tariffs (see "Making the most of post-divestiture tariffs for data network design," DATA COMMUNICATIONS, January 1987, p. 135).

The private-line pricing structure has been evolving since divestiture. With the escalation of installation charges and multiple tariff changes each year, a network manager will think twice before making network changes to take advantage of tariff pricing abnormalities, since they can change again within a few months.

Interaction: Availability, performance, and cost
Network optimization can be viewed as the process of balancing the design factors against one another to reach a network configuration that gives the best combination within the constraints specified. The interaction of the three major network requirements of availability, performance (response time), and cost is shown in Figure 5. The charts show the interactions of the three major design factors when one of the factors is held constant against the others.

Each variable, when held constant, affects all the others. The curve itself is a nonlinear, quasi-step function. Cost will produce a discrete step on a curve, but performance and availability may produce nonlinear steps. This is shown in greater detail in the close-up view in Figure 5C.

Changes in components, hosts, modem speeds, and line topology produce changes on the X and Y axis. If the speed of one network line is increased to 56 kbit/s from 9.6 kbit/s the user will notice the difference. But the change may have a minor effect on the total system performance for a network composed of many lines. The net effect of minor changes may almost appear to produce a continuous curve.

In Figure 5A, the same cost will produce greater avail-

5. Balancing act. *Network optimization is the process of balancing availability, performance (response time), and cost against one another.*

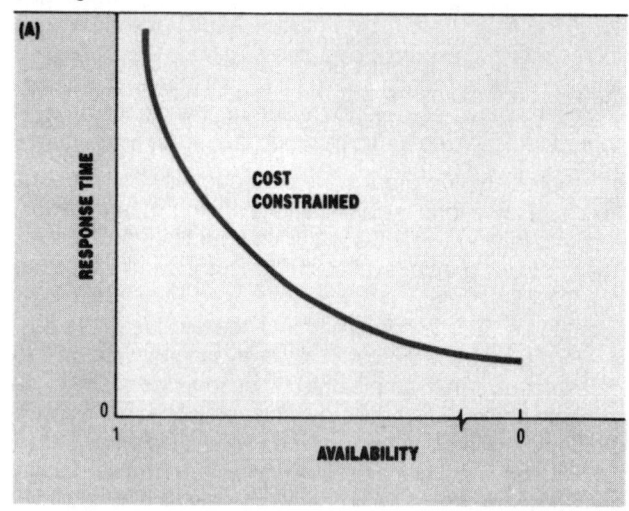

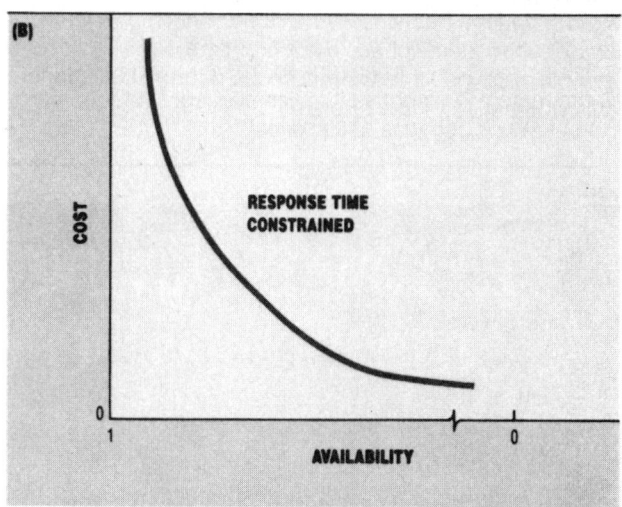

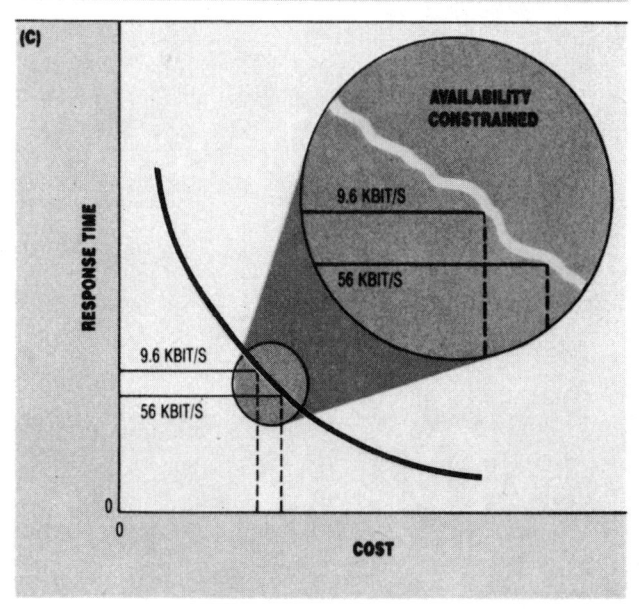

ability with lower performance or lower availability with greater performance. The figure applies to a large backbone system with alternate routes. The failure of one route can force the data to travel over multiple node hops, thus increasing the response-time delay. There is a cost built into the network configuration for a level of availability since the alternate routes have to be sized to handle the newly routed traffic. Voice traffic may not have to be sized because it can be off-loaded to the public switch. The higher the availability, the greater the data capacity that must be held in reserve for possible alternate routing. This will also apply to having two 9.6-kbit/s lines operating in parallel. This provides greater availability but less performance than one 19.2-kbit/s line.

In Figure 5B, the same performance produces greater availability with lower cost or lower availability with decreased cost. An example of a response-time constraint is when an analog line modem is replaced with a modem of the same speed but a newer technology. Cost has increased because of the cost of the new modem; availability should be improved because of the new technology. Response time has stayed the same, since there was no change in modem speed. There may be a slight improvement in response if fewer messages need to be retransmitted because of line errors.

In Figure 5C, the same availability produces greater performance with increased cost or lower performance with decreased cost. If a digital 9.6-kbit/s line is increased to 56 kbit/s, the system cost will increase as a result of the higher tariff, but response time will be decreased (improved performance). We are assuming that the availability of the two digital services is about identical. This is shown in the close-up view of the "availability constrained" chart, Figure 5C. In this example, the new 56-kbit/s line should provide extra capacity to provide greater throughput and improve response times. This assumes a similar workload before and after the installation.

Network availability can be improved through several means: redundancy, alternate routing, using higher-availability digital lines, and minimizing the length of any outage by automating network operations. Network managers will always strive to maximize availability and performance within cost constraints, but in the long term any increase in availability or performance generally requires a corresponding increase in cost. ■

Peter M. Haverlock is a senior instructor at the IBM Telecommunication Education Center in Thornwood, N.Y. He received his B.S.E.E. from the University of Illinois and has served as an adjunct instructor at the U.S. Department of Agriculture Graduate School, Washington, D.C.

James A. Craig and William L. Gewirtz, AT&T Bell Laboratories, Holmdel, N. J.

A network architecture for retail bank networks

Competition in banking breeds innovation: A topology that puts LANs into branches expands data processing and networking options.

Competition is forcing fundamental changes in the banking industry. This in turn is forcing many banks to carefully reevaluate the deployment of their network resources.

With deregulation of the industry, financial services have become an increasingly attractive business opportunity to a broad range of competitors. Brokerage houses offer a supermarket of financial services to upscale clients. Retailers leverage off internal networks to provide point-of-sale and other electronic credit and debit card services. And even traditional manufacturers are providing expanded financial services beyond their own products.

Banks have responded to these competitive challenges in a variety of ways—aiming to both reduce expenses and develop new revenue opportunities. Current trends in the banking industry are having a heavy impact on the network architectures of retail banks. These trends include market and technical forces:

■ *Automation of traditional services.* Branch banking applications traditionally have been segmented into teller and platform operations. Tellers stand behind the counter and process transactions, such as deposits, withdrawals, and account maintenance. Platform officers sit behind the desks and assist customers with opening accounts, loan applications, and financial planning.

Banks will continue to automate traditional banking applications for greater efficiency and productivity. ATMs (Automated Teller Machines) already perform many of the traditional deposit and withdrawal teller functions; paper-intensive platform functions such as opening accounts are increasingly performed on line; and self-service terminals enable users to query account balances. The technological challenge is to reduce the personnel expense associated with routine banking service operations.

■ *Marketing of products and services.* While much of banking automation is oriented toward expense reduction, more significant automation trends focus on the marketing of new products and services to distinct market segments. Banks will use existing client relationships to market a growing range of financial products such as home equity loans, Individual Retirement Accounts, Certificates of Deposit, insurance, mutual funds, and brokerage services. Technology must permit a branch officer to integrate customer information and financial product information and use it in a marketing strategy that targets different sets of products for different demographic segments.

■ *Reconfiguration of branch distribution.* Retail banking's greatest asset (and greatest expense) is the branch network used for delivering products and services. Banks must leverage the "bricks and mortar" expense of the branches to ensure that each acts as a profit center and is differentiated by a product mix designed for the client base it serves. Branches may range from standalone ATMs to minibranches at grocery stores to traditional full-service branches. Branch technology must efficiently support both traditional applications and the evolving marketing applications over the entire distribution of branches.

■ *Mergers and acquisitions.* Banks continue to merge with and acquire other institutions in a drive to achieve a sufficient asset base necessary to compete in a fully deregulated market. The acquisition strategy provides the potential for substantial expense reduction as a result of economies of scale; however, acquisitions also pose the challenge of integrating dissimilar application technologies, networks, and operations structures.

Competitive forces and the banking industry's response to those forces are driving fundamental changes in the technological infrastructure of banks. In essence, banking is gradually changing from a transaction and operational business to an information management and marketing business. This migration will have profound implications for the technological infrastructure of banking in terms of both branch automation and networking architectures.

Banks traditionally have used controller-based technologies for branch automation. Branch controllers were originally developed to provide on-line access for teller transactions. More recently, controllers have been used to support platform functions, such as the opening of new accounts and the downloading of customer files from the host for marketing purposes.

Controller technology
A controller architecture typically supports dumb terminals and printers. The controller provides a multiplicity of functions: device driver, print driver, file storage, network gateway, and local processor.

As a technology base, controllers have proved to be limited and typically suffer from performance and capacity problems. Originally developed to support short and highly efficient teller transactions, controllers may not be able to support the larger information flows (for example, full 3270 data screens) associated with platform or marketing functions. Resource contention within the controller may cause tellers to experience substantial response time degradation while platform officers download customer information files and write them to disk. This can result in the classic bottleneck situation in which multiple users are contending for very limited resources (see Fig. 1A).

Historically, banks have placed multiple controllers in branches or upgraded from one generation of controller to the next as often as every two or three years in an attempt to improve performance. As competition continues to force the banks to develop new applications, this can be an expensive and operationally disruptive process, something banks can ill afford.

Microcomputer technology
A more recent technology trend has been the introduction of microcomputers to branch banks. Microcomputers have been introduced primarily for sales support: to provide the platform officer with the ability to perform quick "what-if" financial calculations or to support the marketing of financial products through graphic displays.

The intelligence of the microcomputer has proved valuable in such marketing applications, and most controllers now connect to microcomputers. Nonetheless, there are basic technology compatibility issues associated with integration of two very different technologies: controllers, with their emphasis on mainframe applications, and microcomputers, with their emphasis on distributed processing and standalone applications.

The current integration of microcomputers and controllers typically consists of providing the microcomputer with terminal emulation or "hot-key" access to host applications through the controller. The microcomputer accesses the host as if it were a dumb terminal connected to the controller. Such rudimentary integration of controller and microcomputer cannot support the ultimate objective: application-level integration. The target architecture for retail

1. Before and after. *In A, a branch controller supports specialized teller transaction processing. In B, a local area network uses microcomputers as universal workstations.*

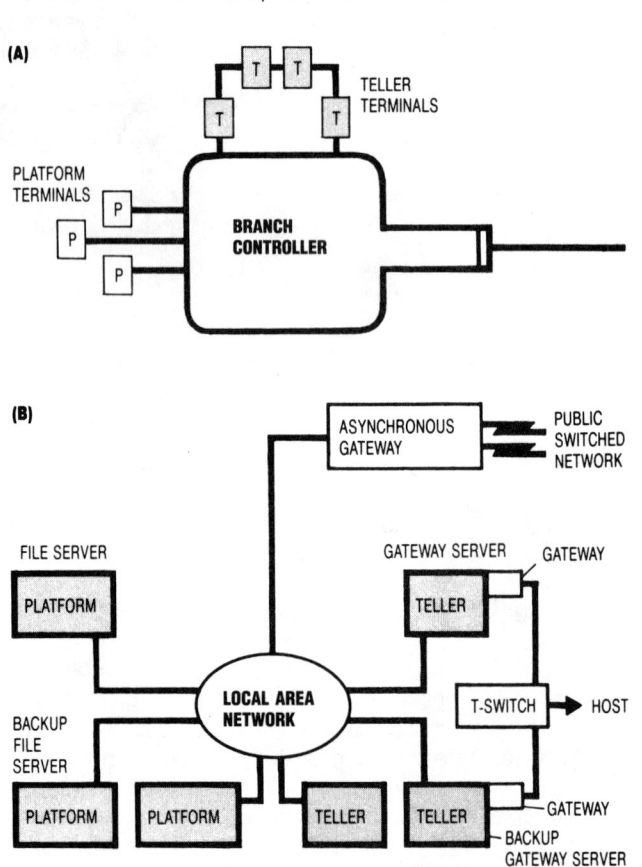

banking would use the flexibility of the microcomputer workstation to access and integrate information from multiple processors and servers distributed over a geographically dispersed network. The architecture that can most effectively address the changing information needs of the banking industry is one in which microcomputer workstations are supported by a local area network.

Microcomputer/LAN technology
LANs provide significant efficiencies through the sharing of both information and resources (disk storage, printers, programs) across a community of users. The high bandwidth of LANs provides users with transparent access to those shared resources, resulting in a branch architecture with greater price/performance than that of controllers.

For branch automation, the hardware would include both client and server workstations in LAN configurations that can be customized for specific price and performance requirements. Client workstations contain resources accessible only to the local user; server workstations contain resources, such as disk storage, that may be shared with other clients on the LAN. Performance considerations dictate whether a server workstation may also support a co-resident client function for greater cost efficiency and use of computing power.

Customer application requirements will drive the software architecture. MS-DOS applications offer the widest options for microcomputer software today, but increasingly sophisticated applications will require the multitasking capabilities of Unix or OS/2. The most efficient application technology may in fact be a hybrid, in which low-cost client workstations operate with MS-DOS, while server workstations use Unix for enhanced multitasking or communications capabilities.

LANs in branches

Figure 1B illustrates the functional components of a typical LAN branch configuration:

■ *Microcomputers as universal workstations*. Microcomputers may be used as both teller and platform workstations and are differentiated only by the peripherals required, such as small display screens or passbook printers for tellers. Microcomputers can provide the routine screen formatting or arithmetic computations at the workstation level without burdening any of the shared LAN resources, such as printers and servers.

■ *LAN linking microcomputer workstations*. The LAN permits functions and resources to be distributed transparently across the branch. A passbook printer may be shared by two teller workstations; an expensive laser printer may be shared by all branch personnel. One client workstation may support a co-resident file server, while another may support a co-resident gateway to the host. By distributing shared

---

On-line loan approval

How do you apply for a loan today? Chances are that you sit down with your local bank officer and fill out a lot of forms, mostly with information that the bank already has about you. The forms are then mailed to a headquarters location where the data is reentered on a mainframe, credit checks are performed, and the loan approval process takes place.

Consider then the potential competitive edge that the bank would have if it could approve your loan in real time, as show in the figure. The loan officer invokes microcomputer-resident software, accesses your current account information through the host gateway, downloads requisite fields, and merges the data into the microcomputer loan application. The bank officer queries you only for any supplementary data.

A dial-up connection may then be established to a credit bureau. Again, predefined credit fields are merged to the loan application and a local microcomputer-based expert system scores the application for approval. The loan officer then routes the forms over the local area network to the branch manager for confirmation of the loan approval. Upon approval, all requisite forms are automatically printed at the LAN print server, and appropriate loan fields are formatted and uploaded to the mainframe.

Fundamentally, a communications-based application architecture allows the branch personnel to streamline a loan process by extracting data in real time from a variety of existing sources, integrating the data at the workstation to deliver a new and more timely product.

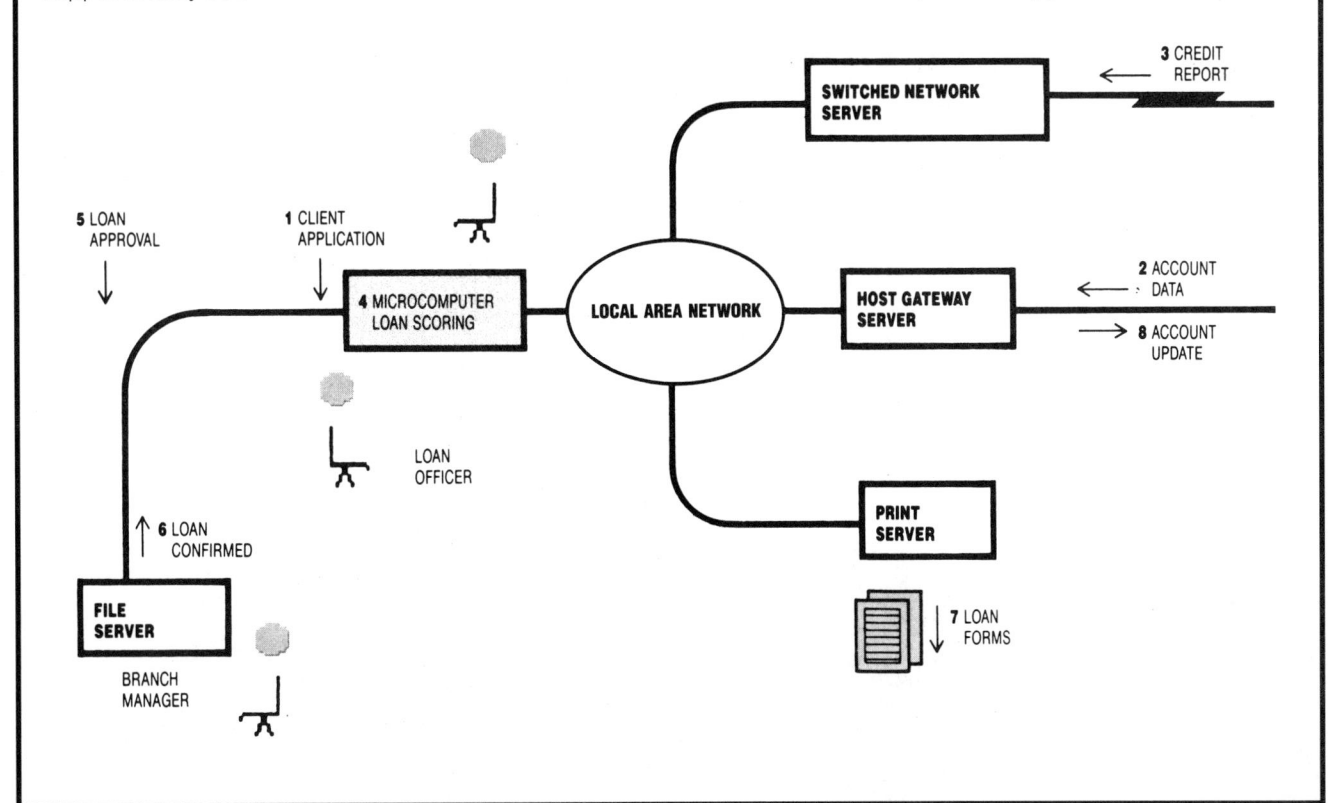

127

resources across the LAN, many of the bottleneck issues encountered with controllers can be avoided or at least minimized.

■ *File and print servers.* File servers support software distribution to the client machines, as well as electronic journaling (or storage) of financial transactions, end-of-day balance reconciliations, and other branch-level administrative functions.

■ *Host gateway server.* The host gateway server manages private-line communications to the host; the gateway server is typically managed by the host as a controller with client workstations on the LAN managed as devices sitting behind the controller. The host may be viewed as the ultimate file server for central storage of all transactions or customer files. However, applications may well be developed at the microcomputer level, relying on the host for data but eliminating the long lead times for application development on the host.

■ *Switched network server.* The switched network server manages switched or dial-up asynchronous communications through value-added networks or the public switched network to information providers (public databases, credit bureaus, or financial service affiliates). The communications capability permits individual branch workstations to access external information sources and integrate the information with local applications or host-based data.

■ *Communications-based application software.* In the migration of banking from a transaction business to an information management business, communications-based applications are fundamental. The software must permit the user to access and integrate data from multiple sources: the local file server, the host via a gateway, and dial-up connections. "On-line loan approval" provides an example of how a communications-based program architecture can operate.

Microcomputer-LAN branch technology brings several advantages to banks:

■ Its open architecture allows microcomputer workstations to be integrated with a vast array of peripherals and software. By contrast, most controllers have proprietary interfaces which may support peripherals only from the same vendor.

■ Standard microcomputer workstations have greater manufacturing economies than specialized financial terminals. Furthermore, the modularity of a microcomputer-LAN configuration makes the cost of branch automation proportional to the number of workstations; this can be particularly attractive in those mini-branches that may require only a few workstations. By contrast, smaller branches pay a penalty for a controller configuration since they must make a substantial investment in the controller regardless of the number of workstations.

■ The LAN can enhance reliability by performing backup functions. The file and gateway servers can act as backups to each other in the event of failure, allowing the branch to continue automated operations, although in a somewhat degraded manner; by contrast, when a branch controller fails, the branch is out of business.

■ Growth options can be managed gracefully through increased segmentation of server functions or through the introduction of more powerful servers as application requirements increase. Since aggregate branch traffic is low, LANs possessing bandwidth in the Mbit/s range are virtually inexhaustible.

■ New applications can easily be developed at the workstation level, avoiding the typical backlog for development of new applications on a mainframe. With communications-based branch software, new programs can be developed quickly on the microcomputer using existing mainframe databases as the ultimate network file servers.

However, as banks use the communications power of such a LAN-based architecture, banking requirements for a wide area network (WAN) architecture will also develop and change whatever is already in place.

Retail banking's network requirements will be driven by the topology of the branch delivery network. As banks continue to acquire and merge, the branch network may well span multiple states and consist of hundreds of branches and several data centers. Evolution of the network architecture will encompass both a backbone network and a branch network.

T1 backbone network

Banks, like other large information users, are deploying backbone T1 networks to integrate voice and data applications as illustrated in Figure 2. Topologically, the backbone network will link T1 multiplexers at hub sites—typically data centers, customer service centers, or large administrative headquarters serving major metropolitan areas. Alternatively, for operational or cost considerations, customers may elect to access shared T1 channel services provided at carrier serving offices. The hub sites, either on customer or carrier premises, serve as concentration points for aggregations of voice and data traffic, originating either at the hub or routed to the hub from smaller branch sites throughout a service area.

The backbone network is implemented through dedicated T1 links, which can provide substantial cost savings relative to multiple low-speed voice and data networks. The backbone network can be used both for intracorporate networks, such as corporate electronic tandem voice networks or dedicated private-line data networks, and for more efficient integrated access to such carrier-switched services as toll-free 800 numbers.

Additional benefits will be realized as T1 networks incorporate Integrated Service Digital Network (ISDN)-based standards. Customers will have substantially greater flexibility in bandwidth management, such as on-demand access to switched carrier services for traffic overflows. ISDN information transfers will also support the development of new network-signaling-based applications that currently exist only on paper (see "An ISDN-based banking application").

Branch bank networks

While T1 networks provide an efficient backbone architecture, the architecture for branch networks will evolve over time to address increasing application and bandwidth

2. T1 backbone. Private T1 networks provide substantial cost reductions by integrating voice and data at hub sites. As Integrated Services Digital Networks become a reality, T1 can be used to access carrier networks for circuit-switched, packet-switched, and bandwidth management services. ISDN backbones provide greater communications flexibility.

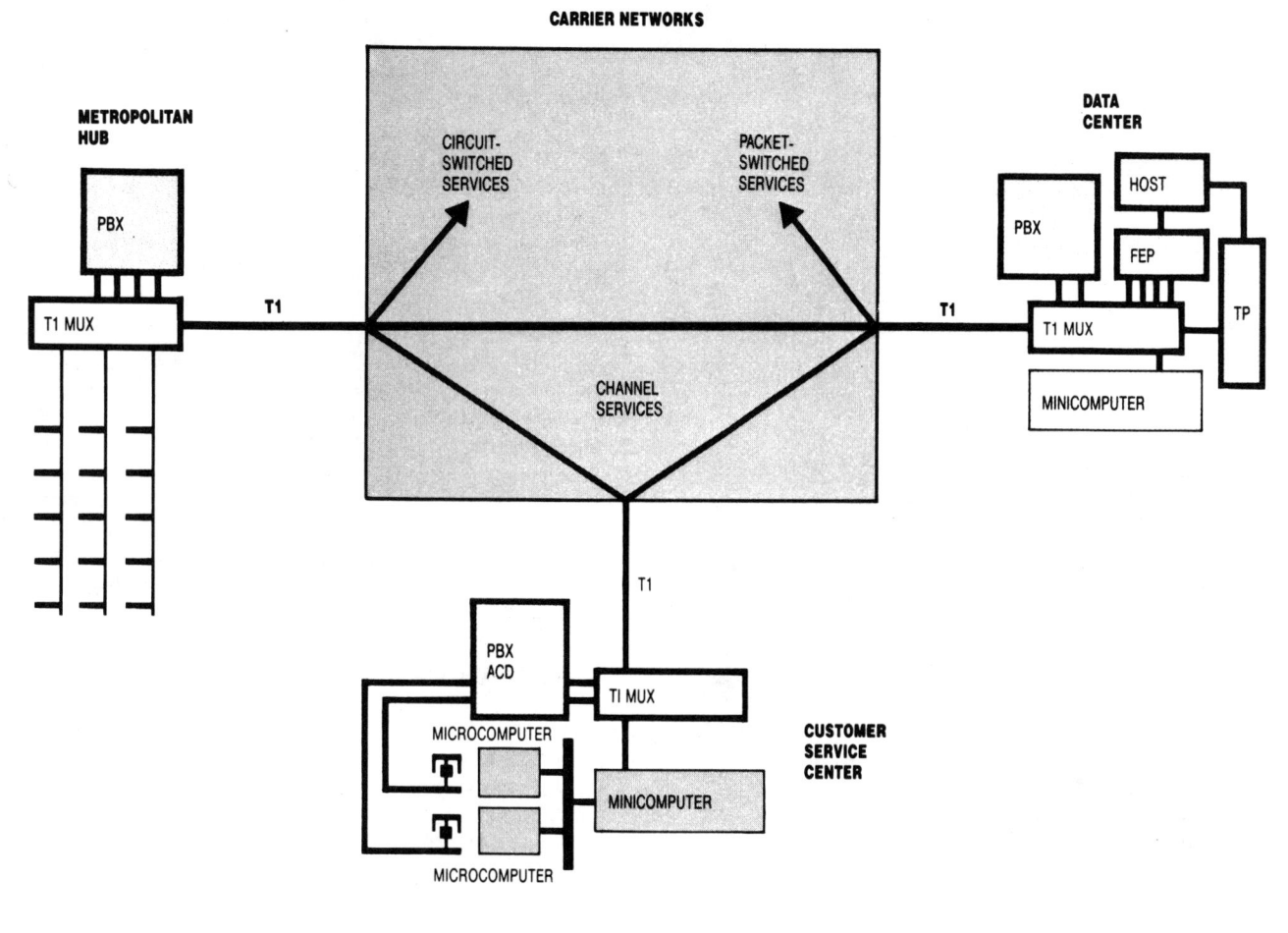

ACD = AUTOMATIC CALL DISTRIBUTION
FEP = FRONT-END PROCESSOR
MUX = MULTIPLEXER
PBX = PRIVATE BRANCH EXCHANGE
TP = TRANSACTION PROCESSOR

requirements. The primary branch network problem today is the multiplicity of data networks often unable to share information in any meaningful or efficient way (see Fig. 3).

It is not unusual for a typical branch to contain three distinct networks: a branch controller network, an ATM network, and a security network. With the potential addition of microcomputer-LAN branch automation, this may result in additional network requirements: a host gateway network and an asynchronous gateway for switched network access. Furthermore, the bank may support a variety of nonbranch networks, such as dial-up, point-of-sale networks or remote microcomputer-based banking services available to small businesses.

Specialized banking applications have evolved over time, and may well use incompatible protocols such as asynchronous, bisynchronous, or Systems Network Architecture's Synchronous Data Link Control. As banks respond to competition, they cannot afford either the cost or the complexity of managing multiple physical networks. Banks may look to either channel-based or virtual circuit-based branch networks, but the objective is to make all branch networks ride over a single transmission link.

Channel networks

Channel-based technology may be extended from the T1 backbone network to the branch. Multiplexing modems, in either point-to-point or multipoint configurations, are routed to the T1 multiplexer hubs and allocated backbone bandwidth to the destination data center. Branch bandwidth is channelized (partitioned into dedicated channels) among distinct branch applications.

A channel-based architecture is generally the least costly option for the integrating of branch networks. It is most efficient for banks with highly centralized mainframe applications and where the role of the network is to provide low-cost transport to the data center.

An ISDN-based banking application

To achieve greater operational efficiency, many banks are establishing telemarketing service centers. The centers can be used to provide routine customer service such as account queries, but they can further be used to market additional banking products.

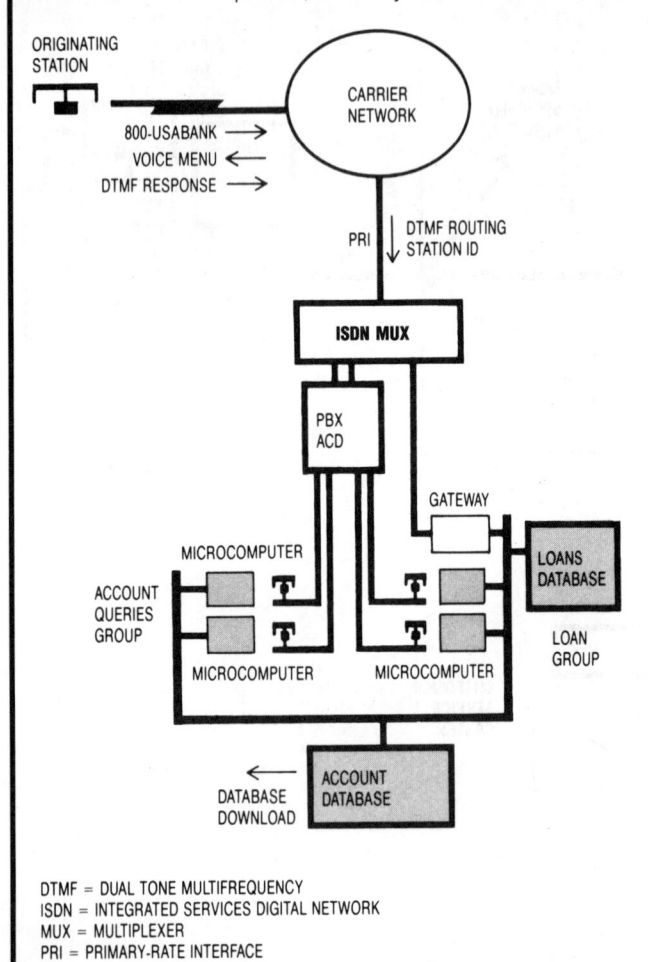

The figure illustrates a bank's telemarketing center. A client requiring a banking service accesses the center through a toll-free service. The client is prompted by speech response queries to dial 1, for example, for customer service, or 2 to apply for a home equity loan. Based upon the DTMF (Dual Tone Multifrequency) input, the call is routed to the appropriate telemarketing group. This has the advantage of allowing specialization of telemarketing functions.

As the voice connection is cut through, the ISDN ANI (Automatic Number Identification) signal from the network identifies the calling party to the telemarketing database. The ANI signal can provide several valuable functions: (1) it can be used to synchronize a client database download to the designated telemarketing representative as the call is cut through; (2) it can provide an additional security screen on callers to ensure confidentiality of financial data; (3) it can be used to track the demographics of incoming calls to assess the efficacy of marketing programs.

ISDN capabilities can similarly be used to enhance outgoing telemarketing applications. The bank may be so successful, for example, in promoting its home equity loans that now it must establish a collections group. Collections may typically consist of a pool of agents working from a list of problem accounts, with the predictable result that many agent call attempts may result in busy signals or no answer.

With ISDN signaling capabilities, call attempts may be driven directly off a collections database. The voice connection is cut through to an agent in synchronization with a database download only when a network signal known as answer supervision indicates that the call attempt has indeed been completed. This can result in substantial productivity improvements in a typical collections application.

DTMF = DUAL TONE MULTIFREQUENCY
ISDN = INTEGRATED SERVICES DIGITAL NETWORK
MUX = MULTIPLEXER
PRI = PRIMARY-RATE INTERFACE

An alternative to channel-based branch networks is a packet-based or virtual circuit switching (VCS) branch network. Branch concentrators are routed in a point-to-point configuration to a virtual switch located at the hub site; virtual circuit switches are interconnected through trunks derived from the T1 backbone network. The VCS architecture can efficiently support both private line and switched traffic to the branch level; since bandwidth is not shared, all branch applications can run at the full trunking speed with potential improvement in response-time performance.

Virtual circuit networks

A VCS architecture may have higher equipment costs, but it will ultimately result in greater operational efficiency and application utility. Its primary advantages are the capability to add, delete, or logically reconfigure virtual networks under software control and the ability to switch every branch workstation to every other endpoint of the VCS network.

VCS technology also provides bandwidth efficiency. It can use bandwidth on a nonchannelized basis, intelligently route multipoint traffic to the designated virtual multipoint drop, and eliminate the broadcast of traffic experienced on a physical multipoint line. It can further allocate bandwidth on demand; branch bandwidth allocated to teller and platform functions during daytime operation may be changed over to download of customer information files at night.

Virtual circuit networks are most efficient for those banks that support distributed processing of applications. VCS networks can support synchronous applications on a transparent virtual private-line basis. Here, the application is managed by the host as if supported on a dedicated private line. VCS further provides switching functions for branch microcomputers, permitting workstation access to

3. Retail banking delivery. Banking applications have developed and grown through strategic evaluations of hardware and software coupled with tactical implementations of least-cost networks. The result has often been a proliferation of networks within and without branch offices. Cost and management of these networks have become prohibitive.

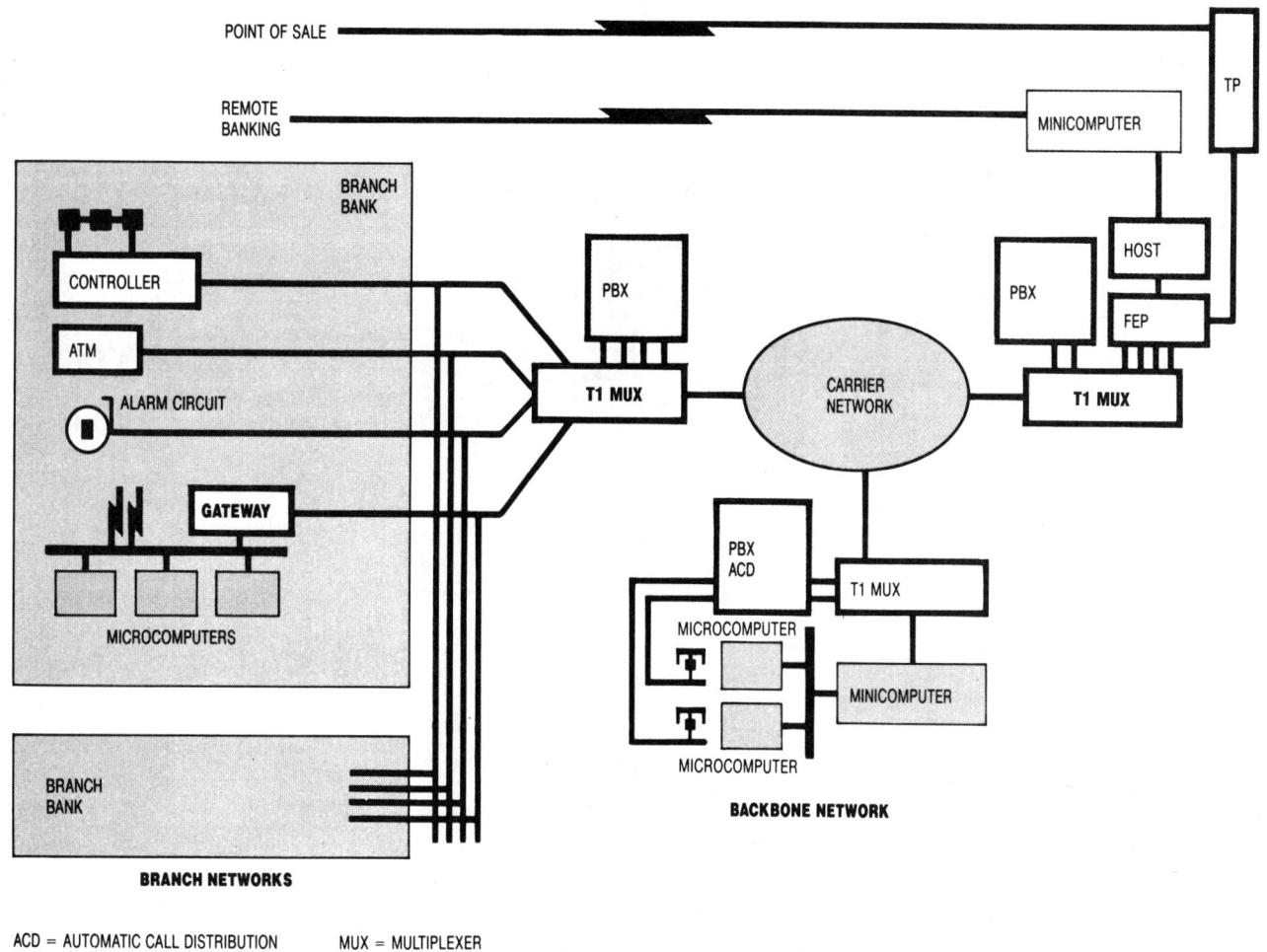

ACD = AUTOMATIC CALL DISTRIBUTION
ATM = AUTOMATED TELLER MACHINE
FEP = FRONT-END PROCESSOR
MUX = MULTIPLEXER
PBX = PRIVATE BRANCH EXCHANGE
TP = TRANSACTION PROCESSOR

any application or information resources on the network. Its primary value is in creating a network infrastructure for delivering new applications. Figure 4 illustrates how a virtual circuit network can be used with the microcomputer-LAN branch automation architecture to deliver network-based banking applications.

Security, ATM, and teller controller applications are supported on a multipoint virtual circuit basis to the host applications at the data center. Since the VCS architecture can intelligently route multipoint traffic only to the destination branch, a single 9.6-kbit/s private line can accommodate a multiplicity of additional data applications as banking needs grow and change.

A microcomputer-LAN platform application at the branch uses both a host gateway and asynchronous gateway into the network. The host gateway is used to access the mainframe data, and the asynchronous gateway can be switched throughout the network or off-net to, for example, a value-added network for access to a credit bureau.

Specialized applications such as electronic mail can be deployed on a dedicated midrange processor deployed at a hub site, providing an implementation alternative to the mainframes. All branches in the network can be served through a menu on branch microcomputers that allows the user to automatically dial and log into the mail server. The mail server can be used for generating form letters or for broadcasting new product updates throughout the branch network. The server can make the day-to-day functioning of the bank smoother and more responsive to customer needs.

The branch can serve as an electronic point of presence for dial-up services. Point-of-sale traffic can be routed to a local-branch hunt group and can ride over the VCS network to a transaction processor. Similarly, small businesses can use microcomputers to dial a local branch and be routed to a midrange processor that contains a daily download of

4. Distributed data switching. *Virtual circuit switching lets banking applications evolve by integrating switched and nonswitched applications. It allows branch-level workstations to access any microcomputer, minicomputer, or mainframe resource at any network endpoint. The planner is freed to put the right software on the right hardware.*

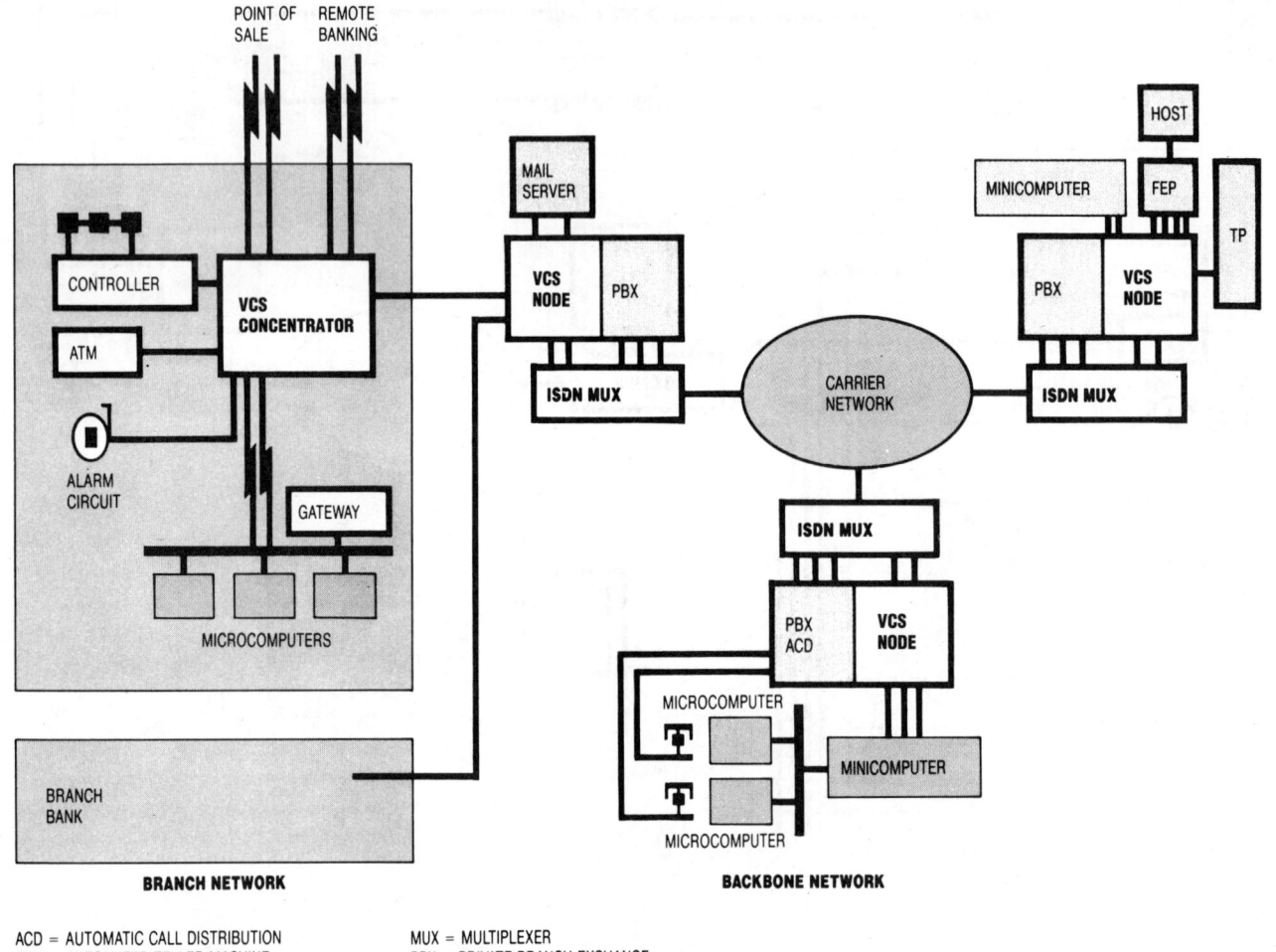

ACD = AUTOMATIC CALL DISTRIBUTION
ATM = AUTOMATED TELLER MACHINE
FEP = FRONT-END PROCESSOR
ISDN = INTEGRATED SERVICES DIGITAL NETWORK
MUX = MULTIPLEXER
PBX = PRIVATE BRANCH EXCHANGE
TP = TRANSACTION PROCESSOR
VCS = VIRTUAL CIRCUIT SWITCHING

account balances. The point of presence not only reduces dial access costs but, more important, provides another link between the small business client and services provided by its local bank branch.

Finally, the branch network can share information with a telemarketing center through the VCS architecture. If a prospective client calls a telemarketing center for a loan application, the client data can be electronically shipped to the nearest bank branch so that local personnel can provide follow-up contact and build on the relationship.

ISDN networks in the branch

In the final analysis, both the channel and virtual circuit technologies provide only a partial solution to the branch network, namely an architecture for integrated data networking. Retail banks will ultimately want to provide integration of voice, data, and image to the branch. Full network integration will be created by the availability of ISDN.

Current transmission services to the branch are limited. Analog and DDS (Dataphone digital service) private-line services do not provide adequate bandwidth, and T1 service may be too expensive. However, with the introduction of local ISDN services, 144-kbit/s bandwidth will be delivered over the local loop. This will provide a high-capacity digital link between the branch LAN and the backbone network, both operating at Mbit/s bandwidths. The branch LAN may ultimately be provisioned with an ISDN interface to act as a high-capacity bandwidth server for the branch.

This technology may also allow the extension of data and voice integration to 144-kbit/s channels that are then linked to T1 or T45 backbone networks acting as hubs. The 144-kbit/s bandwidth would be available for sharing multiple voice channels (32 kbit/s or 64 kbit/s) and multiple data channels or virtual circuits.

High-capacity ISDN digital transport will be provided end

to end between branch workstations and application endpoints, and will open up a whole new range of applications to the bank, including:

■ *Dynamic allocation of bandwidth.* Bandwidth used for voice traffic during the business day can be reallocated for database uploads and downloads during nighttime operation.

■ *LAN bridging over a wide area network.* Images of signature cards may be accessed in real time for more reliable check verification throughout a branch network.

■ *Check truncation.* Electronic images of checks may be captured at the local branch and transported over the digital network. Checks exceeding a certain threshold may be transported immediately, while others would be transported over reallocated bandwidth at night.

Technology directions

Banking applications have evolved over time in response to competitive pressures. In the absence of a fundamental technology direction, application growth has been fragmented, relying on specialized financial devices and application-specific networks.

The LAN-to-ISDN architectural platform will enable banks to integrate the traditional transaction-based banking applications with the evolving information-based marketing applications. Banking applications will be more communications-intensive, requiring integration of multiple data sources at the workstation level. Banking data network architectures will move in two fundamental directions: channel-based architectures, for institutions that elect to maintain mainframe applications, and virtual circuit-based architectures, for banks that migrate to distributed processing. As ISDN capabilities are deployed, high-capacity end-to-end digital transport will open up a range of new applications well suited to an integrated LAN/WAN network architecture. ■

James A. Craig is a technical supervisor at AT&T Bell Laboratories, with responsibility for retail banking architectures. Prior to joining Bell Labs in 1976, he spent a year in Antarctica working for the National Science Foundation. He has an M. S. from Case Western Reserve in systems engineering and a B. S. in electrical engineering from Swarthmore College.

William L. Gewirtz is a division manager for data network planning in the Concept Development Center at AT&T Business Markets Group. He was previously the department head of services and applications planning in the AT&T Bell Labs Architecture Area. He joined AT&T in 1977 after receiving a Ph. D. in computer sciences at the New York University Courant Institute of Mathematical Sciences.

Walter Roehr, Telecommunication Networks Consulting, Reston, Va.

Knocking on users' doors: Signaling System 7

The new standard for signaling between telephone central offices is spreading quickly—and promises a host of new user applications.

What will soon become the highest-volume data network in the world? It is not one of the value-added common carriers like Telenet or Tymnet. And neither is it a government or a military network. Here's a hint: The telephone industry is in the midst of deploying it. The answer is Signaling System Number 7 (SS7), a new, worldwide, packet-switched network specifically designed to handle telephone signaling needs.

In addition to providing the call setup and disconnection traditionally associated with telephone signaling, SS7 will foster a variety of new applications. Credit calls, 800 service, virtual private networks, and a host of other innovative services will be controlled by SS7.

SS7 did not spring from the collective intelligence of one telephone company's marketing department. It is the natural culmination of the evolution of computer-controlled digital switching. The earliest telephone switch signaling techniques emulated the pulses generated by a rotary dial. These pulses were carried on the same wires that carried the user's voice.

Tone signaling came later. It was introduced when multiplexing and carrier techniques made it impossible to signal by interrupting the battery current. Despite the fact that a variety of improved schemes were developed over the years, the tones continued to be carried in the user's speaking channel. It remained a channel-associated signaling scheme.

Moving the signals from a number of user channels onto a single common channel represents a big technical advance. Faster call handling, interdiction of fraudulent "blue-box" tones, and simpler administration are just a few of the many advantages of the common channel arrangement. Such advances could come about only in a computer-controlled switching architecture. In a sense, the common channel becomes a special-purpose data link between two computers that just *happens* to be controlling a telephone circuit.

The first common channel-signaling system to be standardized was Signaling System Number 6 (SS6), which was formulated under the auspices of the International Telegraph and Telephone Consultative Committee (CCITT). Number 6 operates on analog telephone systems using 2,400- or 4,800-bit/s modems. Despite initial worldwide enthusiasm, only North America widely deployed Number 6. In North America, SS6 is known as common channel interswitch signaling (CCIS). The rest of the world impatiently waited for a signaling system tailored for the emerging pulse-code modulation digital facilities: Signaling System Number 7.

The geographically uneven deployment of SS6 set the scene for a dichotomy in the development of SS7. North America has traditionally taken a relatively independent stance vis-a-vis telephone standards. In part, this can be attributed to the size of its network. The North American network represents approximately 40 percent of the world's telephones, and its huge size created special demands. For example, the North American network could not afford to wait for the rest of the world to decide whether common channel signaling was a good idea—it had to proceed with SS6. By deploying SS6 in North America, the pressure to implement SS7 was eased. However, non-North American administrations are not so fortunate. Verbal presentations at the International Switching Symposium (ISS '87) described SS7 introductions that were being accelerated because the existing tone-signaling systems were within months of being seriously overloaded.

The combination of traditional North American independence and different development priorities resulted in two major standards forums developing SS7 recommendations. The CCITT is developing an international version that

1. Out of band. *Both the interoffice signaling arrangement of SS7 and the ISDN access scheme use out-of-band signaling, but each protocol is unique.*

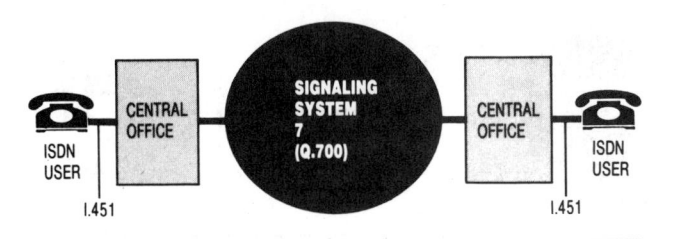

has room to accommodate national versions. In North America, the Exchange Carriers Standards Association T1S1.3 working group is developing its own version. Since ECSA is a recognized American National Standards Institute body, once the T1S1 work is completed, its products become ANSI standards.

In North America the acronym SS7 is used to refer to Signaling System Number 7, generally, and the North American version in particular. If it is necessary to note that the CCITT version is being considered, the acronym SS#7 is used. There are differences between SS7 and SS#7 that, while minor, do prevent direct interoperation.

SS7 is optimized for operation in conjunction with digital telephone systems. While a variety of analog and lower-speed digital transmission paths can be used, SS7 is designed with a 64-kbit/s channel in mind. The 64-kbit/s data links are used to establish a packet-switched network that is optimized for high-capacity, low-delay communications (see "Inside SS No. 7," DATA COMMUNICATIONS, October 1985, p. 120).

SS7 is used only in the backbone network between telephone company switches. One of the primary impetuses for its deployment in North America is the Integrated Services Digital Network; SS7 will support ISDN's enhanced services. However, SS7 is not the protocol users will employ on ISDN access links. An ISDN connection requires the cooperation of two entirely different common channel-signaling systems. Figure 1 shows this arrangement. The users communicate their desires using the ISDN access-signaling protocols, described in the CCITT recommendation I.451.[1] The backbone switches relay these user desires to other switches using the SS7 formats described in the Q.700 series of protocols.

It is interesting to compare these two protocols since, in many ways, they have very similar tasks. The access-signaling protocols in ISDN have a limited task—they only handle user access to the local switch. The signaling protocols do not support independent networks. Rather, they are merely a control mechanism for the affected circuits; they are performing a "network layer" routing function for those circuits. Thus, the I.451 protocol contains all of the required functions in Layer 3. Layers 1 and 2 are stock ISDN: The same Layer 1 is used for all ISDN applications, and the LAPD (Link Access Procedure-D) Layer 2 is used for all packet accesses.

SS7 is conceived as a separate signaling network. It has a seven-layer structure available. Currently, SS7 uses null layers 4, 5, and 6 for circuit-related signaling, but the full seven layers are available, and used, for functions other than circuit-related signaling.

If a PBX is involved in the connection, the user signals to the PBX using the I.451 protocol, and the PBX also uses the I.451 protocols on its public switched network trunks (Fig. 2). It is not necessary for the PBX to make the transformations that allow interworking of I.451 and Q.700—that is a task left to the telephone company switch. The only place that it currently seems likely that SS7 will appear on customer premises equipment is on tandem networks (Fig. 3).

Upper-level protocols

SS7 has a layered protocol structure similar to all modern data communications protocols, though the layers are called levels. The bottom three levels of SS7, which are collectively referred to as the message transfer part (MTP), provide a packet-switched network comparable to an X.25 network. The major differences between X.25 and the MTP are that the MTP provides only a connectionless service and it has a limited (14- or 24-bit) addressing capability. The addressing capability is sufficient for addressing switches themselves, but not for processes or users at the switch.

An SS7 packet switch, which is called a signaling transfer point (STP), is capable of routing the messages using the short addresses. STPs are used to avoid the need for signaling channels on all trunks. A pair of channels replaces the array of signaling channels that would otherwise be required (Fig. 4).

For call-control signaling, the characteristics of the MTP are ideal. The MTP minimizes the processing and transmission overhead and thereby reduces delay. But SS7 gets used for more than just call control. It has many additional uses, including a global address capability and connection-oriented service.

To provide these additional services, the signaling connection control part (SCCP) has been developed. Classic call-related signaling does not use the SCCP. Though the connection-oriented services are still being developed, SSCP global addressing capability is in place and functioning.

In North America, the ISDN user part (ISUP) will provide classic call-handling functions.[2] While ISUP provides functions in addition to the standard call setup and disconnect, it performs call handling extraordinarily well. Interexchange carriers (IXCs) that have deployed ISUP report decreases in call setup time of 3 to 7 seconds. Shaving seconds from a call is valuable if you are paying for switch ports, trunks, and local exchange carrier (LEC) connect time—not to mention enhanced user satisfaction.

But faster call setup is not the only thing ISUP offers. There are additional controls and features of ISDN to support, including: end-to-end signaling, 3.1-kHz speech, 7-kHz speech, 3.1-kHz audio, handling of multimode terminals, changing modes in mid-call, and user-to-user signaling.

The transaction capabilities application part (TCAP) was originally conceived to support two North American database functions: credit card calling and 800 numbers. The

2. PBX talk. *With the exception of tandem switches, SS7 is not being implemented on the customer's premises. If a PBX is involved in the connection, it is not necessary for the PBX to make the transformations that allow interworking of I.451 and Q.700 protocols—that is a task left to the telephone company switch.*

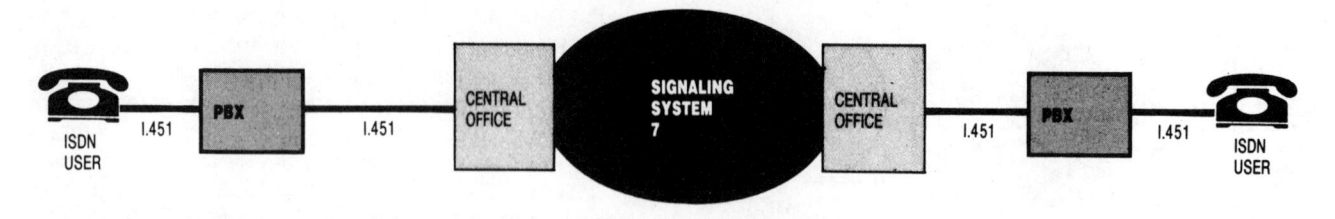

credit card calling database contains the association of telephone numbers and valid four-digit security codes. The contents of this database have been expanded to contain profiles of all lines and billable numbers and dubbed the line information database (LIDB). The 800-number database contains the association of an alias 800 number and the real 10-digit public network number, as well as the billing and interexchange carrier instructions for the number.

The TCAP capabilities address unique North American needs. Calling cards and 800 numbers are not yet widely used outside North America, and TCAP was a product of the North American T1 standards committee. The original TCAP, which is now referred to as Issue 1, was based on the X.409/410 protocols. It was balloted and accepted within the SS7 subcommittee, used as the basis of telephone company procurement specifications, and submitted as a U. S. contribution to the CCITT SS#7 working party.

The working party liked the concept but disliked the U. S. interpretation of X.409/410. CCITT proceeded to rework Issue 1 TCAP for inclusion in the 1988 edition of the CCITT recommendations. Various members of the U. S. delegation then attempted to have the CCITT version adopted as Issue 2 of the ANSI TCAP. This effort bore no fruit.

Currently, the only TCAP with any official status is Issue 1. Work on Issue 2 TCAP has started under an agreement that accepts the following: Issue 1 as the baseline; keeping backward compatibility with Issue 1; and movement toward CCITT TCAP as a worthwhile goal. This agreement ended an extended period during which no progress was made on TCAP because of a bitter disagreement regarding the status of Issue 1 TCAP.

While it is generally accepted that CCITT TCAP (Q.795) is a more elegant protocol, there does not appear to be any function Issue 1 cannot accomplish. There are claims that under some unusual circumstances—such as a loss or delay of a message—Issue 1 can enter a situation that requires a time-out for recovery. However, Issue 1 TCAP has been successfully tested in the field.

It should be noted that TCAP is not used during the actual establishment of a call. There had been some movement toward using TCAP during call forwarding, but that function is now once again firmly positioned within the ISUP. TCAP can be used to establish forwarding arrangements or to determine the suitability of a particular forwarding command, but the actual forwarding is accomplished using ISUP.

The inherent flexibility of the TCAP remote operations syntax readily allows more responsive and complex implementations of the classic database functions. A single 800 number could result in routing to different regional offices, based on the call origin. The 800 service can be easily enhanced with services of the "play announcement, collect digits" type, wherein the caller is queried by a recorded announcement for further information. A caller might hear, "If you are trying to reach Sales press 1, for Service press 2, for all other functions press 3."

Similar automated attendant services are currently available on some high-function PBXs and call-distribution switches, but this version could be embedded in the public network and result in routing the call, via the public network, to different lines and even different cities. While other kludges could be used for providing these features, TCAP provides a simple, straightforward implementation.

Database access for 800 numbers and calling cards provided the initial inspiration for TCAP, but there are myriad new functions, not directly involved in a call step-up, for which TCAP is the ideal vehicle. Some examples are:

- *Camp-on.* A calling party, upon being informed that the called party is busy, can initiate a message indicating that the call should be set up whenever the called party becomes available. All that is kept in the queue is a record of the calling party's number.
- *Call-forwarding management.* Both the updating of a subscriber's current forwarding address and the transfer of that information to a calling switch can be accomplished via TCAP messages.
- *Call gapping.* This is the flow control of remote facilities (overload prevention). The request for call gapping, as well as the interval, is transmitted back to call sources by a TCAP message.
- *Virtual private network management.* Instructions to remote switches that cause the rearrangement of the trunking in virtual private networks can be carried in TCAP message exchanges.

Operations, maintenance, administration

The operations, maintenance, and administration part (OMAP) of SS7 was designed for the care and feeding of the signaling network only, but recent developments have led to its being the leading candidate for the complete management of all aspects of the backbone circuit-switched network. OMAP defines the functional requirements that a management capability should provide,

3. A user SS7? *At the present time the only place that it seems likely that SS7 will appear on customer premises equipment is on tandem networks.*

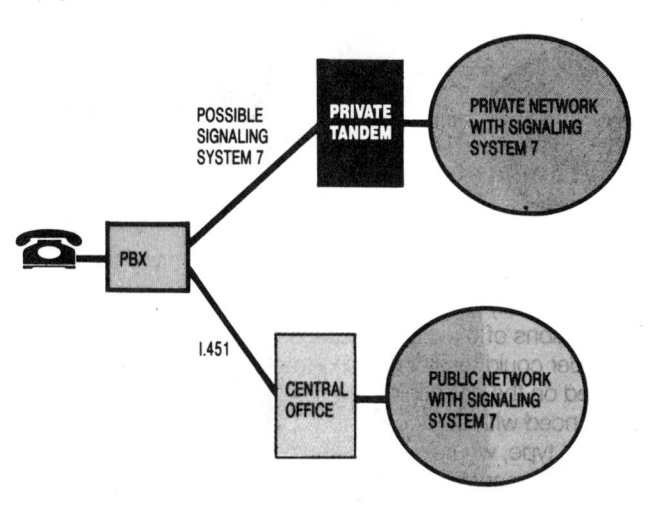

including the transmission of alarms, operational and performance statistics, program dumps, updates of tables and programs, etc. It also provides the means for formatting this information for transmission.

There are several arrangements for the management and control of SS7 networks and for applying SS7 to network management. Most basic management capabilities are built into the individual levels of the various protocols. Thus, Level 2 protocols include a running estimate of the average error rate to monitor link quality while upper levels have extensive reasonableness and format checks.

Any level that detects a problem informs the management entity: the operating system of the local network processor. The original SS7 deployment plans did not include native SS7 management—thinking was dominated by the existing environment in which a manufacturer's proprietary switch-management arrangements were in charge of the switch. First under AT&T patronage and more recently under Bell Communications Research (Bellcore), there have been moves to establish nonproprietary operations system standards using the BX.25 protocols to provide a generic management capability. The initiative led to SEAS, the signaling engineering and administration system. Manufacturers are required to provide a BX.25-based interface to their internal management entity. The regional Bell operating companies are currently going forward with plans to deploy SEAS central processors to manage their SS7 networks.

OMAP depends upon TCAP for the transmission of its messages. The remote operations capability of TCAP is just what OMAP needs for managing distributed processes—there is no need for OMAP to repeat that effort. The history of the parallel development of TCAP and OMAP is interesting. OMAP predates TCAP. There was early appreciation of the need for a network management capability. The initial OMAP efforts centered on functional requirements: What should be measured and what should be controlled. The design of a protocol to accomplish such functions came later. TCAP started as the database query protocol. Once the designers of OMAP started to consider the details of the protocol they needed, it was apparent that it already existed—it was TCAP!

The functions needed for managing the SS7 network are generic—they are needed for the management of any network. Therefore, recent drafts of the CCITT Study Group XI report on Operation Administration and Maintenance (COM XI-R 93-E) on Telecommunications Management Network protocols suggest that OMAP be used for the management of the backbone circuit-switched network. ISDN access networks would be managed using the D-channel protocols. Currently, the D-channel (I.450) protocols do not contain plans for detailed management functions. It is likely that the functionality built into OMAP (and TCAP) will be grafted onto these access protocols.

T1.11X, which specifies the measurements and controls to be implemented, has already gone out for letter ballot, and the July 1988 T1S1 meetings in San Diego incorporated suggested changes. During the July meeting the OMAP bearer protocol (T1.11Y) was approved by the TCAP subcommittee. There was some doubt, in that OMAP has chosen to more closely align itself with the CCITT recommendations than the ANSI/ECSA TCAP. The compromise that won approval for OMAP was the acceptance of the designation "Private TCAP" for OMAP messages. Since OMAP will be reusing some of the same identifiers for different purposes, this designation will allow discrimina-

4. Redundant redundant. *A signal transfer point (STP) handles message routing. A pair of channels replaces the array of signaling channels that would otherwise be required.*

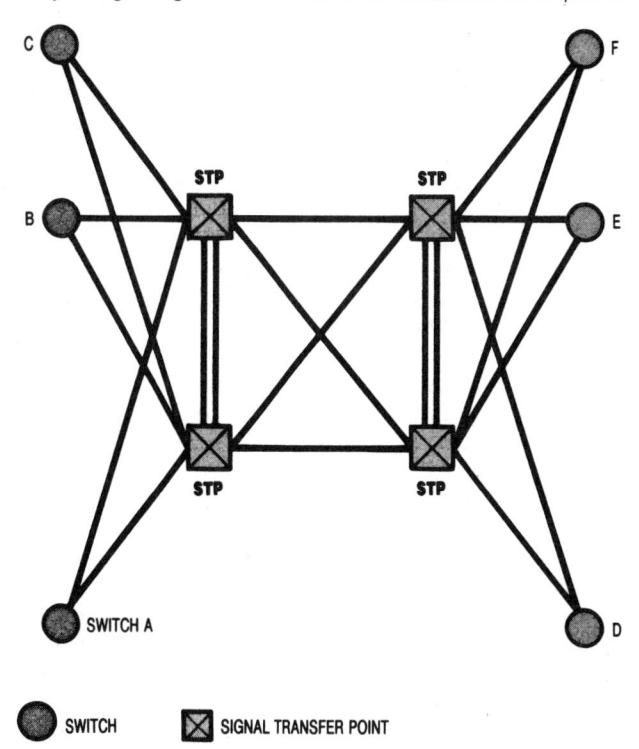

tion, on other than an addressed process basis, between OMAP and existing TCAP applications. Rapid formal approval of T1.11Y looks highly likely. Everyone appears to be pushing hard to keep OMAP rolling—there is a sense of urgency and cooperation.

Non-signaling applications
One of the ways a telephone company can have more control of its services and the switches it buys from vendors is to use simple switches and smart databases. This arrangement is shown in Figure 5. A service switching point (SSP) is any switch equipped with SS7 and TCAP. The database is contained in the service control point (SCP). The essence of the intelligent network is the use of these components to provide for unplanned services.

Just as the local switch need not have any knowledge of the real location of an 800 number or how to handle it, the local switch need not implement a feature the telephone company wishes to introduce. In both cases the switch can refer to a centralized database for instructions. Since the database and the instructions it contains are under the direct control of the telephone company, there is the hope that modifications can be introduced more quickly, uniformly across all switch models, and at a lower cost. The means for accessing the database is SS7 and TCAP—just as in the 800 function.

During 1987, the intelligent network was big news—every telephone company seemed to be rushing to deploy it. Lately, the ardor appears to have cooled. This cooling may be caused by the attention the telephone companies have had to devote to Open Network Architecture. ONA is a judicially mandated initiative that will result in the telephone companies separately tariffing the various components of their services. However, it is still likely that some form of intelligent network will appear.

One of the opportunities that ONA may provide is regulatory approval for providing additional services. With ONA, the telco monopoly advantage is eliminated because all value-added service providers are given access to the service components at the same price the company has to pay to use these services. SS7 plays a key role in several of these new services.

The significant fact about the SS7 network, from a non-signaling viewpoint, is that it has been designed from the ground up as a transaction network; it handles short messages without the overhead of setting up connections. One place this capability could clearly be advantageous is in general credit card validation checks. TCAP was designed to provide this function for calling cards being used to charge telephone calls, but clearly a more generic service is possible. Already telephone companies are establishing generic validation databases so calls can be charged to general-purpose credit cards as well as calling cards. The telcos could sell just transport service to the various credit validation centers, or they could be providing access to their own databases, if regulatory approval were gained.

While it is true that SS7 was designed for backbone signaling on the public network, recent developments

5. Future present. *Reconfigurable components are the essence of the intelligent network. SSPs are equipped with SS7, and databases are housed at SCPs.*

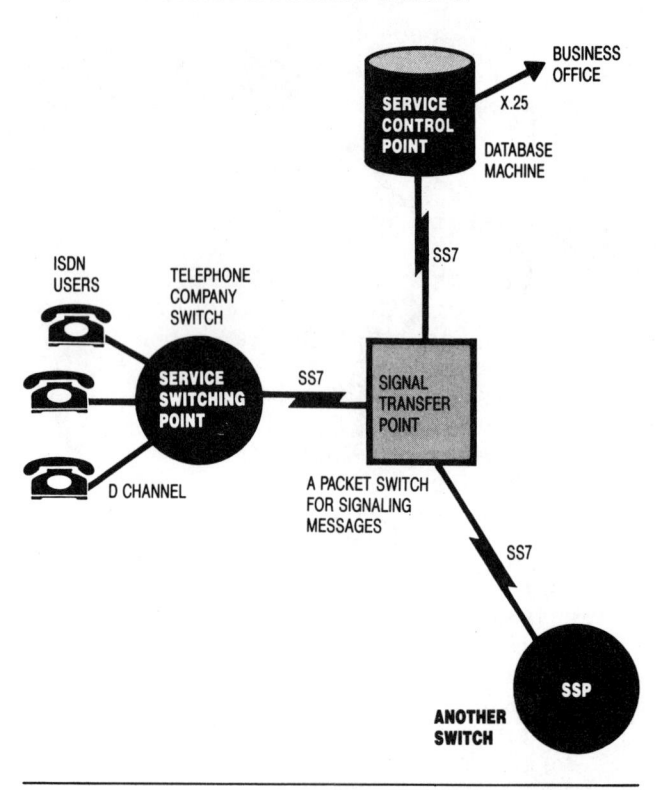

clearly indicate a wider role for SS7 within and between various private and public networks.

The simplest of these extended uses would be in private networks. Large private tandem networks have requirements that parallel those of the public network. Furthermore, private networks do not have the regulatory and operational constraints that can limit the utilization of the signaling network. In a private network, concerns such as capital investment, billing and charge-back, cross-subsidization, and security can be satisfied by executive fiat rather than a regulatory process.

Using a high-performance common channel-signaling system within a private network allows feature transparency across all of the switching nodes in the network. Management capabilities are also enhanced. TCAP and OMAP applications can be readily supported, and a private network could establish its own version of the intelligent network. Several vendors have already announced versions of SS7 for private tandem networks.

NEC offers SS7 as the standard tandem signaling technique on its line of NEAC 2400 PBXs. AT&T has not yet offered SS7 on its System 25, 75, or 85 PBX line but does provide SS7 on 5ESS switches offered to large organizations. The 5ESS is normally a central office switch, but it is offered as a PBX for very large private networks. As a central office switch it must have SS7; the capability is kept when the 5ESS is offered as a private switch. Northern Telecom has announced its Meridian Super-

Node, a private network version of its DMS family of telco switches. Thus, the SuperNode is approximately the equivalent of AT&T's 5ESS private network offering. One of the features Northern is touting on its SuperNode is the capability of providing Feature Group D access. Feature Group D is used by IXCs to provide access to their networks by the general public via the LEC. A private network using this arrangement would effectively be establishing itself as an additional IXC. A similar access by the private network to IXCs, with the SuperNode acting as the LEC, is possible. In both of these arrangements, SS7 would be used to provide signaling.

Not all PBX vendors agree with this arrangement: Both Siemens and IBM, which took over the Rolm product line, have declared that extension of the ISDN access signaling protocol is sufficient for tandem private networks. While this statement is patently true, since tandem networks are currently operated using in-band tone signaling, it is not clear that a signaling system designed for access should be forced to play a role for which it was not designed. It appears that by the time all the extensions are added, the complexity will be greater, the functionality will be less, and the extensions are likely to be nonstandard and proprietary. The larger, more sophisticated PBX customers apparently are aware of this, and there are numerous rumors of all PBX vendors starting to take SS7 more seriously.

SS7 rollout

Signaling System 7 has emerged from the standards committees and test beds and is currently operationally deployed in a number of public and private networks. However, this deployment, like any major software deployment, is not without its setbacks.

There are still occasional reports of bugs and temporary removal from service. Frequently, these bugs do not involve SS7 itself but rather the interaction of other switch software modules, such as trunk testing procedures or the generation of customer information tones with the SS7 modules. These setbacks may delay but will not prevent the ultimate deployment of SS7.

As might be expected, there are significant differences in the various telephone companies' approaches to SS7, but a clustering of similar attitudes falls out according to the LEC-IXC dichotomy. The LECs are most interested in additional services, database, and intelligent network applications of SS7. The IXCs, with the possible exception of AT&T, tend to be most interested in faster call completion.

This leads to very different priorities for establishing internetwork SS7 connections. The IXCs are impatient for the still faster call setups that internetwork connection would allow. The LECs focus on the vulnerability of the large databases that will be accessed and the need for security via a gateway screening function at the interconnection points. AT&T's schedule for releasing the interconnect software for the 2A STP is September 1989. For now, the IXCs have to be satisfied with IXC-to-IXC interconnections.

The IXCs—AT&T, MCI, and U S Sprint, as well as many of the small fiber-based companies—see faster call completion as the main SS7 boon. They are looking for two benefits from faster call setup: greater customer satisfaction and reduced access-time charges by the LECs.

AT&T had CCIS, the North American version of Signaling System Number 6, so it does not feel a great pressure for rapid deployment of SS7. SS7 is perceived as a way of doing things a little bit better, faster, or cheaper. Probably most important, it is a way of supporting ISDN. Deployment of SS7 is slated to follow the availability of the ISDN primary-rate interface at AT&T 4E ESS tandem switches. The schedule calls for 18 switches in 1988 and 68 in 1989. The total count of 4E tandems is 110, but not all need transition in order to allow universal access to SS7 and ISDN. The result is that AT&T will continue with a mix of common channel signaling based on SS6 and SS7 at least through the mid-1990s. Meanwhile, AT&T is experimenting with SS7 to the customer's premises.

MCI had some initial delays and made several false starts on SS7, but it is currently deploying SS7, and the company claims it will complete deployment by the end of 1988.

U S Sprint has been one of the most aggressive deployers of SS7. Sprint has established three pairs of STPs and two SCPs. This complex started supporting operational traffic in 1987. Sprint had problems bringing up SS7 on its DMS 250 switches, but these problems appear to be clearing up. Satisfying improvements in the call setup time within the Sprint network have been obtained. There are a number of smaller IXCs that have deployed SS7, a sampling of which is presented in the table.

If a LEC wishes to provide value-added services such as 800 service or credit card validation, common channel signaling is a necessity. In the pre-divestiture days of the U. S. telephone industry, common channel signaling was used for toll traffic. Toll traffic became largely the province of AT&T after divestiture. Most of the regional Bell operating companies (RBOCs) prefer building a new SS7 infrastructure to rebuilding a CCIS 6 environment. But the various

A sampling of smaller interexchange carriers who have deployed SS7

INTEREXCHANGE CARRIER	SWITCHES W/SS7	REPORTED SETUP SAVINGS (SECS)	INTERCONNECTS NOW	INTERCONNECTS PLANNED	NOTES
ACC	5 DEX	4.5	1	3	1
LITEL TELECOMMUNICATIONS CORP.	3 DMS	3 TO 7			
MIDAMERICAN COMMUNICATIONS CORP.	12 DEX	6		1	
TELECONNECT CO.	6 DEX				2

DEX = DIGITAL SWITCH CORP.
DMS = NORTHERN TELECOM
1 = 800 NUMBER BASE COMING
2 = EACH DEX INCORPORATES A 'MINI-SCP' TO SUPPORT 800 SERVICE. NOT CURRENTLY PLANNING TO DEPLOY SIGNAL TRANSFER POINTS BUT DO SEE AN EVENTUAL NEED.

RBOCs have different opinions about how aggressive the rollout should be.

The two RBOCs with the most aggressive deployments of SS7 are Bell Atlantic and BellSouth. Bell Atlantic has major deployments in New Jersey and substantial deployment in Pennsylvania, Virginia, and West Virginia. BellSouth has STPs in Atlanta, Ga., and Birmingham, Ala., and plans early support of citywide Centrex services. Nynex is at the other extreme; it has yet to commit to any implementation plan and still talks about the number of services that could be provided via CCIS 6. Most of the SS7 action occurs among the major LECs, but there are some smaller independents that are starting to consider implementation and the possibility of shared signaling networks and databases.

BellSouth is proud to be the LEC leader in the deployment of SS7. It is now moving to deploy in all its major metropolitan areas. By the end of 1988 it plans to complete all equal-access tandems. BellSouth has a policy of converting an entire city rather than selectively by switches—a policy it developed during the early days of equal-access conversion.

Bell Atlantic is using AT&T STPs and will be implementing the database access functions, including a LIDB and the 800 number functions, when regulations permit. But the impressive commitment is that of Bell Atlantic, which is aggressively moving ahead with plans to implement SS7 on the local access switches in order to support customized local area signaling services (Class) in northern New Jersey, in the Washington, D. C., metropolitan area, and in a few locations in Pennsylvania. Class is used to identify the calling number to the receiving party.

Southwestern Bell has moved forward on SS7 in order to implement the 800-number database, which was scheduled to go on line during the last half of this year. However, the Federal Communications Commission has put that on hold. LIDB, the next milestone, is planned for 1989. The Southwestern Bell SS7 network currently consists of two pairs of STP/SCPs; the STPs were supplied by AT&T.

U S West is deploying three pairs of Ericsson STPs. It had originally planned to bring up the 800 database as the first use of SS7; with that on FCC hold, the next milestone is LIDB, scheduled for early 1989. Initial tests of the Ericsson STPs were good.

Ameritech is just starting to get serious about SS7. It recently closed the bidding on an STP buy, but things are expected to move along quickly now. Nynex has not yet progressed to committing to deploy SS7.

GTE will start deploying within its local exchange companies by November. It is using 2A STPs from AT&T. The first to go in will be in California. The first service planned is Class on its GTE 5 switches. Eventually, it plans to have a total of four regional STPs (two pairs)—one pair in California and the second in New England and the Midwest.

Vendors
In addition to AT&T and Northern Telecom, the two main vendors of central office equipment, Digital Switch Corp. (DSC) has carved a niche for itself as an SS7 hardware vendor. Initially, DSC specialized in STPs, but it has progressed to combine the SSP's tandem call-switching capability and the SCP's database capability in its MegaHub product. DSC also produces a small SS7 concentrator that is targeted at enhanced service providers. It generally is marketed with several low-speed, 9.6-kbit/s ports and one high-speed port. With 20 low-speed links, this device would sell in the $150,000 to $200,000 range. By way of comparison, a typical rule of thumb for an STP is $20,000 per 56/64-kbit/s port.

The performance of DSC's large STPs is impressive; it is currently rated at up to 80,000 message packets per second. A typical large X.25 packet switch handles 1,000 packets per second. There are two factors that account for the speed difference between SS7 and X.25 packet switches. The first is market pull—there is a clearly perceived need for these high levels of performance. The second is the lower processing load of the connectionless service used for SS7.

Test gear is also crucial to the successful rollout of SS7. Checkout before live-traffic activation, monitoring during operation, and rapid debug in case of failure, are of critical importance to network owners. However, there is significant disagreement about how best to provide this capability.

One means of supporting that need is the built-in capabilities described above under OMAP. In addition, there is agreement that some standalone test capability probably is needed. Bellcore has built a tester that it uses for checkout for its RBOC owners. Similarly, Bell Canada has built one to test equipment entering the Canadian network.

Several test equipment vendors have manufactured standalone equipment, but there is disagreement about how this equipment should be deployed. When compared with X.25, there is universal agreement that SS7 needs much less external testing: the specification is better; there is built-in monitoring capability; and significant redundancy and automatic switch-over are provided. This would argue that the main use of test boxes would be during implementation and troubleshooting.

On the other hand, the task is probably more critical than most X.25 applications. Furthermore, when SS7 is fully deployed, there will be more services and service providers intimately interconnected. X.25 does not have the same level of involvement in the applications. This would argue for wider deployment of test equipment. The initial SS7 installations have been something of test beds, so they have been heavily provided with patch, test, and monitor capability. As field experience accumulates, it is likely that there will be simpler installations. One of the determinants of the need for external test capabilities is the data link interface used. Loopbacks must be externally provided if the older V.35 interface is employed; the built-in loopback capability of the DS0A interface reduces the need for external hardware.

Protocol Technology Inc. (PTI) is currently recognized as the leading SS7 test instrumentation vendor. In July 1988, PTI announced it was planning to be acquired by Tekelec Inc., one of the other leading SS7 test equipment vendors. PTI's boxes have simulation as well as monitor capability and tend to be complex and expensive but very capable.

Tekelec has two SS7 testers available: its older 707 single-function machine and the Chameleon 32, which has dual 64-kbit/s ports. It has implemented MTP, SCCP, TCAP, ISUP, and TUP. The company's machines also have a simulation mode that is programmed with C language commands.

Atlantic Research has an SS7 module for its Interview-750 machine; it tests only the lower layers (MTP) of SS7. Currently, the company does not offer a simulation capability.

CXR Telecom offers a similar low-end test capability in its 845 and 841 digital monitors. The two devices are similar, differing only in the means used to store captured data. CXR Telecom markets its 84Xs as economical devices with limited capability; they cost approximately $8,500. The line does not offer a simulation mode and has not yet implemented protocols above MTP.

Northern Telecom produces an external test device for internal use; it is not offered for general sale. But it is an option on a switch purchase. Northern sees a limited market for the device. Hewlett-Packard, Scotland, has shown a tester built for the European market that implements TUP and MTP. It did not test SCCP, ISUP, TCAP, or OMAP.

In conclusion, SS7 provides the control base upon which a more innovative, responsive, global telecommunications complex can be built. This high-speed, high-volume, low-delay, packet-switched network is already at work in numerous locations and soon will be quite ubiquitous. But, like the adolescent it is, SS7 is still growing, expanding, and evolving. ∎

Walter Roehr is an independent consultant who specializes in communications and computer technology. He has authored numerous articles and reference texts and follows Signaling System 7 developments closely.

Footnotes

[1] The content of I.451 also appears in the Q series of signaling recommendations as Q.931. The dual designation of this protocol reflects its dual parentage—part of ISDN (and therefore included in the I series) but developed by signaling experts in CCITT Study Group 11 (and therefore given a Q-series number).

[2] Outside North America there has been extensive deployment of the SS#7 telephone user part (TUP). This is an earlier, call-handling protocol with limited capability. ISUP provides all of the functions of TUP and a great deal more. There are no plans to deploy TUP in North America. Both TUP and the call-handling portions of ISUP interface directly with the message transfer part, without the intermediacy of the SCCP. TUP+ is being used in Europe to provide limited ISDN service.

Daniel R. Seligman, Codex Corp., Mansfield, Mass.

Mastering SS7 takes a special vocabulary

Forget the computer jargon, SS7 has its own terminology. Concepts familiar in a computer communications environment are often disguised in the SS7 lexicon.

For someone with a computer background, one of the most confusing aspects of Signaling System 7 (SS7) is its terminology. And concepts that are familiar to someone with a computer communications background are often disguised in the SS7 lexicon by unfamiliar names. Here's a handy translation.

In SS7, a network node is referred to as a *signaling point*. There are three types of signaling points. A *service control point* (SCP) supports applications that provide services such as 800 number service. A *service switching point* (SSP) is the point of origin of a request for services. And a *signaling transfer point* (STP) is, for all intents and purposes, a packet switch, capable of accepting a packet on an incoming channel and transmitting it on the appropriate outgoing channel.

The SS7 term most closely associated with a communications line is a *signaling link*. Strictly speaking, *signaling data link* refers to the physical properties of the communications line, while signaling link is reserved for a communications line capable of reliable message exchange between two adjacent signaling points. In general, multiple signaling links connect the same two signaling points for performance and reliability purposes. These multiple signaling links are referred to as a *link set* and correspond roughly to an SNA transmission group.

Similarly, a sequence of signaling points between the origin and destination of a message is referred to as a *signaling route,* and the collection of all routes between the origin and destination, a *signaling route set.*

Protocol Levels

SS7 consists of four layers, or levels, somewhat analogous to the Open Systems Interconnection (OSI) model layers as shown in the table. However, the terminology and the functions of the levels are not strictly in accord with the OSI model. Moreover, the venerable architectural rules that underpin the OSI model are often broken.

Signaling data link functions (Level 1) are concerned with providing a bidirectional communications path between two adjacent signaling points. This level corresponds directly to the OSI physical layer.

Signaling link functions (Level 2) support reliable delivery of messages between two adjacent signaling points and correspond rather closely to the OSI data link layer. Indeed, the signaling link protocol is quite similar to high-level data link control or synchronous data link control protocols. The signaling link functions are often called, collectively, the *link control function.*

Signaling network functions (Level 3) enable data messages and control information concerning outages and congestion to be exchanged between nonadjacent signaling points, corresponding more or less to the OSI network layer. Level 3 is also called the *common transfer function.*

The *signaling connection control part* (Level 4) supports several categories of connectionless and connection-oriented service as well as the addressing of individual applications on a signaling point. Such features are commonly associated with the OSI transport and higher layers.

The lowest three layers of SS7 are collectively called the *message transfer part* (MTP), a carryover from a time when these layers were thought to be sufficient to deliver user data between two remote signaling points. The MTP provides a datagram service between two signaling points.

The signaling connection control part (SCCP) was developed only as it became apparent that SS7 would have to support more than signaling and needed more explicit addressing and more sophisticated services between remote signaling points. The MTP and the SCCP are collectively referred to as the *network services part.*

The choice of SCCP as a name is unfortunate, since

SCCP is invariably confused with SSCP, even though the SNA System Services Control Point bears very little resemblance in function to the SS7 concept.

While SS7 does show the OSI influence, SS7 is characterized by a less than strict adherence to OSI principles. The signaling data link and signaling link functions correspond directly to the OSI physical link and data link layers. But the one-to-one correspondence ends at the link layer.

The signaling network functions are divided into two major categories: *signaling message handling* and the *signaling network management.* The signaling message handling is concerned with routing messages to their appropriate destinations—either to local applications or remote signaling points. It is consistent with the network layer functions of the OSI model. However, the signaling network management manifests some differences, both in terminology and in function, from what experience with computer communications would suggest.

In computer communications, network management is usually a feature that enables a user or an application program to monitor, control, and troubleshoot a network. But in SS7 parlance, the term signaling network management is used to describe how a signaling network functions. It refers to the signaling network's ability to: divert traffic from one signaling link to an alternative *(signaling traffic management);* manage the state of an individual signaling link *(signaling link management);* and exchange control messages to permit the network to adapt to outages and congestion *(signaling route management).*

Signaling link management is concerned with such link-oriented activities as activation, deactivation, and restoration of signaling links. In a conventional computer

SS7 architecture and the Open Systems Interconnection model

ORIGINAL SIGNALING SYSTEM 7 LEVELS	OPEN SYSTEMS INTERCONNECTION LAYERS	REVISED SIGNALING SYSTEM 7 LEVELS
USER AND APPLICATION SERVICE PARTS	APPLICATION	USER AND APPLICATION SERVICE PARTS
	PRESENTATION	
	SESSION	
	TRANSPORT	SIGNALING CONNECTION CONTROL PART
SIGNALING CONNECTION CONTROL PART	NETWORK	SIGNALING NETWORK
SIGNALING NETWORK		
SIGNALING LINK	DATA LINK	SIGNALING LINK
SIGNALING DATA LINK	PHYSICAL	SIGNALING DATA LINK

1. Message signal unit. *Certain fields in an SS7 packet are shared among different levels. This adds efficiency but makes it impossible to replace one protocol with another.*

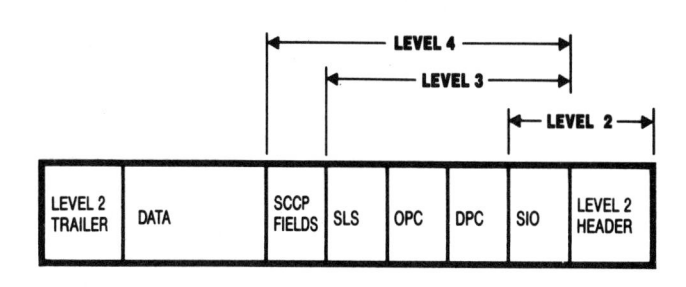

| LEVEL 2 TRAILER | DATA | SCCP FIELDS | SLS | OPC | DPC | SIO | LEVEL 2 HEADER |

LEVEL 4 spans SCCP FIELDS through DATA; LEVEL 3 spans SLS through SCCP FIELDS; LEVEL 2 spans SIO through SLS.

DPC = DESTINATION POINT CODE
OPC = ORIGINATING POINT CODE
SCCP = SIGNALING CONNECTION CONTROL PART
SIO = SERVICE INFORMATION OCTET
SLS = SIGNALING LINK SELECTION

context, these features are more likely to be associated with the data link layer than the network layer.

The SCCP is analogous to the upper half of OSI's network layer, while the signaling network functions comprise the lower half. SCCP is intended to support basic and sequenced connectionless services and three classes of connection-oriented services: basic, flow control, and error recovery with flow control. All grades of service are built upon the datagram service provided by the signaling network level. In one sense, such services seem more characteristic of OSI transport than the network layer.

Another unique feature of SS7 concerns its treatment of the fields in the message headers associated with the communications protocols at the various layers. If SS7 were handling communications in a computer communications context, its use of the fields would be suspect because fields are shared among different protocol levels. This violates the rigid separation characteristic of most computer communications protocols.

Figure 1 shows an example of an SS7 data packet or *message signal unit.* The fields corresponding to the various protocol levels are indicated.

The *service information octet* (SIO) identifies the type of message, its priority level (for congestion control), whether the SS7 network is domestic or international, and a limited address for the MTP user concerned with the message. The very presence of the field is necessary for the signaling link functions to distinguish the message signal unit from other types of signal units; the congestion and message type information is used by the signaling network functions; and the address is needed by SCCP or any other user of the MTP. Thus, a single field is used by three layers.

Routing is handled by the *routing label.* The routing label is made up of the destination address, or the *destination point code,* the source address, called the *originating point code,* and the *signaling link selection* field. The signaling link selection field is a bit configuration that permits load sharing among redundant signaling links. Both the SCCP and signaling network functions utilize the routing label. This is an example of three fields used by two different layers.

Common access to a single field makes it impossible to replace one level with a corresponding level without perturbing other levels. For example, DEC's replacement of its wide area network data link protocol with a protocol more suitable for an Ethernet would be impossible in SS7 without affecting other layers. The rigid structure of SS7 reflects its origins as a self-contained signaling architecture, which was developed prior to general acceptance of OSI layering and OSI principles.

Reliability issues

SS7 is characterized by a pronounced emphasis on reliability. In a computer communications environment, inaccessibility of services or inordinately long response times are occasionally acceptable. In a telephone service environment, on the other hand, major losses in revenue can result from even a few minutes of downtime. As a result, SS7 has highly developed reliability features.

Signaling links between adjacent nodes are generally deployed in groups of as many as eight. SCPs are laid out as mated pairs; identical service nodes are used in different geographic locations. STPs are also deployed in pairs. Numerous alternative routes typically connect SCPs with remote SSPs. The signaling network functions are designed to handle these redundancies. In addition, an elaborate exchange of SCCP messages between an SCP and its mate is required before an application can be taken out of service.

SS7 adapts to degradations and failures by propagating messages designating intermediate signaling points as *transfer-allowed, transfer-prohibited,* or *transfer-restricted* with respect to a given destination. The allowed and prohibited states indicate whether the signaling point is capable of routing traffic to the destination or whether an alternative route must be sought. This type of behavior is rather typical of computer communications environments. The restricted state, however, represents a refinement of this concept and indicates that the signaling point in question is still capable of routing traffic to the destination but the route is in some fashion degraded and that an alternative route should be used if possible. A restricted or prohibited signaling point is tested periodically for changes by an inquiry and response procedure called a *signaling-route-set test*.

SS7 supports a relatively advanced scheme for adaptation to congestion situations, as depicted in Figure 2. Each message is assigned a priority level. Congestion status is determined by the number of occupied transmit buffers associated with an outgoing signaling link. Three congestion statuses are defined for each signaling link, each characterized by three thresholds: *congestion onset, congestion abatement,* and *congestion discard*. Congestion abatement and congestion discard thresholds for a given congestion status are set respectively below and above the corresponding congestion onset threshold.

At congestion onset, warning messages are sent to appropriate signaling points. If the congestion situation does not improve, the number of occupied transmit buffers increases. When the congestion discard threshold is crossed, messages with priorities less than the congestion status are discarded. If the congestion situation improves, the number of occupied transmit buffers decreases, the congestion abatement threshold is crossed, and the congestion status is dropped to the next lower level. The congestion status of a congested signaling route set is periodically tested via an inquiry and response procedure similar to the signaling-route-set test called a *signaling-route-set-congestion test*.

---

2. Signaling link congestion. *SS7 supports a sophisticated scheme for congestion management. For each signaling link, three congestion statuses are defined.*

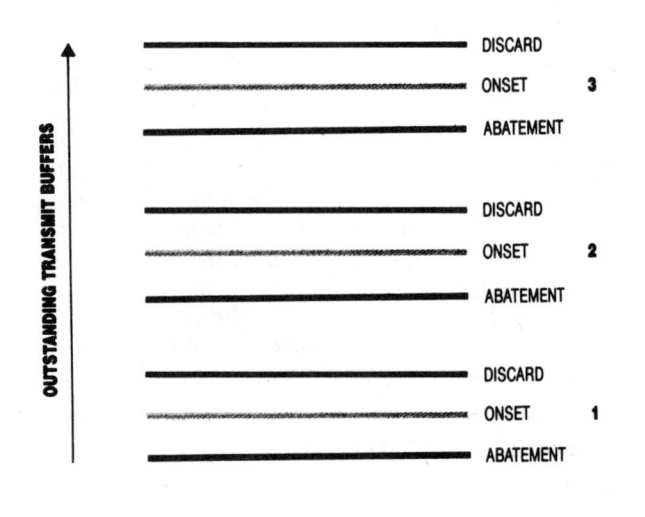

This scheme is far more complex than congestion control procedures in the more common network architectures. Contrast this scheme with, say, DECnet Phase IV where preference is given to messages routed through a node over messages originating at the node. In DECnet Phase IV, outgoing messages are simply discarded when predetermined buffer thresholds are reached. No messages are sent to neighboring nodes to indicate a congested situation.

SS7 remains uncomfortably poised between its telecommunications antecedents and its computer communications future. To the users in the computer world who will inherit it, SS7 presents the challenge of a different worldview and an emphasis on different capabilities than the more conventional computer network architectures. The first step toward mastering it is to understand SS7 terminology. ∎

Daniel R. Seligman is a consulting engineer at Codex Corp. Until recently, he was a senior consultant at Technology Concepts Inc., Sudbury, Mass. He has authored several papers on computer network analysis and design and lectured on various data communications topics. Seligman received his Ph.D. in physics from Yale University in 1976.

Section 2
Design and Operation

James Broughton, Impel Corp., Berkeley, Calif.

Computer-aided engineering enhances network design

Great graphics capabilities are proving powerful as alternatives to traditional techniques for planning with increasing ease.

Today's telecommunications professionals are constantly pressured to enhance, rebuild, or add new communications capabilities to their networks. At the same time, they are also charged with carefully monitoring and managing their network's use. As a result of these dictates, managers are starting to explore alternatives to traditional network design and management procedures. Computer-aided design and engineering devices (CAD/CAE) offer the ability to perform many design and management tasks automatically and with greater ease than do older network management hardware and software. These new CAD/CAE tools combine graphics and database management with automated engineering analysis and report generation to provide computer-driven applications that aid engineers and operators in their daily duties.

Today, computer graphics can be combined with database management software in network design tools. This allows engineers to create design concepts and use graphics to describe those concepts in a CAD/CAE database. From this data, engineers can analyze designs using calculation programs.

Further, as a by-product, CAD/CAE design tools create a facility database that becomes available to operations managers after network installation. In the future, CAD/CAE tools will be used to troubleshoot networks. They will provide operators with tools to perform network diagnostics, using information from the facility database to drive diagnostic hardware and software.

Good CAD/CAE application software is design-oriented. It steps engineers through design tasks in a way similar to manual methods. Using a CAD/CAE design tool, for example, an engineer usually begins by creating a schematic diagram of an equipment configuration with cable connections indicated between device ports. Only the major equipment and its relationship in the network need be shown—not the cable connections. Once created, these schematic diagrams form the basis for design-rule checking when equipment is placed in racks and located on floor-plan drawings.

By means of the schematics, a user can automatically extract a major bill of materials from the CAD/CAE database. Cable connection lists (Fig. 1) can be extracted as well. Each report is created automatically by sorting information that is entered during design into a user-defined report format. Once the schematic design is complete, equipment rack-elevation views can be created. Rack options for each type of gear can be accessed and displayed automatically from parts libraries resident in the CAD/CAE software.

Groups of similar equipment designated in the schematics (such as receivers, transmitters, multiplexers, and amplifiers) can be listed beside the rack options. The user can then select the rack and the corresponding equipment to be placed in it. A rack-elevation view can be created automatically with application software by inserting each piece of equipment in its appropriate location within the rack (Fig. 2). The gear is assembled and labeled in one step without user intervention. This feature enables users to standardize rack configurations and to improve the speed and accuracy of rack design. In addition, a standard bill of materials for equipment-mounting hardware can be attached to the rack.

After the racks are assembled, the CAD/CAE database contains dimensional information describing the size of each equipment bay. Engineers may make layout plans by selecting a bay from elevation drawings and placing it on a floor plan of the facility. To do this,

1. Parts listing. *Using the schematics, the computer-aided design and engineering tool can easily extract the information necessary to provide a fairly complete bill of materials. Rack options for each type of gear can be accessed and displayed automatically from parts libraries resident in the CAD/CAE software.*

CABLE-CONNECTION LIST FOR EQUIPMENT CABLING

FROM	TO	SIGNAL CABLE	EQUIPMENT CONNECTION PORTS	LENGTH (FEET)	CONNECTOR
BAY 101	BAY 104	123	M-124, PORT J-11 PP-71, PORT A-13	41	JARRELL BC-124-A
BAY 101	BAY 105	211	R-1, PORT J-13 PP-74, PORT B-12	67	JARRELL CD-156-G

M = MULTIPLEXER R = RADIO
PP = PATCH PANEL

the application software should be able to display the elevation and floor-plan drawings simultaneously in split windows on the graphics monitor. The user selects a bay by pointing to it in the elevation drawing and then places it on the floor-plan drawing by pointing to the desired location. The application software automatically displays the bay on the floor plan dimensioned to scale. Each bay is placed and arranged by the designer until the layout is complete.

After layout, the bays are given a tag number, for example, Bay 101 or Bay 102. After tagging the bays, the application software should annotate elevation and schematic drawings with the bay number (Bay 101 or Bay 102) by labeling each rack-elevation view and each item of equipment in the schematic diagrams. The user simply provides the bay number in response to a prompt.

After equipment layout, signal cable ladders and power cable ducts can be set in place by selecting paths for cable routing. The paths are displayed in the equipment plan or elevation view (Figs. 2 and 3) and are dimensioned as required. The CAD/CAE application should automatically break the power cable ducts and signal line ladders into standard lengths and correctly place the connectors required for assembly. Each connector and all nuts and bolts are automatically entered in the database so a detailed bill of materials may be generated for purchase orders. Unistrut supports for the ducts and ladders may be designed in the same way.

One of the strengths of CAD/CAE applications is their ability to provide automatic design-rule checking. This is particularly useful in cable routing. CAD/CAE implementations can automatically route signal and power cables according to the design application—once power cable ducts and signal cable ladders are in place. This is possible because the database contains information about all cable connections from the schematic diagram—as well as the equipment locations on the floor plan. The CAD/CAE software will group all signal or power cable that runs between bays, count it, and determine the type of cable required based on the equipment to which it connects.

In addition, CAD/CAE implementations automatically specify and enter cable connectors in the database, based on equipment requirements. Design rules can also be used to perform calculations that specify equipment operating parameters. This capability can

2. Elevation view. *A rack-elevation view can be created with application software by inserting each piece of equipment in its appropriate location within the rack.*

ANNOTATED RACK ELEVATION VIEW OF EQUIPMENT BAYS

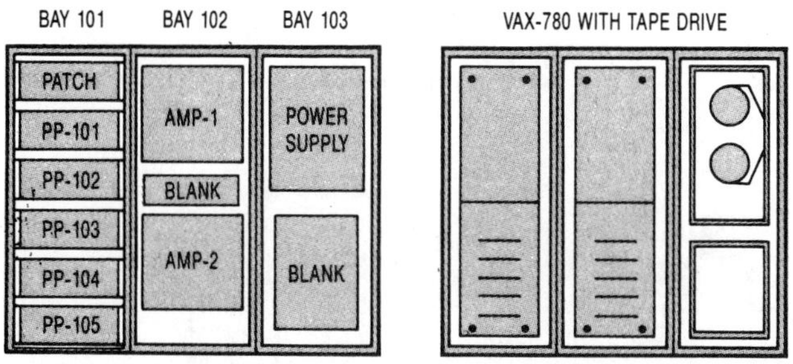

AMP = AMPLIFIER   PP = PATCH PANEL

3. Planned plenums. *Signal cable ladders and power cable ducts can be set in place by selecting paths for cable routing. The paths are displayed as required.*

EQUIPMENT LOCATION PLAN

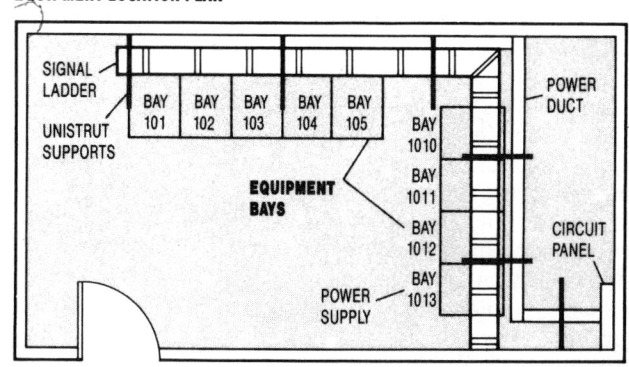

be applied to the design of baseband and broadband local area networks (LANs). In the case of baseband network design, the application can track the length of cable connecting central processing units (CPUs) to terminals or intelligent devices. The designer can then route the cable to each device from a transceiver located on the backbone network. For example, with an Ethernet network, cable lengths are limited to approximately 50 meters. If the designer exceeds that length during cable routing, the application should warn that the 50-meter design rule has been violated, and the connection should be redesigned.

The same procedure can be applied to the backbone network where the cable length, for example, cannot exceed 1,500 meters without amplification. Legal distances between transceivers can also be shown on the backbone. What's helpful to the designer is that this capability also aids in quickly positioning transceivers at the right location.

Broadband design is considerably more complex than baseband design. For broadband networks, the application software must access equipment performance information and make calculations of signal strength in order to specify components correctly as they are placed in the design. Application software can easily specify tap values by determining the signal strength, in decibels (dB), along cables from the headend to the desired tap location in the outbound direction (Fig. 4).

A database containing tap performance data is then accessed, and a tap is selected automatically based on the known dB input and the required dB output to the device being connected to the network. The designer then routes the cable to the next device, and a different tap is automatically specified. This process repeats itself until all devices are connected. Inbound taps and the network headend may be designed using similar procedures.

These capabilities relieve designers of tedious calculations, aid in system balancing, and reduce costly design errors. Equipment specifications can be automatically abstracted into schedules. Also, bill-of-materials reports (Fig. 5) can be generated without user intervention. Because they improve the accuracy of design calculations and construction documents, CAD/CAE systems offer a high measure of return on investment even before network productivity improvements are considered.

Documenting the job

Using report generator utilities, cost estimating can be accomplished accurately by accessing the cost information of each item in a project and deriving the material description and its cost. The report generator will count the number of items in the database, apply the unit cost, and subtotal and total costs for groups of materials. This information can then be printed in hard copy. Purchase orders can be created automatically as well. These capabilities greatly reduce the time-consuming tasks of job costing and procurement. Other reports that can be created easily include bill of materials, cable connection lists, equipment schedules, engineering calculations, and design error logs. A design error log is a hard copy of design rule violations and database inconsistencies in a project.

The CAD/CAE database created by the design engineering team can and should contain complete information about the network being installed. What is more, if the database is carefully created, it can be updated after initial construction and turned over to an operations group to aid in network management. Operators then have the advantage of an accurate description of components and operating parameters for the network they manage. In addition, the engineering design application software can be used to design and manage add-on jobs that occur over the life of the network. A separate application can provide operators with tools required to manage the network daily.

The facility database created by the engineering design team can be updated by the installation team after the network is constructed. This information reflects the network's conditions as it was being built. The database contains an accurate graphic representation of the network, as well as all equipment information and operating characteristics. The network operations group can then use this database to aid in daily facility management.

Facility management software tracks the network changes and generates operations and inventory reports. For example, with broadband LANs, operations managers must frequently make network changes as employees change jobs or as new equipment is added to the network. Facility management software can aid the operations group in reconfiguring the network to accommodate new requirements. For instance, operations managers can project network requirements based on the number and location of employees within a facility. Once design criteria are established for the location, the engineering design application can be used to reconfigure the network. The reconfiguration may require the addition of new taps, which can affect design of the headend, remote amplifiers, and other taps. The CAD/CAE engineering design application can be used to design, engineer, and specify the new

***

4. Broadband. *For broadband networks, the CAD/CAE tool must access equipment performance information and make signal-strength calculations.*

BROADBAND NETWORK SCHEMATIC OF HEADEND AND FIRST DEVICE

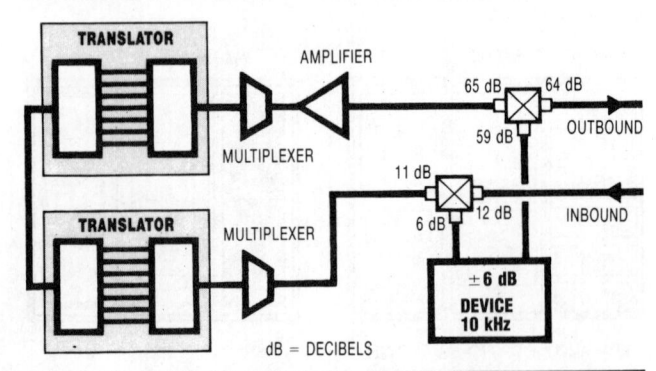

5. Material lists. Because parts lists for each network component are included in the database, a complete bill of materials can be generated automatically without user intervention. Because it improves the accuracy of design calculations and construction documents, CAD/CAE offers a high return on investment.

BILL-OF-MATERIALS REPORT

ITEM NUMBER	MATERIAL DESCRIPTION	QUANTITY	UNIT COST (DOLLARS)	SUBTOTAL COST (DOLLARS)
12792	LADDER MOUNTING BRACKETS, PART 127-I, ALUMINUM WITH 1/2" BOLTS-1 1/2"	67	27.95	1,872.65
48209	POWER CABLE DUCT, PART 1243-Q8, 1/4" STEEL, 4" WIDE × 3" DEEP, LENGTH-FT.	156	16.71	266.76

configuration by adding information to the existing database to describe new operating parameters. Calculation programs can then be used to design and specify equipment operating requirements.

Finally, the CAD/CAE application can be used to issue construction documents, such as drawings, bill of materials reports, and procurement documents. With the addition of this new information, the CAD/CAE facility database accurately describes the new network configuration. As other changes are made to the network, this cycle is repeated. Used in this way, the CAD/CAE application's family of tools becomes the primary working environment in which designers and operators may carry out their daily duties. The result is more accurate communications among the key personnel, reduced errors, and increased productivity.

Making the buy
With an adequately controlled CAD/CAE facility database, companies can better control inventory by using the database as a mechanism for generating inventory reports for management. Application tools, for example, can help make inventory projections based on personnel shifts within the organization. Many facility managers responsible for moving entire corporate divisions now use CAD/CAE applications for such tasks as adjacency analysis, space planning, lease management, and procurement.

These applications can supply a great deal of information to telecommunications personnel responsible for helping implement corporate moves. The facility planners' database establishes design criteria for the engineering and operations groups to use in planning network design and facility management. When the CAD/CAE implementation is used as a tool for managing people and facilities, its importance to the organization increases dramatically.

Like network diagnostics, facility management applications make heavy use of report generation tools. For this reason, the flexibility of report generation utilities should be a high priority when the purchase of a CAD/CAE system is considered. In fact, it is generally wise for purchasers to move carefully when deciding to buy a CAD/CAE system and to be sure of wholehearted management commitment prior to purchase. In addition, careful selection of a reliable vendor who will assist in installing and implementing the CAD/CAE tool is a must for success.

Among the many important facts to consider is the time required to train personnel and develop parts libraries, custom menus, and report formats as well as custom software programs that fit corporate requirements. In general, this can take six to 12 months of hard work by many people before an entire design project can be accomplished with CAD/CAE.

Several tasks are necessary to prepare an organization for use of a CAD/CAE implementation. At minimum, these tasks include the following items:
■ Equipment component libraries must be created that conform to corporate design and operating practices. Graphics representations must be drawn, and performance information must be defined for the parts.
■ Custom report formats must be defined to provide report generation that conforms to standard corporate practices.
■ Menus must be created that provide the user interfaces to standard equipment components, standard report formats, and calculation programs.
■ A bill of materials must be added to equipment components to indicate miscellaneous materials that must be purchased but are not shown on drawings (nuts and bolts for installing equipment, for example).

Each of these tasks should be completed prior to attempting a complete design project using the CAD/CAE design tool. In addition, management must recognize the need to train thoroughly all people who will be involved in the project—including supervisory personnel not involved in the day-to-day use of the CAD/CAE device. In this way, all team members will understand the need to manage the daily use of the CAD/CAE implementation carefully so that it is used to its fullest. ■

James Broughton, now with Amdahl Corp., was product manager with Impel Corp. when he wrote this article. Broughton received a B. S. in architectural engineering and an M. S. in mechanical engineering from the University of Texas at Austin. He has been active in the development of applications for CAD/CAE in the construction and telecommunications industries.

Robert Cameron, Northern Telecom Inc., Richardson, Tex.

The planning and managing of packet-switching networks

A soup-to-nuts recipe for managing a packet network, from forecasting user needs to administering and maintaining the finished product.

Packet-switching networks contain many features that allow network managers to analyze, forecast, engineer, install, maintain, market, provide, and administer packet-switching services efficiently. In contrast to more rigid networking architectures, a packet network is a living, growing entity requiring an operating organization that can fully exploit the benefits of the network. Organizations that have implemented packet networks have, in almost all cases, underestimated both the magnitude of the project and its impact on the organization. However, the benefits to users and the positive impact on the communications budget make a packet-switching network a significant corporate asset.

In the following examination of the key activities in the operation of a packet network, the Canadian Datapac network will be used to illustrate the major points of network management. Suggestions are made for structuring the operating organization in order to maximize the efficiency both of the organization that supports the network and of the network itself.

Telecom Canada, an association of the nine major Canadian telephone companies, and Telesat Canada, Canada's national satellite carrier, introduced Datapac in 1976. Currently more than 40,000 users place more than 72 million calls annually and transfer more than 12 million kilopackets of data (one kilopacket equals 1,000 packets of up to 256 characters each). The network has experienced growth rates in some years of 60 to 100 percent with more than 30,000 additions, deletions, or changes to subscriber service data each year. Datapac, therefore, is a good example of managing a network through massive growth transitions.

Figure 1 identifies the major tasks involved in operating a packet-switching network. The hub of these activities consists of a set of network administration and management features and an efficient operating structure. The manager of a small private network may have sole responsibility for all of these activities. Conversely, management of large public networks may be carried out by several departments, each supporting one function. Regardless, both situations require well-designed network administration and management implementations. While all of these activities may take place at once, the cycle starts logically at the analysis function and proceeds clockwise through engineering to administration and maintenance.

In both private and public networks, the manager must first be able to analyze exactly what is happening on the network. Based on this analysis and other inputs, the manager can forecast network growth and, when appropriate, engineer and install new components. Once installed, the network requires maintenance to ensure the maximum level of service. Network services can then be marketed in a public network or, in a private network, promoted to various departments within the organization. In either case, user ports must be provided. Finally, administrative activities, such as billing and traffic measurement, must be carried out. The efficiency with which these activities can be performed depends heavily on the features of the network administration and management organization, and on the structure of the operating organization.

Terminal use

Prior to installing a packet network, it is not always easy to determine the precise number of terminals within an organization or to know how each terminal is being used. Experience has shown that managers typically underestimate the number of terminals in their organization. This is because the proliferation of microcomputers, many with communications capabilities, is not

1. From the left. Key activities in operating a packet-switching network can be segmented into clearly definable areas of administration and management.

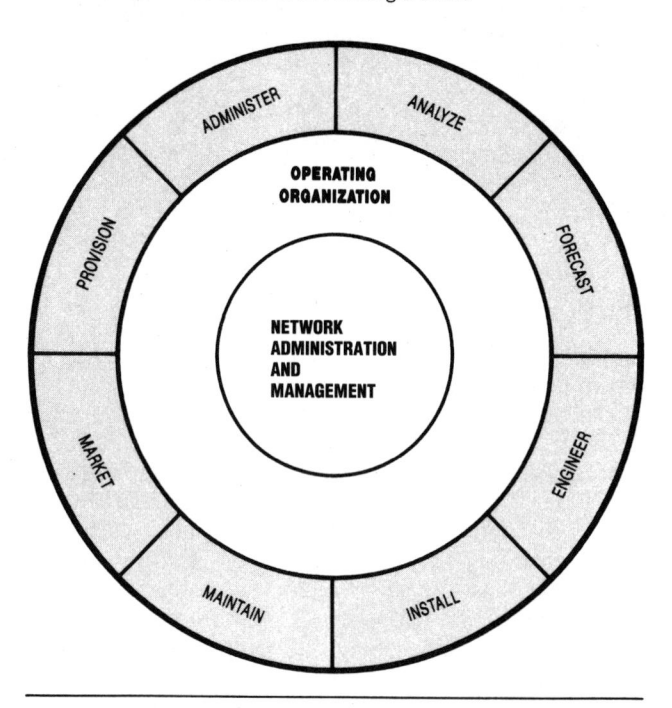

well-controlled in most organizations. Individual departments, for example, purchase many microcomputers and terminals that may not appear on the data processing budgets. Nevertheless, the network manager is frequently required to make an informed guess on the number of terminals, the average message sizes, and the volumes of incoming and outgoing data. However, a great degree of care should be taken in establishing these figures.

For example, on the Datapac network, the typical 1.2-kbit/s asynchronous terminal has an average use outbound to the mainframe of less than one half of 1 percent, and an inbound use of about 6 percent. Average message sizes are approximately 20 characters outbound and 300 characters inbound. These low figures are not surprising, given the relatively low operator typing speed and idle "think time" associated with interactive terminal input. If, on the other hand, the asynchronous port supports a microcomputer with relatively large file transfers to and from a mainframe, the line use and message sizes will be significantly larger.

While traffic volumes must be estimated prior to network installation, once the network is working, the network administration and management provides detailed statistics on traffic volumes, user community of interest (who is talking to whom, and how much), and individual processor, line, and internodal trunk use.

It is very important to analyze community of interest (COI) data in addition to trunk use. The first impulse, after identifying a heavily used trunk, might be to add a parallel trunk. However, on Datapac, this could have the effect of drawing even more traffic onto a particular route, even though examination of the COI information could indicate a more efficient topology change.

The next step in the cycle is the forecast phase. The analysis should have identified trends in COI and component use for input to the forecast. This data alone, however, will not provide sufficient information to forecast future network demand accurately. Managers of private networks must survey user departments to ensure that there will be sufficient capacity available to accommodate new traffic resulting from planned applications. For example, an electronic mail server can be overlaid on a packet network easily. This application could have significant impact on network evolution. Public network operators must gather input from market forecasts of current services and from service planners who may be working on the introduction of new value-added services.

Key to accurate network engineering
In the Datapac organization, for example, the combination of trend analysis, market forecast information, and services planning input results in the production of a Datapac Current Plan. This plan identifies, for the five-year planning period, new equipment locations and capital requirements for each six-month segment. This procedure allows managers to budget and schedule so that the network can be expanded in an efficient manner.

The third phase is engineering. The key to accurate network engineering is the ability to feed network statistics and COI data, as well as forecast growth, into an interactive network design program containing the characteristics of the network components. No network design program available today has the capabilities to produce the optimal design. That is why it is important for the designer or network engineer to be able to interact with the program and to postulate "what if" scenarios. The engineer can change node and concentrator locations and quickly observe the resulting shifts in traffic. Trunks and other network components can be removed from service to simulate failure conditions, and the loads due to alternate routing can be identified in order to ensure that sufficient capacity is available on the secondary paths.

An automated design tool that allows the manager to experiment with such hypothetical scenarios is essential in networks of more than five or six nodes. For instance, additions to Datapac, with more than 52 nodes and 150 full-duplex 56-kbit/s trunks, could not possibly be performed manually. The engineer could not calculate the source and destination of all of the traffic that would move to the new routes, including which path it moved from and which paths it would reroute to in case of a trunk failure.

Installation
In the network management cycle, the installation phase is the most straightforward activity. Most vendors' equipment is modular and fitted with connectors for ease of installation. With Datapac, new nodes are first tested on a standalone basis to ensure basic integrity. Then they are loaded with test software and connected to the network via a single trunk to allow the network control

center (NCC) to monitor performance. The single trunk attachment prevents the node from acting as a tandem switch (handling trunk-to-trunk traffic) and perhaps affecting users in other parts of the network during the test phase.

Datapac network managers use performance verification tools to ensure proper operation of the new components. Only after the node has operated within the network for 48 hours without generating any alarm conditions to the NCC is the operational software loaded. All trunks are then enabled, and the node becomes an integral part of the network. The nodes immediately recognize the identities of new neighbors, as well as the speed of the trunk connections, and automatically update their routing tables to include the new paths. The nodes then send routing table update packets to their neighbors. These updates propagate until a node receiving an update packet does not have to revise its routing table. Thus the NCC is not required to handle topology changes.

Maintenance

The maintenance phase is the next step in the network-management cycle. In a packet network, maintenance activities are based on a philosophy that prevails in the marketing, provisioning, and administration segments of the cycle as well.

Most references to packet switching depict the network as a cloud. The implication is that users need only concern themselves with gaining access to the cloud, not with the procedures that are used within the cloud for call control and routing and for delivery of data. While it might be somewhat unsettling to users to visualize their data disappearing into a cloud, this logical segregation of user access facilities from the backbone network greatly simplifies network maintenance, marketing, provisioning, and administration.

In the backbone network, for example, Datapac instantly detects faults and automatically initiates recovery and fault analysis and reporting procedures. As discussed previously, adaptive routing techniques immediately reroute data packets without affecting service to users. Redundant network components ensure a high level of service. This implies that if network availability is maintained, any problems must be due to incorrect user procedures or caused either by the source or destination access line or by the user equipment.

In Datapac, responsibility for network availability and individual user service rests with two separate departments. The NCC is charged with optimizing network availability and reliability, while individual node-site personnel handle access-line problems. The operators at the nodes also repair any faults of local equipment, working closely with NCC personnel.

Node personnel cannot add or remove any network component without NCC clearance. Weekly availability and reliability reports are generated, and abnormalities are analyzed and corrected. NCC operators log each network failure, noting the action taken, the action to be taken should the fault recur, and the serial numbers of the circuit packs involved.

Datapac node operators have a comprehensive arsenal of diagnostic and monitoring tools to allow them to respond to user problems. Operator commands allow automatic line loopback tests, access to line statistics (such as CRC errors and retransmit counts), and the ability to monitor actual physical, frame, and packet-level states. The operator can also query the user configuration data within the node. In addition, variable thresholds can be set, on a per-line basis, to raise an alarm when the number of events, such as CRC errors, exceeds the preset threshold. Using these tools at the source and destination ports, operators can isolate and correct individual user problems rapidly.

The NCC and node operators are the first level of support in the hierarchical maintenance organization shown in Figure 2. Unresolved network or customer problems are automatically escalated (after a predefined time period) to the appropriate technical support engineer. Faults that cannot be rectified at this level are ultimately referred to the equipment manufacturer or the network developer.

The marketing managers responsible for individual access services and value-added service overlays are presented with the usage-and-trends data required to manage each service offering. The only requirement is that market forecast data be fed into the forecasting phase of the network management cycle. The service provisioning phase is also carried out with minimal interaction with the administrators of the backbone network. However, there must be some relationship between the assignment of a new user to a processor and determining the remaining capacity of the processor.

In the ideal world, "rules of thumb" could be created to indicate whether or not a given processor had enough capacity remaining to accommodate a given line. For example, each processor could be allocated an arbitrary number of load units. Each line being assigned to the processor would reduce the number of units by a specific amount, based on line speed and typical characteristics of the line type (asynchronous/synchronous, polled/contention mode, and so forth). Another method would be to assign lines to a processor until buffer use and delays exceeded acceptable levels.

Load limits

Datapac managers use a more sophisticated technique for assigning lines than the one described above. The network administration software on each node provides the Datapac assigners with a weekly report on assigned and remaining capacity for each line processor on every node in the network. The report generator analyzes peak and average buffer use, used and remaining processor cycles, and any degradation to service of existing users. It thus arrives at guidelines for assignment of new ports. At this point, a rule of thumb is used. Since the assigning function is a clerical process, some method is required to simplify the assigning guidelines. The approach taken in the Datapac network is to produce the assigned and remaining capacity report in terms of "load units." New lines are then assigned, using up load units until the processor is at its maximum assignable capacity.

This method also takes into account the valid assertion that the use of load units, as the sole method of allocating network capacity, is an ineffective method of assigning ports to a network. The load unit is a dimensionless, arbitrary unit. Its only purpose is to provide clerks with a conservative, simple method of assigning new lines. The next runs of the assigned and remaining capacity reports will track the actual load on the processor due to the additions, and it will provide an additional indication of the number of load units remaining. In some cases, the report may indicate that lines may have to be moved to processors that are less heavily used. The objective is maximizing the use of node equipment, while providing responsive service to network users.

With the exception of the assigning process, the access-line design, equipment ordering, and hardware installation activities are similar to the provisioning of any other dial-up or dedicated data circuit. However, with packet-switching network service, there is an added dimension to the service provisioning process: the creation and downline loading of subscriber service data. The network administration and management scheme should contain a subscriber services database. This database should feature user-friendly access so that new subscribers can be added easily and their service data validated and downloaded to the network nodes. It should be possible to add or delete a user or to change port characteristics without affecting service to other network users.

With the hardware installed, and the service data resident in the network nodes, service can then be activated. On Datapac, this activation can be automatic, occur at a future date, or be activated by operator command.

The key criteria, when evaluating subscriber services management software are as follows: the input is user-friendly, the downline loading is nondisruptive to network operations, and users can have their service manually or automatically activated without affecting other users on the network.

Network supervision

The last step in the cycle is administration. With the network operational and users in service, there are a number of administrative functions that must be carried out. The network manager should be sure that the network administration and management implementation provided by the network vendor is capable of supporting these activities.

The manager should assign responsibility for the administration of the customer services, accounting, statistics, and alarms databases. Each department's needs for read/write access or read-only access to the various files should be evaluated. Read/write access should be permitted only when absolutely necessary. Departmental needs should be met by stock or custom application programs that automatically access the required files, retrieve and process the information, and

2. Upping the problem. *The network control center and node operators are the first level of support in the hierarchical maintenance organization.*

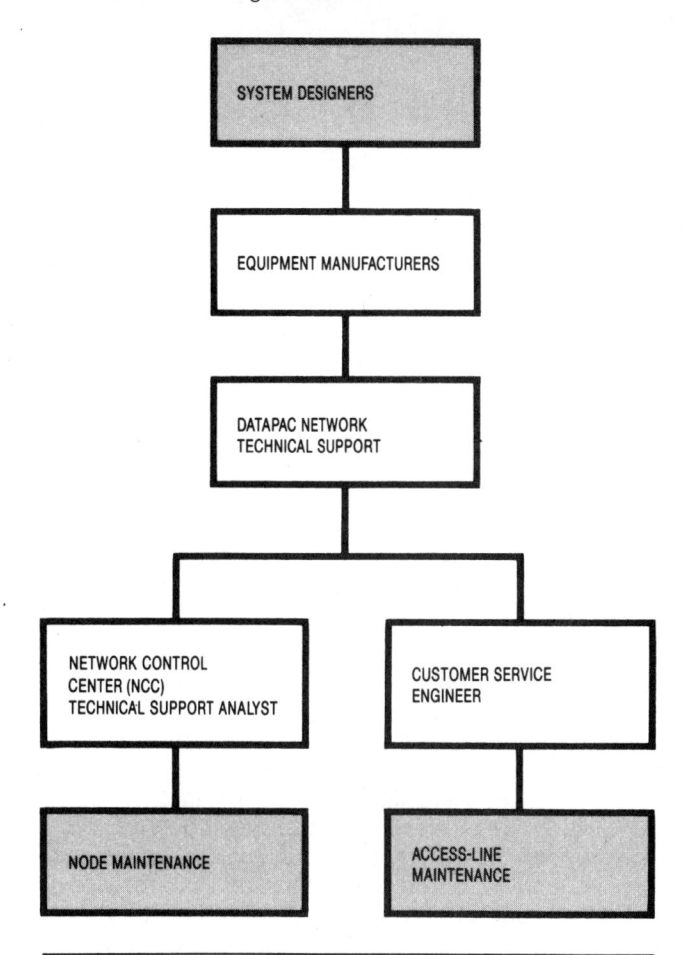

put the results in the format required by the user. File backup features are also necessary.

Datapac network node operators can produce regular reports on usage levels of each network component. As noted earlier, the network manager needs these reports in the analysis and forecast stages for planning network evolution. The reports are also critical to the assignment process in the provisioning phase. An additional administrative function, not directly related to a network administration and management scheme, is nevertheless critical to reliable network operation. There is a direct correlation between network performance and the quantity and quality of training courses attended by operations, maintenance, and administrative personnel. The success of the Datapac network is partially due to the emphasis that Telecom Canada places on internal training. Network operators can take advantage of the wide variety of training courses developed by Telecom Canada on virtually every aspect of network operation. ∎

Mr. Cameron attended Ryerson University, Toronto, Can. Prior to working for Northern Telecom in the United States, he was a systems engineer in the company's Canadian headquarters.

James H. Green, Pacific Netcom Inc., Portland, Ore.

Microcomputer programs can be adapted for data network design

Without buying costly simulation programs, a network designer can harness the microcomputer to perform network modeling.

A data network designer's job is partly intuitive and partly analytic. While the only way to gain the former is through experience and training, acquiring the analytic tools for the job requires merely learning how to apply mathematical formulas. Using such formulas is not overly complicated and can be done with a hand calculator, but this is a tedious and time-consuming task, subject to error. Moreover, data network design is a matter of optimizing cost and service, which often requires reiterative calculations—a particular specialty of computers. Clearly, experienced designers can multiply their effectiveness with computerized design aids.

Several programs for network design are available from time-sharing services or as software packages for mainframe and personal computers. Although these programs differ technically, they share one characteristic—they are expensive. For someone who regularly prepares network designs, these programs are well worth the cost, but for those who prepare designs only occasionally, the cost is difficult to justify.

Fortunately, other types of microcomputer packages can be adapted to network design. Two of these, SmartForecasts from Smart Software Inc., Belmont, Mass., and Formula/One from Alloy Computer Products Inc., Framingham, Mass., can be particularly useful.

Networks are designed by one of three methods: analytic modeling, simulation, or a time-honored method that could be called "seat of the pants." This latter method, performed by adding traffic load until service degrades to an unacceptable level, is probably the most prevalent because it can be quickly applied and requires no special equipment or knowledge. When the load exceeds capacity, however, response time degenerates and costs increase.

The simulation method uses a computer to represent the number of connection attempts, the length of messages, and other network characteristics. If the model is correct, a computer can accurately simulate network performance. However, simulations are costly, either in programming effort or in out-of-pocket expense.

The modeling technique uses formulas to predict the performance of the network components. Although no network conforms exactly to a modeling formula, the results in practice are close enough that the network will perform well if the modeling design assumptions are reasonably accurate.

The effectiveness of a data communications network is generally measured by two factors: throughput and response time. Throughput is defined as the number of information bits that the network can transport per unit of time. Response time is variously defined but can be considered the elapsed time from pressing the enter key until the first character of the response is registered at the receiving terminal. An ideal data communications network does not impair throughput or response time, but, of course, such an objective is never actually realized. The task of a designer is to configure the network so that the components between the sending and receiving ends impair overall network throughput as little as possible.

Data network design requires, first, selecting a topology: that is, the general pattern or configuration in which terminals are connected to the host. The simplest topology is a point-to-point circuit. For batch applications, the design problem is to maximize throughput. For interactive applications, the objective is to minimize response time. In the star topology, communications lines radiate from a central host to individual terminals; typically, these lines are designed as individual point-to-point circuits.

Computing the response time of a point-to-point circuit is fairly straightforward. Simply put, a calculation of the response time involves accumulating the delay in gaining network access, transmission time of a data block, and host response time. This latter function is a given to the network designer and is not a function of the network design.

The tree topology, shown in Figure 1a, presents a different design problem because the circuit is shared by multiple terminals in order to reduce circuit costs. The penalty is the overhead imposed by a circuit-sharing strategy. The most common method of sharing is polling, in which the host addresses a polling message to each terminal sequentially. The terminal responds with a message or an end-of-text (EOT) signal indicating it has no traffic to send. In general, the throughput of a polled network is lower than a point-to-point network because the polling messages add overhead (noninformation) bits. Because polling networks are used primarily for interactive applications, measurements of response time are of the utmost importance.

The design of polled data networks is further complicated by the variability of message lengths. To calculate response time, the designer must acquire statistical information about overheads of polling messages, the distribution of length of messages, and the probability of occurrence of each message type. The statistical mean and variance are used in response-time formulas (as will be illustrated).

A third type of topology, hierarchical, is shown in Figure 1b. To conserve circuit costs, circuits from the terminals are concentrated at branching locations. Terminals are assigned to concentrators or multiplexers and share a backbone circuit to the host. Concentrator locations are selected to minimize circuit costs while meeting response-time objectives.

(The number of possible topologies in a complex hierarchical network is too great to handle by formula. Determining concentrator location—called topological optimization—is accomplished empirically. Two common methods, called add and drop heuristics, are applied manually and are beyond the scope of this article.)

A different kind of problem faces the designer who must decide how to size shared appartus. Common problems include:
- Determining the number of dial-up ports on a time-shared computer.
- Determining the number of circuits between a concentrator or multiplexer and a host.
- Determining the number of devices in a PBX modem pool.

The techniques for making these calculations are based on queuing theory, which has long been applied by telephone engineers to determine the number of circuits required between switching machines, the number of operator service positions needed, and the quantity of line circuits required to serve a common group of users. Formulas for these applications are quite complex and somewhat impractical to solve by hand. Circuit designers typically use standard tables to

1. Polling. *Each terminal is polled for data in networks with a tree topology (A). A hierarchical setup (B) uses multiplexers and avoids polling overhead.*

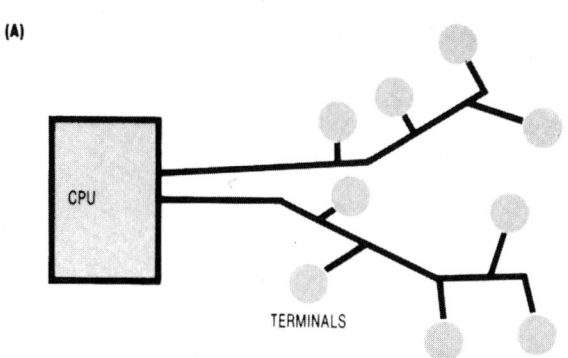

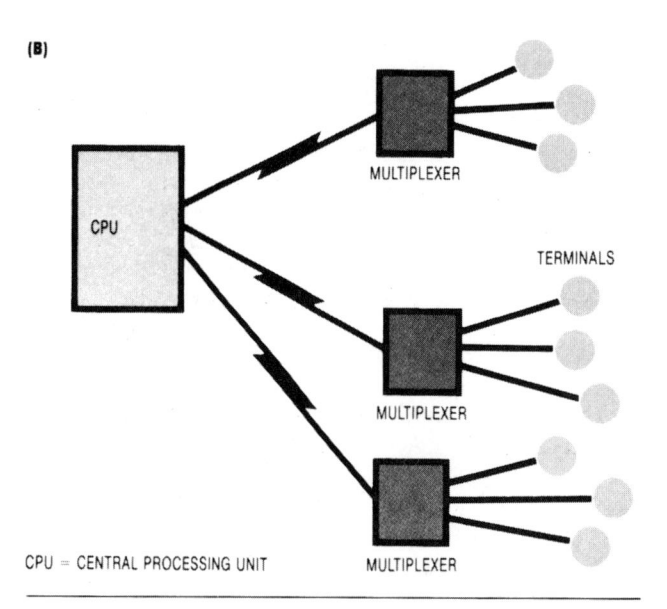

perform computations such as these.

To apply the above techniques, a designer must have access to a computer, tables, or curves plotted from the formulas. The software packages described here have the versatility to handle design problems with ease and accuracy without requiring expensive programming. The primary shortcoming of these packages—compared with a more expensive custom-made program—is the lack of tariff information. This is not a serious drawback to the occasional designer, who can optimize the network on the basis of distance and can obtain cost data from common carriers.

Forecasting with SmartForecasts

The starting point for all network designs is a valid forecast. Forecasting is probably the most difficult task the designer faces and is undoubtedly responsible for more design failures than any other factor. A forecast begins with a measurement of current usage and projects future usage by using statistical techniques. The difficulty of collecting information should not be underestimated. In a new application, for example, usage information may be unavailable, meaning that the designer must correlate expected usage with some other factor.

The designer has two primary strategies for developing a forecast. The first is using a curve-fitting technique to project historical information into the future. Several techniques, such as linear regression, exponential smoothing, and moving averages, are typically used.

The second strategy is to correlate usage information with another factor of known or assumed accuracy. A statistical technique called regression analysis is used to project one factor from another. For example, a grocery chain might correlate electronic cash register transactions with growth in sales volume. A credit union could correlate data transactions with members' deposits and withdrawals.

SmartForecasts is a statistical package available for the IBM PC and compatible machines. Its curve-fitting programs are particularly useful in forecasting because the program automatically determines which of five forecasting methods yields an answer that best fits a table of data. The program is designed to accept weekly, monthly, quarterly, or annual figures and to project them through several time periods.

The forecast period depends on the degree of control the network manager has over network reconfigurations. If the circuits can be dynamically rearranged under software control, a forecast of daily or weekly demand is an invaluable tool for administering changes. Longer-term forecasts are used to project circuit or equipment additions and rearrangements and would most likely use figures that were collected on a monthly or quarterly basis.

For example, consider the weekly transaction volumes shown in Table 1. This table is entered into SmartForecasts' input form and is projected for several periods into the future. The more historical information that is available, the more accurate the forecast will be (provid-

2. Seer. *There is a 90 percent chance that the future values in (A) will lie within the indicated range. The regression forecast in (B) appears a better predictor.*

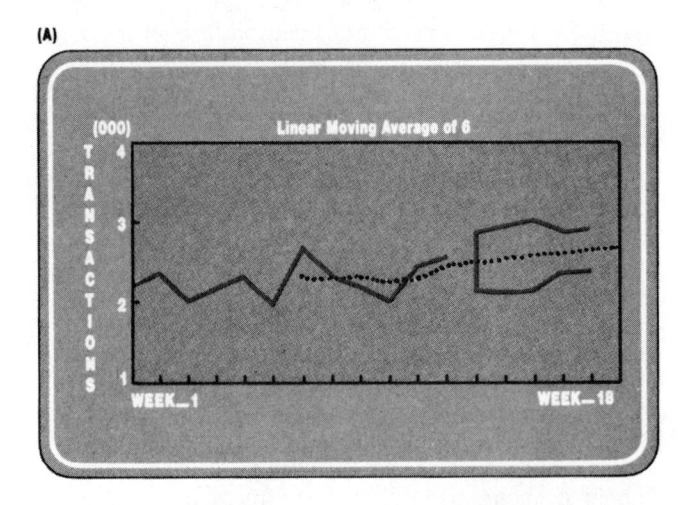

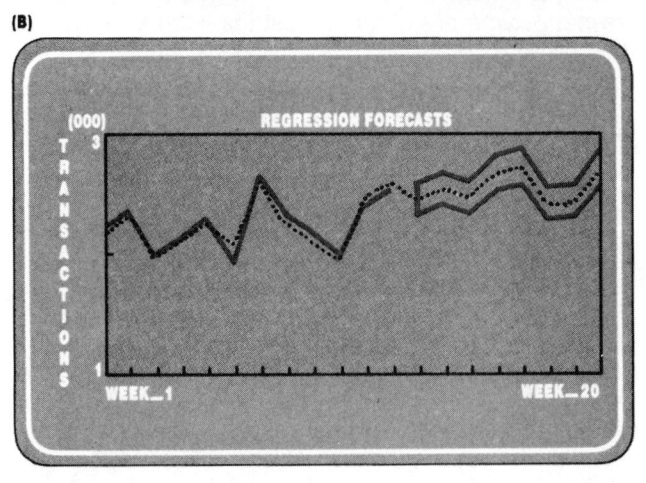

ing an unforeseen change does not occur). The possibility of such changes makes it inadvisable to forecast too far into the future. As an aid, the SmartForecasts input screen cautions the user on the probable limits of accuracy of future figures.

SmartForecasts can allow the operator to choose one of the five methods of forecasting; the program can be instructed to hold a "tournament" among the five methods and to choose the one with the best result. Table 1 also shows the results of the forecast data for six time periods, with upper and lower limits of accuracy shown. Of course, there is no guarantee that some unforeseen event will not force the actual results outside the forecast. If the designer knows of such initially unanticipated events, such as a clearance sale, the program provides for manual adjustments to the forecast. SmartForecasts provides graphic as well as tabular output. Figure 2a shows a graph of the data listed in Table 1.

Another effective way of forecasting is to correlate transaction volumes with some other forecast of business activity that the company regularly produces. Forecasts of business activity produced by trade as-

Table 1: Weekly transactions and sales

WEEK	TRANSACTIONS	SALES
1	2,200	330,000
2	2,364	345,000
3	1,995	319,230
4	2,134	325,760
5	2,321	339,700
6	1,944	325,650
7	2,664	365,250
8	2,325	340,300
9	2,186	327,600
10	1,998	315,200
11	2,439	355,900
12	2,560	366,000
PROJECTED:		
13	*	355,000
14	*	362,000
15	*	356,000
16	*	372,000
17	*	377,000
18	*	352,000
19	*	354,000
20	*	375,000

Table 2: Forecast of transactions using sales data

WEEK	APPROXIMATE 90 PERCENT FORECAST INTERVAL		
	LOWER LIMIT	FORECAST	UPPER LIMIT
13	2,345.88	2,476.91	2,607.94
14	2,429.83	2,565.71	2,701.59
15	2,357.97	2,489.60	2,621.23
16	2,547.37	2,692.56	2,837.75
17	2,605.24	2,755.99	2,906.73
18	2,309.44	2,438.86	2,568.28
19	2,333.77	2,464.23	2,594.69
20	2,582.16	2,730.62	2,879.07

sociations or government agencies may also correlate with a company's data transactions. SmartForecasts uses regression analysis to project from one forecast to another. For example, Table 1 also presents sales volume as well as data transactions. The projected sales for weeks 13 to 20 are added to the table. The program is instructed to regress transactions against sales. If desired, as many as 10 different variables can be used for this analysis.

The results of the regression can be displayed in both graphics and tabular form. Table 2 and Figure 2b show the results of the regression forecast. The results are similar but not identical to the linear moving average forecast of Figure 2a. In this example, the forecast limits are much narrower with regression because data transactions correlate well with sales. The advantage of using regression analysis is that it is an effective way to take advantage of forecasts of business activity made by people whose particular specialty or responsibility requires them to stay in touch with expected changes.

Figure 3 shows another SmartForecasts option, a display of the degree of fit between transactions and sales. The tight cluster of dots about the diagonal line indicates a close correlation between sales and transactions. The R-square figure in the lower right corner indicates the degree of the fit. The closer the fit is to 100 percent, the higher the correlation between the variables.

These two examples give only a brief indication of the capabilities of SmartForecasts. The program also includes the capability of selecting the leading indicators for a specified variable and making adjustments because of seasonal fluctuations. The program also allows the final results of the forecast to be plotted in a variety of forms, including histogram, scattergram, or time series graphs.

Network design with Formula/One

With a forecast of demand and the network topology chosen, the designer is ready to begin configuring the network. Formula/One is a general-purpose problem-solving program that insulates the user from the mechanics of mathematical manipulation. The program is admirably suited to solving the formulas needed to optimize a data network. It accepts algebraic formulas with any degree of complexity; it also accepts formulas with many different parts, solving each formula in turn for an unknown and supplying the result to a subsequent formula. The answer is displayed as a value, table, or curve, depending on how the user constructs the problem. "What if" questions are answered by changing values and recalculating or by giving the program a range of values to compute. Although Formula/One is not designed specifically as a forecasting program, it is also capable of regression analysis and curve fitting. The same forecasting techniques described for SmartForecasts above can also be applied by Formula/One.

(The example discussed below uses half-duplex circuits; full-duplex circuits can be designed by making minor modifications to the formulas.)

In a point-to-point circuit, the designer's objective is to maximize throughput and minimize response time. Both can be calculated quickly with Formula/One. Throughput is a function of several variables:
- Modem reversal time.
- Propagation delay of the transmission medium.
- Line error rate.
- Number of overhead or noninformation bits.
- Transmission speed.
- Data block length.

The designer has little control over some of these factors. The propagation delay and error rate are a function of the type of circuit selected. The number of overhead bits and the bits per character are functions of the protocol. Modem reversal time varies for each modem. While the designers have some control over these factors, their usual task is to select the modem speed and data block length that maximize throughput. If the block length is too short, throughput drops because of the number of overhead bits. If it is too long, throughput drops because of time spent in resending faulty blocks.

One formula for estimating throughput on a point-to-point circuit is:

$$\text{Throughput} = \frac{K_1(M - C)(1 - K_2 K_3 E)^M}{(M/R) + dT}$$

3. Tight fit. *The degree of fit is a function of how closely the dots are clustered around the diagonal line and the R-square figure in the graph.*

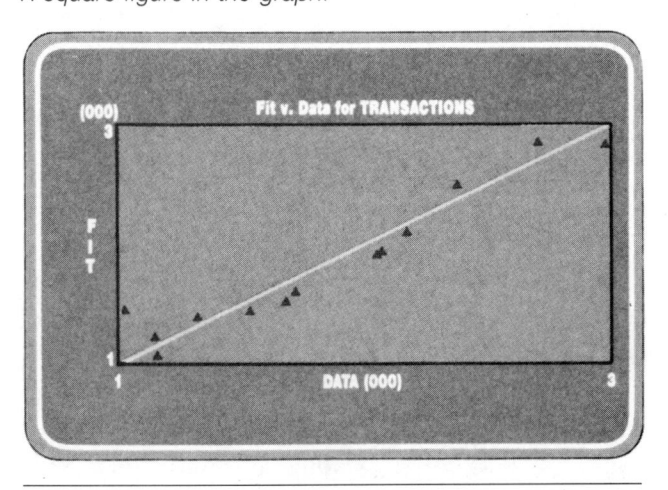

In this equation the variables are:
K_1 = Information bits per character.
M = Message block length, in characters.
C = Average number of noninformation characters per block.
K_2 = Bits per character.
K_3 = Nonretransmission constant.
E = Bit error rate of the circuit.
R = Line transmission rate in characters per second.
dT = Time between blocks in seconds.

The variable K_3 is a factor less than one that is used to discount the effect of multiple bit errors in a single character and errors that do not cause retransmission in a block (such as errors in the parity bit). If this factor is not known, it can be set at one without greatly affecting accuracy.

The factor dT is the time in seconds between blocks. This is a function of the modem reversal time, the propagation delay of the circuit, and the time to transmit acknowledgment characters. These are designated as RTS, PD, and ACK, respectively, in the following formula:

dT = 2RTS + 2PD + ACK

The factors in this formula are constant for a given protocol, modem, and transmission medium, but the designer may want to change them to observe the difference with a satellite circuit, a DDS circuit, or a fast turnaround modem. In a full-duplex circuit, modem reversal time is zero. The elements of dT should be treated as variables in order to observe the changes resulting from different design assumptions.

Response time (T_r) is the sum of the average times required to send an input message from the terminal to the host (T_{in}), CPU processing time (T_{cpu}), and the time to send an output message from the host to the terminal (Tout). In terms of the CPU processing time and the variables used to compute throughput, response time is:

$T_r = W_{in} + T_{in} + T_{cpu} + W_{out} + T_{out}$

The Formula/One package provides several types of input sheets. The primary ones used in circuit design are equation, variable, display, and plot sheets. The first step in using Formula/One is to enter the formulas in the equation sheet using standard mathematical notation. As symbols are entered in the equation sheets, the program enters corresponding symbols in the variable sheet. If a value is assigned to a variable by the user, the program treats it as an input variable; otherwise, the program treats the value as an output and the equation is solved for its value.

With its 17 variables, the variable sheet for throughput and response-time calculations is rather imposing. The program provides a display sheet to shield the user from all the variables except those that will be changed frequently. In this application, the display sheet shows block length, throughput, response time, bit error rate, and modem speed. By pressing the escape (ESC) key, the user can gain access to all other variables. After all variables are entered, the user instructs the program to solve the formula by pressing the recalculation key.

Optimizing block length requires repetitive calculations. To ease this task, Formula/One provides an input list that is accessible from the variable sheet. The designer enters a range of block lengths and instructs the program to fill the list. After recalculating, the designer selects the plot sheet and tells the program which variables are to go on the X and Y axes. The result is displayed in a graph similar to that shown in Figure 4. The results can be displayed in a table, although choosing the optimum block length is easier from a graphic representation. A family of curves can be developed to show the results of different bit error rates, modem speeds, propagation delays, and other such variables.

Queuing

The problem of determining the quantity of shared data communications gear is similar to the problem of determining how many servers are needed to accommodate customers in a waiting line at a bank or hamburger stand. For such problems, the discipline of queuing theory is very useful. To solve the queuing problem, three variables must be known or assumed:

■ The arrival process—how frequently customers arrive at the serving process.
■ The serving process—how long it takes to serve arriving customers.
■ The queue discipline—how customers behave when they encounter delays.

Many telecommunications design problems are identical to the waiting line model. Transactions that arrive at dial-up input ports of a time-shared computer serve as a fruitful example. The designer's objective is to provide exactly the right number of ports. Too many ports result in excessive cost; two few result in poor service. The problem is to determine the time distribution of calls arriving at the computer and the time distribution of the computer's handling of the transaction after it has been accepted. Also, the designer must know how the users react when a busy signal is encountered. For example, do they immediately redial or wait

4. Plateau. *The output of the Formula/One package shows the throughput in bit/s. The optimum block length is chosen where throughput reaches a peak.*

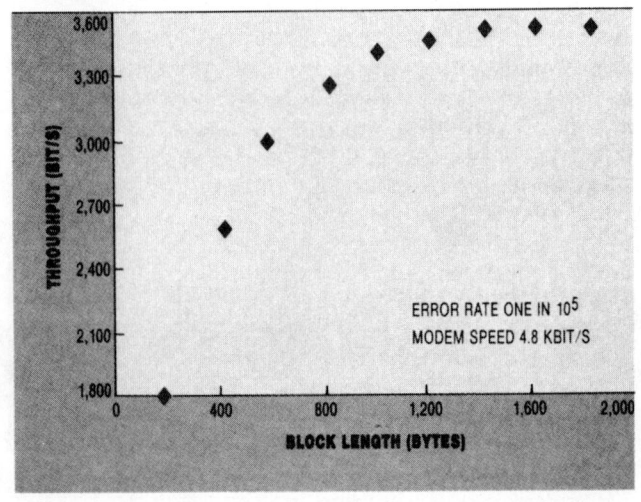

5. How long? The graph in (A) shows the probability that the service time will be T units long. The graph of (B) shows the response-time curve of a polled circuit.

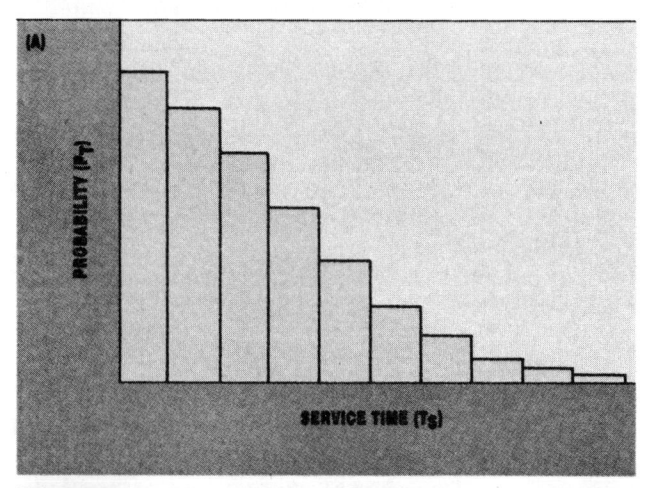

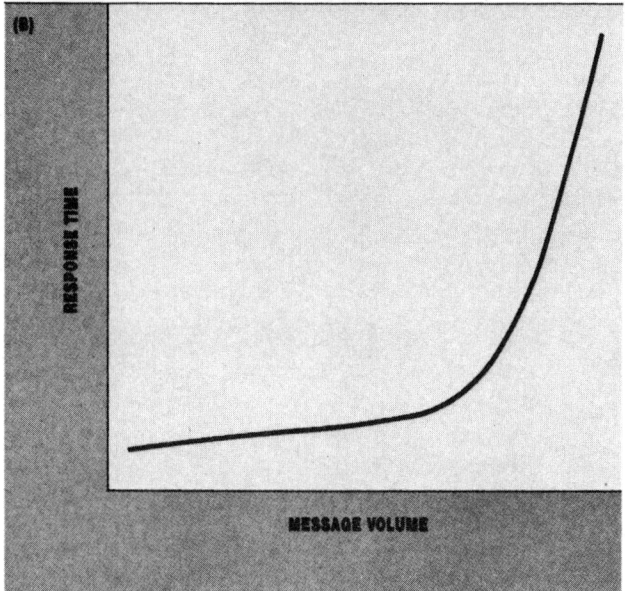

a certain length of time before redialing? Or does the computer have the ability to hold them in queue until a port is idle?

In telecommunications, the amount of usage of shared circuits or apparatus is a function of the number of times it is accessed (call attempts) and the holding time of the attempts. Two units of usage are used in design formulas: hundred call seconds (CCS) and Erlangs. These units measure the amount of occupancy of circuits or apparatus. One Erlang is equal to 36 CCS; each of these quantities represents 100 percent occupancy of a single circuit or server.

The first variable in queuing theory is the arrival rate. In most applications where users arrive at a point to be served by multiple servers, their arrival rate falls into a predictable pattern. If the number of attempts and holding time are plotted, they characteristically fall into a Poisson distribution. Circuit design formulas assume Poisson distribution of arrivals at the serving mechanism.

The second variable in queuing theory is service time, or how long it requires the server to satisfy the demands of the transaction. The network design formulas assume an exponential service time, such as the curve shown in Figure 5a. It is important to understand that these assumed arrival and service times are valid only as long as some external force does not interfere. For example, if a high level of blockage or unexpected loads occur, the distribution of attempts and service times will no longer resemble these models.

With the assumption of Poisson arrivals and exponential service times, the next consideration is the queue discipline, or what happens when the user encounters blockage. When blockage is encountered, the user can hang up and immediately reattempt, wait before reattempting, or be held in a queue until a port becomes idle. These three disciplines are called blocked calls held (BCH), blocked calls cleared (BCC), or blocked calls delayed (BCD), respectively. Table 3 shows the forumlas used for each of the three different disciplines.

Formulas as complex as these are far too unwieldy to solve by hand. Telephone designers have long used tables to size circuit groups. These tables are available for Poisson, Erlang B, and Erlang C formulas. The list the quantity of circuits, traffic volume, and grade of service in terms of percent blockage. To select the proper number of ports, a designer must know the amount of traffic volume and have an objective for the amount of blockage that can be tolerated in the network.

While traffic tables are satisfactory for data communications networks designers, if a microcomputer is available, design problems can be solved with Formula/One. This eliminates the need for books of tables, reduces the possibility of error, and provides a means for obtaining intermediate values that is easier than interpolating between rows and columns on a table. Finally, if the computer and program are already available, the cost of providing tables for designers is avoided.

Table 3: Queuing formulas

QUEUE DISCIPLINE	FORMULA	
BLOCKED CALLS CLEARED (BCC)	ERLANG B	$P = \dfrac{A^n/n!}{\sum_{X=0}^{n} \dfrac{A^x}{x!}}$
BLOCKED CALLS HELD (BCH)	POISSON	$P = e^{-A} \sum_{x=n}^{\infty} \dfrac{A^x}{x!}$
BLOCKED CALLS DELAYED (BCD)	ERLANG	$P = \dfrac{\dfrac{A^n}{n!} \dfrac{n}{n-A}}{\sum_{n=0}^{n=1} \dfrac{A^x}{x!} + \dfrac{A^n}{n!} \dfrac{n}{n-A}}$

P = PROBABILITY OF BLOCKAGE
A = TRAFFIC DENSITY IN ERLANGS
n = NUMBER OF SERVERS
e = NAPERIAN LOGARITHM BASE (2.718+)
X = NUMBER OF BUSY CHANNELS

Traffic tables are not easily programmed into Formula/One. While the previous design formulas can be entered after a few hours practice with the program, the traffic tables require programming Boolean logic, which requires more experience with the program. Alloy Computer Products Inc. will furnish the queuing programs free of charge to all registered users of Formula/One.

Polled multidrop lines

As terminals and message volume are added to a polled data circuit, performance is degraded because both the extra overhead and the message traffic use circuit time. Polled circuits exhibit a response-time curve like the one in Figure 5b. The designer's objective is to calculate response time versus message volume and to configure the network for the best cost/service balance. A secondary objective is to determine how many terminals the circuit will support. However, the design is more sensitive to message volume than to the quantity of terminals. The steps in designing a polled circuit are:

1. Calculate circuit overhead.
2. Calculate message transmission time.
3. Calculate terminal response time.

The overhead in a polled circuit is a function of the protocol. The formulas can easily be converted to full duplex by eliminating modem turnaround time and changing the quanity of overhead bits. A typical half-duplex protocol is shown in Table 4. Three transmission sequences must be examined:

- Negative poll.
- Input transmission (terminal to host).
- Output transmission (host to terminal).

The time to transmit a polling sequence is a function of both the quantity and length of input and output messages and of the time to send the overhead bits from negative polls, input messages, and output messages. With this protocol, the polling overhead is 10 characters, input overhead is 19 characters, and output overhead is 22 characters. The designer must determine the number of overhead characters, modem reversals, as well as propagation delays for the protocol that is in used.

In the formulas in Table 4, CH is the time in milliseconds to transmit a character, which is a function of the line transmission speed. Assume that the line speed is 4.8 kbits/s, or 600 characters per second (cps). Character time is 1.67 ms (derived from 1/600 cps). Assuming modem reversal time (RTA) to be 150 ms and propagation delay (PD) to be 10 ms, the overheads would be: Negative poll: $1(150) + 2(10) + 10(1.67) = 187$ ms. Input overhead: $2(150) + 4(10) + 19(1.67) = 372$ ms. Output overhead: $2(150) + 5(10) + 22(1.67) = 387$ ms.

The second part of the problem is to add message transmission times to the polling overheads. The time required to transmit an input block is the sum of the overhead plus the message size divided by the line speed.

For example, a 100-character input block at 4.8 kbit/s would have a transmission time of $0.372 + 100/600 = 0.5386$ sec. $= 539$ ms. Not all blocks are the same size, however. Computing response time requires calculating transmission time of the average message, which is determined by multiplying the block transmission for a given block length by the probability of that length block's occurring. Message statistics must be collected to provide this information. The transmission time for each block length is multiplied by the probability of that time occurring. The sum of this list is the average transmission time for a block (T_{in}). The variance is computed by multiplying the square of the transmission time by the probability of occurrence. Variance (V_{in}) is the sum of the squares of transmission times minus the average transmission time. The output message statistics, T_{out} and V_{out}, are calculated in the same manner. The results are fed into the response-time model.

As mentioned, response time is the sum of waiting time on input, input transmission, CPU processing, waiting time on output, and output transmission time. However, separate models are required for the three line disciplines: half-duplex line held (polling ceases while the CPU processes the message); half-duplex line released (polling continues during processing but further input messages are queued); and full duplex.

For the half-duplex line held discipline the first formula computes waiting time (W_{in}) from the mean poll cycle time (T_c) and the variance in the poll cycle (V_c).

$$W_{in} = 1/2(1 + V_c/T_c^2)T_c$$

To determine poll cycle time, it is necessary to measure line utilization on input messages (R_{in}), CPU processing time (R_{cpu}), and output messages (R_{out}). Poll

Table 4: Sample half-duplex protocols and overhead formulas

SEQUENCE	OVERHEAD
NEGATIVE POLL CPU: SYN SYN CUA CUA DVA DVA ENQ TERM: SYN SYN EOT	$(R_{np})(RTS) + (P_{np})(PD) + (C_{np})(CH)$
INPUT MESSAGE CPU: SYN SYN CUA CUA DVA DVA ENQ TERM: SYN SYN STX --TEXT--ETX BCC BCC CPU: SYN SYN ACK TERM: SYN SYN EOT	$(R_{in})(RTS) + (P_{in})(PD) + (C_{in})(CH)$
OUTPUT MESSAGE CPU: SYN SYN CUA CUA DVA DVA ENQ TERM: SYN SYN ACK CPU: SYN SYN STX --TEXT--ETX BCC BCC TERM: SYN SYN ACK CPU: SYN SYN EOT	$(R_{out})(RTS) + (P_{out})(PD) + (C_{out})(CH)$

CPU = CENTRAL PROCESSING UNIT
TERM = TERMINAL
SYN = SYNCHRONIZING CHARACTERS
ACK = ACKNOWLEDGMENT CHARACTER
EOT = END-OF-TEXT CHARACTER
CUA = CONTROL UNIT ADDRESS
DVA = DEVICE ADDRESS
ENQ = INQUIRY
RTS = MODEM REVERSAL TIME
PD = PROPAGATION DELAY
CH = CHARACTER TIME
R() = NUMBER OF MODEM REVERSALS
P() = NUMBER OF PROPAGATION DELAYS
C() = NUMBER OF OVERHEAD CHARACTERS

cycle time is a function of message volume (M) and overhead of a negative poll (O_{np}). Note that as line utilization approaches 100 percent, poll cycle time approaches infinity.

$$T_c = M\, O_{np} / 1 - (R_{in} + R_{cpu} + R_{out})$$

The variance in poll cycle time is a function of the message statistics computed in the previous section and the mean and variance of CPU processing time, which are obtained from the computer manager.

$$V_c = IT_c(V_{in} + V_{cpu} + V_{out}) + IT_c(I - IT_c / M)(T_{in} + T_{cpu} + T_{out})^2$$

(In this equation, "I" is equal to the number of messages that arrive per minute.)

With these three forumlas in the Formula/One equation sheet, response time can be computed with a fourth formula. The output waiting time T_{out} is zero in this line discipline because transmission is suspended during processing.

$$T_r = W_{in} + T_{in} + T_{cpu} + W_{out} + T_{out}$$

Clearly, manual calculations of this type are both time-consuming and subject to error. Formula/One is particularly adaptable to this type of computation because all four formulas can be entered in its equation sheet. When the calculate key is pressed, the program solves the equations in the proper order to yield response time (T_r). ∎

Further reading

Chou, Wushow. *Computer Communications,* vol. 1: Principles. Englewood Cliffs, N. J.: Prentice-Hall Inc., 1983.

Doll, Dixon. *Data Communications*. New York: John Wiley & Sons, 1978.

Freeman, Roger L. *Telecommunication System Engineering*. New York: John Wiley and Sons, 1980.

Held, Gil. "No more guesswork for sizing network components" DATA COMMUNICATIONS, vol. 12, no. 10, October 1983.

Sharma, Roshan La., Paulo T. deSousa, and Ashok D. Ingle. *Network Systems*. New York: Van Nostrand Reinhold Co., 1982.

Tannenbaum, Andrew S. *Computer Networks*. Englewood Cliffs, N. J.: Prentice-Hall Inc., 1981.

James Green is president of Pacific Netcom, a telecommunications consulting firm based in Portland, Ore. He has a B. A. in agricultural engineering from Oregon State University and an M. B. A. from the City University of Seattle. He has written three books on communications, including: Automating Your Office *(McGraw-Hill, 1984),* Local Area Networks *(Scott-Foresman, 1985), and* Handbook of Telecommunications *(Dow Jones-Irwin, March 1986).*

Larry Cynar, Cynar & Associates, Thousand Oaks, Calif., Don Mueller, Mueller & Associates, Torrance, Calif., and Andrew Paroczai, TRW, El Segundo, Calif.

Computers design networks by imitating the experts

By using the right knowledge base, an expert system can locate an unplugged cable or design an entire network.

How many times have you seen a field technician working on a tough problem by talking on the phone to the network support engineer back in the plant? What if that expert could be right there in the room? How much faster would the job go?

Today, with advanced microprocessor design and advances in artificial intelligence (AI), increasing amounts of knowledge can be stored and used for applications by people with relatively little expertise in the application area. With an expert system (ES), the expert's knowledge can accompany the technician.

An expert system (see "A summary of expert system terms") is simply a computerized interpretation—into a set of logical rules—of the expert's years of experience. These computer programs are frequently called artificially intelligent expert systems (AIES).

Most early AIES applications were in university or research and development environments. But now much of this early research can be applied more practically with more immediate benefit. Some of the specific AIES applications can be applied to data communications networks. For example:

Configuration control. AIES implementations can be used to configure the components of a data communications network. A statistical multiplexer, for example, needs to have terminal speeds configured, flow-control protocol and interfaces defined, and trunk speeds determined before it can be properly installed in a network. Large switching systems or network control centers need to have all the data pertinent to the network in its configuration, such as alternate routing, queue priorities, tandem switching, least-cost routing, port allocation, passwords, and so forth. These and other factors are all necessary inputs that must be considered when designing large networks.

Fault isolation. Common symptoms of network failure can be entered in the computer, and the AIES will guide a troubleshooter through even the most complex problem as if an expert were there.

The AIES is constantly learning by having its knowledge base improved from problem-resolution feedback. If a course of action proves itself incorrect, the knowledge engineer, or programmer, can change the rule that chooses that action to include another needed input or fact. Thus, knowledge acquired by the AIES is obtained from the best source, one or more experts having skills in the subject.

This article looks at the architecture of a feasibility model AIES used to manage communications networks. This prototype has the capability of designing a network from user requirements, diagnosing network problems, and acting as a network design and troubleshooting training vehicle.

Network design

When human experts design a network, the activity in each of the steps is as follows:

User requirements. The user's needs are reviewed for feasibility and consistency. The result is a set of network requirements.

Traffic requirements. The designer takes the realizable requirements and calculates the traffic required, based upon the number of channels, type of data, data rate, and usage.

Network topology. The designer interconnects the nodes into the network based on nodal location, traffic, or alternate paths required.

Trunk-line selection. The designer selects the most cost-effective trunking facility between the interconnected nodes. This selection is based on traffic requirements between nodes, nodal location, and availability of facilities.

A summary of expert system terms

expert A person who has acquired special skill in, or knowledge of, a particular subject.

expert system A computer implementation that achieves high levels of performance at task areas that, for experts, require years of special education and training. An expert system includes software that achieves this performance. The software consists of a knowledge base and a control strategy.

knowledge This entity consists of heuristics, facts, and beliefs.

heuristic A rule, or a fundamental law, or a procedural tip.

facts and beliefs Propositions or data whose validity is accepted.

knowledge base That of an expert system is the repository of knowledge in a computer. The knowledge base is built by extracting knowledge from experts and accumulating factual data from text books and manufacturers' data sheets. The knowledge base may be organized into rules and frames.

rule Asserts a fact about an area of expertise. A rule is expressed in the form *if it is raining, then wear your raincoat*. The phrase *if it is raining* is the rule's condition.

condition The left side of a rule, or the rule's hypothesis. If the condition is true, then the consequent is true.

consequent The right side of the rule, or the rule's conclusion. The consequent is true only if the condition is true. Conditions and consequents may be made complex, using the conjunctions "and" and "or," as in "if it is raining *and* you do not have a raincoat, then buy a raincoat *or* stay home."

control strategy A computer program for making inferences from the knowledge base. The control strategy in an artificially intelligent expert system uses forward-chaining to make inferences.

forward-chaining A control strategy that produces new decisions by affirming the consequent propositions of a rule with conditions that are currently believed. As newly affirmed propositions become conditions, they change the current set of beliefs and additional rules are applied recursively until no beliefs are added.

frame A knowledge representation scheme. A frame contains the collection of attributes that an object possesses. In a frame, the object's attributes are described in terms of various slots and particular slot values. The attributes of the object are arranged in slots of the frame. An attribute can be a specific property, or it can be a relationship to another object. For example, a multiplexer is a kind of nodal equipment.

slot in a frame, a slot may identify another frame as the attribute of the object. This allows frames to be organized hierarchically, in a structure called a frame diagram. Slots may correspond to intrinsic features, such as name, definition, or creator. They may also represent derived attributes, such as value, significance, or analogous objects.

Nodal-equipment selection. Nodal equipment (modems, multiplexers, switches, statistical multiplexers, and so forth) is selected based on bandwidth requirements, functional requirements (contention, switching, protocols), special feature requirements (auto fallback, redundant logic, reserve bandwidth), and the designer's knowledge of product availability.

Design presentation. The designer presents the design to the client. This is done by itemizing the services and equipment to be acquired, calculating the start-up of overall and monthly costs for various combinations of lease/purchase options and by generating a network diagram.

The design process consists of at least all of the above stages. In each stage, a set of input facts is presented, and the designer uses his knowledge and available data to synthesize a set of output facts. This process can be modeled in a computer by using a set of production rules to represent the designer's knowledge. These input facts are then presented to what is known as a forward-chaining inference engine. This is basically a data-driven, inference mechanism, or program designed to discover the maximum amount of information based on the original input facts.

The input facts are stored in short-term memory (the fact base); the forward-chaining inference engine presents the facts to the production rules. If a rule fires—that is, if all the conditions for that rule prove true—an additional fact is put in the fact base, and the new fact base is again presented to the production rules. This process continues until no more rules fire. The resulting fact base will contain the original input facts as well as all the facts that were deduced by the knowledge base. These deductions can then act as inputs to the next stage. Figure 1 shows a block diagram of this inference procedure.

Knowledge representation

The AIES goes through the same processing stages as the human expert. In each stage a new knowledge base (rule set) is brought in, and the facts deduced from the previous stage are used as inputs (Fig. 2).

Two forms of knowledge representation are used in the AIES: production rules and frames. Production rules are

163

1. First pass. *Facts are put into a knowledge base and go through various stages until a rule is inferred. The rule is added to a short-term memory bank.*

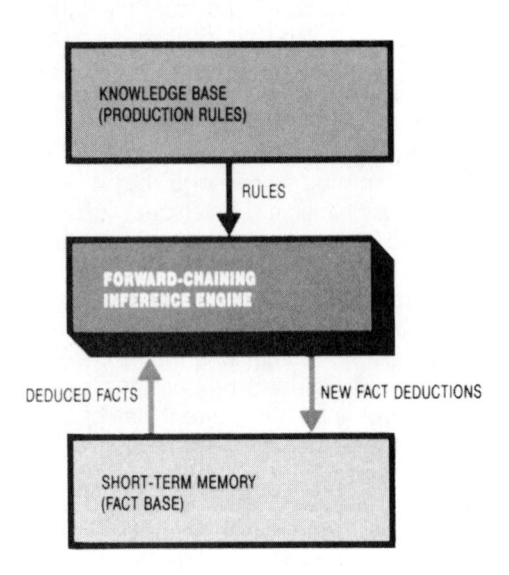

a series of "if-then" rules that deduce new facts. Following are a few of the many trunk-selection rules to illustrate the relationships between production rules and frames. Finally, a set of input facts is given, and the forward-chaining deduction process is tried. The purpose is to decide what type of trunk facility to provide based upon requirements:

Rule Bandwidth 1. *IF bandwidth is less that 4.8 kbit/s, THEN bandwidth is low.*

Rule Bandwidth 2. *IF bandwidth is between 14.8 and 28.8 kbit/s, THEN bandwidth is medium.*

Rule Bandwidth 3. *IF bandwidth is between 28.8 and 56 kbit/s, THEN bandwidth is high.*

Rule Bandwidth 4. *IF bandwidth is between 56 and 250 kbit/s, THEN bandwidth is very high.*

Rule Bandwidth 5. *IF bandwidth is greater than 250 kbit/s, THEN bandwidth is ultra high.*

Rule Dial-up 1. *IF a trunk is not selected, and dial-up is allowed, and terminal rate is less than 300 bit/s, and bandwidth is low, THEN provide Bell 103-based dial-up modems and select a truck.*

Rule Dial-up 2. *IF a trunk is not selected, and dial-up is allowed, and terminal rate is up to 1.2 kbit/s, and bandwidth is low, THEN provide Bell 212A dial-up modem connections and select a trunk.*

Rule Voice grade. *IF a trunk is not selected, and the bandwidth is not high, very high, or ultra high, THEN provide a voice-grade circuit, and select a trunk, and so forth.*

The trunk selection rules are run for each trunk, as determined by the topology design stage. The topology design stage outputs user requirements, traffic requirements, and topology (trunks and their bandwidths) to the trunk selection rules. For example, suppose that one of the trunks has the following facts associated with it:

- Bandwidth is 14.4 kbit/s.
- Dial-up is allowed.
- Terminal rates are less than 1.2 kbit/s.
- DDS service is available.

The following sequence of events would occur:

2. Mock thought. *The artificially intelligent expert system goes through "thought-processing" stages much the same way a human expert would.*

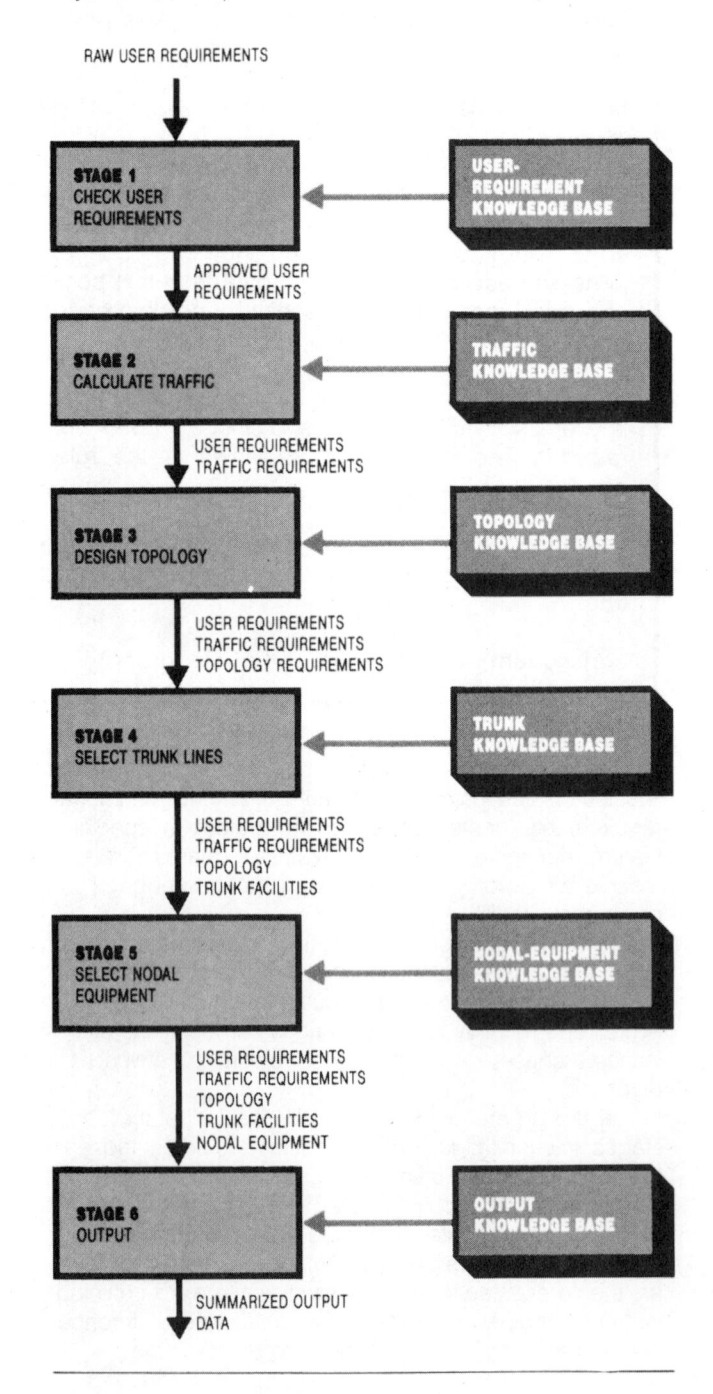

1. During the first pass through the rules, Rule Bandwidth 2 fires. "Bandwidth is medium" is added to the fact base.

2. During the second pass through the rules, Rule Voice grade fires. "Provide voice-grade circuit" and "trunk is selected" are added to the fact base.

3. During the third pass through the rules, no rules are fired, the deduction is complete, and the trunk will be provided with voice-grade circuits.

Note that the fact "trunk is selected" inhibited all other trunk-selection rules from firing. This is used as a rule condition to halt the processing. The AIES will next clear the fact base and present the next trunk and its facts for trunk selection. This process continues until all trunks in the network have been selected.

The rules presented simply illustrate the AIES principles and are by no means complete. A real algorithm would need on the order of 100 trunk selection rules.

Note that altering or adding new rules can change and improve the performance of the AIES. A rule can be added to take care of a special case without affecting the other rules. This modularity allows the knowledge base to grow and improve over time. Frames may be used to specify the fundamental relationships between objects. In the AIES, the network topology, the trunk services, and nodal equipment choices are represented using frames. Using the nodal equipment as an example, the highest-level classification is the network. Subclasses of network include nodal equipment, trunk facilities, and network topology. Subclasses of nodal equipment are modem, multiplexer, statistical multiplexer, protocol converter, concentrator, switch, and so forth.

Each of the subclasses has specific instances, such as different equipment types from different manufacturers (Fig. 3). The actual frame representation of the nodal equipment with slot values assigned is as shown in the table. The particular types of equipment which satisfy the requirements are obtained by searching through the frame-based representation.

Knowledge engineering facilities

The AIES has extensive knowledge-editing capabilities. The rules and frames may be listed and edited using an on-screen editor. There is a trace facility which allows the internal workings of the inference engine to be displayed in a special window during a session. As deductions are made they are displayed. The rule being used is displayed. This allows the knowledge engineer to view the internal processes during a session and make appropriate corrections to the knowledge base in order to obtain desired results. This is essential to provide efficient knowledge-base updates.

The user enters network requirements on a template displayed on the screen in a full-screen editing mode. The interface is similar to a spreadsheet; the user tabs through the entry form using the cursor positioning keys to enter requirements.

After data entry, if it is valid and consistent, the system proceeds to design the network. If the data is not valid, the system explains to the user the inconsistency and asks for corrected input. At the end of each

Design criteria

THE PARTICULAR TYPES OF EQUIPMENT THAT SATISFY REQUIREMENTS PREVIOUSLY DETERMINED ARE OBTAINED BY SEARCHING THROUGH THIS FRAME-BASED REPRESENTATION.

NODAL EQUIPMENT

RESIDES IN	NODE
MAY BE	PURCHASED
MAY BE	LEASED

MODEM

IS	NODAL EQUIPMENT
HAS	ANALOG TRUNK INTERFACE
HAS	TERMINAL INTERFACE
CONNECTS TO	TELEPHONE LINE

MULTIPLEXER

IS	NODAL EQUIPMENT
HAS	DIGITAL TRUNK INTERFACE
HAS	MULTIPLE TERMINAL INTERFACES
CONNECTS TO	MODEM

MULTIPLEXER A

IS A	MULTIPLEXER
LOW-SPEED RATES	100 BIT/S TO 1.2 KBIT/S
HIGH-SPEED RATES	4.8 KBIT/S TO 9.6 KBIT/S
LOW-SPEED RATES	RS-232-C, MIL188B (MILITARY SPECIFICATION)
HIGH-SPEED INTERFACES	RS-232-C, MIL188B, V.35 (INTERNATIONAL SPECIFICATION)
COST	$5,000
FEATURES	REDUNDANT COMMON LOGIC
POWER REQUIRED	500 WATTS
MODE	IBM BIT INTERLEAVED
SYNC TIME	100 MILLISECONDS (MS)

MULTIPLEXER B

IS A	MULTIPLEXER
LOW-SPEED RATES	1.2 KBIT/S TO 4.8 KBIT/S
HIGH-SPEED RATES	56 KBIT/S TO 250 KBIT/S
LOW-SPEED INTERFACE	RS-232-C
HIGH-SPEED INTERFACE	FIBER OPTIC
COST	$50,000
MODE	TIME-DIVISION MULTIPLEXED (TDM) CHARACTER-INTERLEAVED
SYNC TIME	100 MS

stage of the design, the AIES displays the deductions it made.

The user may, at any time, stop the AIES and ask for an explanation, modify the design, or change a requirement. This allows the user to understand why the design is evolving along a certain direction and to modify the design process if desired. In many cases, the user will think of new requirements as the design is presented, or he or she may reduce or eliminate requirements as the cost of that requirement becomes apparent. This is exactly analogous to the negotiations that go on between users and human designers as a design unfolds.

A human expert has the ability to explain how a

conclusion was drawn and why a conclusion was not drawn. A good expert system must also have this capability.

Explanation facilities
The prototype AIES used in this article explains how a fact was deduced by looking backward through rules that deduced the fact. For example, the first pass through the rules in our sample trunk selection session deduced the fact "bandwidth is medium." If the AIES were asked how that fact was deduced, it would respond "bandwidth is medium" was deduced by Rule Bandwidth 2, which states:

IF bandwidth is between 14.8 and 28.8 kbit/s, THEN bandwidth is medium.

The facility to answer how a fact was established is very powerful. Not only can the user see what requirements are causing certain design directions to be taken, but the user is getting trained in the design process as well.

If AIES were asked why "Provide Bell 212A dial-up" was not concluded, it would answer, Rule Dial-up 2 could have been used. The following conditions pass:
1. Trunk is not selected.
2. Dial-up is allowed.
3. Terminal rate is less than 1.2 kbit/s.

But the bandwidth is high.

The "why-not" facility is a powerful aid to the knowledge engineer in debugging the rule set as well as to the user for training purposes.

Computational requirements
AIES implementations consume a lot of memory and computation time, particularly as the knowledge base increases. A full-blown AIES with perhaps 1,000 rules and 1,000 more nodal equipment types defined would require a DEC VAX 750 or equivalent computer. The prototype AIES was developed on an IBM PC, which proved to be adequate because the knowledge base was held to less than 100 rules and only a small sampling of nodal equipment types was used. However, the prototype can be easily transferred to a larger computer because it is written in common LISP, the standard AI programming language. ∎

Larry Cynar, president of Cynar & Associates, received a B. A. from the University of Redlands, Redlands, Calif. Don Mueller received a B. S. E. E. from the University of Iowa, Iowa City. Andrew Paroczai received a B. A. from the University of California at Berkeley, and an M. S. in computer science from the University of California at Los Angeles.

This article is part one of a two-part article.

C. R. Abbruscato, Racal-Milgo, Sunrise, Fla.

Choosing a key management style that suits the application

Message authentication and schemes for encryption range in complexity. If the use warrants the cost, pick the newest high-security protocol.

One of the most difficult aspects of encrypting a data communications network is key management. Poorly done, it can undermine network security. Overdone, it can present an ongoing, burdensome expense. There is no single solution for key management that might apply to all data networks. The chosen method of key management must be consistent with the overall security objectives and the architecture of the network.

When devising a key management scheme, the selection of an encryption algorithm is a relatively easy first decision. For U. S. companies, the choice is usually the Data Encryption Standard (DES), which has been endorsed by the National Bureau of Standards (NBS) and the American National Standards Institute (ANSI). Once that security-related issue has been resolved, questions of physical security must then be addressed. Specifically, the cryptographic equipment needs to be secured against improper handling. Selecting cryptographic equipment and making it secure present an opportunity for cost/security trade-offs. As such, it is also a juncture at which the security of the network itself may inadvertently be undermined.

Physical security becomes a key management issue in cases where the key is stored on a small peripheral module that is not itself physically secured to the cryptographic unit. A bold individual (not necessarily a professional thief) can do more damage to the network's security by "borrowing" the key module for a short period than a cryptanalyst can do with a Cray supercomputer over an extended period. Because of their cost and inconvenience, it is tempting to overlook the need for physical access controls. But they should be considered, for the chain in this instance is truly only as strong as its weakest link.

Before keys can be distributed and used, they must first be created—and they must be random. Keys must be generated so that there is an equal probability of any key or vector in the 2^{56} key space being created. (While DES keys are 64 bits long, 8 of them are parity bits, leaving 56 information bits.) Generating true random numbers can be accomplished in a number of ways. One method uses the inherent noise from a diode as a source. With this approach, the diode noise is amplified and sampled, which sounds simple enough but requires careful design.

For example, if samples are taken at zero crossings of the amplified noise, any voltage offset will create a bias in favor of one polarity producing keys from only a portion of all the theoretical possibilities. Pseudorandom generators are an attractive alternative, but they are just as easy to implement incorrectly as are hardware designs. A simple, practical method for generating random keys is to use the DES algorithm to encrypt a unique, but not necessarily random, seed number. For example, a digital representation of the day, date, and time can be encrypted; the result is then combined with some arbitrarily chosen number and encrypted again under DES. The resultant number has all the characteristics of a randomly produced number.

Key entry

Federal Standard 1027 "Telecommunications: General Security Requirements for Equipment Using the DES" and ANSI X9.17 "Financial Institution Key Management (Wholesale)" both call for dual control during the performance of such manual activities as loading in new keys. The most common method of dual control is via two physical locks on cryptographic equipment. The dual control occurs when the physical keys for the two locks are controlled by two individuals, both of whom must unlock their respective locks to permit any manual key management operations. In this way the network will not be compromised if one individual is compromised.

Manual key entry can also be from printed or written form. Since the security offered by the DES algorithm is no greater than the protection given to the DES keys, special precautions should be taken when working with keys printed out in unencrypted form. When manual key entry into a cryptographic device is from printed form, a technique called split knowledge can be used. Two or more people are given unique full-length key vectors (64 bits) in printed form. They do not show the key vectors to each other. These key vectors are individually entered into the cryptographic gear, where they are combined with binary arithmetic known as exclusive-OR to form the actual key. Again, if the key vector held by one person is compromised, it does not necessarily compromise the network.

Key entry is typically made through a connector on the cryptographic device, although some equipment may have a keypad on each unit to enter keys from printed form. The emerging standard for key entry is a 9-pin connector specified in Federal Standard 1027. This connector is a good choice even for encryption hardware integrated into a computer. Manual key entry from a computer keyboard is not as secure or convenient a method as it first appears to be, mainly because the key should not be displayed on a CRT. The emissions from a CRT can be picked up from outside an unshielded building, and the display can then be reconstructed. Typing in 16 hexadecimal-character keys without a display makes an awkward method of key management even more awkward. Other emissions from the encryption equipment are usually sufficiently low due to compliance with the electromagnetic interference requirement of FCC Code of Federal Regulations No. 47 (CFR47) Part 15J.

The National Security Agency (NSA) supplies DES keys for government use. They produce keys with unimpeachable randomness. However, they provide the keys in printed form or on paper tape. To facilitate key entry from paper tape, the government has a special paper tape reader, called the KOI-18. This manual device is widely used in the government and will soon be made available for some private sector use. Federal Standard 1027 was written with the KOI-18 in mind and specifies its electrical interface requirements for key entry to be compatible. The physical connector on the KOI-18 is different from the 9-pin D-subminiature connector specified in Federal Standard 1027, but the conversion is easily made in the cable linking the KOI-18 with the cryptographic device.

Manual key distribution
Before discussing manual key distribution techniques, an even more primitive method should be mentioned: cryptographic equipment with factory-installed keys. This requires essentially no key management and has very limited applications. For low-risk applications where the users are merely looking for an impediment to intruders, DES encryption units with factory-installed keys may be an adequate, low-cost option to provide some data security. However, the inability to change keys periodically is such a severe restriction that this method cannot be recommended.

Manual key distribution involves the form in which the keys are prepared and the means to transport the keys to the communicating parties who must share them. Keys can be prepared in printed form, on paper tape, or in electronic form (e.g., random-access memory or EEPROM [Electronically Erasable Programmable ROM]) for transport to the cryptographic devices.

A way of implementing dual control with split knowledge during key transport is to encrypt the keys with a special key that is unknown by the individual transporting the keys. This is easily done for keys that are in an electronic form and are stored and transported in key transport modules. A module containing the security key is manually loaded into the equipment that generates the user keys; it is also manually loaded into the user cryptographic equipment. Before the keys are loaded into the key transport module, they are encrypted under the module security key. The key transport module is taken to the user's cryptographic equipment, where its contents are read into the unit, which then decrypts the keys with the module security key. Knowledge of the module security key by itself does not compromise any user data and possession of the encrypted key transport module by itself does not reveal any secret information.

In regard to key management, NSA has interpreted Federal Standard 1027 to require that keys be manually loaded directly into each cryptographic unit and that those keys be the ones used to encrypt the data. While the federal government requires dual control, it does not require split knowledge for any key entry. Apparently, it is felt that the government's ability to qualify and screen its personnel provides sufficient protection.

Electronic key distribution
The private sector expects and demands a level of convenience and sophistication that is not addressed in Federal Standard 1027. To distribute keys manually every time the data-encrypting key is changed is inconvenient and expensive to the point of being burdensome. Fortunately, there are techniques available that offer more convenience at lower operating cost while still providing for satisfactory security.

Figure 1a shows the one-key method of key management used in Federal Standard 1027. A single key is manually loaded into both ends of the communications line. Since that key is used to encrypt the data going over the line, it is the data key, labeled KD in the figure. To lessen the demands of frequent manual key distribution required by this one-key method, most of the commercial cryptographic equipment provides for a two-key method of key distribution. This is illustrated in Figure 1b. As with the one-key method, there is a data key, KD, to encrypt the data. However, this key is not entered manually into both ends of the line. Instead, the data key is transmitted down the communications link, from one cryptographic unit to the other, encrypted under a key-encrypting key (denoted by KK in the

1. One key vs. two. *The one-key method of key management (A) requires manual loading into both ends of the communications line. In (B), the data key is transmitted down the line permitting highly automated key management. The key-encrypting key must still be entered manually at both ends.*

(A) ONE-KEY METHOD

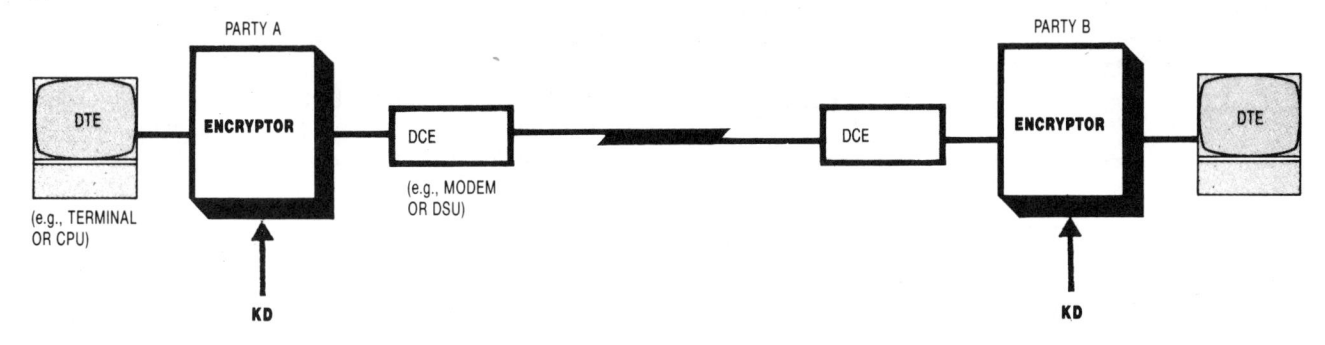

PARTY A — DTE (e.g., TERMINAL OR CPU) → ENCRYPTOR → DCE (e.g., MODEM OR DSU) → ... → DCE → ENCRYPTOR → DTE — PARTY B

↑ KD   ↑ KD

(B) TWO-KEY METHOD

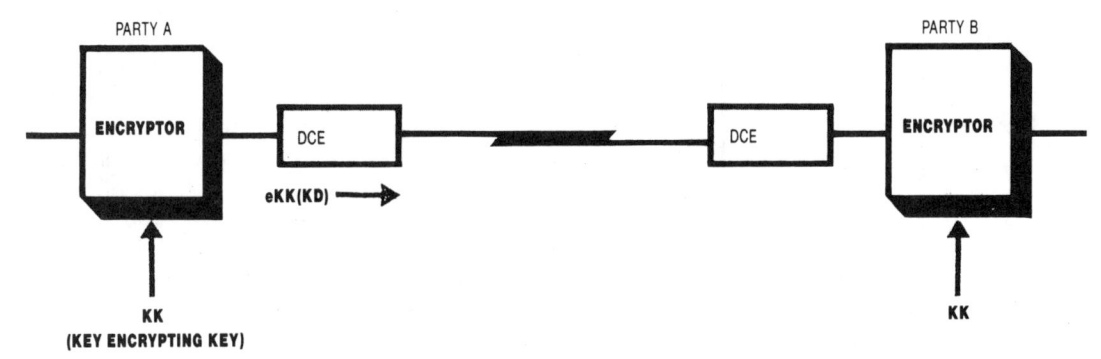

PARTY A — ENCRYPTOR → DCE — eKK(KD) → ... → DCE → ENCRYPTOR — PARTY B

↑ KK (KEY ENCRYPTING KEY)   ↑ KK

CPU = CENTRAL PROCESSING UNIT
DCE = DATA CIRCUIT-TERMINATING EQUIPMENT
DSU = DATA SERVICE UNIT
DTE = DATA TERMINAL EQUIPMENT
KD = DATA KEY

NOTE: eKK(KD) MEANS THAT KD IS ENCRYPTED UNDER KEY, KK.

---

figure). This downline loading of data keys permits highly automated key management with relatively little penalty for changing data keys. The automation stops, however, at the point of the key-encrypting key. For the two-key scheme, the key-encrypting key must be entered manually into the cryptographic units at both ends. Since the key-encrypting key's exposure to human contact is relatively brief, it is deemed to have a relatively long crypto-life; even though it must be manually entered, the time between entries is much longer than with the one-key technique. The net result is more frequent automated data key changes with less frequent manual key entries. Typically, one end of the link has responsibility for generating keys, although it is certainly permissible for the data keys to have been loaded manually prior to the downline transfer.

A more automated key management strategy is the three-key method. It uses the public key technique to distribute the initial key-encrypting key electronically. Unlike the DES, which uses the same key for encrypting and decrypting, public key schemes use two different but mathematically related keys: one for encrypting, the other for decrypting. The encrypting key is made public, while the decrypting key is kept secret. Knowledge of the encrypting key alone is insufficient to determine the secret decrypting key except by mathematical brute force, which presents a prohibitive work factor.

Three-key method (public key)

If Party A wishes to send a secret message to Party B, he obtains Party B's public encrypting key, E_B, from the public file; he then uses E_B to encrypt the message. This message can be decrypted only by the decryption key, D_B, which is known only to Party B.

Similarly, Party B uses Party A's public encrypting key, E_A, to send secure messages to Party A. Note that a public key file is not necessarily required. The system can be designed so that each party merely asks the other party to send its public encrypting key over the communications link.

In the three-key scheme, the public key is used to

2. Operational threat. *A spoofer would join the communications link and try to imitate the traffic. The parties communicating must have a way to identify each other.*

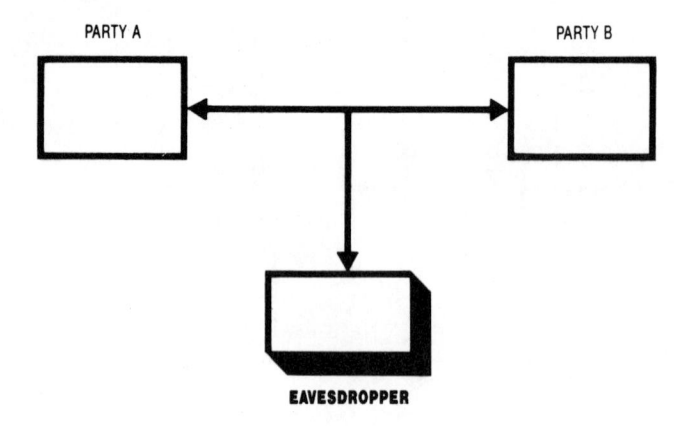

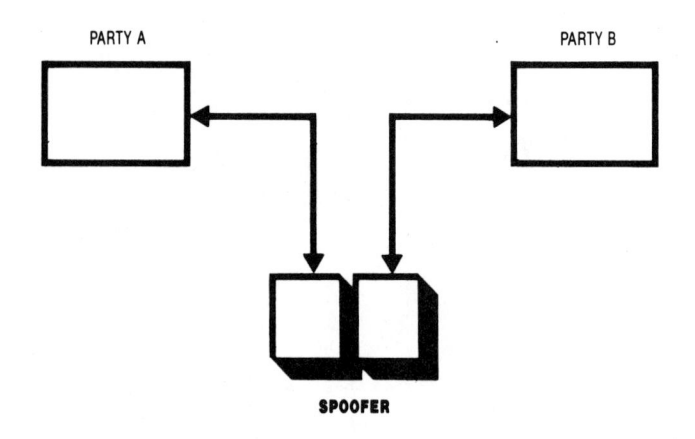

encrypt a key-encrypting key for downline loading. This eliminates the need for manual distribution of key encrypting keys. As with the two-key method, once a key-encrypting key is shared between the two parties, subsequent data keys can be electronically transferred.

Spoofers

An operational difficulty must be overcome when public key methods are used: the threat of a "spoofer." While an eavesdropper would simply monitor and observe traffic between two cryptographic devices, a spoofer would actually insert himself into the communications link and attempt to imitate the traffic. Thus, while the devices think they are talking to each other, they are actually talking to the spoofer. This can be seen in Figure 2.

In the case of a three-key scheme, how does Party A know that the public key it receives from Party B actually came from Party B and not from a spoofer? Furthermore, how does Party B know that the DES key received from Party A actually came from that location and not from a spoofer?

Clearly, there must be something or some piece of knowledge that Parties A and B share that can be used to identify each to the other. This verification can take many forms and can have varying degrees of practicality. In analyzing these methods, it is important to remember the nature of the threat. In distributing DES keys manually, it is critical that the secrecy of the keys be maintained. With the public key technique, since the encrypting key can be made public, the main threat is from substitution, not from loss of secrecy.

While DES encryption provides secrecy, it also provides inherent protection against substitution. But using a previously distributed DES key to protect the public key seems at first to negate the benefit of the three-key method. During initial installation this is true, but for subsequent key transfers there is an advantage.

The purpose of the original DES key, which will be called here the verification key, is to verify the integrity of the transmission path so that the system cannot be spoofed during the first public transaction (see "Authentication"). Under the protection of the verification key, that transaction transfers a key-encrypting key, which is under the protection of the public key, from the cryptographic equipment at one end of the link to the cryptographic equipment at the other end. Along with the key-encrypting key, a new verification key is transmitted, to be used to protect against substitution on the next public key operation. By creating a chain of verification keys in this manner, the spoofer is effectively shut out. The attacker must break the verification key in real time during the transaction in order to make a substitution. This is impossible with today's technology.

The verification key strategy can still be effective without the chaining. In that case, the same verification key is used continuously. If the verification key is "broken," then a spoofer can perform a public key substitution, so the effective life of the verification key must be established and the key replaced accordingly.

The distribution of the first verification key need not be done in secret prior to the first public key transaction. A nonsecret, default verification key can be used during the first key transaction in combination with a simple manual authentication technique to guard against spoofing. This strategy calls for the computation and display of a message authentication code (MAC). The MAC is computed with the default key using the DES algorithm in the cipher block chaining (CBC) mode of operation. The MAC is the residue of the last DES operation on the message. If any bit is altered in that message, then at the time the MAC is computed at the receive end of the link it will not match the originally computed MAC.

When Party B sends the message containing the public key, he appends the MAC to the message. In addition, he displays the MAC so that it can be read by the installer or operator. When Party A receives the message he also computes the MAC and displays it. Similarly, MACs are computed and displayed for the message containing the key-encrypting key and a new, secret verification key encrypted under the public key from Party A to Party B. If the two parties can verify that the MACs match, then they have verified that they are talking to each other and not to a spoofer; if a spoofer had substituted his own key, then that message would have a different MAC. Even though the default key is

Authentication

One valuable application of the DES (Data Encryption Standard) algorithm is in authenticating, or verifying, the integrity of messages. For this purpose, the DES is used in the cipherblock chaining (CBC) mode of operation (see figure).

To start the process, the first block of 64 bits of data (D_1) is processed through the DES algorithm. The result is exclusive-OR'ed with the second block of 64 bits of data (D_2). The output of the exclusive-OR operation is fed back through the DES algorithm and then combined with the next block of 64 bits of data. This continues until the entire message to be authenticated has gone through this process. Since each pass through the DES computation includes information from the previous block, the residue at the output of the last DES process is a direct function of the entire message and the key. If any bit of any of those components changes, then the output of the last DES computation will change. The 32 most significant (or left-most) bits of that last computation are called the message authentication code (MAC).

When a message is transmitted, the MAC is calculated and appended to the message. The authorized receiver of that message should have already obtained the correct DES key. Upon receiving the message, the receiver also computes a MAC from the message and compares his computed MAC with the MAC sent with the message. If the MACs are identical, then the message (as well as the key) has not been altered. The procedures for message authentication are described in detail in ANSI X9.9 "Financial Institution Message Authentication (Wholesale)."

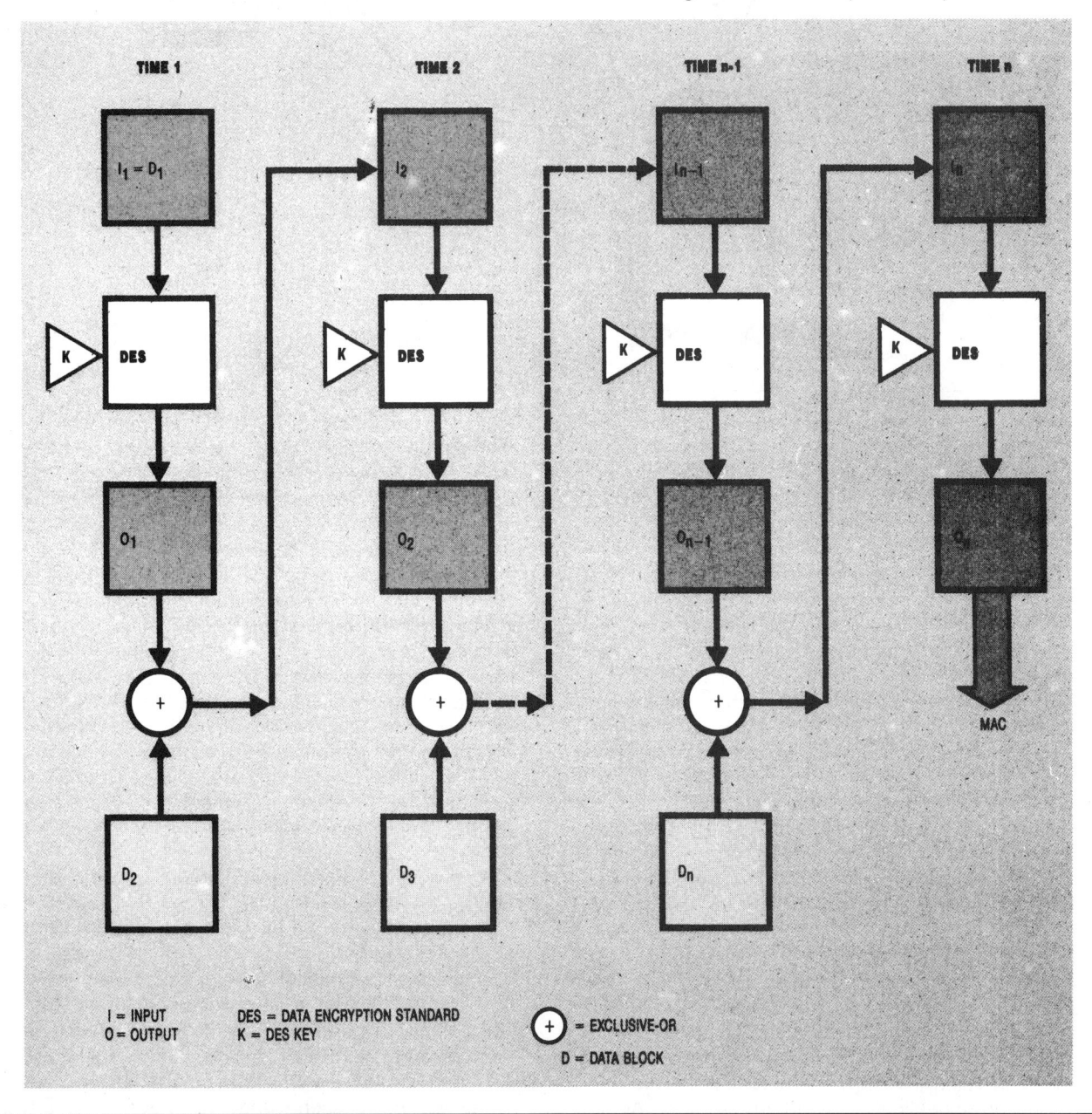

I = INPUT
O = OUTPUT
DES = DATA ENCRYPTION STANDARD
K = DES KEY
⊕ = EXCLUSIVE-OR
D = DATA BLOCK

3. Introductions all around. *The Key Translation Center (CKT) allows two or more parties who do not share a key with one another, but who do individually share keys with the CKT, to establish a keying relationship. The asterisk indicates that the keys are double the length of other keys.*

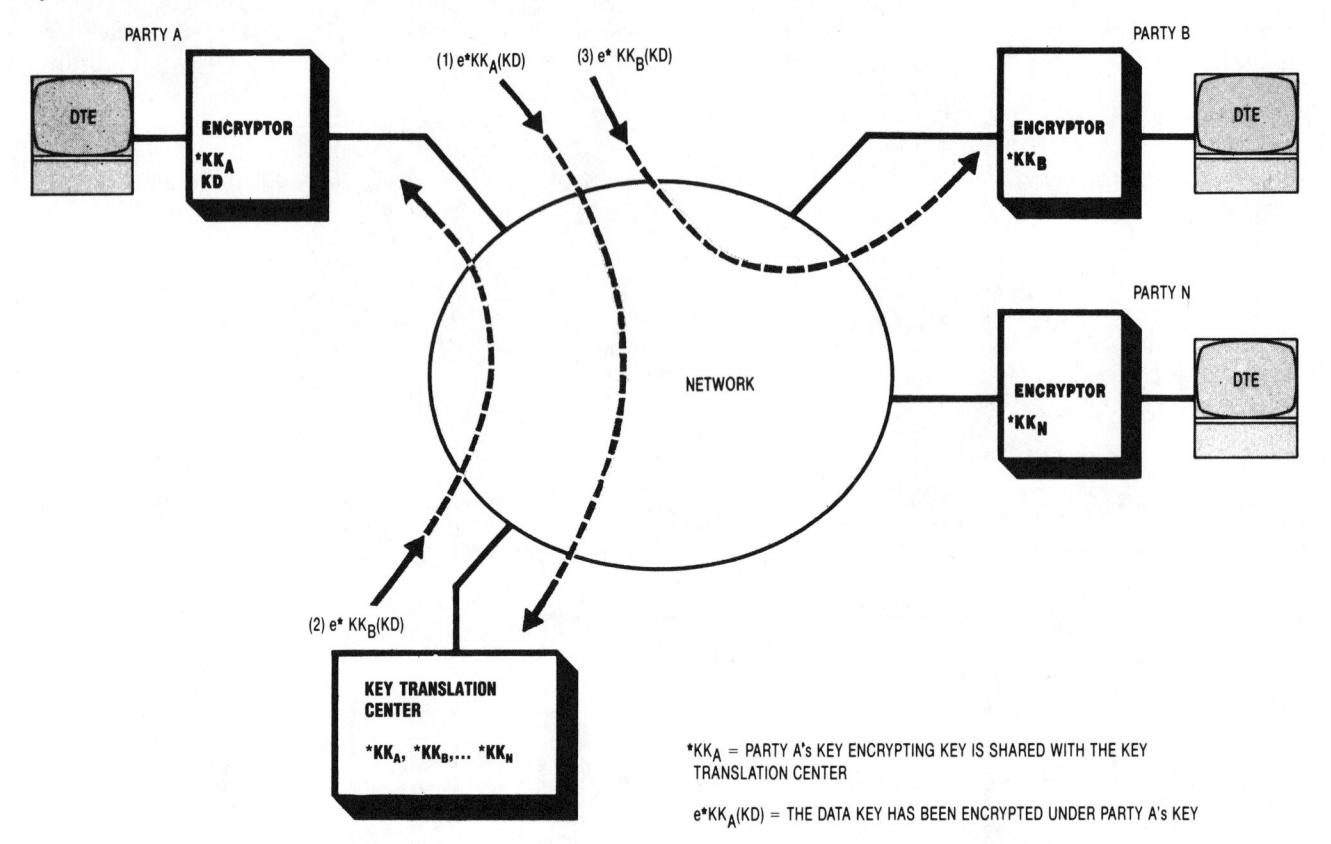

$*KK_A$ = PARTY A's KEY ENCRYPTING KEY IS SHARED WITH THE KEY TRANSLATION CENTER

$e*KK_A(KD)$ = THE DATA KEY HAS BEEN ENCRYPTED UNDER PARTY A's KEY

not necessarily kept secret, the authentication of the messages still works.

What remains is figuring out a way for Parties A and B to verify those MACs. The most positive verification method is to have another secured channel between the two sites. This occasionally happens when communications lines are added between established facilities, but is not the norm. For a more typical case, a simple means of verification is by a telephone call between the sites where the operators or installers know each other. If they are strangers, then they can exchange information or phrases previously exchanged by another convenient medium, such as the mail. The bottom line is that there must be some means of verifying the individual parties since if the spoofer can break into one line, he is certainly capable of breaking into a second one. Of course, one piece of information that the two parties can exchange is a DES verification key. Enough methods for ensuring verification exist to permit various levels of security over a range of costs.

Centralized key management

Control, maintenance, and operation of data communications networks are becoming increasingly centralized, with a small cadre of personnel directing the entire network. A continuation of such centralized control can be expected when cryptographic capability (either encryption or authentication) is added. With the April 1985 approval of ANSI X9.17 "Financial Institution Key Management (Wholesale)," that expectation can be fulfilled. X9.17 provides a protocol to permit secure automated key transfers between cryptographic facilities or between devices in a point-to-point configuration or involving a centralized facility. X9.17 describes two types of key centers—a Key Translation Center (CKT) and a Key Distribution Center (CKD).

■ **Key Translation Center.** The CKT allows two parties who do not share a key with one another, but who individually share keys with the CKT, to establish a keying relationship. Figure 3 shows Party A establishing secure communications with Party B. They do not share a key, but Party A shares a key, denoted by $*KK_A$, with the CKT, and Party B shares a key, denoted by $*KK_B$, with the CKT. The asterisk indicates that the keys are double-length keys (see "Building blocks for the X9.17 protocol").

Party A, the initiator, generates or acquires a key that will be used with Party B. The figure shows a data key, KD, being used, but translating a key-encrypting key is also permitted. Party A encrypts KD under $*KK_A$ and sends it to the CKT as part of a message that instructs the CKT to encrypt KD under Party B's key, $*KK_B$. The CKT performs that task and sends the newly encrypted key back to Party A. In that form, only Party B (in addition to the CKT) can decrypt that message to obtain KD. Party A sends the encrypted key to Party B,

Building blocks for the X9.17 protocol

ANSI X9.17 "Financial Institution Key Management (Wholesale)" combines a number of simple techniques to create a secure protocol for key management. Its building blocks include message authentication, key pairs, multiple encryption, offsetting of keys, notarization of keys, and key usage counters. Some convenient notation follows:
- Operators are shown by the lowercase letters:
  "a" for authenticate.
  "e" for electronic code book (ECB) encryption.
  "d" for ECB decryption.
- Exclusive-OR is indicated by " + "
- Concatenation is indicated by ||.
- The first letter of a key designator is "K." The second letter denotes the type of key so that:
  "KD" is a data key.
  "KK" is a key-encrypting key.
  "KN" is a notarizing key.
- The third letter further defines the key as follows:
  "o" means the key has been offset.
  "l" means the left key of a key pair.
  "r" means the right key of a key pair.

An asterisk "*" in front of a key character sequence indicates that the key is a DES key pair. "(*)" in front of a key character sequence indicates that the key is a DES key pair.

Keys. A key-encrypting key pair consisting of two 64-bit variables with a left part, l, and a right part, r, is denoted by:
$$*KK = KKl \parallel KKr$$

Encryption and decryption. Encryption and decryption of KD_1, by KKy is shown by:
$$KD_1 \text{ encryption} = e\,KKy\,(KD_1)$$
and
$$KD_1 \text{ decryption} = d\,KKy\,(KD_1)$$
If the encryption and decryption of KD_1 is performed by the key pair *KKy, then that is shown by:
$$KD_1 \text{ encryption} = ede\,*KKy\,(KD_1)$$
$$= eKKly\,(dKKry[eKKly(KD_1)])$$
and
$$KD_1 \text{ decryption} = ded\,*KKy(KD_1)$$
$$= dKKly\,\{eKKry[dKKly(KD_1)]\}$$
Going a step further, the encryption and decryption of key pair $*KK_1$ by key pair *KKy is:
$$*KK_1 \text{ encryption} = ede\,*KKy\,(*KK_1)$$
$$= eKKly\,\{dKKry[eKKly(KKl_1)]\} \parallel eKKly\,\{dKKry[eKKly(KKr_1)]\}$$
and
$$*KK_1 \text{decryption} = ded\,*KKy\,(*KK_1)$$
$$= dKKly\,\{eKKry[dKKly(KKl_1)]\} \parallel dKKly\,\{eKKry[dKKly(KKr_1)]\}$$

Note that the key pair algorithm can be used with a single key. In that case:
$$KKly = KKry = KKy \text{ and}$$
$$KD_1 \text{ encryption} = ded\,*KKy(KD_1) = eKKy(KD_1)$$
$$KD_1 \text{ decryption} = ded\,*KKy(KD_1) = dKKy(KD_1)$$

Authentication. When Cryptographic Service Messages (CSMs) are authenticated in X9.17, the message authentication code (MAC) is computed using the technique defined in ANSI X9.9, Section 4.0, using a data key.
$$MAC = a\,KD(\text{data})$$
A CSM containing a single KD is authenticated using that KD. If two KDs are sent in a CSM, then the key used for the MAC computation is the exclusive-OR of the two data keys.

Counters. Counters are used to indicate the number of times a KK or *KK has been used for encrypting other keys in CSMs. The parties using a particular (*)KK synchronize their counters and use those counters as a means to protect against message replay and for use in offsetting keys. The counters are incremented (never decremented) each time a (*)KK is used. The recipient of a message keeps an expected count and compares that with the actual count received in the message. If the count is a repeat, the message is thrown out and an Error Service Message is sent back to the originator of the message. If the received count matches or is greater than the expected count, then the message is accepted (assuming the MAC checks), and the recipient's expected counter is readjusted to synchronize it with the originator's count.

Offsetting of keys. Since keys are encrypted using the electronic code book mode, it is cryptographically advantageous to include some variable element in the key-encrypting key. This is done by offsetting the (*)KK with a counter. That is, the (*)KK is exclusive-OR'ed with a counter. In X9.17, offsetting is always used to transform a (*)KK prior to encryption of a key by that (*)KK. If CT is the value of the counter, then:
$$KKo = KK + CT$$
and
$$*KKo = (KKl + CT) \parallel (KKr + CT).$$
The exclusive-OR of a 64-bit key KK, with a 56-bit counter, is performed as follows: Combine the first byte of KK formed from the first seven high-order bits of the counter. The eighth bit is the parity bit for that byte and is readjusted to give the byte odd parity. Then the second byte of KK is exclusive-OR'ed with the counter byte formed from the second seven bits of the counter, and so on.

Notarization of keys. Notarization is a method for sealing keys with the identities of the communicating pair. Once sealed, or notarized, keys can only be recovered with knowledge of the key used to perform notarization and the identities of the communicating pair. A KD or a KK can be notarized by encryption with a notarizing key, (*)KN, which is formed by exclusive-OR'ing (*)KK with a notary seal, NS.

Suppose that Party A wants to send a key to Party B. Let $FROM_1$ be the first eight characters of Party A's identity and $FROM_2$ be the second eight characters of Party A's identity. Similarly, let TO_1 and TO_2 represent the first and second halves of Party B's identity. If necessary, the identities are replicated to form 16-character identifiers. For example, if Party A's identity is "PARTYA", then $FROM_1$ is "PARTYAPA" and $FROM_2$ is "RTYAPART". Let *KK be the key to be used to compute the notarizing key. Then
$$*KK = KKl \parallel KKr$$
$$KKl = KKl + FROM_1$$
$$KKr = KKr + TO_1$$
From this we now compute the components of the notary seal, NS, where,
$$NS = NSl + NSr$$
$$NSl = eKKR(TO_2) + CT$$
$$NSr = eKKL(FROM_2) + CT$$
where CT is a counter.

Finally, the notarizing key:
$$*KN = (KKl + NSl) \parallel (KKr + NSr).$$
The above expression is then used to encrypt (or notarize) either a KD or a (*)KK.

Now, if you think this is hard to compute, try breaking it.

4. Dealing out the keys. *The sole job of the Key Distribution Center (CKD) is to distribute data keys (A). This is accomplished at the request of the communicating parties or as part of the CKD's own key management program. In (B), the data network is expanded to show transmission lines with link encryption.*

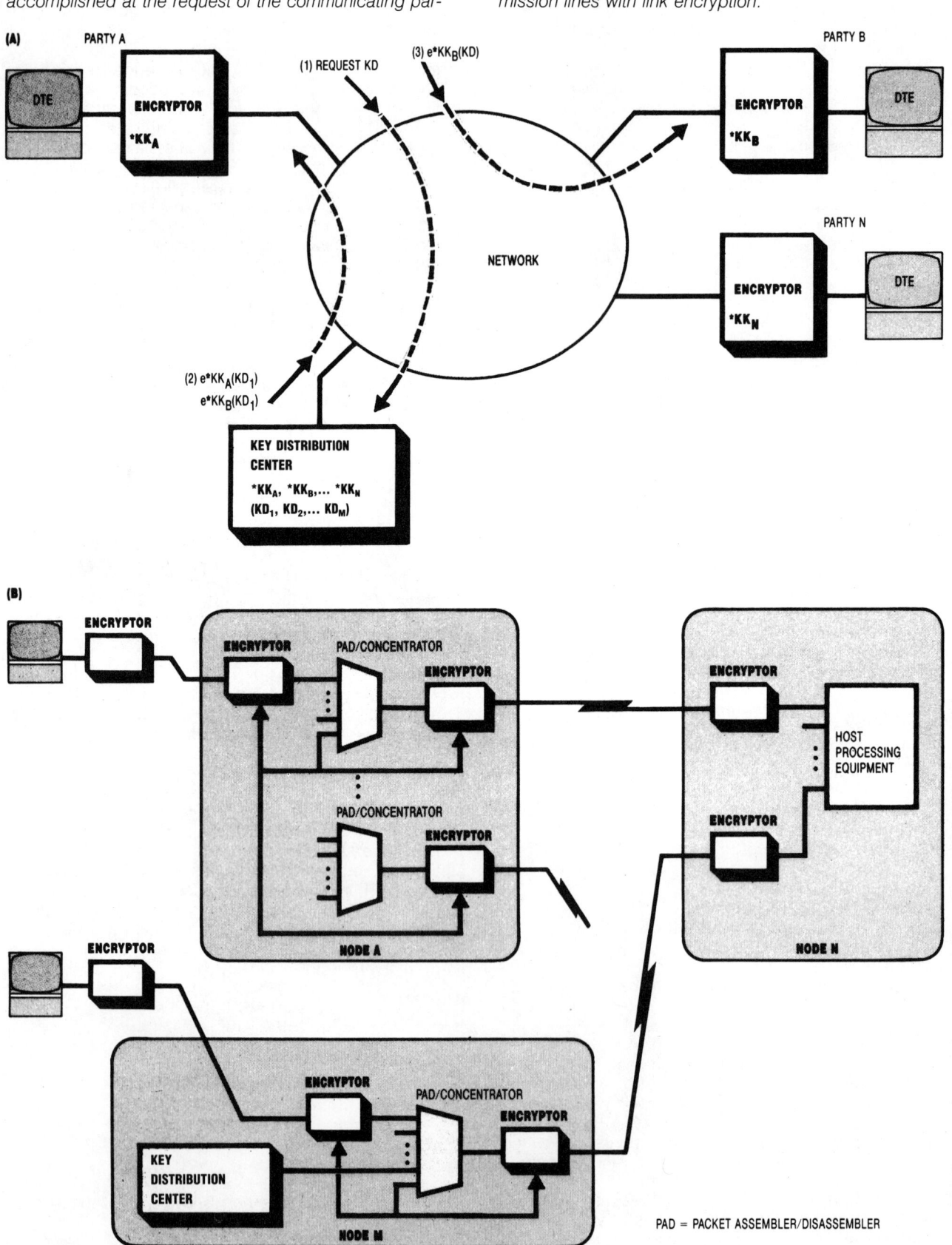

who decrypts it. Both parties now have KD and can begin data communications using that key. A KK could have been transferred instead. With Parties A and B sharing a key-encrypting key, they can downline load data keys without the involvement of the CKT.

The value of a CKT can be fully appreciated in a switched network, where any party has the capability to talk to any other party. Each possible pair of communicating parties must share a key, but that key must be different from the other keys for all the other possible communicating pairs in the network. Thus, a 10-party network would require each party to store 45 keys, a 100-party network requires 4,950 keys, and a 1,000-party network requires 499,500 keys. This is the number of combinations from n objects taken two at a time. $C(n,2)$ $n(n-1)P$ 2. With the CKT, each party need only store one key—the one he shares with the CKT. The CKT must store only as many keys as there are communicating parties in the network. Clearly, without centralized key management even medium-sized switched networks would be unmanageable.

■ **Key Distribution Center.** The CKD can also whittle down the number of keys required in a network, but it operates somewhat differently from the Key Translation Center. The CKT does not generate keys, but it can translate either data keys or key-encrypting keys. The sole job of the CKD is to distribute data keys, either at the request of one of the communicating parties or as part of its own key management program. Figure 4a shows a CKD configuration with Party A initiating a key transaction so that a keying relationship with Party B can be set up. Party A does not send any keys but, rather, requests a new data key. The CKD sends a data key (either generated or acquired by the CKD) in two forms—one encrypted by $*KK_A$ and the other encrypted by $*KK_B$). Party A, receiving the message containing the encrypted data keys, decrypts the one encrypted by $*KK_A$ and sends the one encrypted by $*KK_B$ to Party B—who decrypts the data key and can then encrypt or authenticate data to Party A.

The CKD can be used for key management in other types of data networks including networks employing link-by-link encryption. Figure 4b shows the data network expanded to demonstrate transmission lines having link encryption. The network itself is used to route the CKD messages to the individual encryptors. As shown in the figure, a port off the packet assembler/disassembler (PAD) is connected to the encryptors on at least one end of each link. Since each link can coordinate its own key transfer, only one end of the link need be accessed by the CKD. Figure 4B implies an X.25 packet network, but the concept works just as well if the PAD/concentrators are replaced with multiplexers and modems as in an analog switched network.

In a network like this, it is desirable to control and coordinate key management from a central point. That is where the CKD is located. This arrangement permits all data key generation to be performed by the CKD. In addition to determining what data keys are used, the CKD can determine when key changes are made.

The benefits of electronic key transfers are obvious, but how does one implement those transfers in a secure manner? How does one guard against inadvertent trapdoors in the protocol or against message modification, substitution, deletion, or replay? With ANSI X9.17, the only public standard available dealing with DES key management.

ANSI X9.17 protocol

The security of the X9.17 protocol for electronically transferring keys is predicated on the Cryptographic Service Messages (CSMs) that are defined, the flow of those messages, and, most importantly, the construction of those messages. Together they form a secure protocol that preserves and maintains the high level of security required of keying material. The CSMs for point-to-point key management are:

■ RSI (Request Service Initiation Message). An optional message class that requests that a new keying relationship be initiated. The originator of an RSI might not have a key generation capability.

■ KSM (Key Service Message). A required message class that transfers a key from an originator to a recipient.

■ RSM (Response Service Message). A required message class that provides an authenticated response to a KSM.

■ ESM (Error Service Message). A required message class that reports an error in a CSM.

■ DSM (Disconnect Service Message). An optional message class that is used to discontinue keys.

When using these CSMs, Party A and Party B must first share either a key-encrypting key (KK) or a key-encrypting key pair (*KK). For convenience in this article, (*)KK denotes either a key-encrypting key or key pair. Take a simple case where Party A doesn't have the capability (or permission by the Network Supervisor) to generate or acquire keys but wants to communicate with Party B under a new key. Party B sends an RSI message to Party A requesting that Party A send a key (or keys). If the RSI message has an error in it, Party A returns an ESM to Party B.

Party A sends the requested keys in a KSM. If a (*)KK is sent, it is encrypted under the (*)KK shared with Party B. The accompanying data keys (KDs) in the KSM are encrypted under the new (*)KK sent in that message. If a (*)KK is not sent, then the KDs sent in the KSM are encrypted under the (*)KK shared with Party B. If an encrypted initialization vector (a starting sequence of bits to initialize the DES process) is sent in the message, it is encrypted under the KD (the second KD if two KDs are sent) in that KSM. The KSM is authenticated using the KD or KDs sent in the message. If two KDs are sent, then they are exclusive-OR'ed together with the result used as the authentication key.

Party B responds with an RSM if the KSM is received correctly, or with an ESM if there is an error in the received KSM. If Party A sees an error in the received RSM, he returns an ESM to Party B.

Party A may resend a KSM to Party B an arbitrary number of times, but Party A will not send a new KSM (that is, involving new keys or a new count for the (*)KK specified in the message) until the old KSM is acknowledged by an RSM or an ESM.

Using the DSM requires some special considerations. When either party wishes to terminate a keying relationship, that party sends a DSM. If the DSM is received correctly, the other party returns an RSM; otherwise an ESM is returned. The key named as the authentication key (or the only data key shared by the two parties if no authentication key is explicitly identified) in the DSM is retained to authenticate the subsequent RSM. It is then discontinued. When a (*)KK is discontinued, all keys sent that are encrypted under that (*)KK are also discontinued without being named in the DSM.

So far in this discussion, whenever a CSM has been received in error, an ESM has been sent. However, when an RSM responding to a DSM contains an error, no ESM is sent. The party sending the RSM has discontinued his keys, and manual recovery procedures are required. Besides the CSMs used for point-to-point applications, three more CSMs are used with Key Translation Centers and Key Distribution Centers:
- RFS (Request for Service Message). A required message class in a Key Translation Center environment that sends keys to a CKT to be translated for the ultimate recipient.
- RTR (Response to Requester Message). A required message class in either a CKT or CKD environment that responds to an RFS, an ERS, or an RSI to the center. An RTR may be initiated by a Key Distribution Center.
- ERS (Error Recovery Service Message). A required message class that reports count and key errors to the CKT or CKD and requests resynchronization of the count fields and reinitiation of service.

When a Key Translation Center is used, the Message flow between Parties A and B uses the same CSMs as those used in the point-to-point example, but the messages are constructed somewhat differently. Party A requests a key translation with an RFS message containing new key(s) to be translated and eventually sent to Party B, the ultimate recipient. If a (*)KK is sent, then one KD must be sent. The (*)KK is encrypted under the *KK shared between Party A and the CKT; the KD is encrypted under the (*)KK sent in the message. If a (*)KK is not sent, then at least one and at most two KDs are sent, encrypted under the *KK shared between Party A and the CKT. The RFS message is authenticated using the KD(s) sent in the message. The CKT responds with an RTR if the RFS is received correctly; otherwise, with an ESM.

After receiving the RFS, the CKT decrypts the keys sent in the message. The decrypted keys are then notarized using the *KK shared between the CKT and Party B, the count associated with that *KK, and the identities of Parties A and B. The notarized keys are put into the RTR message, which is sent to Party A. Party A assembles a KSM, which includes the notarized keys as received in the RTR, and sends it to Party B, who responds with an RSM. While the message construction is slightly different, the KSM, RSM, and ESM are handled as in the point-to-point case. The message flow in a CKD environment is similar to that of the CKT except that an RSI is used from Party A to the CKD to initiate service. The CKD provides the KDs for the key exchange. The RTR message from the CKD contains two identical sets of KD(s). One set has the KD notarized with the *KK shared between Party A and the CKD, while the other set has the KD notarized with the *KK shared between Party B and the CKD. This second set is denoted by KDU (U for ultimate recipient) and is passed on to Party B in a KSM. Again, the KSM, RSM, and ESM are handled as they were in the other cases.

Levels of security

How much of this protocol procedure is necessary? The answer is as simple as it is imprecise: It depends. Nothing more definitive can be said unless the question defines the security requirements of the application. Since X9.17 is a standard, it has an obligation to be correct and complete in achieving its objectives. The skeleton of the X9.17 protocol is simple; the complexity enters into it when tailoring the protocol to deal with specific, frequently subtle, issues. This protocol would be used when the security requirements are such that it is not sufficient to guard only against the obvious attacks that can "break" a security arrangement. Attacks that may make breaking the system only slightly easier must also be prevented.

Not every application requires the protection offered by X9.17, which was specifically designed for use in wholesale banking, where a single electronic fund transfer (EFT) can involve millions of dollars. EFT transactions in that dollar range certainly justify substantial protection. But what about EFT transactions for retail banking with only hundreds or thousands of dollars involved? Or what about non-EFT networks where the intent is to provide privacy in corporate or government networks? To answer, the overall security requirements must be known. Even then, the choice of security equipment and features frequently rests with the judgment of a few individuals, and that judgment is typically affected by necessary compromises.

Not very many corporate security officers have the budget to implement the level of security that a rigorous security analysis might dictate. Moreover, the security requirements themselves may defy quantification. The value or worth of maintaining privacy may be difficult to quantify and in the end becomes a business decision based on the available budget. While security officers may not control the security budget, they are expected to implement data communications security consistently. Key management is particularly vulnerable, not only because its importance can be misunderstood, but also because it can be compromised due to inadequate budget or to its operational inconvenience, or to a combination of the two. A reduction in overall security to match the value of that which is being protected and the resource allotted to that protection is the reality facing most security officers. The need to provide an overall consistent level of security, at whatever level they expect to achieve, is their most important and most difficult requirement. ∎

Charles Richard Abbruscato is manager of encryption and T1 products at Racal-Milgo. He received a B.S.E.E. from the University of Connecticut and an M.S.E.E. from the Georgia Institute of Technology.

David M. Piscitello, Burroughs Corp., Southeastern, Pa., Alan J. Weissberger, Teledimensions Inc., Santa Clara, Calif., Scott A. Stein, Honeywell Information Systems, Phoenix, Ariz., and A. Lyman Chapin, Data General Corp., Westborough, Mass.

Internetworking in an OSI environment

The authors, who are active in OSI standards groups, provide authoritative information on protocols and connection methods.

Standards bodies concerned with fleshing out the Open Systems Interconnection (OSI) model are also tackling the associated problems of internetworking—that is, communications between an interconnected set of networks. The seven-layer model from the International Organization for Standardization (ISO) has become familiar within the industry. OSI provides a framework for the interaction of users and applications in a distributed data processing environment that may include a wide variety of both computer and terminal equipment as well as many different communications technologies.

The term OSI refers to the seven-layer architectural reference model and to a set of standards that describe how to provide communications among computers and terminals. To examine internetworking, one must understand OSI terminology. For each of the seven layers, a layer service is defined that identifies the set of functions that the layer provides. Layer services in OSI are of two general types: connection-oriented, which allow the service users (entities in the next higher layer) to establish and use logical connections; and connectionless, which allow the service users to exchange information without having to establish a connection (see "Connections vs. connectionless?").

Within each layer, protocols operate to provide the services defined for that layer. As a number of protocol-selection options exist in some layers of the reference model (for example, the Transport Layer defines five distinct connection-oriented protocol classes), conformance requirements are specified by each of the layer-protocol standards. When in compliance with the required suite of standard protocols prescribed for OSI, configurations are considered to be open.

The lowest two layers, Physical and Data Link, provide technology-specific access to the media that is used to interconnect the network equipment. The third layer (Network) performs data routing and relaying.

The fourth layer (Transport) provides for the reliable, error-free, end-to-end delivery of user data. An equally important and often overlooked function of the Transport Layer is to determine the most cost-effective means of providing data-transport service. Given specific quality-of-service (QOS) constraints by the higher layers, the Transport Layer matches a Transport protocol to the Network Layer service provided. (The QOS is defined by the characteristics of a connection-oriented or connectionless transmission as observed between the end points.) Thus the QOS requested is satisfied in the most expedient and cost-effective manner. Note that any meaningful analysis of the problems related to network interconnection often requires consideration of the Transport Layer as well as the Network Layer.

Internetworking is important
To be a successful standard, OSI must provide a homogeneous environment in which information may be accessed and exchanged independent of the immediate network to which corresponding users are attached. Network interconnection, or internetworking, has thus become a most important issue.

The advent of local area network (LAN) technologies has lent new urgency to this effort. Part of the world treats LANs as a sophisticated, multipoint data link; the remainder, as merely one more type of subnetwork (OSI parlance for a network as part of an internetwork) that must be accommodated by OSI. Therefore, it is inevitable that internetworking solutions involving LANs would be similarly divided.

The problem of interconnecting networks to form a single, "global" network is an inevitable consequence of the recent explosion of network technology. Network designers have investigated and implemented a number of interconnection strategies that attempt to facilitate communications among computers and terminals connected to different networks. While all create a homogeneous networking environment by resolving differences in network technology, access method, address structure, and administration, two appear to be most applicable to an OSI global network.

Strategy 1. In this case, the networks to be connected:
- Offer predominantly connection-oriented services.
- Exist where close cooperation among the network administrations can be achieved and enforced.
- Exist where the extent to which the individual network services differ is limited.

With this approach, connection-oriented internetworking may be achieved by relaying the services of one network directly onto corresponding services of the other networks.

An underlying assumption of this network interconnection strategy is that it is easier to solve the problems associated with subnetwork interconnection when the services that the networks offer are the same than when they are different. Hence, to ensure that the services presented on each side of a relay point are sufficiently similar so that a direct mapping of one set of services onto the other is possible, enhancement of individual networks on a "hop-by-hop" basis (see below) may be necessary. This approach is particularly attractive with public data networks and in countries that centralize provision of network services under a single government-controlled administration, such as a Postal, Telegraph, and Telephone (PTT) organization. In such environments, strong regulatory constraints operate to limit network diversity.

Strategy 2. Here, the networks to be connected offer a mix of connection-oriented and connectionless services and exist where network administration is largely autonomous. The extent to which the individual network services differ cannot be predicted or controlled.

In such configurations, connectionless internetworking preserves individual network autonomy and service characteristics of the networks to be connected. This is achieved by conveying the information necessary to support a uniform network service in an explicit Internetwork Protocol (IP). This protocol makes minimal assumptions about the services available from each of the interconnected networks.

One important observation about these two network interconnection strategies: The service that is ultimately provided to the user at the Transport service boundary is the same in both cases. The only real difference between them is where the end-to-end reliability functions are performed. Hop-by-hop enhancement tries to combine both internetwork routing and end-to-end reliability functions in the Network Layer, thereby making each network connection (hop) a miniature end-to-end transport connection. An internetworking protocol, on the other hand, concentrates on dynamic routing (the proper province of a Network Layer protocol) and leaves to the Transport protocol the responsibility for ensuring end-to-end reliability (the proper province of a Transport Layer protocol).

In practice, hop-by-hop enhancement may provide a reliable Network Layer service. However, the users most concerned with reliability, security, and data integrity—most notably the military and the banking industry—require a Transport service that guarantees those properties from end system to end system. (An end system is defined as a seven-layer configuration. For more detail on this and other OSI topics, see DATA COMMUNICATIONS, "The status and direction of open

Connections vs. connectionless?

In the earliest work on Open Systems Interconnection, communications was modeled exclusively in terms of connection-based interactions. These interactions proceeded through the three familiar phases of connection establishment, data transfer, and connection release. Traditional monolithic networks—where applications are the users—are the source of this connection model. The classic example is the voice telephone network, which is operated directly by human users who establish connections (call), transfer data (talk), and release connections (hang up the telephone).

An alternative model of connectionless interactions, which begin and end with the transmission of a single self-contained data unit (often called a datagram), was developed as an extension to the OSI architecture. Standards developers, working in today's much more complex multiple-network environment, encountered interconnection scenarios that could not be modeled satisfactorily in entirely connection-oriented terms. Both models have been applied successfully to the specification of OSI layer services and protocols.

The essential difference between these two models is that a connection preserves the state of peer-to-peer communications from one data transfer to the next, storing and distributing information regarding the connection within the service provider; connectionless transmission does not. The components of a connection-oriented network—such as gateways, switches, and interface processors—operate collectively over time to create and maintain state information for each connection that is established between one point on the network and another. (Examples of state information include the locations of the communicating peers, the sequence number of the last data unit forwarded, and the status of peer-to-peer flow control.) This information relates each new data transfer to previous transfers on the same connection, so that the network deals with data units containing shorthand "pointers" to the information base (such as a connection identifier) rather than the information itself (such as a full destination address).

A connectionless network, on the other hand, does not establish or maintain any relationship between individual data transfers. All of the addressing and other information needed to convey data from source to destination is included explicitly in each data unit. Connection-oriented networks recover from errors through global-state resynchronization; connectionless networks, with no state to preserve, use timeouts and retransmissions. Intuitively, connection-oriented networks are "deterministic"; connectionless, "probabilistic."

The obvious question that arises from the existence of two different interaction models in the OSI environment is how to choose between them. Unfortunately, the connections vs. connectionless debate has often been carried on as if the question were simply, "Which model is 'better' "—without regard to the context in which it is applied. Those arguing in such terms would have one believe that because a connection seems to be the best model for a particular application, it must also be the best model for the entire data communications structure that supports the application. Or they argue that because connectionless transmission is the most efficient mode of operation for local area networks that use a particular contention-based access method, it must also be the best model for the operation of all types of networks.

A less dogmatic—and much more practical—approach would consider the individual characteristics of the two types of operation in terms of the applicability of one model or the other to specific interconnection scenarios. This approach recognizes that OSI can be viewed from a number of different perspectives, and that it is often necessary to apply different techniques to the solution of different problems.

In general, when a service that processes a large number of data units in essentially the same way—while keeping track of the data transfer state on behalf of the communicating peers—is desired, the connection model is more useful. When the dynamic flexibility of a service that processes data units independently—or the simplicity of a service that does not maintain state on behalf of the communicating peers—is desired, the connectionless model is more useful. In the real world, of course, there are almost

---

systems interconnection," February 1985, p. 177.) In any configuration in which end-to-end reliability is important, the most reliable Transport Layer protocol available must be used, and any effort to enhance the subnetworks is largely wasted (although the user, of course, is expected to pay for the enhancement).

Hop-by-hop: For simplicity?

In strategy 1, gateways (Network Layer relays) perform a mapping of the service offered by one network onto another. In general, the gateways do not add services. Rather, they perform the relaying and switching functions necessary to bind the individual subnetworks into a unified or global network. A consequence of this approach is that either all of the subnetworks must inherently provide equivalent services, or each must be enhanced to some common level of service.

The ISO has included this interconnection strategy in its "Internal Organization of the Network Layer" standard (ISO 8648). Called hop-by-hop (subnetwork) enhancement, the approach may be summarized as follows (Fig. 1): All subnetworks that are to be interconnected must provide exactly the Network Layer service (usually, connection-oriented), either directly or through enhancement. Any subnetwork that does not provide this service must be enhanced or modified to do so. Relays are used

always practical trade-offs rather than absolute positions, and they must be evaluated as such.

From the perspective of the OSI user, the only relevant concern is whether the application's operation is connectionless or connection-oriented. From the perspective of the OSI builder, the important issue is the way in which a particular combination of connectionless and connection-oriented layer services and protocols solves the technical, administrative, and economic problems of interconnecting OSI networks. These two perspectives are essentially independent. The user's choice is based on the characteristics of the application. The builder's choices depend on the available network technologies, an overall strategy for achieving network interconnection, and the economic and administrative constraints of a particular operating environment. The clearest illustration of these choices is found in the nature of applications (the ultimate users of OSI) and the different ways in which multiple networks can be interconnected.

Beneficial examples
Some interactions between peer application entities involve the exchange of many related data units. These interactions must be performed in an explicit application context that defines the intent of the entities' exchange. Such applications can benefit from a connection-oriented service. Examples include bulk file transfer (particularly when checkpoint/recovery features are implemented); virtual terminal usage (long-term attachment of a terminal, workstation, or other interactive device to a host computer, for which the security established during an initial log-on procedure is an important part of the context associated with a connection); and access to distributed network components, such as print servers and remote-job-entry stations.

Other application interactions are either entirely self-contained or do not benefit from the context characteristics of a connection. Examples include "inward" data collection (periodic sampling of a large number of data sources, as in a sensor network); "outward" data dissemination (the distribution of a single piece of information to a large number of destinations); broadcast and multicast (group-addressed) communications; and a variety of other request/response applications (such as directory and time-of-day servers) in which a single request is followed by a single response, with no significant relationship between one message and the next.

Interconnection realities
If all networks were of the same type, used for the same purposes, administered by the same organization, and operated under the same tariffs and regulations, interconnecting them would pose few major technical problems. Clearly, not many of these conditions—much less all of them—are met in the real world.

There are excellent reasons for the providers of public network services to prefer that their interactions with users and with each other be connection-based. These service providers must deal with unpredictable and widely fluctuating loads; must limit variations in quality of service to a relatively narrow range, according to the terms of a legal contract; and must charge for their services on a fair and auditable basis in accordance with that contract. Deterministic global resource allocation is of paramount importance. On the other hand, private networks, such as LANs, tend to be owned and used by the same organization. Their operating costs are generally recovered in ways that are only indirectly related to individual instances of use.

Public-network administrators exercise greater global control over their configurations than do private-network administrators, to the extent that public networks are operated (and perceived by their users) as one global network. Private networks, however, usually consist of a number of individual, interconnected smaller networks, forming an internetwork topology in which boundaries persist that are related to administration, local control, and mode of use. The managers of private internetwork topologies are concerned with the flexible and reconfigurable interconnection of a variety of individual networks, and they are correspondingly reluctant to make too many assumptions about the nature of individual underlying services. These networks tend to be connectionless.

---

to passively map the connection establishment, data transfer, and connection release facilities of one subnetwork onto another whenever network connections cross subnetwork boundaries.

The enhancement of subnetworks that do not provide the OSI network service is usually accomplished in either of two ways: through direct modification of the protocol used to access the subnetwork, as is the case for the 1984 version of CCITT Recommendation X.25, or through the use of what is called a subnetwork dependent convergence protocol (SNDCP). An SNDCP operates on top of a subnetwork access protocol (SNACP), such as X.25, to provide the elements of the OSI network service that are missing from the access protocol. Such a protocol (ISO 8878) has been developed for operation with X.25's 1980 version.

One of the consequences of the hop-by-hop approach is that when a subnetwork administration revises its access protocol to provide the OSI network service, all data terminal equipment (DTE) attached to that subnetwork must often be revised. If the SNACP is modified principally to accommodate internetworking, the user and the DTE manufacturer must absorb the migration expense regardless of whether the DTE is to be used for internetworking—or even whether it will be used for OSI at all. Often such a revision does not result in any signif-

1. Enhancing hop-by-hop. *All subnetworks to be interconnected that do not provide the Network Layer service (usually connection-oriented) must be enhanced to do so. Gateways are then used to map the connection establishment, data transfer, and connection release facilities of one subnetwork onto another.*

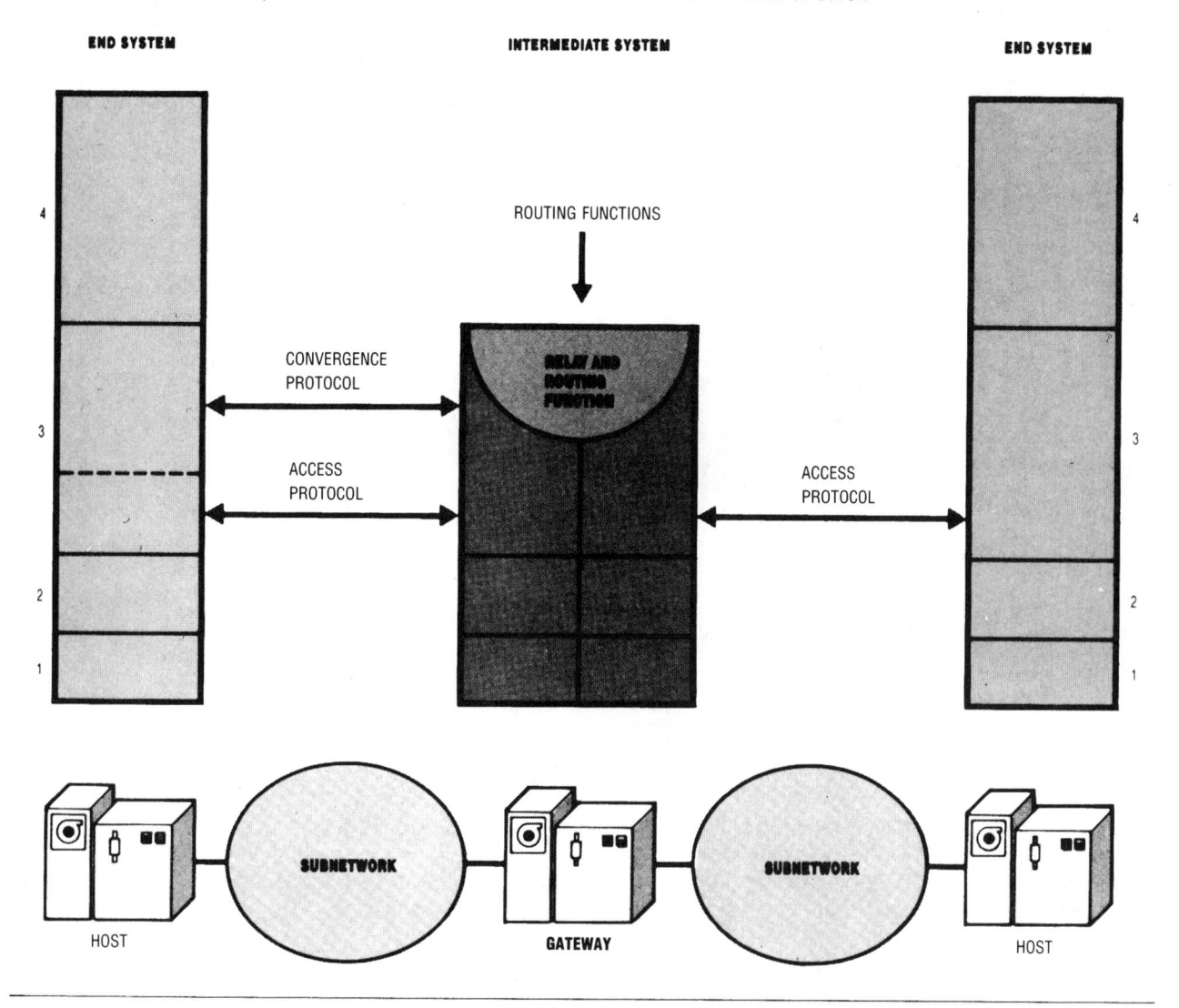

icant improvement in the service provided in the local environment. In fact, accommodations for connection-oriented internetworking often result in lower performance (due to, for example, additional protocol overhead) for local information exchanges.

This is both unreasonable and inefficient. In most subnetworks, local traffic predominates; hence, most subnetwork designs justifiably emphasize efficiency for local traffic. Protocols and addressing are often tailored for the local topology, traffic requirements, and underlying technologies (such as LANs) employed. It should not be necessary to change intranetwork operation—in particular, to modify its access protocol—solely for the purpose of internetworking. The imposition of additional overhead on intranetwork communications just to accommodate the less-frequent instances of internetworking is a poor bargain for most users. Moreover, problems of network interconnection cannot always be resolved by unilaterally imposing a single SNACP on all subnetworks. A more pragmatic approach to internetworking is to examine the interconnection scenarios, then base configuration design on those scenarios that are to be accommodated.

The use of an SNDCP permits existing terminal equipment to coexist in the intranetwork environment with equipment that supports the full OSI network service. However, since an SNDCP is designed to operate in conjunction with a specific SNACP, one must be developed for each subnetwork type to which a particular device may be attached. Such a proliferation of convergence protocols is hardly desirable.

Hop-by-hop subnetwork enhancement, then, is a practicable choice only where the type of service offered by the majority of the subnetworks to be interconnected corresponds very closely to the OSI network service. It is a poor choice for configurations in which the subnetworks provide dissimilar services, provide different or unpredictable qualities of service, or are managed and administered in different ways. In these types of configurations, an alternative method of

2. Connectionless internetworking. *The networks to be connected offer a mix of connection-oriented and connectionless services. The connectionless internetworking strategy is based on the use of a single end-to-end protocol to provide the Network Layer service over any combination of subnetworks.*

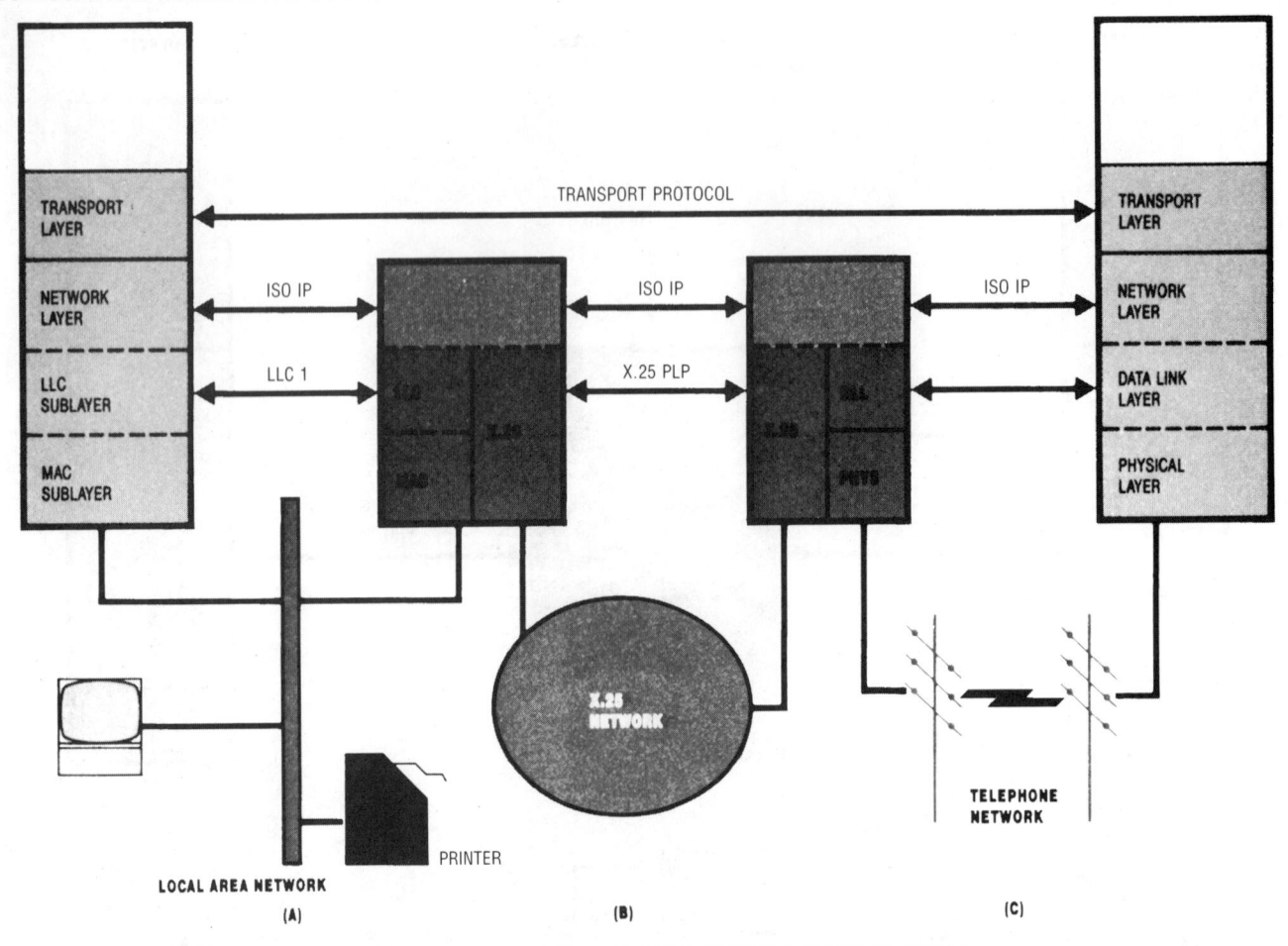

DLL = DATA LINK LAYER
IP = INTERNETWORK PROTOCOL
ISO = INTERNATIONAL ORGANIZATION FOR STANDARDIZATION
LLC = LOGICAL LINK CONTROL PROCEDURE
LLC 1 = LOGICAL LINK CONTROL PROCEDURE, CLASS 1
MAC = MEDIUM-ACCESS COMPONENT
PHYS = PHYSICAL LAYER
PLP = PACKET-LEVEL PROTOCOL

network interconnection should be considered.

Strategy 2 is based on the use of a single end-to-end protocol to provide Network Layer service over any combination of subnetworks (Fig. 2). This IP is operated in a sublayer above the subnetwork-specific protocols in both the hosts and the gateways (see "Protocol relationships"). It performs the addressing and routing functions necessary for end-to-end communications independent of the subnetwork-specific functions operating in each of the interconnected subnetworks. The underlying subnetworks should provide only a data transmission service. No subnetwork enhancement is necessary; an IP can be operated directly over the Data Link Layer.

This second approach to network interconnection was successfully applied by a number of user communities and computer manufacturers long before the ISO began to examine the problems of internetworking. The Defense Data Network (DDN) of the Department of Defense (DOD) and such private networking archi-

Protocol relationships

The relationship of the ISO Internetwork Protocol to the Transport protocol, to others in the Network Layer, and to the Data Link Layer is illustrated in Figure 2 of the accompanying story. The ISO IP may operate (a) over a LAN that uses the IEEE 802.2 LLC (Logical Link Control) Type 1 Class 1 Data Link procedures; (b) over a Public Data Network that uses the X.25 Packet Level Protocol; or (c) directly over a leased or switched telephone line that provides the OSI Data Link service through data link procedures, such as ISO's HDLC or IBM's SDLC. An addendum to the ISO IP, a Draft International Standard, describes how to operate the IP over (a) and (b) above. The U. S. technical subcommittee on the Network Layer (ASC X3S3.3) introduced a description of (c) in April of this year.

tectures as Xerox's XNS (Xerox Network Systems), Burroughs Corporation's BNA (Burroughs Network Architecture), and Digital Equipment's Decnet use internetwork protocols. These protocols provide an efficient and flexible means of interconnecting networks that differ in their types and qualities of service, but they do not impose unnecessary constraints and complexity in the local environment.

The ISO has formalized the second, or internetworking protocol, approach in its "Protocol for Providing the Connectionless-mode Network Service" (ISO 8473). It is designed to operate entirely within the computers (gateways and hosts) and terminal equipment attached to public-switched and private packet data networks, LANs, or other subnetworks. Like forerunners of its kind, the protocol ignores the idiosyncratic characteristics of the individual subnetworks, extracting from each subnetwork a data transmission service with no quality of service or other essential properties. The protocol demands so little from the underlying service (in the subnetwork sublayer or the Data Link Layer) that it may be operated directly over leased and switched telephone lines that provide the OSI Data Link service.

Here is how the ISO IP operates in the OSI environment. Computer and terminal equipment attached to different networks exchange data packets—called Internetwork Protocol Data Units (IPDUs)—in a connectionless (datagram) mode of operation. Each IPDU is routed independently. The route an IPDU takes is determined by the current state of the network links, rather than by the state that existed at some previous connection-establishment time.

Much of the information necessary to determine a route (such as source and destination addresses and quality of service) is conveyed in the protocol control information (PCI) of each IPDU. Additional information, which enables end and intermediate nodes to cooperate dynamically in determining the best route for each IPDU, is distributed and maintained by Network Layer management functions and protocols. In most cases, routing decisions are made independently by each forwarding node. However, the network entity (an active element within the Network Layer) of the end system that generates an IPDU may specify the route the IPDU should take by employing the protocol's optional "source routing" function.

The ISO IP assumes only that each subnetwork traversed is capable of serving as a "data pipe." The IP also harmonizes the service offered by each subnetwork into a uniform connectionless network service. Because subnetworks are required only to provide a data pipe, the mapping of the service required by the IP onto that provided by most SNACPs is extremely simple. For example, in the case of a LAN station operating under the procedures of IEEE 802.2 Type 1 Class 1 (unacknowledged connectionless), the mapping is a relatively trivial one-to-one process.

The ISO IP provides a connectionless network service to the Transport Layer. Unlike a connection-oriented service, a connectionless service dynamically allocates such resources as CPU, buffers, and data link availability on an as-needed basis rather than statically for the duration of a connection. This method of resource allocation is extremely efficient for normally encountered traffic conditions. Coupled with a deterministic or adaptive routing algorithm, it generally results in optimal use of network resources.

It is possible, however, that unusually high traffic volume (involving considerable delay), a temporary shortage of buffer space, the loss of a data link, or some other transient condition will cause an IPDU to be lost. In addition, the Network Layer may not be able to deliver IPDUs in the same order in which they were generated since alternate or parallel routes may be used for the transmission of a particular sequence of IPDUs. (These other routes would be used to enhance throughput or service quality, or in response to changes in the topology or service characteristics of the underlying networks.)

Thus where end-to-end reliability is important, the Transport Layer must provide reliable end-to-end connections that preserve data integrity and sequence. This is achieved by operating a Transport protocol that is able to recover from the loss, duplication, corruption, or reordering of data. In most internetworking configurations, service interruptions such as those mentioned above are relatively infrequent and short-lived. The Transport protocol design must also ensure that any penalty associated with the reliability function is assessed only when the function is invoked. That is, the additional overhead for normal (error-free) operation should be negligible. The ISO Class 4 Transport Protocol (TP 4) was designed precisely for this purpose.

Of the five Transport protocol classes developed by OSI (see Table), Class 4 ensures data integrity and data-transfer reliability when there is a possibility that any of the underlying networks will lose, corrupt, or reorder data. TP 4 utilizes a number of sophisticated error-detection and recovery mechanisms. Detailed below, they include transport protocol data unit (TPDU) numbering, retention of TPDUs until acknowledgment, retransmission of TPDUs following a time-

Comparing transport protocol classes

FUNCTION	CLASS				
	0	1	2	3	4
ERROR RECOVERY	NO	YES	NO	YES	YES
EXPEDITED DATA TRANSFER	NO	YES	YES	YES	YES
EXPLICIT FLOW CONTROL	NO	NO	YES	YES	YES
MULTIPLEXING	NO	NO	YES	YES	YES
DETECTION AND RECOVERY FROM:					
LOST TPDUs	NO	NO	NO	NO	YES
DUPLICATED TPDUs	NO	NO	NO	NO	YES
DISORDERED TPDUs	NO	NO	NO	NO	YES
CORRUPTED TPDUs	NO	NO	NO	NO	YES

TPDU = TRANSPORT PROTOCOL DATA UNIT

out, resequencing of TPDUs, and TPDU checksums. TP 4 provides a reliable, end-to-end transport service no matter what happens in the underlying networks.

All TPDUs that contain user data (DT TPDUs) and are forwarded to a given destination are marked with a sequence number. TPDU sequence numbering is used by flow control, resequencing, and recovery mechanisms to ensure that each transmitted DT TPDU is acknowledged by its destination within a given time interval. If this time interval elapses without receipt of an acknowledgment TPDU, the DT TPDU is retransmitted with the same sequence number as the original transmission. The sending transport entity (an active software element in the Transport Layer) retains a copy of each transmitted DT TPDU until an acknowledgment TPDU is received from the destination transport entity. Once the acknowledgment is received, the resources used to hold the copy are released.

As indicated earlier, misordered (out of sequence) DT TPDUs may occur within the Network Layer. To ensure that data submitted by the source session entity is delivered in sequence to the destination session entity, the receiving transport entity uses the sequence numbers of the DT TPDUs to reconstruct the original user-data sequence. Only TP 4 has this capability.

TP 4 makes use of a checksum mechanism to detect the corruption of TPDUs that are not detected by the Network service. If the receiving transport entity determines that a TPDU has been corrupted, the TPDU is discarded. The retransmission mechanisms described earlier will cause the TPDU to be retransmitted by the sending transport entity.

TP 4 provides explicit flow control. Cooperating transport entities determine a maximum number of outstanding, unacknowledged TPDUs (called the window size) for both communications directions. During the data transfer phase, neither transport entity is permitted to transmit or retransmit more than this number of DT TPDUs until an explicit acknowledgment of previously transmitted DT TPDUs is received. This mechanism, coupled with congestion control procedures in the Network Layer, provides a means of maintaining a uniform service quality for users of the transport service.

Much ado about nothing

One of the most misunderstood features of TP 4 is its extensive use of timers and retransmission mechanisms. These mechanisms are widely recognized to be necessary when a Transport connection is constructed over one or more error-prone subnetworks. However, it is frequently assumed that the presence of the mechanisms results in inefficient operation when the underlying networks are more reliable than the error-prone subnetworks. This is not true. The protocol overhead associated with TP 4 is no greater than that associated with TP 2 and TP 3 during data transfer. In fact, the DT TPDU format is exactly the same for TPs 2, 3, and 4.

From a procedural standpoint, if TP 4's reliability mechanisms are not invoked (that is, TPDUs arrive in sequence, uncorrupted, without loss of data), TP 4 has no more overhead than does any other Transport protocol class. Careful implementation of the protocol and adjustment of the retransmission timers will ensure that no unnecessary delays are introduced during normal operation. When the underlying networks are highly reliable, therefore, the procedural cost of running TP 4 is the same as the cost of running any other Transport protocol class. However, if errors do occur—TPDUs could be lost, duplicated, or corrupted by the underlying networks—then clearly the reliability mechanisms invoked by TP 4 to recover are necessary. This can hardly be classified as overhead.

Proponents of the hop-by-hop approach claim that there will be simplification in Transport protocol operation as a result of the high degree of reliability provided by the enhanced subnetworks. The Transport protocol class of choice in this environment is usually TP 1. However, in the absence of errors, TP 1 and TP 4 operations are essentially equivalent. If an error does occur, the situation is quite different. TP 1 can recover after a failure is correctly signaled by the Network service but cannot detect or recover from errors that are unsignaled or incorrectly signaled, such as corruption of user data. Hence, any user community concerned with data integrity must provide recovery mechanisms at the user application. It is difficult to understand how this can be called simplification.

The protocol's advantages

Because the ISO IP operates in a sublayer above each subnetwork, all the subnetwork sees is data, which is precisely what it saw before there was a need for internetworking. Subnetworks, therefore, are completely unaffected by the presence of the IP (and, for that matter, of OSI). Subnetwork protocols are left intact, and users can continue to employ existing terminal and computer equipment and software. Since the subnetwork protocols continue to be concerned only with efficient operation in the specific environment for which they were optimized, local traffic is not penalized by the introduction of internetworking.

The simple data transmission service required by the ISO IP is readily obtained from the vast majority of subnetworks in existence today. The ISO IP provides segmentation and reassembly mechanisms to accommodate different subnetwork packet sizes. Each subnetwork can continue to use the packet sizes for which it is best suited. Note that the ISO IP was designed specifically to satisfy the OSI connectionless-mode Network service. Therefore, the Transport Layer need not rely upon the quality of service characteristics of any individual subnetwork to provide the QOS requested.

One of the benefits of using the Internetwork Protocol is that the changes required to provide communications beyond the local subnetwork environment are restricted to the terminal equipment and gateways involved in internetworking. In all cases, subnetwork autonomy and local-traffic service characteristics are preserved. Terminal equipment can readily be designed so that the processing overhead associated with internetworking is incurred only when internetworking is actually performed. Routing functions within the Network Layer determine whether the destination is in the local subnetwork. If so, the Inactive Network Layer Protocol (INLP) is used.

The INLP is actually a "null IP subset." It is used strictly to enhance performance, and it serves as an indicator to the destination end system that the complete ISO IP is not present, that the IPDU originator was on the local subnetwork, and that the source and destination subnetwork addresses have been mapped directly onto the corresponding OSI Network addresses for the purpose of expediting the transmission. Contrary to what has been stated in an earlier article (DATA COMMUNICATIONS, "Of local networks, protocols, and the OSI reference model," November 1984, p. 129), the routing functions that determine when to use the INLP are performed entirely within the Network Layer, and its use is completely transparent to the Transport Layer or to a user.

Finally, from the standpoint of product development, there is one significant advantage of using the ISO IP instead of a hodgepodge of subnetwork dependent convergence protocols. Since the IP can be operated over any subnetwork access or data link protocol, users and DTE manufacturers seeking OSI compatibility can limit the number of convergence protocols they must provide to one.

The tailor-made solution
The nature of most LAN applications ensures that a large proportion of the information exchange takes place within the confines of a single LAN or bridged-LAN topology. For other applications, however, data exchange beyond the local area must be made possible. There is an urgent need for the industry to consider a way to interconnect LANs to other LANs—and to other networks such as packet- or circuit-switched types. Primarily, the impact of providing interconnection of these subnetworks should be minimal. While nonlocal traffic may be generated infrequently, the information exchanged over long (nonlocal) distances is likely to be some of the most important. The ISO IP offers a tailor-made solution to this complex set of LAN interconnection requirements.

Since both the IP and the LLC1 (Logical Link Control Class 1) have relatively simple "state [of operation] machines" (essentially "send" and "idle"), Network Layer design in LAN workstations, as well as in LAN-to-LAN gateways, is quite simple. Each Network Layer service user request-to-send-data results in the generation of one or more IPDUs. The IPDUs are passed to the LLC sublayer along with the destination's LAN station address. The LLC sublayer encapsulates the IPDU into "Unnumbered Information" frames and submits them to the "Medium Access Component [device]" for transmission. When the frame is delivered to the destination LAN station, the receiving LLC sublayer strips the control information from the LLC1 frame and passes the IPDU to the internetwork sublayer for forwarding by the routing function (at a gateway) or for delivery to the transport layer (the "user" of the Network Layer service at the LAN station). The procedures for operating the ISO IP over LLC1 are described in greater detail in an addendum to the ISO IP: ISO 8473/DAD1.

Gateways between connection-oriented networks (particularly X.25 types) and LANs are the only devices that require the complexity of a connection-oriented state machine. Typical operations involve a call, reset, or disconnect that is pending or awaiting confirmation. However, the complexity is confined to the gateways: End systems on the LANs do not have to operate a connection-oriented state machine at the Network Layer. Also, in many scenarios a wide area network, such as an X.25 type, is but one "hop" among many. Particularly for internetworking scenarios involving many LANs, the number of hops on which the overhead of an additional protocol over a wide area network is incurred is relatively small.

A significant benefit of using the ISO IP is that gateway complexity is greatly reduced because of the way subnetwork connections—in particular, X.25 virtual circuits—are managed. ISO 8473/DAD1 identifies a set of mechanisms and timers for opening and closing X.25 logical channels. This greatly reduces the number of occasions on which state transitions associated with maintaining a connection interfere with the efficient operation of the gateway.

The data transmission service realized through the manipulation of X.25 virtual circuits is essentially the same as that offered by an OSI data link. The establishment and release of X.25 virtual circuits is governed by timer mechanisms. Logical channels are left open (available) to transmit IPDUs for as long as is economically and administratively efficient.

Some concrete examples
The ISO IP should be used where LANs are involved in internetworking; benefits are derived from resource optimization, throughput enhancement (through the use of load-splitting techniques), and redundancy and resiliency (the ability to adapt to redundancy).

The ways in which the ISO IP enables Network service users to get more for their money are considered next.

Resource optimization. Using the hop-by-hop approach (Fig. 3), each LAN station (end system), A, must establish a separate connection to each X.25-network device (end system), B, in order to transfer data. For each connection, resources (such as buffers, a connection-state-information base, and CPU) must be reserved in both end systems as well as in the relay or intermediate system, I_1, for the duration of the call. In particular, I_1 must have ample capacity to maintain all of these connections, even if no traffic is passed. Clearly, if connections remain idle for long periods of time, valuable network resources are wasted.

In contrast, if the ISO IP is used, the sending end system may free resources as soon as the data unit's transmission is completed. Normally, the receiving end system reassembles the IPDU prior to indicating its delivery to the destination Transport entity. I_1 processes each data unit separately, allocating buffer space as required and maintaining a simple state machine (send or idle). Any communicating pair of Network service

3. Hopping between networks. *Using the hop-by-hop approach, each station, A, must establish a separate connection to each X.25-network device, B, in order to transfer data. For each connection, resources must be reserved in both end systems (A and B) as well as in the intermediate system, I_1, for the call's duration.*

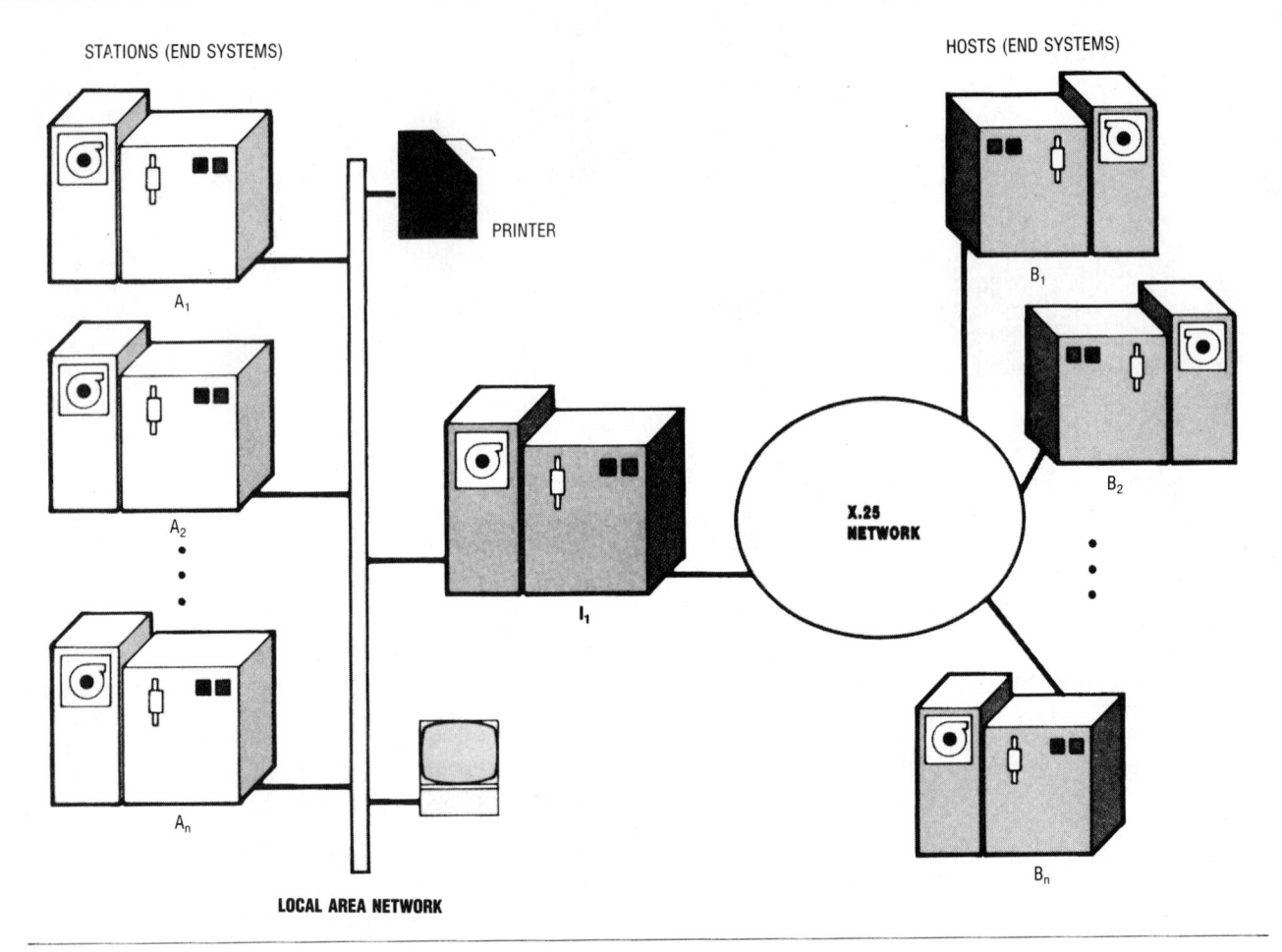

users (hereafter identified as A,B) that has long periods of inactivity between transmissions imposes no overhead. Therefore, the ability of the intermediate system to process transmission requests from any other communicating pair (A,B) remains unaffected. This typically results in highly efficient use of resources.

Throughput enhancement. In many internetworking scenarios, the ability to route IPDUs independently is particularly useful. Data exchanged between hosts attached to one subnetwork can be routed to hosts on a different (remote) subnetwork without the constraint that all data must be routed down the same path (and hence, through the same gateway). Using multiple paths to transmit data to the same destination typically improves throughput and response time.

The ISO IP provides an excellent means of equalizing a potentially vast throughput disparity through the use of load splitting techniques. Consider those configurations (Fig. 4) in which an IEEE 802-compatible LAN (operating at 4, 5, or 10 Mbit/s) must work with a lower-speed wide area network (the maximum throughput of which is normally 48 or 56 kbit/s). In a connection-oriented environment, a separate connection must be established to each network to transfer data. In addition, this separate connection must be relayed through either I_1 or I_2 relay systems.

A number of constraints can immediately be identified.

■ Some knowledge of which intermediate system (I_1 or I_2) is to be used for which connections—as well as how many connections each intermediate system can maintain—must be known prior to establishing connections, to prevent intermediate-system overloading.

■ Once a connection is established via one intermediate system, the resources of the other cannot be utilized for this connection, as no mechanisms exist for preserving the sequence of data to be transferred if routed via parallel paths.

■ If an intermediate system experiences a failure, all of its connections are broken and must be re-established through a different intermediate system.

In contrast, use of the ISO IP resolves all of the above constraints. Each IPDU contains all of the information necessary to uniquely identify it, and each is processed and routed independent of all other IPDUs. Hence, either or both of the intermediate systems can be used to transmit data between end systems (A,B).

Routing functions performed as part of the ISO IP's operation make it possible to use any available intermediate system to transmit IPDUs This is extremely

4. Splitting the load. Here, an IEEE 802-compatible LAN (operating at 4, 5, or 10 Mbit/s) must work with a lower-speed wide area network (whose maximum throughput is normally 48 or 56 kbit/s). With load splitting, multiple paths that exist between end systems and gateways are used instead of a single path.

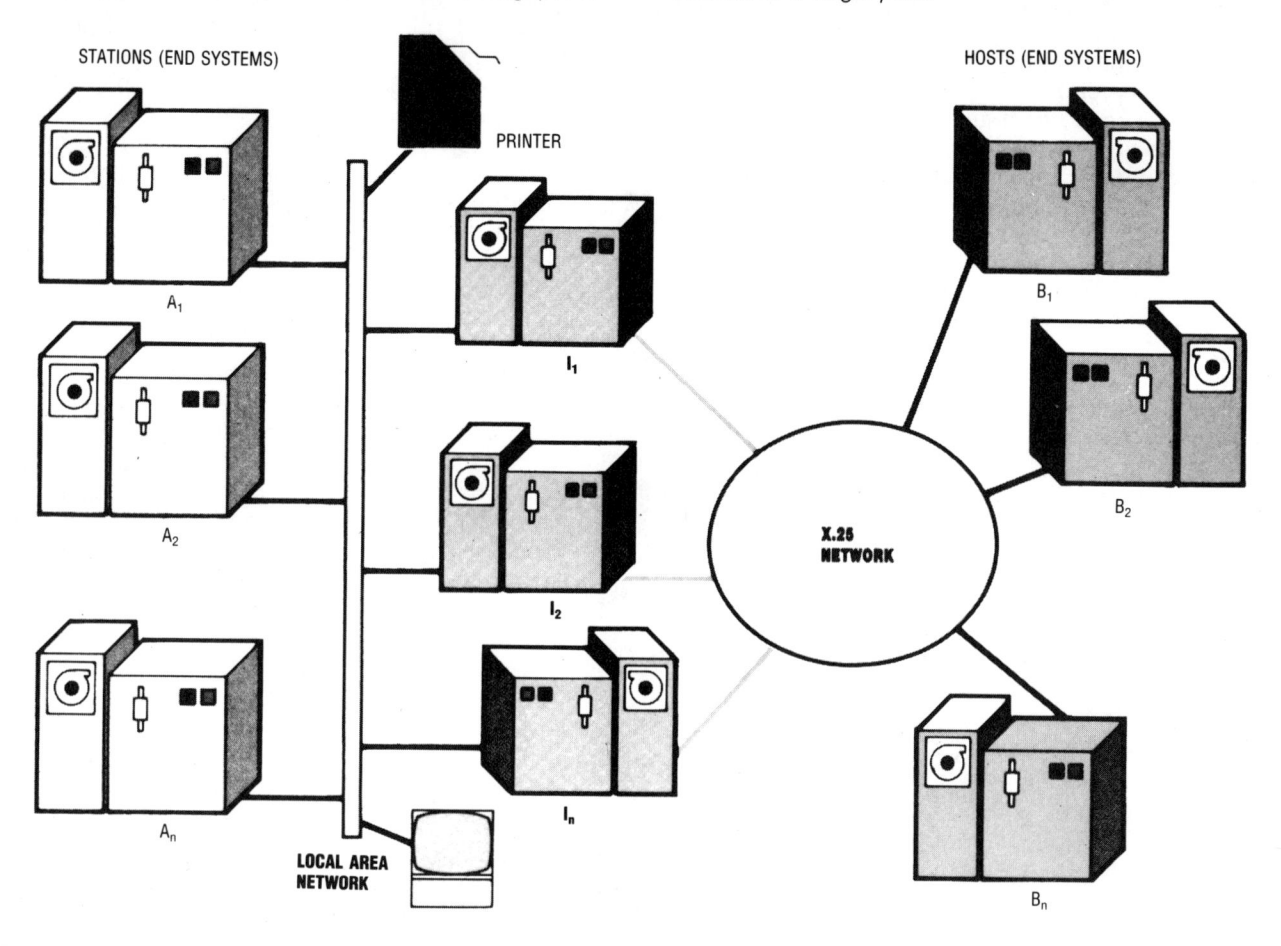

beneficial: Throughput can be effectively improved by a factor n, where n is the number of intermediate systems linking the subnetworks. This technique, called loadsplitting (see "What is load splitting?") is particularly important in internetworking applications.

Consider the case where the subscribed throughput class for a single DTE is 9.6 or 56 kbit/s. Most LAN stations can be expected to operate several orders of magnitude faster. Even if one were to attempt to resolve this throughput disparity through a single DTE using multilink procedures, there would be other problems. One of them: Most data circuit-terminating equipment (DCE) would be unable to respond to such throughput requirements and would impose serious flow control restrictions on the user.

Failure is not final
Another important benefit realized by using the ISO IP is the ability to provide redundancy and timely response to loss or temporary failure of an intermediate system. If an intermediate system fails, or if connectivity to a subnetwork or data link is lost, all communications is not necessarily broken between end systems and other intermediate systems utilizing this intermediate system. In fact, in nearly all routing techniques used in connectionless networks today (and certainly in those anticipated for the OSI environment), traffic can be redirected to other available intermediate systems without a lengthy communications interruption.

If an intermediate system fails, or if connectivity to a subnetwork is lost, most routing algorithms used in connectionless networks today (especially those in the DOD's DDN and in some private networks) respond quickly. Through an exchange of management protocol data units, gateways and end systems are informed of the failure, and traffic is redirected to an alternate available gateway or link. Later, when the failure is corrected, a similar management exchange is used to inform the gateways and end systems.

While the network is responding to such a failure, some IPDUs may be lost. (Upon expiration of a lifetime value encoded in each IPDU, the IPDU is automatically discarded.) Retransmission mechanisms operating in the Transport Layer ensure that no user data is lost or duplicated as a result of the failure.

A similar error-recovery function could be provided in the hop-by-hop approach, in which case the network signals a RESET. Typically, though, such recovery mechanisms are not provided. Normally, a human operator must intervene when calls must be re-estab-

What is load splitting?

One benefit of using a connectionless Network Layer service to support a connection-oriented Transport service is particularly prominent. The functions associated with maintaining user data integrity (sequence preservation, error detection, and error correction) can be separated from functions associated with the exchange of data (routing, Network Layer addressing, and network resource utilization). Networks that use the ISO IP (Internetwork Protocol) are likely to take advantage of this crossover of services by employing load splitting in their routing mechanisms. The reasons: to increase traffic-handling efficiency and to improve overall network performance.

When load-splitting techniques are applied, multiple paths that exist between end systems and gateways are used instead of a single path. IPDUs (Internetwork Protocol Data Units) are routed over these paths in such a way that the traffic load is distributed more evenly throughout the network (see figure).

Load-splitting techniques require participation by Network Layer entities in both end systems and gateways in the routing of IPDUs. This may be perceived as additional overhead. However, the benefits to be had by load splitting substantially outweigh the expense. In the case of a LAN/wide-area-network configuration, significantly higher station-to-station bandwidth may be realized between a LAN station and an X.25 DTE by utilizing more than one DTE (in perhaps more than one gateway) over the slower X.25 network. The same effect could not be accomplished using a multilink procedure under the X.25 packet level protocol. Here, to maintain the sequence of data packets, all traffic must be restricted to the same DTE.

When load-splitting techniques are employed, a higher degree of "resiliency" is also provided. If one of three paths in use suddenly becomes inoperative, the remaining two may support the traffic load. And as a result of a balanced distribution of traffic, individual paths in the network will be less susceptible to overloading. When load splitting is coupled with the flow control features available in TP 4, the network is likely to experience fewer incidents of congestion.

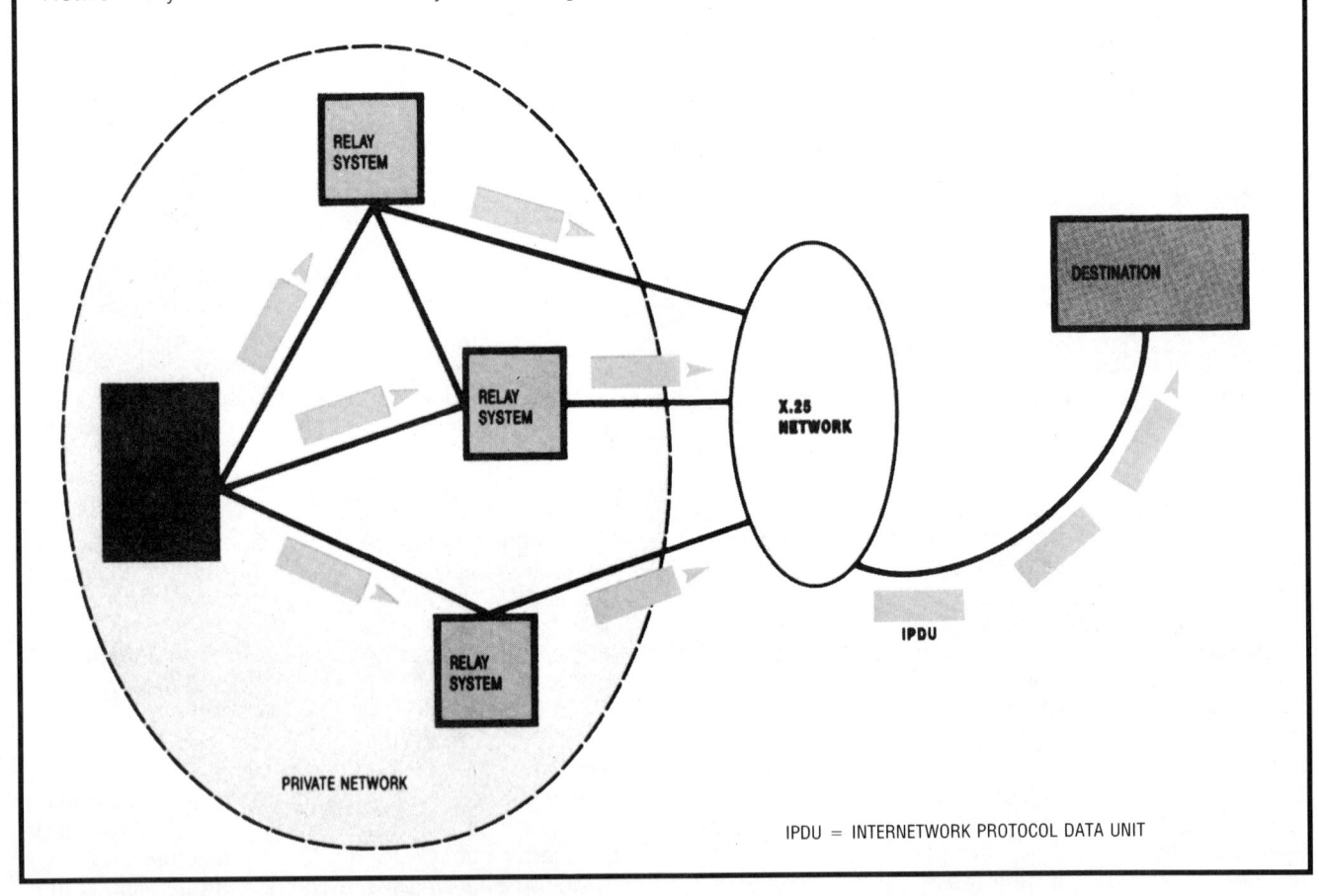

IPDU = INTERNETWORK PROTOCOL DATA UNIT

lished to a different DTE as a result of a gateway loss.

The ISO IP standard is in the final publication stage. Enthusiastic support for the protocol and the connectionless internetworking strategy it supports already exists. The European Computer Manufacturers Association (ECMA) has already issued a standard that describes a subset of the protocol suitable for operation over LAN clusters. (These are physically interconnected LANs that offer a uniform service-data-unit size and, hence, do not require the segmentation and reassembly functions defined in the ISO IP.) ECMA intends to reissue its standard, to be fully compatible with the ISO IP.

The National Bureau of Standards (NBS) will issue the

ISO IP as a Federal Information Processing Standard (FIPS). As a FIPS, the IP can be cited in procurement statements by any federal agency.

A significant endorsement for the ISO IP has come from the National Research Council (NRC) of the National Academy of Sciences. Asked by the Defense Communications Agency and the NBS to investigate the suitability of ISO TP 4 and IP in military environments, the NRC conducted a detailed analysis of the ISO protocols. The members unanimously agreed that the DOD should adopt the ISO standards ". . . as costandards with TCP/IP and move toward [their] eventual exclusive use." (TCP/IP is the DOD's Transmission Control Protocol/Internet Protocol).

Support for IP goes beyond the standards arena. General Motors (GM) has selected the ISO IP for its Manufacturing Automation Protocol (MAP) specification. This standard for factory automation is supported by over 50 vendors, including IBM, AT&T, DEC, and Honeywell Information Systems. GM sponsored a demonstration of MAP at the Autofact '85 conference, which included a successful implementation of the ISO IP. GM has installed MAP at its Saginaw Steering Gear's "factory of the future" and intends to continue such installations throughout 1986.

NBS is developing what is hoped to be a worldwide concatenation of public and private networks using OSI protocols, called OSINet. The ISO IP is prominent in its suite of protocols. The purpose of OSINet is to establish continuity of implementations using the OSI protocols. OSINet will undoubtedly encourage users and vendors to adopt the ISO IP as the solution to network interconnection. ■

David Piscitello is vice chairman of Accredited Standards Committee (ASC) X3S3.3, the task group that oversees the development of OSI Network Layer service and protocol standards. He is involved with network architecture design at Burroughs and holds a B. S. in mathematics from Villanova University. Alan Weissberger participates in ASC and IEEE committees dealing with ISDN, public data networks, and LAN standards. He is a consultant specializing in implementing data communications standards and is an adjunct professor at the University of Santa Clara. Scott Stein participates in ASC and ISO committees that are defining OSI standards. He is a consulting engineer within Honeywell and received a B. S. from the University of Michigan. Lyman Chapin is chairman of ASC's X3S3.3 task group and vice chairman of the ACM SIGComm (Association for Computing Machinery Special Interest Group on Communications). He designs and implements network products at Data General and holds a B. A. in mathematics from Cornell University.

Robert W. Junke, Computer Task Group, Buffalo, N.Y.

A user implements a T1 network

It included overcoming timing problems and using part of the network to beta-test two new multiplexer features: Bypass and automatic alternate routing.

As business competition increases, network users are demanding the ability to rapidly address their changing needs, to perform much higher speed transmission, to provide new functions (such as full-motion video teleconferencing), and to guarantee greater reliability. These factors, combined with a competitive telecommunications industry and tremendous technological progress, have spawned a new generation of cost-effective networks.

One example of this network evolution exists at the Marine Midland Bank N. A. Transmission specialists at the bank designed and installed a network comprised of several T1 (1.544 Mbit/s) circuits across New York State. These circuits provide a conduit through which various types of information—such as voice and data—can be combined and transmitted. AT&T Communications is the primary supplier of Marine's T1 circuits. At each end of the conduit is multiplexing equipment that assembles all of the inputs into a format suitable for transmission. Information received at the conduit's other end is disassembled and distributed to the appropriate outputs.

Selection of multiplexers is the most important decision to be made when installing a T1 network. The multiplexer products of five vendors were reviewed in great detail over a three-month period. Table 1 summarizes these requirements—along with a vendor point-scoring matrix to facilitate the decision-making process. The matrix was used to quantify the relative importance of the requirements and the capability of the vendors to meet each of them.

Vendor E (Timeplex, of Woodcliff Lake, N. J.) and its Link/l multiplexer achieved the highest point score and was selected. The company had also made significant progress toward developing two very important features: bypass and automatic alternate routing (AAR),
both defined below. Marine Midland was Timeplex's beta-test site for these two features. Before detailing the bank's network operations and the beta-test results, it would appear beneficial to first briefly review the conventional means by which voice and data communications have been accomplished, and to demonstrate why T1 networks can be a superior alternative.

Separate networks are required to transmit voice and data in the "conventional" communications environment. Each business center has a private branch exchange (PBX) that provides intrasite voice communications by connecting local telephone users to each other for the duration of the call.

These PBXs also provide the connections for *inter*site communications. For example, if a user at site A must talk to a user at site B, the PBXs at each site will set up the required end-to-end connection. In the conventional network, leased analog lines tie the PBXs together. Not surprisingly, the circuits are commonly referred to as voice tie lines.

The number of tie lines between PBXs is, of course, dependent on the amount of voice traffic between the sites. (Although it is possible to multiplex voice traffic via frequency-division multiplexing, this technique is being phased out in favor of less-expensive methods, such as those noted below.) Voice engineers must provide enough tie lines to minimize the probability of a user at site A being prevented from talking to a user at site B because all tie lines are in use.

Installation of T1 circuits and associated multiplexing equipment represents a significant departure from previous schemes. To take advantage of T1's capacity, multiplexers at each end of the circuit combine individual voice and/or data input channels for transmission. Nondigital inputs, such as analog-voice signals, must be digitized via codecs (coder/decoders) prior to the multiplexing and

Table 1: Vendor scores

REQUIREMENT	MAXIMUM POINTS	VENDORS				
		A	B	C	D	E
PRODUCT RELIABILITY	10	6.5	7.5	8	9	9
QUALITY OF 32-KBIT/S VOICE	9	9	2.5	4.5	8	8
BYPASS, AAR, AND OTHER FEATURES	8	1	1	1.5	4.5	6.5
EASE OF OPERATION	8	1	2	3	6.5	8
OVERALL LEVEL OF SOPHISTICATION/ FUTURE DIRECTION OF PRODUCT	5	1	1.5	2	5	5
FIELD SERVICE SUPPORT	5	2.5	2.5	3	4	4
PRICE	4	4	4	1	2	2
EASE OF UPGRADING	4	1	1.5	1.5	4	4
TOTAL SCORE	53	26	22.5	24.5	43	46.5

AAR = AUTOMATIC ALTERNATE ROUTE

transmission operations.

The pulse-code modulation (PCM) technique requires 64 kbit/s for codec operation. A more recent voice compression technique—utilized by Timeplex—requires only 32 kbit/s of the T1 circuit. As a consequence, up to 44 voice tie lines can be replaced using one T1 circuit. Further, data communications lines connected to the multiplexers no longer require modems, since transmission takes place over a digital facility. (Comparatively inexpensive channel-port cards are used instead.) The cost savings: Replacement of approximately 10 to 15 voice lines will reach the break-even point on the cost of the T1 facility.

Bank basics
The Marine Midland network interconnects its major business centers throughout New York State. In Figure 1, the numbered circles represent multiplexers as network nodes. Each business center in the network has a PBX. Units from AT&T, Northern Telecom, and Rolm make up Marine's multivendor PBX environment.

Buffalo, with nodes 1, 2, and 3, is also the location of Marine's IBM mainframe processors (one 3084, two 3081s, and one 3033AP) and front-end processors (two 3725s and two 3705s). Users from the other business locations access these computers utilizing a variety of devices, such as IBM 3270s (at all centers), IBM System 36s, IBM 3780 remote-job-entry workstations, and Diebold automatic teller machines.

Also in Buffalo, connected to node 1, is a microcomputer (IBM PC-XT), used to supervise the entire network. The network manager uses this workstation to perform software-controlled diagnostics, channel additions, and most network modifications—such as topology and channel speed.

Node 4, in North Syracuse, connects to IBM mainframe processors (one 3081 and one 3033UP) and a 3725 front-end processor, used for application program development. Programmers use these processors from Buffalo and New York City. Another use for the North Syracuse processors is as a contingency production center in the event that a serious problem arises with the processors in Buffalo.

Since New York City is one of the world's leading financial centers, Marine has significant operations there: nodes 6, 7, 8, and 10. And, due to the critical nature of financial transactions originating there, communications between Buffalo and New York City must *always* be maintained. Nodes 9, 11, and 12 are located in Albany, Melville, and Rochester, respectively.

A better way
Bypass allows digital information originating at one node to be transmitted through one or more intermediate nodes to a destination node. Information can then be transmitted between locations not directly connected by a point-to-point T1 circuit.

Significantly, no demultiplexing and remultiplexing (also known as "drop and insert") must occur at intermediate nodes. (Refer to "The anatomy and application of the T1 multiplexer," DATA COMMUNICATIONS, March 1984, p. 183, for a more detailed explanation of drop and insert.)

An example of the benefits of bypass is the traffic pattern between Buffalo (node 1) and Albany (node 9). Were it not for the bypass function at node 5 in Syracuse, either an additional T1 circuit would have to be installed directly between nodes 1 and 9, or extra hardware would be needed at node 5 to use the drop-and-insert method.

Another bypass application: the programmers at node 8 (New York City) communicating with the processors in node 4 (North Syracuse). It allows node 8 traffic to be routed over the internodal paths 4-1-2-8 and 4-5-6-8. Perhaps more importantly, if a disaster in Buffalo forces the backup use of the processors at node 4, bypass enables traffic from nodes 6, 7, 8, and 10 to be readily rerouted to node 4.

As to AAR, consider what occurs when a circuit failure severs, in effect, T1 communications between two sites. AAR guarantees network availability for the affected sites by re-establishing voice and data communications over

1. Bank business. *Multiplexers and PBXs at network nodes interconnect Marine Midland's major business centers throughout New York State.*

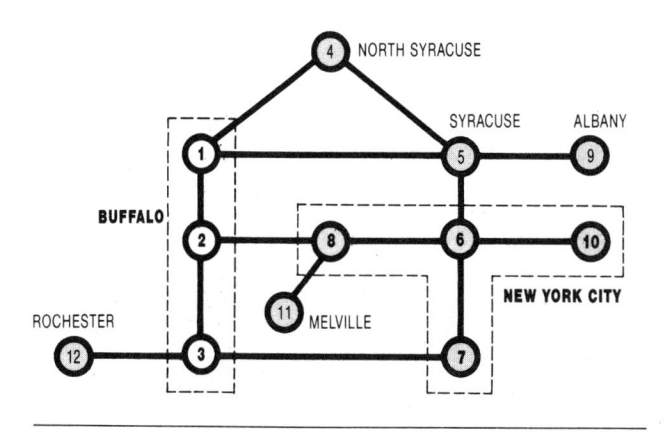

2. The test network. Nodes 2, 3, 6, and 7 were separated from the production network to conduct beta testing of bypass and automatic alternate routing.

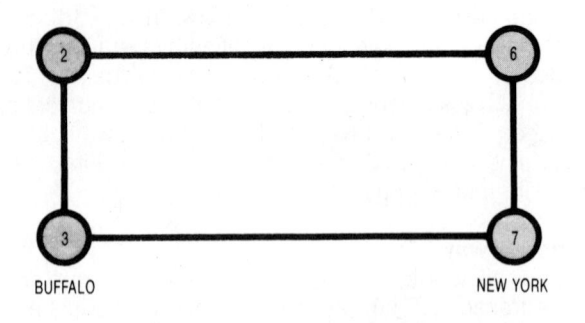

alternate T1 network pathways.

In Marine's network, transactions amounting to tens of billions of dollars are transmitted daily between Buffalo and New York City alone. If a T1 circuit between these centers fails (for example, the circuit between nodes 3 and 7), alternate paths through the network must be found to re-establish communications. A logical alternate route, for example, would be via nodes 3-2-8-6-7.

When Timeplex completed its development of bypass and AAR and needed a beta-test site, a portion of Marine's network was found to be available. Marine had earlier installed multiplexers in Buffalo, North Syracuse, Syracuse, and Albany (nodes 1, 4, 5, and 9—the "production network"). The voice and data circuits between these sites had successfully migrated to the T1 network.

Marine had also installed the remaining Buffalo multiplexers (nodes 2 and 3) and two in New York City (nodes 6 and 7), but had not yet migrated voice or data circuits onto that portion of the network. Nodes 2, 3, 6, and 7 were therefore made available as a network for beta-testing bypass and AAR. A test plan was designed, and voice channels were set up on the test network. Analyzing equipment (Digilog's DLM V), capable of transmitting various bit patterns, was installed at each node to test data channels.

The connections to the four available nodes were physically separated from the production network. New hardware—including the prime and backup bypass modules—and firmware was installed on the test nodes, and testing began (Fig. 2).

The first priority was to test bypass. The voice channels were defined (parameters established) at 32 kbit/s from a supervisory station (an IBM PC-XT) at node 2. After the channels were connected and tested point-to-point (via nodes 2-6 and 3-7), the first bypass channels were connected—by keyboard commands—via nodes 2-6-7. A second set of channels was connected via the 3-2-6 route. Telephone calls were then made. E and M control signaling worked satisfactorily, and conversations over the test channels were of excellent quality. (E and M refers to a commonly used signaling interface whose purpose is to control call setup. The letters are derived from the original implementation: "ear and mouth" signals.)

Data channel speeds ranging from 2.4 to 56 kbit/s were defined. A variety of routes was established after point-to-point data channels were successfully tested. Initial routes were via three nodes, with the middle one being the intermediate bypass node. These routes included 2-3-7, 2-6-7, 3-2-6, and 3-7-6. All tests were successful. Next, routes using all four nodes were tested. These included 2-6-7-3 and 6-2-3-7. Again, performance was satisfactory.

The above tests were repeated several times with consistently positive results. Therefore, the bypass portion of the beta test was considered successful. It was also found that setting up and connecting channels was quite simple. The network manager needed only to define the channel at the originating and terminating nodes, not at any of the intermediate ones.

The multiplexers established the connection over the most efficient route, as defined by a vendor-supplied algorithm. The network manager does have the option of specifying a route if desired. Marine makes frequent use of this option to balance the loading across the network. Also, intermediate nodes were interrogated from the supervisory port to determine the status—such as the defined parameters—of channels being bypassed through them. The same channels were also "looped" at a node to perform loopback testing.

Playing it safe

As indicated earlier, AAR is an extremely important safety feature. When a T1 circuit fails, the network must reconnect channels over alternate routes. Note that AAR is limited by the available bandwidth of alternate transmission routes. Also, transmissions to and from "isolated" nodes (those connected to the network by a single circuit, such as nodes 9 through 12 in Figure 1) have no available alternate route. In those cases, consideration was given to providing dial backup for critical data circuits, and the invocation of a voice contingency plan that, for example, would allow expanded direct-distance dialing.

During AAR testing, T1 circuit failures were caused by disconnecting the cable between the multiplexer and channel service unit (a repeater-signal monitoring and channel-interface unit). Every possible AAR path was tested using both voice and data channels. Each time a circuit failed, an alternate route was successfully established. For example, data channels were connected directly between nodes 2 and 3. When this circuit was disconnected, the alternate route (2-6-7-3) was established and transmission resumed. The mainframe's protocol would cause a repeat of any "lost" transmission.

On the test network, data channels began to reconnect over alternate routes in approximately 30 seconds. After 90 to 120 seconds, all data channels had successfully completed reconnections. Once determined, the maximum reconnect time can be used to "fine-tune" the data network's performance. For example, remote devices cannot respond to polls from the front-end processor until their respective channels have reconnected over an alternate path. If the front-end processor receives no response to its polls, it will—after a preprogrammed

3. Setting up the call. *Separate paths carry control and voice signals. The M control lead transmits ground when the channel is not in use and voltage—called "battery"— when it is in use. The E lead receives signals from the multiplexer when the channel is not in use; when the channel is in use, E is grounded.*

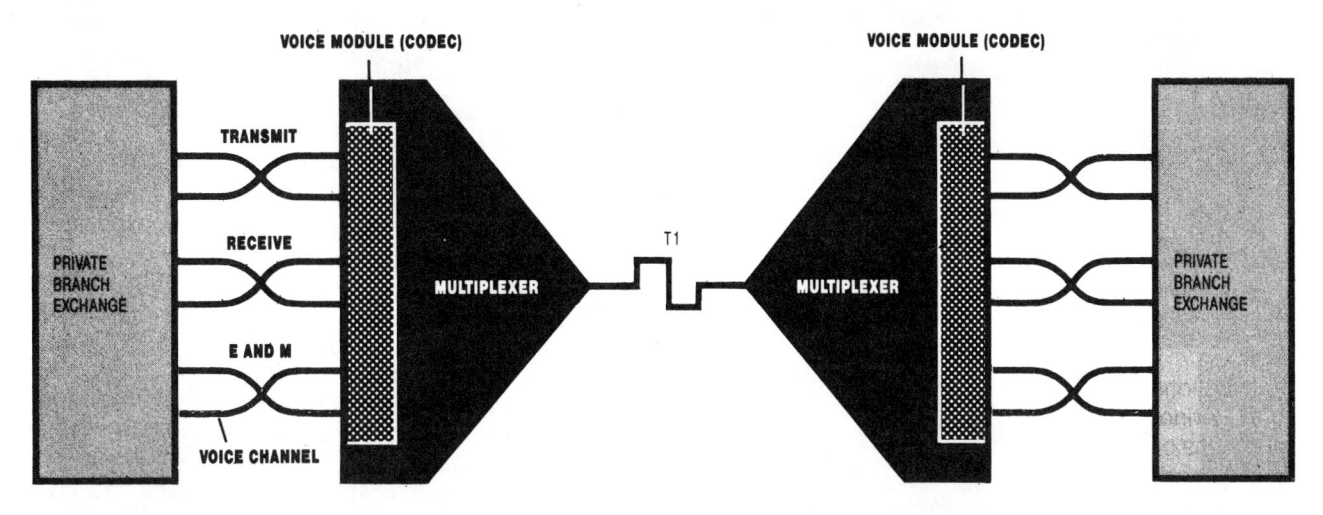

time—deactivate the device. Reactivation is accomplished by manual procedures.

Front-end processors (via the Network Control Program on an IBM 375) should be programmed to continue polling remote devices for a period exceeding the maximum time required to reroute the channel through the network. This prevents deactivation, eliminates the need for manual recovery, and prevents a lengthy service outage. With proper planning, users at remote locations will probably view the outage simply as a short period of poor response time.

Marine was concerned about the operation of voice channels. Under the previous firmware release, during the period of a T1 circuit failure a problem resulted if users attempted to place a call. The PBX would select a T1 channel but, since the T1 circuit was inoperable, the call could not be completed. The caller would wait for an indefinite time, then finally hang up.

Marine requested a "fix": When a T1 circuit fails, the multiplexer should send busy signals to all the PBX channels connected over that circuit. The PBX would then select an alternate means of completing the call. Timeplex modified the operation of the E and M control signals on the voice channels between the PBX and the multiplexer to resolve the problem.

The fix

Separate paths are used for control (E & M) signals and voice signals (Fig. 3). The M lead transmits ground when the channel is not in use and voltage—called "battery"—(supplied by the PBX) when it is in use. The E lead receives incoming signals from the multiplexer when the channel is not in use ("open" condition). But when the E lead is grounded, it signifies to the PBX that the channel is in use. With this in mind, Timeplex modified the operation of the E lead so that when a T1 circuit failed, the E leads on all affected voice channels were grounded by the multiplexer until the voice channels had been rerouted or the T1 circuit was operational again.

Timeplex also developed a method to reconnect channels in a priority sequence. When channels are defined, one of 16 priority levels is assigned. When a failed T1 circuit is operational again, the channels with the highest priorities reconnect first, thus minimizing the outages on high-priority circuits. If the extra bandwidth on the alternate pathway(s) becomes exhausted before all channels have reconnected, only the lower priority channels would remain disconnected.

The network manager could manually disconnect lower priority channels whose primary route happens to be on a portion of the alternate route. This would free up additional bandwidth. For example, suppose the circuit between nodes 3 and 7 failed, and all affected channels reconnected over the 3-2-6-7 route, except for four 9.6-kbit/s data channels. The data channels could not reconnect because no bandwidth was left on the T1 circuit between nodes 2 and 6. But if one 32-kbit/s voice channel were disconnected between nodes 2 and 6, three of the four data channels would reconnect.

The following test demonstrated the network's ability to recover from a T1 circuit failure requiring complex rerouting. The test involved 60 channels and required five distinct primary routes to automatically reroute. Also, the amount of bandwidth the network would attempt to reroute over one of the T1 circuits would exceed the 1.544 Mbit/s allowed. (Table 2 outlines the test parameters.)

The circuit between nodes 2 and 6 had 1 Mbit/s of bandwidth allocated at the start of the test. The circuit between nodes 3 and 7—the common one for the five primary routes—was disconnected. Upon circuit failure the network attempted to reroute 0.6 Mbit/s of bandwidth between nodes 2 and 6, exceeding the 1.544-Mbit/s limit.

Upon failure of the common circuit, all channels between nodes 2 and 6 reconnected except for four 9.6-kbit/s low-priority data channels (over the 2-6-7 path). The circuit between nodes 2 and 6 had almost all of its 1.544-Mbit/s

Table 2: Test parameters

ROUTE NUMBER	ROUTE TAKEN (NODES)	PRE-AAR BANDWIDTH ALLOCATED (KBIT/S)	ALTERNATE PATH (NODES)	POST-AAR BANDWIDTH ADDED TO CIRCUIT 2-6 (KBIT/S)
1	2-6-7-3	260	2-3	0
2	2-3-7-6	160	2-6	160
3	2-3-7	80	2-6-7	80
4	3-7-6	160	3-2-6	160
5	3-7	200	3-2-6-7	200
		860 TOTAL		600 TOTAL

AAR = AUTOMATIC ALTERNATE ROUTE

channel allocated (99 percent utilization). The four data channels became connected either when other channels on the circuit between nodes 2 and 6 were disconnected, or when the circuit between nodes 3 and 7 was reconnected.

Staying in synch

In a time-division multiplexed network, one clock source must provide timing to all nodes. Marine's network uses the 6-MHz internal clock of one of the multiplexers to provide a common clock for all nodes.

In the four-node test network, the 6-MHz clock in node 2—known as the master clocking node—provided internodal timing to all nodes. The signal transmitted from node 2 to node 3 was used by node 3 to derive its clock. Similarly, the signal sent from node 2 to node 6 was used by the latter to derive its clock. And finally, node 7 derived its clock from the signal transmitted to it from node 6. Ultimately then, node 7 received its clock from node 2.

Additional timing sources can be defined within each node. Then, if a node lost its primary clock source, it would have a secondary clock source for fallback, and, if need be, a tertiary source. For example, if the circuit between nodes 2 and 3 failed, node 3 would fall back to its secondary clock source (node 7). In this case, node 3 would then derive its clock from the incoming signal from node 7.

Testing uncovered a problem here. Node 6 was defined so that if the circuit between nodes 2 and 6 failed, node 6 would fall back to its secondary clock, which meant it would derive its timing from node 7. But what would happen if node 6 fell back to derive its timing from node 7, and node 7 stayed on its own primary clock (from node 6)? Nodes 6 and 7 would successfully clock off (supply timing to) each other and would lose synchronization with nodes 2 and 3.

Timeplex resolved the problem by requiring the node that lost its primary clock source to broadcast a loss of clock message to all other nodes. Any node that derived its clock from this node would then fall back to its secondary clock source. Therefore, when node 6 lost its primary-clock source, node 7 would be notified. Node 7 would fall back to its secondary clock source and derive its timing from node 3. This would allow node 6 to derive its clock from node 7 without loss of synchronization to nodes 2 and 3.

Putting it all together

With testing completed, node 8 and associated T1 circuits were installed. Also, bypass and AAR were installed on the production nodes (1, 4, 5, and 9). The production and test networks were then combined, yielding the present configuration. The effort quickly paid off. Shortly after the installation, an intermittent problem developed on the circuit between Buffalo and Syracuse (nodes 1 and 5). The problem was not bad enough to cause the circuit to lose synchronization and invoke AAR. But it degraded the voice quality and caused bit errors on the data lines. Until the problem was resolved, the network manager simply bypassed the voice and data channels over three routes: 1-4-5, 1-2-8-6-5, and 1-2-3-7-6-5.

Some of the Marine network's benefits are:

- *Substantial cost savings.* The bank estimates its savings to be over $2 million per year when compared with conventional analog communications.
- *Rapid addition of circuits to meet user needs.* New channels can be configured in minutes from and to any point in the network from a supervisory workstation. A similar installation using analog facilities from a common carrier takes several months.
- *Use of higher transmission speeds.* Marine is upgrading its base of 9.6-kbit/s 3270 controllers to 19.2 kbit/s and migrating from binary synchronous communications to synchronous data link control. On the T1 network, the upgrade is readily accomplished via keyboard input by changing the speed parameter on the channel definition of the affected nodes at the supervisory station in Buffalo. Response times are being measured in the 1-to-2-second range, at least half that of the lower-speed controllers.
- *Flexible network routing.* Bypass allows programmers on any of the New York City nodes to communicate with processors in North Syracuse. And, in the event of a disaster at the Buffalo center, production-user communications will be resumed by simply routing required channels to the North Syracuse contingency production center.
- *Priority recovery with AAR.* Channel recovery from a T1 circuit outage is rapid (30 to 120 seconds) and does not require manual intervention to restart polling. Were it not for AAR, all of the benefits derived from a T1 network may not have outweighed the potential risks of a lengthy network outage. ∎

Bob Junke is now with the Computer Task Group as manager of its telecommunications consulting services. At Marine Midland, he was the project manager of the T1 network described above. He has taught at SUNY-Buffalo and Niagara University, and holds B. S. and M. S. degrees from Purdue University.

Marc A. Wernli, Northwestern Bell Telephone Co., Omaha, Neb.

The choices in designing a fiber-optic network

When do you use an LED as a light source? When a laser? And what about multimode versus single-mode fiber? Here are one expert's answers.

The many fiber-optic networks that have recently been installed use different combinations of light sources and fiber-cable types. A lot of questions have been raised concerning the advantages and disadvantages of these various combinations. This article attempts to answer some of these questions and to identify some of the applications for each combination.

Two types of light sources are used for fiber-optic networks. The most common, found in the first fiber networks, is the light-emitting diode (LED). The other that has gained popularity, enabling both the bandwidth and distance of fiber-optic networks to increase, is the laser.

Both the laser and LED are semiconductor devices that emit a beam of light when a voltage is applied. The two devices display different optical characteristics and have different sensitivities. Most of the devices are made from gallium arsenide (GaAs). They are also "doped" with other materials, such as indium, aluminum, and phosphide.

Figure 1a shows the optical power output of a typical LED as the modulating current through the device is increased. Note that the LED has almost a linear characteristic, emitting light pulses (the digital data) proportional to the current as it is changed from near zero to about 220 milliamperes (ma). A further current increase would, typically, result in nonlinearity or LED failure.

Compare the LED plot to that of a typical laser (Fig. 1b), which also shows the optical power output as the current through the device is increased. The laser has almost a switching characteristic—because of the plot's steepness—as the current is increased above the threshold level at "a," here about 28 ma. Lasers normally have a fixed "bias" current, just above the threshold level, that causes the device to emit a low light level. The additional modulating current resulting in a change from "a" to "b" results in the laser's light output changing from the threshold to a high level— that is, a light pulse. A laser is normally able to operate at bit rates higher than an LED's because a comparatively small change in current (from "a" to "b") causes a larger increase—almost two orders of magnitude greater—in optical output power than an LED's range.

The output power of an LED is spread over a wide range of wavelengths, resulting in the device having a spectral width of approximately l00 nanometers (100 10^{-9} meters—abbreviated nm). The laser has a much more narrow spectral width, in the range of 4 nm. (Figure 2 displays a comparison of the two spectral widths.) Since different wavelengths propagate at different speeds in a fiber cable, a device with a narrow spectral width will have less pulse dispersion than a device with a wide spectral width. This characteristic results in a laser able to operate at higher bit rates than an LED.

Below visibility

Fiber-optic light sources normally operate at the 850-, 1,300-, or l,500-nm wavelength, which is in the infrared portion of the frequency spectrum. This is below (lower frequency) the visible-light portion, which is 400 to 700 nm. Most fiber-optic networks transmit digitally, with the network "bandwidth" indicated by the maximum data rate; most video transmissions are analog, with an increasing trend to digital.

The usable power of a laser or LED is the power that can be coupled into the fiber "pigtail" (connecting strand) from the light-source module. A fiber has a limited acceptance angle, and any light from an LED or a laser that is outside this angle will not be coupled into the fiber. Figure 3 shows the acceptance angle of a fiber—typically 30 degrees—and the output patterns of both an LED and a laser—typically 90 and 30 degrees, respectively. The acceptance angle and output patterns are cone shaped and can be made more compatible by having a lens on the end of the fiber. The coupling losses for a laser range from approximately 3 to 6 decibels (dB); for an LED, from 10 to 30 dB, depending on the type of fiber with which the LED is being coupled.

There are pronounced differences in the characteristics of LEDs and lasers. These differences turn into advantages and disadvantages for each device, some of which are noted below.

The advantages of an LED are:
- Low cost—approximately $150; a laser costs about $1,500, but the price is dropping each year.
- Long life—20 to 100 years; a laser lasts about five to 20 years.
- Less sensitivity to environmental changes—a temperature increase of 30 degrees Celsius (54 degrees Fahrenheit) decreases an LED's lifetime by a factor of 2; the same temperature increase decreases a laser's lifetime by a factor of 10.

The disadvantages of an LED are:
- Low output power—approximately -10 to -15 dB; a laser's output power is about 0 dB.
- Narrow bandwidth—normally less than 135 Mbit/s; a laser's bandwidth is greater than 1 Gbit/s.
- Ability to couple only to multimode fiber; a laser couples to either multimode or single-mode fiber (see below for a discussion of the fiber types).
- Wide beam of light; a laser produces a narrow beam of light.
- Wide spectral width—typically 100 nm; a laser's is much narrower—typically 4 nm.

It should now be apparent that each light-source device has its own suitable areas of application. If the wrong device is used in a particular application, the result will be either unacceptable performance or excessive cost.

Suitable sites

The LED light source can be used in fiber applications where the distance between network end points is no more than about 12 km (about 7.5 miles) and the data rate needed is less than 135 Mbit/s. (Repeaters—called regenerators—are generally used when longer distances are required; see below.) Multiple-fiber cable is used, with individual fibers spliced to new ones for additional distance where practicable. The maximum allowable distance between end points will vary, since it is dependent on the type of fiber used and the number of splices that are expected to be made. The LED's major application areas have been in campus environments and local area networks (LANs).

The laser should be used in applications where the end-to-end network distance is longer than 12 km—up to about 45 to 110 km (about 30 to 70 miles), after which regeneration is required—or the data rate required is greater than 135 Mbit/s. The laser is used in metropolitan area networks or in networks that have end points in different towns.

1. Optical power. *Note in A that the light-emitting diode has almost a linear characteristic, emitting light pulses proportional to the changing modulating current. This contrasts with the laser's response in B, which displays almost a switching characteristic as the current is increased above the threshold level.*

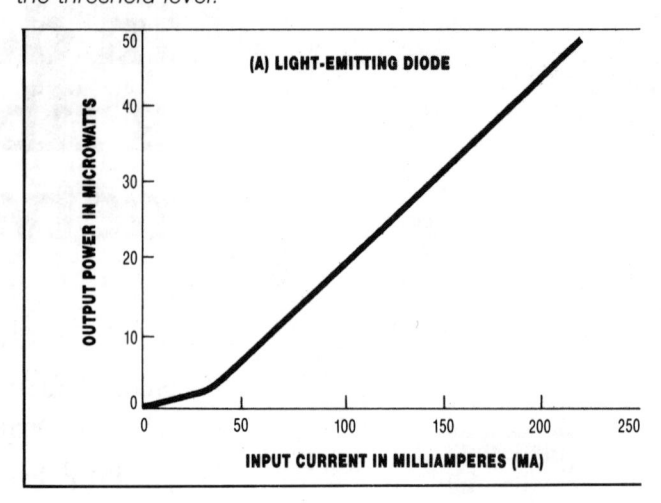

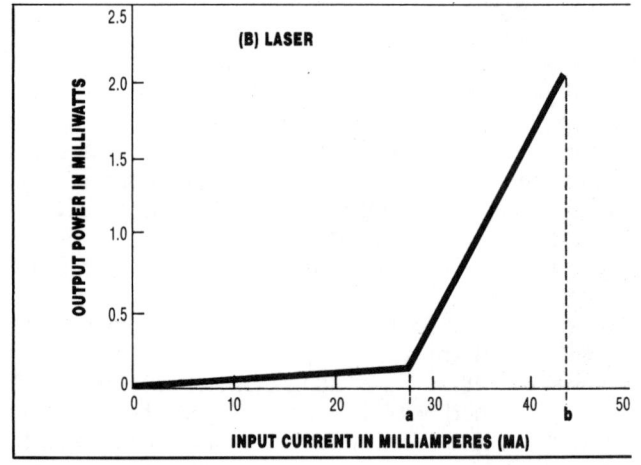

At present, the success of coupling LEDs to single-mode fiber has been limited. A slotted-aperture device, called the edge light emitting diode (in contrast to the older "surface" LED), is being tested for this coupling. The application has a high coupling loss (approximately 15 db), but can be used on single-mode fiber networks that extend less than 7 km (about 4.4 miles). While the conic emitted-light pattern cross section of the surface LED is circular, that of an edge LED is elliptical.

Cheaper, too

The perceived trend is to develop lasers that have longer lifetimes and are not sensitive to environmental changes. As these lasers are produced in quantity, their cost should drop to the $200-to-$500 price range. An additional capability of these lasers is a data rate beyond 10 Gbit/s. There is also an effort under way to develop LEDs that have a narrower light beam and will not have high coupling losses with single-mode fiber. The first practical versions of such products will probably reach the market by early 1987.

Fiber cable is divided into two major categories based on the number of modes (paths) that light propagates along its glass core. The fiber cable that had the greatest usage prior to 1983 contained multimode fibers. This cable has an outer diameter of 125 microns and the core that the light travels in is 50 microns in diameter. (A micron is one millionth of an inch.) The core of this cable allows several modes of the light frequency to exist concurrently.

Most of the cable that is being used today still has a 125-micron outer diameter, but the core diameter is only 7 to 9 microns. This small core only allows one light path or mode to exist; therefore, the cable is called single mode. Note that multimode cable is easier to splice because of the larger diameter core.

Multimode fiber undergoes a light-power loss in the range of 2 dB/km and a data rate limited to approximately 135 Mbit/s. The loss associated with this type of fiber is a function of both distance and bandwidth: the distance-bandwidth multiple. Therefore, the greater the distance between network end points, the narrower the bandwidth (the more limited the data rate). The bandwidths of multimode fiber are in the range of 400 to 1,000 MHz-km—which accommodates a range of 45 to 135 Mbit/s—at an 850-nm spectral width, and 400 to 1,500 MHz-km (accommodating a proportionately higher data rate) at a 1,300-nm. spectral width. The losses in the fiber are due primarily to cable impurities absorbing light. (Incidentally, plastic-core optical-fiber cable has recently become available, but it generally has a reduced-distance capability compared to the more-dominant glass types. This performance inferiorty is due primarily to its greater incidence of impurities.)

Single-mode fiber has a loss in the range of 0.5 dB/km and a data rate limitation in the gigabits-per-second range. The data rate limitation is normally due to pulse dispersion, which is a function of distance. This dispersion is caused by the different wavelengths of light propagating at different speeds in the fiber, causing the pulses to broaden.

Presently, there is a greater demand for single-mode than multimode cable. (One exception: Some LAN applications use multimode 62.5-micron-core fiber cable. Multimode fiber normally has a 50-micron core.) This is because most fiber networks being installed today cover long distances (greater than 10 km) and carry high data rates (above 90 Mbit/s) that require single-mode cable.

Of course, multimode fiber should only be used where its distance and bandwidth limitations do not cause a problem. This would be in distribution networks in buildings, campus environments, and LANs.

Faster and farther
Single-mode fiber is best applied where there is a present or future need for data rates higher than 135 Mbit/s or a transmission distance greater than 10 km between end points within the network. Single-mode fiber applications would include metropolitan area networks and full-motion video transmission, with network end points in different towns.

Most of the research and development effort is directed toward single-mode fiber to reduce both pulse dispersion and loss. This includes better splicing materials and equipment. The two main methods of joining fiber cable are fusion splicing and fiber-optic connectors. Fusion splicing is accomplished by aligning the two fiber ends through a microscope and fusing the fibers together with an electric arc. This generally results in a splice with less than 0.1 dB of loss.

Optimizing the junction
Fiber connectors use various methods of aligning fibers, along with a refraction-index-matching gel between the ends of the fibers, to reduce losses due to an optical mismatch (refraction differences). Fiber connectors have losses in the range of 0.1 to 0.5 dB. Research and development on multimode fiber will probably be limited, because future improvements in light sources and wider bandwidth requirements will limit its uses—except in LAN applications.

The cost of fiber-optic cable has gone through a dramatic reduction—an order of magnitude—in the past few years and there is only a small difference (less than 10 percent) in price between single-mode and multimode fiber. The cost could continue to fall in the future depending on how long the present production volume continues. A more likely development is that future fiber will support much higher transmission rates for approximately the same price as today's fiber. (For information about fiber optics standards activities, see "Brewing standards conflict blurs fiber optics' future," DATA COMMUNICATIONS, February, p. 55.)

Not a static technology
Fiber-optic networks provide significant advantages over other types of data communications networks. To fully obtain these advantages, however, requires extensive planning and design to fit an organization's unique requirements.

In the past few years, fiber-optic networks have gone through many changes in maximum data rate, distance, and cost. During this same period, the design, installation, and maintenance of these networks have also gone through changes, to the point where many organizations are now offering services in the networks' design and installation.

Most fiber networks that were installed before 1983 had a maximum data rate of 45 to 90 Mbit/s and were limited to a metropolitan area. Some of the networks required regeneration every few miles. These networks used multimode-fiber cable and both LED and laser light sources. Some of the locations had alternative facilities utilizing a different technology in case of a fiber network failure. Many of these fiber networks had marginal economic advantages over other technologies; fiber was used where it was extremely expensive to accommodate certain data's additional bandwidth requirements with a different technology.

It is anticipated that the networks being installed today will have a service-failure rate of less than one every five years due to equipment, *if* a redundant

2. Spectral width. *The laser has a much narrower spectral width—therefore less dispersion—than the light-emitting diode: about 4 nanometers versus about 100.*

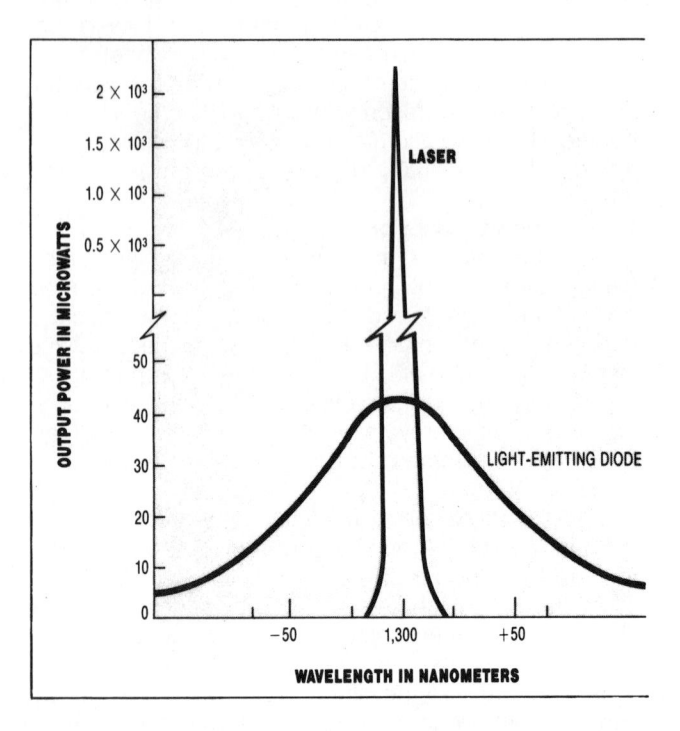

("protection") channel is provided. At present, the records of many networks show service failures occurring from every six months to a year due to fiber-cable cuts and human error.

In the past several years there have been many fiber-optic networks installed, and there are many more that are in the planning process. If all announced fiber networks are built, the result will be a transmission capacity that could not have been imagined a few years ago. A majority of the networks currently being installed are utilizing single-mode fiber cable with laser light sources.

The capacity of these networks is in multiples of 45 Mbit/s and ranges up to 1.2 to 1.7 Gbit/s. The distances range from a few kilometers to networks that cover several states. These networks have excellent economic advantages over other technologies (operating costs are lower because of less-costly maintenance), provide the primary facilities, and have no—or limited—alternative facilities for backup.

The future of many fiber networks exists in higher data rates (1 Gbit/s or more) and longer distances—greater than 50 km (about 30 miles)—without requiring regeneration. These areas are where much of the research and development is being directed today. The resulting networks will probably use improved types of single-mode fiber cable and laser light sources and be capable of carrying huge quantities of video and data.

Needs more attention
Future local fiber networks must also be able to economically distribute data signals from the large-capacity networks to various buildings, LANs, and individual locations. This area has not had nearly as much effort directed toward it as the high-capacity portions have. The smaller fiber networks and the portions of these networks that will distribute the signals from the large-capacity ones have more options for types of fiber and light sources, because of the reduced bandwidth and distance requirements.

The first decisions that have to be made about a small fiber network are the type of fiber to be used and the bandwidth required. The next step is to choose a light source based on the type of fiber chosen, the bandwidth, and the size of the network. Besides the network's initial bandwidth, future bandwidth requirements will have to be determined before any design can be completed. It may be difficult to estimate future bandwidth requirements. But the effort *must* be made because the findings can have a significant effect on the network design.

Economic decisions
Once the initial and future bandwidth requirements are quantified, the number of fibers to be used in the various sections of the network will have to be determined. It may actually be more economical to employ a larger number of fibers with lower individual bandwidths than to place the total bandwidth requirement on fewer fibers with higher individual bandwidths. The choice will depend on the distance between network termination points and how much of the total bandwidth requirement is for the future. A major decision factor will be based on the costs of high-speed (above 45 Mbit/s) versus T1 multiplexing equipment, with the latter having more fibers in the cable. The incremental cost of additional fibers in a cable is not as high as that of high-speed multiplexing equipment on a short (less than 12 km or about 7.5 miles) fiber network. An example of this is a network where new fiber-optic cable is being placed and the data rate (bandwidth) requirements are in multiples ranging from 1.5 to 45 Mbit/s. If the network is shorter than 5 to 10 km (3.1 to 6.2 miles), it is probably cheaper to increase the size of the cable and use individual multiplexers in the 1.5- to 45-Mbit/s range.

After the bandwidth and distance requirements of the individual fibers are identified, the types of fiber and light source can be determined. If multimode fiber is used, the light source can be either an LED or a laser, depending on the network length. An LED can be used if the network is small—no longer than the aforementioned 12 km; otherwise a laser may be required. If single-mode fiber is used, the light source will probably have to be a laser, for now.

Who does what
There are many organizations that sell fiber-optic services, which range from bandwidth on their existing networks to completely installed new networks. There are other options, one of which is leasing fibers, with the customer providing the light source and multiplexing equipment for the leased portion of the cable. A data-transmitting organization should go through a thorough analysis to determine what options are available, and

3. Outputs and angles. *Any light from a light-emitting diode or a laser that is outside a fiber's limited acceptance angle will not be coupled into the fiber.*

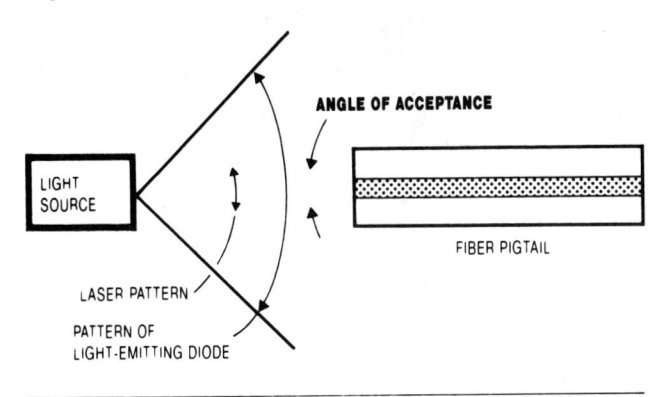

which of those are most suitable.

As is usually true of any network, the decision to build and maintain a fiber-optic network involves the commitment of capital dollars and the issues of how the network will be maintained and who will do the maintaining. A too conservative or an overly optimistic forecast of bandwidth requirements could cause a waste of capital dollars or a later expensive replacement or reinforcement of the network. If the decision is made to lease the fibers and own the multiplexing equipment, additional problems may be encountered. These include: trying to locate and clear troubles via several organizations involved with different portions of the network.

The decision to lease bandwidth conserves capital dollars and places the burden of maintenance and the provisioning of future bandwidth on someone else. With this burden placement comes the possible problems of bandwidth pricing and the reliability of the leasing organization to be responsive to maintenance requirements. The price of bandwidth can vary from a few hundred dollars up to several thousand dollars per megabit per second per month, depending on various factors. Some of these factors include shared versus unshared facilities, standard versus nonstandard bandwidths, transmission distances, and how quickly the initial investment is recovered. A data communications manager must have confidence in the abilities of any organization from which bandwidth is to be leased. ■

Marc Wernli is a staff manager in the engineering planning department at Northwestern Bell. He holds a B. S. and an M. S. in electrical engineering from the South Dakota School of Mines and Technology in Rapid City. He is also a member of the National Society of Professional Engineers.

Hal B. Becker, Consultant, Phoenix, Ariz.

Can users really absorb data at today's rates? Tomorrow's?

Some say no, that many have reached their limit, fostering faulty judgment. But new tools promise to relieve the overload.

In 1775, news of the "shot heard 'round the world" took four days to travel the 180 miles from Lexington, Mass., to New York City, and another 11 days to reach Charleston, S. C. Today, news of a significant event circles the globe in less than a minute, and the impact is observed in the stock and commodities markets no more than 30 seconds later. The average speed of transmission in the first case was approximately 2.25 miles per hour. Now it is about 1.5 million miles per hour.

Data communications and related computer technologies present people with so much information so fast that the number of decisions being made often becomes a substitute for their quality. Gone forever are the days when there was time for contemplation and reflection concerning information received and the course of action it might dictate.

Symptoms of this transition can be seen in the record number of post-Depression bank failures that occurred in 1984, and the continuing increase in the failure rate of other corporations. Another new post-Depression record was set in early 1985 when seven banks in four states failed in a single day.

Computer and data communications technologies, however, have evolved far more rapidly than the comprehension abilities of the people who use them. The bad decisions that led to the recent bank failures may be tied to the inability of people to assimilate exponentially increasing volumes of information presented to them in rapidly diminishing time periods, to sort through it and separate the useful from the "noise," and to make quality decisions. While the symptoms of the ailment are troubling, there is a positive prognosis. But first, let us explore some of the background.

The speed at which news could be transmitted progressed slowly and erratically over the centuries. In 490 B. C., the Greek runner Pheidippides ran 22 miles in just over three hours to bring news of the Greek victory over the Persians at the town of Marathon. He then fell dead of exhaustion. His average speed was slightly over seven miles per hour. For several centuries, communications transmission rates were limited to the speed at which a person could run.

The decision-making speed of humans may be related to their maximum running speed. Evolving over millennia, human data acquisition and decision-making processes may have been genetically selected, derived from the skills necessary both to avoid capture by predators and to capture prey. Human decision-making speed may reach a peak during chases of this type, as constantly changing information, received by the senses, is processed and decisions are made either to avoid or achieve capture. It is questionable how much this decision-making speed has increased in the 15,000 years since the last major Ice Age receded.

Data by foot

In 1690 A. D., almost 22 centuries after Pheidippides, express runners were maintained by armies to carry news from battlefields to command posts. These runners could travel a maximum of 20 leagues (about 60 miles) in 24 hours, averaging 2.5 miles per hour.

In 1775, the aforementioned news of the shot heard 'round the world was carried on horseback at an average rate of about 2.25 miles per hour. Under ideal conditions, a horse could carry information at a maximum of 300 miles in approximately 52 hours, for an average speed of 5.77 miles per hour.

By 1795, mechanical semaphore signaling was being tested. In one instance, a message was transmitted—in relays—the 500 miles between London and Plymouth, England, in three minutes. The average speed for this distance was 10,000 miles per hour. The cost of building, staffing, and maintaining such mechanisms precluded their wide use.

In 1815, bankers in London, England, made a killing in the stock markets using the news of Napoleon's defeat at Waterloo by the Duke of Wellington. They received the news by carrier pigeon and thus had it before the public did. Another of the Duke's pigeons held a record for long-distance communications, traveling 7,000 miles from West Africa to England in 55 days, averaging 5.3 miles per hour. In the long-established tradition of great communications feats, the pigeon died after delivering its message. [In 1941, a pigeon traveled 803 miles in 24 hours for an average speed of 33.46 miles per hour.]

In 1844, Samuel F. B. Morse invented the telegraph and the Morse code. The first message, "What hath God wrought?" was carried on a line between Washington, D. C., and Baltimore, Md. Soon, a skilled telegraph operator could achieve transmission speeds of 50 words per minute. By 1866, the many small telegraph companies that sprang up were consolidated into the Western Union Telegraph Co., which owned over 2,000 offices and 10,000 miles of lines.

At this time, news could travel rapidly around the United States. Overseas communications still relied on ships, which took many days to travel the Atlantic Ocean to Europe. Thus an event occurring on one continent would be old news by the time the message was received on the other.

In 1866, the first transatlantic telegraph line was laid by the then-largest steamship in the world, the *Great Eastern*. Not counting the propagation delay involved, a message could be sent across the Atlantic in the time it took an expert telegrapher to tap it out on the key. A 50-word message could be sent and received in one minute. News was now much more current when received, and an immediate response was possible.

On March 3, 1876, the patent for what would become the telephone was issued to Alexander Graham Bell. The telephone quickly began to replace the telegraph in the United States. People did not have to be skilled telegraphers. They simply picked up the instrument, spoke into it, and listened. By 1928, the first transatlantic telephone line had been laid. Thus did the world's shrinking accelerate.

Chatter speed

A rapid, easy-to-use instrument of communications was available to the masses. News of an event traveled quickly. Issues could be discussed and questions answered promptly. The ease of use quickly made the telephone an indispensable element of every culture that adopted it. The place of the telephone in American culture was best summarized by Marshall McLuhan, who said: "The telephone began as a novelty, became a necessity, and is now regarded as an absolute right" (from *The Telephone Book*, by H. M. Boettinger, Riverwood, 1977).

By the mid-1970s, the AT&T Long Lines Department had coaxial cables between Pittsburgh, Pa., and St. Louis, Mo., capable of handling in excess of 100,000 simultaneous conversations. During the same period, Americans placed an average of 12 million interstate telephone calls per day. Later, waveguide technology was capable of handling 500,000 simultaneous conversations.

The invention of the vacuum tube produced a similarly dramatic advance in communications. Starting with simple crystal radios, powerful receivers evolved that were built around the new invention. By the mid-1930s, an estimated 13.5 million radio sets were in use in the United States. While not cheap (over $100 each), they fulfilled the growing desire to obtain increased amounts of information in shorter time periods and, hence, were purchased in large numbers.

Television produced an even greater impact on an information- and entertainment-hungry public. From the earliest public demonstration by Bell Laboratories in 1927 to the first antenna installed on New York's Empire State Building in 1931, the industry has, to date, produced millions of high-quality, relatively inexpensive sets.

The early telegraph, slow as it was, had one significant advantage over the telephone, radio, and television that followed: The sender presented the telegrapher with a copy of the message to be sent. The telegrapher at the other end presented the recipient with a copy of the message received. Thus both sender and recipient had a reasonable facsimile of the message, which could be examined and retained for future reference. For several decades, subsequent technologies lacked this feature.

Few people can write nearly as fast as most people can talk. And recording gear was not yet available for the masses. As communications speeds increased, therefore, the relative amount of information retained decreased. Continuing progress in computer and printer technologies, despite accelerating communications speeds (both discussed below), offered a method of coping with this problem.

Figure 1 represents the increasing communications rates provided by continuing advances in technology. The example used is the time required to send 250 words 3,000 miles.

Notable events

Concurrent with the advances in communications speeds, three major revolutions occurred that produced profound and long-lasting changes in cultures exposed to them: the Agricultural Revolution, begun about 14,000 years ago, probably resisted by few; the Industrial Revolution, begun in the mid-eighteenth century, not readily beneficial to all (particularly those who lost their jobs to automation); and the Information Revolution, begun in the mid-twentieth century and still accelerating.

Several generations of information elements have come and gone: vacuum tubes, replaced with transistors, replaced, in turn, with third-generation integrated circuits. The current fourth-generation elements, consisting of very-large-scale integrated (VLSI) circuits, will be replaced by a fifth generation that has yet to be defined but is clearly on the horizon.

Computer and communications technologies are providing exponentially increasing volumes of information to exponentially increasing numbers of users in exponentially decreasing time periods. It is not yet clear whether this is a blessing, a nightmare, or a mix of the two. Are human assimilation and comprehension abilities evolving at the same rate as information production and transmission rates? The answer appears to be no.

1. Increasingly rapid transit. *Shown is the time required—through the ages—to send 250 words 3,000 miles. Printers help to cope with the mass of data.*

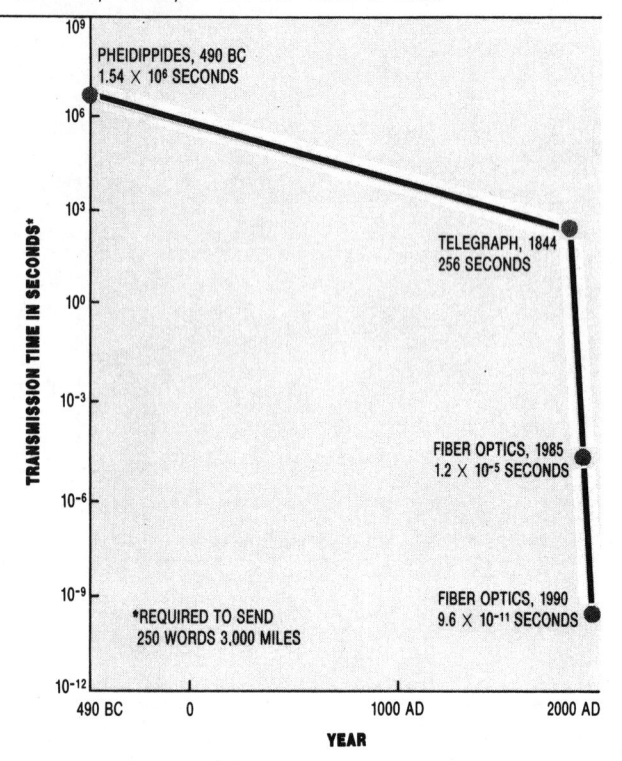

[Graph: Transmission time in seconds vs. Year, showing:
- PHEIDIPPIDES, 490 BC: 1.54×10^6 SECONDS
- TELEGRAPH, 1844: 256 SECONDS
- FIBER OPTICS, 1985: 1.2×10^{-5} SECONDS
- FIBER OPTICS, 1990: 9.6×10^{-11} SECONDS
*Required to send 250 words 3,000 miles]

In the Sumerian era, 4,000 B. C., information was recorded by using sharpened sticks on wet clay tablets to write cuneiform symbols representing the desired message. After they were dried, the tablets provided a relatively permanent record, which could be stored or carried as needed. They contained approximately one symbol per cubic inch of clay. Thus a person could convey to others as much information as could be lifted. Since writing in clay tablets took so long, it can be assumed that literate people could read and acquire knowledge much faster than it could be produced and transmitted. This situation was to prevail for several thousand years.

The Egyptian invention of papyrus—later to be known as paper—produced another dramatic change in the recording density of communications media. A single sheet of paper could contain considerably more information than could a clay tablet, and, of course, it was more convenient to carry. Early documents were prepared one at a time; if more were required, they were laboriously copied. People could still read and acquire knowledge faster than it could be prepared and transmitted.

The invention of devices for printing multiple copies of information, while known for centuries, did not greatly accelerate the production rate until the mid-fifteenth century, when Johannes Gutenberg invented a printing press that used movable type. The rate at which information could be produced, copied, and transmitted began to soar to the exponential heights observed today, while the human ability to assimilate and comprehend such volumes continued at its relatively linear and leisurely rate.

The recording density achieved by Gutenberg was approximately 500 symbols (characters) per cubic inch—500 times the density of the earlier clay tablets. Figure 2 illustrates the increasing recording density of various media. By the year 2000, semiconductor random access memory should be storing 1.25×10^{11} bytes per cubic inch.

In Europe at about 1501, the incunabula (total number of books printed before that year) equaled about 20 million, and the population was approximately 70 million. Thus there was an average of slightly more than one book for every three people. By 1600, just 150 years following the introduction of Gutenberg's press, an estimated 140 million to 200 million books (representing 140 thousand to 200 thousand titles) were in print. By this time, the population of Europe was up to 100 million, which meant there were approximately two books per person. Considering the literacy rates of the time, the information-production curve had already soared well beyond the comprehension rate of the masses.

Modern computers equipped with laser printers can produce information at the rate of 20,000 lines per minute, well beyond the human reading rate (Fig. 3). Fiber optic data communications facilities will soon be transmitting information at the rate of one billion bits per second. Consider the following, possibly absurd, example:

Twenty thousand lines per minute produce 9.6 million lines of print in a single eight-hour shift. Assuming 55 lines per page (60 characters per line), this rate produces 349 books of 500 pages each, which totals almost 576 million bytes of information. A fiber optic data communications link, transmitting one billion bits per second will require just 4.6 seconds to transmit the entire eight-hour production of the single laser printer.

Assuming an average of 550 words per page, and a reading rate of five words per second (a slower rate usually equates to a higher comprehension and retention), a total

2. Package compaction. *Recording density continues to increase. Gutenberg achieved about 500 symbols per cubic inch—500 times that of the earlier clay tablets.*

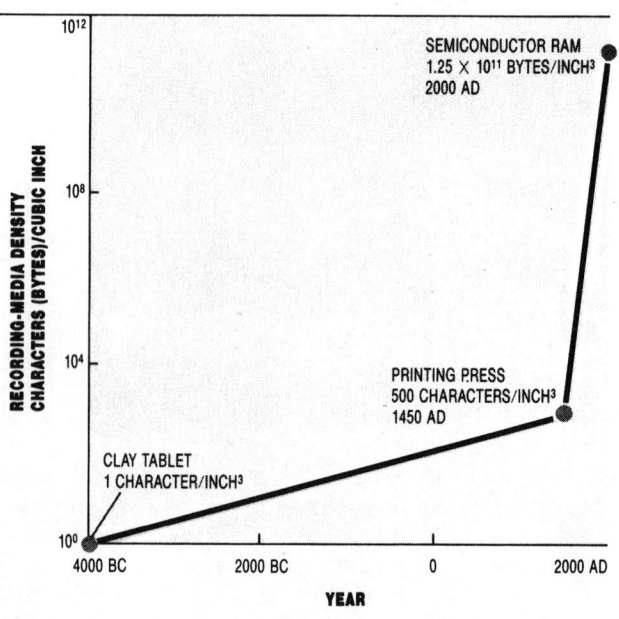

[Graph: Recording-media density (characters/bytes per cubic inch) vs. Year:
- CLAY TABLET, 4000 BC: 1 CHARACTER/INCH³
- PRINTING PRESS, 1450 AD: 500 CHARACTERS/INCH³
- SEMICONDUCTOR RAM, 2000 AD: 1.25×10^{11} BYTES/INCH³]

RAM = RANDOM ACCESS MEMORY

of almost 5,332 hours are required for one person to read a single night-shift's production from a single laser printer. Since the printer will produce another 349 volumes the next night, one can assume that each company owning such a device employs enough people to read and act on each day's production before the next batch arrives. This requires a total of 666 people reading steadily for eight hours.

Furthermore, companies will soon be equipped with the aforementioned billion-bit-per-second fiber optic communications links. Thus they could send an entire copy—several, if desired—of each night's production to their outlying regions and divisions. With little imagination, it is clear that this technology, used effectively, can immobilize an entire army of people every 24 hours. What is being done with all this information?

As the Information Revolution continues, and computer and communications technologies spread among the business, professional, and work environments, a growing number of people are beginning to exhibit several common symptoms of information overload. So much information is being presented in such a short time that people are incapable of assimilating it all before the next batch appears. In many instances, the rush to get the current task finished is accelerated before the next batch of information arrives in the in-basket or pops up on the terminal screen, demanding attention. As a result, the number of decisions made has, in many cases, become a substitute for their quality.

Faster than desired?

The "float" time associated with information is diminishing, perhaps at a faster rate than that associated with checking accounts and credit cards. A trip of two weeks or more will often result in the bill with the current credit card charges beating the traveler home. When news of a major event reached a neighboring town 200 years ago, it was often too late to act. Today, news of an invasion or a new war is flashed around the globe in a minute or less, and the reaction can be seen 30 seconds later in the price of gold on the world market. Communications media of all types have reduced the information float time to the point that reactions taken to new information are often driven more by the emotion of the moment than by reasoning, logic, and contemplation.

Stock market investors appear to be driven more by mob psychology than by the worth of either a particular company or its products. How much thought and deliberation have occurred when the recommendation of one analyst, flashed through an investor network, causes thousands of people to call their brokers and issue frantic "sell everything" orders?

Or what happens when instant communications indicates that the market has plunged seven points—or soared 11 points—in the first trading hour. This causes a number of investors to leap to get in on the action. Seven points on a base of 1,800 points is about four-tenths of 1 percent. Before the days of instant communications, moves of this nature took considerably longer; today, they are commonplace. If the market is not plunging or soaring, it is stagnating (meaning no change), which apparently invokes other investor reactions.

One of the basic tenets of the Information Revolution is: "Next to people, information is an organization's most valuable resource." To many computer vendors and users alike, this apparently means: "If a little information is good, more is better." Coupled with the adage, "Time is money," the two concepts have produced some interesting results.

Since time is a constant resource and information appears to be unlimited, many users are impatiently awaiting the next generation of computer and data communications technology. They will use it to produce even greater volumes of information in a given time period. For many of these users, the most visible result will be their ability to make even larger mistakes faster (see "Two disasters").

Computer-generated and -transmitted information acquires an aura of authenticity that most people do not question. The explosive proliferation of microcomputers has added considerably to the confusion. Newspaper articles relate stories of analysts who have programmed their latest sure-fire algorithms into their microcomputers. They become entranced by the output and continually recommend the purchase of losing stocks, while regularly passing over winners. Perhaps, in the market vernacular, a "technical adjustment" is necessary.

Employees in corporations of all sizes are victims of the same myth of omniscience. Having finally been sold—after many long and trying years—on the value of computers and their work-saving attributes, many employees do not question the accuracy and value of the output data.

The anonymity of computers perhaps offers a partial explanation for the employees' indifference. A report or set of figures prepared by a human can usually be traced, in a few easy steps, to the originating individual. Questioning elements of the report is thus perceived as questioning the individual or group that prepared it.

But questioning a computer-prepared report appears to be another matter. The swift, anonymous, electronic device

3. Outpacing comprehension. *Computers and lasers are producing 20,000 printed lines per minute. Optical fiber will soon transmit one billion bits per second.*

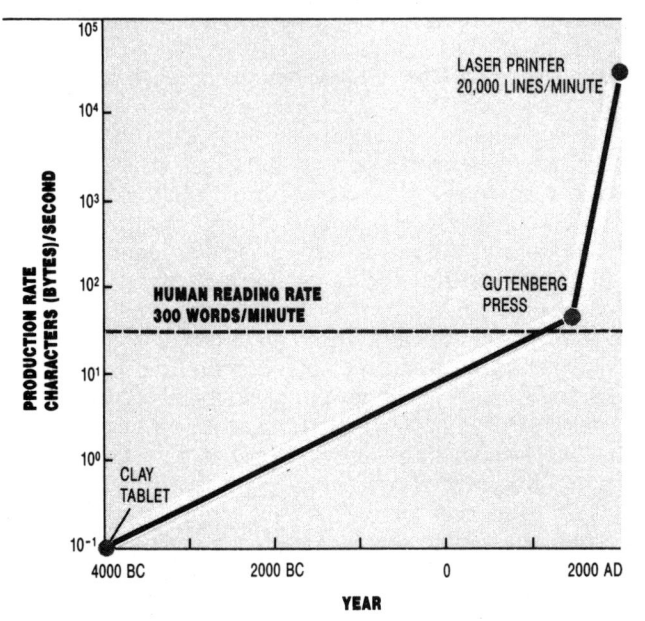

> **Two disasters**
>
> The availability of increasing volumes of information raises some critical questions. How often do fatal flaws in computer-generated and -transmitted designs remain undetected until too late? How many engineering projects fail because the simulator models were faulty? Or the appropriate questions were not asked? Or the answers were incorrectly interpreted? Furthermore, how many decisions to proceed are based on a lack of any negative response transmitted from the computers?
>
> **Substitute for humans?**
>
> Have computer outputs been ignored in some areas where they could help? Have they become a premature substitute for human intelligence in other areas? The collapse of a Kansas City hotel skywalk in 1981 and the recent space shuttle disaster may be two very visible examples of these phenomena. In the former, design changes made in the skywalk suspension were, apparently, not properly evaluated—the right questions were not asked. In the latter, the computers had not detected any reason to halt the countdown, so the launch proceeded. (Only later were factors publicly revealed—such as low temperature and its effects—that were not part of the automatic-halt process.) In both cases, early clues—part of the masses of available data—apparently were not brought to a high enough decision level or were ignored.

can sort through huge local and remote databases in minutes, extract selected information, manipulate it in a variety of ways, and present it simultaneously to several people in widely separated locations. The workings and the outputs of such mechanisms are beyond the comprehension and questioning ability of most users.

Computers are not often treated with the same anthropomorphic tenderness given other inert devices, such as ships, aircraft, and automobiles. The reluctance to accord computers this status may be attributable to the anonymity discussed earlier. "Blame-the-computer" is another aspect of this perceived anonymity. Examples include: "We're sorry the check didn't get mailed last week; our computer was down." Sales banners advertising: "Gigantic overstocked inventory clearance—our computer goofed!" "I'm sorry your order wasn't shipped; our computer was down." It is somehow easy to place the blame for these actions on an inanimate computer, whether or not it was involved in the problem. Many people presented with these excuses accept them readily.

The situation is slowly changing for the better. But current application development methodologies are horribly slow. A large, complex program can take 12 to 18 months to design, debug, and implement. All too often, by the time the application is up and running, the problem has changed so much that it is time to start modifying the program—or to scrap it entirely and start over. The "vanishing-problem" syndrome is alive and well. In many situations, the user opts to live with the less-than-adequate solution that is no longer completely applicable and tries to recover some of the investment. Thus users today are often forced to fit their problems to the solution.

Another possible effect of the Information Revolution: Both newly formed and well-established corporations are failing in increasing numbers. Many new ones are "high technology" companies that enjoy a wild ride as the darling of the investor community—only to discover, usually too late, that the ride was a short one. The scenario is now familiar: Information obtained from the many research and analysis firms indicates that the market that their product addresses will be $8.1 billion, say, in another year. Using their spreadsheet programs, these firms calculate that with only one-tenth of 1 percent of that market, the company will have gross sales of $8.1 million.

The lure is often irresistible, and a number of now-classic traps await. Two of the biggest: an overestimation of the ability to produce and successfully market the product in the face of intense competition, and an underestimation of the complexities and dynamics of the user environment.

Banks represent another visible effect of the Information Revolution: More banks—65 of them—failed in the first 10 months of 1984 than during any full year since the Depression. By mid-1984, the Federal Deposit Insurance Corp. (FDIC) had covered bank losses of $2.4 billion—more than four times the amount paid in the preceding 47 years. In 1985, another record was set as 120 banks failed. And FDIC Chairman L. William Seidman does not foresee any decline in the bank-failure rate this year.

Bankers are obviously making a lot of bad loans, bad decisions, or both. What role do computers play in this scenario? A few bankers will admit, off the record, that they are impressed with a loan applicant who presents a well-thought-out, computer-generated financial plan when requesting a loan. So long as the cash-flow projections, anticipated operating figures, selected ratios (such as debt-to-equity), and trends meet their requirements, bankers find it hard to question the proposal. A few are aware of just how easy it is to prepare such a proposal using a microcomputer and a spreadsheet program; many are not.

Are the bankers to blame? Partially. Are the loan applicants to blame? Partially. Are the computer and software vendors to blame? Partially. Each contributes a piece of the problem. Bankers will have to learn to look beyond the slick, computer-generated proposals and question applicants in more depth. Applicants will have to learn that the one-tenth of 1 percent projections and a business plan based on them are risky business. They should also learn to scrutinize the role that computer and data communications technologies play in their decision-making process. Many entrepreneurs appear to be so myopically entranced with their microcomputers, spreadsheets, and remote databases that they begin accepting them as "givens." Finally, vendors have been selling computers and related hardware and software as nothing short of magic for years—this will likely never change.

The recent breakup of AT&T and the deregulation that followed contribute another twist to the communications environment. A comparison with the computer industry illustrates this complication. For 25 years, computer vendors fought to retain their individual, incompatible postures. The goal was simple: If my terminals are incompatible with every other vendor's terminals, my users will have to buy my terminals. For several years it worked.

The efforts to change this thesis moved slowly, until the recent microcomputer explosion. Suddenly, everyone had to be compatible with IBM, and the race was on. Most successful look-alike microcomputers *are* compatible. A number of vendors who tried to remain independent have failed. At least at the mainframe, terminal, and microcomputer level, compatibility seems to be the watchword. The data communications world is another matter.

Deviations
At least six international organizations are hard at work trying to derive a workable set of data communications standards. It has not been easy. Every time a new standard, or piece of an existing one, is announced, it is often a matter of weeks before some vendor announces: "Yes, we are 90 percent compatible with the new standard. However, we chose to implement a certain feature of it slightly differently." The scope of the world's data communications networks and the multiplicity of vested interests will likely keep this problem recurring. While there is admitted progress in many standards areas, new applications continue to appear that are as yet without standards of any kind.

Back to deregulation: Perhaps the earliest and most visible indication of things to come is represented by the telephones provided in airports for travelers. During predivestiture, to make a telephone call from an airport—be it voice, or data via acoustic coupler—was only a matter of finding an instrument and picking it up. They all worked the same, and it was a simple matter to drop in a coin and make the call. If so desired, the call could be charged to a home or office number.

Today, placing a telephone call from an airport is considerably more complicated. The instruments are more easily located, however, because there are so many of them. Arranged neatly in a row, in little soundproof enclosures, can be found:
■ A telephone with no dial or pushbuttons, to be used only to answer airport paging.
■ Another telephone with a set of pushbuttons, but no place to insert a coin. The instructions indicate that this phone is to be used only for calls on specified, non-Bell (or is it AT&T?) carriers, which require an access code to use.
■ Another telephone with pushbuttons but no place to insert coins. The instructions indicate that this phone is to be used for credit card calls only, which will be charged through AT&T (or is it Bell?), providing you have an acceptable code number.
■ Another telephone that looks more like a computer terminal with a small CRT screen. It takes several minutes to decipher just what can be done with this instrument. Kids like to play with it. Adults like to ignore it.
■ Finally, with the user having received a short, on-the-spot lesson in the advantages of deregulation, a small, "plain-vanilla" telephone with pushbuttons and a slot for coins. It is usually the busy one. This is the instrument that must be used to place a simple local call.

Communications implies a two-way exchange of information. Successful communications occurs when the sender's information or knowledge, after being prepared and readily transmitted, is acquired and understood by the recipient. Now that computer companies are making an effort to allow computers to interact freely, telephone companies seem determined to revert to the posture maintained by the computer vendors 25 years ago. George Santayana was right: "Those who cannot remember the past are condemned to repeat it" (from *The Life of Reason,* volume 1, 1905-1906).

The problem goes far beyond airport telephones. The good news is the countless inexpensive, computer-driven telephones available today presenting truly staggering arrays of features and options. The bad news is that no two of them work alike. The pleasures of the money saved by switching to the new equipment and by the use of the new features are often quickly replaced with the frustration of lost or misrouted calls.

A hopeful sign
The existing incompatibilities in computer, data communications, and voice facilities appear to be contributing to another phenomenon: the increasing entropy of the information with which these mechanisms deal. (Entropy may be defined here as a measure of the randomness or disorder within information.) Viewed globally, from the perspective of the rapidly expanding national and international communications networks and their inconsistencies and incompatibilities, it appears that entropy is increasing. There is hope, though: Standards bodies have several efforts under way (Integrated Services Digital Network and Open Systems Interconnection are two) that are intended to eventually ease the situation.

If the growing number of decisions discussed earlier are based on information that is unverified, received late, or inaccurate, the confidence level in these decisions must also decrease. While difficult to quantify, this aspect of the information issue makes for interesting speculation.

The three revolutions discussed earlier occurred over progressively shrinking time periods. In both of the first two revolutions (Agricultural and Industrial), the benefits were achieved over considerable time periods, with succeeding generations becoming familiar with the new technologies and adapting to them. The changes in any one lifetime were so gradual that they could usually be accepted with proper motivation and training.

The Information Revolution has been evolving within considerably less than one lifetime. The changes in the last 25 years have been—and promise to continue to be—dramatic. But the benefits provided by this revolution remain a mystery to many.

A candidate for a fourth revolution has recently appeared: artificial intelligence. While still in its relative infancy, and considered by some to be poorly named, artificial intelligence may, in time, provide the means to cope with the vast amounts of information produced and transmitted by the computers of the third revolution. The first 10 years of the revolution will likely provide dramatic insight to the process called intelligence, whether exhibited by a human or a machine. (Figure 4 presents the shrinking time frames associated with major technology revolutions.)

To date, in spite of news articles to the contrary, artificial intelligence is not very intelligent. Clever, yes. Brilliant, occasionally. Intelligent, not yet. In many cases, the harder the goal of artificial intelligence is pursued, the more willing the machines become to display genuine ignorance.

Small companies are springing up rapidly and presenting products labeled, in one way or another, artificial intelligence. Articles and books abound describing "thinking machines," and seminars and conferences on the subject are rapidly increasing.

One fundamental problem continues to defy solution: an acceptable definition of the term "intelligence." Describing intelligence is not necessarily the same as defining it. The resolution of this problem may be a long time in coming.

Meanwhile, efforts in two areas related to artificial intelligence bear watching: human-to-machine interfaces and expert systems. They may offer the initial means to cope with increasing amounts of computer-generated and computer-transmitted information.

Visible progress

Human-to-machine interfaces are improving steadily. Although natural-language interfaces are not here yet, progress in that direction is visible. Major remaining obstacles include: content and context analysis, the resolution of ambiguities inherent in natural languages, and the derivation of directions or instructions contained in a statement that can be presented several ways.

Expert systems are developing rapidly in a number of disciplines, including medical diagnosis, failure analysis, and geophysical exploration. These systems are being developed in two areas that will have a significant impact on the Information Revolution: operating systems and data communications network control.

Operating systems, which control computer resources, are becoming increasingly complex. The problems of efficiently scheduling and allocating these resources in response to rapidly changing user demands are starting to exceed the capacity of human operators.

In the days of large, centrally located configurations connected to a network of remote terminals, the operator's problems were much simpler. Today, commonplace distributed configurations contain many sizable processing centers that are interconnected through megabit-per-second networks that span continents and oceans. Further, these installations maintain databases in the gigabyte range that are accessible 24 hours a day.

Operating systems are being developed that increasingly resemble evolving expert systems. The goal is to achieve the most efficient scheduling and allocation of resources in response to increasingly dynamic user demands. Ultimately, combinations of three basic procedures (see below) will be applied, and all three are totally dependent on the rapidly increasing bandwidth available in communications networks.

In some situations, with one procedure the process (application software) will be transmitted for execution to the location of the relevant database. In a second procedure, the database, or portions thereof, will be moved to the process. In other situations, the process and the database will be moved to a third location (the third procedure), where processing hardware is idle, or where excess capacity exists.

Another use of the third option will be to reconfigure around device or network failures. Fault-tolerant proces-

4. Shrinking time periods. *Major technology revolutions are occurring more frequently, and their durations are growing progressively smaller.*

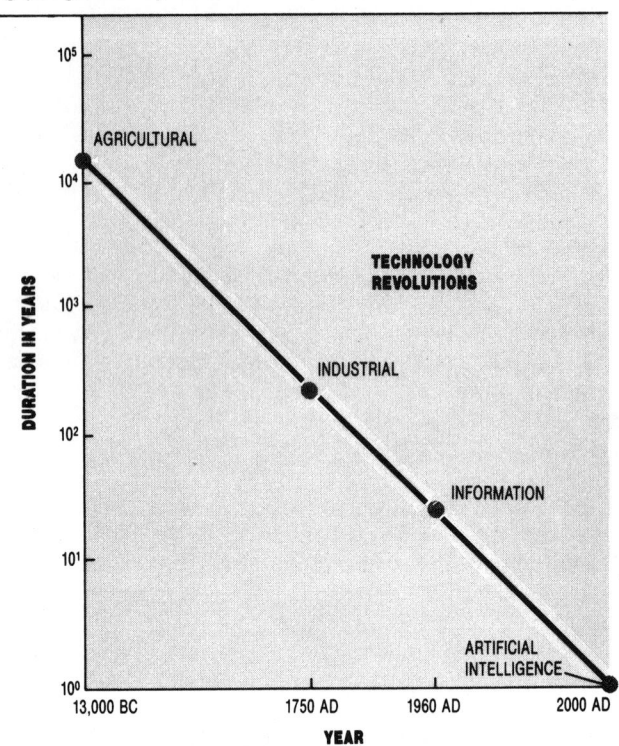

sors, previously limited to somewhat specialized applications (such as airline-reservation and process-control networks), will become increasingly common. Computer users of all types will recognize how dependent they have become on information processing technologies and will demand higher availability and reliability.

The control of network communications is also exceeding the capacity of human operators. As voice becomes digitized and integrated within networks already handling a variety of other types of traffic, the dynamic allocation of bandwidth requires decision-making at rates beyond human abilities. This level of control, exercised when all resources are functioning properly, is one of several communications areas requiring attention.

A second area involves the selection of the least-expensive route through a network. Considerations include the urgency levels of traffic, the load on the network, and, possibly, the time of day. The ultimate goal is the shortest propagation delay at the lowest possible cost. Computer-driven private branch exchanges are already exhibiting primitive versions of this logic.

A third area involves the diagnosis of network-component failures, followed by dynamically reconfiguring around them to maintain the desired throughput and availability levels. Many failures today result in an interruption of service to users, followed by manual location and diagnosis of the fault by human operators using sophisticated equipment at network control centers. User demand for continuous, trouble-free service coupled with the desire to keep control-center-personnel costs down will result in the application of expert systems to the task.

The major problems involved in applying expert systems to these three examples include: extracting the decision-making rules from "expert" human operators and controllers and building them into the "inference engines" that drive the expert systems; designing and implementing the data-acquisition logic required to build and maintain the necessary databases; building in the equivalent of human, intuitive tests of reasonableness; and testing the developing expert systems in realistic environments to raise the confidence in them to acceptable levels.

Speed it up
The use of expert systems to develop application software is another area receiving considerable attention. The major goal is to shorten the current development time such that applications can be developed and modified more readily, in response to the increasing rate of change observed in the user environment. In addition to the major problems discussed above, this development task includes the evaluation of techniques for verifying deterministic answers to questions while using probabilistic procedures.

Beyond these areas lies the enticing realm of applying expert and intelligence-emulating mechanisms that will be largely self-programming. At the core of this research, again, is the question of the definition of intelligence. The search for this definition appears similar to that of particle physicists seeking to understand the fundamental nature of matter. In spite of several centuries of speculation, intelligence researchers are still seeking the identity of their "atom": intelligence itself.

Considerable progress has been made in understanding and emulating the sensory mechanisms that contribute to the development of intelligence in humans. Visual, aural, tactile, and olfactory mechanisms are being produced that begin to approximate their human models. Speech synthesis is commonplace in a variety of devices. While these sensory and speech schemes are clearly a part of the quest for artificial intelligence, they do not yet provide much of an insight to intelligence itself. They are, in a sense, clues to a still unsolved puzzle.

Stated simply, the goal of intelligence may be the survival of a species. If that is true, then the question of the origin of the survival goal must be asked. Why is it necessary for a species, any species, to survive? Where did the survival goal originate, and why?

Consider bacterial or viral species. As soon as humans develop a drug that is effective, these species often evolve to a different, resistant form. Is this behavior perhaps a form of intelligence applied to the survival goal? Much research activity in the field of artificial intelligence is centered around the task of "teaching" computers knowledge, as if they have already acquired the ability to "learn." Without first instilling a goal (perhaps survival) in computers, thus motivating them to learn, it is doubtful that they will ever exhibit true intelligence, regardless of the amount of knowledge they have acquired.

Within the next few years, it may be possible to produce a rough approximation of the human brain with semiconductor technology. When equipped with the sensory mechanisms discussed earlier, it could then be exposed to a variety of external stimuli. Without a survival goal, the behavior of such a device is questionable.

Instilling a survival goal in such a device poses a number of interesting questions. How long would it take for the rudiments of survival, or any other goal, to appear? Could the process be accelerated, or would it take as long to develop as it did in humans (tens or hundreds of thousands of years)? Should it appear, what end would it serve? What form of natural selection would appear that would retain useful characteristics and discard the rest?

It may be that many of the questions being asked about creating artificial intelligence in computers are the wrong ones. Time will tell. If artificial intelligence should become a fourth revolution, it may be accelerated out of the necessity to cope with the computers of the Information Revolution. Two outcomes appear to exist: The problems of the Information Revolution will be solved, or simply moved to an even more abstract, incomprehensible realm. Given the current headlong rush into the artificial intelligence environment observed in the industry, the latter seems, for the moment, more likely.

Social and cultural implications
The Western World's fascination with computers and related technologies presents an interesting contrast to what is observed in Japan, now several years into its well-publicized fifth-generation project. In the Western World, new technologies are often rushed through development, placed in production, and introduced to the market at an ever-increasing rate. Before the social and cultural impact of one revolution (computers) is identified and evaluated, most researchers appear to be on to the next one without so much as an occasional backward glance.

The Japanese have approached the problem from a considerably different perspective. They began with an evaluation of the social and cultural needs of their society. Then they asked: "How can technology help us in meeting these needs?" A number of these needs were evaluated in approaching their fifth-generation project, including:
■ The successful integration of 100 million people—in a country about the size of California—with an increasingly information-oriented culture.
■ Culturally acceptable methods of adapting to a society that is becoming highly educated and older.
■ Greater efficiencies in managing scarce natural resources and energy.
■ Enhancing Japan's ability to become an internationalized culture.
■ Improvements in fields currently exhibiting low productivity.

These are, by any standards, ambitious goals. It is significant that the Japanese started with the social and cultural goals and then asked how technology could help. The Japanese also evaluated the current fourth-generation computer technologies. The conclusion: that generation would not take them where they wanted to go.

Stated another way, much of the Western World appears determined to seek change for the sake of change alone. As a rule, the Japanese, on the other hand, choose to select desired social and cultural enhancements and then apply technology to achieving them. Most Western cultures continue to surround themselves with increasingly sophisticated examples of technology, without first evaluating their cultural and social impacts. With current production techniques, the products of a relatively few are made available, at affordable prices, to the many. Are the data acquisition and processing skills of the many—as facilitated by today's increasingly ubiquitous microcomputers—up to the challenges represented by these devices' copious outputs?

Automobiles and telephones are necessities. It is not clear that computers fall into the same category. While many organizations are entirely dependent on computers for their survival in competitive environments, their overall impact on society has yet to be determined.

The computer industry continues its shakeout. IBM retains its dominant position while hundreds of others, including the telephone companies, pursue a niche in the market. The recently observed sluggishness in the computer industry and the closely related semiconductor industry is attributed to many things, including the shakeout process itself. The slowdown may reflect the initial stages of a general misunderstanding by the masses. Surveys abound suggesting that 40 percent of all microcomputers delivered to date are gathering dust or are used for inconsequential tasks, having failed to provide the benefits promised when purchased. Many first-time users have discovered that they do not possess anything approaching the computer literacy required to apply the machines to their tasks, in spite of vendor assurances that it would not be required.

What role will computer and data communications technologies play in coming years? Stafford Beer, in his book *Brain of the Firm* (John Wiley & Sons Ltd., 1981) poses the question elegantly: "The question which asks how to use the computer in the enterprise, is, in short, the wrong question. A better formulation is to ask how the enterprise should be run given that computers exist. The best version of all is the question asking what, given computers, the enterprise now is."

Many—perhaps most—organizations can be observed pursuing answers to the first two versions of the above question. Few, if any, are seeking answers to the third version. The third question cannot be answered until a sufficiently high level of computer literacy exists in a sufficiently large number of people. Herein lies the problem.

A majority of middle- and senior-level managers in companies of all sizes are people in the 35-to-65 age bracket. Many management techniques they use were developed long before computers and data communications technologies began their explosive evolution. While surrounded by examples of these technologies on all sides, they have little understanding of how the technology works and claim: "I am not computer literate."

The computer-user community supports a growing number of consultants who can apply computer and communications technology to business problems, thus relieving the user of the task. Many of these consultants, like their counterparts employed by the users, have become specialists in narrowly defined areas like operating systems, database architectures, data communications, and specialized application package development. While providing quality work in their chosen specialty areas, many lack the "renaissance" perspective required to step back, view the enterprise as a whole, and ask: "What is the nature of the enterprise I am dealing with, now that com-

5. Convergence salvation. *While humans are becoming more computer literate, computers—through "artificial intelligence"—are becoming more human literate.*

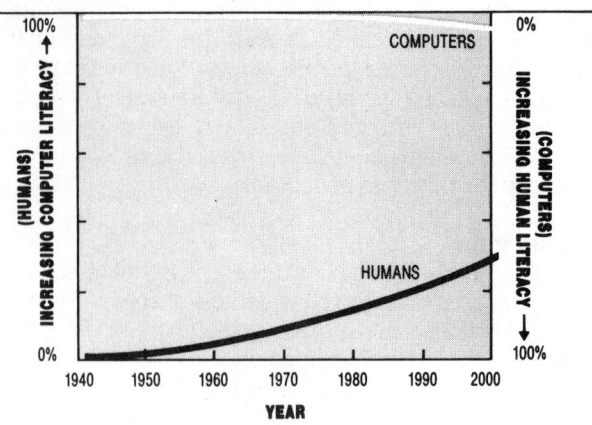

puters play such an integral role in it. How is it changing? What is the rate of change observed?"

From another perspective, the computer industry continues to pursue the goal of "user friendly" equipment. Like the definition of artificial intelligence, the harder it is pursued, the farther away it seems to be. Much of this activity has obscured the fact that the first generation of "friendly users" is already here, and their average age is about 12 to 15. (It is estimated that 92 percent of U. S. public schools now have computers for student use; this compares with just 30 percent three years ago. And at the public schools of at least one state—Texas—computer literacy is a graduation requirement.)

Those friendly users (noted above), introduced to information technology at an early age, have no inhibitions about approaching a keyboard, pressing a key, observing a response, and pressing another key. Soon a dialogue is established, and they are launched effortlessly and painlessly into a realm largely unknown to those just one generation removed. When these people assume the middle and senior management positions in their chosen fields, computer literacy will be an integral element of their management skills. Many of them will look back and wonder: "What was it like back in the old B. C. [Before Computer] days?"

These new managers will inherit the legacy left by the present generation. They will bring a Renaissance perspective to the problems that the earlier generation failed to develop and apply. They will be able to answer the third of Stafford Beer's questions and proceed accordingly.

While the next few decades pass and the friendly users achieve their management goals and increase their computer literacy, another evolutionary process will be occurring: Computers, through their enhanced capabilities and, possibly, intelligence, will become increasingly human literate. Figure 5 presents this convergence. Just when the two will converge remains to be seen. ∎

Following 20 years with GE and Honeywell, Hal Becker started a career as an information management consultant. He has authored two books as well as many articles and makes presentations at major industry conferences.

Anthony Badalato and Paul Drimmer, Coopers & Lybrand, New York, N.Y.

Ways to avoid the stumbling blocks that foil T1 network success

Proper planning includes considerations of network growth, deadlines and how to meet them, and vendor-choice criteria.

High-capacity digital transport facilities—also known as T-carrier circuits, 1.544 Mbit/s and up—have become popular as an overall communications solution where information flow is dense among a few buildings. They are economical, accommodate current requirements, and allow for extensive future expandability.

But for all the T-carrier's strong points, many stumbling blocks potentially impede successful network installations: The number of alternative configurations and equipment choices can be confounding; design, organizational, and financial questions abound; and projects can easily go astray without leadership that balances broad technical perspective with hands-on experience. No organization will be able to exploit T-carriers properly unless it matches the technology with its needs, formulates a thorough and realistic implementation plan, and dedicates the appropriate resources—both technical and personnel—to the implementation project.

Most often, T-carrier installations are chosen to replace multiple circuits between two or more concentration points because they are more economical. For instance, as shown in Figure 1, a New York organization with 20 voice tie lines, 25 data communications lines, and a circuit for remote security surveillance between its uptown and downtown offices would pay roughly $142,800 in annual operating costs. Without accounting for any tariff or equipment increases, the company would pay approximately $700,000 over five years for leased voice and data circuits and such support equipment as modems and network management/diagnostic devices (located in the modems and included in their cost).

Besides tangible costs, other factors and considerations may have a greater influence on the rising cost of the cited network architecture. Some of these factors are:
- The rate of circuit (or application) growth or decline between the concentration points and the consequential incremental costs of adding or deleting circuits and support equipment.
- The ability to provide communications links responsively between the concentration points (taking into account the impact on the organization if deadlines are missed).
- Trends in private-line and, if applicable, public switched network tariffs, along with the potential impact of legislation, such as private-line access charges.

Even excluding these factors from the cost calculations, the organization would be paying out, over five years, almost half a million dollars more than the $220,000 price tag to purchase, install, and run a T1 circuit of 1.544 Mbit/s capacity (Fig. 2). Most of the T1 cost is represented by the leases of the transmission facility and channel bank (multiplexing) equipment.

Who's going to do it?

No wonder, then, that T-carriers—and the vendors offering them—are proliferating (see "T1's time has come"). For instance, in the New York metropolitan area, companies such as Manhattan Cable, Local Area Telecommunications, Teleport, and Damanet have joined New York Telephone in offering T-carrier facilities.

In addition, some companies are privately installing bypass links to achieve T-carrier capacities. For example:
- Citibank has obtained the necessary right-of-way licenses from New York City to install a fiber optic network in Manhattan, connecting its midtown headquarters with offices in three downtown locations. This arrangement bypasses telephone company central-office switching.
- Izod Ltd., a subsidiary of General Mills, has turned to Local Area Telecommunications Inc. to provide microwave facilities to transmit voice and data between two buildings that are only five blocks apart.
- Morgan Guaranty Trust Co. uses Gemlink microwave radio links, originally provided by General Electric, to back

1. An economical alternative. A New York organization with 20 voice tie lines, 25 data communications lines, and a circuit for remote security surveillance would pay roughly $142,800 in annual operating costs—over five years, almost half a million dollars more than the $220,000 to purchase, install, and run a T1 circuit.

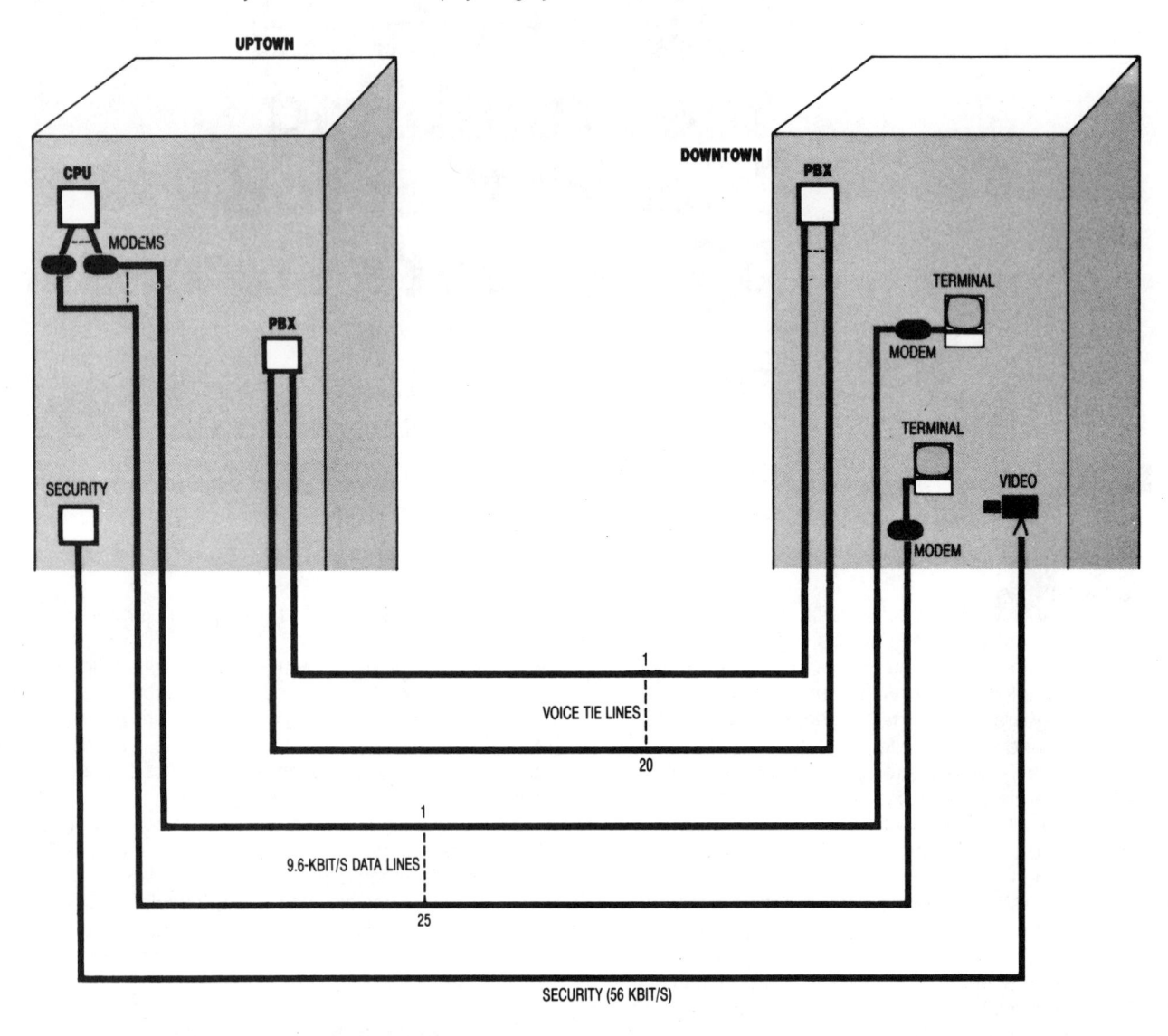

CPU = CENTRAL PROCESSING UNIT   PBX = PRIVATE BRANCH EXCHANGE

up private facilities installed by Manhattan Cable to connect three buildings in close proximity in lower Manhattan.

Although many high-speed (Mbit/s-range) communications alternatives are available, each should first be assessed in the context of the particular organization. To begin, a thorough examination should be made of all pricing and performance factors—most directly from a concise request for proposal (RFP).

Among the cost factors to be considered are the one-time (as for construction) and recurring costs related to the option—as well as the potential derived therefrom—to regulate or stabilize price increases over a five-year period. The network planner must determine the one-time costs (such as for construction) and recurring costs associated with this price stabilization. To help in this determination, the planner should obtain circuit-availability rates, mean-time-to-repair statistics, and the degree of backup and redundancy offered.

Also to be considered is the impact that the high-speed network would have on other corporate networking plans, such as the ability to communicate with a disaster recovery (backup) site. In an analog world, the user would dial up the site via modems, thereby initiating the required backup connections. With T1 multiplexers, there are no modems. One possible solution is an offering by AT&T called Accunet Reserve 1.5. Here, the distressed user telephones AT&T, which could take up to an hour to restore digital service.

One planning task involves evaluating the potential supplier's ability to provide initial and subsequent transmission facilities within an acceptable time frame. The vendor's network-monitoring capabilities should be investigated; if a private-network alternative is also being considered, the cost and availability of interrogative tools necessary to troubleshoot and oversee the network should be explored. Not to be overlooked is an integrity assessment of the routing paths between and within concentration points. Examples include paths between microwave-repeater hops and alternative cable routing within a building. Finally, the type of transmission facility chosen should match the user organization's ability to operate and maintain it.

Choose your mux
Perhaps the most crucial component for performance is the channel bank or time-division multiplexing equipment. Complicating the choice of equipment is what seems to be the almost-daily entrance of new vendors into the field. For example, Infotron Systems, Spectrum Digital, and Cohesive Network have joined more-seasoned vendors, such as Avanti Communications, Timeplex, Datatel, and ATTIS (AT&T Information Systems), in providing channel-bank equipment.

The basic no-frills channel bank simply subdivides a T1 trunk into 24 independent channels at a data rate of 64 kbit/s each. They typically also provide some rudimentary monitoring and diagnostic capabilities, all for a base unit price of about $5,000 to $10,000. Beyond that, certain vendors offer more-sophisticated equipment with complex applications.

Common examples of complexity are advanced voice-encoding algorithms. Typical are continuous variable slope delta (CVSD) modulation and adaptive differential pulse code modulation (ADPCM), both providing economic voice communications. (CVSD uses a one-bit sample to encode the difference between two successive signal levels; the per-channel sampling rate is usually 32,000 times a second. ADPCM, standardized by the CCITT [International Telegraph and Telephone Consultative Committee], uses three or four bits to describe each sample, which represents the difference between two adjacent samples; sampling is done at 8,000 times a second.)

Currently available network-control technology enables a technician to monitor, accumulate channel-usage statistics on, and run diagnostics for various network points from a centrally located console. Another capability from a central site is that of reconfiguring the network without dispatching personnel to remote locations. This reconfiguration may be initiated either manually or automatically. If automatically, the "trigger" may be based on a preset condition, such as time of day.

Another useful digital-network feature is the ability to interconnect multiple T1 circuits, eliminating the need for additional channel banks for backup and multipoint-networking applications. Duplicate crucial network elements—such as processors, modems, and power supplies—are recommended to minimize the threat of an overall failure. Additional features are emerging every day. However, with the more-sophisticated multiplexers costing from $40,000 to $50,000 each, feature packages should be chosen carefully on the basis of current and foreseeable requirements.

Check out the vendor
Finally, the supplier's reputation for delivery, long-term support, and responsive and economical service also should be weighed in the selection process. With a limited base of qualified technical personnel to select the elements for, and to implement, a T-carrier installation, many vendors are providing a turnkey service. This includes not only the installation of their channel bank equipment, but also implementation of support equipment like switches and line drivers from other vendors, as well as overall project

2. Beneficial solution. *T1 networking requires—in this case—just one circuit as the functional equivalent of the multiline approach of Figure 1. Most of the T1 expenses are represented by the user's lease costs for the transmission facility and the associated channel-bank (multiplexing) equipment, and for maintenance.*

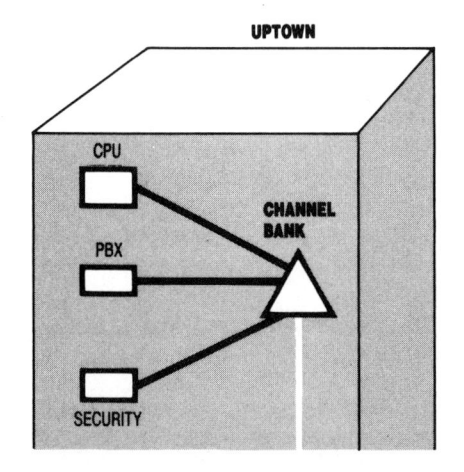

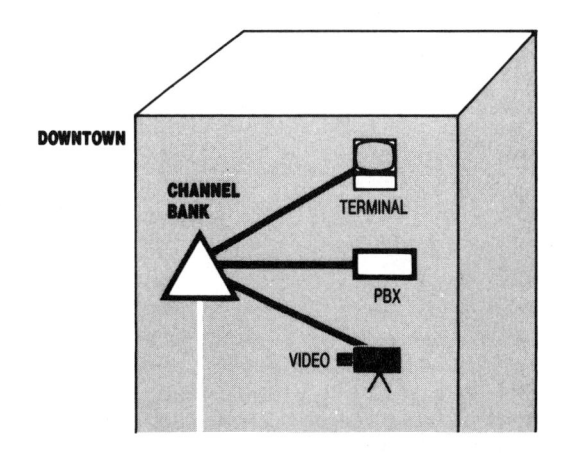

CPU = CENTRAL PROCESSING UNIT   PBX = PRIVATE BRANCH EXCHANGE

T1's time has come

The key reason for the sudden growth of T-carrier service is its progressively lower installation and operating costs and its greater capacity per channel compared with conventional cable pairs. Combined with a spiraling user demand for immediate access to information, T-carrier installations are now proliferating in the commercial environment. But the basic technology has been evolving within the former Bell System and other carrier networks during the past two decades. Its typical application has been to link central switching offices.

Bit by bit

Inside the carrier network, user information is propagated by either analog or digital transmission. (Analog refers to the use of continuously varying signals. Digital transmission involves discrete signal levels.) Whenever digital transmission is employed to carry voice messages, the analog signal must be converted into a pulse-coded stream. As shown in the figure, this pulse code modulation (PCM) represents an eight-bit code. These eight bits appear as one sequence in a frame and show the state of the channel at that instant. Since there are 24 channels and eight bits per channel in a frame, there are 192 (24 × 8) bits carrying customer information in a frame. An additional bit, called a framing bit, is added to every sample for timing and control. Thus, there are 193 bits sent in each frame. To faithfully reproduce voice frequencies of 4 kHz, a sampling rate of 8,000 frames per second must be maintained. The equation reduces to [(192 information bits one control bit) per frame] 8,000 frames per second = 1.544 Mbit/s.

The digital hierarchy is not limited to T1 signaling alone. The table, "Digital hierarchy," (derived from the second reference cited in the accompanying story) shows the North American digital multiplexing hierarchies based on a 24-channel group, the number of 64-kbit/s voice channels that can be derived, and the transmission scheme that is typically used to carry the signal.

Digital hierarchy

LEVEL	BIT RATE	EQUIVALENT NUMBER OF VOICE CIRCUITS	TRANSMISSION MEDIA
DS-0	64 KBIT/S	1	
DS-1	1.544 MBIT/S	24	CABLE PAIR: T1 RADIO
DS-1C	3.152 MBIT/S	48	CABLE PAIR: T1C, T1D
DS-2	6.312 MBIT/S	96	LOCAP CABLE: T2 RADIO: 2 GHz
DS-3	44.736 MBIT/S	672	RADIO: 6 GHz, 11 GHz FIBER: T3
DS-4	274.176 MBIT/S	4,032	RADIO: 18 GHz FIBER: DEVELOPMENTAL COAX: T4

DS = DIGITAL SIGNAL   LOCAP = LOW CAPACITANCE

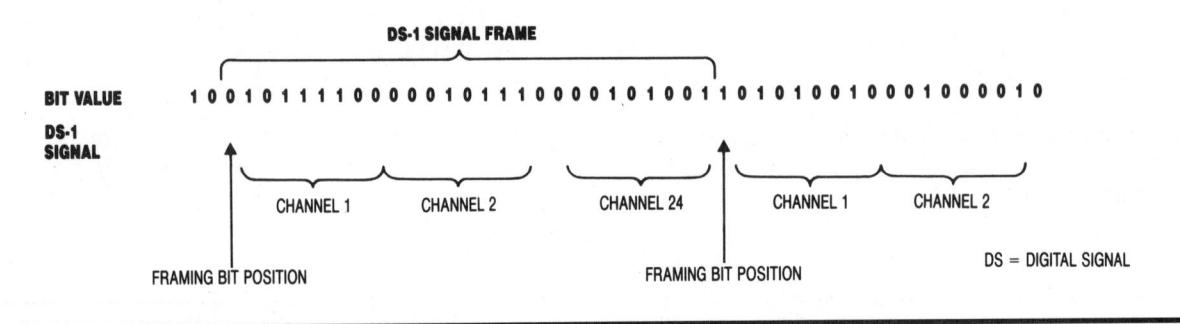

```
                        DS-1 SIGNAL FRAME
BIT VALUE   1 0 0 1 0 1 1 1 1 0 0 0 0 0 1 0 1 1 1 0 0 0 0 1 0 1 0 0 1 1 0 1 0 1 0 0 1 0 0 0 1 0 0 0 0 1 0
DS-1
SIGNAL
              ↑                                                     ↑
            CHANNEL 1    CHANNEL 2          CHANNEL 24    CHANNEL 1    CHANNEL 2
         FRAMING BIT POSITION            FRAMING BIT POSITION       DS = DIGITAL SIGNAL
```

management. The quality of these offerings—as well as the need for them—should be factored into the buying equation.

A cautionary note: Some provision must be made to back up the transmission facility in case of a major failure. In conventional networking, if a company has several low-speed—such as 9.6-kbit/s—circuits connecting two concentration points, one or two circuits can fail at the same time—only under extraordinary conditions would all circuits be out of service simultaneously. With standard multiplexing techniques, however, a T1 circuit between two points might carry as many as 240 data channels of 4.8 kbit/s each. If the T1 circuit or multiplexer failed, all 240 channels would be lost.

Companies that bypass local telephone exchanges must realize that it is difficult to marshal the resources in manpower and alternative routing that a local telephone company can gather in case of failure. It is clearly critical to include a reliable backup plan and its costs in any proposed network strategy.

Therefore, a dual T-carrier installation is highly recommended. The backup circuit, at minimum, requires an interbuilding and intrabuilding route different from the primary T-carrier route. A second vendor might also be considered for the backup.

For fallback purposes, dial backup components can be added to the modems formerly used with the deleted dedicated circuits. The advantage here is that, after the initial purchase of dial backup equipment, costs are limited only to telephone usage. However, this application requires modems, dial backup units, and switching gear; and it carries a high incremental expansion cost. To contain expenses, dial backup should be used only when a limited number of vital applications need to be supported with a backup operation.

The network planner should understand that the

technology behind T-carriers is completely different from that of analog circuits. As a result, technicians' skills must be updated to match new requirements demanded by the high-speed network.

The need for an overall implementation plan soon becomes apparent. Properly staffing the project is the first priority. Implementation activities are technologically demanding but, more important, require abundant time and attention to detail—what an organization oriented toward operations usually cannot spare. It is absolutely pivotal to get the right people—through training, recruitment, or consultants—to usher the project to completion.

Brain trust at work
At the start, planners must identify the project steps, their relationship to each other, and their duration. Meetings among users, vendors, designers, and other contributing parties should amplify and refine the project plan. A seasoned project manager can plot the schedule while questioning potentially overlooked subjects. The latter include the availability of sufficient air-conditioning and electrical interfaces and the speed at which local craftsmen can install cabling and other components.

The manager should also be able to gauge the speed at which an organization works to get things done. For example, staffs can debate the fine points of contract terms and conditions for weeks, but the manager's goal is to develop a tight, realistic, and well-synchronized plan that also considers potential problems.

Several software packages are available to assist in formulating a solid implementation plan. These tools, using some form of the program evaluation review technique (better known as PERT), prompt users to look at the entire project as a series of logical steps. PERT is also an aid in weighing the relative importance of project duration, prerequisites to certain tasks, milestones, manpower levels, and costs. One of PERT's highlights is making apparent the project's critical path, where delay in a vital task would delay the entire schedule. Planners must, therefore, pay particular attention to the critical path and, as much as possible, try to reduce the risks.

Once the go-ahead is given to order equipment and circuitry, everything may look good on paper. However, what happens when the T1 circuit is not available when other components arrive on schedule? Equipment sits idle, and the project payback is either delayed or diluted.

Prepare for contingencies
These pitfalls can be avoided by developing a thorough project schedule, identifying critical items, and planning an alternative course of action in advance. For instance, contingency measures should be provided once it is realized that a delay in T1-circuit delivery could bring the project to a standstill. An additional circuit from another vendor could be ordered, along with the primary circuit, or a 56-kbit/s circuit could be used, resulting in a more limited amount of traffic. (Note that many T1 multiplexers can also operate at slower speeds.) By building in contingencies for potential pitfalls, planners attack implementation tasks rather than possibly reacting haphazardly to them.

A strictly realistic schedule should be developed. For instance, management is often overanxious to benefit from the new network operation, and it is easy to be pressured into developing a highly optimistic schedule. But with equipment delays, damages, and losses in shipment, every event cannot be predicted, nor can planners build each and every contingency into a plan.

Once an implementation plan has been developed, meetings involving all project participants should be scheduled regularly. However, to ensure that these meetings are most productive:
- The project manager should chair the meeting, even if outranked by another participant.
- Discussion should be documented in terms of major activities, progress and problems, the responsible persons, and delivery dates.
- Complex technical issues should be noted and handled separately in a smaller group; schedule another meeting only for those who can contribute to solutions.

Also, keep management advised of proceedings. The meeting notes, if organized and limited to essentials, can serve as a management report.

Perhaps the most difficult aspect of the entire project is cutover. As a guide during this phase, when applications are transferred from the old to the new architecture, a simple adage may help: "Don't burn bridges behind you." More specifically, do not dismantle the old network until all new components have been fully tested and are up and running for at least a month.

In addition, use noncritical information when first testing the new network. When critical information is first entered onto the new network, be sure users can revert quickly to the old one if problems arise. And finally, when an application has been brought onto the new configuration, monitor it for any changes in performance, such as variations in response time and poor voice quality.

A short time after the new network is running, convene a post-implementation review to document the lessons learned. Ideally, this newly gained knowledge can be carried into the next venture, and on to new communications technology. ■

Anthony Badalato, a senior telecommunications consultant with Coopers & Lybrand, specializes in data and voice communications networks. Badalato has been involved with the analysis, design, and implementation of several networking projects for financial and other institutions. He received a B.A. in economics and business administration from Lehman College in New York City and an M.B.A. from Iona College in New Rochelle, N.Y. Paul Drimmer, a manager with Coopers & Lybrand Management Consulting Services, is a 20-year veteran of the data communications industry. Drimmer has completed several domestic and international strategic planning and network-implementation assignments. He holds a B.S.E.E. from the Milwaukee School of Engineering in Wisconsin.

Additional reading
Goldberg, Hugh M., "The why, how, and what of digital data transmission," DATA COMMUNICATIONS, November 1984, p. 169.
Junke, Robert W., "A user implements a T1 network," DATA COMMUNICATIONS, June 1986, p. 155.
Mier, Edwin E., "Long overdue, T1 takes off—but where is it heading?" DATA COMMUNICATIONS, June 1985, p. 120.

Nicholas Papadopoulos, President, Thrysos Consulting Inc., San Francisco, Calif.

Combining data and voice network management

Sharing transmission but separating switching are key to optimum integrated operation of mixed voice and data networks.

Increasingly, communications managers are asked to plan and manage both data and voice networks. Some companies have added significant data communications requirements, requesting that their telecommunications managers manage data as well as voice networks. In other cases, voice networks have been handled by the local telephone company. And since divestiture, some companies have brought voice network management inhouse; in these cases, data managers are asked to take on the voice network also.

Integrated Services Digital Network is becoming a key concept (see "ISDN: Users think it's a distant prospect. Wrong," DATA COMMUNICATIONS, December 1985, p. 193). Both ISDN and the underlying similarity in voice and data transmission mean that the telecommunications manager must at least be aware of both data and voice network management. The manager needs this awareness to talk, at the very least, to colleagues who control the other part of the network.

Comparisons and contrasts
ISDN is based on the observation that both data and voice can be represented and transmitted by bit streams. Modems show the converse: Data and voice can be represented and transmitted by analog signals in the voice band. Of course, there are significant differences in the characteristics of data and voice calls. Bandwidth, accuracy, delay requirements, and error detection and correction are significantly different for voice and data transmission. Peer protocols between partners and network elements are well established in voice networks; they are a mass of confusion in data communications.

The combination of differences and similarities in voice and data transmission has interesting implications for network managers, who are confronted with creating a cost-effective, flexible, and reliable network that meets the organization's needs. Fulfillment of this mission translates to the need to share network components where feasible and to separate functions where necessary. Network elements are shared where the data and voice requirements are close; they are separated where the requirements conflict. Strategies for data network and voice network management can be examined so that appropriate philosophies and practices developed for each can be applied to the other.

Although the basic transmission medium for data and voice is the same, there are some startling differences in voice and data requirements, as can be seen in Table 1.
■ *Accuracy.* Data transmission requires high accuracy, preferably better than one part in 1 million; worse than 0.1 percent introduces significant overhead and possible errors in the ultimate data transfer.

Voice transmission is insensitive to error rates that would destroy the possibility of data transmission. One percent error is barely perceptible on a voice call; 10 percent error is acceptable.
■ *Bandwidth.* Bandwidth requirements can vary from one or two bits per second (fire and security alarms, for example) to several million (high-speed file transfer). Channel occupancy requirements vary widely and can be extremely sporadic. An alarm network has very low channel occupancy, using the channel only when an event occurs. Depending on protocol, file transfer can take up to 80 to 90 percent of the channel. A terminal session is bursty—long silences interspersed with rapid chunks of data transmission. The terminal user thinks and keys, and the computer responds.

Voice bandwidth and channel occupancy requirements are fairly constant—about 300 to 3,400 Hz and an average 40 to 45 percent occupancy in each direction.
■ *Delay.* Data transmission is insensitive to widely varying delays in the same conversation. Certain traffic, such as an

Data and voice characteristics

CHARACTERISTIC	DATA	VOICE
BANDWIDTH	WIDELY VARYING 1 BIT/S TO MBIT/S	CONSTANT 300- TO 400-Hz CHANNEL
CHANNEL OCCUPANCY	WIDELY VARYING 1 PERCENT TO 90 PERCENT BURSTY	FAIRLY CONSTANT
DELAY REQUIREMENTS	INSENSITIVE TO INCONSTANT DELAY	MUST BE CONSTANT AND LOW (0.5 SEC)
ERROR SENSITIVITY	PREFER LESS THAN ONE IN 1,000,000 BIT/S OVER 0.1 PERCENT; INTOLERABLE	INSENSITIVE TO 1 PERCENT; 10 PERCENT ACCEPTABLE
ERROR DETECTION AND CORRECTION	MUST BE PROGRAMMED	PARTICIPANTS MANAGE
PROTOCOLS	WIDELY VARYING, INCOMPATIBLE	STANDARD, ALMOST COMPLETE CONNECTIVITY

interactive terminal session, is sensitive to the amount of delay. However, this delay need not be held constant as long as it is less than a certain amount. Data can be buffered or flow-controlled to handle momentary congestion in the network or a device.

Although voice transmission is insensitive to error, it is extremely sensitive to delay: Delay must be small. Satellite feasibility tests by Bell Labs determined that delay, for example, cannot be greater than 0.25 to 0.50 seconds, after which point voice communications is nearly impossible. Constant delay is the key to intelligibility.

■ *Protocols.* Protocols for data conversations vary widely; even two devices that nominally speak the same protocol—for example, X.25—may not be able to communicate. Protocol technology has advanced substantially since its inception; there are a number of old protocols, such as Telex and 3270 bisychronous, that need to be supported and are too expensive to modernize. Furthermore, high-level protocols, such as transport and session, are not yet standardized.

On the other hand, voice networks demonstrate full connectivity. Almost any phone can be connected to almost any phone in the world and if the partners can understand each other, voice communications can take place.

■ *Error correction.* Protocols are necessary to provide error correction in data transmission. Often, the underlying transmission media do not provide the end-to-end reliability required for data communications, and sophisticated error detection and correction techniques are required. Compared to people who are partners on voice calls, computers are ill-equipped to detect and correct errors.

Because voice calls have two intelligent beings as partners, the transmission system does not have to provide for error recovery or for sophisticated protocols. Humans can ask to have phrases repeated, if necessary.

Although the goal for network management is the same for voice and data networks, the optimal solutions differ. The solutions vary depending on the characteristics of data and voice. Nevertheless, significant cooperative planning is possible, and significant sharing of facilities is desirable.

Voice networks are inherently circuit switches—a transmission path connecting the two partners is dedicated to the call. This characteristic follows because constant delay is critical to voice calls. In expensive transmission media, such as transoceanic cables, the average 40 percent occupancy in each direction on the circuit is exploited to carry more circuits by such techniques as time-assigned speech interpolation (TASI). Usually, however, a constant 300- to 3,400-Hz path is dedicated for the duration of the call. With the emergence of cheap fiber optics, there is no compelling economic reason to abandon this practice.

Generally, circuit switches make inefficient data switches. Data switching uses various buffering schemes to make optimal use of network elements. Insensitivity to nonconstant delay means that data in a call can be queued during periods of momentary activity peaks. Because of the synergy between data computation and communications, advances in computer technology can be exploited to create better data communications switches. Lower memory costs mean that buffering can be exploited throughout the network to reduce sensitivity to overload. Lower computation costs mean that more sophisticated communications protocols that increase reliability and throughput can be put in place.

X.25 could not have been implemented 15 years ago, for example, because the packet assemblers/disassemblers on network ends require memory and buffering. Within the nodes themselves, buffering is used to reduce sensitivity to overload. High-level data link control and Advanced Data Communications Control Procedure satellites use sophisticated transmission schemes, maintaining packets, windowing, and checksums to ensure integrity of link. These newer protocols need the computing horsepower for increased reliability and more sophisticated error detection. Cheap memory is used to buffer packets to allow frames to remain outstanding. Protocol conversion and local area network drivers are other examples of communications techniques that were dependent on the evolution of cheap, powerful computation.

Some pitfalls

Mixing data and voice on the same network can be catastrophic. Data calls can cause near-fatal congestion on voice networks optimized for voice traffic. Conversational terminal traffic, which is commonly carried over telephone networks and modems, can tie up capacity so voice traffic is not sustainable. Excessive dial-tone delay was noted at a Bell Labs location when significant terminal traffic was placed over the voice Private Branch Exchange using modems. Local telephone networks have been expanded beyond standard voice engineering guidelines because of increased modem traffic. Voice calls have short holding (conversation) times, usually about five minutes. Data calls can go on for hours. Expanding the voice network to carry the additional data traffic can be cost-prohibitive. A better solution is the creation of a separate data-switching network to carry the terminal traffic.

Carrying voice over data networks is a worse idea. The usual digitizing scheme for voice takes 56 kbit/s; a modern standard is 32 kbit/s (see "Third-generation codecs pave way for future digital networks," DATA COMMUNICATIONS, September 1984, p. 173). This bandwidth is huge by data network standards. There are schemes, such as linear-predictive coding (LPC), to reduce data bandwidth for voice, but they tend to be expensive and inflexible. LPC usually breaks down when more than one person is talking on one end of a connection; the method is highly sensitive to human speech. Because voice and data are different, one switch handling both kinds of communications ends up saddled with an unnecessary complication of function, increasing possibilities for failure through unaccounted-for interactions between its two different modalities.

Additionally, acrobatics are necessary to ensure that delay is kept constant to provide intelligible voice communications. These acrobatics play havoc with schemes, such as buffering and flow control, that are used to increase network capacity. For example, packetized voice can force significant and expensive enhancements to a packet-switched network used to carry the voice call (see "Leased-line tariffs, services restructured," DATA COMMUNICATIONS, February 1985, p. 45).

Although wholesale and indiscriminate mixing of voice and data networks is not desirable, a certain amount of facility sharing is both possible and desirable. Essentially, transmission facilities can be shared, whereas switching capabilities cannot be shared. Sharing of transmission facilities can occur at several levels, from local area wiring to long-distance trunk lines. Limited sharing of switching facilities can be effectively used for timely trials of communications services, for low use, for off-peak re-use of facilities, and for backup.

The first recommendation for integrating voice and data network management is to build and maintain separate voice and data switch networks while searching for the means to share transmission facilities. Consider the use of

1. Five levels. *The traditional North American voice network hierarchy demonstrates an application of the gravity model of calling. High-usage trunks connect various offices.*

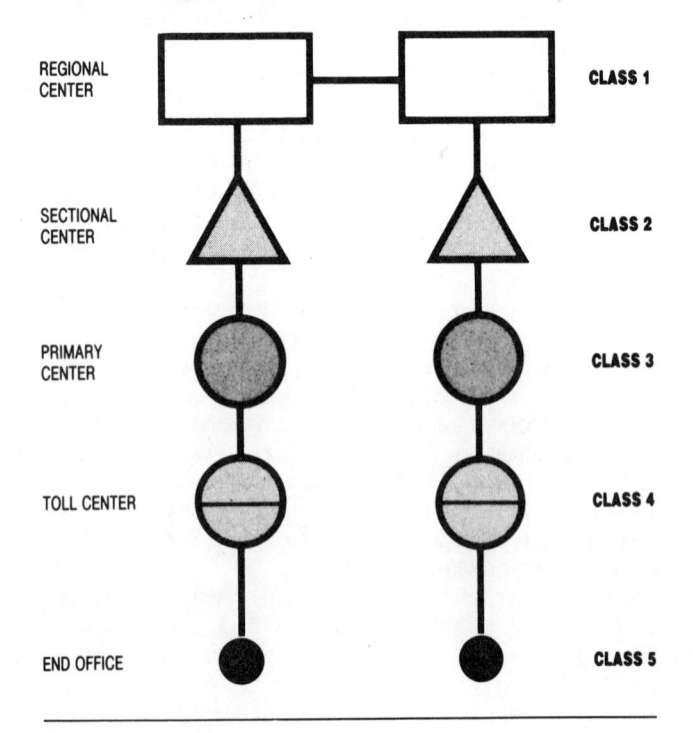

2. Typical voice. *This corporate voice network makes use of tandeming for interlocation switching. This design is most appropriate for large private-line networks.*

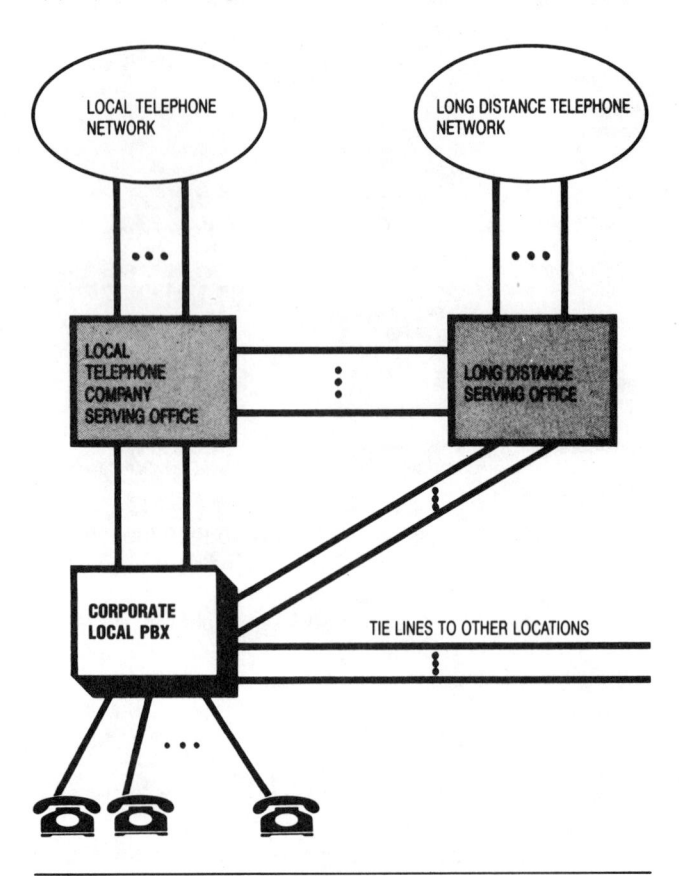

several data networks with interconnections between them (see "The middle ground between public and private networks," DATA COMMUNICATIONS, July 1985, p. 115). Often a particular network, say Systems Network Architecture, is optimized for specific applications. It is sometimes better to gateway separate networks than to attempt to put everybody's traffic on the same switching network. It may be better, for example, to maintain an X.25 network for Digital Equipment Corp. VAX machines with a gateway to host SNA, and not create one monster network to carry all the traffic an organization generates.

A major lesson from voice networks is the gravity model of calling: Calls are more frequent between parties that are geographically close. This observation leads to a hierarchical structure of voice switching and transmission networks—from local key equipment, PBX or Centrex, central offices, and an interoffice hierarchy in the traditional telephone network. Figure 1 shows the traditional five-level hierarchy in the North American Network; Figure 2 shows a typical corporate voice network.

In data networks, this hierarchical approach might lead to several LANs tied together with a data PBX that afforded access between the LANs and to connections outside, such as commercial data networks or circuits to other locations. The LANs are used for immediate, in-office communications. The data PBX provides for tandeming, that is, connecting LANs for local, interdepartmental communications, as well as to the outside world. The data PBX

3. Recommended data. *This corporate data network should include tie lines and local data PBXs. They can coexist with other networks.*

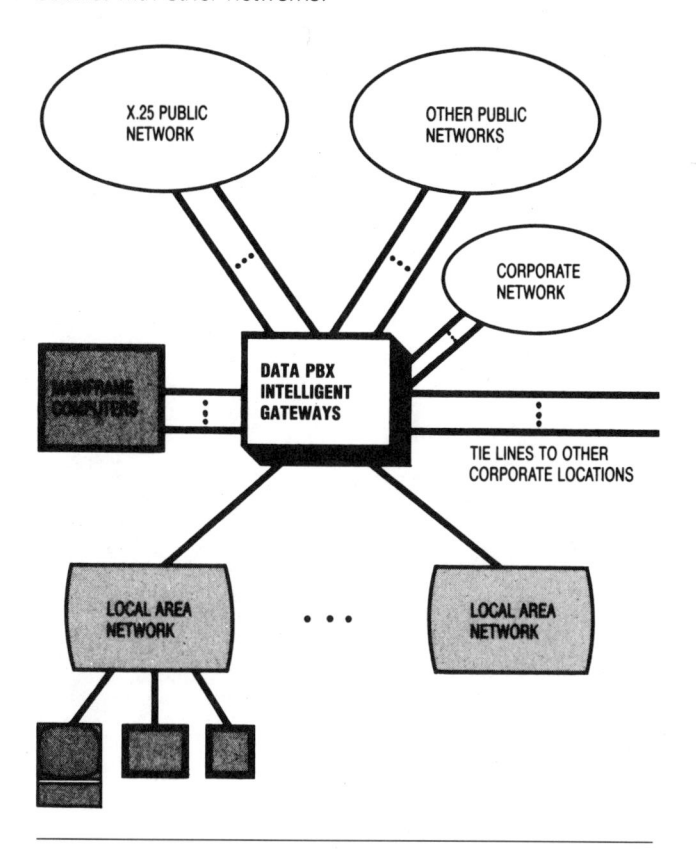

[Figure 3: Diagram showing a central "DATA PBX INTELLIGENT GATEWAYS" node connected to X.25 PUBLIC NETWORK, OTHER PUBLIC NETWORKS, CORPORATE NETWORK, MAINFRAME COMPUTERS, TIE LINES TO OTHER CORPORATE LOCATIONS, and two LOCAL AREA NETWORKs.]

can be expanded into a collection of intelligent gateways that allow internetworking between diverse LANs and larger networks (see "Gateways link long-haul and local networks," DATA COMMUNICATIONS, July 1984, p. 111). Figure 3 shows a proposed corporate data network structure that makes use of tandeming.

Another observation is that it is cheaper to buy bandwidth in bulk. For example, a T1 channel (1.544 Mbit/s) can provide a number of voice channels and data circuits at a substantially lower cost than buying the circuits individually (see "Long overdue, T1 takes off—but where is it heading?" DATA COMMUNICATIONS, June 1985, p. 120). Furthermore, with appropriate terminating equipment, it is possible to shift the allocation of voice and data circuits as needs change over time, be it over the course of a day, a month, or a year. A large-capacity connection could be used during the day to provide circuits for inter-office communications; during the night it could be reconfigured to provide bulk file transfer for data processing at a back-office host located off-site.

Sharing of transmission facilities extends to office wiring. Short-haul wiring can usually be done by twisted pair for both voice and data networks. Telephone-like wiring makes use of twisted pair for both voice and data networks. Similar cabling uses twisted pairs for connecting RS-232-C from terminals to computers or switches. The IBM Cabling System uses twisted pair for connecting stations to the Master Access Unit. An implication for voice networks is that digitization should occur closer to the telephone set:

Channel banks could be placed where the Master Access Units are located, and T1 used to carry to the PBX or the Central Office. If digitization is brought to the wiring closet, fewer wires are needed to go to the PBX. This tactic also creates an interchangeability of wiring between the data and voice PBXs, so cabling can be adjusted as loads require; in case of outages, spares are automatically built into the network.

Figure 4 shows a proposed integrated local corporate data/voice network. Both data terminals and voice telephones are connected by twisted pair to twisted-pair punch-down blocks in wiring closets. LAN controllers or cluster controllers are wired on the other side of the block for data networking. The local data controllers and the voice twisted pair are carried over campus wiring to their respective PBXs or switches. Note that a distributed voice PBX could have channel adapters distributed throughout the campus, possibly at the same location as the LAN controllers.

The voice and data PBXs provide campus interconnections and access to the off-campus services through data- and voice-specific connections as well as through cross-connection to broadband carrier networks. The broadband carrier network, possibly T1, could connect to other locations as well as to an integrated voice and data local office. Note that other patch panels and cross-connection devices are possible and may be desirable. For example, a cross-connection might allow termination of PBX circuits on either voice-specific or broadband equipment. The idea is to build as much modularity and interchangeability into the network as possible, which will provide flexibility for backup and load balancing.

Another data network concept is being implemented in voice networks. Traditionally, a telephone has been thought of as the termination of a twisted copper pair that extends from the PBX or the central office to the set. Increasingly, however, the telephone is viewed as a voice terminal, similar to a data terminal with abilities to signal the switch to perform such functions as hold or three-way calling, which were performed electromechanically by outboard key system units.

A paramount consideration in planning networks is backup planning for failure. Network problem management consists of these phases: detection, isolation, work-around, and restoration. A network fault must first be detected, implying a need for telemetry of network performance. Both voice and data network suppliers are guilty of not providing an integrated, standard means of reporting network outages. For example, there is no standard for terminals controlling electronic telephone switches. When the fault is detected, the failing network elements must be identified and removed from service. Also, a clearly stated backup plan must be in place for a work-around while the network element is out of service. In some cases, noncritical work can be deferred. In other cases, dial backup over public networks is appropriate. Finally, restoral of service must take place.

Effective backup planning will dictate the use of interchangeable network elements as well as patch-panel or cross-connect facilities to bypass failed elements. Major advances are being made in both the voice and data arenas to provide centrally controllable cross-connection

4. Integration. *This proposal for a local voice and data network uses cross-connection to form links to broadband networks. It is a design for corporations.*

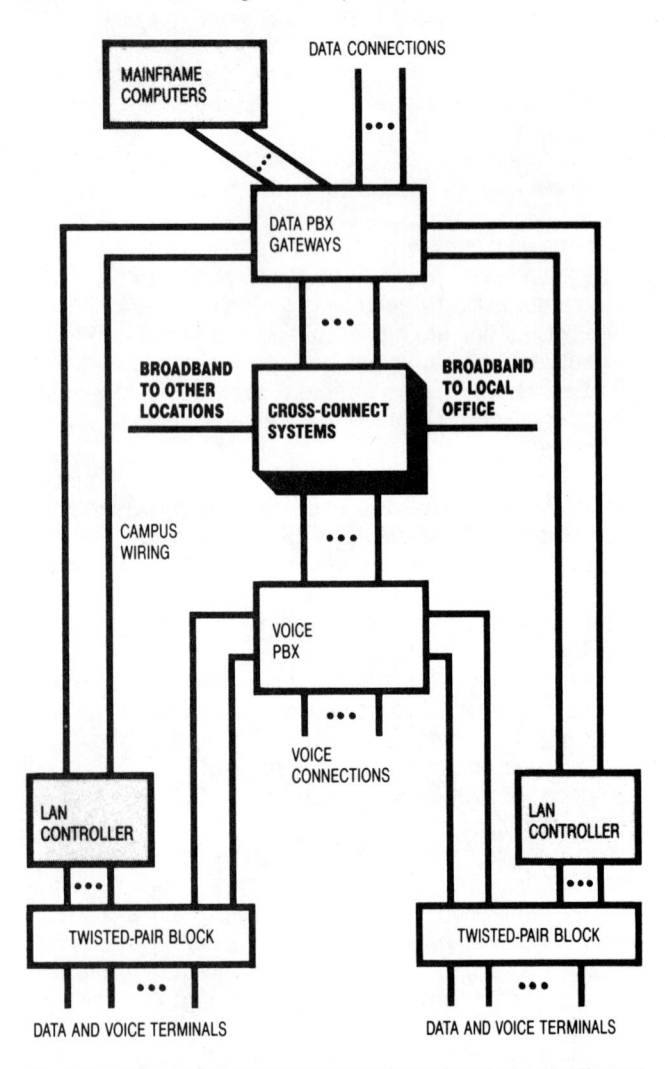

devices to meet this need. Digital Access and Cross-connect System, T1 cross-connect devices mixing data and voice centrally, and remotely controlled PADs are examples are existing cross-connect devices. The manager of an integrated voice and data network should be alert to the possibility of interchanging voice and data transmission facilities.

The key to successful integration of voice and data networks is knowing when to use the same facility for voice and data and when to use different ones. Use of the same components can lead to reduced cost through load sharing, and better and cheaper backup provision. Indiscriminate mixture of voice and data on the same facilities can result in substandard performance obtained at great expense. ∎

Nicholas Papadopoulos has a bachelor's degree in engineering sciences from the University of California at San Diego and master's and doctorate degrees in engineering and economics from Harvard University. Before starting Thyrsos Consulting, Papadopoulos worked at Bell Laboratories in Murray Hill, N. J., and at Data Architects.

John T. Becker and Curtis G. Gray, Network Synergies Inc., West Lafayette, Ind.

Network designers face shifting industry givens

Faster transmission is no longer always better. The traits of modems, messages, and terminal devices must be seen as a unit, and the logical solution may not be today's best.

Growth and expansion of data communications networks, coupled with changes in the common carrier industry, are giving network designers new concerns to ponder. Among other things, they must reevaluate several assumptions on which design efforts and design tools have been based. Under scrutiny are the following industry givens:

- The least-distance network will be the least-cost network.
- Network backbone links and certain high-volume links (3725 transmission groupings, computer-to-computer links, X.25 node connections, circuits to large terminal cluster controllers) can be defined and designed in the same manner, using the same formulas, under which single-station multidrop lines are defined.
- Faster modems will provide both better response times for users and less costly networks.
- Network design, as a job description, deals in a large, static environment.
- Networks can be economically designed—not by using data-modeling techniques at design time, but by engaging in data and performance measurement and evaluation at a later stage.

Moving into the new world of network design let's see where the future concerns will be and why we must challenge the old assumptions.

Least distance least money?

Before the divestiture of AT&T, when long-distance rates subsidized local telephone costs, the mileage-dependent portion of a leased-line environment was a substantial factor in the overall cost. Rates at that time could be $6 or more per mile for interstate lines. Today the same rates have dropped to $2 per mile and lower. How does this affect the designer?

The following table describes the costs associated with a small network under the old rate structure and under the new rate structure.

DESCRIPTION	OLD RATES	NEW RATES
COST PER MILE	$6.00	$2.00
TERMINATION CHARGES (LOCAL LOOPS)	36.00	140.00
MODEMS (9.6 KBIT/S)	150.00	150.00

Using these costs we can design hypothetical examples of a least-distance network and a least-cost network. Figure 1 shows the former.

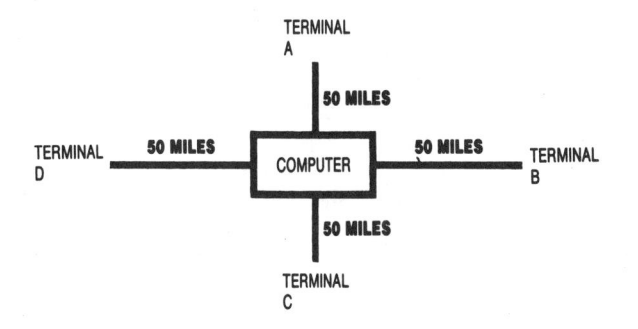

```
                TERMINAL
                   A
                   |
                50 MILES
                   |
TERMINAL  50 MILES  ┌────────┐  50 MILES  TERMINAL
   D ──────────────│COMPUTER│──────────────   B
                   └────────┘
                   |
                50 MILES
                   |
                TERMINAL
                   C
```

In this simplified example it would appear that four lines from each terminal to the computer would be the least-distance arrangement.

Under the old rate structure the cost would be as follows:

LINES, 200 MILES, @ $2/MILE	= $1,200.00
8 TERMINATION CHARGES @ $36.05 EACH	= 288.40
8 MODEMS @ $150.00 EACH	= 1,200.00
TOTAL COSTS:	**$2,688.00**

219

Under the new rate structure the cost increases slightly, as follows:

LINES, 200 MILES, @ $2/MILE	=	$400.00
8 TERMINATION CHARGES AND LOCAL LOOPS @ $140.00 EACH	=	1,120.00
8 MODEMS @ 150.00 EACH	=	1,200.00
TOTAL COSTS:		**$2,720.00**

Next assume that a least-cost network is the goal. Figure 2 shows a network with a total of 263 miles compared with the least-distance network of 200 miles.

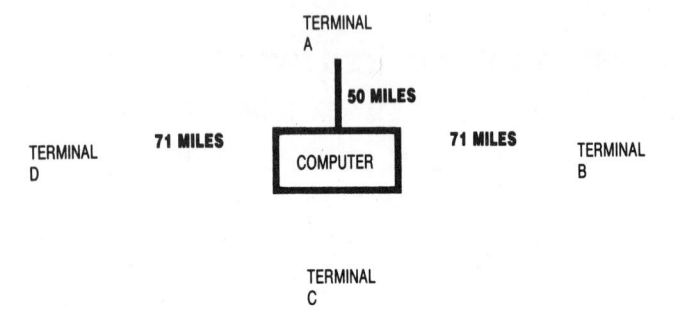

```
                    TERMINAL
                       A
                       |
                    50 MILES
                       |
TERMINAL  71 MILES  +--------+  71 MILES  TERMINAL
   D   ------------| COMPUTER|------------    B
                   +--------+
                       |
                    TERMINAL
                       C
```

Under the old tariff structure, the cost would be:

263 MILES @ $6/MILE	=	$1,578.00
5 TERMINATION CHARGES @ $36.05 EACH	=	180.25
5 MODEMS @ $150.00 EACH	=	750.00
TOTAL COSTS:		**$2,505.25**

The least-cost network, even under the old structure of tariffs, would be 7 percent less expensive than the least-distance network. However, under the newer tariff structure, costs are as follows:

263 MILES @ $2/MILE	=	$526.00
5 TERMINATION (LOCAL LOOPS) @ $140.00 EACH	=	700.00
5 MODEMS @ $150.00 EACH	=	750.00
TOTAL COSTS:		**$1,976.00**

The 263-mile network costs 27 percent less than the 200-mile configuration. Although we had to assume that traffic volumes at the terminal locations could be handled in a multidrop environment, the example nonetheless illustrates an important point. Designing to least distance rather than least cost is expensive, and switching the cost emphasis away from distance and toward fixed-site costs is changing the way networks should be designed. Clearly, focusing on cost instead of distance can make significant differences.

One company found that by approaching the design of its 103-terminal network from the perspective of least distance, it needed 16 circuits; but by looking at least cost, it needed only nine circuits. The savings amounted to $3,000 per month.

While the least-distance network had $183 less in mileage-dependent line charges, the least-cost network had seven fewer modems and 14 fewer communications parts. In addition there were reductions in related support needs that resulted in significant overall cost savings.

The time-honored rule that at over 60 percent line or circuit utilization, response times to users will degrade to an unacceptable level is no longer true in all cases. While this maxim was somewhat reliable in the days before backbone networks, remote transmission controllers, and cluster terminal controllers (that will support at least 20 terminal devices) were common, it is even less reliable now. Networks that are designed according to this old rule may require spending more money than is necessary on circuitry, as opposed to spending it on design.

Designing according to the 60 percent rule assumes two mathematical characteristics of a data communications circuit:

1. Traffic arrival volumes in the critical time frame (busy hour) will be distributed in a random, or Poisson, manner.

2. As a result of that distribution, queue depth and traffic load will develop on the network, which will cause transaction service rates to become exponentially distributed in nature.

Assuming a Poisson distribution of communications traffic arrival rate, however, is assuming a worst-case condition (see "Network-design formulas"). Similiarly, the assumption that the service rate of the communications line is exponentially distributed is assuming yet another worst-case scenario. Coupled, these assumptions can lead to the overdesign of certain types of communication links.

To see how the 60 percent rule would yield an over-

Network-design formulas

One approach to network design makes use of Poisson formulas, another use exponential formulas.

Poisson mathematical formulas represents a method of defining the random nature of events on a communications line.

As an example, if traffic at a location is said to average 60 transcactions per hour, a network designer would not consider the network as having one transaction per second. Instead he must recognize that in some minutes there will be no traffic and at other times there may be multiple transactions in a given minute. Poisson formulas make it possible for this variation to be addressed when designing data networks.

Longer periods

Exponential formulas also allow network designers to address the fact that some transactions will take longer to service than others. An example would be that if it takes half a second to handle a transaction on a circuit—when no other request to use that circuit is made—it will take longer periods of time to process some transcactions if multiple requests for the circuit are made at the same time. At one-half seconds per transaction, when three requests are made at exactly the same moment, the first request will take one-half second to process, the second will require one full second, and the last transaction will require one and one-half seconds.

Exponential formulas allow the network designer to recognize the differences in transaction service times that occur as a result of Poisson (random) nature of certain circuit types.

designed network, take a simplistic example of a remote transmission controller located in Boston, supported by some multiple of 9.6-kbit/s lines from a host in Chicago. The traffic assumptions will be as follows:

Traffic volume to Boston 11,000 messages per hour.

Traffic volume from Boston 11,000 messages per hour.

Average message length to Boston 400 characters.

Average message length from Boston 150 characters.

Average response time required less than 2.0 seconds.

The 95th percentile response time required less than 5.0 seconds.

(A 95th percentile response time says that 95 percent of transactions will be completed within the time frame specified.)

In configuring this network the designer must find out how many 9.6-kbit/s land circuits will be needed both to handle the volumes and the message sizes involved and to meet the performance goals.

Performing at 19.2 kbit/s (two 9.6-kbit/s rims), with the Poisson and exponential assumptions and with a full-duplex line protocol, yields the following results:

CHICAGO-TO-BOSTON LINE UTILIZATION	= 52.05 PERCENT
BOSTON-TO-CHICAGO LINE UTILIZATION	= 87.45 PERCENT
AVERAGE RESPONSE TIME	= 2.64 SECONDS
95TH PERCENTILE RESPONSE TIME	= 6.99 SECONDS

This does not meet the performance goals of 2.0 and 5.0 seconds. Adding another 9.6-kbit/s circuit to bring the total line capacity to 28.8 kbit/s results in the following:

CHICAGO-TO-BOSTON CIRCUIT UTILIZATION	= 45.01 PERCENT
BOSTON-TO-CHICAGO CIRCUIT UTILIZATION	= 68.59 PERCENT
AVERAGE RESPONSE TIME	= .98 SECONDS
95TH PERCENTILE RESPONSE TIME	= 2.60 SECONDS

This scenario meets the service goals. But the design is far from economical. In the real world, the Poisson distribution changes, which can be seen by applying a statistical leveling of the traffic resulting from the large number of individual events occurring at the remote transmission controller.

When a designer assumes that data traffic arrives at communications lines following a Poisson distribution, he can also assume that the calculated response times will represent a worst case under the given traffic load, characteristics, and service rate. Worst-case assumptions can lead to overdesign and more costly networks, especially when the assumption is cascaded. In other words, it is incorrect to assume that the arrival and service rates will still follow a Poisson distribution as data traffic passes from terminals to local controllers to remote transmission controllers. The arrival rate of traffic is usually shaped at each point to a progressively more deterministic distribution: that is, to be less random in nature. Also, the service rate, or time to transmit a transaction or transmission frame, becomes more centered about a given value rather than exponentially distributed about an average.

Thus the arrival of transactions as they pass from terminal to controller to remote controller begin to follow a more deterministic or regular rate. In a like manner, the rate at which each transmission frame or transaction is serviced also begins to follow more of a set rate rather than a rate centered about a value that can vary. This makes possible the use of queuing formulas and response-time formulas that yield more realistic values than does the assumption of a worst case at each point in the network.

Queuing formulas and response-time formulas that use the assumption of Poisson arrival rates and exponential distributed service rates generally yield larger queue and service time delays than do formulas assuming deterministic arrival rates and service times. The key is to know when and how to apply the correct formulas. Current data communications networks require the use of both types of formulas: Poisson distributions and deterministic distributions of both arrival rates and service times.

Generally, Poisson distribution is applicable when dealing with multipoint terminals or terminals directly connected to a communications line or device. The reason is simple. Terminals, by their nature, generally exhibit random transaction generation. To be safe in the design of the terminal portion of the network, a Poisson arrival rate should be assumed. However, when designing backbone links or links between X.25 packet nodes, for instance, the distribution of traffic and certainly the link service time becomes more deterministic and less distributed about an average value. This means that designs for these links that assume Poisson distributions may actually lead to overdesign by trying to get bandwidth to meet certain response time criteria. Compounding the problem is that these backbone links are generally of a higher bandwidth and, hence, more expensive; so overdesigns tend to invoke a much larger cost penalty.

When this effect is recognized, and the design modeling is performed, the result is that at a total line capacity of 19.2 kbit/s the performance goals are met. The old 60 percent guidelines are invalidated, and a significant amount of money is saved.

CHICAGO-TO-BOSTON CIRCUIT UTILIZATION	= 52.08 PERCENT
BOSTON-TO-CHICAGO CIRCUIT UTILIZATION	= 87.45 PERCENT
AVERAGE RESPONSE TIME	= 1.55 SECONDS
95TH PERCENTILE RESPONSE TIME	= 4.10 SECONDS

In leased line charges alone we have saved $10,520 per year, and we have eliminated the need for: two modems, two local loop charges, and four input/output controllers. The total cost savings will easily exceed $20,000 annually. Later measurement can verify that the two circuits at 9.6 kbit/s meet the response times that were predicted.

The use of communications lines by large numbers of users has resulted in a change in the way the designer must plan networks if cost is to be considered. The designer must recognize when the various mathematical design assumptions may be used and must have the ability to use the mathematical and statistical formulas available.

While it is often appropriate to use the Poisson and Exponential method for design, and such assumptions are certainly the safest to make, the cost effect of this approach must be recognized so that business opportunities are not lost due to the projected cost of the communications environment needed to support them.

Everyone has seen network situations where increased transaction volumes resulted in longer response times and poor customer service. Installing higher-speed modems is often thought to be the solution.

Faster is better?
In one case the following occurred on a set of 2.4-kbit/s data transmission lines (the network had synchronous protocol, two half-duplex circuits, 20 nodes, and exponential service time):

CIRCUIT	NODES	MODEM SPEED	AVERAGE TRANSACTION SIZE		AVERAGE TRANSACTION ARRIVAL RATE/SECOND	
			SEND	RECEIVE	SEND	RECEIVE
1	10	2.4 KBIT/S	30.	30.	.69	.69
2	10	2.4 KBIT/S	30.	30.	.69	.69

CIRCUIT	NODES	RESPONSE TIME		CIRCUIT UTILIZATION
		AVERAGE	95TH	
1	10	2.25	5.96	60.96%
2	10	2.25	5.96	60.96%

The two circuits have small message sizes, which is characteristic of many configurations that are used for financial and point-of-sale applications. Reasonable response time is achieved for the average and the 95th percentile objectives, but still better user service was desired.

According to the initial plan the approach was to get faster modems, an expensive but apparently logical course of action from an intuitive point of view. The same network was measured using both 4.8- and 9.6-kbit/s modems where the transmitted message sizes and volumes remained at a constant level.

At a transmission speed of 4.8 kbit/s, the following results were seen (the network had synchronous protocol, two half-duplex circuits, 20 nodes, and exponential service time):

CIRCUIT	NODES	MODEM SPEED	AVERAGE TRANSACTION SIZE		AVERAGE TRANSACTION ARRIVAL RATE/SECOND	
			SEND	RECEIVE	SEND	RECEIVE
1	10	4.8 KBIT/S	30.	30.	.69	.69
2	10	4.8 KBIT/S	30.	30.	.69	.69

CIRCUIT	NODES	RESPONSE TIME		CIRCUIT UTILIZATION
		AVERAGE	95TH	
1	10	2.76	7.32	65.75%
2	10	2.76	7.32	65.75%

Using 9.6 kbit/s yielded the following (the network had synchronous protocol, two half-duplex circuits, 20 nodes, and exponential service time):

CIRCUIT	NODES	MODEM SPEED	AVERAGE TRANSACTION SIZE		AVERAGE TRANSACTION ARRIVAL RATE/SECOND	
			SEND	RECEIVE	SEND	RECEIVE
1	10	9.6 KBIT/S	30.	30.	.69	.69
2	10	9.6 KBIT/S	30.	30.	.69	.69

CIRCUIT	NODES	RESPONSE TIME		CIRCUIT UTILIZATION
		AVERAGE	95TH	
1	10	3.08	8.16	68.14%
2	10	3.08	8.16	68.14%

Although in the preceding case the messages are able to go over the lines at a faster speed, the higher-speed modems, with their longer training (equalization) times, took longer to get to the user than did the slower modems.

Faster is not always better. The characteristics of the messages, the modems, and the terminal devices must all be viewed as an integrated whole if the proper network is to be designed and the proper equipment installed. The modern network designer must see through what appears to be the logical solution and understand how equipment and applications will operate in the real world.

Design is a static function?
Once, long ago, the network designer's job could be viewed as a static function. The first large networks—for airlines, banks, and financial lending institutions—had certain characteristics that allowed for a one-time design effort. New airports were not opened every day, and the initial networks were designed to support only a limited number of applications.

With the increased use of information in the business world, networks have become dynamic entities. Today the network designer must not only address the initial design of the network but must also be concerned with the following questions:
- What if a new application is added to the network?
- Will an acceptable level of user or customer service be maintained?
- What if I change the equipment in the network?
- Will service be improved or degraded?
- Will my network be able to absorb the volume increases that the business is experiencing?
- Should the new locations be added to existing circuits or given their own circuits?
- What effect will these new locations have on service levels if they are added to existing circuits?
- What will be the least expensive circuit that a new location can be added to without degrading service to an unacceptable level?

Today's network designer must recognize the fluid nature of the data communications environment and be prepared to answer the questions asked. The simple answer to achieving a one-time economical network design will no longer meet the needs of business.

As organizations begin to rely more on the ability to move information between locations, the locations involved, the volume of the information, and the dependence that people have on that information will continue to grow.

Perishable commodities—those that lose value rapidly, like empty airline seats after take-off or futures sell orders issued after the market has reached the down limit—illustrate the need to maintain timely information flow.

Just as the financial planner must have the ability to view different financial services and possibilities, the network designer must be prepared to live with, and operate within, the dynamic environment of modern networks.

Design techniques

In the new world of network design assuming that all circuits may be treated the same is no longer fruitful. Traditional methods of designing networks, using line holding time (the amount of a line capacity that will be used by the data and control characters often referred to as line utilization percentage), must by definition assume that all circuits will function in the same manner. Operating under this given, the data communications designer becomes responsible for deciding when the design tools or programs have achieved their optimal level of design.

Today one approach to network design is by using design tools that perform data modeling directly at design time. By modeling how the data will perform as circuits are being constructed, the designer can achieve the needed performance goals (response time needs) in a single modeling run, and by modeling at design time the appropriate service time models can be used to ensure a least-cost environment.

An example of the advantages of data modeling at design time follows:

Service model A will assume the traditional Poisson-exponential service time assumptions, and service model B will recognize the statistical leveling that occurs on many network backbone links.

Looking at a typical communications environment we can see that intelligent remote transmission control units in a possible network diagram may appear as shown in Figure 3.

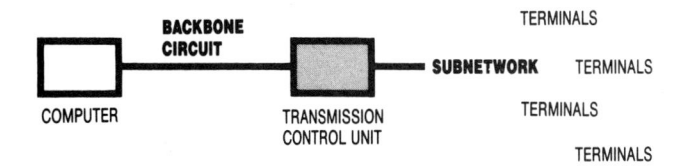

By performing initial data modeling we can design the above network, on a multilevel basis, using data model A for the subnetwork circuits, data model B for the backbone data circuits, and incorporating into our network any delay time in our remote transmission controller. ■

John Becker is a principal and founding officer of Network Strategies Inc. He also serves as a panel member of the American Arbitration Association. Prior to founding NSI, he was a founding officer and vice president of Mid-America Computer Corp.

Curtis Gray is a principal in Network Strategies and the primary author of the STAR (Simulated Telecommunications Analysis Reporting) network design and financial planning software. He has an M. A. in electrical engineering and computer science from the University of Wisconsin.

William E. Bracker, Technology Research Associates, Tucson, Ariz.,
and Benn R. Konsynski III and Timothy W. Smith, University of Arizona, Tucson

Metropolitan area networking: Past, present, and future

MANs have sprung up from CATV technology, established for some 35 years and considered a fairly stable operating environment.

Community-antenna television, or CATV, networks were initially constructed to bring entertainment to rural areas that could not otherwise receive television signals. However, it was not long before someone came up with the idea of putting an antenna on a hilltop or a building and linking it via a cable to an entire city, thus allowing everyone to receive the same-quality signal. When the revenue potentials of such broadcasting became apparent, CATV networks were built in most major metropolitan areas.

The first cable systems usually could carry five channels, but new methods and hardware soon increased this capability to over 70 channels on a single cable. Then dual-cable systems (54 channels per cable, 108 channels total) were installed in some locations.

By the early 1970s, it was apparent that using CATV networks for data communications as well as video had a huge revenue potential. The key to these revenues was that heavy users of the local telephone companies, most of them owned at the time by AT&T and known as the Bell operating companies (BOCs), could bypass the local loop with CATV to send data between their divisions or to access long-haul networks and save substantial communications costs. Remember, however, that CATV systems were originally intended for one-way transmission of video signals for entertainment purposes.

Cable threatens to absorb some of the voice communications business as well, and where cable operators can strike a deal with long-distance carriers, the local loop and AT&T will be bypassed altogether. Needless to say, the telephone operating companies and others who stand to lose markets or potential markets have raised both political and technical objections to these CATV intrusions, leading to many controversial issues.

In 1970, a popular article titled "The Wired Nation" described cable systems and their potential in terms that the public could appreciate. Many feel that this article (and a subsequent book) painted such a glowing picture of the money that could be made with two-way CATV services that it initiated several private and public research projects to investigate the area.

CATV is a controversial topic by its very nature, as some have noted, since CATV networks could cross over or erase well-established lines. These lines include those laid out by local telephone companies, broadcasters, press, publishers, educational institutions, software producers, telecommunications networks, corporate networks, and local area networks (LANs).

The controversies immediately took on political and technical implications, prompting several agencies to step in with suggestions, standards, and directives. In 1972 the Federal Communications Commission published "CATV Report and Order" and "Operational Rules and Standards for CATV." The former required that local government and community service channels and two-way capabilities be provided on new systems; the latter was concerned with the details of technical standards required for a wideband two-way CATV system.

On the political end, the BOCs took the fight to Congress and the state and local regulatory bodies. Their argument paralleled the antitrust arguments of AT&T to the federal government—that heavy-duty trunk service and long-distance links subsidized local universal telephone service and that without this business local rates would surely rise.

Legislation was introduced to keep CATV out of the data and voice transmission business and the BOCs out of entertainment. The cable companies encountered a slump in the mid-1970s, but nonetheless went ahead with the installation of two-way cable capability. Congress split between the House and Senate over regulation, with the House supporting the BOCs. Today the cable operators are permitted to contract for franchised data transmission services in metropolitan areas. But uncertainty over rate regulation and bandwidth-leasing arrangements still cloud the issue.

Locally, the cable companies have successfully challenged the jurisdiction of local regulatory commissions, but there have been mixed court rulings over all aspects of the controversy. Many of these are yet to be resolved, and the future is still uncertain.

From the standpoint of technical standards, there is similar controversy. Arguments rage over access methods, contention, ingress incursion, upstream cascading, and other subjects. The one major attempt to standardize beyond the FCC's efforts has been undertaken by the Institute of Electrical and Electronic Engineers. The Metropolitan Area Network Working Group IEEE 802.6 was formed in April 1981 to investigate an expanded scope of the local network standard. Its purpose was to specify the major functional capabilities needed for MANs and to highlight how these capabilities differ from those already spelled out in the LAN standards. The goal was to address the major implications and thus set up a migration path to accommodate existing and emerging technologies. Since MANs had not proliferated widely (at least commercially), the time seemed ideal to set up standards.

Under the IEEE's outlook, the LAN standards are limited to 2 kilometers, whereas the MAN standards are intended to span a distance of greater than 5 km. Unlike LANs, which are designed mainly for data transmission only, the MAN specifications support data, voice, and video transmission.

The committee recognized that a network spanning a metropolitan area may extend across multiple transmission media—for example, satellite, microwave, and optical fiber. Because of the distance and the multiple media, the previous access methods for LANs—CSMA/CD (carrier sense multiple access with collision detection), token bus, and token ring—were viewed as having serious deficiencies. Time-critical applications such as voice and video cannot tolerate long access delays.

Entire bandwidth available
As a result, a new access protocol will be developed for the stations with-in the MAN to gain access to the shared medium. The working group is moving toward a time-division multiple-access (TDMA) protocol using reservations. With TDMA, the time continuum is divided into continuous, discrete time segments called frames. Each frame is subdivided into N slots, and each station is allowed to transmit at channel speed for the duration of the slot it is allocated. In other words, the station has the entire channel bandwidth available to it during transmission. The station can send data, voice, or video information that is digitally encoded.

Since broadband coaxial cable is the most common medium currently available, it will likely be the first medium. Here are some of the key points that coax-based MANs present, as noted by Custeret Cheung:
1. Coexistence of conventional commercial CATV channels and MAN channels—the cable operator can have MAN service residing on one or more of the channels and still have the remaining channels to offer CATV services.

Table 1: Radio service

SERVICE	BANDWIDTH, 5 TO 30 MHz
AMATEUR RADIO	7.00-7.30
	14.00-14.25
	21.00-21.45
	28.00-29.70
CITIZEN'S BAND	26.96-27.41
HIGH-POWER INTERNATIONAL BROADCASTING	5.95-6.20
	9.50-9.78
	11.70-11.98
	15.10-15.45
	17.70-17.90
	21.45-21.75
	25.60-26.10

NOTE: MARINE AND AERONAUTICAL, FIXED AND MOBILE SERVICES OCCUPY ALMOST ALL THE REST OF THE SPECTRUM BETWEEN 5 AND 30 MHz.

2. Types of service—the MAN offers a high-bandwidth channel in the megabit range, to service data, voice, video, and trunking.
3. Cost-effectiveness—the acceptance of this standard will be highly dependent on the cost-effectiveness of the system.

The important point to note is that the cable operators themselves are not imposing or prescribing any standards. The user is choosing his own protocols for access contention and error checking, and the methods most commonly employed have been CSMA/CD and polling techniques in a master-slave implementation.

Technical and other considerations
The cable companies usually use 75-ohm coaxial cable with splits, depending upon how many cables there are in the system. Interactive telecommunication requires bidirectional (two-way) transmissions. If there is only one cable, the approach is to split the cable into downstream away from the head end) and upstream (toward the head end) paths.

There are two common approaches to dividing up CATV cable capacity: sub- and mid-split. By far the most used by commercial cable operators is the sub-split system, which covers the band of 50 to 450 MHz in the forward direction and 5 to 30 MHz in the reverse direction. The mid-split system covers 250 to 450 MHz in the forward direction and 5 to 200 MHz in the reverse direction. A third approach, the high-split system, has not been used, since the guardband between the two splits covers a bandwidth over which the major networks transmit. Two or more cables obviate the need for splitting for data can be placed on a separate cable from commercial video.

The common method has been to use a combination of frequency- and time-division multiplexing with pulse-code modulation. For the time-division multiplexing (TDM), the standard almost universally adhered to has been the T1, an old Bell standard for mixed voice and data.

With time-division multiplexing, each transmission signal is spread over a given frequency by interleaving portions of that signal over time. The sequence of time slots dedicated to a certain source is a channel, and one cycle of time slots is a frame.

In T1, data is taken from each source in samples (7 bits plus an eighth for supervisory functions), and 24 sources are multiplexed, for a total of 192 bits per frame plus 1 for control. Sources are sampled 8,000 times per second, so that the required data rate for T1 is 8000 times 193, or 1.544 Mbit/s.

The sources that constitute the various channels are themselves frequency-division-multiplexed to make maximum use of the bandwidth.

Problems encountered with CATV data transmissions include ingress and egress incursion and reverse cascade. Ingress incursion is noise penetration into the cable, an especially prevalent occurrence in the lower bands, which is a very crowded portion of the frequency spectrum (Table 1). Naturally, the sub-split approach, although easier to accommodate in more primitive cable operations, is particularly susceptible to this phenomenon.

Egress incursion is caused by data transmissions on the MAN at high frequencies. In these high bands, the cable emits its own interference to other transmissions, including airline broadcasts. Regulation concerning this phenomenon is already in the works.

Unlike ingress and egress incursion, which deal with noise entering or exiting the cable, reverse cascade is damage to the transmission from within the MAN. It was discovered when downstream-only CATV was changed to bi-directional by order of the FCC that amplified noise collects at the head end. The longer the cable and the higher the number of terminal devices, the more noise on the cable. Indeed, this is the major technical problem that cable operations must solve.

The incursion problems can be addressed with good equipment, clean lines and error protocols. Also, if upstream transmissions were reallocated to frequencies above 400 MHz, they would be far less vulnerable to ingress.

As for reverse cascade, some believe that it is a phenomenon whose effect is transparent (or at least easier to live with) for MANs of under 200 miles. Hence, by modularizing the MAN into sections of less than 200 miles one can control (or at least reduce) the effect. At the hubs, the signal can be cleaned, so that there is no unity gain on the head end.

Radio-signal leakage

Other problems include those that normal maintenance should avert. For example, if any one of the hundreds of thousands of coaxial connectors in a typical urban system loosens, or cracks, or becomes corroded, the integrity of the coaxial shield is breached, allowing radio signals to leak into the system. Once the system allows any leakage, it is subject to impulsive noise from automobile ignition; corona around electric power substations, industrial electrical machinery or heating; or arcing at loose electrical connections.

Thomas F. Baldwin (see further reading) presented a four-generational model for the development of two-way cable systems, the first two of which were operational. The goal of his model was to provide successive, cost-feasible steps toward achievement of a full-service broadband communications system where independent stages of technical development were matched with service applications.

Briefly, Baldwin's four-generational model for two-way communications is:

1. The first generation, per-program pay television, has been operational for subscribers since 1973 in Columbus, Ohio, by Telecinema.

2. The second generation adds interactive response capability to the terminals. This capability permits responses to questions presented through the signal received by means of a simple pushbutton-type converter for digital responses. This stage has been in operation since 1977 in Rockford, Ill., by Broadband Technologies Inc.

3. The third generation adds a microcomputer chip and read-only memory.

4. The fourth generation adds low-cost memory storage in the terminal.

The principal difference between the first and second generations is not in hardware, but in software. It is important to note that the computer system must process the response data in real time, returning us to the old contention problem, which is still unclear.

Cable operators seem confident that no technical problem is insurmountable. However, until a protocol from the IEEE or some other source is adopted and worked out in an application, doubts will remain about the CATV MAN's ability to run in real time.

The local phone companies

The local telephone companies have two possibilities for a transmission medium. If they build the plant, it will probably be fiber-optic cable. If they successfully push the cable operators out of the business, it is likely that they will offer to take over and run second cables where installed, so that they will be coax operators.

The phone companies have a switched system in place. This is a big advantage, and they have emphasized it publicly and in the courtroom. The fiber-optic systems will no doubt take advantage of this in-place switched system, while coax implementations will probably follow the cable industry's lead toward TDM along the T1 standard.

The local phone companies, however, have a huge network of twisted-pair lines that are every day becoming more antiquated in a ditial world. Many local loops are burdened with loading coils that restrict the bandwidth, or capacity, of the loop to 4 KHz for long-distance transmission.

Either case—optical fiber or coax—will require money, and the question is, Where will it come from? The phone companies have asserted that they can raise the capital from normal operations, which raises two questions: What advantages the metropolitan governments expect from the local phone companies like those cable operators have bestowed (or at least promised), such as studios, free air time, and no-charge city loops? Does this claimed ability imply that the phone service is subsidizing the data transmission business, which the local phone companies have said all along is just the opposite?

Another approach that is quite common is to use both CATV and telephone lines. This hybrid approach is frequently used with Videotex applications. It is useful when relatively infrequent responses are desired from the user end, but it does not offer "true" two-way data communication capabilities.

Table 2: Social issues

ISSUE	SOLUTION
PUBLIC HAS TOO-HIGH EXPECTATIONS	EDUCATION—OPERATORS NEED TO TONE DOWN EXTRAVAGANT CLAIMS FOR CABLE/MAN SYSTEMS
FEDERAL COMMUNICATIONS COMMISSION POLICIES	STANDARDS, DEREGULATION—GOVERNMENT MUST RESPOND FASTER TO SOCIAL REQUIREMENTS AND TECHNOLOGY CHANGES
THEORY VS. PRACTICE	MAN/CABLE OPERATORS MUST IMPLEMENT A REASONABLE SUBSET OF CAPABILITIES; TOO MUCH TOO SOON MOST LIKELY WILL FAIL
BOUNDARIES AND FRANCHISE	REGULATION, STANDARDS (ISSUES: WHO REGULATES? WHO CONTROLS? MONOPOLY LICENSE FOR SERVICE?)
INTERCONNECTION	REGULATION AND STANDARDS

With this approach, audio/video transmissions are generated at the head end and transmitted over the CATV cable to the user. The user also establishes a telephone connection with the source to transmit responses back to the head end.

Political considerations

Overshadowing and embroiling the technical considerations of using CATV in MANs is the heated political context. The functional word in the entire discussion is "bypass." With CATV, some of local telephone companies' most lucrative customers could abandon the local loop. That would not be so expensive at present, but it could prove to be in the future, so the local phone companies are acting to stop this practice.

There is also a threat to the local phone companies' voice business with CATV, but it does not appear too strong, nor does CATV seem inclined to pursue it except where some major customer should want a mixed voice-data medium.

Rather, the telephone companies have centered their arguments on the threat to local rates should CATV steal some of their heavy users in the dedicated data transmission business. They claim that the heavy users subsidize local service, and should those users defect to CATV, individual monthly bills will surely rise, even to the point of threatening the universal affordable service that the local phone companies are supposed to provide.

For example, Pacific Bell estimates that 1.2 percent of its customers generate about $760 million in revenue every year, an amount that presumably goes a long way toward covering the huge fixed costs that are endemic to this business. Cable operator supporters deny this, stating that data transmissions account for barely 1 percent of the telephone operators' revenues and that the loss of this income would not or should not result in increased rates to individual users.

Furthermore, say CATV supporters, the local phone companies now have 98 percent of the data transmission market and microwave companies have another 1 percent, with only 0.15 percent to cable companies. True competition does not exist and should be made to, say these proponents, or else the cable companies should be free of regulation. This regulation is the critical issue, and it has done more to stifle and upset the industry's progress than anything else.

The local telephone companies have asked local regulatory commissions to step in and control or reject the attempts of cable operators to enter the data transmission field, arguing that the cable operators are promoting themselves as public carriers just as the phone companies are and should therefore be regulated just as the phone companies are.

The cable operators point out that MCI, GTE Sprint, and other alternative long-distance carriers are essentially unregulated, since their market share is so minuscule compared with AT&T's. In its relationship to the local telephone companies at the local level, CATV is in the same boat as these carriers: they do not threaten the mainstream of public traffic, they only threaten a highly lucrative source of the phone companies' revenues.

At present, it would seem that the cable operators are beginning to make their point. Two recent court challenges to the authority of local regulatory commissions in Omaha and New Orleans seem as though they will be upheld. And the argument that the heavy users are subsidizing local phone service seems untrue, at least in some cases. For example, in California in 1982 the telephone operator's cost of providing data transmissions exceeded revenues from that source; that is, without data transmissions, the cost of local service would actually have gone down.

In any case, no one expects subsidizing of service to continue, and it is the future that is really at stake here. One result of all the suits and countersuits has been to slow the CATV companies from aggressively marketing their product. In fact, the largest single complaint of would-be users of the cable for data transmissions is that the cable operators seem unsure about their own commitment to this direction.

The cable operators are people who have made their success in entertainment. In many cases they neither know nor seem to care to learn anything about data transmission. Where service has been installed, it has often been tentative and poorly supported.

What seems certain is that the cable operators have arguments strong enough to pursue a way out of the legal jungle they find themselves in, but there must be a stronger show of determination before they will be able to demonstrate credibility as a bona fide carrier.

In terms of social considerations (see Table 2), the Cable Television Information Center raised the issue of the actual layout of the cable distribution system and its effects on the community. The center contended that the design layout of the system could arbitrarily bring together or segment existing neighborhoods and communities. It suggested that attention should be paid when deciding on the boundaries, so that they are defined to coincide with neighborhood boundaries, school districts, or political jurisdictions.

Fundamentally, there are two kinds of decisions involved:

1. Can (and should) the boundaries for a single hub be defined to coincide with other political and social boundaries?

2. Should hubs be defined to permit each hub service area to be franchised to a different owner?

The question of boundaries is not one of technology, but rather of local community feelings and expectations about education and public access programming, and of data accessibility. This problem may be helped somewhat once interconnection issues between two adjacent cable operations have been addressed.

These issues arise when hubs within a multihub system serving a single city are interconnected or when systems serving several adjacent communities are interconnected.

An interdisciplinary approach

Observers like E. Witte addressed the types of problems that arise as a result of the introduction of new communications technologies and suggests that special attention should be paid to the human, social, economic, and political problems that result.

In 1974 a supranational association, the Munchner Kreis, was founded (based in West Germany) with the purpose of bringing together leading figures in the fields of science, politics, industry, commerce, and communications for communications research. The group believes that this interdisciplinary approach is necessary if the new technologies provided by the scientific community are to be accepted and integrated into existing society.

The Munchner Kreis is primarily concerned with the problem of gaining people's acceptance for new communications technology. Since the impact of communications is so broad and important, the social and economic issues must be considered from the standpoints of the nation, state, local community, organization, and individual. In other words, serious attention must be paid to the interconnections between communications and democracy. To carry this point one step further, if it is our desire to someday develop a worldwide infrastructure that will guarantee international communications, that infrastructure must be accompanied by homogeneity in the legal, organizational, and commercial realms.

An organization like the Munchner Kreis is very important, since it serves as a type of megastructure for organizing current technology and as an indicator of what is to come. To prepare the general public for the possibilities of communication and to predict and study people's reactions to those possibilities, the association aims at an international membership drawn from the widest variety of fields of specialization so that it will be able to deal with the problems of communication on as comprehensive a basis as possible.

As we move toward the 1990s and beyond, it is obvious that there will be major and fundamental changes in the way organizations and individuals do business and interact with each other. Exciting new possibilities will become available as technological innovations reach the commercial marketplace. However, it is important that we keep our expectations aligned with reality and remember that our future is as much in the hands of the politicians and law-makers as it is with the software and hardware engineers. ■

Further reading

Baldwin, Thomas F. "A Systematic Plan for Realization of Full-Way Cable Systems: Four Generations of Technology and Applications," *Two-Way Cable Television: Proceedings of a Symposium Held in Munich, April 27-29, 1977* (New York: Springer-Verlag, 1977).

Brownstein, Charles N. "Two-Way Cable Television Applied to Non-Entertainment Services: The National Science Foundation Experiments," *Two-Way Cable Television*.

Cheung, Casteret, "Metropolitan Area Network Standards—IEEE 802," *Digital Communications on Commercial Cable, Digest of Papers*, Spring Compcon 83, Session 38, Feb. 28-March 3, 1983 (San Francisco IEEE Transactions on Cable Television, 1983), pp. 479-483.

Mason, W. F. "Overview of CATV Developments in the U. S.," *Two-Way Cable Television*.

Smith, Ralph Lee, *"The Wired Nation," The Nation,* Vol. 210, No. 19, May 18, 1970.

Smith, Ralph Lee, "The Wired Nation: Cable TV—The Electronic Communications Highway" (New York: Harper & Row, 1972).

Taylor, Archer S. "Cable Television Encounters Computers," *Digital Communications on Commercial Cable, Digest of Papers,* Spring Compcon 83, Session 38, Feb. 28-March 3, 1983 (San Francisco IEEE Transactions on Cable Television, l983), pp. 474-478.

Cable Television Information Center, Technology of Cable Television, Washington, D. C., 1973.

Walters, Sylvane. "Two-Way CATV Systems," *Two-Way Cable Television*.

Witte, E. *The Munchner Kreis*—A Supranational Association for Communications Research, *Two-Way Cable Television*.

William E. Bracker, Jr., is vice president, research, Technology Research Associates, a Tucson-based consulting firm. Bracker holds a Ph. D. in management information systems from the University of Arizona.

Benn R. Konsynski, professor in the Management Information Systems Department at the University of Arizona, teaches courses in advanced systems analysis and design, office automation, data communications, and distributed processing.

Timothy W. Smith is a candidate for the Ph. D. in management information systems, with a minor in electrical and computer engineering, at the University of Arizona. Smith received a master's degree in business administration from the University of Miami and a master's in science from the University of Arizona.

Nathan J. Muller, Telecom Planning & Analysis, Huntsville, Ala.

ADPCM offers practical method for doubling T1 capacity

Network managers who have long wrestled with balancing cost and quality in private T1 networks now have a solution.

Escalating equipment costs and the frustratingly long lead times needed to order, install, and cut over new T1 facilities are forcing communications managers to turn to adaptive differential pulse code modulation, or ADPCM. When pressure builds to upgrade the transmission capabilities of private voice and data networks, ADPCM gear can double the number of voice conversations that can be carried over T1 facilities.

By implementing ADPCM gear on T1 channel banks, substantial network economies can be realized with only a modest increase in hardware costs—and at virtually no sacrifice in transmission quality. Besides being economically attractive, ADPCM adds a high degree of flexibility to network planners since it limits reliance on the local carriers.

Until recently, the easiest way to keep pace with increasing voice and data traffic was by ordering more line termination equipment and leasing additional T1 lines from local carriers—all at substantial expense. In T1 equipment, voice signals are converted to digital pulses for transmission at 1.544 Mbit/s. At this speed, a single T1 line usually accommodates up to 24 channels or as many as 24 simultaneous voice conversations.

ADPCM halves the number of bits required to encode a voice signal accurately, so T1 transmission capacity is doubled from the original 24 channels, to 48. This provides users with a 2-for-1 cost savings on monthly charges for leased T1 lines, as well as reducing the capital outlay for line termination equipment. And since many ADPCM devices handle network management signaling inband, all 48 channels may be used for voice or data transmission.

The conventional encoding technique used in T1 lines is called pulse code modulation. In PCM line equipment the voice signals are sampled at the minimum rate of twice the highest frequency level. This translates to a rate of 8,000 samples per second since the telephone industry's standard voice signals range from 200 to 4,000 cycles per second (200 Hz to 4 kHz). The amplitudes of the samples are then encoded to binary form using enough bits per sample to keep the quantizing noise low while still maintaining an adequate signal-to-noise ratio.

In a PCM-based application, this means eight binary bits per sample are required, which allows up to 256 discrete amplitude values (255 amplitude values are used in North America), or frequency levels. The conversion of a voice signal to digital pulses is performed by the coder-decoder, or codec, which is a key component of D4 channel banks. The codec then multiplexes 24 channels together to form a 1.544-Mbit/s signal suitable for transmission over digital facilities, like the T1 leased line.

ADPCM does not shortcut the 8,000-sample-per-second requirement for encoding 4-kHz voice signals. Instead, the ADPCM coding device, called the transcoder, accepts the sampling rate, then applies a special algorithm to reduce the 8-bit samples to 4-bit words using only 15 quantizing levels. These 4-bit words no longer fully represent sample amplitudes; instead, they contain only enough information to reconstruct the amplitudes at the distant end.

A crucial element of the transcoder circuitry is its adaptive predictor feature, which predicts the value of the next signal based only on the level of the previously sampled signal. Since the human voice changes little from one sample interval to the next, prediction accuracy for voice tends to be very high.

A feedback loop built into the predictor circuitry ensures that voice variations are followed with minimal deviation. The deviation of the predicted value measured against the actual signal value tends to be very small and can be encoded with only four bits, rather than the eight bits used in PCM. In the unlikely event that successive samples vary widely, the algorithm adapts by increasing the range represented by the four bits though at a slight increase in the noise level over normal signals.

At the other end of the T1 line, another transcoder performs the process in reverse. Predictor circuitry reinserts the deleted bits and restores the original 8-bit code. Since the receiver's decoder must be able to track the transmitter's encoded signal, both must utilize the same algorithm. This locks the user into a brand of transcoder.

Since voice and data sampling patterns differ greatly, separate adaptive predictors are dedicated to voice and data applications. A speech-data detector is used in the transcoder to decide which prediction algorithm to employ.

Other schemes

Other circuit multiplication schemes are used over T1 facilities, such as Continuously Variable Slope Delta Modulation (CVSD) and Time Assigned Speech Interpolation (TASI). But the encoding schemes employed in these techniques are effective only within a relatively narrow range of T1 applications.

CVSD modulation uses a sampling rate as high as 32,000 samples per second. Since the difference between each CVSD sample tends to be very small, a single bit can be used to represent the change in the slope of the analog curve. The big advantage of CVSD is that it can be used to encode voice signals at a lower bandwidth than ADPCM. CVSD will support voice data rates of 16, 24, 32, and 64 kbit/s. The drawback is that the quality of the CVSD-modulated speech is not as good as an ADPCM-encoded signal at 32 kbit/s.

TASI takes advantage of the pauses in human speech to interleave several conversations together over the same channel. Since most speakers cannot talk and listen simultaneously, the effective utilization of the network in most voice conversations is typically somewhat less than 50 percent—and the natural pauses between utterances drops the efficiency an additional 10 percent. TASI-based schemes seek out and detect the active speech on a line and assign only active talkers to the T1 facility. Thus, TASI makes more efficient utilization of time to double T1 capacity.

Although TASI increases T1 capacity, it also may cause significant delay. TASI equipment has also been found to cause "clipping" in small tie line configurations. Clipping occurs when speech signals are deformed by the cutting off of initial or final syllables.

Relative merits

PCM, ADPCM, and CVSD encoding schemes each have advantages and disadvantages depending on their application (see table). PCM-encoded voice quality is good and relatively simple to implement; its drawback is that it delivers only enough bandwidth for 24 channels.

ADPCM also delivers toll-quality voice, and it doubles T1 transmission capacity for voice. Though it passes modem traffic at just 4.8 kbit/s or less, this limitation concerns only those users with a very large installed base of modems.

Businesses frustrated by having to lease separate private-line circuits for voice and data can garner substantial cost savings by incorporating ADPCM technology into their networks and integrating their voice and data traffic. For example, in channels that carry data at speeds greater than 4.8 kbit/s, the ADPCM function can be disabled locally at a front control panel or remotely via an administrative

	BENEFITS	DRAWBACKS
PULSE CODE MODULATION	INTEGRATES VOICE AND DATA AT 1.544 MBIT/S	LIMITED TO 24 CHANNELS
	SINGLE CIRCUIT IS MORE EFFICIENT THAN 24 STANDARD VOICE-GRADE CIRCUITS	INSTALLATION AND CUTOVER MAY TAKE AS LONG AS 18 MONTHS
	MORE ECONOMICAL THAN EQUIVALENT NUMBER OF VOICE CHANNELS	
	MAINTAINS TOLL QUALITY VOICE, SUITABLE FOR TANDEM APPLICATIONS	
	SIMPLE TO IMPLEMENT	
CONTINUOUSLY VARIABLE SLOPE DELTA MODULATION	DOUBLES TRANSMISSION CAPACITY OF T1 CIRCUIT	DOES NOT EXHIBIT TOLL-QUALITY VOICE AT 32 KBIT/S
	INEXPENSIVE AND RELATIVELY SIMPLE TO IMPLEMENT	DOES NOT RELIABLY PASS 4.8-KBIT/S MODEM TRAFFIC
	SUPPORTS VOICE DATA AT 16, 24, 32, AND 64 KBIT/S	HIGHLY SUSCEPTIBLE TO NOISE INTERFERENCE
		LIMITED TO POINT-TO-POINT APPLICATIONS
ADAPTIVE DIFFERENTIAL PULSE-CODE MODULATION	DOUBLES TRANSMISSION CAPACITY OF T1 CIRCUIT	MORE COMPLEX, MORE EXPENSIVE TO IMPLEMENT THAN CVSD
	EXHIBITS TOLL-QUALITY VOICE AT 32 KBIT/S; RELIABLY PASSES 4.8-KBIT/S MODEM TRAFFIC	LIMITED TO SUPPORTING 32 KBIT/S
	INCORPORATES EXTENSIVE TEST FEATURES; OPTIONALLY PROVIDES IN-BAND SIGNALING	LIMITED TO PASSING 4.8-KBIT/S TRAFFIC
	PROVIDES NETWORK PLANNING FLEXIBILITY AND COST CONTROL	

terminal connected to an RS-232-C/V.24 supervisory port. Either arrangement allows noncompressible signals to be passed at their normal rate. Since these instructions are typically saved in nonvolatile electronically programmable read-only memory, there is no need to reset or reprogram them after power outages or network failures. Though ADPCM is slightly more complex to implement than PCM or CVSD, its benefits override its limitations.

CVSD falls between PCM and ADPCM in terms of implementation complexity, but it does not exhibit toll-quality voice at 32 kbit/s and will not reliably carry even 4.8-kbit/s modem traffic. CVSD is highly susceptible to noise interference, which results in a relatively poor-quality signal. Thus it performs best in point-to-point applications, where low cost is the overriding factor. Unlike many ADPCM devices, most CVSD devices do not offer the option of selecting dedicated signaling channels. Not only is CVSD unsuitable for multinode applications, but because

CVSD implementations are vendor-specific, CVSD does not work with all AT&T carrier facilities.

Considering the high monthly cost of T1 lines, ADPCM devices provide substantial savings to businesses operating private networks. For example, the current charge for a single 712-mile Accunet circuit from New York to Chicago runs about $23,000 per month, with an additional charge of about $4,000 for installation. Some circuits between New York and Los Angeles can cost close to $1 million a year. Compare these costs with the one-time charge of about $16,500 for a complete ADPCM network, which can double the T1 channel capacity between these cities.

Bottom-line benefits
Further savings can be won by using higher order digital multiplexers between ADPCM devices (Fig. 1). For example, the output of two ADPCM devices can be multiplexed together over a single T1C carrier (3.152 Mbit/s), increasing channel capacity by four times. With a T2 microwave link (6.312 Mbit/s), channel capacity may be increased eightfold to 192 channels and, over a T3 satellite link (44.738 Mbit/s), may be increased 58 times to provide 1,344 channels.

ADPCM transcoders do not require special repeaters to regenerate signals over long distances nor do they require ancillary test equipment, so benefits also come in the form of direct hardware savings. And when ADPCM gear is added to the network, it does not complicate the testing of T1 facilities.

Determining whether the source of a fault lies in the line itself or with the added equipment is simplified by the "loopback" points found in most ADPCM circuitry. Loopbacks allow operational verification of individual T1 channels or the entire T1 link (Fig. 2).

For example, using a specialized test mode, the user may verify that the entire near-end ADPCM scheme and individual channels are operational. In another test mode, the user may verify near-end operation as well as the operation of the entire transmission link.

Still another test mode may allow verification of both near and distant transcoders and the entire link. End-to-end voice quality may be verified through an analog-digital-analog voice port.

Another feature of most transcoders is an automatic internal bypass, which ensures that the original 24 channels of a T1 line remain available for use in the event of a network failure. This feature is especially important when considering the revenues and productive working hours that are lost during downtime and the confusion that often results when the transmission of critical information is disrupted.

Cost savings aside, ADPCM devices are easy to order and install, as well as to operate and maintain. Some T1 circuits may take as long as 18 months to order and put into service, compared with 30-day off-the-shelf delivery for most ADPCM devices.

Installation of the ADPCM device may be as simple as mounting an additional shelf onto the line termination equipment rack, inserting the required ADPCM modules, and wiring these modules to the channel bank assembly — an easy task for an experienced technician. For an extra charge, some vendors will pre-wire an equipment rack when customers order both the D4 line termination equipment and the transcoders, and then pre-test the entire configuration before shipment.

1. Savers. Using higher-order multiplexers, ADPCM devices increase capacity of T1C to 96 channels, T2 microwave to 192 channels, and T3 satellite to 1,344 channels.

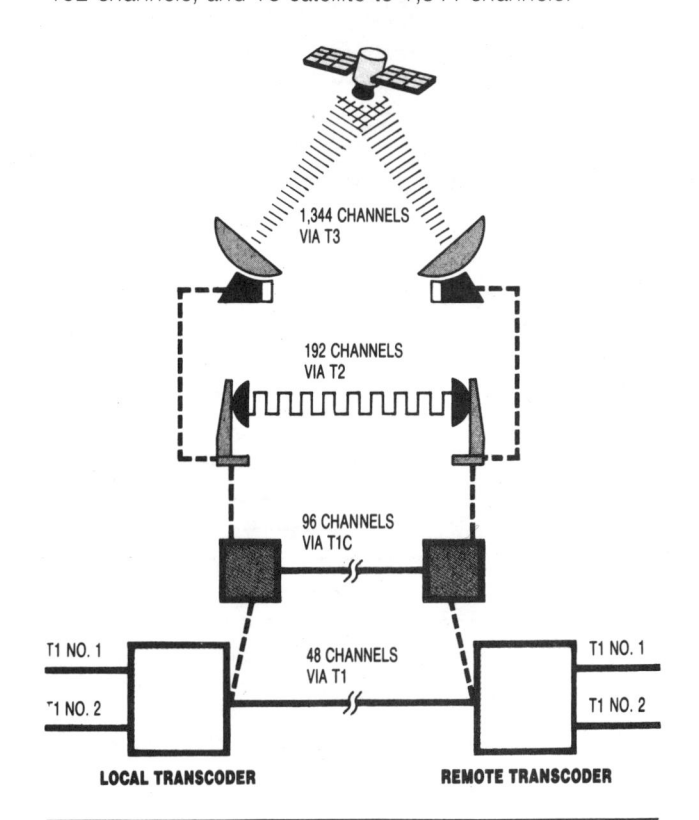

The use of ADPCM devices also permits greater latitude in designing or expanding T1 facilities. Because most ADPCM networks are modular, T1 capacity can be increased as needed simply by adding modules. This eliminates the long lead times required to order T1 circuits to meet forecast requirements.

If future transmission requirements fail to materialize as forecast, ADPCM cushions the impact. T1 tarrifs allow the carrier to levy a penalty charge if lines are canceled prior to the start of service. ADPCM gives the user the ability to increase the traffic-carrying capability of the network without central office assistance.

Questions that may arise during the purchase of ADPCM gear are what will happen to the investment when an Integrated Systems Digital Network becomes available, and will the bandwidth glut created by fiber optic networks negate the need for circuit multiplication techniques? ISDN promises intelligent, end-to-end digital connectivity on a network that will support simultaneous voice and data. Bell operating company field trials are scheduled through 1988. Various manufacturers have announced the availability of so-called ISDN chips so that off-the-shelf products may be on the market by the end of the decade.

Under AT&T's concept of nodal architecture, ISDN primary rate PBX-oriented services (23 BD) will be available to users through its No. 4 ESS toll switching centers accessible via T1 links. Each of the 24 channels can be defined

2. Maintenance mode. *Loopback features in most ADPCM transcoders simplify device and network test procedures. Tests include the near-end ADPCM scheme and individual channels, the entire transmission link, and both near and distant transcoders. End-to-end voice quality may be verified through an analog-digital-analog voice port.*

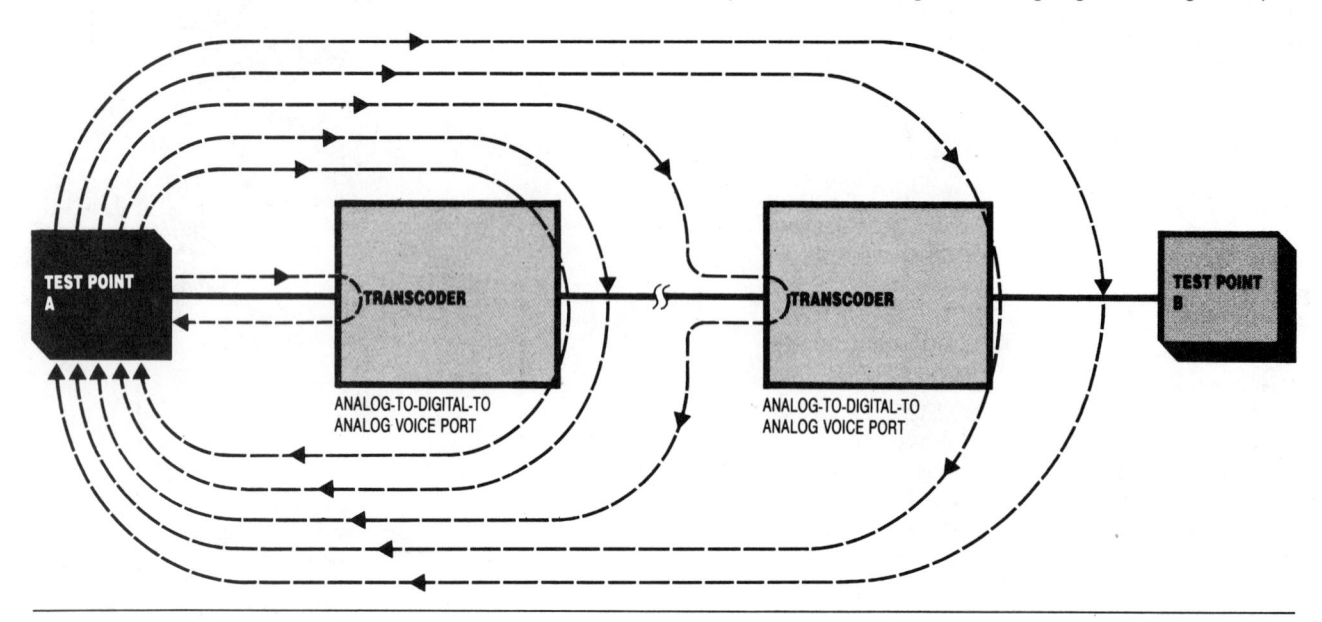

as a special type of circuit to suit the constantly changing needs of individual users, giving customers not only total control over bandwidth allocation, but also service definition—all with minimal reliance on the telephone company. Thus some channels may be defined on demand as AT&T Megacom 800 circuits, while others may be defined as Accunet switched 56-kbit/s, Accunet 1.544-Mbit/s, or Dataphone digital service.

The Bell operating companies plan to offer ISDN services initially through their Centrex exchanges once field trials prove that there is a genuine demand. These ISDN services will be brought up on switched facilities under the carrier's control.

T1 already allows private network users to allocate bandwidth any way they choose—and ADPCM enhances the economic flexibility of network planning. So while ISDN will become a powerful tool for managing corporate voice and data communications, it should be viewed, at least initially, as just another option with which to design, manage, and maintain a high-volume digital backbone network. The initial rollout of ISDN services is in the future. In the beginning, ISDN will exist only as a pilot technology and probably will remain a limited offering at least until the mid-1990s. It may also involve high up-front costs.

Although ISDN promises eventually to become economical, it may force users to give up some of their freedom. ISDN is a service provided largely by telephone companies. In using ISDN, T1 users may have to give up some of the operational latitude that they have grown accustomed to with T1 and be at the mercy of AT&T once again for timely and reliable installation, repair, and maintenance services, as well as features and network options.

Users opting for ISDN services may find themselves locked into a single-vendor solution and lose the bargaining power they now have with private networks. As with T1 circuits, corporations would then be at the mercy of both AT&T and the telephone companies for priorities and schedules for ISDN circuit installation and cutover.

Becoming dependent once again on the telephone company is a prospect few T1 users would welcome. It would not be the wisest decision, therefore, to suddenly abandon T1 in favor of ISDN, even if that were to become possible tomorrow. The reality of ISDN is that such services will evolve gradually, taking into consideration the available capital investments of AT&T and the telephone companies.

Although AT&T could conceivably make ISDN so attractive that it makes T1 less affordable, the emergence of fiber optic networks makes this highly unlikely. As fiber optic networks continue to evolve and become operational, the number of T1 circuits sending voice, data, and video around the country will skyrocket. This portends lower rates for long-haul interLATA T1 circuits and may lengthen the payback period of circuit multiplication networks that were put into service along high-density routes. But the fiber optic networks, as currently laid out, are only backbone networks linking major cities. Such networks may not coincide with the backbone facilities of private networks; not even fiber is economical for reaching into every nook and cranny of the country. Circuit multiplication schemes such as ADPCM will continue to be cost-justifiable for far-flung private networks for the foreseeable future and may play an expanded role in providing economical short-haul access to fiber facilities.

For businesses burdened with the escalating costs of operating their own private networks, the technology embodied in ADPCM represents a viable, economical alternative for increasing transmission capacity. What's more, ADPCM technology adds a high degree of flexibility to network planning—with only minimal reliance on the telephone company. ∎

Nathan Muller, who specializes in hardware and software evaluation and selection, is an independent consultant in Huntsville, Ala. He holds a graduate degree in management from George Washington University.

David W. Campt, University of California, Berkeley, Calif.

Interconnectivity: Risks may be substantial, but the benefits are great

Nowadays, with various options available, users do not have to wait for official standards to link dissimilar equipment.

As the general business environment gets more competitive, computer and communications departments feel the heat. Driven by corporate desires to simultaneously cut costs and accelerate services, communications managers are realizing that they cannot rely on the general advance of technology to allow that to happen; instead, they must consider other strategies.

Adopting a new approach can range from changing the network topology, acquiring more up-to-date technology, or breaking away from a firm's historic computer and communications suppliers. Users do this to take advantage of the many new opportunities in the more competitive computer industry. In addition to established companies encroaching on others' established turf, new entrants are constantly entering the market, trying to leapfrog over the veterans with new technologies.

When contemplating such a move, of course, a communications manager must look at networking issues. Will new processing equipment fit into the network and be accessible to all the users who eventually may want the machine? Will the new computer be able to share data with the extant devices? Will state-of-the-art communications devices offer better price-performance characteristics but perhaps limit networking flexibility?

Without question, many users are not even contemplating such issues. Even though they may be limiting their networking options and not reaping the benefits of new technology, many managers are content to stick with the computer or communications vendors their firms have historically used. This is not unreasonable. Companies that remain loyal to one vendor may find that their suppliers are more attentive to problems, and certainly will not encounter the situation of many vendors pointing at each other as the cause of the user's problems.

What follows are four stories about users who have dared to venture down the multivendor road. Even though all of the companies involved are satisfied customers, it is clear that networking between vendors does create challenges. Finding a product that can link different vendors' equipment can be a major task itself, and after the link is installed, the user may have to patiently work with the supplier to iron out the kinks.

Perhaps most importantly, a tool that flexibly links different vendors' products is likely to create work. Communications staff may have to invest substantial amounts of energy to make sure the new connectivity options are easy to use and properly controlled. The multivendor environment is, after all, more complex (a fact that makes some managers decide to return to a simpler configuration).

Still, if the new, mixed-vendor setup is properly handled, the benefits to the organization can be great. In many cases, interconnectivity products solve the problem of multiple terminals on people's desks, and can lead to a more cost-effective use of processing capability. As one user found, links between different vendors can expand a company's options in selecting equipment, because it has fewer concerns about connectivity.

Similar to widely accepted communications standards—which will eventually replace all ad hoc multivendor links—easy connections between different vendors expand the arena of competition. Thus, the benefits of interconnectivity extend to all users, whether a particular company pursues the multivendor environment or not.

Securities Industry Automation Corp.
For several years, SIAC (Securities Industry Automation Corp.) has used a Sperry (now Unisys) mainframe to support its trading and market data reporting for the New York Stock Exchange bond market.

A subsidiary jointly owned by the American and New York Stock Exchanges, SIAC provides computer services to the American Stock Exchange, the New York Stock Exchange, and the securities industry nationwide. On the mainframe resides the SIAC Automatic Bond System, a piece of software that matches available bonds to bond traders' requests. Hundreds of traders and order-room clerks access the mainframe via 3270 terminals, which historically have been connected to the Sperry host via a Collins Radio Corp. (a subsidiary of Rockwell) 8561 front-end processor. About 60 lines were connected, operating at speeds of 2.4 kbit/s or 4.8 kbit/s.

Several years ago, it became clear that the Collins machine, in the words of Steven Oliphant, a member of the SIAC development team, was "rapidly reaching the end of its useful life." Maintaining a more than 10-year-old machine was becoming increasingly difficult, the device had reached its maximum capacity, and there was no upgrade path available from Collins. SIAC then began a plan to upgrade the Collins equipment.

In addition to migrating to more current front-end processing techniques, SIAC had other goals in mind. A good percentage of terminal usage is comprised of inquiries into a database, so there was a desire to speed up terminal response time by reducing the path length to the host. One way of doing this was to use a front-end processor that could actually handle database inquiries, thus leaving database updates to be handled by the mainframe. Such a solution would not only reduce the response times, but also some of the load on the host.

As do all mainframe vendors, Sperry offered a front-end processor, the DCP product line, and SIAC considered using the company's products for its solution. SIAC eventually decided to upgrade the Collins front-end processor to a machine from Tandem Computer, a company that markets a number of fault-tolerant machines that can double as front-end processors and host computers. SIAC has a number of Tandem machines in-house, and has a lot of experience with the company's equipment.

After choosing the Tandem equipment, SIAC still had to find a means of connecting the Tandem and Sperry hosts. This task was relatively unusual at the time, since SIAC was one of the first to make the attempt, and there were few alternatives.

After evaluating the alternatives that existed, the organization decided to use Network Systems Corp.'s Hyperchannel to implement the link. There were some problems, however. The initial version of the Tandem-to-Sperry link had a number of bugs, something any user of a newly developed product should expect. "One might say we did some development work for NSC," Oliphant says, "in terms of finding some of the problems with the equipment. With any new communications software, there are going to be things to shake out in the user environment that the vendor cannot find." NSC was responsive to the problems, Oliphant adds, saying its hardware was very reliable (Fig. 1).

Although the SIAC's next major network improvement is still being pondered, SIAC plans to continue upgrading and expanding its network. Future expansion toward a more distributed configuration was, in fact, part of the reason SIAC wanted an intelligent front-end processor. "There are a number of different things you can do with a Hyperchannel," says Oliphant. One possibility is using Hyperchannel

1. SIAC. *As much as possible, the SIAC network is modeled after the OSI model. The Sage 6100 communications subsystem offloads all 3270 polling from the Tandem front-end processor. NSC Hyperchannel software provides Tandem-to-Unisys connections and may be used to support network enhancements.*

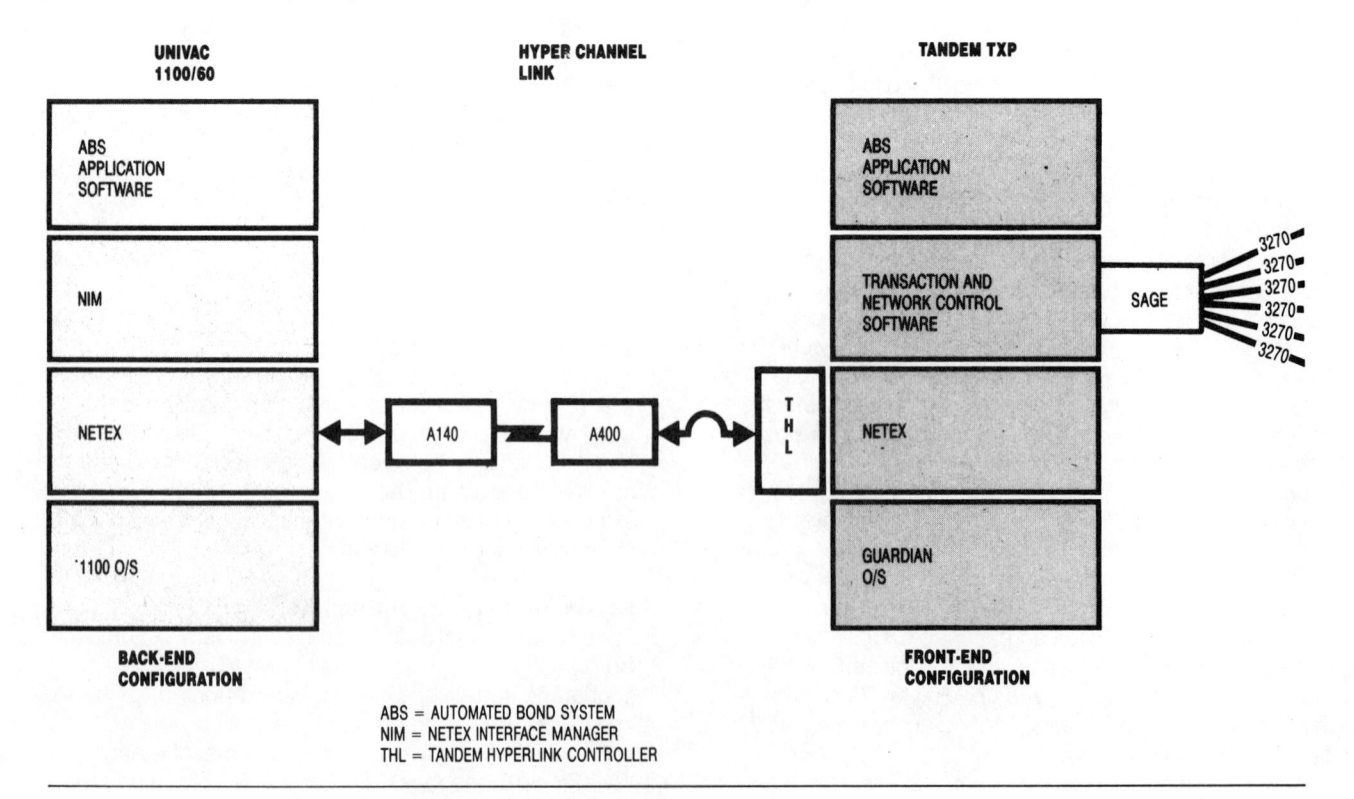

to link the Sperry and Tandem machines and an in-house IBM mainframe. Currently, the only way for the machines to share data is by outputting tape to a tape drive and reading that data into the other machine.

Also, a number of SIAC employees need to connect to both the Tandem and Sperry machines. Instead of putting multiple terminals on a desk, SIAC equips those users' microcomputers with communications cards for both mainframes. This still requires that multiple lines be run from one office, however.

"Obviously, in a full networking solution you want to interconnect all of the hosts together and have one terminal that can connect to each one," says Oliphant. "I am blue-skying with this, but these are the kinds of things that are possible when you have a Hyperchannel in-house," he adds. "SIAC is also currently using the Hyperchannel hardware connecting its remote peripherals to its IBM mainframes."

Westinghouse

Given IBM's dominance of the mainframe computer market, almost every vendor in the industry has had to form some type of strategy to ensure that its equipment is compatible with Big Blue. These offerings, as well as third-party connectivity equipment, can give users the opportunity to acquire other vendors' processing equipment without forming a segregated networking topology.

In fact, many users have turned toward non-IBM communications gear not because the equipment provided more versatile connectivity, but because of better prices or performance. But little in the computer industry is permanent, including users' preferences. Some users have ventured down the alternate vendor route and, for one reason or another, later gone back to Big Blue's fold.

Several years ago, some of Westinghouse's 34 business divisions began to replace IBM 2701 and 2703 front-end processors (these products were predecessors of the 3705) with Comten front-end processors of various sizes. At the time, Comten's equipment included a few features that IBM's did not, one of the most important being the ability for both batch and interactive terminals to select the target host in realtime.

Because of these advantages, Comten became the standard front-end processor supplier for several of Westinghouse's large computer centers. This continued for a few years until about 1984, when IBM released the 3725. After that product became available, the Comten base at Westinghouse began to undergo a gradual erosion, as some computer centers let Comten leases expire and picked up new leases for 3725s.

Westinghouse Communications Systems in Pittsburgh, Pa., is charged with providing telecommunications and data services to Westinghouse business units. According to Charles Winschel, manager of Data Communications Systems, the users who switched were by no means dissatisfied customers. "Most of the users who replaced the Comten did not view the equipment as a major source of any kind of problem," he says. Winschel has the task of providing some coordination between the 34 largely autonomous computer and communications departments. One of the primary reasons for the erosion of the Comten base, says Winschel, was the additional functions provided by the 3725 and the Network Control Program. As a result, some managers no longer saw enhanced value in the Comten gear.

Users had other motives for switching to IBM as well. Because Comten equipment must completely meet the specifications of any new release of SNA, there is usually a several month lag between the availability of new SNA capabilities on IBM's equipment and on Comten's. "I believe there has generally been a 10-to-14-month lag in the availability of new releases," says Winschel. "In some environments, this can be a problem."

Moreover, many managers did not want to face the potential hazards of the multiple-vendor environment, Winschel adds. "One manager simply said, 'With the network growth and increasing complexity, I would rather be in a single-vendor situation.'" Although Westinghouse uses multiple vendors extensively, he continues, the well-known vendor finger-pointing issue has not been a major problem.

"Any time you get finger pointing at a low level, you can always go to a higher level. Eventually, you can always bring vendors to the table. If each of the vendors claims 'It's not my problem,' you just have to say to them, 'Look, I have a problem. I am the customer, and we are are going to work together until we find my problem.'" Vendors have usually responded to Westinghouse, he says, and not deflected responsibility.

While some computer centers have switched from Comten to IBM front-end processors, others have stayed with the smaller vendor, in part to remain insulated from each new wrinkle of SNA. Some managers think being on the front lines of using a new IBM product is not desirable. "No one wants to be the pioneer," says Winschel, "because he is the one with the arrows."

Another reason for the loyalty some computer center managers pay to Comten is the effectiveness of its equipment. "The hardware has been extremely reliable," says Winschel. "Our analysis of the two hardwares says that hardware is not an issue in either company's equipment."

While Comten's support has been generally good, Winschel points out that IBM's support is more consistent across different locations: "Local support from Comten vacillates; it is really situation- and area-dependent. At any time you might find that Pittsburgh has excellent support, while it might be less excellent in Chicago because of some recent promotion or something." This variance should not be surprising, Winschel says, given Comten's relatively small size. With a large company such as IBM, he points out, an expert can be found very quickly.

Beyond the realm of front-end processor hardware, Westinghouse has also ventured into the world of IBM alternatives. More than 50 percent of the company's mainframes are based on IBM architecture, and a number of these machines, such as NAS, are plug-compatibles. In addition, the broadcast and credit companies have historically used Burroughs (pre-Unisys) mainframes, although the credit division has, within the past 18 months, explored the use of an IBM mainframe.

This movement toward IBM does not necessarily represent a movement away from Burroughs, but rather emerges from the fact that the company wants to maximize its options for buying software packages, and more is available on IBM equipment. "For any particular need in the credit environment, you can fill books on commercially developed software for IBM, and that simply isn't true for Burroughs," says Winschel.

Since it did not have an IBM machine, the Pittsburgh-based credit company contracted with a Westinghouse subsidiary in New Jersey to use some of its excess capacity on its NAS mainframe. Until the credit company gets its own IBM machine, the credit company must make sure all of its remote offices around the country have reliable links to both the Burroughs and IBM machines. Conceivably, the design of these links should reflect the fact that a terminal user will spend more than 80 percent of terminal time accessing the Burroughs machine.

To implement the links, the credit company is relying on a nationwide T1 network that Westinghouse is building for all of its subsidiaries. To any user in the credit department, it appears that there are two separate paths to the machines. In actuality, a local multiplexer routes the user over a long-haul link, and a multiplexer in Pittsburgh routes the user either to the local Burroughs mainframe or to the IBM machine in New Jersey.

The T1 network is only one aspect of Westinghouse's strategy for consolidating its various internal networks. "On the first level of the ISO [International Organization for Standardization] model, we are building a physical network to serve the corporation's voice and data needs," says Winschel. As a result, the corporation has an active program to integrate disparate networks' transmission media under T1.

The other part of Westinghouse strategy operates at the network level. About two years ago, the corporation implemented a private packet-switched network, Wespac, that is designed to serve the general connectivity needs of the corporation's computers. "Penetration is growing very, very rapidly. The X.25 interface to these computers is a tremendous capability," says Winschel. This network is Westinghouse's principle strategy for providing more integrated communications between the corporation's computers, the ranks of which include a number of minicomputers from a variety of vendors, such as Prime, Hewlett-Packard, and Digital Equipment Corp.

The main hub for Wespac is located in Pittsburgh, with main U. S. switching nodes in Livingston, N. J., near New York, Baltimore, Atlanta, Charlotte, N. C., Orlando, Houston, Sunnyvale, Calif., and Chicago, plus Brussels and London internationally (Fig. 2). The network supports hundreds of access circuits, the majority of which are not dialed. Small minicomputers or groups of PADs (packet assemblers/disassemblers) reside at about 400 access locations, split equally between terminal PADs, used for input only, and addressable host PADs. Corporate electronic mail is run through four Prime 850 machines, and a CDC computer acts as a front-end processor to a Cray in Pittsburgh that supports Westinghouse engineering and scientific research around the world.

While this network consolidation is taking place, Westinghouse is continuing the development of its many IBM-based networks, which handle much of the corporation's current traffic. "We are moving IBM traffic to the packet-switched network, although only a very small percentage

2. Wespac 1986. *The U. S. backbone is shown with international connections. Small minicomputers and groups of packet assembler/disassemblers reside at about 400 remote locations. Terminal PADs are used for input only. Electronic mail and scientific computing are carried over the packet-switch network.*

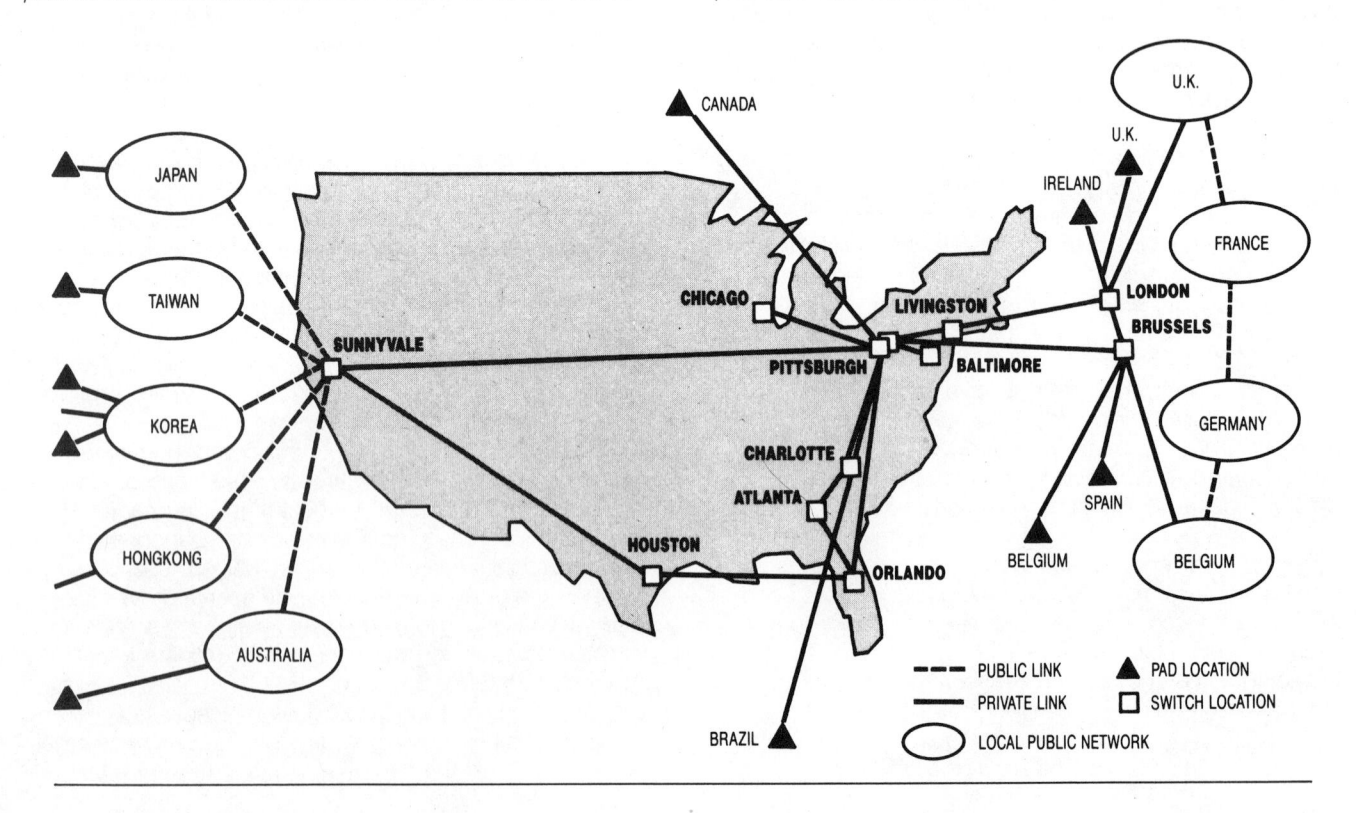

[has been migrated] at the moment," says Winschel. There is no intention to move batch traffic or large file transfers from SNA onto the packet network, although "the PC Network is a natural for X.25." The dual network strategy leans toward putting traffic as much as possible onto Wespac, for the X.25 network is easier to maintain.

Occasionally, Westinghouse will use a minicomputer vendor's SNA products for linking to IBM mainframes, although generally speaking the corporation prefers the X.25 links. "The minicomputer products typically provide one-way connectivity to IBM, meaning that they can look like an IBM PU 2 in an SNA network, for example," says Winschel. "That's great for getting them into the IBM world, but it doesn't do a darn thing to get people from the IBM world to them." Winschel looks forward to the advancement of IBM features such as LU 6.2 to help alleviate this problem.

Down in Maine

While some organizations have only recently begun to address multivendor connectivity, others have faced the challenge supporting diverse vendor's equipment for quite some time.

Prior to 1977, the government of the State of Maine used all IBM mainframes. But starting in 1972, the network used by that state government began to be based primarily around a Honeywell mainframe, the current incarnation of which is a DPS 852 triple processor configuration with about 600 terminals. A few years ago, a reorganization caused the state's data processing department in the capitol of Augusta to incorporate an IBM mainframe as well;

3. Maine in the early days. *Terminals were specific to an application, and some users were required to use more than one terminal to access the same machine.*

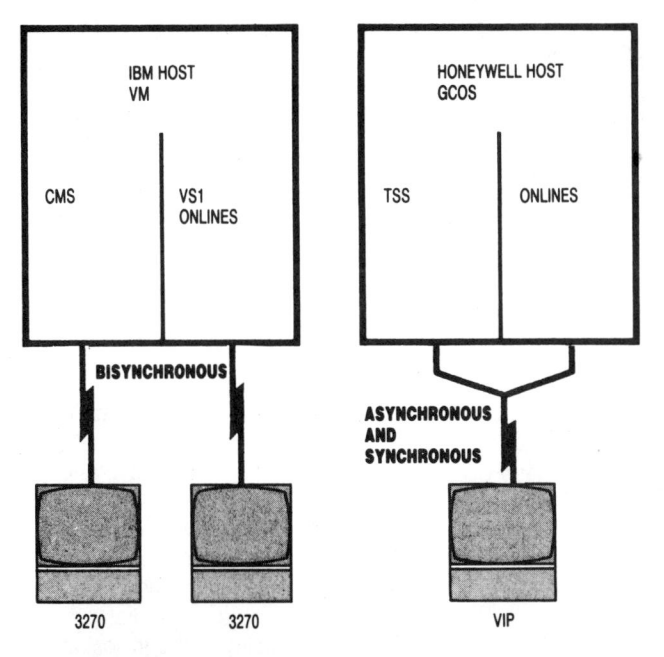

GCOS = GENERAL COMPREHENSIVE OPERATING SYSTEM
TSS = TIMESHARING SUBSYSTEM
VM = VIRTUAL MACHINE
VS1 = VIRTUAL STORAGE 1

the state's Department of Transportation moved and left behind its IBM model 370 135 mainframe, with the proviso that the Bureau of Data Processing continue to provide IBM support.

Initially, it seemed as though the Honeywell time-sharing services and user facilities were stronger. Eventually, though, that changed. "It seemed that during the late 1970s IBM poured a lot of money into CMS (one of the company's operating systems, the Conversational Monitor System) and its facilities, and opened up the operating system to third-party developers," says Carl Weston III, the state's deputy director of central computer services. "The pendulum swung toward IBM in terms of ease of development."

As a result, the computer services group, which supplies computing services to 70 to 80 state agencies, acquired a variety of applications for the IBM machine from a number of sources. As the range of IBM-compatible software advanced, management noticed that an increasing number of agencies kept their databases on the Honeywell host, but developed their new user tools on the IBM machine (Fig. 3).

As more and more users' desks became the residence of more than one terminal, it was clear that a terminal standard was necessary. Even though it was considered far from perfect, the 3274 terminal was chosen as a standard because of the wide support it gets in the industry. Since the state must put most of its contracts out for bidding, the computer services group did not want to greatly limit its future options by standardizing on a Honeywell terminal and the associated protocols.

Choosing the IBM terminal did not, however, resolve the standardization issue. The Honeywell mainframe supported IBM's bisynchronous 3270 protocol, but the IBM was configured to support SDLC (synchronous data link control). "It looked like we would have one terminal on each desk, but we would still have to run two lines," says Weston. "We needed a gateway that would take data from each terminal and point it to the right machine."

Finding a gateway that could accomplish this was far from easy. When the department started looking for this equipment, many vendors were marketing equipment that could perform this routing function for asynchronous protocols, but most of the gear could not handle more sophisticated protocols. The department investigated a variety of major players, including Honeywell, IBM, and Sperry. Different bids were taken from IBM, Burroughs, Univac (then selling Varian equipment), but all had problems. It was not that the hardware could not do the job; instead, the software investment required was beyond the resources of the data processing department.

One important criterion was that the equipment keep response low by minimizing "double buffering," which Weston describes as what happens when a front-end processor is line-attached to another front-end processor. "The Honeywell could do the switching, for example, but it could not channel-attach to either device," says Weston. "It would have been basically a front-end to a front-end."

Although protocol converters, such as the IBM Series/1, were also evaluated, and some equipment brought in and tested directly, response times were not adequate, in some cases twice as much. The computer services department also considered building the facility in-house, but did not want to shoulder the entire development burden. Says Weston: "We wanted a situation where the vendor was also in bed with us and had a commitment to develop this capability."

After five years of fruitless searching, the department became intrigued when Comten claimed its Model 3690 front-end processor could do the job. Following discussions with Comten's technical staff, Maine's processing professionals decided to install the company's equipment for a 90-day trial. "In our contract," says Weston, "we listed the things we wanted the product to do, specified how we would measure the performance, and agreed that Comten had 90 days to make it all work" (see "Populations in the trial period"). It took Comten engineers about four days to provide the switching capability to the initial test set of 14 lines, Weston adds, and the purchase was completed in about six weeks.

Now the Comten machine is channel-attached to the IBM 4381, and is line-attached to the Honeywell front-end processors (Fig. 4). All of the terminals that are linked directly to the Comten machine or to the IBM host (either directly or through IBM 3705 front-end processors) can go through the Comten machine and access the Honeywell DPS 852, and most recently Model 88, applications. Currently, users at Honeywell terminals cannot link to IBM applications.

Users select the environment they want through menus. They can choose the VM (Virtual Machine) or MVS (Multiple Virtual Storage) operating systems on the IBM machine, or the Honeywell machine. When servicing users who want Honeywell applications, the Comten front-end processor essentially functions as a protocol converter between the IBM SDLC protocol used by the terminals and the IBM bisynch protocol that the Honeywell host can understand.

As originally planned, the ability for users to switch between the two mainframe environments has allowed them to conveniently access the productivity tools on the IBM host and connect to the Honeywell databases. As is often the case, however, once a communications capability is in place, new uses for it seem to multiply rapidly. The department plans to incorporate an enhancement to the Comten front-end processor called the integrated protocol converter, which gives the front-end the ability to handle

Populations in the trial period
1) Six 3270 bisynchronous and SNA terminals on point-to-point and multipoint lines operating half-duplex.
2) One 3270 3777 SNA multipoint line operating full-duplex.
3) Two bisynchronous RJE (Remote Job Entry) point-to-point connections operating half-duplex.
4) One asynchronous start-stop line (TTY line) operating half-duplex.
5) VM (Virtual Machine) Passthrough software applications operating half-duplex, passing from an IBM machine through the Comten equipment to Honeywell.
6) Three MAF/RHO bisynchronous lines connected to the Honeywell Datanet.

4. Progress in Maine. *The IBM 3270 and PC compatibles, plus any distributed processor capable of emulating a 3270 device (such as a Wang or IBM Series 1 or 8100) use SNA.*

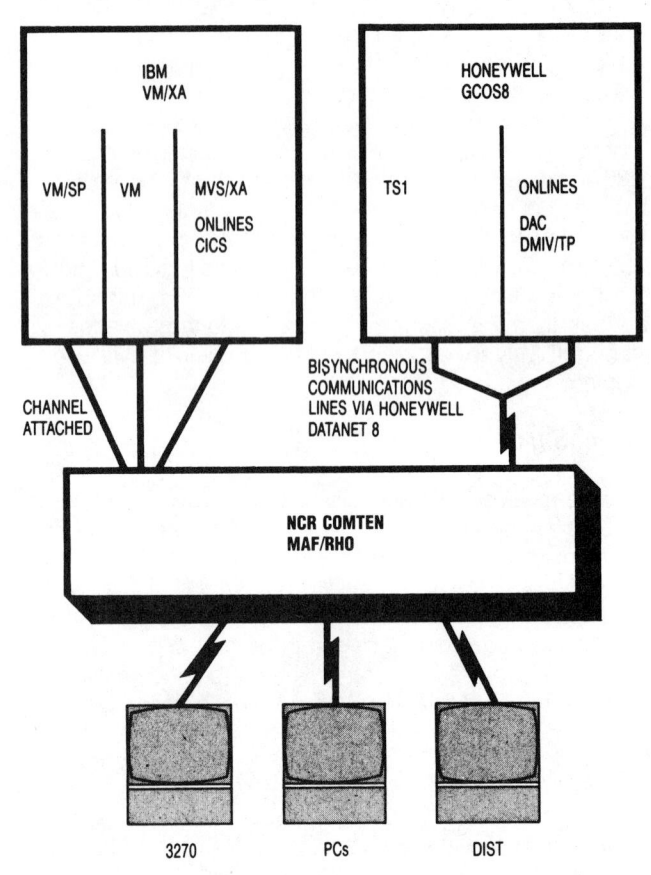

CICS = CUSTOMER INFORMATION MANAGEMENT SYSTEM
DAC = DIRECT ACCESS
DIST = DISTRIBUTED PROCESSORS
DMIV/TP = DATA MANAGEMENT IV/TELEPROCESSING
MAF/RHO = MULTIPLE ACCESS FACILITY/REMOTE HOST OPTION
MVS/XA = MULTIPLE VIRTUAL STORAGE/EXTENDED ARCHITECTURE
TS1 = TIMESHARING 1
VM/SP = VIRTUAL MACHINE SYSTEM PRODUCT
VM/XA = VM/EXTENDED ARCHITECTURE

asynchronous devices. With this capability, state employees will be able to work at home more easily.

Since the initial installation, Maine's Department of Administration has installed a redundant backup to the network, upgrading the Honeywell front-ends to Datanet 8/30s, and adding Wang IDS, a software product that allows transfer of Wang-formatted documents over System Network Architecture networks.

More importantly, the state can provide a whole range of new services to the general public. "We can set up some sort of network where people can get information about post-secondary educations, for example, on a dial-in basis," says Weston. "The dial-in facility is tough for us, though, because it opens up security problems."

Without a doubt, using a third vendor to implement communications between different mainframe vendors has provided Maine with capabilities it could not have had otherwise. In fact, Weston envisages that the state will continue to replace its IBM 3705s with Comten equipment. His advice for users who want to venture into the multi-vendor environment: "If you are not afraid of the other [non-IBM] companies and don't regard them as unknowns, I think it is easier." Still, he adds, "I don't think anyone would do it [use multiple vendors] if they did not have to."

Chrysler

In any corporation, the task of providing disparate users with somewhat integrated access to information is formidable. But for a company that designs and manufactures complex products, the problems are even thornier: integrating office automation and data processing functions with both computer-aided design/computer-aided manufacturing (CAD/CAM) equipment and the machines running the factory floor. For such manufacturing companies, the diversity of related computerized operations makes the multiple vendor interconnection problem especially important, and quite difficult.

For several years, Chrysler Corporation has slowly nudged its computer environment—dominated by DEC minicomputers and IBM and Control Data mainframes—toward one in which users in varied parts of the company could have easy and quick access to information generated in totally separate divisions. This might allow, for example, a marketing department to have instant access to the latest work of the design department.

Although this kind of interconnectivity has long been a goal of Chrysler's information-processing professionals, a key ingredient of the connectivity was almost stumbled upon, in the sense that it was purchased for another purpose.

Chrysler's engineering center has historically relied mostly on CDC and DEC computers, while the company's less technical sectors have tended to use IBM mainframes. Several years ago, when it became clear that the company needed a means of connecting the engineering group's computers—both to each other and to the rest of the business—an elaborate strategy was adopted.

"Our basic strategy was to have a common database for product design so that we could have a single instance of product information in electronic form," says Walter Weglarz, manager of technical computer center systems and operations-engineering. Once this goal was achieved, different parts of the engineering division would be able to refer to the same automobile part in the same way. At the time, each of its vendors had its own databases which were, of course, incompatible with the other vendors. In order to pursue its goal of a common database, Chrysler had to invent its own graphics and CAD/CAM standards, a task those industries are only recently beginning to address (Fig. 5).

A similar situation was true in communications as well, for CDC, IBM, and DEC have separate communications architectures, and these were used for linking different Chrysler machines within each vendor's product line. For links between different vendors, the company established its own internal communications standard, a variant of the bisynchronous protocol.

The topology of this home-grown link between the company's engineering and business systems was not

5. Cyberman. *The CAD terminal displays a wire-frame representation of an H-body car, used in the LeBaron GTS and Dodge Lancer.*

very sophisticated. "This was done via a store-and-forward-type operation. It wasn't machine-to-machine communications," says Weglarz. "We basically had the classical node in the middle situation where dissimilar processors would connect to that machine with the same standard Chrysler-developed protocol."

Although this solution worked for five to 10 years before the implementation of LCN in 1982, increased processing and communications needs eventually began taking their toll on the link. "Eventually the capability of the store-and-forward operation—even using short-haul cables and internal dedicated lines—was not enough to handle the bandwidth necessary for high volumes of data," says Weglarz. At the time, Chrysler was limited to the 56-kbit/s services offered by AT&T and the local carrier. As a result, data transfers were often done by sending data to the middle switching node, then outputting the data to magnetic tapes. These tapes would then be hand-carried to the target machine.

In addition to these difficulties, the engineering computer center was facing the more typical problems of modern corporate computer centers. The technologies that drive computational speed (for example, the speeds of memory chips and central processing units) have been advancing faster than the technologies that govern input/output capacity (for example, disk retrieval algorithms and channel speeds). As a result, getting faster and faster processors did not necessarily have the expected payoff in increased productivity for computer users. "It was obvious we had to extend the number of computers, not simply have bigger and faster computers that would be starving for I/O," says Weglarz.

LCN to the rescue

As a result, about five years ago Chrysler began looking for a way to permit more data sharing between its more than 20 CDC machines, and the mainframe vendor's Loosely Coupled Network (LCN) provided the means. One of its primary functions is to allow multiple CDC mainframes to be linked together, making it possible for users at one terminal's machine to easily call applications and inspect data on other mainframes. For Chrysler, this would help make the common design database more

CHRYSLER ENGINEERING'S TECHNICAL COMPUTER CENTER HOST COMPUTER NETWORK

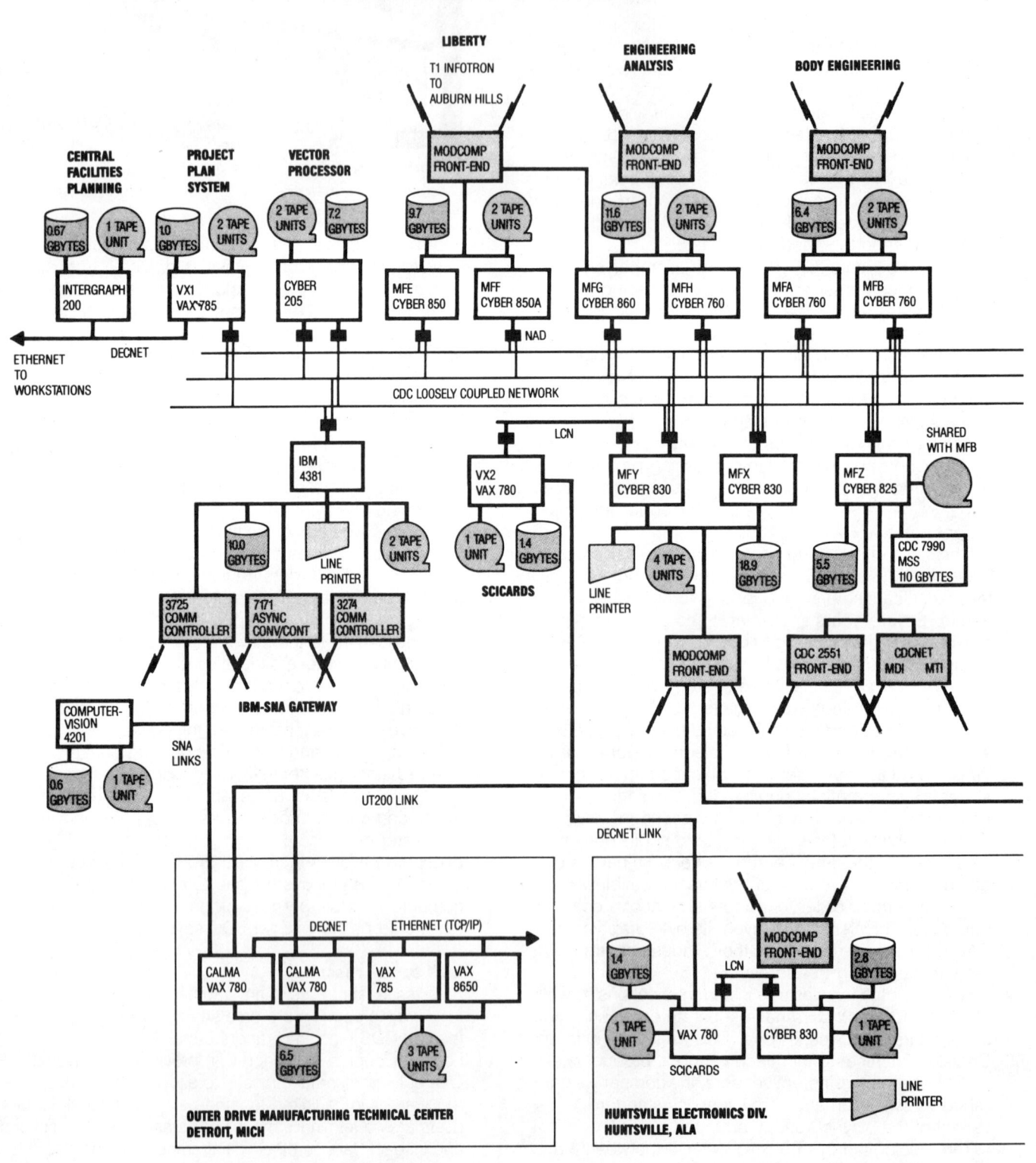

6. Chrysler. *Mainframes numbers X, Y, and Z perform electronic storage. Once data is taken from a sending machine, it is stored only in this archive, until a specified data threshold is reached, where it is sent on to a third level of electronic filing. Engineers can create panels and steering geometry, fitting out vehicles with cargo and components.* ■

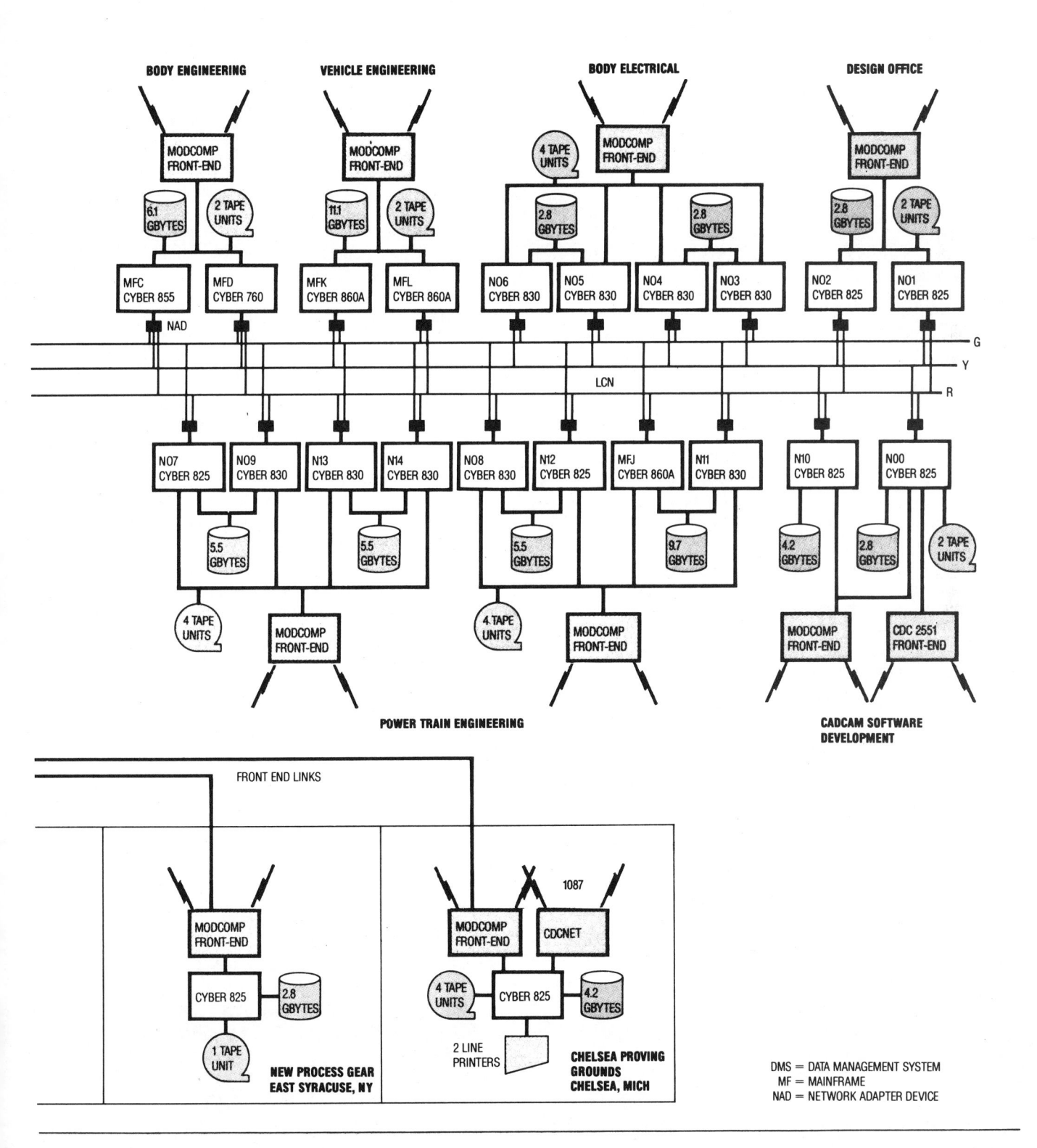

DMS = DATA MANAGEMENT SYSTEM
MF = MAINFRAME
NAD = NETWORK ADAPTER DEVICE

accessible. Thus, for example, an employee in the advance product development group could make a future design concept available to other areas of the company. The manufacturing division might use this information to determine the possible impact of this design on manufacturing operations.

CDC's LCN is a combination hardware-software product operating similarly to NSC's Hyperchannel. Time-driven, at Chrysler the product works over a semi-rigid coaxial network managing remote and local trunks (Fig. 6). The product is heavily weighted toward software, so that data translation functions are microcoded into network access devices strung out along the network, saving processor cycles on both sending and target machines. The RHF software supports primary and secondary data channels with Chrysler achieving actual throughput in the range of six to 10 Mbit/s on multiple trunks in parallel.

Chrysler expected LCN to help provide better management of its CDC-resident data, as well as allow better use of the various computers in the engineering center. LCN would make it easier to run a computational task on the appropriate CDC machine.

As use of LCN continued, communications management discovered that LCN could also accommodate non-CDC machines. LCN provides something akin to a high-speed mainframe channel extension, as well as software that allows a user to transfer data at very high speeds between dissimilar mainframes.

Obviously, this would give the computer center even more flexibility in assigning each task to the optimal computer. "Even though we did not originally expect it," says Weglarz, "we wound up with a multivendor network, with no concerns about data conversions being done twice on different computers. We actually could do [the conversion] in the network access device, and the data was easily transportable from any computer to any other computer." As a result, Weglarz says, it was easier to apply any vendor's advantages and output the data anywhere on the network.

For example, analysts in the automotive design group frequently need finite element analyses. For a very small job, an analyst might compose and run the job on his or her microcomputer. However, it might take a microcomputer hours, or even days, to complete a large model. With LCN in place, the analyst can compose the model locally, then send it to a computer that will accomplish the task in a length of time dictated by the analyst's needs.

Transferring the LCN's potential into reality was not a trivial issue. As with many other highly touted products in the communications field, users who purchase a communications product providing a good deal of flexibility must be willing to invest substantial amounts of effort in customizing the product for their needs.

For starters, Chrysler computer center management had to structure the network so that each work group was directly connected to a computer (or cluster of computers) appropriate for its needs. Doing this properly minimizes the need for network access and decreases response time for the most common transactions.

Perhaps more important, management had to create applications that could give engineers more computational flexibility without requiring them to navigate either the network or the vagaries of each machine. For instance, the previously mentioned automotive designer should (and does) only need to specify the resource by a mnemonic to send data. "The people who are using these machines are engineers, designers, technicians, and clerks. They do not need to be computer experts," says Weglarz.

To insulate the users from the network topology and the idiosyncracies of each computer, computer center programmers wrote the software that takes the analyst's request, builds the appropriate job control language, routes the job to the machine, retrieves the output, and presents it to the user. If desired, the user can specify that the job be run on any available machine, and Chrysler's networking software will choose the machine (Fig. 6).

Although making the LCN tool work well required significant effort, the benefit to the Chrysler Corporation was quick in coming. "We found there were limitations to how well you could effectively interconnect our complex of computers relying on manual intervention," says Weglarz. "With the advent of a single delivery mechanism," he continues, "the staff could be redirected to other activities; [they became] more productive."

Moreover, Chrysler is in a much better position to choose the best software for its needs, with less regard for compatibility considerations. Ironically, the file transfer flexibility that LCN provides gives the automaker more freedom of choice with hardware and software. It has far fewer worries about gear being unavailable to users on their existing machines. Says Weglarz: "The networking capabilities enable us to improve how we spend money on computer resources."

This extends to input/output devices as well. A supplier may have a peripheral device for its machine that is more suitable for a particular task. It might be available today on IBM and not until a future time on CDC, or vice versa. Software will not have to be rewritten for the plotter or peripheral to run on a CDC or a VAX machine.

Weglarz's vision of future improvements focuses more on other vendors than on CDC. "We emphasize the need for wider bandwidth in the 80-Mbit/s range as well as adopting the communications ISO standard. If I want to get brand X computer that has technical prowess for a certain job, I want it to connect to LCN." He adds that as a potential customer, Chrysler would rather get a prospective computer vendor and CDC together to implement that link rather than go around LCN and make a connection some other way.

Even though quite satisfied with the capabilities that LCN has provided, Weglarz is quick to point out that there is at least one other product, Network System Corporation's Hyperchannel, that provides many of the same capabilities as LCN. And while happy with LCN, he looks forward to further industry standardization, such as that which is taking place because of international standards and LU 6.2. This will allow better matching of jobs and computer resources. "Why aim a cannon at a fly on the wall when you can use a flyswatter?" he asks. ■

David W. Campt has a B.S.E.E. from Princeton and worked for several years as software editor for DATA COMMUNICATIONS. *He is now studying at the University of California-Berkeley's School of Public Policy, focusing on issues in technology.*

Gilbert Held, 4-Degree Consulting, Macon, Ga.

A dozen ways to beef up your network with a port selector

The latest port selectors can improve the efficiency of almost any network—particularly those with mixed data rate terminals.

Within the last decade, port selectors—also known as dynamic data switches or contention concentrators—have evolved from simple hardware-based contention devices into sophisticated software-controlled line-switching units that have become the heart of many data communications networks. Now, with new operational characteristics and networking options, the rationale and methods for employing port selectors have taken on a whole new perspective.

A port selector can be seen as a "black box" whose primary job is to provide a number of incoming data sources with contention access to a number of outgoing lines connected to a computer resource, or resources. Port selectors are generally available as standalone units; statistical multiplexers may have switching functions built in.

The motivation behind the development of port selectors was economics. Most large-scale mainframes service a geographically dispersed mixture of terminal devices. In many organizations, multiplexers were installed in larger offices to concentrate data from multiple sources onto a few high-speed lines connected to the central computer. Network managers soon realized that very rarely, if ever, were all the multiplexer ports in use at the same time. This meant that the front-end processor ports of the central computer were also not being fully utilized.

Figure 1 shows a typical mainframe-based network consisting of a mixture of multiplexed data sources, individual data sources connected by leased lines, and locally connected data sources. Note that a total of 74 front-end processor ports are required to service the local and remote data sources, regardless of their activity.

Suppose that, at any one time, a maximum of 40 terminals are communicating. There is then an underutilization of existing front-end processor ports, as well as an excess of ports needed to satisfy the connection requirements of the terminal users. In this case, a port selector designed to permit 74 line-side connections to contend for 40 port-side connections would not only reduce the front-end port requirements, but also increase the use of fewer front-end ports.

Operational overview

In its basic configuration, a port selector consists of either a standalone unit or a cabinet containing rack mounts into which a power supply, common logic, and adapter cards are inserted. Depending on the manufacturer, adapter cards can function as line terminations, port terminations, or both. The network illustrated in Figure 1 does not employ port selectors, so each front-end port must individually service a data source—unless multidrop lines and supporting software are installed.

To successfully support a variety of line terminations, port selectors must also support a wide range of "connect" and "disconnect" control sequences. The most frequently supported connect sequence is the ring indicator signal of the RS-232-C interface, which permits dial-up access to the port selector.

Since data sources directly cabled to the port selector (including terminals that might be connected through a multiplexer) will not provide a ring indicator signal, other connection sequences must also be supported by the port selector. Two other connection procedures supported by many port selectors include recognition of a data terminal ready (DTR) signal and/or activation by data activity.

Once a line-side data source seeking connection is recognized, the port selector central logic—software routines—scan the port-side connectors, searching for an available port. If one is encountered, the port selector initiates a cross connection, establishing a temporary physical circuit between the line-side termination and the port-side termination. This connection is sustained for the duration of the call.

If no ports are available, the port selector can either transmit an appropriate message and issue a disconnect, or it can place the caller seeking access to a port into a "waiting" queue. As Figure 2 illustrates, subsequent action can vary, depending on the operating features of the particular port selector.

Disconnection

Once a cross connection is initiated, the port selector must be able to recognize one or more disconnect sequences, in order to free a port for other users after a session terminates. Typically, a port selector will treat the cessation of the DTR or the clear-to-send (CTS) or received-line signal detect (RLSD) signals for a predefined period of time as the signal to terminate the cross connection and drop the connection. Since directly connected terminals and computer ports normally toggle the DTR signal, such devices are supported by scanning for the drop of that signal for a predefined amount of time.

For data sources connected through modems, either the CTS or RLSD signals are normally used as the indication to drop the connection. In those cases, the port selector will scan for activity by either signal and, if they are absent for a predefined time period, it will terminate the previously established cross connection and perform internal housekeeping functions.

Other disconnect sequence signals supported by some port selectors include: recognition of the BREAK signal that is transmitted when a terminal operator depresses the BREAK key on most keyboards; or the absence of activity on a line for a predefined period of time, known as an Activity Absence Time-out.

Figure 3 shows a port selector performing line-to-port contention, which reduces the number of computer ports required to support all network devices. Note that the contention rate is 74 lines to 40 ports, or 1.85:1. Users will typically want a contention ratio of 2:1 or less. When data sources are scattered over many time zones, however, which means whole areas of the country are likely to be inactive during certain times of day, ratios as high as 4:1 have been successfully employed.

With the configuration shown in Figure 3, 34 front-end processor ports can be eliminated—or used for other purposes. The cost of a front-end processor can approach $1,000 per port. The per-line, per-port cost of a port selector, on the other hand, typically ranges between $100 and $200, so significant savings can be achieved. Still, while the savings in ports and the reduction in front-end processor costs are paramount, the more sophisticated functions of newer port selectors can make them versatile networking devices in their own right.

One example of added functionality is the ability of many

---

1. Unreconstructed. *In a typical network that does not employ port selectors, each front-end processor must service an individual data source unless multidrop lines and software are installed. The drawbacks to network economy inherent in this topology demand another solution. Though unglamorous, port selectors are an efficient alternative.*

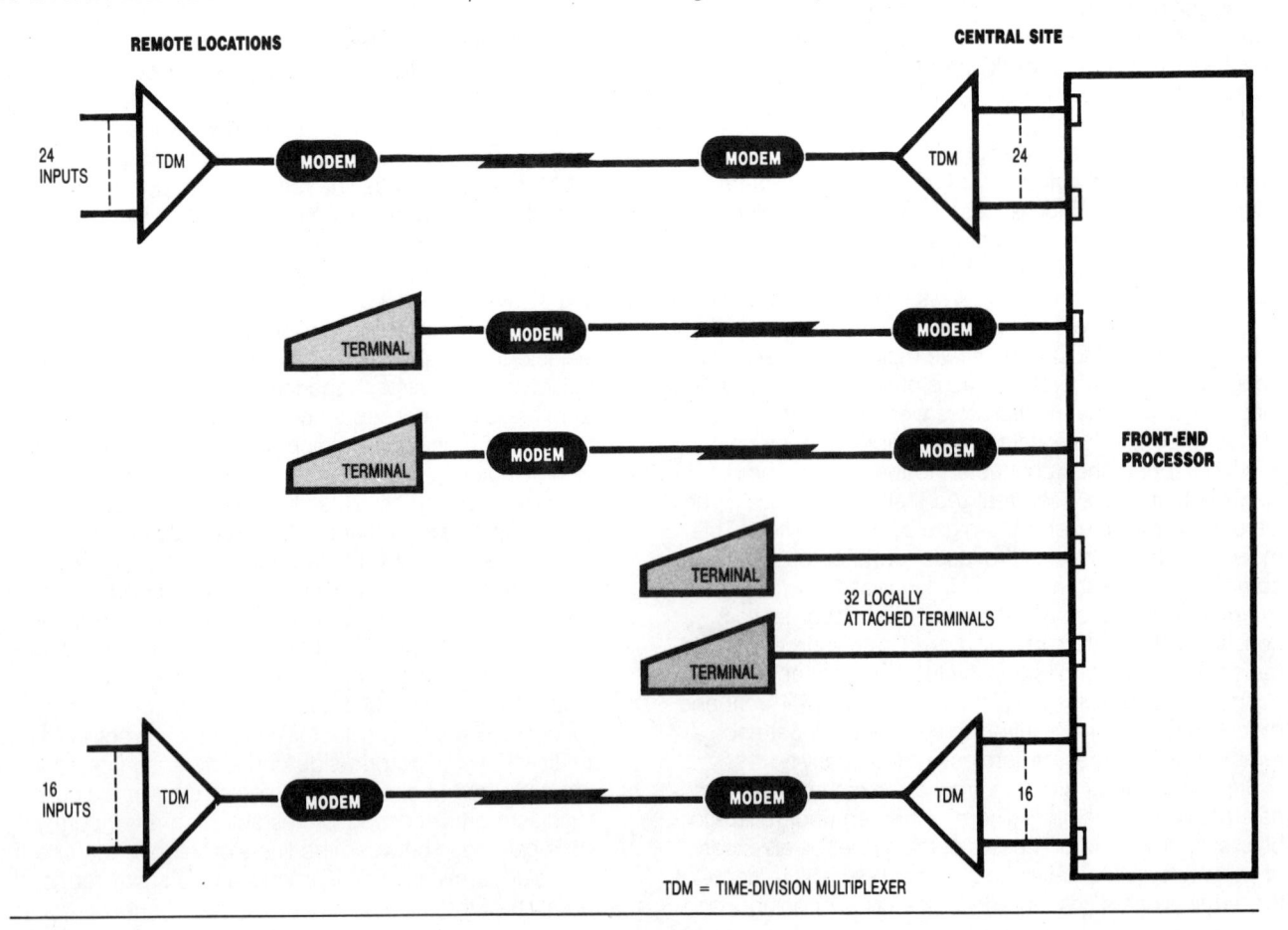

2. Contention. *When performing contention, a port selector allows N inputs to vie for a smaller number (n) of ports, thus saving on port costs.*

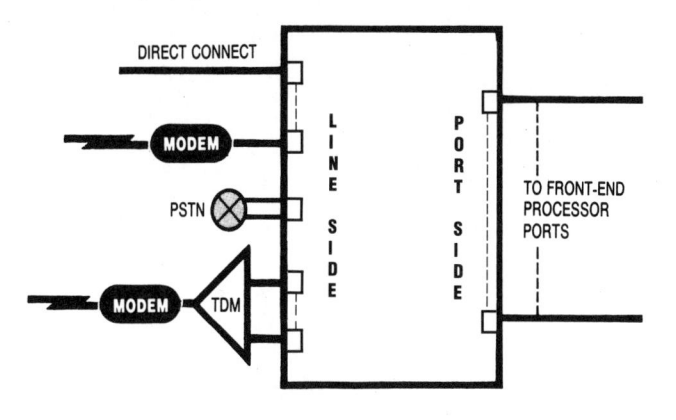

PSTN = PUBLIC SWITCHED TELEPHONE NETWORK
TDM = TIME-DIVISION MULTIPLEXER

port selectors to subdivide their ports into groups—commonly referred to as port classes. This capability permits port selectors to extend automatic data-rate (autobaud) detection in a network to fixed-speed ports on a front-end processor, thus enabling access to many computers over a common network facility.

These added functions can be used for what is called speed-controlled routing. Figure 4 illustrates the use of a port selector with port-class subdivisions providing speed-controlled routing from automatic data rate ports to fixed-speed ports on a front-end processor. Although all modern front-end processors can be configured to support automatic data rates, some processors (such as the IBM 3725 communications controller) may require the user to purchase additional software in order to obtain this capability. Other front-ends may require the acquisition of additional optional hardware; some processors require both additional hardware and software.

Since the additional software may not be capable of running in a fully loaded front-end processor, the resulting memory expansion, if such expansion is even possible, can add considerably to the front-end's cost. Thus, the speed-controlled routing capability of a port selector can be a very attractive alternative for networks featuring a mixture of terminals operating at different data rates, enabling users to access ports of the appropriate speed on the front-end processor.

As illustrated in Figure 4, terminals operating at different data rates at a remote location can dial a rotary connected to a multiplexer with autobaud-detection capability. Here,

3. Modifying the network. *By using a 74-line by 40-port configuration, the port selector eliminates 34 front-end processor ports. Depending on the cost of the ports eliminated, the port selector can pay for itself through the reduction of hardware. Users must determine for themselves what an acceptable proportion of lines to ports should be.*

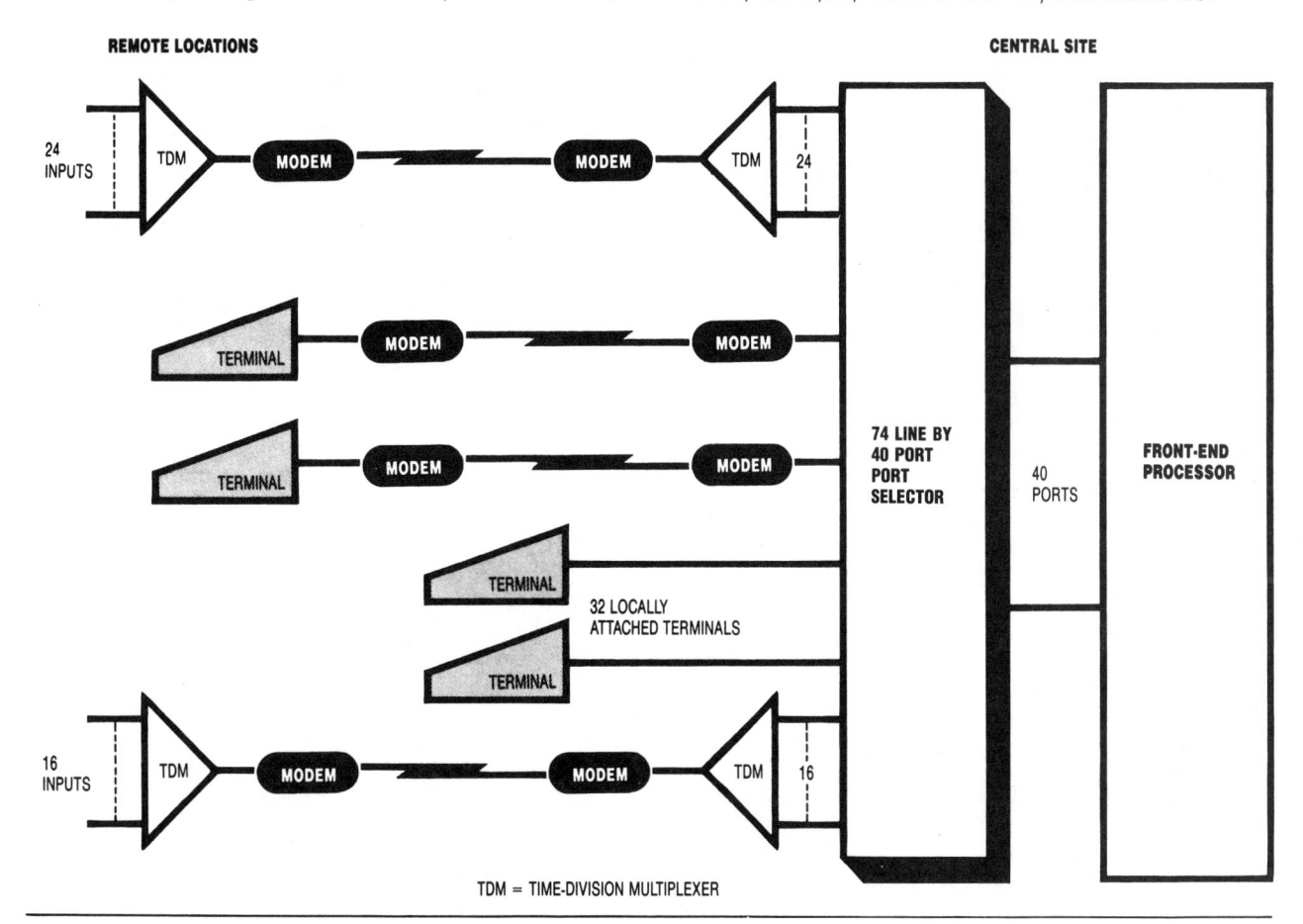

TDM = TIME-DIVISION MULTIPLEXER

245

4. Class divisions. *Port selectors enable speed-controlled routing to govern the input of the appropriate front-end processor ports. This makes it possible to use fixed-speed computer ports dynamically on a network. Port selectors can make allowances for the lack of autobaud functioning either built into or programmed into front-end processors.*

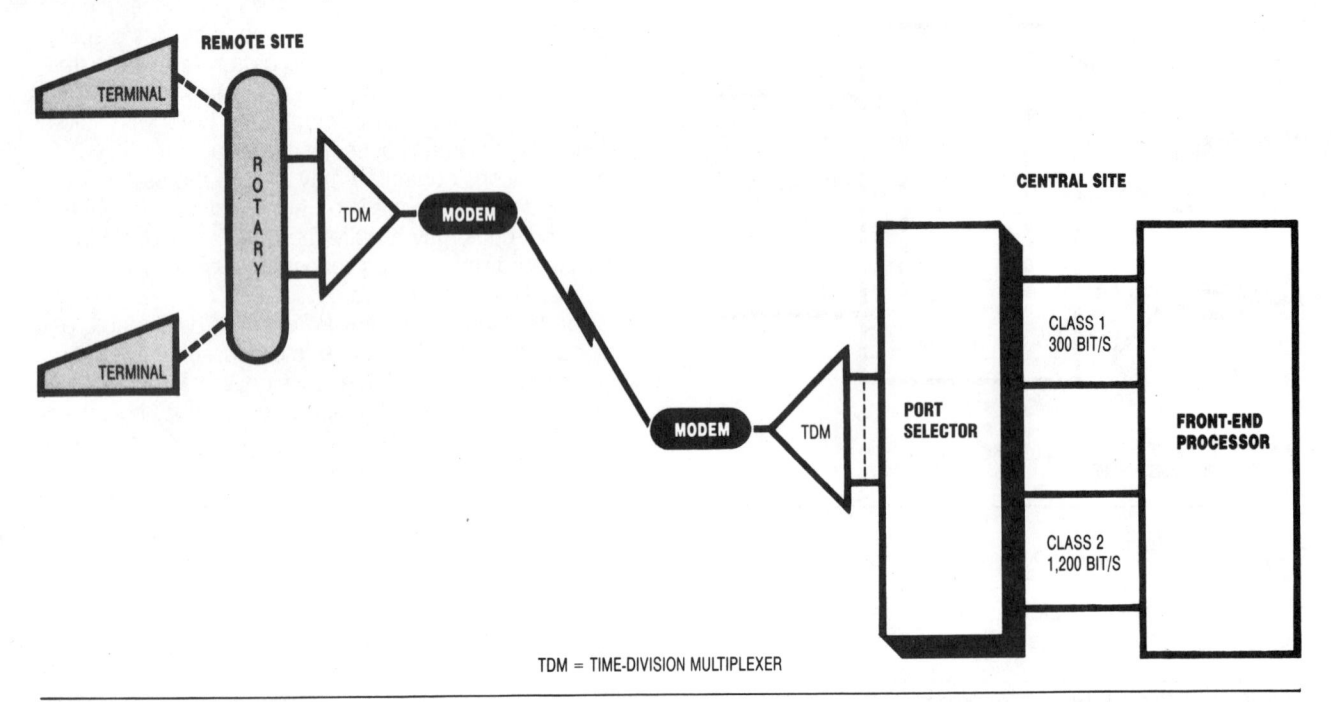

TDM = TIME-DIVISION MULTIPLEXER

the multiplexer measures the pulse width of the incoming signal and adjusts to the operating rate of the terminal.

At the central site, the ports of the port selector have been subdivided into two classes, each based on a different operating data rate of the fixed-speed front-end processor ports. After the demultiplexed data enters the port selector, the selector will prompt terminal users for the class or routing group for which they want a connection established. A user selecting, say, the numeral 1 would initiate a cross-connection search for an available port assigned to class 1; entering a 2 would cause a cross-connection search for an available port assigned to class 2.

Asymmetrical

Although most port selectors performing speed-controlled routing simply enable any incoming transmission of a particular data rate to contend for access to a front-end port supporting the same data rate, some devices permit asymmetrical port access. In such situations, the port selector functions as a speed converter, permitting a data source operating at one speed to be connected to a front-end port supporting a different data rate.

Perhaps the most valuable feature of port selectors is their ability to provide a network or network segment with common access to multiple computers, whether located in the same building or separated by thousands of miles. Figure 5 shows how a port selector can be employed to provide common network access to multiple computers.

As illustrated, all terminals in Boston are connected to a multiplexer, the output of which is routed to New York. In New York, the multiplexed circuit is demultiplexed and then input to a port selector that supports two classes, or routing groups. If the terminal operator selects class 1, a cross connection is established between the line side of the port selector (servicing the multiplexer) and one of the ports assigned to class 1 that connects to the computer in New York.

If class 2 is selected, a cross connection is attempted between the line side of the port selector and a port connected to a multiplexer, which routes data to a computer in Miami. Since users in Boston can access either computer by first being routed to New York over a common circuit, the expenses associated with a separate circuit to connect Boston and Miami are avoided. Also, if terminal users in New York require access to the computer in Miami, they can simply be cabled to the line side of the port selector, enabling them to contend for access to class 2 ports along with the Boston terminal users.

Unlike packet switches, which siphon small portions of data from many users onto a shared line, port selectors permit users to build an on-demand line-switching network. Since many port selectors permit definition of as few as seven to as many as 256 port classes, users can employ this capability to build economical, though complex, line-switching networks.

Wrong number

When a large number of port classes are employed, however, users may be more prone to enter an improper class number and then be routed to the wrong destination. Two features offered with many port selectors can be used to minimize this misrouting: welcome messages and symbolic class names.

On some port selectors, the welcome message is placed into ROM (read-only memory). Since this requires the port selector vendor to "burn in" the message, network managers must carefully consider the composition of the message because it is both time-consuming and costly to

5. Share the wealth. *Common network access to multiple computers can be obtained with the use of port selectors. This creates a line-switching as opposed to a packet-switching network. The savings in long-distance costs can be considerable, as shown in this example set on the Eastern Seaboard. Such line-switching networks can be quite complex.*

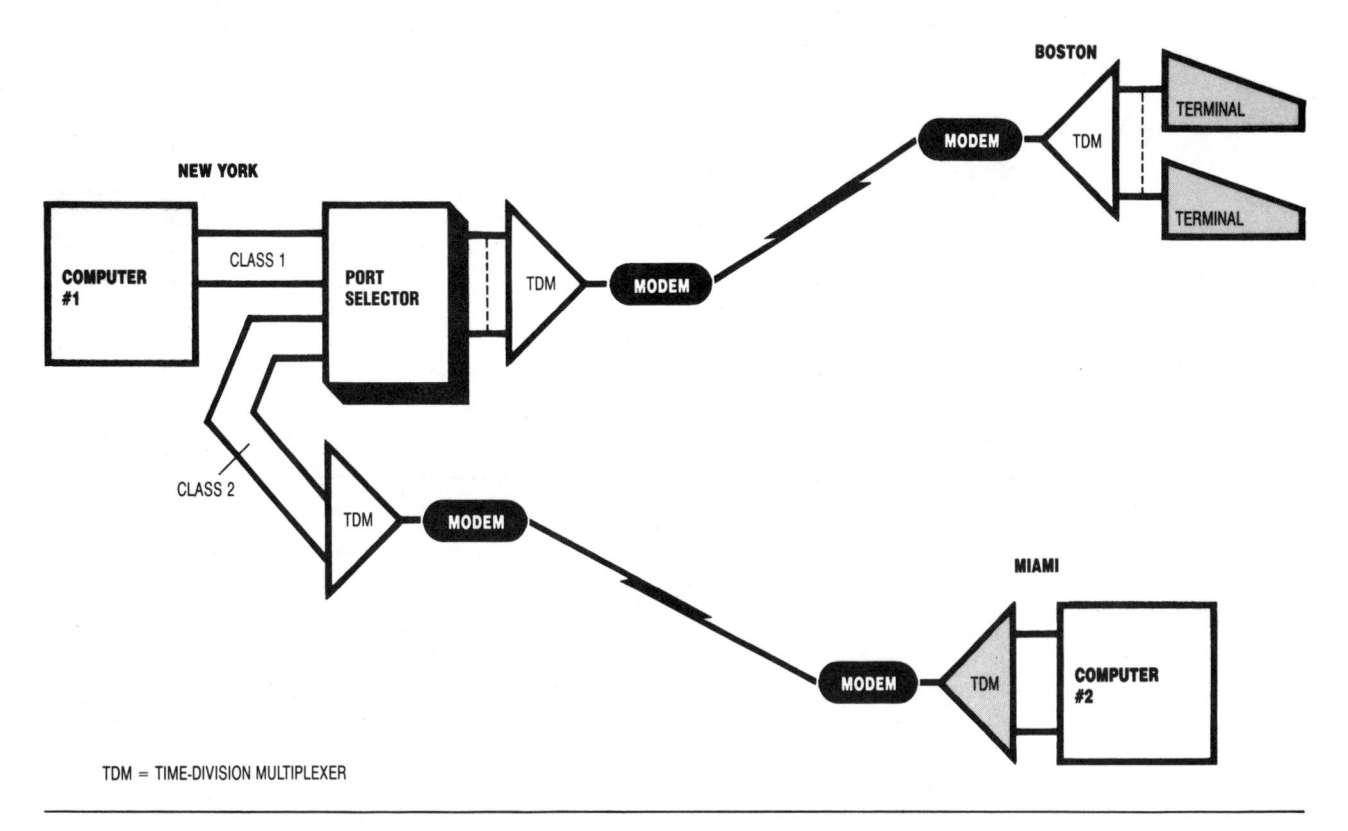

TDM = TIME-DIVISION MULTIPLEXER

change. On other port selectors, the welcome message resides in RAM (random access memory) and can be changed instantaneously through a console attached to the device. The welcome message is immediately sent to each user line after a connection between the user and the port selector is established, and before the user enters a routing class.

By the careful construction of a two-part welcome message, on-line assistance can be provided to network users. Users can, for example, select a port class number that provides interactive help. This capability can be provided by port selectors that support both general welcome messages as well as messages assigned to particular port classes. Figure 6 illustrates a sample help message that could be used to assist forgetful network users by explaining the various routing classes used to access different computers or speed groups on a front-end processor.

Since errors in numeric keyboard entries are easily entered and the codes assigned to port classes may not be meaningful, some port selector vendors support symbolic name classes. This option permits users to define names to each class, permitting, for example, terminal users to enter BOSTON in place of a numeric 1, so that data is routed to the port class that services a computer in Boston. Some port selectors limit symbolic class names to six or eight characters while other devices permit up to 64 characters, so some names may have to be abbreviated or truncated to fit within the character length constraints.

Another valuable feature offered with some port selectors is the ability of the console operator to limit line entry access to one or more specified groups of port classes. As shown in Figure 6, this feature could be used to limit some line inputs to IBM computers and others to Hewlett-Packard or Honeywell units. Since this access security method requires each line input to have a specific set of port-class permissions, difficulties can arise when network input into the port selector comes through telephone rotary equipment, which can dynamically alter the route a data source takes to access the port selector (Fig. 7).

Any terminal user accessing the network through the rotary at the remote site can be routed over line 1 to line N into the port selector. Therefore, implementing access security by the assignment of line inputs to port classes would require N lines to be treated as entities. As the data

6. Where to turn. *This sample help message generated by a port selector aids users who cannot remember the easily forgettable numbers that associate different kinds of lines.*

PORT CLASSES AVAILABLE FOR SELECTION
1 = HEWLETT-PACKARD BOSTON
2 = HEWLETT-PACKARD MIAMI
3 = IBM 300 BIT/S MACON
12 = IBM 1,200 BIT/S MACON
44 = HONEYWELL MACON

247

7. Restriction. *Access security made through the assignment of line inputs to port classes can limit user flexibility. Some vendors also provide password protection. Since it has been said that the greatest threat to national security is microcomputers, port selectors provide a graceful way to control access to sensitive corporate data.*

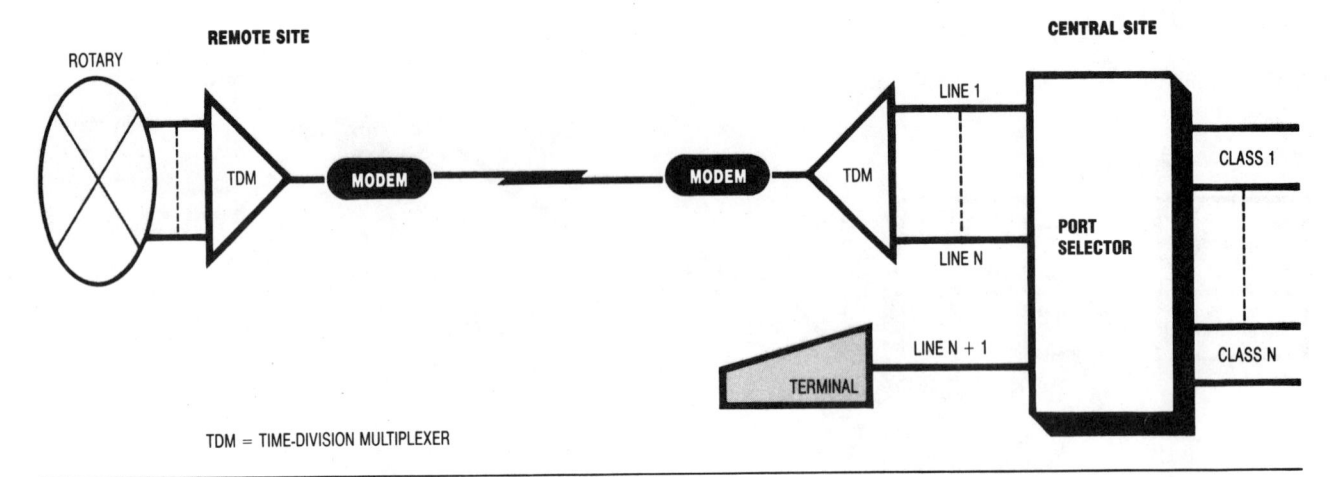

source on line input N1 is fixed, its assignment to a specific port class or group of port classes affects only the terminal connected to that line input. This can occur since the position on the rotary varies.

Thus, if the last four digits of the telephone number of the rotary were 5500, and if 5501 and 5502 were in use when a terminal user dialed 5500, he or she would be rotated to 5503, resulting in a connection to the fourth port on the multiplexer. The data source would then be routed onto line 4 of the port selector. This means that access security implemented by the assignment of port class permissions to lines would require the console operator to treat all lines originating from the rotary as one line group for access security purposes.

Suppose some network users located at the remote site illustrated in Figure 7 required access only to port class 2 while other users were unrestricted in their port class access. Satisfying such a set of circumstances would require the terminals that are only permitted to access port class 2 to be directly cabled to the remote multiplexer. Since this may not be physically possible, some port selector vendors now provide password-controlled permission to port classes. In this case, a terminal can connect to any line of the port selector and be cross-connected to any port class as long as the terminal operator knows and enters the valid password to access that class.

Statistics

Since a port selector is the heart of many networks, any statistical information it provides can be very valuable in determining the flow of traffic through a network. In addition, by analyzing the statistical data generated by some port selectors, users can take corrective action to resolve potential network problems before such problems affect the network's users.

Figure 8 illustrates the statistical log format generated by a Micom 600 port selector, as well as seven representative activity sequences showing how this information can be used to assist in the network operation and design process. The first activity sequence indicates a normal port selector connect (C) from line 1 (L001) to port 1 (P001) on Day one at 7:30 A.M. The second activity sequence denotes a disconnect due to time-out (DT) of the cross connection of port 1 (P001) to line 1 (L001) at the indicated time on Day one. Since statistics can be routed to a predefined port or to an operator's console, a microcomputer can be connected to the port selector to record the activity as it occurs throughout the day. A program can then be developed to analyze the logged data and produce appropriate reports.

One example of an appropriate report based on an analysis of logged port selector statistics might be the frequency of disconnects due to time-outs caused by terminal users reaching a predefined period of inactivity.

8. Formats. *This Micom 600 statistical log depicts the network management information that can be made available through port selectors.*

ACTIVITY SEQUENCE NUMBER	REASON INDICATOR	LINE OR PORT NUMBER	LINE, PORT, OR CLASS NUMBER	DAY	TIME hh:mm:ss

REPRESENTATIVE ACTIVITY SEQUENCES

0001	C	L001	P001	001	07:30:00
0002	DT	P001	L001	001	07:45:00
0003	C	L001	P001	001	07:45:30
0004	C	L002	P002	001	07:47:15
0005	C	L008	P003	001	07:48:42
0006	D	P001	L001	001	07:49:18
0007	C	L014	P001	001	08:01:12

C = CONNECT
D = DISCONNECT
DT = TIMEOUT DISCONNECT

Port selector features to consider

FEATURE	REQUIRE-MENT	VENDOR A	VENDOR B
NUMBER OF LINES SUPPORTED			
NUMBER OF PORTS SUPPORTED			
LINE-SIDE CONNECTIONS SUPPORTED			
RING INDICATOR			
DATA-TERMINAL-READY			
DATA ACTIVITY			
DISCONNECT SEQUENCES SUPPORTED			
CLEAR-TO-SEND DROPPED			
BREAK			
ACTIVITY TIMEOUT			
INPUT DATA RATE SUPPORT			
WELCOME MESSAGE			
FIXED			
VARIABLE			
NUMBER OF PORT CLASSES			
ACCESS SECURITY TO PORT CLASSES			
LINE TO PORT			
PASSWORD			
SYMBOLIC CLASS NAMES			
ASYMMETRICAL CHANNEL SPEEDS			
QUEUING			
STATISTICS			

This information might be used either to change the time-out period or re-emphasize to users that when they take a break they should log off the computer to permit other users to use the organization's computer resources.

Another use of port selector statistics is to examine the traffic intensity offered to both individual ports and port classes. This information can then be used to determine if a reconfiguration of port classes or an addition to the number of existing ports configured into a particular port class is warranted.

Since few, if any, networks are similar, the selection criteria for obtaining port selectors will vary based on the network requirements of the organization. "Port selector features to consider" contains a list of major features that should be considered during the hardware evaluation process. Although this list can be greatly expanded and does not include cost factors, it should provide a firm foundation for an evaluation. ■

Gilbert Held, director of 4-Degree Consulting, is an internationally recognized author and lecturer. Twice a recipient of the Interface Karp award and a winner of the American Association of Publishers' award, Held is the author of 12 books and more than 60 technical articles.

Michael W. Cerruti and Maurice Voce, Intel Corp., Phoenix, Ariz.

Zap data where it really counts — Direct-to-host connections

Attaching directly to a mainframe channel, bypassing the front-end, can pay off for high-speed links to LANs, or to other mainframes.

Connectivity used to mean mainly the interconnection of terminals and host processors. Now, of course, it has evolved into a much grander issue, encompassing not only terminal applications, but also the need to connect diverse islands of automation — consisting of microcomputers, minicomputers, and mainframes, plus various network topologies and a wide range of peripherals.

For increasing numbers of users, a key component of this broad connectivity challenge is making a connection to a mainframe, such as an IBM System/370. The host connection not only opens extensive databases to authorized users, but also provides access to the programs resident in the mainframe. Host access also lets users exploit the mainframe's considerable processing power for performing data-intensive tasks.

In addition, a mainframe connection allows multiple network workstations to simultaneously run the same complex application program resident on a mainframe, and to accomplish tasks cooperatively with it.

For example, to make reservations and seat assignments for individual airline flights, travel agents can use workstations in their offices in cooperation with a remote airline mainframe. After logging onto the network, the agent identifies the customer's flight number, then sends the necessary data to the remote mainframe for processing (Fig. 1).

The mainframe in turn provides information to the agent regarding the type of craft involved and kinds of seats available. Based on that information, the agent can readily book reservations, enter and check locally maintained accounting data, and then return the selection information to the central mainframe where it can be stored and then retrieved by the airline for its use.

This type of cooperative interaction, here between an airline mainframe and processors at travel agencies, is becoming increasingly common. Among the benefits offered, one is to relieve the central mainframe of performing the entire processing task itself — and thus avoid becoming overloaded — while travel agents can be provided with locally maintained data required to run their business. Further, with this cooperative interaction, the various workstations share in accomplishing work efficiently — usually within a more reasonable time frame than is possible if the mainframe did the entire job.

Connectivity crisis
Whatever the application, the need to boost organizational productivity by pooling resources and distributing information effectively between users and the mainframe is becoming an important part of the overall connectivity picture. But at the same time that users are recognizing the potential of the mainframe connection — or perhaps because of this recognition — a phenomenon that some observers have termed a "connectivity crisis" is occurring in many networks.

Many users purchased their computers believing that they would connect easily with mainframes and other processors; others selected their equipment without giving the issue of mainframe connectivity any consideration. The crisis is felt when users realize that they were not sold "instant" connectivity when they invested in their computing devices. Further, when users realize that traditional methods for connecting diverse tools to the mainframe are not always as easy or as efficient as anticipated, the crisis deepens.

The mainframe connection has been traditionally accomplished by linking to a mainframe or I/O (input/output) channel through telecommunications solutions. (As used here, telecommunications refers to remote access.) Principally, these links have been used to connect terminals, remote peripherals, and remote mainframes to the central

1. Cooperative processing. *A mainframe connection allows multiple network workstations to run the same complex application programs resident on the mainframe.*

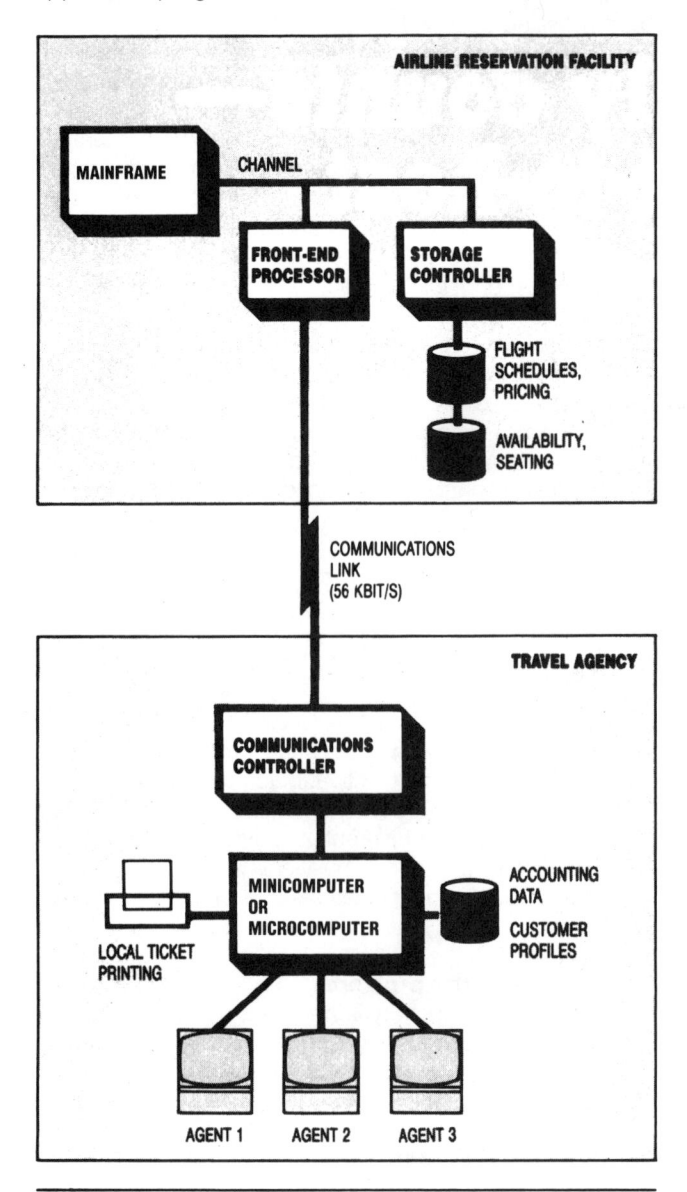

mainframe. In such applications, telecommunications connections allow remote users to move screen images of data interactively. File transfers, however, are handled somewhat less efficiently. Addressing these non-time-critical data transfers with the traditional telecommunications link has nevertheless been quite successful.

Telecommunications connections are optimized for terminal-oriented messages. In today's computing environment, however, a broader orientation is needed if diverse computer networks and peripherals are to be connected in a more symmetrical peer-to-peer relationship with the user's mainframe.

The connection must also be broad enough in scope to support different sizes of data, satisfying diverse users with differing data transfer requirements. In addition, the connection should be easy enough to implement so that the mainframe becomes an accessible server for the various devices, instead of an inflexible center around which all other tools must be configured with comparatively greater effort or difficulty.

As Figure 2 shows, telecommunications solutions use terminal emulation hardware and software and associated communications processors to connect to the mainframe. Generally, they are limited to maximum data transfer rates of 56 kbit/s. This rate is inadequate to sustain acceptable interactive response times (the goal is no more than three seconds) on networks supporting hundreds of users or applications that involve the transfer of large amounts of information in real time.

The bottleneck

The problem with telecommunications solutions is that they require 12 to 13 minutes, on average (in the case of a 56-kbit/s digital facility), to move a 1-Mbyte file between mainframe and microcomputer on a network. Although higher transmission speeds such as T1 are available, they are still limited to transferring data in terminal mode, which limits the transfer block size.

Such high data transmission speeds saturate the processing bandwidth of the front-end communications

2. Old solutions. *Telecommunications uses terminal-emulation hardware and software and associated communications processors to connect to the mainframe.*

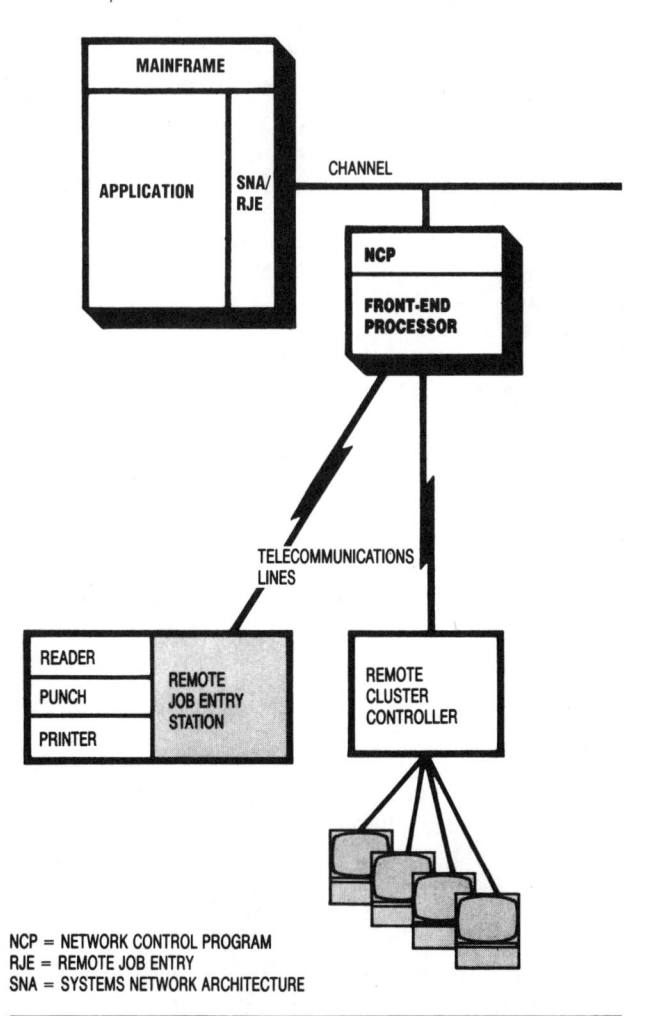

NCP = NETWORK CONTROL PROGRAM
RJE = REMOTE JOB ENTRY
SNA = SYSTEMS NETWORK ARCHITECTURE

3. Connectivity needs. *A chemical company's accounting department transfers to the mainframe via a cluster controller. The processing plant connects via LANs.*

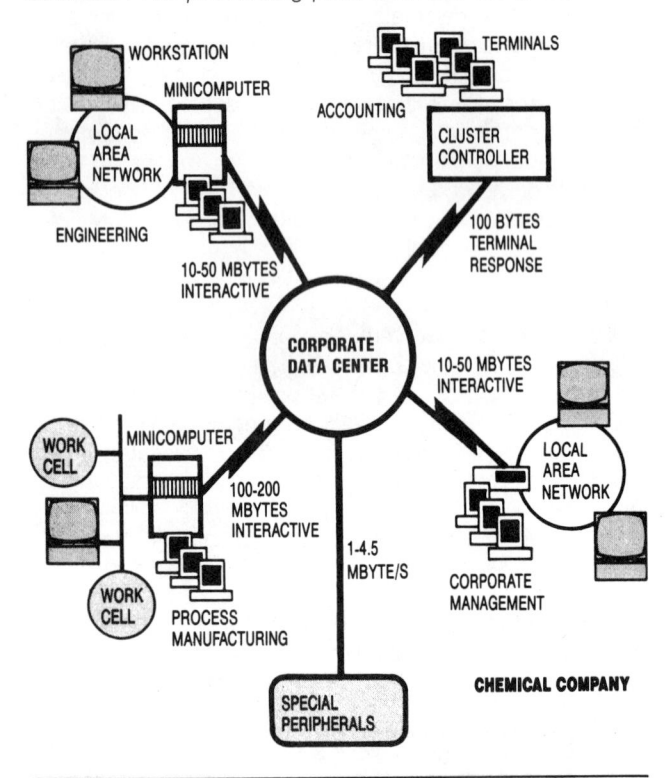

controller, thereby greatly increasing the cost of the connection. As a result, users face a bottleneck that can hamper throughput and, subsequently, productivity. Moreover, the mainframe link must be able to accommodate ever higher transmission speeds such as 100 Mbit/s supported on new fiber-based LANs (local area networks).

Chemistry lesson
To understand the demands placed upon a mainframe link by users with different types of mainframe attachments and varying data transfer requirements, consider a typical chemical company (Fig. 3).

The accounting department is one of the company's major data users. The department is in the same building as the mainframe, but beyond the maximum allowable distance for local attached devices. The department connects to the corporate mainframe primarily by terminals and remote peripherals, and it shares data and applications running on the mainframe.

Accounting, in a transaction mode, typically sends small data blocks (2 kbytes) to the mainframe. It uses standard telecommunications links, such as synchronous data link control over leased lines, for such small data transfers, and its users find this transmission method adequate. When the department has to transfer larger data blocks, however, accounting finds transmission considerably slower (the previously mentioned 12 to 13 minutes per megabyte). Such batch transmissions are typically performed during off-peak hours.

A second company group, those in the processing plant, connect to the corporate mainframe via local area networks, terminals, and other computing devices. In this way, the group can perform real-time processing tasks, access applications on the mainframe, and share data with other groups in the company. This user group represents medium-sized data exchanges of 100- to 200-kbyte blocks.

The company's engineering organization connects to the mainframe chiefly via other computers such as departmental minicomputers used for local data "crunching." The engineering group relies on the mainframe for the temporary or permanent storage of data that later can be extracted for further analysis. The transferred blocks of data can be as large as 50 Mbytes and require data transfers approximating local disk-transfer rates (250 kbyte to 1 Mbyte per second; for these comparisons, one byte equals 8 to 10 bits).

In our scenario, a fourth group, the corporate management team, connects via local area network attachments to the corporate mainframe so that the group can tap into its shared databases and programs. These users, like those of the engineering staff, require the equivalent of disk-transfer rates for large amounts of data: 50 or more Mbytes at a time.

Yet with a telecommunications link to the mainframe, all four groups, which represent very different sets of expectations, connections, and distances from the mainframe, must rely on the same connectivity performance characteristics.

Ultimately, because a telecommunications link favors terminal-type connections, the result is that data transmission can be slowed down to an unacceptable level when bulk file transfers or time-critical batch transmissions are required. Thus this link is likely to fail to efficiently meet the mixed demands of the different user groups in the chemical plant scenario—and in organizations with a similar dichotomy of local communications and mainframe-access configurations.

Reconsidering the problem
It is clear that connectivity is not just a matter of installing cable—it involves much more than fitting a computing tool or new user into a wiring scheme. Instead, connectivity must, by definition, also include the capability of connecting diverse applications with varying data transfer requirements to a range of mainframes. It must also provide organizations with a "comfortable," user-oriented way of interacting with the mainframe.

What has been lacking is a connectivity approach removing the mainframe from its position at the center of the network universe, making the mainframe an accessible server, and featuring direct attachment to the mainframe by varied user communities at channel speeds (3 Mbyte/s maximum).

Connecting to the mainframe means opening the machine's architecture so that the mainframe becomes a productive, integral part of the user's world, not a monolith around which everything else must be designed and configured. What is needed to realize such a connectivity scenario is a bidirectional bridge directly between the IBM mainframe and diverse users that can be incorporated into existing architectures in order to protect the user's investment in already-installed devices. Replacement costs are thus avoided.

4. Meeting the needs. *Shown is a mainframe computer with an open-architecture channel adapter attached directly to a mainframe channel, operating at channel speed. It allows the user to readily connect multiple devices through standard protocols, such as IEEE 796, without changing software at the application level.*

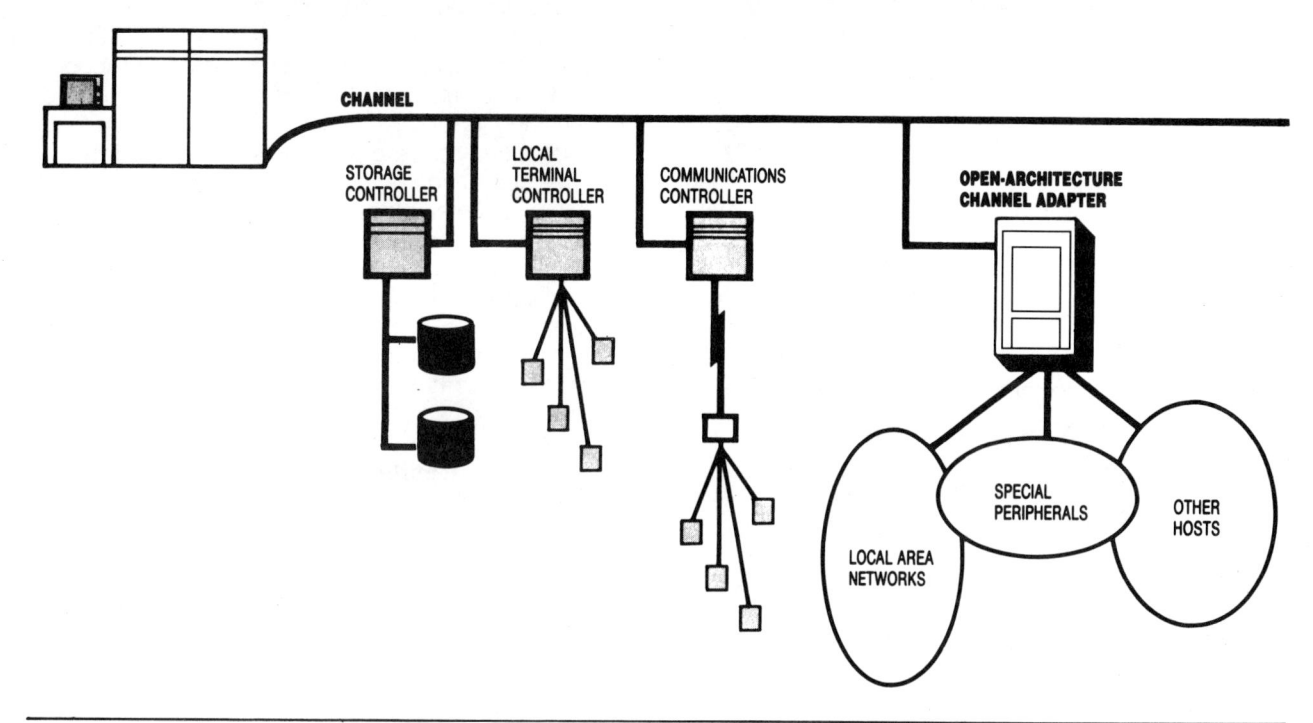

Recognizing the growing need to make the mainframe easier to access as a node on a network, IBM now offers built-in connectivity on its new 9370 computers. The 9370's integrated communications controllers enhance local area network-to-mainframe connectivity by opening the latest IBM mainframe architecture to a broad range of users. Architecturally speaking, that was accomplished by enabling the built-in controllers to interface directly to local area networks and to other computers, without having to go through front-end processors. But these built-in connectivity devices are not available for the entire family of IBM computers. Therefore, they do not meet the connectivity challenges posed by "closed" System/370-type mainframe architectures.

Another product that has attempted to address System/370-type connectivity is the multisource channel adapter, a device that provides a pathway for connecting to the mainframe various computing tools, such as local area networks and high-speed peripherals. Until recently, however, most channel adapters have been designed with proprietary interfaces. That has restricted users to connecting only those products offered by the channel-adapter vendor, thus severely limiting user choice of those computing tools. In addition, most of the adapters have been special-purpose connectivity products capable of handling only one type of connection, and have not supported general attachment for multiple devices.

A nonproprietary open-architecture channel adapter can bring less limiting solutions to users and serve as a connectivity "platform." It can go beyond the terminal orientation of telecommunications links. Shown in Figure 4 is an open-architecture channel adapter, which is an interface device that attaches directly to the mainframe channel at channel speed. It allows the user to readily connect multiple applications through standard protocols, such as IEEE 796, without changing software at the application level.

Channel-attached connectivity devices act as high-speed (3-Mbyte/s) interfaces between various types of IBM-compatible mainframes and local area networks, minicomputers, other mainframes, microcomputers, and specialized peripherals such as optical disks. Such connectivity allows original equipment manufacturers (OEMs) to customize their applications by using programmable interfaces.

Enhancing performance

Channel-attached connectivity opens an IBM mainframe to a wide array of non-IBM devices, networks, and computers. This would be similar to IBM—in a departure from its tradition—"going public" with the bus architecture of its comparatively low-cost microcomputer so that vendors could build applications for a variety of environments (such as LANs and archiving tapes) by using add-in printed-circuit boards.

An immediate benefit of the open-architecture channel adapter application is that peripheral manufacturers who wish to supply products for the mainframe are spared the expense of developing connectivity technology and can focus on product development.

The adapters can also connect mainframes to minicomputers, such as the DEC (Digital Equipment Corp.) VAX, so that VAX users can exploit the resources of both computers. The adapters also allow users to connect various LAN topologies to the mainframe.

The adapter allows multiple connections through a single

channel attachment, minimizing the number of channel attachments. And it handles combinations of applications concurrently, so that disparate elements—such as LANs, optical disks, and minicomputers—can share the mainframe economically. The ability to connect diverse applications concurrently lets users multiplex the bandwidth of the channel for slower devices. Adapters that provide a buffer for data flowing at channel speeds to slower applications offer an effective way to enhance channel throughput. The open-architecture channel adapter allows users to define their application priorities by user or message type—that is especially important when multiple applications coexist in a single adapter.

When evaluating potential adapter vendors, users should be on the lookout for vendor-sponsored development facilities. Where such resources exist, users have access to technically knowledgeable personnel who can help with testing implementations, providing important assistance with application development.

Another vendor-evaluation criterion is vendor experience in providing mainframe-related products and support. The vendor's commitment to ongoing research and development, and its long-term viability, are more obvious points for consideration (see "Connectivity guidelines").

Besides standard connections, the open-architecture channel adapter is designed to deliver tools and methodologies for those users requiring customized applications. The tools include libraries of application routines and development environments that allow simulation of a final configuration during the implementation cycle. Methodologies include protocols that make tasks easier to perform, such as a disciplined methodology for interfacing to the adapter's control unit.

Evolving expectations

The open-architecture channel adapter enables a user to connect several applications concurrently—an efficient approach—which makes the mainframe accessible to many different users and applications. That technology is based on industry standards such as IBM's OEM interface or Federal Information Processing Standard 60.

Attaching to megabyte-per-second channels becomes even more critical as channel speeds steadily increase. Amdahl has already announced a 4.5-Mbyte/s channel. And there is speculation that IBM will move into fiber optics, which should drive channel rates even higher—to at least 12 Mbyte/s.

The speed associated with local area network technology is also on the rise. Standard LAN technology currently offers 1.25 Mbyte/s (the equivalent of 10 Mbit/s) capability. Thus the 3-Mbyte/s data-transfer rate offered by channel-attached adapters is more than adequate for most LAN applications.

However, as LAN speeds increase further, the bottleneck will be moved back to the channel. Users must assure themselves that the connectivity interface between the LAN and the mainframe is adaptable enough to accommodate improved performance on both sides in a constant evolutionary game of catch-up.

The emergence of new peripheral devices requiring connection to the mainframe will also continue. The adapter's channel-speed performance will be a boon for interfacing between the mainframe and advanced devices treated as peripherals, such as aircraft simulators, optical disks, and CAD/CAM (computer-aided design/computer-aided manufacturing) workstations. ■

Mike Cerruti is strategic marketing manager for Intel's System Interconnect Operation. He has worked with mainframe software and channel connect technology for the past 15 years. Maurice Voce is product marketing manager for Intel's System Interconnect Operation. He has had more than seven years' system engineering and marketing experience ranging from mainframes to personal computers and networks. Voce holds a B.S. in mathematics and computer science from UCLA.

Connectivity guidelines

Some general connectivity rules for effecting a high-speed direct-to-mainframe link apply to a number of different applications and to evolving technological changes:

1. The channel adapter should support industry standards such as those adopted by the International Organization for Standardization (ISO) and the IEEE. Examples: the Open Systems Interconnection (OSI) model and token ring and Ethernet networks.

2. Tools for network management should be provided.

3. An important feature to consider is the ability to support multiple device types so that the channel adapter looks to be whatever the mainframe deems is appropriate for the type of function being connected. This avoids the expense and performance inefficiency of multiple layers of protocol conversion software on the mainframe.

4. Look for a channel adapter that can be readily configured with a wide range of adapter boards to support standard applications. This will reduce development time in obtaining specific application solutions. It also enables a simple connectivity to support a number of different connections, such as to a Digital Equipment VAX minicomputer, an Ethernet or token ring LAN, a Manufacturing Automation Protocol application, or an ASCII terminal. These multiple mainframe-connectivity applications could also be run simultaneously.

5. Where there is a time-critical element to the data flow, such as in engineering and scientific environments, look for a high-speed interface of at least 3-Mbyte/s transmission in a data streaming mode (not limiting transmission to a predetermined amount).

6. Choose a vendor carefully, because the vendor's experience, technical knowledge, service, and support are key to the connectivity's success. Pick a vendor whose record shows long-term reliability. Look for a vendor who can help develop and test applications, perhaps by providing a development laboratory. And not least, make sure the vendor can provide training and field support for the entire connection, including the application.

Lee Mantelman, DATA COMMUNICATIONS

Orchestrating people and computers in their networks

Networking is the rage, for both computers and people. Opportunity awaits those who can meld these technical and human processes.

Network managers must relate to two groups of nontechnical people: end users on the one hand, and corporate management on the other. Reciprocally, these groups must also deal with their network managers. Yet there often seems to be a mile-high wall separating the three.

A wealth of management studies has explored the relations between end users (sometimes called "workers") and management. Yet, for network managers, who today may hold their corporation's future in their hands, there are few guidelines for relating to the other groups. This is ironic, as network managers have a unique opportunity to bridge communications gaps between people, a strategic key to corporate growth.

A very simple analogy could serve as an interface, so to speak, between the technical managers and everyone else: the analogy between people and computers in their respective networks.

Workers relate to one another in a variety of interpersonal networks, both those formally defined by the corporation and those informally constructed in day-to-day interaction. The job of corporate management is to orchestrate these networks while thriving within networks of their own. Information networking gurus can wield much more constructive influence by exploring these "people networks"; comparing them to the jumble of processors, software, and linkages that they call home; and using each to explain and understand the other.

Many disciplines, including the social sciences and even the performing arts, can help network managers understand people networks and explain computer networks. At least two conferences, one on Computer Supported Cooperative Work (CSCW), sponsored by MCC (Microlectronics and Computer Corp.) and several scholarly societies, and the Conference on Office Information Systems, sponsored by the Association for Computing Machinery, bring together social and computer scientists concerned with group processes. Networking managers should consider attending. They would not be the only nonacademics; some two-thirds of the 300 participants at the CSCW conference were from industry.

To study networking technology is to study sociology. This idea is not as farfetched as it might first appear: Computer networks evolve to meet the needs of the human networks that spawn them, and they subsequently, and quite naturally, affect their users' social structures. Moreover, just as learning a new program means grappling with the mind that created it, so, too, a firm's experience with a new network is essentially the meeting of two sets of complex social processes. Understanding this can help network managers achieve a good fit.

Network managers need to understand the social processes that they are trying to support, so they can select networks that are harmonious with those processes. An example is the contrast between IBM's Systems Network Architecture (SNA), historically a rigid hierarchy, and Digital Equipment Corporation's (DEC's) Decnet, a peer-to-peer networking scheme. "The architecture of a network must match the organizational philosophy of the company, whether that is centralized, decentralized, or distributed," observes Glenn Habern, a partner with Chicago-based Arthur Young & Co.

In their makers' images
Networks develop in accord with the social structures that surround and influence—indeed breed—them. This is clear in the case of the following packet-switched data networks:
■ Arpanet was built to support military needs. The architecture permitted new nodes to be added and removed rapidly, yielding a flexible topology. This made the network a success with military researchers and with researchers in general, generating so many connections on college

campuses that the military and classified users have since broken away and formed their own subnetwork.

■ Telenet, a commercial spin-off of Arpanet, was built to conform internally to de jure standards such as X.25. Based outside Washington, D. C., Telenet's corporate ambience was predisposed to national and international standards. Setting standards from on high and then building a nationwide network in accordance with them also reflected an East Coast way of doing things.

■ Tymnet, by contrast, grew up on the West Coast as a more ad hoc solution to time-sharing access needs. The network was built on the fly using proprietary internal protocols, which let it conform to the land, so to speak, as it branched eastward. New needs were met with tools developed for the occasion.

Telenet and Tymnet grew more alike with time, but remnants of their different origins linger. Both vendors introduced private-network offerings based on their own public networks, so prospective purchasers should consider the extent to which the private-network architectures still reflect the differing development philosophies and then decide how well each one might fit in the user's setting. Networks most richly support the kinds of interaction that align with their communications philosophy.

Computer architectures have also tended to reinforce the structure of the vendor firm within the user organization. SNA, originally favored by highly centralized companies or divisions, was developed by a monolithic enterprise run with an almost martial rigidity. Decnet, preferred by engineering divisions and firms, grew out of an environment in which all computers, like all engineering professionals, had an equal chance to express themselves.

The center and the challenge
In the meantime, network managers are faced with the decline of centralization, which has been evidenced by a trend toward fewer mainframes and more minicomputers and microcomputers.

IBM, the preeminent mainframe vendor, is still trying to adjust. In network management, a major question mark in decentralized networking, IBM is attempting to restore control to the mainframe with Netview and Netview/PC. It is enhancing its SNA protocol suite to include Logical Unit 6.2 for peer-to-peer communications. And architecturally, it is issuing Low-Entry Networking (LEN), a remake of SNA that more closely resembles Arpanet. Philosophically, IBM seems to recognize that a new day is dawning.

"Intercommunication will change corporate structure from a vertical hierarchy to a broader, more horizontal one," says Dr. Robert Carberry, IBM's director of engineering and technology. "That's one example of change due to computer communications."

Carberry's remark reflects a parallel dilemma facing corporate managers: Individuals and departments have a growing sense of and demand for autonomy, in large part bolstered by powerful, low-cost processors. However, rather than a threat, both technical and organizational managers face what may be their greatest challenge yet. They must try to adapt their command of centralization to the support, guidance, and coordination of more autonomous, distributed intelligences.

The secret to doing this is already being discovered from the bottom up: networking. This type of networking, the kind that people do, is seldom discussed by data communications professionals, but it is increasingly important to their corporate counterparts.

In his ground-breaking book, *Corporate Networking* (Free Press, N. Y., 1986), Robert K. Mueller, former chairman of Arthur D. Little, offers a manifesto for corporate managers that also can be readily adapted by technologists. "Given all the advanced technology, we tend to forget the value of social networking, the informal gossip channels, and verbal and written grapevines that persist in all organizations," writes Mueller.

Turbulent times are prompting the move away from centralization. According to Mueller, "top-down communication can be very effective in a military, religious, political, corporate, or other conventionally structured organization. However, when changes in organizational culture, climate, structure, or process take place in complex social systems, top-down, linear communication paths are often inadequate."

The question is not whether people and computers are networking in all sorts of decentralized ways. They are. The question is how to manage these networks. The old model of command and control is less workable with more autonomous entities. However, managers can have some influence in the process. "The care and feeding of the network requires managers to shift their focus from directing and controlling to supporting," writes Mueller.

One way of doing this is to build, or to adapt, existing hierarchical structures for use as overlays on emerging networks. An example of this is the hierarchical supervisory overlay network, which is used to solve problems in packet networks with a distributed-mesh topology.

IBM wants its Netview-based family of tools to serve as an overlay for control networks of non-SNA equipment. It faces a similar challenge in using subarea SNA networks to integrate LEN networks. Corporate managers will be trying a variant of this: "Use of an overlay permits a manager to realize the advantages of a network mode while still maintaining the strengths of a bureaucratic organization," Mueller writes.

Mueller offers eight steps to designing and setting up a network of people. Data network designers could tap these ideas for their own network-building efforts. At least six of these have obvious computer equivalents, which follow Mueller's steps in parentheses:

■ Clarify the network's purposes. (Know why you need a computer network.)

■ Inventory your own resources, such as knowledge, contacts, and so on. (Each network node takes stock of its local resources.)

■ Identify needed resources unavailable locally. (Determine what resources are needed from other nodes.)

■ Select a suitable network structure. (Software-defined networks can do this, as can the more complex self-designing organizations in distributed artificial intelligence.)

■ Assess what kind of networker each person (node) in the network will be. For example: who is enough of a focus of communications to serve as a steering committee member (which nodes could serve as backbone nodes); who will be less communicative—"isolated" (singly connected nodes); who in your network can span boundaries and link

you to some other set of persons (bridges, gateways, or protocol converters).
- Decide which networking process will be most effective. (Determine which structure-forming interaction style, such as SNA-style system generations or Arpanet-style routing-table exchanges, would best establish the network.)

Other parallels between people networks and computer networks can be found in concepts and terms that apply equally well to either of them. Computer networking lingo is beginning to escape into management and social science usage, as computer jargon did earlier. The following examples, from Mueller's glossary, are suggestive of a text on computer networking:
- Bridge: A member of multiple clusters in the network.
- Clustering: The number of dense regions in a network.
- Density: The number of actual links in a network as a ratio of the number of possible links.
- Gatekeeper: A network member who screens messages (like a filter); a link to external domains (a gateway).
- Openness: The number of actual external links as a ratio of possible external links.

The similarities are no accident. As Mueller points out, "the integration of human networking with computers is part of the social network movement." And Mueller seems to insist that this movement is essential for corporate growth.

Protocols as dialogue

These examples have shown how people's communications can be illuminated with concepts from data networking. The opposite is also true. The protocols in network architectures reflect human communications at each layer.

Network protocols are actually human dialogue embodied in software. When two users or applications use a layered architecture, their interactions are conveyed by messages being passed between and within the layers. These messages are ghostly software trails of the conversations once agreed upon by each layer's designers.

The same is true of computer use in general. Clicking a mouse somewhere on a screen activates a module of code that sends a "mouse-click notice" to a subroutine along with a parameter stating the cursor location. The interface between the calling module and the subroutine is one agreed upon linguistically by the programmers: "When I send you this message, it means mouse-click, and I'll tell you where." Though the original programmers are no longer present, the computer duplicates their interaction.

This relation of human to machine is even clearer in a communications setting. In the physical layer, for example, names of RS-232-C leads and the handshaking sequences that use them suggest people using the telephone. "Ring indicator" means an incoming phone call. The called party picks up: "Data terminal ready." "Carrier detect" equates to the parties' ability to hear each other. "May I say something?" asks one—"Request to send." "Yes, what is it?" replies the other—"Clear to send." The dialogue continues over the transmit and receive leads. This exchange is not the only one possible, of course, and software on each side would have to agree on the sequence. (One can imagine designers thrashing out the details: "When I raise this lead, it will mean this, and you respond like that.")

However, computers have a lot to say to each other, more than can be represented by any number of physical leads. There is no lead, for example, that means "What was that again?" A plethora of link-level protocols, above the physical connection level, arose to support more complex dialogue and to support varied hardware environments. Byte-level messages such as ENQ, ACK, and NAK evolved into bit-level equivalents with the same linguistic meaning. Again, the meanings and proper use of these messages must have been agreed upon.

More complex multidrop link disciplines may also reflect their makers' attitudes about communications. IBM's hierarchies operate in a military fashion, so the 3270 polling protocol resembles a sergeant drilling a row of privates: "Number 2! Anything to send?" "No, sir!" "Number 3! Anything?" "Yes, sir! Here it is, sir!"

In local area networks, token-passing protocols (such as IBM's) resemble a cross between "Dress right, dress! Sound off!" and the game of telephone, where messages are passed from person to person. By contrast, the collision protocols favored by many other vendors are more like crowded family dinners, with everyone trying to shout at once. They are more chaotic but potentially more democratic as well.

Intermediate-layer protocols, such as those in the International Organization for Standardization's Open Systems Interconnection model, tend to be more civil. Each protocol entity deals with its counterpart across the network via its underling, that is, via the entity at the layer below it.

The dialogue may go as follows: Request (entity A asks its underling to convey a request to entity B), Indication (B's underling informs B of the request), Response (B gives its underling a response for delivery to A), and Confirmation (A's underling relays B's answer to A). Such sequences are used for requesting network-layer calls, resetting transport-layer synchronization points, negotiating session-layer services, and so on.

The application layer, however, has a problem like that of the physical and link layers: Individuals and organizations have a lot to say to each other, more than can be handled by one or even several protocols. For this reason, a number of protocol standards at this level are already in the works, and more are expected.

Above application: The people layer

Many of these protocols are being designed for computer-derived uses such as file transfer and virtual terminal. However, some will be able to support interpersonal messaging. An examination of the finalized international standards that are intended to support actual users could reveal the standards setters' underlying view of people in organizations.

There is some question as to how widely accepted the person-to-person standards will be. These methods of communication, developed in committee, will succeed to the extent that they allow users to interact as the standards makers intend them to. Engineers often design networking software with themselves in mind, but the ultimate users turn out to work and think much differently. Thus, people are often forced to "make-do," to work around the technology and to make their organizations do what their networks allow them to do.

One vendor, Action Technologies Inc. (AT) of San

Francisco, has gone in the opposite direction. AT has created a networking software package called the Coordinator, the development of which was driven by social processes—rather than vice versa.

The product is based explicitly on a concept of human interaction within organizations that was spelled out by two of the company's founders, Terry Winograd and Fernando Flores, in their book *Understanding Computers and Cognition* (Ablex Publishing Corp., Norwood, N. J., 1986). In their view, "much of the work that managers do is concerned with initiating, monitoring, and above all coordinating the networks of speech acts that constitute social action."

Talk is dear

Our use of language to do things, the story goes, can be viewed in terms of five types of "speech acts." We can make statements, express emotions, and pronounce declarations, which are made true by virtue of their being said in an appropriate context. We can also make requests of and commitments to each other.

AT's founders were most interested in requests and commitments. They felt that these two types of speech acts underlie the true business of business, namely, the many "conversations for action" each worker conducts with other people. (This differs from the more freestyle, open-ended "conversations for possibilities," which let users deal more in "what ifs" than "would yous.")

The founders derived a number of conversational "moves" or speech acts that can follow each other in a conversation for action. Such a conversation may begin with an offer or a request (Fig. 1). Either of these may be followed by a counteroffer or by a decline, which ends the conversation. They may also be followed by a promise (or an accepted offer), which may be either fulfilled or reneged upon. If fulfilled, the results may be accepted or rejected, ending the matter (renegotiation may also follow an unsatisfactory outcome). If reneged upon, the options are likewise to drop the matter or to renegotiate.

All the moves shown in the diagram change the participants' commitments to each other. Other available moves, such as acknowledging a message without committing, or making a commitment to commit at some future time, are also possible, but they would involve no such change in the state of the conversation.

1. Conversation manager. *The heart of the Coordinator is a model of commitment in conversations. Users are helped along through the flow, with suggested responses to incoming messages guided by where that message is located in the model. Other types of messages can be handled, but the ones shown here are key.*

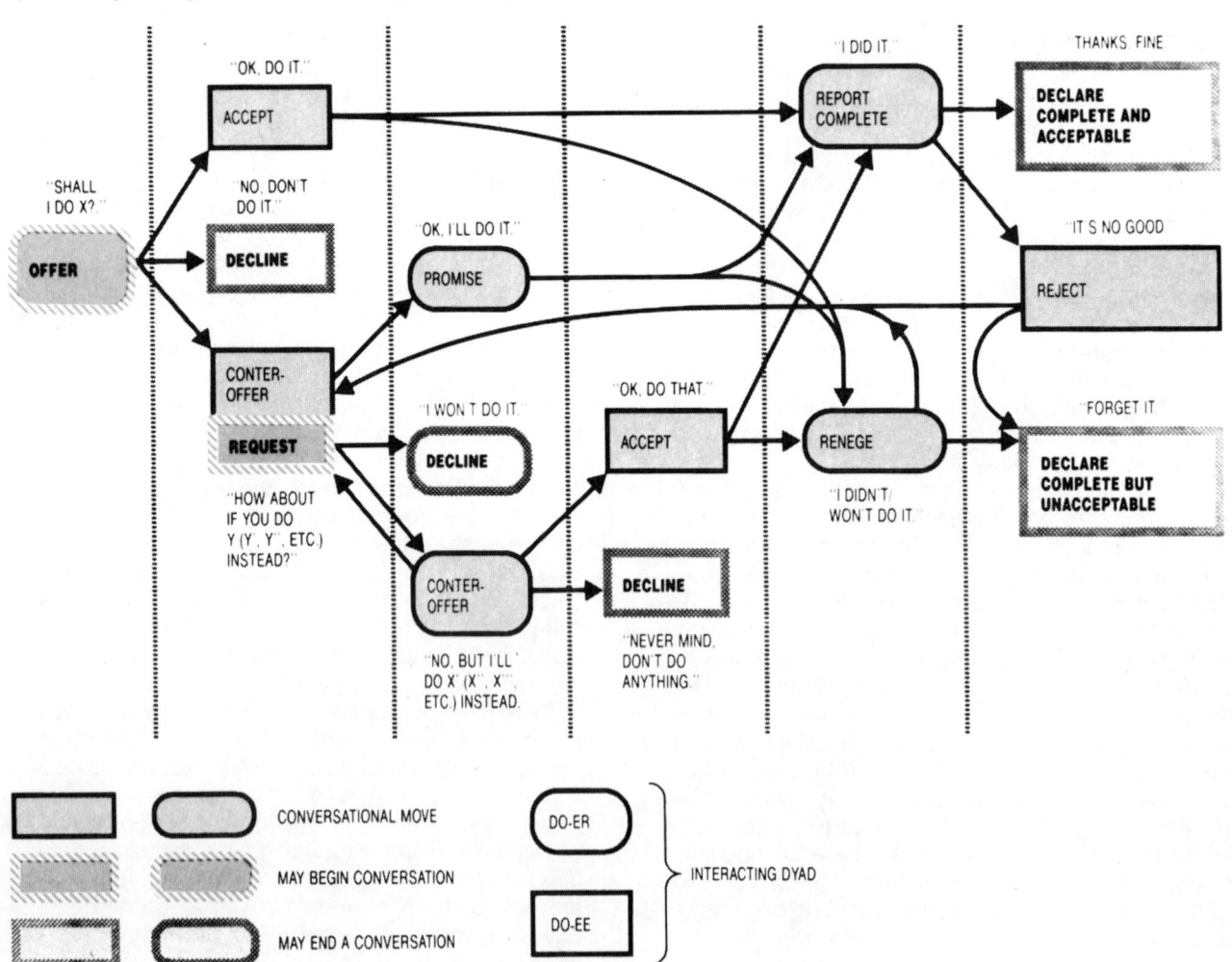

AT's software developers encapsulated the flow of speech acts into a module called the Conversation Manager. In the process, members of the development team, working at considerable distance (some were based in Mexico and others in California), had to solve such networking problems as:
- error-checking with recovery and redialing (for sending releases of code back and forth to each other);
- automatic pickup and delivery of messages;
- network administration and accounting;
- simple user naming conventions despite ever-increasing network complexity; and
- background communications services.

According to Juan Ludlow, the project's leading technical developer, "The hardware and software were an outgrowth of the conversations we were having and needed to support." This networking solution was termed the Communications Manager.

The speech-act view of communications espoused by Winograd and Flores is reproducing itself among thousands of users in more than a hundred firms that have installed the Coordinator since its release in 1985. And more firms are likely to sign up: Novell Inc. (Provo, Utah), a supplier of local area network (LAN) operating systems, is now marketing the Coordinator with AT.

More significantly, AT has distilled the Communications Manager into a development tool called the Message Handling Service (MHS), which Novell is pitching for use in such applications as distributed databases and accounting programs. (The name is curiously chosen. Novell's MHS is likely to, but should not be, confused with the Message Handling System [also MHS] part of the X.400 international standard. Novell claims that its MHS is consistent with X.400's MHS and can produce compatible packets, though the two are not identical.)

Novell is pushing its MHS hard. The package runs on Novell's Netware and several competing network operating systems. These include 3Com's Ethershare, Banyan's Vines, IBM's PC Lan Program, and any DOS 3.2-compatible LAN. The Coordinator is the first application program to use MHS.

Communications manager

The Coordinator and MHS can support as many as ten thousand users. A workstation on a local area network would run the Coordinator software (Fig. 2). This software can access records of the user's conversation and a calendar, as well as DOS (Disk Operating System) files, on the file server.

When a user posts a message, it is packaged by the sealer and shipped via MHS to the file server's holding area. Messages are then sorted, addressed, routed, bundled, and queued by the "MHS server," which may be a dedicated microcomputer. A module called the "router" or the address manager keeps a tree-shaped directory of individual users, where they are, and how to deliver messages to them.

Messages destined for other workstations on the LAN are delivered immediately. Outbound messages are sent via a "transport server" (for message pickup and delivery) either periodically, at certain clock times, or after a given number of them have accumulated. After delivering its own

2. MHS. *Novell's Message Handling Service, derived from Coordinator technology, lets users send local or remote messages by simply specifying the recipient's name.*

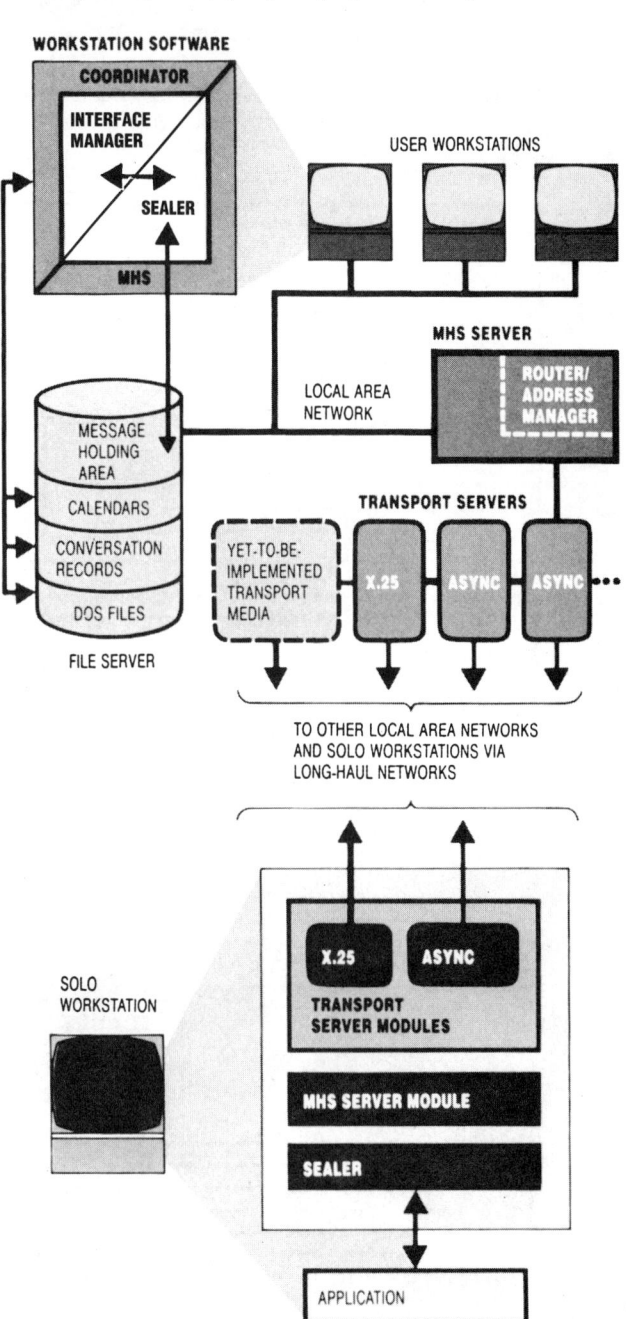

DOS = DISK OPERATING SYSTEM
MHS = MESSAGE HANDLING SERVICE

messages, the originating MHS server checks for mail waiting to be picked up.

The transport servers may reside on the same machine as the "connectivity" (or communications) server or on dedicated hardware. All these modules, along with local software, may reside together on an independent or "solo" workstation. Another module, called a converter, would be needed to converse with IBM's Profs, DEC's All-in-One, MCI Mail, Telenet's Telemail, or other environments. Some

3. Information Lens. *A prototype at MIT lets users sort and view their mail and easily devise new messages. Lens can also take actions on certain types of messages based on rules set by the user. In building a new message, fields can be explained, defaults viewed, and predefined alternatives accessed and inserted.*

INCOMING MESSAGE PANELS　　　　**OUTGOING MESSAGE PANELS**

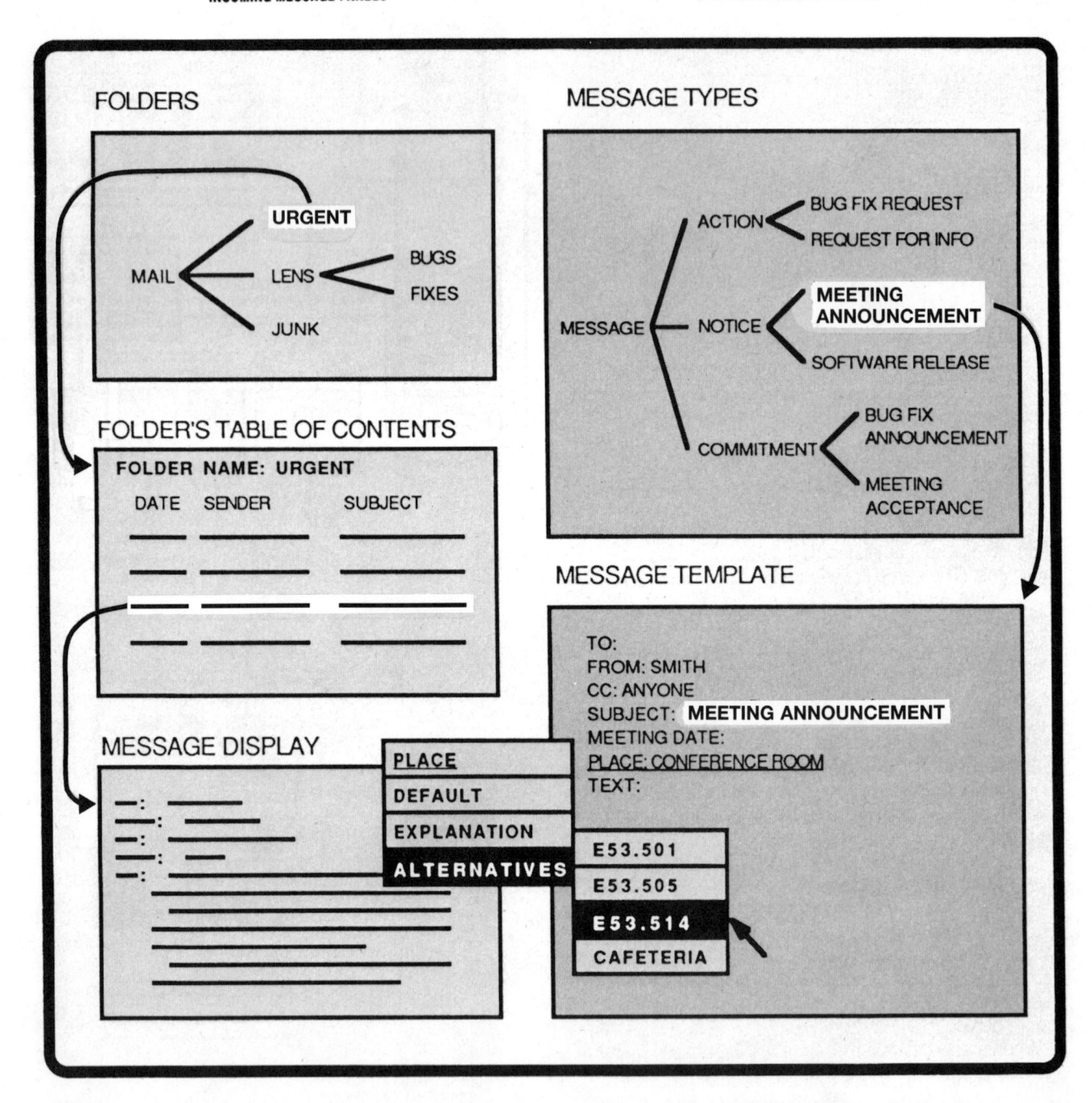

FOLDERS
- MAIL → URGENT, LENS, JUNK
- LENS → BUGS, FIXES

FOLDER'S TABLE OF CONTENTS
- FOLDER NAME: URGENT
- DATE　SENDER　SUBJECT

MESSAGE DISPLAY

MESSAGE TYPES
- MESSAGE → ACTION, NOTICE, COMMITMENT
- ACTION → BUG FIX REQUEST, REQUEST FOR INFO
- NOTICE → MEETING ANNOUNCEMENT, SOFTWARE RELEASE
- COMMITMENT → BUG FIX ANNOUNCEMENT, MEETING ACCEPTANCE

MESSAGE TEMPLATE
- TO:
- FROM: SMITH
- CC: ANYONE
- SUBJECT: MEETING ANNOUNCEMENT
- MEETING DATE:
- PLACE: CONFERENCE ROOM
- TEXT:

Menu: PLACE / DEFAULT / EXPLANATION / ALTERNATIVES
Alternatives: E53.501, E53.505, E53.514, CAFETERIA

---

of these converters have already been built.

As networks grow, userless LANs can be set up to act as communications hubs. For example, Action Technologies supports more than 3,000 workstations with a hub that includes two X.25 transport servers and six asynchronous ones (four at 1.2 kbit/s and two at 2.4 kbit/s). The only processors needed for this hub, which handles more than 50,000 messages a month, are eight IBM Personal Computers (PCs) and two PC/ATs. Companies just starting to use the Coordinator can avail themselves of public hubs set up by Action Technologies.

While the Coordinator is aimed at a broad spectrum of business settings and situations, it concerns only the way people's messages affect their commitments, not the contents of those messages. Researchers at MIT's Sloan School of Management, led by Thomas W. Malone, are more willing to involve computers in the substance of human interplay. They have devised a prototype network called the Information Lens (or simply "Lens") that is more general, yet also more nebulous, than the Coordinator.

Lens provides AI-like support of group information sharing. The prototype runs on Xerox 1100-series workstations over Ethernet. A workstation user would see a tree-shaped directory of folders for messages (Fig. 3). Clicking "Urgent," for example, shows that folder's contents. Any message in the folder may be similarly displayed.

On the outgoing side of the screen, another window shows outgoing message types, each of which describes a semistructured message (a message with certain predefined fields). For example, a meeting might have time, place, and date fields. "Lecture," a subtype of "meeting," inherits its characteristics. Thus, a lecture would have time, place, and date, as well as a field to list the speaker. As shown, actions, notices, and commitments are all subtypes of the type "message," with subtypes of their own.

Clicking on the message type "meeting announcement" brings up a template for that message. The From, Subject, and perhaps Date Sent fields might be already filled in. If, say, the default meeting place was the conference room, the user could click it for a menu explaining the field, stating the default, or listing alternative sites, one of which may be selected with the mouse. All these are editable.

The semistructured approach lets users more easily compose messages (since most fields will already be filled in) and select, sort, and prioritize them (since people already process paper mail that falls into recognizable categories). It also helps them process messages. Lens suggests some likely responses to certain messages or offers to take actions based on them. For example, after presenting a meeting-announcement message, Lens might ask if it should add the meeting to the user's calendar.

Lens can also automatically act on messages based on user-specified rules, such as the following:
- If an action request has a deadline of today or tomorrow, then move it to the Urgent folder.
- If an information request concerns such-and-such a topic, then show it to me.
- If the subject is a meeting, then move the message to the meetings folder.

Messages may also trigger automatic responses. For example, say someone sent out a meeting proposal. A recipient might have a rule that automatically constructs a meeting acceptance message and, upon the recipient's approval, forwards it to the sender. In principle, messages could be bandied about with no human intervention. Of course, this could quickly become pointless, but if rules concerning a group's routine operations were carefully constructed, automatic messaging could be an entree for distributed artificial intelligence.

Users can build their own filters that let them screen information, such as rules for throwing away junk messages. Another type of rule lets users receive information they would not otherwise have gotten, by means of the "Anyone server." This mechanism lets users find messages that they are interested in from a pool of public information.

Specifically, senders may specify Anyone as one of the recipients. A rule such as "If the subject of a message addressed to Anyone concerns such-and-such a topic, then show it to me" brings all messages on that topic from the mail server to the user's attention. The Anyone server is thus a user-driven variation of the old concept of selective dissemination of information.

These features allow Lens to serve as a front-end message-processing assistant for each individual user. Taken together, users' front-ends can augment, and to some extent automate, the flow of information in an organization. The structure is also malleable, evolving as users add, change, and delete rules and message types.

Power to the nodes

The Coordinator embodies an extreme view about the proper use of computer networks to support human ones. People must make their speech acts explicit; computers serve only to amplify and coordinate human commitment but can perform no speech acts of their own. By contrast, the Information Lens borrows ideas from AI and gives computers more of a hand in the content of people's work. Still, the developers of Lens emphasize that they focused not on building intelligent, autonomous computers but instead on supporting the knowledge and processing of human work groups.

Another research effort gives even more autonomy to the network while retaining the primacy of human communication. Workers at the Computer Research Laboratory at Tektronix Inc. (Beaverton, Ore.) have devised an electronic mail prototype that supports a special type of message, called an Envoy. Envoys are not simply passive messages; they are intelligent messengers that perform various tasks in the network.

For example, an Envoy might be deployed to set up a meeting. The Envoy would make the rounds to the workstation of each person on the meeting list, asking a calendar module if that person were available. If so, the Envoy would ask the calendar to pencil the person into the meeting. If not, the Envoy would follow previously set guidelines of what to do. It might scratch the person off the meeting list, if it had been told that the person was nonessential. Otherwise it might try to reschedule the meeting by revisiting its previous stops. If it had no rules about what to do, it might return to its sender and request some. At any rate, when it had completed its rounds as best it could, the Envoy would return and inform its sender of the outcome. Envoys are thus electronic helpmates that can offload from their human users the drudgery of organizational life.

Distributed AI

Taking this idea farther leads to the opposite extreme from the Coordinator: Computer networks that not only support the work of people in organizations but that can even perform some of it. This is one of the potentials of distributed artificial intelligence. DAI gives computers a more active role in creating, manipulating, and conveying meaning.

A small but energetic community of DAI researchers has held annual conferences since 1980 under the aegis of the American Association for Artificial Intelligence. Recent meetings have reflected a split between fine-grained connectionists, who see DAI as AI on parallel processors, and those taking a more coarse-grained approach, in which relatively sophisticated software entities work together to solve problems.

Expert system mise-en-scène

CSI's scenario was as follows: The expert systems work together to solve the jigsaw-puzzle problem of newspaper layout. To do so, they interact much in the manner of people; in fact, the project's architect, Michael Stock, wanted to simulate what the key people at a newspaper would say to each other if they were "glued together at the bellybutton." Although the bit patterns exchanged over the network do not map directly into English, the meaning of the messages equates to the dialog below.

A word about the cast of characters:
- The Edition Manager (EM) expert system oversees and coordinates the entire production process. It uses a detailed economic model to select the most profitable, yet editorially acceptable, way to construct the newspaper.
- Space Reservation (SR) handles requests for space in the newspaper from display-ad salespeople, editors, and the classified ad department.
- Shapeup (SU) does a high-level outline, or shapeup, of a newspaper edition, including how many sections are in the edition, what the sections contain, and so on.
- The Press (PR) expert system configures the printing presses to manufacture the newspaper. It contains numerous rules about press physics and operations.
- Dummy (DU) does the bulk of the actual layout work (final decisions are made by a person at a page-layout terminal). From the Edition Manager to Shapeup to Dummy, the newspaper is constructed in ever more refined steps.

For the most part, these expert systems are based on extensive interviews with human experts. And like their human counterparts, the artificial department heads do not always agree. Overcoming disagreements is crucial, especially in so time-critical an arena as the daily newspaper. Step 23 describes CSI's solution to this problem.

1. The salesperson places an order (see figure).
2. SR-EM: "I have an ad for a new day. Create a new edition."
3. EM-SU: "Shape me up a new edition."
4. SU consults the historical database and
5. does an outline of the new edition.
6. SU-EM: "O.K."
7. EM-PR: "Do a press configuration for this edition."
8. PR looks at SU's outline and
9. does a press configuration.

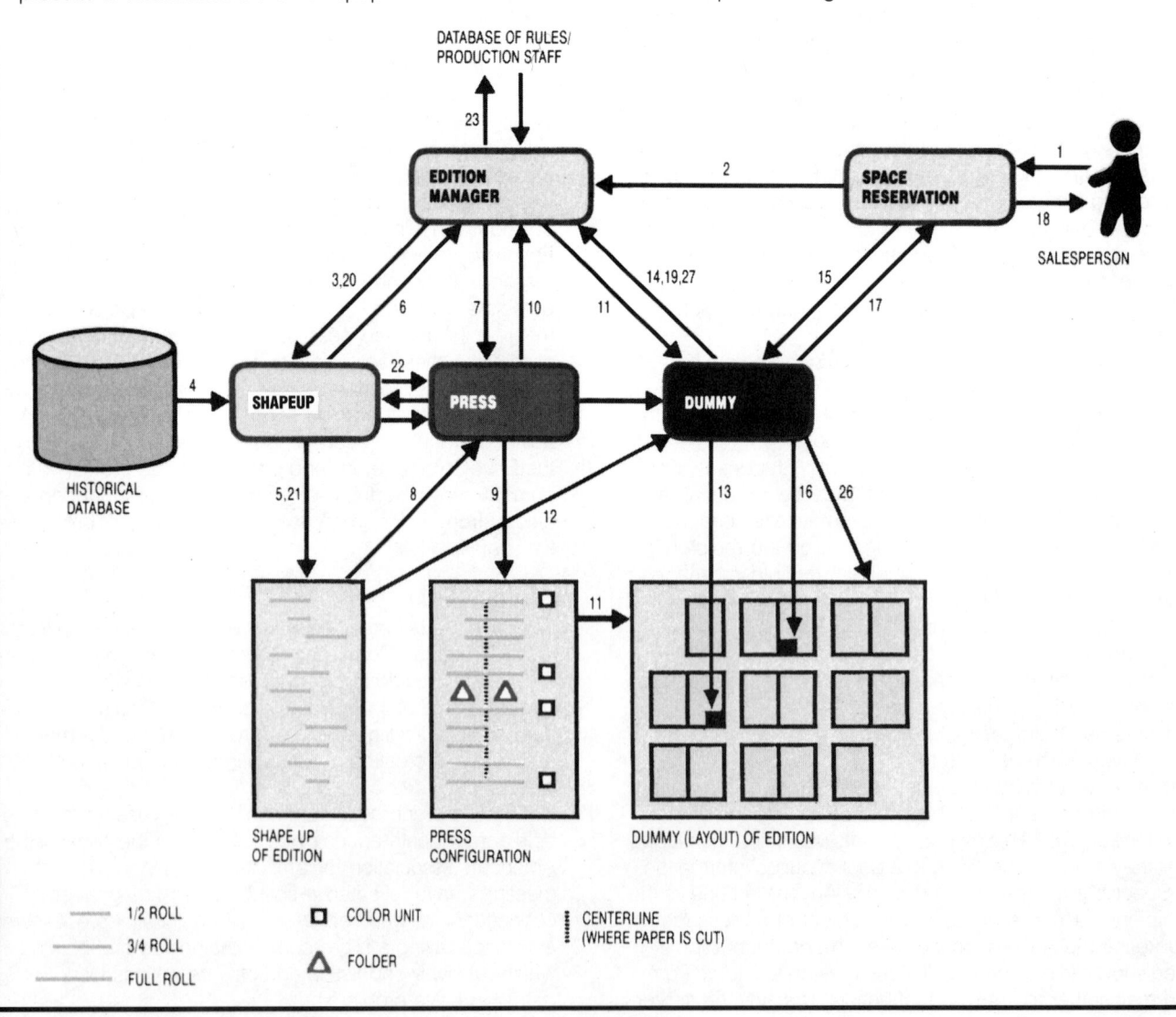

10. PR-EM: "O.K."
11. EM-DU: "Map out this edition."
12. DU gets the outline created by SU and
13. places the first ad in the new edition.
14. DU-EM: "O.K."

Once an edition has been created, SR goes directly to DU to place ads.

15. SR-DU: "Here's another ad."
16. DU attempts to place the ad.

At this point, several things can happen:

17. DU-SR: "O.K." (if placement is successful).

Or, if DU cannot place the ad because it violates an absolute constraint (such as "no ads belong on the editorial page"), it responds:

17. DU-SR: "Unable to place."
18. SR-salesperson: "The ad was rejected" (or, if the newspaper's policy is to take all ads: "O.K., but be forewarned that your ad may not run here").

However, if DU cannot place the ad because it violates a design constraint, it responds:

17. DU-SR: "Stand by."
18. SR-salesperson: "Attempting to place the ad."

Then the following sequence ensues:

19. DU-EM: "We need more space in section X."
20. EM-SU: "Here's a new reqirement from DU. Redesign accordingly."
21. SU creates a new outline (provided there is an acceptable way to accommodate the request).
22. SU-PR: "Here's the new outline of the edition. Reconfigure the presses accordingly."
23. At this point, a conflict might arise between the expert systems, just as it might if people were involved. Trading color, press capacity, and complicated setups against optimum newspaper design is a trial-and-error process, more so than between the other functional areas (although the other expert systems might disagree as well.)

The EM listens to the background chatter between the expert systems. At some point, it determines that a dead-end conflict has arisen (if the constraints are totally contradictory or if the interexpert-system dialog passes some timeout or maximum number of transactions, the limits of which may shrink as press time approaches). The EM then steps in and tries to resolve the dispute based on a special set of rules.

First, the EM looks at allowable perturbations in how the paper is built (such as an alternative manufacturing process). If it finds one, it uses it. If it does not, it then consults another set of rules to determine which (human) member of the production staff should settle the issue (since the decision might involve a high order of business and editorial criteria). The EM notifies that staff person (or the next person in the pecking order set up during the installation) by setting a blinking alarm on that person's screen. At the same time, EM tries out arbitrary constraints as a last-ditch effort.

24. PR reconfigures (if so decided).
25. PR-DU: "Here's new press configuration."
26. DU executes a new layout.
27. DU-EM: "The edition design is now clean."

This line of inquiry not only is more likely to affect network managers, whose networks may someday include these entities, but also mirrors the world of professional interaction that networkers will increasingly be called on to support. In addition, this form of DAI best illustrates the almost human-level dialogues that may be the future's networking protocols.

Proto-publishers

Most DAI research has been purely theoretical. One exception, an ambitious project in applied DAI, was until recently under way at Composition Systems Inc. in Elmsford, N.Y.

CSI has reportedly spent over three years and five million dollars building a decision-support tool for the logistics of newspaper production. The project could have revolutionized the publishing industry and perhaps even the networking industry with its idea of multiple expert systems cooperating with each other as well as with a team of human professionals.

Although a product was said to be within six months of beta delivery, Crossfield Electronics Ltd. of Hemel Hempstead, England, which bought CSI a year ago, pulled the plug on the project last May because of concerns about its marketability, technical feasibility, and return-on-investment potential. Although the project was "worthwhile, exciting, and ahead of its time," according to a Crossfield spokesman, it lost the competition for internal resources against more conventional projects. Even so, the considerable work done on it to date offers an object lesson in how networks may one day relate to their users.

A well-established firm was behind the endeavor. CSI has supplied editorial and advertising departmental networks to newspapers including the Wall Street Journal for more than 20 years. It has been a DEC original equipment manufacturer for nearly as long. CSI's current product consists of PDP/11s and VAXs linked via Ethernet.

The firm was attempting to build a number of cooperating expert systems working on different aspects of the newspaper pagination problem (layout, design, and manufacturability). In the attempt, CSI had become an alpha site for most of DEC's AI products and a beta site for Inference Corp.'s Automated Reasoning Tool, an expert system generation tool. Programming was done in Lisp.

Setting: The real world

Daily newspapers are a chaotic affair. Shifting deadlines, press equipment troubles, and the placement of new advertisements and changes or cancellations to existing ones make the newspaper a classic problem in control and communications. CSI wanted to offer a solution. The standard alternatives of traditional, brute-strength computation and monolithic, centralized AI were rejected because, in such a hectic setting, constant demands and pressure would require quick response, high availability, and fault tolerance. The large memory and processing requirements of AI applications and the need to sell packages to different-sized newspapers led CSI to a distributed, modular approach.

CSI knowledge engineers interviewed experts, such as editors, press operators, and typesetters, at 20 newspapers and determined to write a collection of specialized expert systems. To get them to work together, the develop-

Networking AI for networking's sake

CSI's commercial DAI effort, one of a mere handful, sheds light on a possible new type of network. Another notable endeavor, underway at GTE Laboratories, is exploring new ways of dealing with existing ones. Researchers there are trying to apply distributed AI to communications network management.

GTE is no newcomer to AI, having developed the Central Office Maintenance Printout Analysis and Suggestion System (Compass), an expert system for switch diagnosis (see "AI carves inroads: Network design, testing, and management," DATA COMMUNICATIONS, July 1986, p. 119). Compass, like the other network-management expert systems, is a standalone program. In fact, centralized AI setups have been the sole alternative to conventional network operations strategies. Why, then, is GTE pursuing DAI?

The advantages of distributed problem solving are similar to those of distributed processing in general. They are as follows:
- The computing power of small, inexpensive processors can be harnessed;
- Local processing resources can be applied to local problems; and
- Reliability is enhanced by avoiding a single, centralized problem-solving agent that could fail and take the network down with it.

The researchers believe that applying the techniques of distributed expert systems to network management problems could greatly automate network operations, yielding knowledge-based setups that observe network activity and modify the network to better handle its traffic. They claim that truly autonomous, self-healing networks are the logical end product of their research.

However, users will not benefit in the near future from GTE's DAI efforts, for three reasons. First, most of the DAI work is pure research, not directed toward a product. Second, the research is not expected to bear fruit for some time. And third, work at GTE Labs is aimed at GTE, not directly at user firms. Still, users can glimpse one possible future for their own networks by looking at the direction GTE is pursuing.

In this vision, network management will require a number of different expert systems to handle such areas as the following:
- Tariffs and line quality. To help telephone companies offer users a choice between lower cost and higher quality, an expert system could monitor requests for service and make sure that they can be fulfilled based on what it knows about resources.
- Diagnostics. Each element in a network, such as a trunk or a multiplexer, may require its own expert system. The expert systems, like the networking gear, may come from different manufacturers.
- Maintenance. A maintenance expert system differs from a diagnostic one in that it looks for patterns of failure and suggests maintenance rather than repair action. AT&T's Automated Cable Expertise (ACE) is an example of an expert maintainer.
- Traffic control. Other programs will be in place to manage dynamic routing and exceptional situations such as congestion.

These expert systems are intended to solve localized problems, such as with individual callers, distinct networking components, and regional failures and demands. They can, therefore, take advantage of the distributed approach. However, this approach, for all its merits, has a drawback: It requires what DAI theorists have called "global coherence." That is, the problem-solving entities must work together efficiently and toward a globally optimal solution, but without duplicating each others' work and without being over- or underutilized. Moreover, the expert systems should also be able to distinguish between those situations that require them to work together and those that do not.

It is equally important to decide how the expert systems should be organized to solve the problem at hand. This

(A) FAULT ANALYSIS

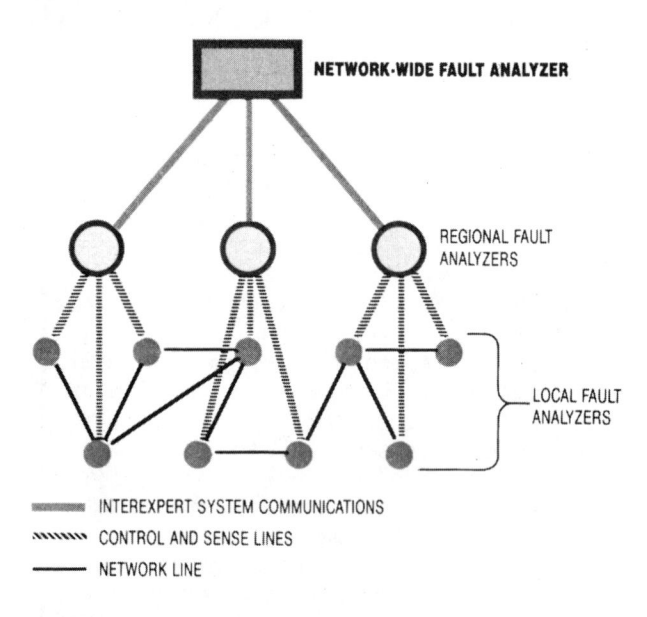

- NETWORK-WIDE FAULT ANALYZER
- REGIONAL FAULT ANALYZERS
- LOCAL FAULT ANALYZERS

— INTEREXPERT SYSTEM COMMUNICATIONS
⋙ CONTROL AND SENSE LINES
— NETWORK LINE

(B) VIRTUAL NETWORK MANAGEMENT

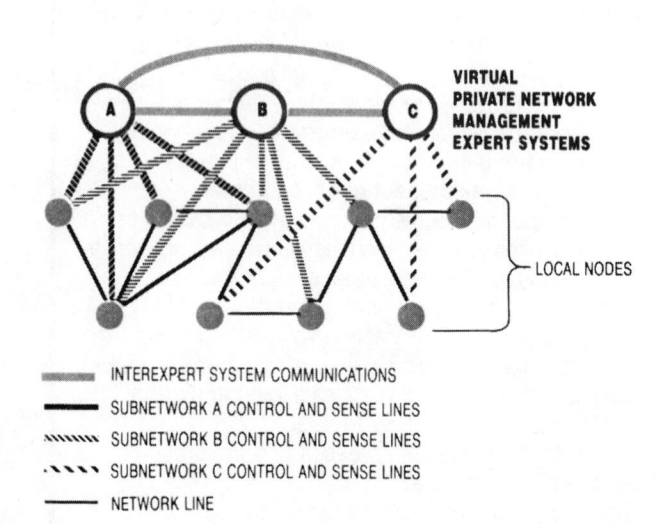

- VIRTUAL PRIVATE NETWORK MANAGEMENT EXPERT SYSTEMS
- LOCAL NODES

— INTEREXPERT SYSTEM COMMUNICATIONS
— SUBNETWORK A CONTROL AND SENSE LINES
⋙ SUBNETWORK B CONTROL AND SENSE LINES
⋙ SUBNETWORK C CONTROL AND SENSE LINES
— NETWORK LINE

decision can be finalized when the expert systems are initially set up, or the organizational design can be managed and updated over time. One ultimate aim of DAI workers is to create problem-solving entities that can organize themselves in optimal ways based on their understanding of the problem, as groups of people come together in different ways to meet different contingencies.

■ **Two examples.** To illustrate how organization relates to task, the GTE researchers cite two networking problems suitable to different organizational designs (see figure): Fault analysis (A), which lends itself to a hierarchical, or "trickle-up," DAI organization and virtual private network management (B), more amenable to expert systems that function peer to peer.

■ Fault analysis. In a network equipped with a number of Compass-like expert systems (local fault analyzers at network nodes), each expert system would receive error messages from local facilities and use them to diagnose local failures. Messages relating to regional failures would be passed to regional expert systems, which in turn would pass network-wide problem messages to a top-level analyzer. This hierarchical arrangement of intelligence is in fact a distributed organization, albeit a particularly rigid one.

■ Virtual networks. In software-defined networks, user firms may be assigned their own virtual private subnetworks, which may one day be managed by distinct expert systems. Unlike the hierarchical fault analyzers, these expert systems may function as peer-level entities. Strictly local changes, such as subnetwork administration, would be handled by each expert system independently. Before making changes affecting other subnetworks, however, an expert system would have to confer with its peers.

For example, suppose each subnetwork had been assigned a set of physical trunks. If subnetwork A had more traffic than it could handle, its management expert system could ask its peer in subnetwork B for unused resources. Moreover, the expert systems could step in for one another in a pinch, boosting reliability.

GTE Labs has created a test-bed, written in InterLisp D and Loops and running on Xerox workstations, to determine when distributed control, in particular a DAI-based mechanism, is better than centralized, operations research-type decision processing. The test-bed simulates a circuit-switched network. Switches and links are defined as objects. Researchers can use it to test different network traffic management schemes.

The work will unfold in stages. The first step will be to construct a standalone network control and traffic management expert system and to benchmark it against scenarios run with no intelligent controls. The next step will be to test hierarchical traffic control, where a single entity coordinates the activities of various local controllers. Since this step is comparable to actual current practice, it could be benchmarked against the performance of a network control person. The final step will be distributed control.

ers would have to solve significant practical networking problems, such as shadowing, load balancing, and performance tuning and monitoring. They also tackled sticky DAI problems. For example, disputes between the expert systems would have to be resolved by another expert system or, if need be, by a person (see "Expert system mise-en-scene").

The expert systems were based on the knowledge of real experts but could be customized by each newspaper. Though many operations are common to all papers, local ways of doing things would have to be part of the AI network. Parameters would have been set from information that the user provided by filling out forms and menus in the installation process. This process itself was to be handled by an expert system.

Newspapers still face the pagination problem. As a recurring problem-solving activity that varies within a pattern, pagination was an ideal candidate for the DAI approach. This approach could also be useful in other settings where specialized individuals work toward a common goal, but each works with their own needs and aesthetics. Some such environments would require new expert systems (see "Networking AI for networking's sake"). Others are currently guided by committees of human experts.

Where is it all heading?

The new messaging technologies presented here combine with DAI principles to offer a glimpse of a fascinating future for data communications, of a world in which the cold logic of computers would be fully and explicitly infused with a human sense of the theatrical. Work groups at all levels could achieve the richness of an ensemble, with people's interaction enhanced by an organically evolving network of people and computers.

As an example, consider the senior management of some hypothetical future corporation. The management (and, ultimately, the corporation) would be decentralized. Each corporate officer would be responsible for finding innovative solutions to some area of concern. Unlike today, however, computer mediation would allow more extensive discussion between the officers. Through these discussions, the consequences of the corporation's decisions could be more accurately simulated. Such a simulation could improve on today's corporate decision-making process, which can be unwieldy or ineffective and may even seem heavy-handed.

The simulation's goal would be to let the officers put forth or try out any number of scenarios. These scenarios would interact with each other and with timely databases. Full-scale, computer-mediated deliberation could be achieved with a technological hybrid of the Coordinator and the Information Lens.

Weaving conversational webs

A scenario would begin as a Coordinator-assisted conversation for possibilities. It would be part of an ongoing stream of conversations that could evolve and continue despite management reorganizations and other personnel changes at the top. At this stage, the medium would be transparent. The conversations among the computer-conferring officers would generate many cross-references, or areas where one officer's ideas relate to or affect the area of concern of another. Such cross-references could be marked or tagged as special attractors for related future messages in an Information Lens-like fashion.

To see how a scenario might actually unfold if implemented, the conversational scenario would have to interact with similar conversations for action and possibilities taking place in other networks, such as those supporting each officer's division. Through these divisional networks, queries and messages could filter to networks of other corporate, governmental, and academic organizations that had been tagged with relevent keywords.

A proposed scenario would also be enhanced by feedback from databases on related topics. These databases would be more accessible if they were encoded in the same object-oriented, message-passing style as the scenario networks. Standards would conceivably govern interfaces from the networks to the databases, as well as between the networks. This could help bring electronic data interchange and other application-level promises to fruition. Scenario networks and the technologies needed to create them appear to be within reach, judging from the progress made in messaging-support and DAI applications. They would help senior management more fully explore corporate strategies and directions.

Augmenting these networks with DAI presents even bolder prospects. Fully implemented, scenarios would require a great deal of attentiveness. Even in normal use, the human interaction-based network described above could result in too many messages for people to keep up with, since all relevant remarks of each participant in each network would be passed to all other interested users. But extending the idea of the Envoy message could help avoid this problem by bringing such networks into the realm of Distributed Artificial Intelligence.

The potential of proxies
Like an Envoy, the conversation that formed a scenario could act on behalf of its participants, if given the autonomy and guidelines to interact with other conversations. It would then be able to front-end for the participants, as CSI's various expert systems would do for newspaper people. In fact, such conversations would arise in a manner similar to CSI's customization process. By interacting with a conversational entity and specifying their needs, users would be molding it to do their bidding.

These possible futures entail a significant challenge, but they also embody a major opportunity. The flourishing of both people networking and computer networking, along with the growing dependence of each on the other, means that corporate and technical managers will have to work much more closely. Fortunately, the emerging tools that are forcing this to happen will also be instrumental in making it happen. ∎

Kevin R. Sharp, Burr-Brown Corp., Tucson, Ariz.

Protecting networks from power transients

Mother Nature is a meanie, and lightning has a particularly shocking effect on communications gear. Forewarned is forearmed.

Fireballs rolling along bare conductors, electrical transformers blown to smithereens by a lightning strike. These and other kinds of electrical equipment hazards are the products of electrical storms—a kind of destruction that is as old as the nation's power grid.

Over the years the communications industry has amassed an enviable record in meeting the threat of outages and spikes caused by the weather. But as communications gear becomes more sophisticated, it becomes more susceptible than ever to "transient" voltage damage, especially the extremes induced by the weather.

In the face of increasing vulnerability, communications managers should understand how such "spikes" are generated—and what steps can be taken to protect computers and networks against them.

Unlike machinery and lighting, communications gear reacts much more quickly to changes in electrical conditions. Large electric motors continue to rotate by inertia for many tenths of a second if power is temporarily disturbed. Likewise, the human eye cannot detect light failures shorter than several hundredths of a second. But when communications equipment and computers encounter an outage of even just one millionth of a second—the kind of disturbances that would go completely unnoticed in motor circuits and lighting applications—the result can be real problems, especially in the high-speed digital circuitry that now is ubiquitous in communications networks.

Since computers and communications facilities consume only about 1 percent of the commercial electrical power generated in the United States, users should not count on the power utility to protect their networks. It is inevitably up to communications equipment users to protect their equipment from power-supply transients. And unless the user takes steps to achieve this protection, the result can be a pattern of equipment damage—and data errors—that may be difficult to trace.

How widespread is this problem? Too often, damage to electronic equipment is not attributed to the spikes and surges that plague the commercial power grid. If a piece of electrical machinery is damaged by power-line surges or voltage spikes, the damage is obvious and sometimes spectacular. In the power-supply circuits of electrical machinery, the total electrical energy that is available is often hundreds of horsepower—even if the actual consumption is less than a fraction of one horsepower. Once the insulation of these circuits is damaged, all of the available power can be released, causing melted wiring, burned insulation, and hazardous fires.

In the digital electronics used in communications networks, however, a combination of deliberate circuit design and fortuitous component characteristics limits the available power inside the equipment to a few tenths or hundredths of a watt, which is less than a tenth of a horsepower. Still, this available energy is more than enough to damage the sensitive semiconductors and integrated circuits that comprise communications equipment, even though it is far less power than is needed to cause the spectacularly obvious damage often associated with electric motor failure. Therefore, tracing communications component failures to power-supply transients is often difficult.

If a pattern of equipment damage or failure emerges over time, electrical transients could be the problem. Users should be particularly sensitive to patterns of failure in power supplies and transmitter/receiver circuits in equipment. The pattern will usually be time-dependent; maybe a company experiences a lot of transmitter/receiver failures in the fall of each year (when thunderstorms are most frequent in many parts of the country). Or maybe most of the communications equipment's power-supply damage occurs on Monday mornings (when heating and air-conditioning systems are most active).

Another suspicious sign is a sudden change in the failure

267

rate, such as an industrial park installation that had no hardware failures for two years and then had four power-supply failures in two weeks. (Maybe a welding contractor just rented the shop next door.) If the user detects any of the previous patterns, he or she should first try to determine where the problem is coming from, and then take measures to protect against it.

How new a problem?

Communications equipment is susceptible to damage and data errors from power sources, communications lines, even the telephone network itself. The telephone industry was the first to protect data and power lines from the influences of electrical transients. But today, semiconductor-based equipment is more transient sensitive than older vacuum tube equipment. And the low-power CMOS (complementary metal oxide semiconductor) circuits are more sensitive than the bipolar semiconductors that were ubiquitous just a couple of years ago.

Because components used in communications equipment are now much more susceptible to transients, what used to be a "pop" during a phone conversation can today cause a funds transfer to be routed into the wrong account or a purchase of 10,000 shares of the wrong stock. It is up to the data communications network installer to protect his or her own equipment and data from a power source designed to work lighting and machinery, and to ensure that data communications lines subject to lightning surges do not damage equipment.

Prudent location of communications lines is the easiest and most economical way to minimize the network's exposure to damage. Some areas, like overhead communications lines, are much more susceptible to lightning-induced spikes than buried conductors. Also, since transients are by nature propagated in either electrical or magnetic conductors, the use of fiber optic cables wherever possible will go a long way toward avoiding the problem altogether. The guiding rule is always to keep a transient from getting into the communications network in the first place rather than rely on protective devices on each piece of equipment to protect the user's investment in expensive communications hardware. After the user has taken reasonable steps to keep transients out of the network, the next tasks are to understand what circuits and devices will protect the equipment at risk and to select commercially available equipment that uses these techniques to greatest advantage.

The first step in protecting equipment is to understand where the damaging transient spikes come from. Most spikes are caused by one of four phenomena: lightning, switched inductive loads, electrostatic discharge (ESD), and nuclear electromagnetic pulse (NEMP). This article will deal only with the first three causes, since hopefully these are the only sources a communications network will need to survive. However, for critical military communications networks the effect of high-altitude nuclear device detonation must be considered.

Whatever their source, transients can enter a communications network three ways: from the power grid before entering the communications installation, from other electrical or electronic equipment attached to the power network inside the installation, and through data lines.

Lightning may be the most studied of the transient generators, but it is also the toughest to guard against because the energy released during a lightning stroke is enormous. Direct strikes may contain voltages of 1,000,000 V and currents of 200,000 A. According to a paper presented to the Institute of Electrical and Electronic Engineers power conversion conference in 1976, this energy would be sufficient to lift the Queen Elizabeth 2 luxury liner 2 feet! It is virtually impossible and definitely not economical to protect delicate communications equipment from these primary effects of lightning if the energy gets into the communications network. The guiding rule must be to not allow these high energy spikes into conductors in the first place.

The best solution for protecting outdoor communications lines in high-lightning areas is to use fiber optic cables with nonmetallic sheaths. These cables will not attract a lightning strike and will be mostly immune to the effects of nearby lightning flashes. If running fiber optic cable is not practical then a lightning-protected zone should be created with lightning rods, lightning conductors, and grounding equipment. The single conductor on the top of most standard electrical utility poles is one example of a lightning conductor. A communications conductor suspended under the power lines on a transmission tower will usually be protected from lightning. If a communications conductor is stretched here, however, the user should watch out for electrical/magnetic interference from the power line and protect equipment and personnel from the possible hazard of a broken power cable resting on a communications conductor.

Secondary effects of lightning contain much less energy than primary effects, so they can be controlled with available products even after the energy gets into communications conductors. When lightning strikes the ground (actually lightning current streams from the ground to the cloud), three effects take place that can cause problems for data communications equipment: induction, bound charge release, and noise radiation.

Large earth currents must flow in order to neutralize the charge deposited by the stroke. These currents can induce voltages in nearby conductors, particularly buried electrical lines (but not fiber optic cables).

Just before the lightning strike, electrostatic fields can charge elevated conductors to more than 30 kV for every meter the cable is suspended above ground. This bound charge is no hazard to communications equipment since the air around the conductor is also "charged" to the same potential.

This electrostatic field of 30 kV per meter will rapidly collapse to about 150 V per meter after a lightning strike. This field change will leave the overhead conductor with a net potential of about 30,000 V for every meter above the ground, effectively generating a voltage spike of over 100,000 V on a power line with 10-meter-high towers. This high voltage is manageable by reasonably sized and priced transient protectors because there is not much energy stored in the charge.

The final secondary lightning effect to be considered is magnetic induction, or noise radiation. A lightning bolt will act as a vertical antenna, radiating high-frequency interference. This is what causes radio and TV static during

1. Thunderstorm days. *The frequency with which lightning will strike close enough to damage communications equipment varies widely across the country. The isoceraunic map shows the average number of days per year that a person standing at a data communications site would be able to hear a thunderstorm.*

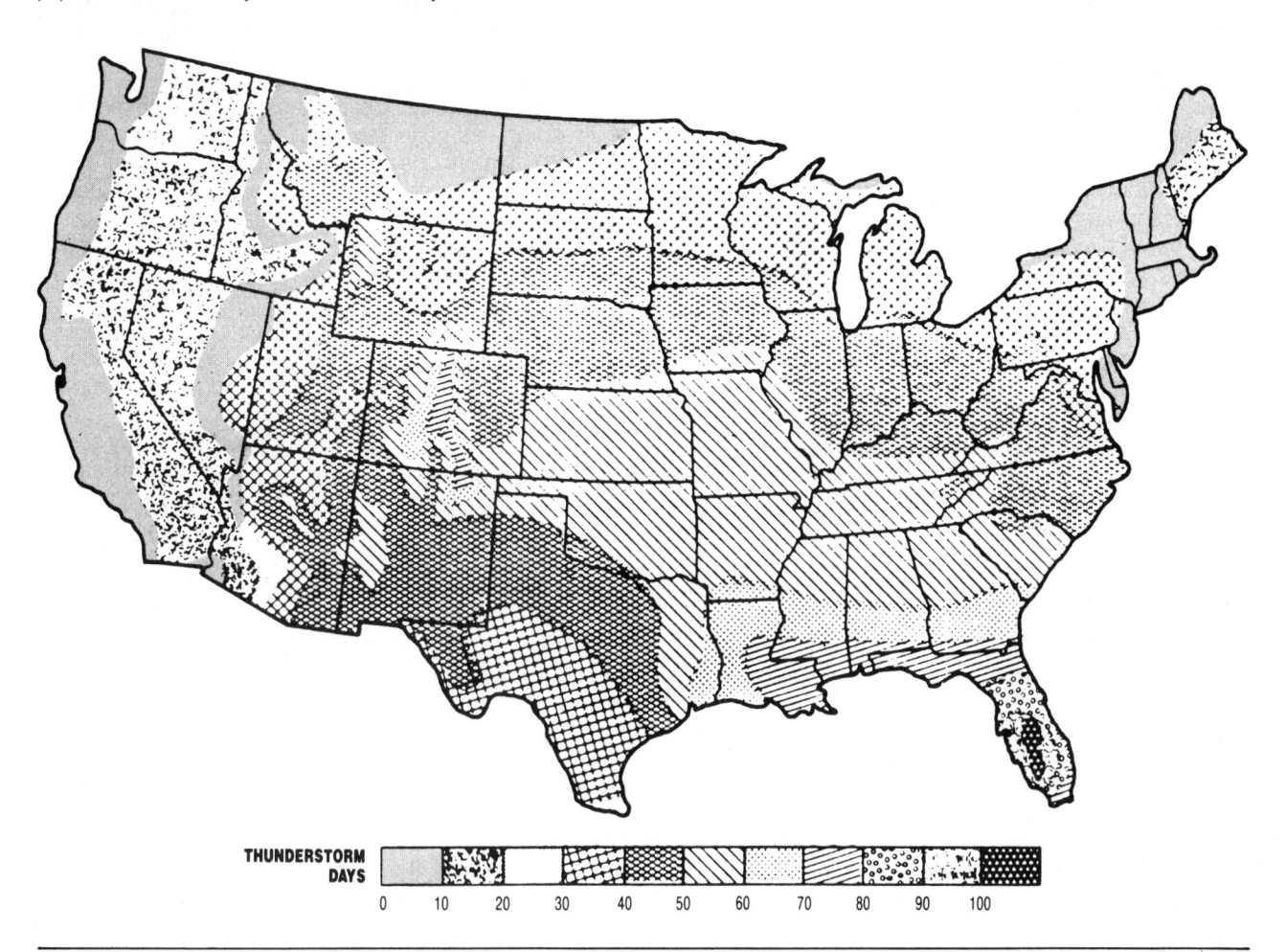

Table 1: Ground lightning strikes vs. thunderstorm days

THUNDERSTORM DAYS PER YEAR	GROUND FLASHES/SQ. KM./YEAR		GROUND FLASHES/SQ. MI./YEAR	
	NOMINAL	RANGE	NOMINAL	RANGE
5	0.2	0.1-0.5	0.5	0.26-1.3
10	0.5	0.15-1	1.3	0.39-2.6
20	1.1	0.3-3	2.8	0.78-7.8
30	1.9	0.6-5	4.9	1.6-12.9
40	2.8	0.8-8	7.3	2.1-20.7
50	3.7	1.2-10	9.6	3.1-25.9
60	4.7	1.8-12	12.2	4.7-31.1
80	6.9	3-17	17.9	7.8-44
100	9.2	4-20	23.8	10.3-51.8

thunderstorms, and it can cause data loss and control errors if it gets into communications networks. Power lines and communications conductors will act as receiving antennas, conducting this interference right into an unprotected piece of equipment. Many commercially available transient suppressors contain filters that help control this problem.

Lightning-induced transients are not the only threat to communications hardware; other damaging transients may originate from inside and outside the building. Switching loads, resonating circuits associated with switching devices, power circuits requiring the action of fuses or circuit breakers, and arcing faults all contribute to an electrical environment that can be damaging to the electronic devices used in communications equipment and the computers processing the data. All of these sources produce transient voltages directly proportional to the rate of current change and contain energy directly proportional to the square of the current (a particular problem during power-supply faults, when the current can be very high before a fuse or circuit breaker opens).

Assessing the threat

After the communications professional knows what can happen to equipment and what causes the damage, it is necessary to estimate the magnitude of the threat to each node in the network. This is done by considering the likelihood of a nearby lightning strike and the frequency of spikes caused by other sources.

The frequency with which lightning will strike close enough to damage communications equipment varies widely, from hundreds of times per year in the southeastern United States to less than 10 times per year on the West Coast. To determine how often lightning will strike near a proposed installation, refer to Figure 1, a map of thunderstorm days per year (isoceraunic map). Find the location on the map and read off the isoceraunic number. This number is the average number of days per year that a person standing at that site will be able to hear thunder at least once during the day (meaning lightning has struck within about 14 miles). To convert this number of thunderstorm days per year to an average number of lightning strikes per year refer to Table 1. The number of expected lightning strikes per year per land area ranges from 1/2 strikes per square mile for an isoceraunic level of 5 to 24 strikes per square mile for an isoceraunic level of 100.

Besides lightning, transients may be generated by the switching of inductive loads, particularly rotating machinery and large relays. When current flows through an inductor, energy is stored in a magnetic field. If the current is removed, the stored energy will be released, often quite suddenly if the current is cut off by a switch or circuit breaker. While load-switching transients are much smaller than lightning-induced ones, they occur much more often. Instead of hitting equipment with one knockout punch like lightning, load-switching transients can take out equipment with a long series of body blows.

Because switching transients and lightning transients account for the majority of the damaging transients in an electrical installation, the user should add to the lightning-threat evaluation done with isoceraunic maps a switching-transient evaluation made much more intuitively; there is no easy way to determine exactly how bad a particular installation will be. Industrial areas present the greatest possibility for switching transients, so the user should be particularly alert for problems if heavy electrical equipment, such as a crane, is in use nearby. Arc welders are another bad transient source.

Even though heavy industrial areas contain the most obvious transient generators, just because an installation is in a nice clean office is no reason to ignore transient problems. A residential air-conditioning unit can produce 6 kV spikes when it shuts off. Residential power circuits also generally have lower impedances because fewer large loads are connected to a circuit at one time. Because the voltage that can be generated by a transient with fixed energy is inversely proportional to impedance (Voltage energy2/impedance), a transient generator can produce larger voltage spikes in a residential setting than in an industrial circuit.

An analysis of switching-transient sources usually shows that most installations are vulnerable. If two or more large switching sources are present at an installation the user should protect the equipment as though a moderate lightning threat was present—even if the location would indicate a low lightning incidence.

Possibly the most brutal transient environments occur in vehicles, both spark-ignited and electrically driven ones. Spark-ignited engines produce problems in several ways. In order to burn fuel most efficiently, high voltages must be applied to the spark plugs at precise times. This means that extremely high transient voltages are purposely generated by the ignition circuitry. These transients produce radio-frequency interference and induce transient voltages in the power supply and in any conductors close to the ignition cables.

If a manager thinks he can get away from the vehicle transient problem by going to a nice clean battery-operated lift truck, he should think again. If anything, these vehicles make it even harder for electronic equipment to survive. The biggest problem with electric trucks is the fact that the electric motors draw large DC currents through highly inductive motors. Often, particulary in dirty factories or humid conditions, the battery cables will not make good contact with the battery posts. When a cable vibrates loose the current to the motor is abruptly cut off, suddenly releasing the energy stored in the motor's magnetic field. The failing battery cable presents an open circuit to the energy released by this field decay, producing voltages high enough to generate an arc between the battery terminal and the battery cable. This arc establishes a conductive path, re-energizing the motor and allowing the transient voltage to decay until a vibration again momentarily jars the troublesome battery cable loose. This process happens so fast that the operator will normally be unaware that it is occurring. Literally thousands of spikes can be generated every operating shift for weeks until the cable finally fails altogether, forcing a repair.

Because of the transient environment and the fact that power-supply voltages in vehicles vary greatly according to the state of the battery and the charging circuit, the survival of any electronic device in a car or on a forklift is a very special design case. Do not expect success if you take a 12-volt DC radio-operated modem and use it on a vehicle unless it was specifically designed for that use.

With so many possible sources of power problems, how can a communications network possibly be protected from all of them? Actually it cannot. The user must gain an insight into what the transients look like, the energy contained in most of them, and then make some rational decisions about what level of protection to provide for the different components in the communications network. Transient protection, like the transients themselves, is statistical in nature. No criterion is worst case and no design is foolproof.

Disturbances that cause equipment problems are divided according to their duration into surges and transients. Transients are generally considered to be less than about 8.4 milliseconds (ms) in duration (one-half cycle of 60-Hz power), surges are longer than a half cycle. Power-line transients normally arise from some electrical disturbance outside of the power grid, a surge is usually the result of a mechanical reaction by the utility power supply to a sudden load change. The designer of protective circuits must consider surges and transients separately since extremely short transients generally only require the ability to conduct large currents with low voltage drops without regard to heating, while dealing with longer transients requires the ability to dissipate heat to avoid damage to the protective device.

Semiconductors are particularly prone to heat dissipation failure from surges since their active region is the atomic junction of two crystalline regions buried inside a silicon block. If the transient endures for too long the heat at this junction can actually vaporize a portion of the crystal, destroying the protective device and exposing equipment to damage. Therefore, just because a device is specified to protect a multiplexer from 3,000 V transients, a failed transmitter that applies 120 VAC to an input for a few seconds may still damage it. In communications lines almost all voltage disturbances will be transients. The only major surge source is a failed transmitter that applies its power-source voltage to its output.

Because transients are sudden releases of stored electrical energy they are almost always unipolar, exponentially decaying current or voltage spikes. They result from a redistribution of electromagnetic energy, either from a sudden shift of charge (as in lightning) or a rapid collapse of a magnetic field (as in an inductive load switch). Quantifying the transient environment is difficult at best since oscillograms of actual transients look like short, high-energy bursts of white noise.

Instead of trying to actually describe the transients occurring, standards have evolved that describe repeatable transient conditions against which equipment can be tested. As the installed base of equipment tested according to a particular standard grows and studies are made about the equipment's ability to survive real-world transient conditions, the standard will be refined and give birth to new revisions or new standards. More than 20 transient standards have been written in the United States and Europe since the science began, and the evolution continues as electrical transmission techniques change and the requirements of the connected equipment develop.

The best course of action for the user is to specify that protective equipment comply with the current most-

2. Protecting formula. *The IEEE-587 standard describes two unipolar impulse waves, one voltage and one current, that approximate the effects of transient conditions.*

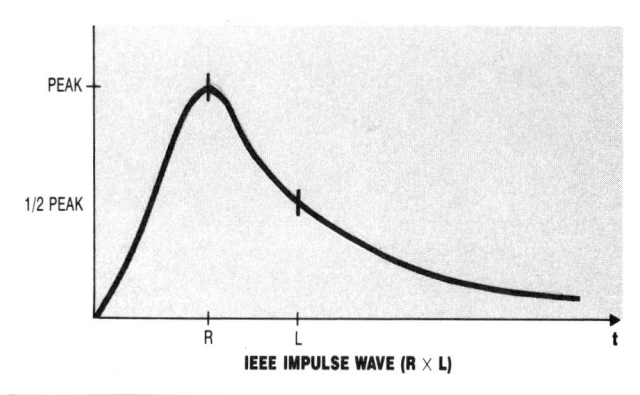

LOCATION CATEGORY	R	L	PEAK
B	1.2 µS	50 µS	6 kV
B	8 µS	20 µS	3 kA

accepted standards and monitor the equipment's ability to protect the gear in the required transient environments. Currently the most widely accepted transient standard in the United States is IEEE-587, which describes two unipolar impulse waves, one voltage and one current, as the best approximation of conditions likely to be experienced outdoors in overhead lines and indoors within 30 feet (10 m) of the electrical service entrance (Fig. 2). The voltage wave rises to a peak in 1.2 microseconds and decays to half that value in 50 microseconds. The peak voltage is dependent on the location, but is commonly 10 kV or greater. The wave is slower-rising and shorter-lasting based on years of observation. One explanation is that characteristic inductance (tending to slow current impulses) in transmission lines is greater than characteristic capacitance (tending to slow voltage changes).

Two things happen as an electrical transient travels through a building's power wiring. First, the unidirectional shape becomes osciliatory because of resonance in the wiring and loads. Second, an installation gets the first level of transient protection—flashover in the wiring devices—for free. Somewhere in every circuit, air is used as an electrical insulator. Overhead power lines are not insulated at all, relying on their distance from the ground to protect people from the thousands of volts applied to the power line.

There is, however, a limit to this air insulation. Given enough voltage, air will eventually break down and the ionized atmosphere left is a very good conductor. This arcing action is what happens during a lightning strike. The amount of voltage required to start this arcing action is proportional to the air-gap distance; the smaller the gap, the lower the voltage required to create an arc. Outdoor 120 V power lines are usually spaced so that the breakdown voltage is about 10 kV, while indoors the flashover voltage is closer to 6 kV. Therefore, just moving a communications device inside reduces by 40 percent the magnitude of the voltage spikes to which it may be subjected.

Unlike power circuits, communications-line transients are

dominated by unidirectional impulse waves indoors and out. These waves are less well defined than those in power lines because fewer studies have been done, but in general they are one-fourth to one-third longer and slower rising, due mostly to the wave "smearing" effects of buried coaxial cable. Older data on communications-line transients are not representative of the problem today, since they were mostly gathered from overhead telecom lines, and show rise times and durations one-fourth to one-third the figures expected with buried coaxial cable.

Older transient protector designs may not work in new coaxial circuits since the coordination of different elements in a protective circuit is often dependent on some assumptions about the rise times of the transient. Designs that use gas tubes or other spark-gap devices as part of the protection circuit are particularly sensitive to the speed with which the transient voltage rises, clamping at voltages that are partially proportional to the incoming transient rise time. The moral of the story when protecting communications lines is to make sure that a protector bought for use in a coaxial circuit was designed for use with the specific coaxial cable type.

Unlike voltage transients, voltage surges normally arise because of some electromechanical action, so they are much longer in duration than transients. If a large load is suddenly removed from a utility grid, a temporary surge may result. Even though surges take much longer to pass than transients they are much smaller in magnitude. Ninety percent of surges will be less than twice the normal voltage, 99 percent less than three times normal.

Preparing a defense
Once the user understands how surges and spikes effect the network, the next step is to protect against them. Most equipment is sold without any specified immunity to transients. It's up to the installer to protect the equipment. The best approach is a combination of good installation, strategic protection, and calculated retrofit.

Good installation means determining where the worst transients will occur and not installing equipment there. Outdoors, whether aboveground or buried, is the worst place to have communications equipment during an electrical storm. Indoors, with as many feet of conductor between the equipment and the outdoors as possible, is the best place to install equipment to avoid lightning transients. Therefore, whenever planning an installation, the user should select a location with at least 100 feet of wiring between it and the electrical service entrance if possible.

When it is not practical to install a device away from transient exposure, then it must either be protected or the user should cross his fingers. Frankly, sometimes it is wisest to cross your fingers. From the isoceraunic map and Table 1, see if a location would be expected to have more than about 12 lightning strikes per year within about a half mile. If so, then it's probably best to protect all equipment. Also, if any local conditions or equipment make a location a moderate to high risk, protect all equipment (Table 2).

What components and circuits are available to protect communications gear against transients and how should they be used? Two transient entry points must be guarded for an installation to be protected: communications lines and power lines.

Table 2: High risk factors

- HIGH LIGHTNING ACTIVITY
- OVERHEAD ELECTRICAL COMMUNICATION CABLES
- LARGE ELECTRICAL MOTORS
- ARC FURNACES
- WELDERS

Most of the transients on communications lines will come from secondary effects of lightning strokes, and therefore will contain considerably more energy than the average power-line transient found inside buildings. Communication-line transients contain so much energy they can cause spikes in nearby cables through induction. Protect data lines at the building entrance to avoid this kind of transient coupling. If it is not practical to protect communications cables exactly at the building entrance then use magnetically shielded cable from the entrance to the protectors, and keep a good distance between the communications cables and any unprotected data lines or power circuits.

Good distance here varies somewhat with installation circumstance. It is important to remember that the goal is to avoid magnetic induction between two conductors and that magnetic induction is most effective between two conductors that are parallel and close together for a long distance. Therefore, never run unprotected cables in the same cable tray or conduit that contains the unprotected communications cable. If the two cables are parallel for more than 10 feet they should be separated by at least 10 feet, if parallel at all they should be separated by at least 3 feet, and if they cross each other at right angles they should be at least 1 foot apart. If the cables are in a grounded conduit system, all the distances may be halved.

After protecting the data lines in an installation, protecting the communications equipment's power supply is the next priority. One approach to power-supply protection is to run a dedicated line, or separate circuit, all the way from the building's electrical service entrance to the computer

3. Transient protection. *The first widely accepted transient-protection standard, was IEEE-472, the test wave used for data and signal lines not exposed to lightning threats.*

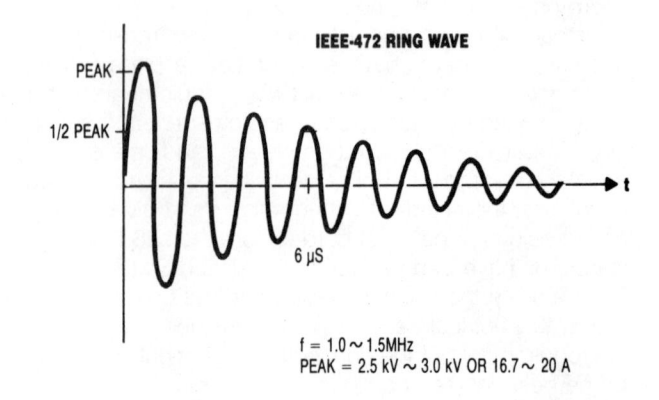

IEEE-472 RING WAVE

f = 1.0 ~ 1.5MHz
PEAK = 2.5 kV ~ 3.0 kV OR 16.7 ~ 20 A

Table 3: Partial list of transient protection devices

POWER CONDITIONING DEVICES

MANUFACTURER	MODEL/SERIES	FILTER	IEEE-587 RATED	PRICE (US $)	PHONE
PILGRIM ELECTRIC	SMART STRIP	YES	YES	175	1-516-420-8990
PILGRIM ELECTRIC PLAINVIEW, N.Y.	VOLTECTOR	YES	YES	174	1-516-420-8990
TII INDUSTRIES	TII 428 MKII	NO	YES	42	1-516-789-5020
TII INDUSTRIES COPIAGUE, N.Y.	TII 439	YES	YES	58	1-516-789-5020
TOPAZ SAN DIEGO, CALIF.	SURGEBUSTER	YES	NO	60	1-619-279-0831
TRIPP LITE	ISOBAR	YES	NO	48-110	1-312-329-1777
TRIPP LITE CHICAGO, ILL.	SPIKE BARS	YES	NO	40-50	1-312-329-1777
S. L. WABER	DATAGARD	NO	NO	16-36	1-800-257-8384
S. L. WABER WESTVILLE, N.J.	DATAGARD	YES	NO	45-160	1-800-257-8384
LEA DYNATECH	SE	NO	NO	97 UP	1-213-944-0916
LEA DYNATECH FOUNTAIN VALLEY, CALIF.	KLEANLINE	YES	NO	209 UP	1-213-944-0916
EFI SALT LAKE CITY, UTAH	DPI	YES	YES	119-129	1-800-221-1174
TRANSTECTOR HAYDEN LAKE, IDAHO	VARIOUS	NO	NO	145 UP	1-800-635-2537
ACCO WHELLING, ILL.	50671-50676	OPTION	NO	30-100	1-800-222-6462
MCG DEER PARK, N.Y.	VARIOUS	OPTION	YES	87 UP	1-516-586-5125
ELECTRONIC SPECIALISTS NATICK, MASS.	ISOLATOR	OPTION	YES	90-425	1-800-225-4876
CONTROL CONCEPTS BINGHAMTON, N.Y.	VARIOUS	OPTION	YES	32 UP	1-607-724-2484

NOTE: IEEE-587 RATING REFERS TO WHETHER OR NOT THE MANUFACTURER INDICATES THE IEEE-587 CATEGORY ON THE DATA SHEET. THOSE PRODUCTS SHOWN AS NOT IEEE-587 RATED ARE NOT NECESSARILY LESS RUGGED THAN THOSE CARRYING THE RATING, JUST HARDER TO COMPARE.

DATA LINE PROTECTION

MANUFACTURER	MODEL/SERIES	COMPATIBILITY	PHONE
TRIPP LITE CHICAGO, ILL.	SPIKE BARS	RJ-11	1-312-329-1777
LEA DYNATECH FOUNTAIN VALLEY, CALIF.	TRANSIENT ELIMINATORS	BARRIER STRIPS, RJ-11 COAX, RS-232	1-213-944-0916
POLYPHASER GARDNERVILLE, NEV.	IS	75/50 OHM COAX BARRIER STRIPS	1-800-325-7170
EFI SALT LAKE CITY, UTAH	DPI	RJ-11	1-800-221-1174
TRANSTECTOR HAYDEN LAKE, IDAHO	DLP, LMP, TSJ, FSP	RJ-11, RJ-45, RS-232	1-800-635-2537
MCG DEER PARK, N.Y.	DLP	BARRIER STRIP, BNC, RS-232, 50 PIN RIBBON, RJ-11, RJ-45	1-516-586-5125

or communications room. This approach, although required by some mainframe computer companies, is not really much help by itself. At best it will somewhat decrease the communications equipment's exposure to load-switching spikes originating in the building, at worst it may lead to the false impression that the best location for a communications/computer room is right next to the service entrance.

A more effective protection technique is to create safe islands of protected power where needed by installing commercially available surge suppressors between sensitive communications gear and the power source. Literally hundreds of companies manufacture transient-suppression equipment, ranging in price from less than $50 to more than $1,000, depending on features. For communications protocols and networks that are intolerant of noise these devices can be purchased with built-in electromagnetic/radio-frequency interference filters. Some equipment will automatically turn on the user's printer, CRT, and modem in the right order when a desktop computer's power is turned on. Since some computer systems expect all the peripherals to be energized and ready to talk when power is applied to the central processing unit, these fancy power-protection strips are quite convenient for unskilled operators to use.

Determine what level of protection a piece of equipment or group of devices require, install some sort of protection device between that equipment and the rest of the power wiring, and prominently label the protected outlets as off limits to any equipment other than authorized equipment. Make sure that the building's wiring is connected to the input side of the suppressor and the protected island is wired to the output side, because some products use circuits that are unidirectional protectors and won't protect equipment well if installed backwards.

How much is enough?

A good place to start in determining the level of protection requires a device to analyze the location in which the equipment will be installed. Two protection standards will help in this analysis: the Institute of Electrical and Electronic Engineer's IEEE-472, and IEEE-587. Both standards define the testing waveforms appropriate for devices to be installed in particular locations. The standards don't dictate the protective device operation, just the transient waves to be expected. When evaluating a product for suitability in an application, find out what the output from the protector will be if it is subjected to the proper standard transient wave, and determine if the communications equipment will survive that voltage.

The first widely accepted transient protection standard was IEEE-472, adopted by the IEEE in 1974. It was designed as a standard transient environment for electrical substation apparatus subjected to transients due to the switching of large inductive loads. The standard is widely accepted as the specified test wave for data lines and signal lines not exposed to lightning threats.

The IEEE-472 standard test wave is an exponentially decaying oscilliatory wave with an original peak of 3,000 V through 150 Ohms source impedance (Fig. 3). The bottom line for protective equipment is that it must survive 20 A spikes, positive and negative, for all signal lines.

Because of increased data and some misinterpretation of the scope of IEEE-472, a new standard was issued in 1980. This standard is IEEE-587, which is now referred to as American National Standards Institute/IEEE C62.41-1980. It is meant to deal with power wiring subjected to lightning influences as well as load-switching spikes. This standard, like IEEE-472, only defines the transient waves to be expected in particular locations, not how to protect against them. This standard is the best one to specify for transient protectors on communications lines and for communications equipment power lines.

IEEE-587 defines three classes of locations. These locations are the starting point in defining the protection requirements of all electronic equipment. The first step in transient protection is to install equipment in the location of lowest possible exposure. If this is cannot be done then make sure the equipment can survive repeated exposure to the test waves defined by the standard.

The harshest environment defined by IEEE-587 is outdoors (location category C). In outdoor locations equipment will be exposed to much higher peak transient voltages because wiring flashover will occur at about 10,000 V instead of the 6,000 V level common for indoor power circuits. The standard transient defined by IEEE-587 for outdoor environments and service entrances is a unipolar exponential spike.

As the power line enters a building the environmental rating decreases and becomes much more definable. The open circuit voltage and short circuit current impulse waves have the same shape as for category C locations, but they now have specific maximum values, 6 kV for the voltage wave and 500 A for the current wave. Any device that includes a crowbar device must be able to instantaneously switch between the voltage and current impulse without damage when the crowbar operates. This location is defined as category B, major feeders and short branch feeders. Devices in these locations also need to handle an oscilliatory wave defined in the standard.

After the power wiring travels far enough inside the building the cumulative wiring impedance limits the short circuit current available to damage protectve equipment. Therefore if the wiring length is at least 10 meters (30 feet) from a category B area and at least 20 meters (60 feet) from a category C area then the current wave peak is defined as 200 A. The voltage wave remains the same.

In general, IEEE-587 allows users to specify the transient protection requirement of their communications equipment by knowing where the equipment is installed. If it is outside, it's category C; indoors but within 60 feet (cable length) of the service entrance, then it's category B, farther than 60 feet from the service entrance, it's category A.

Table 3 lists a few of the more than one hundred manufacturers of power conditioning and data line protection devices, many of whom categorize their equipment according to which IEEE-587 standard wave they are designed to suppress. The products are here rated as neither good nor bad; the suitability of a particular product is dependent on the exact requirements at each location. It is up to the user to determine which products offer the best price/performance combination. ■

Kevin Sharp, a design engineer with Burr-Brown Data Acquisition and Control Division, received his B.S.E.E. from the University of Missouri-Rolla in 1981.

Larry L. Duitsman and Madeline M. Pinelli, Boeing Computer Services, Bellevue, Wash.

An all-purpose model to aid in all phases of network design

With many computer architectures to choose from, managers and users now have a common model that will foster better computer and network planning.

In today's increasingly competitive business environment, a company's ability to establish a unified approach to purchasing and managing information processing resources has become critical. In many corporate environments it is the communications management that most keenly feels the need for a unified approach. This is because the communications team suffers most when the information infrastructure is put together without sufficient forethought and planning.

The development of a unified approach should not fall solely on the shoulders of communications managers, however. Other computer managers, department managers, and users should also consider as vital the adoption of a singular, organization-wide information strategy. Not only would such a plan reduce the organization's expenditures on information resources, but it would also enable the organization to make better strategic decisions about information handling.

Bringing diverse parties into agreement over computer and communications resources is no mean feat. Organizations need a common model that the managers, analysts, and users can employ that will allow them to define network requirements with sufficient specificity and enough compatibility to make integration easier. (The term "model" is defined here as "a description or analogy used to help visualize something that cannot be directly observed.")

Ideally, this model should be based on the idea of managing the many processing and communications resources—including types of data, specific products, as well as architectures—that are usually in the organization (see "What is an architecture, anyway?"). While a number of different network-planning models have been proposed, the Strategic Information Management (SIM) model, detailed here, has been a very useful tool for defining the interactions and interrelationships between heterogeneous computing components.

SIM is a general framework that will allow analysis of all of the diverse processing and data models using a single common technique. Aided by an SIM model, an organization can shorten the length of the development cycle by defining a common discipline to increase the understanding of functional and integration requirements by management, users, and analysts.

Similarities and differences
Understanding the SIM model requires an awareness of major differences and similarities between SIM and the International Organization for Standardization's Open Systems Interconnection (ISO/OSI) architecture and protocols.

The ISO/OSI reference model and the protocols used to implement the model have been developed to directly address the need for information interchange between heterogeneous computing environments. While this is obviously important, users should recognize that the OSI reference model does not address the actual computing hardware or operations required to support information services, but only the exchange of data between a network's two nodes. The ways that the data is accessed, obtained, and manipulated, and how the information is displayed are matters outside of OSI's realm.

The SIM model addresses communications and these other issues in a manner somewhat similar to OSI: The SIM model exploits the advantages of layering of different functions within the architecture. Layering allows the establishment of protocol boundaries to define how information flows from one layer to another.

The Strategic Information Management model also facilitates modularity; that is, the replacement of a product in one layer with minimal or no impact on products in other layers. In a SIM model, the layering concept is applied across all of the information components, not just the communications portion covered in OSI. The same advan-

What is an architecture, anyway?

Users have heard many industry analysts and observers refer to future "information architectures" as the key to more complete integration of the components of current and future computing environments. The prominence of the International Standards Organization's Open Systems Interconnection (ISO/OSI) reference model has done a great deal to popularize the use of architecture in information networks. Unfortunately, a good deal of confusion remains about exactly what an architecture is.

An architecture is defined as the set of rules necessary to build an object or to interface one object with another. An object can be anything, and this definition can be applied to any situation. The ISO/OSI model uses a layering architecture to define a common array of communications protocols. Bakers also use a layering architecture in the construction of certain cakes. There is no conflict with the use of the word architecture in these examples.

Much of the confusion associated with the use of architecture in reference to information processing results from the word "architecture" being used interchangeably with the word "standard." ISO has developed an architecture and a standard at the same time, and OSI is sometimes assumed to be the same thing.

A standard is defined as a convention established from a number of alternatives. Organizations adopt standards or conventions for information technologies to manage their spread, use, and integration. These standards may only exist within an organization, or they may be adopted from de facto standards set by a majority of peers using them, or from international standards set by standards committees.

An information services standard is a convention established from a number of alternative architectures. An architecture can be developed by an organization, a vendor, or a standards committee. Once developed, it may or may not be adopted by any organization as a standard.

The OSI/ISO model is an attempt to develop a networking architecture. The adoption of this architecture by the world's standards organizations would make it an international standard. However, it is still one architecture selected from a number of alternatives. There are other architectures that address the same requirements as the ISO/OSI model. Both the Transport Control Protocol/Internet Protocol (TCP/IP) and the Xerox Network System (XNS) are common architectures that also solve the problems addressed by the ISO/OSI model. When a business decides to use either, instead of ISO, it is still establishing a standard for that organization. In this case the organization's standard is not the same as the ISO's international standard.

Architecture management involves the evaluation of the architectures available in order to recommend certain ones as business standards. The integration problem found in many computer networks does not stem from a lack of architecture; in fact, the opposite is true. There are currently hundreds of mainly proprietary architectures available to control the automated processes of any one task. (IBM's System/370 and DEC's Decnet are well known examples). The challenge of architecture management is to reduce the number of required architectures to one common set of standards.

The process of determining which architectures to use as standards is very similar to the process of selecting hardware or software products. Evaluation criteria can be developed based on business requirements, goals, and objectives. Candidate architectures can then be evaluated against the criteria. There may even be a make or buy decision involved. If an existing architecture does not provide the capabilities needed, a new one may be developed. The MAP/TOP architecture being developed by General Motors and Boeing is an example of a scheme being invented because existing architectures did not meet anticipated needs.

Integration results from the reduction of the number of architectures implemented to perform a task. If an organization focuses on the multiple architectures used in computing environments today, it does not need to wait for tomorrow's standards. Current architectures can be evaluated and selected as standards. As new architectures become available, migration plans can be developed for their adoption.

---

tages of flexibility, modularity, and reduced costs that the ISO/OSI model provides for communications can be realized across the entire landscape of information services.

Solutions that are built on the OSI architecture have successfully addressed the problem of interconnection of computers, which is one aspect of the more general challenge of integrating diverse computing environments. This suggests that similar success might be achieved for architectures of the environments outside of communications. Efforts are currently under way to define an architecture that includes interconnection within the larger scope of distributed computer services. The European Computer Manufacturers Association (ECMA) has recently proposed a model that tries to add structure to a larger piece of the information services environment (see "Another emerging model").

The SIM model is a framework for displaying interconnection, distributed services, and any other architectures that might be involved in a computing environment. An SIM model is meant to highlight information architectures and their use in defining processing interfaces. It is a window to help designers and users visualize the interactions between the information resources. The primary purpose is to allow an organization to evaluate how several types of products, data, and architectures—be they generated by international organizations like ECMA or OSI or by vendors—would interact with current or future environments.

Unlike OSI, an SIM model does not specify portions of the computing environment, even at the architecture level. Instead, it is a tool with which users and managers can describe extant resources and evaluate the effect of changes in data types, products, or architectures.

To become the foundation of a strategic management methodology, an information resources model should:

Another emerging model

The ECMA Distributed Services Framework is a general model for all distributed services. The combination of distributed services would constitute an integrated information service. Three layers—user, client, and service—make up the distributed services framework, which can be used to evaluate architectures. The model recognizes the need for a structure to ensure that each of the various distributed services can communicate with each other as their standards are developed.

Figure 1a shows a nondistributed environment where all of the production and support applications reside on the same computational node. If an application required a distributed solution, a service interface that included communications protocols would have to be created to represent the communications requirement to the remote machine.

The ECMA model requires that the distribution of function be transparent to the user; this implies that the application interface cannot change. Figure 1B shows the creation of a client portion of the application remaining with the user. The application interface has not changed. The service portion of the application has been distributed to another node, but the application interface remains the same. The service interface and the service access protocols are the communications required to perform the separation.

OSI communications occur only in the area inside of the dashed lines. Therefore there is no conflict in the OSI notion of protocols being used only between peer entities. The client and the service are complementary entities that communicate through service access protocols.

In Figure 2, the ECMA model is used to represent the Interpersonal Messaging System of the CCITT X.400 Message Handling System standard. The clients of nodes A and B are the User Agents. They assist the user in dealing with message functions such as sending, receiving, filing, and forwarding. The services of nodes C and D are the Message Transfer Agents (MTA), responsible for relaying the messages to their destinations.

The service access protocol between the client and service would be the submission and delivery protocol (P3) specified in the X.410 and X.411 protocols. The message transfer protocol (P1) between node C and node D would also function as a service access protocol. In this case, one of the services would act as a client to the other service. P1 is also defined in the X.410 and X.411 standards.

The ECMA distributed services framework represents the next step in the evolution of the architectural model. The intent of ECMA is to represent all of the distributed services such as mail, file, retrieval, and print, with one common framework. ECMA has extended the ISO reference model to address architectures outside of the connectivity arena.

More importantly, ECMA Distributed Services Framework has begun to address the integration of the diverse standards being developed by the various national and international committees. This effort must continue to expand in scope until there is an architectural model that references all of the interfaces necessary to automate information processing.

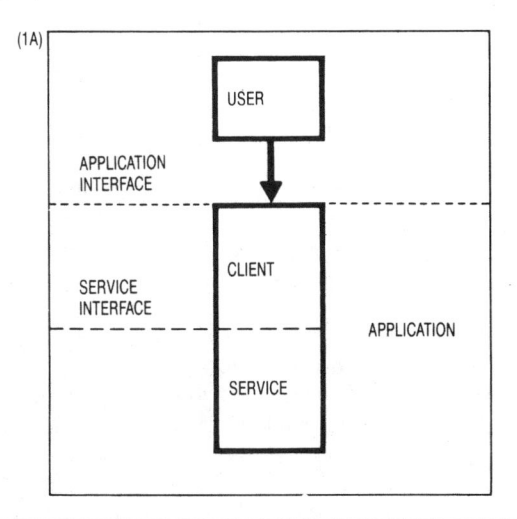

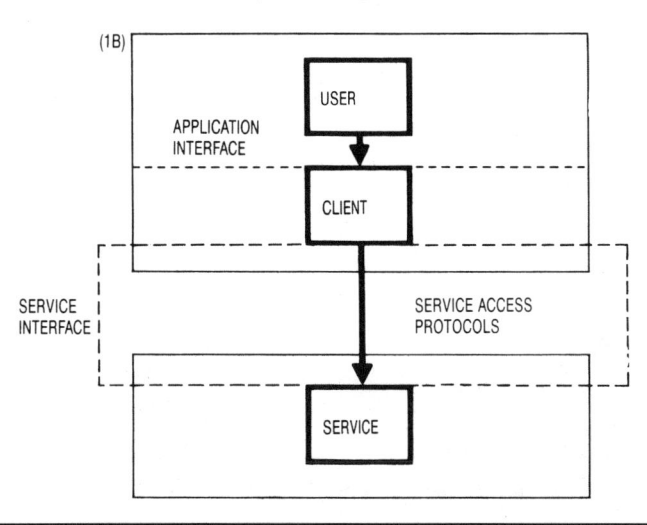

- View the organization's information resources as a whole and as the sum of its parts;
- Be easily understandable to sophisticated users, management, and analysts, and be an educational device for anyone who is in the process of learning about information technology;
- Provide different views for different audiences, while using only one model with one set of data;
- Be easy to sketch to aid visualization of the various information resources;
- Allow the display of varying levels of hierarchical complexity depending on the purpose;
- Have a design that encourages the use of computers in the evaluation of architectures.

The SIM model consists of four layers: the user interface layer, the application layer, the data layer, and the processing layer. Figure 1 shows the ordering of the layers and the categories found in each layer. Each layer may use the services of any other layer at any time. The four layers combine to construct a complete representation of information resources.

The relationship between the SIM model layers is illustrated by the metaphor of the "information vehicle." As will be explained, various layers of the SIM model are analogous to the steering, wheels, fuel system, and engine that make a vehicle operational. The driver (user) of the vehicle is not as concerned with the operation of its components as with reaching a destination. It is up to the designers and mechanics (managers and analysts) to ensure that each component functions correctly and compatibly with the others for the vehicle's smooth operation.

The user interface layer is concerned with presenting information to the user, thereby providing the user with a view of the automated processes. The user interface layer contains the applications with the capability to handle the user's interactions with any information resource.

In the information vehicle analogy, the user interface layer addresses the controls—the steering mechanism, the gas and brake pedals—needed for operation. It also covers information output, or those measurements tantamount to the speed of the vehicle, the amount of fuel available, security (keys and locks), and comfort (the ease of use of seats and gauges).

Similarly, the user interface layer of information processing contains the applications handling the input and output of information to the user. A user-interface support application would typically be responsible for cursor control, screen layout, keyboard functions, and other user input and output devices. The support applications also identify the location of a command line and the specific key to strike for help functions.

Access security is contained in the user interface layer as well, as are the applications supporting the ergonomics of the user interface. These handle the physical attributes of the workstation such as mice, touch screens, and the type of keys on the keyboard.

Tailored

The application layer provides the capabilities that most help the computer users perform their jobs. In the information vehicle, the application layer would address all of the dealer options that tailor the vehicle to a particular individual. Some—wheels, for example—are required. Others—a radio, for one—are intended for convenience, but a radio can also increase productivity if it helps the driver avoid a rush hour traffic jam.

Typical production applications include decision support software, library services, and engineering and manufacturing design tools. Production applications also include personal services software that increase individual productivity. The specific mix of user applications needed for information processing depends on the number and variety of capabilities required.

Production applications use the applications located in the other layers of the SIM model. Depending on the function of the production application, it may employ a user-interface package, a data-access package, and a communications package to perform a single task.

The data layer consists of support applications that manipulate data. The layer is further divided into data-access services, data-interchange services, and data-storage services, a combination of which represents what is needed to manage data flow through the network. This layer contains the fuel used by the information vehicle. Included in the data layer are the fuel lines for accessing the storage tank, the methods of storage, and the delivery and metering mechanisms. If required, there are also methods for locking the gas cap and disabling the fuel pump for improved security.

Data services

In information processing, data-access services are the support applications providing access to the data. They include the programs that address the integrity of the data, database manipulation languages, and the data dictionaries, directories, and encyclopedias.

Data interchange services allow for the interchange of information from one application to another, regardless of their location. These services address data transfers between computers, whether colocated or geographically disperse, and translations between disparate data formats. Data storage services—say, database management systems, data administration, archive and backup, and the

1. Four layers. *The SIM includes all computer processes in its layering scheme, allowing users to more easily keep track of the functions of various components.*

USER INTERFACE SERVICES			USER INTERFACE LAYER
PRODUCTION APPLICATION SERVICES			APPLICATION LAYER
DATA ACCESS SERVICES	DATA INTERCHANGE SERVICES	DATA STORAGE SERVICES	DATA LAYER
NETWORKING SERVICES		OPERATING SYSTEM SERVICES	PROCESSING LAYER

underlying database structure—are concerned with storage.

The data layer also includes data models used to represent the data's type and location. The relationships between the data models and the rest of the SIM model provide the link between the physical data and the applications.

The bottom layer is the processing layer, which is concerned primarily with the interfaces between the rest of the information processing applications and the hardware. It contains operating systems and networking services, and it is the only layer in this model that is concerned with the network's hardware specifics and transmission equipment.

In the information vehicle metaphor, the processing layer would include the engine and the interfaces between the engine and the rest of the vehicle. The engine provides the power to make the information vehicle work. If the vehicle had multiple engines, the processing layer would also contain facilities to coordinate each engine's operation.

The operating system services of the processing layer are the boundary services between the hardware and the rest of the information resources. Operating system support software addresses performance measurement, maintenance utilities, user libraries, device drivers, and computing hardware.

Networking services provide the communications between any two applications, be they production or support applications. This layer includes local area networks, high-speed host-to-host networks, and network management.

Networking services and operating system services are both included in the processing layer because they provide many of the same interface services. Each have functions that are concerned with the sending and receiving of data within one computer or between different machines.

As the information vehicle metaphor suggests, each SIM layer addresses a component crucial for successful operation. Arguing that any layer is more important than another is similar to maintaining that an automobile's engine is more important than its wheels; in fact, removing either would make the vehicle useless. Layering allows the interfaces to be defined so that one engine may be replaced by another to improve performance. The power of the SIM model is that it demonstrates how the components of each layer work with respect to each other. Figure 2 illustrates the relationships between some applications necessary to interconnect a workstation and a host computer. The model could represent interactions between any two or more computers.

The templates on the left and the right represent the workstation and the host, respectively. Each template has been divided into the four layers of an SIM model. Developing a template for each computer shows the hardware connectivity requirements and highlights the integration necessary for both the software and the data.

The smaller boxes inside the templates represent a function to be performed. The model is designed to allow any application to be placed somewhere on a template. On the workstation, the user interface layer contains menus

2. Separation of function. *To provide a clear view of how various component parts work together, SIM templates can facilitate better planning. For example, although the word processor and file storage might be included within a single package today, the organization may want to separate those functions in the future.*

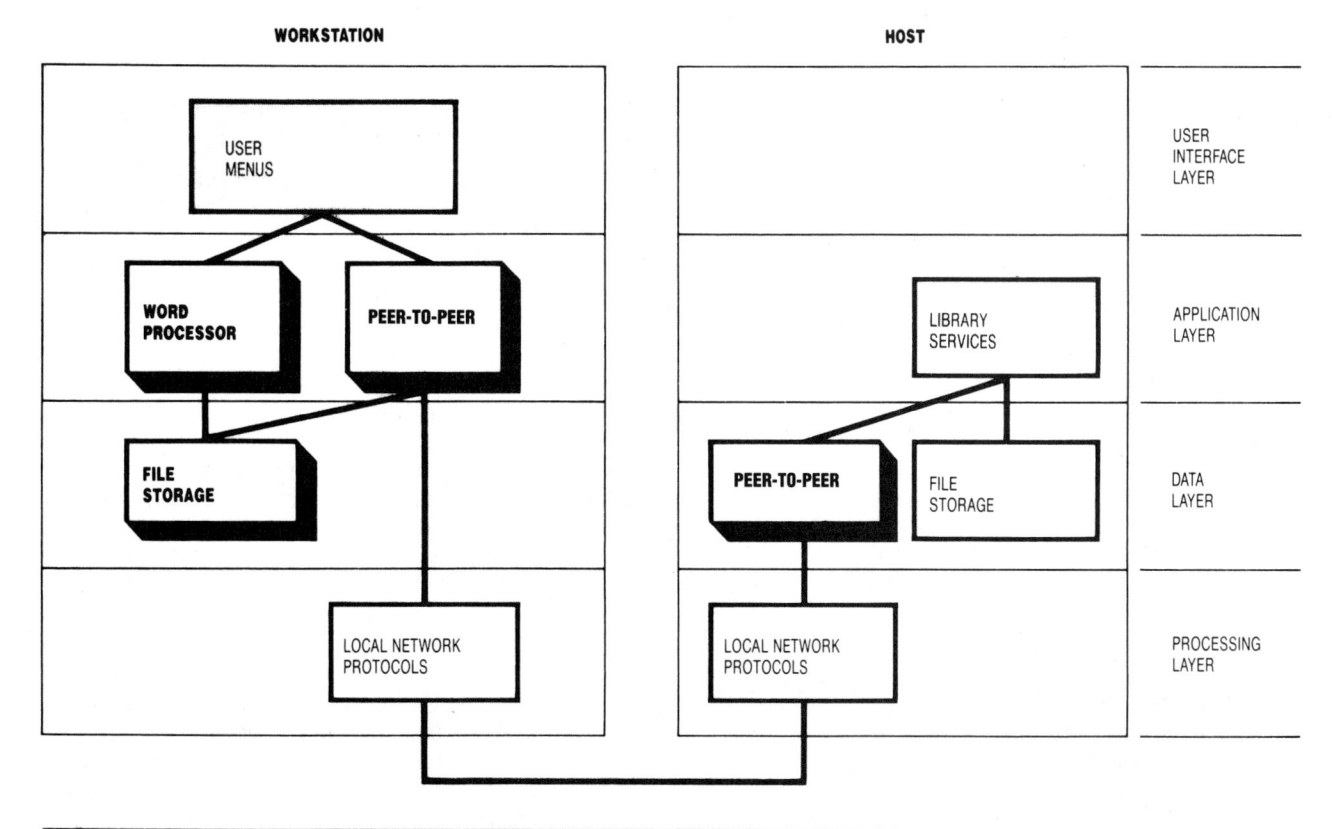

3. Map it out. Before automating any business process, it is vital that the organization's managers clearly understand the relationships between the various steps in that process. In the co-authoring example, managing the actual assignments is the most complex task and is likely to dictate many of the project's software requirements.

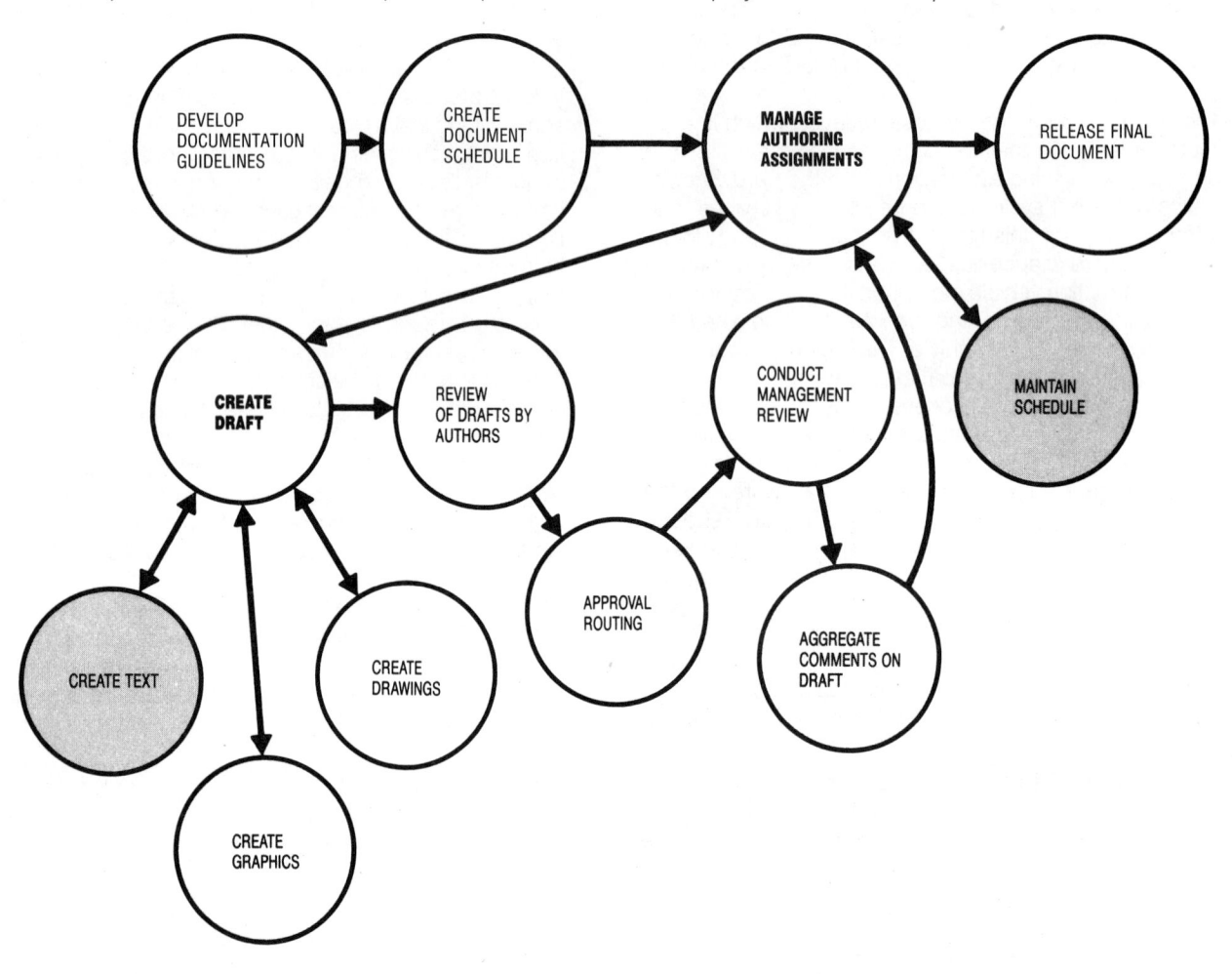

accessing the applications of word processing and the peer-to-peer interface.

The lines between the boxes represent the information flow of the service components. Each line has attributes indicating the type of interface and data shared between the components. The line between the workstation and the host connecting the local network protocol boxes of Figure 2 represents the protocols, wire, and types of data used to communicate between machines. Line attributes between peer-to-peer and network protocols include the protocol boundary and the types of data flowing between the two boxes. If the local network protocols indicated by the box were replaced, the peer-to-peer interface would be affected only if the attributes of the line between them changed. In many product implementations these two functions are combined so the word processor can use its own storage mechanism.

Business scenario
An SIM model is a tool that both users and network analysts can use in the innovative application of information resources. A common model eliminates much of the confusion that is traditionally associated with translating a user's requirements into design specifications.

In automating the information processes, the first step is to determine what the process entails. Suppose the process in question is the co-authoring of an article; Figure 3 is a process flow diagram for this task. "Co-authoring" in this context refers to any technical or management report or study that is jointly worked on by two or more people, who may or may not be in the same work area.

The individuals whose tasks are represented by the bubbles of the process flow diagram should not be concerned with the hardware required to perform their tasks. Their concerns should be ease of use, functionality, and data integration. The user will employ the SIM model to develop a "business view" that will satisfy the requirement of the process flow diagram.

The analysts responsible for implementing the business view have the same concerns for ease of use, functionality, and data integration as the users do. Additionally, they must address the computing and connectivity requirements necessary to support the business tasks. The SIM model can be used by the analysts to develop a "conceptual design view" that will provide a functional design to meet the requirements outlined in the business view.

Using the SIM model requires the iteration of eight steps. To develop the business view of the co-authoring process shown in Figure 4, the steps might take place as follows:

■ **Step 1: Determine the level of detail.** The content of a sketch aided by the SIM model depends on the intended audience. A business view should minimize the detail, while highlighting users' concerns of functionality, data integration, and ease of use. To maintain the same level of detail, SIM users should ask if any object on the sketch is "a type of" another object. For example, in Figure 4, if business graphics were "a type of" CAD (computer-aided design) graphics, the template would include different levels of detail.

■ **Step 2: Draw the SIM template.** The business view addresses only the user interface and application and data layers of the model. It is not concerned with hardware or any of the elements of the processing layer.

■ **Step 3: Add the application functions.** The functions sketched in the application layer are derived from an analysis of the tools and capabilities required to perform each of the individual co-authoring processes as identified by the bubbles in Figure 3. For example, it was determined that electronic mail, the schedule/calendar, and project planning facility were required in order to "Manage Authoring Assignments." All of the other application functions listed here can be directly related to one or more of the co-authoring tasks.

■ **Step 4: Add the user interface.** In this example, as with most environments, the users desired a common user interface.

■ **Step 5: Add the data.** The next step is to determine the data requirements for each of the functions added in Step 3. The need for data integration determines the extent to which storage units are grouped together. In this example, all of the document functions require read/write access to data generated by the other document functions. As a result, in Figure 4 all of the storage for these functions is placed in a single box. The planning and document functions need to import and export data between them, but they do not need direct access, so they are placed in separate storage boxes.

■ **Step 6: Connect the boxes.** The lines drawn between the boxes in Figure 4 indicate the relationships between each of the components. Each line connecting boxes in Figure 4 represents information flow between separate functions in Figure 3. Each line represents the support necessary for a component to perform its job.

■ **Step 7: Check for completeness.** Because it serves as the outline of each task in the larger process, management can use Figure 3 for planning purposes. For example,

4. User's view. *Users will primarily be concerned with building a business view that illustrates the user interface, application, and data layers. The lines between boxes indicate required flows of data. Components needing read/write access to the same type of data are grouped together in the model's data layer.*

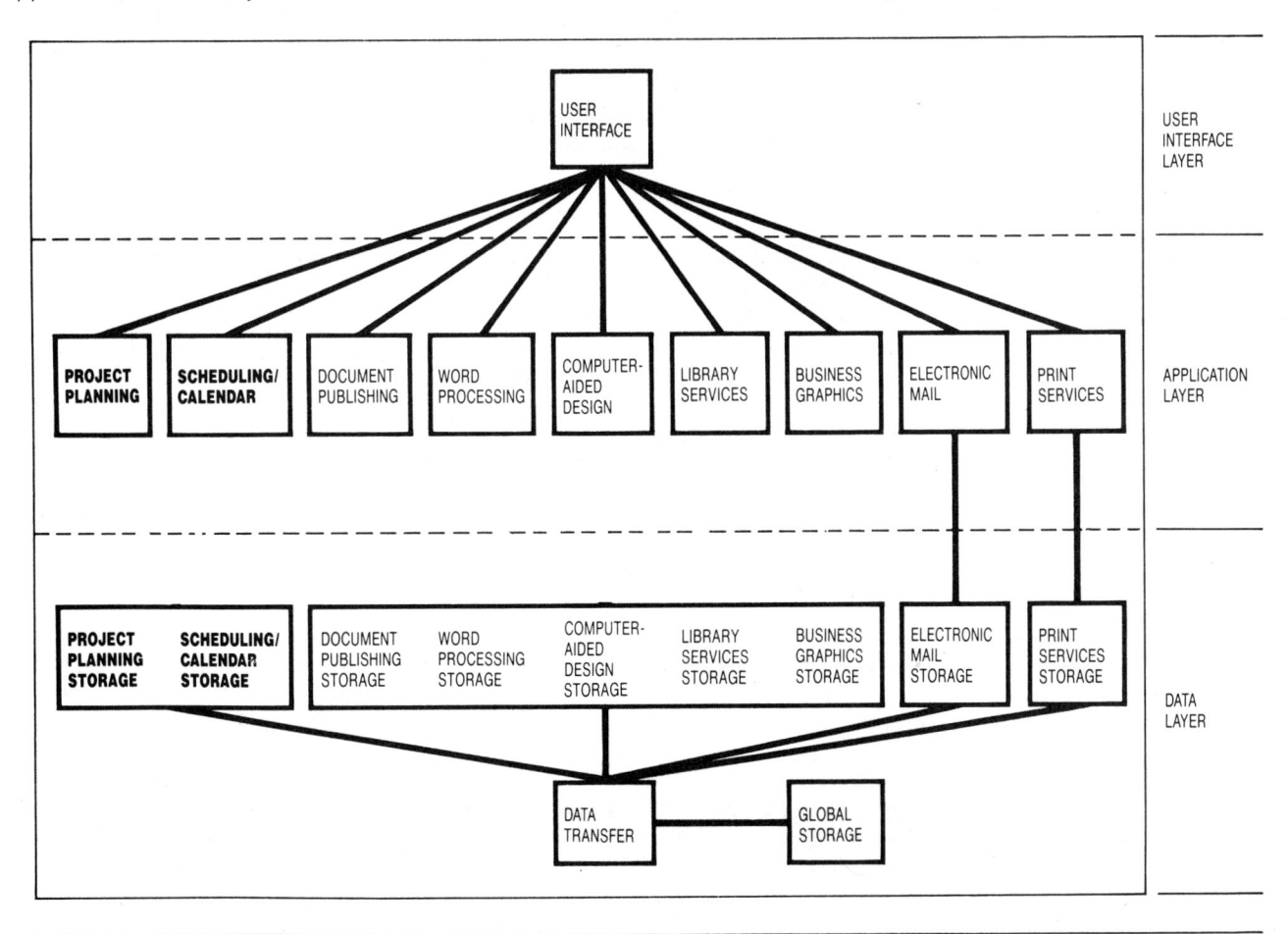

5. Clear relationships. *All four layers of the Strategic Information Management model are illustrated in the conceptual design. By building a conceptual design, managers and analysts can easily see which components need to access each other. This can be done in a general manner without specifying the size or make of the computer.*

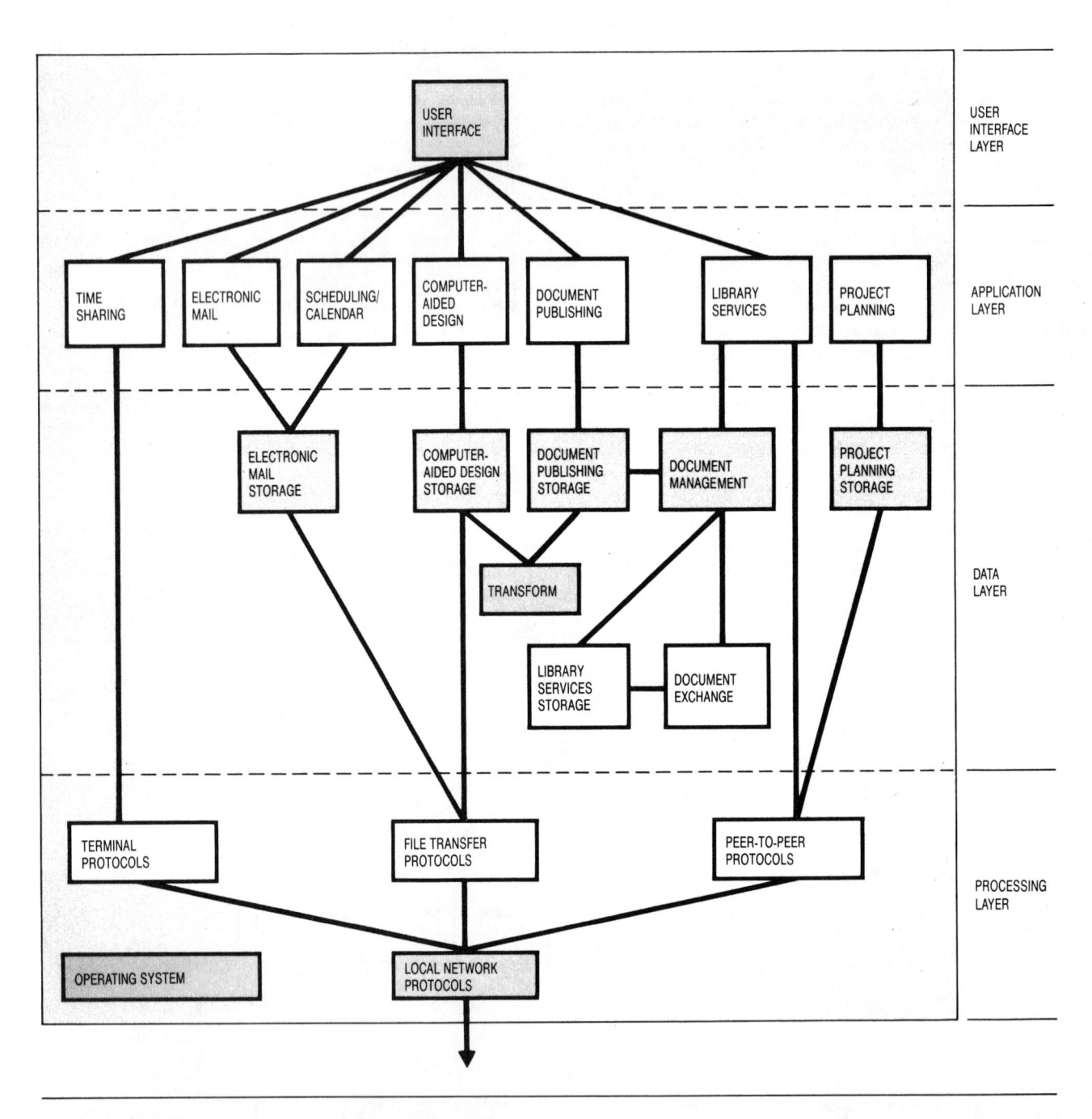

the individual responsible for the "Manage Authoring Assignments" process can prototype how each subtask would be performed using the business view. Each task—from obtaining the authors' current schedules to sending the new assignments—can be evaluated for ease of use, functionality, and data integration.

■ **Step 8: Iterate.** Users of the SIM model must remember that the model is a sketching tool to foster clear thinking about needed information resources. It is not important that the information be displayed correctly or succinctly in the first sketch. In fact, the prototyping of Step 7 is likely to suggest new ways to support a given process. There may be an infinite number of sketch alternatives, each able to support the co-authoring process. SIM users will not serve themselves by becoming overly attached to a particular representation.

Level of detail

As mentioned, the SIM model can also be used by analysts and managers to assess the processing and communications components needed to automate the co-authoring process. These components and the relationships between them will be illuminated in a conceptual design view that the managers will put together. Building a conceptual design view requires many of the same steps as above, although it takes place at a different level.

A design view addresses the computing and connectivity issues in addition to the ease of use, functionality, and data integration issues of the business view. It can be used for highlighting, but it can also portray enough detail to convey how all of the components work together to perform a desired task.

The design view includes all four layers of an SIM model (Fig. 5). One template is created for each type of computer proposed to support the design. For the co-authoring example, a workstation and a host were selected as the typical hardware to perform the co-authoring process. It should be noted, however, that in a conceptual design, the precise size of the machine is not defined. One design could be developed and implemented by many different sizes of organizations, each selecting hardware based on its particular requirements.

An analysis of the business view application layer identifies the capabilities required for the co-authoring process. The components of the conceptual design that directly affect the user are placed on the template in the application layer.

An important consideration in any network design is the allocation of applications across the computing environment. In the conceptual design of Figure 5, the workstation was provided with terminal emulation to allow access to a host's electronic mail facility.

Support applications

Once the application layer has been filled to meet the desired functions, the next step is to add the support necessary for their successful operation. Depending on the type of support provided, the support applications may be located in either the data layer or the processing layer. In Figure 5, the library services application requires additional support from the data layer to exchange, manage, and store documents. Communications support from the processing layer allows the interchange of data between the host and the workstation.

Many of the applications that have been identified on the template require direct action by the user for their operation. These are represented as boxes in the user interface layer. Figure 5 shows two user interfaces, one each for the workstation and host. The project planning function uses only the workstation user interface. When required on behalf of the user, it communicates in a peer-to-peer manner with the host project planning application. Other host applications require the user to invoke a terminal emulation facility and sign on to the host interface.

In this design, separate data files and storage formats were necessary. That required the addition of translation and data loading and extraction facilities to satisfy the requirements of the business view. The host data layer in Figure 5 contains both CAD and document publishing storage. An application has been included to transform one storage format to another.

Once placed in the model, the relationships between the applications can be represented as lines between their respective boxes. The attributes of a line are the protocols used to communicate; the method of communication; and the types, frequency, and amount of data communicated. The line between peer-to-peer and library services may have the attributes of either IBM's LU 6.2 or ISO's CASE protocol boundary. Operating systems have no lines drawn to them, although they obviously have a tie to all of the applications.

One of the strengths of an SIM model is the ability to determine if there are any holes in the design. Each function of the application layer of the business view should match exactly with a capability found in one of the layers of the conceptual design view. The specific flow of tasks developed for the business view can also be evaluated in the design view for ease of use, functionality, and integration. An SIM user can quickly spot any function that is not well supported in the design. For example, in Figure 5 it is evident that the project planning data is not integrated with the rest of the co-authoring data. A scheduling translator would be necessary to meet the requirements of the business view.

Again, sketching is a prototyping tool to help with the iterative process of designing a computing environment. Working with a sketch will suggest alternative ways to perform the same task. Analysts involved in template generation are likely to be heard saying "if only there were a line or a box here," which will suggest new uses of existing products or new opportunities for development.

SIM modeling adds clarity and consistency to the use of information technology by business. The resources required for a particular solution and their application can be displayed against the backdrop of the total business picture. Using a common model provides a direct link between business processes and the conceptual design.

SIM provides a visibility tool to enable the technology transfer of new concepts between individuals of similar or different disciplines. The building-block approach of an SIM model allows each layer to be focused upon individually with varying degrees of complexity. However, if the impact on the global information resource must be determined, the model is capable of an enterprise-wide analysis.

A foundation for analysis

Many forces can influence the effective design and delivery of information services. Resource modeling is the foundation for Strategic Information Management, building the relationships between business information processes and enabling technologies.

The Strategic Information Management model can be applied by an organization throughout the development life cycle of a network to gain consensus among, and bridge the gaps between, an organization's users, analysts, technical staff, and equipment vendors. Moreover, because the SIM model can be easily sketched or automated, it can help all of these participants visualize how the effects of change in one phase of development might ripple through the entire life of a project. ■

Larry L. Duitsman has spent 10 years with Boeing in data communications and systems development. He is responsible for developing a SIM-based information management methodology. Madeline M. Pinelli is responsible for strategic planning for end-user computing for Boeing Commercial Airplane Co.

James G. Herman and Mary A. Johnston, BBN Communications Corp., Cambridge, Mass.

ISDN when? What your firm can do in the interim

Does ISDN fit in with the corporation's networking strategy? The Bell operating companies' trials offer some helpful hints.

As the Integrated Services Digital Network (ISDN) labors toward viability, many corporate and government communications users find themselves in a quandary: invest in existing technologies today or defer current purchases under the assumption that ISDN equipment and services will soon be widely available. In particular, communications managers fear that ISDN will create a generation of prematurely obsolete customer premises equipment (CPE).

Corporate communications managers confront the ISDN dilemma at the same time they are facing increased pressures to use their network resources strategically and lead their companies into new markets. For example, when one firm (such as an airline) introduces a new electronic service or feature (such as an enhanced reservation network), it forces its competitors to do the same. This often means that decision making on new equipment and services must be compressed. In the case of decisions relating to ISDN, this is unfortunate, as the best strategy for the moment may well be a wait-and-see attitude.

For those managers who cannot wait for the technology to unfold because of business pressures, now is the time to define corporate guidelines for ISDN-related procurement decisions. This article makes the case for considering ISDN for the core of a premises architecture because, in many cases, ISDN will first impact functionality at the customer-premises level. Users can control what goes on inside the building walls but not the local telephone company's ISDN deployment schedules.

Figure 1 depicts one scenario for the move from current premises strategies toward ISDN. In the customer-premises environment, ISDN may mean the replacement of complex wiring schemes and separate terminal equipment with universal wiring and integrated voice/data terminals. A look at both historical trends in technology development and current ISDN trial activity indicates that ISDN's importance will not make an impact in the short term.

Today, users can choose between private branch exchanges (PBXs) and local area networks (LANs) as their primary form of data CPE. LANs and PBXs have their roots in vastly different technologies, namely packet and circuit switching, respectively (see "Circuit versus packet switching"). In general, LANs are used for dedicated applications while PBXs support casual users.

ISDN-compatible PBXs are a natural replacement for today's digital PBXs. Vendors of such equipment hope the added ISDN functionality will permit PBXs to take over a larger data-handling role. ISDN PBXs offer the promise of integrated wiring and switching for voice and data. Yet the lagging development of data protocol and networking support in ISDN puts it at a disadvantage in an era when LANs and packet networking are becoming increasingly functional and accepted by the user.

Reacting to an ISDN publicity blitz and swarms of consultants preaching voice/data integration, users currently making a commitment to LANs worry that a data-only premises technology will foreclose integrated options later on. Other users, who are considering today's third- and fourth-generation PBXs as their primary technology for both voice and data, worry that the data-handling capabilities of these technologies and the later ISDN-based generations will continue to fall short of the advances made by technologies optimized for complex data-handling needs.

PBXs used for data switching have yet to deliver on their early promise. Despite claims that PBXs would become the major means of handling local data-switching needs, users have remained unconvinced. On average, less than 5 percent of local data-switching requirements in most organizations are currently handled by PBXs. It is now clear that the evolution of data networking technology will continue well into the next decade. Will ISDN be the overarching framework for the delivery of new data ser-

1. Old and new? *ISDN could turn the customer premises from a web of coax and twisted pair (A) to a single wiring scheme supporting voice/data terminals (B).*

(A) CURRENT PREMISES ARCHITECTURE

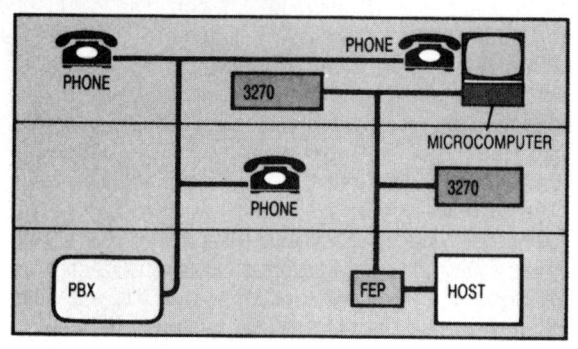

(B) ISDN-BASED PREMISES ARCHITECTURE

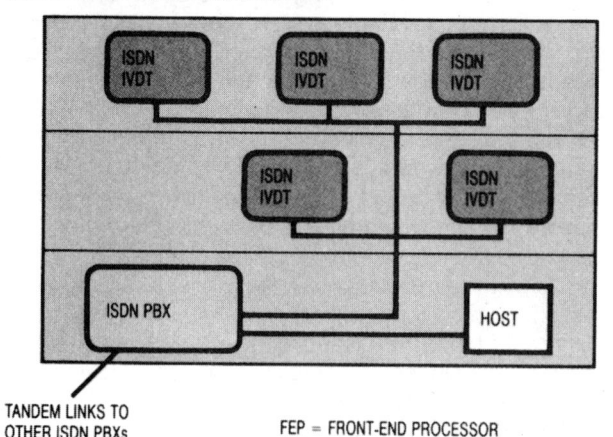

TANDEM LINKS TO OTHER ISDN PBXs

FEP = FRONT-END PROCESSOR
ISDN = INTEGRATED SERVICES DIGITAL NETWORK
IVDT = INTEGRATED VOICE/DATA TERMINAL
PBX = PRIVATE BRANCH EXCHANGE

vices, as the carriers are hoping, or will it be forever catching up with other innovations?

While ISDN will indeed become an important influence in the mid-1990s, most communications managers will be safe in procuring a currently available LAN or PBX today, provided that:
■ the product can be depreciated in seven years or less;
■ the product conforms to the major international standards, such as X.25 and RS-232, that are supported by the ISDN environment; and
■ the vendor has sufficient resources and staying power to migrate smoothly to ISDN when it takes hold.

It is important that planners not grasp at ISDN, LANs, or PBXs as the one true technology, but retain a balanced perspective. ISDN may not succeed as a data technology any more than today's PBXs have. LANs have their disadvantages as well, in that they are not suitable for voice, may require special cabling, are still experiencing changes in standards, and will continue to evolve rapidly in the coming years, forcing periodic equipment upgrades and/or replacements. Technology development proceeds in cycles that include distinct exploratory and maturation phases. It is wise to let new technologies mature somewhat before launching a multimillion-dollar commitment to them that will be expensive to derail should major problems arise.

Each stage in the development of key technologies calls for certain actions on the part of large user organizations (those with users numbering in the thousands). The stages are as follows:
■ During the early phase of a technology, when vendors are still conducting a great deal of research, the appropriate user action is limited experimentation and analysis of prototype offerings.
■ When a set of viable vendor approaches emerges, the technology enters a development phase. During this phase, enrichment of functionality is the main focus. Leading-edge users may begin to adopt the technology as a major network component. They should realize, however, that they are still gambling on its long-term viability.
■ At some point, a successful technology reaches a level of acceptance and standardization that makes it suitable for widespread investment. Vendor and user actions and priorities at this stage focus on operational issues such as efficiency and cost reduction.

In the case of ISDN and LANs, these phases have overlapped; the experimentation phase of ISDN is coinciding with the development phase of LANs (Fig. 2). However, the cycle will not end with ISDN. High-speed fiber LANs based on the 100-Mbit/s Fiber Distributed Data Interface (FDDI) standard are currently in development, as are wideband ISDN services on the order of 135 Mbit/s. These will appear just as narrowband ISDN enters maturity.

To ignore the LAN option in a premises architecture simply because ISDN has appeared on the horizon is a poor decision: LANs can be put to real use today, whereas ISDN is still far from operational. Given a four- to five-year depreciation cycle for LANs, installations made in 1987 will have depreciated in the early to mid-1990s. At that point, ISDN, if it has materialized, can replace other types of customer-premise data networking.

Users lessons from BOC trials

A commitment to ISDN today is not required either to enhance current networks or to avoid equipment obsolescence in the future. However, major users should consider limited investment in ISDN experimentation now. Participation in ISDN trials, particularly those involving a wide range of premises equipment, is one way for some users to determine the applicability of ISDN to their own network architectures and to have an influence on the further development of the standards.

Vendors such as AT&T and Northern Telecom have put forth elaborate scenarios involving extensive integration among home shoppers, store-based telemarketing setups, retail-chain buyers, manufacturers, and business analysts setting sale prices. At its best, ISDN could provide all these parties with instant access to information via existing telephone networks. However, that day is far off. Current ISDN trials are intent on simple replication of existing services. Pricing information, critical for cost/benefit analysis, is unavailable. Clearly, users will not abandon in-place gear for such dubious offerings overnight.

ISDN will be adopted slowly, in an evolutionary manner. For some users, ISDN may eventually provide a flexible premises architecture. For others, dedicated LANs may be equally important. The Bell operating company (BOC) trials provide the best current information on applications and

Circuit versus packet switching

Users making premises equipment decisions for the ISDN era can look to the past for lessons on the successful deployment of new technology. The question of whether to base a dedicated premise communications architecture on existing technology such as LANs or on the new ISDN technology is the latest in a long-running debate between circuit switching and packet switching. Circuit switching is the technology of the public telephone network and has been a major transmission service for over 50 years. Packet switching, on which LANs are based, was introduced in the late 1960s as an alternative to dedicated or dial-up circuits.

- Circuit-switched service, often called dial-up, provided the ability to connect to different destination end-points, one at a time, through the same access line. That is, your telephone is not hard-wired to a single destination but instead can be used to call any endpoint on the public switched telephone network. However, in general, you can use your telephone to carry on only one conversation at a time.
- Packet switching was the first major alternative to dedicated circuits and dial-up service as methods of transmitting data. Unlike circuit switching, which was a voice technology, packet switching was the first network technology designed explicitly for data transmission. It offers the ability to connect to multiple, different destination end-points simultaneously over the same access line. The packet network interprets the data that is sent to it and multiplexes that data over high-speed circuits.

Protocols are conventions for formatting the data sent to the packet network, and the history of packet switching has been largely the story of the development of these protocols. Packet network protocols establish conventions for addressing and code conversion. The protocols also provide error and flow control. These are essential in computer-to-computer communications that do not tolerate errors and can have radically different speeds at each end of the connection. Packet protocols provide a means for the user's device to communicate intelligently with the network concerning the service options desired. They also allow the network to report intelligently on the progress of the transmission.

How ISDN fits in

The goal of ISDN is to integrate access to both circuit and packet-switched services, permitting users to rely on a single user terminal and wiring arrangement. It provides important advances over basic circuit switching. First, it allows more than one circuit-switched connection to occur simultaneously over a single user interface, similar in some ways to the characteristics of packet switching.

Thus, a large mainframe computer can connect to an ISDN switch using the primary rate interface to carry on 23 separate 64-kbit/s circuit switched exchanges at one time. ISDN allows the sophistication of protocol-based control to be extended to circuit-switched services through the D-channel. Again, this is something of a retrofit of the benefits of packet-switching techniques onto circuit-switching frameworks.

At this time, ISDN does not call for any changes to the underlying transmission networks. That is, both packet- and circuit-switched networks will continue to function independently. Research is underway to packetize all communications, including voice. Such technology depends heavily on the high throughput and low error rates associated with fiber optics and is unlikely to be commercially deployed until narrowband ISDN and fiber are widely available (the late 1990s or later).

The dichotomy between circuit-switched and packet-switched services is quite obvious at the premises level:

- PBXs are circuit switches that provide a connection service for data equivalent to dial-up. As they have evolved, PBXs have attempted to provide services that are more tailored to data transmission. In particular, manufacturers have developed non-blocking versions of PBXs, which have sufficient capacity to enable all end-points to use the switch simultaneously. However, this comes at a price, since they still dedicate capacity rather than share resources as in a packet environment. Feature-rich PBXs became a major component of large telecommunications networks with the deregulation of CPE.
- LANs are a form of packet switching specifically adapted to premises communication. They are protocol-based and provide for error and flow control as well as sharing of the communications medium. Like packet networks, they allow for multiple, simultaneous connections and do not block. The first major LAN technology was Ethernet, developed at Xerox's Palo Alto Research Center in the early 1970s.

For many years, there has been much discussion on whether PBXs or LANs would become the dominant form of premises data-handling technology. As the ISDN mythology has flourished, PBX vendors have been positioning themselves to ride its coattails.

It is crucial that users who are trying to develop a coherent network architecture, which is likely to use both technologies, understand how ISDN will impact PBXs and LANs. ISDN is promising to integrate the benefits and features of both packet and circuit switching into a single technology. It relies upon a limited set of protocols, channel types, and equipment interface to provide building blocks for complex network architectures.

Concrete incentives for users to adopt ISDN will evolve slowly over the next five years. Users should watch for ISDN support of data services in general (and packet requirements, including LANs, in particular) before they make widespread commitments to ISDN. In the meantime, however, organizations should begin to prepare for the integrated environment of the late 1990s by beginning to integrate their voice and data management organizations through efforts at internal consensus building.

In the long term, ISDN is likely to fulfill its promise, providing greater integration of voice and data services than is currently available. In particular, it will bring the sophistication of data communications protocols to the world of voice, driving down network maintenance costs via more sophisticated tools and automation. It will also greatly increase the user's ability to manipulate the network.

2. Phases. *Technology matures in distinct phases or cycles. ISDN is following on the trail of LANs but will be followed by higher-speed and -capacity technologies.*

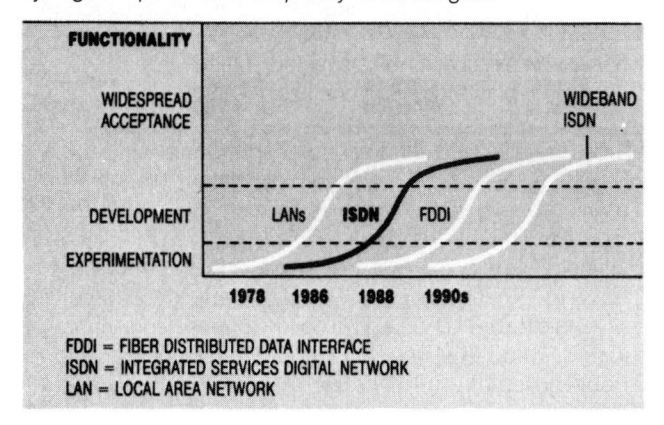

FDDI = FIBER DISTRIBUTED DATA INTERFACE
ISDN = INTEGRATED SERVICES DIGITAL NETWORK
LAN = LOCAL AREA NETWORK

should be carefully monitored by any firm considering the inclusion of ISDN in its network design plans.

In particular, users should watch for demonstration of significant new applications (such as those linking phone numbers and databases) in the area of packet data or integrated voice and data. Innovative, cost-effective features and functions would indicate that ISDN had reached a mature enough stage to be included in user networks. Unfortunately, users who must make premises decisions today have only wide-area ISDN trials from which to learn. Still, they can extrapolate premises strategies by looking at the capabilities of those trials.

Early bias
The first round of BOC-sponsored ISDN trials has demonstrated a natural telephone-company bias toward circuit-switched voice services. This was a logical development, given that the majority of early standards efforts by the CCITT (International Telegraph and Telephone Consultative Committee) focused on the circuit-switched world of public networks. After all, ISDN was first conceived of as a way to upgrade the performance and maintenance of the public common-carrier networks. Many of the countries represented at the CCITT support few, if any, private networks and often enforce restrictions that preclude private purchase of CPE, making such considerations of minor importance to them.

In particular, support for private packet networks and signaling between private PBXs (such as in tandem networks) were placed on the back burner. In recent months, the American T1D1 standards committee has begun to take data and private network requirements more seriously. When the CCITT meets in 1988 to approve ISDN standards for the next four years, the American representatives are expected to push for attention in these areas.

Although the first stage of most ISDN trials has been circuit-switched and voice-oriented, trials due in 1988 and 1989 should test a wider range of functionality. Most of the trials have started off with single "islands," central offices supporting a few dozen ISDN lines. Over time, the BOC trials plan to network together several such central offices. Plans for integration of BOC ISDN services with long-haul networks are still sketchy.

Consequently, users begin to wonder: What good is half an ISDN? That is, why invest in ISDN at one corporate location when it will not be universally available? Why not stick with virtual network services or existing tandem networks instead? The BOCs have few answers to these questions. Instead, they are seeking to segment user markets, targeting ISDN at the group most likely to appreciate network-based enhancements. Users of Centrex are a major target segment. These users are most likely to benefit from half an ISDN network, since they depend on the public telephone company to support their on-premises switching requirements. From the start, enhancement of Centrex (a public network service) has been a major ISDN goal of the BOCs.

In fact, some users seeking PBX solutions have found BOCs' ISDN proposals in the pile of responses. As the ISDN debate has matured, it has become evident that the premises environment will actually benefit from ISDN long before network services demonstrate major ISDN-based advantages. This is because users can buy ISDN switches and develop their own private integrated voice/data environments while the public carrier market is likely to remain fragmented for some time.

A closer look
ISDN relies upon universal availability of digital transmission plants using out-of-band signaling technology, which provides greatly increased intelligence over the public telephone company plant. Trials of ISDN access technology have been scheduled by all of the regional holding companies (see table). From evaluating these trials, it is increasingly clear that the first phase of commercially available ISDN services will focus on improved circuit-switched services. They will provide packet support and access almost as an afterthought.

The first phase of many trials focuses on the basic-rate interface in a Centrex environment (the basic rate is 144 kbit/s divided into two 64-kbit/s information-carrying B channels and one 16-kbit/s signaling and information-carrying D channel). User benefits cited are as follows:
■ voice/data integration using a single access line;
■ coaxial cable elimination, as circuit-switched data in the ISDN environment uses existing Centrex wiring; and
■ 64-kbit/s facsimile.

A few trials also include primary-rate interfaces (1.544 kbit/s divided among 23 64-kbit/s B channels and one 64-kbit/s D channel). Primary-rate functionality is oriented toward PBX-to-central office connectivity and host computer connections to the network.

Circuit-switched data, such as that commonly transmitted via dial-up circuits and modems, will be supported early on, but packet-mode data services connecting to ISDN packet handlers will lag. Packet data will be supported first by circuit-switched connections between the user and existing packet network nodes such as AT&T's Accunet. Later, as standards are hammered out for packet-mode services, users will be able to access ISDN packet transmission on an end-to-end basis.

User organizations involved in the trials range from McDonald's Corp. and the Arizona Department of Transportation to internal users in the Bell companies and the central office vendors themselves. Ameritech's well-publicized McDonald's trial is a good example of early

ISDN applications. Because its early focus is the business office environment, the trial:
- integrates personal computers and voice handsets;
- supports modem pooling;
- offers circuit-switched data;
- demonstrates gateways to packet-switching facilities;
- illustrates how ISDN eliminates coaxial cable requirements for new terminal-to-host connections; and
- replaces key equipment behind Centrex.

The packet portion of the trial routes packet data via a circuit-switched B-channel connection rather than a true packet-switched connection. No provision is made for interconnection with LANs. Since only basic access is being tested, data rates will presumably be limited to 64 kbit/s or less. When primary-rate access becomes available and specifications for the H channels are firmed up (H channels are standard ISDN interfaces, now under development, that range from 384 kbit/s to 135 Mbit/s), the opportunity will exist for higher transmission rates. Other BOCs will demonstrate file transfer, interfaces to mainframes, 64-kbit/s facsimile, and videotex.

The lag between the availability of circuit-switched applications and packet-switched support is 18 months to two years under current plans for trials. Part of this is due to the fact that packet-handling equipment for central offices is still under development by vendors. Available equipment, such as the packet-handling function in the NEC digital adjunct, uses proprietary protocols, which will later need to be redefined to fully comply with CCITT recommendations.

Most of the initial intraLATA ISDN trials will function as islands, with plans for multi-central office internetworking scheduled for early 1988. All 1987 trials, such as the McDonald's trial, are limited to the intraLATA jurisdiction. Southwestern Bell has announced its intention to provide an interLATA trial in 1988 and will serve as the Bellcore test case for such internetworking.

AT&T intends to offer interLATA ISDN in 1987, but it will only be available to customers connecting directly to the AT&T point-of-presence via a primary-rate access line (most likely terminating on a System 85 Primary Rate Interface), bypassing the local telephone company. This option, if delivered on schedule, may provide ISDN signaling to a limited number of very large corporations. However, widespread commercial interLATA ISDN service, supported by BOC access lines, is not anticipated before 1989 or later.

InterLATA services currently available, such as AT&T's Software Defined Network or US Sprint's Virtual Private Network, are often termed pre-ISDN services. This term indicates that they provide more functionality than traditional network-based services but are still not entirely compliant with the international ISDN standards. And again, voice and circuit switching are generally stressed.

The immediate future: What users should do

For users interested in ISDN, patience is a necessary virtue. Like many earlier technologies, implementation of ISDN will take at least a decade. In fact, the initial CCITT planning documents written in 1980 and 1981 indicated a phased implementation taking 10 to 20 years.

As it has unfolded to date, ISDN implementation is fraught with obstacles and contradictions. ISDN cannot succeed unless it is available universally, and this great expense to the common carriers might mean high prices for ISDN-supported services. However, users will not buy ISDN unless it saves them money over existing services.

Moreover, the goal of universal standards and limited interfaces seems out of synch with the pluralistic, competitive telecommunications scene in the United States. The

Selected plans for ISDN trials

REGIONAL HOLDING COMPANY	APPLICATIONS	CENTRAL OFFICE SWITCHES	START DATE
AMERITECH	BASIC ACCESS/CENTREX	AT&T 5ESS	FOURTH-QUARTER 1986
BELL ATLANTIC	BASIC ACCESS/CENTREX	SIEMENS EWSD AT&T 1AESS	MID-1987
BELLSOUTH	INTERNETWORK VOICE, DATA, AND PACKET	AT&T 5ESS NTI DMS-100 NTI SL-10	FOURTH-QUARTER 1987
	CENTREX	SIEMENS EWSD	SECOND-QUARTER 1987
NYNEX	BASIC AND PRIMARY ACCESS	SIEMENS EWSD	THIRD-QUARTER 1987
	BASIC AND PRIMARY ACCESS	AT&T 5ESS	THIRD-QUARTER 1987
PACIFIC TELESIS	THREE-NODE NETWORK LINKING MULTIPLE VENDORS AND PACKET DATA	AT&T 5ESS NTI DMS-100 NEC NEAX 61 NTI SL-10	THIRD-QUARTER 1987
SOUTHWESTERN BELL	THREE PHASES LEADING TO INTERCONNECTION WITH LONG-HAUL CARRIERS	AT&T 5ESS NTI DMS-100 ERICCSON AXE	SECOND-QUARTER 1987
US WEST	MULTIPLE TRIALS BY MOUNTAIN BELL, PACIFIC NORTHWEST BELL, AND NORTHWESTERN BELL	AT&T 5ESS NTI DMS-100 NEC NEAX 61 GTE GTD-5 ERICCSON AXE	FOURTH-QUARTER 1986

multivendor interworking promised by ISDN may not be in the immediate interests of key players. In particular, CPE vendors that gain competitive advantage from proprietary protocols are leary of the ability to use multiple PBXs in private networks. Consequently, these vendors may have little interest or motivation to hurry adoption of ISDN.

Users are particularly hungry for enhancements in the complex areas of data communications and network management. However, this area of ISDN standards has, until recently, received relatively little attention. Given the slow pace of standards approval and the swift pace of technological change, ISDN standards may reach the market too late. Yet, if vendors rush to introduce ISDN products and services before they are fully tested, early disasters and frequent revamping of the specifications could kill support for ISDN as a data technology.

During the next several years, users should practice the principle of caveat emptor regarding purchase of ISDN products and services. Compatibility among different vendor's equipment will not be ensured by the mix of proprietary solutions and international standards. As with any new technology, transition costs and retraining expenses should be carefully estimated and factored into decisions.

ISDN will eventually be widely available, because the long-term survival of the BOCs hinges on widespread customer acceptance of ISDN services. However, BOC agendas that focus on public network cost savings and preservation of installed central office equipment often differ from user priorities for sophisticated services that provide competitive business advantage. To date, the trials have been unable to demonstrate much that can be labeled innovative or revolutionary. In fact, ISDN trials struggle to replicate business services, such as digital Centrex, access to private packet networks, and customer-controlled reconfiguration, which are already supported.

Each user organization must make its own decisions regarding the applicability of ISDN. Business priorities regarding control, growth, and risk avoidance must all be evaluated. Given the slowness with which ISDN will penetrate the public networks, users should focus on assessing ISDN choices at the premises or campus level. In general, premises equipment is purchased with the expectation that it will serve the firm for a number of years (generally four or five for LANs and five to seven for PBXs, depending on individual corporate policies regarding depreciation allowances). Premises equipment procured today may still be in service as ISDN becomes more important in the early to mid-1990s. By comparison, network service offerings from the BOCs or long-haul vendors are often leased by the month and involve less of a long-term corporate commitment. Consequently, planners feel less at risk by deferring ISDN planning in the area of leased services.

Planning strategies

However ISDN is factored into a telecommunications plan, it must complement the firm's broader network strategy. For instance, the first companies to benefit from ISDN may have the following:
■ communications focused in large metropolitan areas (the first geographic areas that will have widespread ISDN);
■ a commitment to telephone company services such as Centrex;

3. ISDN orientation. Five years from now, users favoring ISDN-based networks will likely see ISDN PBXs playing the major role in linking scattered sites.

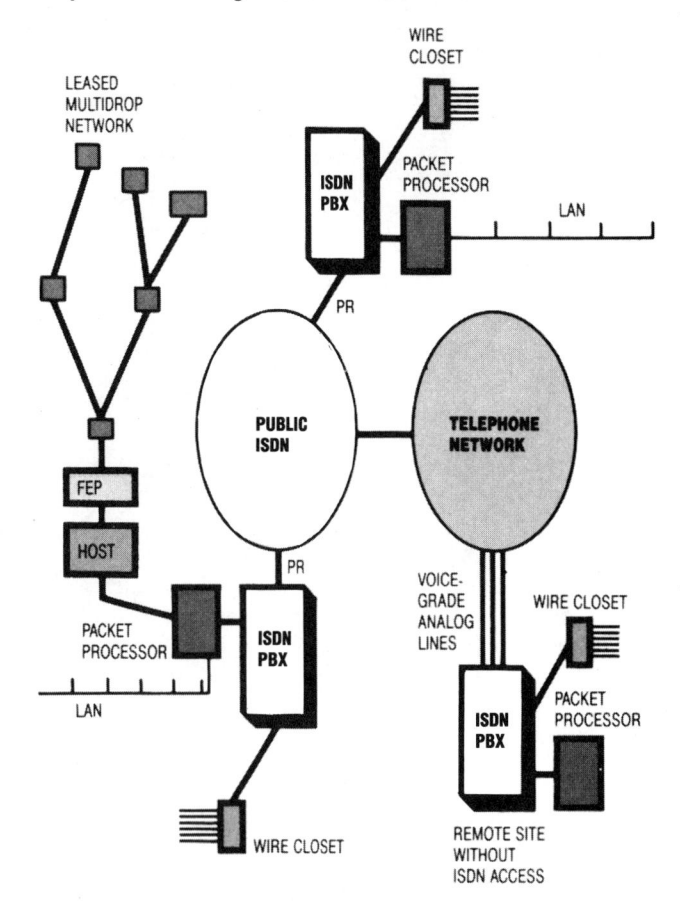

FEP = FRONT-END PROCESSOR
ISDN = INTEGRATED SERVICES DIGITAL NETWORK
LAN = LOCAL AREA NETWORK
PBX = PRIVATE BRANCH EXCHANGE
PR = PRIMARY RATE

■ extensive operations in Europe, where ISDN will be more important and more generally available;
■ voice/data integration as a major goal (hotels and airlines are likely to seek such integration); and
■ too small a scale to support a dedicated private network.

Such companies might develop a network topology akin to that shown in Figure 3. This strategy makes extensive use of telephone company services and ISDN PBXs. LANs and dedicated multidrop networks still exist at the edges of the network, but ISDN PBXs and central office services are the centerpiece. In making premises decisions today, this type of company must demand a smooth evolution to ISDN, buying only from vendors with well-defined commitments to international standards and ISDN deployment.

By comparison, some corporate networks give priority to the following principles:
■ independent support for voice and data;
■ significant commitment to LANs;
■ stable low-speed (under 9.6-kbit/s) data networking or wideband (over 1-Mbit/s) requirements (both of which are outside of current ISDN specifications); and
■ dependence on privately owned resources that are

4. Private-network plan. *If trends continue, private networks of 1992 may be based upon integrated circuit/packet switches linked by high-speed lines. These switches could support data equipment directly for local distribution. The network would also link with public ISDN networks but would not depend upon them.*

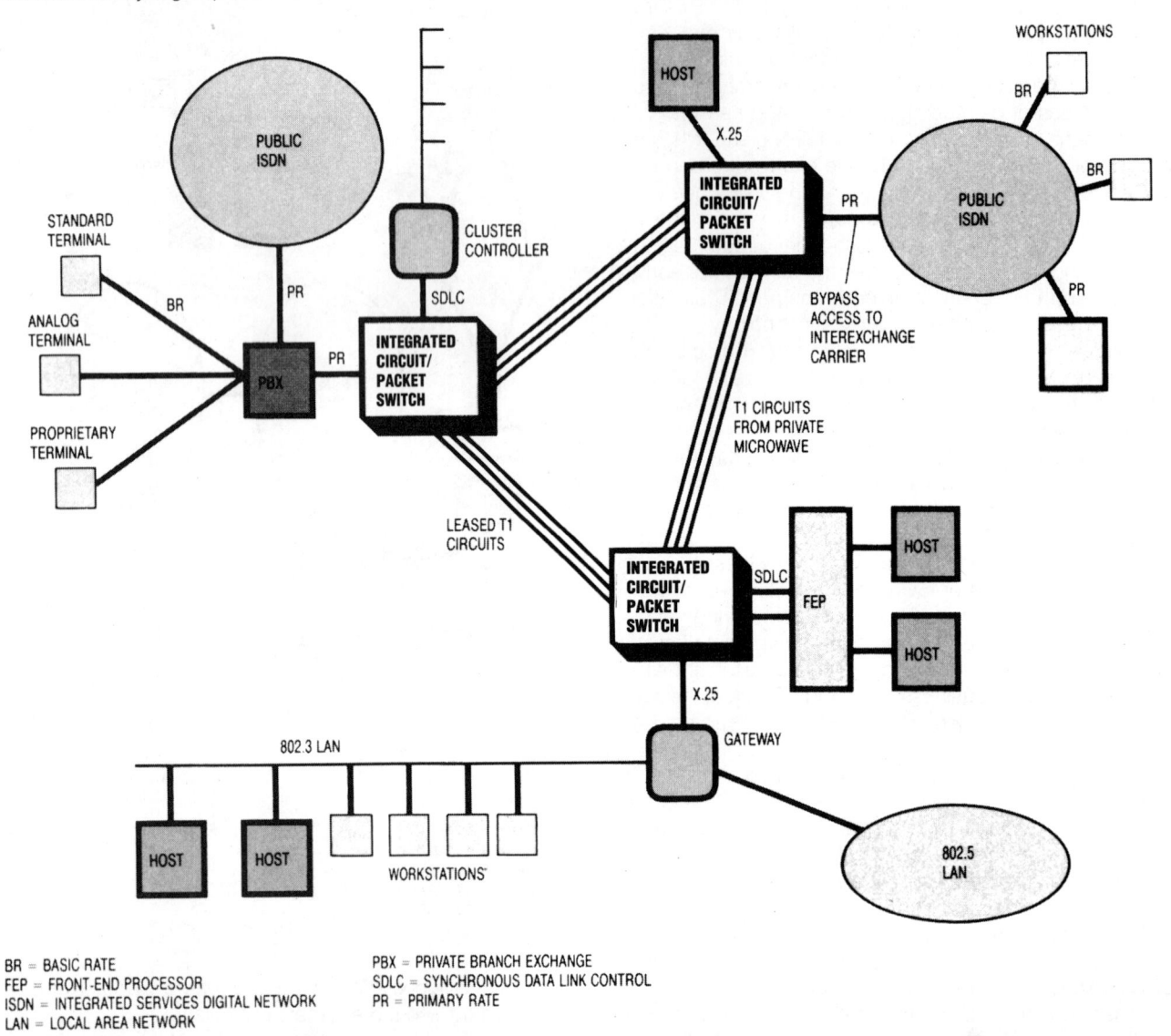

BR = BASIC RATE
FEP = FRONT-END PROCESSOR
ISDN = INTEGRATED SERVICES DIGITAL NETWORK
LAN = LOCAL AREA NETWORK
PBX = PRIVATE BRANCH EXCHANGE
SDLC = SYNCHRONOUS DATA LINK CONTROL
PR = PRIMARY RATE

planned and administered by a large internal staff.

Such organizations, which tend to be data intensive, will prefer a long-term architectural plan like that shown in Figure 4. This strategy treats ISDN as an adjunct to the privately owned core network. A private setup of this sort will make extensive use of T1 backbones, LANs, and bypass of the public networks.

In summation, ISDN capabilities that transcend existing public network capabilities will appear slowly. Truly new services will require time for the technology to mature. However, ISDN will eventually provide the foundation for such new capabilities as protocol-based, circuit-switched services and support of multiple services simultaneously.

Consequently, users should begin to provide for ISDN in their long-term (five years or more) technology plans but not stall decisions that are required by current business demands. Priority should be given to network and premises architectures that support heterogeneous mixes of equipment and adhere to international CCITT and Open Systems Interconnection standards. Compliance to these standards will ensure that networks can grow and mature over time, incorporating ISDN as appropriate. "Working technologies of today are preferable to paper claims of ISDN tomorrow" is a maxim that should guide users as they design their architectures and communications strategies. ■

James G. Herman writes, teaches, and works as an independent consultant in Cambridge, Mass. Formerly, he was director of the Telecommunications Consulting Group at BBN Communications. Herman holds an A. B. in mathematics, summa cum laude, from Boston College and is a member of Phi Beta Kappa.

Mary A. Johnston is a senior consultant in the Telecommunications Consulting Group at BBN Communications. She holds a Masters of Public Policy from the John F. Kennedy School of Government at Harvard University and a B. A., magna cum laude, from Drew University.

Gilbert Held, U.S. Office of Personnel Management, Macon, Ga.

Another use for the versatile protocol analyzer: Capacity planning

Network managers are bringing the workhorse analyzer to bear on identifying—and resolving—line-congestion problems

In the right technician's hand, the protocol analyzer is a troubleshooting tool par excellence, but network managers are taking it up as well and turning it to a slightly different application, network planning.

Technicans have long used the protocol analyzer—also called a data communications analyzer and data communications test set—to monitor and test data communications lines. Most of the newer analyzers, however, are usually more sophisticated than simple line monitors; they can also capture and play back data traffic. When knowledgeable hands exercise the full capability of the protocol analyzer, the tool can also yield useful network-capacity information needed for planning.

Although the capabilities of protocol analyzers vary considerably, there are certain features common to most of them (see table). More sophisticated protocol analyzers can include additional features such as ISDN support, response-time analysis, automatic protocol determination, and line-utilization measurements.

Some sophisticated analyzers may even present line-utilization measurements in easy-to-use graphic form. A few protocol analyzers include a built-in programming language that permits the operator to develop programs that can be executed to perform data trapping, event logging, statistics gathering, and other communications functions.

There are several reputable manufacturers of protocol analyzers, each with many good analyzers to choose from. For example, the Model 400, from Digilog, contains a built-in 3.5-inch diskette drive for recording and playing back data. It is the first model above the entry-level members of the Digilog protocol analyzer family. (More sophisticated members of the family provide a color-display capability, fixed-disk storage, and a line-utilization statistics capability that are not included with the Model 400.)

The Model 400, like similar devices manufactured by other companies, uses a menu-driven operation scheme that minimizes the amount of time required by an operator to master it (Fig. 1). From the menu, the operator can select a specific menu, or two additional master menus, from which other specific menus can be selected.

Setting the trap

For example, selecting Option B calls the Trap Menu to the operator's screen (Fig 2). This menu, like the others, permits the operator to select specific data fields by using four keys on the keyboard. To jump between fields, the operator can use the TAB FWD (Tab Forward) and TAB REV (Tab Reverse) keys. The FIELD SEL FWD (Field Select Forward) and FIELD SEL REV (Field Select Reverse) keys can be used to scroll forward or backward through the options available in each field.

Here is how FIELD SEL FWD and FIELD SEL REV keys work. The Search field for Trap 1 is initially set to OFF. When the Trap Menu is displayed, the protocol analyzer's cursor is automatically positioned on the Search field of Trap 1, since this is the first field into which the operator can enter a selection. By pressing the FIELD SEL FWD key, the operator cycles through the options for the field, which are displayed to correspond to each depression of the key. Options for this field include OFF, SEND, RECEIVE, and BOTH. The FIELD SEL REV key can be used to select the same options in the reverse order. Once an appropriate option is selected, the operator can use the TAB FWD key to position the cursor in subsequent fields.

The operator can define and match two sequences of characters or bits. When a match occurs between defined characters or bits and the data on the line, data capturing can be either started or stopped. Data selected can be sent to the monitor and displayed along with a time function. The operator can also select to increment one or more counters or trip an audible alarm.

Common protocol analyzer features

- LINE MONITORING, RECORDING, AND PLAYBACK
- (BERT/BLERT) TESTING
- EFS TESTING
- EVENT TRAPPING
- DTE/DCE EMULATION CAPABILITY
- DATA DECODE CAPABILITY

BERT = BIT ERROR RATE TEST   DTE = DATA TERMINAL EQUIPMENT
BLERT = BLOCK ERROR RATE TEST   DCE = DATA COMMUNICATIONS EQUIPMENT
EFS = ERROR-FREE SECONDS TEST

The Trap Menu permits the operator to define up to eight characters, or one bitmask. The bitmask is used to match against selected bits of a byte or it can be used to trap a bit-oriented protocol when the use of a character by itself may not be sufficient.

The occurrence of a match between the user-entered characters or bitmask can be used to trigger a number of other operator-defined activities. The operator can, for example, select YES or NO for the Stop on the Trap 1 field. Options for the Trap 1 Display field are NORMAL and REVERSE, which define how data will usually appear in the unit's display buffer. The leading field for Capture On includes the options START and STOP, which are used to define whether the unit will start or stop capturing data on a trap. The trailing Capture On field defines when the protocol analyzer will initiate the action elected in the leading field and includes the options UNUSED, TRAP1, TRAP2, PARITY ERROR, and CRC/LRC ERROR.

Counter/Timer Menu

By itself, the Trap Menu on the Digilog 400 has limited utility for capacity planning. This is because the menu limits the operator to defining the data to be matched and the actions the protocol analyzer should take. However, when the Trap Menu is teamed with the Counter/Timer Menu, the protocol analyzer becomes a potent tool in the capacity-planning process.

The Digilog 400 Counter/Timer Menu has four counters and three timers (Fig. 3). The first field for each counter illustrated in Figure 3 can be set to OFF, PARITY ERROR, CRC/LRC ERROR, TRAP1, or TRAP2. When it is set to TRAP1 or TRAP2, the counter value will reflect the number of times the user-entered data or bitmask in the Trap Menu was encountered. The second field for each counter will display the current value of the counter. The Reset Counters At Run field can be set to either YES or NO. This setting is employed to inform the protocol analyzer whether or not to reset the counters to zero each time an event trapping is executed.

Each of the three time counters has three fields that can be set by the operator, and a fourth field that is used by the protocol analyzer to display the timer's current value. By setting the appropriate values in these fields, the time counters determine the time that elapsed between occurrences of a selected event or a pair of events. Each of the first two time counter fields can have a value of 1 to 4, however, no two fields can have the same value. Thus, 1 to 2, 1 to 3, and 1 to 4 would be valid entries while 1 to 1 would be invalid. The third time counter field, which is initially set to a default value of OFF, can also be set to ON by the operator.

By first setting two trap conditions and then equating each trap to a counter, the protocol analyzer operator can use the time counters to clock the timing between traps. In effect, this capability can be used to determine the duration of predefined events, such as the time duration from sign-on to sign-off.

Capacity planning

How can analyzers be used for capacity planning? Consider a mainframe computer site accessed by remote terminals through the public switched telephone network. The usual means to access the mainframe is to use telephone lines connected to auto-answer modems. Usually the group of lines will be placed onto a common rotary or hunt group so remote users need only dial a single telephone number to get access to the first avail-

---

1. Functional menu. Like many other protocol analyzers, the Digilog Model 400 uses a menu-driven system to simplify its operation by the novice operator.

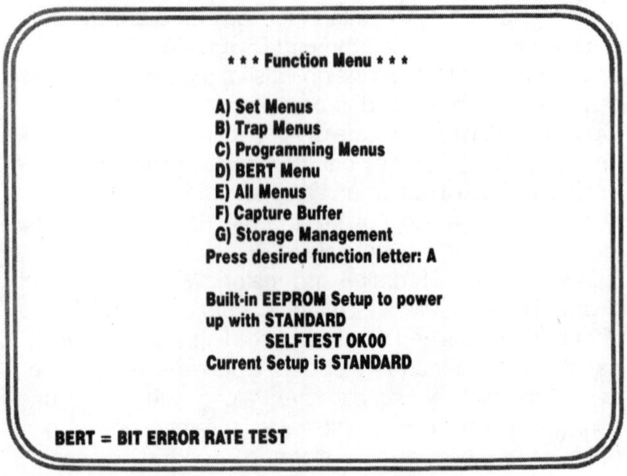

```
* * * Function Menu * * *

A) Set Menus
B) Trap Menus
C) Programming Menus
D) BERT Menu
E) All Menus
F) Capture Buffer
G) Storage Management
Press desired function letter: A

Built-in EEPROM Setup to power
up with STANDARD
              SELFTEST OK00
Current Setup is STANDARD

BERT = BIT ERROR RATE TEST
```

2. Initial trap. Option B selected from the Model 400 main menu calls the Trap Menu. Using four keys, the operator scrolls through the options available in each field.

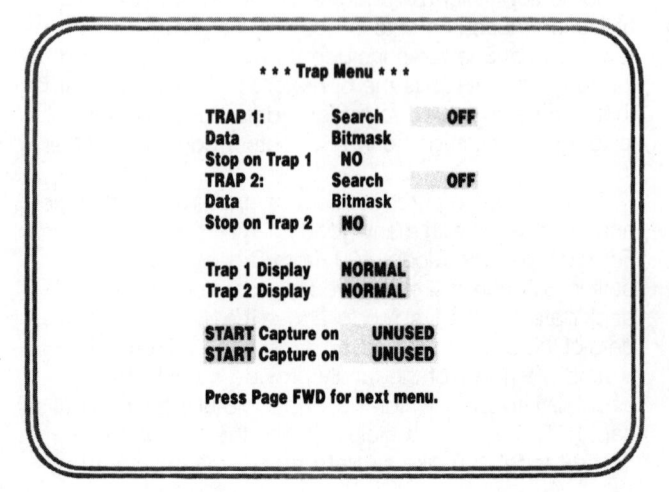

```
* * * Trap Menu * * *

TRAP 1:         Search         OFF
Data            Bitmask
Stop on Trap 1  NO
TRAP 2:         Search         OFF
Data            Bitmask
Stop on Trap 2  NO

Trap 1 Display  NORMAL
Trap 2 Display  NORMAL

START Capture on    UNUSED
START Capture on    UNUSED

Press Page FWD for next menu.
```

3. Counter/Timer. *The Model 400 Counter/Timer Menu has four counter and three timer options. The timers and counters can be used to trap selected bits or bitmasks.*

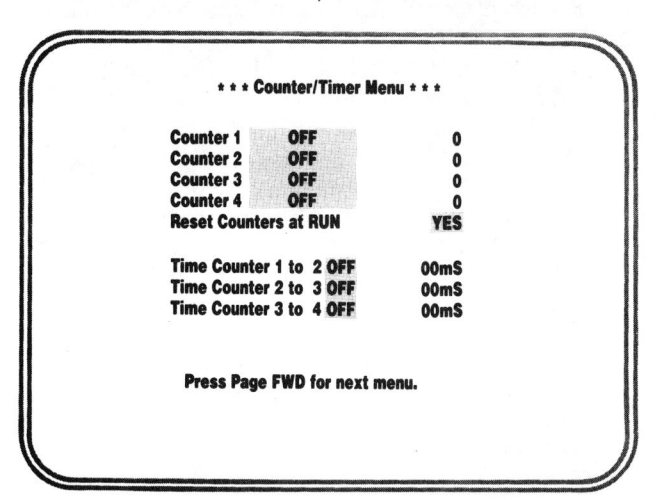

```
* * * Counter/Timer Menu * * *

Counter 1         OFF              0
Counter 2         OFF              0
Counter 3         OFF              0
Counter 4         OFF              0
Reset Counters at RUN              YES

Time Counter 1 to 2 OFF            00mS
Time Counter 2 to 3 OFF            00mS
Time Counter 3 to 4 OFF            00mS

Press Page FWD for next menu.
```

able telephone number in the rotary group (Fig. 4). Thus, if a 10-number rotary uses the number XXX-10000 as the first telephone number in the group, users would be rotored to XXX-10001 if the first number was busy and so on. Only if all 10 numbers were busy would the caller receive a busy signal.

One of the more frequent problems associated with the use of the rotary is selecting an appropriate number of lines to service terminal users at remote sites. After an appropriate number of lines, modems, and computer ports are connected, a secondary problem becomes how to verify the utilization of telecommunications lines. Although managers have no difficulty tracking the growth of usage to the point where frequent busy signals result in end-user complaints, managers find it more difficult to determine if the lines are underused. How to determine that the network has excess capacity is a problem often encountered by communications managers.

Sizing the number of lines needed for a rotary group can be calculated using simple mathematics (DATA COMMUNICATIONS, October 1983, p. 199). Determining if the initial sizing was correct is where the protocol analyzer comes into play. Although many mainframe vendors include comprehensive billing log software with their hardware, these programs typically use USERID, Account Codes, and system usage for billing purposes. These measures are of

4. Rotary access. *A remote user need only dial a single number to access a computer through the first available line in a rotary group.*

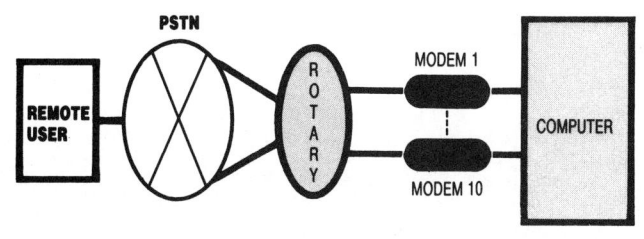

```
         PSTN
                            MODEM 1
REMOTE  (XXX)  ROTARY                 COMPUTER
USER                        MODEM 10

PSTN = PUBLIC SWITCHED TELEPHONE NETWORK
```

5. Setting the trap. *To count individual sessions and their duration, the protocol analyzer must first be installed between a computer port and a modem.*

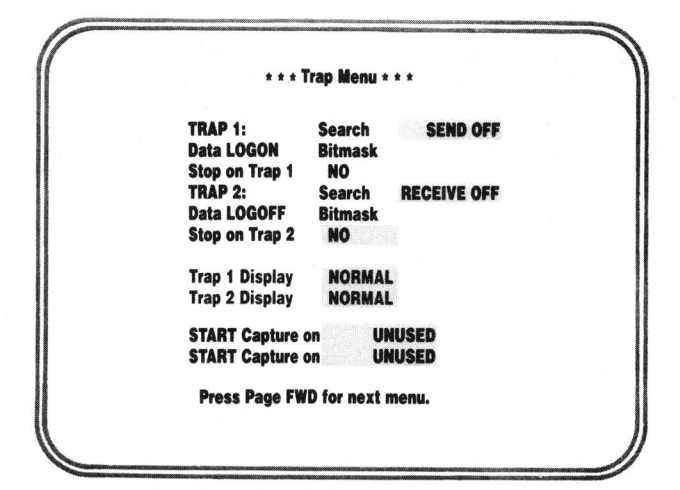

```
* * * Trap Menu * * *

TRAP 1:          Search       SEND OFF
Data LOGON       Bitmask
Stop on Trap 1   NO
TRAP 2:          Search       RECEIVE OFF
Data LOGOFF      Bitmask
Stop on Trap 2   NO

Trap 1 Display   NORMAL
Trap 2 Display   NORMAL

START Capture on              UNUSED
START Capture on              UNUSED

Press Page FWD for next menu.
```

little help when it comes to determining line usage. And computer billing tapes rarely record the physical line used in a session, so another means must be found to measure the activity on individual lines.

Trap setting

Suppose a remote user accessing a mainframe enters a LOGON command and at the end of the session the computer generates a LOGOFF with an XX/YY/ZZ message indicating the time of the LOGOFF. Using the LOGON and LOGOFF as data to be trapped, the Trap Menu on the Digilog 400 could be set as indicated in Figure 5. Since the protocol analyzer will be installed between a computer port and modem, the send side of the line contains data flowing to the computer while the receive side of the line carries computer responses. Thus, the send side of the line would be set to search for LOGON while the recieve side of the line would be set to search for LOGOFF. The remaining fields of the Digilog 400 Trap Menu should be set to their default values: NO for stopping on a trap, NORMAL for

6. Counter/timer settings. *Once the traps are set, counters and timers must be set to measure sessions and durations.*

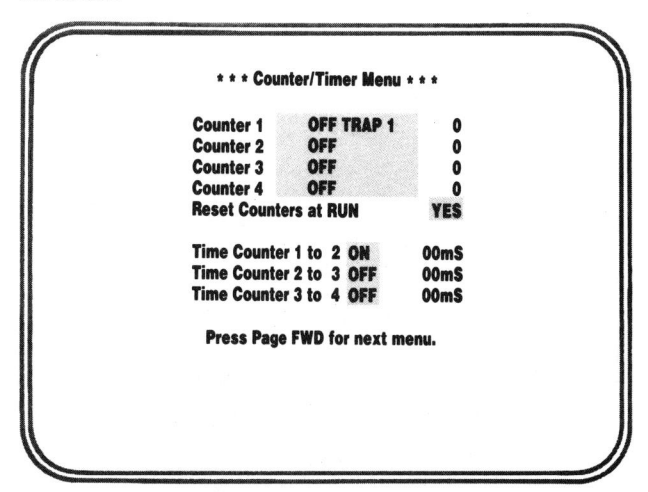

```
* * * Counter/Timer Menu * * *

Counter 1         OFF TRAP 1       0
Counter 2         OFF              0
Counter 3         OFF              0
Counter 4         OFF              0
Reset Counters at RUN              YES

Time Counter 1 to 2 ON             00mS
Time Counter 2 to 3 OFF            00mS
Time Counter 3 to 4 OFF            00mS

Press Page FWD for next menu.
```

display, and UNUSED for data capture, since the purpose is only to count sessions and their duration.

Once the Trap Menu is set, the Counter/Timer Menu must be adjusted from its default field values. Figure 6 illustrates changes made to the menu to count the number of sessions and their duration. The first change to the Counter/Timer Menu involves setting the Counter 1 field value to TRAP1. Then, when a TRAP1 match occurs, the first counter will automatically increment by one. Next, the time counter for Traps 1 to 2 is set to ON, which results in the session duration being measured.

After the Counter/Timer Menu is set, the operator can press the RUN TRAP key to initiate the trapping operation. Thereafter, the operator could periodically display the Status Report Menu, assuming the Reset At Run option was set to NO. This permits cumulative counting even if the analyzer is stopped.

By measuring the traffic on line n, the operator in effect is measuring the traffic that overflowed lines 1 through n − 1 on the rotary. This information can be used to determine both the proportion of lost traffic on a particular line or group of lines, as well as the grade of service. The grade of service, in this case, is defined as the percentage of calls that can be expected to be blocked and encounter a busy signal.

Even without applying traffic engineering mathematics to the sizing process, the network manager may be able to restructure lines and components based upon recorded statistical information—essentially gaining the same end. Suppose the monitoring of lines 9 and 10 during a given time period showed no utilization while line 8 showed two calls occurring and a total duration of 18 minutes. This type of data indicates that the rotary could easily be downsized to a maximum of eight dial-in lines. This change would result in the saving of two lines, two modems, and two computer ports.

The exclusion of control signals from traps and the milliseconds duration of the timer count in the Model 400 creates some trapping problems. Excluding control signals from a trap makes events data dependent. Thus, if LOGON or LOGOFF occurs naturally in the data, it would result in the trap of a false session and session duration. To eliminate such problems, users can develop a program in the Digilog Command Language (DCL) that will respond to different control signals.

Other protocol analyzers, such as the Atlantic Research 3500 Interview, can be used to set counters and timers based upon the state of interference leads. In the 3500, the Ring Indicator can be used to set a counter and timer. If the computer port lowers DTR after a session terminates, the occurrence of this interface activity could be employed to stop the timer. Another interesting feature of the Atlantic Research 3500 is the ability to select timer activity either by milliseconds or seconds. By selecting timing activity to be recorded in seconds, the operator is assured that daily recording will never result in an unobserved timer overflow. ■

Gilbert Held, chief of data communications for the U. S. Office of Personnel Management at Macon, Ga., is this year's Karp Award winner. He is also a recipient of an American Association of Publishers Award and the author of 12 books and over 60 technical articles.

Charlie Bass, Ungermann-Bass, Santa Clara, Calif.

Data networks' endangered and protected species

The world of data communications will converge on SNA and OSI, as the two protocols converge on each other. And proprietary protocols' days are definitely numbered.

Alternative data communications architectures have proliferated in recent years, but only two—IBM's Systems Network Architecture (SNA) and the International Standards Organization's (ISO's) Open Systems Interconnection (OSI) reference model—are destined to survive. All others, including popular vendor-specific architectures such as Digital Equipment Corp.'s Decnet, and open architectures such as TCP/IP (Transmission Control Protocol/Internet Protocol) and XNS (Xerox Network Systems), are an endangered species. For despite the large followings these alternatives have gathered, the world is lining up behind SNA, OSI, or both.

SNA's survival is assured through the enormous presence of IBM in the worldwide marketplace, and IBM's continued commitment to this architecture. More importantly, IBM's customers invariably contribute to SNA's survival through their confidence and investment in SNA, making it the most widely deployed data communications architecture in the industry.

Until recently, critics scoffed at the notion that the OSI reference model would evolve beyond a concept into a concrete, practical protocol suite for multivendor networking. Even fewer took seriously the idea that OSI would rise to such prominence in the networking world that it could stand shoulder to shoulder with IBM's SNA. But that is precisely what is happening.

The rise of OSI
The OSI movement has been steadily picking up momentum and followers since its founding in Europe in the 1970s. Originally, users and vendors banded together under the auspices of ISO to devise an open architectural model for heterogeneous networks. Over the past 10 years, OSI advocacy has evolved from an architectural framework to a set of specific protocols. In the process, it has grown into an international movement with a U. S. following that encompasses large private-sector users as well as government computer users who have come to recognize the importance and potential benefits of an industrywide international standard. Early endorsement contributing to OSI's momentum in the United States came from two of the nation's largest commercial users of computers and communications equipment—General Motors Corp. (GM) and the Boeing Co.

GM became the chief corporate sponsor of the Manufacturing Automation Protocol (MAP), an OSI-based networking standard aimed initially at accelerating GM's internal industrial automation program. Boeing has promoted MAP's twin, the OSI-based Technical and Office Protocol (TOP), a closely related standard for supporting office automation, scientific, and engineering applications.

Both companies have specified MAP/TOP compatibility as a requirement in future computer and communications procurement, and both have invested significant resources in promoting the virtues of OSI. In particular, GM and Boeing were instrumental in founding the MAP/TOP Users' Group (see Table 1).

With prodding from this united and committed user community, computer and communications vendors have come to realize the significance of OSI. Consequently, an impressive and somewhat unlikely collection of such vendors came together in late 1985 to form the Corporation for Open Systems, or COS (see Table 2). COS was chartered to promote the adoption of OSI standards and, in particular, to verify that individual implementations of MAP/TOP products conform to common specifications. COS can also be viewed as a move by vendors to offset the power wielded by the MAP/TOP user community and regain some degree of control in implementation strategies. While COS members recognize the potential benefits of OSI, a consolidated implementation strategy assures vendors that their investment is synchronized with their

295

Table 1: Major MAP/TOP corporate affiliates

3COM CORP.	INTEL CORP.
AEG	KAISER ALUMINUM AND CHEMICAL CORP.
ALUMINUM COMPANY OF AMERICA	LITTON INDUSTRIES AUTOMATION SYSTEMS
ARTHUR ANDERSON & CO.	M.W. KELLOGG
AT&T TECHNOLOGIES	MICHIGAN BELL TELEPHONE CO.
BOEING COMPUTER SERVICES	MINNESOTA MINING & MANUFACTURING CO.
BRIDGE COMMUNICATIONS	MONSANTO CO.
CHRYSLER CORP.	MORTON THIOKOL
CINCINNATI MILACRON	MOTOROLA INC.
DEERE & CO.	NAVISTAR INTERNATIONAL CORP.
EASTMAN KODAK CO.	NORTHERN TELECOM
E.I. DuPONT	NORTHROP CORP.
ELECTRONIC DATA SYSTEMS	ORACLE CORP.
ERICSSON INC.	PHILLIP MORRIS USA
EXXON CHEMICAL CO.	POLAROID CORP.
GENERAL DYNAMICS	PRIME COMPUTER
GENERAL INSTRUMENT CORP.	PROCTOR & GAMBLE CO.
GENERAL MOTORS CORP.	REYNOLDS METALS CO.
GOULD INC.	ROCKWELL INTERNATIONAL
HONEYWELL INC.	SHELL COMPANY OF AUSTRALIA LTD.
HUGHES AIRCRAFT	SQUARE D CO.
INDUSTRIAL NETWORKING INC.	WEYERHAEUSER INFORMATION SYSTEMS
INGERSOLL-RAND CO.	

competitors and partners and that they will not be behind or ahead of one another in the market. While it can be argued that this synchronicity reinforces OSI interoperability objectives, it also allows individual vendors to pace their investment and transition to OSI with minimal risk.

Of the various U. S. government agencies that have contributed to the OSI movement, the most notable are the Department of Defense (DOD) and the Department of Commerce (DOC). As a result of a 1985 internal study, the DOD decided that it would move its vast collection of networks from TCP/IP to OSI. This decision was made based on the conclusion that, over time, off-the-shelf OSI-compatible products would be more readily available than those based on TCP/IP despite TCP/IP's widespread acceptance in the scientific and engineering communities. The DOD is presently determining how it will make the transition from TCP/IP to OSI while maintaining the operational integrity of its networks.

In 1986, under the leadership of the late Department of Commerce Secretary Malcolm Baldridge, 15 federal agencies formed a task force to develop communications specifications for their own internal networking needs (see Table 3). The result was GOSIP (Government OSI Profile), first released in January 1987. GOSIP is essentially a restatement of TOP. Fundamental to a consistent position on OSI is the common reference to U. S. National Bureau of Standards (NBS) Implementors' Agreements as the detailed protocol specification for MAP, TOP, COS, and GOSIP. These agreements, reached in NBS-sponsored workshops, are the underpinnings that loosely hold together an otherwise disparate collection of special interest groups. The prospects for a single industrywide OSI standard would be reduced if the NBS forum did not function as a U. S. network high court. The NBS process was somewhat unstable during 1986 because of federal budget cuts and the uncertainty of the role of the NBS testing versus COS. While there was talk of privatization of the NBS network laboratory, it was generally agreed that testing and certification were the domain of COS, while NBS would play a critical role in bringing the industry together to evaluate and resolve technical issues.

The GOSIPspecification should have a galvanizing effect on the OSI movement in the United States. Traditionally, critics have complained that OSI is nothing more than a paper standard, lacking practical implementation and, more importantly, customers. This opinion is likely to change as the DOD, DOC, and affiliated agencies move to have GOSIP adopted as a Federal Information Processing (FIP) standard. Once GOSIP achieves such status, OSI will become an approved and, in some cases, a required specification for federal computer and communications equipment procurement.

To further reinforce the practical applicability of OSI, COS and the MAP/TOP committee recently announced an event to be held in June 1988 to demonstrate factory, office, and engineering applications executing over MAP-, TOP-, and possibly GOSIP-compatible networks. This event was originally distinguished from previous demonstrations of OSI interoperability by the sponsors' insistence that only readily available products and not prototypes be employed. This

Table 2: Corporation for Open Systems (COS) members

ADC TELECOMMUNICATIONS
AETNA LIFE & CASUALTY
AMDAHL CORP.
APOLLO COMPUTER INC.
APPLE COMPUTER INC.
AMERICAN TELEPHONE & TELEGRAPH CO.
BECHTEL POWER CORP.
BELL COMMUNICATIONS RESEARCH
BOEING COMPUTER SERVICES
BRIDGE COMMUNICATIONS INC.
BURROUGHS CORP.
CCTA, HER MAJESTY'S TREASURY, BRITISH GOVERNMENT
CITICORP
3COM CORP.
CONCURRENT COMPUTER CORP.
CONTROL DATA CORP.
CONVERGENT TECHNOLOGIES
DART & KRAFT INC.
DATA GENERAL CORP.
DIALCOM INC.
DIGITAL EQUIPMENT CORP.
DOW CHEMICAL CO.
E.I. DU PONT DE NEMOURS & COMPANY INC.
EASTMAN KODAK CO.
THE EQUITABLE
EXCELAN
GENERAL ELECTRIC CO.
GENERAL MOTORS CORP.
GOULD INC.
GTE SERVICE CORP./TELEPHONE OPERATIONS
HARRIS CORP.
HEWLETT-PACKARD INC.
HONEYWELL INC.
HUGHES AIRCRAFT CO.
INTERNATIONAL BUSINESS MACHINES CORP.
INTERNATIONAL COMPUTERS LTD.
INTEL CORP.
ITT CORP.
MOTOROLA INC.
NCR CORP.
NATIONAL SEMICONDUCTOR CO.
NETWORK SYSTEMS CORP.
NORTHERN TELECOM INC.
ONTARIO MINISTRY OF GOVERNMENT SERVICES
 COMPUTER AND TELECOMMUNICATIONS DIVISION
PACIFIC BELL
PRIME COMPUTER
PROCTER & GAMBLE CO.
RETIX
ROCKWELL INTERNATIONAL
SPERRY CORP.
SUN MICROSYSTEMS INC.
SYTEK INC.
TANDEM COMPUTERS INC.
TELEX COMPUTER PRODUCTS INC.
TELENET COMMUNICATIONS CORP.
TEXAS INSTRUMENTS INC.
TOUCH COMMUNICATIONS
VISA INTERNATIONAL
WANG LABORATORIES
XEROX CORP.

---

objective has been compromised by the reluctance of several large vendors to participate under such conditions. In keeping with its charter, COS will provide testing tools and services to event participants. (It is notable that COS has committed only to a subset of the full MAP/TOP.)

As further evidence of the momentum of OSI, Digital Equipment Corp. (DEC), arguably the most advanced network supplier among the major vendors, has openly embraced OSI, despite recent criticism of MAP by Chairman Ken Olsen, and it has begun to migrate its proprietary Decnet architecture toward conformance with OSI standards. It should be remembered that Olsen is primarily criticizing MAP's lower two layers, which are based on broadband token-bus technology, not OSI as a whole. His remarks need to be weighed in light of DEC's limited broadband capabilities. The company has not developed broadband technology on its own and has instead relied on outside vendors. DEC has also made an enormous investment in baseband Ethernet and is understandably reluctant to make a similar investment in broadband token-bus.

Finally, DEC now is the market leader in supplying processors for factory automation, and the ascendancy of MAP will create a more level playing field. Conversely, MAP works in IBM's favor since the company has not been a

Table 3: Government OSI Profile (GOSIP) members

DEPARTMENT OF AGRICULTURE
DEPARTMENT OF COMMERCE
DEPARTMENT OF DEFENSE
DEPARTMENT OF ENERGY
ENVIRONMENTAL PROTECTION AGENCY
GENERAL SERVICES ADMINISTRATION
DEPARTMENT OF HEALTH AND HUMAN SERVICES
DEPARTMENT OF HOUSING AND URBAN DEVELOPMENT
DEPARTMENT OF THE INTERIOR
DEPARTMENT OF JUSTICE
DEPARTMENT OF LABOR
NATIONAL AERONAUTICS AND SPACE ADMINISTRATION
NATIONAL SCIENCE FOUNDATION
OFFICE OF MANAGEMENT AND BUDGET
DEPARTMENT OF TRANSPORTATION
DEPARTMENT OF THE TREASURY
NATIONAL COMMUNICATIONS SYSTEM

strong player in the factory. MAP provides an opportunity to increase its factory presence. An obvious consequence of Olsen's comments is that DEC's commitment to MAP is being questioned, putting its sales representatives who deal with GM in an unenviable position.

Regardless, DEC's strategy for OSI involves retaining the external characteristics (such as the user interface) and network-management features of Decnet, while internal layers are brought into OSI conformance. Even though this evolutionary approach to OSI is not consistent with MAP/TOP guidelines, it retains a valuable familiarity and comfort level for Decnet users while leveraging DEC's investment and differentiation in network management. Full MAP/TOP conformance is measured by top-to-bottom adherence to OSI specifications; DEC addresses conformance on a layer-by-layer basis and retains the overall proprietary nature of its architecture. The success of this evolutionary approach will certainly be studied by IBM vis-a-vis its own SNA strategy.

As it becomes clear that SNA and OSI are destined to be the surviving architectures of choice, other vendors will evolve their proprietary offerings to the OSI standard. MAP/TOP purists may frown on this approach, but it allows customer and vendor alike to simultaneously hold on to the past and move into the future.

SNA is an assumption
The long-term survival—and even the preeminence—of SNA is an assumption that most MIS (management information systems) directors and data communications managers have already made because of IBM's position as the world's premiere computer supplier. This, in turn, adds to the momentum of SNA as both IBM and its customers continue their commitment to and investment in SNA and, possibly more important, SNA-based applications.

But it is simplistic to assert that SNA's success relies wholly on the strength of IBM or the blind faith of its users. Remember the PCjr and PC Network? SNA will, in fact, survive for a more important reason: It is a well-conceived, flexible, and robust data communications architecture.

Why is it then that SNA is often criticized for being inefficient, unfriendly, and out of date? Much of that criticism was valid while it strictly reflected SNA's roots in an earlier generation of computing. Introduced in 1974, the architecture supported a hierarchical approach to networking, in which an all-powerful central host lorded over dumb terminals and other attached devices that wished to employ the host in a computing task on their behalf. Though valid, this model of computing became restrictive as ubiquitous intelligent workstations, especially IBM's own PC, created a peer-to-peer, server model of computing and the concomitant local network industry. SNA designers could not have foreseen the microcomputer-to-mainframe phenomenon, but it is a testament to the architecture's excellence that SNA has been extended to embrace today's full spectrum of networking needs, encompassing clusters of PCs, departmental minicomputers, and corporate hosts.

As a result of SNA extensions announced and partially delivered over the past two years, peer-to-peer communications is widely supported; mainframe hosts are not required for routing; dynamic reconfiguration is possible; the location of files is transparent; and network management is open to other vendors. The cumulative effect of these features has been the revitalization of an architecture that was near the brink of becoming obsolete. SNA is now positioned to support current and foreseen network requirements as well as any competitive communications architecture.

SNA and OSI: Comingling and coexistence
SNA's current and expected popularity should not obscure the temptation to meld the richness of SNA with the vendor independence and design economy of OSI. Primarily, IBM is in the business of selling MIPS (million instructions per second) and file storage, which the overhead problems of SNA promoted. To their credit, realistic OSI advocates do not relish being isolated from a ubiquitous SNA infrastructure and its wealth of applications.

Despite its obvious reluctance to embrace a standard it does not control, IBM is pragmatic and often opportunistic. Consequently, while the corporation continues to champion SNA and would prefer to ignore OSI, it has realized that to sell today to MAP factories and to buyers in Europe—and tomorrow to TOP offices and to the U. S. government—it must find ways to counter and accommodate the OSI challenge.

In fact, IBM has participated broadly in the OSI phenomenon. In the United States it has announced and delivered MAP-based products and is an active participant in OSInet, a community of vendors who test the interoperability of their protocol suites across an X.25-based internet. In Europe, IBM has set up centers for research, development, and testing of OSI in Heidelberg, West Germany; La Gaude, France; and Rome, Italy; and has released a special version of SNA called OTSS (Open Transport and Session Support) that is OSI-compatible at the Transport and Session layers.

It is unlikely that DEC's strategy of migrating to OSI while retaining the external characteristics of Decnet has gone unnoticed by IBM. SNA and OSI are not simply destined to compete with one another, but also, as IBM's introduction of OTSS indicates, to coexist and comingle. They may even meld to the point of being indistinguishable.

Many similarities in structure and capabilities already exist between the two architectures, including the fact that each one consists of seven layers (Fig. 1). Starting with the bottom layer, SNA consists of (1) Physical Control, (2) Data Link Control, (3) Path Control, (4) Transmission Control, (5) Data Flow Control, (6) Presentation Services, and (7) Transaction Services. By comparison, the OSI reference model's layers are: (1) Physical, (2) Data Link, (3) Network, (4) Transport, (5) Session, (6) Presentation, and (7) Application.

While specific features differ between the two models, the first two layers of each architecture provide physical connections and data transfer; Layer 3 provides routing functions, Layers 4 and 5 establish and maintain end-to-end connections; Layer 6 enables the proper interpretation of information; and Layer 7 provides the user with application-oriented services. These structural similarities increase the feasibility and probability that they will coexist and comingle over time.

Conceptually, the most straightforward way for SNA and

1. SNA vs. OSI architectures. *Similarities of structure smooth the way for the merger of the two architectures. While details differ, their objectives are similar.*

	SNA		OSI
7	TRANSACTION SERVICES	USER SERVICES	APPLICATIONS
6	PRESENTATION SERVICES	INTERPRETATION	PRESENTATION
5	DATA FLOW CONTROL	ESTABLISH AND MAINTAIN END-TO-END CONNECTIONS	SESSION
4	TRANSMISSION CONTROL		TRANSPORT
3	PATH CONTROL	ROUTING	NETWORK
2	DATA LINK CONTROL	PHYSICAL CONNECTIVITY AND DATA TRANSFER	DATA LINK
1	PHYSICAL CONTROL		PHYSICAL

OSI = OPEN SYSTEMS INTERCONNECTION
SNA = SYSTEMS NETWORK ARCHITECTURE

OSI to coexist is through the application of gateway processors. Rather than merge SNA and OSI, such gateways would provide a functional bridge between the two either by receiving protocol transactions on a layer-by-layer basis from one domain and transforming them into comparable functions of the other or an application process residing concurrently on both protocol stacks (Fig. 2). A layer-by-layer transformation allows end-to-end connections and transparency of the gateway function, but it is far more computationally intensive and limited by the compatibility of features at each layer. Both gateway approaches can localize transformation to a server and provide common network management and access control. However, this approach requires a readily accessible, if not dedicated, resource and implies a disjointed view of the network domain. Implied in this method is that the two architectures remain separate, with crossover limited to application services. Consequently, a gateway server may be impractical or less desirable than a distributed approach that intentionally blurs the distinctions and boundaries between the two domains.

2. Alternate gateways. *The choices are a flow between networks with a layer-by-layer transformation, or an intermediate node able to talk with both architectures.*

(A) LAYER-BY-LAYER TRANSFORMATION

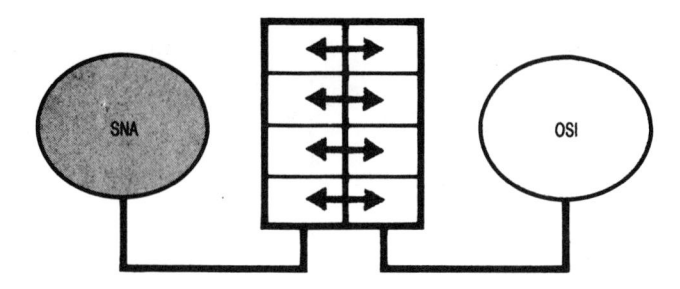

(B) DUAL-STACK APPLICATION

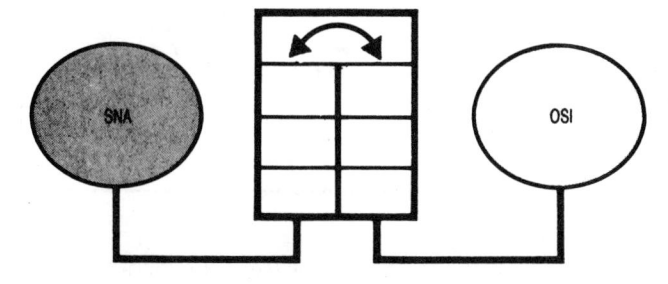

OSI = OPEN SYSTEMS INTERCONNECTION
SNA = SYSTEMS NETWORK ARCHITECTURE

Device emulation

A simple and commonly used approach for allowing devices native to one network to communicate with devices on a foreign network is device emulation. In particular, IBM 327X terminals and controllers are being emulated on virtually all non-SNA networks in order to access IBM host-based applications. In fact microcomputer-to-mainframe communications are dominated by PCs emulating 327X terminal devices linked to an IBM host application. Originally, SNA assumed that this link was a leased line; the connection can now be leased, switched, or, in some cases, an X.25 network.

For microcomputers resident on non-SNA networks, the device emulation protocol is encapsulated and delivered to a gateway server that transmits it to a host through an SNA link protocol such as SDLC (synchronous data link control). Device emulation is limited, however, because the networks are not truly interoperable: Files can be transferred and remote users connected to hosts, but some communications, such as host-to-host, are restricted. Obviously, this device emulation technique can and will be used across OSI networks, providing the same level of SNA-to-OSI communications as alternative protocol suites (Fig. 3).

This particular mingling of OSI and SNA functions may be incidental to the more pressing objective confronting OSI providers: accommodating Netbios, today's dominant network application interface. In principal, Netbios is a Session Layer interface that is not tied to a particular underlying protocol suite. Indeed, a number of vendors support a Netbios interface on top of XNS and TCP. Consequently, it is highly probable that an enterprising vendor will soon incorporate the Netbios interface in an OSI stack, allowing OSI to support a growing number of applications including 327X/SDLC emulation packages.

Netbios is not officially part of either the SNA or OSI protocol suites and, consequently, its continuance is disquieting to much of the IBM organization. Netbios is considered a tactical, not strategic, offering, introduced by Entry Systems Division to support clusters of PCs in the absence of a viable SNA-based solution. As SNA's features are enhanced and strategic application interfaces are identified through SAA (Systems Application Architecture), Netbios is noticeably excluded. Indeed, it is anathema to the MAP/TOP organization because it is not consistent with its strategic direction. However, it has a large and growing base of support that can alleviate the near-term scarcity of OSI-based applications.

3. Device emulation. *Virtually any device protocol can be encapsulated in OSI conventions presentable for an SNA gateway. This technique passes for micro-to-mainframe communications in most networks.*

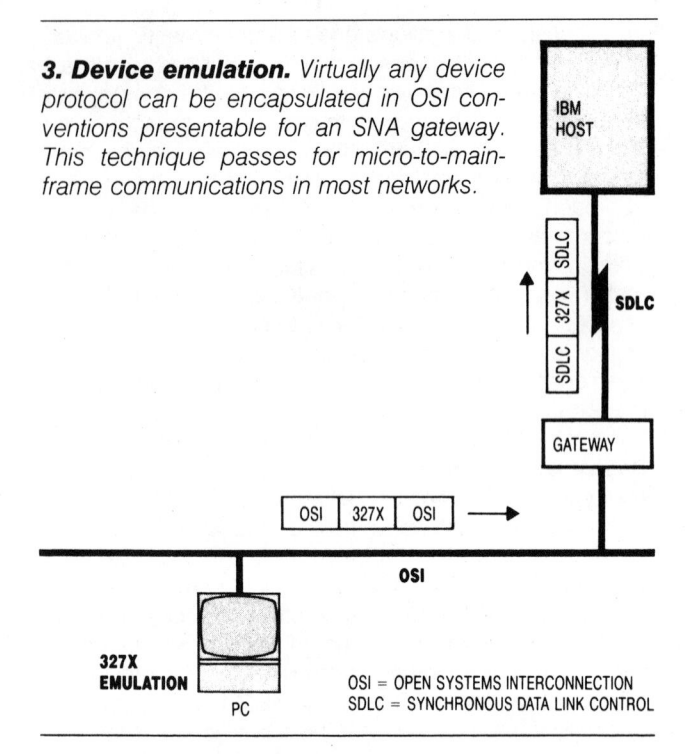

OSI = OPEN SYSTEMS INTERCONNECTION
SDLC = SYNCHRONOUS DATA LINK CONTROL

Complementing the myriad sources of 327X emulation products, IBM now supports the connection of non-SNA devices, namely ASCII terminals, to SNA hosts through protocol conversion options in its 7171 communications controllers. It is easy to imagine a new generation of SNA terminals that can switch between native and ASCII modes of operation. At this point, both SNA and ASCII (OSI) devices can easily connect to both SNA and OSI networks.

Work is under way on a Virtual Terminal Protocol (VTP) for the OSI suite. VTP will establish a uniform and generic set of terminal operations for communicating across an OSI network. Once VTP has been widely adopted, it is realistic to assume that IBM will incorporate VTP protocol conversion in its front-end processors, assuring continued SNA-to-OSI cross-connections.

Internet and interstation transmission

Another straightforward and somewhat mundane instance of SNA and OSI blending is when one acts as a transmission service (layers 1 and 2) for the other. A common example of this approach is when SNA sessions operate across X.25 packet-switched networks. While IBM was comparatively late and limited in its support of X.25, SNA now offers a number of options for taking advantage of X.25 links. This is an instance where SNA and OSI have come together with little fanfare, but with significant benefit in circumstances where long-distance dedicated lines (SNA's original transmission model) are impractical. While the X.25 interface can be functionally viewed as providing Layer 3 functions, generally, SNA treats X.25 as a simple transmission link and retains full control for routing between subarea nodes. A more provocative question is whether or not there will be examples of OSI sessions operating across SNA links? The answer is, yes.

Due to the vast array of SDLC-compatible hardware and software, it is unavoidable that this standard will be used for internet transmission between OSI networks. As before, routing (Layer 3) will remain in the client network and the SNA-to-OSI crossover will occur at the interface between layers 2 and 3.

Comparable to internet transmission services between networks is interstation transmission on a local area network (LAN). In this environment IBM has facilitated an SNA-to-OSI crossover by adhering, in its own Token-Ring PC adapter product, to the Layer 2 interface standard known as Logical Link Control (LLC) or 802.2 adopted by the IEEE-802 committee. By adhering to this LLC interface (Fig.4), a common Layer 3 client can reside on any of the IEEE-802 data link standards: 803.2 CSMA/CD (carrier-sense multiple access with collision detection) or Ethernet; 802.4 token bus; and 802.5 token-ring. The implications of this standard interface and IBM's adherence to it are remarkable. Not only can OSI layers 3 through 7 be easily wedded to IBM's Token-Ring technology, but SNA protocols can reside on token bus (MAP) and Ethernet (TOP). This will undoubtedly result in SNA and OSI protocols residing on common local networks, which will provide additional motivation to provide interoperability between the two. Gateway servers, as discussed earlier, can then be applied to achieve a dialogue between stations that reside on the same LAN but speak different protocols.

4. LLC interface. *The Logical Link Control is a single interface between Layers 2 and 3 that allows mixing and matching of SNA and OSI upper layers with all 802 standards.*

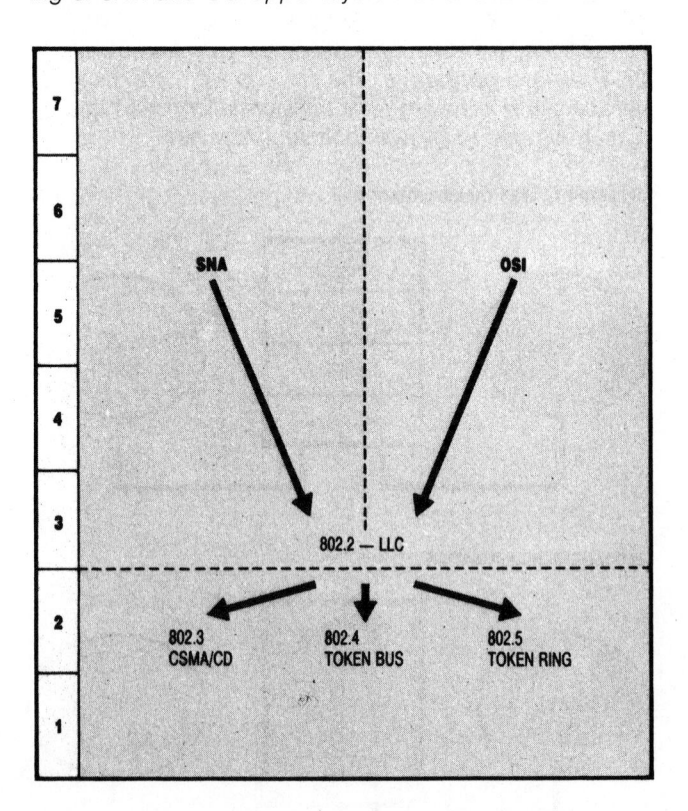

CSMA/CD = CARRIER-SENSE MULTIPLE ACCESS WITH COLLISION DETECTION
OSI = OPEN SYSTEMS INTERCONNECTION
SNA = SYSTEMS NETWORK ARCHITECTURE

Advanced Program-to-Program Communications (APPC) protocol is a recent landmark extension to SNA that supports program-to-program or peer-to-peer conversations between a broad spectrum of computing devices ranging from PCs to large hosts. The significance of APPC is its uniform treatment of any two devices on the microcomputer-to-mainframe spectrum at each end of a conversation. Prior to APPC, a 370-class host was required to manage any conversation and it was the dominant party. APPC conversations consist of a set of commands or verbs defined by an Application Program Interface (API), which are enacted by Logical Unit (LU) 6.2 operations. Consequently, for two programs to engage in a peer-to-peer conversation they must adhere to the APPC API and reside on devices supported by LU 6.2.

APPC/LU 6.2

With the introduction of APPC/LU 6.2, IBM not only set the strategic direction for distributed processing with SNA, it moved ahead of other architectures, including OSI, with provisions for interprocess communications. While Netbios was excluded, APPC was included in the list of SAA interfaces to be supported on all strategic processors.

Subsequent to its introduction, the European Computer Manufacturers Association (ECMA) in 1985 considered recommending that APPC/LU 6.2 be incorporated into the OSI standard roughly at Layer 6 by ISO. Since APPC was built outside the context of OSI, there has been some argument about whether it should be installed at layer 5 or 6. In addition to enhancing its basic functions, the inclusion of APPC/LU 6.2 would make it easier to make a port of APPC-based applications to OSI networks. As IBM migrates strategic offerings such as Disoss (Distributed Office Support System) to APPC, a wealth of applications become available through conformance with APPC. This was, undoubtedly, what ECMA had in mind. However, the other side of this proposition was the inherent advantage that IBM would gain in time and expertise in the context of OSI. This was a political fireball that ECMA preferred to drop rather than defend based on its win-win attributes. It could have been argued that the ECMA community, IBM, and ISO all would have profited from its adoption. ECMA vendors and customers would have gained easy access to IBM applications software, IBM would have made an easy entry into the OSI world, and ISO would have gained a well-defined and critical protocol. But because IBM was on the winning side of the equation, the other players were wary and the opportunity slipped away.

In January, the prospect of incorporating APPC into OSI was raised again by ECMA to a more receptive interna-

5. APPC API. *Like the Logical Link Control interface, this verb set is becoming the interface specification for first joining, then merging, SNA and OSI.*

6. SNA-OSI merger. *As data crosses the SNA-OSI threshold created by interfaces associated with the ISO model, the identity of each architecture is lost and is irrelevant.*

API = APPLICATION PROGRAM INTERFACE
APPC = ADVANCED PROGRAM-TO-PROGRAM COMMUNICATIONS
DISOSS = DISTRIBUTED OFFICE SUPPORT SYSTEM
LU = LOGICAL UNIT
OSI = OPEN SYSTEMS INTERCONNECTION

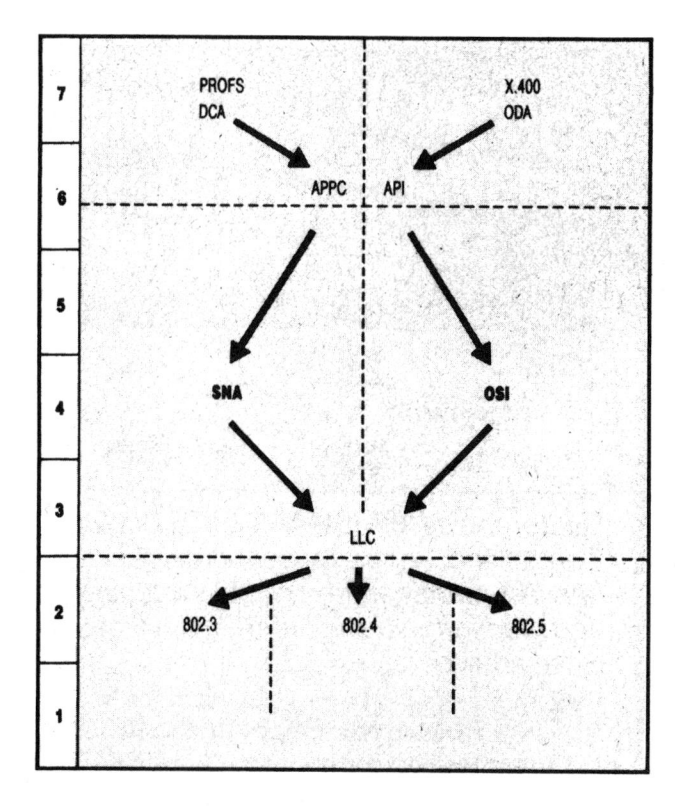

DCA = DOCUMENT CONTENT ARCHITECTURE
ODA = OFFICE DOCUMENT ARCHITECTURE
OSI = OPEN SYSTEMS INTERCONNECTION
PROFS = PROFESSIONAL OFFICE SYSTEM
SNA = SYSTEMS NETWORK ARCHITECTURE

tional audience. In the intervening year, IBM introduced APPC/VM (Virtual Machine) as part of its 9370 announcements operating on Ethernet and bisynchronous links. With its announcement came the realization that, as an interface specification, APPC API could be decoupled from the details of LU 6.2, reducing IBM's advantage (Fig. 5). Consistent with ISO's tradition of massaging a proposal and leaving its imprint, APPC is likely to emerge different than it entered. This may be timely, since IBM is massaging APPC API to resolve minor differences introduced in its different operating systems—APPC/VM is not identical to APPC/PC. By stating that, through SAA, APPC/PC and APPC/VM will converge, users can infer what they were not aware of before: that the two are different.

Applications
The upper reaches of SNA and OSI are where the application protocols most relevent to users are located and where IBM has the advantage of momentum. For document definition and interchange, IBM has defined Document Content Architecture (DCA) and Document Interchange Architecture (DIA), which its electronic messaging service PROFS (Professional Office System) is built upon. In OSI, document definition is provided by Office Document Architecture (ODA) and electronic messaging primitives are contained in the X.400 specification. Will there be a cross-fertilization of these specifications? Certainly. Many vendors have supplied transformation utilities between non-IBM-generated documents and DCA. Such utilities will exist for ODA-to-DCA transformation.

X.400 is one of the most anxiously awaited OSI protocols because it can be the basis for corporatewide messaging between disparate vendors' equipment (the Holy Grail of OSI). Given the existing support of SNA connections by these same vendors, it is likely that X.400-based messaging will occur over SNA circuits. In fact, IBM has announced two such products, the X.400 Disoss Connections and the X.400 Message Transfer Facility.

Indeed, SNA and OSI are destined to intertwine and potentially to merge to the point of being indistinguishable. Consider a distributed electronic mail service that operates between a personal computer and a VAX, based on X.400, that conforms to an APPC interface joined to OSI layers 3 through 5 with the PC residing on an Ethernet and the VAX sitting on a Token-Ring (Fig. 6). Which is SNA and which is OSI? Who cares?

OSI purists may be put off by such scenarios, but between the temptation to access SNA-based applications and IBM's posture vis-a-vis OSI, they cannot, and should not, be avoided. The real issue is the interoperability that comes with open architectures. As IBM defensively opens SNA, the identification of this amalgamation of standards is less important than what they accomplish. ■

Charlie Bass is cofounder and director of Ungermann-Bass Inc., chairman of Touch Communications, director of Sierra Semiconductor, consulting professor of electrical engineering at Stanford University, limited partner of Sequoia Technology Ventures, and editor of Computer Networks *and* IEEE Networks. *Prior to founding Ungermann-Bass, he held a variety of positions at Zilog Inc. from 1975 to 1979, ending up as general manager of its systems division.*

Robert S. Alford and Carl Holmberg, BBN Communications, Cambridge, Mass.

Implementing automatic route selection in T1 networks

Problems like bandwidth constraints and cost containment in T1 networks become tractable when the right routing algorithms are put to work.

Information networking industry pundits have discovered an entirely new segment of the T1 market: the high-end, circuit-switched, private T1 network. Although no consensus has been reached regarding a clear definition of this new class of high-capacity digital network, it can be loosely defined by the following characteristics:
- The size of the network ranges from a minimum of six nodes, upwards to hundreds of nodes.
- A typical node in such a network terminates a minimum of four T1 trunks.
- Dynamic allocation of bandwidth is available between nodes.
- The circuits can be reconstructed upon failure of a T1 link or a switching node.

These high-end networks are not only large, but they also have highly complex topologies. As a result, the art of routing circuits through the network is becoming more complex than it has ever been. To stay ahead of networking's next wave, communications professionals need to know how connections are established, routes defined, and how bandwidth is managed to effectively leverage this exciting new technology.

These topologies challenge the usual categorization of ring or mesh. By strict definition, they are partially connected meshes. Each node can be connected to any number of other nodes. The manner in which nodes are interconnected is dictated by traffic flow and T1 availability. No consideration is given to maintaining network symmetry. The nodes may be connected by one T1 trunk, two T1 trunks, or a dozen T1 trunks. A circuit established between nodes may resemble a point-to-point connection between an originating and terminating node, or it may traverse any number of intermediate nodes to reach its destination. There may be hundreds of potential routes between any two nodes in the network.

An important facet of this complexity is the network's volatility. At any time, a node or a trunk may be added to the network. Similarly, T1 and node failures will occur, instantly changing the dynamics of traffic flow. The combined effects of complexity and volatility in the network make it impossible to use static configuration methods to control routing and bandwidth allocation. Automatic algorithms must be developed.

These algorithms must be designed in such a manner that routes can be defined and connections established in just a second or two. Furthermore, the network must be designed with a means of accommodating changes in the topology without requiring a manual change in the database at each node.

Since the days when the first multipoint and drop-and-insert multiplexers were introduced, the question of how to allocate bandwidth has been an issue. Traditional multiplexer manufacturers do not offer the complexity, expansion capacity, or functionality needed to satisfy the new high end; newer network products are needed. The designers of these wide area network products face the challenge of either adapting old methods of routing to more complex situations or conjuring up wholly new approaches to the problem.

Conventional routing

The science of routing theory has devised a large number of approaches to determining paths through a private network. Two conventional methods are used to provide bandwidth allocation in simpler networks. The first entails nailing up the bandwidth on a full-time basis. In large networks, where many users can contend for bandwidth, the practice of nailing up circuits is very wasteful of costly T1 capacity and invariably results in higher than necessary overall monthly T1 charges. This is particularly true in networks where the majority of the traffic carried is voice.

Voice traffic volumes typically reach a peak at busy hours in the morning and afternoon. The number of tie lines needed between PBXs is configured to carry busy-hour loading. The network management staff that chooses to dedicate bandwidth for voice is guaranteeing that excess bandwidth, which could be used to carry data or other types of traffic, will sit idle during periods of low voice demand.

The second conventional method of bandwidth allocation is table-driven alternate routing. With table-driven alternate routing, the network management staff defines tables specifying a first choice route, a second choice route, and a third choice route. When a circuit becomes active, the node processor attempts to establish the connection via the first choice route. If the first choice is congested, it selects the second choice, and so forth. In this manner, bandwidth can be shared among all circuits and allocated on an on-demand basis. Therefore, table-driven alternate routing overcomes the greatest drawback of the dedicated-bandwidth approach.

There are two approaches to the implementation of table-driven routing. The first makes the originating node responsible for defining the entire route from origin to destination. A list of possible routes for each device connected at that node are maintained in memory. The originating node attempts to set up the first route. If it is blocked at any point, the attempt is abandoned and the second route is tried. Because the originating node has no knowledge of traffic loading across the network, each route must be attempted and must fail before trying the next choice route. As a result, it may take a long time to set up a connection if it is necessary to step down through seven or eight choices.

This approach to alternate routing tables also introduces network management problems. Figure 1a, representing a three-node network, helps illustrate why. There are only two routes from A to B: A-B and A-C-B. By adding one node, the number of routing optionsis more than doubled (Fig. 1b): A-B, A-C-B, A-D-B, and A-C-D-B. With 10 fully connected nodes, there are more than 1,000 possible routes. In practice, few networks will be fully connected; however, most large networks will grow well beyond 10 nodes.

Even if the network management software could accommodate 1,000 alternate routes for each circuit, defining such a table-driven database would be an unwieldy job. In practice, no database will accommodate this many alternate routes. A more typical number is 10 routes per circuit. A typical node could require circuit definitions for more than 200 devices. If 10 routes are to be defined for each, the network management staff would be faced with the task of working out and entering more than 2,000 routes for a single network site.

Worse, each time a node is changed or a new T1 added, most or all table entries would have to be overhauled. Therefore, the network management staff must intelligently select the 10 optimal routes. However, the network management staff cannot predict every network condition and every possible failure. It is inevitable that some circuits will be blocked from time to time even when bandwidth exists to carry the call.

Automating the generation or maintenance of routing tables helps somewhat. Software can be devised to create the tables. The network management staff makes the modifications to the network configuration then initiates a routine that redefines the tables. In this manner, the cumbersome nature of table-driven routing can be lessened, but it can never be eliminated. In addition, automating the generation or maintenance of routing tables does not solve the problem of long connect times and blocking when bandwidth is available.

The second approach to table-driven routing is forwarding. Here each node is assigned the responsibility for only a portion of the eventual route. Each nodal processor only needs to know the paths directly available to it. This simplifies the problem of database management since each node needs to know only the availability of bandwidth in a handful of trunks. In addition, it is not necessary for all nodes to track topological changes or network failures.

The forwarding routing process begins when the originating node recognizes a request to establish a circuit. It then consults a table and makes a selection of one of the T1 trunks connected directly to it. The responsibility for

1. Management nightmare. *A table-driven router stores several alternative routes in memory. Though only two routes are needed in a two node network (a), adding just one node doubles the number of routing options (b). Ten fully connected nodes would require more than 80 possible routes stored in memory.*

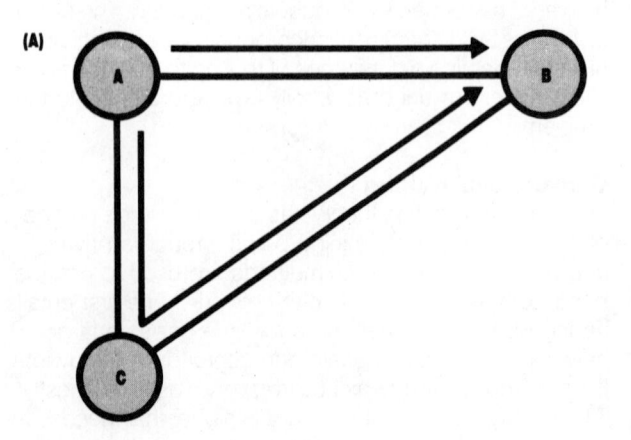

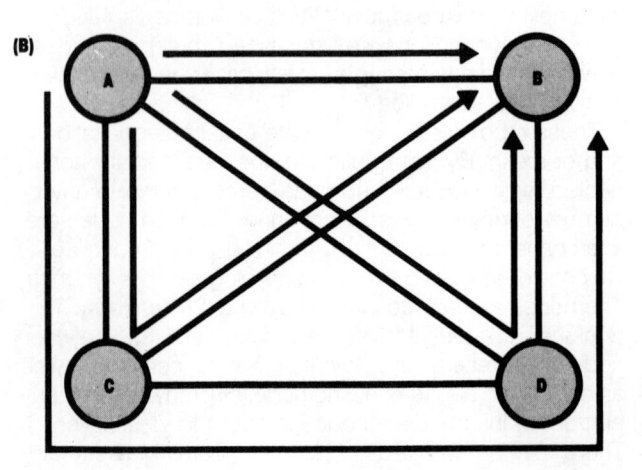

routing is now passed to the next node encountered, which in turn selects a path and forwards the connection request to the next node. Routing continues in relay fashion until a complete path to the final destination is established.

Forwarding does not offer an effective approach to routing in large networks. It offers too little control over the nature of the route selected. Paths may be obtained that tandem through many nodes, unnecessarily tying up bandwidth when more direct routes exist. Also, its trial and error approach can result in long setup intervals.

In complex networks, forwarding tables are particularly prone to loops. Loops occur when routes double back on themselves, never resulting in a completed connection. This problem is particularly acute when the tables are in the process of being updated, with some nodes retaining older tables while other nodes are using the updated tables.

Designing intelligence in
The conventional routing methods offer two choices: either throwing bandwidth at the routing problem or creating a routing method that is overly complicated. Clearly these solutions are not satisfactory; it is more useful to allocate bandwidth on a demand basis. Therefore, designers of the most recent generation of T1 networks have built the routing intelligence into the network itself so that bandwidth allocation can be done on a real-time basis.

As designers, we consider the need for design modifications before considering how intelligence will be built into the network. Any routing technique we use, be it static, table-driven, or more sophisticated, will be subject to revisions as the routing software evolves in parallel with networking needs. Therefore, consideration must be paid to providing for graceful routing software upgrades.

It is not acceptable to take the entire network out of service in order to modify routing software. New and old software must be able to coexist within the network during a phase-in period without disrupting network operation. Furthermore, it should be possible to escape back to the old software if problems arise.

Network intelligence is usually built into the network in one of three ways. Primary routing responsibility can be given to processors located in each node. Or processors strategically located at various nodes in the network can handle all routing requirements on a load-sharing basis. Another way to handle network intelligence is for a single dedicated processor to take over all routing functions.

A single dedicated processor is not desirable because it can become a single point of failure for the entire network. In addition, during periods of high demand, such as the period following a T1 failure when affected circuits must be rerouted, a single processing device can be overburdened. The result is overly long circuit setup times.

Designing processors into each node has the advantage that routing tasks are distributed. However, there are also problems associated with keeping all those processors current. Each routing processor must have the latest information about bandwidth use throughout the network. It must be made aware of what bandwidth has been allocated, what bandwidth has become free, what failures have occurred, and what changes to the network configuration have taken place.

By placing the routing processing task in each node, the question of network-wide modifications has become more complicated. The phase-in of new routing software must now be completed over an extended period because every single node is affected. Should the old software be needed again, it would be necessary to swap out software in dozens of network nodes.

The ideal network design places load-sharing path servers strategically throughout the network. Path server processors are dedicated exclusively to the routing function. This modular path server design offers the flexibility to configure more call-setup capacity as the volume of connection requests grows.

A sufficient number of path servers should be planned to match routing demand; however, the number should not be inflated so that it complicates the task of updating. This design offers the added benefit that each node has access to several remote routing processors so that failure of one such processor does not render a node unable to originate connections. Finally, the load-sharing approach eases the problem of modifiability. It is easier to phase in new software when the change is limited to a few path servers.

Deploying sufficient processing power in the network for routing is the easy part. The challenge is in developing an algorithm for making routing decisions. Some rules seem axiomatic. For example, the ideal number of hops for a point-to-point circuit would seem to be one, however, this is not always true. Furthermore, the decision becomes far more complicated in more extended networks where there may be dozens of routes to the same destination.

Defining the circuit
Before a route can be allocated, the requirements of the circuit that is to be built must be known by the network. These may either be specified by the user at the time of connection request or preconfigured in the originating node's database. User-specified circuits are not generally provided in T1 private network management schemes. When available, it is a welcome feature for offering more flexibility and an easy solution to one-time circuit requirements. An example of a user-specified circuit is the public telephone network. The user dials digits to define the destination and the network routes based on these digits. Another example is a packet network where the user specifies the destination address on the network.

For administrative reasons, the user who provides the address for a connection is not usually extended the privilege of specifying circuit requirements. The user is not expected to have the necessary sophistication in his understanding of the network. Instead, all users are given a standardized set of requirements. For example, no caller on the public telephone network is guaranteed a cleaner or more direct path than anyone else.

Private networks tend to be used for private line replacement—as a result, most circuits can be preconfigured. Since the network itself assumes routing responsibility, the network management staff's job is simply to establish a profile for the circuit in the network's database. We know the circuit's origin because we know where it is connected into the network. The network management staff must identify the destination as part of the circuit profile. Any destination can be the target of multiple origins contending for a particular resource (e.g., terminals contending for a host port). Conversely, any origin may have multiple acceptable destinations, allowing for the establishment of port groups. These multiple destinations may not necessarily be connected to the network at the same node.

Once the origin and destination are known, there is sufficient information from which to select a route. However, the route may not be acceptable for other reasons. The network management staff must lay out other rules for what would be an acceptable route for a particular circuit. Since each application is different, these rules must be defined on a circuit-by-circuit basis. Important characteristics include:

■ *Delay*. Each T1 in the network introduces some amount of propagation delay. The amount of delay varies with the length of the circuit and the transmission technology used. The overall delay of a route is arrived at by summing the delay of each traversed link and intermediate node. The path server is aware of these values. As the path server tries possible routes, it sums delays as it strings links and nodes together. Once a particular route exceeds the acceptable delay limit set down in the circuit definition, it is abandoned as an alternative.

■ *Dollar cost*. The network management staff may choose to assign dollar costs to each trunk in the network. The intent here is to bias traffic away from more costly routes such as measured-use T1s like AT&T's Accunet T1.5 reserved service. As routes are built through the network, the cost of trunks is summed. If a particular path exceeds the allowable limit, it is abandoned.

■ *Error rate*. Some types of transmission are more tolerant of noisy T1 circuits than others. Applications and protocols vary according to how well they handle errors. Digital voice, for example, is highly tolerant of errors since the human ear is incapable of discerning the distortion created by an occasional handful of bad bits. The same volume of errors, however, can cut sharply into the efficiency of a high-speed data application. Error rate of a particular link or as a composite of all links making up a route can be weighed in the routing decision.

■ *Failure rate*. Common carriers do not guarantee 100 percent up time. Just 2 percent down time per year means that a particular T1 will be out of service for 175.2 hours. State-of-the-art T1 network switches overcome this by rerouting circuits upon failure. But some disruption is inevitable when a T1 failure occurs. Brief periods of halted transmission will occur as a result; some or all circuits may not be rebuilt because there is insufficient bandwidth designed into the network. The failure history of a particular route is a function of the failure history of all T1s it traverses. How gracefully a circuit handles disruptions varies from application to application. The network management staff may want to specify different levels of failure tolerance into the circuit profiles.

■ *Encryption*. Some T1 trunks in the network may be outfitted with encryption equipment to scramble transmissions. In networks where security is a requirement on some circuits, the network management staff may specify that some circuits are always sent via encrypted links and other circuits never traverse encrypted links.

■ *Fewest links*. In a sense, this criterion is very similar to dollar cost. The more T1 links allocated for building the circuit, the more expensive the network is. A circuit that tandems through many nodes to reach its destination may block several circuits along the way. For example, Figure 2 represents a six node network. One single wideband circuit routed A-C-E-B-D-F can block nine circuit requests (i.e., from A to E, from A to D, from A to F, from B to E, from B to D, from C to E, from B to F, from C to D, and from C to F). When all other weighing factors are equal, the path that uses the fewest T1 links should be the one that is used. The network management staff may choose to set an up-per limit to the number of links that can be used to establish a circuit.

Together, all of the characteristics noted above make up a class of service. The class of service is defined in a circuit profile on a circuit-by-circuit basis. At the time a circuit request is initiated, the network has less than a few seconds to apply all of the factors within the class of service for the circuit, find a route, and set up the connection.

Volatility is a routing problem because modifications to the network topology affect the availability of routes through the network. It can also be a management problem if updating each path server's knowledge of the network in a timely manner falls upon the shoulders of the network management staff. If the network management staff fails in this responsibility, the efficiency of routing is compromised.

2. Cost-cutter. *The more T1 links allocated to building a circuit, the more expensive the network. When all factors are equal, the path that uses the fewest T1 links is the best.*

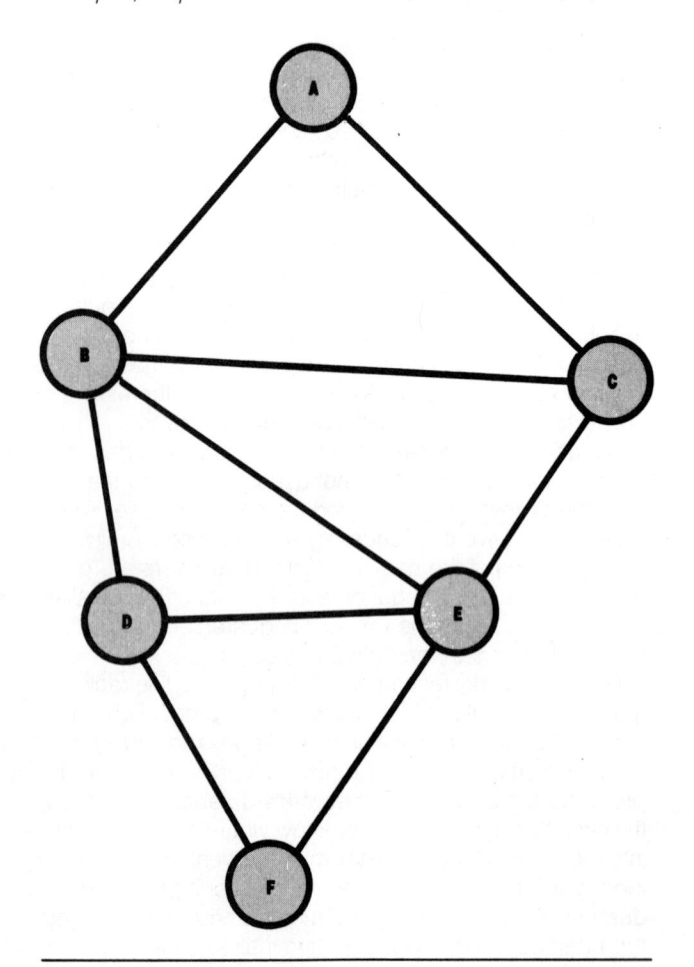

Worse, an error on the part of the person entering the database can severely handicap network throughput.

The network design should foresee these pitfalls. The network should be intelligent enough to handle the task of updating all path server databases automatically. There are a number of processes that can be devised to handle network volatility. One way is to make it the responsibility of new or changed nodes to communicate the changes to all path servers as soon as the modifications are enabled. This is done through what is called a boundary propagation algorithm; here is how it works (Fig. 3).

The process begins as soon as the the T1 trunks connecting the new node to existing network nodes are enabled from the network management terminal. The new node then initiates a handshaking procedure with each of the existing nodes to which it is directly connected. Existing nodes already have virtual communication links to the path servers through which they request routes when devices connected to them become active. Through these virtual links, the existing nodes notify each path server of the existence of a new neighbor node.

The path server now initiates a virtual path of communication to the new node. Through this link it requests details on the T1 resources provided by the new node. The new

3. Keeping up with changes. *As network topology changes, so must the node path servers. As new nodes are added to the network, changes must be communicated to path servers as soon as modifications are enabled. A boundary propagation algorithm can update all path server databases automatically.*

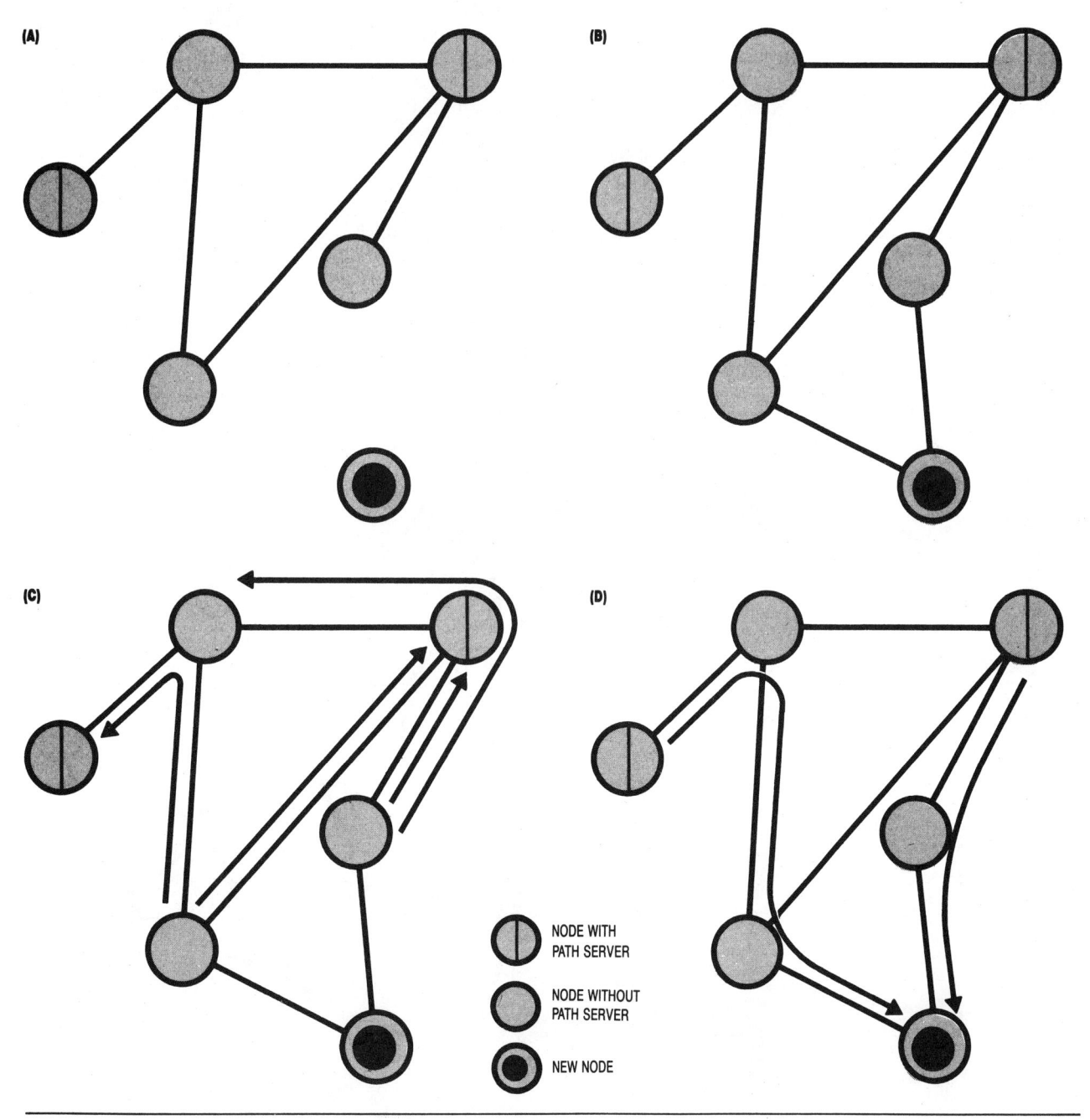

node responds by transmitting the information requested. The new node is now part of the network. It may originate circuits, serve as a termination for circuits, and act as a point through which other circuits can be routed. The entire process of making each path server aware of the new node and its resources varies according to the size of the network, the volume of other demands on the path server, and the total number of path servers configured into the network. In general, the time involved to complete this process is no more than a few seconds. Had this process been handled manually, it would involve at least an hour of the network management staff's time.

Three-step selection

Dynamic routing by a path server involves three distinct activities: resource qualification, tree building, and final path selection and qualification.

Resource qualification involves identifying those resources (i.e., links, switches, and terminations) that are capable of supporting the circuit. For example, if the desired connection is a 19.2-kbit/s terrestrial-only circuit, then we would only consider available ground-speed links with at least 19.2 kbit/s of free bandwidth. Links not satisfying these requirements are eliminated as alternatives. As the restriction to ground-speed links demonstrates, resource qualification can include a number of considerations besides basic bandwidth availability.

The resource qualification activity essentially confines the search for an optimal path to that subset of network resources that independently meet our circuit criteria. It is within this subset of resources that we can incrementally construct a tree of "least-cost" paths, which are then used to determine our optimal path.

Cost is defined as resource cost, which is quite different from dollar cost. Resource costs become highly significant during tree building. A cost is assigned to each T1 and each node in the network. These are used to determine the least-cost path since the cost of a path between two nodes is defined to be the sum of the component costs. Any combination of resource management policies and trade-offs can be represented by using different circuit cost profiles embedded in the class of service. The policy for how cost definitions are assigned can be based upon multiple criteria, including circuit and network requirements.

The tree building process essentially involves incrementally extending a tree of least-cost paths in a branching fashion (Fig. 4). The tree starts from a destination and then, through iteration, works toward the origin. At each stage of

4. Tree building. *To keep costs within limits, a tree building process involves incrementally extending a tree of least-cost paths in a branching fashion. The tree starts from a destination and, through iteration, works toward the origin. At every stage, the tree of paths is extended to each set of adjacent neighbor nodes in a least-cost manner.*

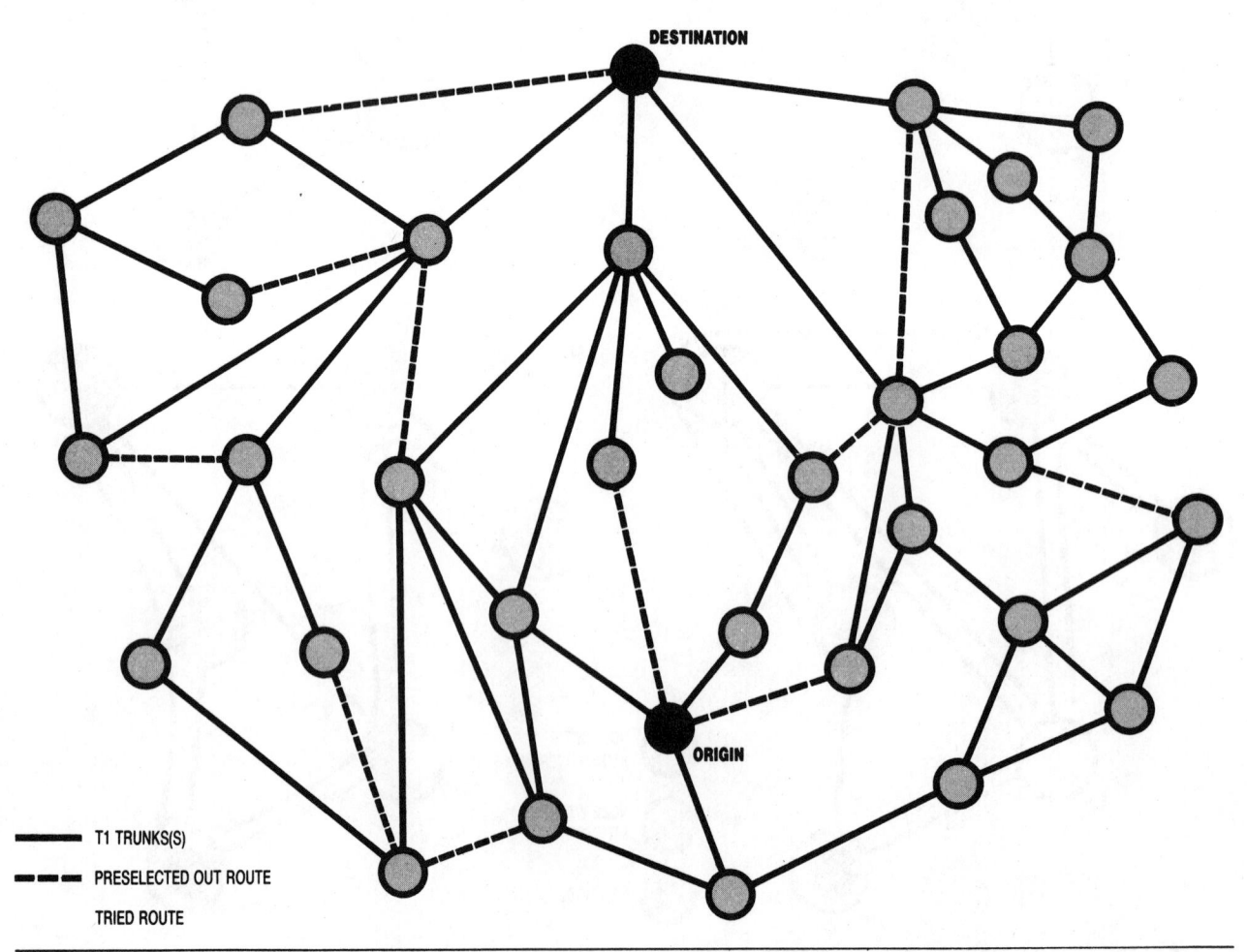

this process, the tree of paths is extended to each set of adjacent neighbor nodes in a least-cost manner. Branches are abandoned if the cost they accumulate as they pass through trunks and nodes exceeds the acceptable cost defined for the circuit.

One of two things will happen to terminate tree building before the origin is reached: Either the incremental extension procedure will exhaust all paths without reaching the origin, or the origin will be reached. Once the origin is reached, unproductive extensions are ignored and the branching process quickly dies out. Frequent blocking indicates that more T1 trunks need to be deployed in the network.

The tree building is completed when a least-cost path is found. The least-cost tree is then traced back from the origin to the destination. The tree building algorithm will directly yield only one least-cost path, even if there are multiple least-cost paths.

The tree building can also start from a collection of destinations rather than from a single destination. This is done for circuits that may terminate in any one of a number of distributed ports that have been identified as a port group. The process will then build a forest of trees. When completed, and if there is no blocking, only one tree will include the origin—and this tree will determine the destination. The destination is selected as part of the same process that selects the least-cost path.

Final path qualification and selection may now be applied. If desired, the least-cost path tree can be processed in order to identify all least-cost paths. One can then apply additional criteria to qualify and select among our set of least-cost paths. For instance, the least-cost tree may be built using dollar cost, but a final qualification criteria may limit the circuit to five or fewer links.

It is desirable to eliminate these final selection and qualification considerations by building them into the cost definition or the tree building process more directly. Post selection represents additional computational overhead. Building constraints into the tree building process may actually reduce overhead by limiting alternatives.

There are basically only two reasons for a company to install a private wide area network: greater reliability and reduced cost. These are the criteria on which a network is judged by the network management staff because these are the same criteria by which the telecommunication manager's performance is evaluated. The network management staff's job is not an easy one. It entails balancing the needs of each individual user with the common need to optimize the network throughput.

An effective routing algorithm helps the network management staff with this balancing act. A reliable routing algorithm also reduces costs in two ways. First, the need for bandwidth is held down to the level of what is necessary, reducing the monthly cost of T1 facilities. Second, the network becomes much easier to manage. As a result, manpower costs can be contained at a time when network management expertise is at a premium. ■

Robert S. Alford is vice president of circuit-switching development at BBN Communications. He holds a B. S. E. E. from the University of Hartford. Carl Holmberg, a divisional engineer at BBN Communications, received his M. S. from Carnegie-Mellon University.

James F. Kearney, Independent Consultant, Andover, N. J.

Making the most of your existing telephone wiring for data traffic

A wide range of components and technologies enable twisted pairs to be used for data access. Why not use what's already there?

Many day-to-day business activities have become dependent on computer terminals for information retrieval. Some companies, the fortunate few, foresaw their data processing requirements during the construction of the building or buildings that would house their host and terminal users and planned accordingly. However, most users are faced with the difficult task of moving into an existing building and retrofitting it to meet their data needs. Such users can save fortunes in special construction and building modification by using the twisted-pair wiring already in place. To do so, they should be aware of the methods by which user terminals may address one or more hosts within a multistory building, the hardware needed for these methods, and the ways in which they can be implemented.

The majority of buildings in use today were constructed with no thought for data traffic. No special duct work was laid, no special wall mounts or receptacles were strategically placed to allow terminal equipment to be connected. Such wall mounts could provide easy access through duct work to a central-site access point for one or more hosts. In most cases, the building was built with consideration given only to voice. In other words, minimal duct work was laid to allow the twisted pairs used for telephones to be located throughout the building. More importantly, little if any duct work is available for future use. Under such constraints, users should consider data-carrying twisted pair as a cost-effective networking option.

What do you need?

Before deciding on a networking solution, however, users should consider their business goals. If all terminals are directly connected to a host on the same floor and no switching or gateways are required, terminal connection is easily satisfied using RS-232-C cabling (for distances up to 50 feet) or twisted-pair cabling using line drivers (for distances greater than 50 feet). These simple solutions are inadequate, however, in multistory buildings with more terminals than host ports, access required to modem pools and/or gateways, the requirement to restrict access on a selective basis, the need to access multiple hosts, and limited cabling space.

Twisted-pair cabling may be used in this more complex situation as well. However, the following must be addressed when designing or optimizing a twisted-pair network:

Design considerations. Each department will have its own goal or goals. All department goals should complement the goals of the corporation. Who actually will use the network and when will it be used? Where will it be used? Define the application in detail and specify why each department needs this connectivity.

User considerations. Determine if there are full-time or intermittent users, as well as their technical background and general attitude toward the method of accessing resources. Determine the types of data to be transmitted along with the volume of data. Is the data to be treated as confidential? Where does the user want the terminals to be located?

Terminal considerations. Define the characteristics of the terminal to be used (block mode/character mode, asynchronous, synchronous, or combination). What terminal speed is required? Desired? Define the number of terminals and printers and terminal-user accessibility to the printers. How many terminals will be direct-connect, dial-up, local, or remote?

Network considerations. What are the transmission rate requirements? Define the type of communications gateways needed. What is the total cost for the transmission media? Can the cost be reduced by clustering or multiplexing? Carefully define issues that deal with reliability, security, and sensitivity of the network.

Enhancements to twisted pair

VENDOR OFFERINGS	MINI LINE DRIVERS	TWISTED-PAIR T1 MULTI-PLEXERS	VOICE-OVER-DATA MULTI-PLEXERS	DATA PBXs
APPROXIMATE PRICING	$85 EACH	$13,000 EACH	????	$250-450 PER LINE*
BYTEC CORP. SOUTHBOROUGH, MASS. (617) 480-0840				✔
CODEX CORP. MANSFIELD, MASS. (617) 364-2000				✔
DEVELCON INC. DUBLIN, CALIF. (800) 423-9210				✔
EQUINOX** MIAMI, FLA. (800) 328-2729			✔	✔
GANDOLPH DATA INC. WHEELING, ILL. (312) 541-6060			✔	✔
GENERAL DATA COMM INC. MIDDLEBURY, CONN. (203) 574-1113			✔	
INFOTRON SYSTEMS CORP. CHERRY HILL, N.J. (800) 257-1688				✔
MUX LAB ST. LAURENT, QUEBEC CANADA (800) 361-1965	✔			
NEVADA WESTERN SUNNYVALE, CALIF. (408) 737-1600	✔			
MICOM SYSTEMS, INC.** SIMI VALLEY, CALIF. (805) 583-8600	✔	✔	✔	✔
TELELABS INC. LISLE, ILL. (312) 969-8800			✔	
TELTONE CORP. KIRKLAND, WA. (206) 827-9626			✔	

*PRICING FOR A DATA PBX DEPENDS ON SIZE AND CONFIGURATION OF OPTIONAL FEATURES. TYPICALLY, PRICE PER LINE DECREASES AS NUMBER OF LINES INCREASES. REDUNDANT COMMON LOGIC, WHERE CONFIGURATION, ACCESS-CONTROL, AND USER CONNECTION INFORMATION HAS HOT BACKUP, IS STRONGLY RECOMMENDED.

**T1 MULTIPLEXERS OFFERED IN STANDALONE AS WELL AS INTEGRAL UNITS (IN WHICH MULTIPLEXER LOGIC RESIDES ON DATA PBX CARD).

Programming considerations. Where should the switching intelligence reside (that is, in the terminals, data switch, or host)? Should the host do switching, queueing, protocol conversion, or statistics gathering? Will software in the host provide adequate security for the users' applications?
Design calculations. What are the response-time requirements? Will the users tolerate queueing delay? Define the number of terminals, lines, and gateways to be used now and in the future. What are the effects of line errors on the network? What is the cost of the network?

When people consider using twisted pair for data and voice or for data only, they immediately think of the telephone companies. For the most part, the telephone utilities are credited with making the public aware that this capability exists. However, over the past five or six years, independent manufacturers of communications equipment have entered the arena with products that use twisted pair for data. These manufacturers offer users an unprecedented array of options (see table).

Three ways to go
There are three methods, each based on a different device, that allow users to transmit data over twisted pair. Common to all three is what is known as line-driver capability. This transmission method, similar to that of a modem, allows data to be sent over twisted pairs, received at the other end, and passed along. The following methods make use of this capability:

■ The miniature (or "mini") line driver is a simple device that attaches directly to the back of the terminal's RS-232-C interface. On the other end of this mini line driver is either four spade lugs (U-shaped wire attachments) or an RJ-11 modular jack for connection to the twisted pair. The purpose of this device is to be able to drive data at speeds up to 19,200 bit/s a distance of up to two miles. This type of device does not require any direct power, but takes its power from RS-232-C control leads on the data terminal equipment.
■ The line driver/multiplexer uses a device similar to the mini line driver on the line side. However, on the data terminal side, the line driver/multiplexer supports as few as four terminals (usually four per channel card) and as many as 128.
■ The voice/data multiplexer allows both voice and data to be transmitted over the same in-house phone line simultaneously. A station unit at the user location merges data and voice over the phone line by means of frequency-division multiplexing. All lines transmitting both voice and data pass through a central-chassis voice/data unit that filters the voice out from the data. The voice is passed on to the voice private branch exchange (PBX), while the data is terminated in an RS-232-C connector on the back of the central chassis. From the connector, the data is then passed to the data terminal equipment.

Consider the three scenarios in terms of the following application: There are 48 users on each floor of an eight-story building. Each user has a phone and also requires a terminal (a total requirement of 384 terminals). Assume also that the host (or hosts) is located to accommodate all of those terminals, say, in the basement. (Note that any of the applications to be considered can also be configured to allow hosts on any floor.)

Envision, for a moment, the internals of the building. On each floor, limited duct space can be used to run cables. Duct space is often restricted by national safety codes that prevent wires being packed too tightly. The ducts may have been small to begin with or they may be presently filled up by other users. These ducts are fed by risers, holes of at least four inches that rise up through each floor on at least one corner of the building. The telephone company has a riser for itself, and there should be at least one more. The telephone company may have riser space that the user could occupy or it may even be able to offer the user access to spare pairs in the riser.

All the terminals may be connected to a host using mini line drivers (Fig. 1a). Each line driver is associated with two twisted pairs (four wires). This is the most cost-effective of the networks in terms of initial dollar layout per user. However, this approach creates a bottleneck at the riser. Descending from the eighth floor, 96 pairs of wire are added at each floor until, upon finally reaching the destination, the user is faced with unbundling 768 pairs of wire. What was saved in hardware cost has been forfeited in flexibility and space-saving.

Alternatively, all terminals may terminate in a time-division multiplexer that incorporates the T1 carrier over two twisted pairs (Fig. 1b). These T1 multiplexers can be configured to support up to 128 individual terminals each. This reduces the number of wires needed within the riser from 768 to 16 pairs, allowing greater flexibility for future growth. Terminals are connected to the time-division multiplexer via twisted pair and mini line drivers.

The third approach incorporates voice/data multiplexers that allow the use of existing phone lines (Fig. 1c). In this example, data from each user terminal is combined with voice and brought down to the central voice PBX. Just before it reaches the PBX, however, the data is stripped from the voice, which is passed on to the PBX, and forwarded to the host. In this example, no new wiring is added, though the price of the station units increases. Conduit is now available for use by any future application or technology.

Each of these approaches has advantages depending on the circumstances. Obviously, if riser space is a problem, the user may not want to wire individual terminals, especially in large numbers. There are then three possibilities:

- Separate data wires can be run from a given location (preferably the phone closet on each floor) to each terminal site.

The terminals may be linked to a line driver/multiplexer, which can accommodate up to 128 host users on only two twisted pairs. At that point, it is demultiplexed and connected to the host.

- If no space is available in the riser, the most effective way to accomplish connectivity is with voice-over-data multiplexing. In this method, voice and data are combined, using frequency-division multiplexing, at a user's station unit. The voice/data signal is then transmitted over the same telephone wire as was originally used for voice alone. Just prior to reaching their destination, the voice and data are again split into two independent streams by a central unit. At this point, voice is passed to the PBX and data is passed either to a host or to another piece of communications gear for relay to a remote location.

Why add a data PBX?

Most current environments that deal with terminal access are faced with the following considerations:

- The number of available host ports may be inadequate compared with the number of terminals required to access multiple mainframes;
- A means must be provided to allow users to dial to outside services; and
- Users in an asynchronous environment must be able to

1. Options. Below, mini line drivers (A), line driver/multiplexers (B), and voice/data units (C) carry data over twisted pair to 384 terminals, 48 on each of eight floors.

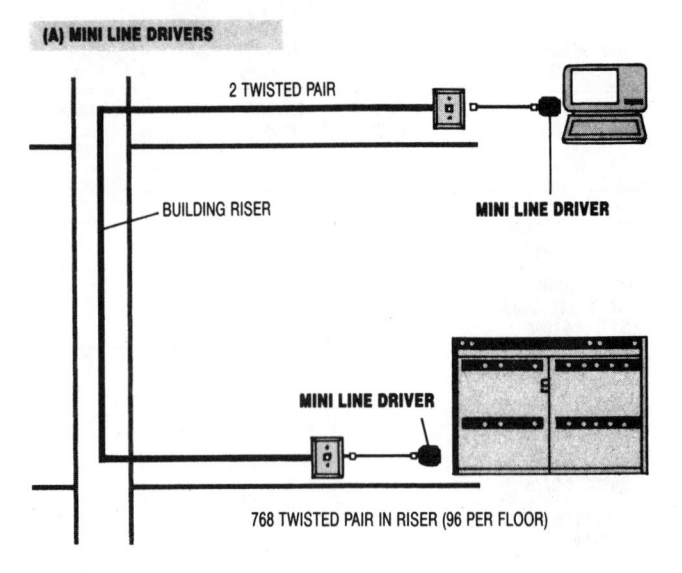

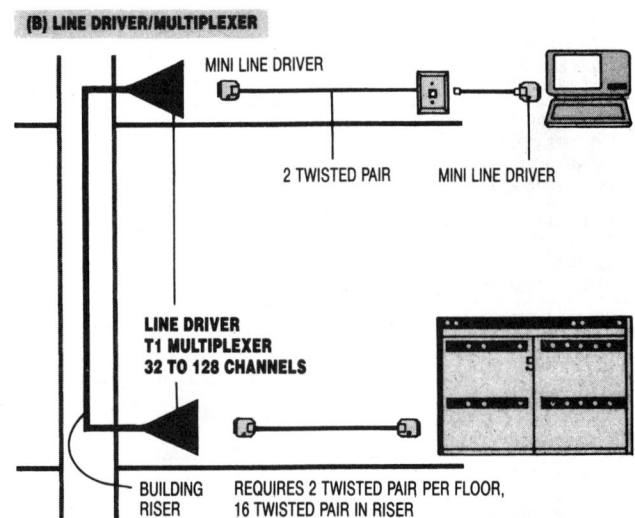

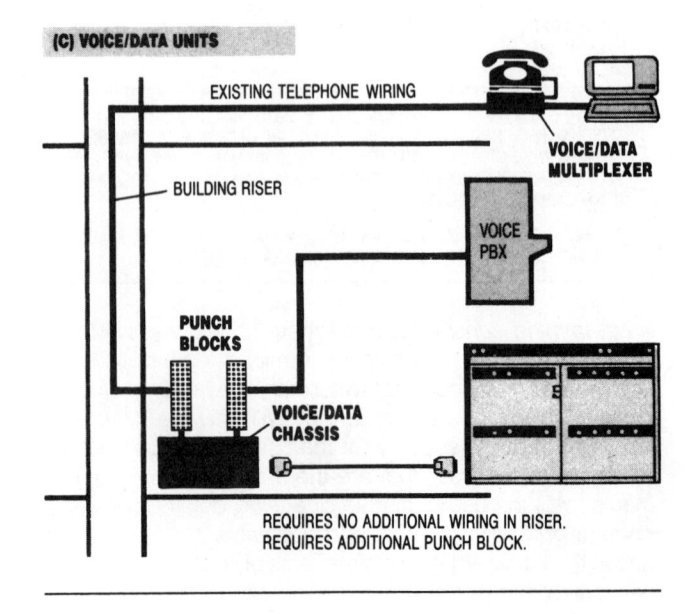

(A) MINI LINE DRIVERS

(B) LINE DRIVER/MULTIPLEXER

(C) VOICE/DATA UNITS

2. Hub. *A data PBX can serve twisted pair-attached terminals with a variety of services, including protocol conversion, multiplexing, and remote access. It provides contention, allows universal access to multiple hosts, remote access through dial modems, and gateway access to packet-switching networks and other environments.*

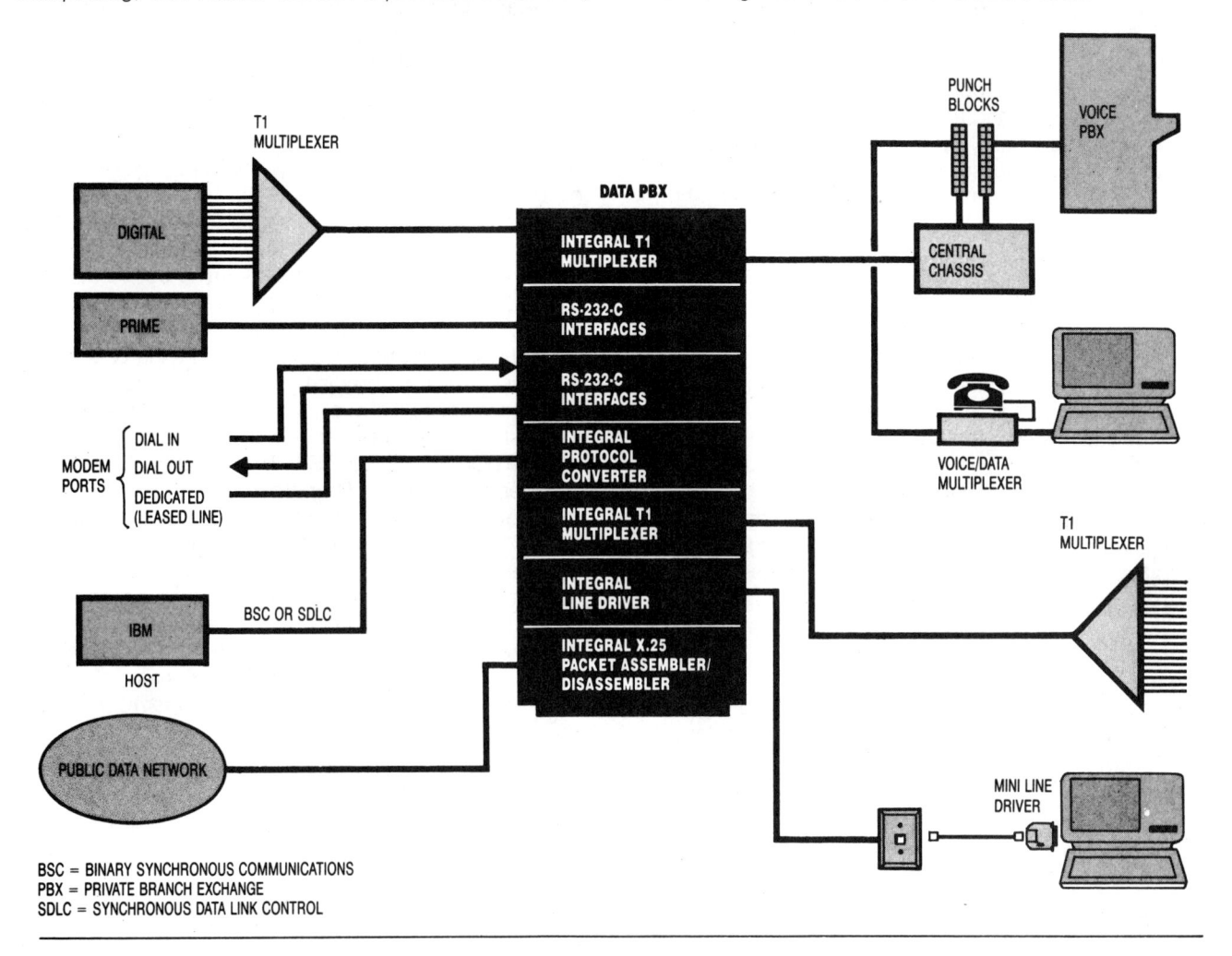

BSC = BINARY SYNCHRONOUS COMMUNICATIONS
PBX = PRIVATE BRANCH EXCHANGE
SDLC = SYNCHRONOUS DATA LINK CONTROL

attach to either an X.25 packet network or through protocol converters for access to an IBM environment.

With any of the three approaches considered in Figure 1, it may be advisable to use a data PBX as the hub of a twisted-pair network (Fig. 2). Most mainframes cannot accommodate hundreds of simultaneous users. The data PBX offers the following benefits:

- It provides contention for interactive users. In the example of 384 users and only 100 host ports, users transmit data only, say, 30 percent of the time. Using one host port for every four data PBX users provides user connectivity while yielding significant savings over the expense of adding mainframe channel cards.
- The data PBX, through its configuration, allows users access to multiple mainframes. With a mainframe on each floor, a user on the eighth floor may easily access a host on the third floor. This access can be provided and/or restricted either by setting up specific class routing or by limiting access through passwords.

(Class selection is the means by which a user requests connection to a given resource. Resources are usually associated with a numeric class or a symbolic name. Users wishing to access the network usually begin by hitting a carriage return upon powering up the terminal. The data PBX responds by generating an ENTER CLASS message. The user then simply enters the numeric class or symbolic name for the resource desired. If a connection is made, the user is prompted GO. If, however, the resource is unavailable, the user will be prompted by a BUSY message. The user can either wait until the resource is available or access other resources, if there are any.)

- Remote access, whether across town or across the country, is usually accomplished through dial modems. However, if everyone in the example above needed access, it would be prohibitively costly to go out and purchase 300 dial-up modems for people who would use them only occasionally. Therefore, a pool of modems could be attached to the data PBX and accessed through class selection.
- Most networks will eventually need to access gateways to X.25 networks and the synchronous IBM world. Today's data PBXs incorporate modules that allow both an X.25 gateway and IBM protocol conversion. This allows any user to access that gateway through class selection.

The data PBX acts as the hub of a star network, a form of a twisted pair-based local area network (LAN). Its main

313

purpose is to provide interactive terminal access to a host. File transfers can be handled over this medium; however, the files transferred should not exceed the megabyte range, or the relatively low speeds available through the switch (often 19.2 kbit/s or less) will jeopardize network performance.

Most networking managers know that they cannot bring up all the processors and terminals designated for all the floors at the same time. They must develop a strategy for bringing users and processors onto the network gradually. As each floor is added to the network, users already on the network must experience little or no down time. In the eight-floor scenario described earlier, the networking manager could proceed as follows:

■ Install the data PBX and one floor of users and processors. While the users would not have the capability of switching to another processor at this time, they would become accustomed to the log-on procedure, thus eliminating the shock associated with redefining the procedure as the network grows.

■ Install additional floors on the network as equipment becomes available and requirements become clearer. This can be done with no down time for existing users.

■ Add gateway modules and modem capability to the data PBX as the applications dictate. This move likewise would not adversely affect existing users.

Conclusion

There are four major alternatives in local area networks: twisted pair, baseband, broadband, and fiber. Fiber is still in its infancy. This is due largely to the fact that no standards have been developed for such issues as fiber width and connection methodologies. Fiber Digital Data Interchange (FDDI), a standard for the handling of data, is just now reaching fruition. However, there is no doubt that, when standards are mature, fiber will become the dominant medium for LANs. The following is a partial list of fiber optic manufacturers users may wish to contact for further information on fiber networks:

■ Codenoll Technology Corp., Yonkers, N. Y., 914-965-6300
■ Fibronics International Inc., Hyannis, Mass., 617-778-0700
■ FiberCom, Roanoke, Va., 800-423-1183
■ Raycom Systems Inc., Boulder, Colo., 303-530-1620

Among the four networking methods, twisted pair is by far the most cost effective for local terminal access. Where twisted pair costs roughly 25 cents a foot, coaxial cabling may cost five times as much. (The economics for broadband and fiber vary because of multiple-channel operation.) The twisted-pair network is transparent to data (having no built-in protocols), so application software/firmware within the host is not required. Any switching that must be done will take place in the data PBX. Twisted-pair LANs actually complement coax and fiber, because they can handle local, low-volume terminal access while the other media handle large, high-volume file transfer and video. ■

James F. Kearney is an independent data communications consultant. He has been in the data processing and communications field for the past 19 years. Prior to starting his own business he worked for such companies as Codex, Timeplex, Wang, and Micom. He has also guest lectured at the United States Military Academy at West Point.

Alan B. Taffel, Telenet Communications Corp., Reston, Va.

Packet-satellite networks: Updating and expanding the hybrid concept

The merging of X.25 and satellite technology into a single network can create a star-mesh topology combining the strengths of both.

Hybrid data communications networks, while by no means new, have nevertheless taken on new dimensions recently. Generally understood to mean a combination of public and private network facilities, the most common type of hybrid is a packet-switching network that blends public value-added network (VAN) functions with private packet-network implementations. With the introduction of low-cost satellite earth stations and economical, high-bandwidth terrestrial transmission, several new forms of hybrids have emerged. Two new twists represent enhancements to the hybrid concept: packet-satellite networks and integrated digital networks.

In order to understand these new forms of hybrids, it is worthwhile to first examine the concept of a combined public/private packet network. Figure 1a depicts the most basic form of hybrid: a single switch on customer premises connected to a VAN. The switch allows devices connected to it to exchange data without incurring either the cost of multiple links to the VAN or usage-sensitive charges normally associated with public networks. Yet these devices may also communicate with hosts or terminals connected directly to the VAN. In this example, network management functions are performed by the VAN's network control center (NCC), and these services extend to the CPE (customer premises equipment) switch as if it were a part of the larger network.

Figure 1b extends the premise of Figure 1a by adding a second data center. A CPE switch at each data center allows a connection between the centers external to the VAN. The two centers may communicate internally and with each other without any traffic or connect-time charges. The switches are linked to the VAN, primarily for reasons of expanded access and to provide a path for alternate routing. Figure 1b also shows a customer taking on some network management responsibility. Here, terminal links with the VAN's NCC and allows customers to monitor and diagnose their own portion of the hybrid network.

Corporations and government agencies may grow to have fairly extensive private packet networks of their own. This would be the case, for example, in a large company that implemented a network to serve as a corporate utility. The various company subsidiaries and their respective applications justify a network that allows sharing of company resources and the advantage of economies of scale. The large company sharing the private network resources is analogous to a VAN being shared by several companies. In both cases, economies accrue by consolidating usage over a single network.

In Figure 1c, a private network is shown fully interconnected with a VAN. Even in a large private network there are likely to be remote sites that do not justify installation of a private node, particularly when the VAN already offers a point of presence that is priced attractively for low-to-moderate traffic volume applications. Thus, the advantages of the hybrid model continue to hold, even in large-scale applications. Further, Figure 1c indicates that a hybrid user may acquire and operate his or her own NCC. In this instance, the VAN's NCC may serve as backup.

These three examples illustrate the advantages of the hybrid configuration to both traditional VAN users and traditional private network customers. For public network users, a hybrid reduces the cost of communications between data centers. Hybrids also allow such users a graceful and cost-effective growth path, undertaking any desired degree of network control. Private network users recognize the benefits of being able to add remote network sites for the cost of standard VAN dial-up rates. Such VAN access ports are typically available on demand, requiring no installation lead time. The VAN can also provide backup routing, backup network management, and international access. The many benefits appear to account for the surge

1. Hybrids. *The simplest type of hybrid network, a single customer switch connected to a VAN (value-added network), is shown in A. In B a second data center has been added to the configuration, and in C the user's network is fully connected. Hybrid forms have evolved over time to take advantage of changing technologies and customer needs.*

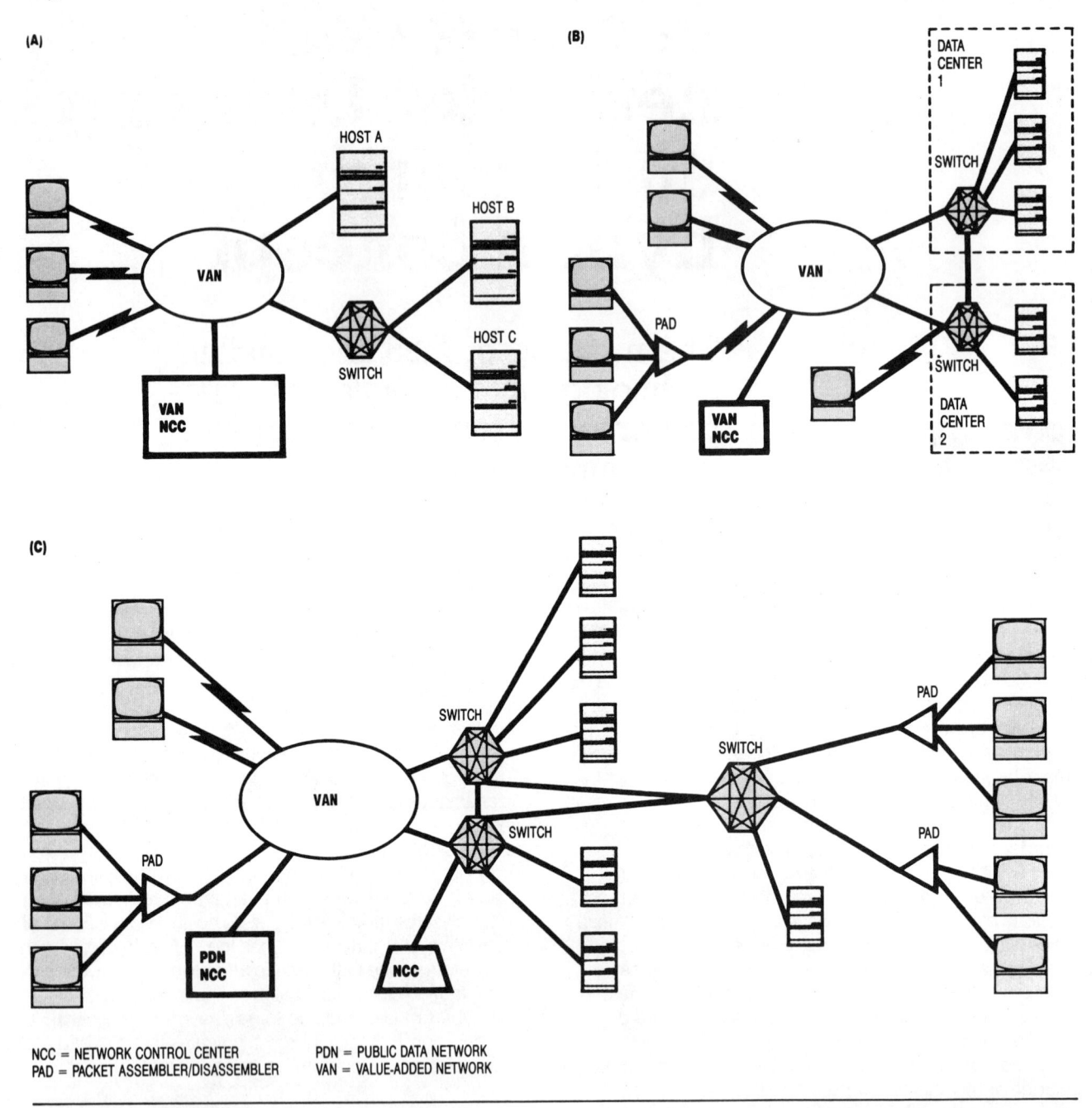

NCC = NETWORK CONTROL CENTER
PAD = PACKET ASSEMBLER/DISASSEMBLER
PDN = PUBLIC DATA NETWORK
VAN = VALUE-ADDED NETWORK

of interest in hybrid networks.

With the expansion of hybrid network usage has come an awareness of a new network requirement: transparency. In a hybrid network, users may connect to a VAN-based port as readily as to a private port. For reasons of training, documentation, and support of traveling users, it is important that a user's contact with the network be the same regardless of access point. That is, the hybrid nature of the network must be transparent. Further, users expect their network management services to be comparable regardless of which portions of the hybrid they are using. Thus, management services must be transparent as well.

In order to achieve transparency, a common network architecture (and perhaps a commonality of network equipment between the public and private elements of the network) is essential. To meet this requirement, users need both public and private network equipment and services that share a common architecture.

Packet-satellite hybrids

One new element that expands the potential applications of the current hybrid model is satellite technology. With the introduction of very small aperture terminals (VSATs), which are small-diameter (1.2- or 1.8-meter) satellite earth sta-

2. Packet layers. A shows elements of the packet architecture. B depicts the access layer with PADs (packet assemblers/disassemblers) and direct X.25 links.

(A)

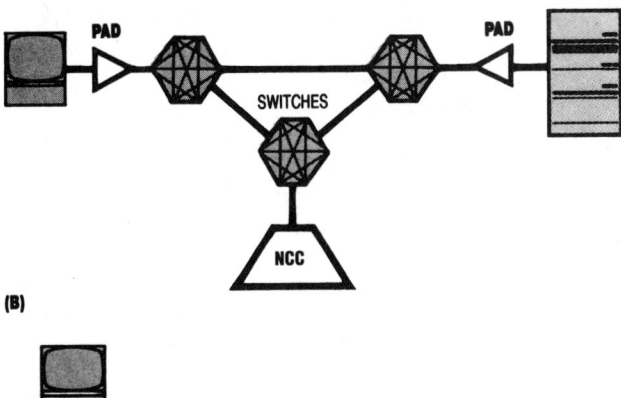

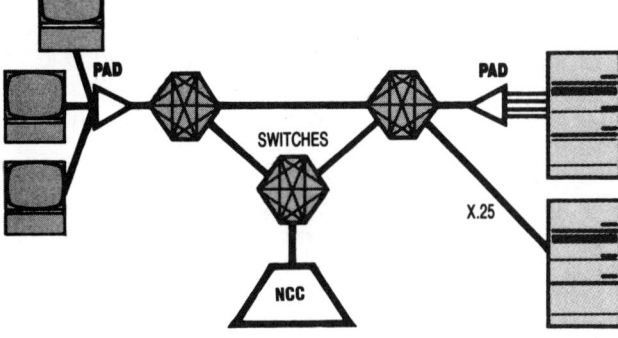

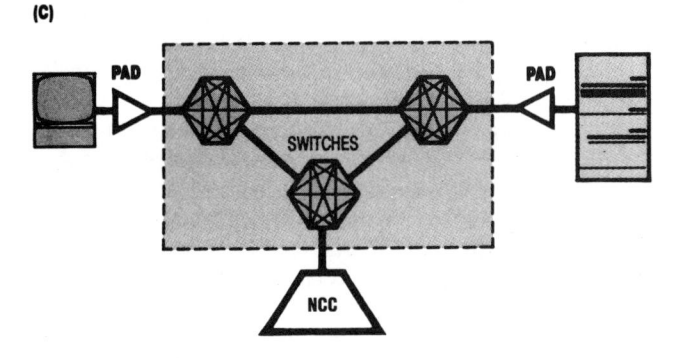

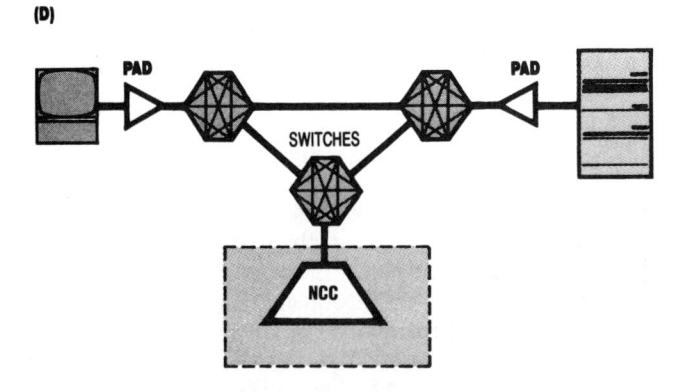

(B)

(C)

(D)

NCC = NETWORK CONTROL CENTER
PAD = PACKET ASSEMBLER/DISASSEMBLER

tions, this once esoteric and costly medium has been made economically viable in a much wider range of situations. In fact, VSATs are now available for about $6,000 to $10,000 per site. Several recent articles have discussed new satellite applications ("Buying satellite services is like opening Pandora's Box," DATA COMMUNICATIONS, April 1985, p. 165); what is new is the merging of satellite and packet technologies to create a new form of hybrid. This expanded hybrid blends the satellite's economy in accessing a central processing site, with the flexibility and manageability of pure packet hybrids.

Packet switching

Today most data communications managers are familiar with the packet-switching technology and its benefits. Architecturally, a packet network may be thought of in three layers: access, switching, and management (see Fig. 2). The access layer enables devices of various protocols—asynchronous, 3270, SNA (Systems Network Architecture), HASP (Houston Automatic Spooling Program), 2780/3780, and so on—to connect to the network. This process is usually accomplished through the use of a packet assembler/disassembler (PAD), which converts the devices' protocols to X.25, the CCITT standard protocol used to create a link with a packet network. It features end-to-end error correction and allows multiple devices to be supported over a single link.

Today, most hosts and front-end processors (FEPs) support X.25 as a native protocol implemented in software. In these cases, host-end PADs are not required, and the host FEP connects directly to the network switching layer. As above, multiple tasks or applications are supported over a single X.25 link.

The switching layer receives traffic from the access layer process, and routes it to the user-indicated destination. Full-function packet switches also generally collect accounting and statistical information and forward it to the management layer processor, often referred to as a network control center. The NCC processes accounting records and statistics, maintains and loads configuration information, controls access to network applications, and allows network monitoring and diagnostic capabilities. In some vendors' architectures, the NCC also is involved with call setup and routing, though this approach leaves a network far more vulnerable to a single point of failure.

The three-layer architecture allows packet networks to be

Technology comparison

TECHNOLOGY	IDEAL TOPOLOGY	SWITCHING	ACCESS CONTROL	PROTOCOL SUPPORT
SATELLITE	STAR; DISPERSED SITES	NO	NO	LIMITED
PACKET	MESH; CLUSTERED SITES	YES	YES	ALL COMMON IBM AND NON-IBM
PACKET/ SATELLITE	STAR TO HUB, THEN MESH; CLUSTERED OR DISPERSED SITES	YES	YES	ALL COMMON IBM AND NON-IBM

3. Packet satellites. *Network control centers can be located at the hub or off-site. The network is bidirectional, but for illustrative purposes, terminal-initiated traffic is shown.*

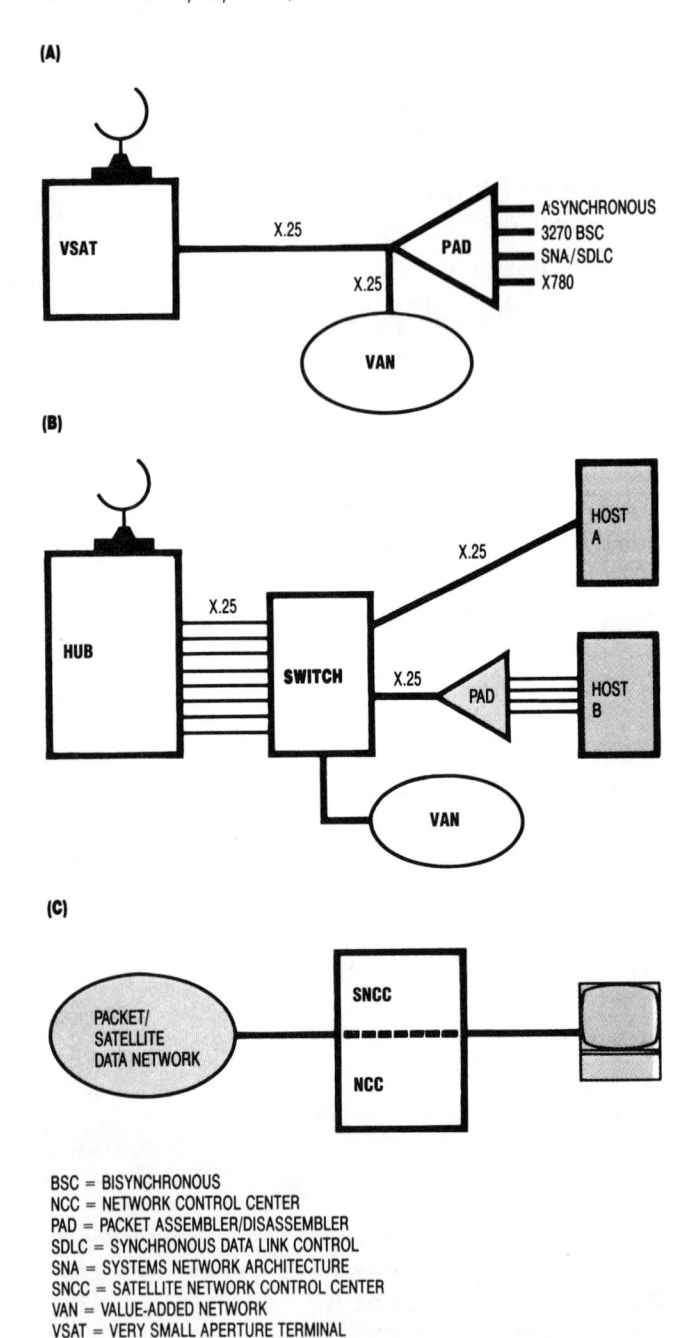

(A)

```
                    ASYNCHRONOUS
VSAT ──X.25── PAD ──3270 BSC
              │     SNA/SDLC
             X.25   X780
              │
             VAN
```

(B)

```
                           ── X.25 ── HOST A
HUB ──X.25── SWITCH
                  ── X.25 ── PAD ── HOST B
                     │
                    VAN
```

(C)

```
PACKET/              SNCC
SATELLITE  ──────  ────── ────── [terminal]
DATA NETWORK         NCC
```

BSC = BISYNCHRONOUS
NCC = NETWORK CONTROL CENTER
PAD = PACKET ASSEMBLER/DISASSEMBLER
SDLC = SYNCHRONOUS DATA LINK CONTROL
SNA = SYSTEMS NETWORK ARCHITECTURE
SNCC = SATELLITE NETWORK CONTROL CENTER
VAN = VALUE-ADDED NETWORK
VSAT = VERY SMALL APERTURE TERMINAL

highly cost-effective when multiple hosts are involved, when terminal devices are geographically clustered, and when a large number of interconnections are desired. In sum, packet networks effectively provide an ideal mesh topological implementation.

Satellite networks

Satellite networks contrast sharply with packet networks. Their optimum topology is a star, whereby multiple distributed remote sites seek access to a centralized data center. At each remote site is an earth station with an

A packet-satellite example

The nation's second largest retail store chain, the K Mart Corp. (Troy, Mich.), is implementing a network. In 1986, K Mart did $23.8 billion in sales. The company maintains more than 2,200 U. S. and Canadian K Mart outlets plus almost 1,500 specialty stores, such as Waldenbooks and Payless Drugstores. The objectives for K Mart's network were to:
■ minimize network costs;
■ provide interactive capabilities for such applications as credit authorization, which can cut costs by lowering bank rates. (Traditional methods of authorization, such as dial-up, are more time-consuming than an interactive network and result in longer customer waits. Speeding credit authorizations alone justified the network, although further economies are foreseen by using the same utility for inventory control and new customer services.);
■ support all existing applications without extensive redesign;
■ create a topology that can easily be upgraded to support additional applications;
■ standardize communications in harmony with international standards; and
■ handle communications needs for the next decade.

■ **Architectural design.** Existing K Mart stores use multiple in-store minicomputers and microprocessors operating under different communications protocols. The K Mart Information Network (KIN) is an IBM Series/1-based store office network for order placement, payroll, invoice processing, and general accounting. Point-of-sale controllers are IBM Series/1s and IBM PC/ATs. K Mart pharmacies run on Onyx PCs, and energy management equipment has its own microcoprocessor. Transmissions were made nightly or in batches as needed; there was no capability for interactive communications.

K Mart decided to use an X.25 packet-switching network to link its in-store equipment because of its international acceptance. Packet switching provides a vendor-independent standard interface for hosts and terminals and allows multiple applications to share common facilities. Message routing occurs in real time, and the entire network can be easily managed from a single point.

■ **Detail design.** Certain applications encompassing 91 percent of K Mart's communications costs, 98 percent of its sites, and 81 percent of the corporation's total traffic were chosen for the K Mart network. KIN was to be included as was sales ticket data for K Mart Apparel (KMA), the $5 billion New Jersey-based subsidiary that merchandises clothing in K Mart stores. Credit authorization was definitely part of the plan, as was a way for regional offices to report to the central office. KMA also needed interactive communications with K Mart International Headquarters. Finally, pharmacy and appliance warehouse terminals plus energy management devices needed to be linked to headquarters.

■ **Request for Proposal.** K Mart prepared a 170-page RFP that was sent out to 17 vendors: AT&T Communications, AT&T Information Systems, BBN Communications, Comsat Technology Products, Equatorial Communications, GTE Telenet/GTE Spacenet, IBM, M/A-Com, NCR,

NEC, Northern Telecom, Paradyne, RCA Cylix, Tandem, Telecom General, Tymnet, and Uninet. While the RFP allowed for either a satellite or terrestrial configuration, vendor responses indicated that costs for a satellite approach would be substantially less and offer more long-term price protection.

■ **Satellite feasibility.** Because of the cost differential between satellite and terrestrial alternatives, K Mart decided to expand its evaluation of satellite vendor offerings. Nine satellite companies (Avantek, Comsat, Fujitsu, GTE Spacenet, Harris, M/A-Com, NEC, SBS, and Scientific Atlanta) were whittled down to three finalists: BBN Communications in conjunction with Harris, GTE Telenet in combination with GTE Spacenet, and M/A-Com.

■ **Vendor selection.** While any of the three provided viable solutions, the joint GTE proposal brought in the lowest cost and had committed $500 million to five new satellites. Ultimately a decision was made based primarily on PAD (packet assembler/disassembler) flexibility, very small aperture transmission (VSAT) versatility, and comprehensive network management tools such as a color graphics network monitoring capability. The solution uses Telenet TP3 PADs, TP4 switches, and a modified TP5 network control center (NCC) that integrates satellite NCC functions. GTE Spacenet is providing a dedicated hub station, 1.8-meter VSATs, and space segments on its G Star satellite network. Due to the inherent efficiencies of the PADs and the VSAT scheme, the entire data application will use only one-half of one satellite transponder.

Telenet itself runs one of the largest public networks in the world, and so seemed to have the experience needed to put together the K Mart packet network.

Certain obstacles had to be overcome in order to fully meet corporate requirements, particularly issues of response time. Since cashiers would be waiting for authorizations for every credit card customer, network round-trip delay should not exceed one second. This was achieved through the use of 56-kbit/s links between the PAD and VSAT, with satellite links at that rate.

The network operations center controls both the packet and satellite network. The entire network or individual components can be observed. Nondisruptive testing can be performed on-line.

As for performance, the K Mart network does suffer from some solar interference—a total of 10 minutes twice a year. Moreover, winds faster than 60 miles per hour, heavy rain, and ice also degrade performance, but 99.6 percent of the time less than one message in 19,500 has to be automatically retried. The life of the G Star satellite is expected to be about 10 years. As the network is configured now, 100 percent network redundancy exists outside the individual stores. If the VSAT link fails at a particular store, K Mart plans a dial-up link to retrieve batch information. K Mart is also planning one-way, full-motion video broadcasts to the stores, regional offices, and distribution centers. —*Walter Bzdok*

Walter Bzdok is senior director of corporate communications and system reliability for the K Mart Corp.

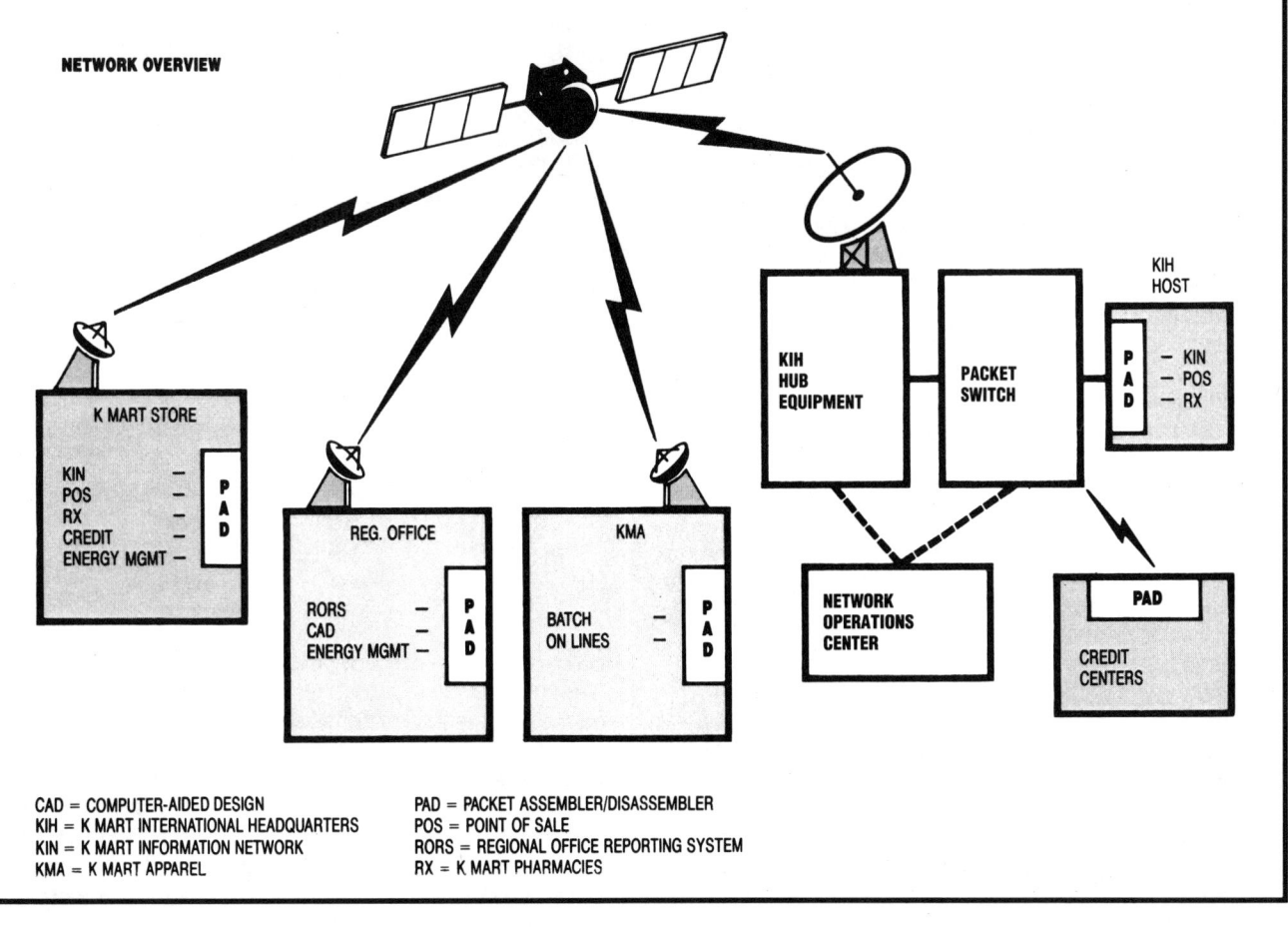

NETWORK OVERVIEW

CAD = COMPUTER-AIDED DESIGN
KIH = K MART INTERNATIONAL HEADQUARTERS
KIN = K MART INFORMATION NETWORK
KMA = K MART APPAREL
PAD = PACKET ASSEMBLER/DISASSEMBLER
POS = POINT OF SALE
RORS = REGIONAL OFFICE REPORTING SYSTEM
RX = K MART PHARMACIES

antenna that communicates with a central hub station. Because the hub is costly to build and install, satellite networks save money over terrestrial solutions only when the hub expenses can be amortized over many remote sites. This is especially true as the cost of remote earth stations continues to decline. In general, an application with less than 100 remote sites is currently not considered a prime candidate for a satellite solution. Interest in satellite solutions has been stirred recently by the introduction of small-dish VSATs. The new ku-band VSATs also allow the easy addition of voice and video applications.

Yet satellites continue to possess several key limitations. Satellites do not switch, but simply focus all traffic to a hub or broadcast it to the remote sites. Thus, linking to multiple host destinations is difficult to implement. In addition, the new VSATs do not yet support many protocols, working primarily with X.25 and SDLC (synchronous data link control). Access control is also lacking. And, of course, the satellite's one-way 270-millisecond propagation delay may be unacceptable in time-critical applications.

Merging two technologies
The "Technology comparison table" shows the disparity between these two technologies, as well as illustrating the benefits of combining them. In a packet-satellite hybrid, inbound traffic is sent to the central hub, but is then routed on to one of what could be many hosts. Thus, there are elements of both the star and mesh topologies in the packet-satellite hybrid. Further, the satellite limitations in switching, protocol support, and access control—none of which applies to a full-featured packet network—are fully overcome in a packet-satellite implementation.

Remote site access in packet-satellite networks is accomplished through both PADs and VSATs. Terminals are connected directly to a PAD, as in standard packet implementation. The PAD is then cabled to the VSAT, using X.25 (see Fig. 3a). The PAD and VSAT need not be separate units, but this approach offers the greatest choice and flexibility. In any case, data from each remote site is beamed to the satellite, and is ultimately received at the data center hub, which consists of RF (radio frequency) equipment and a large dish.

At the hub, data flows to the directly connected packet switch or switches (see Fig. 3b). Since many data centers have multiple hosts, the switch's job is to route the traffic to its destination. In some cases, an off-site host may be the receiver, accommodated through a leased line between the switch and that host site. Alternatively, these remote hosts could reach the data center switches through a PDN (public data network). Either way, all connections to and from the switches are X.25.

Also at the hub, though conceivably at some separate location, would be network management functions. This consists of a processor or processors that monitor the packet elements of the network as well as the satellite elements (Fig 3c). Ideally, the vendors involved will have integrated the two such that operators need not continually alternate between consoles.

At remote sites, the concentrating nature of the PAD means that multiple devices at that site require only one satellite channel and one VSAT port. This frees additional VSAT ports for later use, possibly for voice or video applications. Further, with the aid of the PAD, the VSAT can now support all of the standard data protocols and, depending on the PAD model, several can be supported simultaneously. Some PADs are able to fully emulate the polling of synchronous protocols. This means the remote PADs actually poll the terminals while the host PAD intercepts the normal host poll. The net effect is that traffic only flows between the PADs through the network if there is actual user data to be sent. Thus, bandwidth is used more efficiently and is reserved exclusively for true data transfers. Since polling in some applications can account for upwards of 40 percent of line use, this savings in bandwidth is by no means insignificant.

Packet-satellite characteristics
Packet-satellite networks add a switching component where none existed in a pure satellite approach. Multiple hosts at the hub are easily accommodated, as are hosts that are remotely located. In all cases, the communications protocol is X.25 end-to-end, so data integrity is protected. The use of packet switches also ensures that a disaster backup host can be easily supported. In the event that a primary host fails, the switches can have predefined configuration tables that will route traffic destined for that host to a backup site. This backup routing takes place automatically if the switch senses that it cannot, for any reason, complete a data transfer to the primary host.

Another characteristic of the packet-satellite network is its ability to support routing through a VAN. The PADs may have a terrestrial X.25 link available for permanent or dial-up access to a VAN. Switches at the hub can be likewise connected to the VAN. Thus, if a VSAT or even an entire satellite were to fail, all traffic could be switched through the VAN. This scenario truly illustrates a hybrid configuration that makes effective use of many elements: a VAN, private packet equipment, and satellite transmission. Extending this concept, the VAN could support delay-sensitive applications, routing them to the data center switches through its terrestrial circuits.

Network management is also enhanced in packet-satellite networks. Comprehensive diagnostics, statistics, and accounting are available for the elements of the hybrid network. Additionally, network-based access control is available as an option to those who need high security. This is typically provided by table-driven ID and password screening that inspects calls before they are allowed to be connected. Two additional characteristics of packet-satellite networks are transmission costs that can be fixed for a contracted period and an ability to rapidly deploy or reconfigure sites (since telephone company delays are no longer a factor). See "A packet-satellite example" for an implementation of the technology.

Integrated digital networks
A second new form of hybrid results from the integration of packet-switching and circuit-switching technologies, allowing virtually all forms of digital traffic—digitized voice and video as well as data—to share a common high-speed backbone. This form of hybrid, which may be called an integrated digital network (IDN), contrasts with packet-satellite networks in that it may utilize either a full-mesh or star topology, with backbone transmission consisting of

4. Justifying T1. *The left column shows the number of 56-kbit/s circuits required to equal the cost of a T1 circuit, based on AT&T T1 tariffs for interoffice mileage charges.*

A quick comparison

IF YOUR 56-KBIT/S CIRCUITS TOTAL MORE THAN...	AND YOUR CIRCUIT MILEAGE TOTALS MORE THAN...
NUMBER OF 56-KBIT/S CIRCUITS*	**MILES**
2	25
2	50
3	100
4	500
6	1,000

THEN A 24-CHANNEL T1 CIRCUIT IS LESS EXPENSIVE.

*NUMBER OF 56-KBIT/S DIGITAL CIRCUITS REQUIRED TO EQUAL THE COST OF A T1 CIRCUIT. BASED ON AT&T 56K DDS AND US SPRINT T1 TARIFFS FOR INTEROFFICE MILEAGE CHARGES ONLY.

DDS = DATAPHONE DIGITAL SERVICE

terrestrial T1 circuits rather than a satellite space segment.

Whereas low-cost VSAT devices are responsible for the increased viability of packet-satellite networks, it is the reduction of T1 tariffs that have made IDNs economically feasible. Figure 4 shows that T1 access, with a capacity of 1.544 Mbit/s, can be a cost-effective replacement for as few as three 56-kbit/s circuits in low-mileage applications and as few as seven such circuits over long distances. Thus, T1 links, particularly fiber links with their inherent

5. Typical IDN. *The integrated digital network offers many of the functions of ISDN (Integrated Services Digital Network) in advance of approval by standards bodies.*

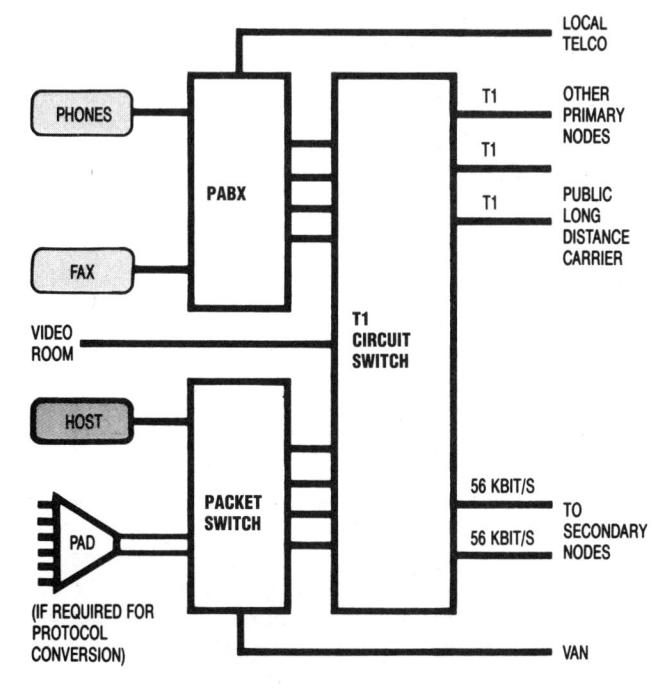

PAD = PACKET ASSEMBLER/DISASSEMBLER
VAN = VALUE-ADDED NETWORK

security and integrity, have become extremely attractive for networks of moderate to high traffic levels.

Other factors have also driven the creation of the IDN hybrid. To manage and use the T1 backbone, the market has seen a proliferation of competitively priced T1 multiplexer/circuit switches. Far more advanced than their pure multiplexer predecessors, these units are more accurately thought of as T1 resource managers, for they allow bandwidth to be intelligently allocated, route traffic around defective links, and feature built-in network management. In short, they may be networked. Perhaps most important, the new T1 switches answer the requirement for shared voice, data, and video over a common backbone.

Comparing circuit and packet switching

While state-of-the-art circuit-switching and packet-switching networks have much in common—including centralized network control, alternate routing, and mesh connections—there are many differences. Packet switching is ideal for a wide variety of data applications, yet its inherent statistical multiplexing makes it most efficient for interactive traffic. Today's T1 networks, on the other hand, use time-division multiplexing and can therefore effectively accommodate very high bandwidth, noninteractive requirements. These might include CAD/CAM (computer-aided design/computer-aided manufacturing) applications and host-to-host file transfer. Additionally, digitized voice (typically in 32-kbit/s or 64-kbit/s channels) and video (768 kbit/s) can be accommodated over the same backbone.

Another difference between these two technologies is the placement of switching control. A packet-switched user can specify data destinations on a per-session basis. In a typical T1-based circuit-switched network, routing from a given point in the network is pre-established by the NCC, and thus the traffic's destination is fixed. However, the NCC can modify routing at any time. Therefore, the true difference between circuit- and packet-switched routing is that control rests with the NCC in circuit-switched networks, and with the user on packet-switched networks. In the next generation of T1 switches, it is expected that user-initiated switching will be available for more and more user devices.

One final contrast is in the area of protocol support. Because a PAD device must emulate a protocol in order to gain such efficiencies as reduced polling overhead and error correction, packet networks support a finite set of standard protocols. Circuit switches, however, are protocol-transparent and not error-correcting. That is, a circuit switch simply passes bits without regard to protocol considerations. While less efficient in many cases, this characteristic does allow such a network to support virtually any protocol.

Configuring an IDN

The access layer of an IDN differs from that of packet-satellite hybrids in that devices may enter the network in a variety of ways. Figure 5 depicts a typical IDN node. Note that most terminals and hosts will continue to access packet equipment as in pure packet networks. However, PBXs or other D4-format voice equipment can connect directly to the T1 switch, as would video sources. High-speed host links and nonstandard protocol devices may

6. IDN private/public hybrid. *This hypothetical network provides economies of scale plus maximum flexibility of configuration. It is served by various public utilities providing backup, overflow, and extended geographic reach. Packet-switching networks, an essential part of the plan, were at one time esoterica used solely by the military.*

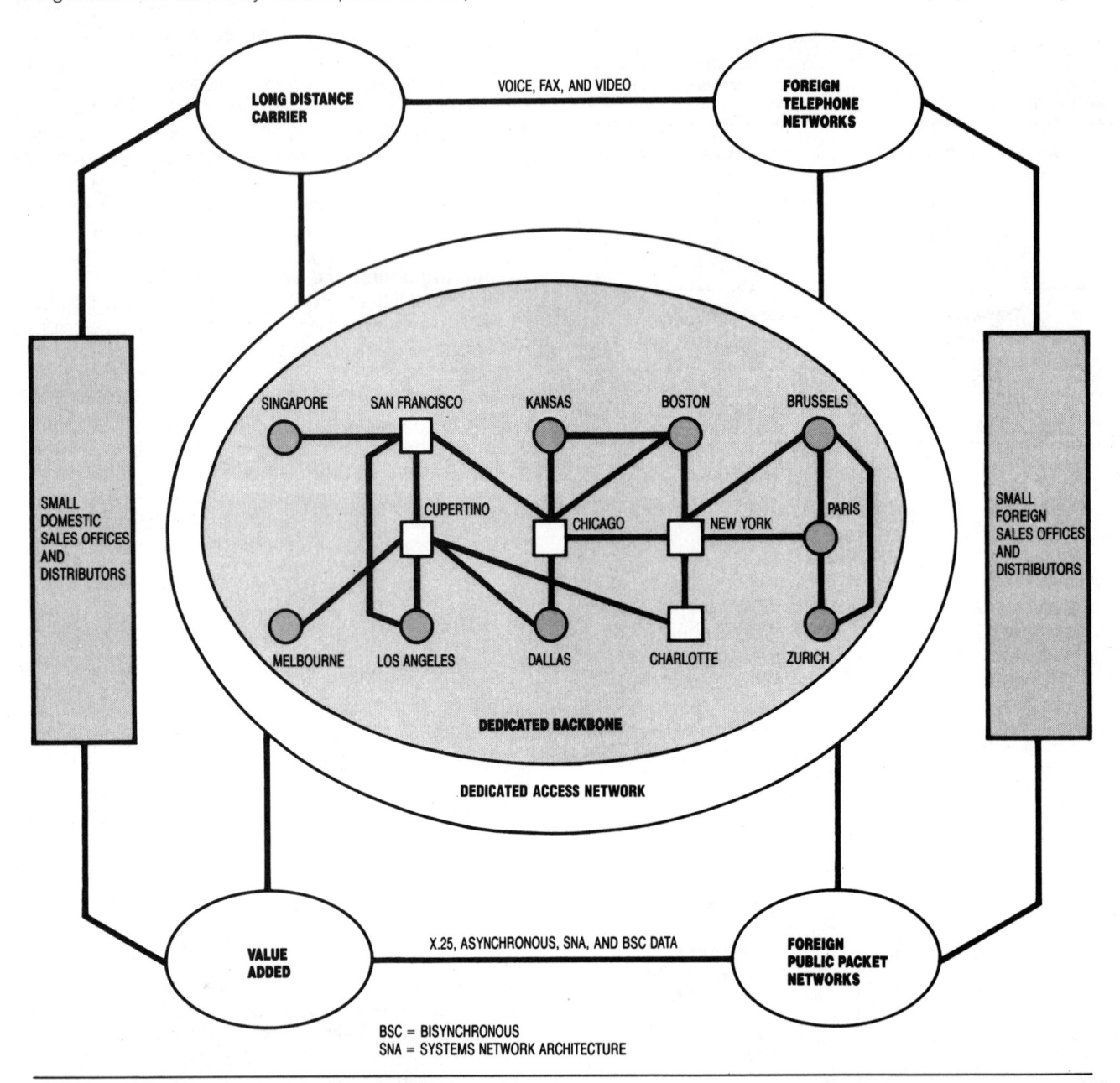

also make direct connections with the circuit switch. Thus, the access layer is really two-tiered, functioning at the packet level and the circuit level, where the tiers are interconnected to permit the best possible connections and maximum flexibility.

The switching layer of an IDN is similarly two-tiered. Packet switches may route traffic to locally connected destinations or through the T1 switch for remote destinations across the backbone. The packet switch will maintain the ability to route international or overflow data to a VAN, preserving an important element of the original public-private hybrid concept. The VAN would also continue to provide access from low-volume sites. At the inner switching level, the circuit switch will route all traffic over T1 (or smaller) links to predetermined destinations. There may also be connections to a public long-distance carrier, used to support overflow voice requirements when the backbone is being heavily used or for access from off-network destinations. In this way, the public-private hybrid concept is extended to the voice world. This expanded hybrid concept is illustrated in Figure 6.

At the management layer, the optimum approach remains an NCC that can handle all kinds of network traffic. Some packet vendors that offer T1 switching technology are in the process of creating such an integration. The results will allow both monitoring and diagnosis of all network elements from a single console. In the meantime, however, a single, centralized network control center, albeit one with dual operator consoles, is available from these vendors at the present time.

Since routing in a T1 circuit-switched network is NCC-controlled, the IDN's management layer must also perform this function. Additionally, the NCC can effect actual topology changes as needed in real time. Because there will typically be spare bandwidth in an IDN network, due to transmission being acquired in the bulk format of T1, the NCC can reserve this excess bandwidth and allocate it as needed. For instance, bandwidth can be reserved at a specified time of day to accommodate a scheduled videoconference or, alternatively, a network manager noticing that traffic has steadily increased between two nodes can simply allocate a larger portion of the T1 link to that traffic. Thus, the IDN NCC affords the network operator an unprecedented degree of flexibility and control.

IDN characteristics

As was the case with packet-satellite networks, the hybrid network that results from combining circuit switching with packet switching incorporates the best characteristics of both technologies. The topology will be a mesh, but can readily be configured to support star applications.

Another characteristic of an IDN is that it allows data communications managers to move immediately in the direction of ISDN. An IDN has in common with ISDN the use of T1 as the primary rate channel. IDN offers many of ISDN's functional characteristics, such as ease of connection, yet, because it is implemented in a private rather than a public network, these capabilities may be realized without waiting for public standards to be finalized.

An IDN is an inherently cost-effective network solution for several reasons. First, the shared digital backbone—consolidating data, voice, and video traffic—obviates the need for separate, geographically overlapping networks. Second, the IDN's ability to control and distribute bandwidth, in effect modifying the network topology as needed, reduces the normal installation fees associated with moving non-T1 circuits. Such flexibility also ensures that network performance for the user is optimized. Productivity of the network's operations staff also improves with the ability to centrally manage a consolidated network.

IDNs and satellite-packet networks are examples of the synergistic nature of hybrid networks. For large corporations or agencies with a requirement to connect many sites to a central data center, the packet-satellite approach is a viable option. For those having ISDN-like needs and requiring greater flexibility with regard to topology, an IDN can satisfy these requirements. In either case, those familiar with and seeking the advantages of packet switching may now find them in an expanded range of cost-effective integrated solutions. ■

Alan B. Taffel, currently Telenet's vice president of U. S. systems sales, began with that firm as a systems engineer in 1979. Since then he has held positions as regional technical specialist, western region systems manager, and director of network systems sales. Prior to joining Telenet, Taffel held several sales and technical positions with minicomputer and mainframe vendors such as Basic Four Corp. and Control Data Corp. He holds a B. S. in information and computer sciences from the University of California, Irvine, where he graduated cum laude.

Nick Berberi, Rockwell International, Newport Beach, Calif.

T1 to ISDN: How to get there from here

In-place T1 products and services, it turns out, will serve the emerging ISDN primary-rate utility very nicely. And so, too, will the DMI PBX-to-computer interface.

While the telecommunications network in the United States appears to be evolving into a network of networks, it is actually in a transitional stage of development—working its way toward an Integrated Services Digital Network. ISDN will be far more than a means to transport information or to create a network of networks. Indeed, it will be an economic engine that will drive and shape the growth of the communications industry worldwide. And ways to enable users of existing digital services and facilities to migrate gracefully, and economically, to ISDN are recognized as essential to ISDN network implementers and vendors alike. As the evolution of ISDN unfolds, computers and communications are becoming indistinguishable. Communications is no longer possible without computers, and computers are considered useless unless they can communicate.

The Digital Multiplexed Interface standard, or DMI, is an example of one development that bridges the gap between these two realms by allowing computers and communications equipment, such as private branch exchanges (PBXs), to communicate effectively and at low cost. The technological strength of DMI is apparent in its D channel signaling, its use of bit-oriented and message-based signaling, and its provision for clear-channel operations.

In the past, signaling information connected through T1 was based on robbed-bit signaling techniques. For example, in the D4 channel bank format, signaling was generally conveyed by robbing bit 8 of every channel from every sixth frame (frames 6 and 12). With DMI, all signaling information is carried in channel 24, thus allowing each of as many as 23 attached users to have full 64-kbit/s capabilities.

Bit-oriented signaling is based on A and B (providing on-hook/off-hook indications, busy-state information, and so on) signaling techniques found in existing D4 equipment. With DMI, signaling occurs in channel 24. This is a stepping stone toward clear-channel functions and provides migration toward message-oriented signaling.

Message-oriented signaling is based on Link Access Procedure-D (LAPD), which allows signaling information exchanged between two entities to be transported in a high-level data link control-type format. When LAPD comes into use, it will handle signaling for establishing, modifying, and tearing down switched calls with a layered protocol.

DMI spreading
The DMI interface standard has met with growing industry acceptance: More than 60 manufacturers are in the DMI users' group. Computer-to-PBX interface (CPI), a competing standard proposal, has no clear-channel capability, is designed to work at 56 kbit/s, and has limited ISDN compatibility.

Some of DMI's benefits are immediately apparent:
- Increased options and capabilities for users and telephone providers in the mixing of both voice and data over a single transmission medium, with access to existing and planned services.
- Cost reduction for both users and service providers by eliminating numbers of twisted pairs and alleviating the administrative complexity associated with large numbers of access points.
- Flexibility: DMI presents an evolutionary standard in step with ISDN, thereby guaranteeing upward compatibility.
- Multivendor compatibility.
- Protection of prior user investments in existing networks.

Role of primary rate
DMI's success stems from its being based on primary-rate technology already in use. Primary-rate-capacity facilities and digital pulse code modulation (PCM) are no longer used exclusively for communications between central offices; they are now used to connect customer premises

equipment (CPE) to the telephone network, too. Primary rate's time-division multiplexing scheme—T1 (North America) and PCM 30 (Europe)—is clearly the fundamental building block in the evolution of ISDN.

Distinguished by its cost-effectiveness and support of ISDN, primary-rate T1 will conceivably play a major role in the office of the future. PCM carriers will coexist with local area networks (LANs) but will provide enhanced services not currently available in local networks. While LANs are widely deployed and can operate at a whole range of megabit/s data rates, they have been limited to data applications and to specific regions. With T1, gateways, video connections, and the integration of voice and data will be available on a single network (see Fig. 1).

Market characterization

Primary-rate T1 technology developed in the early 1960s and has been generally applied to interoffice communications rather than the local-loop network. However, in the mid-1970s, with the development of large-scale integration (LSI) technology, the costs of digital loop carrier networks decreased dramatically, which led to the addition of many new T1 capabilities for the local loop. This resulted in the application of T1 at all three levels of the exchange-carrier hierarchy.

The expansion of T1-based services and the demand for higher bandwidth, better maintainability, and improved cost are enhancing T1's prospects and making it a nearly ubiquitous scheme among CPE and network equipment manufacturers for accessing the public telephone network.

However, because T1 technology justifies itself on cost, traffic volume, and the need to access remote locations through the telephone network, there will be users for whom it will not be a useful option. Those who might not make good use of T1 include users with pure data (as opposed to mixed voice and data) applications, those with LANs that do not need to communicate with or through the outside world, and those who don't need such high speeds.

A number of important changes are now taking place in the structure of the communications and data processing industries that will influence the prospects of using primary-rate T1 at all levels of the digital hierarchy, making primary rate a widely used standard. These changes are as follows.

1. Before and after. *End-user network access is shown before and after Integrated Services Digital Network is in place. ISDN's main advantage is that it allows services such as packet and circuit switching to be accessed from a single point, reducing costs and administrative complexities. Network design and maintenance will also become simpler.*

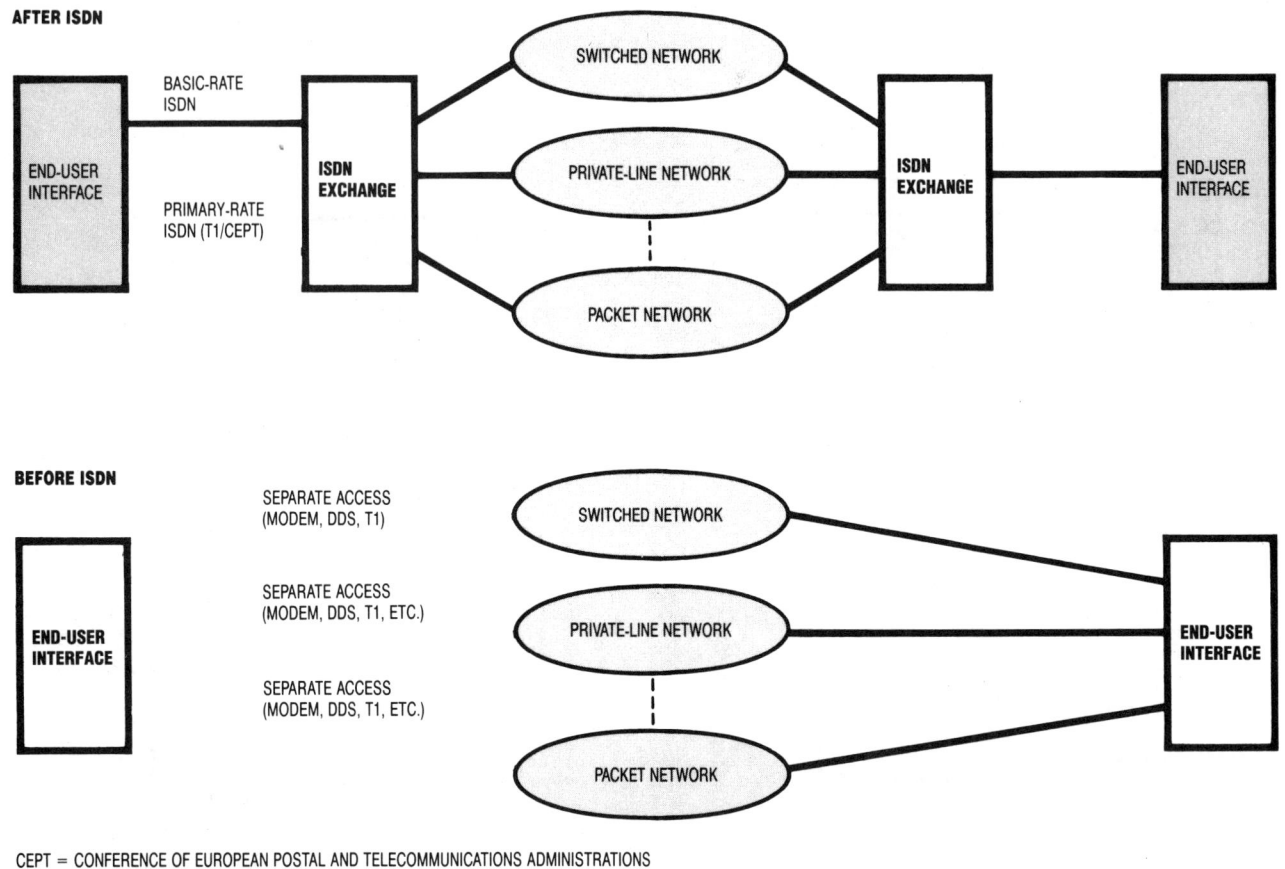

CEPT = CONFERENCE OF EUROPEAN POSTAL AND TELECOMMUNICATIONS ADMINISTRATIONS
DDS = DATAPHONE DIGITAL SERVICE
ISDN = INTEGRATED SERVICES DIGITAL NETWORK

2. Economies. *(A) shows how, by using primary rate, a large volume of voice and data traffic between two locations can be transmitted through a single twisted pair with one point of control. (B) illustrates the costs, hence, economies involved. The example demonstrates the savings resulting from fewer and less complicated network connections.*

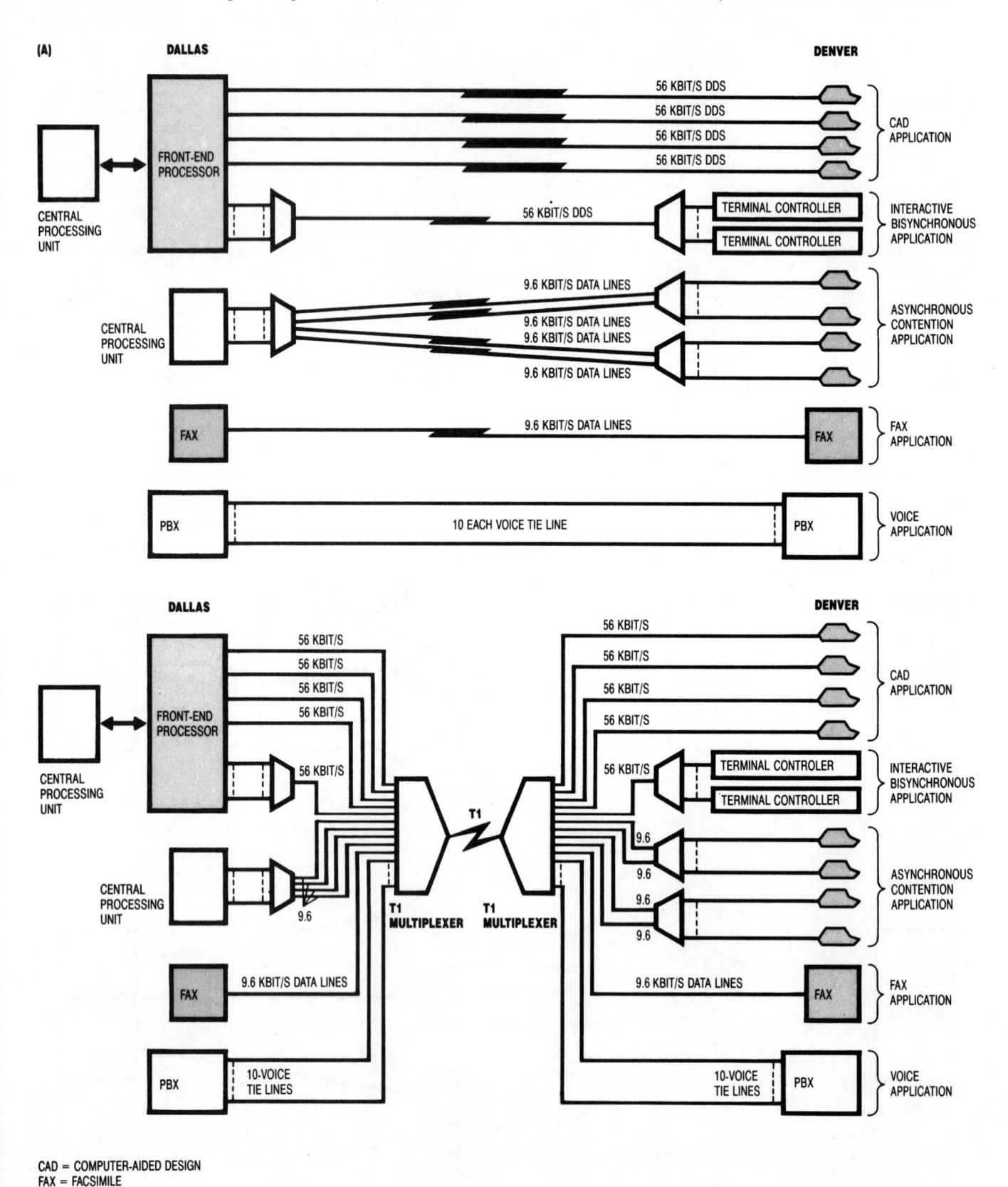

CAD = COMPUTER-AIDED DESIGN
FAX = FACSIMILE

(B)

VOICE GRADE	
INSIDE WIRE: TWO AT $8.00	$16
ACCESS COORDINATION: TWO AT $20.00	$40
LOCAL LOOPS:	
INTRA-LATA: ONE AT $73.00	$73
INTER-LATA: ONE AT $103.00	$103
CENTRAL OFFICE CONNECTIONS: TWO AT $15.50	$31
INTER-OFFICE CHANNEL:	
FIXED CHARGE: $305.00	$305
MILEAGE: 660 AT $.30	$198
TOTAL FOR LINES	**$766**
MODEMS (FOR DATA LINKS): TWO AT $109.00	$218
	$984 PER MONTH
56K DDS:	
INSIDE WIRE: TWO AT $8.00	$16
ACCESS COORDINATION: TWO AT $27.00	$54
LOCAL LOOPS:	
INTRA-LATA: ONE AT $109.00	$109
INTER-LATA: (10 MILES): ONE AT $328.00	$328
CENTRAL OFFICE CONNECTIONS: TWO AT $50.00	$100
INTER-OFFICE CHANNEL:	
FIXED CHARGE:	$1,325
MILEAGE: 660 AT $1.60	$1,056
TOTAL FOR LINES	**$2,988 PER MONTH**
CSU/DSU: TWO AT $35.00	$70
	$3,058 PER MONTH
COST OF CONFIGURATION IN FIGURE 2	
56K DDS: FIVE AT $3,058.00	$15,290
VOICE GRADE DATA LINES: FIVE AT $984.00	$4,920
VOICE TIE LINES: TEN AT $766.00	$7,660
MONTHLY LINE CHARGES	**$27,870**

T1 CIRCUIT	
INSIDE WIRE: TWO AT $8.00	$16
ACCESS COORDINATION: TWO AT $21.00	$42
LOCAL LOOPS:	
INTRA-LATA: ONE AT $556.00	$556
INTER-LATA: (10 MILES): ONE AT $803.00	$803
CENTRAL OFFICE CONNECTIONS: TWO AT $60.00	$120
INTER-OFFICE CHANNEL:	
FIXED CHARGE:	$1,400
MILEAGE: 660 AT $26.00	$17,160
TOTAL MONTHLY LINE CHARGE	**$20,097**
LINE-COST COMPARISON	
MULTI-LINE NETWORK	$27,870 PER MONTH
T1 NETWORK	$20,097 PER MONTH
	$7,773 PER MONTH
	$93,276 PER YEAR

CSU = CHANNEL SERVICE UNIT
DDS = DATAPHONE DIGITAL SERVICE
DSU = DATA SERVICES UNIT
LATA = LOCAL ACCESS AND TRANSPORT AREA

REPRINTED BY PERMISSION FROM 'THE TELECONNECT GUIDE TO T1 NETWORKING', BY WILLIAM FLANAGAN; TELECOM LIBRARY INC., NEW YORK, N.Y., 1987

- Businesses are calling for an industry standard that will allow them to interconnect multivendor communications equipment without resorting to the use of expensive interface devices.
- Primary-rate circuits are now being employed more for wideband access from the customer's premises to the telephone network than for wideband dedicated transmission between customer premises and interoffice transmission, its traditional role.
- Internationally, primary-rate ISDN is seen as strategically important to Postal, Telephone, and Telegraph agencies (PTTs). First, it will allow them to provide advanced network and information services to users, so that as these administrations undergo their own deregulation, they will be able to compete with third-party service offerings and with user-developed private networks. Second, it will assist telecommunications manufacturers in each country by providing advanced domestic markets to stimulate product development for worldwide markets.
- Dataphone digital service (DDS) pricing, especially for multiple 56-kbit/s circuits, already favors primary-rate-based digital services in many cases. For example, six DDS lines (an aggregate 336 kbit/s of transmission bandwidth) equal one primary-rate line (1.544 Mbit/s) in cost but not in bandwidth (Fig. 2).
- In order to avoid rewiring or the installation of expensive LANs, using the convenient medium of twisted pair for high-speed multiplexed interfaces between colocated networks is of growing importance.
- The endorsement of primary-rate-based standards such as DMI by major computer, terminal, and PBX manufacturers clearly underlines the wide acceptance of primary rate as the standard.
- The inability of LANs to support a mix of voice and data services could result in many users maintaining two separate networks—with resultant high administrative costs.
- The need for high-speed transfer of data, including graphics, video, and files between terminals and hosts, without having to rely on separate data networks, is increasing.

Several other significant trends that call for greater bandwidth and speed are evolving in the exchange-carrier and CPE industries. Fiber optics, in just half a decade, has become the dominant technique for very high capacity terrestrial transmission. Today, fiber is gradually being installed in the local loop and CPE environment as a replacement for traditional twisted-pair copper.

Market needs

The most important and immediate advantage of primary-rate technology is its ability to support multiple voice or data connections over a single interface. This can provide a more flexible and substantially less expensive method of accessing information than is possible with conventional interfaces.

Realization of the full potential of primary rate and its benefits can be seen from the perspective of users, network providers, and equipment manufacturers.

From the user's point of view, the primary benefits are:
- Fulfillment of the user's needs at a lower cost than existing alternatives.
- Flexibility resulting from the use of multivendor equipment

3. Device architecture. *The microprocessor controls all the functions in this representative end-user T1-DMI interface device, handling Open Systems Interconnection Layers 0, 1, and the lower portion of 2. Information the microprocessor finds in the channel and address FIFOs tells it where to find the actual line data in shared buffer memory.*

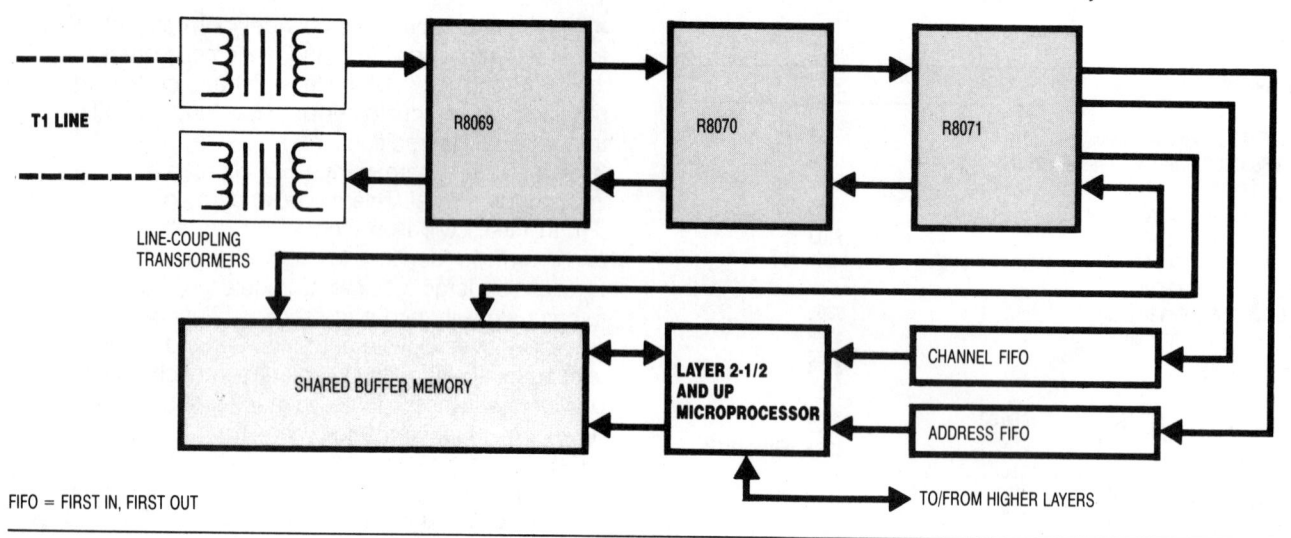

without any consequent degradation of performance.
- Adopting a standard interface allowing higher transmission rates, greater switching capability, and lower cost for network access.
- Ability to access the network services from a single plug.

For network service providers, whether they provide basic or enhanced services, the benefits include:
- Enhanced competitiveness with other private networks.
- Possible cost reduction resulting from the development of less costly digital technology and lower maintenance costs arising from the use of two wires instead of multiple wires.
- Enhancement of services, which should translate into higher revenues.
- Ability to compete with manufacturers, particularly PBX manufacturers, to generate more services.

For equipment manufacturers, benefits will include:
- Access to worldwide markets.
- Increased sales of their own equipment.
- Digital technology that is less expensive and easier to deal with.

Once the Integrated Services Digital Network is in place, users will be presented essentially with three types of connection methods: ISDN primary-rate access to public carrier services from their large locations; ISDN basic access to public carrier services from their smaller locations; and ISDN capabilities encompassing both access schemes for their digital private networks.

Use of ISDN primary rate

ISDN primary-rate access will become the standard means of providing access from larger users' locations to most public networks and services.

Users will have one or more 1.544-Mbit/s or 2.048-Mbit/s access lines between their private networks or local communications setups, as well as to the local ISDN exchange or the network service provider. Users will be able to choose for each individual access channel the services they wish to access. They will also be able to combine channels to provide access to higher than 64-kbit/s bandwidth, so that customers can select bandwidth on a demand basis.

This integrated access will replace today's proliferation of separate access methods involving leased analog data and digital data circuits. End-user organizations will find that switching to primary-rate access for their large locations is very desirable, primarily as a result of the cost advantage and operational benefits. Although large organizations may plan to retain logically separate voice and data networks, integration will take place at the transmission level.

As might be expected, this digital revolution in communications is completely dependent on the development of appropriate high-performance semiconductor devices (see Fig. 3).

It is also clear that T1 is by far a more advantageous alternative than separate connections. Coupled with the cost benefit are simplicity, ease of control and monitoring, and better network management.

Future directions

The advent and wide use of photonics will provide ample bandwidth where primary-rate technology is no longer hindered by certain technological limitations. As the trend progresses and customer requirements for higher speed and lower maintenance cost continue to grow, it is likely that in the near future, T3 (DS-3), which is a multiple of 28 T1s, could become a standard means for connecting the customer premises to the network. Underlying all these predictions is a need to stress that the ultimate success of a product or standard will always be determined by how well it meets customer needs. ■

Nick Berberi obtained a B. S. E. E. from Syracuse University in 1979. He held positions at Litton Data Systems before joining Rockwell, where he has been an applications engineer and product manager and where he is currently manager of system applications for digital network products.

Joseph Fernandez and David E. Liddy, Logica Pty. Ltd., Sydney, Australia

Don't just guess: How to figure delays in private packet networks

Private packet-switching networks pay off, but must they have poor response time? Not if you follow the rules of this simple model.

Some network managers have hailed private packet-switching networks as panaceas for a host of problems. Others, however, worry that such networks could plague users with performance penalties. Both views are understandable: The economies to be gained almost demand private networks, yet users will continue to require adequate responsiveness. A technique is needed that can quell concerns about undue delays.

To determine delays before installing a network, network managers have traditionally been faced with two alternative methods, neither entirely attractive: theoretical techniques, which rely upon statistics that often become unwieldy; and empirical model-building and simulation studies. The simulations may be too simplistic to extrapolate from, the models too expensive to build.

However, a little known hybrid technique for predicting delays is available to network designers. It relies on a combination of theoretically calculated and empirically measured data feeding into a statistical model. This article describes the hybrid technique, defines the concepts that must be understood to use it, and shows how it can predict network response times. A follow-up article will detail two applications of this technique, a simple network that illustrates the workings of the formulas and a more complex example of a network similar to one that might actually be considered by a network planner.

The need
Many large organizations have multiple, separate internal networks. Often, these networks were installed in an ad hoc fashion to provide remote access to computers whose terminals and communications protocols are incompatible with existing computers. However, this approach has led to the following problems:
■ Installing and maintaining trunk lines for multiple networks is unnecessarily expensive, since many of these lines duplicate each other because they run in parallel.
■ Duplicate and costly communications controllers perform similar functions for different computers.
■ Managing a gaggle of networks can be a headache.

Private packet-switching networks, with their ability to flexibly interconnect a wide range of terminals and computers, can rationalize and replace an organization's mix of ad hoc networks. However, many organizations fail to install their own packet networks because of performance concerns. They believe that these networks impose substantial delays on data transport, therefore increasing the response time perceived by users.

This is a myth. A properly designed and dimensioned packet network can support rapid, effective data transfer. In fact, delays imposed by packet networks may pale beside those caused by the heavy loading of host computers or even of the communications controllers that usually interface the hosts with the network.

Methods of estimating delay
Network designers must understand the source and magnitude of these inherent delays. In this way, they can ensure that any additional delays due to the network fall within limits accepted by users. When equipment is selected, the network must be dimensioned to carry the known or projected traffic without unacceptable delay.

The delays imposed by a private packet-switching network can be determined or estimated in several different ways. The methods can be broadly divided into theoretical and experimental techniques.

The experimental methods use measurements of traffic that is injected under controlled settings and traced through the network. Determining the dimensions of the network prior to installation requires that the designer make these measurements without having the whole network in place. One form of measurement involves setting up a miniature test network that duplicates, on a much smaller scale, the behavior of the planned network. A test source is used to inject data into the network, and the data-transit times are measured. The results of these measurements then have to be extrapolated to represent the planned network.

In practice, the process of making sure that the test network duplicates the features of the real network is usually complex and can involve so much labor and equipment as to be considered unworkable. If measurements can only be made on the installed network, the following benefits can still be gained:

■ The results of the measurements can be used to predict how increases in load can be handled.

■ These results can also be useful in pinpointing the locations in the network where long delays are occurring. With this information, an organization can eliminate bottlenecks by upgrading appropriate network components.

Unlike the experimental methods, theoretical estimates of the delays do not require elaborate test networks that can be costly and time-consuming to construct. However, a rigorous, detailed calculation of network delays is at least as complex as setting up the test network.

A calculation of the delays theoretically imposed by a packet-switching network requires consideration of the path taken by data, the detailed working of the software in the node, the processor instruction speeds, the size of delays imposed by software design (such as the operating system), the interprocess communication facility (which handles internal data flow between the processes that usually comprise the code in a packet switch), and the communications line speeds.

The load on a network's components (nodes and communications lines) varies as users create and terminate sessions and enter input or receive output. Thus, the delays are variable and have to be estimated by statistical models rather than being calculated deterministically.

What's taking so long?

Network managers should be clear about the definition of network response time in their evaluation of performance. The response time perceived by a user at a terminal is a combination of several factors, specifically, the time involved in data handling by the terminal, the network transit time, and the host's response time. The first category can be taken by different users to either include or exclude functions such as clearing the screen, unlocking the keyboard, and filling the screen with new information from the host. The host's front-end processor can be considered, in some instances, part of the network, and in others, part of the host. Hence, the FEP response time might be taken as part of the response time associated with either the network or the host.

To isolate the response time of the packet-switching network and compare it fairly with response times of other communications services requires clear definitions of the components of response time as perceived by a user inputting data at a terminal and receiving a reply (a data message) from a host. Response time, in the generic, can be defined as the interval between the user initiating transmission of data through the network by pressing the appropriate key on the terminal and the reception at the user's terminal of the first data character of the reply from the host computer.

The simplest example is a typical session between an asynchronous terminal operating in interactive mode using the standard X.28 protocol and a host computer connected to a packet network. The first event would correspond to the user typing a carriage return to terminate a line of input. The second event would be the display on the terminal of the first character of the response to the line of input by the host computer.

As another example, consider the case where the user is operating in a mode where all input is echoed across the network by the host. This mode of operation is common with DEC host computers, such as VAXs and DEC System-10s and -20s. It is also used where asynchronous terminals are connected through a packet-switched network to external protocol converters that provide emulation of synchronous terminals on the connections to host computers. The protocol converters control all formatting of the terminal screens, so they need to examine each character of input before deciding on the appropriate echo.

In the second example, every character keyed in by a user can be regarded as effecting transmission through the network, and the response is usually the echo of the same character (but not always—it may consist of received data). The response time to the user is then defined as the interval between the input of a character and the display on the screen of its echo (or, in the minority of cases where an echo is not sent, the receipt at the terminal of the first character of the response).

A definition of response time to the user, or *user response time,* that is applicable to many network-planning situations should eliminate terminal-dependent times. Of course, such response time will include transmission times between the terminal and the first node in the network, so it will be dependent on the speed of this link. However, it is reasonable to regard this link as part of the network and therefore to include it. At any rate, as will be shown later, these times are small for the kinds of terminal line speeds that can be (and usually are) used nowadays.

User response time also includes the host and front-end response times. Hence, the part of the response time that is associated solely with the network, that is, the *network response time,* is the user response time minus the host and front-end times (Fig. 1).

The data entered by the user is inserted into one or more data packets and transmitted through the network to the host. Similarly, the response from the host is assembled into a number of packets and transmitted through the network to the terminal.

From the point of view of the network, then, the *network transit time from terminal to host* can be defined as the interval between when the user types the data-forwarding character and the host (or front-end processor) receives

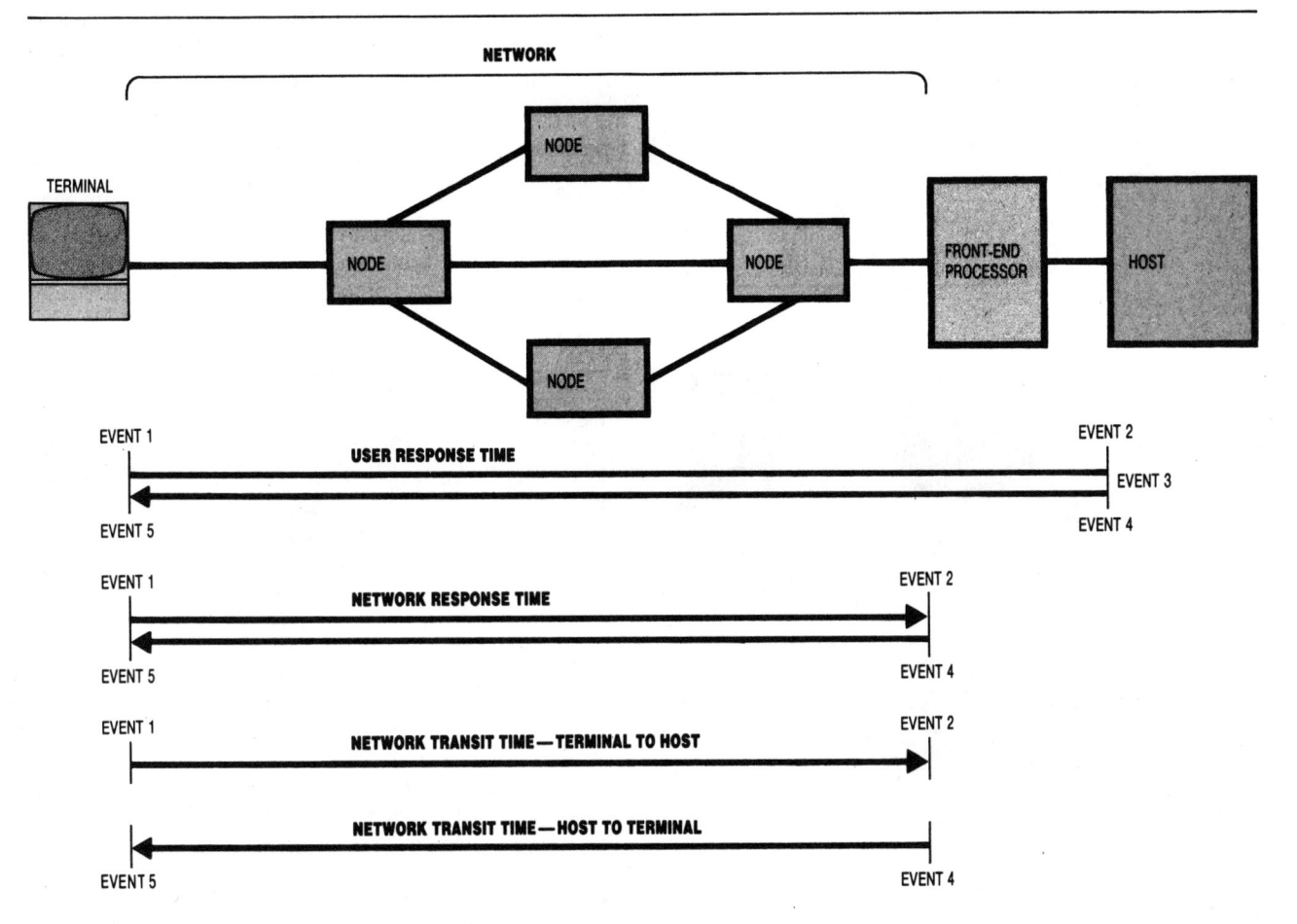

EVENT NO.	EVENT
1	LAST CHARACTER TYPED
2	PACKET RECEIVED
3	DATA PROCESSED AND RESPONSE TIME GENERATED
4	PACKET READY FOR TRANSMISSION
5	FIRST CHARACTER RECEIVED

1. Definitions of delay times. *Before users can measure or calculate response times, they must have clear definitions of the various types of delay that can occur in private packet networks. Network response time is composed of two unidirectional network transit times. However, response time to the user also includes the processing delays imposed by the front-end processor and the host.*

the complete network packet in which the data has been forwarded. Likewise, the *network transit time from host to terminal* is the interval between when the host (or front-end) has the first packet of the response ready for output on the network link and the first character of the packet is displayed on the terminal. The *network response time* is the sum of these two network transit times, covering both directions. This article focuses on transit-time calculation.

■ The basis of this method of predicting response time is to develop a model that describes the network's performance in handling data traffic.

A packet-switching network can be modeled as a *queuing system* (Fig. 2). In a queuing system, customers (data characters or packets) arrive and queue for service provided by one or more servers (nodes or communications lines).

Queuing systems have been studied in great detail, and models have been developed to predict their behavior. A queuing model can be classified as type A/B/m, where A is the probability distribution of arrival intervals (in this case, of data characters or packets), B is the probability distribution of service times, and m is the number of servers.

Most of the calculations that have been done to date use a single-server model. The calculations for multiple-server models are extremely complicated. For this reason, it is not feasible to consider applying them to network-design problems.

Multiple-server situations can arise under various circumstances. For example, a network may perform load balancing and route successive packets to the same destination by different paths, depending upon the load at the time. Thus, each path can be considered a separate

331

2. Packet nets and queues. *The model of a queuing system (A) can be used to study data traffic, either at the level of the network (B) or of the individual node (C).*

(A) QUEUEING SYSTEM

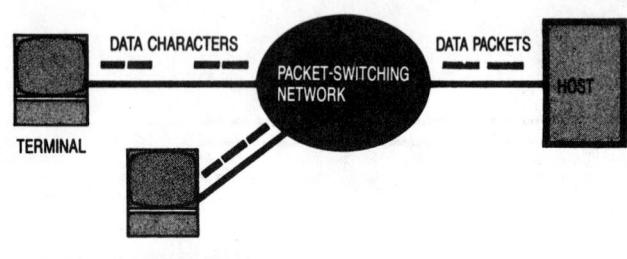

(B) PACKET-SWITCHING NETWORK

(C) PACKET-SWITCHING NODE SERVICING DATA PACKETS

server. Alternatively, each point at which a routing decision is to be made may be regarded as a multiple-server node.

In practice, however, while many networks use dynamic adaptive routing, the vast majority of packets between a source-and-destination pair will follow the same path. Therefore, it is reasonable to regard this situation as one that involves single servers at each point.

A multiple-server situation also arises in nodes with multiprocessor architectures, where the processors are dynamically assigned to process arriving packets. Similarly, there may be multiple communications lines between a pair of adjacent nodes. In either situation, the following approximations may be used to convert to a single-server model:
- One can divide the load proportionally between the servers (nodal processors or communications lines).
- One can aggregate the processing or transmission capacity of the servers and consider them single servers. The first approximation has the effect of overestimating the delays, while the second underestimates them. The network manager must decide for each individual case which option provides the better approximation.

All calculations to be made below assume a single-server model, implying that m in the A/B/m notation of the queuing model is 1. The next step is to decide on the distribution of arrival times of data packets (A in model) and the distribution of service times for node processing and line transmission (B). These quantities determine the type of queuing model to be used.

There are two special distributions for which queuing models have been well developed. One is an exponential distribution, that is, one where the probability of arrivals or the value of service times decreases exponentially as the time increases. The other is where the service times are all the same.

It is reasonable to expect that the second case is a good approximation of the service aspect of a pure packet-switching process. In such a process, the service time should be independent of the packet length, since the task of a node simply consists of examining a fixed-length packet header and rerouting the packet. On the other hand, the transmission time of a packet will be linearly proportional to the packet length. The distribution of transmission times will therefore be the same as that of packet lengths. This also applies to packet assembly and disassembly. These distributions often consist of a cluster around some value, which represents the most common packet size, with roughly equal and diminishing probabilities on either side of this value.

Spreads of this kind typically follow Poisson distribution. Data-packet arrivals at a node can be regarded as independent events. In this instance, the representation of arrival intervals by an exponential function is justified.

The exponential distribution is denoted by M (for Markovian); the distribution with fixed times is known as D (for deterministic). Thus, the M/M/1 model is appropriate for transmission and packet assembly/disassembly times, while the M/D/1 model is suited for modeling the pure packet-switching times between nodes.

Standard formulas

The formulas that these models provide for predicting service and queuing-time distributions are derived from queuing theory. Most are taken from the standard text in this field, Leonard Kleinrock's *Queueing Systems*. In one instance, the formula was derived from an extension of the standard formula using appropriate approximations.

Two kinds of parameters are usually calculated. The most common is the mean value of the relevant quantity (here, the delay time). The other is the so-called 95 percent confidence value of the same quantity. This value represents a limit that the distribution predicts will not be exceeded 95 percent of the time.

Let A = the mean arrival rate and S = the service rate. For an M/M/1 model, the following equations apply:
- Mean service time = $1/S$.
- Mean queuing time = $A/S[(S-A)]$.
- Mean time in queuing system = $1/(S-A)$.
- The 95 percent confidence queuing time = $1/(S-A) \ln(20A/S)$.
- 95 percent confidence time in queuing system = $1/(S-A) \ln(20)$.

For an M/D/1 model:
- Service time = $1/S$.
- Mean queuing time = $A/[2S(S-A)]$.
- Mean time in queuing system = $(2S-A)/[2S(S-A)]$.

An exact formula for the 95 percent confidence queuing time is very difficult to derive, but one can derive approxi-

mations for it in two common cases, as follows:
- When the traffic loads are light and A is much less than S, the 95 percent confidence queuing time = [A ln (20 A/S)]/[2.449(S − A)S].
- When the traffic increases to the point where A is comparable to S, a good approximation of the 95 percent confidence queuing time = [S ln (20)]/[2A(S − A)].

The above formulas represent a fundamental set of equations that can be used to calculate transit times for a packet-switching network. The major outputs are two critical quantities, the mean transit-time delay and the limit for the delay that is obeyed by 95 percent of the traffic.

To calculate response time using the formulas listed above, it is necessary to determine the arrival rates, A, and the service rates, S, for the components of the network. A network consists of nodes and communications lines. The lines connect the nodes with each other and with the terminals and host computers.

The arrival rates will be the data-traffic loads on the nodes and lines. The service rates will be the data-switching and transmission capacities of the nodes and lines. A suitable common unit is needed in which to express these rates; wherever possible, the unit "data packets per second" is used. Note that overhead packets generated by the network are excluded.

A typical network may use several different kinds of nodes, each with a different throughput. Also, the links between terminals, nodes, and hosts will be of different speeds. The service rates are functions of node throughput capacities and line transmission capacities.

There are two basic functions to be performed by the packet-switching network in linking terminals and hosts. One is the assembly of the data stream from the terminals into packets and the corresponding disassembly of received packets into character strings that are sent to the terminals. The other is the switching of the packets.

Some network vendors have implemented these two functions in different types of nodes, known respectively as packet assembler/disassemblers (PADs) and switches; others have a single node type that performs both the PAD and the switching functions (Fig. 3). The calculations below assume separate PADs and switching nodes, but the calculations for the integrated PAD-switching nodes only represent a minor change.

Line capacities and protocol overhead

Communications link speeds are specified in bit/s. Standard speeds in these units must be translated into units of data packets per second, if possible. To do so, the overhead associated with the protocol that is used across the link must be taken into account.

The packet-switching technique results in the generation and transmission of overhead packets that do not include any data. For example, there are acknowledgment packets for data exchange and packets to set up and terminate sessions. These extra packets must be factored into estimates of node performance.

The amount of overhead depends on the communications protocols used. For links between PADs and switch-

3. Implementations. Some packet networks separate the switches from the packet assembly/disassembly (A) while others are built around integrated units (B).

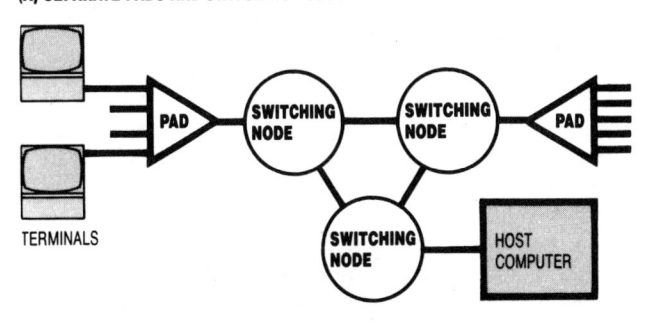

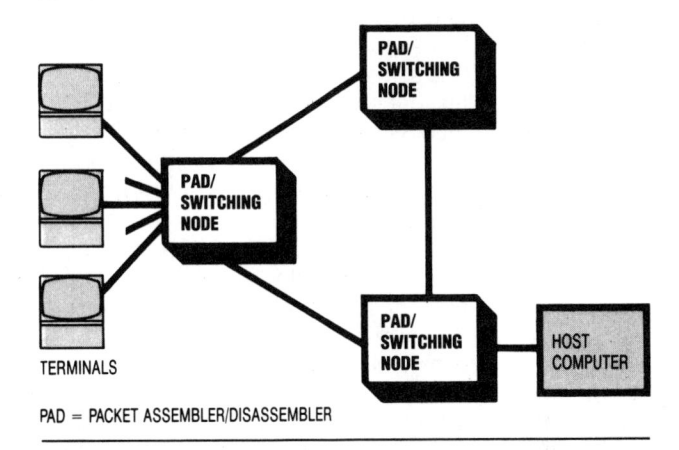

(A) SEPARATE PADS AND SWITCHING NODES

(B) INTEGRATED PAD/SWITCHING NODES

PAD = PACKET ASSEMBLER/DISASSEMBLER

ing nodes, X.25 is most commonly used. However, for links between switching nodes, a proprietary protocol, the details of which are unknown, is often used. In some networks, the proprietary protocol is an extended version of X.75, which therefore closely resembles X.25. In the absence of information on the internal protocols, it is usually reasonable to use X.25 as a fair approximation.

X.25 has a data link layer and a packet layer. The data link layer deals with frames as units of data transmission and sets up a link between two adjacent nodes. The packet layer creates virtual circuits and is responsible for transmission of the data input by terminals and hosts.

The analysis ignores the overhead in initiating links (frame level) and in setting up and clearing virtual circuits (packet level), since these operations occur infrequently compared with transmission of data. The major overhead considered is in the acknowledgment scheme. Here, different considerations apply at the frame and packet level of the X.25 protocol implementation, as follows:
- At the frame level, when data traffic volumes are such that they can fully utilize a link between two nodes in both directions, all acknowledgments can be sent implicitly, as part of the information-frame headers, rather than as explicit acknowledgment frames. This represents a theoret-

ical situation rather than a practical one, since the queuing delays for the link would not be acceptable with full utilization. However, the service rate in the queuing model is the theoretical maximum. Thus, for the purpose of queuing-model calculations, the case of no-explicit-acknowledgments does represent the service capacity of the link.

■ At the protocol's packet level, acknowledgments, in the form of receiver-ready (RR) packets, have to be sent on each logical channel to provide flow control. In theory, the RR packets can also be sent implicitly, although this can only happen if there is sufficient data traffic in the reverse direction on that logical channel. In practice, much more traffic usually flows from a computer to a terminal than in the opposite direction. The analysis uses the pessimistic assumption that an RR packet is generated (in the reverse direction) for every data packet.

It is difficult to treat this situation rigorously with the queuing model. There are now two distributions: In the case of data packets, the exponential distribution of lengths is reasonable, while all RR packets are of the same length. Both are served by the same communications line, but the M/M/1 model applies to the first and the M/D/1 to the second. There is a further complication in that the RR packet overhead is in the opposite direction on the link.

This situation can be handled as follows: A full-duplex communications line may be viewed as two independent servers, one for each direction of traffic flow on the line. When performing queuing-model delay calculations, network transit times are usually computed in a particular direction across the network—terminal to host or host to terminal. For each communication line in the path, only the traffic flowing in the corresponding direction is considered.

Such traffic consists of two sets of packets: data packets and RR packets that acknowledge data packets coming from the opposite direction. It is a reasonable approximation to consider these two sets to be a single set, with an exponential probability distribution of path lengths, provided the mean path length is calculated over the two sets, taking into account their individual mean lengths and occurrence rates.

The arrival rate, A, is now the sum of the data-packet and the RR-packet rates. The service rate, S, is given in units of packets per second by dividing the line's transmission rate by the new calculated mean packet length. Note that the service rate here is no longer in data packets per second, since overhead packets are being included. The M/M/1 model is used to calculate queuing times. However, service times are calculated from the data-packet length (not the mean packet length) and the line transmission rate.

(These rather complicated calculations are made somewhat clearer by the two examples that will be presented in the follow-up article.)

Just as we have defined the service rates, S, of the communications lines, we need to define the service rates of the other servers involved, that is, the network nodes. While communications line speeds are well defined and corresponding capacities can be determined, with node throughput the situation is much less certain. In virtually every instance, we are dependent on the manufacturer's own rating of the capacity of the equipment.

Usually, the manufacturer states the node's capacity in packet/s, but no information is available on how overhead should be taken into account in computing *data* packets per second. The X.25 standard, quite properly, does not specify the way in which the protocol is to be implemented. The relation between rated capacity and true data packets per second is not the same for all manufacturers but, rather, depends on the internal implementation.

For the purposes of the queuing-model calculations, we need to make the best estimate of the relationship between the manufacturer's rating in packet/s and the true throughput of the equipment in terms of data packet/s, giving the model's service rate.

To enable independent verification, it is useful to examine how these service rates can be determined. Basically, this can be done by one of the following methods:

■ A theoretical calculation of the time spent in known processing tasks in the node. Here, a detailed knowledge of the design and structure of the communications and other software (such as the operating system) in the node is required. However, network vendors are usually not willing to release the source code of their software.

■ An empirical measurement of node throughput under controlled conditions. This method is usually the more feasible one. The basic approach is to test a single node in isolation, with a test source and sink of data traffic. A single unit could, in theory, be used as both the source and sink of data. The input rate from the source would be varied and the arrival rate at the sink measured to determine the maximum sustained throughput of the node under these ideal controlled conditions. The source and sink of traffic could be either protocol analyzers or specially programmed microprocessor units, but they must be capable of generating and receiving sufficiently high rates of traffic to adequately test the node.

The calculations of the achievable service rates are worth doing if the information is available. They should certainly be done by the network vendors. The results are useful for two reasons:

■ If they differ significantly from the measured rates, this may indicate a design fault—for example, loops in the code or poor scheduling by the operating system. Such faults can be corrected to produce major improvements in performance.

■ Even if no design faults are discovered, calculations of theoretical performance will point out the existence and location of major sources of delay between the various tasks being performed by the node. This will allow programmers to accurately focus their tuning efforts.

Without detailed knowledge of a particular implementation, we can still make some estimates of service rates by considering the functions to be performed by a node and the time taken to perform them. A node in a packet-switching network can be expected to perform the following functions:

■ Terminal input/output (I/O).

- Reception and transmission of packets (actually frames).
- Assembly and disassembly of packets.
- Switching of packets.
- Synchronous terminal protocol support.
- Asynchronous-to-synchronous terminal protocol conversion.

In addition, the communications software in the node can be expected to include the following functions:
- Network management, including accounting, fault reporting, and assistance with diagnosis.
- Operator interface for control.
- An operating system that controls the operation of all the other software components and which may range from being merely a primitive scheduling mechanism to a sophisticated real-time executive.

Some framework is needed within which to estimate the performance of the network-node software. In early implementations, packet-switching network nodes were invariably programmed in assembly language. The trend nowadays is to use a high-level language such as Pascal or an intermediate-level language such as C. We choose to estimate the number of lines of code in a high-level programming language executed in performing a particular function. We call these units high-level language instructions (HLLIs).

Our strategy is to estimate node performance based on the way the node is probably programmed. To that end, we assume that common contemporary programming practices have been followed by the vendor. We further assume that the code is running on a typical commercially available microprocessor, the node's hardware basis. The instruction times of such a microprocessor allow us to estimate the time required to perform each function. We can expect an HLLI to be translated into about five machine instructions.

Typical execution time for one machine instruction is one microsecond (s). Hence, an HLLI can be expected to take about 5 s. Conversely, the node's central processor can be expected to have a capacity of 200,000 HLLIs. However, the performance of the node as a whole may be greater, since node implementations using multimicroprocessor architectures are now becoming common.

Each of the node functions listed above may be accounted for in an estimate of node performance:
- Terminal I/O. Most node architectures attempt to off-load this function from the central processing unit (CPU) to one or more peripheral units, each with its own microprocessor(s). We therefore assume that the actual character-interrupt handling and the processing of local editing functions are not performed by the CPU and thus do not compete for CPU resources. We also assume that the limits on the node service rates are determined by the capacity of the CPU(s), since extra peripheral units can usually be added relatively easily.
- Packet transmission/reception. Again, most modern node architectures off-load this function to intelligent line-interface boards. Chips that implement the link-level protocol are available. We assume that the processing involved in the framing and transmission of packets does not impose a load on the central processing resources and that the times involved can therefore be neglected in this estimation of service capacities.
- Packet assembly/disassembly. The node has to take the received data stream, create a packet header, and route it to the switching task (or to the switching node). The packet has to be enveloped in a Link Access Procedure-Balanced frame, although some of this may be performed by specialized hardware. We assume that, at a minimum, the hardware performs checksum calculation and verification. If no copying of data is performed, we estimate that the data can be handled in 200 HLLIs.

In practice, the copying of data between buffers is likely to occur at this level, since the terminal input/output process is likely to utilize different buffers than those employed in the packet-assembly and switching processes. We assume only one copy is performed—more than one copy strongly suggests the need for a redesign (which is not to say that this condition will never be encountered with products on the market).

Now things get tricky. A buffer-management scheme, which may be necessary to conserve memory, might have to check buffer boundaries through the copy loop, leading to about 10 HLLIs for each data character. There will be extra overhead at buffer boundaries and at the start and end of the copy. (Readers may be wondering why 10 HLLIs are estimated. They are used for copying a character, advancing each buffer pointer, and checking each buffer pointer for end-of-buffer. The end-of-buffer will require the program to follow buffer links. At copy start and end, buffers will have to be queued and dequeued.)

We estimate that it would require about 1,000 HLLIs to assemble or disassemble an average 50-byte packet that is copied once. Given a processor rating of 200,000 HLLIs as estimated above, a dedicated PAD could handle roughly 200 assemblies/disassemblies of 50-byte packets per second, yielding a total throughput of about 10,000 characters per second.
- Switching of packets. The node has to get a packet from an input queue, scan the packet header, make a routing decision based on the destination address, and insert the packet in an output queue. Dequeuing and queuing is done by changing the values of the appropriate pointers.

Since this is the central function performed by the network nodes, we can assume that the product supplier has put major effort into designing and coding this efficiently. We estimate that this function would be performed in 20 to 100 HLLIs, depending on the sophistication of the routing implementation. A switching node could therefore handle 2,000 to 10,000 packets per second, assuming there was no other overhead.
- Protocol support and conversion. These functions vary widely in their requirements, depending on the sophistication of the protocol used. Consequently, it is difficult to make any estimates that have general applicability.

The protocol functions may be performed in a dedicated PAD-type node or combined with the PAD and switching functions in a single node type. In the former case, extra delays will be introduced in the communication links between the protocol-handling nodes and the switching nodes. In the latter, the raw switching capacity of the nodes will be reduced considerably.

- Other functions. All nodes will participate in the network's management scheme, which usually requires them to gather statistics, respond to probes, and report line outages and other error states. The nodes will also usually provide an operator interface, which can generally be neglected in capacity calculations, since the relatively low level of traffic will not have a significant impact on capacity.

The overhead imposed by the operating system is more important. In practice, a packet-switching node is implemented as a series of processes that require scheduling and interprocess communication. The overhead for these facilities has not yet been taken into account.

It can be assumed that this overhead will roughly halve the capacity estimates calculated thus far. However, if these operating system facilities are implemented in a general way and all the node's software modules are required to use them without recourse to more optimized routines, the overheads may be much higher. Theoretically achievable throughput may be reduced by as much as a factor of 10.

To summarize: Highly optimized, dedicated packet-switching nodes can have capacities in the range of 2,000 to 10,000 packets per second. Overhead caused by the operating system, network management, and other functions can reduce node capacity to 1,000 packets per second or less. A crude estimate of PAD capacity is on the order of 10,000 characters per second.

These estimates, along with principles from queuing theory, can help designers properly allocate network resources. In a coming issue, two detailed examples will illustrate the use of these estimates and principles in designing private packet networks. ∎

Joseph Fernandez is a senior consultant with Logica, an international computer and communications consulting company. He has worked in the field of computer communications for 13 years. Prior to joining Logica, he was a senior research scientist in the division of computing research of CSIRO, the Australian government's research organization. He has a B.Sc. from Makerere College in Uganda and a Ph.D. in physics from Washington University in St. Louis.

David E. Liddy is a consultant with Logica. He has a B. Sc. from LaTrobe University in Melbourne and a Ph.D. in applied mathematics from the University of Sydney. Since joining Logica, he has specialized in the analysis of packet-switching networks and is currently developing a set of computer programs to aid in this task.

337

Harrell J. Van Norman, NCR Corp., Dayton, Ohio

A user's guide to network design tools

Selecting the most cost-effective design from an overwhelming number of options—complicated by ever-changing tariffs—mandates the use of a network design tool.

How does your network grow? Accommodating the expansion and complexity of interconnected networks and accurately forecasting networking requirements is a difficult task. Can your present network configuration adequately handle an increased load? And if so, how much? Are hardware changes required to support additional networking requirements? What part of your hardware configuration should be changed so that resources and services can be shared over the network? How reliable is the network and how will it recover from disaster?

These are just some of the questions facing network managers today. And there are others: How does a data communications manager, faced with expanding networking requirements, select from the many options available to plan and implement an effective network? Do you do networking analysis and design by yourself or depend upon hardware vendors, communications carriers, and consultants for guidance and direction? Are PC-based packages able to provide viable network solutions, or do they need to be tied to more computational power to be effective?

Amplifying the complexity of the situation is the ever-increasing number of vendors offering a wide variety of products, including terminals, protocols, architectures, hardware devices, and network control programs. Amid this proliferation of networking "solutions," it is easy to feel overwhelmed.

Addressing these challenges is often done on an as-needed basis, resulting in network evolution. Networks developed in an evolutionary manner provide incremental solutions to an array of individual networking requirements. With each new requirement, a determination is made for the most cost-effective configuration of equipment and commmunications lines, focusing on each specific need. Design decisions are often made without rigorous analysis of the entire network, since they only address the incremental expansion. This approach seldom provides the best network design for the total communications environment.

Another method—the preferred one—of addressing network requirements is through a structured-design approach. Here, alternative design techniques for the entire network are evaluated in terms of response time, cost, availability, and reliability. While avoiding excessive redesign and allowing for orderly capacity expansion, the process of network engineering evaluates current networking requirements and plans for future network expansion.

Figure 1 represents the total network engineering process. This process begins when you become aware of your current networking requirements and future networking needs and involves monitoring trends in the data communications industry. The trends help determine which state-of-the-art equipment, protocols, and architectures could be successfully used.

An accurate understanding

The process of network awareness includes collecting measurements of current data traffic, taking an inventory of networking equipment, forecasting requirements for network growth, and qualifying the criteria used to evaluate operational trade-offs. Many of these inputs are, by necessity, qualitative or logical in nature; others can be quantified with varying degrees of accuracy. Using a network design tool often requires a greater level of network awareness than most network managers generally possess. Once managers have an accurate understanding of the network—including possible bottlenecks, response-time spikes, and traffic patterns—they can begin the process of network optimization.

Network optimization includes applying network design tools, evaluating alternative network designs through a cost/performance break-even analysis, and optimally

1. Engineering the network. *The design methodology includes network awareness, optimization, and management in a cyclic process of iterative refinement.*

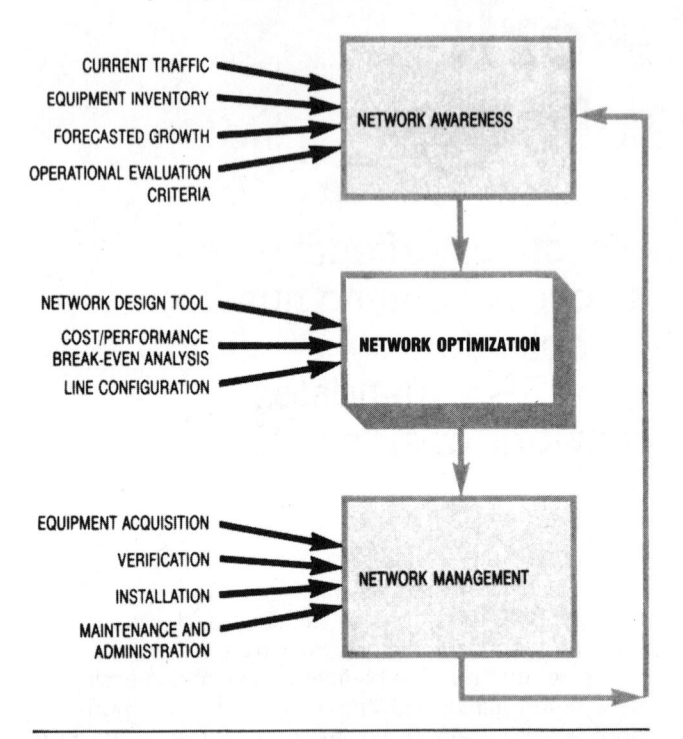

```
CURRENT TRAFFIC ────┐
EQUIPMENT INVENTORY ──→ NETWORK AWARENESS
FORECASTED GROWTH ──┘
OPERATIONAL EVALUATION
         CRITERIA

NETWORK DESIGN TOOL ──┐
COST/PERFORMANCE      ├──→ NETWORK OPTIMIZATION
BREAK-EVEN ANALYSIS   │
LINE CONFIGURATION ───┘

EQUIPMENT ACQUISITION ──┐
VERIFICATION            ├──→ NETWORK MANAGEMENT
INSTALLATION            │
MAINTENANCE AND         │
ADMINISTRATION ─────────┘
```

configuring communications lines. Since there is a nearly infinite number of ways to configure a network and interconnect transmission components, it is not realistic to exhaustively evaluate every possible option. However, by using certain shortcuts, called heuristics (assumptions that are not theoretically justifiable), the number of feasible designs can be scaled down to a manageable few.

Network design tools are developed using heuristic algorithms that abbreviate the optimization process greatly by making a few well-chosen approximations. After an optimized network design is produced, network designers will often use the tool to evaluate different scenarios of transmission components and traffic profiles. Through an interactive process of network refinement, *what if* questions may be addressed concerning various alternative configurations.

Once the network model is in place, proposals for adding or deleting resources (such as controllers, terminals, multiplexers, protocol converters, and remote front-end processors), redesigning application software, or changing routing strategies may be evaluated. Each of the various possible configurations will have associated costs and response times. This produces a family of cost/performance curves that form a solution set of alternative network designs. The final selection of an optimum configuration, based upon operational trade-offs, should be left up to the manager responsible for balancing budget constraints against user service levels.

Network management includes equipment acquisition, verification that the vendor has met the specification, installation, maintenance, and administration. This phase of the network-engineering process feeds information into the network-awareness component, which completes the network-engineering cycle and begins the iterations of design and refinement.

Statistical techniques are employed by network design tools to make certain assumptions about the behavior of a network and the profiles of transactions in the network. These assumptions—in terms of lengths, distributions, and volumes of input and output messages—make the mathematics easier, though usually at the expense of accuracy. The real-world network often violates, to a certain extent, many of these assumptions. Network management is where the optimized network designs meet the realistic world of actual network performance. Detected variances between optimized design and actual performance can provide an improved awareness of host processing requirements, traffic profiles, and protocol characteristics.

Improving network awareness is important throughout the iterations of design and refinement, which progress toward the optimum network. Network management is also where the evaluation of the operational criteria is refined. Selected cost/performance ratios are reviewed in terms of the operational criteria an organization is employing; a determination is made as to the current appropriateness of these ratios.

This all-important phase can avoid crises by doing network management in a preventive rather than reactive manner. The ratios are reviewed before the design is cast in concrete, a better strategy than waiting for design problems to surface that are attributable to poor cost/performance assumptions. To effectively evaluate proposed network design alternatives, one must have a method for predicting network performance and costs. There are four basic methods to forecast network design performance:

■ *Computer simulation.* Traffic is routed through a discrete-event simulator that captures the parameters of the proposed network design. These parameters include host- and transaction-processing times, queuing delays, and buffer capacities. Simulation can be incredibly accurate in timing an individual event—such as message-transit time. However, it generally requires days of processing time to execute. In addition, simulation usually requires specialized expertise to set up the problem, program the simulation, and interpret the results. Simulations overcome problems caused by simplified assumptions. But the cost is significant—not only in computer time but also in the time taken to gather and specify the data needed for generating meaningful network designs.

■ *Computerized analytic models.* Programs employ analytic equations for predicting performance of a design. These equations use statistics and queuing theory under the discipline of traffic engineering. Heuristics are employed to reduce the number of possible network designs to a manageable level. Because all possible configurations are not exhaustively evaluated, this is a potential source of difference between a truly optimal operational network and an optimized network design. Though the network may look

great on paper, it may not operate optimally.
■ *Engineering calculators.* Special-purpose calculators have the standard traffic-engineering equations and tables programmed into them.
■ *Manual calculations.* Here, special-purpose traffic-engineering tables and reference guides are used to determine theoretical performance.

Each of these methods has its benefits and attendant costs. They represent the spectrum, from the very complex and sophisticated to the rudimentary and error-prone.

Establishing the need

Except for networks with less than about 10 interconnections, it is impractical to compute networking calculations manually. If you do not employ a computerized network optimizer, you will not be able to do a very credible job. Attempting to do the work of a computerized optimizer by hand is somewhat presumptuous.

The requisite complex calculations have been incorporated in network-design software tools. However, no network-design program can produce truly optimal designs. This is due partly to imprecise and inaccurate input characteristics and constraints and also to simplifications taken in the mathematical calculations. Nevertheless, evidence does indicate that good heuristic algorithms do produce network designs that are quite close to optimal.

As powerful as network design tools are, they must be appreciated in relation to the total network engineering process. Using network design tools occupies only a small part of the entire business of putting a network together (Fig. 1). The network design process is iterative. The major divisions of the cycle are network awareness, network optimization, and network management. Exercising a network design tool is part of this cycle, not a complete means to an end.

Getting ready

Prior to using network design tools, the process of network engineering requires collecting measurements of current traffic and forecasting requirements for network growth. Deriving precise measurements of current traffic rates is often quite a challenging objective; accurately predicting future network requirements is even more daunting.

Using network design tools is as much an art as a science. Qualifying design considerations and providing accurate values for them is more subjective than objective. It is important to note that the precision of any network design tool is directly related to the correctness of the design criteria upon which the analysis is based. Before undertaking a network design, the designer must know the throughput requirements being placed on the network. Gathering precise data, which establishes these requirements, is perhaps the most difficult task of the predesign activity. Typical data required by network design tools include the following:
■ *Message and traffic profiles of each network application.*
  Lengths and distributions of input and output messages;
  Times and distributions for the host processing required;
  Numbers of application queries per whole transaction;
  Numbers of transactions for each site by application.
■ *Protocol characteristics.*
  Input/output frame size and overhead;
  Polling/select sequence;
  Maximum number of unacknowledged frames.
■ *Network environment.*
  Line speeds;
  Half- and full-duplex line disciplines;
  Input/output propagation-delay and processing times;
  Input/output modem clear-to-send delay times.

Could you provide the above data about your network's performance? Don't be alarmed—most data communications managers cannot. However, when using a network design tool, accurately defining the design criteria is crucial. This is why the integration of a network design tool into a network management mechanism that provides performance-monitoring capabilities is a more credible approach. Without the use of a network performance-monitoring device, much of the required design criteria will result in mere guesses. The end result: The optimization of the network will be proportional to the precision of the design criteria driving the tool.

While using the tool, the network engineer postulates *what if* scenarios through changing node and concentrator

'Precision of any tool is directly related to the correctness of its design criteria'

locations, observing the resulting shifts in traffic. Figure 2 illustrates the application of the network design tool (simplified representation) in the interactive refinement generally produced during an evaluation of various networking alternatives. Iterative improvements are made based on certain cost/performance criteria until the design process ends with an optimized network configuration. Each potential design is validated through a series of feasibility checks based on such constraints as response times, reliability, and availability.

After acceptable conceptual designs are developed, additional phases of network engineering include subsequent circuit reconfiguration, equipment acquisition, verification of specifications, installation, maintenance, and administration. This whole process cycles back with refinements in the network's quantifiable physical characteristics (such as throughput and message profile) and/or the qualitative, logical input constraints (including minimal response times, projected net costs, or reliability measurements).

Iterative improvements are made, evaluating each design composite of network costs and performance against the organization's operational criteria. This process continues until an optimized design is produced that satisfies networking objectives.

Twenty companies supplying network design tools were included in this review (see the table, "Suppliers of network design tools"). Since the focus is primarily on actual designers of software tools, several additional companies that market some of the identified network design tools were not included.

Several significant criteria need to be evaluated before selecting a network design tool suitable for specific environments and objectives:

- Primary objectives. There are two analysis objectives of network design tools—cost and performance. Some tools provide both capabilities; some focus exclusively on performance analysis. These capabilities are applied to three network types or classifications: data, voice, and integrated data and voice.
- Implementation method. There are two general categories of network-performance-prediction techniques: simulation and analytic. Some tools provide a hybrid of both methods; most tools focus primarily on one or the other.
- Processor base. The distinctions between microcomputers, minicomputers, and mainframes are becoming more and more vague with technological advances in VLSI (very large-scale integration) design. An attempt to classify network design tools on the basis of these three categories is provided.
- Source language. The language used to program the software packages that comprise the network design tools is identified. Across the gamut of network design tools, source languages include: Pascal, Fortran, C, Basic, Cobol, Modula 2, PL1, APL, LISP, and Simscript II.5.
- Business arrangements. The procedure required to obtain access to the network design tools is generally either through purchasing a license for the software or obtaining an on-line lease. (On-line, here, means accessing the tools via a timesharing service, for which the user is charged by the time interval.)
- Tariffs. Completeness and accuracy of the tariff database is an important consideration when evaluating network design tools that provide cost-analysis capabilities. The tariffs are changing regularly: Every week there are new filings with the Federal Communications Commission. Tool providers face a significant challenge to offer complete and accurate tariff data.
- Networks supported. A wide variety of network types is supported. They range from private to public, from point-to-point to multipoint, from virtual-switched to circuit-switched.
- Protocols supported. Many link-level protocols are automatically specified in the network design tools. Additionally, most tools provide the capability for user-customized or user-specified protocols.
- Maximum capacity of nodes. Most tools limit the size of the network designs they can handle—the limits are in the hundreds or thousands of devices. Some vendors claim "no maximum."
- Inputs. Network design tools require as input the characteristics of network components—both physical components (like modems, multiplexers, and terminals) and logical components (like performance and cost constraints).
- Outputs. The usefulness of output reports is determined by the understandability, flexibility, and applicability of the network design results.
- Graphics. Some tools support the capability for outputting results graphically. Several software packages allow inputs from graphics devices and display results via graphics and/or plotter devices.
- Help screens. All the network design tools assume some level of networking competence. Varying levels of help and tutorial functions are provided.

Additional factors

There are several other criteria that are not included in the table that should be considered when evaluating a network design tool. Chief among these is cost. However, cost is a difficult item to keep current. It should be obtained directly from the vendor representative listed in the table.

The old adage, "You get what you pay for," generally applies to network design tools. Mainframe-based tools usually provide more rigorous design and analysis and tend to cost an order of magnitude more than microcomputer-based tools.

The table identifies many suppliers of network design tools. (See "Other tools" for some of those omitted.) Many more companies market several of the tools listed. However, only the tool developers, generally, are listed. Provided below are descriptions of some of the major players among the listed suppliers.

(Network design tools are generally classified by the three network types to which they are applied: data and voice, data, and voice.)

Both data and voice

- Aries Group, Rockville, Md. A set of programs is available for the design of both voice and data networks. These tools are claimed to address all aspects of the design problems, such as queuing theory, topological optimization, data communications technology, and tariffs. The tools draw upon an on-line database that contains tariff rates, rate points, and rate structures. The tariffs are for services such as private line (leased) service, WATS (Wide Area Telephone Service), and DDS (Dataphone digital service), covering the range of tariffs that occur in North America. This appears to add up to a very complete tariff database, coupled with a good understanding of the tariff structures.

The tools also provide a variety of report formats for presenting network design results. The tools may be

2. The refining process. Each progressively improved design composite of network costs and performance is evaluated against an organization's operational criteria.

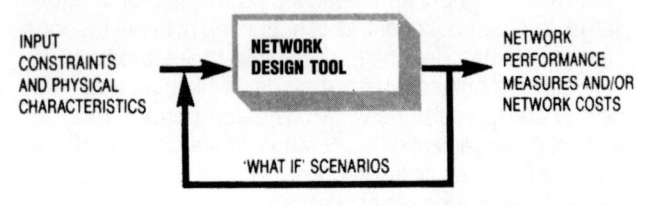

INPUT CONSTRAINTS AND PHYSICAL CHARACTERISTICS → NETWORK DESIGN TOOL → NETWORK PERFORMANCE MEASURES AND/OR NETWORK COSTS

'WHAT IF' SCENARIOS

> ### Other tools
> Three companies not included in the accompanying review are worth noting:
> - IBM, with its Snapshot and Topaz tools, used internally. IBM refused to release any information on its network design tools.
> - CCMI (Delran, N. J.), with its Q-TEL 900 tariff database. This product is used by most tool suppliers to track the ever-changing world of telephone company rates and pricing structures. This tariff information is essential to developing network designs. CCMI does not provide a network design tool.
> - MGT (Waltham, Mass), with its SNA (Systems Network Architecture) Backbone Analyzer. The company had, at one time, intended to market this tool. MGT is presently restricting its use to the company's network-design consulting.
>
> Note that there are many communications consultants possessing and applying network design tools for their internal consulting services, similar to MGT. Those tools are not reviewed in this article.

accessed either by purchasing a license for the software or by obtaining an on-line lease.

■ *AT&T, Bedminster, N. J.* The Enhanced Interactive Network Optimization System (E-INOS) is a tool used by AT&T to design communications networks for its customers as a value-added part of new products and services. E-INOS is a family of interconnected design tools used for voice, data, and integrated voice/image/data networks. This analytic modeling tool has an Integrated Transport Network module to design networks that combine multiple circuit requirements (both data and voice) into common telecommunications facilities.

E-INOS is a network-design package that supports AT&T's customer base. Automated data collection capabilities are provided in the design of voice networks through an interface using SMDR (station message detail recording) tapes. Since this is a tool internal to AT&T, the tariff database for AT&T rates is both current and accurate. Extensive graphics capabilities is another tool feature. AT&T says it is committed to making E-INOS an industry leader in network design and administration. Network designs using E-INOS are a free service provided to AT&T customers through regional customer response centers.

■ *John W. Bridges & Associates, Lewisville, Tex.* The Hybrid Network Design System (HNDS) is a family of integrated design tools addressing voice, packet data, and multipoint data networks. HNDS allows the user to modify theoretical network designs to accommodate real-world situations. In the voice arena, combinations of virtual- and circuit-switched services include foreign-exchange services and 16 different types of WATS services; backbone designs for multinode networks; and complete tariffs for AT&T and OCCs (other common carriers).

Integration of voice and data is handled by HNDS at the transmission level. User screens consider combinations of voice circuits, voice-grade data circuits, DDS-type data circuits, and T1 facilities. HNDS is a hybrid-network design tool with multivendor, multiservice, and multi-application capabilities.

■ *Contel Business Networks, Great Neck, N. Y.* The Modular Interactive Network Designer (MIND) is an integrated set of software tools that design and analyze voice, multipoint data, and packet data networks. MIND-Data is an interactive design tool for private-line, multipoint data communications networks. MIND-Packet addresses the design and analysis of packet-switched or distributed mesh networks. MIND-Voice is used to design least-cost voice networks. MIND-Pricer is an on-line service providing up-to-date tariff information for both data and voice networks. The tools are generally accessed through an on-line lease arrangement.

■ *General Network Corp., New Haven, Conn.* CADnet is a computer-aided design tool using interactive graphics to design, analyze, and optimize voice and data networks. It is targeted for companies with large networks where annual expenses for communications technology exceed $1 million. CADnet permits dynamic simulation modeling of evolving configurations, which includes both network equipment and transmission costs. This is an integrated software tool with a modular architecture that allows users to customize capabilities to their level of networking needs. CADnet accommodates the following networks: voice, centralized and distributed data, packet, local area, and hybrid. CADnet may be accessed through either purchasing a license for the software or obtaining an on-line lease.

■ *Telco Research Corp., Nashville, Tenn.* The Network Pathfinder is a set of tools for designing and refining voice, data, and integrated voice/data networks. Three menu-driven software modules can be used separately or together to answer *what if* questions, assist in determining optimal network changes, improve service, and reduce costs.

The Voice Pathfinder tool provides multinode network design capabilities, including backbone topology, line sizing, and T1- and virtual-network applications. The Exchange Carrier Optimizer II tool provides least-cost design of long-distance services at a single location. The Data Pathfinder tool has an interactive graphics design interface to configure multidrop, multiplexed, and public packet-switched networks.

Integrated voice/data network designs are performed automatically using both the Voice Pathfinder and Data Pathfinder. The tools support error recovery and provide reporting capabilities, including on-line graphics. Access to Pathfinder is obtained through purchasing a license for the software.

Data only

■ *BGS Systems, Waltham, Mass.* Two basic tools, Best/1-SNA (Systems Network Architecture) and Capture/SNA, are the main components of the capacity- and performance-management products. These two tools can operate in an integrated fashion or on an individual basis. Best/1-SNA is an interactive software tool that describes, evaluates, and predicts the performance of the various

Suppliers of network design tools

COMPANY NAME	PACKAGE NAME	PRIMARY OBJECTIVE	IMPLEMENTATION METHOD	PROCESSOR BASE	SOURCE LANGUAGE	BUSINESS ARRANGEMENTS
ACKS COMPUTER APPLICATIONS INC. 1213 WILLOWBROOK DR. CARRY, N.C. 27511 Contact: Wushow Chou (919) 467-7279	ACKS/TOPS (TOPOLOGICAL OPTIMIZATION AND PERFORMANCE SIMULATION)	COST AND PERFORMANCE DESIGN, ANALYSIS, AND OPTIMIZATION	ANALYTIC FOR TOPOLOGY SIMULATION FOR PERFORMANCE	MAINFRAME	FORTRAN	PURCHASE LICENSE
ARIES GROUP INC. 1500 RESEARCH BLVD., SUITE 320 ROCKVILLE, MD. 20850 Contact: Bob Ellis (301) 762-5500	DATA FAMILY: ANALINE, ANXUR, DEXUS, ODEN, QUEUE VOICE FAMILY: JUNO, PROCTOR, QUANDEM	NETWORK COST AND PERFORMANCE EVALUATION, AND CIRCUIT OPTIMIZATION	ANALYTIC	MAINFRAME QUEUE ON MICROPROCESSOR	APL	ON-LINE LEASE OR PURCHASE LICENSE
AT&T COMMUNICATIONS BEDMINSTER, N.J. 07921 Contact: Don Uberroth (201) 234-7505	E-INOS (ENHANCED INTERACTIVE NETWORK OPTIMIZATION SYSTEM)	COST AND PERFORMANCE DESIGN, ANALYSIS, AND OPTIMIZATION OF DATA, VOICE, AND INTEGRATED VOICE/IMAGE/DATA NETWORKS	ANALYTIC HEURISTIC STATISTICAL	MAINFRAME	C LANGUAGE	FREE SERVICE PROVIDED TO CUSTOMERS THROUGH REGIONAL CUSTOMER-RESPONSE CENTER
BBN COMMUNICATIONS CORP. 70 FAWCETT ST. CAMBRIDGE, MASS. 02238 Contact: Debbie Soon (617) 497-2479	DESIGNET	COST AND PERFORMANCE OPTIMIZATION OF DATA PACKET-SWITCHING NETWORKS AND VOICE NETWORKS	EXPERT SYSTEM APPLYING AI AND ANALYTIC TECHNIQUES	SYMBOLICS 3600 SERIES WORKSTATION	LISP	INTERNAL TOOL USED FOR CONSULTING WITH CUSTOMERS
BGS SYSTEMS 128 TECHNOLOGY CENTER WALTHAM, MASS. 02254-9111 Contact: Ottar Glenn (617) 891-0000	BEST-1	PERFORMANCE OPTIMIZATION OF SNA DATA NETWORKS	ANALYTIC	MAINFRAME (IBM AND COMPATIBLE)	FORTRAN	PURCHASE LICENSE

TARIFFS	NETWORKS SUPPORTED	PROTOCOLS SUPPORTED	MAXIMUM CAPACITY OF NODES	INPUTS	OUTPUTS	COMMENTS
THREE UNIQUE SETS OF TARIFFS: AT&T, BOCs, AND OCCs	POLLING POINT-TO-POINT MULTIPOINT PACKET SWITCHING HYBRID (POLLING/ PACKET SWITCHED) TERMINAL ACCESS BACKBONE	BSC SDLC/HDLC ASYNC USER DEFINED	100 TERMINALS PER LINE 40 TERMINALS PER CLUSTER 40 TRUNKS 200 LINKS	TRAFFIC/NETWORK PARAMETERS PERFORMANCE CONSTRINTS LINE TARIFF INFORMATION LINE SPEED CONSIDERATIONS HARDWARE CONSTRAINTS LINE CONFIGURATION LINKS SPECIAL VARIABLES MESSAGE TYPES, SIZES, NUMBERS LINK PROTOCOLS	TRAFFIC AND TERMINAL LOAD FOR POLLED/PACKET LINES TERMINAL RESPONSE TIMES CONCENTRATOR/MULTIPLEXER LOCATIONS AND AMOUNTS EARTH STATION BREAK-EVEN ANALYSIS PACKET DELAYS OPTIMAL ROUTING STRATEGIES GRAPHICS/HELP SCREENS	ENHANCED CUT-SATURATION ALGORITHM
INTERSTATE, INTRASTATE, INTERLATA, INTRALATA, ACCESS. AT&T, BOCs, AND OCCs. U.S. AND CANADA	POLLING POINT-TO-POINT MULTIPOINT PACKET SWITCHING TERMINAL ACCESS BACKBONE LATA STAR	BSC SDLC ASYNC HDLC USER DEFINED	5,600 TERMINALS	TRAFFIC PROFILES LIST OF LOCATIONS OPERATING ENVIRONMENT MESSAGE PARAMETERS CALL PATTERN, DISTRIBUTION, AND VOLUME TARIFF ENVIRONMENT CIRCUIT ORGANIZATION AND END POINTS EQUIPMENT OPTIONS LINK PROTOCOLS	TOPOLOGY RESPONSE TIMES PERFORMANCE MEASUREMENTS LEAST-COST ROUTING TOTAL NETWORK COSTS	COMPREHENSIVE TARIFF DATA GOOD PRESENTATION OF RESULTS INTO REPORT FORMAT GOOD CUSTOMER SUPPORT
CURRENT AT&T TARIFFS LOCAL EXCHANGE CARRIERS FCC NO. 1, NO. 2, NO. 4, NO. 7, NO. 9, NO. 10, NO. 11 BOCs	POLLING POINT-TO-POINT MULTIPOINT PACKET SWITCHING HYBRID (POLLING/ PACKET SWITCHED) TERMINAL ACCESS BACKBONE SATELLITE LANs	BSC SDLC/HDLC ASYNC USER DEFINED	5,000 TERMINAL CLUSTERS 100 FEP LOCATIONS 40 BACKBONE NETWORKS	TRAFFIC COMPOSITIONS EQUIPMENT TYPE AND LOCATIONS LINE SPEED AND LINK PROTOCOLS CRITICAL AND AVERAGE RESPONSE TIMES MAXIMUM LINE UTILIZATION MAXIMUM DROPS PER LINE LOCATION OF DEVICES IN BACKBONE CUSTOMER-SPECIFIC OPERATING CONSTRAINTS TARIFF ENVIRONMENT BACKBONE NETWORK CONSTRAINTS	LEAST-COST ROUTING USER RESPONSE TIME CONCENTRATOR/MULTIPLEXER LOCATIONS, AMOUNTS, TYPES TOTAL NETWORK COSTS LINE UTILIZATION NODE SIZING TOPOLOGY BACKBONE NETWORK CONFIGURATION ISDN BREAK-EVEN ANALYSIS GRAPHICS	ONE FAMILY OF INTERCONNECTED TOOLS PROVIDES VOICE, DATA, AND INTEGRATED VOICE/ IMAGE/DATA NET-WORK OPTIMIZATION GOOD DATA COLLECTION CAPABILITIES CURRENT TARIFF DATA GOOD GRAPHICS
AT&T NO. 9, NO. 10, NO. 11 INTRASTATE INTRALATA OCCs LECs INTERNATIONAL	X.25 PACKET SWITCHING BACKBONE POINT-TO-POINT MULTIPLEXED VOICE NETWORKS	BSC SDLC ASYNC X.25 BURROUGHS POLL SELECT DOD UNISCOPE	2,000 USER LOCATIONS 200,000 TERMINAL 1,700 HOSTS 200 PACKET-SWITCH LOCATIONS	PATTERN, VOLUME, AND DISTRIBUTION OF CALLS EQUIPMENT TYPES AND LOCATIONS RESPONSE TIMES RELIABILITY CONSTRAINTS	TOPOLOGY CONFIGURATION PACKET-NETWORK ACCESS CLUSTERING PACKET-SWITCHING NODE CLUSTERING BACKBONE CONNECTIONS NETWORK COSTS GRAPHICS	USEFUL IN DESIGNING LARGE NETWORKS GOOD EXPERIENCE IN PACKET SWITCHING, NETWORK DESIGN, AND IMPLEMENTATION
NONE	SNA NETWORKS POINT-TO-POINT MULTIPOINT	BSC SDLC	150 SUBAREAS 1,000 TERMINALS PER SUBAREA	TRAFFIC COMPOSITIONS RESPONSE TIMES AUTOMATIC DATA CAPTURE TRHOUGH 'CAPTURE/SNA' TOOL INTERFACE NETWORK DESCRIPTION LINK PROTOCOLS	RESPONSE TIMES UTILIZATIONS THROUGHPUT PERFORMANCE MEASUREMENTS OPTIMAL BUFFER SIZES LINE AND CLUSTER AVAILABILITY HELP SCREENS	GOOD PERFORMANCE PREDICTION FOR SNA NETWORKS GOOD DATA-COLLECTION CAPABILITIES THROUGH 'CAPTURE SNA' TOOL BOTH BOUNDARY NODE AND BACKBONE NETWORK ANALYSIS

SEE LEGEND AT END OF TABLE

Suppliers of network design tools (cont.)

COMPANY NAME	PACKAGE NAME	PRIMARY OBJECTIVE	IMPLEMENTATION METHOD	PROCESSOR BASE	SOURCE LANGUAGE	BUSINESS ARRANGEMENTS
CACI INC.-FEDERAL 3344 NORTH TORREY PINES COURT LA JOLLA, CALIF. 92037 Contact: Paul Gorman (619) 457-9681	COMNET II.5 NETWORK II.5	PERFORAMANCE ANALYSIS AND DESIGN OF COMMUNICAITONS NETWORKS	SIMULATION	MICROCOMPUTER, MINICOMPUTER, MAINFRAME	SIMSCRIPT II.5	PURCHASE LICENSE
CHLAMTAC ELECTRICAL & COMPUTER ENGINEERING UNIVERSITY OF MASSACHUSETTS AMHERST, MASS. 01003 Contact: Prof. Imrich Chlamtac (413) 545-0712	CONSIP II	PERFORMANCE EVALUATION AND OPTIMIZATION OF DATA COMMUNICATIONS NETWORKS AND PROTOCOLS	SIMULATION	MAINFRAME RUNNING ON UNIX OR VMS OPERTAING SYSTEM AND ON IBM PCs	PASCAL	PURCHASE LICENSE
CONNECTIONS TELECOMMUNICATIONS INC. 322 EAST CENTER ST. WEST BRIDGEWATER, MASS. 02379 Contact: Dick Armour (617) 584-8885	MNDS (MULTIPOINT NETWORK DESIGN SYSTEM) DNDS (DISTRIBUTED NETWORK DESIGN SYSTEM)	COST AND PERFORMANCE ANALYSIS AND OPTIMIZATION OF MULTIPOINT AND DISTRIBUTED NETWORKS	ANALYTIC HEURISTIC	MICROPROCESSOR	MNDS - BASIC DNDS - C LANGUAGE	PURCHASE LICENSE
CONTEL BUSINESS NETWORKS 130 STEAMBOAT RD. GREAT NECK, N.Y. 11024 Contact: Zvi Kozicki (516) 829-5900, Ext. 204	MIND (MODULAR INTERACTIVE NETWORK DESIGNER) DATA VOICE PACKET TANDEM INVENTORY PRICER DATA/PC	COST AND PERFORMANCE ANALYSIS AND OPTIMIZATION OF VOICE AND DATA NETWORKS	ANALYTIC HEURISTIC DISCRETE-EVENT SIMULATION FOR DATA PERFORMANCE	MICROPROCESSOR	FORTRAN	ON-LINE LEASE OF PURCHASE LICENSE
DMW COMMERCIAL SYSTEMS 2020 HOGBACK RD. ANN ARBOR, MICH. 48104 Contact: Neill Hollenshead (313) 971-5234	TOP (TELETRAFFIC OPTIMIZER PROGRAM) NODE/1 (NETWORK OPTIMIZATION DESIGN AND ENGINEERING)	TELEPHONE-SWITCH COST AND ROUTING OPTIMIZATION	TOP: SIMULATION, STATISTICAL NODE/1: ANALYTIC	TOP: MAINFRAME/ MINICOMPUTER NODE/1: MICROCOMPUTER	TOP: COBOL AND FORTRAN NODE/1: MODULA 2	PURCHASE LICENSE

TARIFFS	NETWORKS SUPPORTED	PROTOCOLS SUPPORTED	MAXIMUM CAPACITY OF NODES	INPUTS	OUTPUTS	COMMENTS
NONE	POINT-TO-POINT MULTIPOINT POLLING PACKET SWITCHED CIRCUIT SWITCHED VIRTUAL CIRCUIT LAN WAN SATELLITE	CSMA/CD, TOKEN PASSING, PRIORITY, FIFO USER DEFINED	NO MAXIMUM LIMIT	PROCESSING ELEMENTS DATA TRANSFER DEVICES DATA STORAGE DEVICES DATA TRANSFER RATES LINK PROTOCOLS ACCESS TIMES AND METHODS HARDWARE MODULE DESCRIPTION SOFTWARE MODULE DESCRIPTION	NETWORK THROUGHPUT, RESPONSE TIMES CIRCUIT-GROUP UTILIZATION QUEUE STATISTICS BLOCKING PROBABILITIES CALL-QUEUEING AND PACKET DELAYS UTILIZATIONS AND CAPACITIES WAITING-TIME STATISTICS PROCESSOR-LOADING STATISTICS NETWORK BOTTLENECKS HARDWARE TIMING CONFLICTS GRAPHICS; HELP SCREENS	APPLICATIONS PORTABILITY AND EXPANDABILITY (MICRO TO MAINFRAME)
NONE	POINT-TO-POINT MULTIPOINT PACKET SWITCHED RADIO CHANNELS	PROTOTYPES WRITTEN IN PASCAL	10 LINES 100 TERMINALS PER LINE 20 CONCENTRATORS 5 TERMINALS PER CONCENTRATOR	TRAFFIC COMPOSITIONS COMMUNICATIONS LINKS AND NODES TOPOLOGY (STATIC, RECONFIGURABLE, OR MOBILE NETWORKS) WORKLOADS PER NODE NETWORK CHARACTERIZATION PROTOCOL PROTOTYPES	PERFORMANCE EVALUATION	PROTOCOL ANALYSIS THROUGH PROTOCOL PROTOTYPES SOFTWARE EMULATION APPROACH TO SIMULATION, DOES NOT INVOLVE THE MODELING AND PROGRAMMING PHASES AVOIDS DIFFICULT PROBLEM OF MODEL VALIDATION SUPPORTS PROTOCOL TESTING
ALL PUBLISHED TARIFFS ON FILE WITH FCC NO. 9, NO. 10, NO. 11 BOCs OCCs LECs ACCESS AND TRANSPORT	POINT-TO-POINT MULTIPOINT POLLING PACKET SWITCHED TERMINAL ACCESS BACKBONE SATELLITE	BSC SDLC HDLC ASYNC USER CUSTUMIZATION USER DEFINED LAPB	30 HOSTS OR CONCENTRATORS PER NETWORK 270 TERMINAL CLUSTERS PER NETWORK; NUMBER OF TERMINALS LIMITED BY TURNAROUND TIME	TRAFFIC COMPOSITIONS EQUIPMENT TYPES AND LOCATIONS LINE SPEEDS RESPONSE TIMES CIRCUIT LENGTHS AND NUMBERS PATH DELAYS HARDWARE CHARACTERISTICS LINK PROTOCOLS	TOPOLOGY LEAST-COST ROUTING USER RESPONSE TIMES LINE UTILIZATIONS TOTAL NETWORK COSTS PERFORMANCE MEASUREMENTS GRAPHICS; HELP SCREENS	GOOD TRAINING AND CONSULTATION SERVICES
ALL PUBLISHED TARIFFS ON FILE WITH FCC NO. 9, NO. 10, AND NO. 11 BOCs OCCs LECs ACCESS AND TRANSPORT	POINT-TO-POINT MULTIPOINT POLLING PACKET SWITCHED TERMINAL ACCESS BACKBONE SINGLE NODE OR MULTINODE PBX VOICE NETWORK	BSC SDLC HDLC ASYNC USER DEFINED	2,000 TERMINALS 120 NODES IN PACKET-SWITCHED NETWORKS	TRAFFIC RATES AND MIXES EQUIPMENT TYPES AND LOCATIONS LINE SPEEDS RESPONSE TIMES CIRCUIT CONSTRAINTS NODE ADDRESSING LINK PROTOCOLS	TOPOLOGY LEAST-COST ROUTING USER RESPONSE TIMES TOTAL NETWORK COSTS PERFORMANCE MEASUREMENTS GRAPHICS; HELP SCREENS	GOOD SUPPORT OF TOOLS THROUGH CONSULTING SERVICES
AT&T WATS "MEGACOM" BOCs AND DDD MCI PRIM 1 AND 2 (MCI'S VERSIONS OF WATS)	SINGLE NODE OR MULTINODE PBX VOICE NETWORK	NOT RELEVANT	NO MAXIMUM	CALL DETAIL RECORDS TRUNKS ATTACHED TO PBX AVAILABLE ROUTING TRUNKS MAP TELEPHONE NUMBERS TO OPTIMAL ROUTING TRUNKS	TRUNK UTILIZATION NETWORK COSTS TRAFFIC DISTRIBUTIONS OVERFLOW OF TRUNKS	AVERAGE COST REDUCTION OF 10 PERCENT OR GREATER

SEE LEGEND AT END OF TABLE

Suppliers of network design tools (cont.)

COMPANY NAME	PACKAGE NAME	PRIMARY OBJECTIVE	IMPLEMENTATION METHOD	PROCESSOR BASE	SOURCE LANGUAGE	BUSINESS ARRANGEMENTS
ECONOMICS AND TECHNOLOGY INC. 101 TREMONT ST. BOSTON, MASS. 02108 Contact: Mary E. Lynch (800) 225-2496	OPTIMIZER PAK PRIVATE LINE PRICER OCC-PAK	COST ANALYSIS OF BOTH DATA AND VOICE NETWORKS	ANALYTIC	MICROCOMPUTER	C LANGUAGE	PURCHASE LICENSE
GENERAL NETWORK CORP. 25 SCIENCE PARK NEW HAVEN, CONN. 06511 Contact: John Walsh (203) 786-5140	CADNET (COMPUTER-AIDED DESIGN TOOL FOR VOICE AND DATA NETWORK OPTIMIZATION)	COST AND PERFORMANCE DESIGN, ANALYSIS, AND OPTIMIZATION OF DATA, VOICE, ISDN, AND HYBRID (PUBLIC AND PRIVATE) NETWORKS	ANALYTIC FOR TOPOLOGY SIMULATION FOR PERFORMANCE	MAINFRAME	PL 1	PURCHASE LICENSE ON-LINE LEASE
HTL TELEMANAGEMENT LTD. 3901 NATIONAL DR., SUITE 160 BURTONSVILLE, MD. 20866 Contact: Dr. Mike Hills (301) 236-0782	NTD (NETWORK TOOLS FOR DESIGN) NTD-1 LONG-DISTANCE ANALYZER NTD-4A WATS OPTIMIZER NTD-6, -8, -9 SWITCHED-LINE DESIGNER NTD-10 PRIVATE LINE PLANNER	COST AND PERFORMANCE ANALYSIS, DESIGN, AND OPTIMIZATION OF VOICE NETWORKS COST ANALYSIS OF DATA NETWORKS	ANALYTIC	MICROCOMPUTER	BASIC COMPILER	PURCHASE LICENSE
JOHN BRIDGES & ASSOCS. 1660 SOUTH STEMMONS, SUITE 450 LEWISVILLE, TEX. 75067 Contact: John M. Carroll (214) 436-8334	HNDS (HYBRID NETWORK DESIGN SYSTEM) HNDS VOICE HNDS PACKET HNDS MULTIPOINT	COST AND PERFORMANCE ANALYSIS, DESIGN, AND OPTIMIZATION OF BOTH DATA AND VOICE NETWORKS	ANALYTIC	MICROCOMPUTER NDGS VOICE HNDS PACKET HNDS MULTIPOINT MAINFRAME HNDS VOICE	C LANGUAGE A VARIETY OF OTHER LANGUAGES DEPENDING ON THE FUNCTION BEING DONE	PURCHASE LICENSE DESIGN SERVICES

TARIFFS	NETWORKS SUPPORTED	PROTOCOLS SUPPORTED	MAXIMUM CAPACITY OF NODES	INPUTS	OUTPUTS	COMMENTS
AT&T NO. 1, NO. 2, NO. 4, NO. 9, NO. 10, NO. 11 OCCs LECs BOCs INTERSTATE INTRASTATE INTERLATA	POINT-TO-POINT MULTIPOINT	NOT RELEVANT	NOT RELEVANT	NODE LOCATIONS LINE SPEEDS LINE TYPES	LINE COSTS LEAST-DISTANCE ROUTING LEAST-COST ROUTING MONTHLY AND NONRECURRING COSTS HELP SCREENS	GOOD CONSULTING SUPPORT
WATS, DDD, FX PRIVATE LINE SPECIAL ACCESS LOCAL SERVICE RATES INTERSTATE INTRASTATE INTERLATA INTRALATA AT&T, MCI, U.S. SPRINT ALLNET WESTERN UNION ITT BOCs LECs MEGACOM SDN	POINT-TO-POINT MULTIPOINT POLLING PACKET SWITCHED TERMINAL ACCESS BACKBONE SATELLITE SINGLE NODE OR MULTINODE PBX VOICE NETWORK HYBRID (PUBLIC AND PRIVATE) LAN WAN MAN	BSC SDLC HDLC ASYNCRONOUS USER CUSTOMIZED USER DEFINED X.25 ETHERNET DECNET BURROUGHS POLL-SELECT WILL ADD ANY REQUESTED	NO MAXIMUM	TRAFFIC COMPOSITIONS EQUIPMENT TYPES AND LOCATIONS LINE SPEEDS CIRCUIT LENGTHS AND NUMBERS RESPONSE TIMES CDR TAPE INPUT NETWORK ATTRIBUTES USER ATTRIBUTES	TOPOLOGY LEAST-COST ROUTING USER RESPONSE TIMES LINE UTILIZATIONS TOTAL NETWORK COSTS PERFORMANCE MEASUREMENTS CONCENTRATOR/MULTIPLEXER LOCATIONS AND AMOUNTS EARTH STATION BREAK-EVEN ANALYSIS PACKET DELAYS OPTIMAL ROUTING STRATEGIES NODE SIZING ISDN BREAK-EVEN ANALYSIS NETWORK RELIABILITY AND SURVIVABILITY PACKET-NETWORK ACCESS CLUSTERING GRAPHICS; HELP SCREENS	INTERACTIVE WITH GOOD GRAPHICS CAPABILITIES TARGETED AT FORTUNE 200 COMPANIES WITH $1 MILLION IN ANNUAL NETWORKING COSTS MOST COMPREHENSIVE POST-DIVESTITURE TOOL IN THE MARKETPLACE
AT&T WATS FCC NO. 1, NO. 2, NO. 4, NO. 9, NO. 10, NO. 11 OCCs LECs BOCs INTERSTATE INTRASTATE INTERLATA INTRALATA	POINT-TO-POINT MULTIPOINT POLLING PACKET SWITCHED CIRCUIT SWITCHED VIRTUAL CIRCUIT SATELLITE	NOT RELEVANT	500 NODES	NODE LOCATIONS TRAFFIC DISTRIBUTIONS TRAFFIC VOLUMES CIRCUIT LOCATIONS	COSTS CAPACITY UTILIZATIONS	GOOD VOICE NETWORK DESIGN CAPABILITIES GOOD CUSTOMER SUPPORT
AT&T WATS FCC NO. 1, NO. 2, NO. 4, NO. 9, NO. 10, NO. 11 OCCs LECs BOCs INTERSTATE INTRASTATE INTERLATA INTRALATA	POINT-TO-POINT MULTIPOINT POLLING PACKET SWITCHED CIRCUIT SWITCHED VIRTUAL CIRCUIT SATELLITE INTEGRATED VOICE/DATA HYBRID NETWORKS (PRIVATE/PUBLIC), (ACTUAL CIRCUIT/VIRTUAL CIRCUIT)	BSC SDLC ASYNC USER CUSTOMIZED	VOICE: 30 TANDEM-SWITCH NODES, 500 MAIN LOCATIONS PACKET: 45 DATA CONCENTRATORS MULTIPOINT: 45 REMOTE CONCENTRATORS	VOICE: AUTOMATIC DATA CAPTURE FROM SMDR TAPES OR MANUAL INPUT OF LOCATIONS IN NETWORK, TIME-OF-DAY DISTRIBUTIONS, TRAFFIC VOLUMES DATA: TRAFFIC COMPOSITIONS LINE SPEEDS PACKET DELAYS RESPONSE TIMES LINK PROTOCOLS	TRAFFIC SUMMARIES BACKBONE OPTIMIZATIONS VIRTUAL-NETWORK OPTIMIZATIONS POINT-TO-POINT CIRCUIT SUMMARIES END-TO-END DELAYS LEAST-COST ROUTING USER RESPONSE TIMES LINE UTILIZATIONS TOTAL NETWORK COSTS NETWORK RELIABILITY AND SURVIVABILITY COSTS PER UNIT GRAPHICS	GOOD CUSTOMER SUPPORT PACKET/PUBLIC BREAK-EVEN ANALYSIS

SEE LEGEND AT END OF TABLE

Suppliers of network design tools (cont.)

COMPANY NAME	PACKAGE NAME	PRIMARY OBJECTIVE	IMPLEMENTATION METHOD	PROCESSOR BASE	SOURCE LANGUAGE	BUSINESS ARRANGEMENTS
NETWORK SYNERGIES INC. 330 MAIN ST. LAFAYETTE, IND. 47901 Contact: John Becker 1-800-NET-1NSI	STAR (SIMULATED TELECOMMUNICATIONS ANALYSIS REPORTING)	COST OF PERFORMANCE DESIGN, ANALYSIS, AND OPTIMIZATION OF DATA NETWORKS	ANALYTIC	MICROCOMPUTER	PASCAL FORTRAN	PURCHASE LICENSE
PERFORMIX SOFTWARE CORP. 1711 LARKELLEN LA. LOS ALTOS, CALIF. 94022 Contact: Jim Grant (415) 969-3399	NET/PX (NETWORK PERFORMANCE EXPERT)	PERFORMANCE ANALYSIS OF DATA NETWORKS CAPACITY PLANNING	ANALYTIC	MICROCOMPUTER	C LANGUAGE	PURCHASE LICENSE
QUINTESSENTIAL SOLUTIONS (FORMERLY H & S ASSOCIATES) 1335 HOTEL CIRCLE SOUTH, SUITE 206 SAN DIEGO, CALIF. 92108 Contact: Ricki Mundhenke (619) 692-9464	NETWORK PERFORMANCE ANALYZER MODULE NETWORK PRICER MODULE CIRCUIT DESIGNER MODULE NETWORK DESIGNER MODULE PACKET/MESH DESIGNER MODULE	COAT AND PERFORMANCE DESIGN, ANALYSIS, AND OPTIMIZATION OF DATA NETWORKS	ANALYTIC HEURISTICS STATISTICAL	MICROCOMPUTER MINICOMPUTER MAINFRAME	C LANGUAGE	PURCHASE LICENSE OEM SOFTWARE LICENSE DESIGN SERVICES DEVELOPMENT PROJECT
TECHNETRONIC INC. 7927 JONES BRANCH DR., SUITE 400 McLEAN VA. 22102 Contact: G. Jay Lipovich (703) 749-1471	EXPLICIT	PERFORMANCE MEASUREMENT OF IBM DATA NETWORKS CAPACITY PLANNING FOR SNA NETWORKS	ANALYTIC	MICROCOMPUTER WITH MAINFRAME DATA CAPTURE	FORTRAN	PURCHASE ANNUAL LICENSE

TARIFFS	NETWORKS SUPPORTED	PROTOCOLS SUPPORTED	MAXIMUM CAPACITY OF NODES	INPUTS	OUTPUTS	COMMENTS
AT&T AND OCCs FCC NO. 9 AND 10 BOCs LECs	POINT-TO-POINT MULTIPOINT POLLING PACKET SWITCHED MULTIPLEXED TERMINAL ACCESS BACKBONE	BSC SDLC HDLC ASYNC USER DEFINED X.25	3,000 TERMINALS	NETWORK NODE LOCATIONS MESSAGE SIZES RESPONSE TIMES LINK PROTOCOLS LINE SPEEDS DELAY TIMES MAXIMUM DROPS PER CIRCUIT LEC BRIDGING EQUIPMENT TYPES AND LOCATIONS	RESPONSE TIMES CIRCUIT UTILIZATIONS QUEUING ANALYSIS LEAST-COST ROUTING BACKBONE ANALYSIS TOTAL NETWORK COSTS CONCENTRATOR/MULTIPLEXER LOCATIONS AND QUANTITIES NETWORK TOPOLOGY GRAPHICS; HELP SCREENS	ONE-STEP TOTAL NETWORK DESIGN NONDISRUPTIVE NETWORK-LOCATION OPTIMIZER
NONE	POINT-TO-POINT MULTIPOINT POLLING HIERARCHICAL MULTIPLEXED	BSC SDLC HDLC ASYNC USER DEFINED	500 TERMINALS 50 NETWORK ACCESS POINTS 5 HOSTS	CIRCUIT PARAMETERS LINE SPEEDS LINE TYPES END-POINT LOCATIONS MESSAGE SIZES AND VOLUMES LINK PROTOCOLS DELAY TIMES CONTROLLER TYPES	CIRCUIT PERFORMANCE RESPONSE TIMES QUEUING DELAYS TOPOLOGY PERFORMANCE PREDICTION GEOGRAPHIC MAPPING GRAPHICS; HELP SCREENS	GOOD CAPACITY PLANNING GOOD GRAPHICS GOOD REPORTING CAPABILITIES GOOD MAPPING CAPABILITIES
INTERSTATE INTRASTATE INTERLATA INTRALATA OCCs BOCs LECs CROSS VALIDATION THROUGH MULTIPLE SOURCES	POINT-TO-POINT MULTIPOINT (RING, TREE) POLLING PACKET SWITCHED SATELLITE HYBRID (POLLING/PACKET) TERMINAL ACCESS BACKBONE LAN WAN, MAN	BSC SDLC HDLC ASYNC START/STOP WITH FAST SELECT USER CUSTOMIZED USER DEFINED	1,000 USER LOCATIONS PER NETWORK	RESPONSE TIMES MAXIMUM DROPS PER LINE LINE UTILIZATION LINE SPEEDS EQUIPMENT TYPES AND LOCATIONS MESSAGE TYPES, SIZES, QUANTITIES LINK PROTOCOLS BACKBONE NETWORK CONSTRAINTS TARIFF ENVIRONMENT TRAFFIC COMPOSITIONS MAXIMUM ADDRESSABLE DEVICES PER LINE	TOPOLOGY USER RESPONSE TIMES LEAST-COST ROUTING TOTAL NETWORK COSTS LOCAL ACCESS CONNECTIONS CONCENTRATOR/SWITCH LOCATIONS BACKBONE NETWORK DESIGN SATELLITE BREAK-EVEN ANALYSIS GEOGRAPHIC AND LOGICAL MAPS GRAPHICS; HELP SCREENS	UNIX COMPATABILITY GOOD GRAPHICS CAPABILITIES CROSS VALIDATION OF OPTIMIZATION GOOD USER SUPPORT
NONE	POINT-TO-POINT MULTIPOINT POLLING MULTIPLEXED BACKBONE	BSC SDLC ASYNC	100 LINKS 200 NODES	TRAFFIC LOADS NETWORK TOPOLOGY LINK TYPES, SPEEDS, AND MODE ROUTING INFORMATION MANUAL SCREEN INPUT OR AUTOMATIC INPUT FROM HOST GATHERED DATA	LINK UTILIZATIONS PERFORMANCE PREDICTION RESPONSE TIMES QUEUE DISTRIBUTIONS MEMORY MODELING I/O CAPACITIES BACKBONE CHARACTERISTICS WORKLOAD PERFORMANCE GRAPHICS; HELP SCREENS	GOOD AUTOMATIC HOST-DATA CAPTURE GOOD GRAPHICS

SEE LEGEND AT END OF TABLE

Suppliers of network design tools (cont.)

COMPANY NAME	PACKAGE NAME	PRIMARY OBJECTIVE	IMPLEMENTATION METHOD	PROCESSOR BASE	SOURCE LANGUAGE	BUSINESS ARRANGEMENTS
TELCO RESEARCH 1207 17TH AVE. SOUTH NASHVILLE, TENN. 37212 Contact: Laura Westerfield (615) 320-6176	NETWORK PATHFINDER	COST AND PERFORMANCE DESIGN, ANALYSIS, AND OPTIMIZATION OF VOICE, DATA, AND INTEGRATED VOICE/DATA NETWORKS COST AND PERFORMANCE DESIGN, ANALYSIS, AND OPTIMIZATION OF DATA NETWORKS	ANALYTIC HEURISTICS STATISTICAL	MINICOMPUTER MAINFRAME	FORTRAN	PURCHASE LICENSE DESIGN SERVICES
VECTOR SOFTWARE 370 LEXINGTON AVE. NEW YORK, N.Y. 10017 Contact: Anthony Abbot (212) 686-5558	NODE/1 (NETWORK OPTIMIZATION DESIGN ENGINEERING)	DESIGN LEAST-COST LONG-DISTANCE VOICE FACILITIES	ANALYTIC	MICROCOMPUTER	MODULA 2	PURCHASE LICENSE

AI = ARTIFICIAL INTELLIGENCE
APL = A PROGRAMMING LANGUAGE
BOC = BELL OPERATING COMPANY
BSC = BINARY SYNCHRONOUS COMMUNICATIONS
CDR = CALL DATA RECORDING
CSMA/CD = CARRIER-SENSE MULTIPLE ACCESS WITH COLLISION DETECTION
DDD = DIRECT DISTANCE DIALING
DOD = DEPARTMENT OF DEFENSE

FCC = FEDERAL COMMUNICATIONS COMMISSION
FEP = FRONT-END PROCESSOR
FIFO = FIRST IN, FIRST OUT
FX = FOREIGN EXCHANGE
HDLC = HIGH-LEVEL DATA LINK CONTROL
I/O = INPUT/OUTPUT
ISDN = INTEGRATED SERVICES DIGITAL NETWORK
ISO = INTERNATIONAL ORGANIZATION FOR STANDARDIZATION

components within SNA networks. Best/1-SNA can be used to "test-drive" alternatives, allocate workloads, decide when and where to upgrade, and assist with network design and reconfiguration. Capture/SNA is a data-collection and analysis tool that extracts information and aids in managing the performance and capacity of networks based on VTAM (Virtual Telecommunications Access Method). These tools, which run on an IBM (or compatible) mainframe, are accessed through purchasing a license for the software.

■ *Connections Telecommunications, West Bridgewater, Mass.* Two data-network design tools, Multipoint Network Design System (MNDS) and Distributed Network Design System (DNDS), are the main offerings for network optimization. Both tools are microcomputer-based and provide performance modeling, design, and pricing of data networks. MNDS supports multipoint network design under a least-cost algorithm with as many as five different performance "weights."

DNDS is used to design distributed networks—such as X.25 packet switching—and other dynamic-routing, multiple-path configurations. An on-line service supplies tariff information about the telecommunications environment, such as monthly rates for leased-line or dial services, details of the services and boundaries of LATAs (Local Access and Transport Areas), and competitive offerings of OCCs and LECs (local exchange carriers). The company provides training and consulting services to users purchasing a license for its software.

■ *Economics and Technology Inc., Boston, Mass.* The ETI Private Line Pricer is a microcomputer-based tool that provides leased-line tariff information and network analysis. Pricing and routing of point-to-point and multipoint networks are the main features of this network design tool. ETI has also been active in telecommunications public-policy and tariff proceedings. The Private Line Pricer is accessed through purchasing a license for the software.

■ *Quintessential Solutions (Formerly H & S Associates), San Diego, Calif.* Decision-support software tools that run on both microcomputers and Unix-based minicomputers provide design and analysis capabilities for data communications networks. These tools are based upon an architecture that allows the network designer to customize input and output modules. The compute kernel (the lowest level of the operating system) is separated from the input and output modules, and standard interfaces are applied.

TARIFFS	NETWORKS SUPPORTED	PROTOCOLS SUPPORTED	MAXIMUM CAPACITY OF NODES	INPUTS	OUTPUTS	COMMENTS
INTERSTATE INTRASTATE INTERLATA INTRALATA OCCs AND AT&T BOCs LEC BRIDGING WATS, MEGACOM PRIM 1 & 2	POINT-TO-POINT MULTIPOINT POLLING PACKET SWITCHED MULTIPLEXED LATA VIRTUAL/PRIVATE VOICE HYBRID BACKBONE TERMINAL ACCESS	BSC SDLC HDLC ASYNC USER DEFINED X.25	3,000 LOCATIONS	DATA TRANSACTION SUMMARY EQUIPMENT TYPES AND LOCATIONS LINE SPEEDS NETWORK TOPOLOGIES RESPONSE TIMES LINK PROTOCOLS CIRCUIT COSTS VOICE CALL RECORDS HARDWARE CONSTRAINTS NODE LOCATIONS	DATA NETWORK CONFIGURATION TOTAL NETWORK OPTIMIZATION LEAST-COST ROUTING TOPOLOGY TOTAL NETWORK COSTS MULTILEVEL SWITCH CLUSTERING NETWORK ACCESS BREAK-EVEN ANALYSIS LINE UTILIZATIONS USER RESPONSE TIMES CONCENTRATOR/MULTIPLEXER LOCATIONS AND AMOUNTS GRAPHICS; HELP SCREENS	GOOD CUSTOMER SUPPORT SOFTWARE AND TARIFF UPDATES CLEAR, INFORMATIVE REPORTS EXTENSIVE ERROR RECOVERY GOOD GRAPHICS CAPABILITIES EXTENSIVE EXPERIENCE IN NETWORK DESIGN
AT&T WATS, AT&T, MCI, AND U.S. SPRINT MTS INTRASTATE THROUGH DIAL-UP	SINGLE NODE AND SOME MULTINODE PBX VOICE NETWORK	NOT RELEVANT	NO MAXIMUM	CALL INFORMATION SUMMARY OR DETAIL RATES AND TARIFFS QUEUING CONSTRAINTS	TRUNK CONFIGURATION PROJECTED COSTS HELP SCREENS	GOOD HUMAN INTERFACE EASE OF USE

LAN = LOCAL AREA NETWORK
LAPB = LINK ACCESS PROCEDURE-BALANCED
LATA = LOCAL ACCESS AND TRANSPORT AREA
LEC = LOCAL-EXCHANGE CARRIER
MAN = METROPOLITAN AREA NETWORK
MTS = MESSAGE TRANSFER SYSTEM
OCC = OTHER COMMON CARRIER
OEM = ORIGINAL EQUIPMENT MANUFACTURER

SDLC = SYNCHRONOUS DATA LINK CONTROL
SDN = SOFTWARE-DEFINED NETWORK
SMDR = STATION MESSAGE DETAIL RECORDING
SNA = SYSTEMS NETWORK ARCHITECTURE
VMS = VIRTUAL MEMORY OPERATING SYSTEM (DEC)
WAN = WIDE AREA NETWORK
WATS = WIDE AREA TELEPHONE SERVICE

Thereby, capabilities exist to read input-configuration information directly from a database and to read input-performance constraints directly from performance-monitoring devices.

The same standard interfaces are available with capabilities for output into databases, spreadsheets, or standard report formats. The compute kernel includes modules for multipoint network designs, circuit designs, and mesh/packet designs. Complete tariff information is included in a database design. Cross-validation of the tariff database is provided through multiple sourcing. Cross-validation of network optimizations is obtained through interconnection of the family of design tools. Access to the design tools is through purchasing a license for the software.

■ *Technetronic Inc., McLean, Va.* The performance-measurement and capacity-planning software tools for mainframes and networks is called Explicit. The various modules include mainframe performance analysis, MVS (Multiple Virtual Storage) and VM (Virtual Memory) capacity planning, and SNA network modeling and performance/planning. The performance analyzer provides performance measurement, tracking, tuning, and reporting capability, with most functions off-loaded from the host to a microcomputer. Reporting capabilities are also available, including graphics of utilizations, response times, and capacities. Explicit is accessed through purchasing an annual license for the software.

Voice only

■ *HTL Telemanagement, Burtonsville, Md.* Two series of microcomputer-based tools are offered for voice networking: Network Tools for Design (NTD) and Master (Manage and Simplify the Telecommunications Environment Reliably). NTD chooses the optimum mix of long-distance voice communications services for least-cost routing. Master is used for telecommunications resource management. In the NTD series, the tools include modules for long-distance analysis, switch and automatic-call-distribution designing, WATS optimization, and private-line planning. Tariff data is provided. The tools are accessed through purchasing a license for the software. ■

Harrell J. Van Norman is a principal data communications analyst with the NCR Corporation. He earned a B. S. in Systems Science Engineering from Michigan State University.

Kerry Kosak, Desience Corp., Malibu, Calif.

Designing network control centers for greater productivity

A little ergonomics in network control center design goes a long way toward improving output and boosting NCC staff morale

Despite the fact that networks play a key part in corporate activity, little attention is being paid to the nerve center of this function: the network control center. The corporate approach to control center design today runs the gamut from models of efficiency to routinely backlogged, poorly utilized resources. Although a properly designed environment is not the final answer to maximizing human and hardware productivity, recent studies have confirmed that it can go a long way toward improving a less-than-ideal situation.

Part of the problem is that private networks are still novel at many companies, so the concept of a specialized place dedicated to network operations tends to get short shrift. Another factor is that there are few historical studies of optimal command center design.

At the heart of the problem, though, is the intensity of the human/machine interface. The problem is not without precedent. Consider recent technological history: The same problem has already been faced in the scientific and financial communities, both of which provide an excellent starting point for deciding how to build a better data/voice network control center.

Perhaps one of the most vivid images of the last two

1. Form and function. Functional requirements should dictate network control center design. Financial trading requires quick and constant access to two primary sets of equipment: on-line monitors and a telephone. Since traders often work in groups, a systematic means of clustering is needed. And the trading floor aesthetics are also important.

decades is the rows of NASA controllers monitoring individual data terminals during a spacecraft launch. With at least one monitor per person and numerous monitors per spacecraft function, there was need for a well-organized environment that allowed both individual monitor attention and attention to the project's full progression. This was accomplished by arranging low banks of monitors, one per analyst, facing wall displays providing an overview of the entire project.

If the ergonomics of control center design seems to be overstatement of the obvious, consider the following specific design parameters:
- What is the optimum height of the monitor so that it can be the primary focus but still be looked over if needed?
- How deep should the counter in front of it be?
- How close can the monitors be to one another?
- How close can the people be?
- What accommodations should there be for telephones, lighting, and wire management?
- How do you provide for equipment service and changes?

All of these issues contribute to formulating preliminary mechanical-design needs, but equally important is an understanding of each job function and specific needs associated with carrying out that function. To illustrate the importance of function, consider a well-designed trading module used by financial institutions.

The job of trading requires quick and constant access to two primary sets of equipment, the on-line monitors and the telephone. Further, because traders often work in groups, a systematic means of clustering people in either an open floor plan or in rows is a prerequisite, and equipment changes are constant.

Therefore, a well-designed trading module requires flexibility for differing floor plans, modularity to handle diverse and changing equipment, a high level of equipment density, and adequate space for documentation and personal storage (Fig. 1).

To add a little more sophistication to the design equation, factor in the trend toward using the trading floor as a showplace in many financial institutions. Suddenly, aesthetics become relatively important and hidden-wire management, as well as data, telephone-line, and power access must all be juggled without jeopardizing crucial mechanical and ergonomic design elements.

Organization

Designers of specialized environments, such as those seen in financial institutions, have amassed a body of experience that other on-line applications can use. When the lessons learned on the trading floor are combined with the ergonomic work done by many companies and institutions, a substantial pool of data is available that can be applied to the specific needs of the network control environment.

The first thing that is obvious when entering a network

2. Pain in the neck? *Viewing tiers of monitors can often lead to fatigue and neck strain, but when the screens are angled 10 degrees from the vertical, they become easy to see from either a sitting or a standing position.*

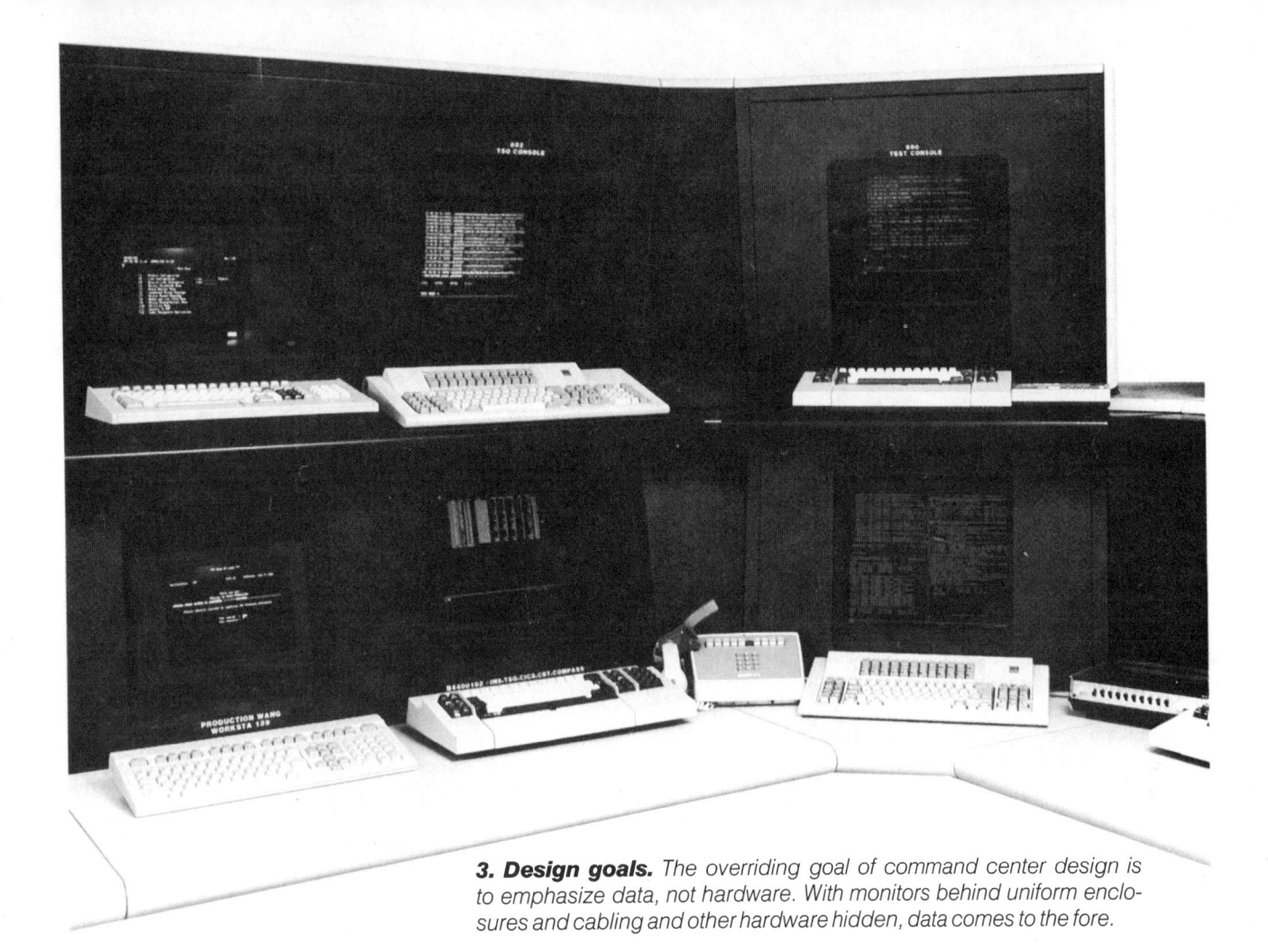

3. Design goals. *The overriding goal of command center design is to emphasize data, not hardware. With monitors behind uniform enclosures and cabling and other hardware hidden, data comes to the fore.*

control center is the ratio of monitors to personnel. In the financial environment, there are generally three or four monitors per person in a defined workstation. In the network control center, it is more common to find several people interacting with common monitors.

This arrangement is further complicated by the fact that different staff operations often share the same space. For example, it is not uncommon to see systems analysts, technicians, and a help desk all in the control area, all sharing the same hardware to one degree or another. Therefore, organization, ease of presentation, and access to a high density of monitors become critical concerns — problems that can be easily complicated when space is restricted or at a premium.

The design solution is vertical stacking. However, since a person may view a stacked monitor from either a sitting or a standing position, optimizing the dimensions and layout of the stacks is critical. Setting the design dimensions compatible with people from five feet to six feet, two inches tall will cover approximately 95 percent of the population. Beginning with a standard desktop height of 29 inches supporting the first tier, the second monitor tier will top out at just about six feet. This height precludes realistic options for a third tier, since it would be almost impossible to access and difficult to see from a sitting position.

Designing the monitor location in the second tier for viewing from a seated position creates other problems. First, because every monitor generally requires that a keyboard be placed in front of it, there should always be a shelf with enough depth to handle the largest keyboard in use (approximately 12 inches deep). Therefore, the viewer will always be looking up and over the shelf. A straight vertical presentation of the second tier monitor can be easily viewed by someone in a standing position but is difficult to see when sitting.

The optimum relationship between an upper and a lower screen is to have both fall within a 20-degree cone of vision of the seated analyst. This requires that each monitor screen be angled approximately 10 degrees from the vertical, which facilitates seeing both screens from either position (Fig. 2).

At the network control site for one of the largest insurance companies in the United States, these design techniques permitted the consolidation of three staff groups into a control area 40 percent smaller than the area previously reserved for the systems analysts. With the addition of two more groups to the same work area, management reported that the ability to quickly pull together people from diverse disciplines for any given problem resulted in a higher level of efficiency and improved customer service.

Accentuating the functional

Beyond organizational and staffing aspects, the functional and mechanical needs of the control center hardware must be addressed. The mechanical layout must provide easy access to monitoring equipment in both tiers, and such things as data lines, power, and air-conditioning access are all essential aspects of the well-designed network operations center. The design of the screen-support stack should be adaptable enough to handle a diverse range of monitor sizes and types and be structurally solid enough to avoid any chance of tipping over when loaded with monitors. These are primary concerns in any design that is freestanding in an open control room space.

Lighting and glare reduction

The most overlooked aspect of control room design is lighting. Generally speaking, most control rooms have too much light and the wrong type. At the center of the problem is screen glare. The reflective nature of the monitor-screen surface in a highly lit environment can become a major distraction to an operator at a terminal. Part of the problem can be alleviated by using a nonreflective surface to cover the monitor, although high levels of light can still wash out the data.

Another aspect of the lighting problem is that despite the fact that it is easier to read a screen in a low-light environment, analysts have reading, writing, and keying tasks to perform on an ongoing basis. The optimum solution for these seemingly conflicting needs is not as complicated as one might imagine. The key is simply to provide controlled light only where needed. This can be accomplished by providing adequate task illumination on the horizontal work surface, supplemented with adjustable indirect ambient lighting kept at minimal levels.

The best solution appears to be built-in direct lighting from a position overhanging the work surface. By using louvered grilles, light can actually be channeled to spill on the horizontal plane of the desktop in front of a monitor. This will provide adequate task lighting for operators without creating a wash or reflection on the monitor screen, while allowing ambient lighting to be reduced significantly.

What happens when area lighting is poorly planned? The experience of a large East Coast securities company sheds some light on this type of problem. The company's new command center was designed to be situated in the room with the mainframe and the support hardware. It was placed on a platform six inches above the computer floor. The additional height, combined with a generally high level of fluorescent lighting, immediately caused glare problems, even though nonreflective screens were in place.

Although the network-operations personnel were looking forward to an integrated control environment, the lighting conditions caused problems. Most of the personnel were frustrated, a few openly complained that they could not work there. The lighting problem added more stress to the settling in and bringing network operations on-line.

The lighting problem was eventually corrected by reducing the overhead light directly above the control environment. But it did require electricians to reopen ceiling access for a period of days, thereby delaying the start-up of a fully functioning installation.

In the grand scheme of things, glare may seem an oversight, but only because it was easily correctable. One only had to hear the complaints of the analysts to understand how critical the proper design of lighting is to the entire command center.

The lighting issue underscores the single overriding goal in control center design: Emphasize the data, not the hardware. In fact, in the perfect environment, all data is contrast-enhanced and well presented; all hardware is practically invisible and almost totally silent (Fig. 3).

A well-designed network control room eliminates all visual distractions. Irregular monitor shapes, cabling, and other hardware are obscured by enclosing the monitors in a dark, nonreflective cabinet. However, enclosing monitors requires forethought to provide functional solutions that facilitate accessibility, ventilation, and wire management.

Another major distraction is noise. Noise is important when deciding on the proximity of the command center to printers and other distractions associated with network hardware. Many companies have used glass enclosures contiguous to the actual hardware area to set off the command center. This helps organize space as well as cut down the noise level in the command environment.

To analysts with so many critical responsibilities, providing an area secure from hardware noise is a definite advantage. In this environment, enclosures can reduce the low-frequency hum of the monitors. There is also a variety of enclosure setups that can reduce printer noise by as much as 90 percent.

Advance planning: the critical criteria

How can networking professionals plan for the future in operations center design? In most instances, the impetus for designing a new command environment is the projected growth of the network. At the command center, growth is compounded by other concerns, such as the phasing in and out of computer hardware and telecommunications equipment. All of these factors have an obvious impact on proper design.

One way to get a handle on these issues is to examine an individual company and follow its planning and design decisions. For security reasons, the company will be called the ABC Corp. It is a holding company for firms in a variety of related industries.

Several years ago, following a change in management, ABC began an aggressive program of acquisition. Some companies remained self-contained subsidiaries, while others were folded into the parent company for economic reasons.

Under this regime, the company's revenues have increased more than fivefold, while management information systems (MIS) capabilities have increased proportionately. At the beginning of the acquisition process, the MIS function was performed in a single large room, but it soon became a series of installations spread throughout a midwestern city. Before MIS consolidation began, three mainframes were located in different parts of the city.

One of the first steps for integrating MIS functions was to build an advanced network consisting of both fiber optic and leased lines. The experience gained from this venture gave MIS operations personnel the confidence to assume responsibility for monitoring and maintaining telecommunications throughout the company.

Three years ago, ABC management realized that it was approaching critical mass and began the planning process

4. Workin' on the railroad. *As the network control center becomes equivalent to the nerve cneter of an entire company, it becomes the paradigm of corporate culture and capabilities. The layout of Burlington Northern's network operations center made monitoring and control operations easier, while the improved asthetics made the command center a company showplace.*

for a centralized command environment that could match projected company growth into the next decade. Three overriding concerns directed their efforts:
- Accommodating all present and future mainframe needs.
- Consolidating the monitoring setups for telecommunications and MIS environmental and support services.
- Planning a three-phase expansion to facilitate growth.

ABC's new data center, only recently completed, is an impressive example of the importance of advance planning. The command center is the geographic center of an installation that takes up the entire floor of a medium-size office building. As such, the command center serves as a functional hub for all the diverse operations it monitors and maintains. The semicircular, double-tiered monitor banks built of modular enclosures, the back wall containing all security, power, and environmental monitoring equipment, and the nearby lounge area for the analysts are all the result of comprehensive planning.

Within the security area, taking up roughly half the floor, are all printing, mainframe, disk-drive, telecommunications, and power resources. The other side of the floor houses offices for the management and support staff. In each of the areas, color schemes, furniture, wall treatment, and lighting have been coordinated for continuity as well as for the needs of specific working environments.

Despite the striking visual impact, an equal amount of precise planning lies behind the walls to accommodate expansion. First, power, cooling, and environmental monitoring and control have all been overbuilt, allowing the raised floor area to be expanded by more than 20 percent with easy access to needed services. Non-weight-bearing modular walls have been used extensively to allow expansion and reconfiguration of the existing center. All of these elements are geared for anticipated MIS growth, which is expected to double again by the end of the decade.

ABC hired a consulting contractor to both plan and execute the work. According to the network-operations manager, this decision was made early in the discussion phase, since ABC realized the importance of support services and that control center design was one area where it lacked the proper expertise.

ABC knew its specific operational needs, but it relied upon the consultants to provide recommendations about power and environmental monitoring—including water detection, main and backup air-conditioning, and Halon fire suppression—as well as security.

A majority of the support-function monitors has been integrated into the back wall of the center. This was done to make the command center a completely functional nerve center in the event of an emergency, an operational consideration often overlooked in the design phase of new centers. Such features as multizoned fire suppression, two-tiered security, and redundant cooling are added dimensions to traditional plans that can become important for both expansion and emergency contingencies.

Human elements

Hardware is not the only important issue, however. As the example of area lighting demonstrated, the needs of the personnel must also be given high priority. Once the explicit design criteria have been met, attention should be paid to the specific needs of the individuals.

The personnel staffing the center should have adequate storage for personal effects and basic supplies. Additionally, storage space for hardware and software documentation should be both plentiful and accessible.

Where monitor banks are stacked two tiers high on a desktop base, adequate work space should be accounted for in the design. However, desktops that are part of the monitor tiers should be set back deeply enough to accommodate either two keyboards (front and back) or a single keyboard and standard binders of documentation. This allows the analyst to use the documentation and keyboard simultaneously on the work surface without requiring cumbersome under-counter pull-out trays or using the analyst's lap as an extension of the work surface. This need is apparent when a software analyst works for several hours without adequate surface area.

Consideration should also be given to where the analysts can spend their time when not specifically working in the control environment. In highly secure areas, one might follow the example of ABC, which built an analyst lounge right off the main secure corridor between the disk-drive operations area and the command center.

Operators' performance can also be enhanced by providing safety features such as structural stability, fireproof construction, abatement of potential radiation, and rounded edges on all exposed surfaces.

The corporate image

As the network control center becomes equivalent to the nerve center of a company, it becomes the showplace of corporate capabilities. So, beyond space, mechanical, and ergonomic needs, aesthetics is playing an increasing role in the location and design of the command center. There is no question that an aesthetic environment can become an integral part of a company's identity, but like all other aspects of the optimum design, it must be factored into the design equation without detracting from other needs, as in the Burlington, N.H., network control center of Burlington Northern Railroad (Fig. 4).

Finally, under all circumstances, the command center must be designed to be redesigned. It must be able not only to account for an ever-changing array of monitoring and technical setups but also be adaptable to changing space configurations and provide hardware and service access. Telecommunications, power, and ventilation also must be considered, as should cable management.

For example, at one IBM manufacturing plant, after just 18 months, the increasing responsibilities of network operations forced a 40 percent expansion of the network control center. By using a modular design that allowed for easy expansion and reconfiguration, the expansion could be made cost-effective and carried out in a fraction of the time required for an entire custom-built design.

In the final analysis, command center design is not simply a function of space allocation and furniture. An enormous amount of advance planning is required for centralized control centers, especially in departments that have significant growth projections. To be effective, questions of morale, hardware, peripherals, and even the building's architecture have to be considered. Eliminating any single aspect can substantially decrease the center's effectiveness and longevity. ■

Kerry Kosak, President of Desience Corp. in Malibu, Calif., received a Bachelor of Architecture degree from the University of Oregon in 1969. He also completed graduate courses at the Harvard School of Design.

Warren Cohen and Nicholas John Lippis III, Digital Equipment Corp., Littleton, Mass.

Building a private wide-area, fiber backbone network

Thoughtful planning, design, and project management ensures plenty of spare capacity in the years ahead for Digital Equipment Corp.'s all-fiber corporate network.

Fiber is still considered by many users to be something of a leading-edge technology, at least as a medium for a private long-distance network. Today, most network managers have not yet reached the stage where they feel as comfortable building a backbone with fiber optics as they would with a commodity item such as T1 circuits. Even so, Massachusetts-based Digital Equipment Corp. (DEC) began laying plans three years ago for constructing its own fiber optic network. Two years later, that network has proved itself not only cost-effective but also technically viable and is presently undergoing an expansion.

Before DEC began building its fiber network, a whole host of business, financial, network-design, and project-management issues had to be resolved, several of which focused the company's attention on a fiber-based network. Planners predicted that the price instabilities of tariffed leased lines would continue for some time, making planning problematic.

The company also realized it would need a high-bandwidth technology with plenty of spare capacity to respond to present and future business needs. Reduced installation time and the opportunity to realize significant cost savings were also important factors favoring the fiber-based network alternative.

The DEC digital lightwave network (DLN), the company's private digital network, is based on fiber optic technology and 45-Mbit/s multiplexing. It initially spanned 80 unrepeated miles, delivering 45-Mbit/s or 90-Mbit/s service to 23 DEC facilities in New England. In 1987, DEC began expanding the DLN.

Phase 2 of the project will add 250 more miles, with backbone links, operating at 90 Mbit/s, connecting an additional 18 existing and 15 planned DEC facilities. Automatic-protection switching via diverse-route looping (that is, alternate-path routing) was added as an enhancement to the original design.

DLN planning began in 1984, when DEC conducted a feasibility study that revealed the positive economics of replacing two large analog Centrex telephone setups and several large analog private branch exchanges (PBXs) with a multiple-host digital switching arrangement. The equipment was to be interconnected by a digital transmission network.

Weighing alternatives

The feasibility study also analyzed the cash outflow of tariffed T1 service. These alternatives were then compared with two options: leasing or purchasing a private, fiber optic, digital transmission network. The fiber optic purchase option was determined to be the least expensive over a 10-year period, resulting in a significant cost savings.

Once the numbers were shown to management, the development work began in earnest on a comprehensive proposal for purchasing the DLN. A request for proposals was issued to a number of competing vendors, and after the responses were analyzed, a contractual arrangement for implementing the DLN was reached with Nynex Business Information Systems (located in Waltham, Mass.). Nynex BIS was appointed the general contractor. It would be the single point of contact with DEC for the design, engineering, provisioning, implementation, and maintenance of the DLN.

Nynex BIS subcontracted to AT&T Technologies for terminal electronics, fiber optic transmission cable, and network-design engineering. Another Nynex BIS subcontractor, Henkels and McCoy of Boston, Mass., would be the principal construction company and would provide outside plant maintenance of the DLN. Nynex BIS would also be responsible for procuring rights-of-way from

1. Distributed star. During Phase 1 development, Digital's long-haul fiber network was configured as a distributed star. Host operates at 90 Mbit/s, 19 remotes operate at 45 Mbit/s.

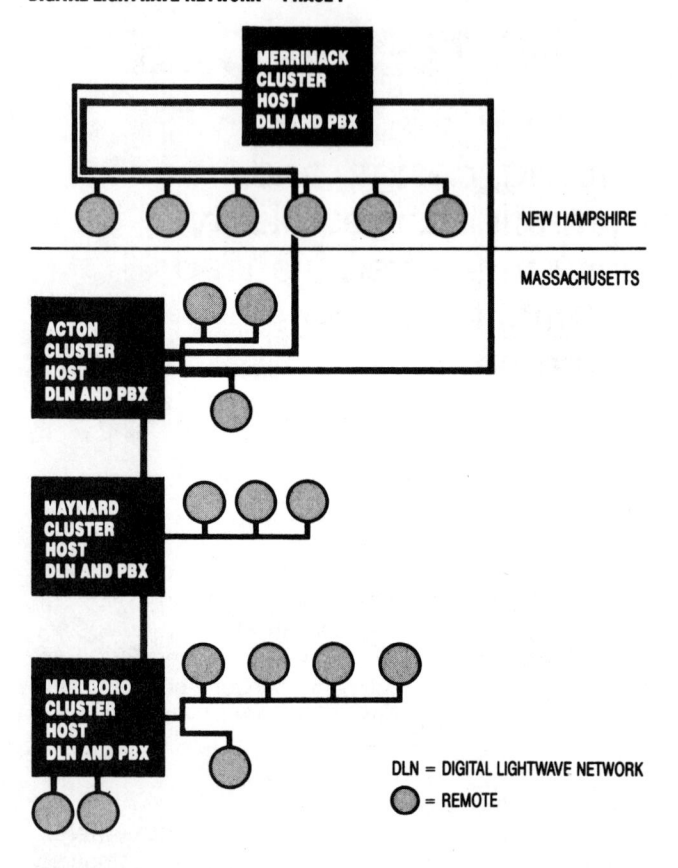

municipalities and coordinating the procurement of pole licenses, interfacing with the utility companies on DEC's behalf for pole and conduit make-ready, or final site preparation, arrangements.

Phase 1

The DLN Phase 1 design was based on Nynex BIS recommendations and DEC modifications of route selections, terminal equipment, topology design, and fiber optic transmission cable. Phase 1 began operating in 1986 and was completed at the end of 1987.

Phase 1 links 23 DEC facilities through a fiber optic telecommunications network spanning 80 miles in Massachusetts and New Hampshire. The DLN network provides DEC with large amounts of bandwidth (that is, 45 to 90 Mbit/s) and the potential for significant additional capacity without changing the fiber medium. The network can accept all forms of digitally encoded information formatted to DS1 specifications for integration of voice, data, and video services.

The topology of the network is a distributed star (Fig. 1).

There are four clusters collecting information from 19 remotes, or concentrators. The four clusters are tied together via AT&T DDM-1000, a dual-DS3 multiplexer. These DDM-1000s are operating at 90 Mbit/s over multimode 50/125-micron fiber at the 1,300-nanometer wavelength. These links define the backbone. The remote facilities are linked to a cluster host via lightguide terminal multiplex assemblies (LTMAs) from AT&T. The LTMAs are termination-circuitry components that operate at 45 Mbit/s over multimode 50/125-micron fiber in the 820- or 1,300- nanometer wavelength.

Should it be necessary to increase throughput, any remote data-collection site can be upgraded to a cluster running at the 90-Mbit/s rate by stepping up to the higher-speed DDM-1000 dual-DS3 multiplexer. Since the distance from host to remote determines the operational wavelength of the link, that wavelength may be modified simply by changing the optical regenerator within the lightguide terminal multiplex assembly or DDM-1000.

To control the network, an on-line, 24-hour-a-day network control center (NCC) has been established. It is built around AT&T's Acorn unit, which provides centralized control and performance monitoring of LTMAs and DDM-1000s via leased 3002 telephone-company circuits. Reporting from remotes to the hosts and from hosts to the NCC is achieved over physically diverse routes.

Figure 2 details the Acorn's hierarchical topology. The network broadcast bridge broadcasts polling information to each hub. The hubs, which are located at host sites, provide data concentration to and from the satellites. The satellites act as the electrical interface to the terminal-electronics components. They also interrogate and monitor the terminal electronics and pass telemetry from them back to the hubs.

If an alarm condition is reported to the NCC, extensive network trouble-reporting, diagnostic, and repair procedures are quickly activated. The DLN has service technicians standing by with lift-bucket trucks, spare fiber cable, and terminal and testing electronics.

There are at least three other network management and control centers within DEC's internal network. One is for the worldwide Decnet operations, another is for the voice digital-switching setup, and a third is for a terminal-switching network. When one part of the network fails, coordination between the various network management and control centers is imperative.

Coordination between the network operations groups is facilitated by formal network trouble-reporting procedures. The DLN network control center plays a major part in these procedures, since it provides physical transport for both the voice and data networks.

Monitor and control

Through the NCC, technicians are able to troubleshoot the DLN and interact with the other network management groups to isolate network problems using sectionalization techniques. The NCC has access to more than a dozen different categories of equipment-failure reports. These alarm-and-control features greatly assist the NCC techni-

2. Control. *Network control is built around AT&T's Acorn setup, a contact-and-closure-based arrangement providing centralized control and performance monitoring of LTMAs and DDM-1000s via leased 3002 circuits. Reporting from remotes to the hosts and from hosts to the NCC is spread over physically diverse routes.*

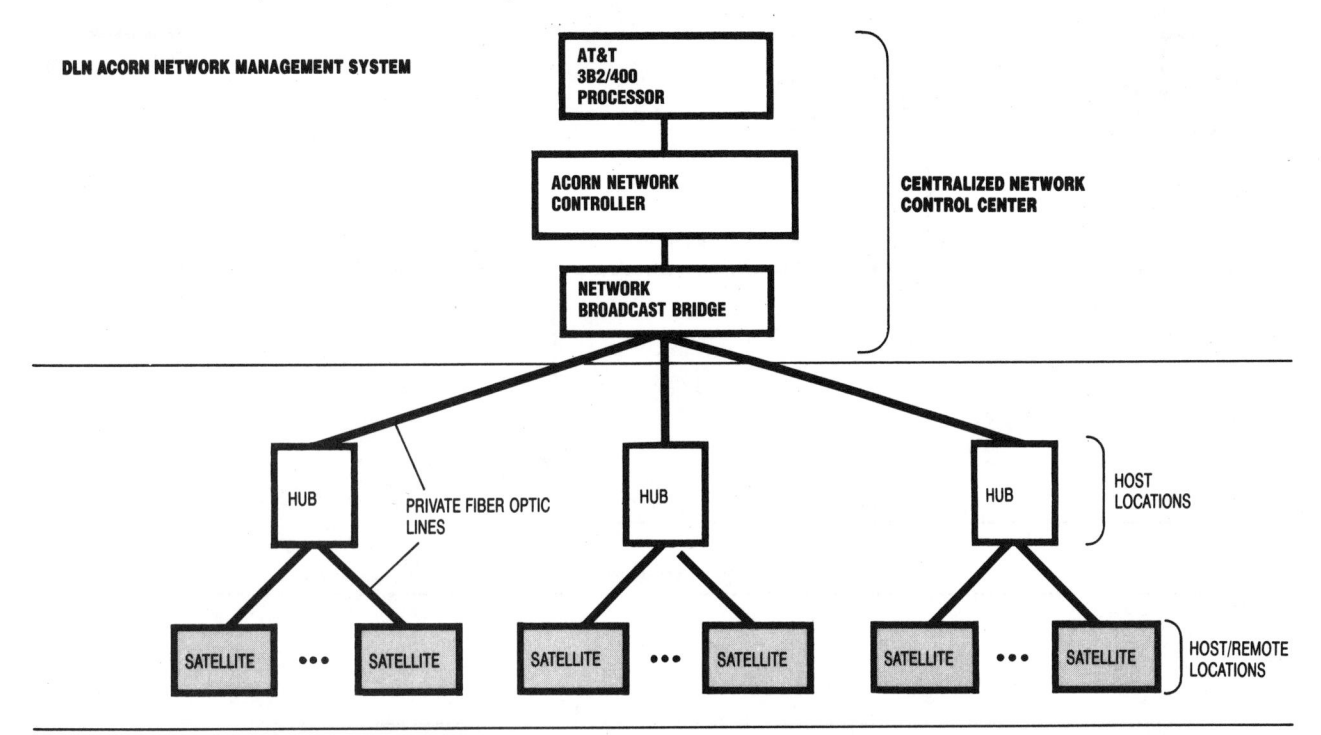

DLN ACORN NETWORK MANAGEMENT SYSTEM

cians in troubleshooting network failures; thus, they help keep the voice and data networks up and running.

Disaster planning was also considered an important part of the network design. During the implementation of the DLN, areas of vulnerability were identified. The problem was sectionalized according to whether vulnerabilities were inside or outside of the plant.

The inside plant was defined as the electronics within a switch room that provides physical transport for services. Typical electronics of the inside plant included the LTMA and DDM-1000 terminal electronics, Acorn hubs and satellites, and maintenance equipment such as optical time-domain reflectometers, T1 bit-error-rate test sets, and optical-power meters.

Vulnerability was assessed in terms of inadequate spare equipment available during a disaster situation. Nynex BIS recommended a fully equipped spare LTMA multiplexer at each cluster host. Other spares, consisting of one Acorn hub and satellite and two optical line-interface units, were also warehoused at the host site. These spares freed DEC from its dependence on the manufacturer's ability to provide emergency replacement equipment.

The outside plant was defined as the fiber routes connecting the various DEC sites. Vulnerability was defined in terms of the location of the utility poles on the DLN fiber routes and the topology of the distributed star network.

Since approximately 93 percent of the Phase 1 fiber installation was to be routed over poles, it was imperative to pay attention to outside plant vulnerability. DEC analyzed utility-pole vulnerability in terms of the location, congestion,

and condition of a given pole. An assessment was also made of the pole's accessibility for maintenance and splicing. A survey of the 80-mile DLN Phase 1 route was undertaken and these specific vulnerabilities noted.

The utility companies conducted their own survey and specified pole replacement and some rearrangement of telephone, power, CATV, and fire-alarm cable to accommodate the DLN cable. The DEC survey was then compared with the various utility surveys. Significant areas of difference between the two surveys usually meant a new route when DEC's offer to replace the poles was not approved by the pole owners.

Distributed star vulnerability

The vulnerability of the distributed star topology was assessed in terms of single points of failure that would cause loss of service and the number of fiber strands within a cable (which vary according to the number of sites served by a given route). The more strands in a cable, the greater the possible impact of a cable break.

To calculate the amount and type of backup resources needed to offset this vulnerability, it was necessary to take into account the existing spare-fiber reels already warehoused, which would be needed to replace failed sections, and the manufacturer's optical power budget per span, which would determine the limit of how many splices a cable would bear. The power-budget calculations note the available loss margins due to the distance between transmitter/receivers and the number of required installation splices. The results revealed the maximum allowable

3. Spares. *To limit the vulnerability that is inherent in the distributed star topology, DEC provisioned spare fiber cable at each network site. A minimum of four fibers were routed to each of the network remotes, one pair of fibers acting as operation transmit/receive, the other pair as a standby transmit/receive.*

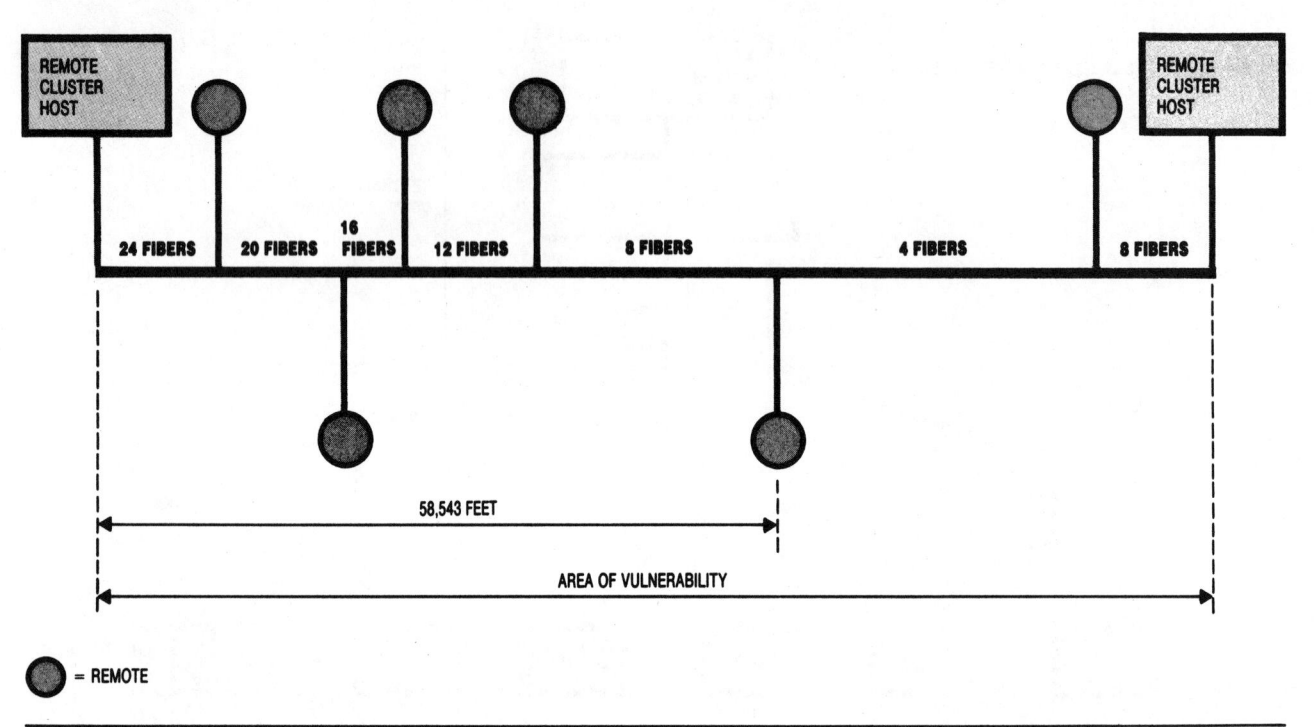

[Figure 3: Diagram showing remote cluster hosts connected via fiber backbone. From left to right: REMOTE CLUSTER HOST — 24 FIBERS — 20 FIBERS — 16 FIBERS — 12 FIBERS — 8 FIBERS — 4 FIBERS — 8 FIBERS — REMOTE CLUSTER HOST. Area of vulnerability spans 58,543 feet. Circles = REMOTE.]

future breaks and splices that could be introduced without degrading signal integrity on that particular fiber link below specified levels. This was calculated using the formula $B = PM/(2 \times TS)$, where B = the number of breaks, PM = power margin available, and TS = typical splice loss.

The fiber-strand count in a cable varies as the route traverses different remote sites. A minimum of four fibers was installed at each remote (Fig. 3). Before the vulnerability-identification and disaster-planning project was initiated, 1,000 feet of spare fiber cable was provisioned for each site on the DLN, with the fiber-strand count equaling the maximum number of fibers entering the switch room.

For example, a host switch room may have 32 fibers entering or leaving; therefore, a 1,000-foot reel of 32-count spare fiber is provisioned. To eliminate this type of vulnerability, 1,000-foot reels of the various fiber-strand counts were required at each facility.

Phase 2

An inherent drawback of distributed star topologies is that single points of failure can bring parts of the network down. If one fiber cable was severed or shattered, a remote site would lose communications with the cluster host. A worse case would be a cluster-to-cluster backbone-link break, resulting in lost intercluster communications from hosts and remotes. To lessen this vulnerability, diverse routing of backbone routes between hosts, as well as diverse routing within the host-to-remote clusters, became an essential element of DLN during the Phase 2 design stage.

The first-phase DLN was designed and cost-justified as a voice-network solution for DEC's headquarters area. DLN Phase 2 was designed as a data-and-voice network that would provide diverse-route protection to all 23 Phase 1 sites as well as 18 new sites. By providing backbone-loop closure between major host locations, the bandwidth, performance, and cost benefits of DLN Phase 2 construction was considered as important as that of Phase 1. A $15 million capital investment to construct and equip the Phase 2 DLN would provide a 10-year pretax data- and voice-circuit savings of $77 million.

During the design, vulnerability-identification, and disaster-planning period of the DLN Phase 1 project, DEC increased its knowledge and experience with lightwave networks. The expertise gave DEC the confidence to work more closely with Nynex BIS in designing the topology of the Phase 2 network.

DEC was now in a position to specify solutions that met business requirements and would no longer rely solely on the contractor to supply recommendations. Most of the DLN Phase 2 topology was designed and cost estimates calculated by DEC from metrics obtained from the Phase 1 experience.

The network topology was designed by selecting routes based on the optimum combination of where the poles are located and what kind of shape they are in as well as on estimates of required rearrangements of existing cable. Road access for maintenance was an important design consideration, as was the availability of conduit. Great pains were taken to select the shortest possible route to service the sites in the area.

4. Strength through diversity. *Adding diversifed routes to DLN changed the topology from a distributed star network into a fully diversified network transport utility.*

DIGITAL LIGHTWAVE NETWORK — PHASE I AND PHASE II

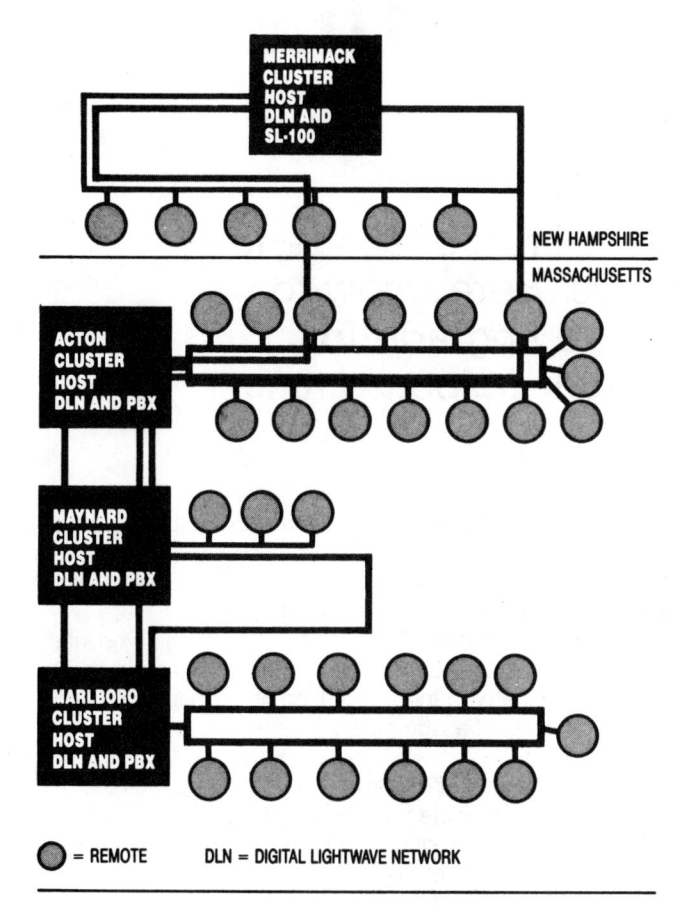

○ = REMOTE   DLN = DIGITAL LIGHTWAVE NETWORK

Other topology design decisions were made by DEC and Nynex BIS on the basis of the following:
- The number of fiber strands needed to support the sites on that route/loop.
- The optical line-interface unit specification, which defines the terminal-equipments' power budgets (how much power is needed to make the terminal gear work), and distance of selected route(s).
- Terminal-electronics specification.

The optimum (minimal) number and location of inside-DEC repeaters based on route distances and power-budget specification.
- Number of spare fibers built into the design.
- Capacity of the DLN network on the first day of its operational life.
- Quantity of spare terminal electronics and DS2 boards.

DEC's Phase 1 experience gave the company enough familiarity with the design and cost implications of various alternative layouts to be able to work more closely with Nynex BIS. This collaborative effort contributed to the success of a final, comprehensive topology design.

DLN Phase 2, currently in progress, was designed to provide 90-Mbit/s service via DDM-1000 terminal electronics. All sites will be connected by 8/125-micron, single-mode, fiber optic cable spanning 250 miles in Massachusetts and New Hampshire. Spare fiber on main routes was added for the 15 future sites, as was subrate or T1 multiplexing gear. Service-restoration capability was increased between the NCC and the outside plant.

The most significant change in the DLN Phase 2 design is the incorporation of diversified routes or loops to all DEC facilities. There are 13 loops built into the Phase 2 design to provide millisecond protection switching via diversified routing.

Adding diversifed routes means that the DLN is no longer a distributed star network. Now it is a fully route-diversified transport utility for metropolitan and wide-area voice and data services (Fig. 4). It provides raw bandwidth to digital PBX switching equipment, DEC computers, T1 data multiplexers, and LAN data bridges operating at T1 rates for extended LAN connectivity. Moreover, it is positioned to transport future T1 and T3 digital video services.

DLN Phase 1 and 2 delivers raw bandwidth to 41 DEC facilities at speeds ranging from 45 Mbit/s to 90 Mbit/s in T1 increments. The topology of the typical link at a remote site includes a digital source that delivers a DS1 bit stream to a digital signal cross-connect (DSX) panel. The DSX panel is where the voice, data, and video services are integrated for transport over the wide area fiber backbone.

The DS1 inputs are then time-division multiplexed up to the DS3 rates (44.736 Mbit/s). After the signal is scrambled, a laser is used to modulate it onto the optical fiber. In the reverse direction, the optical signal is detected by an avalanche photodiode, descrambled, and demultiplexed to the 28 or 56 DS1 channels that are wire-wrapped to the DSX panel.

When Phase 2 construction is completed, it will create a bandwidth bank capable of supporting current as well as future voice, data, and video transport services.

By the end of the decade, the DLN will be functioning as DEC's telecommunications metropolitan and wide area network transport service for digitized voice, data, and video. It will be the link between every major DEC facility in the eastern Massachusetts and southern New Hampshire areas. By "banking" a large percentage of the network's bandwidth capacity for future use, DEC will be well positioned to immediately satisfy increased network-utilization demands. This will provide network connectivity to future unplanned locations in the concentrated transport corridors of DEC's headquarters area. ■

Warren Cohen is network systems manager for DEC's headquarters telecommunications and bandwidth management organization. He received his B. A. in political science from California State University, Northridge, Calif.

Nicholas John Lippis III is strategic analyst with DEC's networks and communications distributed systems strategic planning group. He received his B. S. E. E. from Boston University and is currently pursuing a joint M. S. E. E. at Boston University and the Massachusetts Institute of Technology.

Joseph I. Fernandez and David E. Liddy, Logica Pty. Ltd., Sydney, Australia

Is your packet network a 'model' of response-time efficiency?

Communications lines often are to blame for delays. Here's how to tell which lines to upgrade for improved network response time.

Firms that are designing their own private packet-switching networks must be able to ensure potential users of adequate response time. Beyond any processing delay imposed by the host, the network designer must make certain that no unacceptable delays are caused by the network itself. As described in the article "Don't just guess: How to figure delays in private packet networks" (DATA COMMUNICATIONS, March, p. 197), such delays can be estimated by using queuing-model formulas and an understanding of network-component capacities and loading.

The delay-estimation technique can be learned by applying the model in two sample applications: a simple example, to provide an overview of the essential concepts; and a more complex one, representing a real user's network.

Consider a very simple network as the first example. The network consists of a single switching node connecting a host computer and five packet assembler/disassemblers (PADs), each with 16 attached terminals (Fig. 1). Obviously, this is not a realistic example of a packet-switching network, but it does serve to illustrate the method of calculation that can be applied to more complex cases (such as actual user networks).

The goal is to calculate network transit times for a user at a terminal connected to a PAD via a 2.4-kbit/s line. Each PAD is connected to the switching node by a 19.2-kbit/s link. The link between the host computer and the switching node runs at 48 kbit/s.

The first step is to gather information on the node's load levels and capacities and on the traffic patterns that determine packet sizes. Assume that the manufacturer rates the switching node at a capacity of 800 packets per second and the PAD as capable of handling 4,000 characters per second.

The traffic load from the terminals can vary over a wide range, depending on usage patterns within the organization. In practice, measurements of this usage must therefore be made to provide the appropriate data entered into the equations. These measurements will figure more dramatically in the more detailed example considered later in this article. However, to keep the present example simple and concentrate on the application of the queuing theory formulas, several assumptions are needed to derive the traffic load.

Consider data traffic during the peak period. Of the 16 terminals connected to each PAD, assume 12 are in session with the host during this period. Assume that, at a given moment, half of these have no terminal input/output going on (that is, the users are thinking), a quarter are performing input at a mean rate of five characters per second, and the remaining quarter are receiving screen output at the terminal speed, that is, at 2.4 kbit/s or 300 characters per second. This produces a total input and output rate of 915 characters per second. Thus, during the peak hour, the arrival rate A for the PAD is 915 data characters per second.

The PAD performs packet assembly/disassembly, so that all traffic between it and the switching node is in the form of packets. The next step is to calculate the load on the communications link between the PAD and the switching node.

Assuming a full-duplex link, treat communications in each direction as independent and apply the model to each in turn. For input, the data rate is approximately 15 characters per second. Assume a mean input packet size of 20 bytes. This yields a traffic rate of 0.75 data packets per second. Similarly, for output, using a data rate of 900 characters per second and assuming a mean packet size of 100 characters, one can derive a traffic rate of 9 data packets per second.

1. Simple example. *The problem in the network shown here is to determine user delay by looking at the performance of each of the components and links, in each direction, and then totaling them to find the total network delay from end to end. This example is given to illustrate the model and its application.*

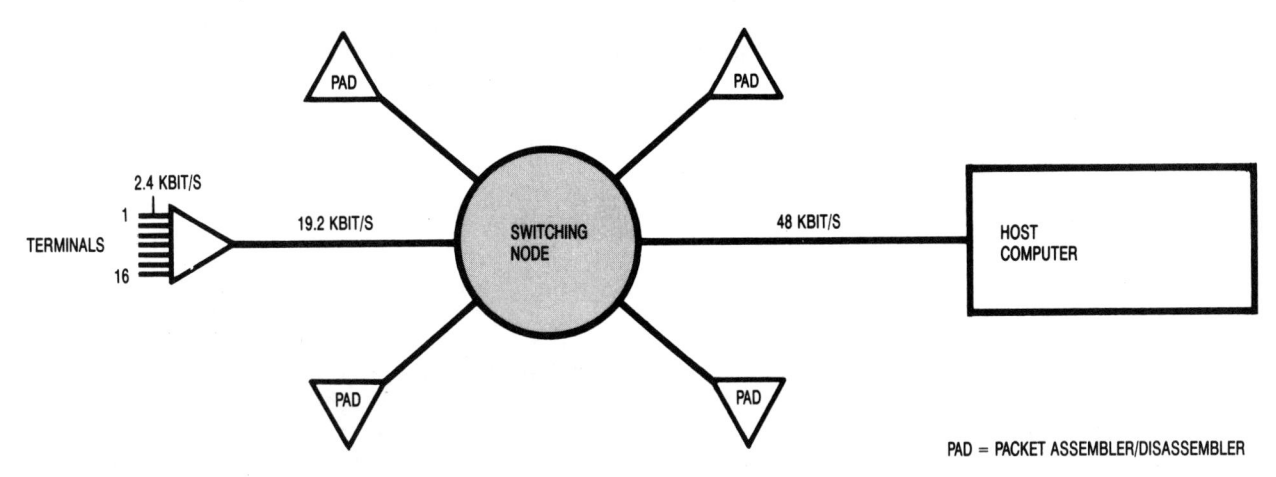

There are five PADs connected to the switching node; assume that all the PADs have the same traffic pattern. Then, the load on the switching node is 3.75 data packets per second from the PADs and 45 data packets per second to the PADs, a total of 48.75 data packets per second to be switched.

Since all the traffic is to a single host computer, the same load is carried by the communications link between the switching node and the host computer.

Refinement

The above values represent the raw figures for traffic loads and equipment capacities. To apply the queuing formulas, it is necessary to translate these values into appropriate arrival rates A and service rates S in consistent units.

There are three types of network components for which this has to be done: the switching nodes, the PADs, and the communications lines. The switching node is rated at a capacity of 800 packets per second. Assume that the node sends a Receiver Ready (RR) packet for each data packet and, thus, that it is handling equal numbers of each of these two kinds of packets in this network. The relatively low number of other kinds of packets may also be ignored. Therefore, the service rates of this node can be taken to be 400 data packets per second. The arrival rate A for this node is the total number of data packets being switched per second. Take it to be 48.75 data packet per second, as noted above.

The PAD throughput is expressed by the manufacturer in characters per second. It is helpful to translate this into a service rate in data packets per second. This is complicated by the fact that the data packet-size distribution for packets from a terminal is very different from that of packets to the terminal. A mean data-packet size may be determined, taking into account packets in each direction, and expressed as the service rate in terms of packets of this size.

Packets with an average size of 20 data bytes arrive at a rate of 0.75 data packets per second from the terminals; packets averaging 100 data bytes in size arrive at a rate of 9 data packets per second for the terminals. Therefore, the total arrival rate A is 9.75 data packets per second. The mean length of arriving packets is 93.85 data characters. A capacity of 4,000 characters per second translates to a service rate S of 42.62 data packets per second for packets of this mean size.

Communications-line bandwidths are expressed in bit/s. As in the case of the PAD, distribution of packet lengths is very different for packets going up the line (PAD to switching node, switching node to host) from the distribution for packets going down the line (host to switching node, switching node to PAD).

However, there is an additional complication. A significant amount of traffic on the same line consists of fixed-length RR packets. As noted earlier, one can treat the line separately in each direction and work out an average packet length in each direction and corresponding arrival and service rates for that direction.

Working it out

So far, this article has merely presented the assumptions of the model and some logical consequences of those assumptions. To derive numbers for anticipated delay, one must perform a set of calculations concerning the performance of the links in the example (terminal to PAD, PAD to node, and node to host) in each direction, as well as for each hardware component.

For insight into how these calculations work, consider the upstream direction (from the terminal) for the link between the packet assembler/disassembler and the switching node. The mean length of data packets in this direction is 20 data bytes. The X.25 protocol adds an overhead of 9 bytes to each packet. For the PAD-to-switching-node link, the input rate is 0.75 data packets per second. Traffic in the other direction is at the rate of 9 data packets per second and will generate RR packets at the same rate in the upstream direction. An RR packet is 9 bytes long.

These two traffic flows (data and RR) give a mean packet length (more correctly, a frame length) of 10.54 bytes. The arrival rate A for the line is 9.75 packets per second. The service rate S for a 19.2-kbit/s line for packets averaging 10.54 bytes in length is 227.7 packets per second.

Once the arrival and service rates are calculated, they are used in either the M/M/1 or the M/D/1 model, as appropriate, to calculate queuing times. The mean service times are also calculated. These calculations are simple, since the service times are just the transmission, assembly, or switching times of packets of known length.

The calculations for each component in the path between terminal and host are as follows:

■ **The terminal-to-PAD link.** Given that terminal input is a sequential process and that the input rate is significantly less than the capacity of the line, no queuing delay will be experienced. Only the service time of the first character needs to be accounted for. In this case, the service rate S = 300 characters per second, so the service time 1/S = 3.33ms.

■ **The PAD.** If the PAD rating = 4,000 characters per second and the packet length is 20 data characters, then the service time = 20/4,0000 = 5 ms.

Assume that one central processor in the PAD handles traffic in both directions. To calculate the queuing time, which will be the same in both directions, requires the mean length of all packets processed. From the terminals, 0.75 packets arrive each second at 20 data characters per packet; from the host, 9 packets arrive each second at 100 data characters per packet.

The average packet length is 93.85 characters, so S = 4,000/93.85 = 42.62 data packets per second, and A = 9.75 data packets per second. Also, the M/M/1 queuing time = A/[S(S − A)] = 6.96 ms.

■ **The PAD-to-switching node link.** The line speed is 19.2 kbit/s. To calculate the service time, assume 8 bits per character. The mean length of a data packet is 20 data bytes, plus 9 bytes of overhead, giving a length of 29 bytes. Thus, the mean service time for data packets = (29 × 8)/19,200 = 12.08 ms.

This line will be carrying two kinds of traffic from the PAD to the node: The 29-byte data frames arriving at the rate of 0.75 frames per second and the 9-byte Ready Received packets, acknowledging data flowing in the opposite direction, at the rate of 9 packets per second.

The mean frame length = [(29 × .75) + (9 × 9)]/9.75 = 10.54 bytes. The mean service rate S = 19,200/(10.54 × 8) = 227.7 packets per second. The arrival rate A = 9.75 packets per second, so the M/M/1 queuing time A/[S(S − A)] = .20 ms.

■ **The switching node.** The switching node's service rate S = 400 data packets per second; the arrival rate = 48.75 data packets per second; and the service time 1/S = 2.5 ms. Therefore, the M/D/1 queuing time A/[2S(S − A)] = 0.17 ms.

■ **The switching node-to-host link.** Since the line speed = 48 kbit/s, the mean service time for data packets = (29 × 8)/48,000 = 4.83 ms. The mean frame length = [(29 × 3.75) + (9 × 45)]/48.5 = 10.54 bytes.

The mean service rate S = 48,000/(10.54 × 8) = 569.3 packets per second and the arrival rate A = 48.75 packets per second. It follows that the M/M/1 queuing time A/[S(S − A) = 0.16 ms.

■ **The host-to-switching node link.** Again, the line speed = 4.8 kbit/s, but in this case, the mean service time for data packets = (109 × 8)/48,000 = 18.1 ms. The mean frame length = [(109 × 45) + (9 × 3.75)]/48.75 = 101.83 bytes.

The mean service rate S = 48,000/(101.83 × 8) = 58.92 packets per second and the arrival rate A = 48.75 packets per second, so the M/M/1 queuing time A[S(S − A)] = 81.34 ms.

■ **The switching node-to-PAD link.** Here, the line speed = 19.2 kbit/s, so the mean service time for data packets = (109 × 8)/19,200 = 42.43 ms.

The mean frame length = 101.83 bytes, so the mean service rate S = 19,200/(101.83 × 8) = 23.57 packets per second. Since the arrival rate A = 9.75 packets per second, the M/M/1 queuing time A[S(S − A)] = 29.94 ms.

Results

The results of the above calculations are summarized in Tables 1 and 2. Note that no queuing times are listed for the terminal-to-PAD lines. For transmission from the terminal to the PAD, the interarrival time is determined by the human typist's speed. Assuming a maximum of 10 characters per second, this time is at least 100 ms. For a 2.4-kbit/s line, the service time is 3.33 ms. Hence, no queuing occurs. For the reverse direction, other considerations produce the same result.

Our definition of transit time is related to receipt of the first character of the response packet. This means that the terminal is idle at the time of arrival.

Totaling of the component delays shows that the mean transit delay for a terminal-to-host data packet is 35 ms, while the delay for a host-to-terminal packet is 210 ms. Of course, these results are based on the assumptions about mean packet size noted above. Calculation of the delays also points out that the largest contributions to delay are made by the communication lines. Thus, if the delays are not acceptable, effort should be directed at upgrading the communication lines rather than the nodes.

Stepping up ... to the real world

A more realistic example of a small packet-switching network is shown in Figure 2. The backbone network has four switching nodes connected in a mesh topology. Each node has a number of PADs attached. Multiple host computers are supported, connected to the switching nodes. The internodal trunks all run at 48 kbit/s, as do the lines between the switching nodes and the host computers. The PADs are connected to the switching nodes by 19.2-kbit/s links. Terminals are connected to the PADs by 9.6-kbit/s lines.

The calculations in this example are much more complex

Table 1: Network transit time for terminal-to-host path

PATH COMPONENT (NODE OR COMMUNICATION LINK)	ARRIVAL RATE	SERVICE RATE	DELAY (MS)		
			SERVICE	QUEUING	TOTAL
TERMINAL-TO-PAD LINK		2.4 KBIT/S = 300 C/S	3.33	—	3.33
PAD	9.75 DP/S	4,000 C/S = 42.62 P/S	5	6.96	11.96
PAD-TO-SWITCHING NODE LINK	9.75 P/S	19.2 KBIT/S = 227.7 P/S	12.08	0.20	12.28
SWITCHING NODE	48.75 DP/S	400 DP/S	2.5	0.17	2.67
SWITCHING NODE-TO-HOST LINK	48.75 P/S	48 KBIT/S = 569.3 P/S	4.83	0.16	4.99
END TO END					35.23

C/S = CHARACTERS PER SECOND
P/S = PACKETS PER SECOND
DP/S = DATA PACKETS PER SECOND
PAD = PACKET ASSEMBLER/DISASSEMBLER

than those in the previous case. One reason is that terminals at different locations have different path lengths through the network to a particular host, so the response time depends on terminal and host location. Another is that any particular terminal has more than one possible route through the network available to its data stream to and from a particular host. The network may perform load balancing and, if it is a datagram network, may send successive packets from a particular terminal to a particular host along different routes. In this case, the response time for a particular terminal is not just given by the transit time over a single path through the network.

Make the following assumptions in carrying out the transit time calculations:

- The network does not use dynamic load balancing but instead sends all the traffic between a particular terminal-host pair along the same route, chosen to contain the least number of hops. If there are multiple paths with the same number of hops, the traffic is split equally between them.
- The transit time is calculated for the above best-route case. If a link goes down, traffic will be routed via an alternate path, which may be longer. The calculations are not done for this case, but it will be obvious to the reader how these calculations could be done.
- As before, each switching node has a capacity of 400 data packets per second and each PAD has a capacity of 4,000 characters per second.
- The mean input (that is, terminal-to-host) packet length is 20 data bytes and the mean output (host-to-terminal) packet length is 100 data bytes.

Estimates and aggregates

To perform our queuing-model calculations for the delay in any component of the network, it is necessary to derive the service and arrival rates for that component. The service rates for the switching nodes and PADs are given above. The service rates for the communications lines are calculated using a mean packet length for each direction of

Table 2: Network transit time for host-to-terminal path

PATH COMPONENT (NODE OR COMMUNICATION LINK)	ARRIVAL RATE	SERVICE RATE	DELAY (MS)		
			SERVICE	QUEUING	TOTAL
HOST-TO-SWITCHING NODE LINK	48.75 P/S	48 KBIT/S = 58.92 P/S	18.10	81.34	99.44
SWITCHING NODE	48.75 DP/S	400 DP/S	2.5	0.17	2.67
SWITCHING NODE-TO-PAD LINK	9.75 P/S	19.2 KBIT/S = 23.57 P/S	42.43	29.94	72.37
PAD	9.75 DP/S	4,000 C/S = 42.62 DP/S	25	6.96	31.96
PAD-TO-TERMINAL LINK		2.4 KBIT/S = 300 C/S	3.33	—	3.33
END TO END					209.77

C/S = CHARACTERS PER SECOND
P/S = PACKETS PER SECOND
DP/S = DATA PACKETS PER SECOND
PAD = PACKET ASSEMBLER/DISASSEMBLER

2. Real-worldesque. *This small packet network, typical of a user's actual installation, shows that the delay calculations are still tractable as the network grows.*

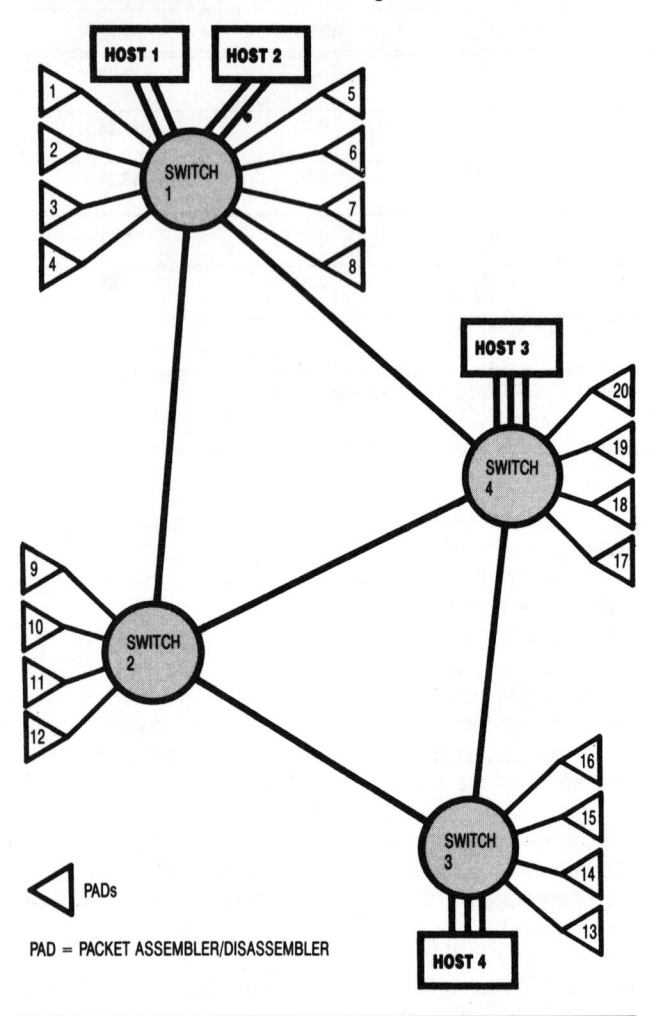

each communications line, derived from taking into account the data-packet and RR-packet rates and lengths, as before.

Since the goal is to estimate the transit times during the peak period, the arrival rates are taken to be the peak data-traffic loads. For the sake of realism, assume that this determination of transit times is being carried out before the network is installed, so no live measurements of data-traffic volumes can be made.

Instead, the designer would have to carry out a survey of user applications and the traffic that they are expected to generate. This survey yields an estimate of daily peak transaction volumes for terminals and hosts and of the average expected transaction-message sizes. From these estimates, one can work out peak input and output data-traffic rates in characters per second for the terminals to each of the hosts (Table 3).

Assume for simplicity that the traffic patterns and volumes are uniform across all the terminals. In practice, different groups of terminals may have different usage patterns, but this can be taken into account by considering each group of terminals separately in doing the traffic-load calculations. For this example, it is best to make the simplifying assumption of uniformity.

Aggregate the data volumes from the terminals at each PAD to obtain arrival rates for the PADs. This information is presented in Table 4. Aggregate also the PAD and host traffic into each switch, as shown in Table 5. Here, one must take into account transit traffic switched by a node and traffic originating or terminating at a node (in a host or a PAD). Finally, aggregate the traffic rates for each of the five backbone communications links, again taking into account transit and terminating traffic, as presented in Table 6.

The arrival rate for any PAD-to-switching node link is the same as the arrival rate for the PAD, since these links are not duplicated. There are multiple links between switching nodes and hosts. Divide the host load equally between these links and treat it as a single-server situation.

At this point, all the information needed for the queuing-model calculations is present in the assumptions stated above and in Tables 4, 5, and 6. It is now possible to calculate the transit delay for any path in the network using this information and the queuing-model formulas.

Roll up your sleeves

As an example, calculate the transit times for data transmitted between host 2 and a terminal connected to PAD 14. The calculations show that the main contributions to delay come from the communications lines rather than the nodes. The largest single culprit is the line between switching node 3 and the PAD. Thus, if a decrease in delay is desired, the speed of this line should be upgraded.

The full details of the calculations are as follows:

■ **PAD 14.** From the terminals, eight 20-character packets arrive per second, so $S = 4,000/20 = 200$, and $1/S = 5$ ms. From the hosts, sixteen 100-character packets arrive per second, so $S = 4,000/100 = 40$ and $1/S = 25$ ms. The average traffic is 24 73.33-character packets per second, so $S = 4,000/73.33 = 54.55$ packets per second. And the M/M/1 queuing time $A/[S(S - A)] = 24/[54.55(54.55 - 24)] = 14.4$ ms.

■ **The PAD 14-to-switch 3 link.** The line is 19.2 kbit/s. Eight 29-byte frames arrive per second from the terminals, so $S = 19,200/(8 \times 29) = 82.76$ and $1/S = 12.08$ ms. RR-packet traffic consists of sixteen 9-byte frames, and the average traffic is 24 15.26-byte frames, so $S = 19,200/(8 \times 15.67) = 153.19$ packets per second. The M/M/1 queuing time $A/[S(S - A)] = 24/[153.19(153.19 - 24)] = 1.21$ ms.

■ **Switch 3.** $S = 400$ packets per second, and $1/S = 2.5$ ms. $A = 193.95$ packets per second, so the M/D/1 queuing time $A/[2S(S - A)] = 193.95/[2 \times 400(400 - 193.95)] = 1.18$ ms.

■ **The switch 3-to-switch 2 link.** The line is 48 kbit/s. Traffic from the terminals consists of six 29-byte frames; $1/S = (8 \times 29)/48,000 = 4.83$ ms. Traffic from the hosts consists of 33.9 109-byte frames, and the RR-packet traffic is 29.05 9-byte frames. The average traffic is 68.95 59.91-

Table 3: Results of traffic survey

	PEAK DATA RATES PER TERMINAL	
	CHARACTERS PER SECOND	DATA PACKETS PER SECOND
TO HOST 1	2	.1
TO HOST 2	2	.1
TO HOST 3	3	.15
TO HOST 4	3	.15
SUBTOTAL	10	.5
FROM HOST 1	20	.2
FROM HOST 2	20	.2
FROM HOST 3	30	.3
FROM HOST 4	30	.3
SUBTOTAL	100	1.0
TOTAL	110	1.5

Table 4: PAD loading

			PEAK LOAD		
			PACKETS PER SECOND		
PACKET ASSEMBLER/ DISASSEMBLER	NUMBER OF TERMINALS	CHARACTERS PER SECOND	FROM TERMINAL	TO TERMINAL	TOTAL
1	16	160 1,600	8	16	24
2	14	140 1,400	7	14	21
3	16	160 1,600	8	16	24
4	15	150 1,500	7.5	15	22.5
5	16	160 1,600	8	16	24
6	12	120 1,200	6	12	18
7	14	140 1,400	7	14	21
8	15	150 1,500	7.5	15	22.5
SUBTOTAL	118		59	118	177
9	14	140 1,400	7	14	21
10	16	160 1,600	8	16	24
11	11	110 1,100	5.5	11	16.5
12	13	130 1,300	6.5	13	19.5
SUBTOTAL	54		27	54	81
13	13	130 1,300	6.5	13	19.5
14	16	160 1,600	8	16	24
15	15	150 1,500	7.5	15	22.5
16	16	160 1,600	8	16	24
SUBTOTAL	60		30	60	90
17	15	150 1,500	7.5	15	22.5
18	13	130 1,300	6.5	13	19.5
19	16	160 1,600	8	16	24
20	15	150 1,500	7.5	15	22.5
SUBTOTAL	59		29.5	59	88.5

byte frames. $S = 48,000/(8 \times 59.91) = 100.2$ packets per second. And the M/M/1 queuing time $A/[S(S - A)] = 68.95/[100.2(100.2 - 68.95)] = 22.07$ ms.

■ **Switch 2.** $S = 400$ packets per second, and $A = 125.55$ packets per second. The M/D/1 queuing time $A/[2S(S - A)] = 125.55/[2 \times 400(400 - 125.55)] = 0.57$ ms.

■ **The switch 2-to-switch 1 link.** The line is 48 kbit/s. Traffic from the terminals consists of 16.8 29-byte frames; $1/S = 4.83$ ms. Traffic from the hosts consists of 11.8 109-byte frames, and the RR-packet traffic is 42.45 9-byte frames, so the average traffic is 71.05 30.34-byte frames. $S = 197.8$ packets per second. And the M/M/1 queuing time $A/[S(S - A)] = 71.05/[197.8(197.8 - 71.05)] = 2.83$ ms.

■ **Switch 1.** $S = 400$ packets per second, and $A = 280.8$ packets per second. The M/D/1 queuing time $A/[2S(S - A)] = 280.8/[2 \times 400(400 - 280.8)] = 2.94$ ms.

■ **The switch 1-to-host 2 link.** The line is 48 kbit/s.

Table 5: Switching-node loads

NODE NUMBER	SOURCE PAD	DESTINATION HOST	VOLUME (DATA PACKETS PER SECOND)	PROPORTION THROUGH NODE	CONTRIBUTION TO LOAD (DATA PACKETS PER SECOND)
1	1-8	All	177	1.0	177
	9-20	1	51.9	1.0	51.9
	9-20	2	51.9	1.0	51.9
TOTAL					280.8
2	1-8	4	53.1	0.5	26.55
	9-12	All	81	1.0	81
	13-16	1	18	0.5	9
	13-16	2	18	0.5	9
TOTAL					125.55
3	1-12	4	77.4	1.0	77.4
	13-16	All	90	1.0	90
	17-20	4	26.55	1.0	26.55
TOTAL					193.95
4	1-16	3	104.4	1.0	104.4
	1-8	4	53.1	0.5	26.55
	13-16	1	18	0.5	9
	13-16	2	18	0.5	9
	17-20	All	88.5	1.0	88.5
TOTAL					237.45

PAD = PACKET ASSEMBLER/DISASSEMBLER

Table 6: Backbone network trunk loads

TRUNK	SOURCE	DESTINATION	NUMBER OF TERMINALS	DATA RATE PER TERMINAL		VOLUME (DATA PACKETS PER SECOND)	PROPORTION ON THIS TRUNK	CONTRIBUTION TO LOAD (DATA PACKETS PER SECOND)
				CHARACTERS PER SECOND	DATA PACKETS PER SECOND			
1-2	PADs 1-8	HOST 4	118	3	.15	17.7	0.5	8.85
	HOST 1	PADs 9-12	54	20	.2	10.8	1.0	10.8
	HOST 1	PADs 9-12	54	20	.2	10.8	1.0	10.8
	HOST 2	PADs 13-16	60	20	.2	12	0.5	6.0
	HOST 2	PADs 13-16	60	20	.2	12	0.5	6.0
	TOTAL							42.45
2-1	PADs 9-12	HOST 1	54	2	.1	5.4	1.0	5.4
	PADs 9-12	HOST 2	54	2	.1	5.4	1.0	5.4
	PADs 13-16	HOST 1	60	2	.1	6	0.5	3.0
	PADs 13-16	HOST 2	60	2	.1	6	0.5	3.0
	HOST 4	PADs 1-8	118	20	.2	23.6	0.5	11.8
	TOTAL							28.6
2-3	PADs 1-8	HOST 4	118	3	.15	17.9	0.5	8.95
	PADs 9-12	HOST 4	54	3	.15	8.1	1.0	8.1
	HOST 1	PADs 13-16	60	20	.2	12	0.5	6
	HOST 2	PADs 13-16	60	20	.2	12	0.5	6
	TOTAL							29.05
3-2	PADs 13-16	HOST 1	60	2	.1	6	0.5	3
	PADs 13-16	HOST 2	60	2	.1	6	0.5	3
	HOST 4	PADs 1-8	118	30	.3	35.4	0.5	17.7
	HOST 4	PADs 9-12	54	30	.3	16.2	1.0	16.2
	TOTAL							39.9
2-4	PADs 9-12	HOST 3	54	3	.15	8.1	1.0	8.1
	TOTAL							8.1
4-2	HOST 3	PADs 9-12	54	30	.3	16.2	1.0	16.2
	TOTAL							16.2
3-4	PADs 13-16	HOST 1	60	2	.1	6	0.5	3
	PADs 13-16	HOST 2	60	2	.1	6	0.5	3
	PADs 13-16	HOST 3	60	3	.15	9	1.0	9
	HOST 4	PADs 1-8	118	30	.3	35.4	0.5	17.7
	HOST 4	PADs 17-20	59	30	.3	17.7	1.0	17.7
	TOTAL							50.4
4-3	PADs 1-8	HOST 4	118	3	.15	17.7	0.5	8.85
	PADs 17-20	HOST 4	59	3	.15	8.85	1.0	8.85
	HOST 1	PADs 13-16	60	20	.2	12	0.5	6
	HOST 2	PADs 13-16	60	20	.2	12	0.5	6
	HOST 3	PADs 13-16	60	30	.3	18	1.0	18
	TOTAL							47.7
4-1	PADs 13-16	HOST 1	60	2	.1	6	0.5	3
	PADs 13-16	HOST 2	60	2	.1	6	0.5	3
	PADs 17-20	HOST 1	59	2	.1	5.9	1.0	5.9
	PADs 17-20	HOST 2	59	2	.1	5.9	1.0	5.9
	HOST 3	PADs 1-8	118	30	.3	35.4	1.0	35.4
	HOST 4	PADs 1-8	118	30	.3	35.4	0.5	17.7
	TOTAL							70.9
1-4	PADs 1-8	HOST 3	118	3	.15	17.7	1.0	17.7
	PADs 1-8	HOST 4	118	3	.15	17.7	0.5	8.85
	HOST 1	PADs 13-16	60	20	.2	12	0.5	6
	HOST 1	PADs 17-20	59	20	.2	11.8	1.0	11.8
	HOST 2	PADs 13-16	60	20	.2	12	0.5	6
	HOST 2	PADs 17-20	59	20	.2	11.8	1.0	11.8
	TOTAL							62.15

Table 7: Transit time from terminal on PAD 14 to host 2

PATH COMPONENT (NODE OR COMMUNICATION LINK)	ARRIVAL RATE	SERVICE RATE	DELAY (MS)		
			SERVICE	QUEUING	TOTAL
TERMINAL-TO-PAD LINK	5 C/S	9.6 KBIT/S =1,200 C/S	0.83	—	0.83
PAD	24 P/S	4,000 C/S =54.55 P/S	5	14.4	19.4
PAD-14-TO SWITCHING NODE 3 LINK	24 P/S	19.2 KBIT/S =153.19 P/S	12.08	1.21	13.29
SWITCHING NODE 3	193.95 DP/S	400 DP/S	2.5	1.18	3.68
SWITCHING NODE 3-TO-SWITCHING NODE 2 LINK	68.95 P/S	48 KBIT/S = 100.2 P/S	4.83	22.07	26.9
SWITCHING NODE 2	125.55 DP/S	400 DP/S	2.5	0.57	3.07
SWITCHING NODE 2-TO-SWITCHING NODE 1 LINK	71.05 P/S	48 KBIT/S = 197.8 P/S	4.83	2.83	7.66
SWITCHING NODE 1	280.8 DP/S	400 DP/S	2.5	2.94	5.44
SWITCHING NODE 1-TO-HOST 2 LINK	75.45 P/S	48 KBIT/S = 281.9 P/S	4.83	1.30	6.13
END TO END					86.40

C/S = CHARACTERS PER SECOND
P/S = PACKETS PER SECOND
DP/S = DATA PACKETS PER SECOND
PAD = PACKET ASSEMBLER/DISASSEMBLER

Traffic from the terminals consists of 92.7 29-byte frames; $1/S = 4.83$ ms. The RR-packet traffic is 58.2 9-byte frames, and the average traffic is 150.9 21.29-byte frames. $S = 281.9$ packets per second. There are two lines, so $A = 150.9/2 = 75.45$ packets per second. The M/M/1 queuing time $A/[S(S - A)] = 75.45/[281.9(281.9 - 75.45)] = 1.30$ ms.

Now, repeat the calculations for the links going in the opposite direction (the components have already been accounted for):

■ **The host 2-to-switch 1 link.** The line is 48 kbit/s. Traffic from the host consists of 58.2 109-byte frames; $1/S = 18.17$ ms. The RR-packet traffic is 92.7 9-byte frames, and the average traffic is 150.9 47.57-byte frames. $S = 126.1$ packets per second.

It can be seen that the arrival rate (150.9) is greater than the mean service rate (126.1). This is why two lines have been placed between host 2 and switch 1. Since this discussion has assumed that logical circuits use only one physical pathway for the duration of a session, the best way to accommodate this multiple-server queue is to divide the load between the servers. The arrival rate is thus 75.45 packets per second and the M/M/1 queuing time $A[S(S - A)] = 75.45/[126.1(126.1 - 75.45)] = 11.81$ ms.

■ **The switch 1-to-switch 2 link.** The line is 48 kbit/s. Traffic from the terminals consists of 8.85 29-byte frames, and traffic from the hosts consists of 33.6 109-byte frames. $1/S = 18.17$ ms. The RR-packet traffic is 28.6 9-byte frames, and the average traffic is 71.05 58.78-byte frames. $S = 102.07$ packets per second. The M/M/1 queuing time $A/[S(S - A)] = 71.05/[102.07(102.07 - 71.05)] = 22.44$ ms.

Note the large increase in queuing time over the host 2-to-switch 1 line, which has 60 percent utilization, compared with 70 percent for this line.

■ **The switch 2-to-switch 3 link.** The line is 48 kbit/s. Traffic from the terminals consists of 17.05 29-byte frames, and traffic from the hosts consists of twelve 109-byte frames. $1/S = 18.17$ ms. The RR-packet traffic is 39.9 9-byte frames, and the average traffic is 68/95 31/35-byte frames. $S = 191.39$ packets per second. The M/M/1 queuing time $A/[S(S - A)] = 68.95/[191.39(191.39 - 68.95)] = 2.94$ ms.

■ **The switch 3-to-PAD 14 link.** Here, the line is 19.2

Table 8: Transit time from host 2 to terminal on PAD 14

PATH COMPONENT (NODE OR COMMUNICATION LINK)	ARRIVAL RATE	SERVICE RATE	DELAY (MS)		
			SERVICE	QUEUING	TOTAL
SWITCHING NODE 1-TO-HOST 2 LINK	75.45 P/S	48 KBIT/S = 126.1 P/S	18.17	11.81	29.98
SWITCHING NODE 1	280.8 DP/S	400 DP/S	2.5	2.94	5.44
SWITCHING NODE 2-TO-SWITCHING NODE 1 LINK	71.05 P/S	48 KBIT/S = 102.07 P/S	18.17	22.44	40.61
SWITCHING NODE 2	125.55 DP/S	400 DP/S	2.5	0.57	3.07
SWITCHING NODE 3-TO-SWITCHING NODE 2 LINK	68.95 P/S	48 KBIT/S = 191.39 DP/S	18.17	2.94	21.11
SWITCHING NODE 3	193.95 DP/S	400 DP/S	2.5	1.18	3.68
PAD 14-TO-SWITCHING 3 NODE LINK	24 P/S	19.2 KBIT/S = 31.72 P/S	45.42	98.04	143.46
PAD	24 P/S	4,000 C/S = 54.55 P/S	25	14.40	39.40
TERMINAL-TO-PAD LINK		9.6 KBIT/S = 1,200 C/S	0.83	—	0.83
END TO END					287.58

C/S = CHARACTERS PER SECOND
P/S = PACKETS PER SECOND
DP/S = DATA PACKETS PER SECOND

PAD = PACKET ASSEMBLER/DISASSEMBLER

kbit/s. Traffic from the hosts consists of sixteen 109-byte frames; 1/S = 45.42 ms. The RR-packet traffic is eight 9-byte frames, and the average traffic is 24 75.67-byte frames. S = 31.72 packets per second. The M/M/1 queuing time A/[S(S − A)] = 24/[31.72(31.72 − 24)] = 98.04 ms. This queuing time is high due to the high utilization — 76 percent.

Conclusion

Tables 7 and 8 summarize the results. Note that the mean transit time for a data packet from the terminal to the host is 86 ms, while that for a host-to-terminal data packet is 288 ms. Hence, the additional round-trip delay imposed by the network is 375 ms.

Put another way, a terminal user of this network would see an additional delay of 375 ms in transaction response, compared with a user directly connected to host 2. These delays can be expected to be short compared to the transaction-processing delays. Therefore, they would not be a significant source of user dissatisfaction.

Provided that packet-switching networks are built to the proper specifications, the transit delays that they create can be kept low. For example, limits of half a second can be achieved relatively easily. With the use of high-performance nodes and high-speed trunk lines, further reduction could be achieved. The delays can be reduced to the point where they are insignificant compared with host-processing delays. Then, the user would be unable to perceive significant degradation in performance compared with direct connection to the host. ■

Joseph I. Fernandez is a senior consultant with Logica, an international computer and communications consulting company. He has worked in the field of computer communications for 13 years. Prior to joining Logica, he was a senior researcher for CSIRO, the Australian government's research organization. He has a B.Sc. from Makerere College in Uganda and a Ph.D. in physics from Washington University in St. Louis.

David E. Liddy is a consultant with Logica. He has a B.Sc. from LaTrobe University in Melbourne and a Ph.D. in applied mathematics from the University of Sydney. Since joining Logica, he has specialized in the analysis of packet-switching networks and is currently developing a set of computer programs to aid in this task.

Alvaro Andraca, McGraw-Hill Information Management Co., New York, N.Y.

Five years after deregulation, FM subcarrier comes of age

Satellite backbone networks and FM radio broadcast facilities are creating real-time links between producers and consumers of news.

The pace of modern business is making real-time information services increasingly attractive to many business executives. The tempo of critical business decisions, which used to be measured in days or hours, is today scored in a matter of minutes. To meet this need, McGraw-Hill Inc., the publisher of DATA COMMUNICATIONS and an information provider to industry for 100 years, began searching for the best method of providing cost-effective, time-critical information to a widely dispersed set of users.

Though the traditional media and electronic on-line services all provide information, McGraw-Hill envisioned building a new service that would beat the competition by providing real-time access to information from a variety of sources at a much lower cost than direct link-up. The winning distribution technique turned out to be a revamped version of the well-known FM subcarrier technology combined with direct satellite transmission.

Frequency modulation (FM) subcarrier is a transmission technique used to modulate an analog or digital signal onto the upper or subcarrier portion of an FM radio station's assigned bandwidth. Deregulation of FM subcarrier broadcasting by the Federal Communications Commission in 1983 breathed new life into this technology and applications started popping almost as soon as the deregulation order was approved (see "Surfacing: FM radio provides a new alternative for carrying data," DATA COMMUNICATIONS, July 1985, p. 173).

Now a hybrid form of FM subcarrier, which combines FM subcarrier broadcasting with direct satellite transmission, is proving itself as a cost-effective method to transmit time-critical information to multiple locations.

McGraw-Hill Inc., in New York, created the McGraw-Hill Information Management Company last year to develop and market new electronic-based information-distribution services based on the company's many news and editorial products. The first of these, McGraw-Hill's Executive One, is a business-information delivery service that offers real-time data to subscribers' personal computers or terminals from a number of sources. The data is transmitted via satellites and FM subcarrier. As part of its service, Executive One provides filtering tools that allow the user to discard automatically whatever portion of the data stream is not pertinent.

Executive One's sources for general news include PR Newswire and AP Online from Associated Press. News summaries and reports are received from Business Week magazine and other New York-based McGraw-Hill sources, as well as from Standard & Poors (S&P), which also provides other financial data. Securities quotes are updated hourly from the major stock exchanges.

Though this information is already available through a variety of media, ranging from radio, television, newspapers, and magazines to direct electronic links and on-line database-retrieval services, the distinctive characteristic of Executive One is its ability to provide real-time access to information from a variety of sources at a much lower cost than direct link-up. This data is compiled and broadcast as it becomes available throughout the business day. Updates from AP Online are broadcast around the clock.

The entire McGraw-Hill Executive One data stream is broadcast to each subscriber location. There, the user selects the information he or she wishes to receive according to categories designated for each source of information. Business Week, for example, offers Daily Reports and a Weekly Selected Article Summary, while PR Newswire categories include News Summary, Newswire, and S&P Creditwire. The user can further refine the data stream by placing keyword filters on each of these categories. Commonly chosen keywords include proper names of individ-

375

Point-to-multipoint alternatives

	LEASED LOCAL LOOPS	ONLINE INQUIRY	DIRECT SATELLITE	SATELLITE/FM SUBCARRIER
REAL-TIME DISTRIBUTION	YES	NO	YES	YES
UP-FRONT USER COSTS (HARDWARE AND INSTALLATION)	$200-$1,000	0	$3,000-$6,000	$800
MONTHLY USER COSTS INCLUDING AMORTIZATION OF MODEM COST	$50-$130	INCREASES WITH USAGE	0	0

uals and companies as well as industry-specific terms and issues.

The selected data can be directed to the screen, printer, or disk drive. (The only exception is AP Online, which requires a separate licensing agreement to permit storage.) A user can be alerted immediately to the reception of especially time-sensitive information by directing the output of data associated with particular categories or keywords to the printer.

McGraw-Hill's primary concerns in selecting a point-to-multipoint data distribution method were real-time distribution capability, low user start-up fees, and low long-term user costs. The table compares the four distribution alternatives that were considered. According to these criteria, FM subcarrier combines the technical and cost advantages of direct satellite transmission with the relatively low reception costs of FM broadcasting.

The Mainstream network

After evaluating the products offered by several network providers, McGraw-Hill opted for a network provided by Mainstream Data (Salt Lake City, Utah). The Mainstream network, illustrated in the figure, combines a satellite backbone network owned and operated by Equatorial Communications Inc. (Mountain View, Calif.) with major FM radio-broadcast facilities around the country to create a real-time link between the information providers and their subscribers.

Executive One data is transmitted to Mainstream Data's network control center in Salt Lake City. This information is received at 300 bit/s to 9.6 kbit/s via dedicated or dial-up telephone circuits, value-added networks, or dedicated satellite channels. All incoming land-based transmissions are protected by error-correcting modems.

Network control is managed by a pair of Stratus XA440 fault-tolerant minicomputers from Stratus Computer Inc. (Marlboro, Mass.). One computer handles network traffic, while the other is used for backup and development. The control center is protected by two independent air-con-

Cost-effective. When McGraw-Hill set out to create a real-time link between the information providers and their subscribers, it turned to a network architecture developed by Mainstream Data. It combined a satellite backbone network owned and operated by Equatorial Communications Inc. with major FM radio-broadcast facilities around the country.

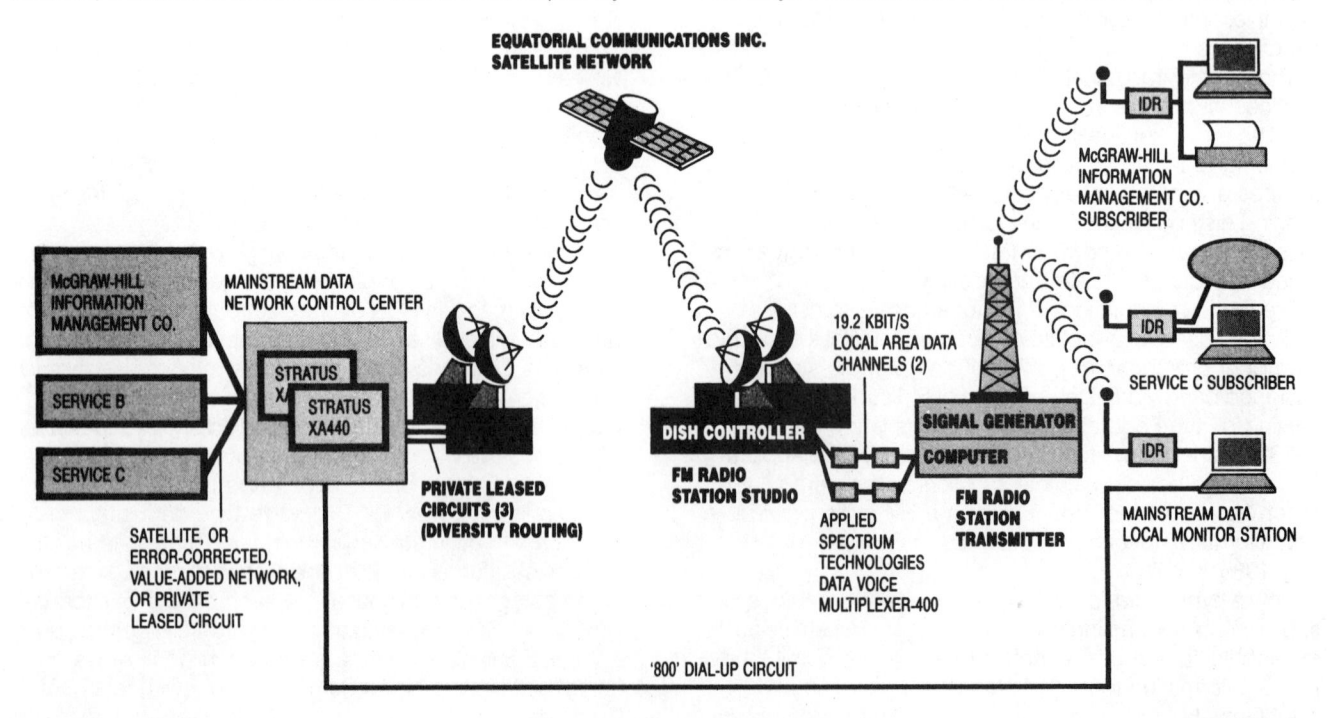

Block those errors

When McGraw-Hill wanted to evaluate the network offerings of various vendors, the critical technical issue was how to judge error performance and network reliability. Because FM broadcasting is not a full-duplex transmission method, no characteristic bit error rate can be assigned. Ideally, a properly designed, installed, and maintained FM broadcasting installation will never deliver a single erroneous bit. Errors that do occur are generally caused by equipment failure, such as a downed transmitter or satellite-transmission faults that are caused by terrestrial interference.

A common approach to error correction in simplex-transmission applications is something called "send-it-multiple-times error correction," or Simtec. Simtec schemes use multiple bandwidths to transmit two or more redundant, parallel, in-band data streams. The receiving unit compares the bit streams to detect errors and deletes redundant data.

Mainstream Data's two-layer error-correction technique eliminates the need in most cases for multiple sends. The first level, a double-bit corrector, detects Gaussian-type errors of up to two bits. The second level is a Reed-Solomon code that corrects bursts as long as 16 bytes.

Network performance is also affected by the signal-modulation technique selected. Schemes that rely on a phase shift in the waveform, such as phase-shift keying and quadrature phase-shift keying, are susceptible to multipath-induced phase distortions, resulting in bit errors. Frequency-shift keying, on the other hand, requires relatively high bandwidth. Mainstream Data utilizes a proprietary amplitude-modulation technique that limits sensitivity to multipath transmission while operating on a relatively narrow bandwidth.

An FM radio station's assigned bandwidth includes two 24-kHz subcarriers located above the audio baseband. The issue of data rate versus bandwidth affects both the efficiency and the viability of a subcarrier data-distribution technique. Vendors of FM-subcarrier networks have developed a number of approaches to this issue, ranging from those that provide a 38.4-kbit/s data rate using the full 48-kHz subcarrier baseband to those that operate on a single subcarrier at rates up to 9.6 kbit/s.

Transmission schemes that require both subcarriers may be difficult to implement, since in many cases one of these carriers is already devoted to other uses, such as internal radio station operations. A common example is the monitoring of a remote, unattended transmitter, in which case a subcarrier is used to broadcast operational data from the transmitter to the studio.

Mainstream Data generates a throughput data rate of 19.2 kbit/s on a single subcarrier. A raw data rate of about 24 kbit/s includes overhead for forward error correction.

ditioning setups and two backup power supplies, one battery-powered and the other gasoline-generator-powered. The network utilizes two parallel but interconnected data paths from the control center through FM transmitters. This prevents interruptions in service caused by hardware failures or environmental interference.

Information received at the control center is assigned a destination address and forward error-correction codes and is packet-interleaved with data from other Mainstream customers onto a 19.2-kbit/s packet-switched signal using statistical time-division multiplexing (see "Block those errors"). The multiplexed signal is transmitted from the control center over conditioned, redundant, private leased circuits to Equatorial's satellite uplink in northern California. Tertiary backup is provided by a dial-up circuit. From there, the signal is broadcast to dual receiving dishes at each of 12 participating FM radio stations across the country.

At each radio station, the signal passes from the dish to a dish controller to a minicomputer located at the transmitter. Here, the signal is reformatted for FM transmission. From the computer, the signal goes to a digital subcarrier generator that uses a crystal oscillator and microprocessor to produce a very stable, digitally generated waveform.

Earlier analog signal generators used coils and capacitors to produce a frequency. Aging and temperature fluctuation could cause these components to generate a signal that drifts into the main audio channel or into a second subcarrier region. Radio station management is particularly concerned with such potential interference, since it can have a direct impact on the quality of the station's signal.

Mainstream's subcarrier signal is modulated directly onto the station's main audio signal for local broadcast. The signal is received at subscriber sites using an FM antenna and Mainstream Data's Intelligent Data Receiver (IDR). Mainstream's single-user data receiver has an RS-232/RS-422 serial interface for connection to a PC, along with a Centronics-type parallel port for hook-up to a printer. It also includes a 256-Kbyte RAM buffer for temporary queuing of data being sent to the printer.

The receiving unit can be configured to operate in either a node or a packet assembler/disassembler (PAD) mode. In the PAD mode, the IDR breaks apart the interleaved data for local processing by a single user. The node mode allows the IDR to act as a network server, passing the data to the network in packetized form. Setup, network status, and troubleshooting instructions as well as data are indicated on the unit's LCD.

The Mainstream Data network is monitored continuously by a series of remote-monitoring stations that report regularly to the network control center. Located in each of the metropolitan areas served by the network is an automatic monitoring station, consisting of an FM antenna, a personal computer with special Mainstream Data software, an IDR, and an autodial modem.

Each station continuously records and analyzes a number of parameters, including error correction and signal quality and strength. The unit receives additional information from the local dish controller regarding satellite signal

Local area data channels: The critical link

In many cases, the receiving site's satellite dish cannot be collocated with the radio station's FM transmitter when a station is located in a downtown area. The downtown areas of cities tend to have heavy radio-frequency activity, which often produces signal interference.

To provide a reliable, low-cost, and easy-to-implement link from studio to transmitter, McGraw-Hill chose local area data channels (LADCs) leased from the local telephone operating company.

An LADC is a continuous metallic pair used primarily for data transmissions. It is cheaper than common 3002 voice-grade circuits; it is also more flexible in application. Tariffs on four-wire LADCs are typically 20 to 50 percent lower than for two-wire 3002 circuits. Two-wire LADCs are even cheaper—only half the cost of comparable four-wire channels.

Whereas transmission over a 3002 line is limited to 300 Hz to 3 kHz, LADC transmissions are constrained only by energy-density limitations. Federal Communications Commission requirements on energy density are designed to prevent the introduction of crosstalk between wire pairs by customer premises modems.

Transmission across an LADC requires the use of a limited-distance modem (LDM) at each end. McGraw-Hill began by employing standard four-wire, non-forward-error-correcting LDMs, but soon detected data errors across the local area data channel.

A unit from Applied Spectrum Technologies Inc. (Minneapolis, Minn.) was considered to help solve this problem. The DVM-400 Data Voice Multiplexer provides forward error correction (FEC) and operates at 19.2 kbit/s at distances up to 4.5 miles on a two-wire circuit.

FEC techniques do not rely on acknowledge/resend procedures to correct data errors. For this reason, FEC modems are often used in simplex applications, such as the Mainstream Data network.

The Applied Spectrum unit combines bit interleaving and forward error-correction techniques to improve the bit error rate of the underlying line condition by a factor of 10^4. A typical LADC link, for example, generates a bit error rate of 10^{-5}; the DVM-400 will typically improve performance to 10^{-9}.

The DVM-400's secondary in-band signaling channel provides diagnostic capabilities such as automatic line equalization, continuous data-line performance monitoring, and remote loopback activated from either end.

The DVM-400 operates above the voice frequency, leaving the baseband open for other applications, as shown in the figure. Mainstream Data utilizes this voice band to transmit a continuous signal from the dish controller to the transmitter, indicating the quality of service of satellite transmission. This information is read and analyzed by the remote monitoring unity as part of the total network-quality profile.

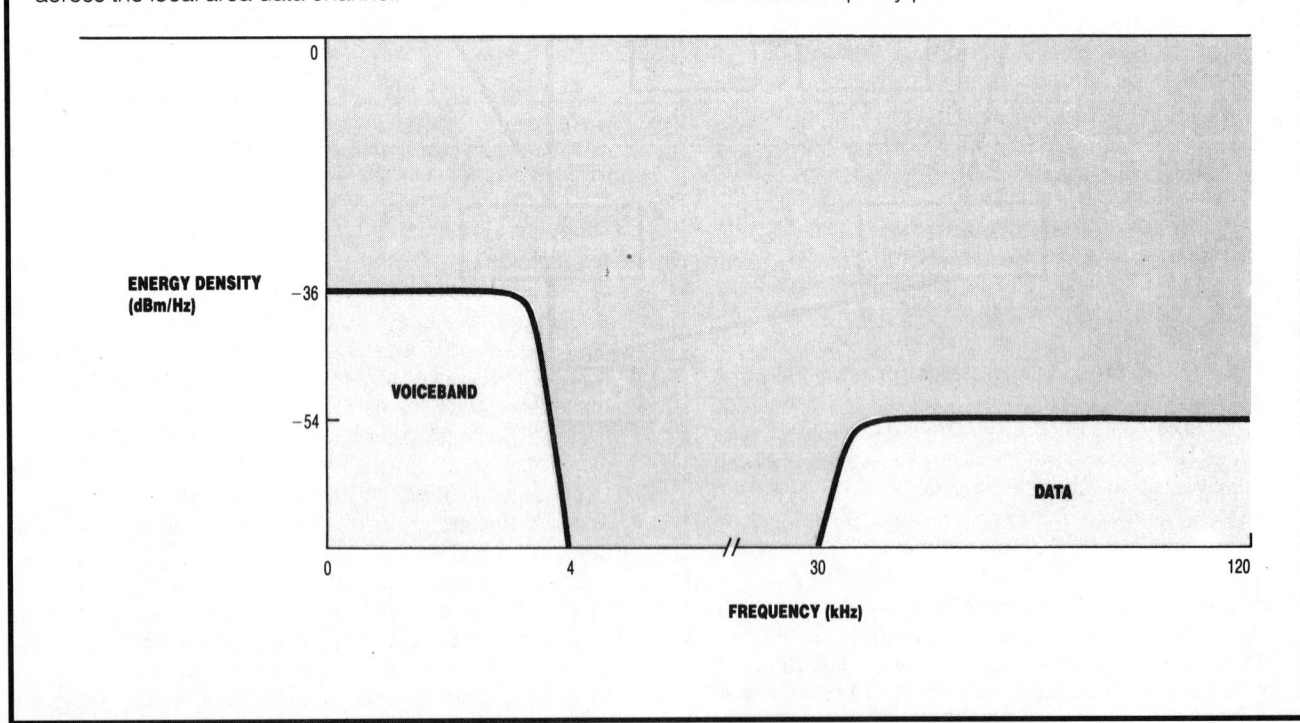

quality (see "Local area data channels: The critical link").

A summary analysis of the data is forwarded to the network control center at regular intervals via a dial-up circuit. If one or more of these parameters is found to exceed predefined limits, the center is notified immediately.

Emergency information appears as a flashing message on one of the network manager's video display. ■

Alvaro Andraca, McGraw-Hill's product manager for Executive One, received his B. A. in finance from Fairfield University in Fairfield, Conn.

Dennis Rose, VLSI Technology Inc., San Jose, Calif.

Wide area network: Sewing a patchwork of pieces together

A growing chip manufacturer met its expanding communications needs through the innovative use of IEEE 802 LANs chained nationwide through satellite technology.

Rapidly growing high-technology companies, dependent upon data communications for delivering their products or services, face some unique problems. Technical managers must supply the necessary computing and data communications resources while still providing for simple and rapid expansion of the network to accommodate the company's dynamic growth. VLSI Technology Inc. exemplifies this dilemma. Our network managers have had to integrate workstations and terminal servers, local area networks (LANs), and satellites.

We have used a number of approaches and technologies to meet our communications needs over the years; each was adequate at the time, but none could keep pace with our rapidly changing requirements. The wide area communications network we have recently installed is an excellent solution to our current needs and gives every indication of being capable of easy, cost-effective expansion as we grow in the future.

VLSI Technology Inc. was founded in 1979 to fill the growing need for very large-scale integrated (VLSI) circuits used in computers, communications gear, and electronics equipment. The firm has been growing by leaps and bounds because of the success of application-specific integrated circuits (ASICs).

ASICs contain numerous transistors, sometimes hundreds of thousands or more, interconnected to provide logic circuits that are designed to meet an equipment manufacturer's applications. Because of this customization, ASICs are having a major impact on the design of all types of electronics equipment. And their success has also had a major impact on the growth of VLSI Technology and on its computers and communications systems.

Customer demand for its products has been such that VLSI Technology has grown to its present sales rate of $150 million a year in just over six years. This extraordinary growth has greatly challenged the firm's computing and communications staff. From the original single building in Silicon Valley, we now have a campus of buildings in San Jose, Calif., two divisions in Phoenix, Ariz., software development centers in San Jose and in France, plus engineering-design centers in the following cities: Boston, Chicago, Dallas, Los Angeles, San Jose, Calif., Fort Lauderdale, Fla., and Princeton, N. J., as well as Munich, Paris, and Tokyo. In addition, VLSI Technology software development and design library centers are located in San Jose and Sophia-Antipolis, France. A new wafer-fabrication plant is currently being built in San Antonio, Tex. This rapid rate of growth has required a constant increase in computing power, communications, and personnel. Since 1982, the number of computers in use has increased from two to more than 300 and the number of terminals has increased from fewer than 100 to more than 700.

Growth and decentralization

The company has established remote design centers that enable our design engineers to be close to their customers, with whom they develop special-purpose ASIC chips. Each VLSI Technology site has its own computers and workstations, has access to a variety of special-purpose computers in the San Jose headquarters, and shares both technical and administrative information with the other company locations. This decentralized computing and resource- and data-sharing capability is essential to our continued success in the ASIC business. We must be located close to our customers, and every site must have easy access to all of VLSI Technology's computing power.

To continue offering a high level of customer support as the company grew, VLSI Technology's management decided to pursue an aggressive strategy of decentralizing the company's proprietary design tools and technical-support processes to the technology centers where key customers are located.

The design centers have become the focal point for practically all sales, circuit-design, customer-support, and training activities. To most VLSI Technology customers, their local design center is the company. As a result, each center must be able to provide the same level of technical assistance and design support as the original headquarters center. In order to remain competitive, each design center must have productive, cost-effective technology in computers, engineering workstations, and data communications.

The centers make it possible for customers to buy or license software and design tools. These can be used to create products at their own companies, at the VLSI Technology design centers, or through joint efforts. Once the chip is designed, it can be turned into a finished product at CMOS (complementary metal oxide semiconductor) and HMOS (high-performance metal oxide semiconductor) wafer-fabrication factories geared for quick ASIC turnaround. Time to market is often the single most important factor in the success of a new product, both for VLSI Technology and for its customers.

Each design center has a complement of engineering computers and workstations, plotters, printers, a Digital Equipment Corp. (DEC) VAX for administrative work, and associated terminals. Each design center has an Ethernet backbone for communications between computers and between terminals and computers. It is essential that the centers be able to share computing resources to make efficient use of the equipment and to handle the spikes in work loads.

Design-center managers and engineering support staff make arrangements in advance for users to access computers at other centers as necessary. However, the design engineer or programmer at a workstation does have access to any device on the network. The transparency of the network makes all computers appear to be on a LAN in the user's building, even though they may be thousands of miles away.

The corporate data communications network that supports the design and manufacturing activities has the added task of providing transmission for the growing volume of traffic for sales-order entry, accounting, electronic mail (VAXmail), and computer-aided manufacturing (CAM) data collection. Over 500 administrative terminals are scattered throughout the company's offices and plants. The network has more than 120,000 unique connections over which data can be transferred: Any of the 300 computers can be connected to any of the others, and the more than 100 terminal servers can connect to any of the computers.

The networks evolve

Of course, such total interconnectibility did not come about overnight. The company's first remote communications network was implemented in 1982 and used dial-up X.25 links to tie the domestic sales offices to corporate headquarters. The first two design centers, located in Boston and Dallas, were opened in 1984. The X.25 network costs rapidly grew from about $1,000 per month to more than $4,000 per month, and the delays in transferring large files imposed a serious burden on the engineers. It quickly became apparent that a faster, higher-capacity data communications solution was needed. The escalating monthly costs of the X.25 network prompted the company to investigate the use of leased lines for the higher-traffic routes.

After some searching, we tied San Jose to Boston and Dallas with leased satellite links connected to our VAXes through tail circuits and modems at each end. Operating at 14.4 kbit/s, the network provided greater capacity than our X.25 service. Unfortunately, however, it also gave us the worst of both the terrestrial and satellite worlds, due to the errors generated by the tail circuits and the transmission delays of the satellites. Large data transfers were difficult and unreliable. Transferring a file from a remote workstation to a central batch machine took considerable time and usually required three individual transfers—if all went well. The user had to initiate a transfer from the workstation to a local VAX. A second transfer got the data from the local VAX across an analog leased line to a second VAX at the remote location. The third transfer moved the data from the remote VAX to the intended remote workstation or computer where the file would be used.

And when all didn't go well, network problem isolation and resolution were often slow and difficult because of finger pointing between the satellite-service vendor, the tail-circuit providers, and the modem vendors.

In the San Jose headquarters, the VAXes in four buildings were tied together with Decnet using some Ethernet segments and some interbuilding RS-232-C Decnet links. By early 1985, the proliferating DEC computers and Apollo workstations clearly needed a more efficient way to talk to each other. To solve this local communications problem, the company extended the Ethernet local area network across the campus of three closely clustered buildings. The San Jose design center, approximately one-half mile away from the other three buildings, remained tied to the campus with the existing RS-232-C Decnet links. Two Hewlett-Packard 3000/70s, supporting over 100 terminals in San Jose, were also tied to the local area network using Ungermann-Bass network interface units, NIU-180s). The NIU-180s allow any terminal to talk to either VAX or HP computers.

The Decnet protocol was used for communications between VAXes, while the transfer of large files between workstations (Apollo, Ridge, HP, and Sun) and batch-processing machines (ELXSIs, VAX 785s, and VAX 8650s) was handled by standard Transmission Control Protocol/Internet Protocol (TCP/IP). Connecting the administrative terminals (DEC VT 100s and their equivalents) to the LAN with the NIU-180s required using the Xerox Network System (XNS) protocol. And the local area transport protocol LAT 11 was used by a variety of terminal servers. Our main campus LAN now connected many computers and hundreds of terminals and workstations located in four buildings and operating with four different protocols.

Our experience with the leased-line/satellite links to

Data link bridges and global addressing

The IEEE 802 standards define the data link layer of the ISO (International Organization for Standardization) model (Fig. A). The 802.3, 802.4, and 802.5 standards describe the three different methods of data transport; aspects of network management and control that are common to all three are defined by IEEE standards 802.1 and 802.2.

At the heart of the 802 standard is a concept called global addressing, which ensures that every device manufactured has a unique identifying address. In each 802 frame, a source address and destination address identify where the packet came from and where it is going. Because every device on an 802-based network has a unique address, the network can ensure that every data packet reaches the correct location. This global addressing concept is contained in the 802.2 standard and applies to 802.3 implementations such as Ethernet, 802.4 token bus networks, and 802.5 token ring networks. The addressing scheme is constant for all 802 standards, although the media access methods may differ and present other kinds of problems.

Figure B shows an elementary network configuration of a small 802 wide area network consisting of two Ethernet local area networks (LAN) connected by TransLAN bridges. Each bridge is "promiscuous," that is, it listens to every data packet transmitted on its LAN. If the packet is originated by a station on the local LAN and its destination is another station on the same LAN, the bridge filters the packet and does not forward it across the transmission link to the remote LAN. If the packet originates on the local LAN and its destination is a station on the remote LAN, the local bridge forwards the packet across the link.

Every packet transmitted onto an Ethernet LAN contains two addresses: a source address that uniquely identifies the sending station and a destination address that defines those stations that will receive the packet. The bridge uses a software algorithm developed by Digital Equipment Corp. that creates a routing table by listening to the addresses in actual data traffic. It does this by looking at the source address in each packet it receives, creating an entry in the routing table, and identifying the network port on which it heard that address. In this way, it learns in just a few seconds of operation the addresses of stations that are on its local LAN and those that are on the remote LAN communicating with the local LAN.

In Figure B, the bridge attached to LAN A learns that stations 1, 2, and 3 are on LAN A by listening to the source addresses of traffic between these stations. It likewise learns that stations 4 and 5 are remote by listening to traffic coming from the bridge on LAN B. This initial learning and table-building process occurs automatically during the first few seconds of operation. Likewise, the bridge on LAN B has built a table identifying stations 4 and 5 as local.

When the bridge on LAN A hears packets being sent from station 1 to station 2, it filters (ignores) them. When it hears a packet being sent from station 1 to station 4, it forwards the packet. To station 1, station 4 then appears to be on LAN A; to station 4, station 1 appears to be on LAN B.

Each bridge continues to learn from each packet even as it is filtering and forwarding, allowing it to relearn the new location of stations as they change physical location within the network. As stations, or even whole LANs, are moved and added, the bridges recognize the new configurations and adapt to them; there is no need for lengthy reconfiguration routines or manual table entries. New configurations and devices are operational as soon as they are installed, saving both labor and time.

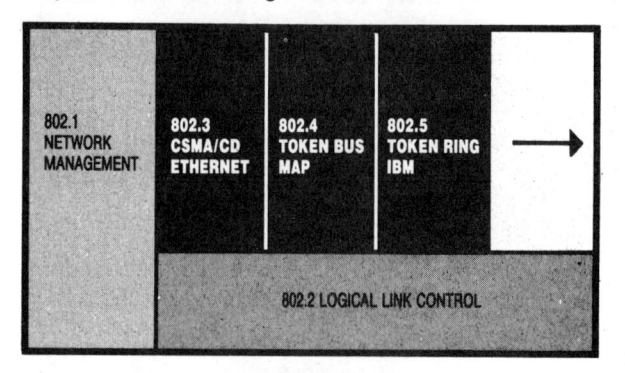

CSMA/CD = CARRIER-SENSE MULTIPLE ACCESS/COLLISION DETECTION

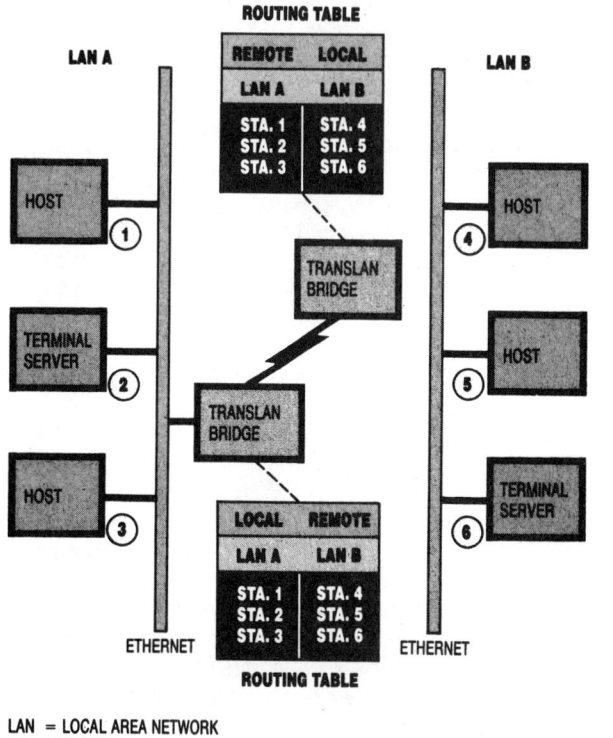

LAN = LOCAL AREA NETWORK
STA = STATION

Boston and Dallas and the Decnet/Ethernet in San Jose had convinced us that we needed a way of tying together the LANs planned for each design center with a more reliable, multiprotocol network that would meet at least four important criteria:

- *Integrity*. Provide maximum file integrity through fast, error-free transmission of large data files to and from any point on the network. These design files, which can be larger than 5 megabytes each, contain every detail of a complex ASIC chip. Any errors in transmission could cause expensive problems for VLSI Technology and its customers.
- *Easy connections*. Provide full, transparent connections to enable sharing of design and computing resources across the network for optimum terminal and computer use 24 hours a day, seven days a week. The wide variety of computers, terminals, and workstations that share the LAN at each location must be able to share the overall network just as easily.
- *Expandability*. Provide a readily expandable network that is easily managed end to end, regardless of geographic location.
- *Economic feasibility*.Provide a stable, predictable, and economically feasible cost per drop to simplify planning, installation, and operations over the long term.

These four broad criteria covered the critical needs of our users and were determined by examining all the problems we had with the existing links. Other important but less significant criteria included:

- *Ease of use*. Everyone ought to be able to use the network easily.
- *Serviceability*. Service personnel must be able to troubleshoot and repair the network easily.
- *Ease of administration*. The network must be easy to manage with minimal extra labor.

Hurdles and solutions

The search for solutions to our network problem led us to consider further use of leased lines, satellite links, and fully cross-connected routes with redundant links. A packet-switching approach using AT&T Accunet was considered but could not provide the direct geographical coverage we needed. We also looked at the feasibility of our own dedicated satellite network using conventional very small aperture terminal (VSAT) technology from a variety of vendors.

At this point, it looked doubtful that we could build a network that would really meet our needs, for the following reasons:

- Leased ground lines were very expensive, provided only point-to-point service, and had data error rates that were marginal for our application.
- Leased vendor satellite links provided the higher data integrity inherent in satellite transmission, but the unreliable tail circuits from the service provider's earth station to VLSI Technology sites virtually eliminated that advantage. Also, the long-term stability of some of the service providers was questionable (as evidenced by RCA Satcom quitting the long-line business during the selection process).

- VSAT technology wasn't well suited to our heavy load of two-way, interactive communications.
- We had found nothing that offered the same protocol transparency across a wide area as our Ethernets provided in the local area networks. (The DEC LAN Bridge 100 could transparently interconnect two Ethernets, but not with the variety of transmission media we needed.)

However, the DEC representative suggested that we talk to Vitalink Communications Corp. about its TransLAN bridge. We found that the TransLAN bridge was protocol transparent and could interconnect LANs over long distances using a variety of transmission media. In our discussions with them, Vitalink representatives proposed a solution that appeared to solve all the elements of our rather demanding network problem.

Vitalink proposed interconnecting the local area networks at each of our plants into a single wide area network by using its TransLAN bridges, employing the most appropriate transmission medium for each link. The long-distance hops between our San Jose headquarters hub and the design centers would be made through a satellite link, using a Vitalink ku-band earth station at each location. (Vitalink calls the combination of its TransLAN software and its ku-band earth stations SatLAN.) Two shorter links, between Boston and Princeton and between Los Angeles and San Jose, would be made through leased lines, and the half-mile distance between our San Jose headquarters and the San Jose design center would be handled by a T1 microwave link.

The concept of a wide area network based on the 802 standard fit well with our positive experience using Ethernet LANs. The TransLAN data link-layer bridge transparently interconnects Ethernets into a single, integrated wide area network based on the IEEE 802.3 standard. The bridge operates at the two lowest protocol layers in the Open Systems Interconnection (OSI) model, the data link and physical layers. The resulting network is transparent to all 802.3-compatible protocols, including the Decnet, XNS, LAT 11, and TCP/IP (Transmission Control Protocol/Internet Protocol) protocols we were using on our LANs. It is also accessible through a variety of protocol-specific interfaces, such as gateways, packet assembler/disassemblers, and terminal servers. As a result, all of our computers, workstations, and terminals can communicate across the single network.

Because the bridges use the 802 concept of global addressing, bridge-configuration efforts are minimal. In the first few seconds of operation, each bridge learns its location in the network—that is, which devices are physically located on its local LAN and which are located on remote LANs. Each bridge then transmits across the serial link only those packets addressed to stations on remote LANs. As terminals, stations, or whole local area networks are moved or added, the bridges recognize the new configuration and adapt to it; there is no need for performing lengthy reconfiguration routines or system generations for the bridges (see "Data link bridges and global addressing").

We had some reservations about using ku band rather

than C band for the satellite links because of the susceptibility of the shorter wavelength signals to fading under heavy rain conditions. The positive side of using ku-band equipment was the much smaller dish sizes with attendant lower installation costs and a faster FCC licensing process. Since SatLAN is designed specifically to support critical applications such as ours, it includes a method of continuously monitoring the quality of each link and automatically adjusting the output power of the individual ground station to compensate for rain effects. In addition, since the network nodes can support satellite or terrestrial transmission paths, parallel or switched circuits can be used as alternate paths on critical links as necessary.

Life with a wide area network

In October 1986, we contracted with Vitalink to provide a self-contained nationwide communications network to link VLSI Technology's domestic design centers with corporate headquarters. The first SatLAN link between Boston and San Jose was up and running in January of 1987, and by the end of February, Dallas and Chicago were on line. The Fort Lauderdale link became operational in June, and the Phoenix operation joined the network in the fourth quarter of 1987. The current network is shown in Figure 1.

The installation process was relatively trouble-free, as can be noted from the fact that all the links were operational in less than a year. Each station was in full operation within one day after its antenna was installed. Antenna installation itself was also relatively simple, but we did learn that, in the South, the building roof may have to be reinforced prior to installing the antenna. In general, roofs in the North are stronger than those in the South to accommodate snow loading. We also had some intermittent microwave radio-frequency equipment problems on the link between San Jose headquarters and the San Jose design center; Vitalink quickly resolved these difficulties using a remote diagnostic capability.

For strategic reasons, the San Jose, Dallas, and Boston design centers are designated hubs, with smaller centers in the same part of the country linked to them for management, administration, and sharing of computer resources. The Los Angeles design center, for example, is connected to the San Jose center 400 miles to the north by a leased 19.2-kbit/s lines. The Princeton design center is connected to Boston by a 19.2-kbit/s leased line. The centers in Chicago and Fort Lauderdale communicate with the Dallas

1. VLSI network. *A combination of ku-band satellite hops, T1 (1.544-Mbit/s) microwave links, and leased lines provides the right bandwidth, performance, and economics. With this setup, new locations can be brought on line or bandwidths changed easily as the company's data communications requirements evolve.*

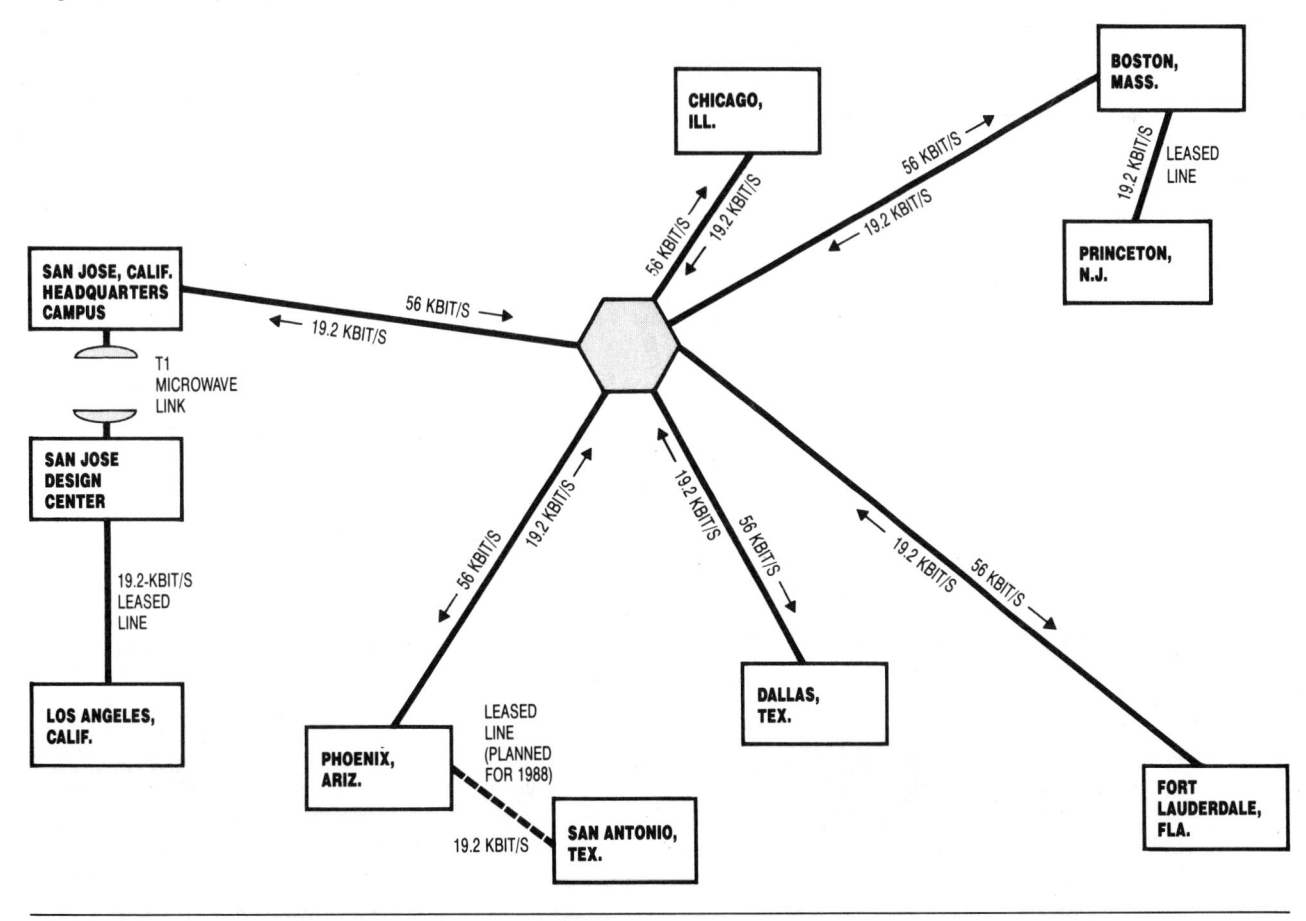

2. San Jose campus. *Multiple resources are linked together at the corporate headquarters location in San Jose, Calif. A variety of computers and workstations are hooked up to an Ethernet network, which extends to other company sites over ku-band satellite connections and T1 circuits as well as fiber optic links.*

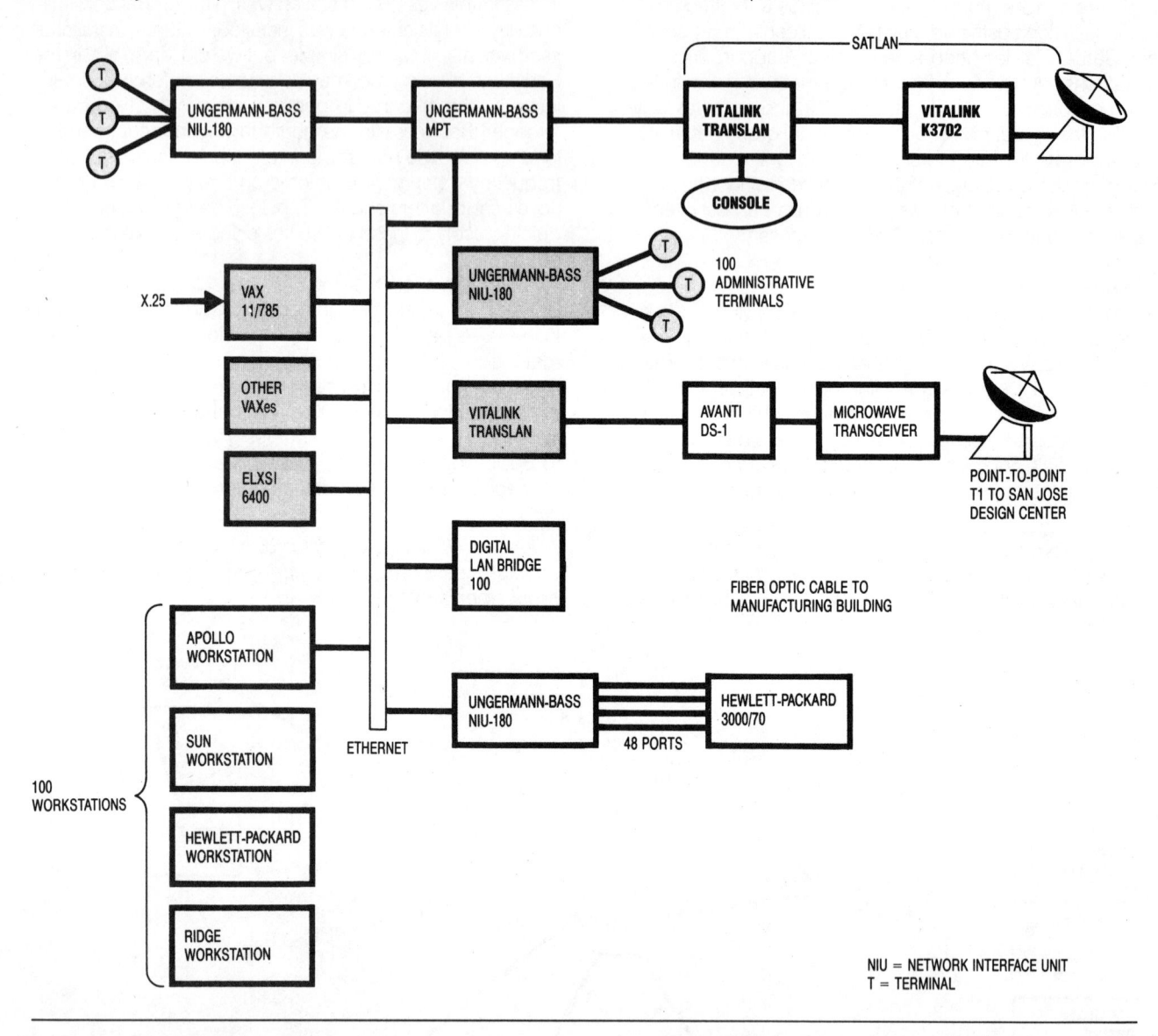

NIU = NETWORK INTERFACE UNIT
T = TERMINAL

center and San Jose through their own satellite dishes.

The individual design centers each have small 1.8-meter dishes, while larger 3.7-meter antennas are employed at the San Jose headquarters and in Phoenix. San Jose transmits in a broadcast mode to all network locations at 56 kbit/s; each design center transmits back to San Jose on a 19.2-kbit/s return channel. The small antenna size at the design centers can provide return data rates as high as 224 kbit/s, which leaves plenty of bandwidth for future expansion as our requirements evolve. The larger antenna size is well suited to data rates up to 1.544 Mbit/s.

The design centers in Munich, Paris, and Tokyo are a different situation, due to the need to route all communications through the national telecommunications services of the host countries. At present, these locations operate more autonomously and communicate with San Jose by means of X.25 packet-switching networks. We plan to install TransLAN bridge nodes in these locations, but the communications links will continue to be provided by the government-controlled agencies.

We are planning a new satellite link to our European development center in Sophia-Antipolis on the French Riviera. This link will use a TransLAN bridge at either end to integrate this office's LAN into the overall 802 wide area network. The actual satellite service will be provided by the French Telecommunications Company (FTCC). We have programs that automatically update changes to the ASIC software and design libraries from one development center to the other; this new link will keep developers on both sides of the Atlantic in sync with each other.

The headquarters computer center at VLSI Technology provides batch computing for the programmers doing chip-design software development and support for the semiconductor engineers who design the physical cells, gate arrays, and interconnects that are used to transform a logic diagram into a chip design. As shown in Figure 2, the headquarters center now has a mixture of mini-supercomputers, including a DEC VAX 8650 and other VAX models, Sun 3/280s, and ELXSI 6400s. A Mach 1000 computer from Silicon Solutions Inc. is a simulation accelerator providing fast confirmation of our circuit designs. A VAX 11/785, in addition to being a regular computing node, also acts as a communications processor for the X.25 portions of the network.

As at all of our major hubs, the computers, workstations, and terminals in Boston are on a LAN coupled to the local TransLAN node, which in turn transmits and receives through the SatLAN satellite ground station. The San Jose design center is the largest of the design centers and provides design services to our customers in Silicon Valley. This center now has two VAXes and dozens of Apollo workstations and terminal ports interconnected by an 802.3 LAN. Vitalink bridges join this LAN with the campus network (and thereby with the 802 wide area network) via a T1 point-to-point microwave link. The San Jose center also supports the 19.2-kbit/s leased line from Los Angeles. The Boston design center is the largest of our centers in the East. Its equipment includes DEC VAX 11/785 and ELXSI 6400 computers, multiple Apollo graphics workstations, and three NIUs providing 24 terminal ports.

The network processors at each of the design centers and in San Jose have a management console to enter network commands and display network statistics. Because each bridge sees every packet on its LAN and all associated links, we can get complete traffic information. We can determine traffic patterns and line use and identify potential areas of congestion.

Vitalink provides continuous on-line network supervision and management as part of its contract. Its network-operations center continuously monitors the satellite transponders, individual data links, and transmission hardware. It also monitors network conditions and logs transmission errors so that preventive maintenance can be used to avoid unscheduled network downtime.

The result is a single, nationwide data communications network that permits our workstation and terminal users at any location to access any computer on the network, transfer files between computers, and share resources among all locations served by the network. The operation of the 802 wide area network has proved to be totally transparant to network users and makes all computing resources on the network appear local.

Benefits

We have realized benefits from the installation of our wide area network in the following areas:

■ The network provides a single, transparent backbone for all groups in the company: engineering, sales administration, financial management, and customer support.
■ The simplicity of the network and the use of the TCP/IP, XNS, LAT 11, and Decnet error-correcting protocols have provided the data integrity necessary for the routine, error-free transfer of large design files from one side of the country to the other.
■ The operation of the 802 wide area network is totally transparent to its users. Accessing a computer or terminal anywhere in the country is done as easily and in the same manner as connecting to a local computer. Our users have not needed any additional training or special commands to use the network.
■ The ease with which the network can be expanded to include new ground stations has been demonstrated as each new location has been brought into the network fold. The lack of interdependency between nodes allows us to add new locations and make network changes without interrupting service to existing locations.
■ Vitalink's network-operations center has provided quick response to any network problems. We are advised in advance of preventive maintenance schedules and often receive advance notice of impending network problems. Network availability has averaged close to 98 percent since the first link was installed and has been over 99 percent for the past three months.
■ The network has provided higher performance for equivalent dollars when compared with leased satellite lines. And it is less expensive and provides better service than leasing multiple, high-performance, point-to-point AT&T terrestrial lines. At present, the monthly line charges and depreciation of the capital investment are virtually identical to the amount we would be paying for leased satellite lines and modems. Three years from now, when the network is fully depreciated, the monthly costs will be substantially less than they are now.

At an August 1987 meeting of all our design center managers, we conducted a survey to determine the level of user satisfaction with our new network. There were no negative comments. Our former communications problems had disappeared.

Our selection of a protocol-independent wide area network with integral satellite transmission facilities provided a smooth expansion of our engineering and administrative data communications capabilities. Data integrity, which had formerly been a serious problem, was no longer an issue. We demonstrated to ourselves, our users, and company management that this wide area network architecture, using the IEEE 802.3 standard, enabled us to interconnect our growing number of locations into an economical and reliable, nationwide network providing excellent performance and simplicity of use with minimal network management problems. ■

Dennis Rose is the engineering support manager at VLSI Technology Inc. in San Jose, Calif. He previously worked for GE Calma for four years as a senior diagnostic engineer and as North American technical operations coordinator. He has a B. S. in chemistry from Westminster College Fulton, Mo.), has done advanced work at the Massachusetts Institute of Technology, and has an M. B. A. from the University of Phoenix.

Bryan F. Gearing, Digital Equipment Corp., Merrimack, N. H.

Coming soon: A cabling standard for buildings

Finally! After a three-year wait, users are about to get a standard for fiber and copper cabling in commercial buildings.

If your company is planning to build a new facility, the question you have probably already been asked is: What's the best way to wire it for voice and data? Well, there will soon be a definitive answer: A new standard for fiber optic and copper cabling in commercial building wiring is almost finished. The Electronic Industries Association (EIA) TR-41.8.1 Ad-Hoc Working Group on Building Wiring, after laboring for three years with dozens of vendors and users, is recommending that 62.5/125-micron multimode fiber become the standard for data and voice backbone applications in commercial buildings and campus-size local area networks.

Final approval by the EIA and the American National Standards Institute (ANSI) could come at any time. Once the standard is approved, it will have a far-reaching effect: Architects and engineers will likely start specifying the new standard fiber optic medium whenever local area data distribution systems are called for. What kind of an impact could such a standard have in the real world? Look at the case of unshielded twisted pair: The RJ connector has become ubiquitous in the office and in the home as well.

No single answer

From a user's perspective, the ideal wiring standard would specify a single medium with a standard connector. But the EIA committee discovered it could not limit its work to fiber; it found a role for unshielded and shielded twisted pair, as well as thin coaxial cable, in building wiring.

The important issues addressed by the fiber and copper wiring standard—media, topology, distances, number of cross connects, and connector loss—ultimately affects the operation and performance of the communications network. It is in the realm of operation and performance that fiber shows its superiority over copper.

The standard targets the commercial office building. The building's internal data-transport wiring is broken down into several elements, including the office, horizontal, backbone, and administration wiring elements. When combined, these elements make up the total office network (Fig. 1).

The office element is the wiring from the workstation to the wall outlet and consists primarily of the line cord. This element is not part of the standard because it is not permanent building wiring.

The horizontal element represents the wiring connection between the serving closet and the outlet on the wall of the office. The length of these cable runs is estimated to be a maximum of 295 feet in commercial offices. Each user requires a separate horizontal element.

The backbone element represents the connection between the local wiring closet and the building closet (building backbone) or between buildings (the campus backbone). The building and campus backbones are functionally the same, though the type of cable used can vary, depending on whether an indoor or outdoor application is envisioned.

The administration element addresses the moves, changes, and rearrangements that invariably take place in a network. It consists of the hardware, labeling, and documentation needed to manage the individual connections. Since the wiring in a commercial building has to accommodate a large number of variables, the creation of a single standard to satisfy the entire user community was difficult to achieve. Building codes, construction materials, and office styles have an impact on the types of cable that can be used and the cable paths. Building styles and dimensions, along with the use of the building, also affect where closets can be placed and the lengths of cable runs.

The star topology works best with the fiber optic standard (Fig. 2). A physical star is versatile: It can be configured in a wiring closet to perform as a logical ring,

1. Wiring elements. *The proposed Electronic Industries Association's standard for commercial office building wiring identifies the internal office, horizontal, backbone, and administration wiring elements. When combined, these elements make up the total office network. The standard presupposes the coexistance of copper with fiber.*

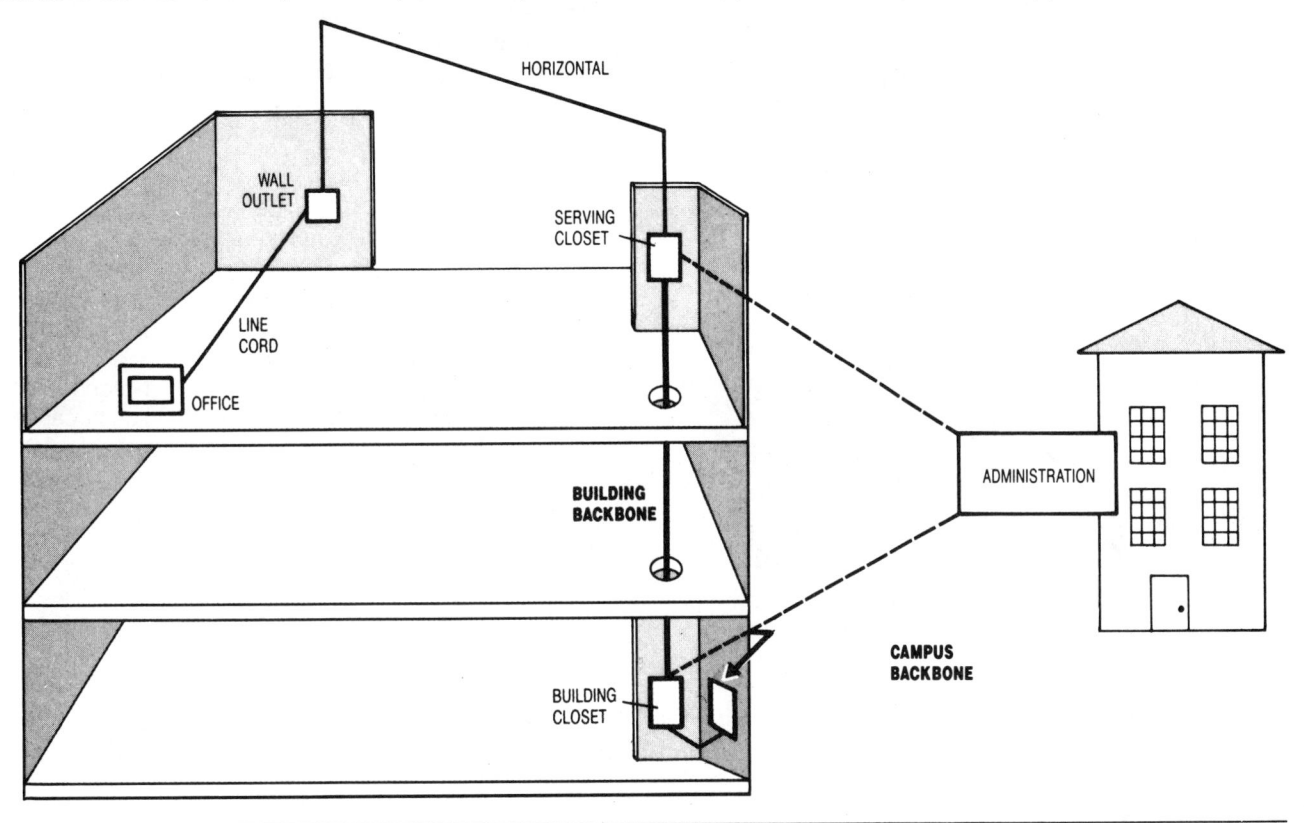

a logical bus, or a logical tree in either a centralized or distributed communications network. In addition, the star has other advantages: It provides central reconfiguration points, for example, which come in handy for maintenance, traffic balancing, or when new technology is introduced.

The flexibility provided by the star topology supports the longevity requirement for building wiring, as well as product and vendor independence. It is also application-independent. This means that the same closets can be used for housing active and passive equipment for many voice, data, and video applications.

Some copper-based products may have distance constraints that conflict with the hierarchical star topology. This is true, for example, of the coaxial backbone for Ethernet—today's most extensively installed local area network (LAN). Not so the fiber-based products. For example, the proposed internode distance for the Fiber Distributed Data Interface (FDDI) is 6,560 feet. This distance is unlikely to affect intrabuilding cable plant requirements and will satisfy most interbuilding requirements.

The two predominant types of media used in wiring today are optical fiber and copper. The optical fiber medium has several intrinsic advantages over copper: It is immune to ground potential differences and to electrical overloads. In addition, its small diameter means it can be threaded into existing ducting, cable trays, and risers—thus avoiding disruptive and expensive installation procedures.

Fiber connections, which include connectors as well as the splices, are still expensive and labor-intensive compared to copper. But this is a dynamic area where significant cost reductions can be expected. For example, a move from ceramic to plastic connectors is already under way, and considerable research is being undertaken to find a way to mate fiber ends without significant polishing. If both efforts succeed, the parts counts and the labor costs associated with installing fiber will drop.

Moreover, light-emitting diode-driven multimode fiber backbones are inexpensive enough to compete with copper. Fiber can be shared by many users, and its distance and bandwidth are superior to copper.

In new installations, fiber generally wins out over copper because of its intrinsic benefits and its future performance potential. At 295 feet, rates on the order of 500 Mbit/s can be achieved using baseband schemes—and multiplexing extends the information-carrying capacity of the cable into the gigabit-per-second transmission range.

If multiplexing cannot provide enough capacity, multimode fiber backbones can be upgraded to single-mode fiber. The cost of replacing multimode with a single-mode backbone to get the added performance can be justified since the bulk of the building cable plant cost remains fixed

2. Top star. *The flexibility of the hierarchical star topology gives the longevity needed for commercial building wiring. A star can be configured as a logical ring, bus, or tree.*

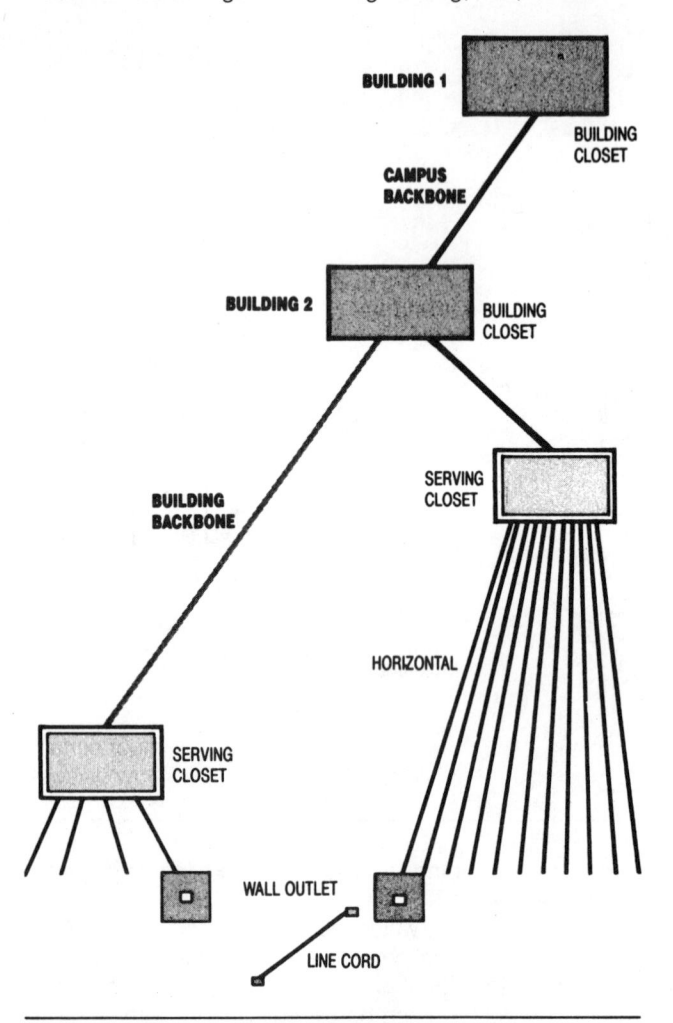

in the horizontal, multimode-fiber plant.

LAN installations in commercial buildings have gone hand in hand with deployment of shielded twisted pair and inexpensive coaxial cable in the office. Three of the major office-automation suppliers have a big stake in copper media:
- AT&T's 1-Mbit/s and 10-Mbit/s Starlan use 100-ohm unshielded twisted pair.
- IBM designed its 4-Mbit/s Token Ring around a 150-ohm shielded twisted pair.
- DEC's 10-Mbit/s Ethernet uses an inexpensive thin coaxial cable.

The standard recognizes each type of medium:
- 100-ohm unshielded twisted pair in four-pair cable.
- 150-ohm shielded twisted pair in two-pair cable.
- 50-ohm thin coaxial cable.
- 62.5/125-micron "dual window" multimode fiber.

The standard provides users with guidance in selecting from among the different media. If a standard that specifies many kinds of media seems far from the ideal, at least it may halt proliferation of different wiring schemes in commercial buildings. The unshielded twisted-pair wiring satisfies current and future voice requirements of the Integrated Services Digital Network (ISDN) and it can be used with most LANs. Coaxial cable compares favorably with unshielded twisted-pair wires in size and flexibility, and it has a standard connector. Though fiber as a horizontal element in building wiring is still rare, in time it likely will supplant copper.

Picking the fiber

The performance characteristics of different fiber proposals were not significant enough to determine the standard fiber size. A proposal for 100/140-micron fiber was eliminated because it did not conform to the commonly accepted external diameter. Nearly all multimode and single-mode fiber cables have an external diameter of 125 microns. An alternative, adopting an 85/125 fiber, was rejected because it has seen little use, and a 50/125-micron proposal was turned back because the 62.5/125 proposal has superior coupling characteristics.

How was the committee able to settle on just one standard size? It was timing. Since none of the major equipment vendors were sufficiently committed to a particular size, they were open to change. The 62.5/125 proposal prevailed.

The copper connectors at the wall outlets and in the cross connects have also been standardized (Table 1). However, the fiber connector is currently undecided. The FDDI committee is standardizing on an ST-based duplex connector; it is polarized and is keyed for the different classes of connection. A smaller duplex connector, which is polarized and unkeyed, will likely be adopted for commercial building wiring at some later date.

All two-wire transmission schemes require a crossover, or a lead transposition, so that transmit at one end of the link goes to receive at the other end. One way to achieve this is by using a crossover in the equipment (Fig. 3). This

Table 1: Commercial building cabling standards

MEDIUM	OUTLET	CROSS CONNECT
100-OHM UNSHIELDED TWISTED PAIR	8-PIN MODULAR; ISDN OUTLET	PUNCH-DOWN OR OUTLET CONNECTOR
150-OHM SHIELDED TWISTED PAIR	IBM CONNECTOR	IBM CONNECTOR
50-OHM THIN COAXIAL	STANDARD BNC	STANDARD BNC
50-OHM THICK COAXIAL	NONINTRUSIVE TAP	?
62.5/125-MICRON MULTIMODE FIBER	?	?

ISDN = INTEGRATED SERVICES DIGITAL NETWORK

3. Crossover simplified. *An equipment-based lead transposition makes it possible to assign the same pin numbers to equipment on both ends of the link.*

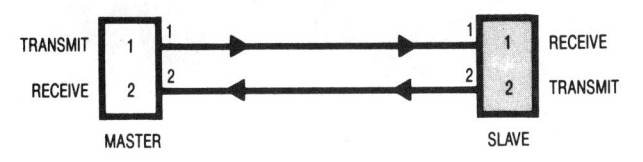

```
TRANSMIT  1 ─────────────►  1  RECEIVE
RECEIVE   2 ◄─────────────  2  TRANSMIT
          MASTER                SLAVE
```

makes it possible to assign the same pin numbers to equipment on both ends of the link. Doing so satisfies cases that define equipment as from and to, but it does not satisfy peer connections—connections between identical equipment that uses the same pin numbers.

In peer-to-peer communications, crossover wiring is needed. If the equipment pin numbers are the same at both ends of the link, simple crossover wiring will suffice (Fig. 4).

When the crossover wiring and extension cables are used, an odd total of crossovers is required to ensure an overall end-to-end crossover. That can be achieved transparently by using couplers that have crossovers built in (Fig. 5). Such an arrangement makes it possible to extend a wiring run by adding couplers and a cable—provided it results in an odd total of crossovers. In this approach, all cables are the same and have the same connector at both ends.

In a building, straight-through cabling is easiest to install and maintain. But the crossover approach is clearly superior for point-to-point interconnections between equipment. With the crossover approach, a single cable type can be used for all wiring, and the user need have no knowledge of lead transpositions.

The crossover wiring poses some minor management concerns, however, since connectors have to be installed in opposition to each other at each end of the cable. This means that knowledge of the remote end is required to install the near end. One way to overcome that is by using directional indicators or arrows along the individual fibers. It simplifies installations, and the arrows on the connectors solve the maintenance and administration problems. At installation time, just line up the arrows on a cable with the arrows on a connector—no pin numbers are needed (Fig. 6).

This solution presupposes adoption of a crossover standard for commercial building multimode cable. If such a standard were finally accepted, it would encourage system designers to follow the crossover concept.

Currently, simplex fiber connectors are used nearly everywhere. This leaves end-to-end crossovers in the hands of the installer. But the FDDI standard calls for a duplex connector with crossover in the wiring, so there is a precedent for the crossover approach. This may help it win acceptance for the fiber commercial building standard as well. However, where straight connections in building wiring are the norm, it is unlikely that a crossover scheme will be adopted for copper.

When planning a new building, the standard should be consulted for medium specifications, topology, distances, and connection losses. Fiber optic cables should meet the published specification; using cables below the specification limitations may work in the initial installation, but when the wiring is reconfigured or a new communications system is added, problems are likely to occur.

By the same token, there are no benefits to installing cables that offer better performance than the specification because future communications systems will not be designed to utilize your cable's superior performance characteristics.

The highs and the lows

The runaway adoption of unshielded twisted-pair wiring presents another potential class of problems—and another reason to consult the standard. Heavy user pressure to operate many established communications networks over unshielded twisted pair has led to the appearance of dozens of kinds of adapters in the marketplace. The picture is further complicated by a large variation in the types of cables being offered.

Amid such confusion, the low-bandwidth applications tend to be very forgiving. However, the high-bandwidth applications may exhibit poor performance if they are not specified for the correct cable. If unshielded twisted-pair wiring is specified incorrectly, it may unfavorably affect such factors as error rate, distance, and the number of nodes that can be put on the network. Also, under certain conditions, unshielded twisted pair does not allow daisy-chaining. Understanding these limitations is important. When the medium is being extended to its limit, operating margins are in jeopardy.

In the horizontal wiring elements of a building, the recommendation calls for two separate cables: one cable of four pairs of unshielded twisted copper and either an additional four-pair unshielded or a two-pair shielded or thin coax is recommended. Mixing signaling schemes within a cable sheath should be avoided in the horizontal runs, since crosstalk can become a problem.

Recommendations for backbone wiring tend to be application-dependent. If telephone traffic is envisioned, 50 pair or more of copper is acceptable. If an Ethernet LAN backbone will run up through the building risers, thick coaxial cable can be specified. Shielded twisted pair works fine for a Token Ring LAN backbone.

But fiber may be the best option. Though cost tradeoffs between copper and fiber must be evaluated, fiber

4. Peer crossover. *In peer-to-peer communications, crossover wiring is usually required. If pin numbers are the same at both ends, crossover wiring will suffice.*

```
TRANSMIT  1 ┐         ┌ 2  RECEIVE
            │ ───────►│
RECEIVE   2 ┘         └ 1  TRANSMIT
          HOST              TERMINAL
```

5. Crossover extensions. *Using couplers with built-in crossovers and extension cable will extend a wiring run, though an odd number of crossovers is needed.*

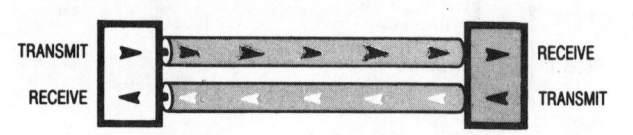

Table 2: Performance parameters of dual window fiber cable

WINDOW	MODAL BANDWIDTH (PER KILOMETER)	ATTENUATION (MAXIMUM PER KILOMETER)
850 NANOMETERS	160 MHz	4.5 dB
1,300 NANOMETERS	400 MHz	2.0 dB

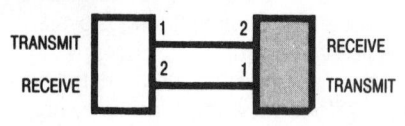

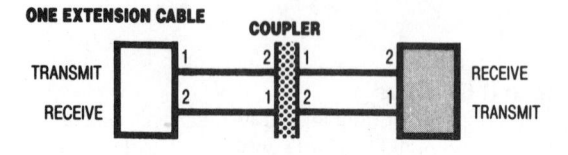

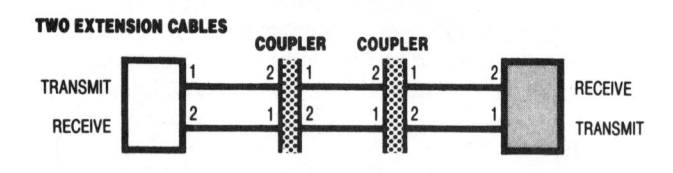

has great potential. And the existence of a standard for fiber will minimize any future risks involved with installing fiber today.

Fiber has the ability to become the first true universal communications utility, and the 62.5/125 size is expected to become the formal North American standard in the near future. The performance parameters of the dual-window fiber cable are shown in Table 2. But adding a fiber into a cable can increase the attenuation, so parameters must always be checked for the entire cable, and not just for each fiber.

To minimize risks and maximize the longevity of the cabling installed in a commercial building, all optical media should:

■ Conform to the specification for dual-window 62.5/125 fiber.

■ Use a hierarchical star topology with two levels of cross connects. The loss of a connector pair is about 1 dB. Thus, a cross connect uses up 2 dB of the link budget.

Generally, the backbone cables between intrabuilding closets should not exceed 1,640 feet and should contain six or more fibers. The distance between closets is dependent on system internode distance and is affected by the number of passive closets between nodes. The suggested distance of 1,640 feet is offered as a conservative number that is unlikely to be exceeded in most buildings.

A dual-ring network topology requires four fibers. But there are other factors to consider when determining the number of fibers in a cable: Will systems not presently sharing the media be added during the life of the cable plant; are there several of the same kinds of equipment in a given closet; and are there enough spare fibers in the cable to replace failed fibers?

Extra fibers should generally be installed because the cost of the cable sheath and the cost of pulling additional cable are offset by putting in spare capacity. The number of fibers should be higher between buildings to provide a more flexible cable plant.

Where the backbone cables between buildings exceed the internode distances of the network, repeaters can be used to extend the distance, provided an appropriate place to house and power them is available.

Horizontal element runs should be a maximum of 295 feet long and contain two fibers. The cost of installing the horizontal plant is orders of magnitude higher than installing the backbone plant, so providing an additional fiber per drop now saves installation costs later.

Upgrading a backbone cable is a far less costly affair. Pulling high-fiber-count cables today, or including single-mode fibers in a backbone, will minimize the need to add capacity in the future.

Adding connectors to a run should always be avoided. Connectors are expensive, they add loss, and they reduce reliability. They should be installed only when there is an application for the cable. Though there is no specific recommendation for connectors today, the ST, in both its simplex and duplex configurations, is becoming a de facto standard.

Deciding which medium to pull to wire a building poses difficult choices. Commercial construction is now in transition from copper to fiber media—though copper will retain a major position for many years to come, especially in the horizontal plant. Fiber is the technology of tomorrow, however, since it makes available the high data rates needed for voice, video, and high-speed data in an office environment. ■

Bryan F. Gearing is a principal engineer in the Network Systems Group at Digital Equipment Corp.'s Merrimack, N. H., facility. He is a member of the Electronic Industries Association's Ad-Hoc Working Group on Building Wiring and holds several patents in telephone switching systems.

6. Cable marking. *Directional indicators or arrows applied directly to the fibers simplify installation, maintenance, and administration problems.*

Ranjana Sharma, Telesyne Consulting Inc., San Jose, Calif.

T1 network design and planning made easier

A few simple steps, with just a little math, can greatly simplify the design of a reliable and cost-efficient T1 network.

A T1 link provides network designers with a DS-1 (digital signal Level 1), 1.544-Mbit/s transmission path, which can be subdivided to carry a variety of different services. The DS-1 signal on a T1 pipe is usually segmented into 24 DS-0 (64-kbit/s) subchannels, which can be grouped or further submultiplexed to transport the following special services:
- Forty-four compressed 32-kbit/s ADPCM (adaptive differential pulse code modulated) voice channels (four bundles of six contiguous DS-0s, each bundle carrying 11 voice channels plus one signaling channel).
- Subrate data channels, consisting of multiple data channels of four rates — 2.4, 4.8, 9.6, and 56 kbit/s — which fit into one DS-0 subchannel.
- Compressed video, which typically occupies one bundle (384 kbit/s), equivalent to six PCM or 11 ADPCM channels.

A physical T1 backbone supports a logical network of lower bandwidth links. This mapping of the logical network onto the physical network actually determines the number of T1 links necessary to support network services, as well as the best physical topology. T1 network design includes all the "classical" network-design steps plus some additional steps required for determining the T1 topology and mapping the logical connectivity to the physical topology.

Classical network design consists of several steps:
1. Determining the type of applications the network will serve.
2. Creating traffic matrices that show expected traffic volumes between source (traffic origination) and sink (destination) pairs.
3. Creating a network structure that defines traffic concentration nodes.
4. Calculating the point-to-point bandwidth required, corresponding to the traffic matrices and the network structure selected.
5. Careful mapping of the required point-to-point bandwidth onto a segmented T1 topology, taking into account user requirements for response time and reliability.

Frequently, these steps must be repeated until requirements are met at the lowest cost.

Network applications and traffic requirements

The first step in the network-design process defines the applications the network will support. This step is important in determining the ultimate bandwidth required between two points, as well as the segmentation of the T1 pipes.

The bandwidth calculations for different application types vary and use different performance criteria. For example, voice-traffic bandwidth computation uses Erlang traffic formulas, while data-traffic bandwidth computation is derived from the principles of queuing theory. Other types of applications, such as videoconferencing, may demand a fixed amount of bandwidth, but only very occasionally, and can therefore steal the bandwidth from other applications on a reserved basis.

Once network applications have been identified, the next essential step is to define the traffic volumes (average and maximum) for which the network must be designed. A traffic matrix presents the traffic origination and destination points, along with the expected traffic between them.

The network-design problem must have some form of estimated traffic matrix as an input. Since traffic patterns vary by time of day (and possibly by day of the month), it is wise to create both an average-traffic matrix and a peak-traffic matrix. Here, peak implies the worst-case traffic-volume conditions that the network must handle. The network should be designed to handle the average peak of daily traffic, while allowing some graceful degradation of service on the worst days of the month.

Traffic volumes must be characterized in some uniform

1. Data-bandwidth sizing. *This four-line topology computes average delays over 14.4-kbit/s links. Traffic matrix numbers are in messages per second.*

TRAFFIC MATRIX (MESSAGES/SECOND)

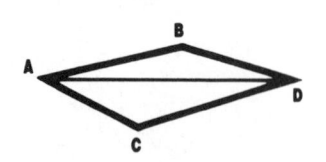

	A	B	C	D
A	0	5	5	10
B	5	0	5	8
C	5	5	0	5
D	10	8	5	0

DESIGN CALCULATIONS

LINK i	CAPACITY C_i (KBIT/S)	TRAFFIC CONTRIBUTION (MESSAGES/SECOND)	TOTAL (λ_i) (MESSAGES/SECOND)	DELAY T_i* (SECONDS)
AB	14.4	AB = 5		
		AD = 2		
		BC = 3	10	0.23
AC	14.4	AC = 5		
		BC = 3	8	0.16
AD	14.4	AD = 8	8	0.16
BD	14.4	BD = 8		
		AD = 2		
		BC = 2	12	0.42
CD	14.4	CD = 5		
		BC = 2	7	0.14

*AVERAGE MESSAGE SIZE 1,000 BITS (μ)

AVERAGE HOPS (PATHS) PER MESSAGE n = $\frac{\text{TOTAL TRAFFIC CARRIED}}{\text{TOTAL TRAFFIC OFFERED}}$

= 45/38 = 1.18

AVERAGE MESSAGE DELAY (SECONDS) = 1.18 (10.88)/45 = 0.29

manner. For data, the usual measure of traffic is given by the average number of messages per second per message source, and the average message length in bits or characters per message. For voice traffic, the measure is usually given in Erlangs of traffic (or in units of hundred call-seconds per hour, abbreviated ccs). These traffic units are derived from the average number of call attempts per hour and the average length per call in seconds.

To derive a detailed traffic matrix, it is usually necessary to provide some form of traffic monitoring capability at the various traffic sources. Such a capability may be present in an operating network, but when a new network is being designed, traffic can only be estimated.

Traffic "community-of-interest" models are sometimes valid as starting points for network design. These assume that traffic between two sites is proportional to the product of their populations, divided by the distance between them. A robust network design should be fairly insensitive to fluctuations of 25 percent in predicted traffic.

Network topologies are generally selected to minimize cost while maintaining a specified degree of reliability. A highly connected mesh structure incorporates direct point-to-point connections between traffic origination and destination points. A hierarchical topology concentrates traffic over fewer high-bandwidth lines, with traffic between several origination and destination pairs switched and cross-connected at tandem-switching locations.

Network topology

A non-hierarchical topology inherently provides richer connections between node pairs and is consequently more reliable. Line and node failures can disrupt traffic from fewer sources. With a hierarchical network, a single tandem node or link can carry traffic between several nodes, so that failures can be more catastrophic. But for a large, geographically dispersed network, hierarchical network design usually results in substantial cost savings because of the concentration of traffic over fewer long-distance links.

Network designs with high-speed T1 backbones are inherently hierarchical, since traffic between several node pairs is combined over fewer physical high-bandwidth links. There is a minimum of two tiers in the hierarchy—the local access level and the backbone—but three or four levels can exist in large networks, depending on the geographical concentrations and distribution of traffic.

The highest-level nodes in the network hierarchy are generally the locations with the largest concentration of network customers; cross-connect devices and T1 node processors are located here. Besides serving collocated network customers, these nodal processors serve as tandem points for pass-through traffic. The next lower level in the hierarchy generally features concentrators or point-to-point multiplexers.

Hierarchical network design consists of determining where to locate major points of traffic concentration, how to connect traffic sources to the points of concentration, and how to interconnect these concentration points. Ultimate traffic sources are terminals, hosts, and telephones.

The traffic volumes from these sources can be concentrated on a per-site or per-location basis, with network equipment such as concentrators, multiplexers, and PBXs concentrating the collocated network traffic. The summed traffic from these sites can be further concentrated into cities that serve as the central concentration point for an area. Each higher level of traffic concentration requires correspondingly higher-capacity PBXs and processors.

The T1 backbone will serve the highest-level traffic matrix. The physical T1 topology is designed after the backbone bandwidth requirement has been determined.

Design calculations

Once one or more traffic matrices are created, the next step is to create corresponding bandwidth matrices. The derivation of capacity from traffic volumes requires design calculations. First, users need to state their desired performance criteria. With interactive data transmission, response time is important and can be defined as the need for the network to provide a specified maximum delay for at least 99 percent of all messages. For voice- or circuit-switched calls, the performance criterion is specified in terms of the percentage of calls that will be given busy signals because of the unavailability of network resources. When selecting bandwidths, there is usually a clear distinction between the design criteria for a backbone as opposed to access lines.

A commonly used formula for sizing data-network link bandwidths is the well-known M/M/1 queuing model. The notation A/B/m is widely used in queuing literature, where A stands for the probability density function for message arrivals, B denotes the probability density function for message service time, and m denotes the number of message servers.

Probability density functions offer randomly occurring events that show the relative frequency with which all possible values of the event will occur. The M/M/1 queuing formula models the time between arrivals and time taken to process a message with an exponential probability density function.

Delay performance can be modeled by:

$$T_i = \frac{1}{\mu C_i - \lambda_i}$$

where

C_i = Link capacity in bit/s
T_i = Queuing + transmission time (seconds) for a communication link i of capacity C_i (bit/s)
λ_i = Average message arrival rate/second on link i (Poisson probability distribution)
$1/\mu$ = Average message size (bits/message, exponential probability distribution)

The mean delay for a message that must travel over several hops before arriving at its destination is approximated by:

$$T = n \sum_{i=1}^{m} \frac{\lambda_i T_i}{\lambda}$$

where

m = Number of communication links in the network

$$\lambda = \sum_{i=1}^{m} \lambda_i$$

n = Average number of hops (or links traversed per message)

In data-network design, the cost of the network is minimized subject to a specified design constraint for average delay T. Figure 1 is a simple example of the use of the above expressions for a four-node network structure consisting of 14.4-kbit/s links.

The routing of messages will be controlled by the data network routing algorithms, so the static calculations shown in Figure 1 are initial estimates for network-design purposes only. In an operational data network, the actual results of changing routing can be quite different, causing instantaneous delays that are not predictable at this stage.

As shown in Figure 2, an important phenomenon to keep in mind when bandwidth-sizing data circuits is: As throughput is increased (more traffic is offered to a network with a fixed capacity) beyond a certain threshold, message delays become uncontrollable.

For voice networks, the design calculations for bandwidth-sizing use different criteria. Voice-channel capacity is based on determining the number of trunks necessary within the design constraint of a specified maximum probability of calls blocked. The number of trunks is calculated by converting voice traffic to Erlangs as follows:

Load per source (telephone) in Erlangs =
(Average calls/source/seconds) x (Average seconds/call)

The resulting calculated number, multiplied by the total number of traffic sources, provides the total Erlang load at a site. Published Erlang B tables and graphs can then be used to estimate the number of trunks required at a site to service the voice traffic at an acceptable grade of service. Here the measure of grade of service is the average ratio of blocked calls to total calls placed.

With digitized voice, the number of trunks computed above is converted to equivalent 64-kbit/s bandwidth PCM links or 32-kbit/s bandwidth ADPCM links. Figure 3 estimates the voice-traffic channel requirement using Erlang load and traffic formulas.

When selecting discrete bandwidth lines to satisfy the results of such design calculations, it is important to remember that better bandwidth utilization is achievable with higher-bandwidth paths. With current tariffs, higher-bandwidth links can also be more cost-effective.

For example, with five 9.6-kbit/s links, with an average message arrival rate of four messages per second per link and an average message length of 1,000 bits per message, the average delay per message derived from the above formula is 0.18 seconds. If the same equivalent traffic of 20 messages per second (five links × four messages per link) is offered instead to a 56-kbit/s link, then the average message delay is 0.03 seconds.

Similar concepts apply to voice traffic. Referring to standard Erlang tables, it turns out that a trunk group size of 10 trunks can serve 4.46 Erlangs of voice traffic at 1 percent blocking. This means that five groups of 10 trunks can serve 22.3 Erlangs of traffic. Again, referring to the Erlang B tables, the same number of total trunks — when combined into a single group of 50 trunks — can serve 37.9 Erlangs of traffic at 1 percent blocking. The drawback is that, with certain network structures and traffic dispersion patterns, it is not always feasible to consolidate traffic over fewer links.

T1 backbone topologies

The network at this stage may logically look like the example shown in Figure 4, with several tree-like structures accessing a defined backbone. The tree-like access lines can be voice tie lines or data lines with various capacities, including point-to-point T1 links.

The backbone itself is assumed to be a mesh configuration interconnected with high-bandwidth links requiring several DS-0 channels. For such a configuration, a physical channelized T1 topology using T1 nodal processors can be quite cost-effective. Since the T1 signal is normally

2. Mean delay. *Increasing traffic loads at fixed capacities can introduce uncontrollable delays as throughput is increased. This creates intolerable waiting periods for users.*

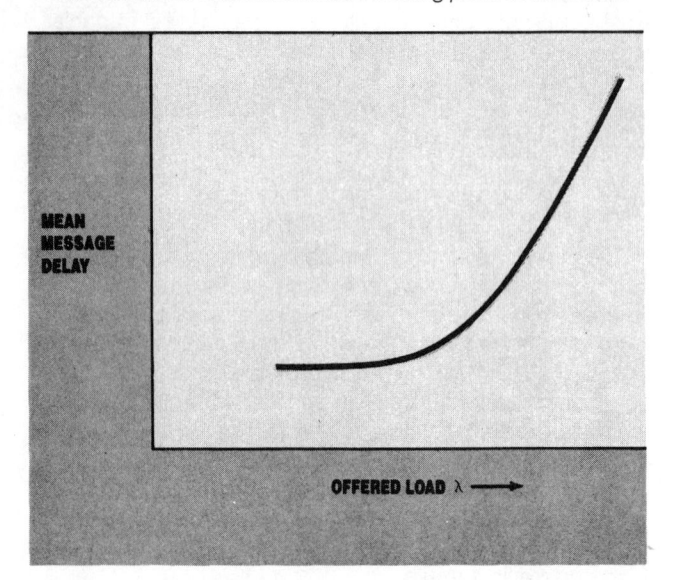

[Graph: MEAN MESSAGE DELAY vs OFFERED LOAD λ]

segmented into lower-capacity DS-0 channels, throughput and delay are still limited by the capacity of a DS-0 channel. But a T1 topology with current tariffs can be more cost-effective for the mesh backbone illustrated in Figure 4.

The next step is to select the T1-backbone topology that can accommodate the required backbone bandwidth. Topology choices are: ring, mesh, tree, or a combination.

The selected topology needs to achieve several objectives:

■ The minimum number of T1 lines terminating at a node must be able to distribute the total DS-0 channel requirement for the node. Referring back to Figure 4, if node 1 requires X number of DS-0 channels stretching to node 2, Y number of DS-0 channels to node 4, and Z DS-0 channels to node 5, then the number of T1 lines terminating at node 1 must together be able to provide at least X + Y + Z DS-0 channels.

■ Every node pair has at least two alternate paths for reliability. The network customer may desire additional redundancy with more alternate paths.

■ The maximum length of the primary path between each node pair is less than a specified number of hops. For example, the network customer may specify that channels between two nodes should pass through no more than two additional T1 nodes.

■ The number of T1 links in the network needs to be able to accommodate both primary and alternate route bandwidths.

In order to ensure that adequate T1 bandwidth is available in the selected network structure, the following steps can be followed:

1. Since standard T1 links are segmented into 24 DS-0 channels, convert the point-to-point backbone bandwidth requirement to equivalent DS-0 64-kbit/s channels.

2. Assign DS-0 channels between nodal pairs using a least-number-of-hops-type algorithm. Such a process is summarized in "DS-0 mapping on T1." This process is tedious if performed manually, but it is easily implemented via a computer program.

3. If adequate bandwidth is not available to meet all the above criteria, restructure the network and remap DS-0 channels.

Figure 5 gives an example of bandwidth requirements mapped into a physical T1 topology. First, create the DS-0 channel requirement matrix between node pairs for the backbone of Figure 4. The topology shown in Figure 5 can be an initial T1 network selection for connecting the six backbone nodes. Using a channel-mapping algorithm, the least-hop primary paths for distributing DS-0 channels over this T1 topology can be found.

As shown in Figure 5, the initial topology selection can accommodate the required DS-0 distribution, but most links are saturated; therefore, not enough alternate routes are available should lines fail. Depending on the level of reliability necessary (the redundancy needed to accommodate single failures or multiple failures), more T1 links should be added.

For this example, adding a second T1 link between nodes 1 and 3 will provide the necessary redundancy by freeing up the capacity on several of the other links. Usually, redundancy analysis can be carried out by deleting one T1 line at a time from the topology and redistributing the DS-0 channels to see if the remaining bandwidth is still adequate.

This sequence of steps, though tedious, is necessary for cost-effective and efficient T1-backbone-network planning. Commercially available network-design tools offer some of the above steps built in, but the network designer still

3. Voice-bandwidth sizing. *Using the four-node topology in Figure 1 to transmit voice, 10 calls per hour are sent between nodes A and B. These are converted into Erlangs.*

TRAFFIC MATRIX (CALLS/HOUR)					EQUIVALENT ERLANGS*			
	A	B	C	D		A	B	C	D
A	0	10	10	15	A	0	.5	.5	.75
B	10	0	12	5	B	.5	0	.6	.25
C	10	12	0	8	C	.5	.6	0	.4
D	15	5	8	0	D	.75	.25	.4	0

*AVERAGE HOLDING TIME 180 SECONDS/CALL

TRUNKS REQUIRED FOR 1 PERCENT BLOCKING PROBABILITY MAXIMUM

	A	B	C	D	TOTAL
A	0	4	4	4	12
B	4	0	4	3	11
C	4	4	0	3	11
D	4	3	3	0	10

4. Sample network. *The hierarchical network design here consists of multiple 56-kbit/s data lines with either 64-kbit/s or 32-kbit/s voice lines.*

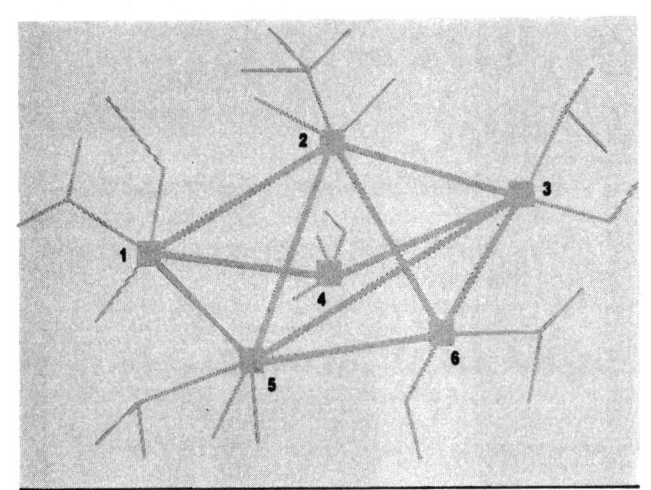

5. DS-0 channel routing. *The channel matrix in the middle of the figure is distributed over the six-node T1 backbone through the distribution table.*

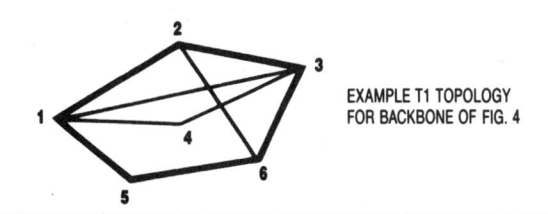

EXAMPLE T1 TOPOLOGY FOR BACKBONE OF FIG. 4

MATRIX OF DS-0 CHANNEL REQUIREMENT DS0

	1	2	3	4	5	6	TOTAL	MINIMUM T1 REQUIRED
1	0	11	56	8	5	5	85	4
2	11	0	10	0	12	16	49	3
3	56	10	0	10	8	15	99	5
4	8	0	10	0	0	0	30	2
5	5	12	8	0	0	15	40	2
6	5	16	15	0	15	0	51	3

PRIMARY DS-0 CHANNEL ROUTING ACHIEVABLE

T1 LINK / NODE PAIR	1-2	1-3	1-4	1-5	2-3	2-6	3-4	3-5	3-6	5-6	TOTAL CHANNEL DISTRIBUTION
1-2	11										11
1-3		8	24	14	10	8		14	10		56
1-4			8								8
1-5				5							5
1-6	1		4		1				4		5
2-3					10						10
2-5	1			1	6	5		6		5	12
2-6						16					16
3-4							10				10
3-5								8			8
3-6									15		15
5-6										15	15
LINK CAP IN USE	21	24	22	20	24	22	24	24	15	24	
LINK CAP AVAILABLE	3	0	2	2	0	2	0	0	9	0	

DS = DIGITAL SIGNAL

DS-0 mapping on T1

The following is the procedure for mapping DS-0 channels on a DS-1 framed T1 link:

1. The total number of DS-0 channels, nodal capacity (NC), which can be served by a node (including pass-through channels), is equal to the sum of the DS-0 channels available over N number of T1 links terminating at the node. For a DS-1 framed signal, the initial value of NC is, therefore, 24 × N. As channels are distributed through a node, its value of NC will correspondingly decrease from its initial value of 24 × N.

2. The initial capacity, NL, of each T1 link is 24. As channels are distributed, the value NL will correspondingly decrease for the affected links.

3. With these definitions in mind, distribute the DS-0 channels required between two nodes, I and J, through a single T1 link; that is, wherever a T1 link directly connects the two nodes. Repeat for all such I and J nodal pairs.

4. Update values for NC and NL for all affected nodes and links.

5. Distribute DS-0 channels between nodes I and J through one intermediate node, M, over two T1 links. The number of channels that can be distributed through node M is the minimum of half the available nodal capacity, NC, at node M, plus link capacity, NL, of the link from node I to node M, plus link capacity, NL, of the link from M to J. Repeat for all intermediate nodes M and nodal pairs I and J.

6. Distribute DS-0 channels between nodes I and J through multiple intermediate nodes M(n); that is, distributed over more than two T1 links. This involves tracing paths between node pairs I and J, each time checking the capacity available at intermediate nodes and intermediate links, and then distributing the maximum number of channels feasible over each path.

---

needs to understand the processes and interact closely with the tool to guide it toward a cost-optimum design. Several iterations of this sequence of steps may be necessary before a satisfactory answer is reached. ■

Ranjana Sharma is president of Telesyne, a consulting firm specializing in telecommunications engineering and network design. She has a B. S. E. E. and an M. S. E. E. from the University of Washington (Seattle) and a Ph. D. in electrical engineering from Stanford University. She has worked in the communications industry for more than 14 years, holding professional and managerial positions at Tymnet, Ford Aerospace, and Bell Northern Research.

Terence D. Todd, McMaster University, Hamilton, Ontario, Canada

Designing gateways into metropolitan area networks

New packet-scheduling algorithms may dramatically increase the capacity of LAN-to-MAN gateways. These techniques work particularly well with broadband cable networks.

At McMaster University in Hamilton, Ontario, some of the issues associated with connecting local broadband coaxial cable networks into metropolitan area networks (MANs) are being investigated. In particular, it has been found in some existing and proposed networks that if the metropolitan area gateway (MAG) is implemented in certain ways, overall traffic-scheduling efficiency can be dramatically improved. (In this context, performance is measured by the mean packet delay, the average time it takes for a network station to transmit a newly generated data packet.) This is true, for example, in networks employing CATV (community antenna television)-based broadband cable backbones. In fact, under some conditions, it is possible to approach a doubling of the traffic-carrying capability of the gateway. The methods used to do this have been named Swift (switch-filtering technique) algorithms.

Broadband CATV backbone internetworks
In many medium-to-large corporate and university networks, a CATV-based backbone is used in order to provide connections between different local area networks (Fig. 1). Typically, the connection between the backbone and the LANs is implemented by MAC (Media Access Control)-level bridges, which operate by bridging packets at Layer 2 of the Open Systems Interconnection reference model. Thus, the MAC-level bridges may be connected to the broadband cable using normal community antenna television data-station modems.

This solution is particularly attractive because it allows for a large and expandable internetwork while permitting higher-level protocol coexistence and transparency. Bridging is accomplished by making use of the global administration of MAC addresses, ensuring that all addresses are unique. In addition, the use of broadband cable technology allows the backbone to support non-data applications such as closed-circuit television.

CATV networks are unique in that they operate using unidirectional data-signal propagation. The CATV data stations are attached to the cable through unidirectional cable taps, assigned to transmit on a specific uplink frequency channel. Connections on the broadband network are then achieved by having the stations listen on a corresponding downlink channel frequency. A head end (Fig. 1) is then responsible for translating the data received on an uplink channel to the correct downlink channel for reception by the stations. Figure 2 shows a cable head end and an uplink/downlink channel pair.

In many CATV networks, there are a number of separate uplink/downlink channel pairs. In Figure 2, the channels are shown as being physically separate (a dual cable configuration), though in practice they may belong to different frequency bands on a single cable backbone.

The most common media access method used on CATV backbones is carrier-sense multiple access with collision detection (CSMA/CD). When CSMA/CD is used on CATV networks, data stations that have packets to transmit sense (listen for) transmission activity on the downlink channel. If the downlink is idle, the station transmits its packet on the uplink channel while performing collision detection. Alternatively, if the downlink is sensed to be busy, indicating a possible transmission in progress, the station waits until the downlink is sensed idle. At that point, it proceeds to transmit on the uplink channel.

When CSMA/CD is used as a local access technique, a transmitting station performs collision detection by comparing the received downlink transmission, bit by bit, with the data transmitted on the uplink. This is done in real time as packet transmission proceeds and is normally performed for a time period known as the collision window. A bit

1. CATV cable backbone internet. *A typical CATV-based broadband internetwork consists of a broadband cable plant used to interconnect smaller local area networks.*

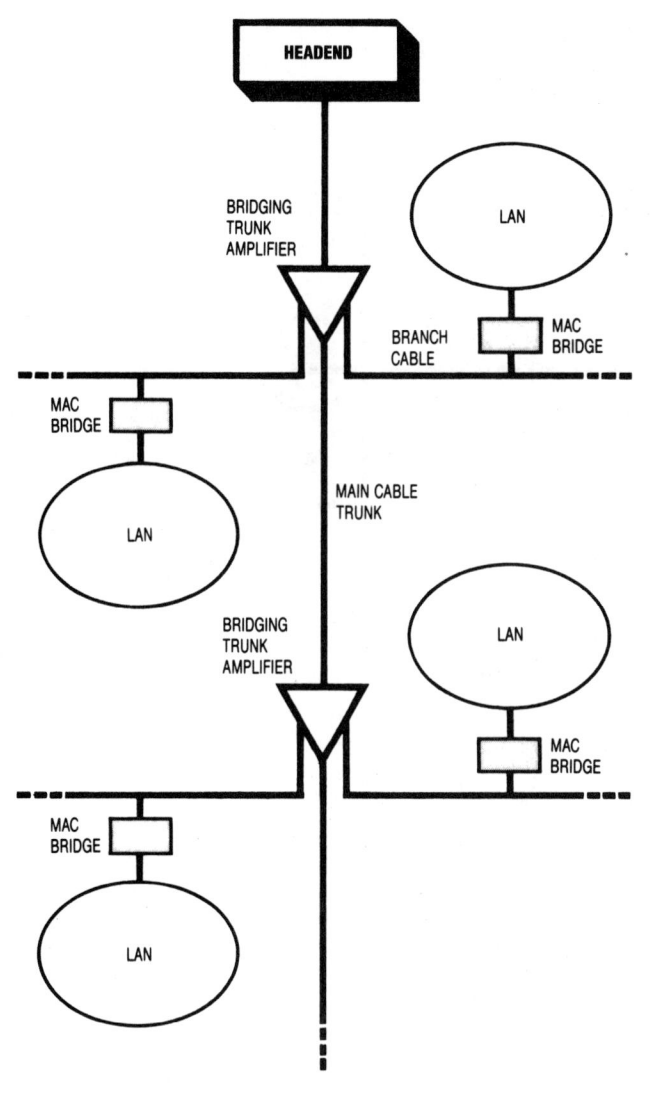

LAN = LOCAL AREA NETWORK
MAC = MEDIA ACCESS CONTROL

mismatch detected anywhere within the collision window signifies to a transmitting station that a collision has occurred, since a simultaneously transmitting station must have garbled its data.

If a collision is detected, the stations involved back off following a collision-reinforcement interval and attempt to randomize subsequent accesses to the channel. The random back-off consists of a time interval whose average value increases exponentially with the number of collisions suffered by a given packet. In accordance with the CSMA/CD algorithm, a station attempts to use the transmission channel again after the back-off interval.

When such a network is connected to a MAN, it is reasonable that the metropolitan area gateway be located on the CATV backbone itself. There are possible advantages to locating the MAG at the head end, as opposed to treating the gateway as a normal CATV data station. Before exploring this possibility, a conventional LAN-MAN gateway design should be examined.

A conventional CATV-MAN interface

In Figure 2, a simple interface between the CATV backbone and a metropolitan area network is shown. Note that attention is restricted to the CATV backbone and its interface to the MAN. Also illustrated is a downlink/uplink channel pair to which local station adapters and the gateway are connected.

In this case, the MAG has been implemented using a standard CATV network station adapter, which connects to an uplink/downlink channel pair as usual.

The functions of the MAG in this instance are conceptually quite simple (Fig. 3). When packets are transmitted on the uplink/downlink channels, the MAG listening on the downlink identifies the packets that must be transmitted into the metropolitan subnetwork and copies them into a buffer, denoted Local/Remote, for further processing. Likewise, packets that arrive from sources elsewhere in the metropolitan area network are buffered in the Remote (or R) buffer. These packets are transmitted by the MAG onto the uplink/downlink channel pair using CSMA/CD. The MAG thus accesses the CATV network in the same fashion as a standard user station.

In Figure 3, consider the packet data traffic flowing within the MAG. The figure shows two buffers. The Remote buffer holds remote packets that have arrived from the metropolitan area subnetwork and the Local/Remote buffer holds packets that have been locally generated by adapters attached to the uplink/downlink channels. Locally generated traffic with remote destinations is called Local/Remote traffic. In this scenario, the cable head end is static, since it only performs the usual task of transferring bits from the uplink to downlink channels.

Close consideration of Figure 2 shows that there are actually three traffic classes: Remote, Local/Remote, and Local/Local. Local/Local packets correspond to locally generated packets (generated by stations on the uplink/downlink channel pair) that have local destinations. Local/Local packets can be transmitted directly to their destinations on the CATV network and obviously do not require the services of the MAG.

The above illustration depicts what would be possible today using existing off-the-shelf devices and technology. However, it is possible for the traffic-handling efficiency of the network to be improved.

Under normal conditions, each packet transmitted by a station (including the MAG) must propagate along both the uplink and downlink channels through the head end. However, if a station is transmitting packets solely to a destination station elsewhere in the metropolitan area, then there is no need for those packets to appear on the downlink channel. (Remember that the packets are relayed by the head end on the downlink channel to allow the destination station to receive them. If the destination station is not

397

attached to this CATV network, then there is no need to do so. However, note that the downlink channel is needed by the local stations to perform collision detection.)

Swift methods attempt to recover or reuse network bandwidth by having remote-bound packets (those that are traveling into the metropolitan subnetwork and thus do not have destinations in the local CATV backbone) transmit only on the uplink channel. While this is occurring, the downlink channel is idle and can therefore be used to transmit packets that have arrived from remote locations in the metropolitan subnetwork. This is the basic objective of Swift operating in a broadband CSMA/CD environment.

There are versions of Swift algorithms that can be implemented transparently to existing CSMA/CD-based network station adapters. This makes it possible to implement a MAG in a head end that is compatible with existing installed local area network adapters.

Swift algorithms

When a MAN interface is designed using Swift, the gateway must be implemented at the CATV head end, referred to as a MAG/head end. Figure 4 shows a single uplink/downlink channel pair that is connected through the head end. Also shown in a simplified form is a switching function (denoted by switch S) performed within the MAG/head end. Note that there are three positions to the switch, which is controlled by the MAG/head end in response to network conditions. The Local/Remote and Remote buffer functions are also included in the figure. The figure illustrates the functional abilities of the MAG/head end arrangement and is not intended to suggest actual implementations.

When the switch (S) is in position S = L (Local), the MAG is connected in the usual way, where the head end acts as a simple bit repeater from the uplink channel to the downlink channel. When this happens, the stations attached to this particular uplink/downlink pair can use CSMA/CD to transmit their packets as usual. Also, the MAG/head end can monitor the packets as they pass by to see which ones must be copied to the Local/Remote buffer for delivery to the metropolitan area network.

When the switch is set to S = R (Remote), the MAG/head end can transmit packets from the Remote buffer onto the downlink channel for reception by their destination stations. During this time, the uplink and downlink channels are disconnected, although it is possible to have a packet being transmitted simultaneously to the MAG/head end on the uplink. Switch position S = C (Carrier) simply places a carrier signal on the downlink channel.

The motivation for Swift is based on the simple observation that packets generated by adapters on a given uplink/downlink need not appear on the downlink channel if the packet is destined for the metropolitan area network (that is, is a Local/Remote packet). In this case, the packet's destination is elsewhere in the metropolitan network, so the downlink may be put to much better use by transmitting packets that have arrived to the Remote buffer. The objective of the algorithm is, where possible, to arrange for

2. Conventional MAN/CATV interface. *The MAG is implemented here as a conventional CATV data station on a specific uplink/downlink channel pair.*

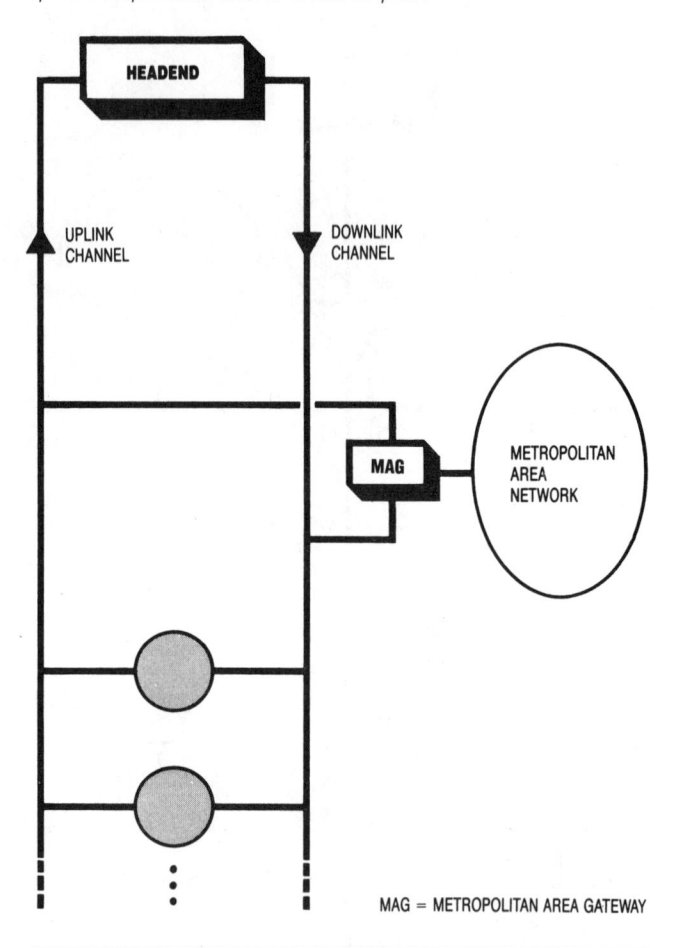

3. Buffering in a metropolitan area gateway. *Packets arriving from the local CATV network are queued in the Local/Remote buffer for transmission into the MAN.*

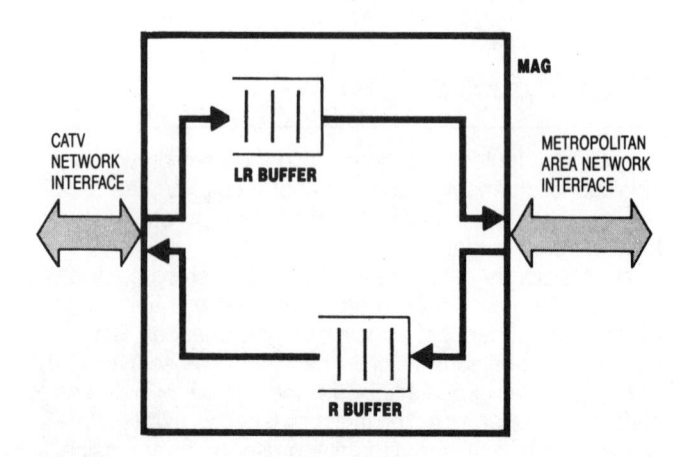

CATV = COMMUNITY ANTENNA TELEVISION MAG = METROPOLITAN AREA GATEWAY
LR = LOCAL REMOTE   R = REMOTE

4. MAG/head end switching. *The Local/Remote and Remote (R) buffers interact continuously with the uplink and downlink channels in a MAG/head end switch. By reading the uplink packet headers and setting the switch to S = R, an intelligent MAG/head end can reuse a large amount of bandwidth, improving network performance.*

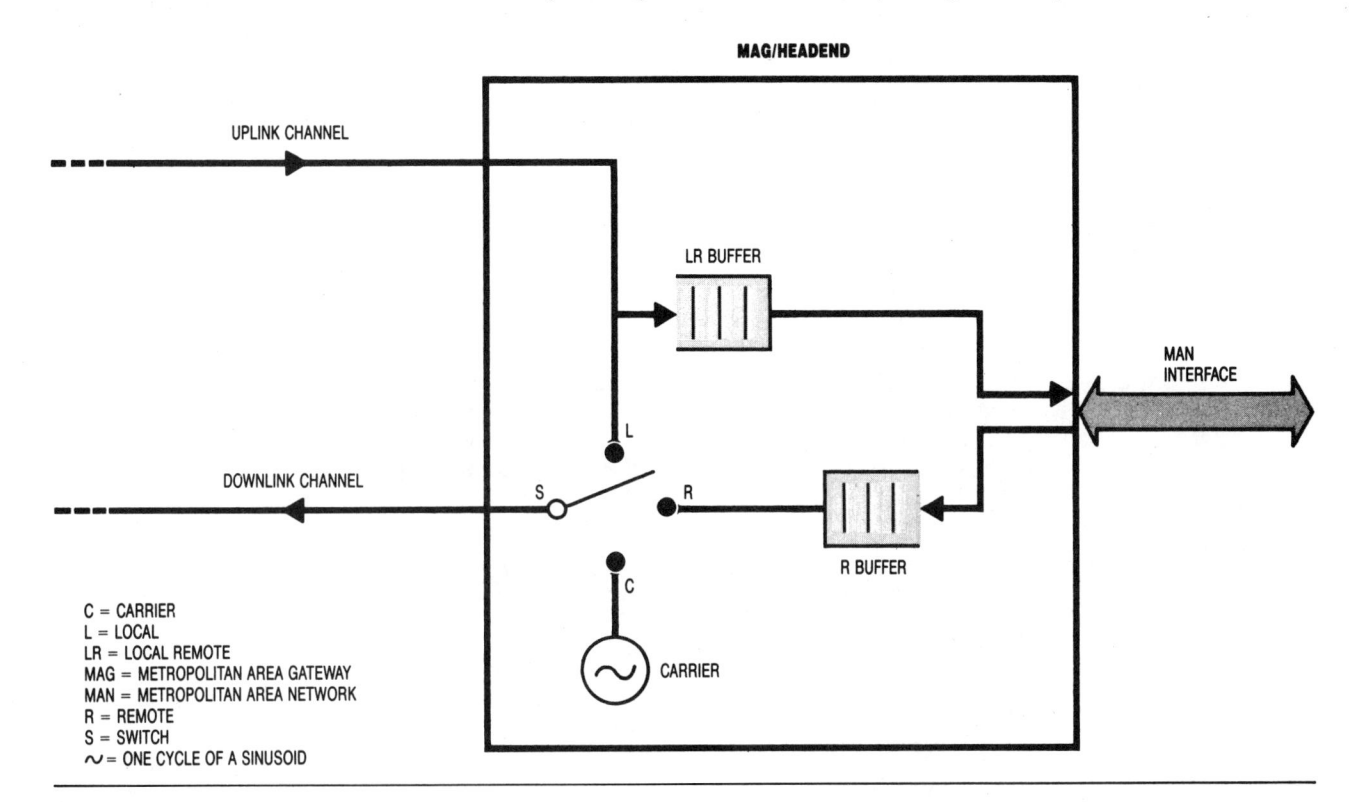

C = CARRIER
L = LOCAL
LR = LOCAL REMOTE
MAG = METROPOLITAN AREA GATEWAY
MAN = METROPOLITAN AREA NETWORK
R = REMOTE
S = SWITCH
∿ = ONE CYCLE OF A SINUSOID

simultaneous local uplink transmissions and remote packet downlink transmissions. This, of course, is only possible when the local packet transmitted is Local/Remote (that is, is not destined for this particular downlink) because the transmission of a Local/Local packet requires the downlink channel if it is to reach its destination.

If the MAG/head end can arrange for simultaneous Local/Remote packet transmissions on the uplink channel and Remote packet transmissions on the downlink channel for a large fraction of the time, the effective bandwidth of the network can approach twice its normal level. With Swift, the uplink and downlink channels can be viewed as two separate transmission channels processing packets in parallel—instead of appearing to users as one logical 5-Mbit/s channel, they may appear as two 5-Mbit/s channels. Under these circumstances, the performance experienced can be improved dramatically.

Swift operation

Assume that the switch (S) has just been set to S = L. The local stations listening on the downlink channel will eventually sense that the downlink channel is idle. Any stations that have packets to send are then free to transmit in accordance with the CSMA/CD algorithm.

Assume that eventually, perhaps after some collisions, a station begins to transmit a packet successfully. The packet, while being transmitted, propagates along the uplink channel, through the MAG/head end, and out along the downlink channel. If the packet is sufficiently long, the packet header containing the destination MAC address will have passed through the MAG/head end well before the complete packet has been transmitted.

Thus, the MAG/head end will be aware if the packet is Local/Local or Local/Remote before the transmission is completed. If the packet is Local/Local (meaning that its destination is a station listening on the downlink channel), then the MAG/head end will allow the packet to completely propagate to its destination on the downlink channel as usual. This is accomplished by maintaining S = L for the duration of the Local/Local packet transmission. If, however, the packet is Local/Remote, then the MAG/head end will set the switch to S = R and transmit a Remote packet (if one is available) on the downlink channel while continuing to receive the Local/Remote packet on the uplink channel.

The Remote packet must be introduced onto the downlink channel in such a way as to prevent other nontransmitting local stations from detecting an idle downlink channel and colliding with the Local/Remote uplink transmission in progress. This can be accomplished easily in CATV networks by having the MAG/head end maintain carrier on the downlink during switching.

Also note that this particular version of Swift is not compatible with implementations of CSMA/CD that employ late collision detection. That is, the local station hardware must not require that the entire packet appear at the receiver of the transmitting station. This is not a problem in broadband cable networks, since collision detection normally extends to the end of a collision window, which includes the MAC header.

In the IEEE 802.3 10Broad36 standard, the collision window runs to the end of the transmitted source MAC address. However, for switch-filtering to be performed in this way, the MAG/head end must be careful not to introduce a Remote packet onto the downlink before the complete collision window period of the local packet has elapsed. Failure to do so would lead to the local station erroneously detecting a collision.

The MAG/head end must also decide what to do if there is a Remote packet backlog and there are no local stations with Local/Remote packets to transmit. Under these conditions, the Remote packets may be delayed for an unreasonable period of time in the MAG/head end. This situation is managed by employing the concept of a scheduling window. During such a window, the MAG/head end makes the uplink/downlink channels available to the local stations by setting $S = L$. At this time, the stations may use CSMA/CD to attempt to generate a successful packet transmission as usual.

If a packet transmission commences during a scheduling window, then the MAG interrogates its header to determine its destination and proceeds normally. If, however, the local stations fail to generate a successful packet transmission within the scheduling-window time period, then the MAG/head end will respond by transmitting a Remote packet onto the downlink. The uplink channel will be unused at this time. Following a transmission (Local/Local, Local/Remote, or Remote), the MAG/head end repeats the process by opening another scheduling window for local stations. In this way, the MAG/head end can set the length of a scheduling window so that the scheduling of remote packets is performed efficiently even when current traffic flows do not consist of Local/Remote packets.

Figure 5 shows an example of how network operation proceeds with Swift. A sequence of events, including packet transmissions, is shown on both the uplink and downlink CATV channels. The figure starts with the MAG/head end setting $S = L$ and opening a scheduling window. During the window, a successful Local/Local packet has commenced transmission. Since the MAG/head end detects that it is Local/Local, the packet is transmitted totally on both channels as shown. This occurs because the MAG/head end maintains $S = L$ for the full duration of the packet transmission. For simplicity, the network is shown with zero propagation delay.

After the Local/Local packet is finished, a new scheduling window is opened. This time, a Local/Remote packet is generated by the local station adapters. At first, the transmission appears on both the uplink and downlink channels ($S = L$), but once the MAG/head end learns that the packet is Local/Remote and that the collision window is expired, it sets $S = R$ and transmits a Remote packet onto the downlink channel. At this point, the network is processing two packets simultaneously.

The Remote packet finishes transmitting first so that the MAG/head end may transmit another Remote packet onto the downlink channel. If the Remote buffer is empty, however, the MAG/head end must set $S = C$ on the downlink channel. The reason for this is that there is still a Local/Remote packet being transmitted on the uplink channel. If the downlink channel were allowed to go idle, other local stations might assume that the uplink was clear and collide with the uplink transmission shown. This happens because the data stations sense the downlink channel to determine whether they can transmit or not.

MAG/head end switching functions can be implemented on existing CSMA/CD CATV networks. Existing local user stations simply operate the CSMA/CD algorithm as usual and may be unaware that the head end is performing the functions indicated. This means that Swift can be implemented in an existing LAN where station hardware is already deployed. Here, it will only be necessary to modify the head end to perform Swift functions.

When should Swift be used?

When considering a Swift implementation, the level and distribution of traffic flow should be considered. Obviously, there is little to be gained if only a negligible portion of traffic flowing on a downlink/uplink channel pair is destined for outside that specific pair of channels (Local/Remote traffic). On the other hand, there must be an ability to interconnect to external networks in any case.

In general, putting a MAG into operation at the head end is a good idea when CSMA/CD is used as the media access protocol. This is true because remote packets can be transmitted onto the downlink in a conflict-free fashion using the head end's total control over the downlink channel. That is, the head end is the only source of packet transmission on the downlink channel. Thus, in a MAG/head end implementation, Remote packets need never generate collisions on the channels.

In a typical multichannel situation, it is likely that certain channels would be designated as having metropolitan-interconnect capabilities. These channels would tend to concentrate outbound MAN traffic and could be reached either through station frequency agility or by conventional bridging. However, when the total traffic levels fluctuate to medium or large values and if the fraction of Local/Remote traffic to total local traffic (including Local/Local packets) is medium to large, then Swift techniques can achieve sizable improvements in performance.

A limitation on the performance improvements possible with Swift is imposed by the mean transmission times of the packets. This happens because Swift relies on simultaneous uplink Local/Remote and downlink Remote packet transmissions in real time. If the packets involved are very short relative to the length of the collision window of the local CSMA/CD stations, then it may be difficult for the algorithm to generate simultaneous transmissions.

Studies have been performed with mean packet lengths as low as a factor of 10 times the collision window length with excellent results; if the collision window were 100 bits in length, this would correspond to 125-byte packets. Results obtained under these conditions with variable

5. Swift example. Local/Local packets appear on both the uplink and downlink channels, as would normally be the case. However, Local/Remote packets may be interrupted on the downlink channel and replaced by a Remote packet. This sequence also shows the use of switch position S = C when the Remote buffer is emptied.

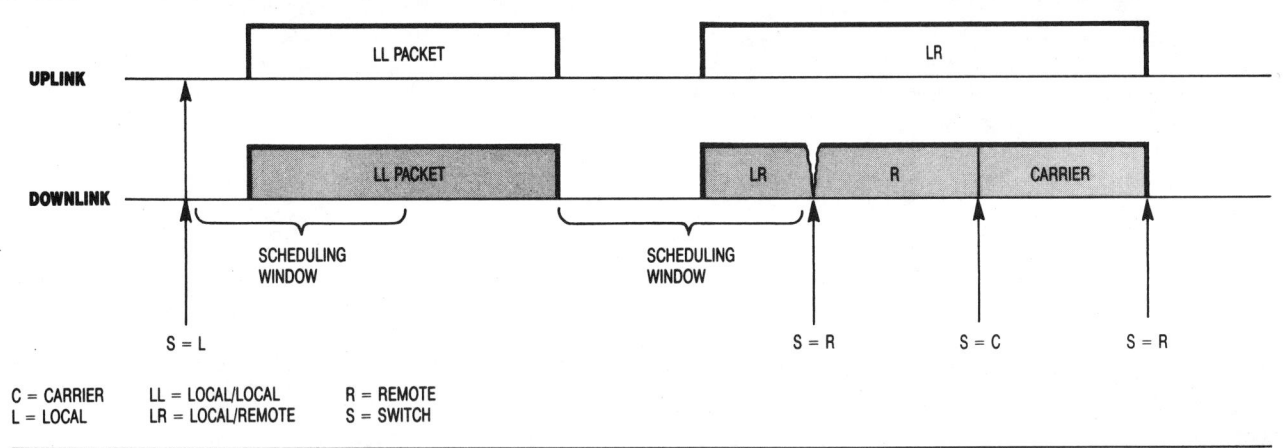

C = CARRIER   LL = LOCAL/LOCAL   R = REMOTE
L = LOCAL   LR = LOCAL/REMOTE S = SWITCH

packet lengths show that, under moderate local station loads, a MAG running Swift can easily achieve equivalent overall mean delays when the remote load is increased by 100 percent over a conventional setup. It is not yet known, however, how small the packet lengths can be and still obtain reasonable benefits.

Figure 6 shows a Swift-based network under heavy traffic flow conditions with packets of fixed duration T seconds. Again, for simplicity, the network is given a small propagation delay relative to T. Also, all the local traffic being transmitted is destined for the MAN (all Local/Remote). In this case—and assuming a large scheduling window—the channel activity goes through cycles.

A cycle begins when the MAG/head end sets S = L and the local stations generate a successful Local/Remote packet transmission. This occurs after a mean CSMA/CD scheduling delay of T_s (scheduling time) seconds. Following this, the Local/Remote packet appears on the uplink and downlink channels for a time period T_{cw} (the CSMA/CD collision window). At this time, the MAG/head end sets S = R and introduces a Remote packet on the downlink channel. Once the Remote packet transmission is complete, the cycle repeats.

The maximum utilization or throughput can be found in this case by dividing the mean length of a cycle into the mean time spent per cycle transmitting packets. The former is given by $T_s + T_{cw} + T$ and the latter by $2T$. Thus, the maximum throughput is given by:

$$\text{Utilization} = \frac{2}{1 + \frac{(T_s + T_{cw})}{T}}$$

When T is large in relation to $T_s + T_{cw}$, then total throughput approaches 2. This value is double the maximum throughput that would be attainable using a conventional solution. In a network without Swift, the maximum throughput is bounded by 1, since no simultaneous processing of packets can be achieved. It is interesting that the requirement that T be large in Swift algorithms is consistent with the general requirement that T be large (in relation to the end-to-end propagation delay) to obtain good performance in CSMA/CD itself.

Parameter optimization

In the basic Swift algorithm outlined above, the length of scheduling window is important. As has been stated, the scheduling window length is the amount of time that the MAG/head end allows the local stations (running CSMA/CD) to generate a successful packet transmission on the uplink/downling channel pair. If the scheduling window expires before this happens, the MAG/head end takes control of the downlink channel and transmits a buffered Remote packet.

The selection of the scheduling window length involves an optimization process basd upon the traffic flow characteristics of the network. For example, if the local traffic flow is large, with a significant percentage of Local/Remote packets, then it is better to select longer scheduling windows. This is true because, under these conditions, it is worthwhile to invest time in waiting for a local station transmission if the chances are good that it will occur soon and be Local/Remote. If the local packet is Local/Remote, then a Remote packet can be transmitted as soon as the local collision window has expired.

It is clear, however, that windows used must be selected by the algorithm on dynamic basis. Now under development are versions of Swift that have these capabilities and obtain excellent performance over a much wider range of operating conditions than before. Methods have also been developed that will allow synchronous Integrated Services Digital Network (ISDN)-compatible voice streams to coexist with data traffic in CATV networks. This method also uses a variant of Swift that can accommodate much higher levels of voice traffic while providing better mean delay performance for existing data stations.

It should be reiterated that, when traffic flows are moderate to heavy, there are many advantages to implementing

6. Heavy channel activity. *Under heavy local and remote loads, a Swift-based network with a large scheduling window will obtain bandwidth reuse.*

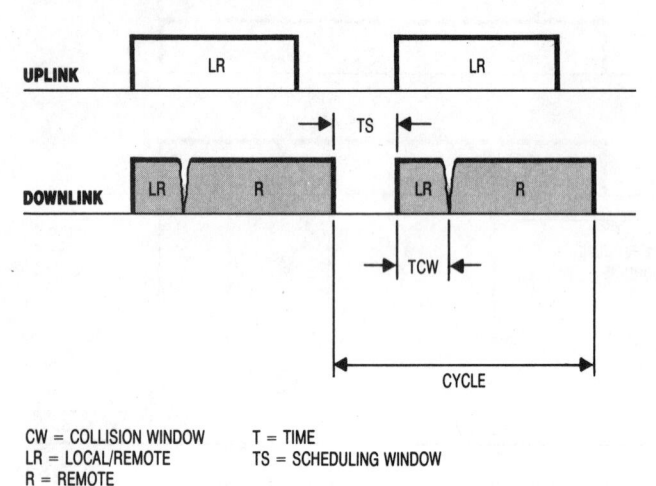

CW = COLLISION WINDOW
LR = LOCAL/REMOTE
R = REMOTE
T = TIME
TS = SCHEDULING WINDOW

a metropolitan area gateway at the cable head end. In any implementation, the decision to employ such an algorithm must be based on a careful examination of the cost and performance trade-offs. Such techniques are applicable to many other non-CATV networks. For example, some architectures use a head end that is based upon fiber optic media intended for manufacturing environments. Swift techniques may be applied here as well. ∎

Terence Todd received B. A. Sc., M. A. Sc., and Ph. D. degrees in electrical engineering from the University of Waterloo (Ontario, Canada). He is an assistant professor of electrical and computer engineering at McMaster University and is a consultant with Sentinel Computer Consultants Inc. in Fergus, Ontario, Canada.

References:

W. Stallings, *Local Networks: An Introduction*, second edition, Macmillan Publishing Co., 1987.

N. Q. Duc and E. K. Chew, "ISDN Protocol Architecture," *IEEE Communications Magazine*, Vol. 23, No. 3, pp. 15-22, March 1985.

G. Ennis and P. Filice, "Overview of a Broadband Local Area Network Protocol Architecture," *IEEE Journal on Selected Areas in Communications*, Vol. SAC-1, No. 5, pp. 832-841, November 1983.

M. B. Akgun and P. Parkinson, "The Development of Cable Data Communications Standards," *IEEE Journal on Selected Areas in Communications*, Vol. SAC-3, No. 2, pp. 273-285, March 1985.

ANSI/IEEE Std. 802.3: Carrier Sense Multiple Access with Collision Detection, Institute of Electrical and Electronics Engineers Inc., 1985.

T. D. Todd, "A Traffic Scheduling Technique For Metropolitan Area Networks," *IEEE Journal on Selected Areas in Communications*, Issue on Interconnection of Local Area Networks, Vol. SAC-5, No. 9, pp. 1391-1402, December 1987.

William Stallings, Comp-Comm Consulting, London, England

Digital signaling: Which techniques are best—and why it matters to you

The NRZ codes common to DP equipment interfaces are not appropriate for LANs and ISDN. But bipolar techniques are.

The emergence of local area networks (LANs) and the ongoing evolution of public and private wide-area telecommunications networks toward digital technology and services have led to a confrontation of sorts over digital signaling techniques. The past—the way computer communications has traditionally functioned—seems on a collision course with the future. And informed users, caught in the middle as usual, may ultimately name the winner: Soon, the choice of transmission equipment may well depend on the digital signaling technique incorporated in that equipment.

Digital signaling may be illustrated by considering the case of digital data being generated by a source, such as a data processing device or a voice digitizer. In either case, the data is typically represented as discrete voltage pulses, using one voltage level for binary 0 and another for binary 1. This traditional encoding technique is known as non-return-to-zero (NRZ) and is common in physical interfaces such as RS-232-C.

A common means of transmitting digital data is to pass it through a modem, then transmit it as analog signals. There are a number of cases where this is not done. For example:
- Baseband local area networks, such as Ethernet and Token Ring.
- Digital PBX connections for terminals, hosts, and digital telephones.
- Digital access to public telecommunications networks over a digital local loop.

In all of these cases, the digital data is transmitted using voltage pulses—called digital signaling. Although it is possible to use NRZ signaling directly, NRZ's form is not compatible with the data rates and/or distances of LANs and digital long-haul networks (including ISDN—Integrated Services Digital Network). Instead, the NRZ signal is encoded in such a way as to optimize performance with these applications (more on NRZ's limitations later).

The data communicator's increasing reliance on digital communications and the multitude of implemented networking applications require a variety of solutions for the problems associated with digital signaling.

Described and compared here are the various encoding approaches used in LANs, long-haul networks, and ISDN. Two of the major issues are signaling-rate requirements and quality and performance.

There are two important tasks in interpreting digital signals at the receiver. First, the receiver must know the timing of each bit. That is, the receiver must know when a bit begins and ends, so that the receiver may sample the incoming signal once per bit time to recognize the value of each bit. Second, the receiver must determine whether the signal level for each voltage pulse is high or low.

A number of factors determine how successful the receiver will be in interpreting the incoming signal: the signal-to-noise ratio (S/N), the data rate, and the bandwidth of the signal. With other factors, such as type and length of transmission medium, held constant:
- an increase in data rate increases bit error rate—that is, increases the probability that a bit is received in error.
- an increase in S/N decreases bit error rate.
- increased bandwidth allows increased data rate.

There is another factor that can be used to improve performance: the encoding scheme. This is simply the mapping from data bits to signal elements. A variety of approaches have been tried. Before describing some of them, let us consider the ways of evaluating and comparing the various techniques. Among the important factors are signal spectrum, signal synchronization capability, error-detection capability, cost, and complexity.

As to the signal spectrum, a lack of high-frequency (high relative to the data rate) components means that less

403

bandwidth is required for transmission. In addition, lack of a direct-current (d.c.) component is also desirable. If the signal has a d.c. component, there must be direct physical attachment of transmission components; with no d.c. component, transformer coupling is possible. This provides excellent electrical isolation, reducing interference.

Finally, the magnitude of the effects of signal distortion and interference depends on the spectral properties of the transmitted signal. In practice, the transmission fidelity of a channel is usually worse near the band edges. Therefore, a good signal design should concentrate the transmitted power in the middle of the transmission bandwidth. This results in lower distortion in the received signal. To meet this objective, codes can be designed to shape the spectrum of the transmitted signal.

For successful reception of digital data, there must be some signal synchronization capability between transmitter and receiver. Some drift is inevitable between the clocks of the transmitter and receiver, so some separate synchronization mechanism is needed. One approach is to provide a separate clock lead to synchronize the transmitter and receiver. This approach is rather expensive since it requires an extra line, plus an extra transmitter and receiver. The alternative is to provide some synchronization mechanism that is based on the transmitted signal. This can be achieved with suitable encoding.

Error detection is the responsibility of a data link protocol that is executed on top of the physical signaling level. However, it is useful to have some error detection capability built into the physical signaling scheme. This permits errors to be detected more quickly. Many signaling schemes have such an inherent error-detection capability.

Finally, although digital logic continues to drop in price, the cost and complexity of the signaling scheme is a factor that should not be ignored.

The most common way to transmit digital signals is to use two different voltage levels for the two binary digits. For example, the absence of voltage can represent binary 0, with a positive voltage level representing binary 1. More commonly, a negative voltage represents one binary value and a positive voltage represents the other (Fig. 1). This latter code, as noted earlier, is known as NRZ. NRZ is generally the code used to generate or interpret digital data by terminals and other EDP devices. If a different code is to be used for transmission, it is typically generated from an NRZ signal by the transmitting device.

A variation of NRZ is known as NRZI (nonreturn-to-zero-inverted [inverted on ones]). As with NRZ, NRZI maintains a constant voltage pulse for the duration of a bit time. The data itself is encoded as the presence or absence of a signal transition at the beginning of the bit time. A transition (low to high or high to low) at the beginning of a bit time denotes a binary 1 for that bit time; no transition, binary 0.

NRZI is an example of differential encoding. In differential encoding, the signal is decoded by comparing the polarity of adjacent signal elements rather than determining the absolute value of a signal element. One benefit of this scheme is that it is usually more reliable to detect a transition in the presence of noise than to compare an absolute value to a threshold value.

Differential encoding is also beneficial in a complex transmission environment, where it is easy to lose the sense of the signal's polarity. For example, on a multidrop line, if the leads from an attached device to the twisted pair are accidentally reversed, all 1s and 0s for NRZ will be inverted. This cannot happen with differential encoding.

The NRZ codes are readily engineered and make efficient use of bandwidth. This latter property is illustrated in Figure 2, which compares the spectra of various encoding schemes. In the figure, frequency is normalized to the data rate. As can be seen, most of the energy in NRZ and NRZI signals is between d.c. and half the bit rate. For example, if an NRZ code is used to generate a signal with a data rate of 9.6 kbit/s, most of the energy in the signal is concentrated between d.c. and 4.8 kHz. (The other codes mentioned are discussed later.)

The main limitations of NRZ signals are the presence of a d.c. component and the lack of synchronization capability. For the latter, consider that, with a long string of 1s or 0s (10 or more) for NRZ, or a long string of 0s for NRZI, the output is a constant voltage over a long period of time (10 or more bit times). Under these circumstances, any drift between the timing of transmitter and receiver will result in the loss of synchronization between the two.

Because of their simplicity and relatively low frequency response characteristics (Fig. 2), NRZ codes are commonly used for digital magnetic recording. But their limitations make these codes unattractive for signal transmission.

1. The formats. *The most common way to transmit digital signals is to use two different voltage levels. For example, a negative voltage represents one binary value, and a positive voltage represents the other.*

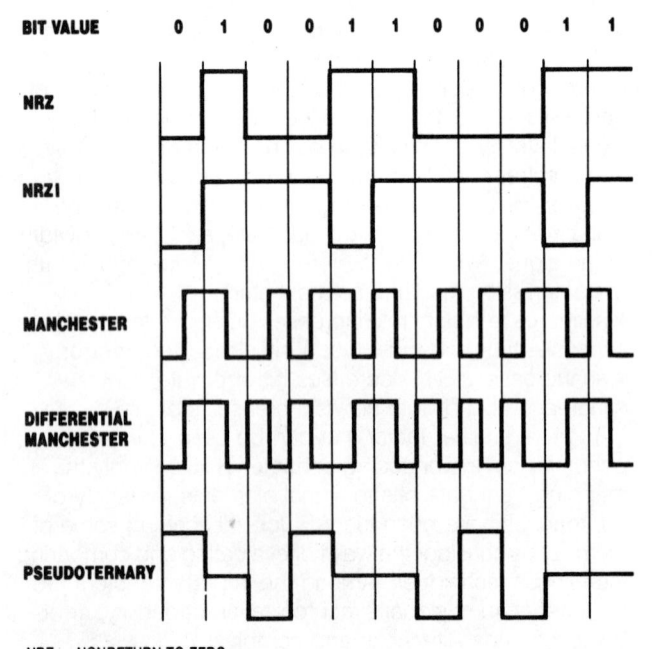

NRZ = NONRETURN TO ZERO
NRZI = NONRETURN TO ZERO INVERTED

2. The spectra. *Most of the energy in NRZ and NRZI is between d.c. and half the bit rate. Most of the energy in biphase is between one-half and one times the bit rate. Thus, in biphase—such as Manchester encoding—the bandwidth is reasonably narrow (there are no high-frequency components) and contains no d.c. component.*

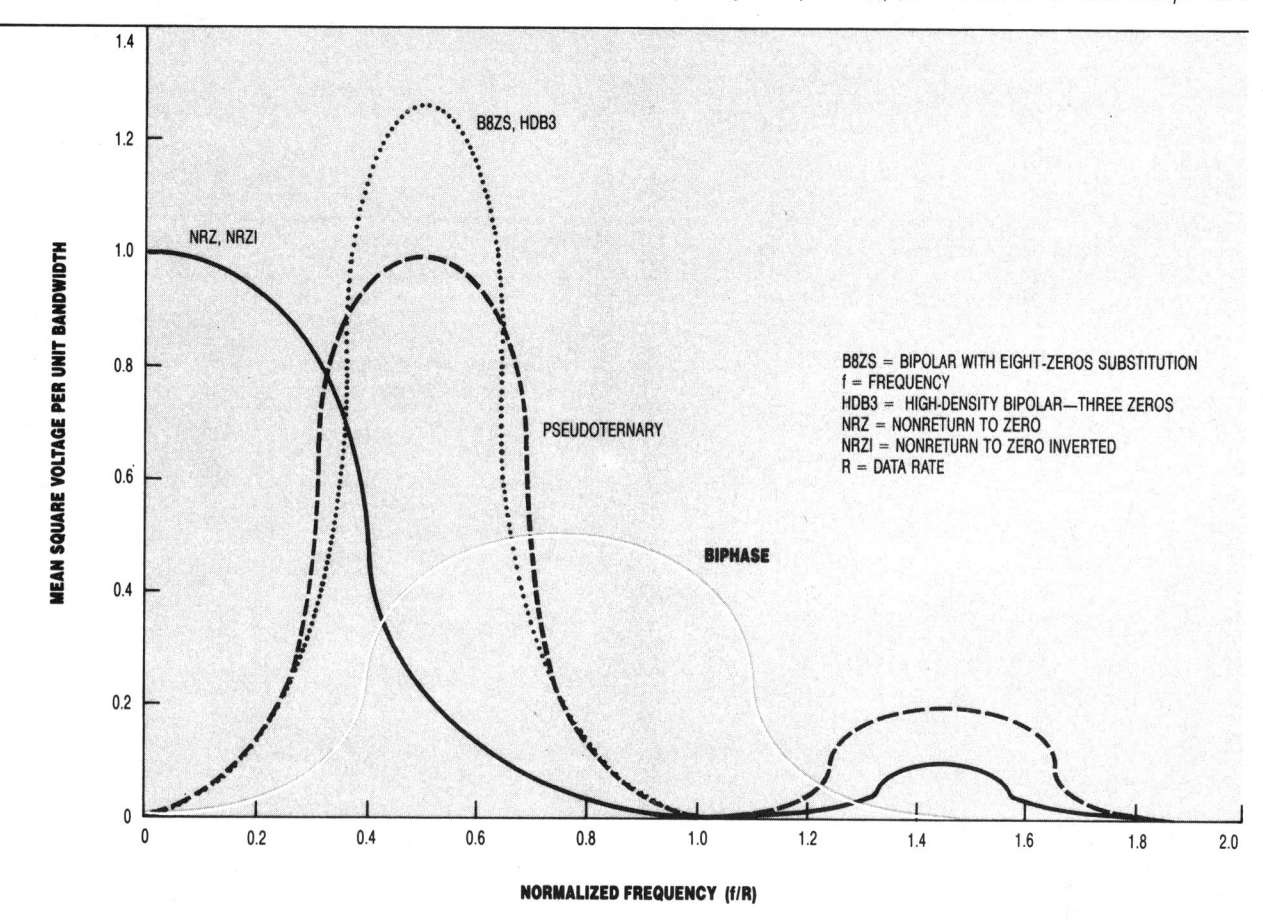

B8ZS = BIPOLAR WITH EIGHT-ZEROS SUBSTITUTION
f = FREQUENCY
HDB3 = HIGH-DENSITY BIPOLAR—THREE ZEROS
NRZ = NONRETURN TO ZERO
NRZI = NONRETURN TO ZERO INVERTED
R = DATA RATE

There is a set of alternative coding techniques, called biphase, which overcomes the limitations of NRZ codes. Two of these techniques, Manchester and Differential Manchester (Fig. 1), are in common use in LANs.

In the Manchester code, there is a transition at the middle of each bit period. The mid-bit transition serves as a clocking mechanism and also as data: A high-to-low transition represents a 1, and a low-to-high transition represents a 0. In Differential Manchester, the mid-bit transition is used only to provide clocking. The encoding of a 0 is represented by the presence of a transition at the beginning of a bit period, and a 1 is represented by the absence of a transition at the beginning of a bit period. Differential Manchester has the added advantage of employing differential encoding (described earlier).

All of the biphase techniques require at least one transition per bit time (transition rate) and may have as many as two (Table 1). Thus, the maximum modulation (digitization) rate—equal to the transition rate—is twice that for NRZ; this means that the bandwidth required is correspondingly greater. To compensate for this, the biphase schemes have several distinct advantages:

■ *Synchronization.* Because there is a predictable transition during each bit time, the receiver can synchronize on that transition. For this reason, the biphase codes are known as self-clocking codes.
■ *d.c.* Biphase codes have no d.c. component.
■ *Error detection.* The absence of an expected transition can be used to detect errors. Noise on the line would have to invert both the signal before and after the expected transition to cause an undetected error (Fig. 1).

The bulk of the energy in biphase codes is between one-half and one times the bit rate (Fig. 2). Thus, the bandwidth is reasonably narrow (no high-frequency components) and contains no d.c. component.

Biphase codes are popular techniques for data transmission. The more common Manchester code has been specified for the IEEE 802.3 standard for baseband coaxial cable and twisted-pair CSMA/CD (carrier-sense multiple access with collision detection) bus LANs. It has also been used for MIL-STD-1553B, which is a shielded twisted-pair bus LAN designed for high-noise environments. Differential

Table 1: Signal transition rate

	MINIMUM	101010...	MAXIMUM
NRZ	0 (ALL 0s OR 1s)	1.0	1.0 (1010...)
NRZI	0 (ALL 0s)	0.5	1.0 (ALL 1s)
MANCHESTER	1.0 (1010...)	1.0	2.0 (ALL 0s OR 1s)
DIFFERENTIAL MANCHESTER	1.0 (ALL 1s)	1.5	2.0 (ALL 0s)
PSEUDOTERNARY	0 (ALL 1s)	1.0	1.0

NRZ = NONRETURN TO ZERO
NRZI = NONRETURN TO ZERO INVERTED

Manchester has been specified for the IEEE 802.5 Token Ring LAN, using shielded twisted-pair wire.

The biphase codes are well-suited for digital signaling on baseband IEEE 802 LANs. In principle, they could also be adapted for the most recent LAN standard, the Fiber Distributed Data Interface (FDDI), which specifies a 100-Mbit/s optical-fiber ring.

Synchronous, but inefficient
For example, for Manchester encoding, a pulse of light would comprise the first or second half of the bit time to represent 1 or 0, respectively, with the absence of light in the other half of the bit time. This would provide the Manchester synchronization benefit. The disadvantage of this approach is that the efficiency is only 50 percent. That is, because there can be as many as two transitions per bit time, a signaling rate of 200 million signaling elements per second (200 Mbaud) is needed to achieve a data rate of 100 Mbit/s. At FDDI's data rate, this approach represents an unnecessary cost and technical burden.

To overcome the data rate burden imposed by Manchester, the FDDI standard specifies a code referred to as 4B/5B. In this scheme, encoding is done four bits at a time; each four data bits are mapped into a five-bit code. Each bit of the code is transmitted as a single signal element (presence or absence of a light pulse). The efficiency is thus raised to 80 percent: FDDI's 100 Mbit/s, for example, is achieved with 125 Mbaud. The resulting cost savings are substantial: A 200-Mbaud optical transmitter/receiver can cost from five to 10 times that of a 125-Mbaud pair.

In order to achieve synchronization, there is actually a second stage of encoding for FDDI. Each element of the 4B/5B stream is treated as a binary value and encoded using NRZI. The use of NRZI, which is differential encoding, aids the ultimate decoding of the signal after it has been converted back from optical to the electrical realm.

Table 2 shows the symbol encoding used in FDDI. There are 32 five-bit codes; 16 are used to encode all of the possible four-bit blocks of data. The codes selected to represent the 16 four-bit data groups are such that a transition is present at least twice for each five-bit code pattern on the medium. Since NRZI is being used, this is equivalent to requiring that there be at least two ones in each five-bit code (recall that in NRZI, a one is encoded by a transition). As can be seen (Table 2), all 16 of the codes that represent data contain at least two ones.

To summarize the FDDI encoding scheme:
- A simple on/off encoding is rejected because it does not provide synchronization; a string of 1s or 0s would have no transitions with which to synchronize.
- The 4B/5B code is chosen over Manchester because it is more efficient.
- The 4B/5B coded data is further encoded using the NRZI technique. This enables the resulting differential encoding

Table 2: 4B/5B Code

CODE GROUP	ASSIGNMENT	
LINE STATE SYMBOLS		
00000	QUIET	
11111	IDLE	
00100	HALT	
STARTING DELIMITER (SD)		
11000	1ST OF SEQUENTIAL SD PAIR	
10001	2ND OF SEQUENTIAL SD PAIR	
DATA SYMBOLS		
	HEX	BINARY
11110	0	0000
01001	1	0001
10100	2	0010
10101	3	0011
01010	4	0100
01011	5	0101
01110	6	0110
01111	7	0111
10010	8	1000
10011	9	1001
10110	A	1010
10111	B	1011
11010	C	1100
11011	D	1101
11100	E	1110
11101	F	1111
ENDING DELIMITER		
01101	USED TO TERMINATE THE DATA STREAM	
CONTROL INDICATORS		
00111	DENOTING LOGICAL ZERO (RESET)	
11001	DENOTING LOGICAL ONE (SET)	
INVALID CODE ASSIGNMENTS		
00001* 00010* 00011 00101 00110 01000* 01100 10000*	THESE CODE PATTERNS SHALL NOT BE TRANSMITTED BECAUSE THEY VIOLATE CONSECUTIVE CODE-BIT ZEROS OR DUTY-CYCLE REQUIREMENTS. CODES MARKED WITH AN ASTERISK, HOWEVER, SHALL BE INTERPRETED AS HALT WHEN RECEIVED.	

3. Two techniques. For the 1.544-Mbit/s primary-rate interface, the coding scheme is known as bipolar with eight-zeros substitution (B8ZS). For the 2.048 Mbit/s interface, the coding scheme is known as high-density bipolar-three zeros (HDB3) code (Table 3). In each case, the fourth zero is replaced with a code violation.

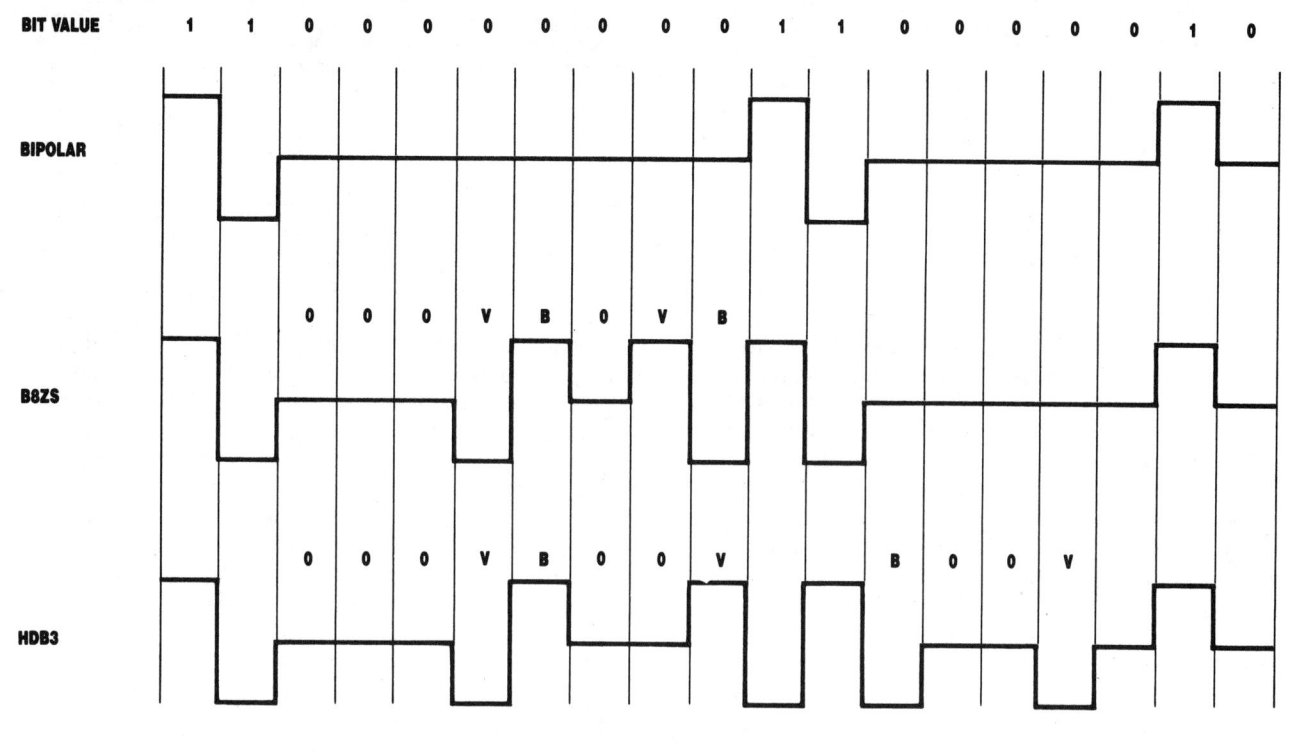

B = VALID BIPOLAR SIGNAL
B8ZS = BIPOLAR WITH EIGHT-ZEROS SUBSTITUTION
HDB3 = HIGH-DENSITY BIPOLAR—THREE ZEROS
V = BIPOLAR VIOLATION

to significantly improve reception reliability.
■ The specific codes chosen for encoding the 16 four-bit data groups guarantee at least two ones—hence, at least two transitions; this provides adequate synchronization.

Only 16 of the 32 possible code patterns are required to represent the input data. The remaining symbols are either declared invalid or assigned special meaning as control symbols. For example, two of the patterns (codes 11000 and 10001) always occur in pairs and act as start delimiters for a frame.

Undesirable
For ISDN, two forms of access are specified: basic and primary. The basic rate provides two user channels and a control-signaling channel at a total data rate—including overhead bits—of 192 kbit/s. The primary interface offers rates of 1.544 and 2.048 Mbit/s.

As with the interface to a LAN, the use of NRZ codes is undesirable for the ISDN basic access interface because of the lack of synchronization and the presence of a d.c. component. To overcome these problems, the encoding chosen for the basic-rate interface is pseudoternary coding. In this scheme, a binary 1 is represented by no line signal; a binary 0, by a positive or negative pulse. The binary 0 pulses must alternate in polarity. The term pseudoternary arises from the use of three encoded signal levels (positive, negative, and zero) to represent two-level (binary) data.

There are several advantages to this approach. First, there will be no loss of synchronization if a long string of 0s occurs. Each 0 introduces a transition, and the receiver can resynchronize on that transition. A long string of 1s would still be a problem, but the basic interface-framing structure includes extra 0s to avoid this problem. Second, since the 0 signals alternate in voltage from positive to negative, there is no net d.c. component. Also, the bandwidth of the resulting signal is considerably less than the bandwidth for NRZ. Finally, the pulse alternation property provides a simple means of error detection. Any isolated error, whether it deletes a pulse or adds a pulse, causes a violation of this property.

Thus, pseudoternary coding has a significant advantage over NRZ. Of course, as with any engineering design decision, there is a trade-off. With pseudoternary coding, the line signal may take on one of three levels. But each signal element, which could represent $\log_2 3 = 1.58$ bits of information, bears only one bit of information. Thus, pseudoternary is not as efficient as NRZ coding.

Another way to state this is that the receiver of pseudoternary signals has to distinguish between three levels (+A, −A, 0) instead of just two levels in the other signaling formats previously discussed. Because of this, the pseudoternary signal requires approximately 3 db more signal power than a two-valued signal for the same probability of bit error.

A second disadvantage of pseudoternary coding is that it lacks a synchronization capability. If there is a long string of ones, there are no transitions, and it is easy for the receiver to get out of synchronization with the transmitter. As mentioned, this is overcome in the basic interface with the use of special framing bits that guarantee a minimum number of transitions. These bits add overhead, of course, so that the full data rate of the interface is not available for user data.

Primary-access interface

As stated earlier, to overcome this lack-of-transition problem in the basic interface, the framing structure includes some zeros. In the primary interface, to make maximum efficient use of the high data rates provided, there are no bits available for such balancing, so another approach is needed. A biphase approach is not desirable because this will raise the signal transition rate. At the data rates of the primary interface, this is an expensive alternative.

Another approach is to make use of some sort of scrambling scheme. The idea behind this is simple: Sequences that would result in a constant voltage level on the line are replaced by "filling" sequences that provide sufficient transitions for the receiver's clock to maintain synchronization. The filling sequence must be recognized by the receiver and replaced with the original data sequence. The filling sequence is the same length as the original sequence, so there is no data rate increase. The design goals for this approach can be summarized as:
- No d.c. component.
- No long sequences of zero-level line signals.
- No reduction in data rate.
- Error detection capability.

Two techniques are specified for the primary-rate interface (Fig. 3), depending on the data rate used (1.544 or 2.048 Mbit/s). The coding scheme used with the 1.544-

4. B8ZS. *If a bipolar code has a long string of zeros, a loss of synchronization may result. To overcome this in B8ZS, two otherwise unlikely code violations are forced.*

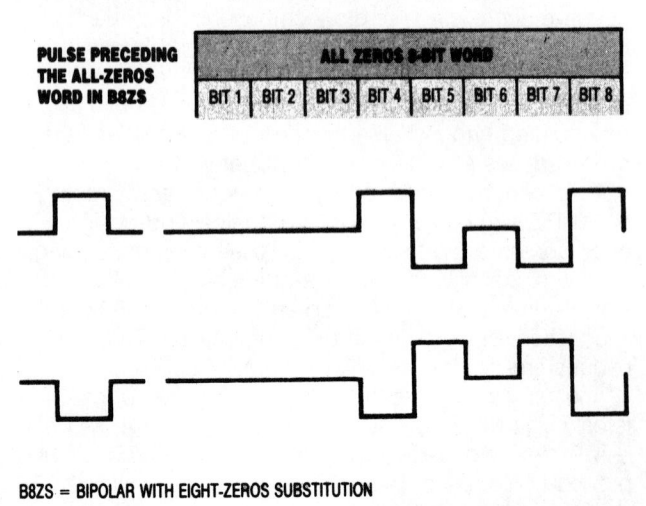

Table 3: HDB3 substitution rules

POLARITY OF PRECEDING PULSE	NUMBER OF BIPOLAR PULSES (ONES) SINCE LAST SUBSTITUTION	
	ODD	EVEN
−	000−	+00+
+	000+	−00−

Mbit/s interface is known as bipolar with eight zeros substitution (B8ZS). As with the basic interface, the coding scheme is based on a pseudoternary code. In this case, the code, referred to as bipolar, is as follows: Binary 0 is represented by no line signal; binary 1, by a positive or negative pulse. The binary 1 pulses must alternate in polarity. Note that the assignment of codes to binary 0 and 1 is the reverse of that for the basic interface. The drawback of the bipolar code is that a long string of zeros may result in loss of synchronization.

To overcome this problem, the bipolar encoding is amended with the following rules (Fig. 4):
- If an octet of all zeros occurs and the last voltage pulse preceding this octet was positive, then the eight 0s of the octet are encoded as 000+ −0− +.
- If an octet of all zeros occurs and the last voltage pulse preceding this octet was negative, then the eight 0s of the octet are encoded as 000− +0+ −.

This technique forces two code violations of the bipolar code, an event unlikely to be caused by noise or other transmission impairment. The receiver recognizes the pattern and interprets the octet as consisting of all zeros.

The coding scheme used with the 2.048-Mbit/s interface is known as the high-density bipolar-three-0s (HDB3) code (Table 3). As before, the approach is based on the use of bipolar encoding. In this case, the scheme replaces strings of four 0s with sequences containing one or two pulses. In each case, the fourth zero is replaced with a code violation. In addition, a rule is needed to ensure that successive violations are of alternate polarity so that no d.c. component is introduced. Thus, if the last violation was positive, this violation must be negative, and vice versa. Table 3 shows that this condition is tested by knowing whether the number of pulses (ones) since the last violation is even or odd and knowing the polarity of the last pulse before the occurrence of the four 0s.

Figure 2 shows the spectral properties of the B8ZS and HDB3 codes. Neither has a d.c. component. Most of the energy is concentrated in a relatively sharp spectrum around a frequency equal to one-half the data rate. Thus, these codes are well-suited to the high-data-rate transmission of the primary ISDN interface. ∎

William Stallings is a consultant and president of Comp-Comm Consulting, London, England. This article is based on material from his Data and Computer Communications, *Second Edition (Macmillan, 1988) and from the recently published* ISDN: An Introduction *(Macmillan, 1989). Stallings holds a Ph. D. in computer science from M. I. T.*

B8ZS = BIPOLAR WITH EIGHT-ZEROS SUBSTITUTION

Fred Yarusso, Merrill Lynch & Co., New York, N.Y.

Moving to new quarters? Don't trip over the cabling issue

A 44-story financial tower runs three cables to each workstation; a relational database helps track and manage the cabling maze.

Cable. Modern industry wouldn't last long without it. But installing the stuff may require the careful orchestration of an 18th-century impresario. Here's how one large financial firm solved its cabling problems while moving more than 5,000 employees into its new location in New York's Lower Manhattan.

Merrill Lynch & Co. began occupying the 44-story South Tower of the World Financial Center (WFC) during the second half of 1987; its move-in schedule continues through 1988. As is typical with office relocation plans, occupancy was scheduled to take place on weekends—sometimes one floor at a time, but often several floors were scheduled for move-in during the same time period, giving the installation crew two days (see table).

This demanding schedule is being met by the concerted efforts of all the members of a dedicated project team. The Communications Department, a part of the team, gives special credit for the success of the move to two main factors: the communications cable that was chosen, and the methods that have been put in place to manage the cable plant.

Planning for the move began in earnest during 1986. At that time a number of subcontractors, architects, and engineers were actively engaged in moving Merrill Lynch into a WFC building that had been completed earlier. They were able to learn from that experience, in which they used a combination of conventional AT&T inside wiring for the voice instruments and native cabling for the wide range of differing data terminals. (The phrase "native cabling" refers to the existing, or vendor-prescribed, cable for a device.) On the positive side of this arrangement, it allowed the project team to forecast budgeting requirements and to meet the users' demands for initial service.

On the negative side, however, this combination of wiring proved to be inflexible; it could not accommodate last-minute changes in user requirements and user relocations. To effect such changes, it was often necessary to run additional cable. During the planning process for the second of the two buildings in the WFC, therefore, Merrill Lynch charged DP Communications Corp. (one of the organizations engaged on the project and managed by the Communications Department) with the task of solving its cabling problems.

DP Communications Project Manager Fred Grover recommended a standard cabling scheme for all user/workstation locations throughout the building. Also recommended was a relational database package to enable computerized tracking of the cable plant during installation and to manage the cable thereafter.

These recommendations, which were eventually adopted by Merrill Lynch, were greeted with some skepticism by many on the project team. Accommodating the diverse types of terminals with a standard set of cables seemed extremely difficult.

The business departments of the company are relatively free to select whatever technology they need to suit their applications, and the potential for connectivity problems is high. After all, the scope of the project encompassed 30,000 square feet at the upper levels of the building (accommodating 250 people) to 70,000 square feet at the lower levels (accommodating 500 people).

The project team knew it would have to accommodate the large number of terminal-based services that provide financial market data and that it would need to support the burgeoning variety of local area networks (LANs). One way to accomplish this is by providing an identical cable set to each workstation throughout a building. A summary of IBM's cabling alternative appeared in the May 1987 issue of DATA COMMUNICATIONS ("How to design and build a token ring LAN," p. 213), and AT&T had previously announced its

Move-in schedule

FLOOR	POPULATION	MOVE-IN DATE
5	360	7/24/87
14/16	343	9/18/87
15	170	10/2/87
12	180	10/16/87
7	380	10/30/87
4	415	11/20/87
6	302	12/11/87
8	484	1/15/88
11/17	268	1/29/88

cabling product, PDS (Premises Distribution System).

Running a single type of cable set to each workstation means that when personnel are relocated throughout the building, their terminals and computers need not be rewired, saving both time and expense. Furthermore, the computerized database of the cable plant offers the capability to perform rapid troubleshooting and to overcome problems during the testing periods (which are notoriously brief on projects of this nature). These planning decisions were agreed on by the project team and have turned out well.

Shopping for cable

Choosing the specific cables to satisfy present and projected communications requirements is never an easy task. It is influenced by the rapid evolution of technology, the long lead times involved in specifying cables for construction projects, and the constantly changing communications requirements of users.

After deciding to wire each user/workstation with the same three cables, Merrill Lynch's project team settled on the following combination:
- IBM Type 1, which is a two-pair, 22-gauge shielded cable.
- AT&T standard inside wiring D-type (which is a four-pair, 24-gauge unshielded cable) for data.
- AT&T D-type for voice.

The IBM Type 1 cable terminates in the user area with an IBM universal connector, the standard connector for this type of cable. The AT&T four-pair cable for data and the AT&T four-pair cable for voice both terminate in the same AT&T 104A block. This is a standard cable terminating block that has two modular connector ports: one for plugging in the voice cable, the other for the data cable.

The termination method was largely dictated by the under-floor cell structure in the building and the type of partitions used. The cables are routed from the nearest communications closet to the user/workstation under-floor. They protrude into the workstation area through a hole in the floor, providing eight feet of slack, and are then terminated.

This arrangement provided a flexible termination method and easy troubleshooting. For this type of application, it compared favorably with other installation methods, such as an initial wall termination. Approximately one week after the user has been settled in the space, when the placement of furniture and terminals is relatively certain, the cables are tied down and neatly dressed.

In choosing the type of cables, the project team decided against fiber for the lateral and the riser cables. Riser sleeves for fiber and spaces for related electronics, however, have been allocated in the communications closets; so, fiber could be added at any time, but it was not specified for the initial intrabuilding communications. The decision, which was made in 1986, was based on several observations: The applications could be supported with twisted-pair wiring, and there was a lack of standardization at the time for particular fiber types and the electronics associated with fiber. Obviously, anyone planning a cable plant for a building to be occupied 18 months from now might reach different conclusions. The Merrill Lynch project team is currently testing fiber products and will gradually phase them into the WFC complex.

In considering whether to use coaxial cable, the team decided it could get along without RG-62 coax (which is the native cable for the IBM 3270 family of terminals) because, although Merrill Lynch is a large IBM user, there are many products that support these devices on twisted pair.

It was harder for the project team to decide whether to use RG-59 coax. This is the native cable used by many terminal-based financial information services. Any given floor of the WFC complex might have hundreds of these terminals for market data services. Baluns that would operate for distances of more than 1,000 feet were required but were not available from any known source. A balun (short for balanced/unbalanced) is a transformer that matches the impedance of one cable type to another. Baluns allow the use of twisted-pair cable between two devices that would ordinarily use coax cable.

Merrill Lynch had been using baluns by StarTek Corp. (Worcester, Mass.) to support some of these services for distances of more than 200 feet. StarTek was encouraged to enhance its products, and it successfully developed a broader line of baluns to operate over the longer distances.

There was also some concern about whether or not the installation could get along without Ethernet cable—specifically, thin Ethernet cable (RG-58)—as lateral cable. Digital Equipment Corp. (DEC) specifies the thin Ethernet cable as one of the components of its lateral set in a premises distribution system. A few Ethernet LANs had already been scheduled for installation, and that number could have grown considerably. The project team was aware of several vendors bringing out products to support Ethernet on twisted pair and decided to visit users and to test these products. It was determined that products would be available in time to accommodate the move-in schedule. After a thorough evaluation of the requirements and related technology, it was decided that the building could be wired with twisted-pair cables—shielded and unshielded.

Choosing IBM Type 1 as the basic shielded cable was

1. The technology room. Each floor has a room set aside for technology equipment. The gear in this area is kept to a minimum and typically consists of file servers, which allow users to share common files, and gateways, which provide access to other networks or services. A simplified representation is shown.

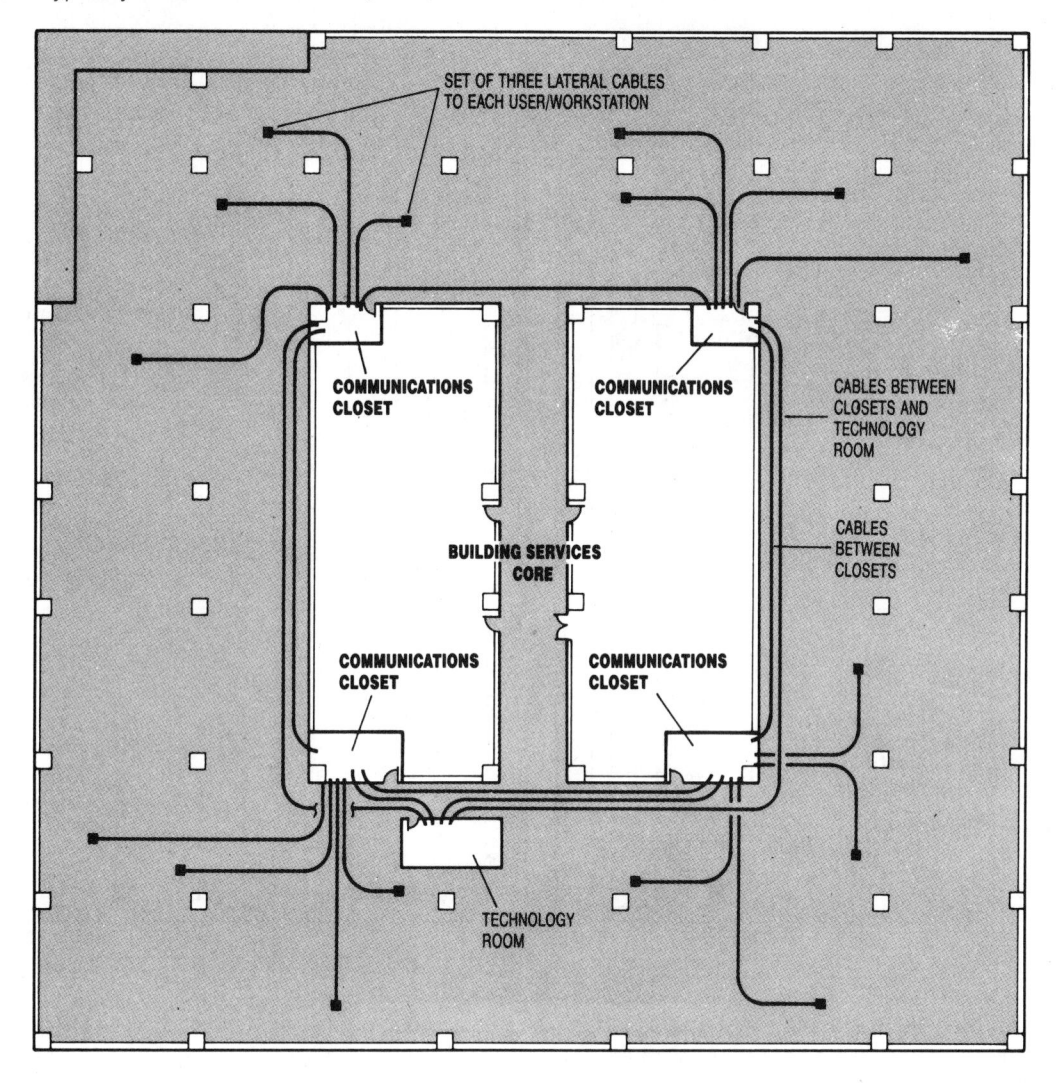

not done casually. The equipment associated with the IBM Cable System, racks, panels, and connectors adds to the cost of the cable, but the project team felt the cost was justified by the following applications:
- A significant number of financial market data terminals, which require shielded cable to operate with baluns over distances of 500 to 1,000 feet.
- Novell/Proteon 10-Mbit/s Token Ring LANs.
- 10-Mbit/s Ethernet LANs.

In addition, there was the potential IBM 16-Mbit/s Token Ring LAN to consider.

The IBM equipment associated with the cables (such as racks, panels, connectors, and patch cords) presents advantages and disadvantages. On the one hand, this gear adds to the cost and requires closet space; on the other, it acts as a standard to which other vendors must adapt their products. For instance, many vendors of LAN hardware provide products that are readily mounted in the IBM Cabling System equipment racks.

Similarly, the unshielded cable for data that the project team chose was AT&T's inside-wiring standard four-pair cable. Recommended by AT&T for its PDS, this cable is in widespread use and, consequently, relatively inexpensive. It offers excellent transmission for a wide variety of communications devices. The cable and associated equipment (such as blocks, frames, and modular adapters) also offer vendors a direction in which they can guide their product development. Merrill Lynch Communications, like many large user organizations, can urge vendors to develop products. As a result of its testing and operational experience, it offers valuable guidance to vendors while not being an R&D organization per se. However, in this case, the company wanted to use off-the-shelf solutions.

Merrill Lynch chose the twisted-pair cabling types of

2. Cabling from floor to floor. *The intrabuilding communications riser cables are shown here. Shielded and unshielded data risers terminate on frames, which provide access to controllers by means of patch cables. Voice riser cables also terminate on frames and access the private branch exchange. Risers interface with lateral cable sets.*

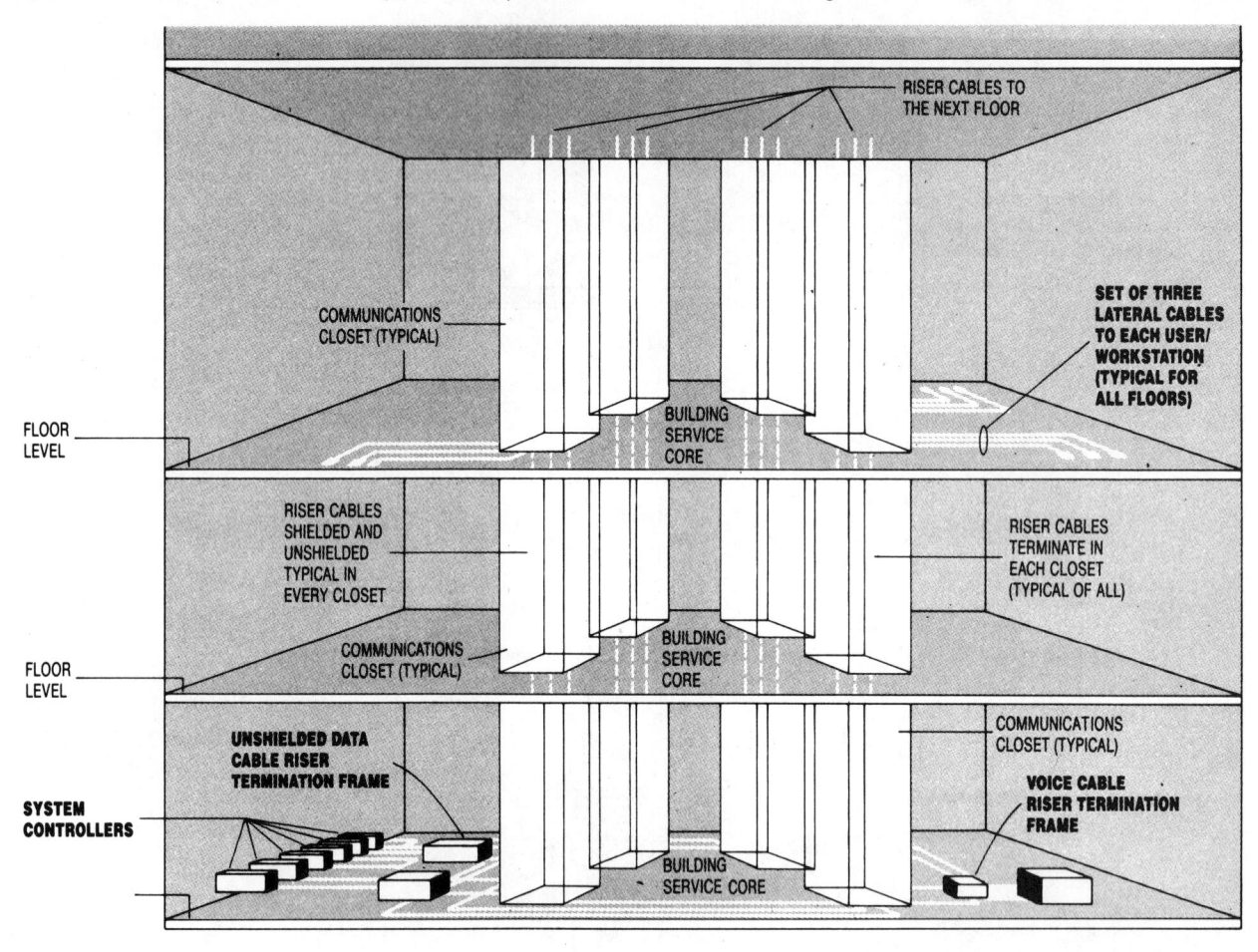

IBM's and AT&T's premises distribution systems because it seemed that the combination could support all of the company's user requirements at the time the new building would be occupied. This is proving to be the case.

The cable plant

The following summary refers to the cable installed for intrabuilding communications. Cables installed for access to external services and locations are not discussed. Merrill Lynch has an extensive and comprehensive network that links all business locations. The transmission media, equipment, and communications carriers associated with this network are beyond the scope of the present article.

The building is cabled with the IBM and AT&T cables previously mentioned. The lateral cables are installed in a star configuration from the serving communications closet to the user/workstation area. The IBM Type 1 cable is terminated in closets on IBM cabling racks and panels. The AT&T four-pair cables are terminated on standard telephone industry cable termination blocks.

Each floor has a room designated for technology. The equipment in the technology rooms is kept to a minimum and usually consists of servers associated with LANs on that floor (Fig. 1). This includes such equipment as file servers, which allow users to share common files, and gateways, which provide access to other networks or services.

The closets and technology room are interconnected with cables: six IBM Type 1 cables plus a 50-pair cable between adjacent closets and between each closet and the technology room. Most of the communications equipment is housed on a specific floor assigned for this equipment. Merrill Lynch chooses to specify a particular floor, or in some cases two floors, for communications equipment in an office tower setting. This is partly because it allows for local concentration of expensive power backup and external communications backup; the service personnel for the numerous Financial Market communications controllers can be more easily monitored; and it is easier for the controllers to be shared by terminal users throughout the building. The AT&T PBX and the intrabuilding voice cable termination frame is housed on this floor. Similarly, an intrabuilding data cable termination frame is housed on this floor. IBM Type 1 riser cables are also terminated here.

Riser cables emanate from this floor to all the communications closets in the building. The voice riser consists of a 400-pair cable to each closet. Three pairs of this riser are assigned to each telephone. Based on the square footage of the floors and number of closets, each wiring closet will support equipment for approximately 125 people.

The data cable riser consists of a 200-pair cable from the data cable frame on the technology floor to each closet. IBM Type 1 risers, quantity 10, also run to each closet in the building. In addition, some IBM Type 1 cables, quantity 4, are run from floor to floor to link up local area networks. A Wangnet riser was installed to support the Wang terminals in the building and to provide a general-purpose broadband riser. A broadband riser was advisable to serve as a backup against the company's falling short on riser count or failing to account for a particular service. The maturity of this technology makes it suitable to carry a variety of data communications services. For instance, it can carry Ethernet service or T1 service.

Figure 2 depicts a summary of the intrabuilding communications riser cables. Note the technology floor shown on the lower portion of the figure. On the left side of the floor, shielded and unshielded data riser cables are terminated on frames. The frames provide access to various controllers by means of patch cables on the frames. Similarly, on the right side of the floor, the voice riser cables are terminated on a frame and access the PBX. The riser cables are shown serving each floor vertically in communications closets. On the uppermost floor in the figure, the risers are shown interfacing with a typical lateral cable set.

Managing the maze
A key factor in meeting strict move-in schedules is the method chosen for documenting and managing the cable plant. The project team at Merrill Lynch set up a computerized database, which included the lateral and riser cables.

The instructions to the electrical contractors to install the cables are issued as computer printouts. The basic riser-cabling work is scheduled to be finished many months before any floors are occupied, and the lateral cables are installed two to three months before occupancy. Two to three months before move-in, the project team makes a final review of all users' communications requirements. Any additional lateral cables and the necessary cross-connections in the serving closet and on the communications floor are fed into the computer database. A variety of reports are then issued from the computer database, and instructions to the electrical contractor are issued. Reports that trace the cable and specific conductors associated with particular user terminals are also issued (Fig. 3).

The trace report provides detailed information about every segment linking the terminal to the controller. The columns on the computer printout under the segment identified as A provide detailed information about the cable termination in the user area, such as terminal, device type, floor, and room. The columns on the printout under segment B provide details of the lateral cable, such as cable number and pair assignment. Segment C details the lateral cable termination in the serving closet, such as closet name and termination block assignment. Columns under segments D, E, and F provide similar details pertaining to the riser cable, such as cable pair assignment and termination. Columns G and H provide terminations for the final segment in the circuit to the control unit.

These trace reports are a very effective tool in troubleshooting problems in the brief time before a scheduled move. Inevitably, last-minute changes occur, but they can be accommodated by the flexibility of the installed cable plant and the capability to trace very rapidly any or all circuits.

Although many services can be supported by either the IBM Type 1 or the AT&T four-pair cable, the Merrill Lynch project team assigned particular services to one or the other data cable for ease of maintenance. Because the four-pair cable is more common and less expensive than the IBM Type 1, the company plans to support devices on the four-pair cable where possible and to reserve the IBM Type 1 for applications that require it. To date, the following devices are being supported on the IBM Type 1:
- Quotron Market Data Service—The monitor and the keyboard on one Type 1 cable.
- Bloomberg L. P. Market Data Service—The monitor only; the keyboard is supported by the four-pair unshielded data cable.
- Reuters Monitor 1 Market Data Service—The monitor only.
- Reuters Equity 2000 Market Data Service—The monitor only; the keyboard is supported by the four-pair unshielded data cable.
- Novell/Proteon 10-Mbit/s Token Ring LANs.
- Novell/Proteon, IBM, and Ungermann Bass 4-Mbit/s Token Ring LANs. (Merrill Lynch has begun supporting these on unshielded twisted pair.)
- Ethernet 10-Mbit/s LANs. (Products are now emerging to support these networks with unshielded cable.)
- Twinaxial Cable Applications, such as IBM system 3X and IBM 5520. (Products are now emerging to support these computers with unshielded cable.)

The remaining devices are supported on the four-pair unshielded lateral cable. These devices are too numerous to itemize, but the project team has adopted a method of allocating particular cable pairs to particular services. Assigning certain pairs to specific devices, where such wiring is possible, results in the partial adoption of a de facto standard for the building; this helps in troubleshooting and maintenance. Furthermore, such allocation provides a method for sharing the cable among two, three, and in rare cases, four different devices. The four pairs of a four-pair telephone cable are specifically identified by the color of the conductors. When they are terminated in a standard 104A termination block, the pairs are assigned to specific pins (Fig. 4).

The termination for the four-pair voice cable is the same as for the four-pair data cable. Note, in the pair assignments for Port 1 "Voice," that Pair 1 comes from two conductors in the cable, the blue/white and the white/blue.

Pair 2 comes from two conductors in the cable, the orange/white and the white/orange. Pair 3 comes from two conductors in the cable: the green/white and the white/green. Pair 4 comes from two conductors in the cable: the brown/white and the white/brown. IBM 3270 devices use Pair 1 with baluns. Pairs 2 and 3 are used by the numerous devices that require two pairs. This includes RS-232-C devices that require four conductors or fewer to operate. It also includes RS-232-C devices that require more than four conductors, and in those cases, the signal is brought onto two pairs by an RS-232-C multiplexer.

During this move, Merrill Lynch has not used more than two pairs for a device. The project team chose to play it safe by using an RS-232-C multiplexer with distance rated at 1,000 feet when there was any doubt about whether a given device could operate over the required distances. The fourth pair of the four-pair lateral data cable is used to support Wang terminals with baluns. This allows the device to access Wangnet, which is available in all closets. If a modem for a data terminal is required, that device is assigned to the fourth pair of the data cable if it is available. The fourth pair, if used for modem support, is cross-connected in the serving closet to the voice riser.

Figure 5 depicts an example of more than one device sharing the four-pair data cable. The upper portion of the figure shows the user floor. To the right is a user area with an IBM 3270 Terminal and a Digital Equipment Corp. VT220 Terminal. To the left of the upper portion of the figure is the serving communications closet. The two terminals in the user area are shown connected to the 104A block in the user area. A four-pair splitting adapter is shown that allows the terminals to be connected to particular conductor pairs. The IBM 3270 terminal is connected to the splitting adapter with a balun, it is assigned Pair 1 of the four-pair lateral cable. The VT220 terminal is assigned Pairs 2 and 3 of the four-pair lateral cable. This RS-232-C-type connection does not require a balun.

In the communications closet, the four-pair lateral cable is shown terminated on the lateral cable termination block. The 200-pair unshielded riser cable for data is also shown terminated on a block. A cross-connection between the lateral cable and the riser cable is shown for each terminal. The lower portion of the figure shows the technology floor where the risers terminate and where the controllers are located. The risers terminate on the Data Distribution Frame. This frame is also connected to the system controllers on the technology floor. The figure shows a patch cable on the frame for each terminal (IBM and DEC) between the riser cable termination on the frame and the controller access cable termination on the frame. The figure also depicts the balun at the controller end of the IBM terminal circuit.

This sharing of the cable is another technique that permits quick adaptation to a change in user needs. Unused pairs of an existing lateral cable can be connected to a riser cable in an hour or two. It would take many hours to pull in a new lateral cable. Not all conductors of a given lateral cable will necessarily be used. When unplanned additional terminals must be supported, it helps to have a

3. Tracking the cable. Reports that trace the cable and specific conductors that are associated with particular user terminals provide detailed information about every

TERMINAL					LATERAL		
DV	FL	ROOM	PRST	SEQ.	NUMBER	PAIR	COLOR
B7	16	114	8/5	1	16BD2E01	0001	WH/BL
B7	16	110	7/9	1	16BD2D17	0001	WH/BL
5T	16	108	7/8	1	16BD2D13	0001	WH/BL
5T	16	106	7/1	1	16BD2C09	0001	WH/BL
5T	16	104		1	16BD2C05	0001	WH/BL
5T	16	103-AB	6/9	1	16BD3D13	0001	WH/BL
5T	16	167-JE	12/9	1	16CD4C21	0001	WH/BL
D2	16	163-KA	11/7	1	16AD3B21	0001	WH/BL
D2	16	163-KB	11/9	1	16AD3C01	0001	WH/BL
D2	16	163-KC	12/2	1	16AD3C05	0001	WH/BL
5T	16	174		1	16CD2C05	0001	WH/BL
B7	16	203-FG		1	16CD3E01	0001	WH/BL
B7	16	127	9/9	1	16BD3A13	0001	WH/BL

DV = DEVICE CODE
FL = FLOOR
PRST = PRESET A CABLE ACCESS HOLE IN THE CELLULAR FLOOR STRUCTURE
SEQ = SEQUENCE

segment linking the terminal to the controller. The reports can be especially helpful when trying to troubleshoot problems in the brief time before a scheduled move. The last-minute changes that will inevitably occur are accommodated by the flexibility of the installed cable plant and the ability to trace circuits rapidly.

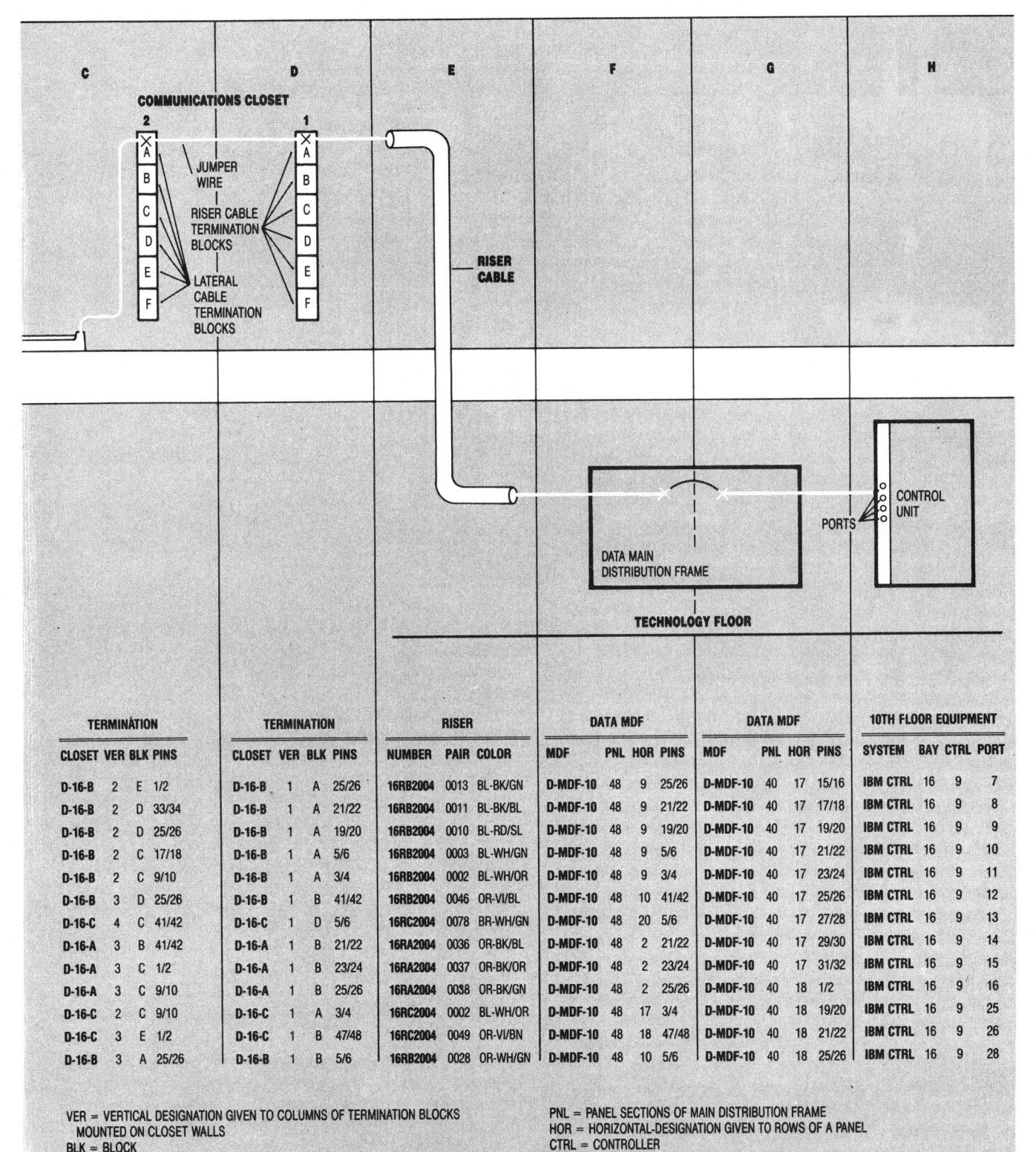

TERMINATION				TERMINATION				RISER			DATA MDF				DATA MDF				10TH FLOOR EQUIPMENT			
CLOSET	VER	BLK	PINS	CLOSET	VER	BLK	PINS	NUMBER	PAIR	COLOR	MDF	PNL	HOR	PINS	MDF	PNL	HOR	PINS	SYSTEM	BAY	CTRL	PORT
D-16-B	2	E	1/2	D-16-B	1	A	25/26	16RB2004	0013	BL-BK/GN	D-MDF-10	48	9	25/26	D-MDF-10	40	17	15/16	IBM CTRL	16	9	7
D-16-B	2	D	33/34	D-16-B	1	A	21/22	16RB2004	0011	BL-BK/BL	D-MDF-10	48	9	21/22	D-MDF-10	40	17	17/18	IBM CTRL	16	9	8
D-16-B	2	D	25/26	D-16-B	1	A	19/20	16RB2004	0010	BL-RD/SL	D-MDF-10	48	9	19/20	D-MDF-10	40	17	19/20	IBM CTRL	16	9	9
D-16-B	2	C	17/18	D-16-B	1	A	5/6	16RB2004	0003	BL-WH/GN	D-MDF-10	48	9	5/6	D-MDF-10	40	17	21/22	IBM CTRL	16	9	10
D-16-B	2	C	9/10	D-16-B	1	A	3/4	16RB2004	0002	BL-WH/OR	D-MDF-10	48	9	3/4	D-MDF-10	40	17	23/24	IBM CTRL	16	9	11
D-16-B	3	D	25/26	D-16-B	1	B	41/42	16RB2004	0046	OR-VI/BL	D-MDF-10	48	10	41/42	D-MDF-10	40	17	25/26	IBM CTRL	16	9	12
D-16-C	4	C	41/42	D-16-C	1	D	5/6	16RC2004	0078	BR-WH/GN	D-MDF-10	48	20	5/6	D-MDF-10	40	17	27/28	IBM CTRL	16	9	13
D-16-A	3	B	41/42	D-16-A	1	B	21/22	16RA2004	0036	OR-BK/BL	D-MDF-10	48	2	21/22	D-MDF-10	40	17	29/30	IBM CTRL	16	9	14
D-16-A	3	C	1/2	D-16-A	1	B	23/24	16RA2004	0037	OR-BK/OR	D-MDF-10	48	2	23/24	D-MDF-10	40	17	31/32	IBM CTRL	16	9	15
D-16-A	3	C	9/10	D-16-A	1	B	25/26	16RA2004	0038	OR-BK/GN	D-MDF-10	48	2	25/26	D-MDF-10	40	18	1/2	IBM CTRL	16	9	16
D-16-C	2	C	9/10	D-16-C	1	A	3/4	16RC2004	0002	BL-WH/OR	D-MDF-10	48	17	3/4	D-MDF-10	40	18	19/20	IBM CTRL	16	9	25
D-16-C	3	E	1/2	D-16-C	1	B	47/48	16RC2004	0049	OR-VI/BN	D-MDF-10	48	18	47/48	D-MDF-10	40	18	21/22	IBM CTRL	16	9	26
D-16-B	3	A	25/26	D-16-B	1	B	5/6	16RB2004	0028	OR-WH/GN	D-MDF-10	48	10	5/6	D-MDF-10	40	18	25/26	IBM CTRL	16	9	28

VER = VERTICAL DESIGNATION GIVEN TO COLUMNS OF TERMINATION BLOCKS MOUNTED ON CLOSET WALLS
BLK = BLOCK
PNL = PANEL SECTIONS OF MAIN DISTRIBUTION FRAME
HOR = HORIZONTAL-DESIGNATION GIVEN TO ROWS OF A PANEL
CTRL = CONTROLLER

4. Allocating cable pairs. *Certain pairs of cable are directed to specific devices. This aids in troubleshooting and maintenance. It also provides a way to share the cable among up to four devices. Note, in the pair assignments for Port 1 "Voice," Pair 1 comes from two conductors in the cable: the "blue/white" and the "white/blue."*

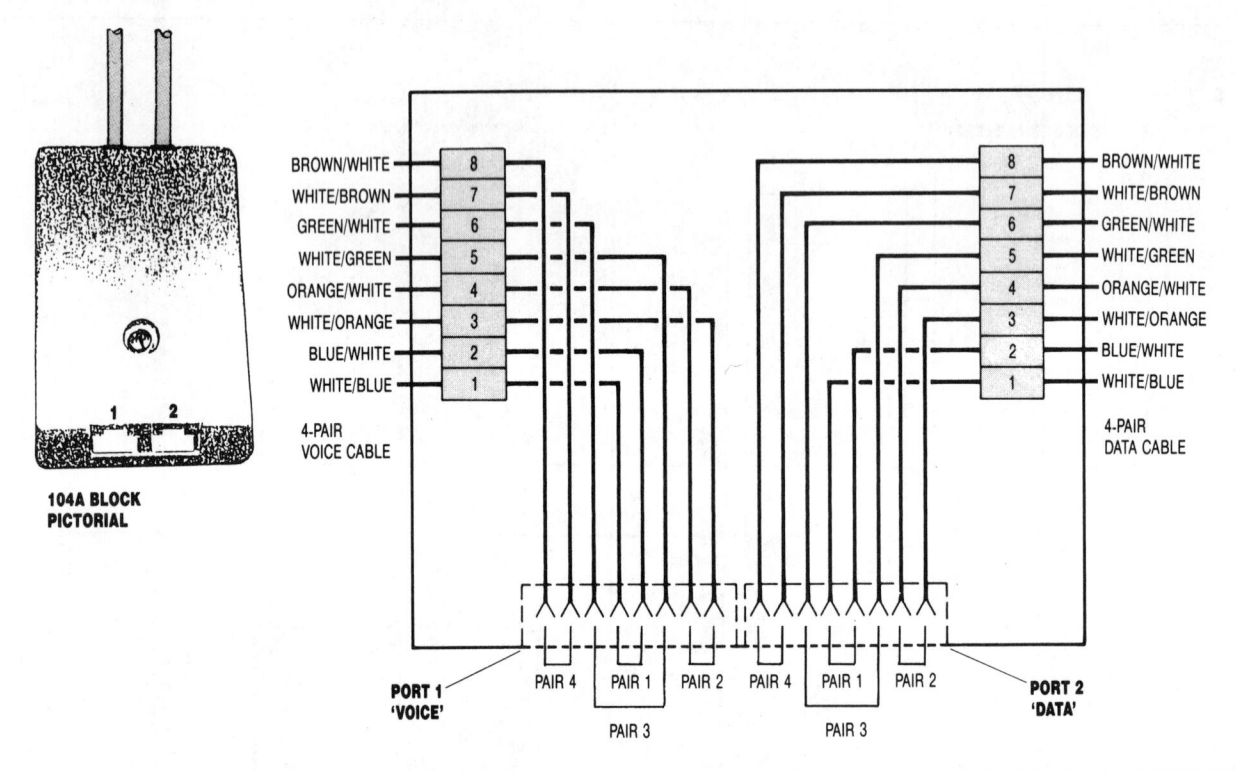

scheme of cable pair assignments to add these terminals in an orderly and predefined way. Extensive testing was done to ensure the successful operation of multiple devices on the same cable.

It is important to test users' applications well in advance of the scheduled occupancy date. To that end, users must cooperate in providing equipment that can be tested. The project team at Merrill Lynch established a test area on the technology floor so that applications could be tested in the environment they would eventually occupy. Occasionally, the team had to conduct tests in the users' old environment so that actual operational demands could be experienced. They also used a small laboratory, which had been previously set up by user organizations to test new local area network products.

Testing a particular application is carried out on varying levels, from users' perceptions to a detailed analysis of the electronic signals being transmitted. On a general level, the operator observes the terminal screen, keyboard, and/or printer operation. Obvious problems of screen intensity, clarity, and keyboard, and/or printer problems will be detected simply during operation.

At a more detailed level, a datascope is extensively used. This is a device that provides for bit error rate testing and can be set up to generate data through the test environment over extended periods and with various data rates. For testing at the lowest level and for analyzing video signals that go to the market data terminal screens, an oscilloscope is used. This device detects signal degradation and allows detailed analysis of the electronic signals being transmitted.

Indeed, testing was and continues to be an integral part of the planning process. All products brought into the Merrill Lynch communications environment as part of this project were tested and continue to be monitored. For future premises wiring and retrofits of existing buildings, the company is evaluating the use of optical fiber and related fiber multiplexers. The high data rate capability of these media will undoubtedly yield benefits. Also, testing will continue to enable the integration of new products that effectively use twisted-pair cable.

Besides testing cable plant technology, the project team will continue to test and validate its cable management procedures. Its computerized cable management techniques are being enhanced to include graphic capability. Given the rapidly evolving state of the art of communications technology, communications managers will have to live with and maintain cable plants consisting of diverse media. The ability to manage these plants and adapt them to serve users will be a continuing challenge.

While there are benefits to be gained from a more standardized cable plant and from well-planned cable management techniques, there are also potential disadvantages: The particular "standard" for each organization and building premises depends on the communications services required; and, of course, the optimal cable plant

5. Sharing cable. Two terminals are connected to more than one device sharing the four-pair cable. A splitting adapter lets terminals be connected to particular pairs.

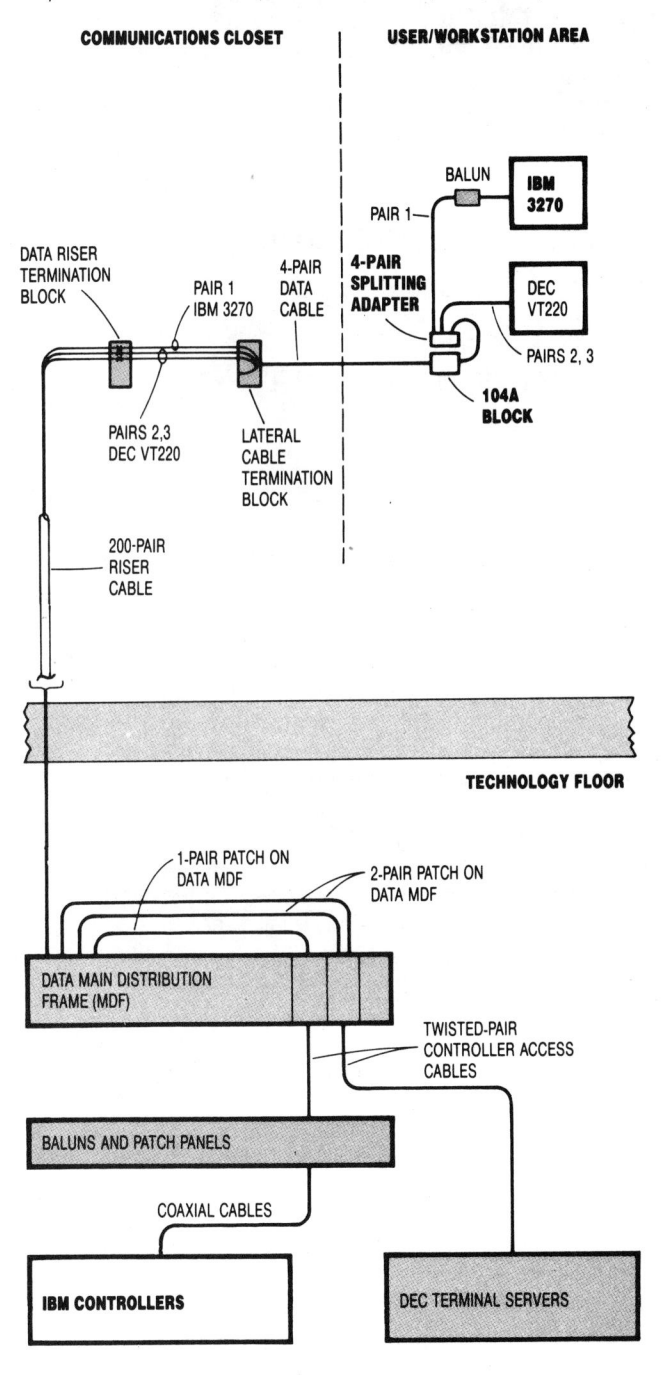

for any organization is constantly evolving. ∎

Fred Yarusso is an assistant vice president and senior systems manager at Merrill Lynch. He joined the company in 1985 and developed the cable design and cable management departments. Yarusso is a 19-year veteran of AT&T, where he had extensive experience in operations, engineering, and sales.

Robert Rosenberg, DATA COMMUNICATIONS

The shots seen round the world

Videoconferencing and facsimile are just the start. Keep an eye on your network traffic: An imaging revolution is under way.

A store in Brooklyn is robbed. The cops pick up a suspect, but if the shopkeeper takes the time to go downtown to file a complaint, he must close the grocery and lose another day's receipts. Elsewhere, a high school student has a chance to earn college credits—but only if she can take a course at another school miles from her home.

The dilemma these people face is familiar to corporate officers and professionals: How can you be in two places at once? Videoconferencing technology got the student and shopkeeper out of the binds they were in, but the technology is not exactly new. Corporations have been using videoconferencing for years to conduct sales meetings and seminars. So where's the imaging revolution?

Look again. Shopkeepers and high school students now have access to technology that a few years ago cost megabucks. Image communication is no longer just for the *Fortune* 100 companies. From credit card processing to video telephones, medical diagnostics to movie making, image traffic is unquestionably growing as a percentage of overall data traffic—both in local area networks and in the public switched telephone network.

By far the largest percentage of imaging traffic today is sent over the dial-up network by inexpensive facsimile devices. In the last four years, Japanese manufacturers have pushed fax out of its corporate niche and created the first mass market for image technology (see "Getting the fax straight"). A New York City delicatessen that takes customer orders by facsimile may not be typical, but in today's business environment, it's hard to imagine life without a fax.

How much fax traffic is moving across the public network is hard to quantify, but experts say it has already affected how the network is used. "The studies I heard say that more than one-half of the traffic between the United States and Japan is fax," says David Vlack, head of advanced hardware architecture at AT&T Bell Laboratories, Naperville, Ill.

While fax is first among the imaging technologies, there are other inexpensive devices aimed at consumer markets that could pump up the amount of image network traffic even more. If you want to *see* how grandma is doing, pick up the phone; video telephones that cost a few hundred dollars are showing up in consumer catalogs and on department store shelves (Fig. 1).

Mitsubishi Electric Co., for example, is marketing two videophones. The Luma videophone is aimed at business users and costs about $1,000 retail. It consists of a handset, camera, and a tiny 96-by-96 pixel, black-and-white display. The unit, which is only about 8 inches by 11 inches in size, takes approximately five seconds to transmit an image after the user presses a button on the set. During that time, voice communications is interrupted.

Then there's the Visitel, a consumer unit that can be cabled to a standard handset, much like a telephone answering machine. It costs $400 retail, according to the manufacturer.

Or have grandmother send a snaphot. A new generation of electronic still cameras that operate without film can send high-quality pictures around the world in a few minutes. These are available from several manufacturers and are priced under $1,000 (Fig. 2).

TV by phone?
Collectively, all the imaging traffic now on the network—conferencing, fax, and the rest—probably still represents but a small portion of the total traffic volume. But if Congress gives local telephone companies the green light to compete with cable TV companies, imaging over the public network could take on a whole new meaning.

In June, the government's National Telecommunications

and Information Administration said telephone companies should be allowed to provide a "video dial tone"—a wideband transport capability under the regulations that apply to common carriers. And in July, the Federal Communications Commission went a step further and called for the removal of cross-ownership restrictions imposed by the 1984 Cable Communications Policy Act (see "FCC gives telcos nod for cable TV business," Washington Newsletter, DATA COMMUNICATIONS, August 1988, p. 25). Removing cross-ownership restrictions would allow the local telephone companies to provide programming, not just the wire transmission medium.

"Good news for consumers, good news for the telephone companies, bad news for cable TV," trumpeted *Forbes* magazine, which said getting into the TV business could be the incentive telephone companies need to start pushing fiber the last mile to the consumer.

"We have a firm belief that Congress and the FCC will change the cross-ownership rules," says Don Marsh, Contel Corp.'s vice president of technology and the corporate point man leading a series of TV-over-fiber tests. Apparently, other large telephone companies, including GTE and Southwestern Bell, are optimistic as well; each is demonstrating fiber as a TV distribution method (Fig. 3). The telephone companies, which had been planning right along to provide image-transmission capabilities as part of broadband ISDN (Integrated Services Digital Network) services, are now revising their schedules to quicken the pace of fiber deployment in light of pending congressional action.

"In our opinion, the information age is coming to America, but its execution is being thwarted by the twisted pair," says Tom Gillett, director of advanced operations testing, GTE Corp., Stamford, Conn. "Regardless of the potential of twisted pair, it is costly to repair and costly in terms of the need to eliminate noise from a copper network. We

Getting the fax straight

In an exhaustive study of image-processing technology, Probe Research of Ceder Knolls, N. J., traced the modern facsimile market all the way back to a fax device patented in 1843, but it wasn't until the mid-1980s that the devices became consumer items.

The International Telegraph and Telephone Consultative Committee (CCITT) Group 1 technology was standardized by that organization in 1968, and a Group 2 version was agreed on in 1976. Group 3 fax technology did not come along until 1980, but its improved coding techniques reduced transmission requirements while scanning resolution doubled over what was then available. In the real world this meant that a single page on a Group 3 facsimile could be transmitted in about a minute compared with the three to six minutes needed by the earlier Group 1 and Group 2 technologies—and the page would be more legible.

"Group 3 is the mainstream now. About 99.9 percent of the faxes that are sold are Group 3 [units] that operate at about 9.6 kbit/s. They use V.29 modem technology and will transmit a page in about 15 to 25 seconds," says Steven Joerg, vice president of sales for Rico Corp., West Caldwell, N. J. (see table).

But Group 3 will eventually be supplanted. "Group 4 is the ISDN [Integrated Services Digital Network] fax and there is not a hell of a lot of them out there right now. But there you are talking about three to five seconds per page," explains Joerg.

The Group 4 machine, which was standardized by the CCITT in 1984, is expected to come into its own when the 64-kbit/s packet network envisioned for ISDN becomes widespread. The Group 4 fax has twice the resolution of today's Group 3 and needs just seconds to transmit a document because it implements a coding scheme that assumes wider network bandwidths and few network-induced errors.

Instead of designing a coding scheme to forgive the errors induced by noise propagating in an analog environment, when the CCITT designed a fax standard for ISDN, it went to "a scheme with almost two times Group 3's compression efficiency," Probe researchers found. Some of the hard-won efficiencies were gained by switching from a one-dimensional to a two-dimensional compression scheme.

"Most of Group 3 is one-dimensional; it compresses only one line at a time. Group 4 fax is two-dimensional. It compares a scanned line to the line above and below. If there are less than 3 bits of change, the fax can map the two lines as if it were a single line. In fact, it can map up to eight lines in this way and use only a small amount of code to do it," says Richard D. Fisher, an independent consultant in Los Gatos, Calif.

Just the fax, ma'am

YEAR	YEARLY REVENUE (FAX BOARDS AND FAX)	STANDALONE FAX PLACEMENTS	COMPUTER-BASED PLACEMENTS
1983	—	50,000	0
1984	—	89,000	0
1985	$828,000	145,000	1,000
1986	$712,000	191,000	2,500
1987	$1,265,000	475,000	11,500
1988	$2,109,000	910,000	33,070
1989	$2,386,000	1,130,000	70,250
1990	$2,753,000	1,320,000	151,000
1991	$2,882,000	1,450,000	227,500

Source: CAP International

---

have 11 years of experience with fiber. In terms of the quality and cost reasons alone, fiber will pay for itself," he says.

GTE, which received a waiver from the FCC in April to operate a coax network in Cerritos, Calif., is also wiring part of that Los Angeles suburb with a companion fiber optic network to learn how to engineer and market broadband services.

Discovering what the consumer expects from a broadband network is a big part of what the telephone companies will be doing in the months ahead.

"We have these studies that tell us the average call lasts about three minutes, that 3 kHz is adequate for voice—though we actually talk at about 15 kHz—and that you encounter about 3 CCS [centi-call seconds] busy-hour traffic per line," says Stagg Newman, division manager of international network and broadband services at Bell Communications Research. "But we don't have this kind of information for broadband. We have to study what the service requirements are that humans will need," he adds.

The experts say the current network switching architecture can be upgraded to meet the initial demands for broadband services. Fiber adjuncts, which are already available from several of the big switch vendors, can be grafted to the current switch hierarchy to provide services over local fiber loops.

"The current switches will have to be upgraded to interface to fiber," says Bell Labs' Vlack. "Our prototype module—the digital adjunct switch—which we demonstrated at the 1987 International Communication Conference in Phoenix, Ariz., has evolved," he says. "The bit rate we showed at Phoenix was 139 Mbit/s. The Sonet standard calls for slightly over 150 Mbit/s, and that's where we are now."

Adding a fiber adjunct to the network is the first plateau the telephone companies will have to reach as they revamp the network hierarchy, planners say.

"We envision it happening in a phased evolution," says Newman. "The first phase would have a separate network for video. The second phase would be to put voice, data, and video on single fiber, but the central office would terminate voice in the current-generation switch while data went to a fast-packet switch. The video would go either to a fast-packet switch or a video switch. The final phase would be to integrate voice, data, and video switching into a fabric that would terminate in a high-speed packet switch or a photonic switch."

Photonic switching is very much a laboratory phenomenon, and it's likely to stay there for some time to come. But there is still plenty of room for image information in today's network. Take the case of the Brooklyn shopkeeper, for example.

Before the advent of the video teleconferencing hookup in Brooklyn, the shopkeeper could count on spending hours traveling downtown to police headquarters to identify the suspect and swear out a complaint. Or, as the Brooklyn District Attorney's office too often observed, the shopkeeper might decide to cut his losses and forget the whole business. Officials at the Brooklyn DA's office recognized they had a problem and found a relatively inexpensive conferencing system could provide a solution.

In a pilot study of videoconferencing conducted last year, the DA's office and the police downtown were linked by codecs, facsimile devices, and two 56-kbit/s lines to a local precinct house. Prosecutors liked what they saw.

Instead of going downtown, "the guy tells you his story at the local precinct house. We make up the complaint and send it back to him on a fax. The whole deal takes an hour—and the guy didn't have to lose a day [of business]," explains Zachary Tumin, special assistant to Brooklyn District Attorney Elizabeth Holtzman.

Image transmission worked well in Brooklyn because the application put few demands on the Picturetel Corp. (Peabody, Mass.) codec used in the experiment. Moreover, system designers could count on relatively little motion during the interview process, so just two digital switched circuits were needed, and they were available from the local telephone company.

Improvements in codec technology, declining equipment costs, and increasing availability of switched digital services is making videoconferencing more attractive for an increasing number of applications.

When first commercialized more than a decade ago, videoconferencing required a dedicated room, costly equipment, and several megabits per second of network bandwidth. The situation today is decidedly changed. The dedicated room with special acoustics and bulky cameras has been replaced by equipment carts that hold all the system components—codec, video camera and monitor, audio subsystem, and control system. It can be wheeled to where it's needed, and the performance is much better.

"The image quality at transmission speeds six times the multiple of 64 kbit/s [384 kbit/s] is appreciably better than even a 6-Mbit/s [system of] a few years ago," says market researcher Scott Douglas of Telemanagement Resources International (Lake Wylie, S. C.). According to industry insider John E. Tyson, president of Compression Labs Inc. of San Jose, Calif., 80 percent of the world's videoconferencing market is now doing its full-motion conferencing on circuits ranging from 384 to 768 kbit/s.

The cost of purchasing a videoconferencing machine has also changed dramatically. "We introduced our first product in 1986 at about $100,000. Since that time, we've seen a two-to-one performance improvement, and we've cut the cost to $60,000," says Picturetel Corp.'s Randy Smith, vice president of sales. Tyson, at Compression Labs, charts the same downward curves in cost: "Prices have dropped by 20 percent to 25 percent per year over the last two years."

More to come

Despite the drop in price, full-motion videoconferencing is still awaiting its golden age. Telemanagement Resources International in its study, *Image Communications,* pegs the total number of motion videoconferencing systems at about 1,000 worldwide.

Industry executives want to pin slow growth on the dearth of digital tail circuits in many parts of the country. But a bigger problem for the industry is the lack of standards that foster compatibility. Compatibility is a prerequisite for building user confidence in the technology and fashioning teleconferencing into a true worldwide network service (see "Eyeing a compression standard").

Piggybacking a videoconferencing system on existing network resources ought to be another way to win user confidence. But the Minx videoconferencing system, which was introduced by Datapoint Corp. about three years ago as a broadband adjunct to its Arcnet local area network product line, so far has had only a lukewarm reception.

Another conferencing outfit, Videotelecom Corp. of Austin, Tex., took a slightly different route to win user confidence by making the familiar PC the fulcrum of its videoconferencing system.

Ken Dickinson, director of video products at the San Antonio, Tex.-based Datapoint, says about 600 workstations have been sold to about 50 customers. "We've been doubling revenues every year, but the revenues are still modest," Dickinson concedes.

The Minx workstation consists of a camera and display in a single terminal that accepts both narrowband LAN and the wideband video connections. Cluster servers, which serve as controllers, switch video calls between workstations via the dedicated broadband network and also act as gateways to the narrowband LAN. To communicate outside the local network environment, Datapoint has built the interfaces needed to link to many of the codecs currently on the market.

Videotelecom Corp. put the PC in the middle of its strategy for videoconferencing. By mounting a video camera-speakerphone assembly on top of a PC monitor and dropping a video interface card into the backplane, the computer becomes a videoconference workstation in a broadband LAN environment. Switching requires an intelligent hub, so Videotelecom developed one built around the PC bus architecture.

To go outside the local broadband environment, the

1. I see. Frame-capture video telephones that transmit low-resolution black-and-white images over dial-up lines are being introduced by several Japanese manufacturers. The commercial version and consumer version (inset) from Mitsubishi Electric can be interworked with units expected to be offered by several other manufacturers.

company also came up with a codec on the PC bus. The codec provides all the compression needed for long-distance communications at speeds ranging from 56 kbit/s to 768 kbit/s.

But not every application calls for full-motion capability. Many organizations need conferencing systems but cannot afford to implement full-motion video (see table). And while telephone companies plan for broadband networks, it is sobering to realize that switched 56-kbit/s service is still not available in many parts of the United States.

Less than full motion

For example, in the rural town of Hancock, N. Y., 16-year-old Melissa is the only student in her high school who placed high enough in standardized tests to take advanced-placement English for college credit. But the classes were only given at a high school many miles away. Melissa is not old enough to drive, and her parents could not afford private transportation.

The Delaware-Chenango Valley school district hit on a way to link Melissa and her teachers using voice-grade lines. Students and teachers can talk together and share the images written on a real-time PC-based chalkboard.

The system is built around proprietary software and modem technology from Optel Communications Inc., New York, N. Y., and a personal computer fitted out as a graphics workstation with a graphics board and a digitizing tablet. The Optel hardware and software divides a dial-up link into two channels, leaving analog voice communications intact and PSK (phase-shift key) modulating the data.

As the teacher talks to the student, the teacher makes notes on a PC graphics tablet just as he would on a blackboard. The modem's notch filter takes a 800-Hz slice out of the voice channel and stands ready to put PSK data into the notch at 1,200 bit/s. As the teacher writes on the tablet, the local graphics processing board translates the handwriting into X and Y coordinates and puts the coordinates into local memory. Optel software takes that information and then compresses and transmits it as binary data to the remote computer.

At the remote end of the link, the data is decompressed by the program, and the XY coordinates are handed off to the local graphics board where they are translated into pixels. Melissa sees the notes in real time.

Besides the role they are playing in conferencing, personal computers have assumed other pivotal roles in moving images on the network. When they are linked with scanners and facsimile boards, PCs become image servers. When they are outfitted with frame-capture boards, they become mini movie studios.

Small video-production outfits are using personal computers, frame-capture boards, and graphics-editing packages to turn out video that mimics what big-time broadcasters do with multimillion-dollar machines.

"Our clients are doing frame-by-frame, high-resolution animation in 2-D and 3-D. They do corporate flying logos, medical illustrations, everything. And the quality of what they do is comparable to what you see on HBO," says Alan Waxenberg, a vice president at Personal Computer Services Inc., a New York City-based video system integrator.

"In the PC world there is a variety of third-party software that transmits captured video," says Waxenberg. For example, a package called Send>it! from Xenas Communications Corp., Cincinnati, Ohio, transmits color images over voice-grade lines at speeds up to 19.2 kbit/s using its proprietary software and modem technology.

Standalone faxes remain the shock troops of the image revolution. But the hottest growth area, albeit still a very

2. Pretty as a picture. Electronic still cameras available from several manufacturers record reproduction-quality images on 2-inch magnetic disks. The portrait, taken from a Sony system, was transmitted using a transceiver built around a 9.6-kbit/s modem that conforms to the V.29 specification.

small percentage of the total fax market, is putting a fax board into a PC and attaching a scanner to create an image server. Last year the sales of fax add-in boards more than doubled, and they are expected to more than double again next year.

Here is how they are being used: When a file mixing text and graphics is created on a PC, it has to be printed on paper before it can be scanned for transmission by a standalone fax. At the receiving end, the same document may then have to be rekeyed into another computer. The advantage of a fax board in such a situation is obvious, and by adding a scanner to the PC, it can be turned into an image communications server.

Companies such as Calera Recognition Systems Inc. of Santa Clara, Calif., and Caere Corp. of Los Gatos, Calif., have developed page-recognition packages that go well beyond the limited optical character recognition commonly associated with desktop scanners. The software can take the bit-mapped output from a scanner or fax board and convert the character portion of the data stream into text files that can be edited by many popular word processing packages.

But dropping a fax board into a card slot next to a modem board runs against the grain of many users. So late this summer, market researcher CAP International brought together modem, fax, and PC add-in board manufacturers to begin developing a binary file transfer standard that would handle both ASCII and image data.

Bell 212 modems, such as the popular Hayes modems, use block error correction on ASCII code with the burden for detection falling on the receiver. When a block of code that has errors is received, the sender must retransmit until a successful transmission is made. A fax scans a document only once, and error correction is a part of the compression scheme. Only when a block is received successfully is it decompressed and printed.

The committee, which is seeking recognition by the Electronic Industries Association, has decided to implement the T.30 and T.4 error-correction method so that a single modem board can be used to transmit both image and ASCII binary data.

The electronic still cameras being introduced by Japanese manufacturers make fast work of processing and transmitting snapshots. Instead of recording an image on film, the cameras use color charge-couple device technology to record reproduction-quality images on 2-inch magnetic disks. Canon, Sony, Sharp, and dozens of others have standardized around a recording format for a 2-inch magnetic disk. Up to 50 analog images can be recorded per disk.

Getting the picture
Using the Sony technology, for example, a reporter fighting a deadline can drop a disk into a recorder and spool the image to a local transceiver. Using dial-up lines, it takes about three minutes to transmit a color image to a remote transceiver. Once it is received, the image can be spooled to a thermal printer for reproduction-quality output.

The transceiver is built around a 9.6-kbit/s modem conforming to the V.29 specification and has fallback capability. It can also be connected to external modems for international calling. According to Jeff Kashinsky, director of system engineering at Sony Corp. (Park Ridge, N. J.), Sony will probably come out with a version with a high-speed (ISDN-compatible) modem in the future.

The resolution of the transceiver's frame grabber is 768 × 480 pixels. Each image is processed in two steps: Sub-Nyquist sampling is performed to reduce the pixel count by half. After the picture is sampled, it is coded using

Networking images

THE MOTION NETWORK

SUPPLIER	KILOBITS PER SECOND											
	56	64	112	128	168	192	224	256	280	320	336	384
COMPRESSION LABS INC.	X	X	X	X	X	X	X	X	X	X	X	X
CONCEPT COMMUNICATIONS[1]	X	X	X	X	X	X	X	X	X	X	X	X
EYETEL COMMUNICATIONS	X	X										
GPT VIDEO SYSTEMS	X	X	X	X	X	X	X	X	X	X	X	X
MITSUBISHI ELECTRONICS	X	X										
NEC AMERICA[2]	X	X	X	X	X	X	X	X	X	X	X	X
PHILIPS KOMMUNIKATIONS INDUSTRIE		X		X								
PICTURETEL CORP.[3]	X	X	X	X	X	X	X	X	X	X	X	X
TANDBERG TELECOM	X	X										
VIDEOCOM INTERNATIONAL CORP.	X	X										
VIDEOTELECOM CORP.[4]	X	X	X	X	X	X	X	X	X	X	X	X
VISTACOM INDUSTRIES	X	X										

1. IF MULTIPLES OF 56 KBIT/S ARE DESIRED, A STEPPED-DOWN T1 LINE, NOT MULTIPLE 56-KBIT/S LINES, MUST BE USED. DEVELOPMENT WORK IS PROGRESSING TO OVERCOME THIS SHORTFALL.
2. INFORMATION ON MOTION REFRESH RATE NOT MADE AVAILABLE BY NEC AMERICA; HOWEVER, QUALITY IS QUITE GOOD (SUBJECTIVE OBSERVATION).

THE CAPTURED IMAGE-COMPATIBLE PERSONAL COMPUTERS

SUPPLIER	IBM PC	IBM XT	IBM AT	IBM PS/2	NOT PC-BASED, BUT COMPATIBLE	NOT PC-COMPATIBLE
AETHRA	X	X	X			
AT&T	X	X				
COLOR VIDEO FAX					X	
COLORADO VIDEO				X		
DATABEAM			X			
DISCOVERY SYSTEMS						X
DISCRETE TIMESYSTEMS[1]	X	X	X			
EASTMAN KODAK CO.					X	
IMAGE DATA					X	
INTERAND						X
MITSUBISHI						X
OPTEL COMMUNICATIONS[2]	X	X	X	X		
STARSIGNAL	X	X	X			
TSI — HORSHAM					X	
VIDEOPHONE[3]						X
VUTEK SYSTEMS[4]	X	X	X			
WAWASEE ELECTRONICS						X

1. IMAGE TRANSMISSION INFORMATION NOT SUPPLIED BY DISCRETE TIMESYSTEMS.
2. OPTEL'S TRANSMISSION TIMES ARE BASED ON RESOLUTIONS OF 256 X 200 PIXELS (16,000 COLORS) AND 512 X 480 PIXELS (8-BIT GRAY SCALE). COMPARABLE TIMES WITH 512 X 480 (WITH 32,000 COLORS) ARE: 1.2 KBIT/S — 6.45 MIN.; 2.4 KBIT/S — 2.8 MIN.; 4.8 KBIT/S — 1.4 MIN.; 9.6 KBIT/S — 52 SEC.; 19.2 KBIT/S — 47 SEC.
3. VIDEOPHONE IS DEPENDENT ON PICTURE ACTIVITY AND GRAY SCALE RESOLUTION RATHER THAN DIFFERENT SPEEDS.
4. THE VUTEK SYSTEM ALLOWS THE IMAGE TO BE SQUEEZED DOWN TO A QUADRANT. WHILE IN THE QUADRANT MODE, THE IMAGE RESOLUTION CAN BE INCREASED TO 640 X 480 PIXELS AND TEXT OR GRAPHICS CAN BE VIEWED ON THE REST OF THE SCREEN.

RESOLUTION OF IMAGES (PIXELS X LINES)			REFRESH RATE (FRAMES PER SECOND)			
			KILOBIT RANGE			
NTSC	PAL	RGB	56-112 (NTSC)	>112 (NTSC)	64-128 (PAL)	>128 (PAL)
256 X 240	256 X 288	512 X 576 PAL; 512 X 480 NTSC	10	15	8.33	12.5
512 X 400		512 X 400	4-30	4-30		
256 X 240	256 X 240	256 X 240	12-30		12-30	
352 X 288	352 X 288	352 X 288	6-30	10-30	6-30	10-30
160 X 120; 320 X 240		480 X 480; 640 X 480	10			
360 X 288	360 X 288					
258 X 288	258 X 288	258 X 288			8.33	
256 X 240	256 X 240		10	10-15		
175 X 119	175 X 140	175 X 140	4-12.5		4-12.5	
160 X 120; 320 X 240		480 X 480; 640 X 480	10			
256 X 240			5-15	5-15		
256 X 240	256 X 240	256 X 240	12-30		12-30	

3. PICTURETEL'S CODEC OPERATES AT SPEEDS DOWN TO 1.2 KBIT/S. AT SPEEDS LESS THAN 56 KBIT/S, THE REFRESH RATE SLOWS DOWN UNTIL IT DELIVERS FROZEN IMAGES AT 1.2 KBIT/S. THE CODEC OPERATES ON EXTERNAL CLOCKING (FROM THE NETWORK), WHICH MEANS THE CODEC OPERATES AT ANY SPEED BETWEEN 56 KBIT/S AND 384 KBIT/S.

4. THE FULL RANGE OF THE VIDEOTELECOM CODEC IS 56 KBIT/S TO 768 KBIT/S.

TRANSMISSION IN SECONDS											RESOLUTION				
				KBIT/S							MONO-CHROME GRAY SCALE[5]	PIXELS (HORIZONTAL)	LINES (VERTICAL)	REFRESH MODE (INTERLACED)	REFRESH MODE (NON-INTERLACED)
1.2	2.4	4.8	9.6	14.4	19.2	56	112	448	560						
	360	180	90								C,H	256	240	X	
60	55	50	45	40	30	20					D,F	350-480	640		X
	112	56	28								H	512	480	X	
			170	110	80	28	14	3.5	2.8		F,H	512	485	X	
240	120	60	30	25	20	10	5	1.3			D,H	1,280	1,024		X
	128	64	32								H	512	480		X
											H	768	488		X
	200	100	50								H	512	480	X	
	40	25	10								G,H	592	440	X	
			16								A,D	320-1,280	240-480	X	X
			5.5								D	96	96	X	
80	40	20	13	12							H	512	480	X	
	10	6	5	3								512	480	X	X
						7	7	7			A,D	1,024	1,024	X	
											D-F	256	256	X	
210	110	50	30								H	512	256		X
	30	10	1								D-F	128	128	X	X

5. MONOCHROME GRAY SCALE IS BASED ON THE FOLLOWING:
A = 1-BIT GRAY SCALE (1/2 LEVEL)
B = 2-BIT GRAY SCALE (3/4 LEVEL)
C = 3-BIT GRAY SCALE (5-8 LEVEL)
D = 4-BIT GRAY SCALE (9-16 LEVEL)
E = 5-BIT GRAY SCALE (17-32 LEVEL)
F = 6-BIT GRAY SCALE (33-64 LEVEL)
G = 7-BIT GRAY SCALE (65-128 LEVEL)
H = 8-BIT GRAY SCALE (129-256 LEVEL)

NTSC = NATIONAL TELEVISION STANDARD CODE
PAL = PHASE ALTERNATE LINE
RGB = RED, GREEN, BLUE

Source: Telemanagement Resources International

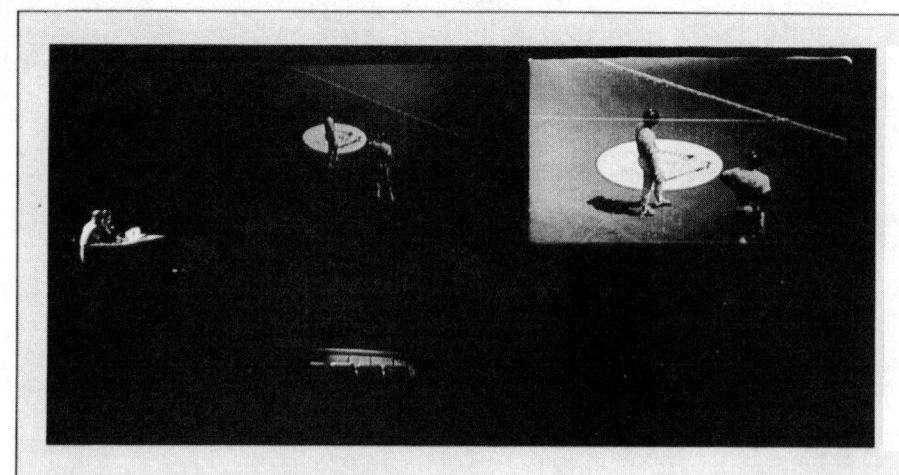

3. Stealing home. Telephone companies see TV as the wedge that will open residential markets to the broadband capability of fiber. Southwestern Bell showcased fiber by putting projections of a high-definition TV signal (left) and a broadcast-quality image (right) side-by-side during a St. Louis Cardinals vs. Philadelphia Phillies ball game.

differential pulse-code modulation. The coding compresses each 24 bits to 13 bits.

Filing photos on deadline has an exotic ring, but it is the more mundane applications—such as integrating the images with database information to speed production of time-critical information—that will ultimately prove the technology's value.

For example, the Board of Realty Information Systems (Boris), East Lansing, Mich., is marketing a turnkey multiple-listing production system to local realty boards. The Photo-List system, which integrates images from the electronic cameras with information contained on a remote database, is being used by about 10,000 agents around the country.

Once a house has been recorded on the 2-inch disk, an agent drops the disk into a video player and spools the image to a frame-grabbing terminal developed by Image Data Corp. of San Antonio, Tex. Boris is using a low-cost version of Image Data's Photophone. The terminal performs about a 10-to-1 compression on the video image while maintaining up to 256 shades of gray. After it's digitized, the terminal's proprietary 9.6-kbit/s modem needs anywhere from 16 to 45 seconds (depending on resolution) to transmit the information to the host. At the host, the digitized image is integrated with database information, then shipped to a printer for publication.

Integrating the digitized images of hundreds of homes with the appropriate data is no small problem, but building a system capable of capturing millions of images, storing them, and manipulating them as records in a database is a problem of a different order. Companies such as Filenet, Kodak, Wang, and, more recently, IBM have introduced systems based on write-once optical disk technology.

Large corporations, banks, and insurance companies are taking the first tentative steps in the direction of image capture and management. For example, early this summer, USAA, a large insurance cooperative formed by retired military personnel, cut over the first stage of a large image network that will eventually give 1,000 terminals generating 25,000 pages of information per day access to a massive image database.

USAA had some definite ideas about the kind of imaging system needed to handle the job. "We did not want a custom system," explains Charles A. Plesums, manager of image systems at USAA's Information Services Company at San Antonio, Tex. So Plesums began casting about for a vendor who would do the job using as little custom equipment as possible. "We tried to get IBM and other companies interested in the image system we needed," Plesums says. Eventually, IBM took the bait.

The USAA system Plesums specified became the prototype for IBM's mainframe-based ImagePlus system. The system is built around the System/370 architecture and a gang of token ring networks.

The integration of images with data in a common database is handled in the mainframe by partitioning the job into routing and database management functions, a server for the imaging subsystem, and a storage subsystem controller.

The routing and database management functions are handled by the folder application facility. It coordinates indexing and management of the magnetic and optical records. The server for the imaging subsystem, called the object distribution manager, arbitrates between the folder-application facility and the storage subsystem controller. It controls communications resources whether they are local or remote and is the interface to the Systems Network Architecture backbone. The storage subsystem controller, called the object access method by IBM, controls access to magnetic and optical storage devices.

When the full system comes on-line, Plesums says upwards of 30 token ring LANs will link the mainframe to workstations and more than 3.5 terabytes of optical disk storage.

American Express, which went with a custom-designed image management system, conquered its massive image management requirements using a variation of the standard Ethernet. The image system, which was built by TRW Financial Systems, Berkeley, Calif., handles about 1.4 million credit card slips per day. The system uses high-speed scanners to lift the image from the credit slips and then store the images on optical disk jukeboxes. The stored images become just another database element and can be printed out as part of credit card users' monthly statements.

As the daily torrent of credit card slips arrives at processing centers in Miami, Fla., and Phoenix, Ariz., they are processed and routed through the system on four Ethernet

Eyeing a compression standard

Recently the videoconferencing industry got a boost from the International Telegraph and Telephone Consultative Committee (CCITT), which is taking dramatic steps to attack the compatibility problem in videoconferencing gear. An experts group on visual telephony in Study Group 15 is developing a single standard for video compression that will be able to handle bit streams ranging from 64 and 56 kbit/s to 2 Mbit/s. The standard should be in solid draft form by the end of next year and could be adopted as early as the latter part of 1990.

The standard, P × 64, will spell out how to implement discrete cosine transforms (DCT) for video telephony. DCT, as it is being implemented for video compression, works on 16-by-16 or 8-by-8 pixel squares; it determines the motion and complexity in each block and codes the block accordingly. The algorithms will provide on-the-fly optimization for regions that require greater definition while providing less pixel data to regions such as background or a still image that requires less information to refresh.

A companion effort, one to develop a standard for freeze-frame compression, is under way in the International Organization for Standardization (ISO). The photo experts of ISO Working Group 8 are also using DCT-based algorithms, but it is unlikely there will be any mapping to the video telephony standard, says Barry Haskell, a member of the visual communications research department at AT&T Bell Laboratories.

"The freeze-frame people have a broader objective than the video telephony group," Haskell says. "The requirements they must satisfy range from how to compensate for poor resolution to the precise bit mapping of images. In video telephony, the only requirement is that you satisfy the viewer. You don't have to go beyond that."

LANs running under TCP/IP. According to Steven C. Grant, director of technology at American Express, nearly 80 percent of the available network bandwidth is being utilized.

TRW was reluctant to provide details about how Ethernet was optimized. However, Philip Schaadt, vice president of product development at TRW, said special link-level boards were built.

"What we did at Level 2 Ethernet is build a very fast networking board and increase the packet upward to 16 kbits per packet. Our Ethernet implementation is standard to the point it can be mounted in a PC and used with standard communications protocols."

While LANs are serviceable in some imaging network applications, there are limits to the amount of information a LAN can transport. Witness the situation in the modern hospital.

In addition to conventional film X-ray and its digital replacement, nuclear magnetic resonance, positron emission tomography (PET), computer-assisted tomography (CAT), and ultrasound—the modalities of imaging used in medicine—are important clinical tools for physicians and surgeons.

The ideal scenario gives physicians random access to the different imaging modalities stored on optical disks through a network. Earlier this year, at a meeting about image processing, Gregory J. Barone, president of Siemens Gammasonics, the medical imaging subsidiary of the German electronics giant, painted a rosy picture of what's ahead: "The hospital of the future will be filmless, and more important, physicians will have the ability to move the images about."

Barone envisions the physician, seated before a bank of displays, able to use wideband networking to compare the information available from different modalities. "The physician will have the ability to move ultrasound next to CAT scans and put them both next to PET," he predicts. "This will enhance the ability of the physician to diagnose."

But such a network does not yet exist. Says Paul Fenster, a visiting professor of electrical engineering at City College of New York: "Every conference talks about this, but it doesn't exist in the real world. It is the problem of bandwidth; that is the biggest problem."

The hard facts are that medical imaging technology is outrunning the network's ability to handle the data. For example, the current generation of CAT scanners has a 512-by-512 resolution. Each pixel in a 512-by-512 array requires 12 bits of data and at least 4 bits of overlay. Thus a single image is built with about one-half megabyte of data. The physician ordering CAT scans for diagnosis does not want to see just one or two views; typically, the doctor orders 20 or more images. And CATs do not have the largest data requirements; a single chest X-ray requires about 8 Mbytes of information.

"The question," says Edward Barnes, manager of multimodality systems at GE Medical Systems, Waukesha, Wis., "is how do you assemble the images in a reasonable amount of time?" Barnes says painting the one-half megabyte of data for a single CAT image on a standard Ethernet takes about four seconds. Since throughput on Ethernet degrades as loading increases, several physicians calling for images from a central location would soon bring the network to its knees. Current practice, say Barnes and Barone, calls for a radiologist or other medical imaging specialist to review all the images stored at a local site and to use the network to transmit only a subset to a consulting physician's workstation for review.

Thus, in medical applications, the network tends to slow the flow of diagnostic information needed by physicians. But in many other applications, such as video teleconferencing, image storage and retrieval, and image communications, there is no network bottleneck. The network is already handling image information—and if the FCC removes the cross-ownership barriers imposed on telephone companies, the quantity of image information moved across the network will become huge. One day, users may be able to end a phone call by saying, "See you later"—and literally mean it. ∎

Jack Covert, Tom Nakamura, and Nilo A. Niccolai, Hughes Aircraft Co., Long Beach, Calif.

High-flying Hughes earns its wide area networking wings

To get the rapid CAD file transfers it needed, Hughes Aircraft created an EDEN, a high-speed wide area network that supports multiple protocols

Electronic computer-aided design (ECAD) and mechanical computer-aided design (MCAD) are most productive and cost-effective when engineering groups can freely transfer design data and share computing resources. These groups are likely to communicate through a local area network (LAN) or computer cluster at sites where they occupy a single floor, building, or building complex. More widely dispersed groups may require wide area networks (WANs), which can extend hundreds of miles or worldwide.

In early 1986, a large engineering design network went on line at Hughes Aircraft Company. Today, the Engineering Design Network (EDEN) interconnects 12 major design centers and more than 2,000 individual nodes (Fig. 1). Most of the network community operates within the 900-square mile area around metropolitan Los Angeles. EDEN is believed to be the largest totally Ethernet-based WAN in the world, both in terms of the WAN's geographical area and the number of nodes it includes.

In the last few years, two major trends in ECAD and MCAD have made it far more difficult to design and operate networks like the one at Hughes. First, design files have been rapidly increasing in size; network file transfers that previously took only a few minutes can now last several hours at the same line speed. Second, the increase in the number of CAD workstations supplied by multiple vendors that use different protocols requires that engineering networks be more flexible in their handling of protocols. At Hughes, the need to offer very high-speed transfer while accommodating multiple communications protocols resulted in the EDEN network.

EDEN was primarily intended to support CAD-based VLSI (very large-scale integrated) chip design by six Hughes computer groups in metropolitan Los Angeles and Tucson, Ariz. The number of nodes on the network supporting VLSI chip design has since continued to grow and now includes myriad workstations and microcomputers. Using Ethernet to tie the network together allows a large number of nodes to be tapped into EDEN through relatively inexpensive transceivers. Each subgroup within the network is isolated by bridges that limit local traffic to the subgroup and make communications possible between groups. The flexibility provided by the network protocol has encouraged other Hughes engineers, such as those performing structural analysis, to use the network and further contribute to the demands placed on it.

EDEN network services

The EDEN network was to provide two principal services to the design groups: a shared central computing hub and high-speed transfer of design files over a highly reliable, highly available communications network. The central computing site was to provide the ability to offload large simulations onto high-performance computers, enable sharing of expensive special-purpose devices such as high-speed simulation engines and plotters, and provide technical training and software tool support. Both the central computing operation and communications were to be developed and operated by the Corporate VLSI Support Organization (CVSO) of the Hughes Communications and Data Processing Division.

By more closely integrating the efforts of the design groups, these services would enhance the interactive environment of computer-aided design; CAD works because engineers iteratively check their designs against technical specifications contained on a mainframe. Progressive increases in the sizes of design files and logic simulations had begun to cause the long turnarounds characteristic of batch computing. More and more of the simulations on small and midsize computers were taking

1. EDEN. *Hughes's Engineering Design network today serves 12 independent design centers, each using an Ethernet. Its LANs connect more than 2,000 nodes. Copper, fiber optic, or microwave links each design center to the central computing facility, which provides shared power for logic-fault simulation and other computer-intensive tasks.*

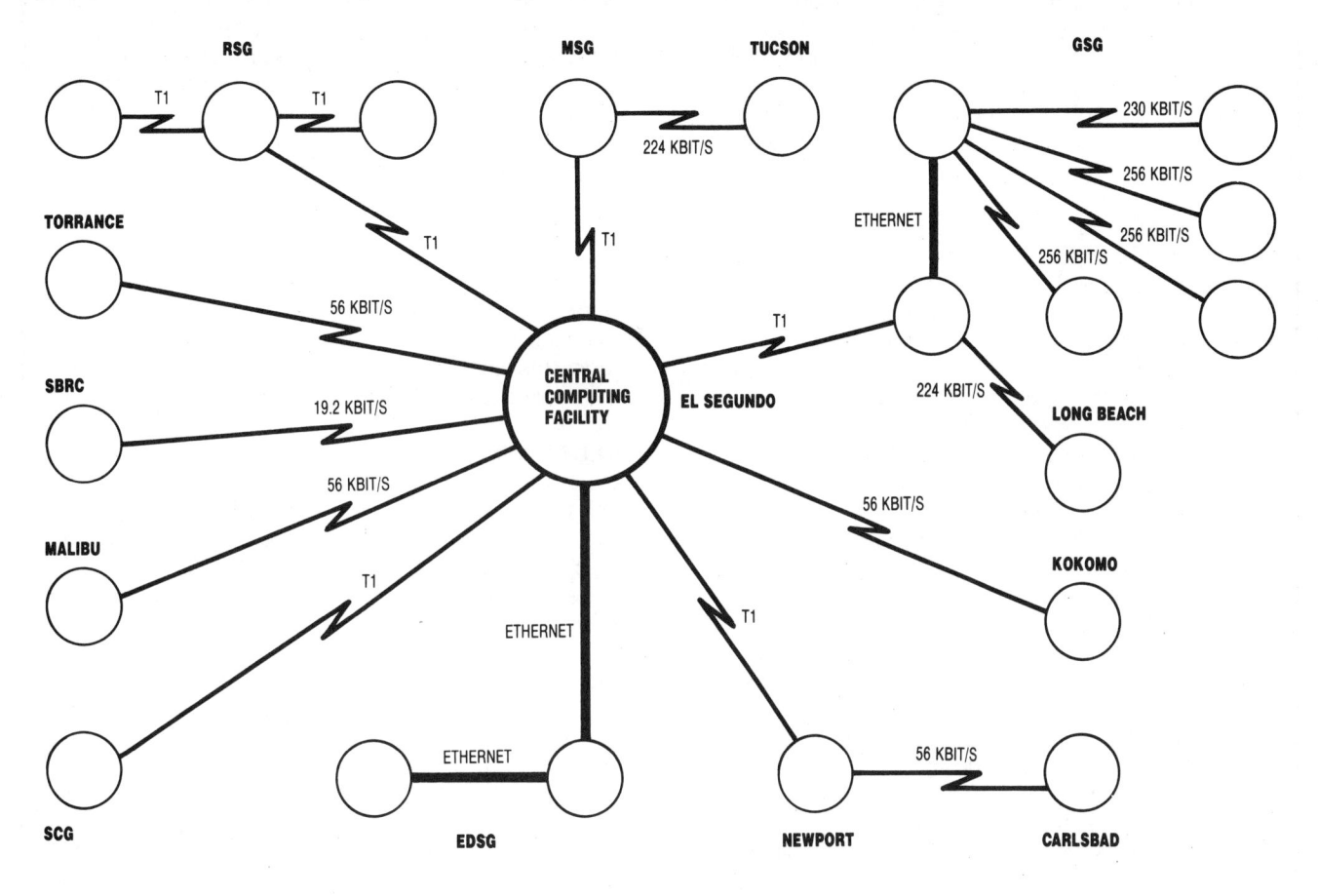

EDSG = ELECTRO-OPTICAL DATA SYSTEMS GROUP
GSG = GROUND SYSTEMS GROUP
MSG = MISSILE SYSTEMS GROUP
RSG = RADAR SYSTEMS GROUP
SBRC = SANTA BARBARA RESEARCH CENTER
SCG = SPACE AND COMMUNICATIONS GROUP

hours, as were file transfers over low-speed lines. As a matter of fact, transporting tape reels by automobile had become the preferred method of transferring files from site to site.

Even before the installation of the communications network, it had been demonstrated that computer-intensive simulations were substantially speeded up by linking special-purpose simulation engines to large central computers. For example, TEGAS5 logic simulations of LSI (large-scale integrated) circuits having 4,000 to 8,000 gates took four to 10 hours to run on a midsize computer but were executed in less than two seconds on a large machine supported by a simulation engine. Later, even with VLSI circuits with 40,000 to 50,000 gates, simulation engines still kept execution time down to a matter of minutes.

As the number of gates in a VLSI design increased, design file size increased by as much as five times. The speed of the few existing communications lines was typically between 1.2 kbit/s and 9.6 kbit/s, and the resulting transfer times often meant turnarounds of several hours. In addition, line errors are more likely in modem-to-modem communications links.

By the time planning began on the EDEN network (see "EDEN network architecture"), several design groups had already acquired CAD workstations. Others were preparing to do so to improve productivity.

Coping with workstations

While most of the early workstations were standalone machines, it was clear that their operation, too, would benefit from the EDEN network's distributed computing. Although a workstation's local computer power could handle much of the designer's ECAD needs, there were many tasks that took too long on small machines (even with newer high-performance workstations) and that were more efficiently performed on larger design-center computers.

Design groups had generally selected either Digital Equipment Corp. (DEC) VAX computers or IBM mainframes for these computer-intensive tasks. The VAX computers were typically owned and operated locally by

EDEN network architecture

The EDEN network's 12 principal communications lines connect the central computing facility to multiple design centers operated by Hughes's six design groups and R&D laboratories in Santa Barbara, Calif., and Malibu, Calif. The CVSO is located both in El Segundo, where it manages the central computing facility, and in Long Beach, where most of the support staff works. Excepting the special link to General Motors Hughes Electronic/Delco in Kokomo, Ind., the most remote site in the WAN is Tucson, Ariz., 400 miles from the network hub.

A star configuration has been used because most of the large design file transfers were between the remote design centers and the central computing facility. The star provided most efficient connections at the least cost for an initial high-speed implementation. Most transfers are round-trips to the central facility for simulation runs. Other transfers include downline loading of CAD libraries to design centers that prefer to store large files at the central facility; the engineers upload the libraries after they have finished working with them.

The speeds of the principal lines currently range from 19.2 kbit/s (for the Santa Barbara Research Center only) to 1.544 Mbit/s for T1 transmission to five design centers. The 19.2-kbit/s rate, the maximum that could be achieved with special modems over existing analog circuits, will be upgraded as soon as new lines can be installed. Singly or in combination, the communications medium may be copper, fiber optic, or microwave. The line speed is matched to the needs of the design center, so one or two lines are likely to be upgraded every year. If the available bandwidth is adequate, first choice is usually Hughes Communications Inc., Hughes's microwave-based corporate common carrier.

■ **Design centers.** The physical extent of the baseband Ethernet LANs at the design centers and their number of active nodes vary widely according to the design group's needs. While the CVSO is responsible for maintaining network integrity and performance outside the LANs, the design groups build and operate their own LANs independently.

A smaller design center such as Torrance now requires only 10 nodes. At the other extreme, GSG has gradually expanded its LAN to what are now 10 subnetworks at five sites stretching from Fullerton's five-mile-wide campus down to San Diego, 90 miles away. Except for a 224-kbit/s microwave link to Buena Park, all the subnetworks are connected through bridges and fiber optics links. GSG reports a current total of 450 nodes, including computers of many sizes, workstations, and terminal servers providing network access to more than 7,300 user terminals.

A more typical baseband Ethernet LAN supports the Missile Systems Group's design center in Canoga Park. There, Ethernet segments in multiple buildings are connected through bridges and a fiber optic line. A 224-kbit/s terrestrial link connects Canoga Park with a Missile Systems factory in Tucson. Most of the computer nodes are DECnet-oriented: VAX general-purpose computers, PDP-11 minicomputers, and VAXstation workstations. Eighty-five workstations, mainly Apollo, Sun Microsystems, and Hewlett-Packard models, are connected directly to the baseband coaxial cable.

Groups of up to eight asynchronous video terminals are connected to the Ethernet cable at various design centers through LAT (Local Area Transport) terminal servers. A terminal can access any computer in the design center's LAN and, with authorization, any other computer in the EDEN network that's running network-supported protocols. ASCII terminals access IBM computers asynchronously by means of reverse LAT servers. At these nodes, the eight terminal ports of the same LAT hardware (with only a slightly modified configuration) are instead connected to eight IBM input/output ports.

■ **EDEN central facility.** The central computing facility at the network hub in El Segundo consists principally of a VAX 8600 computer and an IBM 3090 mainframe connected to an Ethernet LAN segment. A Zycad LE1008 logic simulation engine and a Zycad FE 1004 fault simulation engine are connected to the VAX 8600. In addition to simulations, the two large machines perform other computer-intensive tasks such as design-rule checking and test-vector generation for electronic design. A VAX-11/780 processes administrative applications so that the VAX 8600 can be dedicated to the more demanding computational tasks.

The central IBM 3090 can be accessed over the network with either the TCP/IP or DECnet protocol. The Spartacus controller from Fibronics International (Hyannis, Mass.) and the Interlink controller from Interlink Computer Sciences (Fremont, Calif.) provide TCP/IP and DECnet services. Both controllers are channel-attached. Standard IBM communications is available through 3270 controllers, an NCR Comten communications processor, and connection to the Hughes NJES (Network Job Entry System)/SNA (Systems Network Architecture) network.

The 56-kbit/s DECnet link to Kokomo is connected to the network hub's LAN through a MicroVAX II. In addition to its routing function, the MicroVAX II acts as a security gateway between the EDEN network and GMHE/Delco. Security software checks source, destination, and user IDs before permitting transfers across the gateway. The security procedures provided by the operating systems at the source and destination nodes are also in effect.

Among other communications services provided by the EDEN network is electronic mail. The VMS operating system's VAXmail is supported throughout the network, including IBM nodes and the Kokomo link. Similarly, IBM's Profs (Professional Office System) electronic mail can be sent to Digital computer nodes and to Kokomo. Profs mail between IBM nodes uses Hughes's independent NJES network. Any IBM computer in the NJES network can be accessed by any VAX computer in the EDEN network.

VAX-oriented organizations within Hughes; other groups tended to share IBM computers. The groups' computer managers needed the freedom to add connections to their portions of the network and do so without the consent (and perhaps knowledge) of network managers. With bridges, the computers, workstations, terminals, and other devices within a design center would have to be able to communicate among themselves and do so without affecting network operation.

The basic EDEN network architecture manifested itself: Each design center's computing resources would be connected into a LAN, which would in turn be linked to other LANs through a WAN.

Handling multiple protocols
While many workstation vendors and third-party suppliers recognized the value of distributed computing for workstations, they offered no solid multivendor communications interfaces. Therefore, it was necessary to select for the network those communications protocols that could accommodate current and future workstations plus peripherals that were likely to be acquired for design centers. Nodes using these protocols had to be able to share the same communications media: the baseband coaxial cable in the LANs and the copper, fiber optic, and microwave links in the WAN.

Selecting Ethernet as the common physical transmission protocol was not difficult: it is essentially the same as the IEEE 802.3 standard, supported by most workstation vendors at that time and including some now no longer in business. The higher-level protocols needed to implement network communications had to be compatible with Ethernet.

One clear choice was DECnet. The design groups' VMS-based VAX general-purpose computers and VAXstation workstations would be linked into the network via communications protocols defined by Digital's DECnet networking architecture. in addition to Ethernet, DECnet supports two other physical transmission protocols that were to be useful in the network: CCITT recommendation X.25 and Digital's DDCMP (Digital Data Communications Message Protocol). Before the creation of EDEN, as a matter of fact, three design groups had DECnet/Ethernet LANs, and several others were operating low-speed DECnet/DDCMP lines.

Apollo Computer Inc. (Chelmsford, Mass.), Sun Microsystems Inc. (Mountain View, Calif.), Hewlett-Packard Co. (Palo Alto, Calif.), and other makes of workstations running versions of the Berkeley Unix 4.2 operating system communicate through the TCP/IP (Transmission Control Protocol/Internet Protocol) networking software. VAXstation workstations running on Digital's Ultrix-32 version of Unix also use the TCP/IP protocol. Finally, XNS (Xerox Network Systems) was the protocol specification used by several additional workstation vendors and, besides, was widely used in small office-automation LANs.

EDEN would therefore need to support the Ethernet physical transmission protocol and the DECnet, TCP/IP, and XNS higher-level protocols. Beyond that, the design groups were free to network microcomputers and other devices with nonstandard protocols as long as the protocols were implemented for use within their design centers alone.

The three network-supported, higher-level protocols had the added virtue of reasonably good correspondence with the Open Systems Interconnection (OSI) seven-layer reference model proposed by the International Standards Organization, as shown in Figure 2. This correspondence with OSI helps ensure a smoother transition to OSI in the future as well as reduced dependence on the whims of a single vendor. Although all OSI layers have not been fully defined, software implementing protocols above Ethernet at Layer 1 (physical) and Layer 2 (data link) was already available for DECnet, TCP/IP, and XNS.

The major suppliers of TCP/IP at Hughes are Excelan Inc. (San Jose, Calif.) and the Wollongong Group (Palo Alto, Calif.). Excelan provides OSI Layers 1 through 4 in its Ethernet controller and Layers 5 and 6 for the host computer. Wollongong, on the other hand, provides Layers 1 and

2. EDEN and OSI. *Most layered software implementing DECnet, TCP/IP, and XNS protocols in the EDEN network corresponds with the networking functions defined by OSI.*

ISO/OSI MODEL	DECNET			TCP/IP	XNS
APPLICATION	USER				(UNDEFINED)
	NETWORK MANAGEMENT				
PRESENTATION	NETWORK APPLICATION			TELNET FTP SMTP	
SESSION	SESSION CONTROL				COURIER
TRANSPORT	END-TO-END COMMUNICATIONS			TCP	SPP
NETWORK	ROUTING			IP	IDP
DATA LINK	ETHERNET	DDCMP	X.25	ETHERNET	ETHERNET
PHYSICAL					

DDCMP = DIGITAL DATA COMMUNICATIONS MESSAGE PROTOCOL
FTP = FILE TRANSFER PROTOCOL
IDP = INTERNETWORK DATAGRAM PROTOCOL
IP = INTERNETWORK PROTOCOL
ISO = INTERNATIONAL ORGANIZATION FOR STANDARDIZATION
OSI = OPEN SYSTEMS INTERCONNECTION
SMTP = SIMPLE MAIL TRANSFER PROTOCOL
SPP = SEQUENCE PACKET PROTOCOL
TCP = TRANSMISSION CONTROL PROTOCOL
TELNET = VIRTUAL TERMINAL PROTOCOL
XNS = XEROX NETWORK SYSTEMS

2 in the Ethernet controller and Layers 3 through 6 in the host-resident software. The Ethernet controller is supplied by Excelan, or, for the Wollongong software, by either Wollongong or DEC.

TCP/IP communications between nodes consists of IP in Layer 3 and TCP in Layer 4. Software modules for Layers 5 through 7 are application-dependent, such as Telnet for terminal traffic, File Transfer Protocol for file transfer, and Simple Message Transfer Protocol for electronic mail.

For XNS, Xerox has defined Internet Datagram Protocol, Sequenced Packet Protocol, and Courier. Third-party vendors have written software to the Xerox definitions. The functions in Layers 6 and 7 have not been defined but have been approximated by several vendors. At Hughes, the CVSO has had to write additional software to deal with certain incompatibilities between XNS nodes. CVSO had an outside vendor write additional software to solve an XNS network-numbering problem. Some XNS software implementations do not permit more than one network number, obviously a problem when confronted with multiple Ethernets.

In order to more effectively back up DECnet nodes, the X.25 protocol has been implemented at Layers 1 and 2 on several network nodes. This uses the separate Hughes X.25 terminal network. The Kokomo WAN link uses DECnet's DDCMP to provide Layer 1 and 2 functions.

Extending Ethernet range

When EDEN was planned, the highest copper speed at Hughes was 56 kbit/s, and fiber optic links were to be installed wherever 1.544W-Mbit/s T1 speed was needed. Although well below Ethernet's speed of 10 Mbit/s on baseband coaxial cable, T1 transmission was a substantial improvement over 56 kbit/s—21 seconds versus 357 seconds for a 1-Mbyte file transfer, for example.

In late 1985, the only source of Ethernet LAN bridges offering both T1 speed and the necessary networking software was Vitalink Communications Corp. (Fremont, Calif.), which already had a 224-kbit/s model and expected to announce a T1 model in mid-1986. In order to get the network on line as soon as possible, 56-kbit/s lines were installed immediately and later converted to T1 as bridges and T1 lines became available. As shown in Figure 1, T1 speeds are currently being provided on five of the 12 connections to the central computing site.

Vitalink bridges connect the high-speed communications lines at the design centers and central computing site (Figs. 3 and 4). There are many more bridges in the network; Hughes's Ground Systems Group, or GSG, alone has more than 30 Vitalink and Digital bridges linking its various subnetworks. Bridges operate only at the data link level (Layer 2 in Figure 2) and, in the EDEN network, only with the Ethernet IEEE 802.3 messaging protocol.

As a network filter, the bridge connected to a communications line passes along all Ethernet traffic with addresses outside the design center, whatever its higher-level protocol might be. Conversely, addresses within a center are blocked from the WAN and so are restricted to the design center's LAN. Filtering permits the design center to fully use network services and yet introduce in its LAN any computers, workstations, and other devices that it wishes to support locally without danger of affecting network operation.

Bridges also keep track of the names and locations of active nodes in the network, which is important because changes are constantly being made. They do so by dynamically building and updating node-name tables. Among the few rules that the CVSO has established for design centers are conventions for node names that can be understood for supported protocols throughout the network.

Whenever an existing design-center node transmits a message to a new node in its own LAN and identifies the destination node with a user-oriented name, the bridge adds that node name to its node-name table. When a design-center node addresses a new node not in its own LAN, the bridge passes the node name along for checking to other bridges until the location of the node is identified. The identifying bridge then tells all intermediate bridges what the new node's name and location are.

The Vitalink bridges provide CVSO network managers with performance data on traffic flowing through communications lines. The Vitalink and Digital bridges in the GSG LAN do the same for the GSG computer managers. The data includes the number of message packets transmitted over given periods or at fixed intervals; the addresses of devices in its node-name table; and the number and types of errors.

The CVSO EDEN network managers' principal tasks are to maintain high performance and availability in the backbone or shared portion of the network, defined as all bridges and devices on the network sides of the design center bridges (Fig. 3). Recognizing the independence of the design groups, the network managers monitor the design centers' activities but have no control over them.

Network performance is measured in a number of ways in order to anticipate, isolate, and some operating problems. It also guides the CVSO in optimizing the network's configuration as it expands and changes.

While high line use has been generally viewed as a measure of good network effectiveness, low transfer time is far more relevant in improving engineering productivity. Since transfer time increases with line use, maximum utilization may not be a valid goal in designing a network. This is particularly true today, when installation and operating costs for a 1.544-Mbit/s T1 line are only marginally higher than for a 56-kbit/s line.

The CVSO is developing a network performance benchmarking procedure that will give typical maximum and minimum end-to-end transfer times for specified types and sizes of files. It will then be possible to model performance of potential configurations in selecting optimum alternatives. The files represented in Figure 5 are among those that Hughes has found can present transmission problems.

The plotted performance measurements in Figure 5 compare data throughput (in kbit/s) and transfer time for a 1-Mbyte file (in seconds) for 224-kbit/s and 1.544-Mbit/s (T1) lines.

In Figure 5A, the throughput of the 224-kbit/s line drops

3. In the San Fernando Valley. *The Missile Systems Group's design center in two buildings in Canoga Park is supported by two Ethernet LAN segments connected by a fiber optic link. In addition to the T1 line to the central computing facility, a 224-kbit/s terrestrial link reaches a manufacturing site in Tucson, Ariz.*

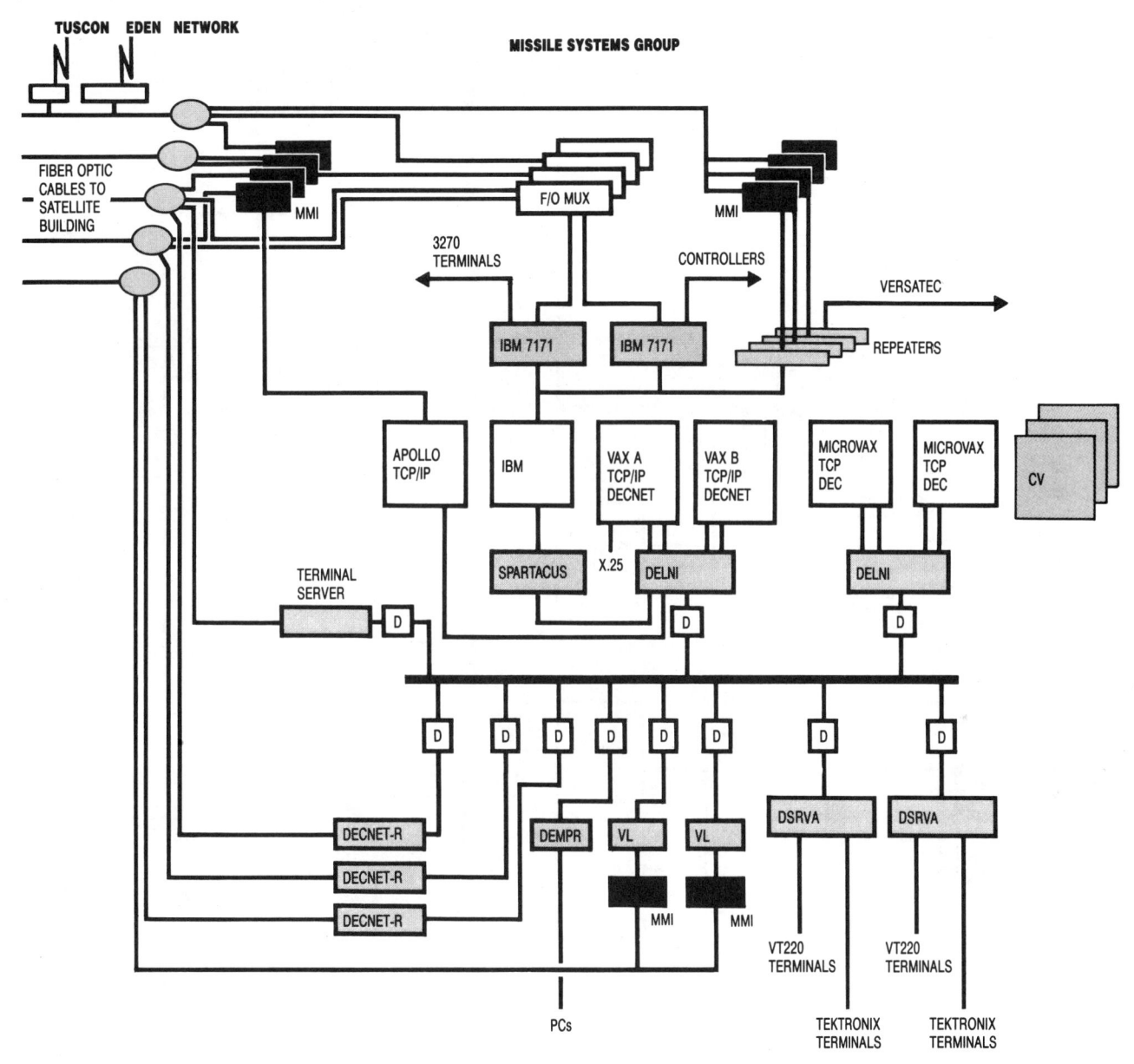

CV = COMPUTERVISION WORKSTATION
D = DESTA (DEC TRANSCEIVER)
DECNET-R = DECNET REPEATER
DELNI = DECNET/ETHERNET LOCAL NETWORK INTERCONNECT
DEMPR = THINWIRE ETHERNET MULTIPORT REPEATER
DSRVA = ETHERNET LOCAL AREA TERMINAL SERVER
F/O MUX = FIBER OPTIC MULTIPLEXER
MMI = FIBRONICS FIBER OPTIC MODEM
TCP/IP = TRANSMISSION CONTROL PROTOCOL/INTERNET PROTOCOL
VL = VITALINK

substantially for both two and three concurrent transfers, while throughput of the T1 line holds quite steady at about 350 kbit/s. This is particularly significant with respect to the output speed of the source node, which for a VAX on an Ethernet is typically between 350 and 450 kbit/s. Since the computer and line are both being used at nearly their full-speed capabilities, the network is operating efficiently for one, two, and three concurrent transfers over the T1 line.

In Figure 5B, the T1 line maintains a steady transfer time of less than 25 seconds for at least three concurrent transfers of a 1-Mbyte file, while the transfer time for the 224-kbit/s line rises sharply from 44 to 123 seconds in going from one to three transfers. If it is assumed a user will not tolerate waiting more than a minute for a file transfer, two concurrent transfers become unacceptable for a 224-kbit/s line. In most cases, transfer time for a T1 line will

4. Central computing. EDEN's central computing facility in El Segundo consists mainly of a VAX 8600 computer and IBM 3090 mainframe connected to an Ethernet segment. The machines are accessed by the Hughes design centers for simulation and other computer-intensive tasks. The Zycad simulation engines are connected to the VAX 8600.

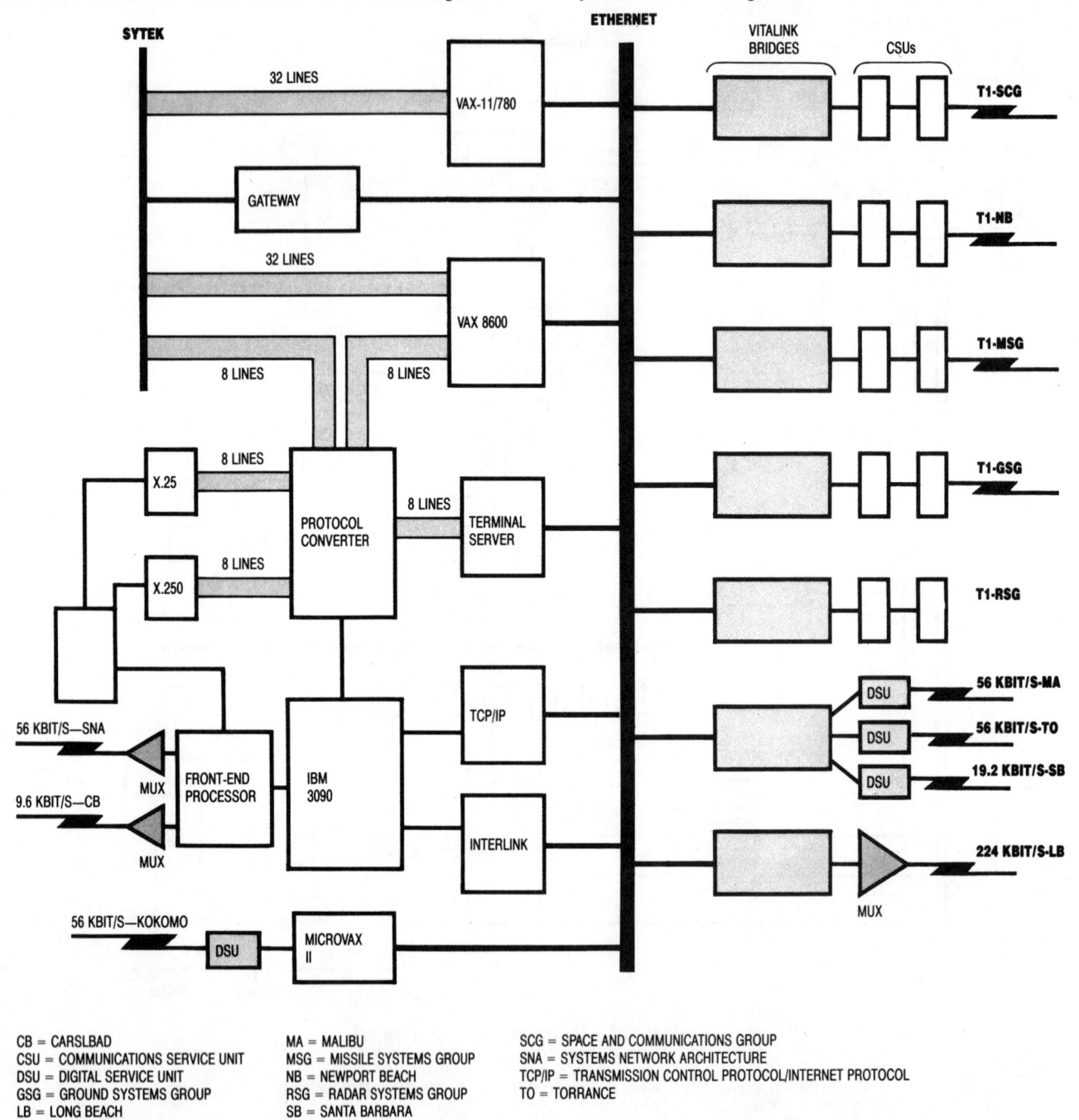

CB = CARSLBAD
CSU = COMMUNICATIONS SERVICE UNIT
DSU = DIGITAL SERVICE UNIT
GSG = GROUND SYSTEMS GROUP
LB = LONG BEACH
MA = MALIBU
MSG = MISSILE SYSTEMS GROUP
NB = NEWPORT BEACH
RSG = RADAR SYSTEMS GROUP
SB = SANTA BARBARA
SCG = SPACE AND COMMUNICATIONS GROUP
SNA = SYSTEMS NETWORK ARCHITECTURE
TCP/IP = TRANSMISSION CONTROL PROTOCOL/INTERNET PROTOCOL
TO = TORRANCE

be acceptable until there are five concurrent transfers. Even with the large number of nodes on EDEN, statistical analysis has predicted that five concurrent large transfers will occur across the backbone less than 1 percent of the time.

The most widely used network performance measuring instrument at Hughes is the Lanalyzer, an Excelan LAN analyzer integrated with a Compaq PC. It measures Ethernet and TCP/IP traffic at a selected point in the network in several ways: total number of message packets; number of short packets (below minimum length); number of Cyclic Redundancy Check and abort errors; location of cable breaks using a built-in time domain reflectometer; and line use in bytes or frames per second. The Lanalyzer's snapshot of network use helps in evaluating the need for upgrading line speed. The main user of the Lanalyzer is Communications and Data Processing's Telecommunication Group, which is responsible for maintaining the network's physical transmission capability.

There are several network maintenance application programs that are useful both for monitoring availability of network resources and for fine-tuning performance. VAX ETHERnim (Ethernet network integrity monitor), which is a layered software module on the VMS operating system, tests paths and generates reports on the relative availability of nodes, expressed as "can you see the nodes."

5. Metrics. *(A) shows throughput as a function of number of concurrent transfers on 224-kbit/s and 1.544-Mbit/s lines. (B) shows transfer time for a 1-Mbyte file.*

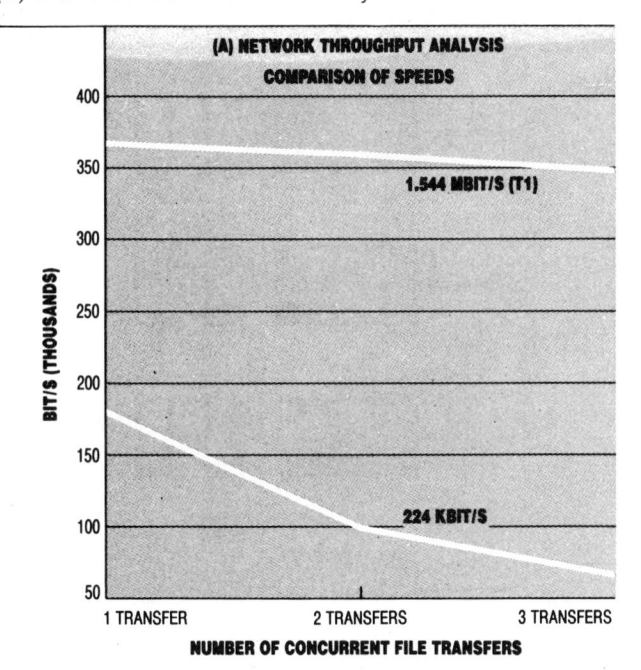

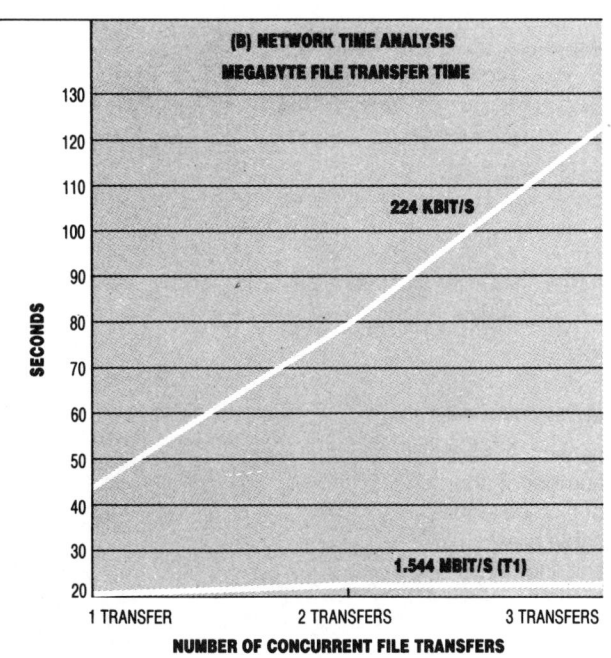

The DECnet Monitor maintains statistics on the operation of DECnet nodes, including how often nodes go down and for how long. This software helps in predicting saturation of lines and evaluating rerouting alternatives for speeding up transfers. The LAN Traffic Monitor listens to message packets on the cable and captures packets for examining traffic patterns. Given the volume of data, good display graphics are essential elements for any network analysis tool, although none currently has gone beyond the test phase for Ethernet WANs. Older protocols such as Systems Network Architecture, X.25, and BSC (bisynchronous communications) have good management tools. But since graphics have only come into common use in the last few years, and Ethernet WANs have only been in existence since 1986, Hughes is still working on finding the right network management display.

In the future

In addition to performing large simulations (access to shared computer power will probably always be valuable to design groups), there will be expanded network services in such areas as engineering and business databases, videotex, and document transfer. Design files that are now typically 1 Mbyte to 2 Mbytes will increase to 10 to 20 Mbytes in size, but higher speeds should be available to maintain acceptable transfer times.

As its first step in this direction, Hughes is planning to install fiber optic links between the central hub and the Radar Systems Group and Space and Communications Group design centers. These links permit the circuits to operate at the full Ethernet speed of 10 Mbit/s, resulting in transfer times as short as one-sixth that of T1 lines.

With higher line speeds, there will also be major changes in network configuration since it will then be possible to distribute shared services and still maintain good network response. What is now physically a star network will become a peer network both physically and logically. In this way, there will be more flexibility for optimizing both individual node-to-node transactions and overall network efficiency and performance. ■

Jack Covert is head of network support services in CVSO. He was previously computer security manager at Northrop Aircraft Company and, before that, an operating system programmer at Rockwell International. He has a B.S. degree in accounting and computer science from Brigham Young University (Salt Lake City, Utah).

Tom Nakamura is manager of application services in CVSO. Prior to joining Hughes, he was a thermal analysis engineer with the space shuttle project at Rockwell International's Space Division. He has a B.A. in physics and mathematics from Occidental College (Los Angeles, Calif.).

Nilo A. Niccolai is manager of CVSO. Prior to joining Hughes, he was a professor of computer science at the University of North Carolina, Charlotte. He consulted with IBM and has served with the U.S. Army in the Chief of Staff's office in the Pentagon. He holds a B.S. in physics and an M.S. in mathematics from the Carnegie Institute of Technology (Pittsburgh, Pa.) and a Ph.D. in mathematics from Carnegie-Mellon University.

Jill Ann Huntington, Datapro Research Corp., Delran, N. J.

OSI-based net management: Is it too early, or too late?

AT&T's latest salvo manages data plus voice; interworks with SNA applications, network devices, and public switched networks.

An old battle is heating up again in the data communications arena. The players: IBM and AT&T. The playing field: the network management market. Not exactly an even match, considering IBM's commanding lead. But this time it's not just one network management scheme versus another: It's OSI-based network management, which most agree holds out to users the last best hope of achieving true end-to-end control over multivendor networks.

AT&T last month announced several new products, including the key component of its OSI-based network management strategy—Accumaster Integrator. The Accumaster Integrator uses SNA Management Application, developed by Cincom Systems Inc. (Cincinnati, Ohio), to obtain logical network management data on SNA networks from either IBM's NetView or Cincom's Net/Master. The Accumaster Integrator is in beta test and will be generally available in the fourth quarter of this year.

AT&T concurrently announced Accumaster Management Services, a network management service (typically off-site) administered by AT&T staff. Each AT&T staff team will be dedicated solely to one customer. Also announced is Release 2 of the Accumaster Consolidated Workstation, featuring an X Window-based display.

The distinguishing feature of the Accumaster network management product family is its OSI-based implementation, built on AT&T's Unified Network Management Architecture (UNMA). Up to this point, UNMA's OSI-based approach has appeared to be more of an idea than a solution. Indeed, some users have viewed OSI as an esoteric subject—until last September, when IBM announced that its NetView would support OSI by mid-1990.

OSI is definitely moving into the mainstream. In the short term, however, due to IBM's massive installed base of SNA networks, NetView's market presence will continue to dwarf OSI-based management systems. SNA will never go away, and AT&T does not expect that it will.

AT&T is not aiming to dislodge SNA, nor does it want to replace NetView with its own SNA network management system. Rather, AT&T's goal is to set the standard in the industry for OSI-based network management—by providing a graphics-based integrating package that will send and receive commands and information to and from hosts, network management subsystems, and various devices in a multivendor network environment.

To make the Accumaster Integrator viable in the SNA market, AT&T is providing a mechanism for capturing SNA management data from NetView (or Net/Master) and correlating that to alerts from various physical devices monitored on the network. The Integrator's Alarm Correlation feature isolates the most likely source of multiple alarm faults—including those cases where the fault may be a logical error (such as a modem malfunction). The Integrator draws on a network device profile stored in a comprehensive configuration database to provide the Alarm Correlation feature. AT&T is clearly targeting multivendor network management, and the Accumaster Integrator presents the first tangible step toward supplying a product that is capable of addressing the growing OSI market.

Both NetView and the Accumaster Integrator provide integrated network management, but their approaches, designs, strengths, and limitations are different (Table 1). The Accumaster Integrator provides several advantages over NetView: immediate OSI support, superior physical network management capability, and comprehensive management for voice networks.

On the other hand, NetView provides superior logical network management—particularly of SNA networks. Most of all, NetView has a three-year lead over the Accumaster Integrator and has evolved to become a de

437

Table 1: Comparing network management

FEATURE	ACCUMASTER INTEGRATOR	NETVIEW
HARDWARE	AT&T 3B2 (SUPERMICRO-BASED)	SYSTEM/370 (HOST-BASED)
OPERATING SYSTEM	UNIX	MVS
UNDERLYING PROTOCOLS	NMP (AN IMPLEMENTATION OF OSI MANAGEMENT PROTOCOLS)	LU6.2 (PROPRIETARY)
VENDOR SUPPORT	17 VENDORS HAVE PLEDGED TO DEVELOP NMP INTERFACES TO THEIR PRODUCTS. MORE THAN 40 VENDORS HAVE JOINED THE OSI/NM FORUM IN LESS THAN SIX MONTHS.	MORE THAN 40 HAVE PLEDGED TO DEVELOP NETVIEW/PC INTERFACES OVER THE PAST TWO YEARS. NUMBERS HAVE NOT INCREASED SIGNIFICANTLY IN RECENT MONTHS. FEW VENDORS HAVE ACTUALLY RELEASED NETVIEW/PC PRODUCTS.
USER CUSTOMIZATION	USING C LANGUAGE.	USING CLISTS AND REXX (A STRING-PROCESSING LANGUAGE).
CONSOLE DISPLAY	GRAPHICS-BASED WITH WINDOWING CAPABILITY, RUNNING ON A SUN WORKSTATION.	TEXT-BASED, RUNNING ON AN IBM PC OR PS/2.
TERMINAL EMULATION ACCESS TO OTHER PHYSICAL DEVICE MANAGEMENT SYSTEMS	YES	NO
SAA COMPLIANT	NO	YES
OSI COMPLIANT	YES	WILL SUPPORT OSI BY MID-1990.
PRIMARY SOURCES OF REVENUE FOR VENDOR	AT&T EXPECTS SUBSTANTIAL REVENUE FROM ITS ACCUMASTER MANAGEMENT SERVICES AND FROM PROVIDING DATA TO LECs ON THE CUSTOMER-ALLOCATED PORTION OF THE PUBLIC NETWORK.	IBM DERIVES SUBSTANTIAL REVENUE FROM BOTH SOFTWARE LICENSES AND SELLING ADDITIONAL MAINFRAMES IN THE LONG RUN.
PRICE	$300,000*	MONTHLY LICENSE FEE $705-$1,255. (NETVIEW/PC HAS ONE-TIME COST OF ABOUT $2,000)

*TYPICAL CONFIGURATION INCLUDES INTEGRATOR SOFTWARE, ONE 3B2-600, TWO INTEGRATOR WORKSTATIONS AND CABLE, PLUS ACCESS TO THREE ELEMENT MANAGEMENT SYSTEMS ONE OF WHICH IS THE SNA MANAGEMENT APPLICATION. THE UNMA APPLICATION, A NET/MASTER OPTION, IS MARKETED BY CINCOM AND PRICED SEPARATELY.

facto standard. Users know that NetView will be around for a long time.

One of NetView's weak spots is its method for linking with non-IBM devices and network management systems—namely, NetView/PC. When the product was first introduced in 1986, IBM boasted that NetView/PC would become a de facto standard for consolidating network management information from devices on a multivendor network.

"The approach we used for NetView/PC could be accepted by the standards people," said Jack Drescher, then IBM's product manager for NetView/PC in Research Triangle Park, N. C. (see "An analysis of IBM's NetView/PC: Potential, potential, potential," DATA COMMUNICATIONS, November 1986, p. 81). Now, more than two years later, NetView/PC's drawbacks have dispelled that hopeful prediction and hindered IBM's efforts to win support from an overwhelming majority of vendors.

Chief among these drawbacks is that NetView/PC is awkward and expensive to implement. Another major limitation has been the inability of NetView to respond to NetView/PC alerts. Version 1.2 of NetView/PC, with a scheduled availability date of May 1989, is designed to provide this support.

IBM also hoped that NetView/PC would extend NetView's reach into managing voice networks. Concurrent with NetView/PC's introduction, IBM announced two Rolm products that used NetView/PC to pass call detail information and alerts from PBXs to NetView. However, with IBM's December 1988 sale of Rolm comes doubt about Big Blue's ability to comprehend the voice market and deal with voice management effectively.

These two gaps in the NetView scheme—capturing data from other vendors' devices and managing voice communications—are weaknesses that AT&T hopes to exploit with the Accumaster Integrator. Both multivendor network management and voice management were primary considerations when AT&T designed UNMA and the Accumaster Integrator.

The new Accumaster Integrator, described in the context of AT&T's Unified Network Management Architecture, lends itself to several comparisons with NetView. Although the comparisons that follow here are not intended to be exhaustive, they should furnish a starting point for evaluating the Accumaster product announcements (see "Glossary" for an explanation of terms).

Three-tiered architecture

AT&T's Unified Network Management Architecture aims to provide end-to-end management of voice and data networks in multivendor environments. It is a three-tiered architecture that follows the ISO/CCITT Open Systems Interconnection (OSI) standards and management framework—to the extent they are defined (Fig. 1).

This standards-based approach contrasts sharply with the direction IBM took several years ago in developing NetView. IBM instead chose to create a de facto standard, designing NetView (and its predecessors, Network Communication Control Facility, or NCCF, and Network Problem Determination Application, or NPDA) to manage networks conforming to SNA's proprietary architecture developed primarily to connect IBM equipment.

NetView and UNMA each provide some network management integration by allowing the user to monitor both logical and physical aspects of the network from one central interface. (Network management products that provide a "logical network view" measure the network as represented by actual traffic passing over it. Products controlling the physical network monitor and control the actual circuits and network nodes.)

UNMA starts with the Accumaster product line on the physical side and (in its initial implementation) enables the user to view actual displays from components on the logical side, such as IBM's NetView or Cincom's Net/Master. This feature is sometimes referred to as a cut-through capability.

To provide the cut-through capability, AT&T has included a 3279 emulation package as part of its Accumaster Integrator. Thus, the Accumaster Integrator console will have the capability of displaying information just as it appears on the NetView console. The Accumaster console (a Sun workstation) will be a terminal defined in SNA and, as such, will have the capability of accessing SNA applications, such as Cincom's Net/Master or IBM's NetView. At present, AT&T has no plans to develop its own logical management package for SNA networks.

In contrast, IBM begins with NetView on the logical side and integrates physical network management through its Link Problem Determination Aid (LPDA) and NetView/PC. Similarly, Enterprise Management Architecture (EMA), from Digital Equipment Corp. (DEC, Concord, Mass.), starts with the Executive control program on the logical side and provides Access Modules that communicate with remote hardware using OSI-based protocols.

Both data networks and voice networks can be managed via UNMA. More important, however, UNMA supports the customer-allocated portion of the public network. It lets users tie together three network management domains: the customer premises, the local exchange network, and the interexchange network.

AT&T can provide UNMA customers with the means to integrate information from public networks with data from their private network management systems. IBM's NetView and DEC's EMA stop short of the public network and offer instead much more sophisticated logical network management capabilities, particularly at the applications level.

Since OSI management standards have not yet solidified, how soon can an integrated solution for managing multivendor networks appear? IBM has decided to wait until mid-1990, when most OSI management standards will have reached International Standard status (Table 2).

AT&T is not waiting—the Accumaster Integrator uses Network Management Protocol (NMP), AT&T's implementation of OSI standards as they exist today. AT&T pledges to modify NMP and its products to conform to any changes that the OSI committees may dictate. In the interim, AT&T is proactively seeking support for NMP and is heavily involved in the OSI Network Management/Forum.

A Unified User Interface is the focus of UNMA's three-

Glossary

The following terms are useful in understanding the new AT&T network management product. They are not presented in alphabetical order because, in part, each term builds upon an understanding of its predecessor in the list.

OSI (Open Systems Interconnection): an architectural model designed to enable computerized systems in multivendor environments to share information. The OSI model defines an open system as one that obeys OSI standards in its communications with other systems. This contrasts with proprietary architectures, such as IBM's Systems Network Architecture (SNA), which are designed primarily to support one vendor's equipment.

DIS (Draft International Standard): the last stage in the OSI standardization process before final approval. Draft International Standards are usually stable enough to allow vendors to commence implementation, although they are still subject to minor modifications before becoming International Standards (ISs).

DP (Draft Proposal): the stage in the OSI standardization process before DIS. Draft Proposals define the scope of the standard and, while subject to modification, are usually stable enough to guide vendors in developing product architectures.

UNMA (Unified Network Management Architecture): a three-tiered architecture that serves as AT&T's blueprint for future network management products and services. UNMA is an open architecture that employs interfaces based on OSI standards.

Network elements: network equipment and services that comprise an organization's network. These include physical devices on the network, such as modems, multiplexers, PBXs, LANs, and hosts. Also included are local exchange carrier (LEC) networks, interexchange services, Postal, Telephone, and Telegraph agencies (PTTs), and international network services. Network elements form the first tier of AT&T's UNMA.

EMS (Element Management System): a system that manage network elements. The term encompasses the plethora of devices on the market that provide network management for certain pieces of the network. Large networks typically employ many separate EMSs to control different vendors' products and services. Examples of EMSs include systems that manage modems, multiplexers, and DSUs (such as the Codex 9800 INMS, Racal-Milgo's CMS 2000, and AT&T's Dataphone II Level IV).

AT&T also uses the term *element management system* to describe certain operations embedded in the phone company's networks that analyze traffic over telco switches, support trouble ticket facilities, and provide related services. EMSs form the second tier of AT&T's UNMA.

NMP (Network Management Protocol): the protocol used in UNMA to communicate information from various vendors' element management systems to the Accumaster Integrating System. NMP is fully compliant with the OSI seven-layer model and the currect definitions of OSI management framework and management information services. NMP performs the same function as do LU 6.2 and SSCP-LU for IBM's NetView/PC.

Accumaster Integrator: AT&T's Unix-based system capable of automatically uploading information from other EMSs via the NMP interface. AT&T announced this product on January 31. According to AT&T, the product will eventually provide overall, end-to-end network management of all network elements in an organization's network.

ACW (Accumaster Consolidated Workstation): an MS-DOS based system that displays a consolidated network view by providing separate windowing sessions to various AT&T EMSs. Release 2 of ACW, announced in January, will allow users to communicate with IBM host applications (using NetView or Cincom's Net/Master) while maintaining active sessions with other EMSs.

NCCF (Network Communication Control Facility): IBM's host-based network management software that provides the operator interface and network logging facilities. NCCF operates as an application program under ACF/VTAM. IBM enhanced NCCF (renamed "The Command Facility") and incorporated it into NetView in 1986.

NPDA (Network Problem Determination Application): IBM's network management product that uses a series of panels to provide operations with information needed to perform problem determination and resolution. An enhanced version NPDA (renamed "The Hardware Monitor") was incorporated into NetView in 1986.

EMA (Enterprise Management Architecture): Digital Equipment Corp.'s OSI-based network management strategy. Like UNMA, EMA provides for multivendor network management using an interface based on OSI standards. Seven vendors have pledged to support EMA. Digital has yet to publish its interface implementation.

---

tiered architecture. Machine-to-machine interaction occurs between the three tiers; human-to-machine interaction occurs at the Unified User Interface.

Tier 1 is made up of network elements, which may include customer premises equipment (CPE), such as modems, multiplexers, LANs, hosts, and PBXs. Local exchange carrier (LEC) networks, interexchange services, Postal, Telephone, and Telegraph agencies (PTTs), and other international network services are also considered network elements.

Tier 2 consists of Element Management Systems (EMSs), which manage network elements. An EMS provides what may be called "local" management capabilities—operations, administration, maintenance, and provisioning functions of managing a particular network element group. Today, most large networks include multiple EMSs, since vendors have traditionally supplied multiple network management products for different devices and services. It is not unusual to find different EMSs for computer hosts, matrix switches, T1 resource managers, and Ethernet

1. Three tiers. *UNMA's Tier 1 consists of Network Elements; Tier 2 has Element Management Systems; and Tier 3 is the Accumaster Integrator.*

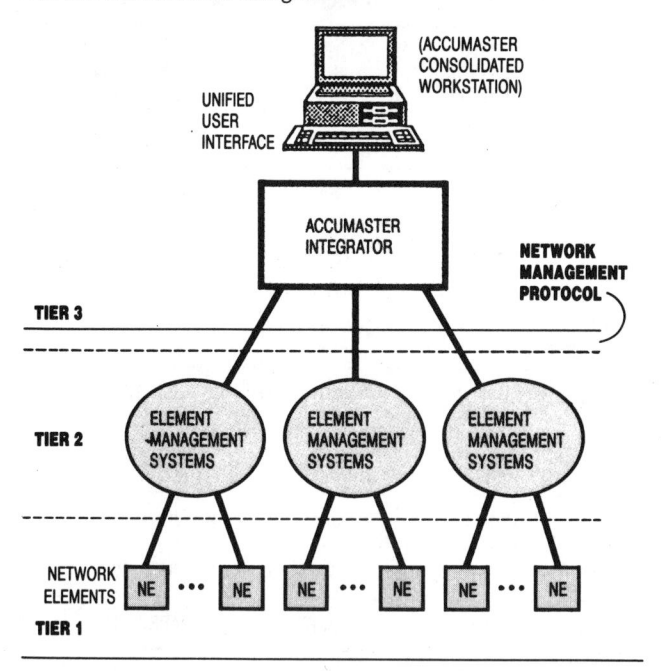

LANs—all in the same corporate network. Each EMS is, all too often, an island unto itself. The inability of disparate EMSs to share information prevents the user from obtaining an end-to-end view of the network.

Customers have access to about 20 AT&T EMSs. An additional 100 network-based EMSs deployed by AT&T are not currently customer-accessible. Over the next decade, AT&T plans to make more of these systems accessible, providing customers with additional trouble ticketing information as well as alarm and performance information on AT&T facilities.

Tier 3 aims to provide the end-to-end view by supporting communications between EMSs and the Integrating Network Management System. This integrating mechanism, the Accumaster Integrator, is the heart of the UNMA strategy. The Accumaster integrating system will communicate with individual EMSs through a common protocol stack. This stack (Network Management Protocol), is AT&T's implementation of OSI management specifications as they exist today. Under UNMA, management integrating packages (such as the Accumaster Integrator) also communicate among themselves via the standard protocol stack (Tier 2-to-Tier 3). UNMA allows communications based on native (proprietary) protocols between the Network Elements and their respective EMSs (Tier 1-to-Tier 2). Communications based on standard protocols between Tiers 1 and 2, however, may evolve in the future.

The Accumaster Integrator provides cut-through capabilities to the EMSs, allowing users to exercise full capabilities of each EMS. This feature gives AT&T's product an edge over IBM's NetView.

As part of the Accumaster product announcements, AT&T unveiled a special application it developed under joint agreement with Cincom Systems. This application, called the SNA Management Application, provides logical information about SNA networks to the Accumaster Integrator.

Also at Tier 3, the Unified User Interface creates the image of a virtual network, so to speak. The Accumaster Consolidated Workstation's enhancements found in Release 2 support this concept by bringing screens from 10 element management systems into one central screen (up to six at any one time, using windowing capabilities). These element management systems include NetView, Net/Master, several AT&T voice management systems, and any element management systems that support VT100 terminal emulation.

Implementing OSI net management

As of January 1989, AT&T had published four Network Management Protocol documents (TR 54004 through 007) to guide other vendors in implementing the interface between their proprietary Element Management System products and the Accumaster Integrator. AT&T's work on NMP, as described in these documents, provides the best insight available to date regarding the practical mechanics of implementing OSI network management.

Based on the OSI seven-layer reference model, NMP is also consistent with the OSI Management Framework and Management Information Services to the extent to which these are defined. NMP is implemented in Layers 4 through 7 of the OSI model and is independent of specific implementations of Layers 1, 2, and 3. NMP is subject to some modification because it depends on some OSI standards that are not yet finalized (see "NMP Status").

AT&T divides its application layer services into two categories: transaction services and file transfer services. Transaction services depend on Common Management Information Service (CMIS), Remote Operations Service Elements (ROSE), and Association Control Service Elements (ACSE). Enhanced transaction services, which provide for two-phase commitment and chaining plus similar sophisticated facilities, require Commitment, Concurrency, and Recovery (CCR) in addition to CMIS, ROSE, and ACSE. File transfer services require ACSE and FTAM. AT&T's implementation of these for NMP, are outlined in document TR54004.

In addition to NMP's implementation across Layers 4 through 7, AT&T has also published NMP application message sets for configuration management and fault management. OSI specifications covering these two functions, as well as for security management, are closer to final approval than are those for performance and accounting management. AT&T is currently working on message sets for the latter as well as for the other remaining UNMA network management functions, although progress to some extent depends on OSI committee progress.

AT&T, however, sees no reason to wait for these functions to attain Draft International Standard (DIS) status before embarking on implementation. The company is developing products that conform to OSI Draft Proposals (DPs), even though DPs are subject to modification. AT&T

Table 2: OSI management status

TITLE	REFERENCE DOCUMENT	WORKING DOCUMENT	DRAFT PROPOSAL	DRAFT INTERNATIONAL STANDARD	INTERNATIONAL STANDARD
OSI MANAGEMENT FRAMEWORK	DP 7498-4	COMPLETE	COMPLETE	COMPLETE	OCTOBER '88
OSI MANAGEMENT INFORMATION SERVICE OVERVIEW	N 2683	COMPLETE	COMPLETE	JULY '89	JULY '90
STRUCTURE OF MANAGEMENT INFORMATION	N 2684	COMPLETE	COMPLETE	MAY '89	JULY '90
COMMON MANAGEMENT INFORMATION SERVICE (CMIS)	DP 9595	COMPLETE	COMPLETE	COMPLETE	SEPTEMBER '89
COMMON MANAGEMENT INFORMATION PROTOCOL (CMIP)	DP 9596	COMPLETE	COMPLETE	COMPLETE	SEPTEMBER '89
CONFIGURATION MANAGEMENT	N 2686	COMPLETE	COMPLETE	JULY '89	JULY '90
FAULT MANAGEMENT	N 2687	COMPLETE	COMPLETE	JULY '89	JULY '90
SECURITY MANAGEMENT	N 2688	COMPLETE	SEPTEMBER '89	JULY '89	JULY '90
ACCOUNTING MANAGEMENT	N 2689	COMPLETE	SEPTEMBER '89	APRIL '90	APRIL '91
PERFORMANCE MANAGEMENT	N 2673	COMPLETE	SEPTEMBER '89	APRIL '90	APRIL '91

pledges to modify its products to comply with any OSI changes, but it expects those changes to be minor. Despite this aggressive approach to product development, it will take a number of years for AT&T to develop full functionality in all its targeted functional areas.

Although AT&T has pledged that it will modify NMP to conform to final OSI specifications, users must realize that the OSI guidelines leave room for differences in vendor implementations—differences that can tarnish the lure of interoperability. Within each OSI layer are not only mandatory services but also optional ones that a vendor may or may not choose to implement. Vendors may opt to extend protocol definitions. For example, AT&T adds a parameter constraint in the ACSE Protocol for NMP that is not required in the ISO standard. Thus, users must be aware that while OSI-based implementations are *open,* this does not mean that they are identical.

Developing interfaces

Since multivendor connectivity under UNMA depends on vendor acceptance of NMP, AT&T proposes that other vendors use the published NMP specification and message sets to develop UNMA interfaces to their products. DEC, Hewlett-Packard, and a few other pioneers, however, are also developing different OSI implementations. Equipment vendors do not have unlimited resources, and many will be forced to choose between AT&T's NMP and other alternatives.

Industry analysts have viewed the lack of vendor support for NMP as a major hindrance to user acceptance of UNMA. AT&T has recently pledged to take a more proactive role in promoting NMP among equipment vendors for LECs and PTTs. AT&T lists 17 equipment vendors supporting NMP (Table 3). This list will undoubtedly grow before early 1990, when the OSI/Network Management (NM) Forum plans to stage an interoperability demonstration.

Before this demonstration becomes a reality, however, the Forum must select a protocol-stack and message-set implementation. This will probably amount to choosing a slightly modified version of AT&T's NMP. While Forum membership does not indicate full-fledged support of AT&T's NMP, each vendor in the Forum has made a substantial commitment to an OSI-based approach to network management—an approach that is much more amenable to AT&T's scheme than IBM's or even DEC's.

It is interesting to note that at the April 1988 IBM Telecommunications Consultants Conference, IBM was proudly pointing out that "maybe two or three vendors" had pledged support for AT&T's UNMA, compared with about 40 vendors supporting NetView/PC. While support for NMP and membership in the OSI/NM Forum has increased substantially over the intervening months, the momentum behind NetView/PC support has remained stationary, at best.

Consensus setting

AT&T played an instrumental part in the July 1988 formation of the OSI/NM Forum. The founding members included Amdahl, British Telecom, Hewlett-Packard, Northern Telecom, Telecom Canada, STC PLC (UK), and Unisys. Forum membership has increased to more than 40, including five new voting members. New voting members include Digital

NMP status

The following section briefly describes the OSI layers and protocols in which Network Management Protocol is implemented. The status of each OSI component is listed to provide a general picture of the degree to which NMP is subject to change.

OSI Layer 4—International Standard Status (finalized). The transport layer establishes (node-to-node) network connections, provides end-to-end data acknowledgment, terminates network connections, and performs related tasks.

OSI Layer 5—International Standard Status (finalized). The session layer establishes endpoint-to-endpoint sessions, specifies duplex or half-duplex service between session users, and coordinates session termination.

OSI Layer 6—International Standard Status (finalized). The presentation layer ensures compatibility between incoming file, record, and data formats, as well as the format requirements of the receiving systems.

OSI Layer 7—Portions have attained International Standard Status (partially finalized). Within the application layer are several sublayers used for network management, including:

- ASCE (Association Control Service Elements): International Standard (finalized)
- FTAM (File Transfer, Access, and Management): International Standard (finalized)
- ROSE (Remote Operations Service Elements): Draft International Standard (relatively stable)
- CCR (Commitment, Concurrency, and Recovery): Draft International Standard (relatively stable)
- CMIS (Common Management Information Service): Draft International Standard (relatively stable)
- CMIP (Common Management Information Protocol): Draft International Standard (relatively stable)

Communications Associates Inc., MCI Communications Corp., and Microtel Ltd. New vendors pledging associate membership include Avant-Garde, Contel Technology Center, Fujitsu America, Hekimian Laboratories, Infotron, NCR Corp., Network Equipment Technologies, Prime Computer, Racal-Milgo, Siemens A.G., and Telwatch.

The OSI/NM Forum's main goal is to accelerate delivery of OSI management-based products by forming a consensus on protocol options, message sets, and management object definitions. Forum members anticipate that such a consensus will not only promote product implementation development but also influence international bodies to adopt standards supporting (or, at least not conflicting with) those implementations. Forum members have pledged to promote Forum-adopted platforms within the international standards bodies.

The Forum plans to stage an interoperability demonstration in mid-1990 and is in the process of defining the messages that will allow that demonstration to take place. These messages address *event management,* which includes fault management and configuration management. The closure date for a Forum consensus on messages is planned for mid- to late 1989. At the same time, the Forum will be working toward a consensus on management architecture, including object definitions that name and define management data.

In January, the Forum achieved its first major milestone when it defined a seven-layer protocol stack for conveying management information. The protocol stack includes X.25 and IEEE 802.3 for the transport layers and is fairly close to AT&T's published definition of NMP.

AT&T has indicated that it will modify NMP to comply with whatever the OSI/NM Forum adopts and, ultimately, with whatever OSI committees adopt. Clearly, each vendor represented in the OSI/NM Forum has individual reasons for joining and will vote for an implementation that promotes those goals. Thus, there is no guarantee that OSI/NM Forum members will automatically adopt NMP without modifications.

On the other hand, regarding anticipated OSI/NM Forum decisions, AT&T has stated that the technical work done on NMP will not change. Furthermore, AT&T's John Miller, director of network management, has stated that "the Forum's goals are totally aligned with AT&T's goals."

DEC is moving full steam ahead in garnering support for its own OSI-based approach—Enterprise Management Architecture (EMA). Announced in Cannes, France, in September 1988, EMA employs a director-entity approach that allows multiple "directors" (digital network management systems) to manage "entities" (analogous to AT&T's Network Elements) on multiple domains.

Directors can exchange management information with, and even take over management duties of, other directors in different domains (in the event of a malfunction). This exchange is possible since EMA specifies a standard OSI-based format for information exchange. At this time, however, DEC has not announced peer-to-peer exchange with its directors and NetView or the Accumaster Integrator. According to DEC, EMA was two and a half years in development.

DEC's EMA has support from seven vendors, including Codex, DCA, Stratacom, Timeplex, and other equipment and software providers. Several of them have pledged support for both EMA and NMP. While the momentum behind the OSI/NM Forum is undoubtedly pushing DEC to reconsider its declination, the company has not announced intentions of joining to date.

DEC has stated that it will deliver OSI-based management products by the end of 1989. It could thus damage AT&T's bid to wield primary influence over the direction of OSI-based implementations. If AT&T's supporters continue to increase in number and if DEC announces no viable strategy for linking to NetView, that damage will probably be minimal.

Adding logical data

Last summer, AT&T and Cincom Systems Inc. announced an agreement to jointly develop an application that will provide logical data to the UNMA integrating system. Cincom currently produces Net/Master, widely accredited as the only real competition to IBM's NetView.

Table 3: Vendor support for NMP

- AVANTE-GARDE COMPUTING INC.
- AVANTI COMMUNICATIONS CORP.
- COASTCOM T1 NETWORKING PRODUCTS
- DIGITAL COMMUNICATIONS SYSTEMS INC.
- DYNATECH COMMUNICATIONS INC.
- EMCOM CORP.
- GENERAL DATACOMM INC.
- HEKIMIAN LABORATORIES INC.
- INFOTRON SYSTEMS CORP.
- INTEGRATED TELECOM CORP.
- KAPTRONIX INC.
- NEWBRIDGE NETWORKS
- PARADYNE CORP.
- RACAL-MILGO
- SYNC RESEARCH INC.
- TELINQ SYSTEMS
- TRIDOM CORP.

Benefiting from Cincom's expertise in SNA, the resulting application, called the SNA Management Application (SMA), resides in the IBM host and allows the Accumaster Integrator to interface with either IBM's NetView or Cincom's Net/Master. The application will support SNA customers by extracting logical information (messages and screens) from IBM systems and displaying them on the Accumaster Integrator Console (Fig. 2). Using this approach, the Accumaster Integrator could include peer-to-peer interfaces with other logical network management systems, such as DEC's EMA.

The SMA builds on the functionality that Net/Master currently provides, according to Cincom, but will not preclude UNMA customers from using IBM's NetView. Furthermore, the SMA does not depend on IBM's Network Performance Monitor (NPM), adds Cincom. The current Net/Master product requires access to the NPM, which IBM could bundle with NetView at a future date. (There have been no indications so far that IBM intends to do that, however, and Datapro believes that this is unlikely.)

The UNMA roll call

1. The Accumaster Integrator. Featuring an X Window-based interface, the product takes logical network management information from the SNA Management Application (an optional feature) and correlates it with physical device data transmitted (via an NMP interface) from various vendors' EMSs. One of the Integrator's most striking aspects is its Alarm Correlation feature. While not infallible, this feature provides a distinct advantage over NetView's implementation of capturing physical device alerts through NetView/PC—particularly when the network administrator must resolve multiple simultaneous alarms.

The Alarm Correlation feature compares alarms as they are received, determines the most likely cause of the alarm, and suppresses secondary alarms linked to the original problem. To accomplish this, the Integrator uses internal logic in combination with network device profiles stored in a configuration database. The feature can help operators determine what to do first and whether the alarms are related by isolating the most likely source of the alarm.

As impressive as this feature is, it is obviously not perfect, since the Accumaster Integrator chooses the *most likely* cause. In the future, says AT&T's Miller, the company will expand this area and will continue to enhance the algorithms that drive the feature. Also, at present, the Alarm Correlation is not user-programmable in the sense that it includes user exits for extensive customization.

The marketability of the Accumaster Integrator hinges on UNMA's open NMP interface, which allows the user to integrate data from other vendors' element management systems. Codex, Racal-Milgo, and other vendors offer modem and multiplexer management products that far outsell AT&T's Dataphone II, for example. Of course, the Accumaster Integrator can only provide information from other EMSs if the manufacturers develop the requisite NMP interface. Seventeen major vendors have already pledged to develop NMP interfaces. Four of those vendors, including Racal-Milgo and Infotron, have already implemented NMP interfaces on their products. The number of vendors supporting NetView/PC is about twice AT&T's figure; however, the momentum behind NetView/PC is surprisingly low, given IBM's installed base of SNA networks. If NetView/PC supporters numbered in the hundreds, AT&T would not have poured so much into the UNMA/NMP development effort.

The *new* Accumaster Integrator not only consolidates but also integrates by processing information. Initially, the product provides control in two clearly related areas: fault management (alarm integration) and configuration management. The Integrator includes a configuration database of all network components. Device profiles include device types, names, locations, and how these devices are interconnected. The product draws on this information to produce real-time graphic displays of network configurations. The Integrator's graphics capabilities also support fault management—including the use of icons to represent network devices and color codes to indicate fault severity and importance. More important, the product allows the user to track down logical or physical faults from one console.

For example, if an alarm comes in from the SMA, the operator can view data from the NPDA Hardware Monitor and query to get a picture of the route from the host through each of the SNA service nodes. The operator can then examine the status of each to see whether or not there is a physical problem correlating to the logical alarm. Also, the Integrator enables an operator to claim "ownership" of

2. Accumaster and SNA. *The initial implementation of the Accumaster Integrator will communicate with other vendors' network management systems via AT&T-supplied gateways.* *AT&T and Cincom Systems jointly developed an extraction application, the SNA Management Application, that feeds the Integrator with data from the SNA focal-point program.*

UNMA-SNA ARCHITECTURAL RELATIONSHIP

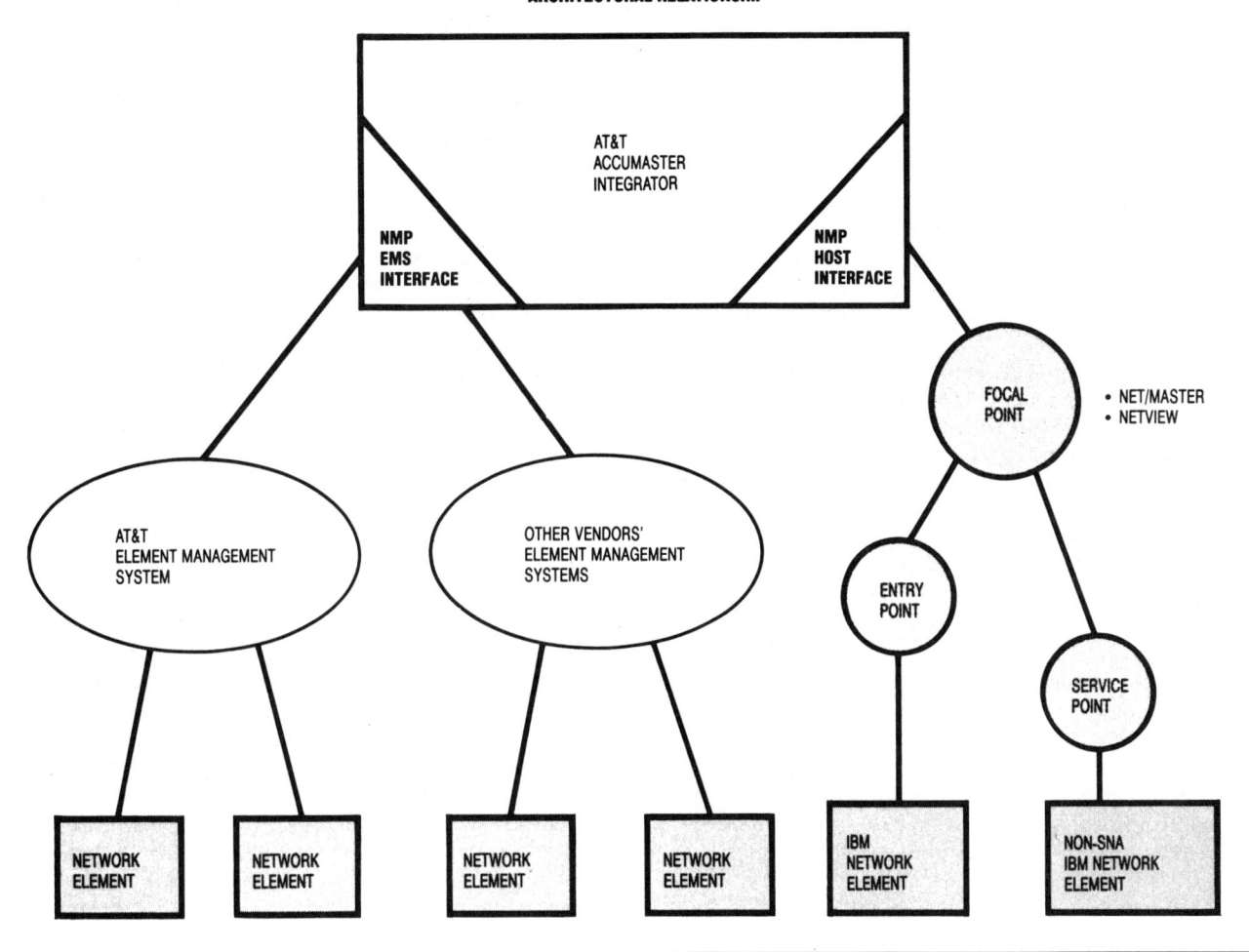

an alarm by setting a flag to indicate to other network operators that the alarm is being handled. An electronic mail feature enables network operators and administrators to communicate information about faults and problem resolution on line.

Besides processing management data through the NMP interface, the Accumaster Integrator includes terminal emulation capabilities to enable the user to cut-through to the individual EMSs. That important feature gives the user the full effect of operating at the EMS console—such as a StarKeeper console. This provides an advantage over the NetView/PC implementation, which presents generic alerts or character strings (extracted from the other physical network devices using code points) rather than the actual EMS console display.

To facilitate an end-to-end view of the network, the Accumaster Integrator creates a unified, "virtual network" composed of both private and public network elements. It does so by combining information from customer premises EMSs with information from EMSs controlling those portions of the public network partially allocated to the customer. Under UNMA, this integrating system may reside either on the customer premises *or* in the public network, or in both simultaneously.

2. Accumaster Management Services. Besides the Accumaster Integrator, AT&T now offers an integrating service called Accumaster Management Services. This service will be administered from various Network Operations Centers (NOCs) that physically reside either in AT&T's network or on the customer's premises. NOCs will be staffed by AT&T personnel teams dedicated solely to a specific customer. When under contract for NOC services, the customer will receive three categories of services: provisioning, fault management, and analysis. Provisioning includes procurement and follow-through on all equipment and service orders, including station moves, changes, and relocations. Fault management includes call receipt, trouble tracking, repair verification, restoration planning, and implementation services. Analysis includes regular, detailed reports on numerous aspects, including response time and meantime to restore.

There are roughly 10 NOCs already in service. About half of these are on the customer's premises. Pricing is on a contract basis and depends on options chosen. Contracts can be designated as five days per week (normal, or expanded business hours) or seven days per week, up to 24 hours per day. This service is available immediately.

3. *Accumaster Consolidated Workstation, Release 2*. The first release of the ACW was announced in September 1987 and became available in mid-1988. Release 2 of the ACW runs on an AT&T 6312 Work Group System (WGS) and uses multitasking applications software to monitor multiple Element Management Systems simultaneously. The 6312 WGS supports the MS-DOS operating system.

The ACW's 3270 terminal emulation facility provides IBM host access from NetView or Cincom's Net/Master. Users can communicate with host applications while maintaining active sessions with other EMSs. ACW Release 2 now supports up to 10 EMSs, displaying the status of six of these simultaneously using the windowing feature. In this release, AT&T has also added VT100 terminal emulation and 513 emulation. The 513 emulation feature gives users the capability to access the AT&T System 75/85's (PBX) Centralized System Management and other vendors' EMSs. The VT100 capability enables users to access network management systems that support VT100 terminal emulation. This includes AT&T network-based information, such as the Management Information Systems Report. MISR is a network administration report system designed to support Software Defined Networks.

ACW operators can observe network changes in real time and interact with the appropriate system to perform testing, problem diagnosis, troubleshooting, and reconfiguration.

4. *Accunet T1.5 Information Manager (AIM)*. This is the first AT&T product adapted to UNMA standards. AIM works either as a standalone EMS or as an EMS under the Integrator. AIM features audible signals and graphics displays, and it focuses on configuration management, alarms, and error performance. The product does not display real-time utilization of T1 lines. AIM keeps various records on failed circuits for 31 days.

The AIM software package costs about $3,000 and will be available to Accunet T1.5 customers in April. There will also be a onetime fee of approximately $1,000 per circuit accessed to initiate service. To implement AIM as a standalone system, the customer must provide a PC and a 9.6-kbit/s private line link to the AIM system.

5. *NetPartner Network Management System*. AT&T's integrating system for LECs is the NetPartner Network Management System (NMS). AT&T introduced NetPartner NMS in 1988, although there are no figures available on how many units have been sold. This product is a hardware/software combination that gives Centrex customers restricted access to the phone company operations embedded in the phone company's network. Specifically, NetPartner translates information that is currently within the phone company's operations into a form that is easier for customers to understand, manipulate, track, and alter. The phone company maintains control over what the customer can access.

NetPartner NMS features a three part architecture, comprised of host equipment at the phone company, customer premises equipment, and operations at various phone company sites.

NetPartner also provides a 3270 terminal emulation capability that provides access to IBM's NetView or NCCF/NPDA or Cincom's Net/Master. In this configuration, NetPartner appears like a cluster controller to the IBM host computer. NetPartner has a menu-driven, graphics-based user interface. The phone company customizes each menu to show only those functions that the customer's company has purchased. Currently, one NetPartner NMS can support five to 10 customers — up to 20 users simultaneously. The phone company defines the number of users per customer. Future NetPartner enhancements are planned, including an ISO-based interface to specific customer premises-based and interexchange carrier products. Some artificial intelligence-based capability is also planned.

AT&T is directing its NetPartner marketing efforts at both phone companies and end-users. AT&T is seeking to convince LECs that end-users are willing to pay extra for the control facilities that a NetPartner-equipped LEC can offer. AT&T is trying to persuade end-users to demand NetPartner capability from their local telcos. AT&T is particularly pursuing those users interested in ISDN capabilities. Pricing information on NetPartner is not yet available.

Nine ways to support users

Network management systems are used to help obtain real-time information on network performance and traffic characteristics, diagnose problems, and reconfigure to meet changing needs. In the past, network management was characterized by separate management systems devoted to providing these services for a particular vendor's product or group of products.

AT&T, however, proposes with UNMA to integrate data from disparate management systems and collectively provide critical network management functions. Specifically, UNMA defines nine functions to support users in managing their networks. The first five of these functions are included in the OSI functional model of management.

■ *Fault management*. The goal of fault management is to maintain network availability at an acceptable level. On a day-to-day basis, this means quick and accurate problem detection and problem determination.

The UNMA management integrator correlates alarm information from various devices to pinpoint the event or fault that may have caused multiple alarms. UNMA calls for complete audit trail of the fault management process, supported by trouble tracking.

In April 1988, AT&T published the Network Management Protocol Fault Management Message Set Specification. This document outlines how NMP enables users to control alarm reporting, manipulate alarm information, and set alarm reporting parameters.

> No OSI-based product supports all nine functional areas defined by UNMA.

■ *Configuration management.* The goal of configuration management is to manipulate network configurations to adapt to changing needs and traffic patterns or to isolate problems. To support this, the network management product must collect information on the current state of the network, noting changes; modify network attributes; and change configuration.

The UNMA definition of this function has four basic aspects: network inventory management, change management and provisioning, name management, and actual connections (relating inventory items to their physical layout). Network inventory management involves tracking all devices and services on the network. Change management and provisioning supports both scheduled and unscheduled movement of telephones, modems, terminals, circuits, and other network components. Name management governs the network directory. In April 1988, AT&T published the NMP Configuration Management Message Set Specification.

■ *Performance management.* The goal of performance management is to identify and correct potential problems before they cause a fault. To accomplish this, the management system must collect data on current network and resource performance levels and maintain performance logs.

UNMA provides the capability to correlate information from multiple EMSs to assist the users in identifying network performance trends. UNMA users will have the capability to monitor selected network components via user-definable measures and thresholds. UNMA will provide both recent history and current performance data, which the user can analyze in order to identify network performance trends.

■ *Accounting management.* This function informs users of costs incurred and enables users to set accounting limits.

Under UNMA, users can compare usage and billing information across related voice networks and subnetworks and potentially obtain more complete billing, verification, and chargeback information. Chargebacks may include charges for fixed-cost items such as telephones or terminals as well as usage; UNMA supports comparison of vendor bill verification and comparison of vendor bills with inventory and internal measures.

■ *Security management.* Security management encompasses access control, authorization facilities, and partitioning the network. The OSI definition of security management also includes support for encryption and key management and the maintenance and manipulation of security logs.

UNMA supports tracking log-on attempts and violations to prevent unauthorized network access. UNMA also supports multiple network management permission levels. Under UNMA, network administrators can manage the network from either one or several different network management operation centers by partitioning the network.

■ *Planning.* While not defined as an OSI management function, planning is widely accepted as a major network management subsystem. The goal of planning is to design and optimize models that describe potential changes to the network. Consolidating usage trends and performance data supports this. Planning typically involves collecting performance data, consolidating usage trends, and analyzing future requirements.

UNMA outlines three common types of planning: capacity planning (day-to-day fine-tuning, such as adding or rearranging trunks); contingency planning (backup and disaster recovery, including estimated costs); and strategic planning (new applications, growth plans, acquisitions, reorganizations).

■ *Operations support.* This encompasses managing the staffing and operation of a network management center.

UNMA defines four aspects of operations support: creating network management center procedures (for trouble logs, maintenance fixes, shift changes, etc.); analyzing work and information flow at the center; analyzing network management center staff requirements; and preparing user training and development plans.

■ *Programmability.* The goal of programmability is to customize the network management package to meet corporate needs. There are no "off-the-shelf" solutions in network management. Programmability is critical because each network is different and networks are characteristically in a state of flux.

UNMA provides parameterization of key characteristics and provides flexible report capabilities, customizable scripts, and custom programming options. It supports programmability in C language.

■ *Integrated control.* The goal of integrated control is to create the image of a single virtual network, even though it may actually comprise diverse, separate management systems. UNMA currently provides consolidation of multiple network control screens into a single, windowing workstation—the Accumaster Consolidated Workstation.

These nine generic function descriptions provide a useful framework for evaluating integrated network management products. Currently, there is no single OSI-based product or service that provides comprehensive, integrated support in all of these functional areas. Over the next decade, AT&T plans to evolve UNMA and the Accumaster product line to fill this gap. ■

Portions of this article were adapted from the January 1989 issue of "Datapro Reports on Communications Software."

Jill Huntington is an associate editor/analyst with Datapro. She holds a B. S. in business and English from Wake Forest University, Winston-Salem, N. C., and a master's degree in Computer Science from the University of Virginia in Charlottesville. Within the computer industry, she has held positions in programming, teaching, and technical writing. Prior to joining Datapro, Huntington was employed as a programmer/analyst with NCR Corp. (Dayton, Ohio).

Remote-control software: PCs apart, but not alone

Packages now give users faster screen updates, LAN versions, more diversified hardware support, even PC-to-Mac connectivity.

Software that allows a PC operator not only to access but also to control another PC is flourishing. Some packages even extend this remote-control capability so that an operator at a distant PC can view and change files in use on any LAN workstation.

But control is a double-edged sword in any context, personal computing not excepted. While these communications-intensive programs may be just what's needed in certain networking applications, yielding control of any processing resource to a remote user can also cause problems for a network manager.

As new releases of remote-control software are enhanced, they grow increasingly more complex. To accomplish the appearance of mutual PC activity, remote-control software today requires considerably more programming effort than do general-purpose communications packages (see "Capturing screens, commandeering keyboards"). It is important that networkers understand some of these program machinations so they may better select the right package and optimize their network configurations.

Remote-control software usually consists of two modules: One, referred to here as the application module, sits on the PC that also runs the application programs (spreadsheets, word processing, and so forth); the other, called the support program, runs on the "remote" PC, from which the user views and changes the application file.

While the user on the application side is working with, say, Lotus 1-2-3, the remote-control software operates in the background as what's called a "terminate and stay resident" program. This means that the program loads on command, installs, and then returns the user to the DOS command level of the machine. The program remains dormant but constantly looks at the communications port; it takes 44 to 67 kbytes of resident memory, depending on the vendor's implementation.

The support-side user may then call in to view the application that is running. Once connected, both users can enter data from either keyboard. The support-side software is considerably larger—typically 122 to 177 kbytes of code—but it is called into memory only when needed.

Such a communications arrangement can be used, for example, to provide remote off-site support to a customer, thereby eliminating travel time to solve problems at remote sites. Or it may be used as a way to access applications and data housed in desktop PCs from laptops or terminals at home or on the road. It may also provide customers with around-the-clock access to on-line order entry/data retrieval, or it can automate the transfer of, say, sales data between far-flung branch offices.

Remote-control programs can also provide a way to emulate the terminals of the most widely used mainframes. They can enable, for instance, an IBM PC emulating a DEC VT 100 terminal to connect to a DEC VAX. The remote-control package could be programmed to access the host automatically and download files that are needed by remote terminals in branch offices around the country. Remote PCs that are running the companion module of the software could then access the PC that is acting as the gateway to the remote host. In this way, files can be accessed, downloaded, or scanned during off-hours.

One of the most recently updated remote-control communications packages is also the one that came out first. Remote2, which was updated to Version 1.1 in February, was originally released in 1979 as Remote. The package's coauthor, Les Freed, had just formed Microstuf Inc. and was simultaneously working on what was to develop into a popular general-purpose communications program: Crosstalk.

The competition started in 1985 with Carbon Copy and Close-Up, and continued with pcAnywhere in 1986 and

Capturing screens, commandeering keyboards

Remote-control communications software has to be flexible. A successful package will accommodate lots of different keyboard types, all the popular brands of monitors, the slew of various modems—and it will be able to work with a majority of popular applications software.

To communicate keyboard strokes and screen images back and forth between two different machines, this type of software has to be prepared for whatever it might meet at the other end. That's why the programming effort for remote-control software is more intensive than for a general-purpose PC communications package. According to Ed Girou, a consultant with R. A. Kottmeier and Associates Inc. (St. Louis, Mo.), remote communications software is more complicated by a factor of three.

At Dynamic Microprocessor Associates Inc., vice president of marketing Peter Byer says that it took DMA nine man-years to write the code for pcAnywhere, but three man-years for the company's well-received general-purpose communications package, Ascom 4.

Part of the difficulty stems from the way different application programs work. Some use the PC's disk operating system (DOS) to access the keyboard and monitor; others go through the basic input/output system (Bios) of the machine. Still others transmit their commands directly to the video memory.

Depending on how the application program sends its commands, the remote-control program may or may not be able to intercept these instructions and redirect them over the communications port and out to a remote user. Early releases of remote-control software had more difficulty than the current packages, however, and most now make an effort to work with the video memory.

"Co/Session is designed around monitoring the video memory directly," says its program developer, Floyd F. Roberts, who is also vice president of Triton Technologies Inc. "Whether an application goes through the Bios to update video memory, or whether an application itself writes directly to video memory, Co/Session will work with it because the software is hooked into it at that level."

What the remote program does to the information it gets from memory is now the frontier for programmers. Co/Session, says Roberts, scans the video memory and checks for changes; it then does a "highly optimized compression of these changes. Because of the effect of compression, we are much faster [than others] in screen update time."

At least one longtime user of remote-control software packages agrees that Co/Session excels at screen updates. "Originally with the remote packages," says Rick Segal, technical adviser for the commercial insurance division of Aetna Life and Casualty Insurance Co. (Hartford, Conn.), "the screen update happened every time there was a keystroke. Now they have migrated to transmitting only that which is updated." —S. J. L.

Co/Session in 1987. In the spring of 1988, Crosstalk Communications, a division of Digital Communications Associates Inc. (Roswell, Ga.), completely reworked Remote and announced Remote2.

Remote2 scores high in the areas that matter most to software reviewers (see "Rating the remotes: A hands-on evaluation"). Its features now include voice/data switching, compatibility with DESQview (from Quarterdeck Office Systems, Santa Monica, Calif.), improved file security, and accommodation of higher transmission speeds. As for program load time, it was found to be significantly faster than that of the competition (Table 1).

When operating in the DESQview mode, Remote2, 1.1, lets the caller (on the support side) see a full-screen display of the single DESQview session; the user at the application side sees a windowed DESQview screen. By running Remote2's application module (which DCA calls the host program) in a number of DESQview sessions, it is possible for a PC to appear as a multi-user host. With Remote2's revised security features, the program can be set to give varying levels of access to each caller: send and receive, which lets the caller upload and download any files; send only, which permits the caller only to upload files; receive only, which allows downloading; and none.

While most remote-control communications packages are designed to be used via dial-up or leased phone lines (generally employing asynchronous transmission using start and stop bits), there are versions of this software that will run on LANs. Because the LAN medium allows for synchronizing the sending and receiving devices, the data can be organized in a synchronous fashion and sent at a fixed—and much faster—rate than the 19.2-kbit/s top end of the dial-up world.

A LAN version of Remote2, R2LAN, announced at last fall's Comdex trade show, is now in beta test. R2LAN runs on any IBM Netbios-compatible link, including DCA's 10net, Novell, and IBM's Token Ring. It provides a LAN with all of the features that freestanding microcomputers can have via Remote2. While the standalone remote-control package gives access to a LAN via modems and serial ports, the LAN version makes use of the higher transmission speeds found on the network. By using the network's asynchronous communications server, users have a variety of mix- and-match access possibilities. A network license of the LAN version costs $795.

But DCA is not the first to announce a LAN version of remote-control communications software. Brightwork Development Inc. (Red Bank, N. J.) announced its NETremote in 1987. Like standalone remote-control packages, it lets the PC user have control of the keyboard and screen of another machine. But it extends this access to any workstation on the network. As a result, any user can access any of the files and peripherals on the LAN, including facsimile cards and modems that might be installed on only one of the PCs.

NETremote supports Novell's Advanced NetWare 2.0a

Table 1: Remote control offerings

	MAXIMUM TRANSMISSION RATE (KBIT/S)	MODEMS SUPPORTED[1]	FILE TRANSFER PROTOCOLS
CARBON COPY PLUS VERSION 5.0 MERIDIAN TECHNOLOGIES INC. IRVINE, CALIF.	38.4	AT&T, USER DEFINABLE	PROPRIETARY, XMODEM, YMODEM, KERMIT
CLOSE-UP VERSION 3.00A NORTON-LAMBERT CORP. SANTA BARBARA, CALIF.	38.4	USER DEFINABLE	PROPRIETARY, XMODEM
CO/SESSION VERSION 3.1C TRITON TECHNOLOGIES INC. ISELIN, N.J.	19.2	PREDEFINED COMMAND STRINGS INCLUDE TELEBIT TRAILBLAZER, TELCOR, VEN-TEL, ETC.	PROPRIETARY, XMODEM
pcANYWHERE VERSION 3.00 DYNAMIC MICROPROCESSOR ASSOCIATES INC. NEW YORK, N.Y.	57.6	PREDEFINED COMMAND STRINGS INCLUDE UDS, ROCKWELL, US ROBOTICS, ETC.	PROPRIETARY, XMODEM
REMOTE2 CROSSTALK COMMUNICATIONS, VERSION 1.1 DIGITAL COMMUNICATIONS ASSOCIATES INC. ROSWELL, GA.	38.4 (115.2 WITH 8-MHz PC/AT)	PREDEFINED COMMAND STRINGS INCLUDE US ROBOTICS, TELEBIT, VEN-TEL, ETC.	PROPRIETARY, XMODEM, YMODEM

1. BESIDES HAYES-COMPATIBLE MODEMS
2. 'APPLICATION' = PROGRAM MODULE RUNNING ON THE PC THAT ALSO HAS THE APPLICATION PROGRAM. 'SUPPORT' = PROGRAM MODULE RUNNING ON THE PC CALLING INTO THE APPLICATION. TSR = TERMINATE AND STAY RESIDENT.

and 2.1 via a direct interface; the IBM Network Program; 3Com's 3Plus; AT&T's Starlan; Banyan's Vines; and Sun Microsystems' TOPS/DOS 2.1.

The network license for use on up to four file servers costs $695. A single-server version, which supports a NetWare LAN, costs $295. When used in conjunction with the asynchronous remote-control packages, NETremote supports distant networks.

A variation on this approach is found in LAN Assist, first announced in the fall of 1987 by Fresh Technology Group (Mesa, Ariz.) and recently updated as LAN Assist Plus 2.0. It requires only about 2 kbytes of memory and supports Novell NetWare, Versions 2.00x and 2.1x.

LAN Assist, which costs $199, lets users monitor problems on the LAN, transfer files to the server from the viewed workstation, maintain software, and conduct training sessions. And it supports other workstations across bridged networks. Used in conjunction with one of the asynchronous remote-control packages, it provides mutual keyboard and screen activity across the LAN.

Power user

While remote-control software is viewed by some as merely a technical support tool (clearly one of its more popular uses), at the international headquarters for the broadcasting division of the Associated Press, Deputy Director Lee Perryman puts Carbon Copy Plus to work in a wide variety of ways (see figure).

Perryman's division handles radio, TV, and cable TV, plus most of AP's corporate and government services. In addition, members of the AP wire service depend on this hub of electronic activity for their news transmissions. It is a business in which the efficient dissemination of information is the product.

AP's Washington, D. C.-based administrative operation has 30 microcomputers permanently wired in an IBM Token

TERMINAL EMULATIONS	NUMBER OF COMMUNICATIONS PORTS ACCOMMODATED	MEMORY SIZE (KBYTE)[2]		COST	
		APPLICATION MODULE (TSR)	SUPPORT MODULE (NOT TSR)	APPLICATION	SUPPORT
VT-52, VT-100, TVI-920, IBM3101, DEBUG	2	61	177	$195	$195
TTY	2	44-67, DEPENDING ON CONFIGURATION	134	$245	$195
TTY	4	51	125	$249 SESSION/XL (A SCRIPT LANGUAGE) = $225	$125
26, INCLUDING VT-52, VT-100, HAZELTINE, PCMACTERM	3	45 (47-50 FOR VERSION 3.1, WHICH ACCOMMODATES EGA, VGA, HERCULES GRAPHICS)	128 (NOT NECESSARY TO USE SUPPORT MODULE)	$145	INCLUDED WITH 'APPLICATION MODULE'
19, INCLUDING VT-100, VT-220, WYSE, HAZELTINE	8	57 (COLOR) 45 (MONOCHROME)	122	$129 BUNDLED = $195	$89

Ring LAN. There are no minicomputers or mainframes at this location, but a partial T1 line gives it a 2.4-kbit/s link to a DEC VAX at AP's main headquarters in New York City. The AP newsroom's terminals are connected via Vitalink to a communications center two blocks away.

AP first installed remote-control communications software in late 1985 and, for the following year, was a beta test site for Meridian Technology Inc. (the makers of Carbon Copy Plus). The software is now used for PC-to-PC, PC-to-LAN, and PC-to-mainframe communications. Applications include remote maintenance, technical support, file transfer, contract service, billing, word processing, spreadsheet work, and selective access of news relay transmissions.

The application version (also called "host" version) of Carbon Copy Plus runs on six PCs in the same Washington building, and the support version runs on 11 PCs at remote sites (one of which is for a networking consultant who dials in to troubleshoot). Perryman also has a copy of the remote version on a PC at home; there is another copy on a PC at the New York site. None of these local PCs need more than occasional access to the LAN.

"If we had hard-wired them," says Perryman, "we would have spent about $1,000 per workstation for network interface cards and cabling. Plus there would have been the added problem of having more stations on the ring to diagnose and fix."

Besides the cost savings inherent in using the PC communications software approach—rather than wiring all these stations directly into the LAN—there is an added benefit in transmission line costs. If AP had to transmit directly to the mainframe in New York rather than dial up through the PC, it would need to use at least 19.2 kbit/s of the T1, says Perryman. "A dedicated 2.4-kbit/s circuit to New York, purchased through the phone company," he adds, "costs about $500 to $600 per month. A full T1, by

Remote-control options. *The Associated Press uses Carbon Copy for point-to-point (A), PC-to-LAN (B), PC-throughout-LAN (C), and PC-to-mainframe (D) applications.*

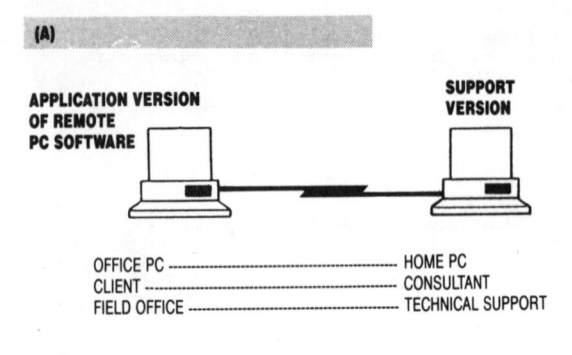

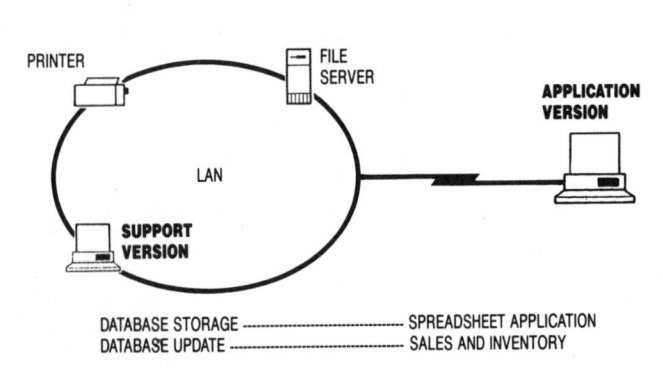

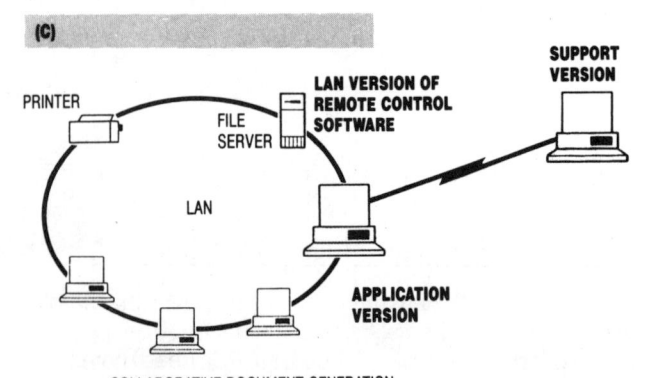

(A)

APPLICATION VERSION OF REMOTE PC SOFTWARE — SUPPORT VERSION

- OFFICE PC — HOME PC
- CLIENT — CONSULTANT
- FIELD OFFICE — TECHNICAL SUPPORT

(B)

PRINTER, FILE SERVER, LAN, SUPPORT VERSION — APPLICATION VERSION

- DATABASE STORAGE — SPREADSHEET APPLICATION
- DATABASE UPDATE — SALES AND INVENTORY

(C)

PRINTER, FILE SERVER, LAN, LAN VERSION OF REMOTE CONTROL SOFTWARE — SUPPORT VERSION, APPLICATION VERSION

- COLLABORATIVE DOCUMENT GENERATION
- TRAINING SESSIONS

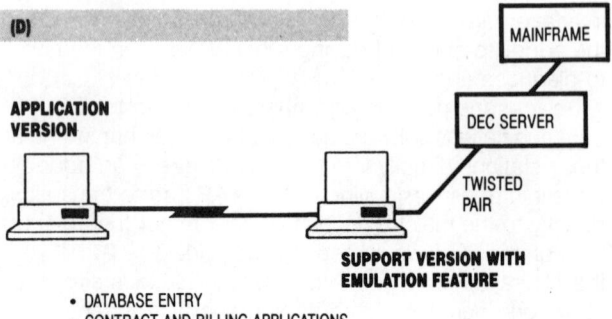

(D)

APPLICATION VERSION — DEC SERVER — TWISTED PAIR — MAINFRAME

SUPPORT VERSION WITH EMULATION FEATURE

- DATABASE ENTRY
- CONTRACT AND BILLING APPLICATIONS

comparison, could cost more than $10,000 per month."

With Carbon Copy Plus on PCs at each site (rather than a network bridge between the mainframe in New York City and the LAN in Washington), the user can choose to have only individual screens of data transmitted, rather than entire files. There are many cases, notes Perryman, where "we might otherwise have to send a 30-kbyte record, even though the user needs only a small portion of it—say, 9 kbytes."

"If entire files had to be sent," says Perryman, "the file and its index record would be locked to all other users during the round-trip transmission, plus the time it takes the user to work on it. If we had network bridges, we'd be moving all of the data. That would slow the response time at the New York end. Why risk having data corrupted when it doesn't need to be sent anyway?"

The AP microcomputers are configured with Telebit Corp.'s Trailblazer modems, which enable dial-up data rates from 9.6 to 19.2 kbit/s, depending on line quality. The PC in New York, an IBM PC/XT compatible from Leading Edge, is directly connected via twisted pair to a DECserver, which provides access to all of AP's computers. It runs the support version of Carbon Copy Plus with its VT100 emulation feature activated. One port is connected to the DECserver and the other to Washington.

On the LAN side, in Washington, D. C., AP uses an Intel 386-based network server. A more high-powered LAN gateway could have been used, but, as Perryman says, "this handles all of the access we need."

Also running on the LAN is Fresh Technology's LAN Assist Plus, which lets designated operators view, maintain, or assist any PC on the network. It can also tie multiple workstations together for training sessions or connect them to one PC that has an attached modem. Perryman also uses the combination of LAN Assist Plus and Carbon Copy to facilitate the collaborative work involved in producing month-end reports.

One of the main, and probably unique, ways that AP uses remote-control software is in selective access to its ongoing information services. The PCs running Carbon Copy Plus have access not only to the LAN but also to a PC that is connected to two different remote services: the Canadian News Relay System, which uses data collected by Broadcast News Ltd. (a Canadian news company based in Toronto); and the Weather Relay System, which gets its data from the National Weather Service.

While Carbon Copy Plus has made Meridian Technologies the leader in this market, Close-Up, from Norton Lambert, is a favorite among original equipment manufacturers (OEMs). Since the first version came out in 1985, the company has sold 300,000 copies. IBM, for example, bought 100,000 copies early last year for use in its support centers. IBM and Norton Lambert later announced that they were codeveloping an OS/2 version of Close-Up.

At the PC Support and Technology group of MONY (Mutual of New York) Financial Service (Teaneck, N. J.), Close-Up has been helping to relieve the telephone hot line since 1986. "We have four 800 [WATS] lines coming in for telephone technical support to the insurance agents," says

Rating the remotes: A hands-on evaluation

To learn how remote computing packages differ, DATA COMMUNICATIONS *commissioned an independent software-testing laboratory to experiment with five of the major offerings. The following are excerpts from the report, which is also partially summarized in the scorecard. The testers exercised the packages using: an IBM PC/XT (640-kbyte RAM, monochrome display, 20-Mbyte hard drive, 1.2-kbit/s Hayes-compatible internal modem, 2.4-kbit/s Hayes external modem); a Toshiba T1200 laptop (1-Mbyte RAM, LCD display, 20-Mbyte hard drive, 1.2-kbit/s Hayes-compatible internal modem, 2.4-kbit/s Hayes external modem); and a Paragon 386E (2-Mbyte RAM, EGA monitor, 100-Mbyte hard drive, 1.2-kbit/s Hayes external modem).*

Carbon Copy Plus, 5.0. Of the five packages reviewed, this is the slowest in program load times, although it is rich in features. It will transfer files as readily (in terms of the mechanics of keystrokes and how to tell the PC what file to send) as any general-purpose communications program — and do so with friendly DOS-like commands. File transfers can also be done in the background at the request of a remote computer even while the user at the host continues to work in an unrelated application. Besides having an excellent terminal emulator, it will accomplish file transfers through Xmodem, Kermit, Ymodem, and a proprietary protocol.

Carbon Copy Plus users at the host [application] and remote [support] sites view the same graphics screen images regardless of whether both systems are similarly configured. A remote computer with a plain Hercules graphics card can see the graphics as displayed at the host equipped with an EGA or VGA card. This function, however, was clumsy at best. Tests between the Toshiba laptop with a monochrome display plus a Hercules graphics card and a 386 IBM-clone with an EGA card were less than satisfactory.

Technical support for the package promised a software patch to make the CGA-EGA connection more livable. The patch was available on Meridian's BBS. Once installed, it helped but did not make the CGA-EGA connection totally acceptable. With this patch, Carbon Copy Plus is the only remote communications program reviewed here that is compatible with a mixture of graphics interfaces.

Carbon Copy Plus is one of the more difficult packages to learn, with an unnecessarily complicated installation procedure. Normally, installation involves a batch file that creates a directory and copies the program files — automatically, with no user intervention. This package requires the user to enter his name and company name to validate the software license. That information is then recorded on the master disk, which in turn will not run until the user has first run the start program.

If users are looking for a program that readily handles graphics applications (like PageMaker) or want to do a significant amount of on-line printing, Carbon Copy Plus is not the package to choose. Graphics packages load and run slowly with this program, and direct printing while on line caused the system to freeze in a variety of tests.

Close-Up 3.00A. Of the packages reviewed here, Close-Up provided the easiest and most efficient method of operating graphics-intensive programs and printing from a remote location. Moreover, installation of Close-Up is the easiest of those reviewed. It transfers only one file to the designated directory, which is also the name of the command typed at the DOS prompt to load the software. This is in stark contrast to the multiple files loaded into directories by other programs after the operator follows a complicated installation procedure.

It is also the most flexible in terms of automating redundant and complex tasks. Besides offering the remote user the ability to access the office PC/Host and all of its applications and files, Close-Up — through its extremely powerful Automated Communications System script language — allows users to preprogram the most complicated of file transmissions and retrievals during off-hours with no human intervention. The big plus here is that you don't have to be a programmer to create automated scripts: Close-Up comes with a "learn" function that will memorize keystrokes for later playback. One of the other packages has a learn feature, but Close-Up's required the least amount of editing of the script.

To write a complicated transfer/retrieval session script, a user need only turn on the learn mode and perform the operation once. During that session, Close-Up remembers all of the keystrokes and the "wait" prompts from the remote system, and records them to a file. From that point forward, the user simply instructs Close-Up to run that script at a specified time, and the exact session is repeated. Users can, of course, add to a script additional sophisticated features that take into account a variety of responses from the remote computer.

For example, if a script were instructed to call a remote site to retrieve financial data and then load that data into a Lotus 1-2-3 spreadsheet at the host [application] end, it could be further instructed with an "If" command telling the host computer to print a copy of the spreadsheet if the figure in the sales column is below a certain point.

Co/Session 3.1. The major drawback to Co/Session is that the script language is not included in the original package. The separately priced script module ($225) adds a great expense to the automation of remote communications.

Called Session/XL, the script program is relatively easy to use and performs the standard automation programs. With it you can automate the transfer of files, automatically run remote programs, monitor the remote screen, print locally or remotely, and more. The language is advanced enough to allow scripts to check conditional statements, branch to other commands, and then return and check for errors. Scripts can be created in any word processor that stores ASCII files (called job files) and are

	PERFORMANCE	DOCUMENTATION	EASE-OF-USE	SUPPORT	OVERALL VALUE	TOTAL RATING
CLOSE-UP, 3.00A	5	5	5	5	5	5
REMOTE2, 1.1	4.5	4.5	4.5	5	4.5	4.6
CARBON COPY PLUS, 5.0	4.5	4	3.5	4	4	4
PCANYWHERE, 3.0	4	4	4	2	3	3.4
CO/SESSION, 3.1C	3	2	3	4	2.5	2.9

0 = POOR; 5 = EXCELLENT

relatively simple to write if you're familiar with the functions of the GOTO, %SET, and LOOP commands. Also, job files can be run from the DOS prompt by typing SXL and the name of the job file.

A major deficit is the script language's inability to "learn" keystrokes during a session for later playback. The program's ability to record a session for later viewing does not create a script file to use in automating future on-line sessions. Moreover, the manual for Session/XL is woefully inadequate for all but the most seasoned users who have had previous script-writing experience. There are no detailed descriptions or how-to's, just lists of commands with truncated definitions.

Session/XL's index does not even contain the word "script." New users will find this intimidating and will most likely not bother to learn the script language or use the automation capability at all.

pcAnywhere, 3. Unlike the other remote communications programs reviewed here, pcAnywhere comes with both the host [application] and remote [support] disks in the same package. Surprisingly, the remote version, called Aterm, is not required to run remote-controlled sessions. Most standard communications programs will interface with pcAnywhere, although users do lose access to some of the upper-end features. The program has no context-sensitive help, but most users will find pcAnywhere's function logical and intuitive.

The documentation borders on excellent, with an unbelievably liberal dose of reproductions of screens and explanations of each element shown. Other manuals could take some pointers here. Unfortunately, there is no separate manual for the remote user to refer to.

The pcAnywhere package provided an adequate measure of error messages. Various attempts to sabotage the communications session were handled smoothly by the program. Tests at 2.4 kbit/s lost their connections on a few occasions, usually during attempts to switch from data to voice. The package appeared to work better with actual Hayes modems than it did with "compatible" versions. As with all but one other program reviewed here, using an EGA system on one end and a CGA on the other produced a mismatch on both displays. There were no error messages to explain the transmission difficulty.

The manufacturer's New York technical support group's phone line (a toll number) was constantly busy; the author never did get through to a person. The company runs a support BBS using its own Chairman bulletin board software, which was easy to access. Although the electronic-mail messages that were left for technical support on software problems were answered, the information was somewhat abbreviated—and the response took a couple of days.

For users who support only one or two remote sites, the $145 suggested list price makes pcAnywhere a good value. If a user will be working remotely with a number of other systems, however, he will have to buy the entire package containing both the local and remote disks.

Remote2, 1.1. The two program modules of Remote2 can be purchased as a bundled package or separately. A big plus for the program is that Crosstalk Mark 4 (and even Crosstalk XVI) can act as the remote module for the host system, eliminating the need to buy the "call" [support] module altogether. Of course, there is a slight trade-off in functions here. The program is not as powerful when not used in conjunction with its remote counterpart, but for some users the trade-off will probably be minimal. Remote2 will even work with a host [application] module and a "dumb" terminal for users who have only that. Although a good deal of the functions go by the wayside with that kind of arrangement, it does show how flexible this program is.

Remote2/Host will run completely in the background, automatically accepting calls and notifying the operator when a remote caller has gained access. It will also operate in a manual mode where the host operator must manually answer phone calls. A third mode, called Restart, runs Remote2 in the foreground, completely taking over the computer and making it a dedicated host unable to perform any functions other than those requested by remote callers.

Remote2 is conspicuously missing some key ingredients that most other packages consider standard options. For example, there's no facility for switching from data to voice, or vice versa. The host module cannot be used as a traditional communications program because you can't dial out with it, except to call the remote module back. There is also no facility for "recording" a session for later playback, although Remote2 does keep track of the number of calls and elapsed time.

The call module of Remote2 will, however, act as a traditional communications program. It, in fact, carries one of the largest phone directories of the packages reviewed here—2,000 entries.

Table 2: Loading the file

	PAGEMAKER	MICROSOFT WORD	LOTUS 1-2-3
CARBON COPY PLUS, 5.0	3:50	3:05	3:01
CLOSE-UP, 3.00A	1:45	1:02	1:17
CO-SESSION, 3.1C	1:20	1:13	1:11
pcANYWHERE, 3.0	1:18	1:17	1:20
REMOTE2, 1.1	:59	:40	:42

THIS TEST WAS DONE BY CALLING THE APPLICATION, OR 'HOST,' MICROCOMPUTER FROM THE SUPPORT, OR 'REMOTE,' MICROCOMPUTER. THE TIME IT TOOK FOR THE APPLICATION PROGRAM TO LOAD IS QUOTED IN MINUTES:SECONDS. SEE 'RATING THE REMOTES' FOR HARDWARE CONFIGURATION. ALL DATA COMPILED AND VERIFIED BY KEN JOY (CANOGA PARK, CALIF.)

Senior Systems Analyst Bruce Ostrover, "but the agents who have PCs and remote-control software get quicker response time. They also get better help."

Before they had the Close-Up installations, says Ostrover, users had to send diskettes to the agents to solve programming problems. "Now when we determine there is a software problem, we can fix it right then via the remote package," he says. Approximately 650 of MONY's 4,000 insurance agents have Close-Up on their PCs, plus 2.4-kbit/s internal/external modems. To cut transmission time, Ostrover says, the files to be transmitted are compressed using public-domain software.

Flexibility
Last August, a review of several remote-control communications packages appeared in two on-line information services: Crosstalk Forum's data library and CompuServe's IBM Communications Forum. Edward L. Girou, a consultant with R. A. Kottmeier and Associates Inc. (St. Louis, Mo.), who wrote the review, believes that the latest development in this software genre is its support for more and more hardware configurations, a feature that varies somewhat among the vendor offerings (Table 2).

One example of this flexibility is the PC-to-Macintosh capability of pcAnywhere, Version 3, from Dynamic Microprocessor Associates Inc. (New York, N. Y.), which was updated to Version 3.1 in mid-February.

PC MacTerm works with pcAnywhere to let users run a PC from a Macintosh, either through a modem, via a direct cable connection, or over an AppleTalk network. But DMA is not just targeting the Mac-to-PC user; one of its large customers is National Data Corp. (Atlanta, Ga.), which has embedded pcAnywhere into its own products. One of these, custom-built for the United States Government, is called the US Treasury Input System.

"We're getting ready to install 200 copies of the US Treasury Input System," says John Forrester, director of microcomputer development at NDC. PCAwhere provides the capability for NDC to dial into stations running it and take control for technical support.

The US Treasury Input System package "records deposits that are made payable to the U. S. Government and transfers the automatic clearinghouse format for the Federal Reserve. The money is then transferred that same day," says Forrester. The software package is being installed, for example, at every U. S. military base that has a banking function, including those in South Korea, Japan, Cuba, and Belgium.

The role that pcAnywhere plays in the US Treasury Input System is in trouble-reporting. There is, in some cases, a 14-hour time-zone difference that NDC technical support personnel must work around. "Using pcAnywhere," says Forrester, "we can see the user's PC environment and watch what is going on as it's happening. Plus, if we had to send out software, we would have to ship it through customs. With pcAnywhere, we transfer it electronically."

Some of the calls for support that Forrester fields are as elementary as how to use the keyboard, what is a backslash, and how to move the cursor on the PC monitor. But Forrester says he can waste as much as three hours on a long-distance call explaining how to "do a directory or display a file." While pcAnywhere does not eliminate the long-distance call, it cuts the time considerably. Another advantage, says Forrester, "is that the customer can go have a coffee break while we do the troubleshooting. When we're done, we can 'warm boot' the machine. The U. S. Government isn't interested in teaching people to be PC literate. These customers only need to be bankers."

Mixing LAN and long-haul
Merging classical phone-line, or freestanding, remote computing software with LAN-based versions is the direction most vendors with products in this area are headed.

At the recent Networld show in Boston, for example, Norton Lambert announced its Close-Up/LAN. It is unique in that it offers one-to-many, rather than just one-to-one, connections. Because the program runs in the background (as a terminate-and-stay-resident module), it allows users to hot-key among the nodes on the LAN and among the applications on the nodes. Such multiple connectivity options, plus promised OS/2 capabilities, further broaden the scope of remote computing software.

In the interest of accommodating a wide range of users, Farallon Computing Software (Berkeley, Calif.) has entered the remote-control arena with a version that runs on Apple Macintoshes. Timbuktu comes in a LAN or a dial-up version (announced last summer), which uses the Xmodem file transfer protocol.

From the range of new offerings, it would seem that remote-control communications software is solving more and more problems for large and small companies. One potential user who has had evaluation packages in-house for several months remains skeptically hopeful.

"We run a support center," says Larry Perlstein, principle consultant/Apple technology at GE Consulting Services Corp. (Bridgeport, Conn.), "and we see a lot of potential in the concept of remote control from a support standpoint. But there are a few concerns: speed, security, and running through bridges, because we have a lot of bridges. I'd like to be able to sit down at the Mac and be able to control the IBM PC, or vice versa. That would be great." And apparently do-able this very minute. ■

Roger L. Koenig, Koenig Communications Consulting, San Jose, Calif.

How to make the PBX-to-ISDN connection

Here are your options for making the big move—from your current, proprietary digital network design to one based on the ISDN standards.

Growth produces change. Your company's sales organization has to move from its current suburban headquarters site to a new office building in the city. In the kickoff planning meeting, your CEO expresses grave concerns about the possible negative impact on revenues that might result from a lapse in communications between the sales organization and its customers. The sales VP makes it clear that he wants no degradation in voice or data services.

The new sales office will need a high level of connectivity with headquarters. From the standpoint of continuity and economy, incoming Direct Inward Dialing (DID) numbers, voice mailboxes, and attendant (private-operator) services should be provided centrally from the headquarters facility. Each salesperson's desktop PC must continue to have 19.2-kbit/s access from the new location to the mainframe's inventory database, price-quoting program, and electronic data interchange and electronic-mail applications.

A tough scenario to implement? Are leased lines to the new facility the only answer? And are they going to add significantly to your telecommunications expenses? What options do you have in equipment selection? Will you need to add another T1-multiplexer node at the new site? How is your maintenance staffing going to be affected?

In today's world of private digital networking, your options are rather limited. Figure 1 shows a current solution for networking the new sales office to headquarters.

To achieve integration of voice-calling features between the two facilities, you would invariably need to buy a new PBX from the same vendor as the one at headquarters. With luck, the PBX vendor's proprietary common-channel networking scheme should, over T1 lines, support the required centralized attendant service, DID integration, and voice-feature transparency (sharing the same features between headquarters and the remote site).

Providing voice-mail service to the sales office from the headquarters' voice processor could be tougher, however. The voice processor needs to receive forwarding-number identification and called-extension status (busy, unavailable, answer supervision) information from the sales office's PBX, through the network. The processor also needs to be able to activate message-waiting displays at the remote-site extensions.

Success in providing voice mail service from the headquarters location will depend on how well the PBX and voice processor share information over their proprietary data link. If a new voice processor is needed at the sales office, the two processors will exchange messages by dialing-up connections through the private network.

Switching and sharing
Asynchronous 19.2-kbit/s data connections from PCs to the headquarters mainframe may be supported economically with an X.25 packet assembler/disassembler (PAD), serving as a statistical multiplexer. The new sales-office PBX, with its proprietary digital telephones, switches data from each PC to the multiplexer channels.

The PBX's data-switching feature also provides for a sharing of local printers and file servers, in addition to concentrating data channels to the mainframe. Providing voice and data to each desktop over a single twisted-pair cable is inexpensive and easy to maintain compared to an overlay of local area networks.

A new T1 multiplexer is needed at the sales office to

1. Proprietary solution. *Here is a current digital solution for networking a new sales office to headquarters. Both PBXs should be from the same vendor. With luck, the PBX vendor's proprietary networking scheme, over T1 lines, will support the integration of the required voice features between the two sites.*

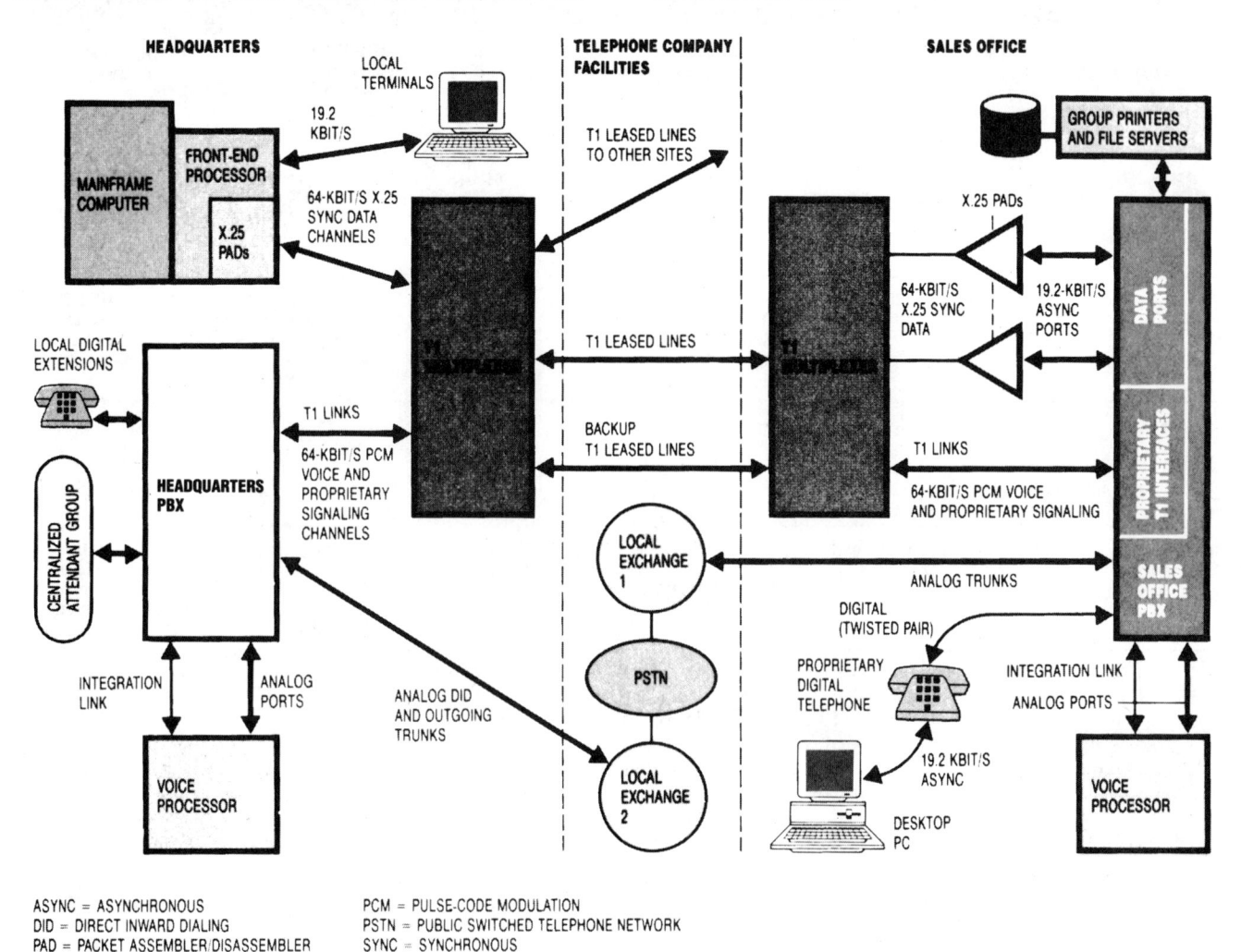

ASYNC = ASYNCHRONOUS
DID = DIRECT INWARD DIALING
PAD = PACKET ASSEMBLER/DISASSEMBLER
PCM = PULSE-CODE MODULATION
PSTN = PUBLIC SWITCHED TELEPHONE NETWORK
SYNC = SYNCHRONOUS

manage the leased T1 lines back to headquarters. Its functions are:
- To allocate DS-0 (64-kbit/s) channels for synchronous data connections between the PADs and the mainframe;
- To combine the PBX's voice and common-channel messaging channels with the data connections; and
- To provide alternative routing in case of T1-link failures.

Redundant leased T1 lines, with separate routing, are needed to ensure continued operation of the sales office in the event of a backhoe disaster.

Will the Integrated Services Digital Network (ISDN) standards make any difference in solving the aforesaid networking problem? Would standardized ISDN networking save money—either in equipment purchases, transmission costs, or in maintenance?

The answer to both questions is yes. ISDN networking promises multivendor-feature integrations, more networking options, and lower costs to users. To understand why, it is helpful to look at how private networks implemented with ISDN standards are different from the proprietary digital networks that are widely implemented now.

During the 1980s, a number of proprietary digital networking techniques were developed by PBX vendors. Most vendors with PBXs that support 500 lines or more now also offer what can be called a "digital networking option." This option uses T1 lines to carry voice, data, and signaling between the PBX nodes in a network.

The first generation of T1 connections to PBXs simply performed the functions of a D3 channel bank—essentially providing tie-line connections with 24 digital PCM (pulse-code modulation) channels. Emulating the analog tie lines that preceded them, these T1 interfaces used "on-hook, off-hook" signaling bits and in-band tones in each channel to establish connections and pass numbering information between networked PBXs.

Subsequently, however, the desire to implement a uniform set of PBX features across networks exceeded the capabilities of in-band signaling to carry this more complex

PBX common-channel T1 networking comparison

PBX	T1-BASED COMMON-CHANNEL NETWORKING	SIGNALING CHANNEL AND LINK ACCESS USED	T1 BEARER CHANNELS	MESSAGE SET AND PROTOCOL USED FOR PRIVATE DIGITAL NETWORKS
AT&T Definity System 75/85	Distributed Communications System (DCS)	A separate, proprietary, data link carries signaling between nodes. An HDLC link access protocol is used.	24 channels per T1 link.	AT&T's DCS uses a proprietary message set between System 75/85 nodes. Messages are transported with an implementation of the X.25 protocol.
Ericsson MD110	T1 links can connect distributed nodes (Line Interface Modules) in a star configuration.	One 64-kbit/s D channel per T1 link is used for signaling. The link access is proprietary.	22 channels per T1 link.	The MD110's distributed architecture uses a proprietary message set and protocol over the D channel established on each T1 or 2.048 Mbit/s connection link.
Fujitsu F9600	'ISDN Network Transparency' available in the first quarter of 1990.	One 64-kbit/s D channel per T1 link is used for signaling. ISDN LAPD access is implemented.	23 channels per T1 link.	Fujitsu's new ISDN networking will use CCITT Q.931 basic call-establishment messages and protocols. 'Supplementary services' messages are based on the Australian PTT extensions, which are similar to evolving ECMA recommendations.
Intecom IBX	InterExchange Link (IXL)	One 64-kbit/s signaling channel is used per T1 link. The link access protocol is proprietary.	InterExchange Links support 21 bearer channels per T1 link.	Intecom's IBX distributed architecture uses a proprietary message set and protocol between IBX nodes.
Mitel SX-2000	Mitel SuperSwitch Digital Network (MSDN)	One 64-kbit/s D channel per T1 link is used for signaling. HDLC link access is implemented.	23 channels per T1 link.	Mitel's MSDN is an implementation of the United Kingdom's Digital Private Network Signalling System (DPNSS). It conforms to British Telecom-recommended specifications. Mitel-specific transparent messages have been added.
NEC NEAX 2400	Common Channel Interoffice Signaling #7 (CCIS #7)	One 64-kbit/s signaling channel is used per T1 link. The link access protocol is proprietary.	23 channels per T1 link.	NEC's CCIS #7 uses a proprietary message set based on a subset of the CCITT recommendation for Signaling System Number 7.
Northern Telecom SL-1 SL-100	Electronic Switched Network (ESN)	One 64-kbit/s signaling channel is used per T1 link. The link access protocol used is LAPD.	23 or 24 channels per T1 link.	Northern's ESN uses a proprietary message set and protocol to network SL-1 and SL-100 nodes. The ESN message set may be carried over primary-rate interfaces, which conform to CCITT standards at Layers 1 and 2.
Rolm 9750	Rolm 9750 Business Communications System (multinode distributed architecture)	The proprietary Control Packet Network (CPN) can use T1 channels dynamically for signaling between nodes.	'Extended Digital Interties' carry up to 24 channels per T1 link.	Rolm's 9750 distributed architecture uses a proprietary message set and protocol for communications between nodes. Signaling messages share bandwidth with bearer channels over T1 or fiber optic links that connect nodes.
Siemens Saturn	Corporate Network (CorNet)	One 64-kbit/s D channel per T1 link is used for signaling. ISDN LAPD.	23 channels per T1 link.	CorNet uses CCITT Q.931-basic call-establishment messages and protocols. 'Supplementary services' messages are defined by Siemens, closely following evolving ECMA recommendations.

ECMA = European Computer Manufacturers Association
HDLC = High-Level Data Link Control
ISDN = Integrated Services Digital Network
LAPD = Link Access Procedure-D
PRI = Primary-Rate Interface
PTT = Postal, Telegraph, and Telephone
SS7 = Signaling System 7

information between network nodes.

Most current T1-networking options use some type of common-channel signaling implementation to carry digital messages between PBX nodes—either over a dedicated separate data connection or one that is shared with voice and data channels. Messages regarding the call-processing and management of a group of connected channels are generated by the software of each networked PBX.

For example, Mary Allen, in the new sales office, has an urgent matter to discuss with the payroll manager at headquarters. She calls, finds the manager's extension busy, and uses the PBX's camp-on (call back when free) feature. The following dialogue of messages might be exchanged over a common signaling channel between the two PBXs:
Sales Office PBX: Request call to extension 2254; extension 6120 is calling with class of service (calling privileges) level 11, text string: "Mary Allen."
Headquarters PBX: Extension 2254 is busy.

ISDN PBX NETWORKING DIRECTIONS	STATUS OF PBX-TO-EXCHANGE ISDN PRIMARY-RATE INTERFACE
AT&T is designing implementations of open ISDN PBX networking based on CCITT Q.931 and evolving ECMA recommendations for supplementary messages. DCS networking capabilities will be migrated to ISDN primary-rate. SS7 support on PBXs will depend on user demand.	Available since 12/87 on System 85. Current features supported are Automatic Number Identification (ANI) and Call-By-Call Selection. New features now in certification.
Ericsson's current implementation of the Digital Private Network Signalling System (DPNSS) in the U.K. will become part of an ISDN networking option in the U.S. during 1990. DPNSS features will be added to Q.931 primary-rate connection capabilities, available at the same time.	PRI-to-exchange connections demonstrated in Australia, Norway, and at AT&T trials. Primary-rate Interface availability in the U.S. is first quarter of 1990.
ISDN primary-rate networking, providing feature transparency between Fujitsu PBXs, will be available within the year on the F9600 and selected Omni products. Fujitsu will respond on an individual basis to requests for an open specification of their ISDN networking implementation.	A primary-rate exchange interface for the F9600 will be available in the third quarter of 1989. Availability for selected Omni PBXs is the first quarter of 1990.
Intecom will adopt ISDN networking standards as they become available. Networking standards based on CCITT Q.931 and Q.932 recommendations are preferred since these would provide a uniform interface for PBX-to-PBX and PBX-to-exchange applications.	A primary-rate exchange interface for the IBX is currently in development and will be available in mid-1990.
Mitel's MSDN currently offers common channel networking functionality equivalent to, and in many cases beyond, CCITT Q.931 and the extensions coming from the ECMA committee. Mitel is also developing Q.931-based SX-2000 networking for future compatibility.	Primary-rate exchange interface available in the U.K., implementing the British Digital Access Signalling System 2 (DASS2). Mitel has not announced an availability date for the U.S.
NEC will migrate its CCIS #7 proprietary networking to an international ISDN standard. This migration will be driven by 'cost-effective user benefits and public exchange capabilities.' The desire is to implement only one international networking standard based on CCITT recommendations.	A primary-rate exchange interface for the NEAX 2400 will be available in the U.S. between the fourth quarter of 1989 and the first quarter of 1990.
Northern Telecom has been a proponent of PBX networking based on a subset of the public exchange Signaling System 7, called the Remote Operation Service Element (ROSE). Northern's exchange-to-PBX ISDN networking and PBX-to-PBX networking are to be based on a common implementation.	Available since 11/88 for the SL-1. Supported features are ANI, Call-By-Call Selection, and NB+D. NB+D supports signaling for up to 16 primary-rate lines (383 B channels) on one D channel.
Rolm declined to comment for this survey. IBM's recent agreement to sell Rolm Systems to Siemens Corporation is expected to influence Rolm's directions in ISDN networking.	Rolm has not announced an availability date for a 9750 primary-rate exchange interface.
CorNet will be available on Saturn PBXs in the third quarter of 1989, and is currently shipping with the Siemens HiCom PBX in Europe. CorNet will continue to be upgraded to ECMA recommendations, leading to CCITT Q.932 and Q.933.	A primary-rate exchange interface for Saturn PBXs will become available at the same time as the CorNet interface: third quarter of 1989.

(Mary Allen hears busy tone from the local PBX and activates the camp-on feature.)
Sales Office PBX: Extension 6120 requests a camp-on to extension 2254.
Headquarters PBX: Camp-on confirmed.
(Mary Allen hears a confirmation tone and hangs up, waiting to be rung back when the payroll manager's line becomes free. The payroll manager receives a message on her displayphone: "Camp-on from Mary Allen, Ext. 6120.")

These messages are usually assembled into packets, along with other messages, and sent over a common data channel between the PBXs. A voice connection between the two locations is never needed during this signaling dialogue.

The table, "PBX common-channel T1 networking comparison," summarizes the common-channel networking implementations using T1-carrier facilities that are being provided by several leading PBX vendors. (The table is not all-inclusive.) Marketing managers from each of the PBX companies were interviewed to assemble the information shown.

Keeping it the same
Today's common-channel networking options offer a high degree of feature transparency between networked PBX nodes. Feature transparency refers to the ability of PBX users to have the same calling and data features across a network of PBXs as they have available at their local PBX. Mary Allen's ability to camp-on to a headquarters extension is an example of feature transparency.

Call-processing information, such as an extension's class of service, is sent across networks to allow remote-site users to share trunking facilities and other calling privileges. A centralized attendant service option is provided over common-channel networks by most PBX vendors. An exception is Northern Telecom's (NTI) SL-1, which currently provides the centralized-attendant feature only over conventional tie lines.

Most common-channel T1 networking in use today is in proprietary, closed implementations. Closed means that no published specification for interworking is available to users, network integrators, or other third-party vendors. PBXs in proprietary networks share most features only with PBXs from the same vendor.

The most widely implemented examples of proprietary T1 networking in the United States are AT&T's Distributed Communications System (DCS), NTI's Electronic Switched Network (ESN), and NEC's Common-Channel Interoffice Signaling Number 7 (CCIS #7).

Despite NEC's choice of names, its CCIS #7 uses a proprietary message set, which is based on a subset of the CCITT recommendation for Signaling System Number 7.

Richard Minthorne, director of product management for NEC America, says that there are no plans to publish an open specification for NEC's CCIS #7.

"This might change in the future if other vendors implemented SS #7 for networking," says Minthorne. NEC's long-term direction, he explains, is to migrate its CCIS #7 networking to the international ISDN standard. This migration will be driven primarily by central office ISDN capabilities. Minthorne stressed that "there need to be cost-effective user benefits to migrate private networks to ISDN."

NTI's ESN is now being offered over the company's SL-1 and SL-100 primary-rate ISDN interfaces. Using the first layer of CCITT ISDN standards, these interfaces structure the T1 bit stream into 24 DS-0 clear (full 64-kbit/s) channels.

In addition to standard ISDN 23 B+D channel use, an implementation option is to provide common-channel signaling for managing up to 16 primary-rate lines (383 B channels — 24 × 16, minus the one D channel) over a single D channel. This N × B + D arrangement provides further bandwidth conservation by using the one D channel for all the channels' signaling.

Proprietary protocol
Consistent with ISDN standards for the data link (Layer 2) of the OSI (Open Systems Interconnection) reference model, Link Access Procedure-D is used to convey messages over common D channels. Still, the actual protocol and structure of messages (corresponding to the upper levels of the OSI reference model) for NTI's ESN remain proprietary to NTI, however. This currently precludes the possibility of interworking through a direct primary-rate connection with an ESN network by any other vendor or service provider.

On the question of networking openness, Bob Hoffman, Product Manager for ISDN Network Services at NTI, responded that the company has already published its first specifications for exchange-to-PBX ISDN primary-rate connections.

"Northern's is a full-product family across public and private networks. I see exchange-to-PBX and PBX-to-PBX ISDN networking as being the same. We need to accommodate customer needs and will continue to talk with other vendors to arrive at common standards," Hoffman says.

AT&T's DCS has been implemented in networks across the United States for roughly six years now. DCS messages are exchanged among System 75, System 85, and Definity nodes over a virtual data circuit using implementations of high-level data link control and X.25 protocols. With a separate common-channel data circuit, all 24 channels of connected T1 links may be used as bearers for voice or data traffic.

Dick Davis, Supervisor of Systems Planning For Definity Generic 2 at Bell Laboratories, says that DCS networking will remain proprietary to AT&T.

"The direction is to create an open [networking] specification with ISDN standards. DCS will retain its current networking features, with new features to be introduced as needed. We intend this to be a smooth transition [to ISDN networks], compatible with the DCS installed base," says Davis.

AT&T development efforts are focused squarely on networking its PBXs with an implementation of the CCITT ISDN recommendations Q.931 and Q.932 (see "When will ISDN networking come together?"). A vendor executive who is participating in AT&T's ISDN trials speculated that we may see a product announcement for ISDN networking of Definity in six to 12 months.

"Networking features such as calling- and connected-party identification are available now [on Definity primary-rate connections]," says Bell Labs' Dick Davis. "It may take two to three years before all of the DCS-enhanced networking features are available on ISDN. In the interim period, we will see hybrid DCS and ISDN networks, with interworking of basic features."

AT&T's recent Definity PBX announcement included primary-rate features for AT&T virtual private network support and load sharing in automatic call distribution applications. N × B + D capabilities were also announced, providing signaling control for up to 20 primary-rate interfaces (479 B channels — 20 × 24, minus one D channel) over a single D channel.

Ericsson's MD110, Intecom's IBX, and IBM/Rolm's 9750 PBXs are designed with what are loosely defined as distributed architectures.

Distributed PBX architectures received a lot of attention in the early 1980s. It was during that period that so-called fourth-generation PBXs were introduced, typified by the CXC Rose, Ztel's PNX, Intecom's IBX, and Rolm's 9000 CBX. Distributed architectures offer the following advantages for network implementers:
- Absolute feature transparency between nodes;
- The ability to share large amounts of bandwidth (connection channels) between nodes; and
- Centralized management and programming.

Similarities predominate
Differences between the capabilities of distributed and networked architectures have blurred in recent years. Both, in fact, use similar principles to exchange call-processing messages between nodes over a common data channel. Distributed software architectures communicate information on network activity to each node on a continual basis, increasing — compared with the networked case — message content and frequency. On the other hand, networked architectures pass information between nodes as the result of an event or query, sending messages to only those nodes involved in a particular transaction.

Comparing feature transparency of networked and distributed architectures, however, is becoming a nonissue. One has to look very hard to find usable features that a networking approach cannot implement.

The main advantage that remains with distributed architectures is their ability to efficiently share large amounts of connection bandwidth over coax or optical-fiber links. High-capacity links are only really useful where "dark fiber" (unmultiplexed optical-fiber cables) can be economically installed, such as within building complexes, in campus environments, or along utility company rights-of-way.

Mark Housley, product marketing manager at Rolm, explains that interconnecting distributed 9750 nodes with

T1 links is an option for customers who are not able to install optical fiber.

"The preferred method of connection is to install internode links, which support 365 channels each via optical fiber, versus 24 channels provided over Extended Digital Interties." Extended Digital Intertie is the name given to Rolm's proprietary T1 interconnection product.

Nodes in a Rolm 9750 installation communicate signaling and management messages over a virtual data circuit, called the Control Packet Network (CPN). CPN uses available connection channels of either internode links or Extended Digital Interties to transport message packets between nodes. The implementation—similar to that of other vendors—is entirely proprietary to Rolm.

Marketing managers at Rolm declined to comment on the anticipated directions for ISDN networking of the 9750. Sources close to the company indicate that after the effects of IBM's sale of Rolm's development and manufacturing facilities to Siemens are more fully assimilated, Rolm is expected to begin development of ISDN networking soft-

When will ISDN networking come together?

When asked about networking with ISDN Centrex, a research manager at one regional Bell operating company (RBOC) comments: "We're still trying to determine the direction internally."

ISDN Centrex (containing PBX-like features), with the ability to integrate with private PBX networks, may still take several years to evolve. A significant problem is that there is currently a lack of ISDN networking standards in ISDN protocol Layers 3 and above. And, unfortunately, arriving at a common standard is hampered by political and commercial interests.

ISDN standards are constructed using the Open Systems Interconnection seven-layer reference model:
- Layer 1 (physical) specifications for primary-rate (DS-1) connections are in CCITT recommendation I.431.
- Layer 2 (data link) protocols for the D channel (called Link Access Procedure-D) are contained in Q.920 and Q.921, which was ratified at the November 1988 CCITT plenary session.

These first two layers are considered stable and have been widely implemented in primary-rate interface chip sets and controller firmware. The trouble begins at Layer 3, the network layer.

CCITT recommendation Q.931 ("ISDN User-Network Interface Layer 3 Specification") covers messages and protocols for basic call set-up and supervision over primary-rate ISDN interfaces. This is the Layer 3 specification that has been implemented to make exchange-to-PBX primary-rate connections in the United States.

Q.931 does not cover the voice and data features that are expected of modern PBX networks, however. Even rudimentary features, such as call transfer, forwarding, and call waiting, are not covered by Q.931.

These so-called supplementary services are the specification's void that both standards committees and manufacturers are attempting to fill. The definition of these supplementary services will eventually be incorporated into CCITT recommendation Q.932, called "Generic Procedures for the Control of ISDN Supplementary Services." Q.932 will be submitted for ratification at the 1992 CCITT plenary session.
- **ECMA and ANSI.** There are two main standards committees at work on the upper layers of ISDN primary-rate networking. The European Computer Manufacturers Association (ECMA) has become the focal point of ISDN networking-standards creation in Europe.

ECMA committee TC32/TG6, which is working on a definition of supplementary services, represents a variety of interests. These interests include those of AT&T, IBM, Northern Telecom (NTI), and Siemens.

The American National Standards Institute's (ANSI) T1S1 is a subcommittee that has grown out of the original T1 committee. The latter was formed five years ago by the Exchange Carriers Standards Association. T1S1 is the U. S. focal point of ISDN networking standards.

A member of the ANSI T1S1 committee explains: "T1S1 is composed of four major groups: regional operating companies, interexchange carriers, manufacturers, and users. Most of the user representation is from government organizations."

Doug Spenser, Data Interfaces Supervisor at Bell Labs, has been a member of ANSI T1S1, ECMA TC32/TG6, and CCITT study groups relating to ISDN networking standards. He describes how the ANSI and ECMA committees go through three stages in the development of a recommendation.

"Stage 1 is the determination of architectures and topologies," he says, "Stage 2 is 'What do you want the protocol to do?' At Stage 3, you actually decide on the message set and protocol. The ANSI and ECMA committees are only at Stage 1."

Another member of T1S1 comments: "The committee is driven in two directions. One direction is with the RBOCs and NTI, who support networking based on Signaling System Number 7 [SS7]. The other direction is set by a group of manufacturers who support networking with extensions to Q.931, the same as the ECMA [TC32/TG6] committee. Bellcore's [Bell Communications Research's] technical recommendations [based on SS7] are in conflict with ECMA recommendations.

"T1S1 is mostly dominated by the interests of the RBOCs. They want an ISDN networking standard that will facilitate their implementation of virtual private networks."
- **Who's ahead?** The consensus of the committee members interviewed was that ECMA TC32/TG6 is farther along in the definition of ISDN networking standards than is T1S1. There is a faction in the ANSI committee (composed of several companies also represented in ECMA) that inputs TC32/TG6 recommendations to T1S1. IBM is one of the supporters of ECMA recommendations

ware based on Siemens's set of CorNet standards. Rolm has already demonstrated ISDN primary-rate interfaces in Europe on the now-withdrawn 8750 PBX.

Ericsson's MD110 is also promoted as a distributed-architecture PBX. Since the MD110's internal switching is built on a European 30 B+D (2.048 Mbit/s, instead of 1.544 Mbit/s) framing structure, the preferred method of connecting MD110 nodes in North America is through media that support 2.048-Mbit/s transmissions.

Optical-fiber, microwave, or copper links are frequently installed to establish groups of 2.048-Mbit/s connections between MD110 nodes. When standard T1 lines (at 1.544 Mbit/s) are used, eight channels per link are restricted by converters, yielding 22 B+D connections (one B channel is used for synchronization). And, again, the protocol for signaling and messages sent on the D channel is proprietary to Ericsson.

Ericsson has been an active participant in international ISDN development. Primary rate exchange-to-PBX connections on the MD110 have been demonstrated in Australia, in the United States. The ECMA committee recommendations are widely expected to become CCITT recommendations in 1992.

Simon Wilders, ISDN product manager at StrataCom (Campbell, Calif.), was a representative to the ECMA TC32/TG6 committee through 1988. He estimates that the networking features of Q.932 will be fairly well defined by the committee by the end of this year.

When asked about the direction of standards activities, Bob Hoffman, product manager for ISDN Network Services at NTI, comments: "It's going to take quite a while for a standard to shake out of the ANSI committee." Hoffman explains that a subset of TCAP [see below], called the Remote Operation Service Element (ROSE), has been defined to implement call-associated messaging for ISDN networking to customer premises.

TCAP stands for Transaction Capabilities Application Part. It is an applications protocol that forms part of the North American public network implementation of CCITT SS7. Current public-network TCAP applications involve database inquiries for credit card calling and 800-number translations (see "Knocking on users' doors: Signaling System 7," DATA COMMUNICATIONS, February, p. 147).

Building on the existing TCAP protocol, applications may be created with SS7 to encompass ISDN networking features, such as:
- Call-forwarding management
- Camp-on to extensions
- Virtual private network management
- Centralized attendant services.

A message set and protocol based on TCAP provides an alternative to ECMA's definition of the supplementary services currently lacking in ISDN-user networking standards. This is the definition desired by many of the RBOCs. Implementation of ISDN features on the public network would be easier if customer premises equipment was able to communicate with public exchanges. This implementation could be facilitated via a subset of the language that the public network is already growing to understand: TCAP.

Bellcore and NTI have been working on ROSE (a TCAP subset), which would satisfy the needs for supplementary services in both PBX-to-PBX and PBX-to-exchange networking applications.

With ROSE as an upper-layer standard to network PBXs, the RBOCs could offer users a smooth migration from PBX-based private networks to virtual private networks and ISDN Centrex. SS7-based supplementary-service definitions bring to the RBOCs less time and a lower cost to implement ISDN networking services.

Development of ROSE-based networking software appears to be well under way. "The Northern SL-1 and DMS-100 have been linked with a ROSE network. Networking based on ROSE is now being discussed with the RBOCs," says NTI's Hoffman.

Bell Labs' Spenser comments: "The RBOCs could offer virtual private networks with either Signaling System 7 or Q.931-based [ECMA] protocols. Q.931 would take more work, but Q.931 and SS7 can be made to interwork. The part of the Q.931 message which can't be handled locally can be sent over SS7."

■ **'Management and maintenance.'** Jim Neigh, head of the Systems Interworking Department at Bell Labs, believes that the ANSI T1S1 committee will recommend two standards based on Q.931 and on SS7.

"Signaling System 7 was designed for the public network. It contains a lot of management and maintenance overhead that customer premises equipment can't make use of," says Neigh, "People are saying that a skinny SS7 could be done, but you need Q.931 for [primary-rate] connections to the exchange anyway. Q.931-based [ECMA] standards have more advantages for private and virtual private networking. AT&T will only support SS7 [PBX networking] if the demand materializes."

Neigh says that AT&T is working on its own implementations of Q.932, staying close to the ECMA standards. "AT&T will publish an open specification for its implementation and evolve its products to be compatible with future standards," he notes, adding: "If you wait for the ANSI standards committee [T1S1], it will be 1992 or '93 before we see products. Vendors can't afford to wait."

Indeed, most PBX vendors are not waiting. First implementations of Q.931-based networks, with vendor-specific extensions, are becoming available this year and in 1990. Neigh says that the implementation of open multivendor ISDN PBX networks in the United States will depend on customer demand. "If a sufficiently large RFP [request for proposal] requiring ISDN interworking comes forward, then we could see a multivendor network in a couple of years. If not, it may be several years."

Norway, and in AT&T Communications trials. The British Telecom pre-ISDN networking standard, Digital Private Network Signalling System (DPNSS), is currently available on the MD110 in the United Kingdom (see "The British DPNSS experience").

Larry Minzey, Ericsson product manager, says that DPNSS will become a T1 networking option on the MD110 in the United States during the first quarter of 1990.

"Basically, the top 10 DPNSS features will be supported, including centralized attendant service," says Minzey. "DPNSS network features will complement Q.931-based MD110-to-exchange and MD110-to-MD110 primary-rate ISDN capabilities, which will be available at the same time."

Only one primary-rate networking technique has so far been used to implement multivendor PBX networks: British Telecom's DPNSS. As a subsidiary of British Telecom, and an active vendor in the U. K. marketplace, Mitel of course offers DPNSS on its SX-2000 PBX. Mitel also offers DPNSS networking in the United States over T1 links between SX-2000 PBXs; it is called the Mitel SuperSwitch Digital Network (MSDN). MSDN is reported to have more than 60 network installations in the United States and Canada.

The British DPNSS experience

In the early 1980s, British Telecom (BT) organized a committee of PBX manufacturers (Plessey, GEC, and Mitel) to work on a common channel signaling standard for PBX digital networking. The committee developed a PBX interconnection standard designed to provide a set of networking features for both voice and data calls. It named its handiwork the Digital Private Network Signaling System, or DPNSS.

Having no accepted ISDN networking standards at the time, the committee built on telecommunications standards that did exist. With the International Organization for Standardization's (ISO) seven-layer Open Systems Interconnection model in hand, the DPNSS was constructed as follows:

- Layer 1: 2.048-Mbit/s digital links with the same channel and framing structure as that used for 30-channel pulse-code modulation. This is defined under CCITT specifications G.703 and G.732.
- Layer 2: High-level data link control, the link access protocol defined under ISO specification 4335.
- Layers 3 through 7: The DPNSS signaling set and compelled protocol (acknowledgment required for each message transmitted) was defined (contained in BT Network Requirements 188) to perform network routing, end-to-end transport, presentation of features, and the exchange of network management information.

The committee's intention was to provide an extensive set of PBX features with DPNSS, but still leave manufacturers with the ability to develop and implement unique networking capabilities. Although DPNSS defines supplementary services for relatively sophisticated applications (such as centralized attendant service and text messaging), manufacturers are still able to implement networking features of their own through "nonspecified information" strings.

DPNSS private networks were first implemented in early 1986. Acceptance since then has been swift. Of 11 U. K. private network managers interviewed in 1988, eight were using DPNSS in their digital networks. Frequently cited reasons for choosing DPNSS were multivendor-PBX feature integration and the ability to use future ISDN offerings from British Telecom to incorporate off-network locations into virtual private networks.

Virtually all RFPs (requests for proposals) for networked PBXs in the United Kingdom now require DPNSS capabilities. DPNSS network users include major government, financial, manufacturing, and transportation networks. For example, the Central Computer and Telecommunications Agency (CCTA) runs the largest DPNSS network in the country for Her Majesty's government. It links 60 PBXs serving over 35,000 extensions in the Westminster area of London.

DPNSS is also being deployed in the broader CCTA network of 170,000 government extensions. CCTA has been the proving ground for DPNSS feature integration between vendors. PBXs from Ericsson, GEC, Plessey, and Mitel are all now a part of the government's DPNSS network.

Plessey, the U. K. market leader, has also been a leader in the implementation of DPNSS on its ISDX PBXs, providing "the feature of the month," as a former Plessey engineer described it. Mitel and GEC are not far behind. Ericsson has only recently entered the DPNSS networking market.

British Telecom's promotion of the DPNSS standard is not totally altruistic. They intend to gain by wooing users from private networks and onto virtual private networks and ISDN Centrex, providing value-added revenue to BT.

To accomplish this, the Digital Access Signaling System 2 (DASS2) specification was created (in 1983) to be a compatible subset of DPNSS. British Telecom began installing DASS2 primary-rate (2.048 Mbit/s) connections between customer PBXs and public exchanges last year.

DPNSS and DASS2 are often termed "pre-ISDN" or "U. K. ISDN" standards. They diverge from CCITT Q.921- and Q.931-specified ISDN primary-rate networking standards at Layer 2 and higher. But with the large installed base of DPNSS networks in the United Kingdom, and BT's current roll-out of DASS2 connections to the public network, it is likely that the island nation will also remain an ISDN island for some time to come.

Indeed, this was found to be a major concern among multinational network managers interviewed in the United Kingdom. With the European continent building ISDN private networks based on Q.931, gateway functions will be needed to connect U. K. sites with those on the continent. Gateways will surely lead to extra complications, feature limitations, and expense for users.

Leo Lax, assistant vice president for ISDN marketing at Mitel, explains that MSDN uses the "transparent capabilities" of DPNSS to exchange Mitel-specific messages (such as text strings for Mitel's on-line telephone directory).

"MSDN is well featured. Users are not satisfied with fewer features [over a network] than they have at a local node," says Lax.

Based on Q.931
Mitel has produced a feature chart that favorably compares MSDN telephony features with those specified by CCITT standards Q.931 (with extensions) and Signaling System #7. Lax adds that Mitel is developing networking capabilities based on Q.931, "since you need to do Q.931 for public exchange connections anyway." He anticipates that the CCITT will adopt "supplementary service" features, similar to those currently used in DPNSS, when the Q.932 specification is fully developed.

Migrating from DPNSS networking to CCITT-standard networking is not seen as particularly difficult. "In addition to differences in the actual format of the messages, DPNSS is a compelled protocol [requiring an acknowledgment for each message transmitted], while Q.931 is not. This will need to be accommodated [by software changes]," Lax notes.

Fujitsu's flagship PBX, the F9600, is billed as "a ground-up ISDN switch." Mike Albers, director of product planning for Fujitsu Business Communications Systems, says general availability of primary-rate interface capabilities is anticipated for late 1989 and early 1990.

Currently dubbed simply ISDN Network Transparency, Fujitsu's PBX-to-PBX networking feature is based on CCITT Q.931. The supplementary service messages (which are needed to implement the majority of PBX features) are designed from Australian PTT (Postal, Telegraph, and Telephone agency) specifications, according to Albers.

Although similar to evolving European Computer Manufacturers Association recommendations for Q.932, the Australian extensions "won't work with other U. S. [networking] implementations," Albers says, explaining that Fujitsu anticipates evolving its ISDN networking implementations as further standards become available.

Will Fujitsu publish an open specification of its ISDN networking implementations? "Fujitsu would respond on an individual basis to requests for a primary-rate networking spec. There are no plans to publish an open specification [as Siemens has done for its CorNet], however," Albers says.

Siemens has taken a bold lead in promulgating a worldwide ISDN PBX networking standard called CorNet (for Corporate Network). Originally created from the German PTT specification, TR6, CorNet has been installed on over 150 Siemens Hicom PBX nodes in Europe. The 700-page CorNet specification was first introduced in February 1988 and is available from Siemens for a small fee.

Scott Augerson, director of product management at Siemens Information Systems, says "CorNet will be available on the Saturn product line in the U. S. during the third quarter of 1989. Primary rate exchange-to-PBX interfacing [Automatic Number Identification and Call-By-Call Selection] will also become available at the same time. Ninety voice and data features will be supported over CorNet in the first release."

Augerson explains that CorNet is a "living" specification. "CorNet implements CCITT Q.931 now, and will be upgraded as Q.932 and Q.933 become available. Siemens has taken its best guess at the Q.932 supplementary service messages. If the messages turn out to be different, we will change CorNet when Q.932 is released," he says.

Augerson is not aware of any multivendor demonstrations being planned for CorNet. He says that the National ISDN Users Group will be soliciting vendor participation for an ISDN demonstration event in the fall of 1990, however.

To date, the majority of PBX vendors have requested copies of CorNet, according to Augerson. None of the vendors in this survey, however, said that they were considering incorporating CorNet compatibility into their products. The most likely U. S. implementer, among large PBX suppliers, is the Rolm Systems division.

What is being gained?
Why are so many PBX vendors putting so much development effort into networking with ISDN standards? What will ISDN networking do that the proprietary implementations don't do now? Consider again the earlier networking problem—connecting the new sales office to headquarters. Figure 2 shows a networking solution that implements ISDN standards.

T1 leased lines still provide the most economical traffic medium between headquarters and the sales office (you pay fixed monthly lease charges rather than measured rates per call). Back-up lines have been replaced by primary-rate interfaces to the local exchange. If connections on the leased lines are lost, all voice and data traffic can be immediately routed over primary-rate connections back to headquarters.

Providing back-up does not mean that the public exchange must understand PBX-feature messages. The sales office PBX (or headquarters T1 multiplexer) simply requests 64-kbit/s clear-channel connections, as needed, from the public network.

This is done with the local D-channel connection to the exchange. Supplementary service messages, which are used to activate PBX features, may be carried as user data over the public network D-channel. This type of network back-up is a near-term solution. Only rudimentary ("pipeline") ISDN transport capabilities are required of the public exchange network. You don't need to wait for your local operating company to offer ISDN features that are on a par with those provided by your PBXs.

In addition to providing back-up, the primary-rate interfaces have consolidated the local exchange connections, replacing a bank of analog trunks. This should make fault isolation and maintenance easier. Having the primary-rate interfaces in place also provides an inexpensive way to handle peak traffic periods—simultaneous with existing T1 lines—between headquarters and the sales office without adding T1 lines. Over all, the transmission economies have been improved.

2. Private ISDN solution. *In this networking solution, which implements ISDN standards, T1 leased lines still provide the most economical traffic medium between headquarters and the sales office. If connections on the leased lines are lost, all voice and data traffic can now be routed over primary-rate connections back to headquarters.*

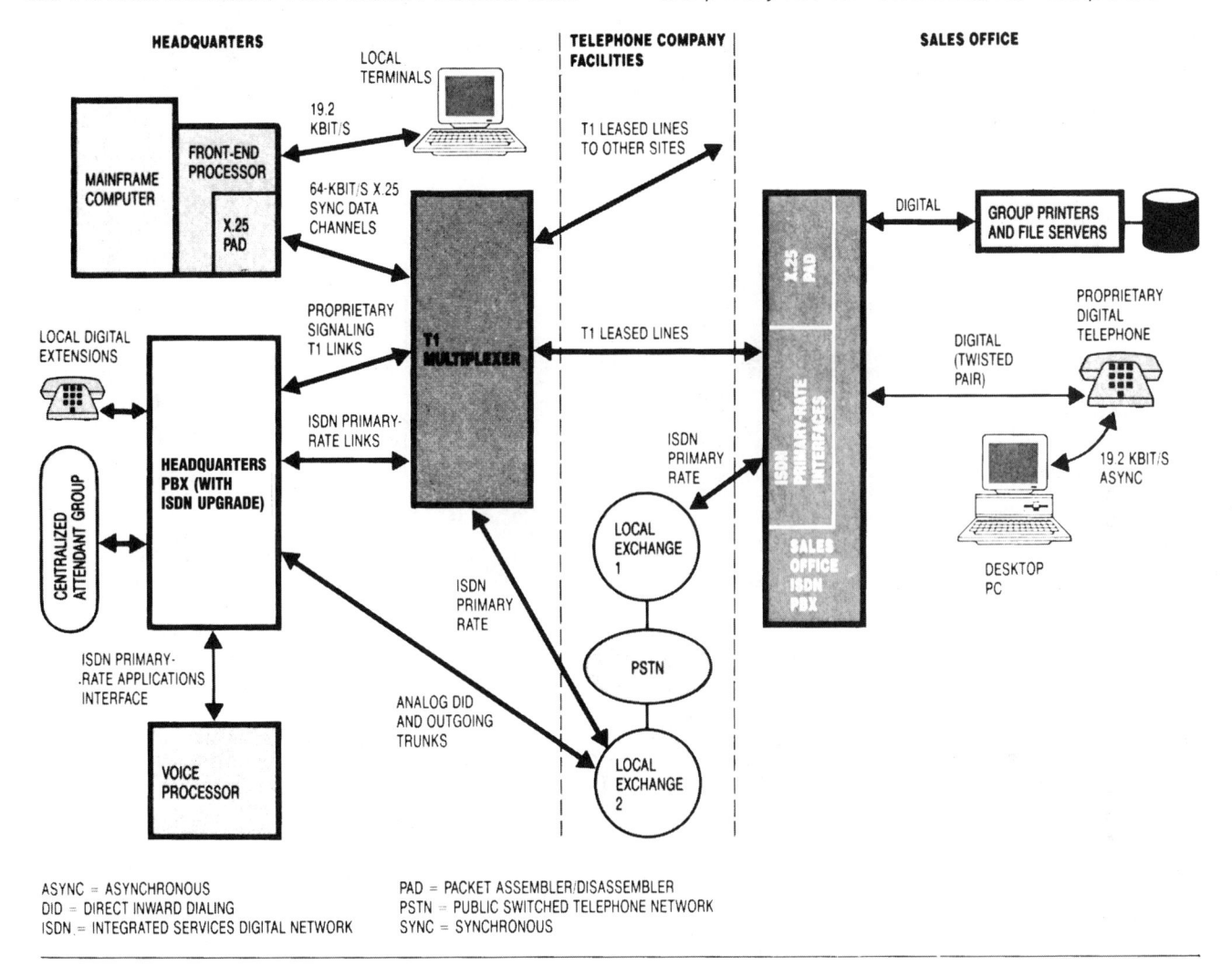

ASYNC = ASYNCHRONOUS
DID = DIRECT INWARD DIALING
ISDN = INTEGRATED SERVICES DIGITAL NETWORK
PAD = PACKET ASSEMBLER/DISASSEMBLER
PSTN = PUBLIC SWITCHED TELEPHONE NETWORK
SYNC = SYNCHRONOUS

The most dramatic difference between the proprietary and ISDN networking solutions is the reduction in equipment needed at the new sales office. For example, you don't need to buy an extra voice processor. With forwarding-number identification, message waiting, and called-extension status messages available in a standard ISDN protocol, the headquarters voice processor can provide a full set of voice-mail features across the network. To accomplish this, the voice processor will need to be upgraded to enable the connection of a primary-rate applications interface to the headquarters PBX.

And there is also no longer a need for an end-node T1 multiplexer at the sales office. Using standard D-channel messages, both the T1 multiplexer at headquarters and the ISDN PBX in the sales office can monitor link performance, enabling connection rerouting from either site.

Now that the PBX can manage the entire data connection back to headquarters, the next logical step is to provide the X.25 PAD function internally. This eliminates the bank of PBX data ports (Fig. 1) that were needed to route data through an external (X.25) multiplexer.

Statistically multiplexing the 19.2-kbit/s channels into 64-kbit/s X.25 packet data provides a much more efficient use of the primary rate and T1 DS-0 channels than subrate speed adaption (padding each 19.2 kbit/s up to 64 kbit/s). Statistical multiplexing is even more efficient than subrate multiplexing (putting three 19.2-kbit/s connections into one 64-kbit/s channel).

With ISDN networking, the PBX has assumed many of a T1 multiplexer's roles at a network end. For the management of private network backbone traffic, the functions of the T1 multiplexer are still very much needed, however.

Using open ISDN networking standards, you will also have a choice of vendors from which to purchase the sales office PBX. You can pick the most cost-effective product without worrying about being locked into a vendor's proprietary networking scheme.

Yet another networking solution is looming on the ISDN

3. ISDN Centrex/private net. Here, a network connecting the sales office to headquarters uses ISDN Centrex. The only added telecommunications equipment at the sales office are ISDN basic-rate telephones. Data connections from the desktop PCs are plugged into the telephones. And the local exchange provides an X.25 PAD facility.

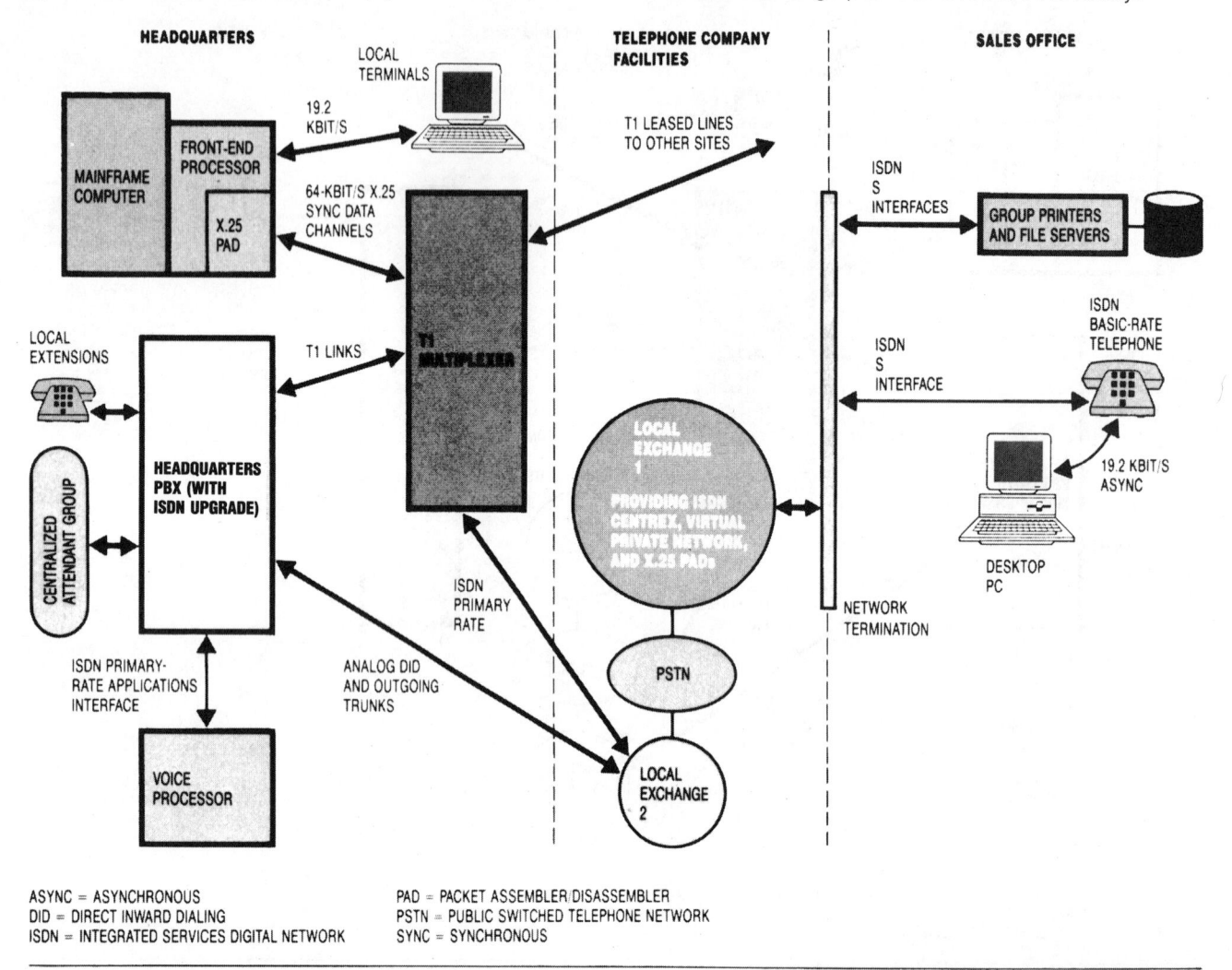

ASYNC = ASYNCHRONOUS
DID = DIRECT INWARD DIALING
ISDN = INTEGRATED SERVICES DIGITAL NETWORK
PAD = PACKET ASSEMBLER/DISASSEMBLER
PSTN = PUBLIC SWITCHED TELEPHONE NETWORK
SYNC = SYNCHRONOUS

horizon, however. Figure 3 shows a network connecting the sales office to headquarters using an ISDN Centrex service. In this implementation, the only telecommunications equipment needed for the sales office are ISDN basic-rate telephones.

The ISDN Centrex alternative

Data connections from the desktop PCs are plugged directly into the back of the telephones. The local exchange provides an X.25 PAD facility for an efficient concentration of data into primary-rate channels. And a numbering plan can be established that gives the sales office a virtual private network to headquarters. There are no leased lines to the new location.

Unlike the ISDN PBX solution, private network integration with ISDN Centrex requires that the local exchange be able to understand, and generate, a full set of PBX feature messages. For your company to achieve the feature transparency and centralization of resources desired, the local exchange's operating company must offer ISDN capabilities equivalent to your PBXs. This will obviously take longer to evolve than the simple ISDN transport and call establishment abilities required of the operating company in Figure 2.

Providing a full set of ISDN features and solutions is definitely a top-agenda item for the Bell operating companies, just as it is with foreign PTTs. ISDN Centrex also provides new opportunities to private network implementers. For example, by using ISDN standards internally, private networks will be able to make use of ISDN Centrex and virtual private networking as a cost-effective means of serving off-network locations. ■

Roger Koenig is an independent consultant providing market-research, technical-specification, and international development planning services to manufacturers of telecommunications products. He previously spent nine years with Rolm, involved with PBX design and development. He holds an M.S. in Engineering Management from Stanford and a B.S.E.E. from Michigan State.

U.S. Companies Involved In ISDN Trials And Service Rollouts

Company	Switch	Access	Carrier	Status	Note
Aetna, Hartford, Conn.	AT&T 5ESS	Basic	Southern New England Telephone	Installation underway	
Alverno College, Milwaukee	Siemens EWSD	Basic	Wisconsin Bell	Started March 1988, ends 1989	Trial.
American Express, Phoenix, Ariz.	AT&T System 85 PBX/4ESS	Primary	AT&T	Started July 1988	First customer of AT&T's Primary Rate.
American Transtech, Jacksonville, Fla.	AT&T System 85 PBX/4ESS	Primary	AT&T	Started Dec. 1987	Beta test site for AT&T's Primary Rate.
Arizona, State of, Phoenix	Northern Telecom DMS-100	Basic	US West Communications	Started Nov. 1986, officially ended	
Boeing Co., Seattle	AT&T 5ESS	Basic	US West Communications	Scheduled start Nov. 1988	
Carnegie Mellon University, Pittsburgh	AT&T 5ESS	Basic		Contract pending	
Chevron Corp., San Francisco	Northern Telecom SL-100/DMS-100	Primary/Basic	Pacific Bell	Installation underway	
Contel Corp., Atlanta	AT&T 5ESS	Basic	Southern Bell	Started April 1988	Paying customer.
Control Data Corp., Minneapolis	NEC NEAX 61E	Basic	US West Communications	Started Nov. 1987, ends Nov. 1988	Trial.
Duke University, Durham, N.C.	AT&T 5ESS	Basic	Southern Bell		
Eastman Kodak Co., Rochester, N.Y.	Northern Telecom SL-100 PBXs	Primary		Started Aug. 1988	First Primary Rate using two SL-100s.
Federal National Mortgage Assoc., Washington	AT&T 5ESS	Basic	C&P Telephone	Started June 1988	
First Data Resources Inc. (American Express subsidiary), Omaha, Neb.	Northern Telecom SL-1 PBX and AT&T 4ESS	Primary	AT&T	Unannounced	
Glaxo Inc., Research Triangle Park, N.C.	SL-1 PBX and DMS-100	Primary/Basic	GTE South	Started June 1988	First Primary Rate/Basic Rate in one trial.
Hardees, Rocky Mount, N.C.	Northern Telecom DMS-100	Basic	Carolina Telephone	Scheduled start Jan. 1989	
Hayes Microcomputer Products Inc., Norcross, Ga.	AT&T 5ESS / AT&T 5ESS	Basic / Basic	Southern Bell / Pacific Bell	Started April 1988 / Sept. 1987 to Sept. 1988	Using ISDN to develop ISDN products.
Hershey Foods Corp., Hershey, Pa.	AT&T 5ESS	Basic	Contel of Pennsylvania	Scheduled start Oct. 1988	Will include ISDN satellite transmission.
Honeywell Information Systems (Honeywell Bull), Minneapolis	Northern Telecom DMS-100	Basic	US West Communications	Started Jan. 87, officially ended	Applications included data/voice transmission between office and employees at home.
Intel Corp., Chandler, Ariz.	AT&T 5ESS	Basic	US West Communications	Started Feb. 1987, ended Aug. 1987	Trial.
Johns Hopkins Medical Center, Baltimore	AT&T 5ESS	Basic	C&P of Maryland	Contract pending	
Lawrence Livermore Laboratory (University of California), Livermore, Calif.	AT&T 5ESS	Basic	AT&T Federal Systems		
Lockheed Missiles and Space Co. Inc., Sunnyvale, Calif.	AT&T 5ESS	Basic	Pacific Bell	Started Sept. 1987, ended Sept. 1988	Trial.
Mass. Institute of Tech., Cambridge	AT&T 5ESS	Basic		Scheduled cutover Oct. 1988	Using AT&T 5ESS as PBX for private network.
Mather Air Force Base, Sacramento, Ca.	AT&T 5ESS	Basic	AT&T Federal Systems	Started Aug. 1988	Model for ISDN deployment at 50 bases.
McDonald's Corp., Oakbrook, Ill.	AT&T 5ESS	Basic	Illinois Bell	Started Dec. 1986	
McDonnell Douglas Corp., St. Louis		Primary	AT&T	Unannounced	
Microcom Inc., Norwood, Mass.		Basic	New England Telephone	Installation underway	Part of centrex contract.
Motorola Inc., Schaumburg, Ill.	Northern Telecom DMS-100	Basic	Illinois Bell	Planning stage	
NASA, Washington	AT&T 5ESS	Basic	AT&T Federal Systems		
Nice Corp., Ogden, Utah.	Northern Telecom SL-1s	Primary			Telemarketing company using private ISDN.
North Carolina State Univ., Raleigh	Northern Telecom DMS-100	Basic	Southern Bell		
Northeast Utilities, Hartford, Ct.	Northern Telecom SL-1s	Primary			PBXs in Rocky Hill Ct. and Meriden, Ct.
Pennsylvania, State of, Harrisburg	Northern Telecom DMS-100	Basic	Bell of Pennsylvania	Contract pending	Statewide network with ISDN in Harrisburg.
Pratt & Whitney, East Hartford, Conn.	AT&T 5ESS	Basic	SNET		
Prime Computer Inc., Natick, Mass.	AT&T 5ESS	Basic	Southern Bell	Started April 1988	Paying customer.
Rockwell Communication Systems, Richardson, Texas.	AT&T 5ESS with two remotes.	Basic	Southwestern Bell	Scheduled start Dec. 1988	40 buildings in a campus environment will be linked via ISDN.
Shearson Lehman Hutton Inc., New York	AT&T 5ESS	Basic	New York Telephone	Started June 1988	Part of 8,000 line centrex contract.
Shell Oil Co., Houston	AT&T 5ESS	Basic	Southwestern Bell	Start Sept. 1988	Plan to use 5,000 ISDN lines.
Southern Methodist University, Dallas	Siemens EWSD	Basic	Southwestern Bell	Started Feb. 1988	
SunTrust Service Corp., Atlanta	AT&T 5ESS	Basic	Southern Bell	Started April 1988	Paying customer.
3M Corp., St. Paul, Minn.	AT&T 5ESS	Basic	Southwestern Bell	Started Aug. 1988	Plan to use 3,165 ISDN lines.
Tenneco Inc., Houston	AT&T 5ESS	Basic	Southwestern Bell	Started June 1988	Plan to use 3,900 ISDN lines.
Texas A&M University, College Station	GTE GTD-5 EAX	Basic	GTE Southwest		
University of Arizona, Tucson	AT&T 5ESS	Basic	US West Information Systems Inc.	Planning stage	Using AT&T 5ESS as PBX in private network.
University of Connecticut, Storrs	AT&T 5ESS	Basic	SNET		
University of Indiana, Bloomington	Northern Telecom DMS-100	Basic	Indiana Bell		
University of Maryland, College Park.				Unannounced	
University of South Florida, Tampa	AT&T 5ESS		GTE South	Started Oct. 1987	
U.S. Dept. of Treasury, Wash.				Contract pending	
U.S. Bank of Oregon, Portland	Northern Telecom DMS-100	Basic	US West Communications	Started March 1987	Trial.
Virginia, State of, Richmond	AT&T 5ESS	Basic	C&P of Virginia	Started April 1988	
West Virginia University, Morgantown	AT&T 5ESS	Basic	C&P of West Virginia	Scheduled start Dec. 1988	Plan to use 660 ISDN lines.

Note: This chart was compiled by *CommunicationsWeek* with information provided by Teleos Communications Inc., Eatontown, N.J., as well as from news releases and published reports. Carriers, switch manufacturers and their affiliated laboratories are not listed.

Lee Mantelman, DATA COMMUNICATIONS INTERNATIONAL

Taking your SNA network abroad

Customs is not the problem when crossing borders with your SNA data. It's how to interconnect and administer a multinational network.

As the business world continues to compress into a single global economy, the effect on data networks—many multinational firms are finding—is just the opposite. To exploit the changing international social and economic landscape, businesses are expanding across national boundaries, oceans, and continents with unprecedented fervor. And it is corporate communications management that is tasked with linking these newly established, or newly acquired, foreign manufacturing and marketing activities into a seamless global network.

Many of these networks are built on IBM mainframes held together with SNA. According to IBM figures, by the end of 1988, users around the world were using SNA to support nearly 40,000 System/370 hosts—more than half of which are outside the United States.

But many pitfalls, as well as potential benefits, await networks that cross familiar national borders. The various PTTs can seem very strange indeed. So network managers should pack the following guidebook of suggestions, offered by users, consultants, and IBM, when planning to "go abroad" with their SNA networks (or with any private networks, for that matter).

Guided tour
■ *Consider granting national autonomy.* On one hand, corporations want to control their network resources. On the other, organizations in different countries often seek some measure of autonomy. Clearly, a compromise is needed.

Like other large IBM users, IBM itself faced a number of technical and managerial problems in building a network that now has more than 3,000 attached mainframes around the world.

"We decided that the best approach was a confederation of networks," says Ray Reardon, head of international networks integration with IBM Europe. Each country manages its own network, with gateways to international networks and applications.

"The trick is, how do you accept sufficient rules to make it all work without cutting across national autonomy?" asks Reardon. One way is for network overseers to set up rules but leave local issues to national data centers. Those rules might include naming conventions, service-level agreements, and when each country will be up and running.

■ *Break out the service center.* Those people in a user organization who fund an international service may not be the ones who benefit from it. To avoid budget battles, Reardon suggests that users establish a service subsidiary to provide and operate international telecommunications.

More simply, this role can be made the mission of one of the country organizations. Of course, users can also opt to turn to a value-added network (VAN) or international managed data network services (MDNS) provider.

"I'm pretty lucky; I've got a critical mass that's enabled me to put together a pretty cohesive international network," says Reardon. "Not all companies have that, which is why there's a VAN/MDNS business."

■ *Explore the role of SNI.* "The main thing about international SNA networking is the number of terminals you're trying to support," says John Wishney, London-based chief executive of communications business development with Electronic Data Systems, another firm with a large multinational SNA network. "As you expand geographically, you tend to add more terminals, and when you get 30,000 or more, you run into severe technical problems."

Time zones can also make life difficult. In a network with Pacific, European, and U.S. hosts, downtime for repair and maintenance is limited. "You'd be busy trying to get the network down and back up again by the time people come

468

1. Three-step process. *No SNI is needed when a remote country only has a single host, but it is useful when there are two or more. Multiple SNI links provide redundancy. Mapping from the original host in the remote country can be translated directly into the SNI tables. In general, the sooner the user deploys SNI, the better.*

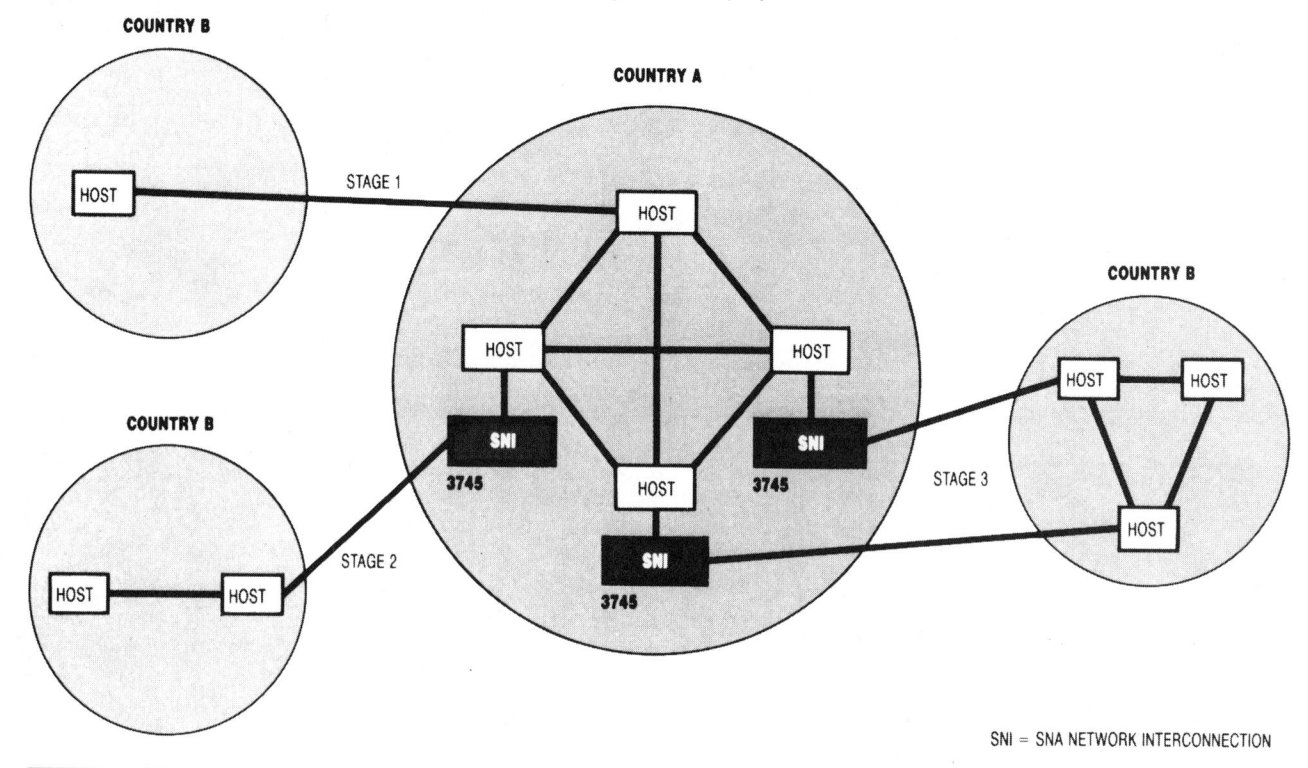

SNI = SNA NETWORK INTERCONNECTION

to work in Japan," says Wishney.

The problem here lies in the SNA system generation (sysgen) tables, which have to be coordinated, synchronized, and loaded throughout an SNA network for any change (such as moving a terminal). "IBM would say, 'You don't have to take the whole network down, just sysgen more devices than are actually active. Then, when new ones come in, activate a spare,'" says Wishney. "But it burns resources to carry all those extra definitions."

One IBM tool can make coping with complex international networks a lot easier: SNA Network Interconnection. SNI, an off-the-shelf software package for the 3745 communications controller, segments a large SNA network into more manageable subnetworks.

SNI functions by separating the address space of one SNI subnetwork from that of another. This requires devices within the subnetworks to be given aliases if they are to be addressed across the SNI gateway.

Originally developed by IBM for its Information Network, "SNI is probably one of the most stable products IBM has ever produced, in that it worked on day one," says Richard Lavender, senior consultant with Logica Financial Systems Ltd., London.

With an SNI gateway in each country, users can localize network administration. "Even with SNI, problems are still there, but you can deal with them on a country-by-country basis rather than struggling with the whole thing," says Wishney.

One disadvantage of SNI might be performance, according to David Welch, senior consultant with BIS Applied Systems in Birmingham, England. "There are limits to the amount of traffic you can handle on an SNI. If traffic between the networks grows, it can cause response times to increase, especially for interactive transmissions," he says.

■ *Plan an orderly growth.* Using SNI also adds a degree of complexity to the administrative task. While two national networks can use the SNI, someone has to administer it and make sure network definitions and route mapping are done and maintained properly.

Welch advises growing the SNI configuration in three stages (see Fig. 1). Each stage calls for local administration (to obtain and manage lines, modems, and multiplexers) and user support (for the local community).

Stage 1—for single-processor sites in foreign countries: Treat the remote hosts as simple extensions to the home network by bringing them into that network's addressing regime. Make sure all procedures used to install and maintain operating software are well documented at the central site and implemented as a management responsibility at the remote sites.

Stage 2—where there is more than one processor in a foreign country: Keep the original network addresses and addressing scheme, but implement them on an SNI and use them as aliases for the new remote network being built. Give the SNI the same address as the original processor.

2. Sneaky ASCII. *With IBM's X.25 Interconnection product, non-SNA data such as ASCII can be slipped across borders, saving money and boosting performance. XI sends local terminal data via one packet network over SNA to the remote host via another. A few other SNA components, namely NSF and NPSI, are also needed.*

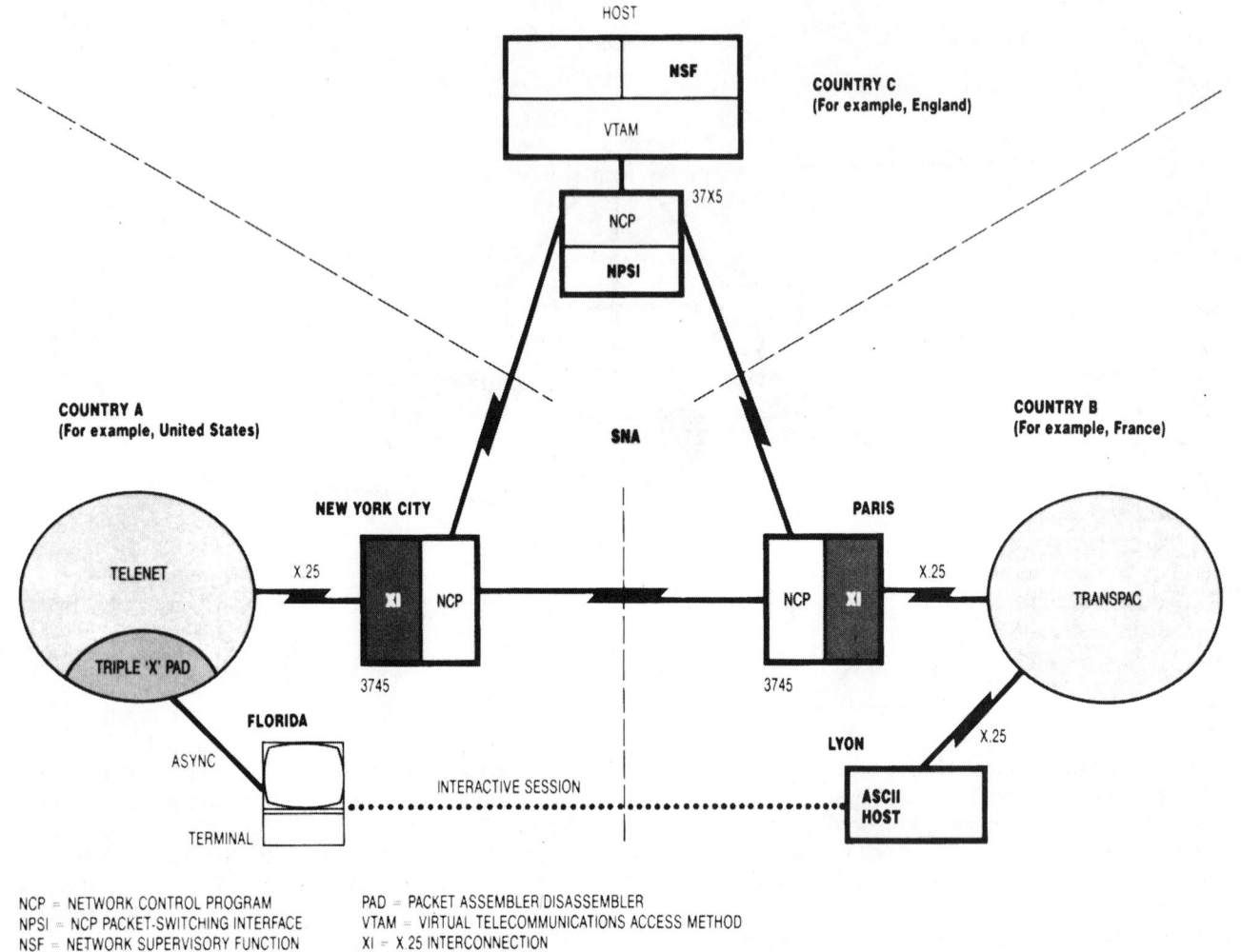

NCP = NETWORK CONTROL PROGRAM
NPSI = NCP PACKET-SWITCHING INTERFACE
NSF = NETWORK SUPERVISORY FUNCTION
PAD = PACKET ASSEMBLER/DISASSEMBLER
VTAM = VIRTUAL TELECOMMUNICATIONS ACCESS METHOD
XI = X.25 INTERCONNECTION

Don't hire armies of software people in each country, just enough for administration and user support.

Stage 3—when the remote network grows and exchanges more traffic with the home network: Install additional SNIs to ensure performance and reliability.

■ *Rationalize software distribution.* With five or more satellite countries, Welch suggests, the network manager might consider providing each remote site with a complete "envelope" of software, including operating system, database management, communications—and possibly even applications—based on the local configuration and requirements. To update a site, then, simply modify a central copy of the envelope and send it out.

This approach is efficient: It's much less effort to generate an envelope and install it in a number of locations than to have each site do its own. Also, the central controlling site then knows what software all the others are using. Standardizing the hardware platform used in the various countries can improve efficiency even more, though this may be politically difficult.

However, Lavender notes, an envelope of compiled code could be a lot of data to transfer over an SNA network, perhaps several Mbytes per node. Rather than generating these load modules at the central site, it might be better to transmit the SNA network definition statements, just a few kbytes, and have the code generated locally. Though this requires the remote-site staff to have some skill, this approach could work for companies with centralized network decision-making.

■ *Let your network do double duty.* Users may not be aware of the potential to send non-SNA data (such as asynchronous ASCII files) over international SNA backbone networks via X.25. This technique, suggested by Logica's Lavender, can save users money and avoid the well-known performance problems of international public packet-switching networks.

The secret ingredient is XI, IBM's X.25 Interconnection product (Fig. 2). XI, which translates information carried in

X.25 packets into an SNA-compatible format, lets users link packet networks using their own SNA backbone as an international transport medium for non-SNA data. In this way, a terminal user only has to pay for a local call to access a host in another country.

Figure 2 shows XI software running on IBM 3745 communications controllers in two countries that are to be connected. Each 3745 is linked either to its local public X.25 network, or directly to an X.25 terminal or packet assembler/disassembler.

This setup also requires the Network Supervisory Function, a NetView subtask, in one of the network hosts to manage the XI nodes and the NCP Packet-Switching Interface that is loaded in that host's 37X5 communications controller.

"Although all the people I know who are doing this are using it within their own companies," says Lavender, "you could almost resell this bandwidth to third parties and no one would be the wiser."

Since the traffic crosses international boundaries, the PTTs could view this as a value-added service if it connected people in different organizations. But, Lavender says, "I'm sure the PTTs don't have a hold of this, because they can't get a hold of it."

■ *Get your staff in sync.* According to Wishney, several staffing issues arise from the vagaries of Europe's telecommunications—and its distance from IBM's U.S. heartland. SNA programmers outside the United States have to accept the fact that they don't have access to the same telecommunications facilities, quality, or availability as their American counterparts.

Generally, whenever something goes wrong in the physical SNA network, it has an impact on the logical SNA network—the realm of IBM's Virtual Telecommunications Access Method (VTAM) and Network Control Program. For this reason, it's important that telecommunications staffers and SNA programmers coordinate closely. "Most human-error problems will come from a breakdown in communications between those two groups," Wishney says.

Another problem is that software release levels may be slightly behind those in the States. IBM-provided patches for software may not always be available. "The problem may be new for Europe but already fixed in the U.S., so it may take longer to find the fix," says Wishney.

■ *Favor multinational vendors.* IBM makes sure that all its offered equipment (communications controllers, modems, multiplexers) is approved in each country in which it is sold. But buying pure Blue is not always possible.

"I can't afford to use IBM everywhere, even if I wanted to," says one user who requested anonymity. "So I stick with larger companies with an international presence, such as Codex, Racal, and Timeplex. We stay with IBM in front-end processors, but NCR Comten should be okay." He adds that smaller, entrepreneurial suppliers, while adequate in one country, might not offer international support. "You might not be able to buy it in another country, and I don't want to mix products in the network." Other users, however, are more loyal to Big Blue.

"IBM will hold people's hands if that's what's needed," says Roger Cornish, European communications manager with Texas Instruments Ltd. in Bedford, England. "They're always very willing to guide their customers."

But users should not be lulled into an international sense of security. "What's remarkable is the variability of support that people get from IBM in various countries," says Ray Northcott, the London director of networking practice with Butler Cox and Partners Ltd.

"I can think of at least one case within the last three years where a customer was poorly served by IBM in one country in Europe but well served in the U.K.," Northcott says, "and requests to IBM U.K. didn't appear to do much in the other country." The reason for this variation, he explains, may be that the user was bigger in England than in the other country.

> **Users working in Timbuktu want the same service from IBM that they get in London.**

■ *Bring in the brass.* Users working in, say, Timbuktu want the same service from IBM that they get in London. And IBM, just like many other vendors, may claim to provide a uniform level of service. "You'll always get reassurance. Every supplier can do everything everywhere," says Northcott, "but words are cheap. You have to get senior management involved."

One user company took an approach that others could also benefit from. It brought in top IBM people to meet with its senior management, who said, "Here's what we want to do, we have these problems, and we want to know what you're going to do about them." The users didn't focus on such technical details as what NetView could do for them today and tomorrow but rather emphasized the commercial and management issues.

"The old adage of no one getting fired for buying IBM has a lot of weight and is a substantial credit to IBM," says Northcott.

■ *Be patient with IBM's central support.* Outside the United States, Northcott says, "IBM seems like a company of national units rather than an international company. So if you put up an SNA network across Europe, it's difficult to get a single point of support."

To address this concern, IBM set up the Global Networking Support Office, a supranational group based in La Gaude, France. GNSO, which has only been active for several months, has offices in France, the United States, Canada, and Japan. Its aim is to help large account teams understand and get answers to such issues as international network design, PTT tariffs, and service availability.

The centralized support group "comes out of user demand for pan-European networking in the 1992 era," says Patrick Abbott, network consultant with the GNSO. Northcott agrees. "For some time, IBM has been asked for and expected to provide this global support. Customers like ICI, Exxon, and Ford operate worldwide."

But it may be too soon to judge GNSO's effectiveness. "The good will and support have been spilling out of IBM verbally, but don't appear to have been established on the ground," says Northcott.

■ *Keep abreast of standards.* Though SNA has been a credible strategy for building international networks, it faces stiffening competition from standardized architectures such as the OSI model.

"Europe has always been very strongly for OSI. As governments start to dictate an OSI strategy, I think they're forcing the issue," says Texas Instruments' Cornish. Wishney adds: "Every man and his brother is now after you to implement standards, and SNA is not an international standard, X.25 is."

Private users now running IBM hosts and SNA networking, Wishney says, "should probably stay SNA when they go international, but I expect to make a different recommendation in about 12 months time" "I wouldn't disagree with that," says Cornish.

One advantage of OSI over SNA in international networks is that there could, in time, be a broader market of people around the world with OSI experience. Wishney says, "VTAM programmers are a fairly esoteric bunch of people. The industry is faced with a serious-enough skills shortage as it is. I can only expect an equal or more acute shortage of SNA people."

But IBM stresses that SNA and OSI are not conflicting approaches. "If you want to connect two SNA networks, use SNI. If you want to connect two dissimilar networks, use the appropriate OSI standard. That's what OSI is supposed to be," says IBM's Reardon.

Though all the users agree that OSI is on its way to becoming more of an international requirement, they bristle at any potential attempt governments might make to mandate its use. "If you're a private user, you should be able to do whatever you like," says Wishney. That includes sticking with a proprietary architecture like SNA, where innovations can be deployed much more quickly. "Remember, standards are a leveling-down process. You could standardize to a point where you start stifling things," he says.

IBM's international database

With all these suggestions firmly in hand, users may decide to plunge right into an international network design, or they may prefer to farm the job out to a consultant. In either case, the designer should take account of service differences between countries.

IBM's Centre d'Etudes et Recherches (CER), in La Gaude, has highlighted 10 such differences (see Table). The ones shown in the table are only examples and illustrations. IBM can, however, provide a specific country-by-country breakdown of each parameter.

"That source is a real database and a real asset, because lots of people would like to know what you can do in this domain," says Etienne P. Gorog, CER's telecommunications center director.

According to Gorog, the reason for discrepancies is that each PTT has been optimizing its services in its own country. Even if several PTTs are also offering international services, they are not necessarily compatible.

Some services provided within a country may not be available between countries. And international facilities that are available may differ in options provided by each country (as with X.25) or in the type of lines available for switching data (only digital lines in

International service variations

PARAMETER	EXAMPLE/EXPLANATION
AVAILABILITY — SPEED: — TYPE: — CONFIGURATION: — TECHNOLOGY:	2.048 MBIT/S IN U.K., 1.920 MBIT/S IN FRANCE ANALOG-FOR-DATA NOT AVAILABLE IN GERMANY DIFFERENT OPTIONS FOR SUCH SERVICES AS X.25 ISDN STATUS VARIES AROUND EUROPE
CONSTRAINTS ON USAGE	VOICE COMPRESSION NOT ALLOWED IN GERMANY BUT PERMITTED (WITH RESTRICTIONS) IN U.K.
CONSTRAINTS ON USER (SINGLE/MULTIPLE)	BANDWIDTH RESALE FORBIDDEN IN MANY COUNTRIES; SWITCHED-TO-LEASED MAY BE FORBIDDEN (ITALY)
TARIFFS	NOT PUBLISHED FOR ALL AVAILABLE SERVICES; TARIFF MAY DIFFER IN COUNTRIES AT EITHER END OF AN INTERNATIONAL SERVICE
LOCAL AVAILABILITY AND DELAYS TO OBTAIN FACILITIES	PTT MAY PROVIDE INTERNATIONAL SERVICE ONLY FROM SPECIFIC NATIONAL LOCATIONS; MAY TAKE SOME TIME TO PROVIDE SERVICE IN AREA WHERE NOT PREVIOUSLY AVAILABLE
EQUIPMENT — APPROVED: — MANDATORY:	PTTs APPROVE NETWORK-ATTACHED EQUIPMENT SOME ALLOW ONLY PTT-PROVIDED EQUIPMENT
COMPATIBILITY OF EQUIPMENT	VERIFICATION TESTS NEEDED TO CHECK SAME INTERPRETATION OF STANDARD AT EACH END
PERFORMANCE OF PTT FACILITIES (RELIABILITY, AVAILABILITY, AND SERVICEABILITY)	NOT ALL PTTS GUARANTEE QUALITY OF SERVICE; INTERNATIONAL LINKS MAY DEGRADE THE PERFORMANCE ACHIEVED AT EACH END
MAINTENANCE	NO AUTOMATED MAINTENANCE INTERNATIONALLY; EACH PTT HAS OWN APPROACH OR IMPLEMENTATION
BILLING	SAME AS MAINTENANCE; ONE-STOP-SHOPPING FOR UNIFIED INTERNATIONAL BILLING AND MAINTENANCE NOT YET AVAILABLE

Source: IBM Centre D'Études et Recherches, La Gaude, France

Germany but switched analog lines as well in the United Kingdom, France, and Italy).

Differences in speed may be due to use of the channel or slot by the PTT for billing or signaling. The PTT may also restrict what the user can do with a given service or who may be allowed to use it. In Switzerland, for example, the monthly cost for a line may vary by two to three times, depending on its use. In Italy, a service that is to be shared among different users must be accessed through switched, and not leased, facilities. Some countries allow no resale at all, while the United Kingdom allows resale of most data facilities, but not voice.

Besides different rates based on local versus long-distance and switched versus leased-line, there's little rhyme or reason to the structure of international tariffs. "There's no case where one country has the same tariff as another," says Gorog. One end of a connection often costs much more than the other. Leased-line tariffs can have ratios as high as one to three.

"Network designers may choose countries as node sites based on these differences," notes Gorog.

Also, not all tariffs are published, especially for high-speed services. In many cases, a PTT has no published tariff for, say, a 2-Mbit/s service, but it can still be obtained. "Just knowing that a service can be or has been negotiated can be critical," says Gorog.

Paying premiums
Another twist is varying transborder premiums. For example, the U.K. tariff to cross the English channel via a leased line can be more than double the cost of covering the same distance within England. France Telecom's cross-channel tariff is about 30 percent more than the French domestic charge (which is itself higher than its English counterpart).

Even if an international service exists in both countries, it may not be available from point A in country X to point B in country Y. If the desired end points are outside the two countries' business centers, "maybe you can't get a link, or you have to wait three years," Gorog says. Planners would be well-advised to get a current local map from each PTT showing precisely where service is available.

The approval process can also be daunting. Each country will approve only certain products for connection to its network. Network planners have to determine whether a PTT requires the use of network-access standards, either voluntary (such as X.25) or mandatory ones (such as specific modulation schemes).

The more countries in a network design, the fewer the designer's equipment options. "We haven't gotten IBM modems approved in Germany because we weren't allowed to," says Gorog. "But in every country where we're allowed, our modems have been approved. Other modem suppliers may be approved in, say, only three out of seven European countries, which is not bad, but you have to be everywhere to respond to an international network," he says. And there may be few international vendors with IBM's clout and perseverance to get its products approved.

"It costs IBM an arm and a leg to get its stuff approved across Europe," confirms Jim Norton, director of industry studies with Butler Cox and Partners Ltd.

Indeed, equipment not only has to be approved (say, a Siemens box in Germany and a Bull machine in France), it also has to be compatible. And even if it works well across the room, the link may perform erratically across an international connection.

The more countries in a network design, the fewer the designer's equipment options.

Getting quality
Another variable is the performance and quality level of PTT facilities, which some countries guarantee and other don't. France is a good case: Its PTT guarantees performance of the switched network at 9.6 kbit/s, while other countries may guarantee it only to a rate of 1.2 kbit/s.

Alternate-routed leased lines and exchange connection may or may not be available in a given country. For example, the PTT may offer high-reliability service customers, for a fairly small charge, an additional link into the local exchange as a hot standby. Such PTT offerings in the Netherlands, the United Kingdom, and France are very worthwhile, according to Abbott.

Crossing a border also makes a big difference here. For example, X.25 connections in the United Kingdom and France each attain a certain level of performance, but that level degrades over the international link.

Finger-pointing can also be a problem in transborder link maintenance. When a line drops, the user normally has to go to both sides to fix the problem. Though several individual countries offer automated network maintenance schemes, not one is the same as or compatible with any other.

Billing problems are as prevalent as maintenance problems: there is no coordination from the two sides of an international line. And users are growing more vocal in demanding a simpler, more modular way of billing. This is an aim of the PTT plan for one-stop shopping and billing called Managed Data Network Service, but, Gorog says, "we don't see that materializing."

Though on a somewhat different level, the services available from value-added suppliers also differ from country to country, as do order-to-install times and the quality of lines and adapters.

Users can try to deal with all these variable themselves, or they can turn to a consultant for help. But with its own experience and the detailed database it has already assembled, IBM should certainly be on users' lists of resources to draw upon in building international SNA networks. ■

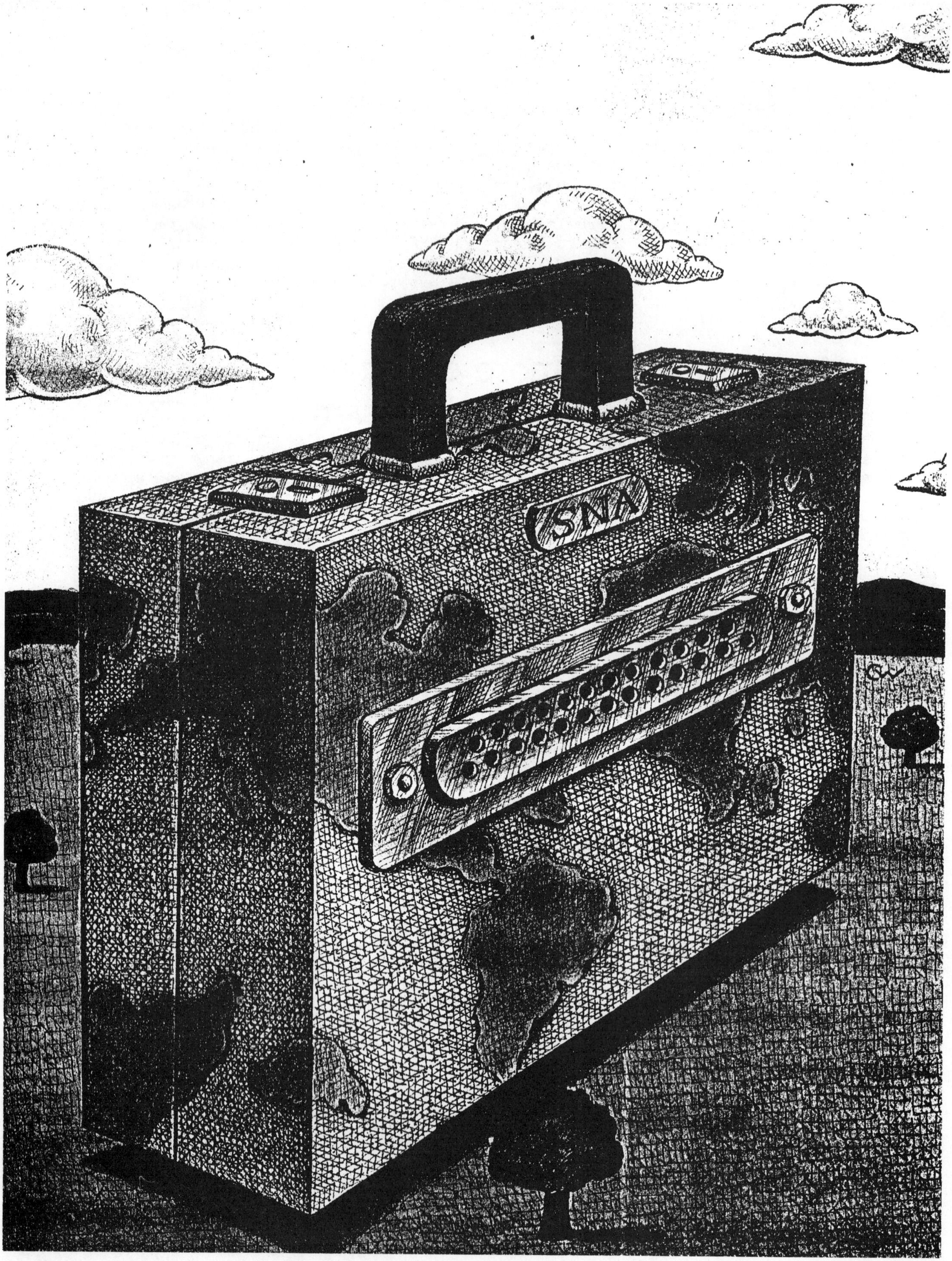
SNA

User survey exclusive

NetView shows highs and lows of net management

More users employ IBM's solution in multivendor nets, but give it a tepid rating

IBM's NetView, the most popular of the installed data communications network management products, continues to represent what's best and worst about how this technology is being implemented in 1989.

A Datapro Research Corp. survey of 157 managers of large networks, shows that 55 percent are employing network management packages in multivendor environments. That's an increase from 43 percent in the 1988 survey. Datapro's numbers indicate that two-thirds of the NetView users are employing the system in multivendor environments (see Table 1). About one-third of all respondents were NetView users, up from 25 percent in

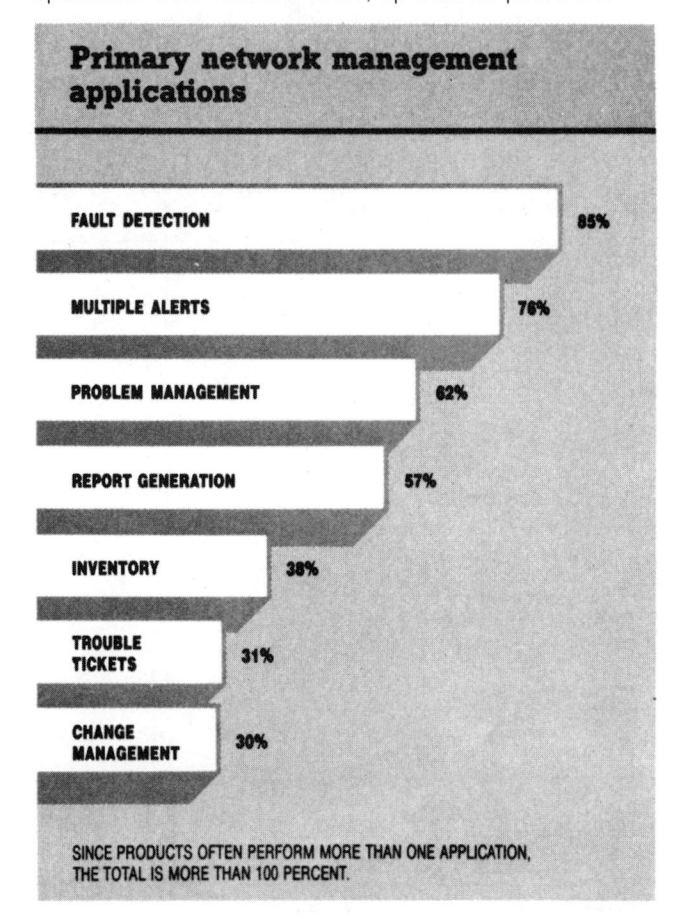

Primary network management applications

Application	%
FAULT DETECTION	85%
MULTIPLE ALERTS	76%
PROBLEM MANAGEMENT	62%
REPORT GENERATION	57%
INVENTORY	38%
TROUBLE TICKETS	31%
CHANGE MANAGEMENT	30%

SINCE PRODUCTS OFTEN PERFORM MORE THAN ONE APPLICATION, THE TOTAL IS MORE THAN 100 PERCENT.

Table 1: Is the package used in a multivendor network?

VENDOR	TOTAL NUMBER OF USERS	YES*	NO	NO ANSWER
AT&T	7	29	57	14
AT&T PARADYNE	13	46	54	—
CINCOM	3	100	—	—
CODEX	12	33	67	—
DIGITAL EQUIPMENT CORP.	3	67	—	33
GENERAL DATACOMM	11	45	55	—
IBM	58	64	34	2
MEMOTEC DATACOM	4	50	50	—
RACAL-MILGO	11	18	82	—
TELENET	3	67	33	—
TIMEPLEX	7	57	43	—
ALL OTHERS	25	68	28	4

*NUMBERS REPRESENT PERCENTAGES

Datapro's 1988 network management survey.

The flip side is that NetView and most of its competitors are still not living up to users' expectations. In fact, users this year gave NetView an overall satisfaction rating of 2.6, down from its 3.0 rating in 1988.

Unrealistic expectations?

On average, the network managers who responded to the survey think that their network management products are between fair and good (see Table 2). This is the same rating (2.9 out of a possible score of 4) that respondents gave last year.

Users generally like the reliability of their network management hardware. But they had mediocre opinions of the software capabilities of the products they use, especially the ability of the network management systems to generate reports or route traffic.

The ratings for installation and operation suggest that vendors' claims of user friendliness are not well enough defined for users. A lack of trained personnel, a common problem in data communications, might explain the 2.8 rating for Ease of Installation. Bringing network management systems on line often involves fairly extensive system programming, a problem most managers cite when discussing the products. Vendors have made their systems easier to use, but apparently only a few have attacked installation problems.

When reviewing these ratings, keep in mind that network management is still a young technology that is constantly changing and experiencing tremendous

User survey exclusive

Table 2: Network management user ratings, on a scale from 1 to 4

MANUFACTURER/MODEL	OVERALL SATISFACTION	EASE OF INSTALLATION	EASE OF OPERATION	EASE OF HARDWARE EXPANSION	EASE OF SOFTWARE EXPANSION	SOFTWARE PERFORMANCE	HARDWARE RELIABILITY
CINCOM—							
NETMASTER	3.7	3.7	4.0	4.0	3.5	4.0	*
DIGITAL EQUIPMENT CORPORATION—							
ALL MODELS	3.7	3.7	3.7	3.7	3.7	3.7	3.7
TELENET—							
TP5	3.7	3.0	3.0	3.0	3.3	3.3	3.7
RACAL-MILGO—							
CMS 2000 SERIES	3.1	3.1	3.1	3.1	2.6	2.9	3.4
OTHERS & UNSPECIFIED	3.3	3.3	3.0	3.0	2.7	2.7	3.3
SUBTOTAL	3.2	3.2	3.1	3.1	2.6	2.8	3.4
TIMEPLEX—							
LINK	2.5	3.0	3.0	3.0	3.0	2.5	3.0
TIMEVIEW	3.5	3.5	4.0	3.5	3.5	3.5	3.5
OTHERS & UNSPECIFIED	3.3	3.3	3.7	3.3	3.3	3.3	3.7
SUBTOTAL	3.1	3.3	3.6	3.3	3.3	3.1	3.4
AT&T PARADYNE—							
ANALYSIS	2.7	2.9	2.9	3.0	2.8	2.5	2.8
ANALYSIS 6510	3.4	3.2	3.4	3.4	3.2	3.0	3.6
SUBTOTAL	3.0	3.0	3.1	3.2	2.9	2.7	3.1
AT&T—							
DATAPHONE II	2.5	3.3	3.0	2.8	3.0	2.5	3.0
OTHERS & UNSPECIFIED	3.5	3.0	3.0	3.5	3.5	3.0	3.5
SUBTOTAL	2.8	3.2	3.0	3.0	3.3	2.7	3.2
CODEX—							
4800 SERIES	2.7	2.3	2.7	2.7	2.3	1.7	2.3
9300 SERIES	2.7	2.7	3.3	3.3	3.0	2.3	3.0
9800 SERIES	3.5	3.5	3.5	3.5	3.5	3.5	4.0
SUBTOTAL	2.8	2.7	3.2	3.2	2.9	2.3	2.9
IBM—							
NETVIEW	2.6	2.5	2.7	2.8	2.7	2.8	3.0
OTHERS & UNSPECIFIED	2.3	2.4	2.0	2.3	2.3	2.4	2.7
SUBTOTAL	2.6	2.5	2.6	2.7	2.6	2.8	3.0
GENERAL DATACOMM—							
NETCON	2.5	2.5	2.6	2.3	2.1	2.5	2.7
MEMOTEC DATACOM—							
ALL MODELS	2.5	3.0	2.8	2.5	2.0	2.5	3.0
ALL OTHERS	3.2	3.3	3.3	3.3	3.0	3.0	3.5
GRAND TOTAL	2.9	2.8	2.9	2.9	2.8	2.8	3.1

*INSUFFICIENT RESPONSE

User survey exclusive

REPORT GENERATION CAPABILITIES	RESTORAL CAPABILITIES	TRAFFIC ROUTING	QUALITY OF VENDOR'S MAINTENANCE/ TECHNICAL SUPPORT
3.5	2.7	2.0	3.0
3.7	3.7	3.7	3.7
3.0	3.7	3.7	3.3
2.8	2.7	2.3	3.4
3.0	2.7	2.5	4.0
2.8	2.7	2.4	3.6
1.5	2.5	2.0	3.5
3.0	2.5	3.5	3.5
3.7	4.0	4.0	4.0
2.9	3.1	3.3	3.7
2.5	2.3	2.2	2.3
2.6	3.2	3.0	3.4
2.5	2.6	2.5	2.8
1.8	2.7	2.0	3.0
2.5	3.5	*	3.0
2.0	3.0	2.5	3.0
1.7	2.3	2.5	2.7
3.2	3.0	2.7	2.7
3.9	3.5	3.0	3.5
2.7	2.9	2.7	2.8
2.2	2.6	2.3	2.9
2.3	1.8	1.5	2.1
2.2	2.5	2.2	2.8
2.1	2.8	1.8	2.3
1.8	2.8	1.0	2.8
2.8	2.9	2.8	3.2
2.5	2.7	2.5	2.9

growth. Yet the fact that little has changed since the 1988 Datapro survey shows that either users—perhaps listening to too many sales pitches—are expecting too much of their systems or vendors are not listening to user's requirements, or both.

How the packages are used

The survey elicited a great deal of information on how the network management packages are being used. As shown in the chart "Primary network management applications," most (85 percent) of the systems were used for fault detection while few (30 percent) employed the products for change management.

Sixty-nine percent of the respondents said their network management products were monitoring modems, 66 percent of the products monitor transmission lines, 33 percent monitor multiplexers, and 24 percent specifically monitor T1 multiplexers.

Eighty-five percent of the rated systems maintained an active database, while 58 percent of the systems support backup-restoral functions. When asked whether response-time monitoring was important to their systems, 56 percent said it was very important, 31 percent said moderately important, 9 percent felt it was not important, and 4 percent did not respond.

Who's rated the best?

This survey was mailed to a cross-section of DATA COMMUNICATION'S readers on a random basis. It is not intended to show market share or be the definitive pronouncement of user satisfaction. It is merely a guide that users may want to consider as part of their purchasing procedures.

The vendors and packages are listed in Table 3 by their average rank in the Overall Satisfaction category, on a scale of one to four. However, each of the top three vendors—Cincom, Digital Equipment Corp., and Telenet—were only rated by three respondents. Others, such as IBM (58), AT&T Paradyne (13), Codex (12), Racal-Milgo (11), and General DataComm (11), were reviewed by substantially more users. Vendors not listed were rated by less than three readers.

Five vendors showed improved ratings in the 1989 survey, including Codex (2.8, up from 2.5) AT&T Paradyne (3.0 from 2.8), and Racal-Milgo (3.2 from 3.0). Two vendors slipped in the overall satisfaction category: IBM and General DataComm (2.5 from 3.1).

Survey methodology

The survey asked users to rate the network management vendors and products as either excellent, good, fair, or poor. Datapro assigned a value of from one to four to each rating, with four being excellent. The tallied numbers for each value were then multiplied by the corresponding weight, and the average taken by dividing the sum of the products by the total number of responses for that category. Table 3 contains these averages.

For more information on the survey, or to obtain a copy, contact Datapro Research Corp., 1805 Underwood Blvd., Delran, N.J. 08075 (609-764-0100). ∎

Robert D. Love and Thomas Toher, IBM, Research Triangle Park, N.C.

How to design and build a 16-Mbit/s Token Ring

Wondering whether to make the move up to a 16-Mbit/s Token Ring? Is new cabling necessary? Here's Big Blue's view on how to get there from here.

Much has happened in the LAN world since our first Token Ring article appeared ("How to design and build a Token Ring LAN," DATA COMMUNICATIONS, May 1987, p. 213). One notable development was IBM's introduction last year of a 16-Mbit/s version of the Token Ring. This is not solely an IBM endeavor, however. The IEEE 802.5 committee of the International Organization for Standardization has voted on an update to the basic TokenRing standard that includes 16-Mbit/s operation. That update is now in the ratification process and on its way to becoming an international standard.

The demands of data transmission at this higher rate (a fourfold increase over the 4 Mbit/s previously allowed) necessitate a fresh look at physical network planning. Indeed, some users may be able to go to 16 Mbit/s using their existing cabling. Others may have to make some changes. And some, depending on their particular topology and traffic requirements, may not need to upgrade to 16 Mbit/s at all.

The issue of data cabling—selecting and installing the right cabling type the first time—has been a bane of data networkers for years. But there is finally some progress being made toward the standardization of data cabling types and installation practices.

The Electronic Industries Association (EIA), for example, is developing a standard for commercial building wiring in its EIA TR41.8.1 committee. This standard, oriented toward local data networking, specifies cable types, distances between wiring closets, and distances from wiring closet to office (the outlet in a user's work area).

In addition, the American National Standards Institute is developing a fiber optic 100-Mbit/s token-passing ring called a Fiber Distributed Data Interface (FDDI) in its X3T9.5 committee. The most important applications for this emerging standard will be direct host connections and high-speed backbones that will interconnect LANs through attaching devices such as bridges and gateways. For these applications, the FDDI attachments will generally be confined to the wiring closets and to the computer center.

These emerging standards are also important during physical network planning for 16-Mbit/s Token Rings because the bulk of data communications cabling in an establishment should be usable over a 10- to 15-year period, with incremental additions made as needed to support new technologies. The specific implementations of LANs will continue to evolve to accommodate changing needs and to incorporate technology enhancements as they become available.

A peek at the future
The emerging EIA building wiring standard calls for copper cabling from the wiring closet to the office. The cable types include: 150-ohm, data-grade media, suitable for 4-Mbit/s and 16-Mbit/s Token Rings; telephone wire, suitable for voice and lower-speed data communications; and coaxial cable, where needed. Maximum wire run lengths from wiring closet to work area should be limited to 90 meters (about 300 feet). For wiring between closets, both copper cabling and optical fiber are recommended.

The optical fiber recommended is 62.5/125-micron multimode optical fiber, the same fiber recommended in the emerging FDDI standard. Multimode fiber of a different size than the recommended 62.5/125-micron fiber (such as 50/125-micron) can also be used for LAN applications. However, connection of different fiber sizes may result in mismatch losses that will decrease maximum transmission distances.

The proposed building wiring standard requires wiring closet placement to adhere to the maximum drop length of 90 meters. It also implies a requirement to provide optical

1. Drive distances. To calculate copper ring segments bounded by optical fiber converters, the ring is divided into segments using optical converters. Allowable drive distances are computed separately for each segment. Physical restrictions limit maximum lobe lengths and the copper wire distances between wiring closets.

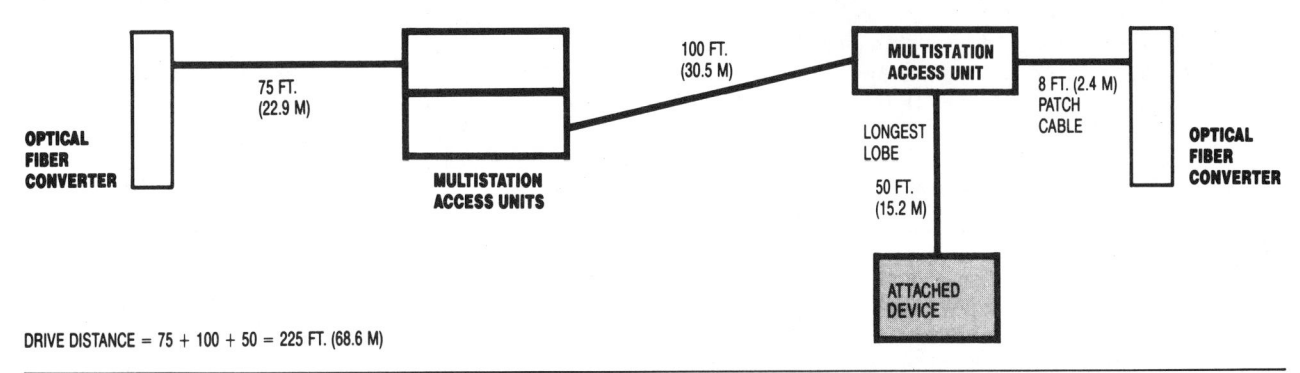

DRIVE DISTANCE = 75 + 100 + 50 = 225 FT. (68.6 M)

connections between wiring closets, or, more economically, to provide the space and capability to add optical fiber as needed. The requirements for copper between wiring closets also fall within the standard.

New restrictions

The fundamental design principles for 16-Mbit/s network planning remain unchanged from those for 4 Mbit/s. However, physical restrictions limit maximum lobe lengths and the copper wire distances between wiring closets (Fig. 1). In addition, telephone wiring is not acceptable for Token Ring operation at 16 Mbit/s (see "Token Ring and twisted pair").

Attenuation of signals on copper wires increases as the square root of frequency, so that for 16-Mbit/s TokenRing transmission, the maximum drive distance from any driver to any receiver is about half of what is allowed at 4 Mbit/s.

Since the drive capability is not significantly reduced on optical fiber links, fiber plays an important role in connecting 16-Mbit/s wiring closets. By using optical fiber between wiring closets, the drive capability is left available for driving signals over the copper lobes.

As the examples below show, copper connection of wiring closets is still effective for the close spacing found in high-rise buildings with vertically stacked closets. The maximum recommended lobe length for a 16-Mbit/s Token

---

Token Ring and twisted pair

The 4-Mbit/s Token Ring was introduced simultaneously on data-grade media and on telephone twisted pair. The major difference in the capabilities of the two transmission media appeared to be the maximum number of stations allowed on a single ring (72 for telephone twisted pair, 260 for data-grade media).

The maximum lobe length allowed for telephone twisted pair was 100 meters (330 feet), which corresponded with the 100-meter maximum recommended lobe length for data-grade media. An examination of the wiring charts for 4-Mbit/s transmission on data-grade media showed a different story—namely, that data-grade media are technically capable of supporting lobe lengths in excess of 300 meters (984 feet).

The reason for the 100-meter maximum recommended length was the projected need to reuse those same wiring runs for 16-Mbit/s Token Rings at a later time. In fact, data-grade media are able to carry Token Ring signals two to three times as far as telephone twisted pair.

When considering the propagation capabilities of data-grade media and telephone wire for 16-Mbit/s Token Rings, the following restrictions must be heeded:

1. Signal attenuation increases as the square root of frequency. Therefore, going from 4 to 16 Mbit/s decreases maximum propagation distances by 50 percent.

2. Interference from external noise is significantly higher for 16-Mbit/s signals on telephone wiring than for 4-Mbit/s signals. In order to compensate for the higher noise levels, transmission distances at 16 Mbit/s are further reduced.

3. The Federal Communications Commission imposes limits on radiated energy above 30 MHz. Since telephone wiring readily radiates energy from the transmitted signals, severe filtering of harmonic frequency content of the 16-Mbit/s transmitted waveform would be required to adhere to FCC limits. This filtering would further restrict drive distances.

Based on the maximum drive distance of 100 meters for 4-Mbit/s Token Rings, and on the limitations imposed by 1, 2, and 3 above, the maximum practical lobe lengths for 16-Mbit/s Token Rings on telephone twisted pair, without resorting to costly coding schemes or specially designed active Multistation Access Units, would be on the order of 30 meters (98 feet). Development of special transmit and receive filters would be required to achieve this capability. Even then, there would be higher error rates associated with the higher noise levels present at these frequencies.

In light of the difficulties and the resulting performance limitations, telephone twisted pair is not a viable medium.

Ring on Type 1 and Type 2 cable (using 22-gauge, 150-ohm, data-grade cable) remains at 330 feet (100 meters). Since this recommended maximum is close to the transmission limits for 16-Mbit/s speeds, the use of Type 9 cable (26-gauge, 150-ohm, data-grade cable) limits useful maximum lobe lengths to 220 feet (67 meters).

For the migration from 4-Mbit/s to 16-Mbit/s Token Ring to be smooth, the lobe cabling must be reusable. If telephone wire was used as lobe cabling at 4 Mbit/s, then either the lobe wiring must be changed before migrating or 16-Mbit/s operation must be restricted to backbone rings that interconnect the 4-Mbit/s rings. If data-grade lobes were used at 4 Mbit/s, with maximum lobe lengths within the stated guidelines, then that wiring can be used to support 16-Mbit/s Token Rings.

In addition, multiple-wiring-closet rings operable at 4 Mbit/s without repeaters may require optical fiber converters at 16 Mbit/s. Multiple-building LANs will definitely require optical fiber links.

The concept of the Adjusted Ring Length (ARL) remains

Optical fiber converters for 16-Mbit/s Token Rings

The IEEE 802.5 standards committee is currently extending the Token Ring standard to include optical fiber propagation. Separate work items address fiber lobe attachment and the cabling within building walls, including the wiring between wiring closets.

Several manufacturers have announced or are developing optical fiber attachments to the Token Ring. One important class of attachments are new optical fiber converters. These converters, now available for 16-Mbit/s Token Rings, are considerably enhanced compared to many of the older models used for 4-Mbit/s ring operation.

The new fiber converters still allow for optical links between wiring closets (separated by at least 2 kilometers) using any good-quality 62.5/125 or 100/140 multimode optical fiber cable. But unlike older repeaters, the new fiber converters often are smart boxes that participate in network management and are able to automatically reconfigure the ring in case of failure of either the converter or the optical fiber.

Two optical converters and the fiber cable between them form an optical fiber subsystem. The subsystem provides two signal paths: one in the primary direction and one in the backup direction. Each converter has a universally administered address, with the upstream (transmitting) converter's address on the primary path and the downstream (receiving) converter's address on the backup path.

During normal operation, the primary path carries the ring traffic and the backup path carries separate ring-maintenance signals that ensure the ring's continued availability. In the event of a failure in the fiber subsystem, such as a broken fiber cable or a power interruption to either converter, the two converters both detach from the ring and wrap the signals at their copper inputs. In this wrapped state, the converters try to establish communications with each other. If they do, then they both reinsert into the ring, reestablishing the primary data communications path.

When the converters are not inserted in the ring, the remainder of the ring operates using the backup path. The removal of the optical fiber converter from the main ring path is observed by network management software. LAN management software sends notification of the physical error condition to the network administrator who then takes repair action. Other LAN users will not be aware that there is a physical failure on the ring.

A second error condition detectable with optical converters is the disconnection of a cable in the main ring path. Because all data connectors also have the automatic wrap capability, the disconnection of a cable on the main ring path, as between MAUs, causes the ring to wrap and use the backup path.

This condition by itself is not cause for alarm, since all stations will continue to operate on the LAN unaffected by the break. However, the condition should be fixed, since a second break or an attempt to wrap the ring to correct for some other failure will now divide the ring into two separate rings. With the optical subsystem in the ring, the downstream converter becomes active in the ring path. Network management uses the appearance of that station as a signal that the ring has entered a wrap state. Before this problem becomes compounded during a ring failure, network management can correct it without having any impact on users of the LAN.

Each optical fiber converter receives, retimes, retransmits, and converts both the electrical signal to optical and the optical signal to electrical. When the converter wraps, the signals on the input copper wires are wrapped at the converter's electrical signal input.

As a consequence, any Token Ring segment bounded by converters must be able to operate as an independent ring if those converters are both in a wrapped state. Therefore, the allowable drive distances represent the entire cabling from the segment's input converter to its output converter, plus the longest lobe length. During a condition with both converters wrapped with only a single attaching device on the longest lobe, the signal path runs from the station's transmitter, along its lobe wiring, along both the forward and return paths between the converters, and back along the lobe wiring to the station's receiver. Since the signal propagates along both the main ring path and the lobe wiring twice, the distances that represent the longest cable path represent one-half the actual drive distance from transmitter to receiver.

Because the converters reamplify and retime both the main ring path and the backup path, and because both those amplifiers and retimers are a part of the signal path during a wrap condition, each converter counts as two stations against the maximum attach limit for the ring.

central to determining the drive distance of 16-Mbit/s rings. For a ring that is wired completely with copper cable, the ARL is the sum of all the interwiring-closet cables, minus the length of the shortest of the interwiring-closet cables. The ARL is then added to the length of the longest lobe on the ring. This total distance is the longest transmission path in the ring.

When optical fiber converters are used, the ring is divided into segments bounded by the converters. Then, as illustrated in Figure 1, the ring-segment drive distance for each copper ring segment consists of the wiring in the main ring path bounded by converters, plus the length of the longest lobe in the segment. (For a description of the optical fiber converter, see "Optical fiber converters for 16-Mbit/s Token Rings.")

When using Table 1 for rings with converters, compute the ring-segment drive distance for each segment, including the number of wiring closets and Multistation Access Units (MAUs) within the segment under consideration.

Our hypothetical building consists of two seven-story wings with a wiring closet in each wing of each floor (Fig. 2A).

Assume that each wiring closet contains two MAUs for a total of 28. The longest lobe is 250 feet (76 meters). The cables between floors of the vertically stacked wiring closets are each 30 feet (9 meters). The cables connecting the two stacks of wiring closets are each 500 feet (152 meters). All cabling is Type 1 or Type 2 wire. Table 1 shows the allowable drive distance for a 16-Mbit/s ring spanning multiple wiring closets or for a ring segment bounded by converters and using Type 1 cable.

Since Table 1 indicates that the maximum number of MAUs allowed in a multiple wiring closet ring without converters is 18, converters must be used to divide the ring into segments. The most obvious initial placement for these converters is on the two 500-foot runs between wiring closets 7 and 8 and wiring closets 1 and 14.

This placement divides the ring into four segments. Two of the segments are the 500-foot runs of optical fiber bounded by converters. The other two segments are the two vertically stacked groups of seven wiring closets each, with the 14 MAUs (2 per wiring closet) and all the associated copper cabling.

Because the two optical fiber segments are only 500 feet long, any good-quality 50/125-, 62.5/125-, 85/125-, or 100/140-micron optical fiber cable will provide satisfactory transmission capability. Since the FDDI and EIA TR41.8.1 standards specify 62.5/125-micron fiber, the most commonly installed size in the United States, we recommend that new installations follow these guidelines.

Now consider the drive distance of the two remaining copper ring segments containing 14 MAUs each. Since these two segments are exactly the same, they can be treated simultaneously. As shown in Figure 1, the ring segment drive distance is the sum of all the lengths of interwiring-closet cable between the repeaters plus the length of the longest lobe. So, for each of these ring segments the drive distance is 30 + 30 + 30 + 30 + 30 + 30 + 250 = 430 feet (131 meters). Using Table 1, we find that for a seven-wiring-closet ring segment containing

Table 1: Drive distances

NUMBER OF MAUs	NUMBER OF WIRING CLOSETS									
	1	2	3	4	5	6	7	8	9	10
1	569*									
2	547	531								
3	525	509	493							
4	503	487	471	454						
5	482	465	449	432	426					
6	460	443	427	411	394	378				
7	438	422	405	389	372	356	340			
8	416	400	383	367	350	334	318	301		
9	394	378	361	345	329	312	296	279	263	
10	372	356	340	323	307	290	274	258	241	225
11	351	334	318	301	285	269	252	236	219	203
12	329	312	296	279	263	247	230	214	197	181
13	253	270	253	237	220	204	188	171	155	138
14	211	227	211	194	178	161	145	129	112	96
15	168	184	168	152	135	119	102	86	69	53
16	125	142	125	109	92	76	60	43	27	10
17	83	99	83	66	50	33	17	—	—	—
18	40	56	40	24	—	—	—	—	—	—

*DISTANCES ARE IN FEET (TYPE 1 OR 2 CABLE) FOR 16-MBIT/S MULTIPLE-WIRING-CLOSET RINGS AND RINGS WITH CONVERTERS.
MAU = MULTISTATION ACCESS UNIT

14 MAUs, the maximum allowable ring segment drive distance is 145 feet (44 meters). Therefore, each segment must be divided again.

Suppose the copper cable between wiring closets 4 and 5 and between 10 and 11 is replaced with optical converters connected by optical fiber cable. Each of the copper segments is subdivided into two copper segments separated by a fiber segment. The ring-segment drive distance for the larger of the new copper segments (wiring closets 1, 2, 3, and 4 in the west wing of the building) is: 30 + 30 + 30 + 250 = 340 feet (104 meters). Table 1 indicates that the maximum allowable drive distance for a ring segment with four wiring closets and eight MAUs is 367 feet (112 meters). Since the other pair of ring segments (wiring closet group 5, 6, and 7, and wiring closet group 8, 9, and 10) are smaller, the ring now meets the 16-Mbit/s wiring rules. All ring segments are capable of independent operation as required by the network problem-determination philosophy, which is to restore the maximum number of ring stations to network operation by removing the malfunction without requiring repair action. The ring now looks like the one in Figure 2B.

There are, of course, other solutions to this problem. If it is impractical to install optical fiber cable between wiring closets 4 and 5 and 10 and 11, a pair of converters can be installed at each output of wiring closets 4 and 11. These converters are connected to each other by an optical fiber patch cable made up with BNC connectors on each end. For the ring segment bounded by the converters in wiring closets 4 and 7 the new drive distance is 30 + 30 + 30 + 250 = 340 feet (104 meters), instead of 30 + 30 + 250 = 310 feet (94 meters), when a converter was placed in wiring closet 5. Since this distance is still much

2. Single building. *A shows the basic layout for single-building 16-Mbit/s tactics as a two-wing, seven-story building. In B, a single-building network includes a 16-Mbit/s intrabuilding backbone. The most obvious initial placement of the optical links is along the two 500-foot (152-meter) runs from wiring closets 14 to 1 and 7 to 8.*

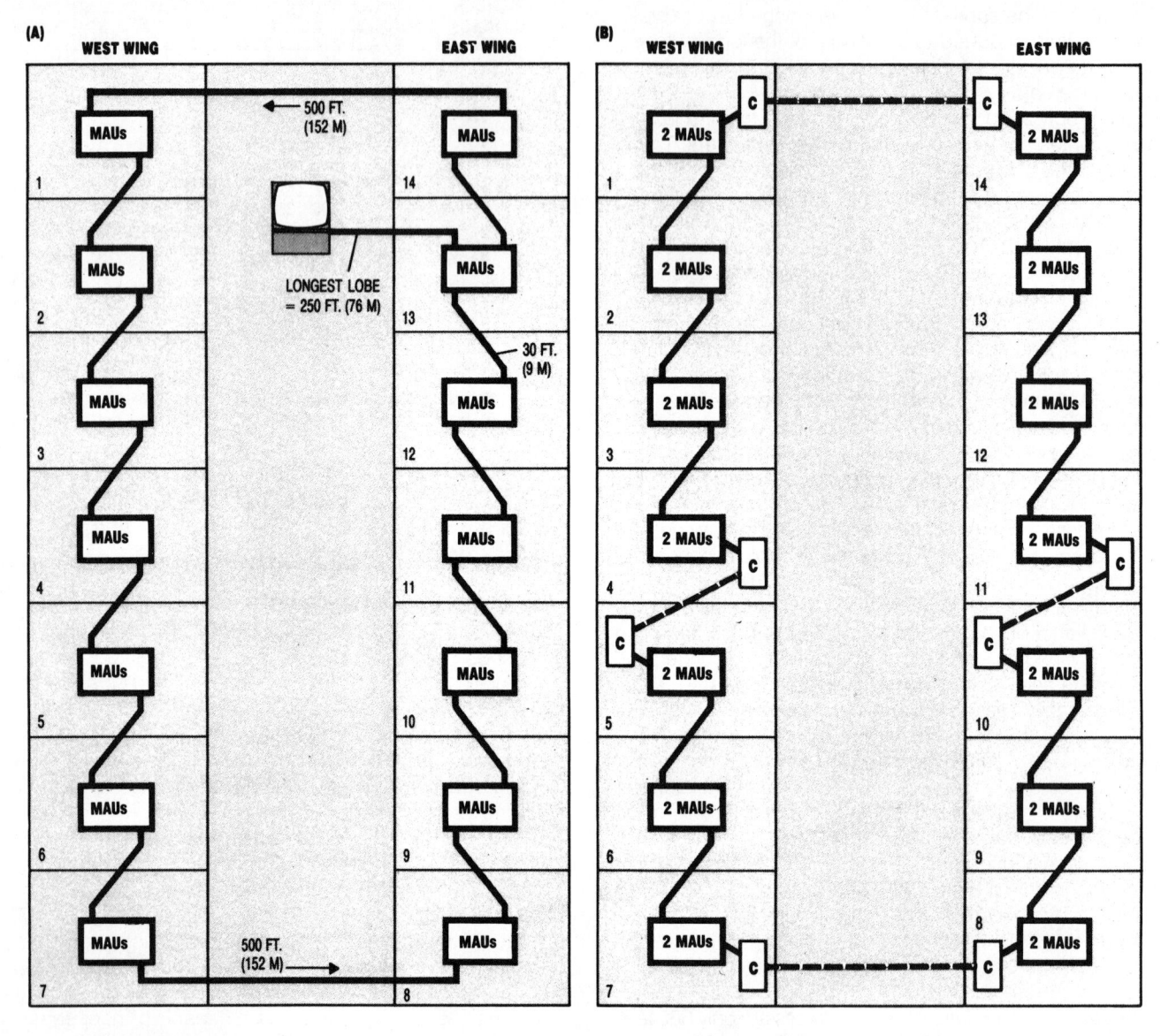

C = OPTICAL CONVERTER
MAU = MULTISTATION ACCESS UNIT

less than the 427 feet (130 meters) allowed for a three-wiring-closet, six-MAU segment in Table 1, the ring segment meets the wiring rules. The tables for ring-segment drive distance count only those wiring closets in the ring segments containing MAUs.

So far, the total number of attaching devices on the sample network has not been increased from the 4-Mbit/s solutions discussed in our earlier article. But it is unrealistic to expect that demand for attachment would not have increased in two years. Now that the network has proven itself, it is time to look at how it can grow.

Each of the 14 wiring closets could be configured to contain a single 4-Mbit/s ring of up to 260 attaching devices (33 MAUs). Table 2 shows that each of the 14 wiring closets could be configured to contain a single 16-Mbit/s ring of up to 168 attaching devices (21 MAUs) with no converters.

The individual rings require no repeaters or converters. Based on Table 3, a 16-Mbit/s Token Ring using Type 9 lobe cables—instead of Type 1—is limited to 104 attaching devices. Note that a 4-Mbit/s ring with 250-foot (76-meter) lobes of Type 9 cable could still be configured with 260 devices on 33 MAUs.

A separate 16-Mbit/s network is needed to link individual

Table 2: Drive distances

NUMBER OF MAUs	\multicolumn{10}{c}{NUMBER OF RACKS}
	1	2	3	4	5	6	7	8	9	10
1	569*									
2	556	523								
3	543	511	494							
4	531	498	481	465						
5	518	485	469	452	436					
6	505	472	456	439	423	407				
7	492	459	443	427	410	394	377			
8	479	447	430	414	397	381	365	348		
9	467	434	417	401	385	368	352	335	319	
10	454	421	405	388	372	355	339	323	306	290
11	441	408	392	375	359	343	326	310	293	277
12	428	395	379	363	346	330	313	297	281	264
13		383	366	350	333	317	301	284	268	251
14		370	353	337	321	304	288	271	255	239
15		357	341	324	308	291	275	259	242	226
16		344	328	311	295	279	262	246	229	213
17		331	315	299	282	266	249	233	217	200
18		319	302	286	269	253	237	220	204	187
19		306	290	273	257	240	224	207	191	175
20		279	263	247	230	214	197	181	165	148
21		253	236	220	204	187	171	154	138	122
22		226	210	193	177	161	144	128	111	95
23		200	183	167	150	134	118	101	85	68
24		173	157	140	124	107	91	75	58	42
25			130	114	97	81	64	48	32	15
26			103	87	71	54	58	21	—	—
27			77	60	44	28	11	—	—	—
28			50	34	1	8	—	—	—	—
29			24	—	—	—	—	—	—	—

*DISTANCES ARE IN FEET (TYPE 1 OR 2 CABLE) FOR 16-MBIT/S SINGLE-WIRING-CLOSET RINGS.   MAU = MULTISTATION ACCESS UNIT

rings together and to provide host attachment. The ideal location for such a backbone ring is often in the computer room to facilitate host connections. In many buildings the computer room is located on the first floor, often in the center of the building, so we will put ours there. The 16-Mbit/s backbone ring will consist of 5 MAUs to allow room for all 14 rings, several host connections, and later expansion.

Bridge works
We will place a bridge (see "Spanning wiring limitations with bridges") in each wiring closet. The longest lobes on the backbone network will run from the bridge workstations in wiring closets 1 and 14. Each of these lobes will be 430 feet (131 meters) of Type 1 cable. Table 2 shows that the maximum lobe length for a single wiring closet ring of 5 MAUs is 518 feet (158 meters). Therefore, the backbone ring will also operate without converters.

As appealing as this solution is, it cannot be replicated for all configurations, nor should it be. It would exceed lobe length limitations if the building were significantly taller or the computer room were not so centrally located. Conduit space may or may not be available for the cabling from each of the bridges to the computer room.

Alternatives abound. For the seven single-floor rings on the west wing, put seven bridges in wiring closet 5. Similarly, put seven bridges for the east wing in wiring closet 10. Provide six cables each from wiring closets 5 and 10 to connect the bridges in those closets to the single-floor rings. Connect the other adapter card in each bridge to a MAU.

The configuration so far is two groups of seven rings each, which are joined together by a backbone ring consisting of seven bridges and a MAU in a single wiring closet. These two groups of rings must be joined together, and the host attachments added.

The first solution connects the single wiring closet backbone rings using optical fiber cable and optical fiber converters. This approach requires two cables connecting wiring closets 5 and 10 with optical fiber converters at each end of each of the cables, totaling four optical fiber converters. Cabling would go from the host to the nearest wiring closet.

A second solution bridges these two backbone rings together. Assume here that the computer room is not centrally located but is in the west wing of the building on the first floor. The ring tying the two wings together will be in the computer room. Place an eighth bridge in each

Table 3: Drive distances

NUMBER OF MAUs	\multicolumn{10}{c}{NUMBER OF RACKS}
	1	2	3	4	5	6	7	8	9	10
1	379*									
2	371	349								
3	362	340	329							
4	354	332	321	310						
5	345	323	312	301	291					
6	337	315	304	293	282	271				
7	328	306	295	284	273	263	252			
8	320	298	287	276	265	254	243	232		
9	311	289	278	267	256	245	235	224	213	
10	303	281	270	259	248	237	226	215	204	193
11	294	272	261	250	239	228	217	207	196	185
12	286	264	253	242	231	220	209	198	187	176
13		255	244	233	222	211	200	190	179	168
14		247	236	225	214	203	192	181	170	159
15		238	227	216	205	194	183	172	162	151
16		230	219	208	197	186	175	164	153	142
17		221	210	199	188	177	166	155	144	134
18		212	202	191	180	169	158	147	136	125
19		204	193	182	171	160	149	138	127	116
20		186	175	164	153	142	132	121	110	99
21		169	158	147	136	125	114	103	92	81
22		151	140	129	118	107	96	85	74	63
23		133	122	111	100	89	78	67	57	46
24		115	104	93	83	72	61	50	39	28
25			87	76	65	54	43	32	21	10
26			69	58	47	36	25	14	—	—
27			51	40	29	18	—	—	—	—
28			34	23	12	—	—	—	—	—
29			16	—	—	—	—	—	—	—

*DISTANCES ARE IN FEET (TYPE 9 CABLE) FOR 16-MBIT/S SINGLE-WIRING-CLOSET RINGS.   MAU = MULTISTATION ACCESS UNIT

Bridging wiring limitations

The IEEE 802.5 committee is in the midst of confirming source routing as a Token Ring standard. Source routing means that an end station is responsible for providing a route through the bridge connected to the target station. This supports a link connection between the two stations. The power of the source-routing algorithm lies in the fact that it allows bridges that connect Token Rings to provide multiple paths between stations, increasing network availability.

Another capability deriving from source routing is advanced network management. All routed frames contain path information, so that a break in the communications path can result in a notification to network management of where the break has occurred. This feature is important in planning for physical maintenance of a Token Ring network.

For smaller networks, including those within single buildings, another bridging strategy is to daisy-chain rings together. Redundancy can be provided here by connecting the first and last rings together.

The number of hops a message can take across bridges limits the daisy-chaining of rings, however. The developing IEEE 802.5 standard limits the maximum number of hops to 14. Some implementations further reduce the maximum hop count to seven for better network performance and control.

Multibuilding networks generally apply a more structured approach to bridging. Local rings, servicing a single building or portion of a building, are bridged to rings covering a segment of the campus. These can be bridged to rings spanning larger segments of the campus, or even the entire campus.

At any level of interconnection, multiple bridges or multiple rings can be used to increase the availability of the network. Using this hierarchical approach with a four-tier structure and a six-hop limit to bridges, a single IEEE 802.5 Token Ring network can support over a million users.

To the extent that communications is primarily confined to local rings, bridges increase the total available bandwidth of the network. Parallel rings and parallel bridges can increase the availability of the network.

In addition, bridges can increase the physical flexibility of the network. By interconnecting both 4- and 16-Mbit/s rings together, the higher speed can be achieved where it is needed in the network, and the better physical design flexibility of 4-Mbit/s Token Rings can be exploited where the higher data rate is not required. Exceptionally long lobe lengths (600 to 1,200 feet [183 to 366 meters]), which are not allowed at 16 Mbit/s, can be supported on 4-Mbit/s Token Rings.

In addition, 4-Mbit/s rings that use telephone twisted pair as the lobe cabling can be bridged to a 16-Mbit/s backbone, thus providing the high-speed backbone to stations that do not have high quality data-grade media available to them.

of the wiring closets 5 and 10, and connect it to the last available lobe connection in the backbone MAU. Connect the other port of the bridges to a MAU located in the computer room. The longest lobe length is from the bridge in wiring closet 10 to the computer room, with a cable distance of about 560 feet (171 meters).

This configuration is within the 569-foot (173-meter) drive distance limit. This solution requires no converters but two additional bridges. The incremental wiring, in addition to the lobe cabling running from the wiring closets to office areas, consists of a cable that runs from wiring closet 10 to the computer room and cable duct space. The duct space allows a maximum of four additional cables between any two floors in either wing. In addition, patch cables are used within the computer room to connect the host to the MAU.

If all the rings in this network operate at 16 Mbit/s and all lobe wiring used was Type 1 or Type 2 cable, this hypothetical network could support up to 2,352 attaching devices without the use of converters (exclusive of host attachment devices). If the lobes were wired with Type 9 cable, up to 1,456 attaching devices could be supported without converters. If the rings on each floor were all operating at 4 Mbit/s, up to 3,640 devices could be supported exclusive of host attachment, again without repeaters or converters.

Further, these are not the outside limits for bridged rings but instead reflect the assumption of one ring per wiring closet with the 250-foot (76-meter) maximum lobe lengths in this building plan.

Campus backbone rings

When multiple buildings are involved, all the buildings should be joined by optical fiber. Existing copper surge suppressors do not support 16-Mbit/s Token Ring operation. Also, joining buildings with optical fiber cabling eliminates lightning and ground-potential-difference problems between the buildings. Finally, the long drive distances provided by the optical fiber converters are especially well suited to campus backbones.

Typically, interbuilding cabling comes into a structure at the same entry point used by telephone cabling, a location appropriate for installing optical fiber converters. The number of networks in a building and the locations of the wiring closets containing those networks will determine the topography of the backbone ring in a particular building.

Figure 3 illustrates the basic layout of the campus. Building A is the same as the one described in the single-building exercise with the installed intrabuilding backbone ring. Building B is a warehouse measuring 1,800 feet (549 meters) by 300 feet (91 meters) and containing 12 rings that are currently not joined together. Building C is a 20-story office tower with a ring on each floor. These rings are not interconnected. The objective is to connect all of the rings to allow peer-to-peer communications and to permit host attachment at the computer room in building A.

In building A, the building utility entry is located approximately 500 feet (152 meters) from the computer room's intrabuilding backbone ring, which will be used to connect to the campus backbone with a bridge. Cables with at least

3. Campus backbone ring. *Token Ring design techniques can be easily extended to encompass multiple buildings, creating a single network. Ultimately, all rings linked together should provide peer-to-peer communications, including attachment to hosts. Building A is a seven-story, two-wing unit, B is a warehouse and C is a 20-story office tower.*

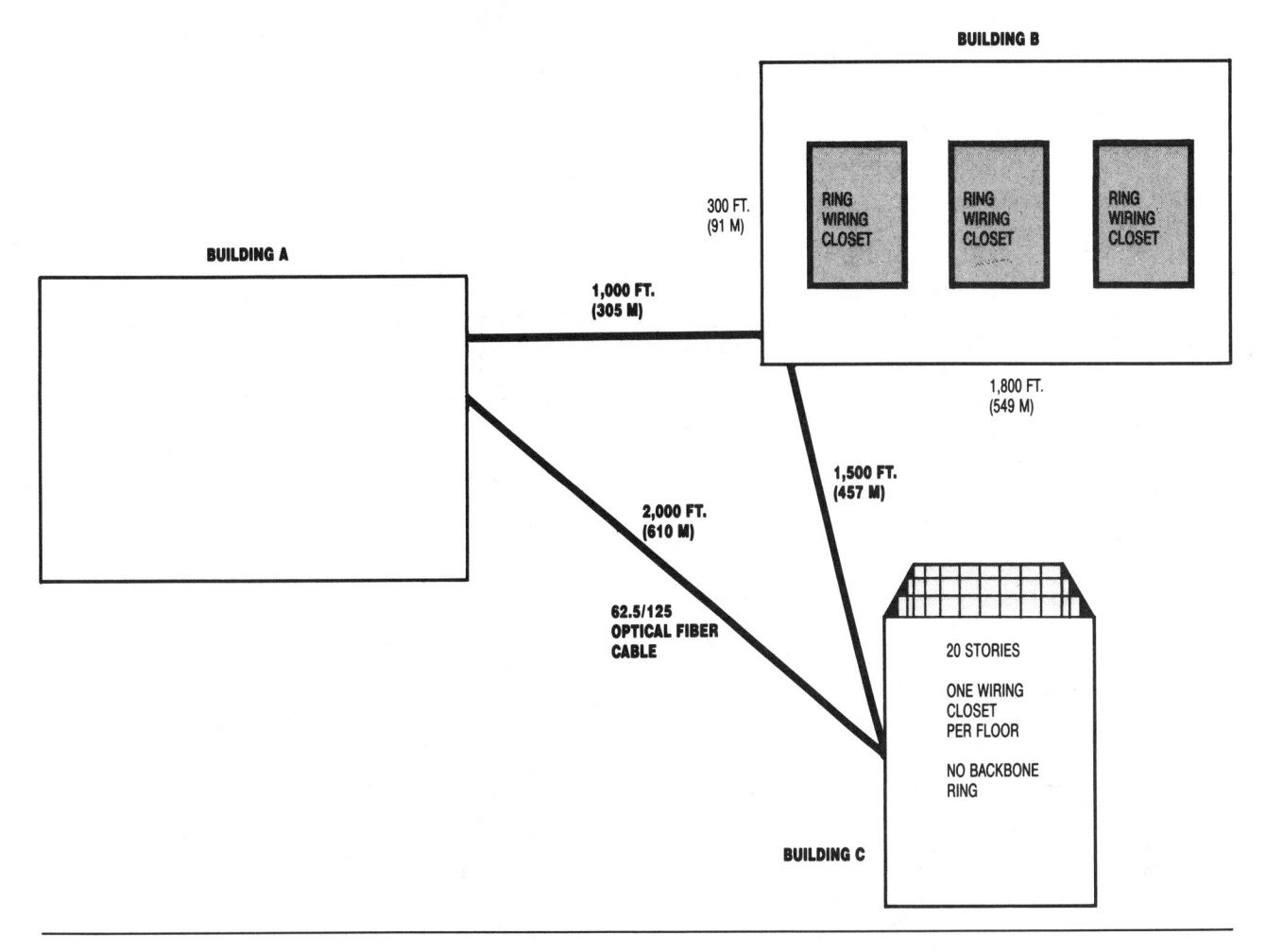

two available optical fiber strands each enter the building from both building B and building C. Since the two cable lengths are well under 6,600 feet (2 kilometers), these four optical fibers should be spliced or connected to intrabuilding optical fiber cable runs between the building entry and the computer room.

Techniques such as fusion splicing or mechanical splicing are commonly used, but where optical loss is not critical, connectors and patch cables are an option. Optical fiber converters can then be placed in a rack with the other network components. A bridge and a single MAU will connect the intrabuilding backbone network to the campus backbone providing both peer-to-peer communications and host attachment for the campus.

Alternatively, the optical fiber converters and the MAU could be installed at the building entry, with a single run of Type 1 copper cable linking the building entry to the computer room. This wire could serve as a lobe connection for the bridge to the intrabuilding backbone located in the computer room.

Building B, because of its large size and the location of its wiring entry point, seems at first glance to present a much different set of design challenges than building A. Because the 12 building-B rings are equally distributed among the three widely separated wiring closets, there are two workable alternatives. One of the design approaches is similar to that used in building A.

The first solution for building B is to build an intrabuilding backbone ring with a single MAU in each wiring closet. All interwiring-closet cabling is optical fiber with an optical fiber converter at each end of the cable run. This intrabuilding backbone ring requires six converters. In addition, a single lobe of Type 1 cable approximately 350 feet (107 meters) long will have to be installed between wiring closet 1 and the building entry point. A MAU, two optical fiber converters, and a bridge at the building entry point provide the connections to the campus backbone. An alternative would be to splice the four optical fibers of the campus backbone to fiber cables run to wiring closet 1. Then the bridge workstation, MAU, and optical fiber converters for the campus backbone could be placed in wiring closet 1.

A second approach eliminates the bridge between the

intrabuilding backbone and the campus backbone by making the three MAUs in building B part of the campus backbone ring. Here, the cabling for the campus ring in this building runs from the building entry, through wiring closets 1, 2, and 3 and back to the building entry.

This approach assumes that virtually all campus network traffic in building B leaves the building and is not transmitted from one ring to another in the building. If that is not the case, the intrabuilding backbone approach would be preferable because local inter-ring traffic would not have to use the campus backbone for communications within the building.

As a general guideline, whenever a ring can be split into two rings and where most of its traffic stays on the local rings, the available communications bandwidth increases. When splitting a ring in two requires most traffic on at least one of those rings to cross the bridge connecting the rings, available bandwidth has not increased. In addition, overall network performance may be somewhat degraded.

All of the wiring closets in building C are stacked, and there is a single ring on each floor of the building. Obviously, either an intrabuilding backbone ring bridged to the campus backbone or the propagation of the campus backbone across the 20 stories is possible. Multiple intrabuilding backbones may also be desirable, depending upon the amount of local traffic being generated. For example, the wiring closets on the first five floors might be joined together by an intrabuilding backbone and all of the other wiring closets by another.

One way to construct an intrabuilding 16-Mbit/s backbone for this building is to place three MAUs in the wiring closet on the tenth floor, while placing the bridges for each of the 20 single-floor rings next to the ring they support. Type 1 cabling is then used for lobes to all of the other wiring closets. Another run of Type 1 cable is required between the wiring closet on the tenth floor and the building entry to accommodate the bridge to the campus backbone. Each of these lobes is on the building C backbone. Since no lobe exceeds 543 feet (166 meters), this ring requires no converters.

Single-closet backbone

There are a number of design principles that make this choice of configurations preferable. By keeping the backbone ring within a single wiring closet, we are able to take advantage of the additional drive distances for single-wiring-closet rings. Also, MAUs decrease available drive distance since they are a source of attenuation in the main ring path. By minimizing the number of MAUs, we can either increase lobe lengths or decrease the number of optical fiber converters used.

The components of the campus backbone in building C are two optical fiber converters connected to the optical fiber cables from buildings A and B and a single MAU.

If all of the buildings on the campus use intrabuilding backbone rings and the distance between any two of the buildings is less than 6,600 feet, the campus backbone will consist of six optical fiber converters and three MAUs. Together they can provide peer-to-peer and host connections for 46 local rings connecting over 7,500 attaching devices. If the local rings are operating at 4 Mbit/s instead of 16 Mbit/s, over 11,500 attaching devices can be supported. In both these cases, no repeaters or converters are required on the local rings.

Points beyond

Token Ring networks can be extended beyond the boundaries of the establishment by employing a remote bridge that allows two rings at some distance from each other to be connected into a single network. A remote bridge connection is made up of two bridges (each attached to both of the rings), two modems, and a communications line operating at rates from 9.6 kbit/s to over 1 Mbit/s.

The configuration just described is hierarchical; between any two rings on the network there is only one path through intermediate rings. This scheme provides a high level of availability in that a single failure does not affect the entire network. If a local ring fails, it affects only those devices on that ring. If an intrabuilding ring fails, the local rings, other intrabuilding rings, and the campus backbone all continue to operate. If the campus backbone fails, local rings and intrabuilding backbones will continue to operate.

Since the campus backbone ring is vital in providing host connection and interbuilding communications, some additional measures can be employed to ensure maximum availability. First, spare pairs of optical fibers should be installed for all interbuilding cable runs. Next, consider installing a dual backbone ring as shown in Figure 4.

Installation of the dual backbone changes the overall configuration to a mesh network, providing more than one path through intermediate rings between two other rings. Under normal operation the campus backbone rings provide an aggregate bandwidth of 32 Mbit/s. More important, if one campus ring fails, the other will still be available so that communications across the entire network is uninterrupted at a data rate of 16 Mbit/s.

The introduction of 16-Mbit/s TokenRing networks and the bridges and gateways that support interconnection of networks to each other and to host computers has greatly enhanced the flexibility of LANs. At the same time, we must remember that greater flexibility and utility, insofar as they make the network accessible to more and more users, demand the maximum amount of reliability.

Because the foundation of the Token Ring network is the cabling on which it is implemented, if it and its relationship to the network have not been carefully planned, reliability may be affected. Indeed, reliability begins with a carefully planned cabling implementation that stays within the boundaries described in this article. Careful implementation pays dividends as well in its flexibility for expansion, ability to handle moves and changes, and potential use with emerging technologies. ∎

Robert D. Love, a senior engineer at IBM's transmission system technology department, has been an active participant in the IEEE 802.5 Token Ring standards committee since 1982. He is responsible for the development of the wiring rules for IBM's Token Ring network. He received his BSEE from Columbia University and an MS in electro-

4. Dual backbone rings. Increased availability and higher traffic handling are achieved with dual-backbone configurations. This technique is extendable to a multitier approach. Dual backbones create mesh networks, implementing the most reliable topology. Communications will continue at an uninterrupted data rate of 16 Mbit/s if one ring fails.

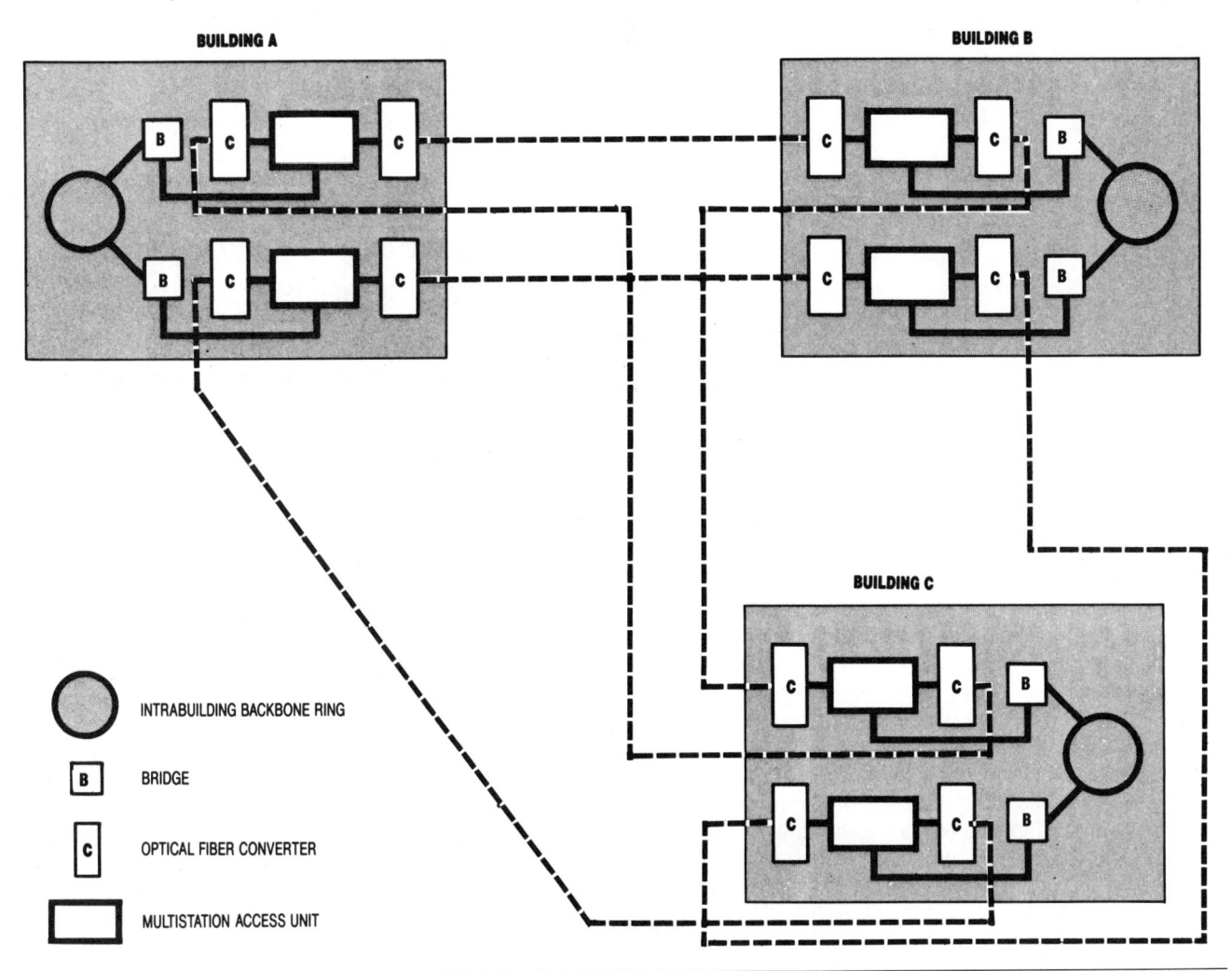

BUILDING A

BUILDING B

BUILDING C

○ INTRABUILDING BACKBONE RING
B BRIDGE
C OPTICAL FIBER CONVERTER
▭ MULTISTATION ACCESS UNIT

physics from the Brooklyn Polytechnic Institute.

Thomas Toher is a staff information developer in IBM's network products information development department. He received a BA from Hobart College (Geneva, N.Y.) and an MA in English from Clark University (Worcester, Mass.).

Suggested reading:

ANSI/IEEE Std 802.5 1985, published by The Institute of Electrical and Electronic Engineers Inc. 1985. The ANSI/IEEE standard for 4-Mbit/s Token Ring operation.

Working documents of the IEEE 802.5 Committee. Local Area Networks Token Ring Media Access Method and Physical Layer. Specifications (802.5X 88/55). 1988. Describes the updates to the standard to support 16-Mbit/s operation. Also, Draft Addendum to ANSI/IEEE Std 802.5 1988 Token Ring MAC and PHY Specification Enhancement for Multiple Ring Networks. (Unapproved draft for comment 2/21/89). Describes source routing.

FDDI and ANSI TR41.8 working papers. FDDI Physical Layer Medium Dependent (PMD) Draft 5/20/88 (X3T9/88 055). Describes the Physical Layer Specification for FDDI. Also, Commercial Building Wiring Standard (PN 1907) (3/89. Draft 6.0). ANSI TR 41.8.1 Draft 3.0. Describes the proposed building wiring rules and defines all allowable cable types

"How to design and build a Token Ring LAN," Robert D. Love and Thomas Toher, DATA COMMUNICATIONS, May 1987, p. 213. Describes 4-Mbit/s Token Ring design.

"Transmission," Richard J. S. Bates, Lee C. Haas, Robert D. Love, and Franc E. Noel, DATA COMMUNICATIONS, March 1986, p. 279. Describes Token Ring transmission over telephone twisted pair.

"Inside Token Ring Version II, according to Big Blue," Norm Strole, DATA COMMUNICATIONS, January 1989, p. 117. Describes IBM's 16-Mbit/s Token Ring offering including bridges and optical converters.

Dean Wolf, Fujitsu Inc., and Steven S. King, DATA COMMUNICATIONS

Making the most of ISDN now

Although reports coming in from ISDN trials praise the high speeds and low error rates, it's apparent that new application planning criteria are needed.

Over the past few years, the noise level from discussions about ISDN theory and internals has grown to an acronym-laden roar—CCITT I.430, SS7, Q.931, 2B1Q, V.120, TE2, and so on. Unfortunately for readers of technical publications, much of this flood of verbiage refers to elements of digital networks that lie outside of a user's domain.

But in the realm of customer data terminal equipment (DTE) and workplace ISDN applications, things are a lot more familiar than the ISDN theory articles would have it. In the course of a user's daily interaction with ISDN, for instance, computers converse with ISDN terminal adapters (the modem-like device that connects non-ISDN computers to ISDN networks) using familiar Hayes AT and X.25-PAD command sets. Q.931 signaling procedures running on the ISDN D channel's Link Access Procedure-D (LAPD) protocol are there too, just as the articles predicted, but they exist well below the level of user operations.

Introductory articles have also given us some very large expectations about what ISDN will deliver. With all that's been said, it is easy to assume that local ISDN services and high-speed terminal adapters (TA) will arrive just in time to meet upcoming demands for increased communications bandwidth, allowing computers to exchange data briskly, at 19.2 kbit/s or even 64 kbit/s. And if that's not fast enough, according to the ISDN gurus, we will be able to order up a couple of ISDN B channels, add some compression, and go for 200 kbit/s, desktop to desktop.

Of course, it's not that simple. Many of the ISDN pilot sites aren't running at 64 kbit/s yet, let alone 200 kbit/s. In fact, most are running at 19.2 kbit/s or less. Given the constraints of existing hardware and software, some initial ISDN applications have been limited to the snail's pace of channels that were designed at the height of the analog era. Let's face it, even today, few communications programmers are coding applications with error-free 64-kbit/s digital lines in mind.

For insight on the implications of ISDN for existing communications setups, consider what happened when the industry first added high-speed analog modems to its arsenal of communications gear. These modems talked to each other at dazzling speeds across voice-grade lines. Unfortunately, the process of goosing throughput exposed previously unconsidered weaknesses in computer-to-modem links, in data rates available on installed ports, and particularly, in production communication software.

In addition to throughput bottlenecks in the non-ISDN elements of their networks, early ISDN users are finding that their application software is not always well suited to a near-error-free environment. These pioneers have discovered that communications software doesn't stop giving time-consuming acknowledgments for every packet just because it's on an ISDN line. In most cases, communications software such as Kermit, Crosstalk, or ProComm doesn't know that it's not talking to an analog modem. This is because the TAs do such a good job of emulating a modem on the DTE side of their interfaces.

Eventually, ISDN ports will be built into every computer, as RS-232 and similar ports are today. But until then, computers will need a TA between them and the switching nodes on a digital network. Consequently, it is the features and limitations of both the TA and the central office switch that determine what services are available to today's ISDN users.

Basic-rate services

Although not available in all possible combinations, the fundamental set of ISDN basic-rate data services consists of:
- B-channel circuit-switched (BCS) service for one or two B channels on an as-needed basis, with data rates up to 64 kbit/s. BCS can be asynchronous or synchronous, and is

protocol-independent, in that it does not assume X.25 or other protocols. Some, but not all, central office switches can provide a permanent B channel circuit as a provisioned service.

- B-channel packet-switched (BPS), an X.25 service on the B channels on a permanent or switched basis, to 64 kbit/s. Some switches allow BPS on one B channel only and may require BPS to be permanent in initial offerings.
- D-channel packet-switched (DPS), an X.25 service on the D channel at sub-16-kbit/s data rates. Circuit switching is not available for current ISDN D-channel implementations.

In spite of all the attention given ISDN's 64-kbit/s data rates and advanced interfaces such as synchronous V.35, much of the initial application work is being conducted with asynchronous equipment at speeds of 19.2 kbit/s or less. This makes sense, considering that currently there are few computers or front-end processors (FEPs) with asynchronous RS-232 ports running at 64 kbit/s. Generally, TAs support the 64-kbit/s rates with synchronous protocols.

Because of the lack of widespread support for 64 kbit/s, TAs use a rate-adaption technique that lets them provide user equipment with speeds of 56 kbit/s, 48 kbit/s, or less on the B channels. TAs come with one or two RS-232 or V.35 ports that can be used to access the BCS services concurrently. At sub-19.2-kbit/s speeds there are typically a variety of TA rate options: 300 bit/s, 1.2 kbit/s, 2.4 kbit/s, 4.8 kbit/s, 9.6 kbit/s, and 19.2 kbit/s.

In the case of X.25 packet switching for the B channels, it is important to distinguish the nominal 64-kbit/s rate from the throughput class of the X.25 virtual circuits, generally 9.6 or 19.2 kbit/s. Multiple virtual circuits can run on a single ISDN channel; their throughput class is independent of the channel speed. If the connecting computer has an X.25 FEP, the TA's internal X.25 PAD can be turned off. When the TA's PAD is on, it can be configured with familiar syntax, such as the Hayes AT or CCITT X.28 commands (see Table 1).

Manufacturers of ISDN-capable central office switches generally provide a B-channel packet-data facility as a provisioned (preassigned) service. This means that the function is set up specifically via a service order, in a

Table 1: Typical ISDN Terminal Adapter Commands

X.28 commands

(All commands are followed by <CR> except <CTRL>P)

PAD command	Function
CLR	Clears a virtual call
INT	Transmits an interrupt request packet
MENU	Switches to off-line command mode
PROFn	Selects a profile
PROF?[n]	Reads a profile (the currently active one if number is omitted)
PAR?n	Reads a parameter value
R	Requests reestablishment of a virtual call
RESET	Transmits a reset request packet
STAT	Requests status of a virtual call connected to the DTE
n...n	Sets up a virtual call
SETmin	Changes or establishes a parameter value
SET?min	Changes or establishes a parameter value and then reads it
<CTRL>P	Escape character

X.28 result codes

COM	Call connected
ERR INC	Error
RESET	Reset
FREE	Call status
ENGAGED	Call status
PAR (list)	Parameter value
CLR CONF	Clear confirmation
CLR ERR	Local procedure error
CLR RPE	Remote procedure error
CLR PAD	Call cleared by PAD
CLR DTE	Call cleared by terminal
CLR OCC	Called number busy
CLR INV	Invalid call
CLR DER	Called number out of order
CLR NC	Network blocked
CLR NA	Access denied
CLR NP	Called number not assigned

AT commands

Command	Parameter	Function
AT		Attention code; precedes all commands except A/ command
A/	None	Repeats last command (Don't use prefix AT)
Dn	n = up to 30 digits	Dials a call (tone dialing only)
En	n = 0	Do not echo commands
	n = 1	Echo commands
H	None	Disconnect a call
O	None	Returns to on-line state
Qn	n = 0	Sends result codes
	n = 1	Suppresses result codes (quiet mode)
Sr=n	r = 0-16	Register number
	n = 0-255	Sets register 'r' to value 'n'
Sr?	r = 0-16	Reads contents of register 'r'
Vn	n = 0	Numeric output of result codes
	n = 1	Verbal output of result codes
Xn	n = 0	Result codes 0-8 and 30-33 (See Result Codes table)
	n = 5	Preceding result codes and 21-23
%I	None	Request to send INTERRUPT packet (for PAD)
%Pn	n = 0-3	Selects specified profile (for PAD)
%P?n	n = 0-3	Reads specified profile (for PAD)
%R	None	Requests to send RESET packet (for PAD)
%S	None	Requests call status (for PAD)
MENU	None	Switches to off-line command mode

AT result codes

Numeric	Verbal (DPS/BPS)	Verbal (BCS)
0	OK	
1	CONNECT	
3	NO CARRIER	
4	ERROR	
7	BUSY	
8		NO ANSWER
21	ENGAGED*	
22	FREE*	
23	TRANSFER*	
30		CALL REJECTED
31		B-CH BUSY
32		INCOMPATIBLE
33		L1 DEACT

*Expanded result codes

manner similar to that of placing a subscription order with a value-added network for X.25 service. Current AT&T 5ESS switch software allows provisioning of a single permanent BPS channel per basic-rate interface (BRI). Switched BPS service will become available with the next 5ESS software release for nondedicated applications. The Northern Telecom DMS-100 switch allows both B channels to be configured as packet-switched, permanent service.

On many ISDN installations, the most attention has been given to the basic rate's packet-switching D channel (DPS), which supports X.25 as the B channel does, only at lower speeds. Part of the reason for this attention is economic, in that the D channel is quite inexpensive compared to the high (but descending) costs for the B channels.

The D channel runs at 16 kbit/s between the TA and the ISDN switch. But because this channel is also used for call set-up and network management signaling, the full 16 kbit/s is not available to the user.

Whenever possible, user equipment is configured to interface with a TA at 19.2 kbit/s, while the TA transfers through the network as fast as current loading conditions allow. The D channel can be shared by multiple devices—on the same BRI—in which case throughput can bog down to unacceptable levels. Trials on a single D-channel attached device on lightly loaded networks have clocked end-to-end throughputs of up to 14 kbit/s. It must be stressed that only ISDN-optimized software can attain these speeds.

Packet heaven

Because of the multiple-virtual-circuit facility of X.25, calls can be set up between multiple DPS- or BPS-connected computers in terminal-to-host configurations (see Fig. 1).

1. Packet-switched. The ISDN central office switch freely mixes circuit-switched and packet-switched channels. The advantage of packet switching is that channel bandwidth is allocated to a call only upon the appearance of data from a terminal device. This data is typically divided into packets of 128 or 256 bytes.

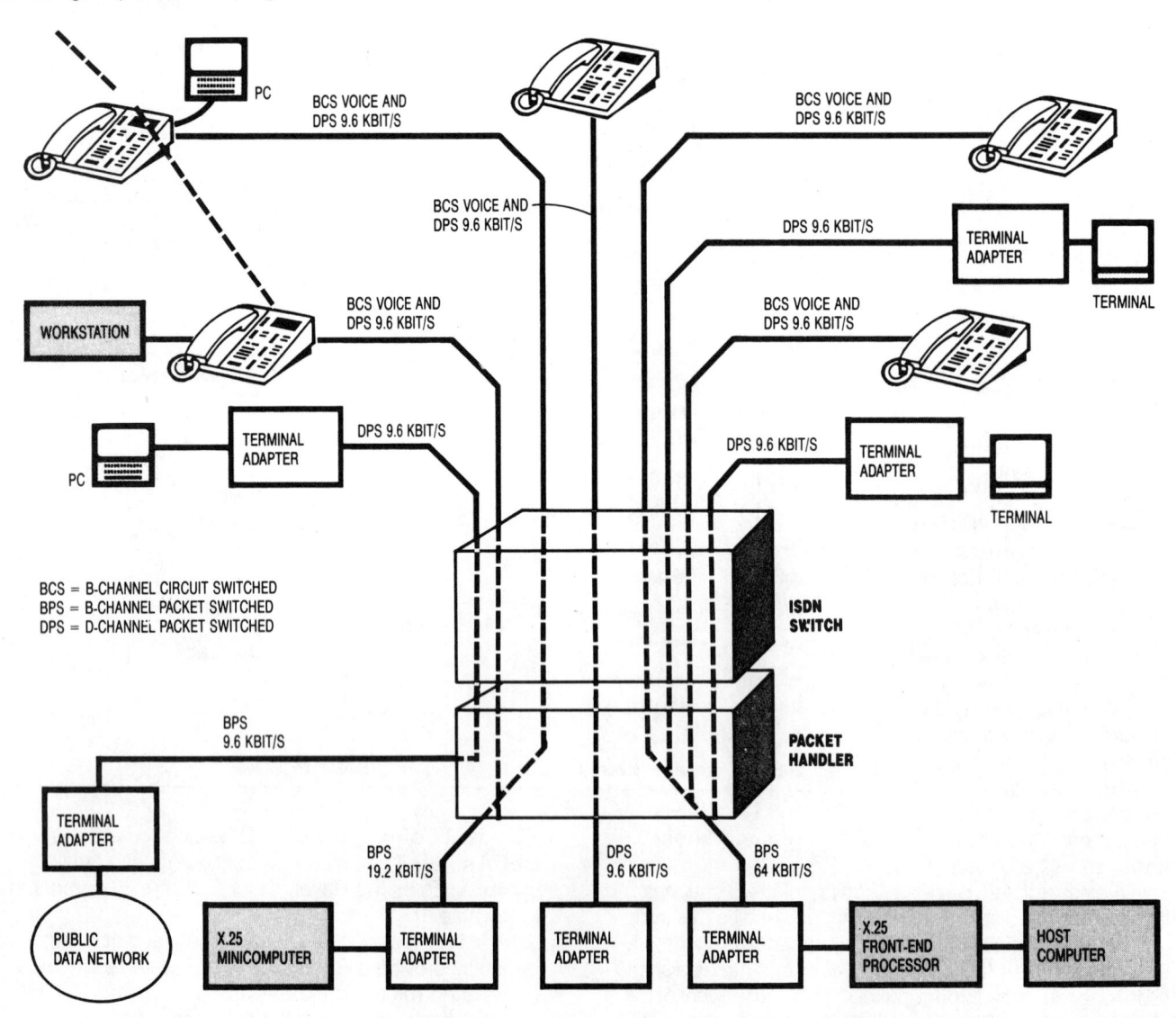

2. Multidropping. *The simplest technique for boosting channel utilization is multidropping. Here, two telephones with integrated TAs share the line with a standalone TA.*

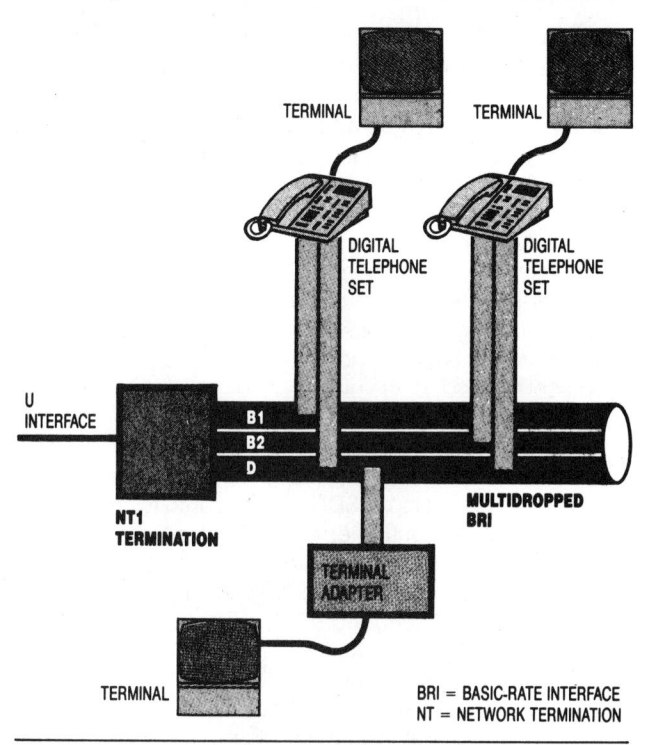

For example, several DPS TAs can create logical channel connections through a central office packet handler to a single BPS-connected computer. BPS service can provide greater throughput than DPS service because of the higher speed of the channel and the ability to request a higher throughput class (an X.25 feature that defines the maximum data rate of a virtual circuit—for instance, 19.2 kbit/s).

Packet switching on the B and D channels brings a number of advantages to X.25, such as the ability to multiplex multiple logical links on the 2B+D channels. The end effect of X.25 on an ISDN is error-free communications from TA to TA. For circuit-switched ISDN, without X.25 error correction, bit-error rates are related to the quality of the local loop, something on the order of one in 10^{-6} bits.

The 5ESS allows abbreviated dialing for X.25 packet-switched calls via a feature called 4-Digit Intercom. Northern Telecom has an automated dial feature called Direct Call, which allows prestoring of the destination address at subscription time.

There will increasingly be alternatives to ISDN's classic one-BRI-per-TA configuration. The availability of these alternatives varies greatly from vendor to vendor, but it usually involves multiple TAs sharing a single basic-rate line. For instance, the ISDN specifications allow up to eight terminals to be multidropped per line.

The advantage of multidropping is greater utilization of the 2B+D line (see Fig. 2). In this configuration, two digital telephones with integrated TAs share the line with a stand-alone DPS TA. The two B channels are fully occupied by the telephone users and the D channel carries the traffic for three terminals. ISDN trials have shown that response times for DPS users can become noticeably slower if more than three or four terminal users simultaneously share the D channel for a typical interactive application such as on-line text editing.

Note that the ability to utilize a BRI in the manner described is dependent on the bandwidth requirements of a given application. For example, if PCs are given D-channel access and the application calls for continuous PC-to-PC or PC-to-host file transfers, it is recommended that the BRI be configured to allow only a single PC to have D-channel access. A BRI configured in this manner could support two voice-only telephones and a PC via a stand-alone DPS TA.

Trials and errors

Given that the majority of in-place software and hardware has been designed to transfer data on the lower-speed, error-prone channels that typify the analog age, a choice that faces users is: either down-size expectations for ISDN, or, upgrade existing components, often at considerable expense and effort. Judging from the trials, there will be a lot of both going on, as ISDN becomes widely available.

To fully comprehend the effects of high-grade digital links, it is necessary to examine the entire end-to-end application, including FEP-to-host communications, packetizing software, port-to-bus interfaces, and file transfer methods. All these components are potential bottleneck areas.

Overengineering the ISDN interface to support a full 64 kbit/s for an application that is throughput-limited to 9.6 kbit/s (such as 3270 protocol conversion) obviously results in excessive cost and wasted bandwidth. This is due to the underutilization of the high-cost TA and FEP with, for example, 64-kbit/s V.35 interfaces.

Suppose you have an application running at 19.2 kbit/s. You feel the application could benefit from a speed upgrade to 64 kbit/s with ISDN BCS service. Before upgrading your network components—such as communications controllers for a host computer—talk to your equipment supplier. Ask for references of users with network environments similar to yours who have installed the application on currently available 56-kbit/s Dataphone Digital Service (DDS) links. Were there any bottlenecks? Was the application able to completely utilize the 56-kbit/s line?

Generally speaking, if an application can fully utilize a 56-kbit/s DDS circuit to obtain a response time or throughput improvement, it is a candidate for ISDN's full 64-kbit/s B channels. If your equipment suppliers have participated in ISDN compatibility tests, you may benefit from their experience and insight about the applications that work best in such an environment. Available ISDN features can vary somewhat, depending on the manufacturer.

The experience of early ISDN users has shown that the penalty for ignoring the basic layered approach of OSI becomes a greater factor with the higher speeds of ISDN. The penalty is failure to achieve full utilization of the 64-kbit/s B channel. Some users were unable to take advantage of the higher speeds because of unnecessary overhead and duplication—both avoidable with the OSI layered approach.

These inefficiencies have been particularly evident at sites where users ran their own software on top of the D channel's X.25, resulting in redundant packetization, flow control, and error checking. Consequently, even if the ISDN is well-tuned, application software duplicates communications functions, with adverse effects on throughput.

IBM network protocols that support X.25 tend to have a different type of bottleneck (see "IBM supports X.25, doesn't it? Clearly, it depends," DATA COMMUNICATIONS, March 1987). IBM's Network Packet Switching Interface (NPSI) contains a great deal of internal overhead, because it is constrained to convert X.25 on the network side to SNA on the host side in both directions.

BCS represents a real opportunity for upgrading SNA links, but for many users of IBM mainframe networks that require X.25 connectivity NPSI must be considered for its effects on throughput. In cases such as these, where the bottleneck is the FEP's communications method, it may be more cost-effective to utilize lower-cost/lower-speed ISDN components to provide connectivity. For example, choose TAs with RS-232 interfaces that support 9.6-kbit/s DPS and 19.2-kbit/s BPS, rather than V.35 interfaces and 64-kbit/s BPS.

Another user experience involved a site that was previously using short-haul modems and an ancient file transfer program with its own custom blocking, error checking, and flow control. This was definitely not a well-layered, OSI-compliant protocol stack. When the short-haul modems were replaced with 64-kbit/s ISDN, with the original communications software was retained, no performance improvement was detected.

With all the problems stemming from running old software on new network links, the question arises: Why not just use a non-error-checking approach that relies on the near-error-free ISDN environment for data integrity?

There have been and will continue to be cases where users move data with non-error-checking procedures on an ISDN. But, keep in mind that, even with the B and D channel X.25 packet services, error correction is only between the TA's—it is not end-to-end.

This means that errors will not be corrected when introduced by the cables and connectors between the TA and DTE, the DTE ports, or the DTE software itself. For instance, if there is a local source of noise interference that is causing bit errors in the DTE's RS-232 port, the most perfect D-channel session in the world can't stop this corruption of data.

That brings us to today's popular error-checking software products. In fairness to existing communication software, it must be said that not all non-ISDN optimized software products are worthless at the new speeds and error rates. End-to-end error correction and flow control is so critical to many applications, existing packages using Kermit, Xmodem, or proprietary transfer protocols will be used widely on ISDN.

Considering that ISDNs will often be employed to connect dissimilar systems, the fact that mature communications products such as Kermit run on so many different platforms is in their favor. And if these mature products have advanced features such as sliding windows, variable packet sizes, and X.25 line-turnaround control, it is likely that they will perform adequately (but not superlatively) on an ISDN.

Packet sizes on ISDNs are generally 128 kbit/s or 256 kbit/s, as specified by the user at provisioning time. On error-free lines such as an ISDN, it is normally recommended that communication software block size be set as high as possible, to reduce protocol overhead. With such good channels, the performance hits from retransmitting one of these long blocks should be few and far between.

Further, the block boundaries of the user software may not relate well to ISDN packet boundaries. This is likely the case when asynchronous software blocks its data in 128-byte increments—not uncommon. After several bytes of protocol are added, the block will no longer fit in the ISDN 128-byte packet. In some cases, two 128-byte packets will be sent for a single 128-byte block of user data.

Very long packets, 2 kbytes or more, are one method used by the makers of ISDN optimized software to maximize throughput. But other ISDN developers will say that long packets are just a Band-Aid, and what is really best for ISDN lines is an error-correcting technique that doesn't use conventional send-a-packet, acknowledge-a-packet algorithms.

Some of the fastest asynchronous software available today for ISDNs does not use standard stop-and-go acknowledging, but instead sends a constant stream of data, with unacknowledged cyclic redunancy checks (CRCs) included every 2 kbit/s, and minimal handshaking at one-minute intervals. With this streamlined approach, an acknowledgment is sent at the end of a file or group of files. In the rare event of a CRC-detected error, subsequent blocks are discarded and retransmission is started at the point of the error.

Besides higher speeds and greater efficiency, ISDN optimized communications software has other advantages over conventional software. (ISDN optimized software includes: HyperAccess, from Hilgraeve Inc., Monroe, Mich.; Excellnet, from ExcellTech Inc., Yankton, S.D.; IS NET, from Newbridge Networks, Herndon, Va.; and Manylink, from Netline Inc., Provo, Utah.) Just as pre-ISDN software has been optimized for the dominant analog modem technologies, ISDN software has been optimized for the emerging terminal adapters. This often involves collaboration between hardware and software makers to develop software drivers, configuration scripts, and user interfaces that are particularly well-suited for TAs.

The down side of this new software is its proprietary nature. Unlike the mature protocols (such as Kermit and Xmodem) that run on hundreds of different hosts and modems, new ISDN software supports a very limited repertoire of platforms. Consequently, to reap the rewards of ISDN-optimized software, it will in most cases be necessary to buy new software for every system in the application.

The correct method of planning for ISDN, while not particularly obvious, is quite similar to traditional network

Table 2: Typical X.25 channel translation settings

X.25 TRANSLATION FUNCTION	VALUE
(A) D-CHANNEL SETTINGS	
PACKET SIZE (SEND OR RECEIVE)	*128 OR 256 BYTES
PACKET WINDOW SIZE (SEND OR RECEIVE)	*2 OR 3 PACKETS
REVERSE CHARGING ACCEPTANCE	YES/*NO
FLOW-CONTROL PACKET NEGOTIATION	YES/*NO
FAST-SELECT ACCEPTANCE	YES/*NO
INTERCOM ADDRESS INDICATOR	FOUR-DIGIT NUMBER
(B) B-CHANNEL SETTINGS	
B1: CHANNEL-PACKET RATE	64 KBIT/S
N2: RETRANSMISSION	2-15
T1: RETRANSMISSION TIMER	20-200
WINDOW SIZE	1-7 PACKETS
PACKET SIZE	*128 OR 256

*DEFAULT

planning methods. The worst thing is to rush out and provision BRI lines without a good idea of communications requirements. Before even thinking about TAs and BRI lines, consider the capabilities and limitations of your application, end-to-end, all layers, hardware, and software. Next, look at TAs.

Do the right thing

Says AT&T senior engineer Harris Barbier: "You really need to know the capabilities of your TA before provisioning." He adds that the user manuals for TAs are a good place to go for specifications and tips on how to set up BRI lines so they will match the features of the TA. Some of the parameters are: packet size, variable sliding window size, number of virtual circuits, and throughput class for packet circuits.

With Centrex service, it is the telephone company's responsibility to enter in the switch the parameters for the ISDN line, but the telco cannot do this without accurate information from the user. Therefore, the user must carefully plan the network—not only how each BRI will be used but also down to specific parameter settings for DPS and BPS. A configuration profile that is established for each digital subscriber line (DSL, same as a BRI) is called a translation. The translation tells the central office how the line is to be used—such as voice only or voice and data—as well as specific voice and data characteristics of the telephone or terminal adapter at the other end. Table 2A gives typical translation options for the AT&T 5ESS D channel. This is a sampling of default values and their interpretations. In some cases, these defaults can be modified from the TA.

■ *Packet size.* The 5ESS allows a packet size of 128 or 256 bytes. Users with heavy terminal-character traffic would select 128, while those with file transfer requirements might choose 256 when their TA supports this value.

■ *Window size.* The switch's D channel window size is two or three packets. For a clean loop set the window to three, for a dirty loop set it to two.

■ *Reverse charging acceptance.* The reverse-charging option tells the switch whether an incoming caller may reverse the charges for the call. This value defaults to no.

■ *Flow-control negotiation.* If the flow-control parameter is set to "yes," TAs may negotiate the window size, packet size, and throughput class for both directions (independently) of the D channel. If this parameter is set to no, the default settings for these parameters will apply.

■ *Fast select acceptance.* This is an X.25 feature for quick calls in applications such as automated credit card approval. It allows data (a credit card number, for instance) to travel with the initial call-request packet. The default for this parameter is "no"—meaning fast select calls are not accepted.

■ *Intercom address indicator.* A four-digit number may be entered for fast dialing inside an AT&T exchange. This is for users of this AT&T service.

X.25 B-channel default settings (see Table 2B) control the parameters that support packet data on the B channel. With the current software on the 5ESS and DMS-100, BPS is a permanent-connection-only service. The X.25 B channel translation functions are similar to those for the D channel, but the values a user chooses for window and packet sizes may differ.

■ *N2 retransmissions.* This is an integer from two to 16 that controls the number of retransmissions for LAPB, level-two frame retransmissions. A value of two is typical for applications on a good loop; a higher value is necessary for a dirty loop.

■ *T1 retransmission time-out.* For this parameter, measured in tenths of a second, a setting of 20 would allow 2 seconds for a LAPB-level retransmission time-out.

■ *Window size.* Unlike X.25 parameters for the D channel, the B channel may have a window size of two to seven packets. Users in areas with problematic loops should keep this number low (two); otherwise, better performance may be gained by increasing it. But high window-size values require more memory in the TA and switch.

■ *Packet size.* The packet size of the ISDN B channel is independent of the D channel. Depending on the capabilities of the TA, this may be set to 256 to lower protocol overhead.

The proper translation settings are a function of application requirements and the capabilities of the terminal adapters and central office switches. TA manufacturers and switch providers are gathering a wealth of implementation experience in this emerging technology area, so the new ISDN user should never be alone when it's time to make things work. ■

Dean Wolf is Manager of Product Marketing with Fujitsu's ISDN Systems Group. He has 13 years experience in the field of data and voice communications. He holds a bachelor's degree from Southern Illinois University, Carbondale. Wolf acknowledges the technical and editorial support of Jim Weldon.

Stephen Fleming, Licom Inc., Herndon, Va.

Get ready for T3 networking

DS-3 will be as commonplace as today's T1 because of coming availability and cost-effectiveness. Here's what to expect and how to implement the new technology.

Many corporate managers are administering T1 networks that are growing at a rate that puts crabgrass to shame, and new users are clamoring for more and more bandwidth. Increasingly, data communications managers are eyeing T3 equipment as a solution to the problems raised by this expansion. Carriers and vendors, aware of the growing interest, are offering equipment and services at the T3 rate (44.736 Mbit/s—commonly called 45 Mbit/s).

As an indication of increasing T3 acceptability, a recent study by Ken Bosomworth of International Resource Development showed the T3 equipment market climbing from $20 million in 1988 to a predicted $330 million in 1994. Similarly, he says, the T3 services market should go from $20 million in 1988 to $900 million in 1994.

To better understand many of the issues related to T1 and T3 networking, one should be familiar with North American regulations (see "The digital hierarchy"). They dictate how digitally multiplexed signals may be transmitted over the public network in the U.S. and Canada. Different hierarchies of signals are used abroad, making direct interchange of voice or data signals with North American networks impossible.

Before it can be justified, a T3 backbone must prove itself economically. In late 1988, AT&T amended its Tariff No. 9 to significantly improve the economics of T3 circuits. The previous tariff involved a complicated scheme of mileage bands. There is now a fixed charge of $6,000 per month and a simple mileage charge based on airline miles between cities. The charge varies from $180 per month for one-year contracts to $150 per month for three years to $130 per month for five years. Carriers other than AT&T offer similar arrangements, often for even lower prices.

A comparison of these charges versus standard T1 charges is shown in the figure. (The T1 calculations assume typical AT&T monthly rates of $2,600 fixed and $14.85 per mile with a 15 percent volume discount; the T3 calculations assume a three-year rate.) As can be seen, a T3 circuit can be cost-justified by as few as four T1s on links of less than 50 miles. (Recall that a T3 represents the equivalent of 28 T1 circuits.) At the other extreme, 10 T1 circuits will always cost more than a T3, regardless of distance. With non-AT&T carriers, the principle remains unchanged, although the break-even points may vary.

Flexibility and control
Implementing a T3 corporate backbone provides users with a measure of flexibility and control over the network. At the low-speed end, circuits may be either 56 kbit/s or 64 kbit/s. These may be used for voice or data services. Clear-channel circuits at 64 kbit/s may also be used for the bearer (B) channels of ISDN.

Intermediate circuits are often referred to as fractional T1 or FT1. These consist of some subset of a full T1 in multiple-DS-0 bundles. For example, one-half of a T1 (768 kbit/s) or even one-quarter of a T1 (384 kbit/s) may be adequate for many videoconferencing applications, without dedicating an entire T1 circuit. Many carriers have begun offering fractional T1 service to capture users who have outgrown 56-kbit/s services but who cannot yet justify a full T1. With the proper switching functions, these may be consolidated within a T3 networking multiplexer.

Full T1 circuits may be switched between sites on demand. The bit rates of these circuits may be 1.544 Mbit/s (the standard T1, including the framing bit), 1.536 Mbit/s (a T1 payload, without the 193rd framing bit), or 1.344 Mbit/s (24 channels of 56 kbit/s each).

The capability to switch this variety of circuits allows the user to put up and take down circuits of differing rates on the T3 backbone as requirements change, without having to work through an external network provider. This can take the form of reducing lead time for a circuit order, time-of-

T3 Networking

The digital hierarchy

Digital transmission techniques were introduced in the former Bell System in the early 1960s for efficient transport of voice signals. At that time, data formed an increasingly smaller percentage of network traffic. Therefore, the basic digital transmission structure is centered around human voice communication.

A plot of amplitude versus time for a speech sample would show significant high-frequency components. Luckily, a successful conversation (involving both understandable speech and a recognizable speaker) requires less than 4 kHz of audio bandwidth. After being processed by a low-pass filter, a speech sample loses its high-frequency components and is ready to be digitized.

The maximum analog frequency to be reproduced is 4 kHz. Therefore, according to the Nyquist theorem (the rate at which data can be transmitted without incurring intersymbol interference cannot be more than twice the bandwidth in Hertz), an 8-kbit/s sampling rate was adopted. By sampling the voice signal every 125 microseconds, its essential information is extracted in analog format and readied for digital encoding (see Fig. A). This pulse-amplitude modulated (PAM) signal contains all the information in the original signal up to approximately 4 kHz. (Because of operational considerations, the actual cutoff in digital telephony is lower than 4 kHz, but the principle remains the same.)

■ **Pulse-code modulation.** The modulation scheme chosen for early digital transmission standards is the easiest to implement: pulse-code modulation. The PAM signal is quantized (see Fig. B) by mapping into discrete amplitude levels, each with a unique binary code. (The example in the figure uses four-bit coding.) Naturally, additional discrete mappings reduce the quantization error and improve the fidelity of the coded signal. The example shows four-bit coding, for a total of 2^4 — or 16 — possible coding levels. Actual devices were designed to implement eight-bit coding, providing 2^8 — 256 — quantization levels. This is sufficient for satisfactory reproduction of the human voice.

Note that by creating eight-bit codes 8,000 times per second (8 kbit/s), the bit rate of a digitized voice signal is eight multiplied by 8,000, or 64,000 bits per second. This is the basic 64-kbit/s channel that is the foundation of much of the digital network. Such a digital channel is often referred to as a DS-0 (digital signal, level 0).

■ **Time-division multiplexing.** The original deployment of digital transmission was driven by a desire to conserve copper pairs outside telephone offices. Each analog voice signal in the existing telephone network consumed a physical pair of copper wires from a subscriber location to a telephone central office. Once the basic building block was digital, it became feasible to multiplex the digital signals together into a higher-order digital signal. The interleaving method used, time-division multiplexing, led to defining the next level of the digital hierarchy as being equivalent to 24 DS-0 signals. This level is referred to as DS-1 or T1.

The multiplexing function to create a DS-1 signal was originally handled by a network element known as a digital channel bank or D-bank. Now, of course, numerous devices offer T1 interfaces.

■ **T1 format.** DS-1 is formed by byte-interleaving 24 DS-0 channels. The per-frame aggregate capacity, deduced from 24 channels of eight bits each, is 192 bits. Repeating the frame 8,000 times per second would result in an aggregate bit rate of $192 \times 8,000 = 1,536,000$ bits/second, or 1.536 Mbit/s.

This, however, is not the DS-1 rate, since a 193rd bit is added to each frame for timing and alignment purposes. This 193rd bit is called the framing bit and brings the aggregate bit rate to the T1 rate of 1.544 Mbit/s.

The framing bit is used to repeat a specific pattern throughout a "superframe" consisting of 12 frames. This pattern is used by receiving terminal equipment to identify and align the incoming bit pattern. The most common framing pattern used defines a D4 framing structure (named after the AT&T D4 channel bank). A newer standard, called extended superframe format (ESF), uses a 24-frame pattern (see "The hidden treasures of ESF," DATA COMMUNICATIONS, September 1986). The extended size of the superframe allows for the transmission of a six-bit cyclical redundancy check and for a 4-kbit/s embedded operations channel.

Note: Each voice circuit in a T1 appears to consist of a 64-kbit/s channel. In actuality, telephone switching re-

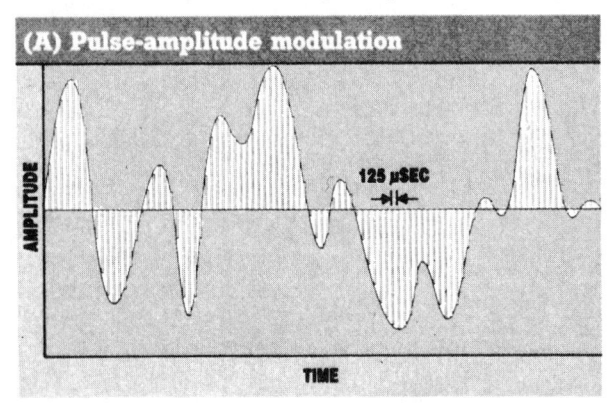

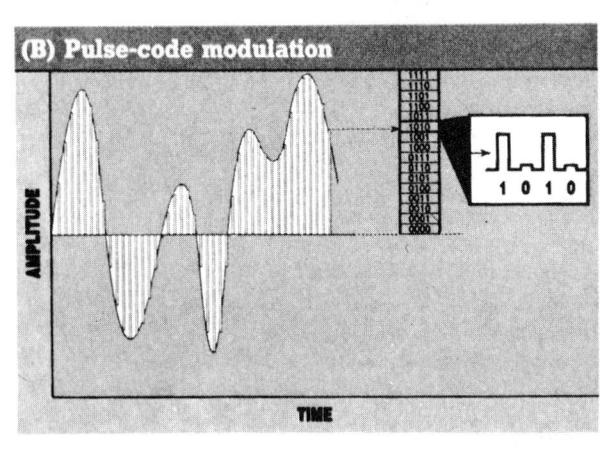

(A) Pulse-amplitude modulation

(B) Pulse-code modulation

T3 Networking

quired a fraction of this bandwidth to handle call-processing functions. In North America, this was handled through robbed-bit signaling: Every sixth frame, the least significant bit is removed from each eight-bit sample and used for signaling. This results in imperceptible degradation for voice traffic. Data traffic, however, can only rely on the untouched seven bits. Instead of a 64-kbit/s clear channel, therefore, DS-0 data circuits have an effective bit rate of only seven bits multiplied by 8,000 per second, or 56 kbit/s.

■ **T3 format.** At a rate of 1.544 Mbit/s, the T1 rate was sufficient for many needs, but not for all. On dense, high-volume routes, higher bit rates were required. The first step in this direction was made with the DS-2 rate, consisting of four DS-1s. (An intermediate rate, DS-1C [3.152 Mbit/s] is sometimes used, but is of no concern to this discussion.) The individual bits from each tributary are bit-interleaved.

DS-2 would appear to have a bit rate four times that of a DS-1, or 6.176 Mbit/s. In actuality, another layer of framing is added at the DS-2 rate. This framing ensures correct byte alignment as before, but also allows for variances in clock rates among tributary DS-1s (asynchronous operation). Special stuffing bits are included, as required, to ensure identical bit rates before bit interleaving. "Fast" signals receive less stuffing; "slow" signals receive more stuffing; adjustments are made continually to achieve a common reference level. These stuffing bits are removed at the far end of the transmission span. A total of 17 framing and stuffing bits are added to each DS-2 frame, leading to an aggregate DS-2 bit rate of 6.312 Mbit/s (8,000 × 17 added to 6.176 Mbit/s).

At this speed, the DS-2 signal can still be carried over copper pair, but the limitations inherent in copper transmission become more restrictive. Specially shielded cable is required to reduce crosstalk and susceptibility to electromagnetic interference. This cable requirement limited the acceptance of DS-2 installations, and it has never reached the wide deployment seen by DS-1 equipment.

By the end of the 1970s, however, a way around the limitations of copper pair was on the horizon. Transmission technologies based on the nearly limitless bandwidth of optical fiber were emerging from the laboratory, and a new layer of the digital hierarchy seemed appropriate. Based on technology available at the time, an asynchronous DS-3 rate was defined as the combination of seven DS-2 signals (equivalent to 28 DS-1s or 672 DS-0s). Framing and stuffing bits, as well as rudimentary error checking and internal communications, were added to the DS-2 tributaries to generate an aggregate bit rate of 44.736 Mbit/s.

Notice that the information content, or payload, of a DS-3 is equal to 672 64 kbit/s = 43.008 Mbit/s. The additional 1.728 Mbit/s of overhead represents the sum of the DS-1 framing bit, the DS-2 framing and stuffing bits, and the variety of DS-3 overhead bits. In terms of network efficiency, the DS-3 format devotes 96 percent of the transmission bandwidth to payload, with approximately 4 percent overhead.

Although the DS-3 was designed to be created from multiple DS-2 signals, there was an obvious inefficiency in using two separate devices for the DS-1-to-DS-2 and DS-2-to-DS-3 multiplexing functions. (The generic names of these devices are M12 and M23, pronounced em-one-two and em-two-three.) A new type of device was created, dubbed the M13 (one-three), which accepted DS-1 inputs and produced DS-3 output (and the reverse).

Note that the DS-3 (or T3) is often referred to as consisting of 28 DS-1s. This is correct, but it is more accurate to describe it as consisting of seven DS-2s. The DS-2 rate has not been eliminated, but simply shifted to an internal rate within the M13. None of the limitations of the DS-1 and DS-2 frame formats are removed by asynchronous DS-3 devices.

■ **North American digital hierarchy.** Coincidentally, as fiber optic technology exploded out of the laboratory, a far-reaching shift took place in the U.S. network. The divestiture of AT&T meant that no single entity would be capable of setting universal standards for the North American network. The technological advances did not slow, however, and higher and higher bit rates became economically feasible. After an abortive attempt at a DS-4 (274.176-Mbit/s) standard, major manufacturers basically went their individual ways with optical fiber equipment operating at a variety of bit rates. Overhead channels, multiplexing schemes, and the number and type of tributary were all decided on a per-manufacturer basis. Optical links between equipment from different vendors became impossible. DS-3, however, was retained as a common denominator and became the standard interconnect for all high-speed fiber gear designed for use in the public network.

Since all high-speed links were asynchronous, bit-stuffing penalties continued to mount. Multiple levels of bit-stuffing (DS-1 to DS-2, DS-2 to DS-3, DS-3 to proprietary) continued to require additional bandwidth for synchronization control. At the same time, the high bandwidth of optical fiber encouraged manufacturers to implement additional overhead functions. Payload efficiency dropped dramatically, but this was acceptable, given the vast capacity of these new transport elements. Hub offices began terminating dozens or, in some cases, hundreds of DS-3 signals.

By the late 1980s, the North American digital hierarchy looked like what is shown in the table.

North American digital hierarchy

LEVEL	BIT RATE	DS-3	DS-2	DS-1	DS-0	EFFI-CIENCY
DS-0	64 KBIT/S	—	—	—	1	100%
DS-1	1.544 MBIT/S	—	—	1	24	99%
DS-2	6.312 MBIT/S	—	1	4	96	97%
DS-3	44.736 MBIT/S	1	7	28	672	96%
3 x DS-3	≈ 139 MBIT/S	3	21	84	2,016	93%
12 x DS-3	≈ 565 MBIT/S	12	84	332	8,064	91%
24 x DS-3	≈ 1.2 GBIT/S	24	176	664	16,128	86%

T3 Networking

day circuit changes, or quick response to a temporary overload condition in some part of the network.

Realistically, T3 networks are not for everyone. Who can justify the investment required to operate a 45-Mbit/s circuit? One group that can is *Fortune* 500 corporations that have already installed nationwide T1 backbones. Smaller organizations with unusually high communications requirements (typically, service companies) also have T1 networks in place today. In any network where multiple T1 circuits have been placed or planned between locations, it may be appropriate to consider T3 service.

Another segment consists of organizations with rights-of-way, such as public utilities, railroads, pipelines, and state and local governments. One of the most cost-effective ways to implement a T3 network is to own the optical fibers required for transmission. Right-of-way organizations have an advantage, since they do not have to go through the negotiations necessary to lay a cable across private property and public thoroughfares. Although the up-front costs of a fiber installation can be significant, they are often offset by eliminating the recurring monthly charges associated with leased T1 or T3 circuits.

T3 technology issues

Once the decision to go to T3 has been made, a number of questions about technology need to be answered.

One of the choices to be made early in the process is whether to lease or purchase T3 circuits. The options involved in each choice are summarized in Table 1.

The first solution is the traditional way of managing communications: leasing the circuit from a service provider. This provider may be the local telephone company, an interexchange carrier (IXC), an urban-bypass organization, or a friendly right-of-way company. In each case, the service provider takes care of bringing a T3 pipe to the customer premises and keeping it working. The fees for such a service are often justifiable if the user organization is not set up for network monitoring, troubleshooting, and repair.

Alternatively, the T3 circuit may be leased only between carrier points-of-presence. Since T1 circuits to the premises are readily available, this combination avoids the problem of providing a dedicated fiber pair to the premises to complete a broadband connection: the "last-mile" problem. This option can be attractive when terminating an entire T3 at a private user's location is not feasible, either because of prohibitive placement cost or traffic patterns.

The alternative to leasing is to own the T3 transmission equipment outright. This implies that the user must have access to one of two facilities: an optical fiber cable or a digital microwave transmission path.

Optical fiber has a much higher transmission capacity than digital microwave. By installing a private cable, a user establishes ownership of transmission resources and, for a one-time cost, avoids the recurring expenses associated with a leasing arrangement. However, it is often impossible to install cable without owning a right-of-way. Placing private fiber cable is normally only practical for short-haul networks within a campus or urban area.

In addition, locating and repairing cable breaks (caused by construction, accident, or malice) can be time-consuming. To get around such difficulties, some users have resorted to leasing "dark fiber" from a telephone company, bypass operation, or IXC. Dark fiber means that the monthly charge pays only for dedicated access to the optical fiber and for cable maintenance; the user is responsible for providing optical multiplexing equipment, network monitoring, and terminal maintenance. This can be especially practical in an urban setting where a private right-of-way would be prohibitively expensive.

Because of the right-of-way and maintenance problems inherent in optical fiber cable, users have turned to digital microwave radio as a T3 transmission medium. Operating basically over line-of-sight paths, these networks provide reliable error-free transmission of a T3 signal across an urban area. Newer units can even be mounted indoors, transmitting through a window, so that tower installation and maintenance is eliminated.

Digital microwave radio has three problems. First, the electromagnetic spectrum is strictly licensed by the Federal Communications Commission. Overcrowding of the airwaves has closed off certain portions of the spectrum in densely populated urban areas. Second, strict spectrum allocations imply that digital microwave cannot be readily upgraded in bandwidth. While a 45-Mbit/s optical fiber can be readily converted to 565 Mbit/s, a 45-Mbit/s digital microwave setup is probably destined to remain at 45 Mbit/s.

Finally, digital microwave devices are inherently limited to line-of-sight distances—sometimes even shorter,

T1 and T3. *Using standard AT&T rates, a tariff comparison shows that leased T3 lines carrying the equivalent of 28 T1s are more economical than multiple T1 lines. For distances of less than 50 miles, only four T1s are required to break even.*

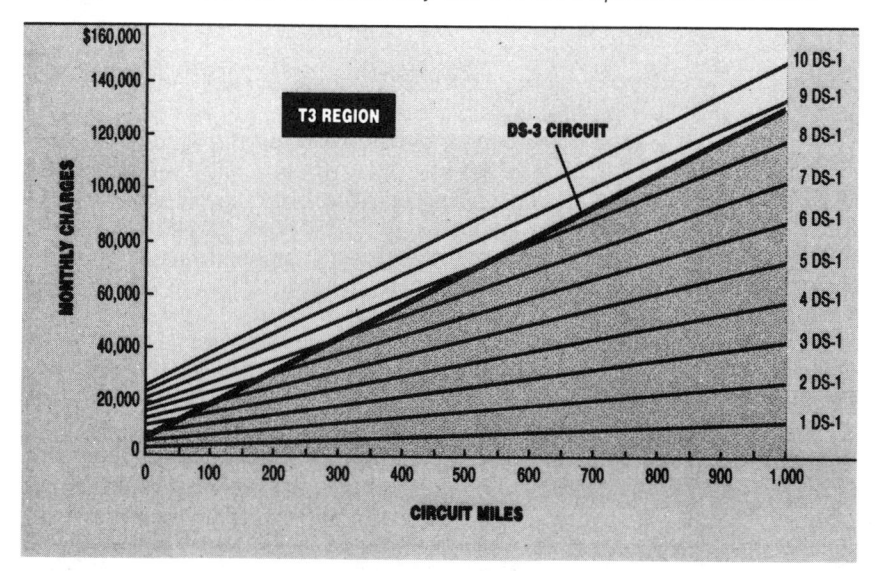

T3 Networking

owing to weather conditions or other hindrances. This is manageable in urban-distance settings, but can cause problems for interstate or national networks.

Many private networks mix optical fiber and digital microwave. The fiber can be used in high-density areas, with microwave hops to lower-density remote locations. Alternatively, the main network can be placed entirely on fiber, with T3 microwave used for emergency restoration in the event of a cable cut. Dual-media transmission is especially attractive to service-critical networks such as banks, airlines, and other on-line transaction-processing companies.

After choosing between leased and owned facilities, the data communications manager must decide what equipment to use to terminate these facilities. With leasing, the terms of the lease may dictate use of a particular vendor's equipment. When using private fiber cable or microwave, the terminal choices are entirely the responsibility of the data communications manager.

Table 1: Facility choices

	LEASED T3 CIRCUITS	OWNED OPTICAL FIBER	OWNED DIGITAL MICROWAVE
ADVANTAGES	SIMPLICITY MINIMUM STAFFING REQUIREMENTS PERFECT FOR LONG-HAUL T3 NETWORKING	ONE-TIME COST 'UNLIMITED' CAPACITY VARIED EQUIPMENT OPTIONS EASY UPGRADES COMPATIBILITY WITH NEW SERVICES PERFECT FOR CAMPUS ENVIRONMENT	ONE-TIME COST RAPID SET-UP PORTABILITY NO RIGHT-OF-WAY ISSUES PERFECT FOR METROPOLITAN AREA NETWORK
DISADVANTAGES	RECURRING MONTHLY COST LIMITED NETWORK CONTROL LIMITED FLEXIBILITY	REQUIRES RIGHT-OF-WAY VULNERABLE TO CABLE CUTS REQUIRES ON-CALL MAINTENANCE	WEATHER DEGRADATION FCC LICENSING RESTRICTIONS LIMITED BANDWIDTH

Equipment choices

All private optical networks share the advantages of near-limitless upgradability. The data rate of modern single-mode fiber reaches many tens of Gbit/s. Whether the fiber cable is owned or leased, users have abundant options when shopping for optical fiber terminals. The 1980s boom in optical installations for the public network has led to dozens of available products from numerous vendors; capacities currently range from a single T3 (45-Mbit/s) to 36 T3 circuits (1.7-Gbit/s). Future installations are expected to operate at even higher speeds. Although these Gbit/s devices are only of interest to telephone operating companies and IXCs, they will certainly be available to private-network users when the need arises.

Once transmission issues have been solved, the user must decide what T3 multiplexing equipment is required for the application. There are three major choices:
- Asynchronous M13s.
- DS-3 cross-connects.
- T3 add/drop multiplexers.

The traditional method of obtaining T3 circuits is via an M13 multiplexer. This device collects up to 28 T1 signals and uses two steps of time-division multiplexing to produce a T3. There are tens of thousands of M13s in service in the public network today; 10 years of manufacturing experience and economies of scale have brought them down to surprisingly low prices ($4,000 to $10,000). The rate and format of the M13 signal are defined in Bellcore document TR-TSY-000009.

These devices have been optimized for bundling T1 circuits for point-to-point transmission in the BOC and IXC markets. They are the T3 equivalent of the "dumb" channel bank—adequate for basic transport, but often inadequate for complex corporate networks. They are usually appropriate only in point-to-point networks that do not require much in the way of flexibility or performance monitoring.

A modification of the T3 standard, known as C-bit parity, was published by AT&T in document PUB 54014, which describes the Accunet T45 service offering. While not adding any flexibility or control to the M13 standard, C-bit parity does provide rudimentary far-end performance monitoring over T3 lines. This standard has not gained the widespread deployment of the traditional M13s.

At the other extreme is the new entry of 3/1 and 3/1/0 digital cross-connect systems (DCS). These are sometimes referred to as DACS (Digital Access and Cross-connect System) products, after the AT&T product line of the same name—or as wideband cross-connects. The types differ in that the 3/1 DCS demultiplexes DS-3 signals to the DS-1 rate, while 3/1/0 machines demultiplex through the DS-1 rate to DS-0, or 64-kbit/s, rate. DCS 3/1 or 3/1/0 devices incorporate an internal switching matrix, providing circuit-switched flexibility and significant network management features. The DCS excels in "hub-and-spoke" networks where T1 circuits from a variety of T3 sources must be interconnected. These products tend to be optimized for dozens or hundreds of T3 ports; although their features may appeal to corporate-network users, their capacity (and price tags) can be overwhelming.

A third multiplexing alternative exists in the form of the T3 add/drop multiplexer (ADM). As the name implies, these units allow the individual T1 circuits to be added and dropped at a particular site. When deployed along a T3 route, these devices can provide the T1 switching functions of large 3/1 cross-connects for a single T3 circuit. Some vendors also provide DS-0-level switching for some or all of the 64-kbit/s circuits within a T3. This provides the user with electronic control of the T3 bandwidth, network surveillance, capital-cost reductions at intermediate sites in a T3 route, and compatibility with the huge installed base of T3-based fiber and radio networks.

Synchronous ADMs can be especially valuable in overcoming the limitations of a T3 microwave network. By

T3 Networking

providing efficient bandwidth allocation, a synchronous T3 ADM can maximize the utilization of a single T3 pipe. With digital microwave, the difference between a single highly filled T3 and two poorly utilized T3s can determine whether a network is economically or technically feasible.

It is worth noting here that a fourth class of multiplexing equipment will be entering the market within the next few years. These devices, based on the Sonet hierarchy (see "Sonet calms choppy waters"), promise to offer a level of flexibility and performance well beyond the traditional M13. Because of their synchronous nature, Sonet networks will provide many of the benefits of the DS-3 cross-connects and T3 add/drop multiplexers for multi-DS-3-rate fiber. Since Sonet networks are incompatible with the installed base of fiber transmission, however, DCS machines or T3 ADMs will still be required to connect Sonet networks with the existing North American network.

T3 ADMs may be directly integrated into intelligent T1 multiplexer networks, acting as a higher-level circuit switch for a broad mixture of T1 tributaries. The same can be said for asynchronous M13 networks.

Although the DS-1 and DS-3 external interfaces are identical, the internal architectures are quite different. The asynchronous M13 bit-interleaves and bit-stuffs DS-1 signals to the DS-2 rate, repeating the process from DS-2 to DS-3. The T3 ADM, on the other hand, synchronously aligns each incoming DS-1, allowing identification of each DS-0 circuit. Therefore, the T3 ADM first demultiplexes the DS-1 tributaries into the component DS-0 signals. After passing through a time slot interchanger (TSI), the signals plus overhead channels are remultiplexed in a single stage from DS-0 to DS-3.

A DS-3 interface may be provided on one or both sides of the TSI. When equipped on a single side, the device acts as a DS-1-to-DS-3 multiplexer, with internal DCS. When equipped on both sides, the device can add and drop individual DS-0 or DS-1 signals to a DS-3 path. Unlike asynchronous designs, there is no limitation on how many DS-0 or DS-1 signals may be added or dropped.

By implementing the DS-0 internal switching matrix, the ADM provides far greater flexibility in circuit arrangement than the hard-wired multiplexing of an M13. The ADM can also switch proprietary or unframed T1 signals at the full 1.544-Mbit/s rate (rather than the 1.536-Mbit/s or 1.344-Mbit/s rates common to other devices). This is accomplished by arbitrarily splitting the signal into 24 bytes of eight bits each, carrying the 193rd bit separately in the T3 overhead, and reuniting the elements in the correct sequence at the receiving terminal. Unframed T1 video signals or non-D4-compatible T1 signals can be carried transparently within such a device.

In addition, the synchronous T3 transmission format (ANSI standard T1.103-87)—also known as Syntran—used in a DS-0 ADM integrates performance monitoring and an embedded control channel into the T3 bit stream. A nine-bit cyclical redundancy check (CRC9) is performed at the DS-3 rate for remote performance monitoring. Also, a 64-kbit/s embedded operations channel permits communications both between individual nodes and between the network and network management devices. For T3 users, the synchronous T3 format provides the same benefits that the extended superframe format (ESF) provides for T1 users.

Compatibility issues

Most data communications managers have learned the importance of compatibility with the public network the hard way. Proprietary interfaces, unless in a pure single-vendor network, often turn out to be more troublesome than useful. When choosing T3 network equipment, there are new network compatibility issues involved.

First, the equipment must be compatible with the existing T3 standards. Luckily, the T3 world consists of well-defined interfaces, so compatibility is not usually a problem here.

One area of concern with T3 is facility compatibility. A number of AT&T and Bellcore publications specify the interface required for a DS-3 signal to be carried over T3 equipment. These specifications make up the Digital Signal Cross-connect, Level 3 (DSX3) standard. The standard includes specifications for line coding, signal pulse shape,

---

Sonet calms choppy waters

No description of high-speed networking would be complete without a discussion of Sonet. This new standard, an acronym for Synchronous Optical Network, is supported by dozens of vendors and public network providers in North America, Europe, and Japan. It operates at multiples of T3 bandwidth. Initially, products will be offered at the following bit rates (OC stands for optical carrier): OC-1, 51.84 Mbit/s; OC-3, 155.52 Mbit/s; OC-12, 622.08 Mbit/s; OC-48, 2.49 Gbit/s.

Sonet brings order to the current chaos of high-speed fiber optics, where each vendor has established independent proprietary bit rates and protocols. In the United States, the first phase of the Sonet standard has been published as ANSI T1.105-1988 and T1.106-1988.

As a synchronous standard, Sonet is well-suited for switching tributary signals within a higher-bandwidth pipe. Initially, switching will be limited to T1 and DS-0 signals, but other service offerings will be defined as time goes on. One future example: a publicly switched Ethernet interface that would allow you to dial a 10-Mbit/s channel cross-country or around the world. Sonet networks operating at OC-3 and higher rates will also be used as the basis for interconnection of broadband services such as Distributed Queue Dual Bus metropolitan area networks and high-definition television.

Sonet is fully backward-compatible with the ANSI synchronous T3 format; it is partially backward-compatible with older asynchronous T3 equipment such as M13s. It does not affect users of digital microwave radio, since there is little need for a synchronous standard higher than T3 for these applications. It will be deployed by the BOCs in the early 1990s. Depending on its success in that arena, other network providers and corporate-network users will follow suit.

T3 Networking

clock rates, frame format, and other parameters of interest to equipment vendors. Any device meeting this DSX3 specification can transport over any T3 equipment, whether optical or microwave, from any vendor.

A second area of concern is terminal compatibility. Once the signal has been successfully transported, issues of terminal compatibility arise. Two major camps of terminal standards exist and have been formalized by Bellcore: the synchronous T3 and the asynchronous T3. These two signals are not directly end-to-end compatible, but must be translated by a third device.

The synchronous standard, being newer, is more fully documented and does not allow much vendor freedom in implementing the specifications. This ensures that any device built to this standard will work successfully with any other, regardless of vendor.

The asynchronous standard is older and has been interpreted differently by vendors. Many of them have added proprietary extensions to provide value-added features or to make up for deficiencies in the format (such as the C-bit parity variation, noted earlier). Their unique implementations of the asynchronous T3 interface can cause compatibility problems ranging from disabling minor features to major network inconsistencies. These solutions dictate that a user employ a single-vendor network.

Since installing T3 equipment in a private network almost always implies connection with new or existing T1 equipment, the issue of T1 compatibility must also be addressed. As with DS-3, AT&T and Bellcore have published details of the T1 interface, collectively referred to as the DSX1 specification. Again, the standards have left room for vendor interpretation, leading to possible inconsistencies.

The most common T1 interface consists of a 1.544-Mbit/s signal channelized according to D4 format. This allows network equipment to identify and switch individual DS-0 (64-kbit/s) signals. This standard was established by AT&T and is nearly universal within the public network. It is normally referred to as D4-channelized or DACS-compatible. The payload data rate of a channelized T1 is 1.536 Mbit/s.

The lockstep limitations of the D4 format led some T1 equipment vendors to ignore the 64-kbit/s boundaries and use a proprietary organization of data within the DS-1 frame. An unchannelized signal requires access to the entire 1.544-Mbit/s T1 format. This allowed more flexibility in transporting varied rates of voice and data, but created incompatibility with the DACS networks installed throughout the public network. Many vendors now offer both options: channelized for public compatibility and unchannelized for maximum bandwidth efficiency. Installations using both types of signals must ensure that their T3 equipment can transport both efficiently.

Table 2: Performance monitoring parameters

PARAMETER	ASYNCHRONOUS T3	SYNCHRONOUS T3
DS1		
BIT ERROR RATE		✓
BIPOLAR VIOLATIONS	✓	✓
SLIPS		✓
CRC6 VIOLATIONS	ESF ONLY	ESF ONLY
AIS DETECT		✓
FRAME LOSSES		✓
ERRORED SECONDS		✓
SEVERE ERRORED SECONDS		✓
DS3		
BIT ERROR RATE	PARITY-BASED	CRC-BASED
PARITY ERRORS	✓	✓
BIPOLAR VIOLATIONS	✓	✓
CRC9 VIOLATIONS		✓
AIS DETECT	✓	✓
FRAME LOSSES	✓	✓
ERRORED SECONDS	✓	✓
SEVERE ERRORED SECONDS	✓	✓

AIS = ALARM INDICATION SIGNAL
CRC6 = 6-BIT CYCLIC REDUNDANCY CHECK

Yet a third variation exists with unframed T1 signals. These normally represent the output of certain T1 video codecs or encryption equipment. Unlike the unchannelized signals, these do not even meet the DSX1 framing standard, but only the clock rate and associated requirements. Again, in a network that must transport unframed T1 signals, it is important to verify that the T3 equipment will be compatible.

Future compatibility

Finally, it is important to note that the DS-3 formats do not represent the final step in network evolution. As higher bit-rate standards evolve, T3 equipment purchased today must be integratable into new networks that will be installed throughout the 1990s.

The user making the leap to T3 will find that all the network management capabilities of T1 products are still available. Indeed, by offering single-point monitoring of all T1 circuits in a network, the addition of intelligent T3 devices can actually make the network operator's life simpler, not more complicated.

Early in the T3 equipment decision-making process, a communications manager will realize that most T3 equipment is optimized for traditional BOC and IXC applications. This can be adequate in some circumstances, but these devices do not lend themselves to the sophisticated management and control requirements of private networks.

A new generation of T3 devices is appearing on the market from a number of sources—from traditional telephone company transmission-equipment suppliers to T1

T3 Networking

multiplexer manufacturers to small start-up companies. All, however, have a common goal: to create intelligent T3 devices for the emerging private marketplace. These devices are a distinct departure from the "dumb" ones used in huge quantities by telephone companies and IXCs.

Communications managers will also have to analyze the T1 circuits currently used in their networks. For seamless operation, the T3 equipment should support all types of T1 circuits likely to be used. Managing different types of T1s on different types of T3 equipment ensures confusion.

Networks with an emphasis on data will tend to use unchannelized T1s in order to pack the maximum number of data circuits into the available bandwidth. This is especially common in networks where one vendor supplies all the T1 multiplexing equipment. These circuits will also be used for T1 video codecs, where channelization is meaningless.

Voice networks will tend to use D4-channelized circuits. Thesecircuits are generated by standard channel banks; they are alsoavailable on most modern T1 multiplexers. Naturally, D4-channelized T1s can also be used for transport of data at any number of bit rates, depending on the capabilities of the T1 equipment.

A possible complication arises with the availability of adaptive differential pulse-code modulation (ADPCM) equipment in private-network multiplexers. ADPCM transmits high-quality voice at 32 kbit/s, 21.3 kbit/s, or even 16 kbit/s, instead of 64 kbit/s. DS-0 switching implies that pairs, triads, or quads of ADPCM signals will be switched as units, which slightly reduces the efficiency of the switching matrix. The problem may be alleviated with the realization that T3 bandwidths bring an end to many of the original arguments for ADPCM. With 28 T1 circuits available, maximum utilization of a single T1 may not be an issue.

Finally, many vendors are considering the use of ESF circuits for enhanced fault isolation and troubleshooting. Although it may not be economical to convert all existing circuits to ESF, it is important to allow new circuits to take advantage of this improved range of functions.

Network surveillance consists of alarm reporting and performance monitoring to continually verify the health of the network. Alarms can consist of equipment failures, carrier failures, or entire node failures. Performance monitoring provides diagnostic information for individual circuits, either before or after a failure. Statistics can be reported for both T1 and T3 levels, as shown in Table 2. Effective use of performance thresholds can encourage preemptive equipment maintenance and minimize network downtime.

Surveillance in the public network is typically handled by multimillion-dollar surveillance centers operating on dedicated mainframes. In the private-network world, a dedicated microcomputer or technical workstation is often a more realistic alternative. The user's workstation is only half the issue. Comprehensive network surveillance requires both intelligent network management devices and intelligent network elements capable of measuring performance statistics such as slips, frame errors, CRC errors, and bipolar violations (see "How to detect frame slips in voice-band PCM channels," DATA COMMUNICATIONS, October 1988). Therefore, even sophisticated network surveillance can obtain and display only elementary alarm information from less sophisticated T3 devices such as M13s.

Many network management products really provide only network surveillance. Reaching into the network to rearrange circuits, either in response to a service request or to a service problem, often requires a technician moving cables on a patch panel.

Newer T3 devices solve these problems by giving the operator direct control over DS-0 and DS-1 connections within the network, resulting in such benefits as time-of-day switching of T1 or T3 or the reserving of bandwidth for an anticipated videoconference. Again, this requires comparable levels of sophistication both for the T3 network elements and for the centralized management software.

Survival

Network disaster recovery is critical at the T3 rate—not many networks can survive the simultaneous loss of 672 circuits. Disasters take many forms. Fiber optic cable cuts are not uncommon; with modern high-bit-rate fiber, a single cable cut can affect a quarter of a million circuits. Both long-haul and local-access circuits are at risk. Central office fires, such as the one at Hinsdale in Chicago, are rare but devastating. Even operator errors, such as disconnecting the wrong cable at the wrong time, can constitute a network disaster.

The first line of defense is battery backup and internal redundancy; T3 equipment built to BOC standards typically has multiple layers of protection switching built in, so that no single module failure can cause loss of service. The BOCs typically specify protection-switching times of 60 milliseconds or better to restore full service over backup modules.

The next defense is network redundancy: By routing redundant T3 circuits over separate physical links, users can protect against even catastrophic carrier-node failures. With excess bandwidth available on unaffected routes, manual or automatic disaster plans may be implemented to restore some or all circuits affected by the disaster. This can take several minutes—longer than the 60 milliseconds specified for equipment protection, but still greatly superior to the hours or days required for manual restoration.

The best defense is autonomous network recovery. For example, survivable T3 ring architectures provide rapid restoration of all circuits in a fraction of a second, without the lag time required to consult central human or computerized network management mechanisms. Such arrangements are now being implemented in the public network. Private users, with their sensitivity to even brief periods of downtime, will almost certainly follow suit. ∎

Stephen Fleming has a BS summa cum laude in physics from Georgia Institute of Technology (Atlanta). He has worked at Bell Laboratories, where he specialized in optical fibers, and at Northern Telecom, where he specialized in multiplexers and cross-connects. He wrote this article when he was Licom's director of marketing for T3 and Sonet products. He has since returned to Northern Telecom.

Ira Brodsky, Datacomm Research Co., Wilmette, Ill.

Dialing data via mobile telephones

An analog service, cellular poses problems for data applications. But special modems, interfaces, and protocols bridge the gap.

Long gone are the days when only corporate executives and movie stars could say, Here's my office number *and* my mobile phone number. Over the past few years, the prices and availability of cellular phone services have decreased to a point where mobile phone technology has joined the mainstream of business and consumer communications. And this is just the beginning: Industry observers expect that by 1990 more than 3 million cellular phone sets will be in use.

What are the data-transmission implications of all of these roving phones? The capability for data-on-cellular is here now; judging by the number of products and development efforts, the demand for data applications over cellular networks is growing.

Although the market for sending data over the cellular telephone network is fairly small today, many industry observers believe that data will account for 10 percent of cellular airtime within the next few years. Others feel that data will eclipse voice in percentage of airtime within 10 years.

Although vendors and industry organizations are engaged in ongoing planning for digital cellular networks, current cellular networks are analog. And, if you think the planned digital cellular network will facilitate data applications, you may be in for a surprise. Unlike ISDN—the nascent all-digital landline network intended to serve both voice and data—digital cellular is motivated mainly by the desire to expand channel capacity, not necessarily to enable user-to-user digital transmission.

With 1991 targeted for the first digital cellular network in the United States, we are told that a wholesale conversion from analog to digital isn't far behind. But don't be so sure (see "Where's the data comm?"). Meanwhile, those who see a near-term benefit to exchanging data with remote mobile stations must contend with the existing analog cellular network.

In the United States and Canada, there are more than 60 cellular service areas that already offer data-specific capabilities. They include networks operated by Ameritech, Nynex, Southwestern Bell, and Compuserve. Applications for digital cellular transmissions are evolving. It is currently used to access databases for such tasks as handling hazardous materials and facilitating field service and sales force automation.

The cost components for a typical configuration include a modem with Microcom Networking Protocol, or MNP (about $300 to $500), the cellular phone (about $1,000), an adapter for the modem-to-cellular connection (about $450), data terminal equipment such as a laptop personal computer (about $1,500), plus the airtime charge. For example, basic service in Chicago from Ameritech is $15 per month plus $0.38 per minute. The setup charge to get onto the network is $35.

Data-specific snags
While RJ-11 jacks are ubiquitous on local loops from the public switched telephone network (PSTN), cellular telephones connect to the cellular network via radio antennas and free space. A standard connector for hooking up data modems and facsimile machines simply isn't found on most cellular telephones.

Besides, communicating over cellular telephone involves interruptions, both intended and unintended. The cellular network was designed not for data communications applications, but to allow a large number of mobile subscribers to share a limited slice of the radio spectrum.

Subscribers to cellular networks are able to share a portion of the radio spectrum because the service area is divided into geographical cells, often represented by hexagons. In the center of each cell is a radio node, or cell site. By restricting the mobile station's transmitter to low power (maximum output equals three watts) and reusing

Cellular

frequencies in nonadjacent cells, the cellular network can serve multiple subscribers over the same channel at the same time. A major drawback: As a mobile station passes from one cell to another, the call must change frequencies. These cell handoffs interrupt the voice channel for periods of 200 to 1,200 milliseconds (typically between 600 and 800 milliseconds) and can cause dial-up modems to retrain, or even hang up. Add natural radio disturbances, and you have what designers of data communications protocols call an adverse environment.

And there's more bad news. Cellular voice channels differ from landline voice channels. At first glance, cellular voice channels appear to offer more bandwidth than do their landline counterparts. However, a large portion of this bandwidth is reserved for out-of-band signaling. An ever-present supervisory audio tone originates from the cell site and is transponded (retransmitted) by the mobile telephone. The network uses this tone, located near 6,000 Hertz, to make certain that the mobile telephone is operating correctly. It is filtered from the audio (speaker's voice or modem signals) and never reaches either party. Audio is muted when a second tone, the signaling tone, located at 10,000 Hertz, is used to confirm cell handoffs or to acknowledge an alert message. (An alert is a message that causes a mobile telephone to "ring.") If a user terminates the call or presses the flash button, the second tone is also sent, which again mutes the audio.

Cellular voice channels differ from landline voice channels in two other ways: cellular allocates 300 to 3,000 Hertz for voice, while landline circuits provide relatively flat response from 300 to roughly 3,400 Hertz; and cellular uses FM with companding and preemphasis. Companding dampens swings in amplitude, while preemphasis is used to boost the high-end frequencies (compensating for FM's characteristic high-end noise).

Finally, a radio impairment (called Rayleigh fading) is encountered in cellular's 800-MHz frequency band that causes an occasional baud (signal interval) to be lost.

Standard 300-bit/s and 1.2-kbit/s modems will work well over cellular telephone as long as adequate steps are taken to make the modem work despite noise and signal interruptions. But medium- and high-speed modems use more complex modulation techniques to squeeze through the limited voice-channel bandwidth; they may not operate reliably over cellular telephone. For example, V.32 modems do not readily operate at 9.6 kbit/s over cellular, but some manufacturers claim to have adapted these units to run in the 4.8-kbit/s fallback mode.

On some networks, however, the link between the cell site and the mobile telephone switching office may employ compression techniques that are acceptable for voice but inhibit data transmissions at more than 1.2 kbit/s. And Group III facsimile modems are intentionally throttled back to 4.8 kbit/s to reduce the bit error rate, turning lost scanning lines into less catastrophic errors like gray spots. (One reason that reduced link speeds work better is that the number of bits a noise spike knocks out is proportional to the bit rate. For example, the same spike that causes four bit errors at 1.2 kbit/s may cause more than 32 bit errors at 9.6 kbit/s.)

How, then, can cellular telephone be adapted for data communications? By meeting these three major requirements:
- A special physical/electrical interface.
- A modem system that can tolerate cellular voice-channel characteristics, including companding, preemphasis, limited bandwidth, and relatively long periods of carrier interruption.

Where's the data comm?

"The two biggest challenges facing the cellular industry are capacity and privacy," says William DeNicolo, president of Telular Inc. (Wilmette, Ill.), a manufacturer of intelligent interface adapters for cellular telephones. According to DeNicolo, these are the issues behind the current push toward creating digital cellular networks to replace the existing analog networks. Note that "digital cellular" does not automatically imply user-to-user digital transmission—it might only refer to the way the network operates internally. This issue has yet to be resolved by the standards bodies.

Nevertheless, DeNicolo believes that the migration to digital cellular networks presents an opportunity for standardizing the way in which cellular networks handle data. "But data isn't a high priority in the digital cellular standards work right now," he adds.

While it was reported late last year that the Telecommunications Industries Association's (formerly the Electronic Industries Association) TR43.5 committee was leaning heavily toward a frequency-division multiplexing scheme—to help keep in line with the FCC's directive stating that digital cellular must not disenfranchise owners of analog units—the Cellular Telephone Industry Association announced earlier this year that U.S. standards makers had, in something of a turnabout, settled on the time-division multiple access (TDMA) approach.

Even if the new digital scheme does not disenfranchise analog users, it still presents problems. DeNicolo points out that owners of the new digital telephones will not be guaranteed service in areas that do not upgrade to digital. And TDMA may pose new obstacles for data applications. Unlike frequency division multiplexing, which grants each user a continuous bit stream, TDMA assigns a time slot to each user. Thus, the sending device must know not only how fast to send the data but also when to send each bit. "Without a standard data interface and guidelines for sending data over a TDMA cellular network, we just don't know how difficult it will be to adapt existing hardware and software to digital cellular," DeNicolo observes.

But all may not be gloomy. If the standards bodies do settle on an interface for data, then users of digital cellular could realize higher speeds, greater reliability, and better security. At least one manufacturer of mobile telephone switching office equipment, Northern Telecom (Richardson, Tex.), has been paying attention. It demonstrated modemless 9.6-kbit/s data transmission over its DMS-MTX switch earlier this year. —I.B.

Cellular

1. Linking cellular and landline nets. While landline users are connected to a local switching office through physical media, cellular subscribers are linked by radio to a cell site. This cell site is linked to the Mobile Telephone Switching Office through metallic or fiber optic cable or microwave radio. The cellular call begins at "power up."

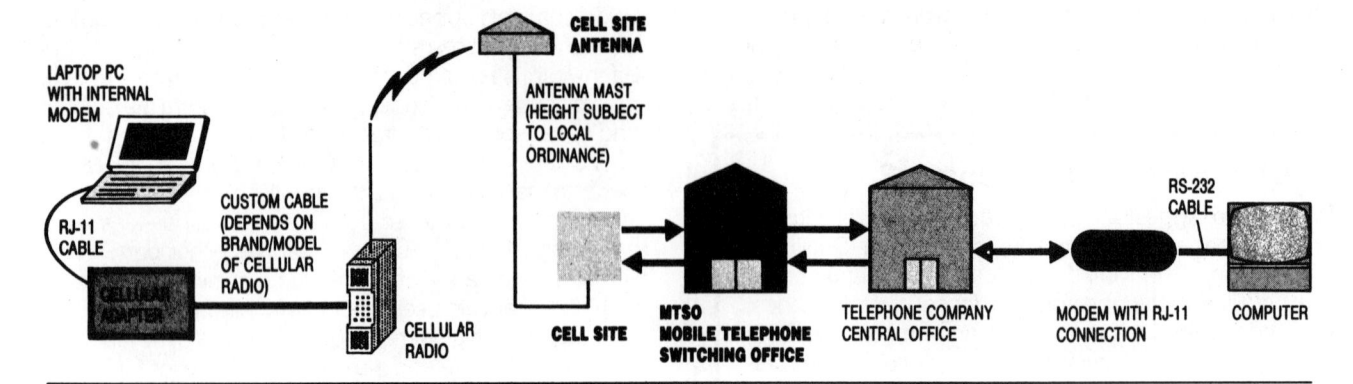

- An error-control protocol that will buffer and retransmit data to counteract noise, flutter (caused by multipath reception of the same signal), fading, dropouts (caused by dead spots in a cell's coverage area), and cell handoffs.

Cell divisions

A closer look at how the cellular network operates will help explain the hurdles associated with sending data over cellular telephone.

The cellular telephone is assigned 825 to 890 MHz, excluding 845 to 870 MHz. The mobile station transmits in the low band (825 to 845 MHz), while the land station transmits in the high band (870 to 890 MHz). In other words, each cellular telephone channel consists of two frequencies: one for sending, one for receiving. And they are offset from each other by 45 MHz.

The Federal Communications Commission (FCC) has decreed that there can be two cellular networks in any given region, and in metropolitan locations there usually are. One is called wireline, because it is licensed to the indigenous telephone company. (The first wireline, in fact, the first commercial cellular telephone network, was placed in service by Ameritech in Chicago in October 1983.) The other is called nonwireline, and the FCC has instituted a lottery to select each nonwireline licensee (because of the huge volume of applicants in most areas).

It seems that the working definition of the nonwireline network is that it can be owned by anyone except the indigenous telephone company. Ironically, Southwestern Bell Telephone and Pacific Telephone are now owners of nonwireline networks in the Ameritech region.

To eliminate interference between the two coexisting cellular networks, frequencies are further divided into "A" and "B" bands. Each uses 333 out of 666 total channels: "A" uses 825 to 835 MHz and 870 to 880 MHz, while "B" uses 835 to 845 MHz and 880 to 890 MHz. Each network provider, wireline and nonwireline, further divides the channels to ensure that no two adjacent cells use the same frequencies. Since each hexagonal cell may have up to six adjacent cells, the 333 channels must be divided seven ways. This allows each cell to handle up to 45 channels. (Actually, current regulations provide optional expansion channels, with even more channels planned for the future. The discussion here is limited to the originally designated channels.)

To minimize "dead spots"—locations where effective radio communication is lost—cell sites are spaced just far enough apart to provide partial overlap in coverage. And cells tend to be smaller in metropolitan areas than in rural areas because of the higher density of users.

Of the 333 channels per network, 21 are reserved as control channels. These are used for network access, call establishment, and call management. In other words, a phone call over the cellular network actually requires simultaneous operation on two duplex channels. The land station transmits on the forward control channel and forward voice channel, while the mobile station transmits on the reverse control channel and reverse voice channel.

Another way in which the cellular network differs from the landline network is in how users are connected to the switching office. While landline subscribers are physically connected directly to a local switching office (the central office), cellular subscribers are linked by radio to a cell site, which in turn is linked by metallic cable, microwave radio, or fiber optic cable to a special switching office designed specifically for cellular telephone. This office is called the mobile telephone switching office (MTSO; see Fig. 1).

Making cellular connections

The typical cellular phone call begins at power up, at which time the mobile telephone reads its internal Numeric Assignment Module memory to determine whether its home network is "A" or "B." Then it scans the relevant control channels, locking on to the strongest receive signal.

Next, the mobile telephone accesses the network. This is done by monitoring the busy/idle bit on the forward control channel and seizing the control channel when this bit indicates the channel is free. The initial transmission over the reverse control channel includes registration information. Registration tells the network which cell the mobile telephone is in, so that incoming calls can be appropriately directed. Registration data, sent at 10 kbit/s, includes the cellular telephone's equipment serial number

Cellular

2. Data formats. *Forward control channel messages include filters to ensure a continuous data stream (A); the reverse control channel (B) returns a coded version of the received digital color code. All networks use these protocols, as specified by the Telecommunications Industries Association (formerly the EIA).*

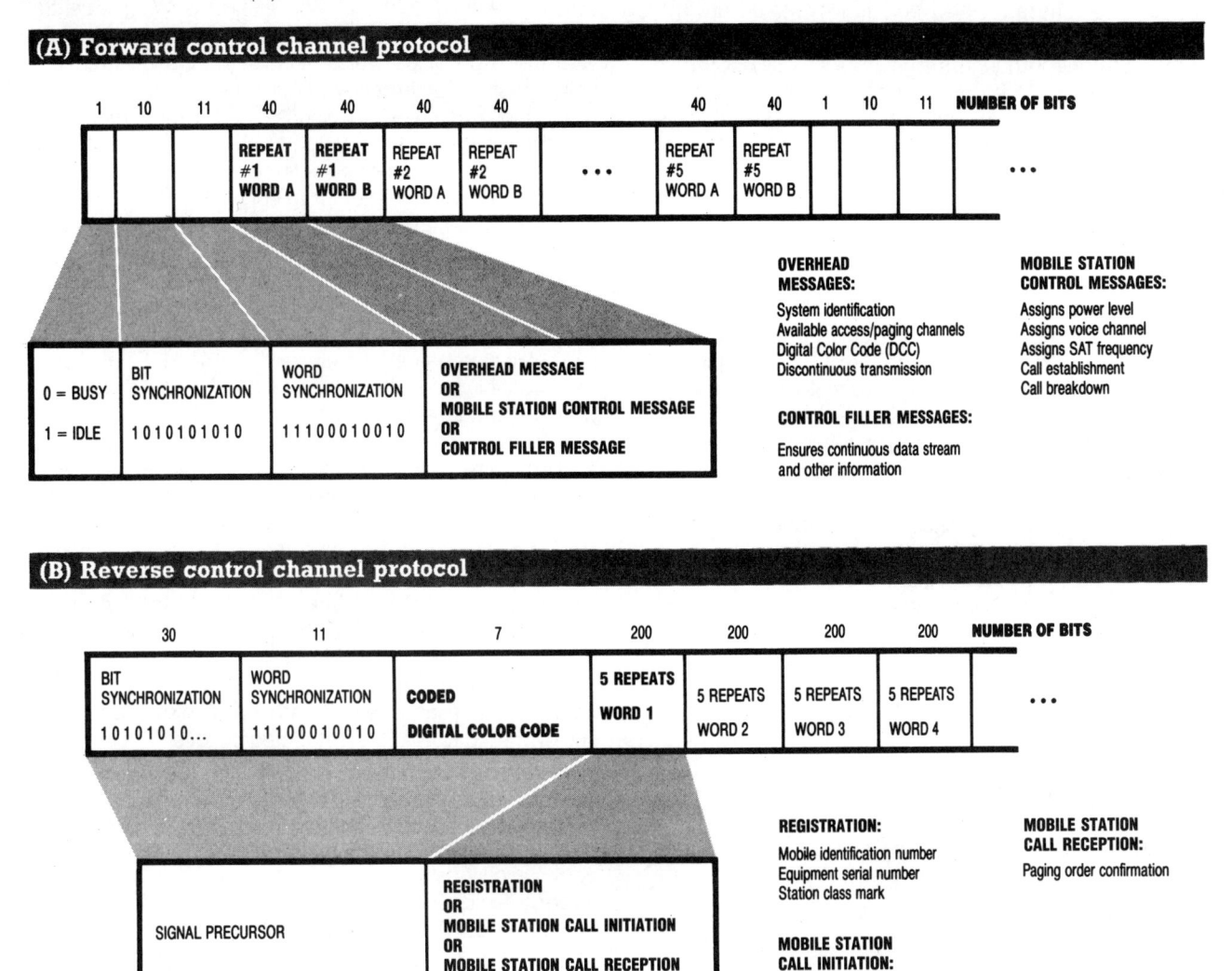

(A) Forward control channel protocol

NUMBER OF BITS	1	10	11	40	40	40	40	...	40	40	1	10	11
				REPEAT #1 WORD A	REPEAT #1 WORD B	REPEAT #2 WORD A	REPEAT #2 WORD B	...	REPEAT #5 WORD A	REPEAT #5 WORD B			

0 = BUSY 1 = IDLE	BIT SYNCHRONIZATION 1010101010	WORD SYNCHRONIZATION 11100010010	OVERHEAD MESSAGE OR MOBILE STATION CONTROL MESSAGE OR CONTROL FILLER MESSAGE

OVERHEAD MESSAGES:
System identification
Available access/paging channels
Digital Color Code (DCC)
Discontinuous transmission

CONTROL FILLER MESSAGES:
Ensures continuous data stream and other information

MOBILE STATION CONTROL MESSAGES:
Assigns power level
Assigns voice channel
Assigns SAT frequency
Call establishment
Call breakdown

(B) Reverse control channel protocol

NUMBER OF BITS	30	11	7	200	200	200	200	...
	BIT SYNCHRONIZATION 10101010...	WORD SYNCHRONIZATION 11100010010	CODED DIGITAL COLOR CODE	5 REPEATS WORD 1	5 REPEATS WORD 2	5 REPEATS WORD 3	5 REPEATS WORD 4	...

SIGNAL PRECURSOR	REGISTRATION OR MOBILE STATION CALL INITIATION OR MOBILE STATION CALL RECEPTION

REGISTRATION:
Mobile identification number
Equipment serial number
Station class mark

MOBILE STATION CALL INITIATION:
Dialed digits

MOBILE STATION CALL RECEPTION:
Paging order confirmation

(ESN; 32 bits), system identification number (SID; 15 bits), and mobile identification number (MIN; 34 bits). Registration is repeated whenever a mobile station switches cells.

The ESN identifies the manufacturer and specific unit. The MIN is a binary encoded version of the mobile unit's telephone number. The SID indicates the "home" network for the mobile station, that is, the network to which the unit subscribes and by which it is billed.

Cellular telephones may operate anywhere in the United States, but when a cellular telephone is used outside of its home network it is said to be roaming. The SID identifies the roaming unit to "foreign" networks, that is, networks other than the user's home network.

In some networks, the control channels are divided into access and paging channels. If the network does not support combined control channels, the mobile station is instructed to retune to a paging channel after registering. The user may then place a call by dialing the desired number and pressing the SEND button. The dialing information is conveyed over the reverse control channel as a data word in what is known as a Mobile Station Call Initiation message. If the "order" (which in the business of cellular telephones is the term for a telephone call) is accepted, the mobile telephone is instructed to go to a specific voice channel.

At this point the caller can speak and listen over the handset. Once the call is in progress, the cellular network continues to exchange control signals with the mobile telephone. Most of this information is sent via the control channel. The land station may use the forward control channel to tell the mobile station to increase or decrease its transmit power, or it may assign a new voice channel prior to a cell handoff.

Figures 2A and 2B illustrate the industry standard data formats used on the forward and reverse control channels. The forward channel uses alternating "Word A"/"Word B"

Cellular

information fields to address mobile stations with even- and odd-numbered mobile identification numbers. Each word is repeated five times to ensure successful reception. The forward channel messages also include fillers to ensure a continuous data stream.

The forward control channel's Digital Color Code (DCC) is used to ensure that the control channel is not confused with a control channel from a nonadjacent cell reusing that frequency. Discontinuous transmission refers to the ability to command the mobile station to change its transmit power.

Reverse control channel messages handle registration, call initiation, and call reception. The mobile telephone reads the land station's DCC and returns a coded version of the DCC, verifying that the phone is locked on to the correct signal. Dialing procedures for cellular networks sometimes differ from landline procedures. For example, if a roaming mobile phone calls another mobile station within the same network, the other station's area code must still be dialed. Anyone wanting to reach the roamer must know the access number of the network the roamer is on.

There is also a major difference between landline and cellular billing. On the landline network you are only billed for completed calls. On the cellular network, you may be billed for access time: from the moment you press the SEND button until you press the END button. This billing method is used because dialing, listening to far-end ringing, and listening to busy signals are actions that tie up one of a limited number of voice channels. The "airtime" approach to billing applies not only to calls you place but also to calls you receive.

When the call is finished the user presses the END button, generating 1.8 seconds of signaling tone over the reverse voice channel. If the user forgets to press the END button, the network will eventually time-out the call.

Modem connections

Dial-up modems are designed to connect to two-wire metallic circuits. For cellular applications, they must somehow connect to the cellular transceiver's audio input/output. This connection is typically found between the handset and the radio. Unfortunately, early attempts at standardizing this interface failed, and most manufacturers have gone to proprietary connections.

Telular Inc. (Wilmette, Ill.) has produced a family of RJ-11 interface converters that tap into many cellular telephone manufacturers' handset-to-transceiver connections. Telular markets this product as the CelJack. Motorola Inc. (Schaumburg, Ill.) also makes a similar box, called the Cellular Connection, for its own line of cellular telephones. Both products provide a modular RJ-11 hookup and simulate a landline PSTN connection. This simulation includes: synthesizing dial tone, accepting keyed tones (Dual-Tone Multifrequency); and pulse dialing, generating ring voltage, responding to off-hook/on-hook condition, and presenting the correct impedance. Not only data modems, but also facsimile machines and standard analog telephones can be connected to cellular transceivers through these interface converters. They also convert two-wire landline devices to the four-wire world of radio transceivers, that is, separate transmit and receive paths.

Telular has found several markets for its CelJack, one of which is linking cellular telephones to PBXs. When a central office is knocked out of service (as happened when fire swept through Illinois Bell's Hinsdale central office), the ability to share emergency cellular links from desktop telephones can save a business—and even human lives. While these products provide the physical and electrical interface necessary in connecting modems to the cellular network, they do not attempt to solve the other problems associated with sending data in the cellular environment. In general, the modem still needs to be modified or configured to tolerate carrier dropouts of up to 1,200 milliseconds. If the modem or fax machine is going to be used in a mobile or portable application, some provision for running on battery power must be made. Finally, an error-control protocol (such as MNP) should be used when sending data over cellular.

Plug-in modems, available for most laptop PCs, are one solution to the battery-power problem. Unfortunately, a standard expansion bus does not exist for laptop computers; each model usually requires a custom modem. These modems are miniaturized and rarely include built-in error control. For such configurations a software package—such as MTE Version 2.1 from MagicSoft (Lombard, Ill.)—will allow a standard laptop PC plug-in modem to communicate with modems supporting MNP.

As with any communications application, end-to-end error-checking software, such as Kermit or Xmodem, is the surest way to guarantee data integrity, even if modems have error-detection/correction features. Unfortunately, the combination of general-purpose communications software and such modem protocols as MNP can introduce redundancies and inefficiencies that will lower throughput. If your communications software program gives you a choice of error-correction protocols, do not choose Kermit or Xmodem because throughput will be slower. Other protocols, such as YmodemG, are better suited for cellular transmission.

Other manufacturers have developed complete solutions for sending and receiving data over cellular networks. These products combine modems, cellular telephone interface adapters, and special error-control protocols to provide reliable data communications under the conditions encountered over cellular, whether in fixed- or mobile-station operation.

The pioneer and leading vendor is SpectrumCellular, headquartered in Dallas. SpectrumCellular has been promoting cellular data applications almost from the inception of cellular telephony and offers a complete line of data products for cellular networks, including special modems for both mobile/portable units and landline/base stations. Other products include a cellular fax machine and a complete portable communications system housed in an aluminum briefcase—including its own laptop computer and printer. The briefcase is primarily intended for hazardous-materials teams—special squads often found in major fire departments.

SpectrumCellular's modem family includes the Bridge,

Cellular

Span, and Spannet, each running a proprietary protocol called SPCL that SpectrumCellular has developed for cellular's adverse environments. The Bridge is always installed at the cellular telephone and can talk to either Span or Spannet. The Bridge can also communicate with standard Bell 212/103 modems, but only in a nonerror controlled fallback mode. The Span is a standalone or rack-mounted unit that is designed for the landline/host connection. Spannet, on the other hand, is intended for installation within the service provider's network. It generally connects back-to-back (RS-232 to RS-232) with a Hayes-compatible dial-up modem. This arrangement makes it possible to meet the special needs of the cellular network,

When customized for cellular, V.32 modems yield 4.8 kbit/s.

while providing connectivity to the majority of installed modems. The only restriction is that the attached landline modem must be an intelligent, speed-compatible unit.

SpectrumCellular has had significant success in convincing cellular network providers and on-line data services to standardize on Spannet. In most cases, the Spannet/standard modem pair is installed in the MTSO, but it can be installed in other "hubs." More than 50 cellular networks—both wireline and nonwireline—in the U.S., Canada, and the U.K.—have selected Spannet.

To place a call through Spannet, the user dials a special data access number. This establishes a link between the Bridge-equipped cellular telephone and the Spannet modem. Now the caller has a transparent, but error-controlled, connection to a landline modem. While it is still necessary to dial out through the landline modem, the process can be automated with a communications script (found on most PC communications packages). But with the exception of their cellular fax and laptop PC-V.22*bis* modem combination, these products are limited to 1.2-kbit/s transmission.

The SpectrumCellular protocol, SPCL, employs both automatic retransmission (ARQ) error detection and retransmission and forward error control (FEC) techniques. According to SpectrumCellular, FEC is necessary to counteract Rayleigh fading. Rayleigh fading causes moderately frequent baud errors: a 1.2-kbit/s link will typically incur 10 to 15 such baud errors per second. (And in the 1.2-kbit/s Bell 212A modulation, each baud represents two bits.) Rather than retransmitting an entire packet of data to correct a single baud error, the FEC technique allows the baud error to be corrected at the receiver. This technique works as long as no more than one in every 12 baud are received in error. But the FEC technique exacts a penalty: of every 24 bits sent, eight must be allocated to FEC.

When large bursts of errors are encountered, the ARQ scheme takes over. A cyclic redundany check is used to verify received data. If data is received in error, a retransmission is requested. The "sliding windows" approach to sending packets reduces delays by allowing the transmitter to kick out packets one after the other, while the receiver sends acknowledgments only for groups of packets. Packets that are not acknowledged within a prescribed time period are automatically retransmitted (negative acknowledgments are not used in this scheme).

SPCL also uses dynamic packet sizing to handle cellular's widely varying conditions. When error rates are low, it is most efficient to send large blocks of data. With larger blocks, the number of bits dedicated to protocol overhead are a smaller percentage of the total bits sent. Under high error rate conditions, it is most efficient to send small blocks, because there is less chance that an error will occur within a small block. Should an error occur, retransmitting a small block is also less time consuming than retransmitting a large block.

SPCL determines the optimal packet size by monitoring the rate at which forward error corrections are performed. When the rate increases, SPCL responds by reducing packet size. Under harsh conditions, adjustments can be made on almost every packet. Packet size may be reduced quite rapidly, but increases are doled out more cautiously.

Maximum packet size is constrained by two factors: space available in the modem's internal memory and time-outs. SpectrumCellular's modems can support packets of up to 30 kbits. But when transmit data consists primarily of keyboard input, a packet will be closed off and sent if the next character is not received within 200 milliseconds. This prevents the "spongy keyboard" effect in networks that rely on far-end echo for their local screen display.

The minimum SPCL packet size is three bytes. When there is no current data to be sent, SPCL repeats the last five or 10 packets. This maintains synchronization and heads off retransmission requests.

Another company, Compuquest Inc. (Schaumburg, Ill.), has announced an alternative approach, based on its Compuquest Communications Protocol (CCP), which achieves link rates of 4.8 kbit/s using V.32 modems that are customized for the cellular environment. Even higher throughputs are claimed for compressible data. Compuquest states that CCP includes an FEC technique powerful enough to allow the receiver to recover up to 72 continuous errored bits without retransmissions. In addition, Compuquest reports that CCP includes both data encryption and a software device for tagging individual modems, features designed to combat cellular's security risks.

The problem, of course, is that cellular telephone transmissions can be picked up by inexpensive, widely available scanners. Fortunately, low transmitter output, split-frequency operation, Supervisory Audio Tone, and cell hand-offs tend to thwart the casual eavesdropper, but they do not pose serious obstacles to determined data thieves. For example, when a data transmission switches cell sites, the frequency that the transmission is on changes. This

Cellular

Suppliers of cellular data products

COMPANY/LOCATION	PRODUCT
AMERITECH MOBILE COMMUNICATIONS SCHAUMBURG, ILL. (312) 706-7851	TURNKEY CELLULAR DATA SERVICE
COMPUQUEST INC. SCHAUMBURG, ILL. (312) 529-2552	TURNKEY CELLULAR MODEMS
MAGICSOFT INC. LOMBARD, ILL. (312) 953-2374	ERROR CONTROL SOFTWARE
MICROCOM INC. NORWOOD, MASS. (617) 762-9310	ERROR CONTROL MODEMS
MOTOROLA COMMUNICATIONS SCHAUMBURG, ILL. (312) 576-6612	CELLULAR RJ-11 INTERFACE
NISSEI ELECTRIC USA INC. CLOSTER, N.J. (201) 768-0085	PORTABLE FAX MACHINES
OMNITEL INC. FREMONT, CALIF. (415) 490-2202	CELLULAR MODEMS
PUBLIC ACCESS CELLULAR TELEPHONY INC. MILFORD, OHIO (513) 248-9258	ELECTRONIC SIGNBOARD PUBLICATION
SECURE TECHNOLOGIES HERNDON, VA. (703) 471-6338	CELLULAR WORKSTATIONS
SOUTHWESTERN BELL TELEPHONE MOBILE SYSTEMS DALLAS, TEX. (214) 733-2169	TURNKEY CELLULAR DATA SERVICE
SPECTRUM CELLULAR CORP. DALLAS, TEX. (214) 630-9825	TURNKEY CELLULAR MODEMS
SUMMARY SYSTEMS DALLAS, TEX. (214) 638-3282	TRUCK FLEET EQUIPMENT
TELEBIT CORP. MOUNTAIN VIEW, CALIF. (415) 969-3800	HIGH-SPEED MODEMS
TELULAR INC. WILMETTE, ILL. (312) 256-8000	INTELLIGENT RJ-11 INTERFACE
TEXAS INSTRUMENTS AUSTIN, TEX. (512) 250-6679	PORTABLE TERMINALS
TOUCHBASE SYSTEMS INC. NORTHPORT, N.Y. (516) 261-0423	BATTERY-POWERED MODEMS
UNIDEN CORP. FORT WORTH, TEX. (817) 267-0521	RJ-11 ADAPTER

means the listener would have to know what cell to retune the receiver to. Even then he would miss some of the transmission.

Modem manufacturers

Although some market forecasters have predicted that data will eventually eclipse voice in the share of cellular airtime, few modem manufacturers have jumped into this fledgling market. Even if some market forecasters suffer from extreme optimism, the sheer number of cellular telephones and portable computers already being used by business suggests a rather large niche.

Telebit Corp. (Mountain View, Calif.) believes that it is uniquely qualified to address high-speed cellular data requirements. Not only does Trailblazer's multicarrier modulation have the ability to adapt to available bandwidth, but its Motorola 68000 processor delivers the extra horsepower needed to meet some of cellular's special requirements. With the exception of providing the RJ-11 jack and battery operation, the Trailblazer comes equipped for cellular today; the user need only enter special set-up commands. One advantage of the Trailblazer is its ability to adjust the number of bits encoded per baud, a feature not found on other modems. Telebit has also discovered, through experimentation, that high-speed connections can be achieved with greater reliability through a minor modification to the modem's handshake sequence.

The Trailblazer has been successfully used in high-speed cellular data applications like linking trauma care vans with hospitals. A paramedic at an accident site can upload X-ray images to a radiologist for diagnosis. Even compressed, these images require large volumes of data and can take several minutes to transfer at low speeds. In early tests, Telebit has achieved average, uncompressed throughputs of five to six kbit/s, an improvement good enough to turn precious minutes into seconds.

Microcom Inc. (Norwood, Mass.), the developer of the MNP protocol, is another modem manufacturer whose products are used in cellular data applications. MNP (through Class 4) is now a required option in the CCITT's V.42 recommendation, granting formal recognition to a protocol that had already achieved the status of a de facto standard. Limited tests suggest that MNP is quite effective in cellular applications in spite of its lack of an FEC technique.

According to Greg Pearson, author of MNP and Microcom's vice president of technology planning, a new MNP service class could meet the special needs of cellular data. "Forward error correcting schemes add horrendous overhead," says Pearson. "Turning FEC on only when needed would be more efficient." Pearson also points out that certain modulation techniques benefit from FEC more than others, another reason to make it a negotiable function.

Other protocol considerations that Pearson sees are:
■ When using dynamic frame sizing in the cellular environment, the frames should start out small (normal MNP Class 4 starts up with large frames).
■ The time-out for consecutive, failed retransmissions must be generous for cellular.
■ Multiple frame selective retransmission, a more efficient ARQ technique found in MNP Class 9, would help maximize throughput when the going gets rough.
■ Finally, the "S" register (which sets the time a Hayes-compatible modem will wait before it hangs up after losing receive carrier) should be set to wait out at least two consecutive cell hand-offs.

Touchbase Systems (Northport, N.Y.) makes a line of portable, battery-operated modems. While Touchbase has

Cellular

not specifically modified its products for cellular telephone, customers report they are using Touchbase's 1.2-kbit/s (Bell 212A/103, V.22/V.21) and 2.4-kbit/s (V.22*bis*, V.22, V.21, Bell 212A/103) modems in cellular applications. The modems run on 9-volt alkaline batteries and can be purchased with MagicSoft's MTE, a diskette version of MNP.

Omnitel Inc. (Fremont, Calif.) is a major supplier of laptop PC modems, including units that implement Spectrum-Cellular's SPCL. This makes certain Omnitel plug-in modems compatible with the dozens of networks supporting Spannet. (For a more complete list of companies that provide products or services useful in sending data via cellular telephone, see table.)

The future of cellular data

Cellular telephone is a young industry, and applications for data transmission are just beginning to blossom. Some of the newer applications are particularly exciting because they provide services that were not technically feasible or cost-effective prior to cellular telephone.

Cellular's promoters see a bright future in combining cellular data links with automatic vehicle locating systems, like Loran C or the satellite-based Global Positioning System (see "Satellite links for the masses: The final frontier?" DATA COMMUNICATIONS, November 1988) to help freight companies manage truck fleets. Fire departments have begun integrating facsimile with cellular telephone, allowing them to download floor plans of burning industrial buildings on the spot. And an Ohio firm, Public Access Cellular Telephony Inc. (Milford, Ohio), is showing mass transit systems how to use cellular telephone to deliver real-time, local advertising to electronic signboards on buses and trains.

While the move to create a digital cellular network holds promise for data applications, it will be up to the pioneers and early adopters to determine just how pervasive cellular data applications become. As Sherry Callahan, vice president of data processing at Carswell Credit Union (Fort Worth), has found, walking into a client's office and demonstrating your company's on-line services without plugging in to the telephone line can grab attention. "Even though computer technology is rampant, people are still impressed when we access our host computer without wires." ■

Ira Brodsky is president of Datacomm Research Co., a data communications marketing consulting firm located in Wilmette, Ill. He has previously held marketing positions with Tektronix Inc. and Gandalf Technologies and was most recently director of product marketing at U.S. Robotics Inc.

John M. McQuillan, McQuillan Consulting, Cambridge, Mass.

Routers as building blocks for robust internetworks

Linking LANs with smart segmentation devices—particularly routers—will provide users with a foundation for the next generation of wide-area networking.

In the past, traffic aggregation and multiplexing were performed primarily by packet switches, multiplexers, and concentrators. Increasingly, however, local area networks serve as the lowest-cost, highest-speed, and most popular traffic multiplexers. As a consequence, many organizations have concluded that their wide-area data network may evolve to become an internetwork of LANs.

Hierarchical networks such as IBM's SNA are built around a multitier structure with micros, minis, and mainframes and extensive distribution of processing services throughout the network. Such networks have been the mainstay of many computing facilities throughout the 1980s (see Fig. 1A).

In the 1990s, a new kind of network architecture will grow rapidly to assume importance at least equal to that of the previous two structures (see Fig. 1B). In a LAN-interconnection network, all the user devices appear to be on a uniform network architecture with equal access to all other clients and servers. Peer-to-peer communication is possible at high speeds through such a network without hierarchical data communications or data processing.

Interconnected LANs replace WANs
The two major computer vendors, IBM and Digital Equipment Corp., have both begun to develop and sell LAN interconnection solutions based on peer network architectures. DEC has been selling Ethernet-based networks to engineering departments for years. DEC's new generation of routers and SNA gateways is intended to provide LAN interconnection solutions for the entire corporation.

IBM has been the leading proponent of hierarchical networks sold to the central processing group. Recently, however, IBM has introduced a broad range of new products based on standards, including TCP/IP and FTAM (File Transfer, Access, and Management) on both MVS and VM. IBM also offers interconnection with Ethernet and 802.4 LANs in addition to its own token ring. The challenge for DEC is to sell LAN-interconnection technology to the MIS manager. The challenge for IBM is to build a new generation of networks. While neither vendor is likely to abandon its traditional market, both will be competing for the emerging market of centrally purchased internetworks.

Thus, the stage is set: Organizations are interconnecting their LANs driven by the decentralization and distribution of processing. Other contributing factors include the widespread availability of low-cost, high-bandwidth fiber optic transmission for local, metropolitan, and wide-area communications. Furthermore, the industry has responded by developing several generations of attractively priced bridges and routers for interconnecting LANs.

For the last 10 years, most corporations have standardized on a small number of wide area network (WAN) architectures. Traditionally, these have been dominated by the major computer suppliers (IBM, DEC, and others) and the X.25 packet-switching standard. We are now recognizing the emergence of a new WAN standard—the extended LAN. Sales of X.25 WANs may reach a peak and begin to decline by the early 1990s. Even IBM and DEC may begin to see their WAN sales slow as a result of competition from their LAN products.

Fast and simple LAN extension
The two main technologies for LAN interconnection are bridges and routers. Bridges are fairly simple devices. They work at Layer 2 in the OSI protocol hierarchy. When segmenting an extended LAN, bridges use the low-level node addresses, seeing only a "flat" address space with no hierarchical grouping represented with logical addresses. That means that they are independent of any higher-level protocols, though these may run in conjunction with a router.

Bridges need to process only the Media Access Control

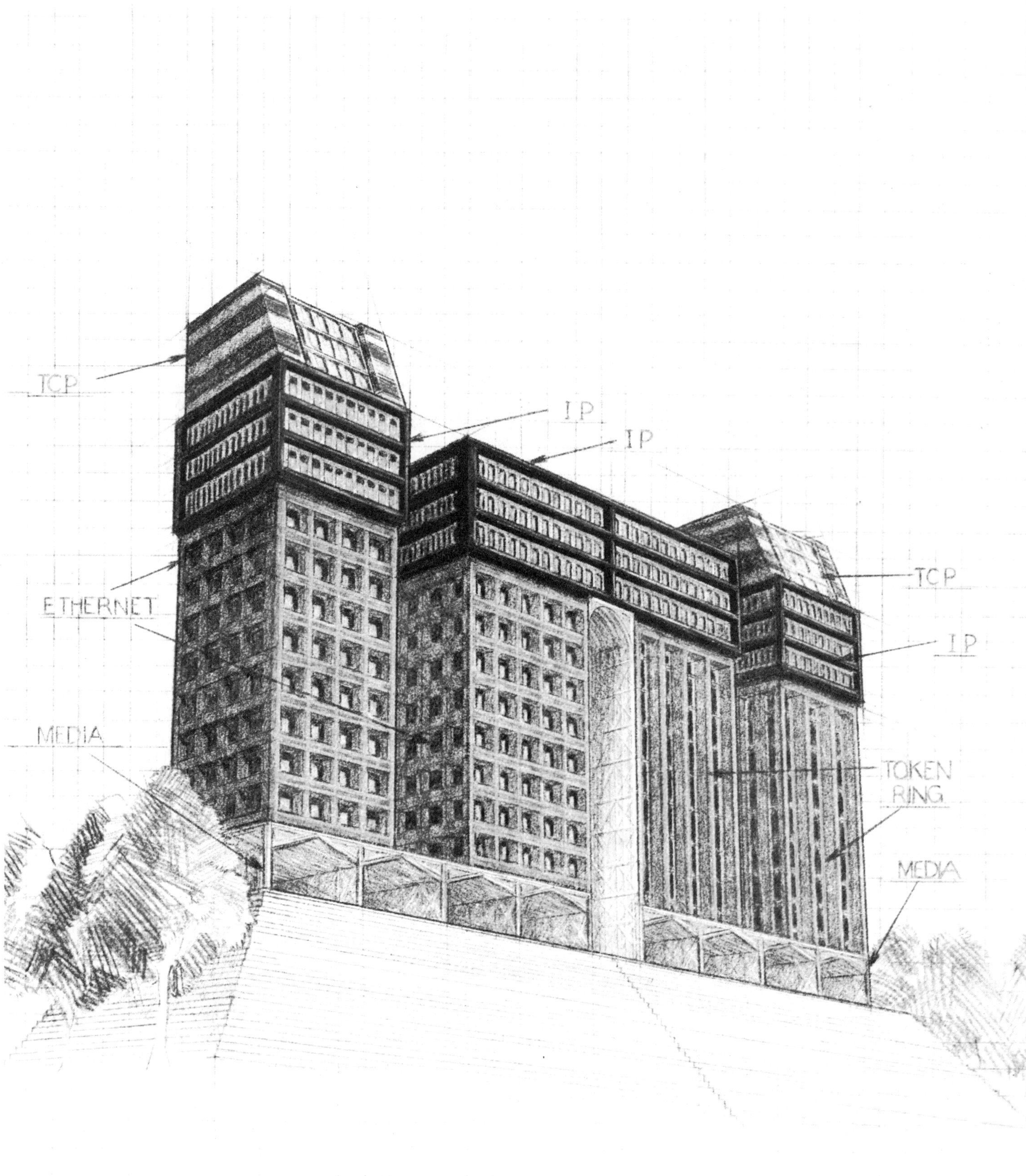
TCP
IP
IP
TCP
IP
ETHERNET
MEDIA
TOKEN RING
MEDIA

Building Blocks

(MAC)-level header on the LAN, a function that can be carried out partly or entirely in hardware. Bridges are, therefore, very fast. They only need to do two things. First, they must filter all the packets on the local area network, detecting which ones can remain local and which ones must be sent to another LAN. The second task is called forwarding, which is transmitting any of these remote packets to one or more other LANs.

These functions may be fixed or adaptive and some bridges can be programmed to carry them out on a selective basis. In general, bridges are used to interconnect similar LANs, but this is not essential. It is possible to bridge between very different LANs as long as both are running a compatible protocol suite at Layer 3 or above (see Fig. 2).

Bridges have a variety of applications, such as connection of coax LANs and twisted-pair LANs or extending local area networks to larger distances and greater numbers of ports. One of the more important applications is to improve the throughput and reduce the congestion on a LAN by dividing it into two pieces and isolating the traffic of each LAN from that of the others. Thus, it is quite common as local area networks grow for bridges to be introduced and for the networks to divide and multiply in a process resembling cell mitosis.

Bridges are very attractive to network managers because they can be inserted in a manner that is completely transparent to the computers and workstations at the source and destination. A customer can operate DEC, Hewlett-Packard, Sun, Apollo, and other computers over the same extended LAN. In contrast, the customer of a WAN technology such as SNA, DECnet, or X.25 has to select a single standard for a wide-area protocol and implement a network of compatible processors.

With bridges, there is no need to renumber any of the computers on the network, or to change the protocols or message headers after the bridges have been introduced. Further, bridges are capable of forwarding many different protocols concurrently so that the network manager does not need to know in advance which protocols are used. Finally, it is possible to bridge LANs at almost any speed and distance. Most organizations start slowly at first, linking a few local networks together.

Begin building bridges

It may be appropriate in the early stages of a network's evolution to install a low-end bridge on a PC platform by adding a couple of LAN interface cards. These imbedded bridges offer a way to get started without much investment or commitment.

As time goes on, most organizations conclude that it is necessary to have special-purpose bridges where traffic levels exceed a few hundred packets per second. Extremely high-speed bridges are available between LANs in the same building. These bridges come with two local area network ports and act as a node on each LAN. Other bridges operate over wide-area distances. These units come with a LAN interface and an interface to a transmission circuit such as a DS-0 or DS-1 line. For intermediate distances, it is possible to link bridges by fiber or other transmission equipment.

There has been some controversy about how bridges should operate. The IEEE 802 group has been working since 1984 on MAC bridging. Two proposals for bridge management protocols have been analyzed in detail: source routing (originally proposed for token ring) and spanning tree (originally proposed for Ethernet).

Spanning tree bridges are designed to be used by any Ethernet-style product, even those developed before 802.3. Source routing bridges add a routing information field to the basic frame format, requiring some participation by the

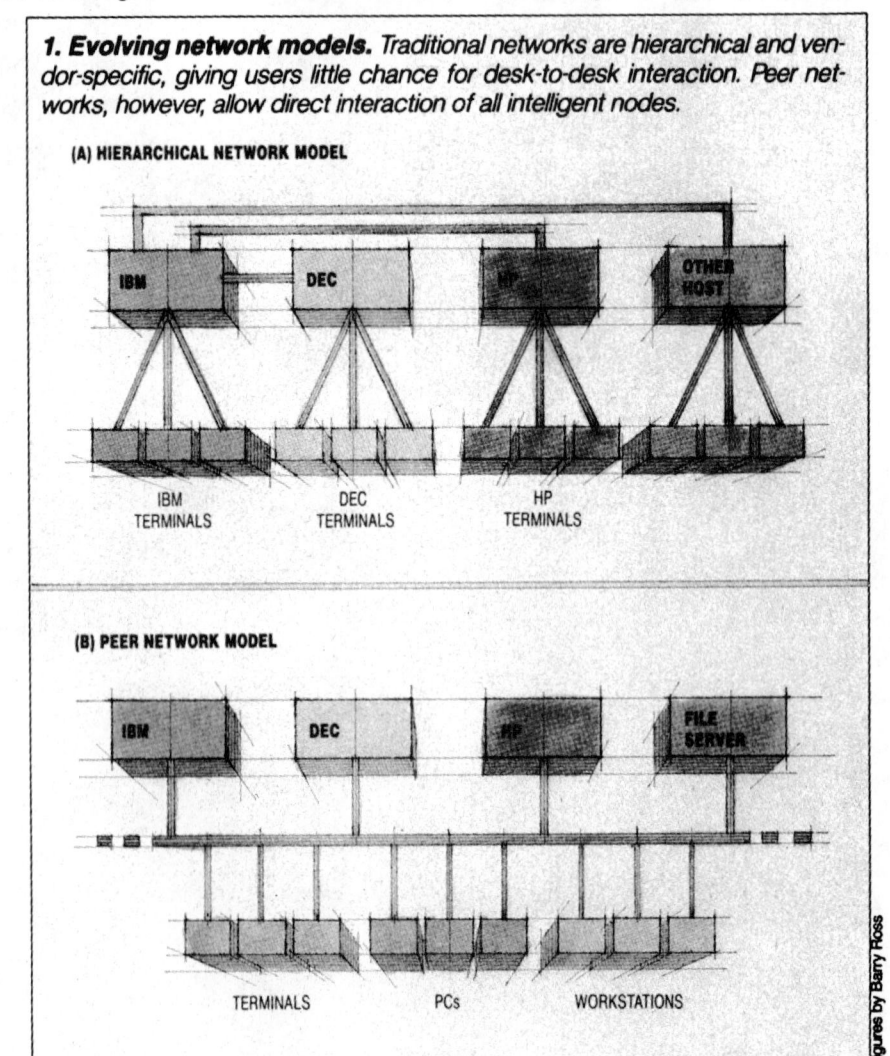

1. Evolving network models. Traditional networks are hierarchical and vendor-specific, giving users little chance for desk-to-desk interaction. Peer networks, however, allow direct interaction of all intelligent nodes.

Building Blocks

end stations. This makes it harder to apply to products already installed in LANs today.

The advocates of source routing claim that it will be applicable to much higher-speed LANs (as much as 100 Mbit/s for the Fiber Distributed Data Interface [FDDI] standard) by avoiding table look-ups in the bridges. Some of the debate concerns whether it is more important to be backward compatible to existing LANs or to plan for future LANs. IEEE 802.1 decided that transparent spanning tree bridges offer the best solution in terms of robustness, compatibility with existing 802 standards, and stability.

The spanning tree protocol has broadened the scope of bridges. Originally, bridges were capable of forwarding on one path only, without any intelligence regarding the network topology. This meant that bridges could not be linked to form a network that contained loops. The spanning tree protocol algorithm manages these loops, keeping certain links inactive to prevent packets from circulating forever. Some bridges are even capable of sending traffic on multiple paths simultaneously.

Another very attractive feature of bridges is that most of these products are intelligent and are capable of learning which nodes are local to each LAN automatically. There is no need to configure a bridge network. It is enough simply to turn the equipment on and let it discover the LAN's configuration and establish its own address table.

State-of-the-art bridges today are capable of filtering 20,000 packets per second and forwarding more than 10,000 packets per second for prices ranging from $2,000 to $20,000. At $20,000 there are advanced WAN bridges capable of alternate routing and good network management. For $2,000 the bridge is local only, with lower packet-processing rates and rudimentary network management. This price performance is an order of magnitude better than conventional packet switches and multiplexers, and two orders of magnitude better than earlier front-end processors in mainframe-oriented network architectures.

All in all, bridges remain the technology of choice for interconnecting small numbers of local networks or those that are in proximity to each other, as well as connecting LANs in applications where extremely high performance is essential.

Enterprise-wide internets

Routers are a complementary technology to bridges. Routers operate at Layer 3 in the OSI hierarchy, which means they are specific to a particular protocol such as TCP/IP, XNS (Xerox Network Systems), DECnet, or OSI. Routers exchange information with each other via a management protocol, in order to establish routes through the networks that interconnect them. In this way, routers form an internet or a "network of LANs." Routers offer improved reliability because they have adaptive routing algorithms, extensive traffic-filtering capabilities, and some security measures (see Fig. 3).

2. Basic bridge architecture. *Bridges are low-level network management devices that can be implemented completely or mostly in hardware. Though they don't offer sophisticated network management, they are fast and cheap.*

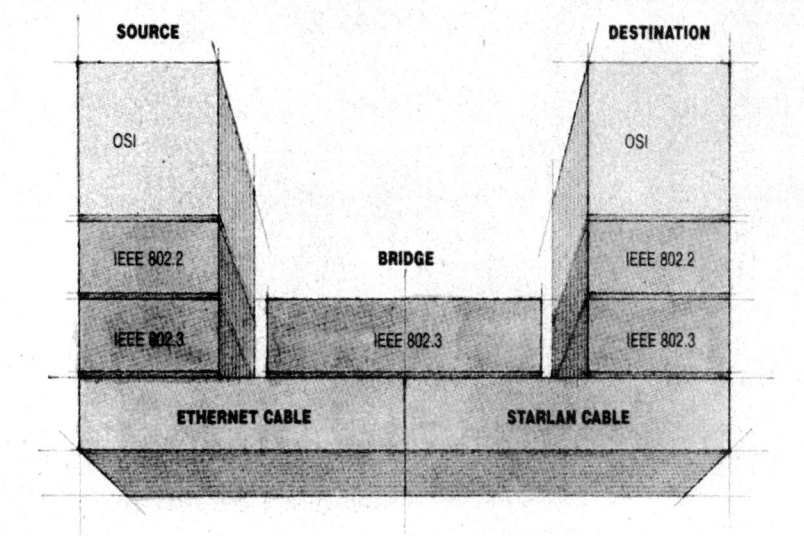

One of the prerequisites to using a router is to standardize on one or more protocols and addressing structures in the network. In order to use routers in a large LAN interconnect environment it is essential to have a protocol that is based on a hierarchical, not flat, address structure.

Usually, router management schemes determine routes based on the top-level address structure (analogous to an area code in the telephone network), then when the packet arrives at the destination router and it is sent into the destination network, that network must take responsibility for delivery to the address within that "area code." Routers are designed to interconnect highly diverse media. The original routers for the Arpanet (called gateways) were TCP/IP devices that could link Ethernet, X.25, radio, and satellite networks.

Along with the advantages of greater reliability, security, and network management information come some drawbacks. Introducing routers in a network means that the end points must all adhere to a uniform address numbering plan. Breaking a LAN into two pieces with a router is not as simple as splitting a LAN with a bridge. In general, bringing in an additional router means renumbering some of the computer systems.

But there are compensations. It is not possible to bridge one LAN segment onto another without limits. After bridging several segments in a row, fundamental network timing limits prevent further bridging. With routers, this problem goes away. Customers can build networks with hundreds of routers using today's technology. Advocates of routers point out that they can serve to protect one network from problems or disturbances on another.

It is quite common for large bridged networks to suffer from "broadcast storms" because some LAN protocols that are broadcast in nature are passed through each bridge,

513

Building Blocks

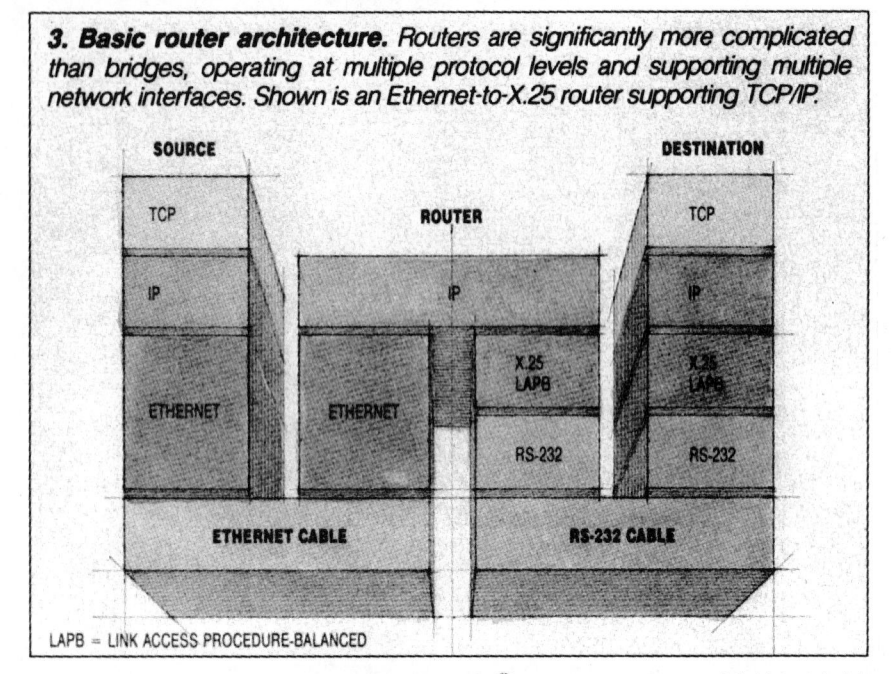

3. Basic router architecture. *Routers are significantly more complicated than bridges, operating at multiple protocol levels and supporting multiple network interfaces. Shown is an Ethernet-to-X.25 router supporting TCP/IP.*

LAPB = LINK ACCESS PROCEDURE-BALANCED

multiplying the traffic level on the extended network enormously. In the case of malfunctions or improperly set parameters, some extended LANs have been rendered unusable because of such broadcast storms. Routers, on the other hand, forward only traffic that is explicitly addressed to them, blocking uncontrolled broadcast traffic.

One of the most important innovations in recent years is the development of multiprotocol routers. Initially, most routers were specialized to a particular protocol suite. Recently, several companies have developed routers that can concurrently route several different protocols such as TCP/IP, DECnet, XNS, and OSI. This is tremendously important for organizations that have not yet been able to standardize on a single protocol. However, there are other protocols (DEC's Local Area Transport and some SNA protocols) which cannot be routed, and so must be bridged. Many routers can also bridge.

In comparison with bridges, routers are somewhat more expensive and often slower. State-of-the-art routers are usually limited to a few thousand packets per second and cost about $10,000 to $20,000. Although a fast router will out-perform a slow bridge, routers are generally slower than bridges because their job requires a fairly significant amount of processing in software.

But speed is relative. A router that is capable of handling 2,000 packets per second is fast enough to connect an Ethernet to a T1 line at full bandwidth. In fact, such a router could interconnect two LANs and two T1s at full efficiency if the packet lengths are sufficiently long. And there are routers available today that are capable of speeds of up to 10,000 packets per second in certain circumstances.

Making the choice

So, which is best—bridges or routers? While many vendors enjoy debating the question, it is often not really relevant for users. Why? Because most customers find that the question has an easy answer for them. If they value an open architecture with standard interfaces and want to preserve flexibility to use any protocol they desire, then the choice is clear—they must use bridges.

On the other hand, some organizations have been able to develop a strongly defined, unified architecture around one or a small number of standard protocols. In this case, where the emphasis is on a clear vision of the future, routers can help to make this vision a reality.

Some people have observed that two-port bridges can, as many LAN segments are interconnected, lead to inefficiencies because traffic from the source segment may have to flow through many intermediate segments on its way to the destination. This reduces some of the traffic isolation advantages that may have motivated the use of bridges in the first place. Multiport bridges that can connect four or more LAN segments are a potential solution to this problem. On the other hand, multiprotocol, multiport routers can fill this product requirement as well.

Whenever there is a difficult choice, such as bridges versus routers, the customer may want to have his cake and eat it too. Now that is possible with the arrival of hybrid bridge/routers, which are available from such companies as Cisco, Halley Systems, and others. In fact, both bridges and routers continue to improve over time, as shown in Figure 4, with a trend toward product consolidation.

Most large customers with many LANs to interconnect are likely to use some bridges locally and some routers remotely and for backbones, favoring hybrids. At the same time, there will still be a market for low-cost bridge-only products and for single-protocol routers.

As networks grow larger and more numerous and as communications processors get faster, customers are tending to use routers for WAN communication. They provide better management and more complete isolation of local communications. On the other hand, as LANs get faster—FDDI and 802.6, for example—bridges may be favored for the highest-speed communication (over 100,000 packets per second) because it is difficult to carry out such high-speed processing in software.

As FDDI technology is used in the local area and T3 circuits become prevalent in wide-area networking, both bridges and routers will be built to interconnect low-speed LANs with FDDI and FDDI with T1 and T3 circuits.

View from above

The movement to a LAN-interconnect architecture has profound consequences throughout any organization in several areas: technical architecture, organizational structure, and network strategy.

A LAN-interconnection network makes it possible to

Building Blocks

unbundle data processing from data communications. Traditionally, minicomputers carried out many different functions, including terminal support, database access, WAN communication, electronic mail, and others. Once a LAN-interconnection architecture is installed, these functions can be split among a number of specialized processors.

Terminal servers are used to interconnect terminals with multiple hosts, not just one. File servers and database servers offload the data-handling function. Specialized electronic-mail servers can run on 386-based processors and specialized servers such as bridges and routers carry out the WAN communications. The minicomputer or other specialized CPU is relegated to the role of a MIPS server.

This is an important transition from the era in which terminals had to be purchased from the same vendor as the host computer. Now PCs serve as universal workstations and are purchased as commodities. In fact, workstations, servers, bridges, and routers are all quickly entering the commodity phase. They are purchased based on price and performance rather than on brand name.

Unfortunately, a LAN-interconnect architecture gives the network management staff a much more difficult job than in the past. Instead of a centralized network with tight configuration control over all workstations, the future lies with extremely flexible and open systems with thousands of users supported by their local departments. Local area network interconnection is a centrifugal force that is leading to a dispersed architecture of microservers instead of more centralized minis and mainframes. It is also leading to an era of local control and management of highly distributed networks.

The second major consequence of LAN-interconnection architectures is a shift in organizational structure in many companies. Traditional organizations are structured hierarchically. This leads in turn to a hierarchical information flow with batched information sent upward in the organization at periodic intervals.

This organization and information flow causes long delays in the decision-making cycle that are familiar and frustrating to everyone. Many companies are trying to flatten their organizational pyramid in an effort to become more responsive to the demands in the marketplace and to shorten their decision cycles. LAN-interconnection networks are a way to assist in this effort.

Decentralization and downsizing

As a company evolves its business practices and flattens its organizational pyramid, it can use nonhierarchical computer networks to send information on a peer-to-peer basis throughout the organization. The information can be streamed from one desk to another instead of being batched from one department to another.

The best early example of such an application is electronic mail. In time, interconnected LANs will make this kind of communication possible for all sorts of business processing. The revolution of network structure will cause—and it must reflect—profound changes in organizational structure.

The third consequence of LAN-interconnection networking is a shift in strategic thinking. The traditional model for competition in business is based on value. In this theory, the relevant equation is: Value = Quality ÷ Price.

The new theory is that there is a third dimension to competition. Companies can not only compete by improving their quality or by dropping their price but also in a third way: by competing in time. The new equation is: Value = Quality ÷ (Price × Delay).

An important new competitive edge is improving the delay the customer sees in working with the supplier. This delay might be in ordering the part, getting satisfactory customer service, upgrading to a new capability, or some other aspect.

Local area network = interconnection technology makes an organization more nimble. Since rapid response is key to future competitiveness, LAN interconnection can play a vital role in improving a company's position in its marketplace.

The interconnection of LANs to form a new kind of WAN is a critical step in the establishment of enterprise networks. Even when the fundamental technologies of bridges and routers become well established, there will still be a number of problems remaining in network management, operations, and security. Installing these network segmentation devices is the first step in a new era of internetworking. ∎

John M. McQuillan is a well known authority on computer networks with 20 years of experience, including the development of the SPF routing algorithm for the Arpanet, which serves as the basis for many LAN routers today, IP and DECnet Phase V among them.

4. A separation that won't last. Though they started out as unrelated devices from different vendors, bridges and routers are in the process of converging.

- SIMPLE BRIDGES
- ↓
- LEARNING BRIDGES
- ↓
- MULTIPORT BRIDGES
- ↓
- BRIDGE/ROUTER
- ↑
- MULTIPROTOCOL ROUTERS
- ↑
- SINGLE PROTOCOL ROUTERS

Edward R. Teja, special to DATA COMMUNICATIONS

Router roundup: Tools for network net segmentation come of age

A head-to-head comparison of the leading router vendors, and the key issues to keep in mind while you are router shopping.

Today, increased competition, evolving user needs, and corporate mergers are forcing network managers to marry previously independent networks. Their task is not to create a network from the ground up, but to aggregate subnets into a viable internetwork that will equally serve the parts and the whole of the organization. When putting networks together, the challenge is to let data—not network faults—flow across internetwork boundaries.

This sounds great, but in many cases separate networks with vastly different networking strategies have been forced into unnatural relationships, in spite of their installed bases of different hardware and software. Integrating diverse systems into a single network, therefore, means accommodating a variety of protocols and user needs without sacrificing overall integrity.

Under such circumstances, it makes sense to build an internetwork that retains the characteristics and autonomy of the individual subnets, while at the same time allowing users to share resources and data without the restrictions introduced by location or protocol incompatibility. The only way that this can happen is through intelligent network segmentation.

The concept of network segmentation reflects an important reality: On a single LAN, every component is a relatively equal partner in the network address space, but this situation can spell disaster in a relatively large network. A single address space in a large inter- and intracorporate network plays havoc with network reliability, management and planning, and flow control.

Here's why. The fallibility of an electronic system is equal to the combined failure rates of all of its components. The more components, the higher the overall failure rate. Furthermore, the reliability of the system can never be any better than that of its least-reliable element. That's a sobering thought when you are mixing PCs, minicomputers, mainframes, and special-purpose servers on a single network. Common sense dictates that an unlimited number of these devices, unrestricted, on the same wire is not an ideal situation.

The best way to manage the interaction of large numbers of devices on a network and also eliminate the single address space is with routers. Segmenting a network with routers inherently builds fire walls between the subnets. A failure in one subnet might disrupt that LAN but won't bring down the entire network. More important, if you create a high-speed backbone and link subnets to the backbone through routers, you can place critical shared resources on the backbone and not worry about the failure of a subnet gaining access to a critical resource (see figure).

Because of the increased demands on segmentation devices, the networking industry is starting to discover that the router is the correct way to segment a network into independent LANs that can easily share data and resources. A router is the only network element that does this job, and it is the only job that a router does best. Bridges are not routers, no matter how much they look like routers, appear to do router functions, or even cost as much as routers (see "Bridges and routers—different tools for different jobs").

Today, there are not as many routers available as there are bridges and repeaters, though this will likely change as internets grow in size and sophistication. The primary third-party router vendors are Cisco Systems (Menlo Park, Calif.), Proteon Inc. (Westborough, Mass.), and Wellfleet Communications Inc. (Bedford, Mass.). These are the aftermarket vendors whose products are not embedded in host computers or file servers, as they are with DEC and Novell routers.

Each third-party router vendor brings a somewhat different approach to routing. The intent of this article is not

Router Roundup

Router isolated backbone. Putting critical resources on the Ethernet or token ring backbone and segmenting the network with routers puts a fire wall between the users and the resources that must not be impaired by faults that may develop on local network segments. Data propagates through the internet, but not faults.

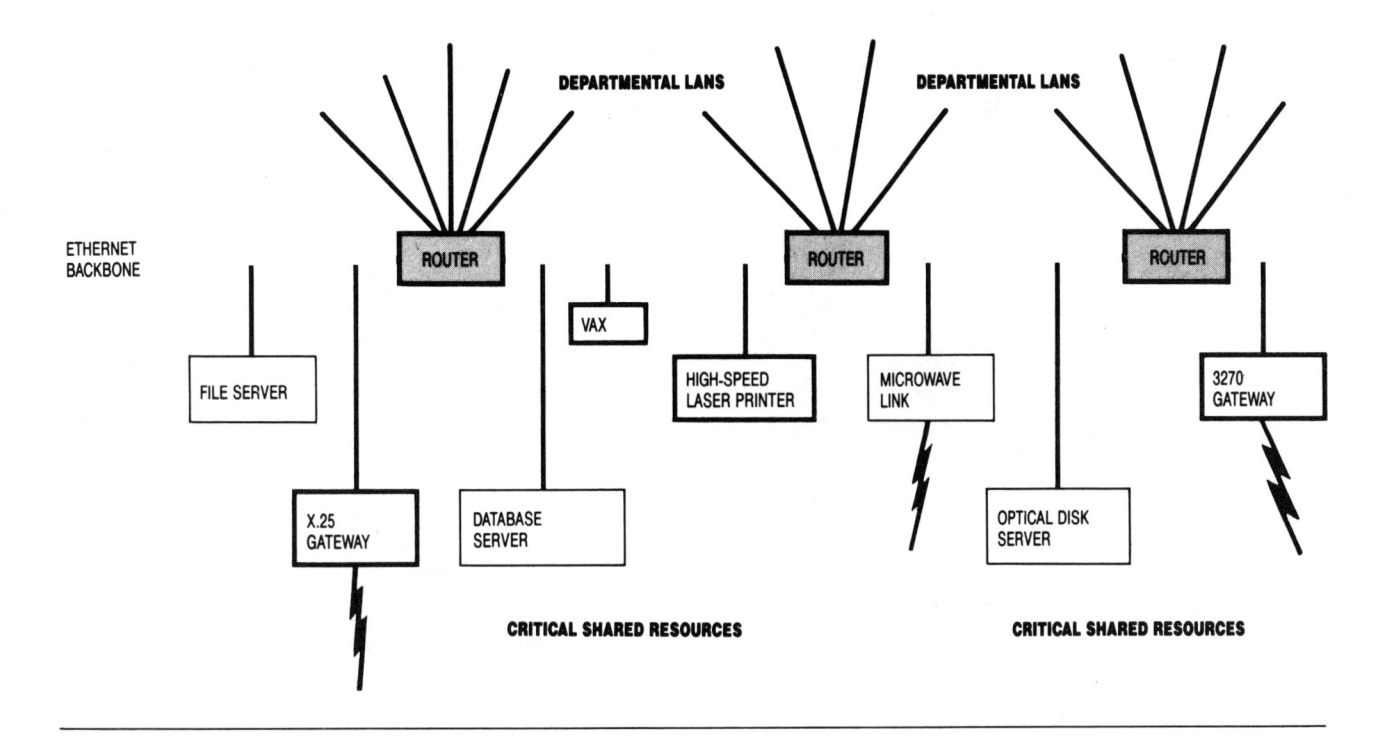

an exhaustive evaluation of all vendor products, but rather an introduction to basic features and functionality available in this emerging industry.

Stacking up the stacks

Each network architecture (SNA, DECnet, NetWare) has its own protocols that define packet formats and the way packets are directed through an internetwork. Many protocols share a single heritage, but there are differences. Each of the third-party router vendors supports one or more network protocols and, in some cases, OSI stack components as well.

Conformance with several network protocols provides interoperability between dissimilar networks. Unlike a bridge, a router can support multiple protocols on multiple protocol layers. This allows interoperability not possible with bridges. For example, a router that has interface and protocol compatibility with both Ethernet and token ring can connect the two, passing messages from one to the other.

Among the well-established protocols are:
- CLNS, the International Organization for Standardization's Connectionless Network Services (as specified by ISO 8743).
- TCP/IP, used extensively both commercially and in military networks.
- XNS (Xerox Network Systems), which is used by many commercial vendors.
- DECnet, from Digital Equipment Corp. One of DEC's important network protocols is Local Area Transport, or LAT, a terminal server protocol.
- IPX/SPX (Internet Packet Exchange/Sequenced Packet Exchange), the protocol used by Novell's Advanced NetWare.
- EGP (Exterior Gateway Protocol), the routing protocol used by gateways on the Defense Data Network (DDN).
- Chaosnet, developed at MIT for the artificial intelligence community, provides routing plus host functions such as status and uptime services.

Some protocols are more important and relevant than others. For example, while Chaosnet represents a specialized and limited market, XNS, IPX/SPX, and TCP/IP are far more common. Still, if you must aggregate subnets that use specialized protocols, it doesn't matter that they may be arcane. Being multilingual is a plus, and some router vendors are more multilingual than others.

Of course, there are different interpretations of every protocol, and standards do not necessarily define all aspects of the protocol well. Thus, there are differences in the robustness of the implementations. The only way to be certain that an implementation is up to par is to look at the vendor's customer base and see if the company has a fair amount of experience in dealing with a particular protocol.

Increasingly, one well-implemented protocol is not enough for a large organization. Flexibility in aggregating networks requires an ability to use multiple protocols simultaneously. Without this, it becomes impossible to effect dependable internetworks.

Network protocols like XNS and TCP/IP are not the only protocols used by a router. To maintain their internal tables and keep abreast of events on the internet, routers talk among themselves using router management protocols.

When a router receives a packet, it must decode the destination address and determine the best path to route it through the internet. This is known as the "policy routing" and also affects how well a router will enact flow control. There are differences in the ways that routers select the best path for a message. Some routers are configured by the network manager, who establishes the cost of each possible path by entering the speed, number of hops, and path length in the routing table. These are termed "static" routers. Most multiprotocol routers are "dynamic" routers, which automatically configure and reconfigure the routing table as needed.

Even though dynamic routing is widely supported, vendors differ on how to go about it. Those differences reflect their philosophies of routing algorithms and the influence of the established protocols they choose to implement.

Basically, there are two competing internal routing algorithms: distance vector and link state. Cisco, Proteon, and Wellfleet currently make use of distance/vector routing protocols derived from Xerox's Routing Information Protocol (RIP).

RIP is a distance vector (Bellman-Ford) protocol that describes the network in terms of "hop count" metrics. In this system, a number between 1 and 15 represents the number of routers between the current node and the destination node, making no distinction between high-speed paths and slower ones.

Some RIP implementations add to the hop count of slower networks to weight them more accurately. But the maximum value of the metric is 15, seriously limiting the effects of such weighting. A router using RIP propagates its metrics throughout the network as part of update messages. These metrics provide a snapshot of the network routing table and its status.

Wellfleet Communications, for instance, uses RIP for TCP/IP environments but features extensions to it that allow its routers to make better use of the network's resources.

Most vendors offer some proprietary extension to RIP, but there is another alternative. The Internet Engineering Task Force of the Internet Activities Board developed a new-generation protocol called the Open Shortest Path First (OSPF), a link state algorithm. Proteon, an early implementer of OSPF, contends that the algorithm elimi-

Bridges and routers—different tools for different jobs

Bridges play an important role in extending networks. They interconnect LANs, passing traffic between them at high rates of speed.

Bridges are Media Access Control (MAC) store-and-forward devices that can run at high packet rates because they ignore network-level (and above) protocols.

Bridges offer more than simple store-and-forward operation, however. They can link dissimilar LANs, for instance. And units such as Vitalink's TransLAN bridges can link local area networks into wide area networks, even interconnecting subnets over T1 links.

Bridges are also significantly less expensive than routers. The price for Proteon's lowest-cost p4100 router, for instance, starts at $3,750, and then one has to buy network interfaces, which range from $500 to $3,900 for a dual-port X.25 interface. CrossComm Corp.'s Integrated Local Area Network (ILAN) bridge costs under $3,000 for token ring networks. And this is not a bargain basement model.

A user gets a great deal of performance from a bridge for less money than the router; what isn't available at that price is intelligent network segmentation and flow control.

Routers provide an intelligent link between networks, using network protocols to examine and then pass only the traffic from one subnet (one side of its interface) that is addressed to a station on the subnet on the other side of the interface. Although bridges such as Vitalink's TransLAN use spanning tree protocols to try to minimize traffic congestion, they are still transparent to network protocols, such as SNA and TCP/IP.

This automatic alternate-path routing by bridges is an invasion of the router's traditional territory. The term for such hybrid status is "brouter." But the router vendors don't take the challenge of brouters seriously. Multiple path-control functions in brouters are proprietary and can only be used with other brouters from the same vendor; it does move bridges into an area of network management. Brouters don't provide the fire wall of segmentation, nor do they reduce cross-network traffic. The functions are different, and what brouters bring to the party is nothing more than a capability of handling higher-peak traffic.

But just when the separation of bridge and router grows crystal clear, along comes Wellfleet. Wellfleet makes both routers and bridges, but rather than offering models of each, the company packages bridging and routing services together. Wellfleet uses its proprietary Communication and Application Protocol Environment to provide both learning bridge and routing services in the same box under selective control of the network manager. Cisco also can bridge and route in the same device. So, with both of these products the answer to the question "What is the difference between a bridge and a router?" for all practical purposes becomes "The configuration you choose."

The advantage of the combined bridge-router approach is that it adapts to any network need. The user pays for all this flexibility, whether all the bridge and router functions are needed or not. Conversely, when a multiple-protocol router is called for, why pay for bridging capability as well? The answer is that some networks don't support Layer 3 (network) protocols, making bridges the only viable solution for interconnection. The combination of bridging and routing in the same chassis supplies the tools to mix and match LANs of all types into a heterogeneous network. —E.R.T.

Router Roundup

nates routing loops and black holes that can crop up with other algorithms.

Rather than broadcasting the entire routing table to all other routers, in OSPF each router transmits a packet with a description of its own links to other routers. Each router receiving the packet acknowledges it to the sender. Routing tables are built from the collected descriptions sent by all the routers. Because link descriptions are small and infrequent, they require little routing traffic, keeping the network free for other messages.

Router selection criteria

Router evaluations must take into account the two protocol touchstones discussed above: router management algorithms and standard network protocols.

In their design process, vendors choose a management protocol, then interpret the routing information contained in the network protocols and use it according to their internal routing algorithms. There is nothing in the packet structure standards that dictates routing policy. So the router can collect routing data from a variety of network routing protocols and send a packet that conforms to yet another protocol. Table 1 shows what network protocols the routers from Cisco, Proteon, and Wellfleet support.

An important criterion in selecting a router is the number and selection of network protocols it supports (TCP/IP, XNS, LAT, and so on). But connecting to a network is more than a question of the right software; it is also a hardware, or physical-level, issue. A router's network interfaces comprise the ports and hardware required for interfacing to existing networks. As with software, there are some obvious standard networks to be accommodated, such as IEEE 802.3 (Ethernet) and 802.5 (token ring). These are so pervasive as to be mandatory, but others, such as wide-area interfaces for X.25, T1, and Dataphone Digital Service, make the router more useful in the average network setting.

In the same vein, it's advisable to find out how many of these interfaces (ports) each vendor's router can accommodate in one box at one time. Much of the router price is in the chassis, power supply, and system intelligence. The more interfaces a single chassis can handle, the more versatile and cost-effective each unit can be. Table 2 lists the maximum number of network interfaces of each type.

Of course, these lists don't tell the entire story. In the case of Cisco, for example, the router chassis has nine slots. Two slots are used by the system CPU card and memory. That leaves seven slots for interfaces, all of which can be used for token ring interfaces. With Ethernet the situation is different.

Cisco supplies two types of Ethernet interface cards: the Multiport Communications Interface (MCI) provides two Ethernet interfaces and two serial (up to 4 Mbit/s) ports; the Serial Communications Interface (SCI) provides four serial and two Ethernet ports. Because of power supply considerations, the chassis accepts only five MCI or SCI cards, limiting a user to 10 Ethernet interfaces per router. With additional "oomph" for the power supply, however, a user can (and several do) use all seven slots for MCI cards, upping the maximum to 14 Ethernet interfaces. This mega-router will route packets between 14 independent LANs from one box.

Another clarification for Table 2 relates to nonstandard interfaces. Proteon, an early innovator in the LAN field, developed its own network interface hardware long ago. So, in addition to the standard IEEE varieties, Proteon provides access to its own LANs, such as the ProNET-80, an 80-Mbit/s token ring network. ProNET-80 allows a user to space adjacent networks 1.24 miles (2 kilometers) apart and use fiber optics for the interconnection. Proteon also offers a complete fiber optic incarnation that lets ProNET-80 systems run over 18.6 miles (30 kilometers). If nonstandard interfaces were included, Table 2 would show a maximum of four ProNET-10 interfaces and three ProNET-80 interfaces.

In general, the fiber optic LAN market is hot. So hot that all three router vendors have wasted no time in promising connections to the Fiber Distributed Data Interface (FDDI). That means, from a practical standpoint, that they'll be available to play with later this year (assuming you are a large customer) and to buy in 1990. The reason for the push is that FDDI provides a 100-Mbit/s network ideal for a backbone in a large system, connection to a high-speed backbone being a major application for routers.

Router roots

Looking at the offerings from Cisco, Proteon, and Wellfleet, a definite pattern emerges, one that points up the heritage of each of three vendors. Including its own implementations, Proteon offers far more LAN interface choices than the other two. A logical circumstance when you consider that Proteon has a history as a LAN interface vendor. Routers were added to the product line to meet the needs of customers building larger networks. This experience in LANs means that Proteon often gets more performance on the LAN side than other vendors, at least with its proprietary products.

In 1986, Wellfleet was founded to develop a product

Table 1: Network protocols supported

Network protocols	Cisco	Proteon	Wellfleet
APPLETALK	YES	YES	NO
APOLLO DOMAIN	YES	Q4-89	NO
CHAOSNET	YES	NO	NO
ISO CLNS	YES	AVAILABLE 2/1/90	NO
DECNET	YES	YES	YES
DECNET (PHASE 5)	FUTURE[1]	FUTURE	
EGP	YES	YES	YES
HELLO	YES	NO	NO
TCP/IP	YES	YES	YES
IPX	YES	YES	NO
XNS	YES[2]	YES	YES

1. Phase 5 should be compatible with OSI.
2. Xerox, 3Com, and Ungermann-Bass implementations

Table 2: Maximum number of interface ports

Interfaces	Cisco	Proteon	Wellfleet Link model	Wellfleet Concentrator model
ETHERNET	10	7	8	26
DDN 1822	7	7	NOT AVAILABLE	
DDN X.25	18	14	NOT AVAILABLE	
DDS	18	14	16	52
PDN X.25	18	14	NOT AVAILABLE	
T1	18	14	8	26
TOKEN RING	7	7	NOT AVAILABLE	

DDN = Defense Data Network
DDS = Dataphone Digital Service
PDN = Public Data Network

that provided both bridge and router functions, on the sound theory that both were needed to solve all networking needs. Coming from Interlan and Codex, the founders brought chip-to-system-software experience to their products, as well as a different way of looking at internetworking. They intended to create a product that would support growth and change. Recently, one customer, testing the success of Wellfleet's intentions, began upgrading a system with four routers to a 24-router network.

Unlike the other two vendors, Cisco started as a router company. Founded in 1984, Cisco came out of Stanford University where its founders had implemented the TCP/IP network under contract to Defense Department's Advanced Research Projects Agency. This work was the basis for Cisco's routers. The firm's experience, therefore, is focused on high-performance, large-scale networks and intercontinental capability. For example, one of Cisco's customers has put together a worldwide network consisting of 300 routers and more than 450 links. A network manager with a similar goal might find comfort in this company's experience.

Router performance: The hard part

Even after a network's protocol and interface requirements are defined, the issue of performance can still make decisions difficult. There are, at present, no viable router performance benchmarks, and a lot of fuzzy spots in the vendor-supplied statistics.

The performance evaluation problem is not limited to routers, by the way. In general, a lack of meaningful performance benchmarks for internetworks is a thorn in the side of the industry. Consider the problem of determining how much traffic a network will support. The network's bandwidth is a measure of the number of bits-per-second the medium will handle. For instance, a token ring interface provides 4-Mbit/s operation.

A more meaningful metric, though, is the number of packets it will handle per second (PPS). This is because there is often a great deal of overhead involved in routing packets. A packet travels at 4 Mbit/s between intermediate nodes in the routing process but is delayed significantly while it is processed by each router in the path. Consequently, the raw bandwidth of a network is no indication of the actual throughput from end node to end node, taking into account the delays. The PPS indication is far more accurate a measurement of performance but far more difficult to achieve.

Unfortunately, when evaluating routers, it isn't easy to discover this specification by looking at the data sheets. For one thing, the vendors don't all measure throughput the same way. Some router vendors count each interface transition (entrance and exit) as a separate passthrough. In other words, if it passes 1,000 packets through both interfaces in one second, this is considered a throughput rate of 2,000 PPS. Other vendors measure the transfer of a packet through the router once, providing a measurement of 1,000 PPS for the exact same performance. To be sure, it's necessary to ask.

During the preparation of this article, Cisco, Proteon, and Wellfleet were invited to provide their packet performance for two particular situations—Ethernet to Ethernet and token ring to token ring. The results, measured in terms of the filtering rate and the forward rate, are presented in Table 3. Filtering rate indicates how many packets per second the router can take off the line. Forwarding rate is the speed the device processes and transfers packets to the output network.

Note that these specifications won't reveal the dropout rate, packet size, delay, or other important factors. They do, however, provide a starting place for discussions on performance evaluation. A user concerned with restricting access to his LAN from some larger network to which it is connected, should ask the router salesperson: What happens to performance when you invoke enhanced access control? In some cases, the performance will degrade. If access control is needed, then the degraded performance figures are the ones to use for comparison.

Another problem with the performance numbers is that they don't usually indicate whether the rates are for a single interface card with multiple ports or for multiple interface cards (performance will often degrade when transfers involve passing through the router's backplane). Similarly, the numbers can represent a sustained data rate over time or a burst-oriented data-forwarding rate. Is the vendor giving both, or just the best the router can do under the best of circumstances? The numbers Wellfleet provided, for example, are for a sustained data rate, across a single interface card, using 64-byte packets (we asked).

Even the seemingly straightforward filtering rate can be a perplexing specification. Proteon contends that it is a bridge specification, not a router specification. That is reasonably true in that a router has to process only those packets that are addressed directly to it. Bridges, by contrast, must make filtering decisions on every packet that hits the wire. In spite of this, other vendors seem concerned with filtering rates for routers, so it becomes a router specification as well.

This measurement nightmare is aggravated by the fact that no two networks are alike. Traffic patterns or peak

Table 3: Head-to-head performance comparison

Interfaces	Cisco	Proteon	Wellfleet
ETHERNET TO ETHERNET			
FILTERING (PPS)[1]	15,000	–	15,000
FORWARDING (PPS)	12,000	1,000[2]	7,200[3]
TOKEN RING TO TOKEN RING			INTERFACE NOT AVAILABLE FROM VENDOR
FILTERING (PPS)	15,000	–	
FORWARDING (PPS)	1,500	1,000	

1. Proteon does not consider the filtering rate a router performance specification.
2. Numbers constant with or without access control.
3. Sustained data rate for 64-byte packets.

PPS = Packets per second

loads that present no hardship to one router can overwhelm another. At this stage in the game, the most valid criterion is experience. The vendor's assurances that the problem won't arise mean something only if the vendor has experience with a network of the scale of—and protocols similar to—what is being put together.

The problems grow as the typical network is growing. At Stanford University, for instance, the campus network includes 51 TCP/IP gateways (routers) and 79 subnets—and it is constantly growing. Most organizations face the same situation. Computational power continues to migrate to the work site, and communications needs grow proportionally.

This trend often puts companies on the leading edge of what is known about router performance. There are no simple solutions, but the facts suggest that the canny user should shop for routers by finding a vendor who has sold to customers with network demands comparable to his own. And today that means much more than a number on a data sheet or benchmark. ∎

Vendor addresses
Cisco Systems Inc., 1350 Willow Rd., Menlo Park, Calif. 94025 (415) 326-1941
CrossComm Corp., Box 699, Marlborough, Mass. 01752 (508) 481-4060
Proteon Inc., Two Technology Dr., Westborough, Mass. 01581 (508) 898-2800
Wellfleet Communications Inc., 12 DeAngelo Dr., Bedford, Mass. 01730-2204 (617) 275-2400
Vitalink Communications Corp., 6607 Kaiser Dr., Fremont Calif. 94555 (415) 794-1100

Edward R. Teja is a free-lance writer who specializes in electronics and communications subjects. His book PC and PS/2 Graphics will be available this fall from Microtrend Books, San Marcos, Calif.

Dale Neibaur, Novell Inc., Provo, Utah

Understanding XNS: The prototypical internetwork protocol

A leading protocol designer conducts a remarkable guided tour to the little-known fountainhead of modern LAN protocols.

In addition to SNA and DECnet, two architectures have had a major impact on the technical direction of network evolution, and played a seminal role in the development of an international layered architecture. Though they are not as well known as IBM's and DEC's architectures, these protocol suites have had an incalculable impact on the marketplace and both evolved from the same prototype architecture, a university research project called PUP.

One of these two influential architectures was specifically designed to continue delivering packets under battlefield conditions, to survive in the face of extraordinary loss of both user devices and internet switches. The second architecture was optimized for typical office, technical, or university campus environments.

Most would recognize TCP/IP as the battle-ready architecture of choice for our vast military computer networks. It's quieter cousin is Xerox Network Systems (XNS).

The legacy of XNS

XNS was born at the Xerox Palo Alto Research Center (PARC), where scientists began modifying PUP (PARC Universal Packet) to create a more robust product. Once introduced, XNS moved beyond Xerox implementations and was adopted for subsequent customization and modification by the leading LAN designers of the early 1980s.

Such LAN industry leaders as 3Com, Novell, and Ungermann-Bass have all used XNS as the cornerstone of their network product offerings, with subsequent profitable results. The architecture has been an active part of the education and work experience of an entire generation of hardware and software engineers.

XNS's ultimate legacy to the history and progress of network architecture may lie in the international arena. The XNS layered approach is known to have been part of the inspiration for the OSI model's designers in the International Organization for Standardization.

The OSI seven-layer model is in many ways an expanded, more complex version of XNS's basic direction. Since no engineering occurs in a vacuum, many designers who worked on the OSI model's functional requirements had worked extensively with XNS and other existing network architectures and they were naturally influenced by the positive features of XNS's architecture when they were developing the world's future interconnection model.

The XNS approach to networking is well-adapted to most office applications, with clean functional layers, clearly defined protocols, and easy-to-implement features. XNS was adopted in the LAN marketplace primarily because of its simplicity and robust structure.

The key to XNS internetworking is datagram delivery, in which every packet is individually routed on a "best effort" basis. It also includes protocols for quick request and response, more extended conversations, error reporting, route-checking, and the constant updating of routing tables. These latter services work together to turn simple datagram transport into a reliable message delivery system. Reliable delivery of entire messages is, of course, the ultimate user need.

Growing an internetwork

A network architecture does not really have an opportunity to show its power, robustness, and dependability until it must serve a complex internetwork with thousands of devices in hundreds of separate LANs. The evolution from a simple LAN to a demanding internet occurs gradually, in response to changing user needs and the productive ways they find to use their network.

Every workgroup or department reaches a point where it has expanded beyond the floor space that a single Ethernet can handle. The network manager's answer is usually to add a second segment and a *repeater* to link them. Since a repeater simply boosts the electrical signal, the second segment has to be identical to the first.

What happens if the user's needs have changed since the first cable was laid? If users want to add a new unshielded twisted-pair cable plant or if their new applications demand the higher bandwidth of an optical fiber, the network manager will put a *bridge* between the two segments. A bridge can link LANs that differ not only in medium but also in topology (a star linked to a bus, for example) or in transmission method (baseband to broadband).

But as the sophistication and number of network nodes increase, bridges become overloaded and network managers look for different solutions. As the sheer number of devices in a particular building climbs, the need for another type of communication between LANs becomes obvious. Now it is time for *routers* to show their value in the internet.

A router's primary task is to manage traffic loads and alternate routing between LANs. As the number of LANs in the extended network grows, the business need for increased management of the LAN-to-LAN communication increases. In this final stage of the growth of the internet, the network has grown into hundreds or thousands of individual servers, hosts, user devices, and peripherals linked by formerly isolated LANs with bridges, gateways, and routers.

The XNS architecture

Unlike the seven-layer OSI model, XNS developed in four levels, beginning with Level 0 (see Fig. 1). As with the OSI model, each layer provides services to the next higher level.

1. OSI and XNS layers. *Though a well-layered architecture, XNS preceded the OSI network model and does not apply the same labels or numbers to its stack components.*

OSI PROTOCOL MODEL	XNS PROTOCOLS
7 APPLICATION	CLEARINGHOUSE GAP — 4
6 PRESENTATION	COURIER — 3
5 SESSION	
4 TRANSPORT	SPP / PEP / RIP — 2
3 NETWORK	IDP — 1
2 DATA LINK	ETHERNET / RS-232 / RS-449 / X.21 — 0
1 PHYSICAL	

GAP = GATEWAY ACCESS PROTOCOL
IDP = INTERNETWORK DATAGRAM PROTOCOL
PEP = PACKET EXCHANGE PROTOCOL
RIP = ROUTING INFORMATION PROTOCOL
SPP = SEQUENCED PACKET PROTOCOL

Although XNS could provide a complete architecture from transmission medium to applications support, the most important and unique aspects of the architecture occur at the network and transport layers, which is where we will spend most of our time.

XNS Level 0 corresponds generally with the first and second layers of the OSI model, referred to as the physical and data link layers. Level 0 protocols physically transmit data from one point to another over a transmission medium, just like their OSI counterparts in the physical and data link layers.

The name "Ethernet" is an obvious candidate at this layer, but another XNS-specific protocol, called the Synchronous Point-to-Point Protocol, is also available. The architecture will accept other common transmission protocols, including the X.21, RS-232-C, and RS-449 standards.

XNS Level 1, the core of its router functionality, corresponds to the internetworking aspects of OSI Layer 3 (the network layer). XNS Level 1 protocols determine where the packets go, including internet source and destination addressing. XNS defines only the Internetwork Datagram Protocol (IDP) for this layer.

Level 2 protocols correspond to OSI's fourth, or transport, layer, focusing on message integrity and multiple classes of service. XNS Level 2 specifies five protocols and five corresponding packet types to give structure to a stream of related packets where required and provide for simple request-response service under other circumstances. These protocols ride on top of IDP services and handle retransmission, sequencing, and flow control and determine how routers share information for route-building.

The Level 3 protocols provide services similar to OSI's fifth (session) and sixth (presentation) layers. These protocols control remote procedure interactions and determine conventions for data structuring, allowing users to access remote resources; print, save, and access files; and communicate with a variety of differently formatted display devices.

XNS also has a Level 4 (OSI's application layer) for application protocols that are implemented for specific platforms. Many applications have been developed on this level that use the underlying XNS internetworking protocols.

Addressing and sockets

To understand the operation of routers and other nodes that use XNS protocols, one must first understand internetwork addressing. Every XNS device has a complete address that includes: a network number corresponding to an individual LAN; a device number corresponding to a physical network card (called a host address); and a socket number relating to different application processes in the same node. In OSI terminology, a device number is the data link layer address and the network number is the network layer address. A "host" in this context is any computer on the network.

Within each network device there can be a number of software entities called sockets. A socket is simply a subaddress within the device's memory space that an application process uses as the sender or receiver of data. Using a socket means that the application process is not bogged down with the details of creating, maintaining, or ending connections for every function call it makes. The application process concentrates on its task and delegates the connection-management tasks to the specialist, in this case the socket. A socket that is regularly used for a specific network function, such as file service or print service, is called a well-known socket.

Since understanding internetwork addresses is the key to understanding IDP's functions, let's take a closer look at the structure of an XNS address. Each XNS address has three parts: A 32-bit network number uniquely identifies the subnetwork (LAN) within the internet, logically; a 48-bit device or "host" number identifies the specific device, physically; the 16-bit socket number identifies the socket that is managing the interaction. The device number is identical to the address of the Ethernet, token ring, or other network card. See Figure 2 for these addresses in the context of an XNS packet.

XNS's designers considered it essential to have absolute device numbers in order to identify an individual device independent of the particular network to which it happened to be attached. The combination of absolute device numbers and logical network numbers allows easy moves and changes as well as concurrent attachment by a device to more than one network.

2. An encapsulated packet. Shown are the IDP fields and their encapsulation by the DLC header and trailer. The DLC layer uses 48-bit devices addresses. Logical network addresses are in the IDP fields of the network layer.

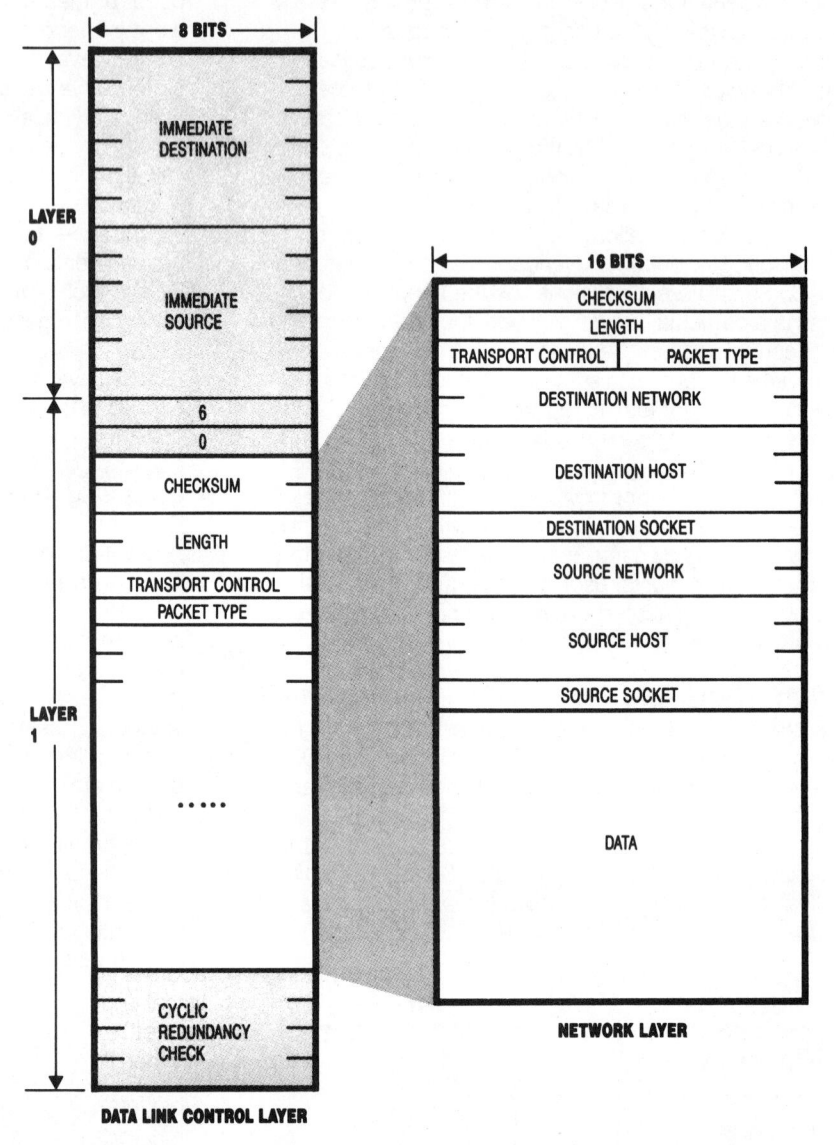

Router revelations

IDP, at the OSI network level, is the workhorse of XNS and all XNS-derived protocols. It resides at the crossroads of the communications stack, where packages of data en route from the application layer are "stamped" with intelligent routing directions and handed to the data link layer for output to the network wire. Understand how IDP and the network layer work and the realm of internetworking will be open to you.

First, an analogy: A router builds routes based on the "next place to send this packet" criterion. Think about how a driver and navigator communicate while en route to an unfamiliar location. The navigator does not give the driver all the landmarks at once . . . the freeway exit, the third light, the left on Main, the right on Second, the railroad tracks, the abandoned farm, and finally, the red house with the white trim. Instead, the driver is told what to do next, and when that landmark is reached, the next one is given.

XNS network routers can only work with each landmark on the particular packet's journey as the previous one is passed.

The process starts when a source node decides to send a packet to a destination node. Before any data is exchanged, the source node must learn the internetwork address of the destination node. The application in the source node most likely knows a high-level symbolic name for the destination node, such as Server5 or VAX23. These names are generally stored in an application table or supplied by the operator. To learn the internetwork address, the source node uses name-service protocols to look up the address in an address directory. Alternatively, the node uses a broadcast facility that can dynamically return the full address of a named network entity, as occurs with Netbios. In either case, this type of name resolution or

XNS

3. Hop-along packeting. *Traffic on a single LAN needs only a 48-bit data link layer device address. When a packet hops across multiple LANs, its DLC address is changed by each router it passes through. By looking up the destination network number in its routing table, the router learns the device address of the next router in the path.*

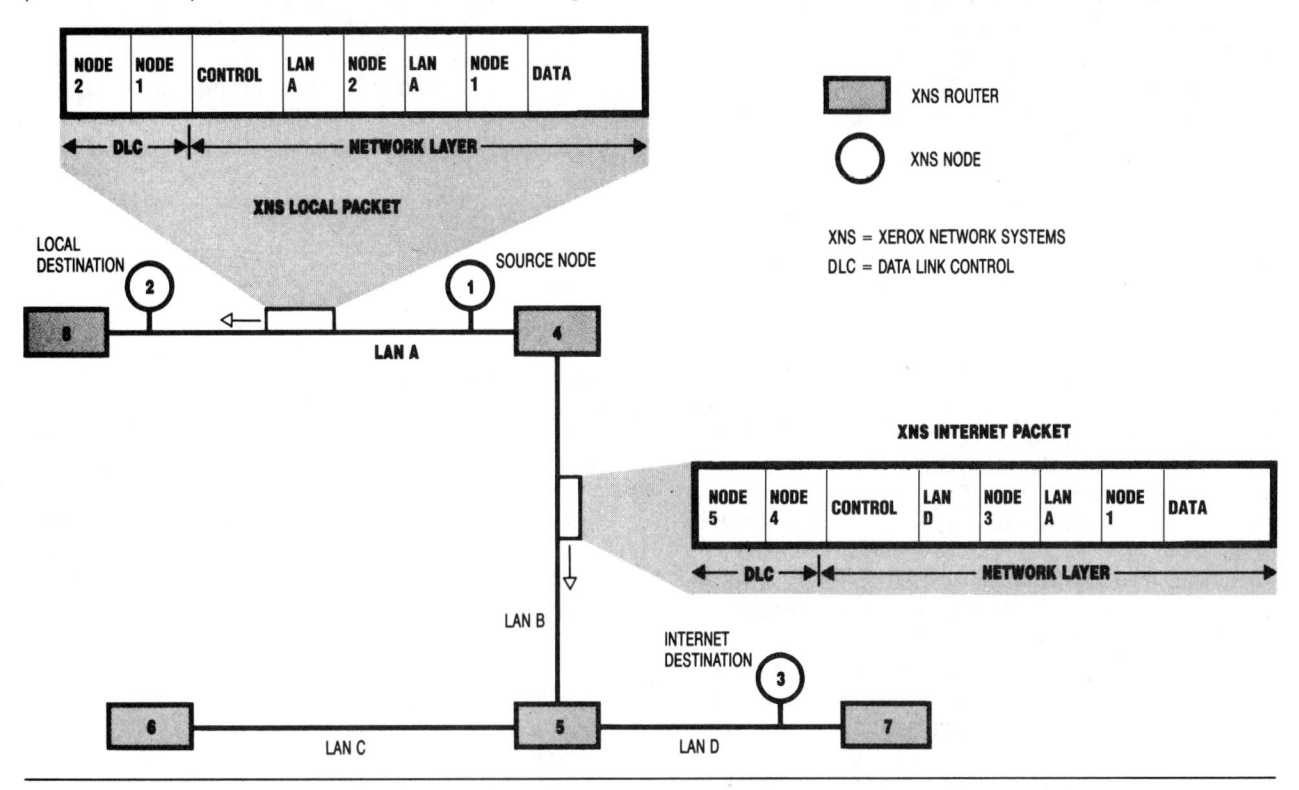

"binding" is above the network layer of IDP, but it must occur before a packet can be sent.

Once the source node has determined the full internetwork address of the destination node, the packet can be sent. IDP, at the network layer, does not understand names, only the network address numbers that it receives, along with the data to be sent, from the higher layers and the application.

Now that the network layer has the user data and an internetwork address, it can make the first routing decision. The network layer software in the source node knows which network (LAN) it is on. If the destination node is on this same network, the source's network layer merely adds its packet header to the user data and hands it to the data link layer with a request to send the packet to the destination node's device number. This is illustrated in the "local packet" example of Figure 3. A simple point-to-point exchange requires no intermediate routing. Much of the traffic on a LAN is this type of local exchange.

Keep in mind that when an internetwork is subdivided by routers, data link layer packets only travel on a single LAN segment that's defined by routers, as with LAN A in Figure 3. Unlike bridges, which pass everything, a router passes only those packets addressed directly to it and that it decides to route.

Now for the internetworking mechanics. When a source node wishes to send to a destination node on another LAN segment, the network layer builds a packet with the destination's full internetwork address inside. But unlike the previous local example, the network layer software tells the data link layer to address the packet to a router's device address, not to the ultimate destination-node device address. The data link layer address is called the immediate address in XNS and is not necessarily the address of the target end node. This is shown in the "internetwork packet" example of Figure 3.

When the addressed router receives the incoming packet, it throws away the immediate address in the DLC header, looks into the packet's destination address field, finds the network number of the target node, and then consults its route tables to decide how best to route the packet.

From its route tables, the router learns the device address of the next router along the best path to the destination node. The device address of this next router is now used for an immediate address, and the packet again is handed to the data link layer for transmission. In this way, a packet may be passed from router to router. Each router in turn strips off the data link layer address and replaces it with the address of the next router along the path to the target node.

Finally, a router on the same LAN segment as the destination node receives the packet and uses the internal destination device address as the immediate DLC address and sends it across the link to the end node.

At this point, the IDP internetwork transmission is complete. If some higher-level software is guaranteeing the

XNS

transmission, an acknowledgment packet could be sent back to the source node now. But without the higher-layer guarantee, IDP just makes a "best effort" to get the packet there, with no retries and no acknowledgments.

IDP assumes that the packet will have a nominal maximum length of 576 bytes. One of the major ways that vendors have deviated from a pure-XNS implementation of IDP has involved a different maximum packet size. Larger-than-standard packets will require that all routers in the network be aware of the correct packet size because routers will almost always discard aberrant-sized packets.

Rest in peace with RIP

At the transport layer, an XNS network designer has five protocols available to achieve the two basic functions of system management and data transfer. XNS transport layer services use the network layer IDP packet for a basic delivery mechanism.

The XNS system management and control protocols include: Routing Information Protocol (RIP) packets for router-to-router communication; Error Protocol packets for routers to report packet failures; and Echo Protocol, which tests a path to an unknown device.

The XNS Data Transfer services include: Sequenced Packet Protocol (SPP) and Packet Exchange Protocol (PEP).

For XNS-derived LANs such as Novell's, the two most important of these transport layer protocols are RIP, which allows routers to exchange route data, and SPP, which provides guaranteed services on top of IDP. Novell's Sequenced Packet Exchange (SPX) is a close adaptation of XNS SPP, just as Novell's Internet Packet Exchange (IPX) is largely derived from XNS IDP.

The key to a router's ability to calculate intermediate addresses on an end-to-end route is in its internal routing tables. In XNS, and Novell's IPX and SPX, the protocol that routers use to update and manage route tables is RIP.

Internet routers use RIP and its special packet format to keep each other informed of the network topology. The basic question routers need to answer is: How do I get to Network X, and how close is it to me? RIP allows internet routers to build and maintain routing tables consisting of network numbers, router addresses, and the number of hops to that router.

4. XNS routing table. *A routing table doesn't show addresses for all the routers in a path. Instead, it tells only the address of the next router in the path and the hop count.*

SUBNET # 4 BYTES	ROUTER PORT 1 BYTE	NODE # 6 BYTES	HOP COUNT 1 BYTE
68A2	A	169A31	2
C43B	A	C6234B	6
1286	B	6789A2	1
D94C	C	7A7B61	4
672A	A	250311	9
⋮	⋮	⋮	⋮

XNS = XEROX NETWORK SYSTEMS

Routing tables (see Fig. 4) contain an entry for each LAN on the internetwork. For each network's entry, the route table gives the device address of an adjacent router that is the next step toward the target network and the port to which the next router's LAN is connected. Also stored in the routing tables are the total number of routers the packet must pass through to arrive at a given destinaton network. This is referred to as the hop count.

Routers are constantly sharing information about themselves and information they have picked up from listening to other routers. In other words, routers gossip. The self-referential information a router offers is usually true; the data on other routers and LANs is as accrate as gossip usually is. Since routers are working with what is essentially hearsay evidence—except when they are talking about themselves—the databases that they are constantly advertising and sharing are only as accurate as the last broadcast they received.

New information propagates through even a very complex internetwork in a relatively short time because routers listen attentively to each other's gossip. A typical RIP implementation will have routers communicating with each other every 30 or 60 seconds. If a router hears what purports to be new information, it will update its table entry. Each entry in the route table has a timer associated with it. If the timer runs out before a given network's existence is reconfirmed, that entry is deleted from the table. In the Novell version of RIP, this timer is four minutes.

The original XNS version of RIP is adequate when everything is working, but the kind of information that does not propagate quickly is the information that a certain network is no longer reachable. One router may realize that it can no longer get packets to a network, but every other router in the internet may have to try and fail before they all conclude that they can't get packets to the network either. In this sense, routers act like incurable optimists.

Standard XNS implementations insist that routers broadcast their RIP information on all LANs they are connected to, including back to the same network from which they originally received the information. This process can use a substantial amount of network bandwidth and is an area targeted for improvement by router vendors. Some vendors' implementations of RIP have added capabilities to counteract these drawbacks, including the ability of routers to broadcast messages immediately when detecting a change.

Practically every implementation includes some special features, customization, and enhancements. With Novell networks, in an effort to avoid spreading outdated and possibly erroneous information, routers do not advertise down the same LAN segment from which they originally received a certain database item. In standard XNS, of course, a router broadcasts its information everywhere. In addition, routers broadcast their RIP packets every 60 seconds rather than every 30 seconds to avoid using up valuable network bandwidth with excessive router traffic.

Also, the Novell implementation added two extra bytes to the RIP packet for a time-to-net parameter, specifying

5. The SPP packet. *Key to guaranteed service, the fields in the SPP transport layer store sliding window dialogues, packet sequence numbers, and acknowledgment information.*

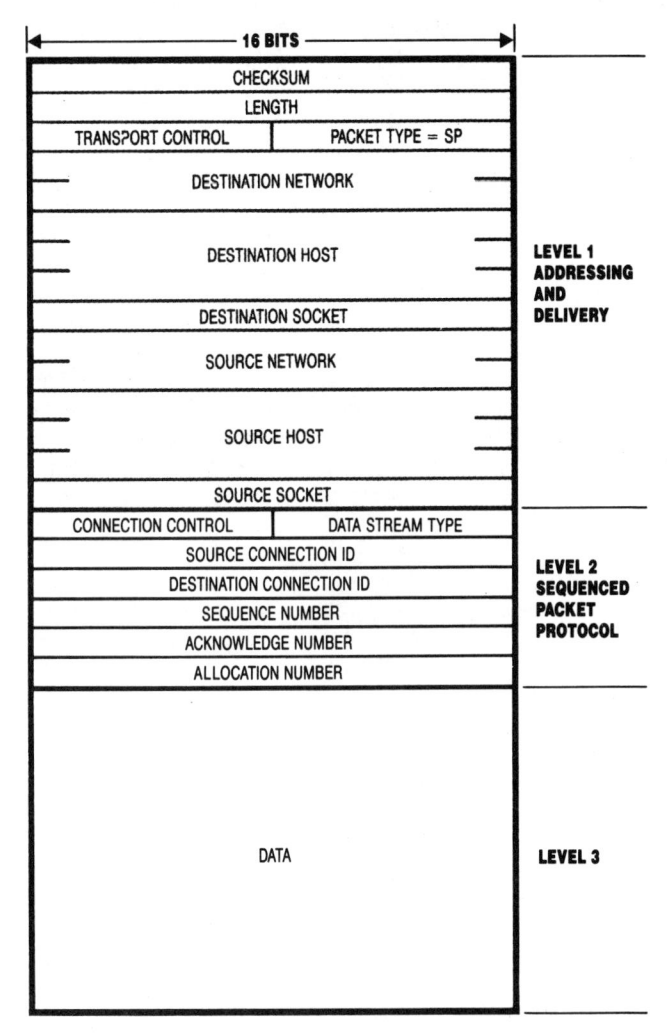

```
|<----------- 16 BITS ----------->|
+---------------------------------+
|            CHECKSUM             |
+---------------------------------+
|             LENGTH              |
+----------------+----------------+
| TRANSPORT CTRL | PACKET TYPE=SP |
+----------------+----------------+
|       DESTINATION NETWORK       |
+---------------------------------+         LEVEL 1
|        DESTINATION HOST         |        ADDRESSING
+---------------------------------+            AND
|       DESTINATION SOCKET        |         DELIVERY
+---------------------------------+
|         SOURCE NETWORK          |
+---------------------------------+
|          SOURCE HOST            |
+---------------------------------+
|         SOURCE SOCKET           |
+----------------+----------------+
| CONN. CONTROL  | DATA STREAM TYP|
+----------------+----------------+
|      SOURCE CONNECTION ID       |        LEVEL 2
+---------------------------------+        SEQUENCED
|    DESTINATION CONNECTION ID    |         PACKET
+---------------------------------+        PROTOCOL
|        SEQUENCE NUMBER          |
+---------------------------------+
|       ACKNOWLEDGE NUMBER        |
+---------------------------------+
|       ALLOCATION NUMBER         |
+---------------------------------+
|                                 |
|              DATA               |         LEVEL 3
|                                 |
+---------------------------------+
```

the expected delay to the end station. This allows a sender to determine an accurate delay to the end station, which avoids both timing-out too early and waiting too long for a packet that is genuinely lost.

Sequenced packet protocol

SPP provides a reliability layer above the simple datagram delivery of IDP. This is similar to OSI's transport level virtual-circuit services. The SPP approach to message integrity involves creating a connection between sender and receiver through two sockets that expect a synchronized stream of packets.

Using sliding window techniques, both sockets start with sequence numbers of zero in the first SPP packet that they send to set up the connection. They continue numbering every data packet that they send out during the course of the connection (see Fig. 5). The receiving socket can put the message back together by linking the user data fields of successively numbered packets.

SPP is a robust, full-duplex service that allows a source to send a number of packets to be acknowledged by a single packet. SPP allows multiple application processes in the same node to simultaneously conduct virtual-circuit sessions with different remote resources.

SSP is key to message integrity in XNS, which becomes an issue in networks because packets can get out of order or be lost entirely. Packets can fail to arrive at their destination for a wide variety of reasons, some of which are application- and network-specific. For instance, when a bridge chokes, it discards frames (packets). Routers using load-balancing algorithms rarely choke, but in a heavily loaded network a momentary buffer overflow with resultant packet loss is possible. Also, if a router finds a checksum error in a packet, it will discard the packet. Finally, the receiver's buffer can overflow, with resultant packet losses.

In addition to these instances of lost packets, some packets are simply delayed and, therefore, out of order because they took a different route from their fellows and encountered longer transmission delays. If they are sufficiently delayed, the sender may time-out on the acknowledgments and generate duplicate packets. In a complex internet with multiple routes available, a network designer can expect that a significant number of packets will arrive out of order and possibly duplicated.

The maximum size of an SPP packet is 576 bytes, the nominal internet packet size, although some vendor implementations vary from this size all the way up to the maximum Ethernet packet size. A pure IDP/SPP implementation is almost impossible to find because of the large number of diffrent vendors who have adopted XNS in their product line.

As user applications evolve in the 1990s, lost and severely delayed packets will become a more significant performance issue in network design. Graphics- and video-intensive workstations are already causing problems on many bridged LANs because of their tendency to hog network bandwidth and clog bridges, locking out other LAN traffic. As digital voice and other time-critical digital applications emerge, these problems will only intensify.

When multitasking, windowing high-performance workstations join their graphics cousins in large numbers in typical LAN installations, the issues of sequencing and message integrity will become a more important part of network protocol design.

Even the best of today's routing and internetwork technologies will not meet the needs of coming LAN applications such as digital voice. Hence, protocol designers must develop new architectures that far exceed the transport capabilities of SPP and similar methods. These future protocols will have to provide classes of service that are suitable for applications that cannot afford the delays.

Other family members

Although IDP, SPP, and RIP are by far the most important XNS protocols, other members of the XNS family have been useful and influential in the evolution of networks.

■ *Packet Exchange Protocol.* PEP was developed to serve transaction-oriented needs in both the general LAN case and the completely connectionless (in OSI terminology) network types. Automated teller machine and credit card verification systems, for example, qualify as entirely transaction-driven, but even the more ordinary LAN environments involve some operations that are exclusively devoted to what XNS calls "request-response."

Since it is designed for simple request-response, the PEP packet has a very simple structure, with only three fields. The ID field lists what action the requester wants, while the Client Type provides a link to a specific transport layer client. The Data field (optional, of course) can have an answer to the request.

For example, a device requests the time of day from a Time of Day server or requests access to a file from an Authentication server. In either case, the server receives a PEP with a request in the ID field and sends back the answer with the same ID and the response in the Data field.

> It may be time to rethink the overall design goals of our networks.

- *Courier.* Subtitled the Remote Procedure Call (RPC) protocol, XNS Courier is similar to session-level services in other architectures. It uses layers below it such as SPP and IDP to provide applications with remote function-execution services. In Courier terms, there is always an active element that is asking for something and a receiving element that is offering a service.

For instance, a client application could use a Courier RPC to make a request to a remote printer, as though the printer was local. To the application, the printer is just a function call away. As with any RPC protocol, Courier takes the local function call, and transfers it to a remote resource (print server, file server, or the like) for execution.

- *Bulk Data Transfer.* Within Courier's world is the Bulk Data Transfer (BDT) protocol, which allows for the transfer of larger blocks of data than Courier can generally handle. The BDT protocol was designed to accommodate file transfers and mainframe-to-mainframe database transfers with bulk data defined as an arbitrarily long sequence of bytes treated as a single Courier data object.
- *Clearinghouse.* The Clearinghouse service is an application-level directory service linking the names and aliases of users to the corresponding network and data link addresses.

Xerox specifies that Clearinghouse databases can, and should be, replicated throughout the network, with a Clearinghouse server on each segment if possible. Since the speed with which a user process can access the Clearinghouse database determines how quickly a user can make a connection to the desired resource, Clearinghouse is generally made resident on a dedicated server, along with another service, called the Authentication service. An authentication process allows specific users to access specific resources while barring others.

- *Error Protocol.* Any device that notices an error in a packet—generally a checksum error, for installation that are still using checksums—and discards it, should send an Error Protocol packet to the sender, specifying the kind of error and the packet that was destroyed. As the underlying hardware becomes more reliable and fewer installations depend on checksums for error control, the Error Protocol is fading from importance in the XNS pantheon. When routers send an Error Protocol packet, they use their well-known router error socket, which does nothing but send and receive error messages.
- *Other applications.* Xerox uses a Gateway Access Protocol to connect XNS to non-XNS systems. In this way, GAP is like OSI's Virtual Terminal protocol, bringing all types of display devices into communication with each other, performing protocol translation service consistent with a Layer 7 protocol gateway. Other services on the application layer include the Filing service, for file access and management; Print service, for printing to remote printers through the network; Document Interchange service, for translation from one document format to another; and the Mail service, for electronic messaging.

The future of XNS

In the non-SNA marketplace, XNS remains very popular, trailing only DECnet and TCP/IP in popularity. We can expect XNS to remain a popular LAN network architecture through the mid-to-late 1990s when TCP and XNS protocols fade from importance and OSI protocol stacks take their place.

XNS's future will be affected by new technical developments and new market directions sweeping the LAN industry. On the semiconductor side, there has been an almost tenfold increase in the accuracy and reliability of the underlying hardware since XNS was first introduced, rendering some of the original design criteria obsolete. The era of XNS's development was a far cry from the brave new world of very-high-bandwidth media, such as today's 100 and 200 Mbit/s Fiber Distributed Data Interface and tomorrow's 10 Gbit/s successors, as well as desktop devices that have exploded with their own hundredfold increase in processing power. New applications, especially desktop graphics and video, will easily use up the rapidly expanding available bandwidth.

These developments will move the performance bottleneck from the medium—where choke points exist now—to the internet devices. The XNS of the 1990s will need new switching technology and bridges, routers, and gateways operating at two orders of magnitude beyond their present capability. In a world where we must have high-performance delivery systems on this scale, it may be time to rethink the overall design goals of our networks. Contrary to the philosophy that protocols can be developed and then forgotten, each wave of new network applications will drive network designers to recast their protocols in increasingly powerful forms. ■

Dale Neibaur, senior system architect, has been with Novell since 1981. As a programmer and member of the Novell "Superset" programming team, he coauthored such fundamental NetWare components as the bindery, IPX/SPX, and much of the NetWare file system.

Gilbert Held, 4-Degree Consulting, Macon, Ga.

Is ISDN an obsolete data network?

Amid all the hoopla, what's getting lost? Here's a comparison of eight applications: How are they served by ISDN versus other data transmission methods?

Although the demand for information about ISDN continues to escalate—and vendors, service providers, and users march to the ISDN cadence—the technology's data transport capabilities are not necessarily better than those of other existing and emerging transmission methods. While it may seem like heresy, developments in other areas of data communications may be rendering ISDN obsolete.

How, for example, does ISDN transmission capacity compare with the transmission rates offered by recent advances in such fields as local and wide area networking, PC video graphics, and voice digitization techniques?

ISDN notwithstanding, a wide variety of technologies exists to satisfy end-user communications requirements. These options yield a range of transmission rates: from fewer than 100 bit/s to millions of bits per second. There are 17 popular categories of such transport mechanisms (see Table 1). Several are specific to particular industry segments. The continuous slope variable delta modulation digitization technique, for example, was originally developed for military applications. This technique encodes each analog voice sample into a single bit based on a comparison of the height of the sample with the height of the previous sample—thus increasing the sampling rate with respect to the data rate. Although voice quality at 8 and 16 kbit/s is marginal, with an increased sampling rate the encoding of 1 bit per sample produces a reasonably acceptable voice quality at 32 kbit/s.

Another digitized voice technology, linear predictive coding (LPC), synthesizes a voice conversation by encoding pitch, energy level, and other voice parameters. Although LPC digitizes a voice conversation into an extremely low data rate, it is very expensive to implement and not suitable for general use.

Two video transmission methods are readily available: Full-motion video, which can be transmitted at about 700 kbit/s, and freeze-frame, which can be transmitted at 64 kbit/s using commercially available equipment that compresses digitized video signals. Although both types of video have been successfully compressed at lower data rates, the current cost of equipment precludes widespread use. A small-band ISDN videophone terminal, according to the Netherlands Foreign Investment Agency, a unit of the Dutch Ministry of Economic Affairs (North American headquarters in New York City), is expected to be available for home and office use in Holland in 1992. The agency says this videophone is expected to provide high-quality moving pictures accompanied by high-quality voice at 64 kbit/s by using a "hybrid method of data compression that combines DPCM [differential pulse code modulation] and transform coding."

Communications applications

Although there are many existing and evolving non-ISDN communications applications, a small subset can be used as a representative base to compare against the data transportation capabilities provided by ISDN (for details of the ISDN architecture, see "ISDN overview"). Eight current and potential communications-related applications that can be expected to occur on an ISDN B channel are listed in Table 2.

Currently, electronic-mail applications are most often satisfied by dial-up access via the switched telephone network at data rates ranging from 110 bit/s to 9.6 kbit/s. Thus, an ISDN B channel should be more than sufficient to serve text-based messaging with this application. If the transmission of graphics images is required as part of an electronic-mail application or as a standalone application, then the type of graphics image, as well as what one considers a reasonable transmission time, will govern whether or not a B channel provides an acceptable level of service.

As for whether ISDN will benefit file transfer operations, at the 64-kbit/s data rate of a B channel, this application is

ISDN

Table 1: Popular transmission options

TRANSMISSION TECHNOLOGY	DATA TRANSPORTATION RATE
ANALOG MODEMS	
SWITCHED NETWORK USE	110 BIT/S–9.6 KBIT/S
LEASED LINE USE	1.2–19.2 KBIT/S
DATAPHONE DIGITAL SERVICE	
SWITCHED NETWORK	56 KBIT/S
LEASED LINE	2.4–56 KBIT/S
DIGITIZED VOICE	
PULSE-CODE MODULATION	64 KBIT/S
ADAPTIVE PULSE-CODE MODULATION	32 KBIT/S
CONTINUOUS SLOPE VARIABLE DELTA MODULATION	8–64 KBIT/S
LINEAR PREDICTIVE CODING	2.4–4.8 KBIT/S
LOCAL AREA NETWORKING	
APPLETALK	.25 MBIT/S
ARCNET	2.5 MBIT/S
ETHERNET	10.0 MBIT/S
TOKEN RING	4/16 MBIT/S
IBM 3270 TWISTED WIRE	2.38 MBIT/S
T1	
NORTH AMERICA	1.544 MBIT/S
EUROPE	2.048 MBIT/S
VIDEO	
FULL MOTION	700 KBIT/S
FREEZE FRAME	64 KBIT/S

served 3.33 times faster than it is with 19.2-kbit/s modems. The number of 8-bit characters that could be transmitted at the ISDN B-channel rate, at 19.2 kbit/s on a high-speed conditioned analog leased line, and at 2.4 kbit/s using a V.22*bis* modem on the switched telephone networkare compared in Table 3.

Applying this data to typical interactive file transfer operations indicates that an ISDN B channel should provide satisfactory performance. For example, consider the standard 360-Kbyte 5.25-inch and 1.44-Mbyte 3.5-inch diskettes commonly used in PCs. Using an ISDN B channel, it would take about six seconds to transfer the contents of a 360-Kbyte diskette and 23 seconds to transfer the contents of a 1.44-Mbyte diskette. Since interactive users rarely transfer the contents of an entire diskette, it appears that a 64-kbit/s data transfer rate should provide an acceptable level of service for interactive file transfer operations.

For distributed computer systems requiring the transfer of large databases, the ISDN 64-kbit/s B channel, however, may not prove satisfactory. As an example of this situation, consider the transfer of a 100-Mbyte file. At 64 kbit/s, the file transfer would require almost four hours!

Text versus graphics

How well ISDN will serve applications involving graphics image transmission is less easily determined than its suitability for electronic-mail and file transfer operations. The time required for the transmission of screen images on PCs varies, as can be seen in an examination of three popular video display modes.

When an IBM PC or compatible is in its text video mode, each ASCII character is displayed in a box defined by a number of vertical and horizontal pixels. The pixels that represent the character are generated by the video circuitry, which interprets the ASCII character. In the computer, each character of the text video image requires 2 bytes of storage. The first byte is the ASCII code that defines the character; the second byte is known as the character attribute and defines such parameters as whether the character is blinking, underlined, highlighted, and so on. Thus, the transmission of a full text screen image would require sending 32,000 bits of data (80-character column by 25 lines by 2 bytes per character—for storage—by 8 bits per character). When transmitted using a V.22*bis* modem operating at 2.4 kbit/s, the screen transfer time is 13.3 seconds. Using an ISDN B channel, the time required to transfer a full text screen image is reduced to 0.5 seconds.

The introduction of IBM's Presentation Manager and the growth in other window-oriented machines—including the Apple Macintosh series, Atari STs, and Sun workstations—will result in an increasing need to transmit and receive computer screens in their graphics mode. In fact, remote computer control has grown in use over the past few years from a single product offering to software marketed by about a dozen vendors, one of which (Norton-Lambert) claimed sales of more than 100,000 units during 1988.

The time required to transmit a graphics screen image depends first on the video graphics mode that is being used. Second, transmission time depends on whether the communications software is capable of performing data compression and/or transmitting only screen changes with respect to a prior transmitted image. In examining the transmission of a graphics screen image, first consider the time required to transmit the screen without compression.

If a PC has the Enhanced Graphics Adapter (EGA) display capability, it can display a window with a resolution of 640-by-350 pixels and one of four possible colors. Such a display requires 224 kbits to transmit a black-and-white image, each bit carrying the pixel image and its color composition: black or white. If an image using any color other than black or white is transmitted, an additional 2 bits per pixel would be used to represent one of the four possible colors, increasing the transmission requirements for a full-screen image to 448 kbits. If a V.22*bis* modem operating at 2.4 kbit/s is used to transmit an EGA screen image, a total of about 187 seconds, or more than three minutes, would be required. Even with the use of an ISDN B channel, seven seconds would be required to transmit an EGA screen.

For a PC with the Video Graphics Adapter (VGA), the higher resolution of the VGA display results in a corresponding increase in the time required to transmit a screen of data. In the VGA graphics mode a resolution of 640-by-480 pixels results in a minimum of 307,200 bits to represent a black-and-white image. Since VGA permits up to 16 possible colors, using 4 bits per pixel to represent color yields a total of 1,228,800 bits. A V.22*bis* modem at 2.4 kbit/s would take about 8.5 minutes to transmit a screen image. Using an ISDN B channel would reduce transmission time to 19.2 seconds.

The preceding computations of transmission time are for a graphics screen without compression, but what effect can

ISDN

be expected through the use of a sophisticated remote-control software package? In the advertising literature for Co/Session 3.0, Triton Technologies Inc. (Islin, N.J.) claims to offer the fastest text and graphics screen updates of any remote-control package. The vendor specifies that the time needed to display a large directory in a window at 2.4 kbit/s is 43 seconds. Since this is approximately a quarter of the time computed for the transmission of a screen under the EGA mode, an interpolation would suggest that transmitting the screen image over an ISDN B channel would require less than two seconds. A similar interpolation for the use of a VGA video mode would reduce the computed transmission time of 19.2 seconds to less than five seconds for transmitting a large directory in a windows environment.

Since late 1988, numerous vendors have introduced register-level superextended VGAs. Downward compatible with conventional VGAs, these cards support the emerging 1,024-by-768 pixel superextended VGA resolution. That resolution requires 786,432 bits to represent a black-and-white screen image. In its color mode, where 4 bits per pixel can be used to represent 16 colors, the transfer of a full screen would require transmitting a total of 3,145,728 bits. At the ISDN B-channel data rate of 64 kbit/s, it would take about 49 seconds to transmit an extended VGA screen. Even if one assumes that remote-control software provides an approximate 4:1 reduction in transmission time (which is what the author experienced using the previously mentioned Co/Session package), the transmission of an extended VGA screen can be expected to take more than 10 seconds on an ISDN B channel. While 10 seconds does not seem particularly long by itself, for a repeating process—such as dialing into a network management system and scrolling through screens—this delay could be irritating at the least and would be unacceptable to organizations that aim to obtain a three- to five-second response time.

Voice versus graphics

There is one communications application that may well be over-served by ISDN: human voice transmission. Since toll-quality voice can be transmitted at 32 kbit/s, the 64-kbit/s transmission rate of an ISDN B channel may actually be excessive. In contrast, graphics applications may pose some difficulty. Although the author is not well versed in the economics of adaptive PCM voice digitization versus conventional PCM, it would appear that having variable-rate channels instead of fixed-rate channels provides a higher level of capability that could alleviate some of the problems associated with transmitting graphics images. As an example, consider the figure, which illustrates two possible modifications that can be made to the data transporting capability of the B channel under a variable-rate channel creation scheme.

The first variable-rate channelization illustrated in the figure could assign 96 kbit/s to a B channel and 32 kbit/s to a B1 channel. This would permit higher data transfers on the B channel and simultaneous transmission of voice, digitized at 32 kbit/s, on the B1 channel. The second variable-rate channelization scheme could merge the transportation capability of two B channels, providing subscribers with the ability to transmit data-intensive applications, such as ex-

---

ISDN overview

Under the evolving ISDN architecture, access to the network is possible via one of two major connection methods: basic access and primary accesss. Basic access defines a multiple channel connection derived by multiplexing data on twisted-pair wiring. This connection method consists of two bearer (B) channels, each operating at 64 kbit/s, and a data (D) channel operating at 16 kbit/s. The ISDN basic access channel format is illustrated in the figure. Because of the composition of the basic access format, it is commonly referred to as a 2B + D service.

The bandwidth of each B channel is structured to carry either one pulse code modulation voice conversation or one 64-kbit/s data transmission. Together, the two B channels permit users to transmit data and, at the same time, conduct a voice conversation using a single telephone line; to converse with one person and to receive a second call; or to place one person on hold while answering a second call.

The D channel provides a transport mechanism for signaling information. Signals on the D channel are designed to control the B channels or to supplement their use by providing subscribers with information not readily available with conventional telephone service. Examples of the former include the carrying of off-hook and dial numbers, while examples of the latter include presenting the calling party or carrying information for home alarm systems or utility meters.

B1 CHANNEL	D	B2 CHANNEL

B CHANNELS OPERATE AT 64 KBIT/S EACH
D CHANNEL OPERATES AT 16 KBIT/S
2B + D SERVICE OPERATES AT 144 KBIT/S

Primary access can be considered a multiplexing arrangement whereby a group of users having basic access share a common line facility. This type of access permits a private automated branch exchange to be directly connected to the ISDN network. In North America, primary access consists of a grouping of 23 B channels and one D channel, with each channel operating at 64 kbit/s. This structure enables the T1 (1.544-Mbit/s) carrier to serve as the transport mechanism for ISDN primary access.

Included in the specifications for ISDN is a provision for packet-switched channels that are designed to carry streams of user information at varying data rates. An H channel is designed to operate at 384 kbit/s, while H11 and H12 channels operate at 1.536 and 1.92 Mbit/s, respectively. Although H channel primary-rate interface provisions exist within ISDN, vendors have not yet designed equipment to operate on H channels. —G.H.

tended VGA screens, in half the time possible with a single B channel.

The Q.931 recommended standard ("ISDN User-Network Interface Layer 3 Specification for Basic Call") includes a provision for the concatenation of two B channels to obtain a 128-kbit/s data transfer capability. In addition, under CCITT I.460 ("Multiplexing, Rate Adaption, and Support of Existing Interfaces") there is a provision that permits a B channel to be split into 8-kbit/s intervals. Unfortunately, neither standard has, to the author's knowledge, been adopted by switch vendors; nor have manufacturers begun work to develop equipment that operates with concatenated or interval B channels. Clearly, variable-rate B channel capability would provide consumers options that are not available under the present ISDN structure.

Videophone and videotex

As previously mentioned, videophones are expected to be commercially available by 1992. The key to the operation of a videophone is the videocodec, which effectively compresses full-motion video images to approximately 1/1,500 of their original information content. Since the videocodecs that are currently marketed sell for around $15,000, it is reasonable to question vendor expectations that mass production will result in economical videophones by 1992. One only has to look at the relatively stable retail price of CD-ROM players between 1986 and 1988 to question expected price declines in high-technology equipment.

One wonders, though, what effect a 128-kbit/s B channel—obtained by merging two B channels—would have on videophone developments. Since the complexity of a compression scheme is proportional to the required compression ratio, the cost of videocodecs that operate at 128 kbit/s could be less than that of videocodecs operating at 64 kbit/s. Although the degree of cost reduction is a matter of

Table 2: Applications versus capacity for ISDN B channel utilization

APPLICATION	CAPACITY
ELECTRONIC MAIL TRANSMISSION	MORE THAN SUFFICIENT
FILE TRANSFER OPERATIONS	VERY GOOD FOR INTERACTIVE WORK; INAPPROPRIATE FOR LARGE DATABASE TRANSFER
GRAPHIC IMAGE TRANSMISSION	REASONABLE NOW BUT LIKELY TO BE UNACCEPTABLE FOR SUPEREXTENDED VGA AND OTHER EVOLVING HIGHER-RESOLUTION SCREEN IMAGES
VOICE CONVERSATION	EXCESSIVE
REMOTE-CONTROL PC SOFTWARE	NOT ACCEPTABLE FOR GRAPHICS IMAGES
VIDEOPHONE TRANSMISSION	TOO SOON TO TELL
VIDEOTEX	ACCEPTABLE FOR TEXT; UNACCEPTABLE WITH HIGHER-RESOLUTION SCREEN-BASED GRAPHICS
INTERCONNECTING LANS AND WANS	BOTTLENECK IN SOME APPLICATIONS

Table 3: Transfer of 8-bit characters

TIME	64 KBIT/S (ISDN B CHANNEL)	19.2 KBIT/S*	2.4 KBIT/S**
1 SECOND	8,000	2,400	300
30 SECONDS	240,000	72,000	9,000
1 MINUTE	480,000	144,000	18,000
1 HOUR	28,800,000	8,640,000	1,080,000

*VIA CONDITIONED ANALOG LEASED LINE.
**VIA V.22bis MODEM ON SWITCHED PHONE NETWORK.

conjecture, its existence provides another rationale for the implementation of a variable-rate set of B channels.

Current videotex implementations create screen images by transmitting character codes that form cells, or blocks, of characters on the screen. Although videotex pictures are rather crude compared with EGA and VGA images, videotex data transmission requirements are relatively modest. Thus, 2.4-kbit/s transmission used on analog facilities will be substantially improved through the use of an ISDN B channel.

If videotex should eventually use pixel graphics, the same problems mentioned for remote-control PC software can be expected. That is, the time it takes to update screen images on an ISDN B channel would increase from a fraction of a second to more than 10 seconds for videotex using pixel graphics.

Interconnecting LANs and WANs

A network curiosity in the 1970s, local area networks have become an important communications mechanism for most organizations. Since the data rates of most LANs are in the megabits-per-second range, the delay encountered when connecting geographically dispersed LANs or connecting LANs to WANs via an ISDN B channel could severely tax a user's ability to do real-time work. This is because interconnection through Direct Distance Dialing or the ISDN B channel would be at 56 kbit/s or 64 kbit/s, compared with a megabits-per-second transfer on a LAN, resulting in delays when moving data between networks.

If data is in the form of high-resolution screen images, using the ISDN B channel will result in a slight bottleneck compared with the data transfer rate that can be achieved on a LAN. For other LAN operations, such as downloading a file contained on the server of one LAN to a workstation on a second LAN, the B channel will probably provide a sufficient data transfer capability. In fact, for many LAN-to-LAN data transfer operations, the key to whether a B channel is sufficient or acts as a bottleneck will be the number of users simultaneously requesting cross-LAN data transfers and the type of data transfer involved. Although broadband ISDN (BISDN) can be expected to provide a much higher data transporta-

ISDN

tion capability than is currently available in ISDN field trials, its implementation is probably five or more years distant. Because of this, ISDN as it now exists will probably be a bottleneck to some LAN and WAN interconnections.

ISDN problem areas

Besides the question of whether ISDN serves current and evolving applications as well as or better than other transmission technologies, there are problem areas that could delay the implementation of ISDN. One of the key features associated with digital ISDN telephone sets is the capability of displaying call progress and calling information. The telephone sets' ISDN capabilities can only be used when both the calling party and the called party are served by ISDN-compatible central office switches that can communicate with one another. This stumbling block may affect how long it takes to get ISDN telephone sets into the marketplace.

The communications between ISDN switches—known as Signaling System Number 7 (SS7)—is also both a key component of ISDN and a major bottleneck to universal ISDN services. When this communications method sets up telephone calls it uses circuits that are separate from those actually used for conducting the calls. Also known as out-of-band signaling, this technique permits the network database to be accessed as part of the call setup process, providing communications carriers with the ability to offer a variety of enhanced services to subscribers. Unfortunately, without a widespread deployment of SS7, ISDN will resemble a series of isolated islands. Thus, users implementing ISDN have no guarantee that they will obtain all or a portion of the advantages associated with its use.

Another area of concern is the data transmission rate between non-ISDN and ISDN devices. Until pre-ISDN equipment is replaced, access to most ISDN facilities will be through the use of terminal adapters, which convert 19.2 kbit/s to 64 kbit/s by appending null bits. Functioning as sophisticated speed converters, these expensive devices will use a rate-adaption process to enable non-ISDN equipment to function at the ISDN data rate. But since the device does not make the data come out of the PC faster, a PC with a serial port would still be limited to an effective data rate of 19.2 kbit/s when attached to an ISDN B channel through a terminal adapter.

Summary

Is ISDN an obsolete data network? In general, an ISDN B channel provides a transmission capacity that is sufficient for most current and emerging applications when viewed from the perspective of data rate capability. For some applications, such as extensive file transfer operations, high-resolution screen image transmission and interactive remote control involving window-oriented screens will result in the ISDN B channel being a data transfer bottleneck. For other applications, such as transmitting data between distant LANs, the bottleneck potential of the ISDN B channel will depend on the number of users simultaneously requesting remote access to another LAN and the applications they intend to perform.

Although BISDN can be expected to alleviate bottlenecks of data transfer on the B channel, its implementation is far too

Modifying the B channel. *Using variable-rate channelization can result in a higher level of transmission capability. Two methods are shown.*

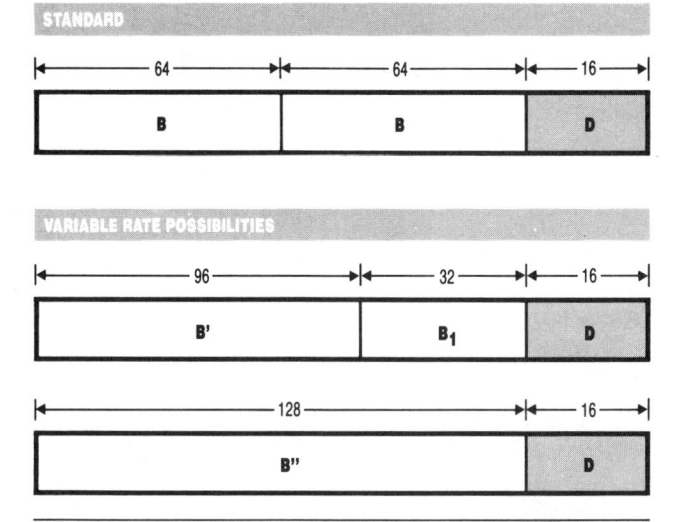

distant to overcome current and pending transmission problems. Bottlenecks at the B channels could be eliminated, in the near term, by using packet-switched H channels and variable-rate B channels. Unfortunately, the use of both H channels and variable-rate B channels requires actions by communications vendors—out of the hands of the user.

Although the capability for high-speed data transfer exists, vendors have not begun developing equipment to use these facilities. Thus, for the near future, the use of an H channel does not appear to be a reasonable way to alleviate B-channel bottlenecks. With respect to variable B channels, the implementation of Q.931 and I.460 also appears to be in the distant rather than the near future. Owing to the preceding, this author expects vendors to come to the rescue, probably developing a device equivalent to a conventional limited-function multiplexer for a 2B + D channel. If this occurs, the ability to combine two 64-kbit/s B channels into a 128-kbit/s data pipe (for applications that do not require voice coordination when simply performing file transfers, sending screen images, or accessing a distant LAN) can be expected to alleviate many of the data transfer problems foreseen in this article. Eventually, with the implementation of equipment supporting I.460, the author expects the splitting of a B channel into 8-kbit/s intervals; voice digitization advances that simultaneously permit quality voice transmission at 16 kbit/s with data transfer occurring at 112 kbit/s; and full data transfers at 128 kbit/s, which will provide considerably more user flexibility and eliminate most B-channel bottlenecks. ∎

Gilbert Held, director of Macon, Ga.-based 4-Degree Consulting, is an internationally recognized author and lecturer on data communications subjects.

References

Frost & Sullivan Inc., Report E1095, Digital Telecommunications/Europe

Netherlands Foreign Investment Agency, Telecommunications Industry Update, 11/15/88

Edwin E. Mier, editor at large, DATA COMMUNICATIONS

New signs of life for packet switching

X.25 packet switching has always had its place. But now that includes linking LANs and hauling SNA data. And with OSI gaining momentum, packet backbones are a way to cover all bases.

What do corporate giants Westinghouse, American Airlines, Manufacturers Hanover Trust, and United Parcel Service have in common? Each of these firms, and many others, have recently decided that their next-generation data networks would be based on X.25 packet-switching backbones.

"The packet backbone is more reliable, more versatile [than existing, proprietary network architectures]," says Bill Jewell, managing director of communications engineering for American Airlines (Fort Worth, Tex.), "so it fits many standards." (By versatile, Jewell means that the new network can handle more multivendor equipment and a greater variety of traffic.) American announced in March that it had contracted with Northern Telecom for a new multimillion-dollar, multiyear packet-switching backbone network. This network replaces a predominantly IBM-based, leased-line network. Initially, the network will be used primarily for internal traffic; later, it will be used for American's SABRE reservation service.

Another recent convert to the packet backbone is the Online Computer Library Center in Dublin, Ohio. OCLC announced in mid-November that it will spend $63 million over the next five years to install and operate a new backbone packet-switching data network. OCLC's major applications are internal traffic and access from various customer devices. The existing network is a mixed bag: asynchronous dial-up customer devices, with internal traffic handled by 3270 and other methods.

Why the dramatic upsurge in new packet networks? According to OCLC and others, there are a host of reasons that involve both current and future network requirements.

"We want to maximize our connectivity, we want to be less vendor-dependent, and we want to accommodate very different kinds of technology at our network end points," says Fred Lauber, OCLC's manager of telecommunications systems engineering and project manager for the new packet-switching backbone.

(The OCLC network is being installed, and packet equipment supplied, by Telenet.)

Packet switching? The old Defense Department data-transport technology developed in the 1960s? Sure, it's served Arpanet well, but isn't it mainly for wide-area TCP/IP networks? For those Unix- and VAX-based academic networks?

Not any more. While the latest and greatest developments in T1 networking (and its downsized derivatives, fractional T1 and switched 56-kbit/s service) have dominated the trade press lately, the packet-switching industry has hardly been idle.

A spate of recent trends and developments, including cheaper and more powerful packet-switching hardware and new software for marrying LANs and SNA traffic with X.25 nets, have significantly bolstered packet switching's capabilities, applicability, and esteem in the eyes of users and vendors.

Indeed, it is the rosy prospects for packet-switching's future that underlie the technology's renewed appeal. And this has caught many industry analysts by surprise. Just a few years ago, analysts were writing off packet switching as a mature and stagnant technology that faced only declining market prospects.

In a major 1987 network-technology overview, the Boston-based Yankee Group had forecast that over a five-year period (1987 to 1992), the number of private global packet networks (those traversing at least one national border) was expected to grow from perhaps 150 to 600. But a few months ago, the Yankee Group revised that forecast, noting that the number was growing much faster than expected.

According to Jack Freeman, senior analyst for data communications at Yankee Group, "it now seems there

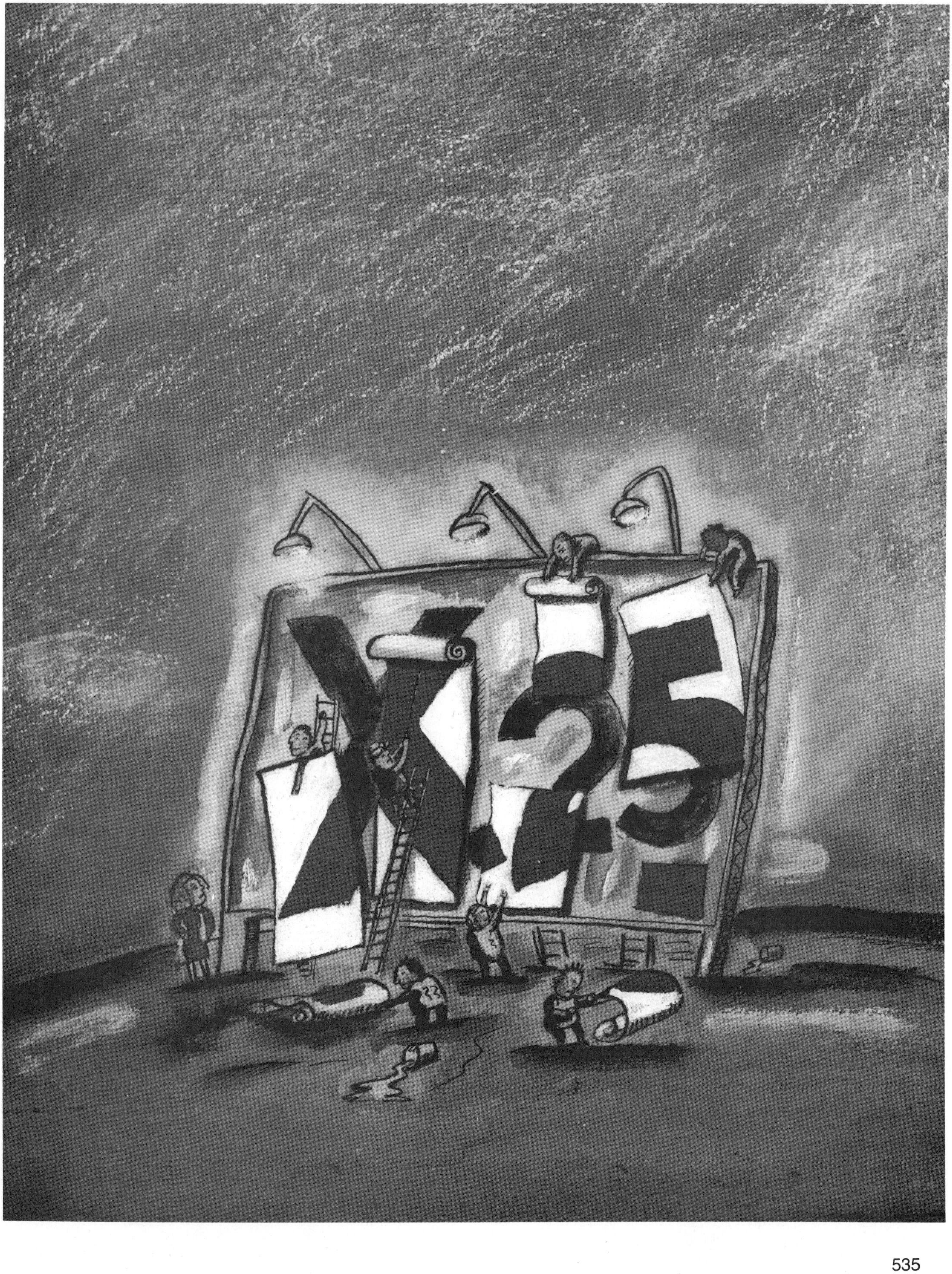

Packet Switching

could easily be over 1,000 such global packet backbones by 1992." He adds that there are currently "from 200 to 250" such networks, "and it's growing very fast."

What's new
Among packet switching's technical developments:
- *Adaptability for SNA.* Though X.25 and SNA have for years been regarded as diametrically opposed networking technologies, IBM terminal-to-host and computer-to-computer traffic can now be carried along with other traffic types over a packet network as quickly—and some say even more efficiently—than over dedicated SNA leased lines.
- *LAN to WAN.* As a technology for linking LANs to WANs, and to remote LANs, X.25 packet gateways and transport backbones are becoming as popular as point-to-point bridges (and are being deployed in ever greater numbers, by some market estimates). Packet-network gateways typically interact with LAN stations using the Netbios (and soon, Named Pipes) software interface and the Server Message Block (SMB) client-server protocol.

Other router-class gateway devices operate on the LAN as TCP/IP nodes and communicate through a wide-area packet network with other TCP/IP nodes. And new LAN-to-packet-network gateways are promising support for other popular LAN protocol stacks and client-server protocols (such as the XNS adaptations of older 3Com LANs, as well as those of Novell and Banyan).
What's more, new public data network services designed expressly for LAN-to-remote-LAN connectivity (such as one from Reston, Va.–based Telenet, due out next month) will debut next year.
- *More packets, lower prices.* During the 1980s, packet-network price-performance ratios improved by an order of magnitude. According to Northern Telecom, a leading supplier of high-capacity, general-purpose backbone packet networks, the capital cost per 100 packets per second (pps) of private-packet-network throughput has dropped from more than $100,000 in 1979 to about $16,000 today (Fig. 1).

And within the last few years, packet-network call-processing rates (the number of virtual-circuit call setups and tear-downs that can be handled per unit of time) have soared. Where a packet network might have processed 50 calls per second two years ago, it might now handle more like 500 calls per second, according to Carlton Rice, Northern Telecom's manager of network engineering. What has mainly brought about this increase in call-processing speed is faster processors and streamlined call setup procedures.
- *X.25 ubiquity.* Every major computer vendor—including IBM—now offers an integral X.25 packet-network interface. These typically consist of a high-speed (typically 64-kbit/s) synchronous line interface, along with software customized for maximum packet input/output directly into the data format of the vendor's computer. Integrated interfaces eliminate the PAD protocol conversion that had been a chokepoint of sorts at the host-connection end of the packet network.

And since 1976, when X.25 was first standardized, it has

1. Plummeting packet prices. *During the 1980s, the cost per 100 pps throughput has dramatically dropped, from more than $100,000 in 1979 to about $16,000 today.*

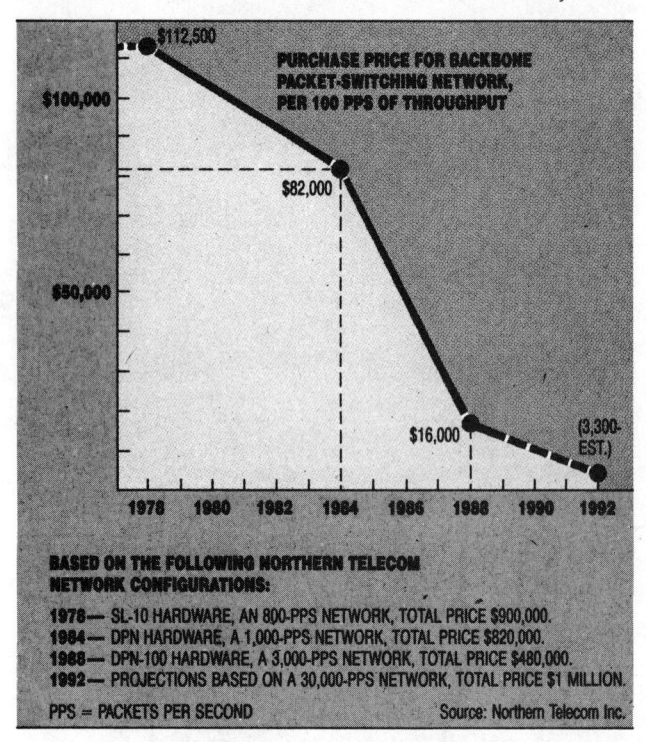

PURCHASE PRICE FOR BACKBONE PACKET-SWITCHING NETWORK, PER 100 PPS OF THROUGHPUT

$112,500 (1978)
$82,000 (1984)
$16,000 (1988)
(3,300-EST.) (1992)

BASED ON THE FOLLOWING NORTHERN TELECOM NETWORK CONFIGURATIONS:
1978— SL-10 HARDWARE, AN 800-PPS NETWORK, TOTAL PRICE $900,000.
1984— DPN HARDWARE, A 1,000-PPS NETWORK, TOTAL PRICE $820,000.
1988— DPN-100 HARDWARE, A 3,000-PPS NETWORK, TOTAL PRICE $480,000.
1992— PROJECTIONS BASED ON A 30,000-PPS NETWORK, TOTAL PRICE $1 MILLION.
PPS = PACKETS PER SECOND   Source: Northern Telecom Inc.

been revised, enhanced, and refined in subsequent 1980, 1984, and 1988 versions. Users and vendors report few problems anymore in mating X.25-based packet equipment from different vendors.
- *Voice and (packet) data?* Where X.25-based packet transport has long been a mainstay in many of the world's developed countries, except the U.S. (largely because of the relative affordability and availability of leased lines, including digital data service and T1), the efficiencies and other benefits of packet switching, including the prospects for implementing mixed-media traffic over the same packet backbone, are winning over many new converts—even in the U.S.

New high-capacity packet switches (30,000+ pps) can now handily fill T1-capacity trunks. And all indications are that real-time, toll-quality voice will routinely ride along with data on packet backbones within two to four years (see "Less processing, higher throughput"). Telenet and others offer this capability today in specially engineered private packet networks. (Telenet inserts software-based fast-packet switches that it OEMs from Campbell, Calif.-based StrataCom between its packet switches.)

The catalyst for mixed-media (voice, data, video, graphics) backbone packet switching will be frame-relay and fast-packet specifications that are now being defined in the standards community. (Frame relay is a streamlined mode of packet switching based on the CCITT's Link Access Procedure-D, which minimizes packet-layer processing. Fast-packet switching is an embryonic packet-transport technology featuring small, fixed-length packets—called cells—and fast, silicon-based logic execution.)

Packet Switching

■ *And then there's OSI.* With the U.S. Government's OSI procurement doctrine (GOSIP) effective in August 1990, X.25 packet-switching support will be required for all new systems and networks involving wide-area data communications that are bid to the government after that time. Since the U.S. Government is the largest purchaser of computer and data communications equipment and services in the world, this will accelerate the widespread deployment of OSI standards-based computer networking, for which X.25 is currently the only wide-area protocol set encompassing Layers 2 and 3 (the data link and network layers, respectively).

Vendors who are now busily implementing OSI's higher-layer protocols, as Retix Communications is, say that the software mechanisms defined for tying in OSI's higher layers, enabling access to and egress from X.25-based packet networks, are straightforward and efficient.

Big Blue packets

Most agree that a fundamental incompatibility between X.25 packet switching and IBM terminal-to-host networking had long stifled the widespread deployment of packet backbones, especially in the U.S. and especially in the IBM-dominated general-business sector. But this is not the case anymore. Indeed, as IBM proceeds to migrate SNA away from the terminal-to-host, hierarchical structure and toward computer-to-computer, peer-to-peer networking, some argue convincingly that packet-switched backbones can handle SNA network traffic even better than IBM's own leased-line-based designs.

Driven by customer demand, first in Europe and more recently in the U.S., IBM has made several key product introductions (some would say concessions) that enable IBM SNA data to be efficiently transported via packet-switching networks. As a result, the IBM-X.25 incompatibility that existed five years ago has largely disappeared.

Five years ago, many IBM networks were still running character-oriented Bisync (Binary Synchronous Communications, or BSC), which does not relate well to the bit-oriented nature of X.25 and its HDLC (high-level data link control). In fact, the public packet carriers developed a special protocol called Display System Protocol (DSP), which significantly improves the flow of Bisync traffic over a packet network. (DSP has become a de facto industry standard and is also now supported by many private packet network equipment makers.)

Even with DSP, however, transporting Bisync traffic via packet network is less than an optimal, or elegant, solution—especially when compared to the IBM polled, cluster-controller-to-front-end, leased-line approach.

This fundamental incompatibility began to dissolve with the spread of synchronous data link control (SDLC) in IBM networks. Yet, while the bit-oriented SDLC could be much more efficiently handled by X.25 packets, there remained the IBM-network requirement for host polling of remote SDLC devices. Passing polls added considerable, unnecessary data packets to an intermediate X.25 transport network, until packet-network designers discovered they didn't have to carry this polling traffic at all: They could identify and intercept the polls at the host end, issue their own at the terminal end, and not have to devote network bandwidth to transporting them.

But the real marriage of IBM's SNA and X.25 packet switching came about just a few years ago with IBM's general introduction in North America of an IBM front-end processor software package called NPSI (Network Packet Switching Interface). IBM had already furnished NPSI to its European customers for a few years, but demands for it in the U.S. became too loud for IBM to ignore.

NPSI lets IBM SNA traffic take on the look and feel of X.25 packets, using an X.25-oriented, IBM-developed frame format called QLLC (qualified logical link control). These frames can then be zipped through packet networks virtually intact. With NPSI in the front-end (along with the 37X5's basic operating software, the Network Control Program), polls are not even issued anymore—they are intercepted by NPSI.

Packets versus leased lines

Using the DSP protocol for transporting older Bisync terminal traffic via a packet network involves considerable protocol conversion. In general, the more processing required to adapt non-X.25 traffic to a packet network, the lower the throughput, the higher the response time, and so on. (A comparison of different terminal types and their adaptability to packet transport is shown in the table.)

The fastest way to get SNA data through a packet network is by encapsulation. In encapsulation, the IBM frames (typically 3270 or 5250 terminal data streams) are kept intact and bundled up within X.25 packets. On leaving the packet network, the X.25 wrapping is stripped off. Encapsulation is usually employed in packet networks for passing IBM QLLC data to a 37X5 front-end processor running NPSI.

In other environments, the packet network needs to do

> **Bisync traffic via packet net is less than an optimal solution.**

more than merely encapsulate the IBM frames. At a minimum, the recognition and interception of polls at the host end is required (along with issuance of polls at the remote, terminal end).

Packet-switching equipment vendors vary in the degree of processing that their gear applies to IBM SNA data as it enters, transits, and then leaves the packet network. Telenet's approach, for example, is to convert SNA data into its own internal frame and packet format. This is done, says Alan Taffel, Telenet's vice president of strategic marketing, so that IBM terminal traffic can be switched to non-IBM processors via the packet network, with possibly substantial protocol-conversion processing.

In other cases, SNA traffic is carried more transparently.

Packet Switching

Less processing, higher throughput

If the remarkable increase in packet-processing rates and throughput can be attributed to any single network-design trend, it is that more and faster processors are being used and that they are performing less software-intensive processing.

According to John McQuillan, president of McQuillan Consulting (Cambridge, Mass.) and a noted expert in network-equipment design and performance, protocol processing (such as is done in the asynchronous-to-X.25 processing of a PAD) is a heavy additional burden on a single-processor packet switch.

McQuillan says that a single-processor packet switch can process from five to 10 times more packets by not also having to perform protocol processing. When multiple protocol conversions are being handled by the same processor, the drain on throughput is magnified even more.

Indeed, based on the performance data available from suppliers of single-processor packet equipment (such as low-end, standalone packet switches and PC plug-in boards for LAN-to-packet-network gateways), packet throughput in the range of 300 to 500 packets per second (pps) can be achieved, given the 16- and 32-bit microprocessors that are typically employed today.

Users should keep in mind that as they add protocol processing to a packet processor, the packet throughput will drop. And depending on the complexity of this additional software-intensive protocol handling, this drop could be significant. A single-processor packet node, theoretically capable of pushing out 500 pps, could be constrained to perhaps only 50 pps by the addition of a lot of protocol processing.

Multiprocessor designs, especially in sizable nodal packet switches (such as those from BBN, Telenet, Northern Telecom) can obviate this throughput degradation, their designers claim. In fact, the architecture of today's large packet switch—consisting of multiple line controller cards, protocol-processing cards, and packet-switching controller cards, all connected via a multimegabit/s bus—is functionally not unlike today's LAN configurations.

Indeed, some experts see the protocol-processing function that used to be integral to packet switches being performed much more efficiently on a LAN.

"PAD functions are now being done more cost effectively by LANs," says McQuillan. He maintains that the traditional appeal of packet switching—"to multiplex a lot of low-speed terminals to high-speed devices"—is "no longer important." Instead, he says that the promise and prospects of packet backbones for efficient, general-purpose, wide-area transport is the new draw and that anything other than raw packet transport and switching is being pulled out of the packet network to achieve greater throughput performance.

Frame relay promises to streamline packet switching even more—and enable at least a doubling of raw-packet throughput from the same amount of processing power. Frame relay gets its name from the frame designation of data units at Layer 2 (data link layer) of the OSI model. In essence, typical Layer 3 (network layer) functions, where most X.25-based packet processing is now performed, are relieved from the frame-relay packet network (see figure and table).

The long-term design trend, most designers agree, is for many software- and memory-intensive functions—such as correct packet sequencing and error correction—to be moved into the devices that are connected by the frame-relay packet network. Frame-relay operation, largely defined in the CCITT's specifications for Link Access Procedure-D (Q.921), still provides for the detection of errors, for example, but bad packets are summarily dropped. It is up to the source and destination devices to correct such errors, usually through retransmission.

Because X.25-based packet networks are centered on the CCITT's X.25 Recommendation, which is only a standardized convention for access to a packet network, packet-equipment and switch suppliers have always used their own proprietary protocols and architectures within their packet networks for internodal traffic transport and management.

Because of this, most believe that frame-relay switches will be able to be conveniently inserted into existing packet networks and that an X.25-to-frame-relay interface will enable the efficient marriage of the two.

Frame relay will still use much the same packet-processing hardware base as today's packet-switching equipment, most agree. This means that a packet node based on, for example, an 80286 chip, could be reworked to become a frame-relay switch instead of an X.25 switch. But where a node might pump out 100 X.25 packets per second, it would be able to do perhaps 200 to 500 frame-relay packets per second.

Existing hardware designs will be inappropriate, however, for the generation of packet switches that come after frame relay (mid- to late 1990s). Prototypes of packet switches based on asynchronous transfer mode (ATM), or so-called fast-packet operation (also called cell relay by Bellcore), have already been demonstrated. These perform almost no software-based processing and instead perform packet-manipulation functions based on silicon-embedded logic, which is much faster.

ATM fast-packet switches will serve as the basis for the switched multimegabit/s data services that are being preannounced by local and long-distance carriers and data-transport service suppliers. These will appear initially for metropolitan-area transport, mainly as a network facility for linking multimegabit/s LANs.

Unlike X.25 and frame-relay packet switching, where packet size is variable, fast-packet switching uses standard 48-byte cells (as distinguished from Layer 2 frames and Layer 3 packets), plus a 5-byte header. Current designs and prototypes are heralding throughputs of millions of packets per second, which leads some to doubt whether private backbone-type packet networks will ever

Packet Switching

have the throughput requirements to justify such high-horsepower switches (with correspondingly high price tags) through the next decade.

Fast packet will unquestionably be capable of mixed-media transport—real-time voice, data, graphics, video. It is still unclear whether frame relay, by comparison, will offer the capability even for packetized, real-time voice transport. New 32-bit processors (80386, 68020) might enable it, the experts say, but they also agree that frame relay, while a significant enhancement over X.25 packet switching, is not designed to support real-time voice traffic.

—E.E.M.

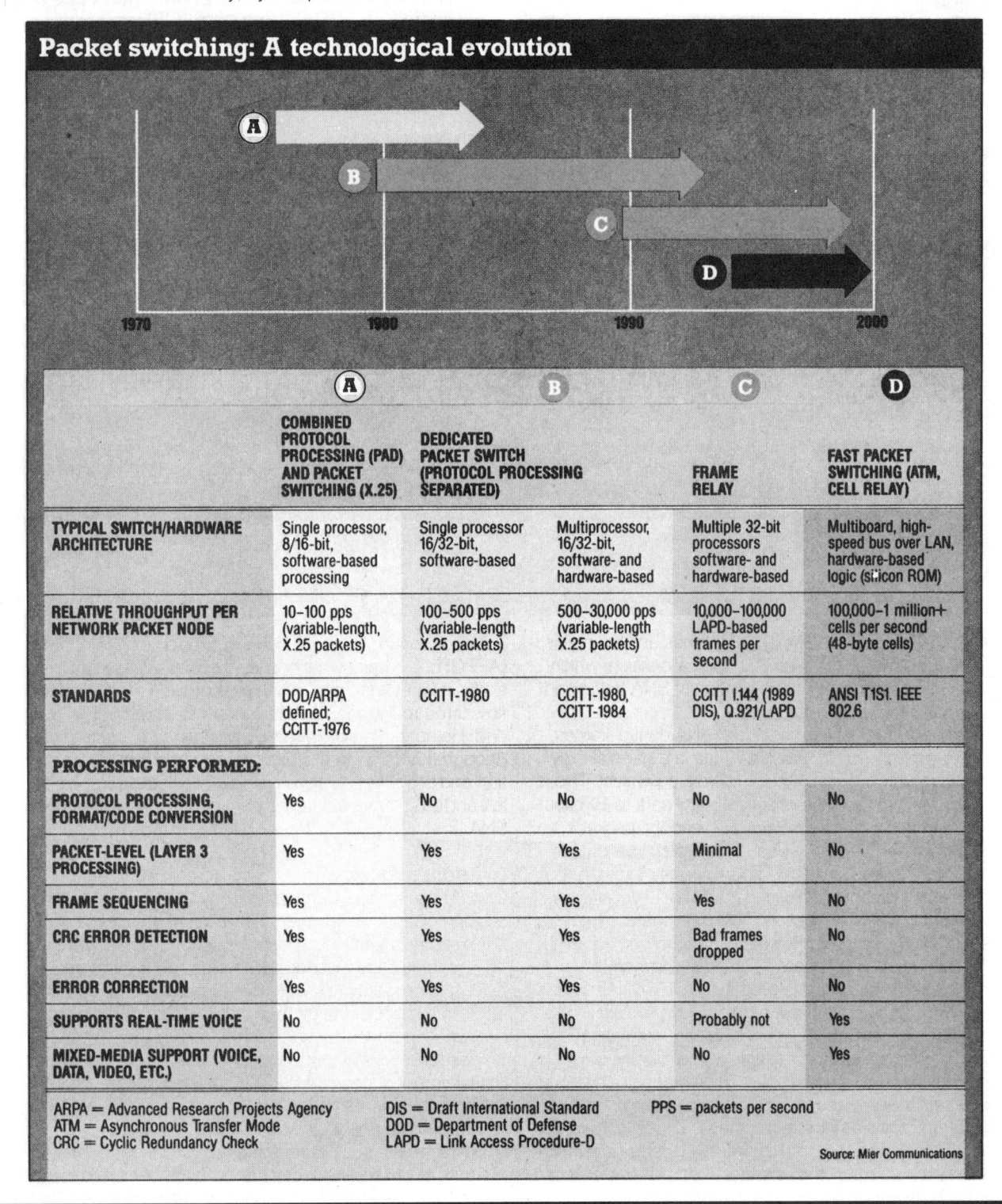

Packet switching: A technological evolution

	A COMBINED PROTOCOL PROCESSING (PAD) AND PACKET SWITCHING (X.25)	**B** DEDICATED PACKET SWITCH (PROTOCOL PROCESSING SEPARATED)		**C** FRAME RELAY	**D** FAST PACKET SWITCHING (ATM, CELL RELAY)
TYPICAL SWITCH/HARDWARE ARCHITECTURE	Single processor, 8/16-bit, software-based processing	Single processor 16/32-bit, software-based	Multiprocessor, 16/32-bit, software- and hardware-based	Multiple 32-bit processors software- and hardware-based	Multiboard, high-speed bus over LAN, hardware-based logic (silicon ROM)
RELATIVE THROUGHPUT PER NETWORK PACKET NODE	10–100 pps (variable-length, X.25 packets)	100–500 pps (variable-length X.25 packets)	500–30,000 pps (variable-length X.25 packets)	10,000–100,000 LAPD-based frames per second	100,000–1 million+ cells per second (48-byte cells)
STANDARDS	DOD/ARPA defined; CCITT-1976	CCITT-1980	CCITT-1980, CCITT-1984	CCITT I.144 (1989 DIS), Q.921/LAPD	ANSI T1S1. IEEE 802.6
PROCESSING PERFORMED:					
PROTOCOL PROCESSING, FORMAT/CODE CONVERSION	Yes	No	No	No	No
PACKET-LEVEL (LAYER 3 PROCESSING)	Yes	Yes	Yes	Minimal	No
FRAME SEQUENCING	Yes	Yes	Yes	Yes	No
CRC ERROR DETECTION	Yes	Yes	Yes	Bad frames dropped	No
ERROR CORRECTION	Yes	Yes	Yes	No	No
SUPPORTS REAL-TIME VOICE	No	No	No	Probably not	Yes
MIXED-MEDIA SUPPORT (VOICE, DATA, VIDEO, ETC.)	No	No	No	No	Yes

ARPA = Advanced Research Projects Agency
ATM = Asynchronous Transfer Mode
CRC = Cyclic Redundancy Check
DIS = Draft International Standard
DOD = Department of Defense
LAPD = Link Access Procedure-D
PPS = packets per second

Source: Mier Communications

Packet Switching

Good and bad fits for packet-backbone transport

TERMINAL COMMUNICATIONS	EXAMPLE	SUITABILITY FOR PACKET TRANSPORT	COMMENT
CHARACTER-MODE, FULL-SCREEN, WITH ECHO-BACK	VT100	POOR	ECHO-BACK GENERATES UNNECESSARY PACKET TRAFFIC. CHARACTER-MODE TRANSMITS SINGLE CHARACTER AT A TIME, MAY REQUIRE A FULL PACKET FOR EACH CHARACTER
TTY-MODE	ANY TTY EMULATION	FAIR	TRANSMITS ONE LINE OF DATA AT A TIME, WHICH PERMITS BETTER PACKET UTILIZATION THAN CHARACTER-MODE
BISYNC CLUSTER CONTROLLER, POINT-TO-POINT	IBM 3274 BSC	FAIR	BLOCK-MODE, GOOD FOR FILLING PACKETS, BUT CHARACTER-ORIENTED FRAME STRUCTURE REQUIRES CONVERSION TO BIT-ORIENTED X.25/HDLC STRUCTURE
BISYNC TERMINALS, MULTIPOINT	IBM 327X BSC	GOOD	POLLING IN MULTIPOINT CONFIGURATION CAN BE ELIMINATED; BEST IF PACKET NETWORK SUPPORTS DSP PROTOCOL FOR BSC DEVICES
SNA/SDLC CLUSTER CONTROLLER	IBM 3174 SNA/SDLC	GOOD	SNA/SDLC BIT-ORIENTED STRUCTURE MEANS FAST PACKET FORMATION; POLLS CAN BE INTERCEPTED/SIMULATED; A LONG PACKET SIZE USUALLY BEST
APPC/LU6.2 DEVICE	PC, PS/2, SYSTEM/36	EXCELLENT	ESPECIALLY EFFICIENT OVER PACKET NETWORK WITH NPSI FRONT-END PROCESSOR AND USING QLLC FRAME STRUCTURE

APPC = ADVANCED PROGRAM-TO-PROGRAM COMMUNICATIONS
BSC, BISYNC = BINARY SYNCHRONOUS COMMUNICATIONS
DSP = DISPLAY SYSTEM PROTOCOL
HDLC = HIGH-LEVEL DATA LINK CONTROL
LU = LOGICAL UNIT
NPSI = NETWORK PACKET-SWITCHING INTERFACE
QLLC = QUALIFIED LOGICAL LINK CONTROL
SDLC = SYNCHRONOUS DATA LINK CONTROL
TTY = TELETYPEWRITER

Source: Mier Communications

Northern Telecom, for example, establishes and maintains an X.25 virtual circuit for each SNA logical session, which provides a clear one-to-one association of SNA traffic and packet-network resource allocation.

In some vendors' packet products, SNA control frames, which may or may not contain user data, are painstakingly converted to their closest packet-network equivalent. This is helpful in tying in management and control data to IBM host-based network management. In other vendors' packet products, all SNA frames—whether control or user data—are passed via encapsulation. This maximizes throughput and response time but makes the packet network effectively transparent to the SNA network. The price paid is usually in a separate network management arrangement, which is required for the packet network, in addition to whatever management arrangement is used for the IBM SNA network.

"Network management is a problem," says Maks Wulkan, executive vice president of Eicon Technology (Montreal). Eicon's products, deployed generally for LAN gateway access to remote IBM processors via packet networks, offer several session-layer interfaces for packet-network transport. For SNA traffic, these include 3270/LU2, 5250/LU7, and APPC/LU6.2. Another handles basic ASCII/VT100 asynchronous communications. And recently, Eicon added a software option for handling IBM's QLLC.

According to Wulkan, Eicon's approach for LU6.2 (APPC) communications is to encapsulate this type of traffic in QLLC packets, which provides the most efficient flow through the packet network and to the IBM NPSI front-end. He notes that by passing SNA traffic transparently (though the Eicon products do take care of poll interception and physical unit device emulation, notably PU Types 2.1 and 2.0), "the packet network knows nothing about SNA."

Different strokes

Using a packet backbone solely to replace a greater number of leased lines in an all-IBM network probably doesn't pay in most instances. A packet network would likely reap savings for the user's organization over an extensive network of point-to-point leased lines in monthly facilities costs, but the payback for the cost of the packet-network components—such as the processors—could take an exceedingly long time. Even if the monthly leased-line bill was reduced from $20,000 to $10,000, the $10,000 saving might have to be spent for packet equipment, amortized over, say, four years.

However, in a network with many multipoint lines, which are handling mainly IBM terminal-to-host communications, a packet backbone network can have considerably more

Packet Switching

appeal. There are several reasons for this:
- In multipoint configurations, IBM host polling of terminals can become an unwieldy portion of the overall traffic. Where polling on a point-to-point circuit usually accounts for only a minuscule portion of the overall line traffic, this can grow in a multipoint environment to a point where individual user response times suffer unacceptable degradation. The elimination of polling through a packet-switching network can ameliorate this.
- Multipoint circuit pricing has become exceedingly complex, with the result in many cases being sharply higher overall monthly costs owing to new per-circuit connection charges levied by the local telephone companies. Packet-network topologies invariably use many fewer leased-line circuits to accommodate many more devices. And if engineered properly, a packet network can usually handle high-peak data-traffic spurts, and even sustain considerable network growth, without having to add transmission facilities (though data-rate upgrades on select links may be needed from time to time).
- A packet-switching twist called fast select enables very short transactions (like credit card authorizations, automated teller machine personal identification number verifications, and similar point-of-sale-type transactions) to be transmitted extremely efficiently. Already supported in Northern Telecom's packet switches, for example, fast select allows a single, short call setup packet to carry a brief data message to the destination and the destination to acknowledge receipt and respond in a single, equally brief call tear-down packet.

For these reasons, packet backbones may be justified as a transport replacement for multipoint environments, even where there is only IBM terminal-to-host traffic. Usually, however, the decision to go with a packet backbone hinges on additional factors, such as the ability to readily switch user's connections.

The right stuff

Many large-network users have a mix of terminals, PCs, minicomputers, and hosts. And one reason packet backbones are in growing demand today is that they readily enable users at any network-attached device to switch from one destination to another, even when different protocols are required, and over the same network link.

The heightened appeal for packet switching in this regard comes from the additional software support many packet-equipment suppliers now offer. Typically, support is now included for the following:
- PAD protocol conversion for asynchronous terminal traffic—usually to asynchronous hosts—may be supplemented with particular terminal-emulation software support (such as for VT100 or 220 terminals, Hewlett-Packard async devices, and Honeywell VIP terminals).
- IBM 3270 Bisync, usually combined with support for DSP or specific Bisync/3270 device emulation.
- IBM 3270 SDLC, emulation of an SNA 3270 cluster controller, or remote standalone 3270 display terminal.
- IBM 5250 SNA/SDLC, emulation of the terminal series used with IBM minicomputers.
- X.25, which will typically be based on the CCITT's 1984 or 1980 versions of the standard. Sometimes connection between one vendor's PAD supporting the 1980 version and another vendor's packet switch supporting the 1984 version can be a problem (especially during call establishment, since the switch offers services and features from the 1984 specification that the 1980 PAD doesn't recognize). To handle such cases, equipment can sometimes be set to run in 1984-suppressed mode, so that communications with 1980-based devices can be accommodated on a line-by-line basis.
- X.32, which defines X.25 features for support over a dial-up (or switched), synchronous connection. This involves user identification and setup negotiation features that typical, leased-line X.25 connections do not require.
- SNA/QLLC, special support for IBM NPSI and the IBM-defined QLLC packet format. A newer version of QLLC, called ELLC (for extended logical link control), has been issued by IBM and will no doubt be supported by packet-equipment suppliers that already handle QLLC. ELLC is the SNA-via-packet-network protocol that IBM is supporting for networking its midrange computers (System/36, System/38, AS/400) via packet networks. It reportedly provides better error-recovery capabilities than its QLLC predecessor, because ELLC keeps more error-recovery decision making at the data link level. By comparison, QLLC reportedly defers many error-recovery decisions to higher-level SNA protocols.
- X.75, for connections between packet networks (private to public, private to private, public to international).
- Any of several leading LAN client-server protocols, such as the SMB of Netbios-compatible LANs. These are supported mainly in the new genre of packet-switching LAN gateways.
- APPC/LU6.2 (IBM's Advanced Program-to-Program Communications/Logical Unit 6.2 protocol), which can be

> **Packet-equipment suppliers now offer greater software support.**

implemented in several ways. One of the most common is where a LAN-to-X.25 gateway or PAD processes the IBM-defined verbs from other LAN workstations (usually PCs). In this way, the PCs don't have to run the full APPC protocol software themselves (requiring from 400+ Kbytes in an MS-DOS PC to more than 2 Mbytes in an OS/2). In this case (supported in such LAN/X.25 gateways as those from Eicon Technology), the PCs send only the APPC verb to the X.25 processor, which then initiates and performs the appropriate APPC protocol function.
- Support for higher-level protocol stacks, on top of X.25, such as TCP/IP. In such products, like those available from Frontier Technologies (Milwaukee, Wis.), the X.25 gateway with TCP/IP support effectively becomes a LAN router.

Packet Switching

- Any combination of the above protocols.

It is unlikely that many networks would need to implement all of these protocol conversions in the same backbone packet network. And it would probably be unwise to do so, since protocol processing is a processor- and memory-intensive operation. As the number of different protocols that need to be supported grows, throughput drops. In most networks, only a few are needed.

"Ninety-five percent of what most are looking for [in packet-network protocol support] is Bisync, 3270/SNA, asynchronous PAD, and X.25," says David Jeanes, manager of data network market analysis with Bell Northern Research (Ottawa). "You need more memory available for multiprotocol systems," he adds.

The LAN connection

The spate of relatively new equipment manufacturers who are specializing in LAN gateways to packet networks seem to reflect the newness of LANs in general. Manufacturers like Eicon and Frontier are recent entrants into the packet-switching marketplace.

But like the LAN industry in general, these specialized-gateway makers are growing rapidly (Eicon, for example, claims 25,000 of its gateways are already installed, and that its products are OEMed to some 40 to 50 resellers and system integrators).

It seems, too, that these LAN gateway specialists are particularly adept at developing efficient software for tying LANs to wide area networks. The traditional packet-switch and PAD manufacturers, by comparison, seem these days to be more focused on hardware development to improve throughput and call-processing performance in backbone packet networks.

Naturally, the LAN-gateway makers are obliged to ensure that their products work with the leading backbone packet-switch suppliers, as well as with the major public data networks of the world. Users should query these gateway suppliers on their connectivity and compatibility experiences with the specific backbone packet equipment they are either already using or considering.

Unfortunately, there are no de jure standards yet that specify how LAN traffic is to be mapped to packet-switching WANs or connections maintained through packet backbones to other, remote LANs. As a result, implementations vary greatly in how this is achieved, what particular features are supported, and in relative performance.

A better approach than IBM's?

To understand the diversity of approaches being taken, consider several of Eicon Technology's SNA/LAN-to-packet-network implementations.

Eicon's appearance to LAN nodes takes several different forms. Currently, communications to the gateway across the LAN uses Netbios; the LAN stations, in the role of clients, use a redirector function to access the gateway (supplied by Eicon, which takes about 10 Kbytes of the workstation's memory). The gateway handles packet call setup and communications through the packet network to the remote IBM host.

The redirector function initiates SMB sessions with the Eicon gateway, which assumes the role of a server. The SMB message informs the gateway to which host and application a virtual circuit through the packet network needs to be established. For 3270 or 5250 terminal communications, the LAN stations need to have software for the terminal emulation; the gateway assumes the identity of the appropriate cluster controller to the host.

The operation is different, however, in the case of APPC/LU6.2 communications (Fig. 3). Eicon supplies software that loads onto the PCs (or PS/2s) needing to establish APPC communications. As opposed to IBM's approach, which is to load APPC/PC on MS-DOS PCs (or the full APPC protocol module from the Communications Manager of OS/2 Extended Edition), the Eicon approach is for the LAN workstations to send only APPC service-request verbs. The Eicon gateway processes these verbs and initiates the appropriate APPC protocol actions and messages.

Eicon boasts of the efficiency of its approach. The memory requirements on the LAN workstations are much smaller than if they handled all APPC processing on their own, and the amount of trans-LAN traffic related to APPC comunications is minimized, says Wulkan.

But even more important to operation over a packet network, the Eicon gateway-based approach involves a single physical unit appearance to the remote IBM host. This is key because the PU definition to the host front-end is the basis for host-issued polls (and not LU existence in SNA networks). And with a single PU in the gateway, polling from the SNA host is minimized. If polls need to be passed to LAN workstations (as when the stations handle all their own APPC processing, and each has both PU and LU entities defined to the host for it), then this additional traffic can unduly burden the packet transport network.

When the Eicon gateway handles cluster-controller processing and emulation across the packet network to a NPSI host, the response time that is achieved is as good as a direct leased-line connection through an IBM 3174 cluster controller.

In fact, according to Wulkan, independent user tests show that response times through an Eicon gateway are, in some cases, even better than through a 3174 cluster controller for clusters of from one to 10 terminals (PCs emulating 3270 terminals). The normalized data (supplied by Eicon) indicates a 4.0-second response time for a 10-node 3174 cluster, and 3.5 seconds for an Eicon LAN configuration supporting 10 terminal-emulation LAN workstations.

"The data shows our technique doesn't degrade performance," says Wulkan. The analysis is not an exact apples-to-apples comparison, however, since the Eicon approach was measured as a cluster-controller gateway on both Ethernet and token ring LANs, communicating using the QLLC encapsulation technique over a packet network to a remote NPSI host. The IBM 3174 can operate as a gateway on a token ring, or it can be configured to handle QLLC communications over a packet network to a NPSI host, but IBM documentation

Packet Switching

clearly points out that the 3174 cannot handle both attachments at the same time.

The spread of OSI presages a rapid proliferation of packet backbones. But the good news for users today is that existing packet networks should be readily adapted to handling high-level OSI communications (in addition to whatever protocols and connectivity are supported currently).

"X.400, for example, is a message-switching network that has been designed to overlay on top of a packet-switching network," says John B. Stephensen, senior vice president of technology at Retix, which has pioneered OSI protocol implementation in a variety of network and processor environments. "For wide-area OSI communications," he adds, "X.25 is the only thing that exists."

According to Stephensen, higher-level OSI protocols will access a packet network through a new OSI protocol standard called ISO-IP (its formal name in OSI circles is the Connectionless Network Protocol, or CLNP; the ISO-IP is used to distinguish it from the Internet Protocol of the TCP/IP protocol suite).

Technically, both X.25 and OSI higher-level addresses are variable in length. In practice, however, an X.25 address (the data terminal equipment address) is usually 14 digits. And the OSI address (called the IP-NSAP—network service access point—address) can be up to 24 bytes in all.

"For OSI, you need to follow any of the eight NSAP address formats," says Stephensen. These formats, for carrying and passing on existing address standards like

2. Two APPC methods. *IBM's APPC approach (A): Load APPC/PC on MS-DOS PCs. With Eicon's approach (B), the LAN workstations send only APPC service-request verbs across the LAN. The PC gateway processes them and initiates the APPC protocol actions and messages. This involves a single physical unit appearance to the IBM host.*

(A) IBM APPC

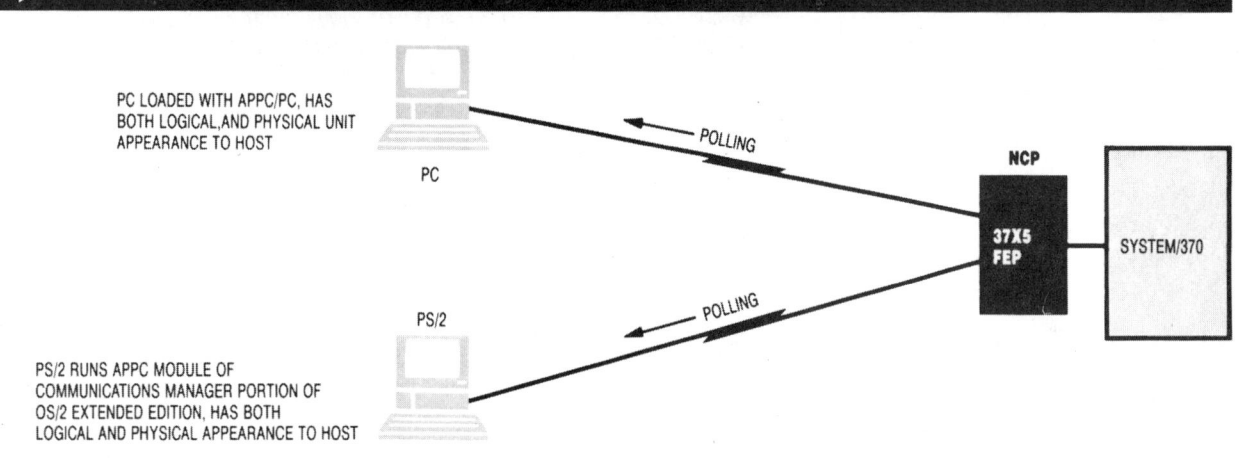

PC LOADED WITH APPC/PC, HAS BOTH LOGICAL AND PHYSICAL UNIT APPEARANCE TO HOST — PC — POLLING → 37X5 FEP (NCP) — SYSTEM/370

PS/2 RUNS APPC MODULE OF COMMUNICATIONS MANAGER PORTION OF OS/2 EXTENDED EDITION, HAS BOTH LOGICAL AND PHYSICAL APPEARANCE TO HOST — PS/2 — POLLING →

(B) Eicon APPC

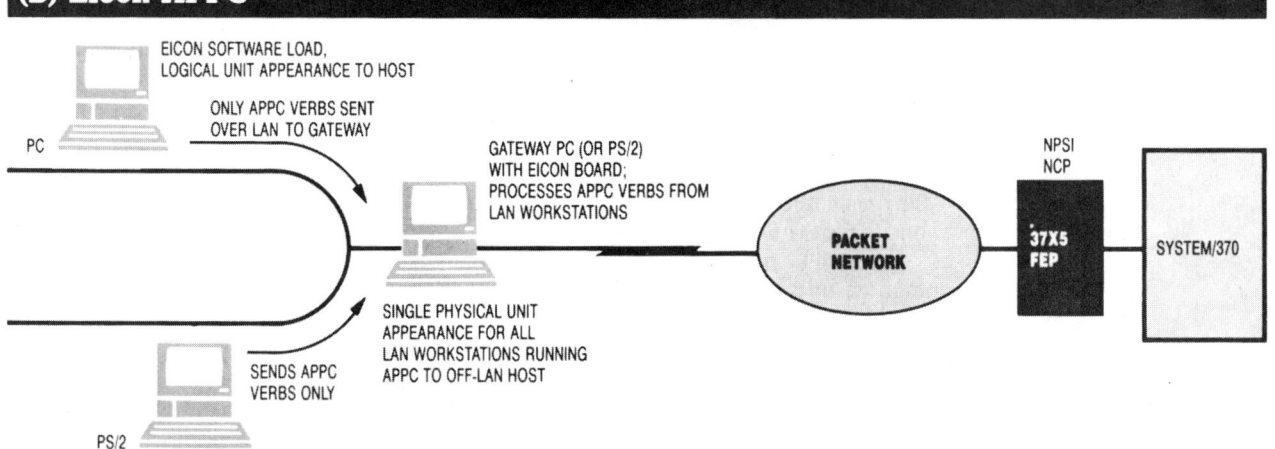

EICON SOFTWARE LOAD, LOGICAL UNIT APPEARANCE TO HOST
ONLY APPC VERBS SENT OVER LAN TO GATEWAY
PC
GATEWAY PC (OR PS/2) WITH EICON BOARD; PROCESSES APPC VERBS FROM LAN WORKSTATIONS → PACKET NETWORK → 37X5 FEP (NPSI NCP) → SYSTEM/370
SINGLE PHYSICAL UNIT APPEARANCE FOR ALL LAN WORKSTATIONS RUNNING APPC TO OFF-LAN HOST
SENDS APPC VERBS ONLY
PS/2

APPC/PC = ADVANCED PROGRAM-TO-PROGRAM COMMUNICATIONS FOR THE PC
FEP = FRONT-END PROCESSOR
NCP = NETWORK CONTROL PROGRAM
NPSI = NETWORK PACKET-SWITCHING INTERFACE

Packet Switching

Telex numbers, telephone numbers, even ISDN addresses, are all clearly delineated in a separate OSI standard: ISO document 8348, Addendum 2.

Forward spin

The state of packet switching is clearly evolving in response to changing times. And while many of the recent developments—packet price-performance ratios, software links, LAN-to-packet gateways—have positioned packet switching to enjoy an unprecedented proliferation in the next decade, other advances are still needed before packet backbones are viewed as the best long-term selection for most users.

More attention needs to be paid to separating packet-network control from transport, for example. Sources, such as packet switch vendors, report that so-called separate-channel control for packet-network operation is being actively studied and explored, both as part of standards activities involving frame relay, for example, and also in private development labs as a technique holding great promise for unheard-of packet throughputs in future generations of packet switches. (In standard packet switching, control packets preempt user-data packets. Separate-channel control, as the term implies, keeps control packets on a channel separate from user-data channels.)

The separation of control and transport is also a key to packet-switching operation over ISDN, which most agree is a marriage that simply must work. Conventional packet switching, characterized by the LAPB link-control protocol, is constrained to operating within the ISDN D channel, which offers only limited bandwidth for data transport. (At this writing, dial-up X.25 over a B channel is not yet a reality.)

LAPD—devised for ISDN connections—is a substantial improvement over LAPB toward achieving separate-channel packet control and enabling full 64-kbit/s, DS-0 packet bandwidth on a dynamic demand basis.

There's little question that packet backbones deployed in the coming decade will be based on 64-kbit/s DS-0 and 1.544-Mbit/s T1 facilities. Most packet gear already supports full-line utilization at 64 kbit/s, and several can also now fully load T1 facilities. (Telenet, for one, uses T1 facilities as part of its packet network.) What's more, the competition in the U.S. for T1 and fractional T1 bandwidth facilities makes their use as trunks in packet backbones inevitable. ■

Edwin E. Mier, editor at large for DATA COMMUNICATIONS, is also president of Mier Communications Inc., a communications and networking consultancy in Princeton Junction, N.J. Mier publishes the CONNECTIONS series of loose-leaf information services and computer-networking guides.

Kumar Shah, Data General Corp., Westboro, Mass.

Managing networks of the '90s

Object-oriented design and standards like OSI, IBM's NetView, and AT&T's UNMA are the keys for controlling distributed networks.

Networks of the 1990s will primarily use the client-server model of computing, also known as cooperative processing, in which applications are distributed between intelligent network nodes. This next generation of networking will require a new way of thinking about network management that groups the attributes of network nodes into modules regardless of how they are implemented on the network. This technique, called object orientation, will allow flexible and timely monitoring of the next decade's heterogeneous networks.

Advances in computing and networking have symbiotically developed to provide greater access to information in the enterprise. The cycles of innovation in the two arenas have ushered in an era in which the dividing line between computing and networking has practically vanished. As the two disciplines merge, the network becomes the computer; users are promised utilitylike access to services; and network management takes center stage. A twofold solution to the problem of managing these incipient networks is emerging.

First there is a growing consolidation around such industrywide de facto standards as IBM's NetView and AT&T's Unified Network Management Architecture (UNMA). In order to provide an integrated enterprise management system, these standards must cooperate with each other and with standards included in the International Organization for Standardization's OSI model. Second, two design technologies have developed that influence network management: client-server processing and object-oriented design. Object orientation often refers to a programming technique in which code and data are contained in the same software module. In this article, however, object orientation refers to the treatment of the network and all the devices, systems, applications, and services within it. Object-oriented programming could be one of the core technologies that would help achieve development of object-oriented networks.

Client-server computing allows OS, OS/2, and Unix-based machines (called clients) to work cooperatively with servers to access computing resources, communications, databases, and other services in the enterprise. The next frontier in client-server computing lies in developing cooperative processing capabilities among the servers.

The key to unlocking the power of heterogeneous server-to-server communications is object orientation, which refers to a way of binding together a set of devices (for example, modems) having common characteristics. This could be accomplished by defining a specific object class for that set of devices. For building network management applications, object orientation has the potential to bring about a true integration of the heterogeneous multivendor networks present in the enterprise today.

Network management architecture has had to change with each stage of computing technology (see "Evolution in computing"). Client-server computing represents the fourth phase in computing technology's evolution, and the corresponding phase of network management revolves around object orientation.

The client-server connection
Computing has now reached the threshold of a new technology: client-server computing. Designed to break down organizational barriers, it emphasizes access to information across the enterprise. This phase of computing aims to bring about an integration of the isolated islands of networks.

Unlike desktop computing, which approached the needs of workgroup users as separate from those of the enterprise itself, client-server computing enables local user groups to be serviced by servers that become access points into the

Network Management

enterprise backbone network. Seamless access to the various remote file, print, communications, and information services is achieved by interactions between the respective servers in the enterprise. This goes one step beyond the capabilities provided by remote bridges, which link heterogeneous LANs at the physical and link layers. The backbone then integrates multiple workgroups and thereby bridges the islands of enterprise computing and workgroup computing (see Fig. 1). Eventually, the artificial barrier separating a LAN from another LAN or a WAN will disappear.

Client-server computing goes one step beyond the sharing of file, print, and communications services on PC LANs; it introduces the concept of cooperative processing. The master-slave relationship between terminals and hosts began to disintegrate with the advent of desktop computing and PC LANs. It gives way even further with client-server computing, where a cooperative kinship exists among all

---

Evolution in computing

Three major waves of innovation have propelled computing technology. The first involved the mainframe; the second distributed computing to departments; and the third provided computing on the desktop via PCs. Simultaneously, managing computer networks has grown ever more difficult.

Computing's first phase dates back to predivestiture, when voice and data networks existed as separate entities that did not interact with each other. The MIS manager oversaw computers and data networks; the telecommunications manager was responsible for voice networks.

Hierarchical, mainframe-centered computing reflected the demand of the time. It emphasized centralized decision-making and presupposed centralized control of corporate computing and database resources. Similarly, the network was managed from a central location: the mainframe. Hence, network management architecture was centralized with the mainframe at its center.

Point-to-point and multipoint switching were the networking technologies that supported mainframe-centered computing. Network management consisted primarily of monitoring the network's physical devices. Since there was a one-to-one relationship between the user of the network (that is, the operator of the computer terminal) and the provider of the network and computing services (that is, the mainframe), the logical networks that could be constructed were point-to-point, which is a relatively easy topology to manage.

When the advantages of a more distributed approach to decision-making became evident, independent mission-critical business units developed within the corporation. This evolution heralded the second wave in the development of computing.

The enabling technology for distributing computing capabilities to individual departments was packet switching. It permitted construction of point-to-multipoint logical networks that provided access to multiple sources of information in the organization. The TCP/IP internetworking protocol and the X.25 packet-switching protocol, for example, incorporated the packet-switching technology to tie together farflung multiple systems.

In its second phase the network consisted of heterogeneous computer systems provided by multiple vendors. This configuration added several orders of magnitude to the complexity of network management.

Since packet switching provided a way for users of any one computer system to connect to more than one system at any given time, it was possible to construct many logical networks overlaid on top of the physical network. Hence, in addition to the management of the physical networks, the network manager also had to manage logical networks.

With T1 transmission and switching technology, voice and data traffic could be treated in a homogeneous fashion. Wide area networks were constructed with statistical multiplexers, low-rate time-division multiplexers, and PBXs feeding into public and private packet-switching networks or into high-speed T1 multiplexers. Remote Ethernet and token ring LANs were linked via bridges, routers, and brouters.

But each WAN/LAN component provided its own management system, and there was an average of 13 different components in the typical *Fortune* 1,000 company.

Since the logical network could use public and private facilities, network management was again made more complicated. Networks included leased as well as switched facilities to connect the customer premises equipment at the local exchange carrier and interexchange carrier, so end-to-end management of a computing session required management of the public facilities.

The trend to decentralize decision-making extended even further, resulting in desktop computing. It built on LAN technology and tied PCs together through PC LANs: for example, Novell's NetWare, LAN Manager, AppleTalk, and token ring (Netbeui). The phrase "workgroup computing" was coined to explain the virtues of PC LAN technology—and the file, print, and communication services that could now be shared among the PCs on a PC LAN.

Management of the hundreds of PCs linked together on the multitude of PC LANs in the organization emerged as the toughest challenge of the era. The MIS director, whose authority had already been wrested away with the increasing use of departmental computing, saw that authority slip away even more as PC LANs continued to proliferate in the organization.

The use of different network management tools for each networking technology made the task of managing the network extremely difficult. The mainframe-based network management system and the management system for PC LANs were completely misaligned. —*K.S.*

Network Management

1. Transparency. Before client-server computing, database links were manually defined; here, the client requests data from a workgroup server, which gets it from other servers.

Client-server or cooperative processing

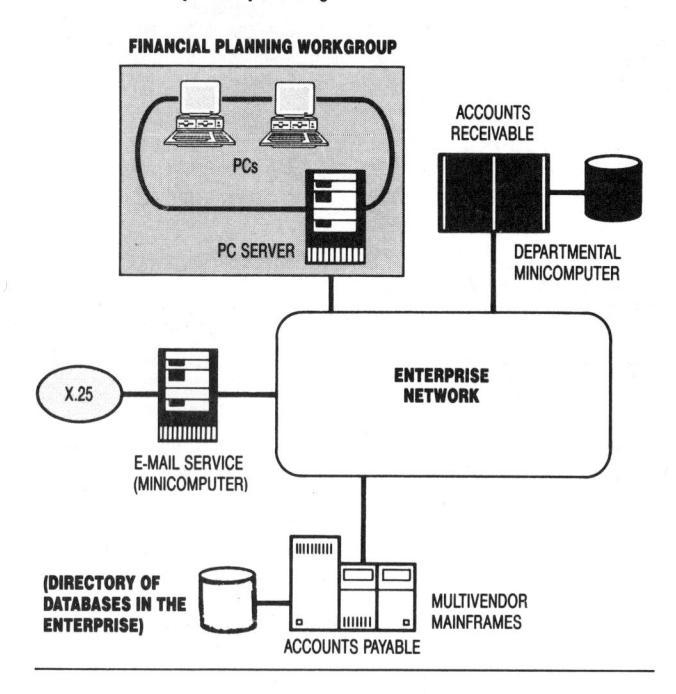

[Diagram: Financial Planning Workgroup with PCs and PC Server connected to Enterprise Network, which connects to Accounts Receivable (Departmental Minicomputer), E-Mail Service (Minicomputer) via X.25, Directory of Databases in the Enterprise, and Accounts Payable (Multivendor Mainframes).]

the computers in the enterprise.

The strength of client-server technology lies in its ability to integrate heterogeneous LAN/WAN and communications protocols (see Fig. 2). Development of enterprisewide electronic mail, directory services, and a variety of other services will be the foundation for developing more sophisticated, enterprisewide office automation, electronic data interchange, electronic funds transfer, and other cooperative data processing applications.

In client-server networks, management has to be distributed—with intelligence built into each system and each island of computing. Only with such cooperation can the network manager have a comprehensive picture of the network. The computer server for a workgroup becomes the access point to the resources and services in the enterprise network for the clients attached to it. Conversely, the server makes the resources within that workgroup available and accessible to the other workgroups in the enterprise.

During the lifetime of the enterpise each workgroup has in the past picked its own computers, operating systems, applications, databases, LAN/WAN technology, communications protocols, and telecommunications devices. The fundamental barrier to development of an integrated network management system is the fact that each resource type has unique characteristics; even in the same class of resources, the resources provided by different vendors have different sets of characteristics.

In lieu of an all-encompassing solution to the problem of enterprisewide management, a few vendors have started to provide short-term solutions to one aspect of it: fault management.

IBM, in its NetView products, has defined an alert Network Management Vector Transport (NMVT) for sending alerts from heterogeneous devices to NetView. AT&T has defined an alert interface to allow other vendors to send alerts to Accumaster, its UNMA-compliant system.

The fault-monitoring capabilities these systems provide for multivendor, heterogeneous networks address one of the fundamental problems of managing the enterprise network. But the solutions do not address the other management functional areas: configuration, performance, security, and accounting management. Alerts are the least common denominator of information provided by the various heterogeneous network resources.

Saved by object orientation

Aiming for a way to reduce the complexity of managing heterogeneous devices, industry experts have developed the object-oriented design paradigm. It helps to manage multivendor devices in a uniform manner, rather than having to customize management for each.

Object orientation builds on the foundation of client-server computing. Its design principles define four basic elements of an object:

- *Object class.* Describes objects as being related to one another and able to inherit attributes and data. A class can be a superclass of another class or an aggregate of other classes.
- *Object instance.* The manifestation of a class.
- *Method.* The behavior or operation supported by classes of objects.
- *Link.* That which connects two classes or instances of classes and defines relationships between them.

Applied to client-server computing, object orientation treats each network resource type as an object class. For instance, an enterprise network would have an object class for modems. The modem object class may consist of a subclass of Hayes modems, a subclass of Codex modems, and a multitude of other vendors' modems. The modem object class has a unique set of characteristics that bind all the modem objects to that class. These characteristics (attributes, methods, and so forth) are defined and encapsulated within the modem object class. Properties specific to a Hayes or Codex modem are defined in their respective subclasses, and properties common to all modems are inherited from a generic modem class.

When a network manager performs a loopback test on any modem in the network, the modem object is referenced by its instance name. Referencing the object also makes available the methods employed to perform the loopback test. Each modem may require a different command to perform the loopback test, but object orientation hides this from the network manager.

By using the object-oriented design paradigm, the network manager can reduce the complexity of network management, since the network now consists of only a handful of generic object classes (modems, time-division multiplexers, statistical multiplexers, T1 multiplexers, PBXs, PC LANs, minicomputers, PCs, and so forth).

Another immense benefit of object orientation that

Network Management

makes it an ideal complement for client-server computing is that the network's complexity is essentially hidden from the user. A compound object or a network resource, such as a corporate database, may consist of many compound departmental database objects. These objects may or may not be on the same server, but the user who accesses a compound object does not have to deal with the process of collecting individual components of the composite data.

The OSI connection

OSI standards committees have recognized the benefits of object-oriented design principles and have started defining a management information base (MIB) that is object-oriented. The MIB will contain information regarding the data attributes for each resource or object in the network.

In order to make the MIB truly open, OSI will act as the registration authority, or clearinghouse, for objects. Once an object is registered with ISO, information related to that object becomes available to other OSI-compliant systems. For such objects as individual layers within the OSI protocol stack, OSI will define a set of public objects, as well as methods and operations to manipulate them. For private objects defined by each vendor, however, the definition of methods and procedures to manipulate the objects is left to the vendor.

To provide access to the objects or services that could be anywhere in the network, OSI has also defined X.500: the standard for directory services. The interactions between the servers for exchanging object-related information and providing access to a variety of services in the network is facilitated by the directory servers.

Attempts made by other standards bodies or vendors (Internet Engineering Task Force for standardizing the MIB for TCP/IP internetworks; IEEE 802.1 for managing link layers; IBM's NetView for managing SNA objects) follow lines similar to the approach adopted by OSI. They lack, however, the object orientation imparted by OSI for the definition and manipulation of the objects.

Since there is no methodology to manipulate all the private objects in a multivendor, heterogeneous network, client-server technology has so far yielded single-vendor solutions. In order to reduce the complexity of client-server computing in a multivendor, heterogeneous network, what is needed is a model that can deal with diverse resources or objects provided by different vendors. Also needed are transparent access to these objects and intelligent interactions between them.

The industry has begun to address the problem of dealing with diverse resources, or objects, attached to the network. The Object Management Group, which was formed earlier this year and is based in Westboro, Mass., has set out to address the need to provide an industrywide architecture for management of objects in general, and distributed objects in particular.

2. Integrating protocols. *The inverted pyramid represents the degree of difficulty in providing the capabilities of each building block of the network, as well as the relative benefits provided by each.*

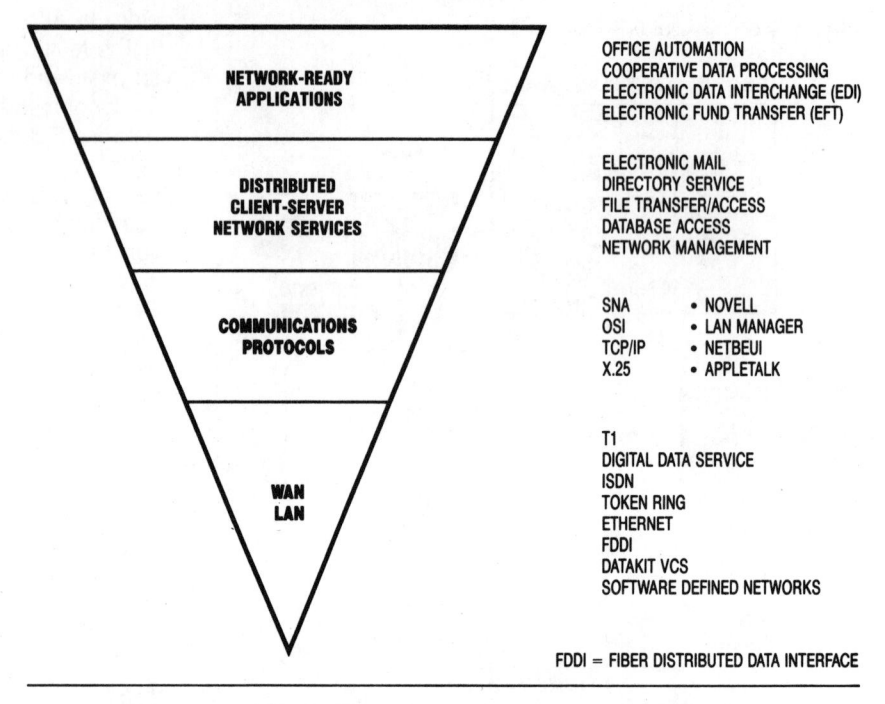

Block	Examples
NETWORK-READY APPLICATIONS	OFFICE AUTOMATION / COOPERATIVE DATA PROCESSING / ELECTRONIC DATA INTERCHANGE (EDI) / ELECTRONIC FUND TRANSFER (EFT)
DISTRIBUTED CLIENT-SERVER NETWORK SERVICES	ELECTRONIC MAIL / DIRECTORY SERVICE / FILE TRANSFER/ACCESS / DATABASE ACCESS / NETWORK MANAGEMENT
COMMUNICATIONS PROTOCOLS	SNA, OSI, TCP/IP, X.25 • NOVELL • LAN MANAGER • NETBEUI • APPLETALK
WAN LAN	T1 / DIGITAL DATA SERVICE / ISDN / TOKEN RING / ETHERNET / FDDI / DATAKIT VCS / SOFTWARE DEFINED NETWORKS

FDDI = FIBER DISTRIBUTED DATA INTERFACE

Distributed object management

With client-server computing, there is no longer a one-to-one relationship between the user and provider of the network services. But this lack of simplicity need not worry the network manager. The solution to managing distributed network computing lies in the client-server and object-oriented technologies themselves.

The four essential components needed to build a distributed object management system are:
- An object model of the network that encompasses the workstations, or clients, and extends across the network to other servers and workstations.
- An object registration and management facility to track the various components of each object.
- Communications methods for interactions between objects across the network.
- Object directories, which function as repositories of information about objects on the network and their relationships to other objects. OSI-defined X.500 standards for directory services would make these directories work in multivendor, heterogeneous networks.

Client-server computing

Applications using the client-server computing model and based on object-oriented design principles are already available. Data General has a blueprint for client-server computing technology called Distributed Applications

Network Management

Architecture; Hewlett-Packard has announced NewWave; IBM recently announced OfficeVision; Digital Equipment Corp. has Network Applications Support; Unisys offers Office Document Architecture; and 3Com has Object Management Platform. They all embody the concepts of object-oriented design and use components of the client-server computing technology.

In existing networks there is a prevalence of certain de facto communications protocols. The strategic MIS systems and networks of the enterprise, for the most part, have a very strong SNA orientation. Mission-critical business units have a variety of different networks: X.25 for access to remote systems through public data networks, and TCP/IP (and more recently OSI) to tie together multivendor, heterogeneous networks. The individual departments have deployed a multitude of PC LANs (Novell, LAN Manager, Netbeui, and so forth).

It is impossible, in a single step, to go from autocracy among the management systems to a completely harmonious and integrated network management system based on principles of client-server computing. An intermediate step to consolidate management systems around de facto and de jure industry standards will be crucial to building an integrated enterprisewide management system.

Major forces in network management

In most enterprises the network control center resembles a retail store for television sets—with a variety of monitors for managing numerous devices, systems, and applications. But some order and consolidation around network management standards are starting to emerge.

■ *IBM's NetView.* Because of the large installed base of SNA networks, IBM's NetView plays a major role in enterprise network management. With its blueprint for integrating non-SNA applications and devices, IBM is also positioning NetView as *the* enterprisewide management system. Numerous vendors have accepted NetView in this role and have provided a way for NetView to manage their devices, systems, and applications. These vendors boast of their products' ability to blend into SNA networks.

While far from a complete solution to the problem of enterprisewide management, NetView provides a level of integration and consolidation that reduces the complexity of enterprise-network management.

■ *OSI.* The International Organization for Standardization (ISO) is providing bridges to future network management possibilities. The ISO aims to let users leverage their investment in various applications found among the heterogeneous islands of computing. So in addition to standardizing the protocols at the bits-on-the-wire level, it has created standards for the exchange of information between applications.

Some of the standards created by the ISO to allow bridging of applications include X.400 for exchanging electronic mail, FTAM (File Transfer, Access, and Management) for exchanging documents, and X.500 for sharing and exchanging directory information.

The ISO's effort to standardize the syntax for exchange of network management information between two heterogeneous systems, or islands of computing, resulted in the definition of CMIS/CMIP (Common Management Information Services/Common Management Information Protocol). The ISO is also in the process of standaridizing the semantics for network mangement information that is exchanged between heterogeneous systems. The ISO has classified the network management information into five distinct categories—configuration, fault, performance, security, and accounting—that mirror the five essential components of managing a network.

What has not been given attention is ISO's attempts to define and treat network and computing resources as objects, and its attempts to standardize them. Object orientation is one of the key enabling technologies that will go a long way toward bridging the islands of computing.

While OSI's standards for physical layer protocols, application layer services, and even certain applications (such as FTAM for file transfer and X.500 for directory services) are in place, standards for network management are not. Because of the way that object orientation treats network resources it can solve the problem of multivendor interoperability. But the lack of OSI standards for the MIB in the short term poses a challenge to both vendors and users.

■ *TCP/IP en route to OSI.* In the absence of a complete set of OSI standards, TCP/IP, the de facto standard for multivendor interoperability, continues to play a significant role in network management. Recognizing that OSI provides a set of standards for interoperability and management that is far more robust than that of TCP/IP, the Internet Engineering Task Force (IEFT, which is the standards body for TCP/IP) has taken a pragmatic stance toward the OSI standards. To facilitate the migration of TCP/IP internet protocols to OSI, IEFT has formulated a set of standards called Netman.

The Netman standards consist of a standard for protocols—CMIP-over-TCP/IP (CMOT)—and a standard MIB

> Netman—a set of standards—aims to east TCP/IP's path toward OSI.

for the objects or network resources to be managed. The CMOT standard was formulated to allow OSI management applications to run on top of the TCP/IP protocol stack unchanged. Because the TCP/IP transport stack could be eventually replaced with the OSI stack without affecting the applications, the users' and vendors' investments in the OSI applications would be protected. While the OSI standards for its MIB are not expected to be available for a couple of years, the MIB for Netman is completely defined and vendors are in the process of implementing it.

TCP/IP standards for interoperability and management are, therefore, ideally positioned to provide a smooth migration to OSI standards. In addition to providing a vehicle for the use of OSI management applications and

Network Management

services, it provides a tool to experiment with the OSI technology without all the OSI standards being available. En route to the deployment of OSI MIB and consolidation of network management systems around the OSI standards, TCP/IP is a perfect facilitator and a catalyst.

■ *AT&T's UNMA.* Neither NetView nor OSI adequately address the management of data communications and telecommunications devices and equipment in the enterprise network. In order to fill this void, AT&T has proposed Unifed Network Management Architecture. Recognizing the virtues of object orientation adopted by OSI for the definition of its MIB and the multivendor interoperability that OSI standards would proffer, AT&T has used OSI standards as the core of UNMA.

With a 60 to 70 percent share of the telecommunications facilities market, the significance of AT&T-provided telecommunications management solutions cannot be overemphasized. Any network management strategy that requires end-to-end control and monitoring has to include the public and private telecommunications management provided by AT&T. Thus, UNMA is a perfect complement to the computer and data network management solutions provided by NetView and OSI.

Step one: Consolidation

What is starting to emerge out of the multiplicity of network management systems is a hierarchy. IBM, with its large installed base of more than 40,000 SNA networks, continues to have a strong influence over the way networks are managed. NetView has emerged as a dominant standard for network management in these networks, with vendors trying to figure out a way to tie into it. Similarly, AT&T continues to be a standard-setter for telecommunications network management. Its blueprint for the management of enterprisewide networks is watched closely by the industry and has been adopted by quite a few vendors. Together, the OSI standards for the protocol (CMIP), the MIB, and the specific management functional areas are a harbinger of solutions to the very interoperability problems that the industry is trying to solve. The picture that is beginning to emerge is that of a two-tiered hierarchy, with IBM's NetView, AT&T's UNMA, and OSI occupying the top tier and the makers of all other components (such as LAN/WAN gear, telecommunications equipment, systems, and applications) providing management systems and solutions that can fit into these three umbrella management systems (see Fig. 3).

NetView, UNMA, and OSI are, in fact, complementary; they are not competing as solutions to the problem of enterprisewide network management. With their heavy investments in the SNA networks, the users are not likely to replace the SNA networks with OSI networks. IBM's NetView, the standard for management of the SNA network, will continue to play a major role in the management of the enterprise network well into the nineties. New networks being put in place are more likely to use the TCP/IP or OSI standards for network management because of the level of multivendor interoperability they can provide. Thus, NetView and OSI standards for managing data communications networks and AT&T's UNMA for managing telecommunication facilities together will provide a more complete solution to the problem of enterprise management.

3. Management troika. *A two-tiered hierarchy is emerging, with IBM's NetView, AT&T's UNMA, and OSI on top of vendors of all other components.*

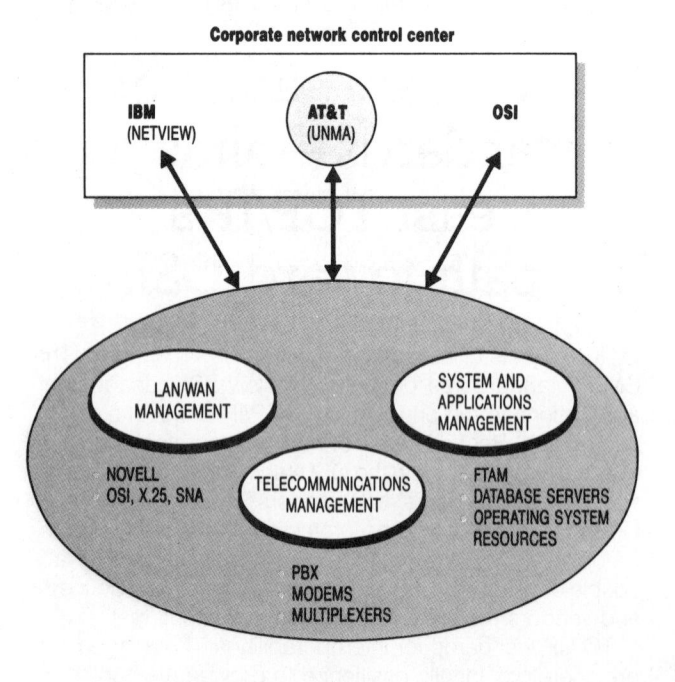

FTAM = FILE TRANSFER, ACCESS, AND MANAGEMENT

Step two: Cooperative management

While the focus during the first three waves of computing was on managing individual devices in the enterprise, in the client-server computing environment the focus has to shift to managing the services and applications in the network.

■ *Common user interface.* Also necessary for managing the networks of the nineties is a common user interface. Adopting industry standards for user interfaces (for example, Open System Foundation/Motif and AT&T's OpenLook) as the enterprisewide standards for user interfaces will facilitate development of a common user environment for the enterprise. Network management, one of the many distributed applications in the enterprise, should adopt the enterprisewide standard for user interfaces as well.

But the user interface must be extendable to provide management for new or modified objects; it must be built using the object-oriented design principles. As new network resources are added to the network, the objects representing those resources will be added to the existing networkwide library of objects. If the presentation services (that is, the user interface services) that are needed to manage the objects are encapsulated within the objects themselves, then the responsibility of representing the differences in object-related information is borne by the objects themselves—not by the user interface.

■ *Distribution of management services.* The ability to distribute network management services will also be important in the coming decade. Cooperative processing between the various computer and communications servers in the enterprise will be the key technology for developing a distributed enterprise management system that serves all the different network management personnel. To provide comprehensive configuration, fault, performance, accounting, and security management services for all the objects in the network, the methods needed to provide those services must be encapsulated within each object.

1. Configuration. By abstracting network resources to a set of specifications that is independent of the implementation of those specifications, distributed object management makes the configuration of networks inherently flexible. When resources or users are added to the network, identity, location, access rights, and other attributes are collected at the nearest sever or directory service. The presence of these resources and users is made known to all the other servers by an exchange of information related to the new resources and users.

2. Fault. One of the biggest challenges facing the network manager today is the multitude of terminals spewing out network failure alert messages. Each device has its own representation of the alert that may or may not resemble the alerts generated by another device of the same type made by another vendor. NetView, with its Alert NMVT, has standardized the format of the alert message to alleviate the problem faced by the network manager. AT&T has similarly defined its own format, which could be used by other vendors. OSI is also defining an alert format.

The problem, however, is that once the basic structure of the alert message is defined, any subsequent changes to its structure would make it inoperable with the existing network management systems. With an object-oriented design approach for the definition of alerts and with an associated object-oriented user interface, the changes made to the structure of the alert would not affect the previous generations of alerts. It would also provide the ability to represent the new alert alongside the old ones.

Additionally, in order to provide expert-system capabilities to correlate alerts and thereby reduce the burden on the network manager, one could make use of the object-oriented technology to build the expert system. At present, as the result of a single physical failure—such as the failure of a T1 line—the network manager is presented with streams of messages originated by each affected device. The expert system, however, could present the network manager with a single alert related to the physical failure.

3. Performance. Treatment of the performance data collected by various network resources as objects allows one to gather and analyze the collective data, which represents the true performance of the network. The computation of the various performance metrics (utilization of links, software components, response time, and so forth) is also facilitated, since each object provides its own method for massaging the data to generate a metric that is universally applicable.

Since performance data is collected by the individual network resources, the process of gathering the data requires the cooperation of multiple resources and servers in the network. Distributed computation of the metrics by the various servers in the network is the key to presenting the performance data to network management personnel—ranging from the LAN administrator to the department network manager to the corporate network planner.

4. Security. Object-oriented security management policies and procedures provide a mechanism that allows objects and groups of objects to be treated consistently. The security access level associated with each object may be kept—as a part of the information about each object—in the directory service. Also, each user-profile object in the directory services may have associated access rights that define the span of access for the user. When making an association between the user object and the service object, the access level required for that operation on a specific object and the access rights assigned to that user are

NetView, UNMA, and OSI are complementary solutions.

matched. This ensures that only the legitimate users are making use of the services.

Moreover, having defined the security access level for one class of objects, the inheritance property of the object-oriented design could ensure a uniform enforcement of security management to all the members of that class and subclasses within the entire network. Similarly, one could assign access rights for a class of users and guarantee that those access rights are enforced uniformly and conistently for all the members of the user group.

5. Accounting. For large corporations where the enterprise network is increasingly being treated as a profit-and-loss entity, it is important to track network resource usage for the individual departments and users. Within distributed, client-server networks, the services used by the clients will be provided through the network; for each transaction several devices, systems, and applications might be involved. Today's accounting systems are far from providing a comprehensive accounting record for the usage of each one of the components. An incremental approach for building accounting systems that can collect and correlate accounting information from the various objects or network resources will facilitate the development of an enterprise-wide accounting system. One solution is a distributed extensible accounting system. It will be able to collect accounting data from multiple databases, which can be provided by the client-server technology and by the communications between the various servers. It will also collect data from the increasing number of heterogeneous databases.

Enterprise architecture
A network management strategy based on client-server and object-oriented design technologies will permit the construction of self-managing domains, which could be workgroups, departments, business units, or even enter-

Network Management

prises. Each domain could be functionally independent of the others in the following ways:
- Maintenance of network configurations.
- Fault collection and analysis, restoral of services.
- Collection and analysis of performance data.
- Enforcement of security mechanisms.
- Generation of accounting policies and procedures.

The construction of these self-managing domains does not, however, preclude cooperation among them; nor does it preclude construction of hierarchical, overlapping, or mutually exclusive domains.

For example, the administrative requirement might call for the services of a PC LAN to be accessible to a particluar workgroup with the additional requirement that the MIS organization be able to generate performance figures for that PC LAN. One could construct a set of security and performance domains that would allow the workgroup to access the services provided by the PC LAN and another set of security and performance domains to allow the MIS organization to collect only the performance data from that PC LAN—but barring access to all the services provided by that PC LAN.

The three key ingredients for successful development of the enterprisewide cooperative network management system for networks of the nineties are:

- Object-oriented user interface to permit the dynamic addition of new and modified objects.
- Network resources built using object-oriented design principles (that is, resources that are self-contained) with their attributes, methods, and behavior encapsulated within the resources themselves.
- OSI-defined Specific Management Functional Areas (that is, configuration, fault, performance, security, and accounting management) built on the client-server, distributed computing model.

One of the most important tools networking and network management personnel need to equip themselves with is knowledge and understanding of the benefits of the two core technologies: client-server computing and object-oriented design. Through effective use and dissemination of their knowledge of these technologies and industry standards, MIS directors and telecommunications managers can better control their enterprise networks. ∎

Kumar Shah is a section manager responsible for developing network management systems for Data General. He has been with the company since 1986. Shah earned an MS in electrical and computer engineering from the University of Massachusetts at Amherst in 1982 and a BS in electrical engineering from the Indian Institute of Technology, Bombay, India, in 1980.

Section 3
Economics

Paul R. Strauss, DATA COMMUNICATIONS

1985 in review

Rapid deregulation spawns conflicts, realignments

Year two of AT&T's divestiture: The FCC allocated users and called Computer Inquiry III; IBM traded SBS for part of MCI; BOCs begged to become gear-makers.

Struggle was the essence of noteworthy events in 1985: the struggle to find the political, scientific, and business conditions needed to survive an age of deregulation. That struggle is by now well understood by information network users. Vendors are only beginning to awaken to a new, nervous era.

There were also events that showed more fundamental tides: the widening use of data communications, the advances of its technology, the solid need for its services. In the depths of the Atlantic, for example, data communications technology was used to help find the sunken luxury liner, R. M. S. *Titanic*.

In Holmdel, N. J., scientists at Bell Laboratories pumped 4 gigabits—62,500 voice telephone calls or about as much information as contained in a 30-volume encyclopedia—through an optical-fiber strand in a single second. And, in Washington, D. C., the Government posted requests for bids on a $4.5 billion private telecommunications network using packet- and circuit-switched data transports, the largest such private network in the world.

Foibles of 2 A. D.
The news had a lunatic, Marx Brothers quality to it. Some examples: IBM disposed of a loss-making subsidiary—and the move was played up as a strategic victory for the computer maker. The Communications Workers of America—which demonstrated in 1985 that it had no power to halt AT&T's 24,000 staff reductions—announced that IBM is now ripe for organizing.

The Justice Department said AT&T had a feature something like protocol conversion in one of its computer programs: a glitch that inexplicably had the effect of changing some letters entered as S-P-R-I-N-T and some entered as M-C-I into the letters and symbol A-T-&-T. During the year, Chairman of the Federal Communications Commission Mark G. Fowler called some of AT&T's long-distance competitors "Chicken Littles"—after he had listened too long to their cries of impending doom.

Meanwhile, signs of real change often got little attention. AT&T, for example, estimated that at the end of 1985 its capacity of 1.03 billion circuit miles was 690 million miles less than that of its combined long-distance competitors.

We're smaller
By saying that it is now, in one sense, relatively small compared with the universe of long-haul carriers, AT&T was looking for a deregulatory break, and before 1985 ended it got several. Another way of looking at the same figures, however, points up the growing clout of data communications users. AT&T's competitors only had 400,000 circuit miles in 1983. With such increased capacity—and with a tidal wave more coming as fiber networks are completed—users will be able to bargain for more attractive rates.

But they haven't been getting them, not yet. Another statistic: By the end of 1985, the Center for Communications Management Information estimates, the average price of a U. S. private line had increased about 40 percent over that of 1983. That price restructuring is touching off trends to private satellite networks and user-owned access facilities that will strengthen throughout 1986 and 1987.

Less substantial, albeit more amusing, were the bickerings of regulators and the squallings of service vendors. In a story bound to get more coverage this year, the Federal Communications Commission (FCC) indicated its growing irritation with the administrator of AT&T's divestiture, Judge Harold H. Greene, by reminding him that a three-judge appeals court, not a lowly

1985 in review

Scene, 73 years later. *Researchers found the sunken liner Titanic by hooking together computers over a four-mile RS-232-C line (its signal boosted and multiplexed together with analog signals). An eight-bit microprocessor on ocean floor turned on cameras and sonar gear. Titanic was a bit more than two miles down, but transmission also went through rolled cable. Next step, fiber links, researchers say.*

Culver Pictures

district court, has jurisdiction over FCC matters.

Judge Greene also was embroiled in disputes with the former Bell operating companies. Throughout the year, he granted scores of deregulation waivers to the BOCs, including, in one case, the right to bypass AT&T. The BOCs, however, seemed not at all satisfied. The most truculent of them, US West, asked that the AT&T-Justice Department settlement be dropped.

Doing that would let AT&T back into the local telephone business, but US West seemed so irritated at Judge Greene that it ignored the logical peril of its request. The other BOCs merely asked for a modification of the Modified Final Judgment.

Judge Greene responded by saying that, with all the new business opportunities he had been granting, the BOCs would be satisfied—after all, he wondered, wasn't their main responsibility decent local telephone service? The chairman of US West retorted with what seemed like splenetic rage. Judge Greene, he said, had no right to consider the quality of local service.

Late in the year, the BOCs settled on another tack, backing legislation that would allow them to manufacture telecommunications equipment. That AT&T is losing money making telecommunications equipment did not seem important to the seven siblings, but they were perhaps irritated when the Republican congressmen who sponsored their bill tacked on the requirement that the gear must be made in the United States. The reason was a growing data communications trade deficit with Japan (see figure). During 1985, AT&T stopped making home telephones in the United States and started manufacturing them in Singapore.

Deregulation decisions

Among the important deregulatory moves were these: In March, the FCC allowed BOCs to launch packet services and to use them to do X.25-to-asynchronous protocol conversion. In June, New Jersey Bell was the first BOC to submit a plan to do just that. But in October, the FCC showed that deregulation often involves more regulation. It rejected that BOC's submission, saying that such a plan would have let local telephone users subsidize data communications.

Throughout the year, a trend toward local deregulation emerged, following the decision of the Massachusetts Public Utility Commission (PUC) in late 1984 to allow intra-LATA rate competition starting in 1987. Oregon, for example, scrapped all regulation of toll service. But other states were not as yielding. So the FCC had to force deregulation on some. In August, the FCC overruled the state public utility commission of Nebraska and allowed a local cable company to offer a data communications service.

The state PUC had barred Cox Cable Communications from offering a high-speed digital service because the service competed with the local telephone company. Precisely, the FCC responded, and we're trying to encourage that competition. The ruling, however, was troubling to states-rights advocates in the Reagan Administration since the cable service operated entirely within the state of Nebraska.

On international matters, the FCC was equally sweeping. Backing a policy decision by the Reagan Administration late last year, the commission ruled that privately owned international satellite carriers are in the public interest. They may not, however, offer links to any public network. An FCC press release said the move would "bring to the world some of the dynamism that characterizes the U. S. domestic data processing telecommunications sector."

Long-distance BOCs

Perhaps as significant a decision was Judge Greene's July ruling that a BOC may build a massive private network—even when that network bypasses AT&T's public long-distance services. Though the breakup decree forbids BOCs from providing interexchange routing, Greene said, "it does not prohibit [BOCs] from

1985 in review

Surface amicability. Last year, feuds broke out between AT&T's deregulation judge Harold H. Greene (left) and the FCC and former Bell subsidiaries. Greene's main 1985 ruling: A BOC may construct a user's private long-distance network that bypasses AT&T.

Less amused. Irritated by steady complaints from AT&T's long-distance competitors, FCC Chairman Mark G. Fowler (right) called them "chicken littles." The FCC's major moves in 1985: letting AT&T sell gear without a separate subsidiary and launching Computer Inquiry III.

providing switching services to customers as end-users for originating and terminating traffic among dedicated interexchange circuits as part of a Centrex or PBX offering." AT&T had asked Greene to block Pacific Northwest Bell from building such a network for the Oregon State Government.

While Judge Greene was offering limited bypass privileges to the BOCs, the FCC was piling them on AT&T. In a move that aroused considerable congressional opposition, the commission approved the Megacom, Megacom 800, and AT&T Software Defined Network tariff proposals, all of which encourage bypass. "Stacking the deck in favor of the largest player," stormed an outraged John D. Dingell, chairman of the House Energy and Commerce Committee.

You've been drafted!

And in another anti-AT&T move, the commission undertook one of the most novel intrusions into U. S. family life since President Gerald Ford advocated flu shots. The reason: By mid-year it had become clear that Americans were bored to tears with the pick-your-long-distance carrier procedure necessitated by the Bell breakup.

Even more fervently than during presidential elections, Americans exercised their right *not* to vote. That wasn't all right with the FCC for several reasons. BOCs, still not recognizing that their former parent is now their competitor, were feeding all unvoting telephone customers to AT&T. Naturally, the other long-distance carriers screamed "unfair."

The FCC attempted to make long-distance competition more fair by forcing Americans to consider the various carriers. But that ruling touched off an election campaign that *Forbes* noted will be more expensive than most presidential contests. AT&T alone spent $200 million in vote-for-us advertising in 1985, and it probably will spend more this year.

The commission forbade the BOCs to assign all the nonvoters to AT&T. Instead, it ruled that BOCs must act like regional draft boards. The names of all telephone users failing to answer the equal-access call must be thrown into a hopper and assigned by lot, in proportion to the other voters in the district.

That ruling offered AT&T's competitors some solice, but it also allowed accidental nonvoters—some, perhaps, who had been out of the country during the equal-access hullabaloo—to be assigned to a vendor that might not even serve one of the places that those voters may frequently call.

In an era where government is supposed to affect the average citizen as little as possible, the sweeping impact of the FCC's move was astonishing. Surprisingly, this unusual bureaucratic maneuver aroused little comment. Similarly, home telephone users were hit by a $1 access fee in June. There was no outcry.

Coming soon: Computer III

There are, however, signs that a bitter AT&T critic may be rising in Congress. Representative Timothy Wirth, the Colorado Democrat who in 1982 sponsored H. R. 5158, a tough divestiture proposal introduced prior to the Modified Final Judgment, has a chance for a seat in the U. S. Senate. Wirth won Senator Gary Hart's backing to run for Hart's seat if Hart quits the Senate to launch a full-time campaign for the presidency.

But legislators come and go; data communications inquiries, by contrast, follow each other like successful movie treatments. The *Rocky* series springs to mind. And like that series, Computer Inquiry III was the 1985 deregulatory proposal offering the greatest chances for drama and sweeping change.

Some idea whose ox might be gored by Computer Inquiry III emerged last year. In September, the FCC allowed AT&T to sell customer premises equipment without using a separate subsidiary (AT&T said that would save it $1 billion a year). However, the commission did insist that AT&T file an accounting plan showing that telephone calls were not subsidizing equipment sales. About a month later, AT&T filed that plan. And, one of the first companies to say that plan really didn't do what it was supposed to do: IBM.

During 1985, IBM asked for its own deregulatory relief: removal of two old antitrust consent decrees, one

1985 in review

Battling for BOCs. Maybe they are just telephone companies, but John D. Dingell, chairman of the House Energy and Commerce Committee, says the BOCs should be allowed to make telephone equipment too. Dingell has been feuding with the FCC, saying the commission has moved too fast on deregulation.

Wide World Photos

of which dates to the 1930s. Although there were dated portions of each decree, some IBM competitors worried that scrapping them both might allow IBM to again sell software solely as part of a hardware transaction. If users could not obtain, say, IBM's Systems Network Architecture (SNA), the plug-compatible manufacturers worried, would they stop buying the lookalikes? Some of AT&T's microcomputers are now IBM PC-compatible—whether it has opposed IBM's request has not yet been made public.

Bad year for Big Blue

With these two exceptions, the giant of Armonk kept above regulatory bickerings. But IBM did not pass through 1985 without some embarrassment. In March, IBM discontinued production of the PC*jr*, though it adamantly denied it was dropping the product.

In February, the company introduced its high-end mainframe family, code-named Sierra, now dubbed the 3090 family. As usual, the new IBM mainframe was praised by most pundits, but Big Blue failed to predict that this year's wait-for-the-latest-mainframe syndrome would be far worse than usual. So many users waited for the 3090 series, which began shipment in October, that IBM's earnings for the first three quarters were down by 12.1 percent.

They would, perhaps, have declined by less if IBM had been able to reduce the losses of Satellite Business Systems, which at the beginning of 1985 was owned 60 percent by IBM and 40 percent by Aetna Life Insurance.

To be fair, the only rational way IBM could have reduced SBS's losses would have been to not launch that joint venture in the first place. SBS, which lost around $100 million a year, became an IBM subsidiary when the Communications Satellite Co. dropped its 33 percent stake in 1984. But, worse than the losses, was the deepening need for investment. SBS's managers were calling for billion-dollar fiber optic networks.

In late May, IBM announced it was buying all of SBS and then trading the company, except for three of its satellites, for 16 percent ownership in AT&T's most aggressive long-distance competitor: MCI Communications.

In October, IBM introduced its long-delayed token-ring local area network. It aroused more questions than it answered. Since the company's first presentation stressed microcomputers alone, observers wondered whether IBM would scrap the similar PC Network—the hardware of which is made by Sytek Corp. Then, some other questions: Why didn't Big Blue show links to its line of minicomputers; and why was no mainframe link displayed? Some analysts think those products are forthcoming, and that regular announcements are planned for this year. Others disagree.

Perhaps the only technical advance the finally-announced token ring had over the versions IBM had displayed in 1983 and earlier was that it was capable of using two-strand telephone wiring for short distances. Starlan, AT&T's local area network, has the ability to use local wiring—one advantage that it had over the local area networks of many competitors.

Whither the wire, and when?

Unfortunately, in the miasmic year 1985 even the benefits of using local wiring became uncertain. Many data communications users think that the wire in their walls is like gold on their property. That is only partially true. The local telephone company owns most wiring installed in commercial premises, but the FCC has ruled that the telephone companies must grant the right to use that wiring, provided that the use causes no problems to the network.

The question is, who will own that wiring in the future? In April, the FCC said it was considering making the local telephone companies transfer that ownership either to the telephone customer on the property or to the building owner—the FCC hasn't yet decided which. When? Why just as soon as the wiring has been amortized.

But the amortization period's length is in doubt. It could be up to nine years from now under the FCC's accounting rules, but it might be shorter or longer if states win their anti-FCC lawsuit that the U. S. Supreme

1985 in review

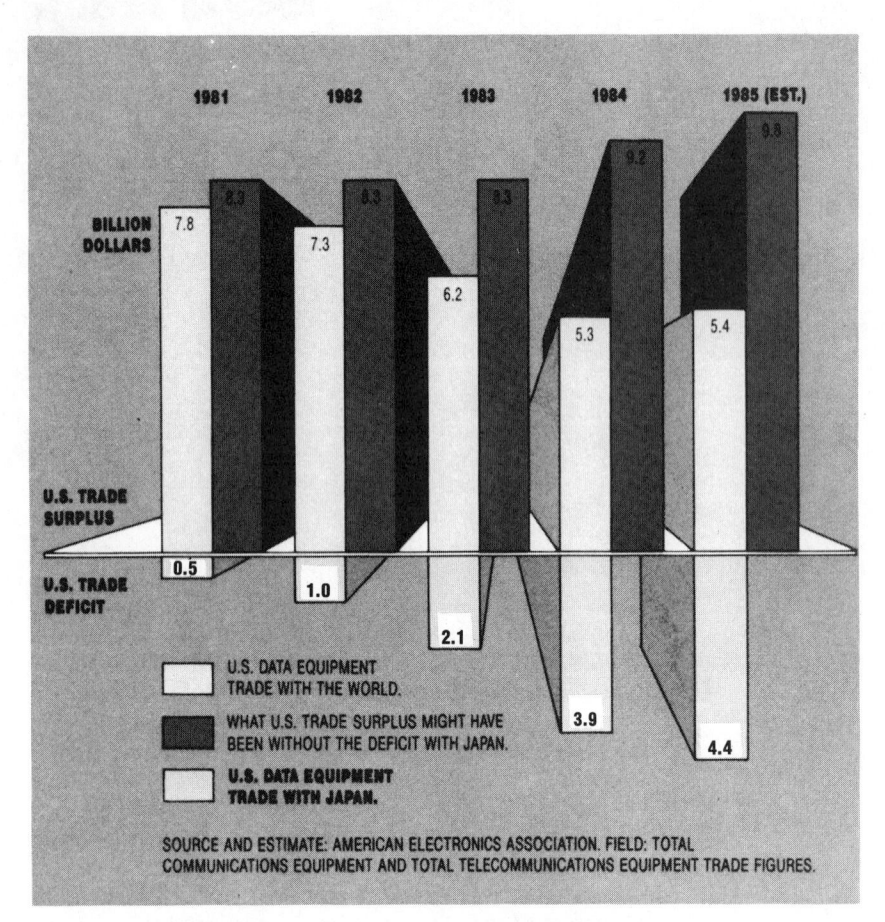

Shrinking data surplus. For the last five years, the United States has exported more computers and telecommunications equipment than it has imported, but that trade surplus has declined steadily while imports from Japan increased by 780 percent. So legislators are talking about putting trade restraints on Japanese data communications gear.

[Chart: U.S. Trade Surplus/Deficit 1981–1985 (Billion Dollars)
- 1981: Surplus 7.8 / 8.3; Deficit 0.5
- 1982: Surplus 7.3 / 8.3; Deficit 1.0
- 1983: Surplus 6.2 / 8.3; Deficit 2.1
- 1984: Surplus 5.3 / 9.2; Deficit 3.9
- 1985 (EST.): Surplus 5.4 / 9.8; Deficit 4.4
Legend: U.S. Data Equipment Trade with the World; What U.S. Trade Surplus Might Have Been Without the Deficit with Japan; U.S. Data Equipment Trade with Japan.
Source and estimate: American Electronics Association. Field: Total Communications Equipment and Total Telecommunications Equipment Trade Figures.]

Court agreed to hear in July. The states claim that the FCC cannot specify the depreciation methods that a telephone company uses. That, they say, should be a state PUC's privilege.

As to the wiring matter, the big legal question is: Can ownership of the wiring be passed to its users rather than to the owner of the building? Some of the advantages of using local wiring would greatly dim if a landlord could make you pay for using it. But that is one of the possibilities.

Troubles of a turbulent year
Such uncertainties were part of the nightmares of 1985. Some other notable disasters:
- DSC Communications Corp. (the former Digital Switch) announced it had revised its earnings for 1984 and the first half of 1985. In each case the revision was downward—from a $32 million profit to a $5 million profit in 1984 and from a $5 million loss to a $14 million loss in the first half of 1985. The reason was that a DSC customer—it turned out to be GTE Sprint, itself troubled by massive losses caused by equal-access competition—had refused to accept a DSC tandem switch worth $72 million.
- Hewlett-Packard, Data General, Racal-Vadic, Paradyne, Apple Computer, Allnet, GTE Sprint, Digital Research, McDonnell Douglas's Communications Industry Systems and Computer Systems divisions, ITT, Wang Laboratories, AT&T, and Sweden's Ericsson Telephone Co. had layoffs or involuntary—usually unpaid—holidays. Mainframe-maker Control Data lost $239 million in the 12 months ending in September.
- In September, the U. S. Postal Service shut down its electronic-mail operation, known familiarly as ECOM. The Post Office has never disclosed how much ECOM lost, but some said it lost as much as $1 each message, and the Post Office said ECOM transmitted as many as a million messages a week. Over the last three years, ECOM probably lost more than $100 million. Exxon Corp. probably lost much more on its Exxon Office Systems unit, which was closed in August.
- An audit team of BOCs chided the BOCs' $872 million-a-year research arm, Bell Communications Research, for spending too much money on such things as testing product quality. The manufacturers really should do this testing, said the audit team, explaining that Bellcore had spent $110 million too much.
- And the Justice Department charged that an AT&T computer program misallocated 87,000 business customers who had asked to be assigned to other long-distance carriers. The cause was error, perhaps a software error, said the department, but AT&T agreed to pay a $1 million-a-day fine if the episode reoccurs (so far, apparently, it hasn't).

Standards used strategically
As always, decisions on standards set the groundwork for new products. Significant in 1985 was the urgency with which companies or nations tried to bend standards or to develop standards that served them best. For instance, the European Computer Manufacturers Association (ECMA) has chosen IBM's peer-to-peer com-

1985 in review

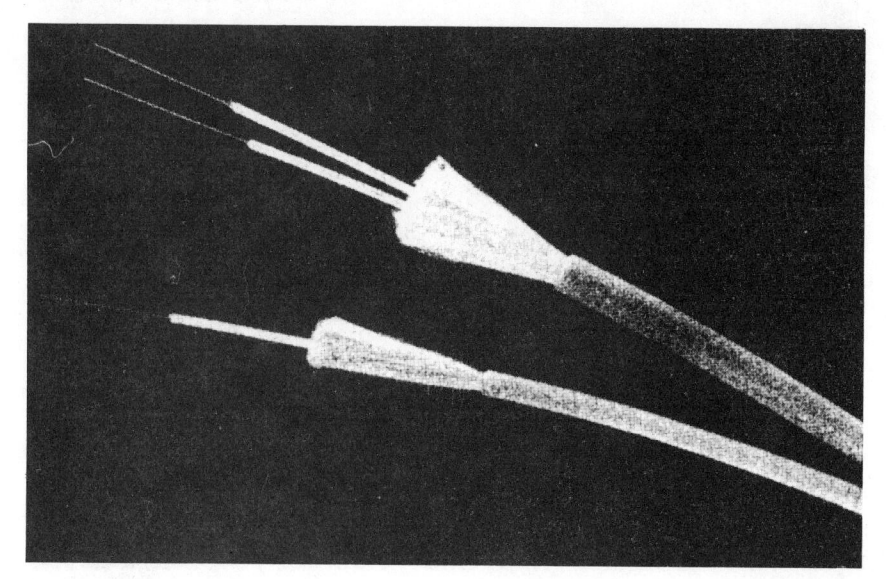

A dozen gigabits. In one major technical breakthrough, Bell Laboratories managed to pour four Gbit/s through a fiber line like one of these. Long-distance telephone companies have been laying thousands of miles of fiber cables, each usually carrying 99 fibers. But only two or three fibers are normally used in each link.

puter-linking interface Logical Unit 6.2 as a basis for the transaction-processing protocol for the Open Systems Interconnection (OSI) model.

Big Blue often has had its products become de facto standards, of course. But this isn't quite the same thing, and ECMA's moving so fast to "universalize" LU 6.2 may set back some of the Armonk giant's plans to sell more computers outside the United States. As part of its European antitrust settlement of 1984, IBM promised to publish LU 6.2 and has done so. It has also told reporters that it intends to make LU 6.2 "open"—in the sense of encouraging links. Still, IBM owns LU 6.2 and might sue a manufacturer that uses it without IBM's permission.

However, the year saw surprisingly rapid agreement on implementing OSI. IBM as well as DEC, Hewlett-Packard, Honeywell, and others have announced commitments to integrate products with the OSI model. IBM has even offered free use of its European Networking Center in Heidelberg, West Germany, to non-commercial researchers who want to test high-level OSI protocols.

OSI and MAP

Of late, IBM has been supporting outsiders' standards quite frequently. Its token ring fits within the IEEE 802.5 standard. It supports General Motors' Manufacturing Automation Protocol (MAP). For that matter, IBM's local area network software, known as Netbios, has become a de facto standard for LAN software.

One significant move that IBM failed to make on standards: IBM, if it had wanted, could have fought to have its wideband 2-Mbit/s LAN, known as PC Network, adopted as a U. S. standard. But Big Blue didn't fight, and the 802.3 committee voted PC Network down. Since PC Network and the token ring serve the same purpose (and compete for the same users), so far, IBM can now back out of the PC Network—if it wishes to—giving the reason that it intends to support universal standards. It is still unclear what IBM really wants: two microcomputer LANs or one.

Less of a political move but vastly more useful is the International Organization for Standardization's proposal for global voice and data address formats, a kind of worldwide zip code for electronic links. The ISO is proposing a 14-digit standard for public data networks, an 8-digit standard for Telex networks; a 12-digit format for public telephone networks; a 15-digit format for the future Integrated Services Digital Network; and some others.

ISDN, however, is having its own standards problems. The race to be first with an ISDN product—such as Siemens' ISDN chip set, announced in October—and the pressure to announce tests of ISDN public switching are making manufacturers select incompatible routes toward an integrated network. Some experts say that the incompatible products already produced may force revisions in the ISDN protocol in the near future. Another deadlock is over the question of whether time-division multiplexing or echo cancellation will be used in the basic ISDN interface.

Playing deregulation catch-up

When it comes to their standards on how much to regulate, and when, governments are usually far less integrated than are data equipment vendors. But one concept does seem to be catching on around the world: deregulation. In recent months, West Germany's Council of Economic Advisors called for a complete restructuring of the Bundespost. A government commission in Holland suggested smashing the Dutch PTT (Postal, Telegraph, and Telephone agency) into three units. In Japan, the Nippon Telegraph and Telephone Corp. ceased being a monopoly and became a publicly owned firm on April 1.

Norway's PTT relinquished its monopoly on most modems, and Italy's telecommunications holding company STET (Societa Finanziaria Telefonica) announced it would sell a large hunk of its stock in Italy's telephone company SIP (Societa per l'Esercizio Telefonica). Other international developments include:

■ Six major European computer makers agreed to support Unix—and probably AT&T's preferred version of Unix, System V—in an attempt to slow IBM's pene-

1985 in review

The sinking of SBS. When two companies announce an $830 million alliance, and the chairman of one of them isn't present, there's often a reason. MCI Chairman William C. McGowan and IBM Vice Chairman John Rizzo (right) announced the swap of Satellite Business Systems for part of MCI. Deal was great for MCI, embarrassing for IBM.

tration of their markets. IBM's response: It is considering making and selling central-office switches in Europe.

- AT&T and British Telecom announced they would sell rooftop-to-rooftop private satellite links using (at least so far) Intelsat's transatlantic satellites.
- Britain moved more rapidly toward a national digital network with British Telecom planning to install digital switches capable of handling 5 million lines by the end of next year.

Mergers and strategic alliances

Although the IBM/SBS/MCI transaction won the lion's share of attention in 1985, the year was notable for many mergers and joint ventures. Most will be discussed in IndustryWatch's summary of the year's business developments (see "Year of the droop, but hardly demise," Industry Watch). The ones of news note include: Ameritech buying software house Applied Data Research for $215 million; Automated Data Processing buying stock quotation service Bunker Ramo for an undisclosed price; and Pacific Telesis's $432 million purchase of the radio-paging firm Communications Industries.

Then there were: 3Com's merger with microcomputer maker Convergent Technologies one month after IBM introduced its token ring LAN; long-distance vendor Lexitel's merger with long-distance vendor Allnet; 3M's sale of its LAN operations to factory-automation specialist Allen-Bradley, which had just been purchased by defense contractor Rockwell International; and British Telecom's near-certain control of Canadian PBX vendor Mitel (in early December, British authorities had approved the purchase, and the two firms were awaiting word from Canadian regulators).

Important joint ventures and marketing alliances included these: AT&T and Electronic Data Systems (General Motors' new subsidiary) formed a joint venture to sell private data networks; Martin Marietta bought 25 percent of Equatorial Communications, and the two firms will together sell satellite networks.

Among the new products: IBM, DEC, Burroughs, Data General, Amdahl, Cray, and other firms announced top-of-the-line processors. AT&T announced its first fiber-optic tariff, called Accunet T45, a 44.73-Mbit/s long-distance service priced at the equivalent of five to eight T1 lines, depending on distance. A short-haul version, called Metrobus, is in the wings.

The virtues of "virtual"

AT&T and MCI introduced "virtual networks," products that replace or supplement private circuit-switched networks by allocating bandwidth on demand rather than by leasing lines. Some outsiders suspected the carriers' rapid increase in leased lines charges were meant to encourage users to move to the virtual networks, where there was a chance of vendors "locking in" users. The reason users might be tied to a single vendor, some analysts suggested, was that much of the information about the networks—its peak periods of utilization, for example—would become the property of the vendor.

However, virtual networks solve many finger-pointing headaches. For that matter, so do network management services. AT&T, Pacific Telesis, and US West have also been introducing these services.

Other notable new products: Dayna Communications' Apple Macintosh-to-IBM PC connector, Macharlie; Proteon Inc.'s 80-Mbit/s token ring LAN; Alliant Computer Systems' superminicomputer using parallel processing techniques; and Channelnet Corp.'s fiber optic extension of an IBM mainframe's byte channel, said to allow transmission at full-channel speeds for up to about 3.1 miles.

What kind of a year was 1985? A year of profit declines and stiffening competition for vendors, tighter budgets for users, and increasing bypass—though not yet on a scale to affect the BOCs' robust earnings. Oddly, since a digital world seems just on the horizon, it was a year of advances in high-speed modems. It was a year of retrenchment, consolidation, and struggle. It was an interesting year. But then, there are two meanings of "interesting." "May you live in interesting times" is no blessing: It is an old Chinese curse. ∎

Edwin E. Mier, DATA COMMUNICATIONS

Comparing the long-distance carriers

Study shows wide differences. Which offers the best signal quality? The quickest connect time? The best bit error rate? Equal access links compared.

The telephone rates of most long-distance carriers are changing so rapidly, and substantially, that it's now nearly impossible for many customers to discern which carrier offers the best prices. With volume discounts, wide disparities between interstate and intrastate rates, distance-based differentials, and incomprehensible time-of-day price permutations, any attempt to compare the carriers serving a particular area is probably not worth the effort—at least not if cost alone is to be the determining factor.

So, in the absence of readily discernible price differences, on what should users base the selection of a long-distance carrier? Why not quality? Do the carriers vary at all in their ability to set up a good connection the first time the caller dials? In the time they take to set up the call? In the number of redials required to get through? How about the quality of the signal that the carrier delivers? And, for those who care about the carriers' ability to transmit data accurately, are some carriers better than others?

In all of these areas, there are, indeed, differences. In some locations, and depending on the measurement criteria, the disparities are major; in others, the differences are narrower, though still noticeable and measurable.

This article presents the results of a recent major study comparing the dial-up quality of the leading long-distance carriers. Measurements were made over hundreds of dial-up connections—all placed through "equal access" central offices—and over very different traffic routes. Among the more significant findings:

■ Despite the connotation of "equal access," AT&T almost always set up calls more quickly than any other carrier— almost twice as quickly as some of its competitors.
■ On average, AT&T put more of its calls through on the first try and required the fewest redials when the first call attempt failed. But MCI had fewer calls blocked due to network "busy" conditions than did AT&T, while Sprint, of the half-dozen carriers tested, had the most.
■ The carrier offering the best bit and block error rate, on the average for all locations, was MCI. AT&T placed third overall and, in fact, offered the worst bit and block error rate over the cross-country route.
■ Sprint offered the strongest receive signal, but its signal was also among the "dirtiest," as measured by background and impulse noise.
■ Despite more than 100 call attempts over several days, it was impossible to set up a modem connection via Allnet over the Northeast-corridor route.
■ The single biggest problem affecting data transmission appeared to be phase jitter and phase hits—at least for the phase-encoded 212A-type modems that were used.
■ The carriers' performance and relative ranking in certain categories, such as signal-loss distortion (loss by frequency), was more or less constant over the different routes, while in other categories, such as bit error rate, performance varied considerably by carrier from one location to another.
■ Of the different routes tested, all carriers offered their best performance—both analog and digital—over the Northeast corridor.
■ When compared with the results of cross-country testing done in 1984, this study indicates that, in several categories, the relative performance of Allnet and Sprint has dropped decidedly, while that of MCI and Western Union showed marked improvement.

Test conditions
DATA COMMUNICATIONS, in cooperation with the Center for Communications Management Information (CCMI, a sister organization within McGraw-Hill Inc. that provides numerous information services covering carriers, their services, and rates) collaborated earlier this year and decided to jointly undertake the project of comparing carriers. In the

early planning stages, several objectives were defined:
- The test would sample the leading long-distance carriers that have constructed and now use their own network transmission facilities, at least to some degree. Specifically excluded would be the "resale" carriers, which offer long-distance service, often in limited or specific geographical areas, and which carry long-distance traffic primarily via "resold" AT&T Wide Area Telecommunications Service (WATS) facilities. The thinking was that, testing resale carriers (even if some were found that served all of the different traffic routes targeted for examination) would really just be a test of AT&T's WATS switching and transmission facilities.
- The same carriers would be tested over several different traffic routes, and variations of quality, by route and by carrier, would be determined.
- For the results of this study to address the current and future state of communications in the United States, all connections tested would be dialed up through local telephone central offices that had already been converted for "equal access."

Route selection

For logistical simplicity, and to minimize the number of test variables, the decision was made to set up a base of operations for all testing at the northern New Jersey headquarters of CCMI. A conventional dial-up business line was ordered and installed by New Jersey Bell, directly linking the Ramsey, N. J., location (in the North Jersey Local Access and Transport Area, or LATA, No. 224) with the serving New Jersey Bell central office. Local digital testing over this circuit showed it to offer reasonably "clean" and virtually error-free performance.

Every connection tested was placed into CCMI over this circuit. Therefore, for every carrier tested and for each dial-up connection, the same local loop at the destination end was always used.

It was decided that, for maximum diversity, testing would be performed over three very different traffic routes:

1. Cross-country, which offered the longest distance possible within the continental United States. A direct-dial business line into the downtown San Francisco editorial offices of DATA COMMUNICATIONS (San Francisco LATA No. 722) was acquired. The serving equal-access central office of Pacific Bell (equipped with a No. 4 ESS) was about a mile away. All cross-country test calls were placed over this access line, to the same destination telephone number at CCMI in Ramsey.

2. The Northeast corridor, which runs from New York to Washington, D. C., reportedly the route having the heaviest telephone traffic in the United States. A conventional residential dial line for the testing was ordered and installed in Gaithersburg, Md. (Washington, D. C., LATA No. 236) by C&P (Chesapeake & Potomac) Telephone. The serving equal-access central office was about a mile and a half away. As with the cross-country testing, all test calls from this mixed light industrial/residential suburb of Washington were placed over the same access line.

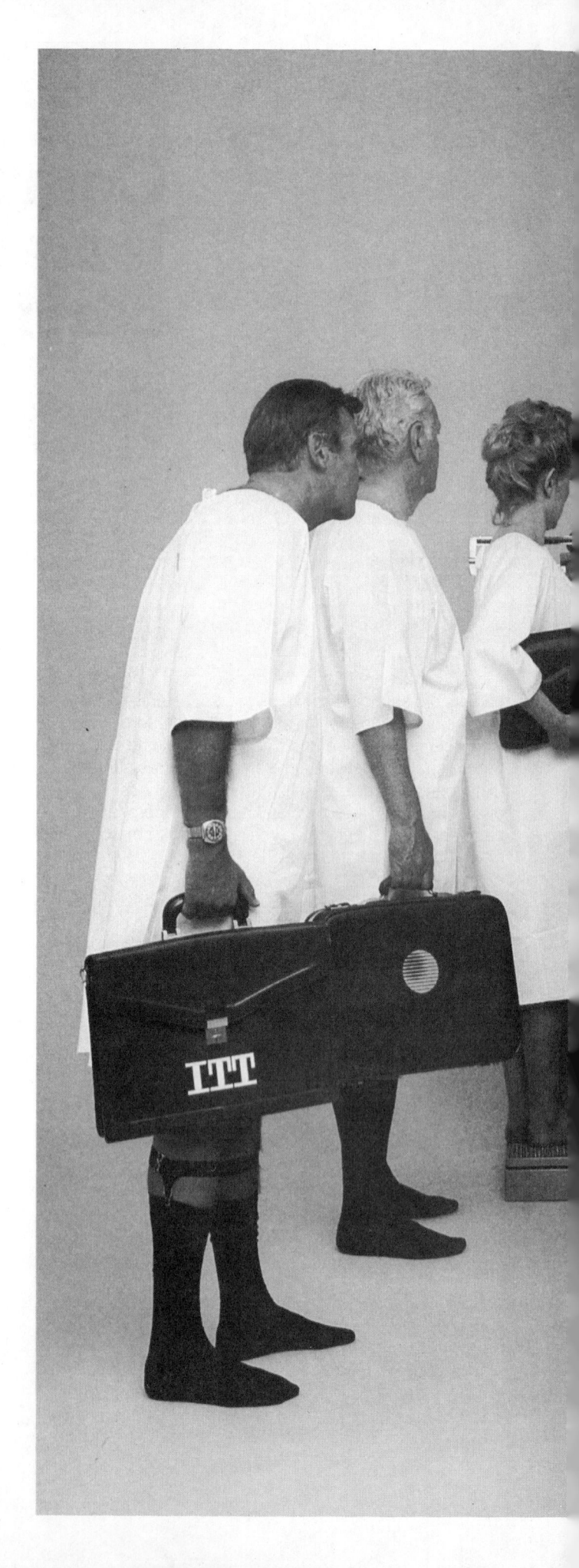

3. Intrastate. To see the variation in quality for a relatively short-distance route, it was decided that tests would also be conducted within a state—in this case, New Jersey. However, to use the facilities of long-distance (or, more appropriately, interexchange) carriers, LATA boundaries have to be crossed. A dial-up access line was ordered and installed in a residence in Mercerville, N. J. (near Trenton in the Delaware Valley LATA No. 222). The equal-access central office (New Jersey Bell) serving this primarily residential neighborhood was about two miles away. As with the other test sites, all connections were placed over this same access line to the northern New Jersey location of CCMI, about 50 miles north in an adjoining LATA.

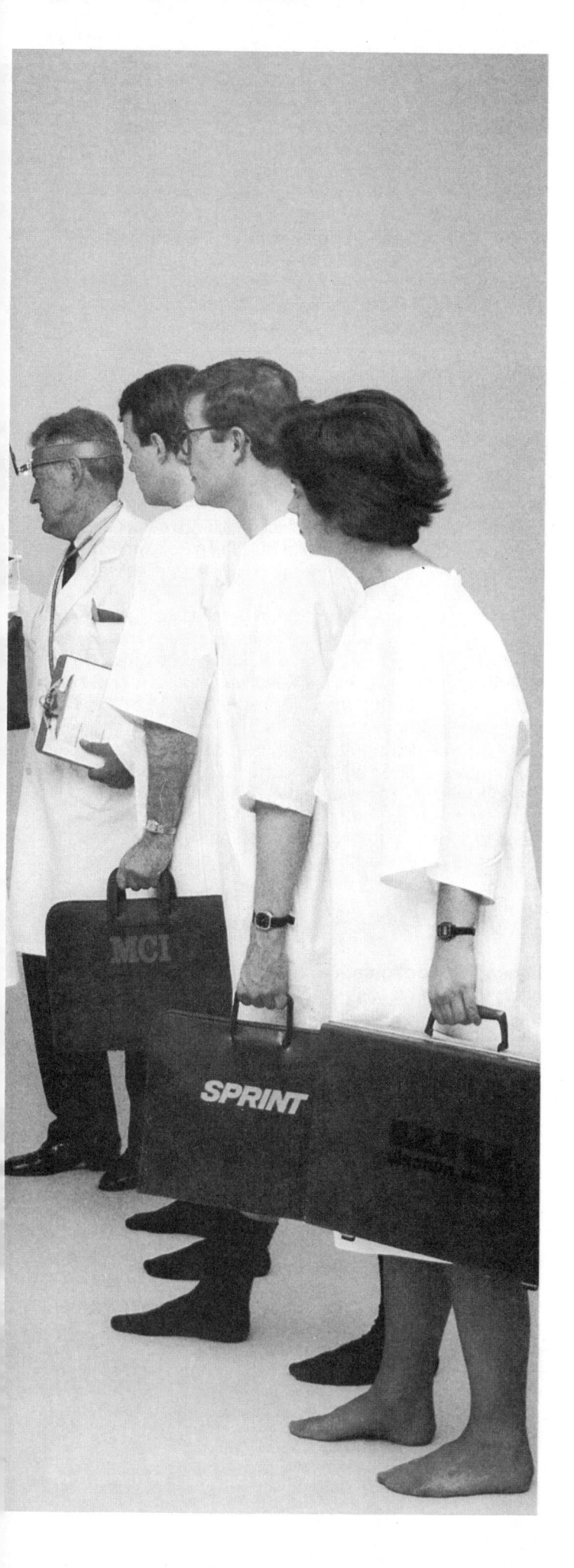

Carrier selection
After settling on the calling locations (all served by equal-access offices), each local Bell operating company, or BOC (Pacific Bell, C&P Telephone, and New Jersey Bell), was contacted in March and asked to supply a list of all long-distance carriers available in those locations. There were about a dozen on the list for each site.

In each location, two or three carriers known to be "resale" carriers were excluded, leaving about nine or ten contenders. (Interestingly, the BOCs regard the carriers differently. New Jersey Bell, for example, clearly distinguishes the carriers as either resale or bona fide carriers, while Pacific Bell refused to make a distinction.)

Because the intent was to compare variations in carrier quality, as well as variations by route, a few carriers were eliminated because they did not serve all three locations. These included RCI, the long-distance affiliate of Rochester Telephone, which served our New Jersey test site but not those in Washington, D. C., or San Francisco. From a remaining list of eight bona fide carriers, all serving the three remote test locations, two more were eliminated: SBS and US Telecom. As of March 1, the acquisition of SBS by MCI had been consummated, and customers could no longer sign up for service from SBS. (We tried but were referred by SBS's salespeople in each location to MCI.)

For different reasons, US Telecom wasn't tested. First, the merger of US Telecom with GTE Sprint was pending regulatory approval at the time, and it seemed likely to be completed by the publication date of this article. (The merger was finalized last month, and the two carriers are now consolidated as US Sprint.) Also, US Telecom refused to allow us to place calls using "10XXX" dialing, which is available only in equal-access locations and which was essential for the test procedure. (GTE Sprint, however, which had the majority of the network facilities, as well as the customer base, of the now-consolidated US Sprint, was one of the carriers tested.)

The six carriers, collectively representing an estimated 98 percent of the U. S. long-distance telephone market, tested were:

1. Allnet (which itself was merged last year with another carrier, Lexitel, and is now the long-distance subsidiary of ALC Communications).
2. AT&T.
3. ITT (marketed in some areas as U. S. Transmission Services, or USTS).
4. MCI Communications.
5. GTE Sprint (now consolidated with US Telecom as US Sprint).
6. Western Union Long Distance (W.U.).

It should also be noted that even these carriers make use of AT&T WATS facilities to some degree, but usually only when their own network capacity is completely filled. We had no way of knowing which of our test calls were being carried by WATS, but neither would a customer subscribing to one of these carriers.

We were therefore willing to accept the fact that the measurements taken, and the qualitative profiles presented here, probably reflect some calls carried by WATS. Of course, customers subscribing to one of these carriers would be equally likely to have some of their calls carried via WATS. Subsequently, the use of WATS, to whatever degree, can reasonably be viewed as a component part and wholly representative of the service these carriers offer their subscribers in general.

Equal access?

As mentioned, a primary objective of the study was to test the carriers exclusively via equal-access central offices. Telephone subscribers in equal-access locations are required to select a primary carrier, although they may still use any number of "secondary" carriers through what is called "10XXX" calling.

With 10XXX access, a subscriber whose primary carrier is, say, MCI may still place calls via AT&T, for example, by first dialing 10288 (which is "10" plus the three-digit carrier identification code for AT&T, "288"). The local telephone central office then routes the call to that particular secondary carrier—assuming, of course, that the carrier is accessible from that particular exchange (central office).

To put all the carriers on an equal basis for test purposes, each carrier that was to be tested was accessed first through the 10XXX secondary-carrier access method. All of the test lines, however, were presubscribed to AT&T as the primary carrier. And assurance was needed that the primary carrier, in this case AT&T, could also still be accessed through 10XXX dialing as a secondary carrier.

(AT&T was selected as the primary solely for expediency: The different BOCs informed us that AT&T could be connected as the primary carrier the same day our local access lines were installed, but that choosing any other carrier as the primary would mean having to wait anywhere from a few weeks to a month or more. This would appear to give AT&T a clear marketing advantage over its competitors in supposedly equal-access areas.)

AT&T was asked if customers could still access their primary carrier by also using the 10XXX procedure. A district manager with AT&T Communications, reportedly an expert in equal-access hardware and software, assured us that any caller could, indeed, alternately access his primary carrier through the 10XXX procedure, and that the equal-access switch would then treat the call in exactly the same way as it would any 10XXX request for a secondary-carrier connection.

We were satisfied, then, that our 10XXX test procedure would work and was valid. We proceeded to sign up with the carriers.

As AT&T access—albeit via 10XXX dialing—was already assured, each of the other five carriers was also then contacted. We explained to each that we wanted to be able to use its services on an interim basis as a secondary carrier from the three different calling locations. We explained that, in each location, we were already pre-subscribed to AT&T and wanted to compare its facilities and prices for a couple of months.

All were delighted to comply, and each furnished us with its 10XXX access code (10444 for Allnet, 10488 for ITT, 10222 for MCI, 10777 for Sprint, and 10220 for Western Union). It is noteworthy that most of these AT&T competitors offer to pay the nominal $5 fee that the BOCs charge customers to switch their primary carriers. What's more, a few even offer to pay the cost for switching you back to AT&T if after a while you are not satisfied.

'Gypsy' calling

All of the carriers told us that they were experiencing billing problems with many of the BOCs in the case of 10XXX dialing, which the carriers affectionately refer to as "gypsy" calling. Under current procedures, the local telephone company keeps track of 10XXX calling, bills the customer along with the regular monthly bill for local service, and then splits the receipts with each long-distance carrier, as appropriate. (The problems with BOC billing became apparent to us as we began receiving bills for our test calls. This is discussed later.)

Two of the carriers, MCI and Sprint, insisted that they bill us directly for the gypsy calls we made over their networks. They apparently were able to collect the call records expeditiously from the local telephone companies, for we received accurate and timely bills from both MCI and Sprint for the test calls they handled.

It should be noted that none of the carriers we got in touch with was given any indication that the purpose for our using their services was to determine their comparative quality and then publish the results. In fact, every effort was made to subscribe to each, on a secondary-carrier basis, just as any business customer might do. Each carrier was told, however, that we intended to use their facilities both for voice calling and for dial-up data transmission. None indicated any concern about our plans for mixed traffic.

Equipment configuration

After reviewing the wide assortment of commercially available telephone-line testing equipment (and there are literally dozens of vendors that offer the type of test gear we needed for our testing—both analog and digital), we narrowed the field to several state-of-the-art devices.

For digital testing, we selected the recently introduced Tektronix 835 digital tester, an improved version of the Tektronix 834, which DATA COMMUNICATIONS staffers used in similar testing two years ago and with which they were already familiar. (Editors are no easier to retrain than dancing bears, and are perhaps a bit more obstinate.)

The Tektronix 835 adds several desirable new features to the predecessor 834, including battery-backed random-access memory. This enables users, who have to perform the same preprogrammed test routines in different locations, as we did, to begin testing without having to reprogram the salient test parameters. The unit also added new software computational features, including a display of the number of errored seconds and the percent of error-free seconds for each data test. In addition, the 835 is packaged the same as the 834, making it compact and easy to carry.

For analog line testing, we selected a newly delivered unit from Hewlett-Packard, the 4945A Transmission Impairment Measuring Set (TIMS). The device performed all of the tests we wanted, plus many more. From our experience with the unit we learned that the wealth of available test routines necessitates considerable practice before the user gains proficiency in the 4945A's operation.

All 4945A tests and results are selected from, and displayed on, a small integral CRT providing simple and easy-to-read output. Regardless of the ease and simplicity, a user has to remember the correct sequence of button pushing for performing each test and must "back out" of each test screen before calling up the next one. It took longer than we would have wished to rotate through the same redundant sequence of tests. The capability to link only the tests we needed, then step through them with a single push button, in the order that we performed them, would have been a great help.

The 4945A devices at both the transmitting and receiving ends worked flawlessly during the months of testing, despite a heavy workout by nearly a dozen different testers.

1.2 kbit/s
All digital testing—bit and block error rate—employed the CCITT (International Telegraph and Telephone Consultative Committee) 511-specified pseudo-random bit pattern, generated by the Tektronix units and then transmitted using Racal-Vadic 212A-compatible modems. These modems were used for two main reasons:

1. We wanted our dial-up data testing to be representative of much of the dial-up data transmission being done today. And 1.2-kbit/s, 212A-compatible, phase-shifted modems are probably the most ubiquitous now used.

2. The Vadic modems also have the ability to store more than a dozen long dialing sequences in battery-backed random-access memory, which was essential for the conduct of this testing. Through front-panel push-button autodialing, we were able to rotate rapidly from one carrier to the next. Also, with the dial sequences stored in the modem, and dialing performed by the modem, it was not necessary to use a connected terminal or personal computer to initiate modem connections. The Tektronix digital test unit, which assumed the role of the data terminal equipment to the modem, could remain connected behind the modem.

An inconsistency in equal-access dialing procedures, however, posed one small problem. For calling from within New Jersey, a "1" has first to be dialed, even in equal-access areas, before any long-distance carrier can be accessed. This is then followed by the 10XXX sequence for a secondary-carrier call. The initial "1" is not required in either the Washington, D.C., or San Francisco areas. As a result, two sets of autodial sequences had to be programmed—one for New Jersey, another for everywhere else.

All the equipment for the test configuration was hooked up locally, and the test procedures worked out and rehearsed, before the equipment and the tester were sent to the actual remote test sites.

In the test configuration, the telephone-line transmit-and-receive wire pair was terminated directly on the HP 4945A analog test unit. Before each test call, the HP unit was configured so that the connection passed transparently through the box to a "handset connection" (which is designed to enable field technicians to attach a handset, dial, and talk over the circuit to be tested). To this connection we attached another telephone transmit-and-receive pair, which terminated with a two-way modular telephone jack.

On one side of the two-way connector was plugged the telephone line that connected the autodialing Vadic modem (which, in turn, connected to the Tektronix tester via an RS-232 interface). The other modular jack position permitted the attachment of a speakerphone monitor, which was used to time the carriers' connection setup.

Because it could affect the transmission characteristics of the line being tested, the speakerphone was activated only during call setup and expressly for the purpose of timing the connection. It was always turned off before any analog or digital testing began.

The caller used a stopwatch: The timer started as the last dual-tone multiple-frequency (DTMF) dial digit was transmitted by the Vadic modem, then stopped after the connection was set up and the return carrier signal was received from the destination Vadic modem. The answering modem, in Ramsey, was set up to always answer an incoming call after a single ring.

The results of the connect-time measurements are shown in Table 1. Keep in mind that the times indicated include one ring cycle at the far end. The ring cycle, generated by the destination central office, is six seconds long (two seconds of ring followed by four seconds of silence). The "ring," therefore, added a minimum of two seconds and as much as six seconds to the total connect time. Each carrier had to contend with the same variable, so about four seconds of the connect time shown is attributable, on average, to the ring cycle. (A voice call, assuming the recipient answered on the first ring, would incur the same amount of connect-time delay.)

Failed calls
Each dialed call attempt was recorded, whether or not it was successfully completed. (The connect times shown in Table 1 are only for successfully completed calls.) In a significant number of cases, calls were not completed on the first attempt. Table 2 summarizes, both overall and by specific route, each carrier's "batting average"—its ability to complete each test call on the first attempt.

From each calling test site, every carrier was tested in rotation. That is, a battery of tests began by attempting a call via Allnet and ended with Western Union (rotating always in the same alphabetical order). Where the first call attempt failed, the failed attempt was recorded, and the same carrier was given up to four more chances to complete a call successfully. If the five-attempt maximum still did not produce a completed call, which was necessary before any digital or analog test measurements could be made, then the testers moved on to the next carrier in the order.

Testing was done in this manner, constantly rotating from one carrier to the next, through 10 such batteries per test day. (Testing began between 6 and 7 A.M. from the calling location and continued through the entire business day.

Table 1: Comparison of connect time

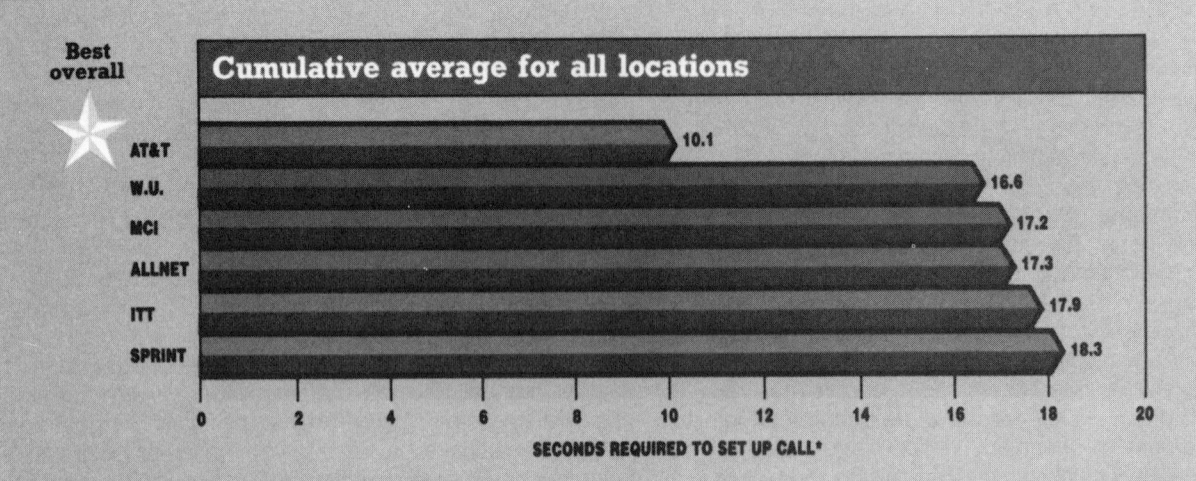

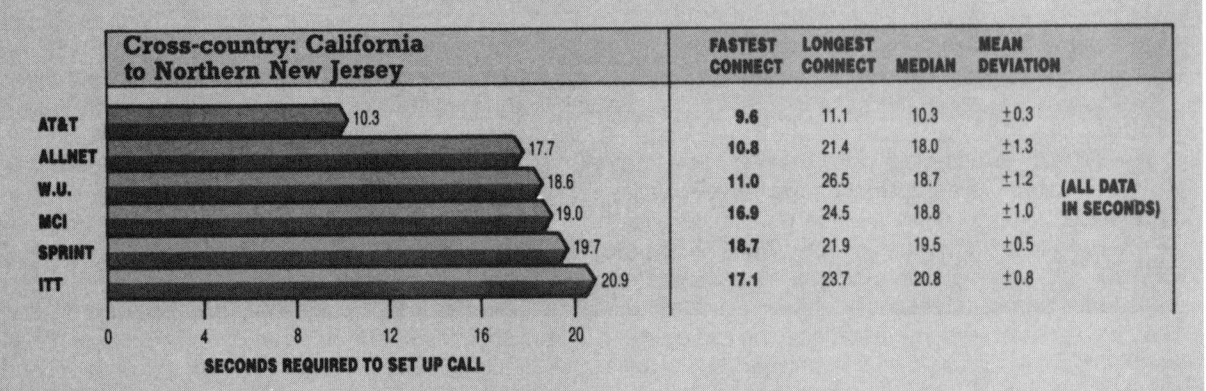

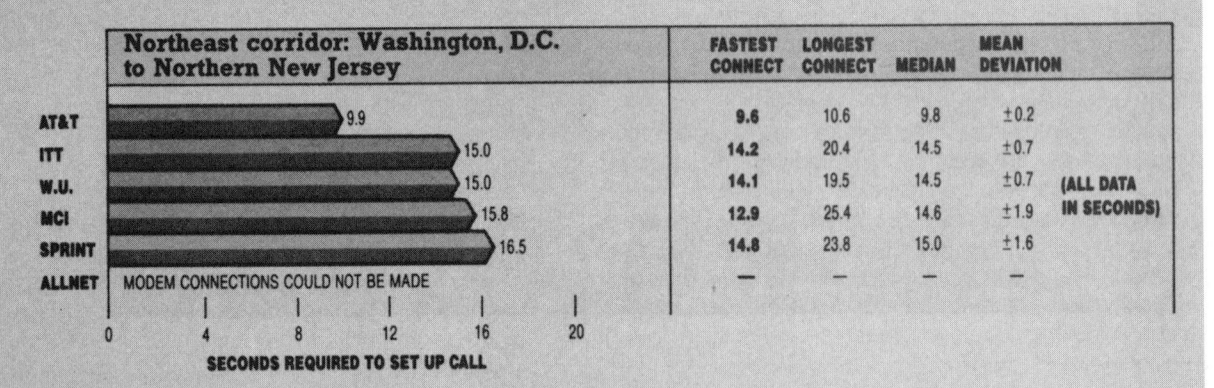

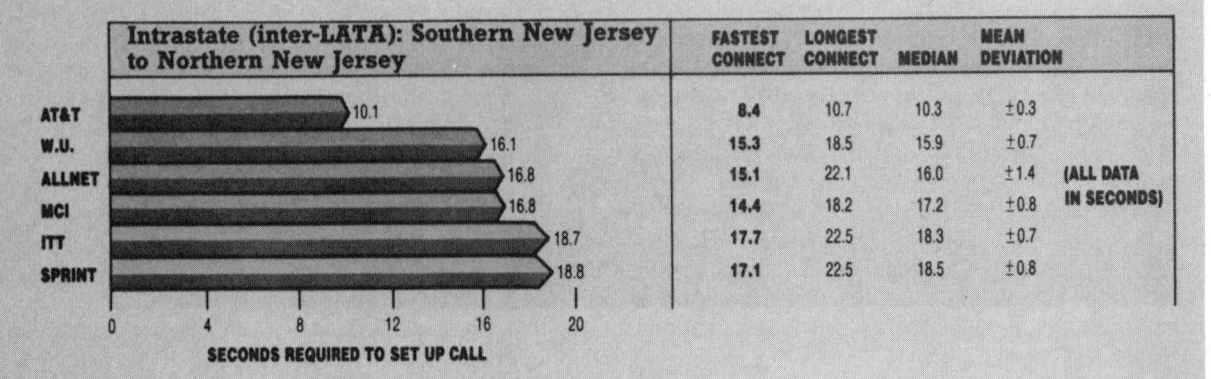

Cumulative average for all locations (Best overall: AT&T ★)

Carrier	Seconds required to set up call*
AT&T	10.1
W.U.	16.6
MCI	17.2
ALLNET	17.3
ITT	17.9
SPRINT	18.3

Cross-country: California to Northern New Jersey

Carrier	Seconds to set up call	Fastest Connect	Longest Connect	Median	Mean Deviation
AT&T	10.3	9.6	11.1	10.3	±0.3
ALLNET	17.7	10.8	21.4	18.0	±1.3
W.U.	18.6	11.0	26.5	18.7	±1.2
MCI	19.0	16.9	24.5	18.8	±1.0
SPRINT	19.7	18.7	21.9	19.5	±0.5
ITT	20.9	17.1	23.7	20.8	±0.8

(ALL DATA IN SECONDS)

Northeast corridor: Washington, D.C. to Northern New Jersey

Carrier	Seconds to set up call	Fastest Connect	Longest Connect	Median	Mean Deviation
AT&T	9.9	9.6	10.6	9.8	±0.2
ITT	15.0	14.2	20.4	14.5	±0.7
W.U.	15.0	14.1	19.5	14.5	±0.7
MCI	15.8	12.9	25.4	14.6	±1.9
SPRINT	16.5	14.8	23.8	15.0	±1.6
ALLNET	MODEM CONNECTIONS COULD NOT BE MADE	—	—	—	—

(ALL DATA IN SECONDS)

Intrastate (inter-LATA): Southern New Jersey to Northern New Jersey

Carrier	Seconds to set up call	Fastest Connect	Longest Connect	Median	Mean Deviation
AT&T	10.1	8.4	10.7	10.3	±0.3
W.U.	16.1	15.3	18.5	15.9	±0.7
ALLNET	16.8	15.1	22.1	16.0	±1.4
MCI	16.8	14.4	18.2	17.2	±0.8
ITT	18.7	17.7	22.5	18.3	±0.7
SPRINT	18.8	17.1	22.5	18.5	±0.8

(ALL DATA IN SECONDS)

*FROM TRANSMISSION OF LAST DUAL TONE MULTIFREQUENCY DIAL DIGIT (10XXX CALLING) TO RECEIPT OF CARRIER TONE FROM DESTINATION MODEM; INCLUDES ONE RING CYCLE (AVERAGE 4 SECONDS).

All test days were scheduled on regular business days, Monday through Friday.)

Readers will note that our definition of a completed call is predicated on the ability of the two modems to connect, and for the connection to remain intact for the duration of digital and analog testing—a period that averaged about 10 minutes. By imposing this modem-connection requirement, we were in all likelihood weeding out a number of marginal connections even before any test measurements were taken.

'Failed connections'

Table 3 shows a summary and general categorization of all the failed call attempts. In the "Network Busy" category, it can be assumed that the carrier lacked sufficient network capacity (either in local access or along the backbone network) to handle the call. These blocked call attempts would not have been completed for either modem connections or voice callers.

In the category "Failed Connections" were all the cases where, for any number of reasons during call setup, the calling modem detected spurious sounds—such as heavy crosstalk or signaling tones—and aborted the connection attempt, usually before the distant end would even ring or answer. Because the calling tester had the speakerphone activated during call setup, these sounds were usually clearly audible.

For failed connections, the calling modem would usually indicate that it encountered what it believed was a voice response at the destination end and then promptly abort the call. It is noteworthy, perhaps, that there were no such instances recorded for AT&T. It is possible that, in a number of the failed-connection cases, a voice caller may have been able to get through. It would seem, however, that an extreme amount of background noise would probably have necessitated hanging up and redialing even for voice callers.

The category "Dropped Connections" includes all the cases where the call went through, rang, and was answered by the destination modem. But then, for reasons that could not be determined, the modems were not able to exchange their handshake signals and disconnected after a time-out of several seconds. In a small percentage of the cases in this category, the connections actually just dropped, inexplicably and for no apparent reason. This generally occurred after line testing had already begun, usually within the first few minutes of the call.

It seemed worthwhile to plot the average number of redial attempts required before a call went through. This data, shown in Table 4, reflects the average number of call attempts needed to obtain a solid modem connection. Except for MCI, every carrier had at least one case where a connection could not be obtained even with five call attempts, which was the maximum. These instances were frequently during the two "busy hours" (mid morning and early afternoon) of the day, which is also when most of the "busy" call failures were encountered.

To more readily compare the connection capabilities of the carriers, thumbnail summaries of the recorded statistics are compiled in Table 5, both for overall averages, as well as by specific route. Note that for the Northeast corridor, most of the carriers completed calls more quickly, with fewer redials and failed calls, than for either of the other two routes.

After a carrier successfully set up a call, the testers at both ends began a two-minute digital test. The test was performed in both directions at the same time and consisted of sending blocks 1,000 bits in length, which carried the CCITT 511 pseudo-random bit pattern.

Bit error

About 116 blocks were sent in each direction during each two-minute test. (Five-minute bit error rate testing was also conducted from time to time, also in both directions at the same time, during which time approximately 289 blocks were sent in each direction.)

The total number of bit errors and block errors was recorded at both ends for each test, before switching to analog line testing. Table 6 shows the results of the bit error rate testing, where the total number of errored bits was divided by the total number of bits sent and then equated to a bit error rate per million bits. The results are shown both by individual route and overall for all locations. Table 7 shows the same comparison data for block errors.

Allnet was not able to set up a single modem connection over the Northeast corridor route, despite more than 100 call attempts that were made over a three-day period. Because modem dialing was the basis for the connection statistics shown, and prerequisite to digital bit and block error rate testing, data for Allnet's poor performance in the Northeast corridor could generally not be included in the overall averages. Tables 1 (Connect Time) and 4 (Call Attempts Required) do not reflect this anomaly in the overall averages, while Allnet's repeated failures are reflected in Table 2 (First-Try Completions) and Table 3 (Why Calls Failed).

Digital performance

Because we had compiled the data, we believed it would also be interesting to see how data-transmission performance looked when viewed in a few other ways.

Table 8 shows two different aspects of data-transmission performance. In the first, the total percentage of two-minute data calls that were free of errors is shown. ote that while AT&T places first, it is almost a three-way tie, with Western Union and MCI each only a percentage point behind.

Also shown is the total percentage of error-free data-transmission seconds, for all data transmissions from all locations. The top four placers appear to be very close, all achieving more than 99 percent error-free transmission. But keep in mind that at 1.2 kbit/s, more than 1,000 bit errors can occur in a single errored second.

All of the data transmission-related measurements—bit and block error rates, percent of error-free connections, and error-free seconds—are summarized in Table 9, both by individual route as well as overall for all locations. The data transmission quality offered by the carriers, relative to each other, changed dramatically from one route to another. This means that, as bit and block error rates seem likely to vary significantly both by carrier and by route, data communications managers might be advised to consider testing several carriers themselves over their high-traffic dial-up data routes.

Table 2: First-try call-completion record

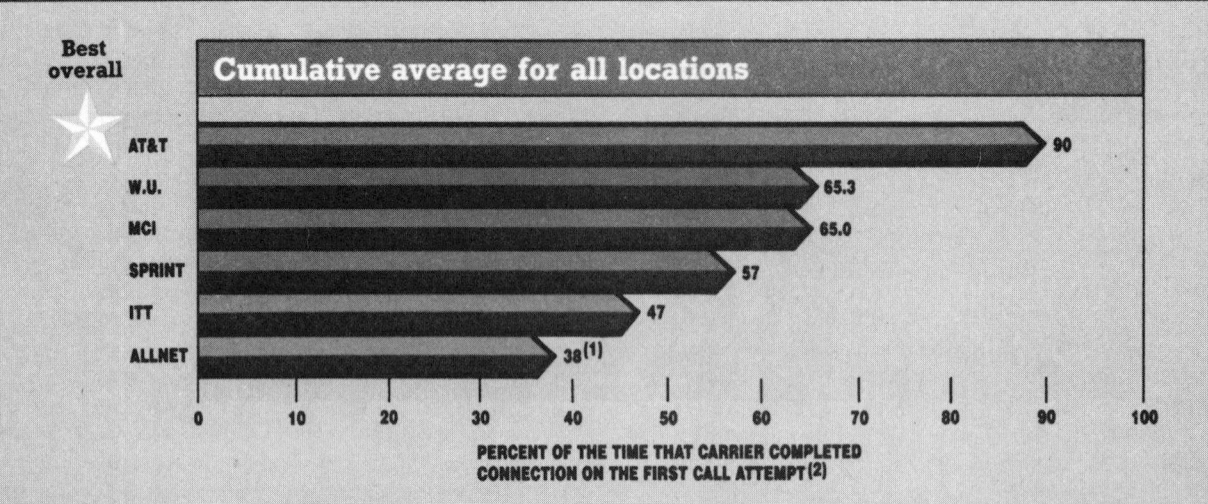

Cumulative average for all locations

Best overall ★

Carrier	%
AT&T	90
W.U.	65.3
MCI	65.0
SPRINT	57
ITT	47
ALLNET	38 [1]

PERCENT OF THE TIME THAT CARRIER COMPLETED CONNECTION ON THE FIRST CALL ATTEMPT [2]

WHAT MEASUREMENT SHOWS:
CARRIER'S ABILITY, COLLECTIVELY AND BY SPECIFIC ROUTE, TO SET UP A QUALITY CONNECTION THE FIRST TIME THE CALLER DIALS

EFFECT ON DATA:
FOR MODEM AND COMPUTER AUTODIALING, A BETTER FIRST-TIME AVERAGE SAVES EQUIPMENT PROCESSING TIME AND FREES UP PORT/MODEM RESOURCES QUICKER

EFFECT ON VOICE:
BETTER FIRST-TIME COMPLETION AVERAGE SAVES TIME ON REDIALS

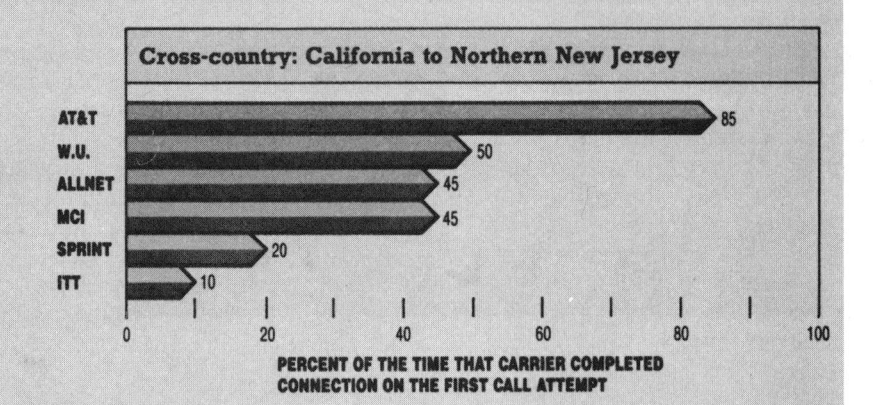

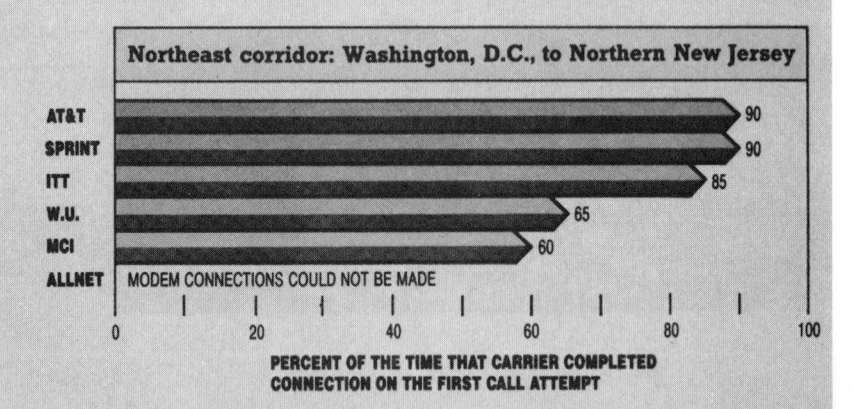

[1] CUMULATIVE AVERAGE REFLECTS THE FACT THAT MODEM CONNECTIONS COULD NOT BE MADE VIA ALLNET ALONG NORTHEAST CORRIDOR ROUTE.

[2] CONNECTION HAD TO ENABLE MODEMS TO CONNECT AND HAD TO REMAIN INTACT FOR FULL ANALOG AND DIGITAL TESTING— A PERIOD OF ABOUT 10 MINUTES. A SUMMARY OF FAILED, BLOCKED, OR DROPPED CONNECTIONS IS MADE IN TABLE 3.

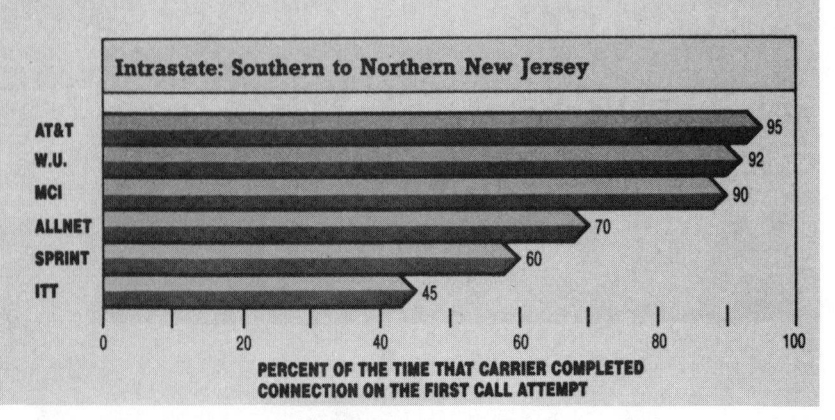

Cross-country: California to Northern New Jersey

Carrier	%
AT&T	85
W.U.	50
ALLNET	45
MCI	45
SPRINT	20
ITT	10

PERCENT OF THE TIME THAT CARRIER COMPLETED CONNECTION ON THE FIRST CALL ATTEMPT

Northeast corridor: Washington, D.C., to Northern New Jersey

Carrier	%
AT&T	90
SPRINT	90
ITT	85
W.U.	65
MCI	60
ALLNET	MODEM CONNECTIONS COULD NOT BE MADE

PERCENT OF THE TIME THAT CARRIER COMPLETED CONNECTION ON THE FIRST CALL ATTEMPT

Intrastate: Southern to Northern New Jersey

Carrier	%
AT&T	95
W.U.	92
MCI	90
ALLNET	70
SPRINT	60
ITT	45

PERCENT OF THE TIME THAT CARRIER COMPLETED CONNECTION ON THE FIRST CALL ATTEMPT

Table 3: Why calls failed

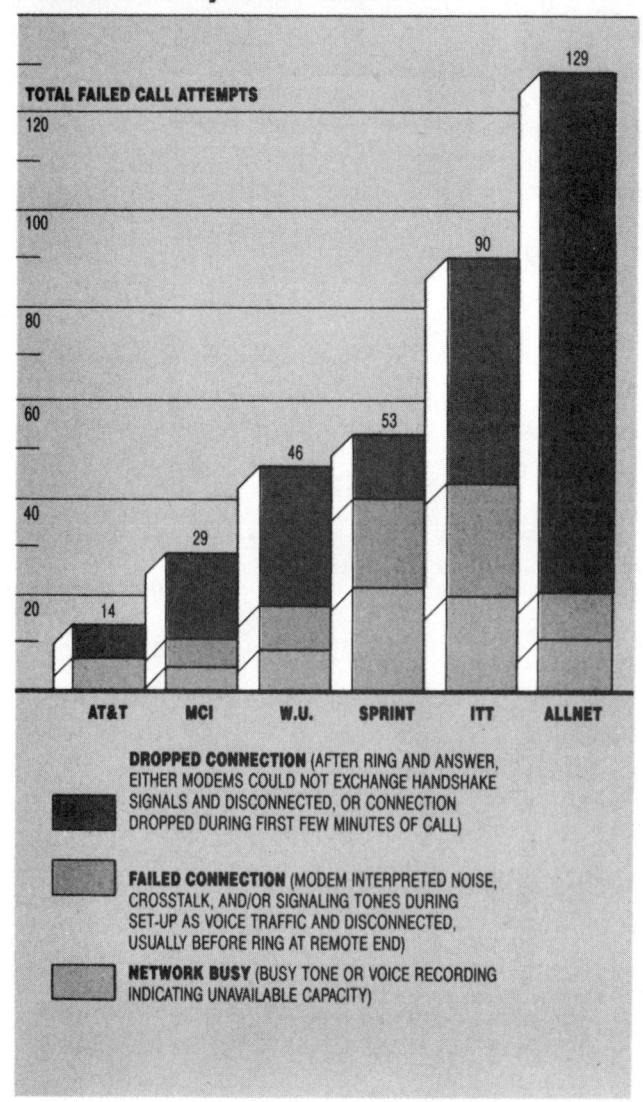

TOTAL FAILED CALL ATTEMPTS

Carrier	Total
AT&T	14
MCI	29
W.U.	46
SPRINT	53
ITT	90
ALLNET	129

- **DROPPED CONNECTION** (AFTER RING AND ANSWER, EITHER MODEMS COULD NOT EXCHANGE HANDSHAKE SIGNALS AND DISCONNECTED, OR CONNECTION DROPPED DURING FIRST FEW MINUTES OF CALL)
- **FAILED CONNECTION** (MODEM INTERPRETED NOISE, CROSSTALK, AND/OR SIGNALING TONES DURING SET-UP AS VOICE TRAFFIC AND DISCONNECTED, USUALLY BEFORE RING AT REMOTE END)
- **NETWORK BUSY** (BUSY TONE OR VOICE RECORDING INDICATING UNAVAILABLE CAPACITY)

On the whole, MCI's data-transmission performance was best, even though it rated first only over the Northeast corridor. Over the cross-country route, Western Union won hands down, with no bit or block errors recorded at all. Western Union placed third over the Northeast corridor and intrastate New Jersey routes and, coupled with its excellent cross-country showing, can be considered a strong second-place contender for overall data-carrying quality.

While AT&T's cross-country data performance was the worst of all the carriers measured, it took top honors in New Jersey (and did just so-so over the Northeast corridor route). A few possible explanations for AT&T's poor cross-country data track record are examined in the following section on analog measurements.

After the two-minute digital tests were completed, the testers "seized" the same line, while the connection was still set up, with the HP 4945A analog test devices. This was accomplished by activating the transmit and receive "hold" coils of the HP unit, which grabbed the line and cut off the "handset connection," to which both the modem and the digital test equipment were connected.

Because each modem is designed to detect the loss of carrier, and then disconnect and drop the connection within a few seconds, it was necessary for the testers at both ends to synchronize their switchover from the digital testing mode to analog. This was done with a countdown procedure. (The testers at either end were always in communication via a separate voice connection.)

Unlike the digital testing, which was conducted in both directions simultaneously, the analog testing (requiring the full bandwidth of the line) could be done only in one direction at a time. The analog test signals were sent first from the remote calling site to the Ramsey base, where the measurements were recorded. Afterward, the line was turned around, with the Ramsey site sending the test signals to the remote site, where measurement data was recorded.

The first measurement taken was the strength of the received signal. This is based on a test tone (or holding signal) sent by the transmitter at 1004 Hz, which is near the middle frequency range of a typical dial-up telephone line. The results are shown in Table 10.

The carrier with the consistently strongest received signal was Sprint (an overall average of about -22 dBm — decibels relative to one milliwatt). Most of the carriers' signal levels averaged between about -23 and -26 dBm. This equates to about 10 to 15 dB of loss overall, since the transmitted signal was always sent at a level of -13 dBm. As the signal level drops, especially beyond about -25 dBm, it becomes hard for the listener to hear the remote speaker. According to old Bell System specifications, a signal level that reaches -30 dBm is unacceptable. (A -30 dBm signal is 10 times weaker than a -20 dBm signal.)

The reader will note that analog measurements were taken for Allnet over the Northeast corridor, even though Allnet could not support a single modem connection over this route. In about every third test battery, in order to get some analog readings for Allnet, the testers disconnected all the digital test gear and manually put through an analog call over Allnet. (The digital gear then had to be reconnected before moving on to the next carrier, which was AT&T.)

Allnet did, therefore, support connections over the Northeast corridor that were manually set up, but not modem connections. We deviated from our normal procedure in this case mainly to ascertain, through analog measurements, why Allnet had such a problem in setting up modem calls. (It would appear to have been due to a combination of factors. As shown in subsequent tables for the Northeast corridor, Allnet had the worst signal-to-noise ratio, the highest amount of phase jitter, the worst peak-to-average [P/AR] reading, and a regular appearance of phase hits.)

Signal-to-noise

The next analog measurement taken was the signal-to-noise ratio (see Table 11), which compares the strength of a transmitted tone with the level of the background noise on the channel. AT&T and MCI both had remarkably "clear" channels within New Jersey and over the Northeast corridor but ranked fourth and sixth, respectively, for noise over the cross-country route.

Consistent with these signal-to-noise ratios is the fact that AT&T and MCI rated sixth and third, respectively, with their cross-country bit error rates. Both fared much better

Table 4: Number of call attempts needed to get through

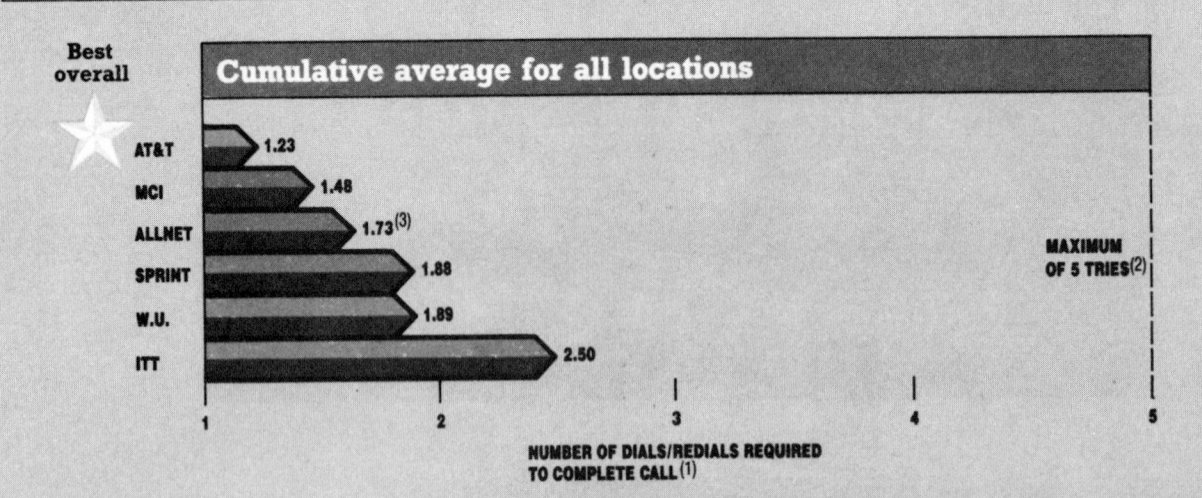

Cumulative average for all locations
Best overall ★

Carrier	Dials/Redials
AT&T	1.23
MCI	1.48
ALLNET	1.73 [3]
SPRINT	1.88
W.U.	1.89
ITT	2.50

MAXIMUM OF 5 TRIES [2]

NUMBER OF DIALS/REDIALS REQUIRED TO COMPLETE CALL [1]

WHAT MEASUREMENT SHOWS:
AVERAGE NUMBER OF DIAL/REDIAL TRIES REQUIRED, BOTH OVERALL AND BY SPECIFIC ROUTE, BEFORE CARRIER COULD ACHIEVE AND MAINTAIN A QUALITY CONNECTION.

EFFECT ON DATA:
LOWER NUMBER OF REQUIRED ATTEMPTS IMPROVES CHANCES OF DATA CONNECTION BEING SET UP QUICKER, AND MAINTAINED WITHOUT DISRUPTION.

EFFECT ON VOICE:
SAME AS FOR DATA.

[1] A VALUE OF 1, THE BEST POSSIBLE AND IDEAL, WOULD MEAN THE CARRIER ALWAYS SET UP A GOOD MODEM CONNECTION ON THE FIRST DIAL ATTEMPT.

[2] CONNECTION ATTEMPTS FOR TESTING WERE MADE IN ROTATION, ALPHABETICALLY BY CARRIER. WHERE THE FIRST ATTEMPT FAILED, UP TO FOUR MORE ATTEMPTS WERE MADE, A MAXIMUM OF FIVE, BEFORE THE NEXT CARRIER WAS TRIED.

[3] DOES NOT REFLECT ALLNET'S INABILITY TO SUPPORT MODEM CALL SET-UP ALONG NORTHEAST CORRIDOR ROUTE.

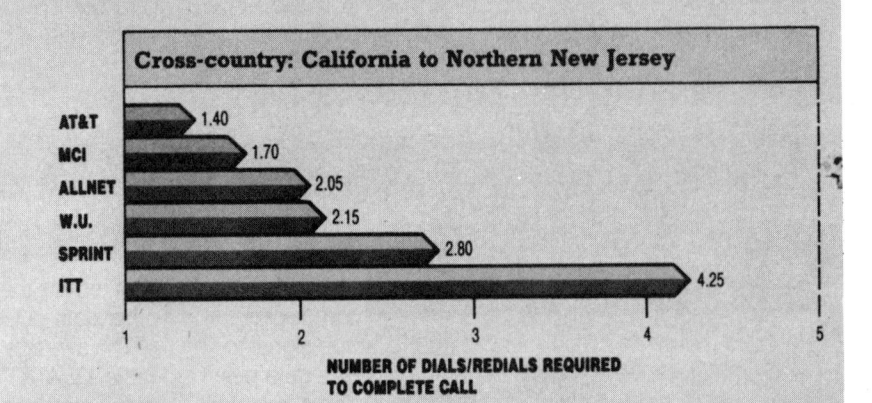

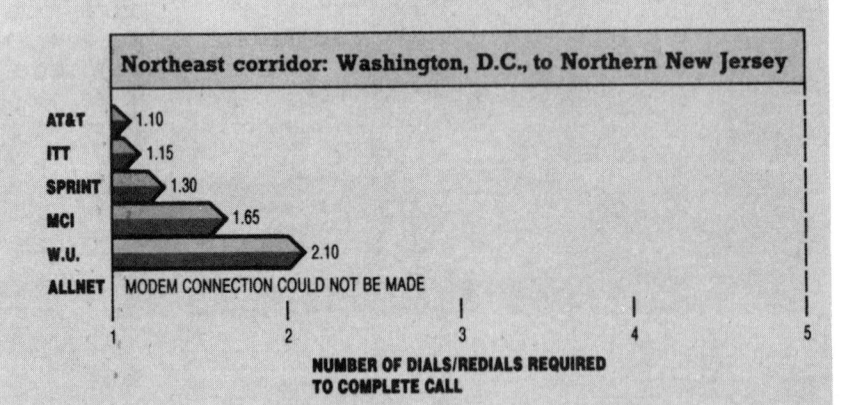

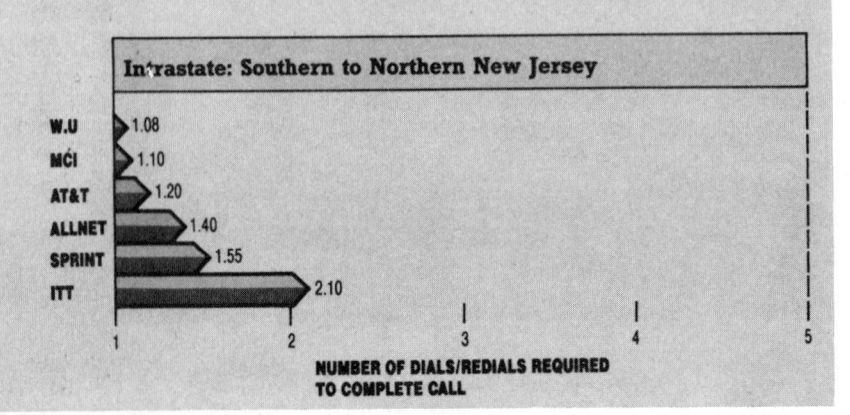

Cross-country: California to Northern New Jersey

Carrier	Dials/Redials
AT&T	1.40
MCI	1.70
ALLNET	2.05
W.U.	2.15
SPRINT	2.80
ITT	4.25

NUMBER OF DIALS/REDIALS REQUIRED TO COMPLETE CALL

Northeast corridor: Washington, D.C., to Northern New Jersey

Carrier	Dials/Redials
AT&T	1.10
ITT	1.15
SPRINT	1.30
MCI	1.65
W.U.	2.10
ALLNET	MODEM CONNECTION COULD NOT BE MADE

NUMBER OF DIALS/REDIALS REQUIRED TO COMPLETE CALL

Intrastate: Southern to Northern New Jersey

Carrier	Dials/Redials
W.U	1.08
MCI	1.10
AT&T	1.20
ALLNET	1.40
SPRINT	1.55
ITT	2.10

NUMBER OF DIALS/REDIALS REQUIRED TO COMPLETE CALL

Table 5: Summary comparison—connection capability

	OVERALL, ALL LOCATIONS	CROSS-COUNTRY	NORTHEAST CORRIDOR	INTRASTATE, NEW JERSEY
AVERAGE CONNECT TIME (SECONDS)	AT&T (10.1) W.U. (16.6) MCI (17.2) ALLNET (17.3) ITT (17.9) SPRINT (18.3)	AT&T (10.3) ALLNET (17.7) W.U. (18.6) MCI (19.0) SPRINT (19.7) ITT (20.9)	AT&T (9.9) ITT (15.0) W.U. (15.0) MCI (15.8) SPRINT (16.5) ALLNET (N/A)	AT&T (10.1) W.U. (16.1) ALLNET (16.8) MCI (16.8) ITT (18.7) SPRINT (18.8)
FIRST-TRY CALL COMPLETION (PERCENTAGE)	AT&T (90) W.U. (65.3) MCI (65) SPRINT (57) ITT (47) ALLNET (38)	AT&T (85) W.U. (50) ALLNET (45) MCI (45) SPRINT (20) ITT (10)	AT&T (90) SPRINT (90) ITT (85) W.U. (65) MCI (60) ALLNET (N/A)	AT&T (95) W.U. (92) MCI (90) ALLNET (70) SPRINT (60) ITT (45)
DIAL/REDIAL ATTEMPTS REQUIRED FOR CONNECTION (AVERAGE NUMBER)	AT&T (1.23) MCI (1.48) ALLNET (1.73) SPRINT (1.88) W.U. (1.885) ITT (2.50)	AT&T (1.40) MCI (1.70) ALLNET (2.05) W.U. (2.15) SPRINT (2.80) ITT (4.25)	AT&T (1.10) ITT (1.15) SPRINT (1.30) MCI (1.65) W.U. (2.10) ALLNET (N/A)	W.U. (1.08) MCI (1.10) AT&T (1.20) ALLNET (1.40) SPRINT (1.55) ITT (2.10)
CALLS BLOCKED; NETWORK 'BUSY' (NUMBER OF 'BUSIES' PER 20 CALL ATTEMPTS)	MCI (1.1) W.U. (1.6) AT&T (1.9) ITT (2.7) ALLNET (3.2) SPRINT (3.9)	MCI (1.8) ALLNET (3.4) AT&T (3.6) SPRINT (3.6) W.U. (3.7) ITT (4.5)	AT&T (0.0) W.U. (0.0) ITT (0.9) MCI (1.2) SPRINT (1.5) ALLNET (N/A)	ITT (0.0) MCI (0.0) W.U. (0.0) AT&T (1.7) ALLNET (2.9) SPRINT (6.5)

over the other two routes, where their signal-to-noise measurements were also much better. Not surprisingly, Western Union, which offered the best average signal-to-noise ratio cross-country, also offered error-free data performance. Overall, the tested carriers averaged a signal-to-noise ratio of at least 30 dB, which is not bad. Old Bell System specifications call for a signal-to-noise ratio of at least 24 dB.

The next analog measurement was signal-loss "distortion," also called the "gain slope," which measures how much weaker the high and low frequencies on the line are compared with a middle-of-the-road tone of 1004 Hz. The results are shown in Table 12.

The measurement of signal-loss distortion was one that remained relatively constant by carrier, regardless of the route tested. AT&T and Sprint usually split the first- and second-place ratings, while ITT was always worst—in fact, consistently much worse than the others. In a few of the measured cases, the lower frequencies were the ones that exhibited the most loss deviation. In the overwhelming majority of the cases, however, it turned out to be the high-frequency component that was either much weaker or missing altogether.

Phase and P/AR

Two other analog measurements are particularly helpful in assessing a telephone line's data-carrying capacity, though the characteristics they measure have a negligible impact on voice quality. They are phase jitter and the peak-to-average ratio.

Phase jitter, usually the result of power-supply fluctuations, causes variations in the relative, and expected, positions of data-carrying signals, or pulses. In short, phase jitter moves the pulses into positions and slots allocated for other data pulses.

The average phase jitter measurements are shown in Table 13. AT&T's was best overall and placed first over two of the routes. It fared worse, however, over the cross-country route (fifth place), where its data performance was also the worst (sixth place). MCI's cross-country bit errors would also appear to be attributable, at least in part, to phase jitter. Most tested carriers averaged four to five degrees of phase jitter, though many had individual measurements near or exceeding 10 degrees (out of phase), which is the worst acceptable under old Bell System specifications.

The peak-to-average ratio is measured by sending a complex signal at 16 different frequencies. The P/AR rating, a number on a scale of 1 to 100 (100 being the best), represents the spreading in time of the different signals. A "bad" P/AR (see Table 14) usually indicates such problems as envelope delay, noise, and loss distortion, though the specific problems are not readily discernible through the P/AR measurement alone. As P/AR value drops, so does the line's available bandwidth for data transmission.

AT&T usually offered an extremely good P/AR; Sprint's was usually among the worst. Allnet's P/AR for the Northeast corridor, where modem connections could not be made, was the worst of the carriers measured.

Table 15 presents a brief summary of all the analog measurements. Overall, AT&T took the most "firsts" (four out of five), though Western Union's analog signal quality was also, by and large, impressive.

MCI, which offered the best data performance overall, captured none of the top places with regard to analog signal quality. Based on the overall averages, MCI main-

Table 6: Bit error rate[1]

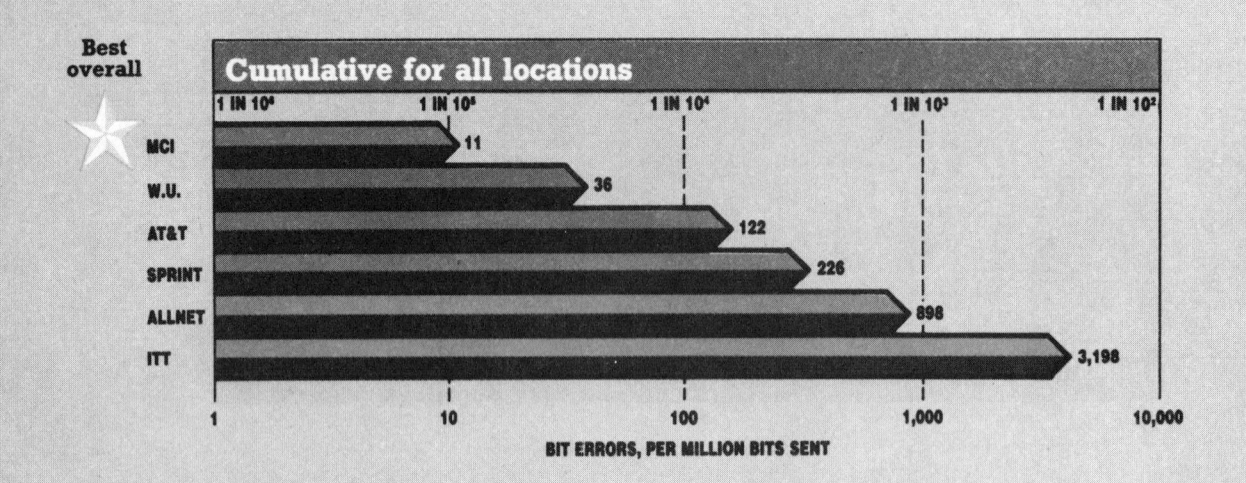

Cumulative for all locations
Best overall: ★ MCI

Bit errors, per million bits sent:
- MCI: 11
- W.U.: 36
- AT&T: 122
- SPRINT: 226
- ALLNET: 898
- ITT: 3,198

WHAT MEASUREMENT SHOWS:
THE ABILITY OF CARRIER, BOTH OVERALL AND BY SPECIFIC ROUTE, TO MAINTAIN THE INTEGRITY OF TRANSMITTED DATA.

EFFECT ON DATA:
HIGH ERROR RATES REQUIRE LONGER CONNECT TIMES FOR THE RETRANSMISSION OF ERRORED DATA.

EFFECT ON VOICE:
NONE.

(1) BASED ON TWO- AND FIVE-MINUTE TRANSMISSION OF CCITT 511 PSEUDO-RANDOM BIT TEST PATTERN. TESTING DONE IN BOTH DIRECTIONS SIMULTANEOUSLY USING FULL-DUPLEX, 1.2-KBIT/S, BELL 212A-COMPATIBLE MODEM TRANSMISSION.

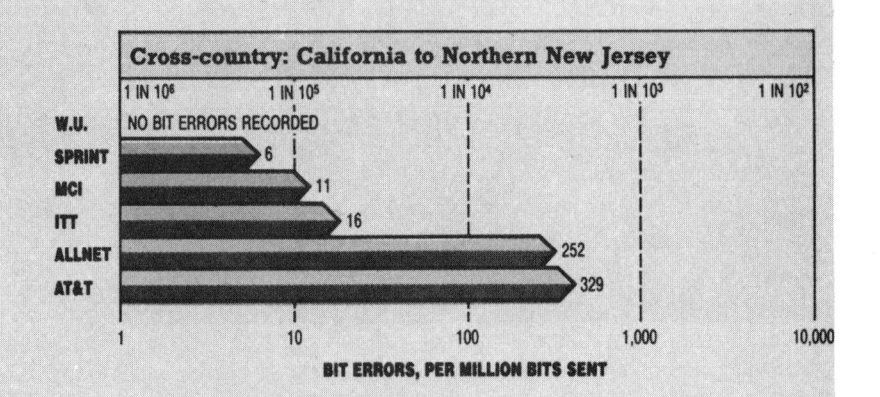

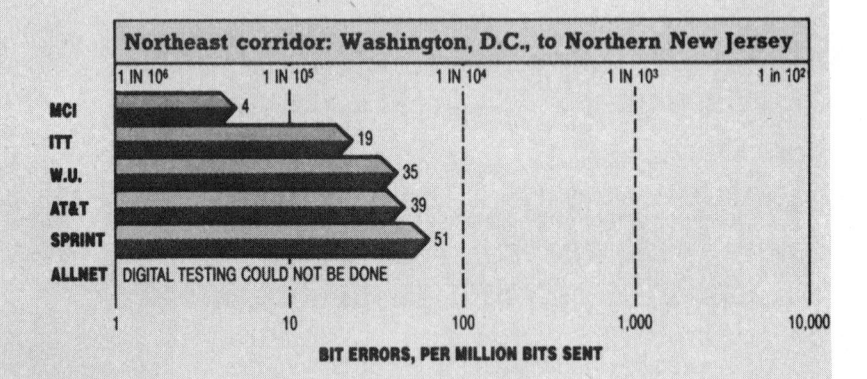

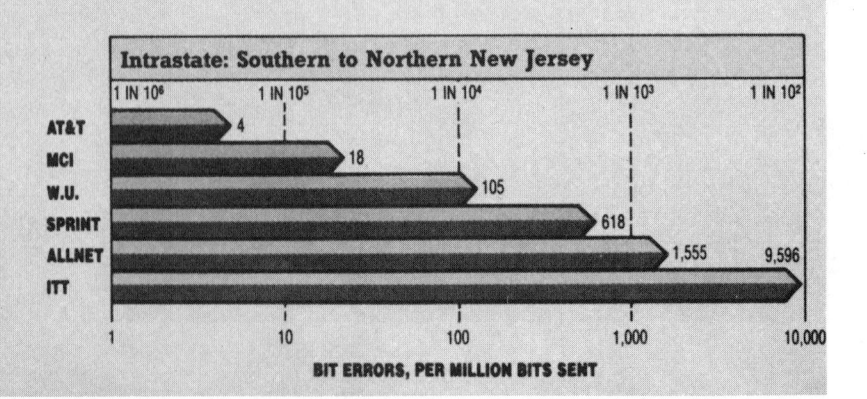

Cross-country: California to Northern New Jersey
- W.U.: NO BIT ERRORS RECORDED
- SPRINT: 6
- MCI: 11
- ITT: 16
- ALLNET: 252
- AT&T: 329

Northeast corridor: Washington, D.C., to Northern New Jersey
- MCI: 4
- ITT: 19
- W.U.: 35
- AT&T: 39
- SPRINT: 51
- ALLNET: DIGITAL TESTING COULD NOT BE DONE

Intrastate: Southern to Northern New Jersey
- AT&T: 4
- MCI: 18
- W.U.: 105
- SPRINT: 618
- ALLNET: 1,555
- ITT: 9,596

Table 7: Block error rate[1]

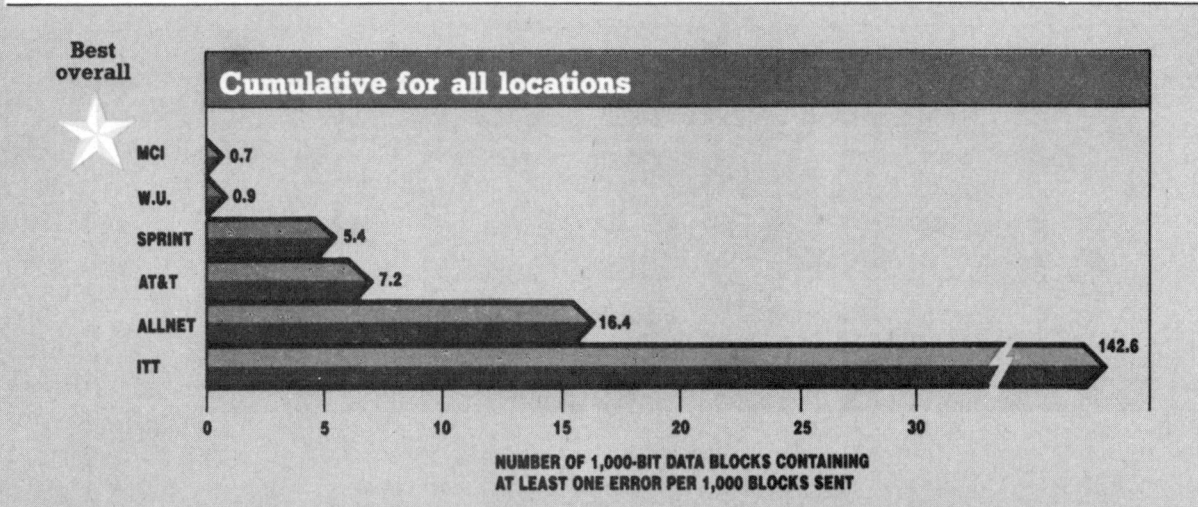

Cumulative for all locations

Best overall ★

MCI	0.7
W.U.	0.9
SPRINT	5.4
AT&T	7.2
ALLNET	16.4
ITT	142.6

NUMBER OF 1,000-BIT DATA BLOCKS CONTAINING AT LEAST ONE ERROR PER 1,000 BLOCKS SENT

WHAT MEASUREMENT SHOWS:
WHEN VIEWED WITH BIT ERROR RATE, REVEALS WHETHER ERROR-CAUSING EVENTS ARE CLUSTERED AND BURSTY OR EVENLY DISTRIBUTED.

EFFECT ON DATA:
HIGH BLOCK ERROR RATE CAN BE ESPECIALLY BAD FOR LONG DATA BLOCKS OR FILE TRANSMISSION AS ERROR LIKELIHOOD, NECESSITATING COMPLETE RETRANSMISSION, GROWS.

EFFECT ON VOICE:
NONE.

(1) BASED ON TWO- AND FIVE-MINUTE TRANSMISSION OF CCITT 511 BIT PATTERN AND FULL-DUPLEX, 1.2-KBIT/S BELL 212A-COMPATIBLE MODEM TRANSMISSION.

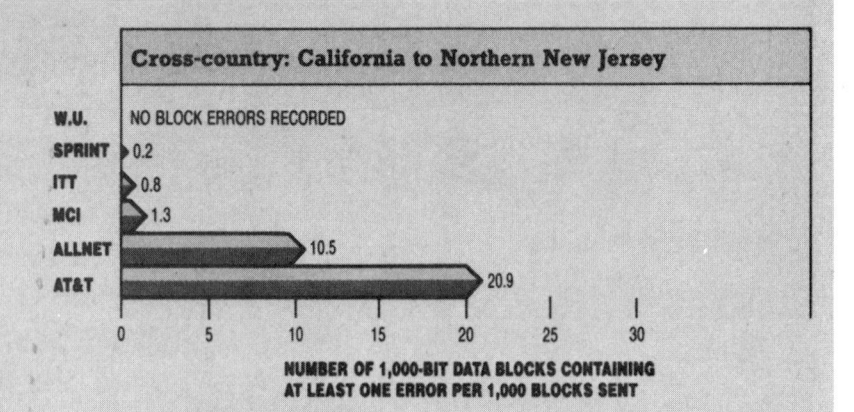

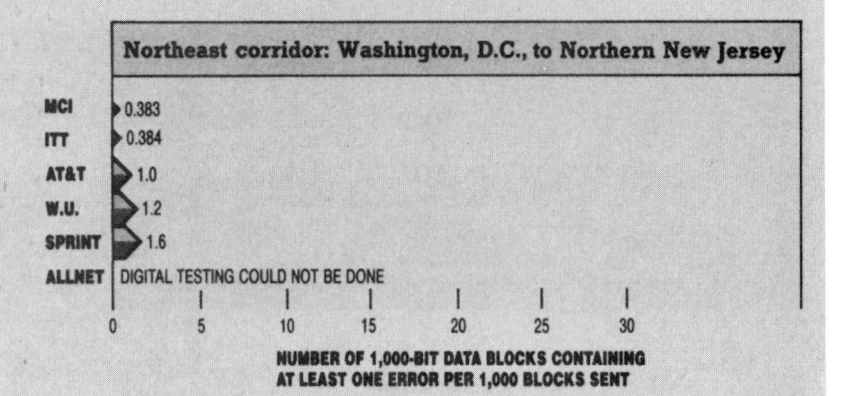

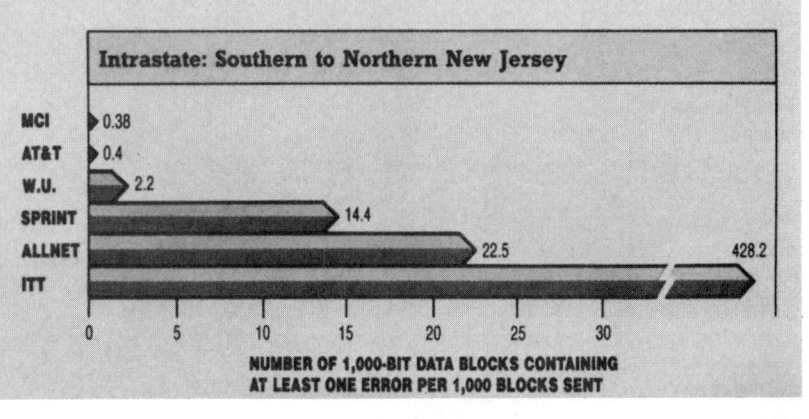

Cross-country: California to Northern New Jersey

W.U.	NO BLOCK ERRORS RECORDED
SPRINT	0.2
ITT	0.8
MCI	1.3
ALLNET	10.5
AT&T	20.9

NUMBER OF 1,000-BIT DATA BLOCKS CONTAINING AT LEAST ONE ERROR PER 1,000 BLOCKS SENT

Northeast corridor: Washington, D.C., to Northern New Jersey

MCI	0.383
ITT	0.384
AT&T	1.0
W.U.	1.2
SPRINT	1.6
ALLNET	DIGITAL TESTING COULD NOT BE DONE

NUMBER OF 1,000-BIT DATA BLOCKS CONTAINING AT LEAST ONE ERROR PER 1,000 BLOCKS SENT

Intrastate: Southern to Northern New Jersey

MCI	0.38
AT&T	0.4
W.U.	2.2
SPRINT	14.4
ALLNET	22.5
ITT	428.2

NUMBER OF 1,000-BIT DATA BLOCKS CONTAINING AT LEAST ONE ERROR PER 1,000 BLOCKS SENT

Table 8: The error-free file

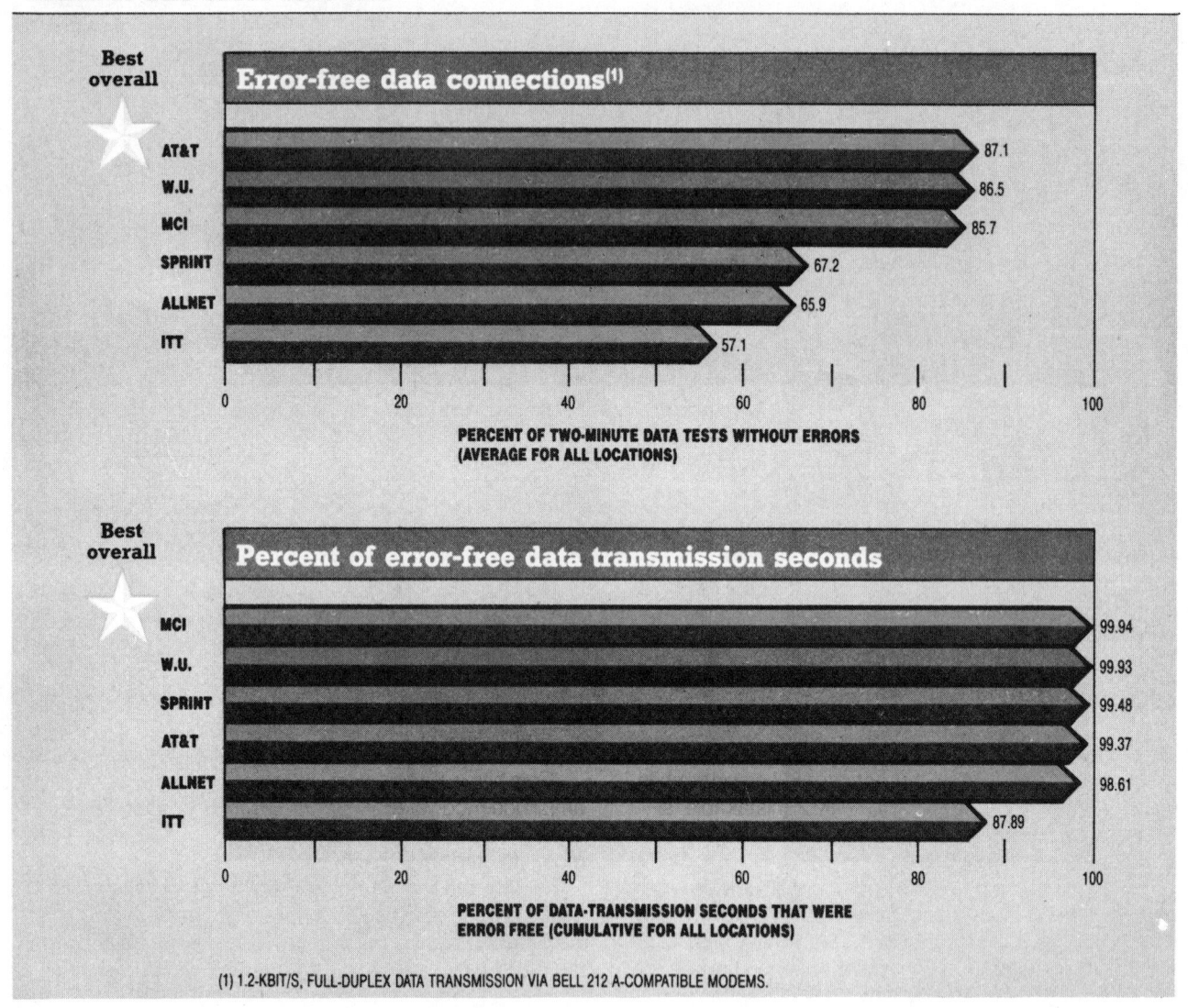

Error-free data connections[1] — Best overall: AT&T

Carrier	Percent
AT&T	87.1
W.U.	86.5
MCI	85.7
SPRINT	67.2
ALLNET	65.9
ITT	57.1

PERCENT OF TWO-MINUTE DATA TESTS WITHOUT ERRORS (AVERAGE FOR ALL LOCATIONS)

Percent of error-free data transmission seconds — Best overall: MCI

Carrier	Percent
MCI	99.94
W.U.	99.93
SPRINT	99.48
AT&T	99.37
ALLNET	98.61
ITT	87.89

PERCENT OF DATA-TRANSMISSION SECONDS THAT WERE ERROR FREE (CUMULATIVE FOR ALL LOCATIONS)

(1) 1.2-KBIT/S, FULL-DUPLEX DATA TRANSMISSION VIA BELL 212 A-COMPATIBLE MODEMS.

---

tained a clearly middle-of-the-road position—not too bad, but not very good either. While its cross-country analog performance could certainly have been better, MCI seems to have demonstrated that a carrier offering a consistent, though mediocre, analog signal quality can be the best choice for carrying data.

There is, however, another measure of line performance that would seem to corroborate MCI's strong data-carrying performance, and AT&T's less-than-optimum data performance despite its many blue ribbons. This is the occurrence of transients.

Transients

Despite AT&T's apparent success with most of the analog measurement criteria, it lost out to MCI for the top data-carrying honors. Why? The reason could be transients, which are transmission impairments of a quite severe nature but which appear infrequently and sporadically. (While the other analog measurements are based on signal sampling at regular intervals and averaging over time, transients are simply counted as they occur—generally over more protracted periods of time.)

(Because transients in our testing were based on one-minute transient counts, done in each direction on each tested dial-up circuit, the number of counts for each transient had to then be equated to a 15-minute period of measured connect time, which is the minimum counting time prescribed for most transient-measurement procedures. Therefore, the transient counts presented in this article are not necessarily the same, statistically speaking, as they would be if each connection had undergone a full 15-minute transient test. Our methodology presumed that, for dial-up quality comparison, it was preferable to take analog measurements over many more connections and then average the results than it would have been to take much longer measurements over far fewer connections.)

One type of transient that clearly has a deleterious effect on the quality of data transmission is phase hits, where, unlike phase jitter, the phase of the signal jumps to more than 20 degrees out of normal phase for a relatively long time—more than four milliseconds.

The relative appearance of phase hits, plotted along with the measured bit and block error rates, is shown in Table 16. Note that MCI had almost no measured phase hits on

Table 9: Summary comparison—data transmission quality

	OVERALL, ALL LOCATIONS	CROSS-COUNTRY	NORTHEAST CORRIDOR	INTRASTATE, NEW JERSEY
BIT ERROR RATE (BIT ERRORS, PER MILLION BITS SENT)	MCI (11) W.U. (36) AT&T (122) SPRINT (226) ALLNET (898) ITT (3,198)	W.U. (0) SPRINT (6) MCI (11) ITT (16) ALLNET (252) AT&T (329)	MCI (4) ITT (19) W.U. (35) AT&T (39) SPRINT (51) ALLNET (N/A)	AT&T (4) MCI (18) W.U. (105) SPRINT (618) ALLNET (1,555) ITT (9,596)
BLOCK ERROR RATE (BLOCKS WITH ERRORS, PER 1,000 BLOCKS SENT)	MCI (0.7) W.U. (0.9) SPRINT (5.4) AT&T (7.2) ALLNET (16.4) ITT (142.6)	W.U. (0) SPRINT (0.2) ITT (0.8) MCI (1.3) ALLNET (10.5) AT&T (20.9)	MCI (0.38) ITT (0.39) AT&T (1.0) W.U. (1.2) SPRINT (1.6) ALLNET (N/A)	MCI (0.38) AT&T (0.4) W.U. (2.2) SPRINT (14.4) ALLNET (22.5) ITT (428.2)
ERROR-FREE, TWO-MINUTE, FULL-DUPLEX DATA CONNECTIONS (PERCENT)	AT&T (87.1) W.U. (86.5) MCI (85.7) SPRINT (67.2) ALLNET (65.9) ITT (57.1)	W.U. (100) SPRINT (95.0) ITT (92.9) AT&T (90.0) MCI (76.2) ALLNET (65.0)	ITT (90.5) MCI (90.5) SPRINT (90.0) AT&T (81.0) W.U. (78.9) ALLNET (N/A)	AT&T (90.5) MCI (90.5) W.U. (76.9) ALLNET (66.6) SPRINT (19.0) ITT (0.0)
ERROR-FREE SECONDS OF FULL-DUPLEX DATA TRANSMISSION (PERCENT OF TOTAL SECONDS)	MCI (99.94) W.U. (99.93) SPRINT (99.47) AT&T (99.27) ALLNET (98.61) ITT (86.89)	W.U. (100) SPRINT (99.98) ITT (99.91) MCI (99.89) ALLNET (98.99) AT&T (97.87)	ITT (99.96) MCI (99.96) AT&T (99.93) W.U. (99.90) SPRINT (99.84) ALLNET (N/A)	AT&T (99.96) MCI (99.96) W.U. (99.87) SPRINT (98.63) ALLNET (98.24) ITT (65.33)

any of the routes, while AT&T did, especially over the cross-country route. The relationship between phase hits and data-transmission performance is unmistakable, although phase hits clearly are not the only cause of problems in data transmission.

A second category of transients is impulse noise—very brief (less than four milliseconds) incidents of very strong noise. These noise "spikes" sound like "pops" and "clicks" to voice listeners. The most common causes are electromechanical switches and relays, installation and maintenance activities, and weather disturbances.

Table 17 summarizes the impulse noise measurements. Note that three different perspectives are presented: overall impulse "hit" counts and high-level impulse "hits," both equated to 15 minutes of measured connect time, and the percentage of all connections where impulse hits were counted.

The percentage of connections with impulse hits shows overall distribution, while the number of hits referenced to 15 minutes of measured connect time shows the relative frequency of this noise. The high-level category shows the number of times, per 15 minutes of connect time, that a high-intensity impulse spike occurred. These are the strongest impulse hits and the ones most likely to mutate the value of transmitted data, since they approach the same power level as the main data-carrying signal. Sprint appears to be the carrier most plagued by high-level impulse noise.

Two other categories of transients—gain hits and dropouts—are not plotted because of their very infrequent appearance. Only a handful of gain hits, where the receive signal level abruptly goes up or down as much as 12 dB, were recorded over the hours of transient counting. This was statistically an insufficient amount to attempt to draw a pattern. At least one gain hit was recorded for each carrier.

(Gain hits will affect data transmission the most for amplitude-modulated modems. Because of the phase-shifted modems we employed, our data transmission was most affected by phase hits.)

The last transient category—dropouts—constitutes cases lasting more than four milliseconds where the signal inexplicably drops by more than 12 dB. Dropouts interrupt the signal flow, which means data is always lost when dropouts occur. Again, an insufficient number of dropouts was recorded for any valid analysis. None was recorded for MCI or Allnet. AT&T logged only one (over the cross-country route); ITT and Sprint each had two (over the Northeast corridor); and Western Union logged four (all over the New Jersey route).

The measurements taken for this study (all testing was done in April and May) and the testing methodology applied are very similar to a study that was conducted in 1984. Equal access was not then a reality, and testing was done only over a single cross-country route (from San Francisco to New York).

To our surprise, after eliminating two carriers that were tested in 1984 but not this year and comparing the relative rankings of the six carriers, we found that the results (shown in Table 18) are remarkably consistent with those of two years ago.

In examining the rankings for eight criteria that were measured both in 1984 and in this year's testing, most of carriers were within two place rankings of their 1984 relative position, and many were unchanged. If one considers as significant only the cases where the carrier changed at least two place rankings, when comparing the 1984 cross-country results with both the cross-country and overall

Table 10: Received signal strength

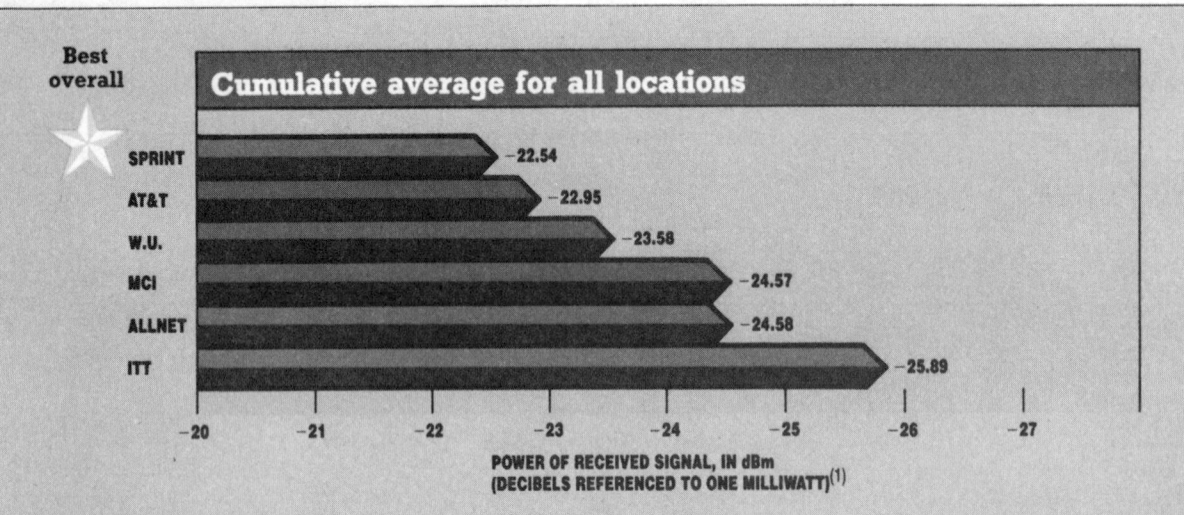

Cumulative average for all locations

Best overall ★

Carrier	Power (dBm)
SPRINT	−22.54
AT&T	−22.95
W.U.	−23.58
MCI	−24.57
ALLNET	−24.58
ITT	−25.89

POWER OF RECEIVED SIGNAL, IN dBm
(DECIBELS REFERENCED TO ONE MILLIWATT)[1]

WHAT MEASUREMENT SHOWS:
RAW STRENGTH, OR POWER, AT THE RECEIVER'S END OF A SIGNAL (TONE) SENT IN THE MID-RANGE TELEPHONE FREQUENCY (AT 1,004 HERTZ; ALL WERE TRANSMITTED AT A POWER LEVEL OF −13 dBm).

EFFECT ON DATA:
GENERALLY NOMINAL, UNLESS THE LEVEL DROPS TO NEAR −30 dBm OR MORE.

EFFECT ON VOICE:
SIGNAL SOUNDS WEAK, BECOMES HARD TO HEAR AS LEVEL DROPS BEYOND ABOUT −25 dBm.

[1] ALL 1,004-HERTZ TEST SIGNALS WERE TRANSMITTED AT A POWER LEVEL OF −13 dBm. THE DIFFERENCE MAY BE CONSIDERED OVERALL SIGNAL LOSS.

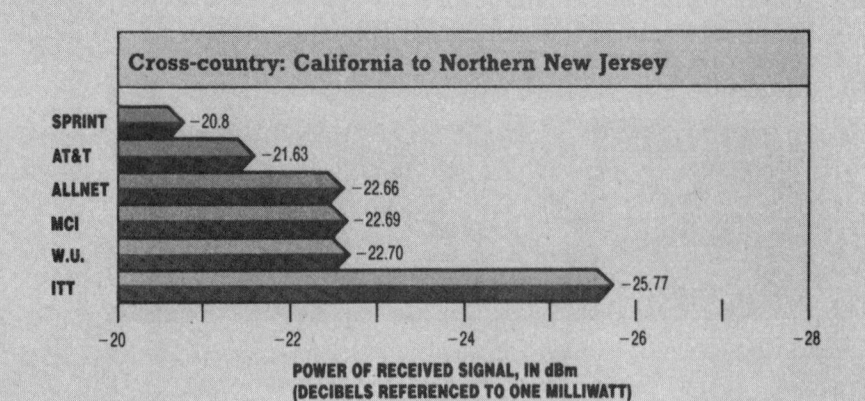

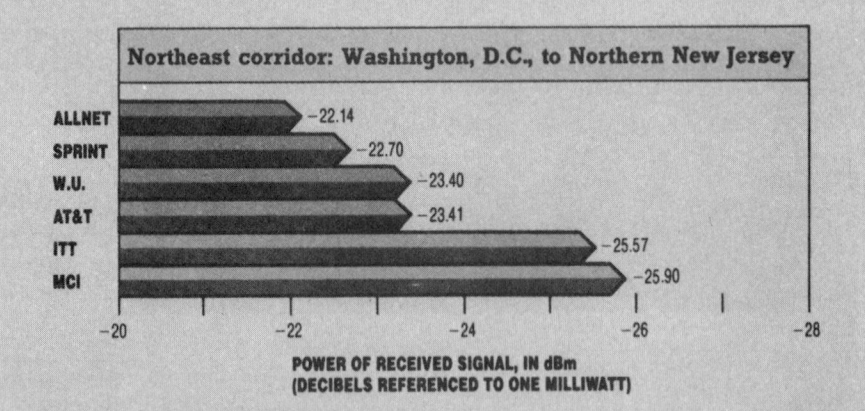

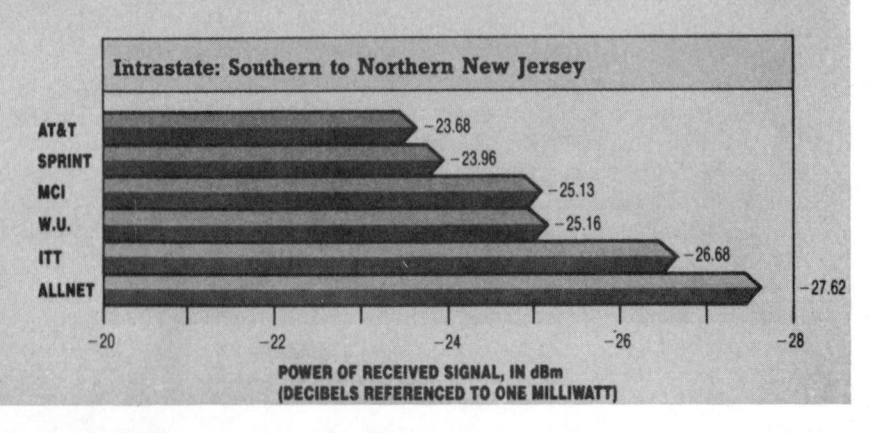

Cross-country: California to Northern New Jersey

Carrier	Power (dBm)
SPRINT	−20.8
AT&T	−21.63
ALLNET	−22.66
MCI	−22.69
W.U.	−22.70
ITT	−25.77

POWER OF RECEIVED SIGNAL, IN dBm
(DECIBELS REFERENCED TO ONE MILLIWATT)

Northeast corridor: Washington, D.C., to Northern New Jersey

Carrier	Power (dBm)
ALLNET	−22.14
SPRINT	−22.70
W.U.	−23.40
AT&T	−23.41
ITT	−25.57
MCI	−25.90

POWER OF RECEIVED SIGNAL, IN dBm
(DECIBELS REFERENCED TO ONE MILLIWATT)

Intrastate: Southern to Northern New Jersey

Carrier	Power (dBm)
AT&T	−23.68
SPRINT	−23.96
MCI	−25.13
W.U.	−25.16
ITT	−26.68
ALLNET	−27.62

POWER OF RECEIVED SIGNAL, IN dBm
(DECIBELS REFERENCED TO ONE MILLIWATT)

Table 11: Signal-to-noise ratio

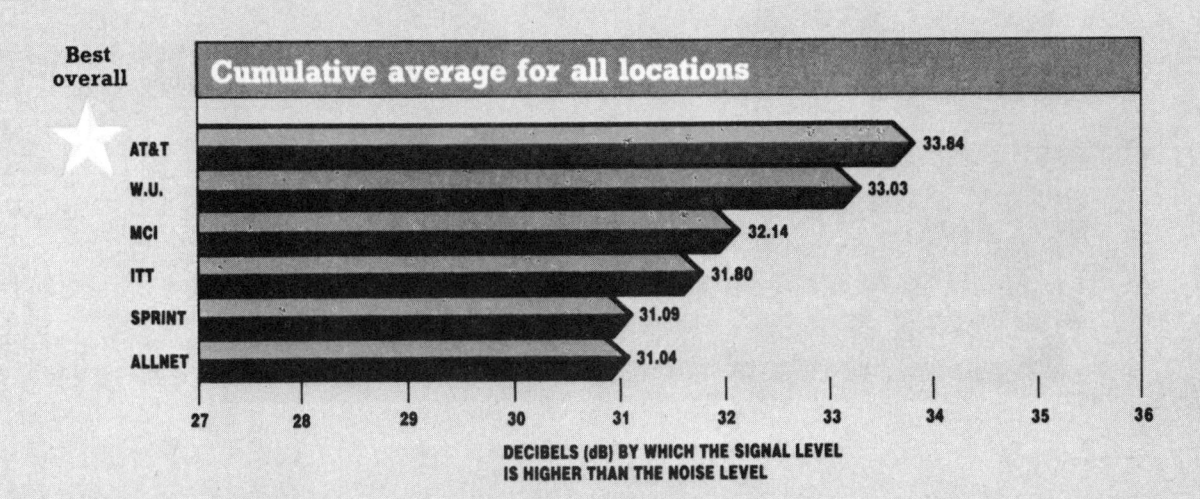

Best overall ★

Cumulative average for all locations

Carrier	dB
AT&T	33.84
W.U.	33.03
MCI	32.14
ITT	31.80
SPRINT	31.09
ALLNET	31.04

DECIBELS (dB) BY WHICH THE SIGNAL LEVEL IS HIGHER THAN THE NOISE LEVEL

WHAT MEASUREMENT SHOWS:
RELATIVE CLARITY OF THE DATA OR VOICE SIGNAL AS COMPARED WITH THE BACKGROUND NOISE ON THE LINE.

EFFECT ON DATA:
SIGNAL-TO-NOISE RATIO TRACKS CLOSELY TO BIT-ERROR PERFORMANCE; AS NOISE LEVEL NEARS SIGNAL LEVEL, NOISE CORRUPTS THE BINARY VALUE OF DATA. AN S/N RATIO OF 30 dB OR BETTER FOR DATA IS GOOD.

EFFECT ON VOICE:
BACKGROUND STATIC AND ELECTRICAL NOISE CAN RENDER EVEN A STRONG VOICE SIGNAL UNINTELLIGIBLE. CLARITY OF SIGNAL ERODES QUICKLY AS S/N RATIO DROPS NEAR OR BELOW ABOUT 25 dB.

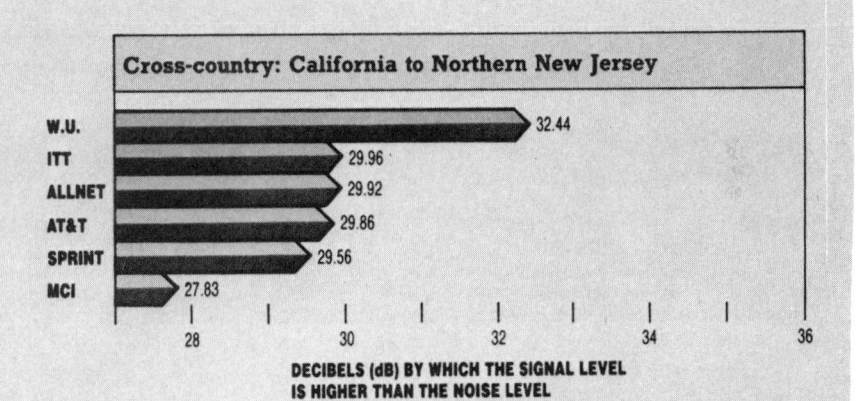

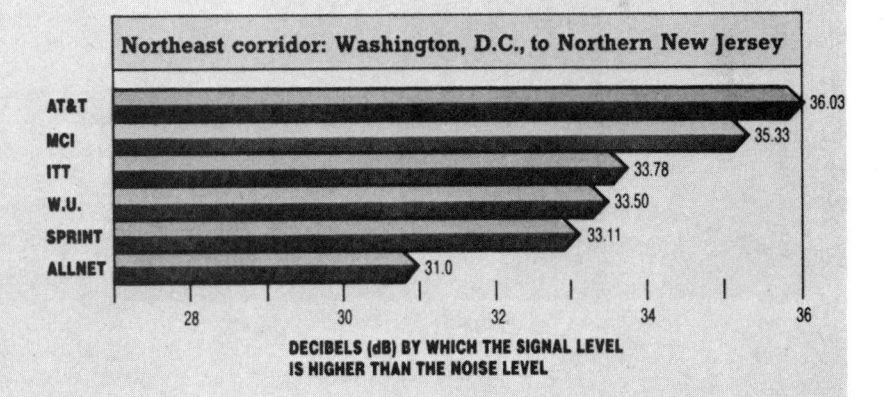

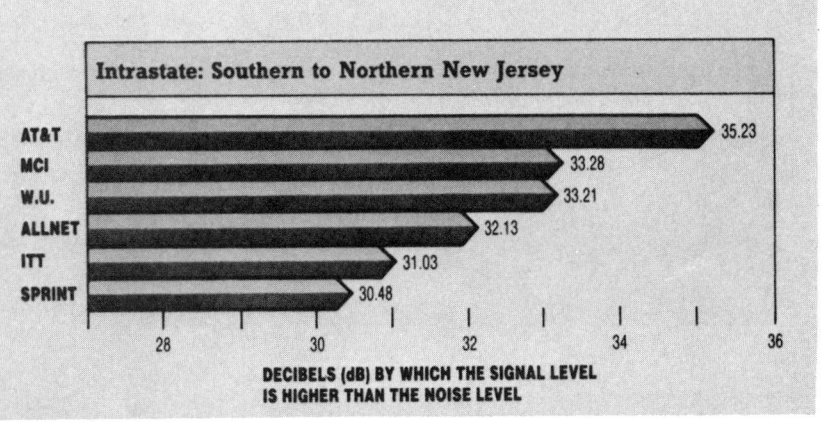

Cross-country: California to Northern New Jersey

Carrier	dB
W.U.	32.44
ITT	29.96
ALLNET	29.92
AT&T	29.86
SPRINT	29.56
MCI	27.83

DECIBELS (dB) BY WHICH THE SIGNAL LEVEL IS HIGHER THAN THE NOISE LEVEL

Northeast corridor: Washington, D.C., to Northern New Jersey

Carrier	dB
AT&T	36.03
MCI	35.33
ITT	33.78
W.U.	33.50
SPRINT	33.11
ALLNET	31.0

DECIBELS (dB) BY WHICH THE SIGNAL LEVEL IS HIGHER THAN THE NOISE LEVEL

Intrastate: Southern to Northern New Jersey

Carrier	dB
AT&T	35.23
MCI	33.28
W.U.	33.21
ALLNET	32.13
ITT	31.03
SPRINT	30.48

DECIBELS (dB) BY WHICH THE SIGNAL LEVEL IS HIGHER THAN THE NOISE LEVEL

Table 12: Signal loss 'distortion'

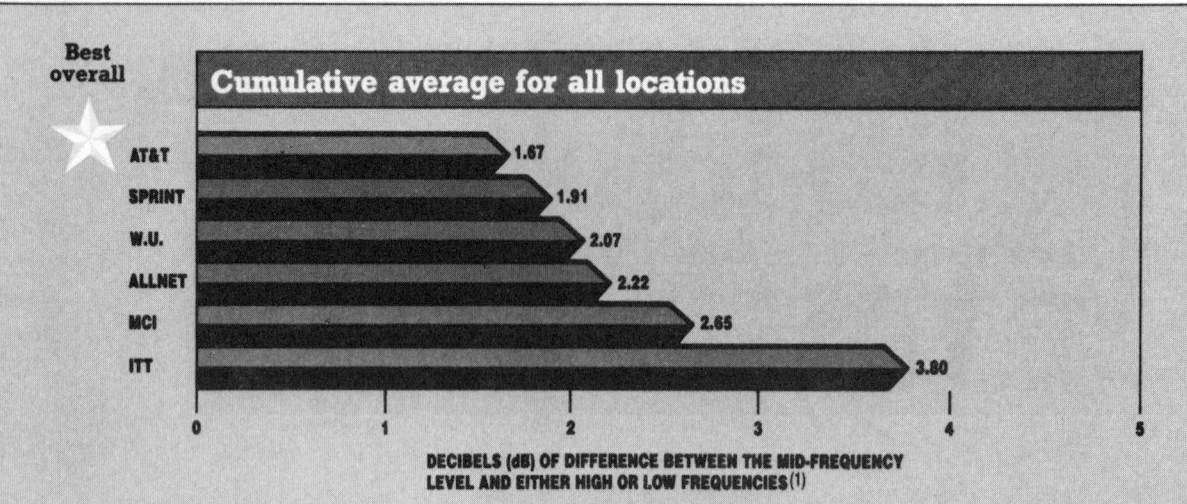

Cumulative average for all locations

Best overall ★

Carrier	dB
AT&T	1.67
SPRINT	1.91
W.U.	2.07
ALLNET	2.22
MCI	2.65
ITT	3.80

DECIBELS (dB) OF DIFFERENCE BETWEEN THE MID-FREQUENCY LEVEL AND EITHER HIGH OR LOW FREQUENCIES[1]

WHAT MEASUREMENT SHOWS:
THE 'CLIPPING OFF' OF EITHER THE HIGH- OR LOW-FREQUENCY PART OF A TELEPHONE SIGNAL; THE ADDED LOSS OF (USUALLY) THE HIGH-FREQUENCY SOUNDS, COMPARED WITH THE MIDDLE FREQUENCY (1,004 HERTZ).

EFFECT ON DATA:
WHERE MODEMS, SUCH AS 212As, USE A HIGH-FREQUENCY CARRIER IN ONE DIRECTION, A LOW FREQUENCY IN THE OTHER, THIS ADDED LOSS CAN PREVENT DATA CONNECTIONS OR SIGNIFICANTLY CORRUPT DATA USING THE HIGH FREQUENCY.

EFFECT ON VOICE:
PARTS OF CONVERSATIONS, WHERE TALKER USES LOW- OR HIGH-FREQUENCY SOUNDS, ARE LOST EN ROUTE TO THE LISTENER.

[1] IN A FEW CASES, USUALLY INVOLVING ITT, THE HEWLETT-PACKARD 4945A TESTER COULD FIND NO HIGH-FREQUENCY COMPONENT TO MEASURE. NONE WAS RECORDED, THEREFORE, AND THIS IS NOT REFLECTED IN THE TABLES. AS A RESULT, THE ACTUAL OVERALL AVERAGE DISTORTION IS WORSE THAN THESE TABLES DEPICT.

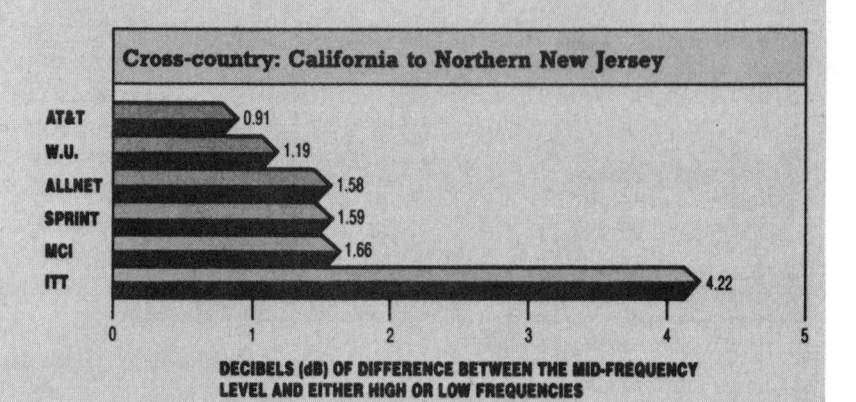

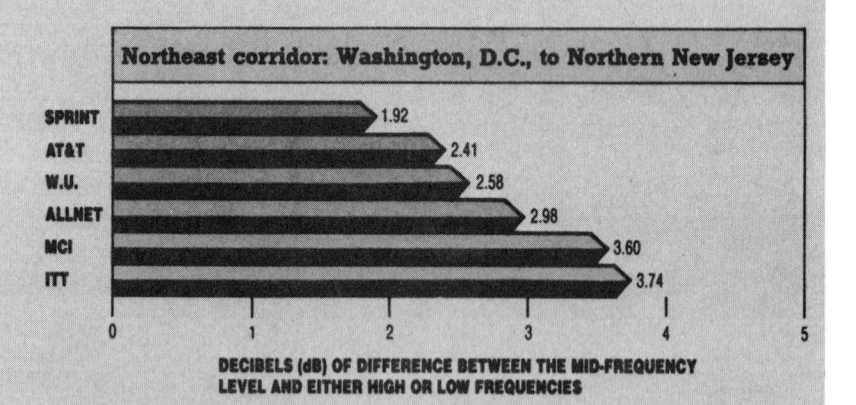

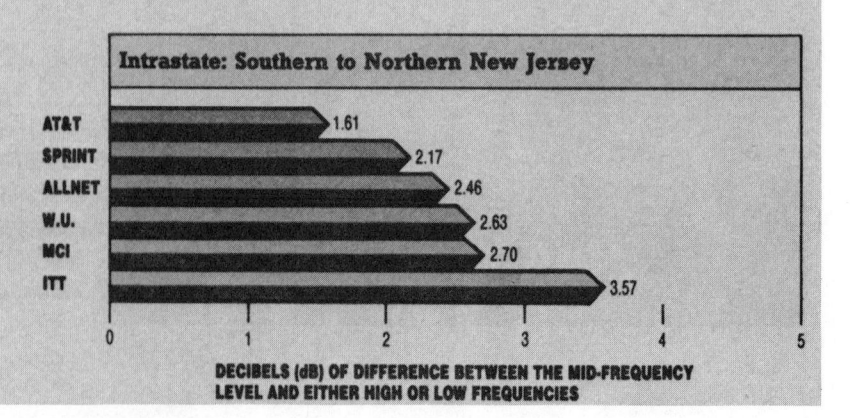

Cross-country: California to Northern New Jersey

Carrier	dB
AT&T	0.91
W.U.	1.19
ALLNET	1.58
SPRINT	1.59
MCI	1.66
ITT	4.22

DECIBELS (dB) OF DIFFERENCE BETWEEN THE MID-FREQUENCY LEVEL AND EITHER HIGH OR LOW FREQUENCIES

Northeast corridor: Washington, D.C., to Northern New Jersey

Carrier	dB
SPRINT	1.92
AT&T	2.41
W.U.	2.58
ALLNET	2.98
MCI	3.60
ITT	3.74

DECIBELS (dB) OF DIFFERENCE BETWEEN THE MID-FREQUENCY LEVEL AND EITHER HIGH OR LOW FREQUENCIES

Intrastate: Southern to Northern New Jersey

Carrier	dB
AT&T	1.61
SPRINT	2.17
ALLNET	2.46
W.U.	2.63
MCI	2.70
ITT	3.57

DECIBELS (dB) OF DIFFERENCE BETWEEN THE MID-FREQUENCY LEVEL AND EITHER HIGH OR LOW FREQUENCIES

Table 13: Phase jitter

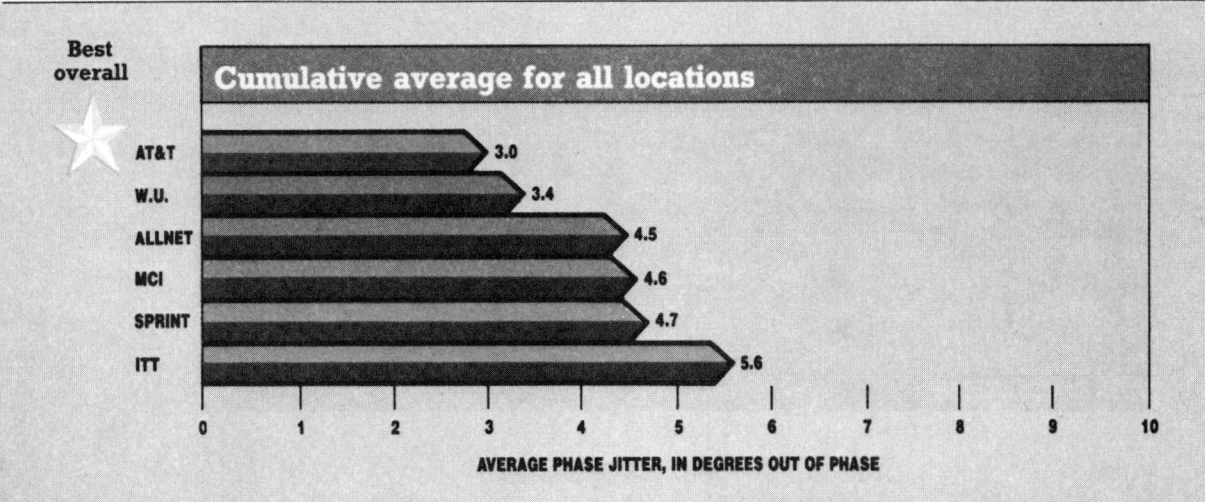

Cumulative average for all locations
Best overall ★

Carrier	Average phase jitter (degrees out of phase)
AT&T	3.0
W.U.	3.4
ALLNET	4.5
MCI	4.6
SPRINT	4.7
ITT	5.6

WHAT MEASUREMENT SHOWS:
AMOUNT OF VARIATION—UNDESIRABLE SHORTENING OR LENGTHENING—OF THE ANTICIPATED WAVEFORM OF THE MAIN TRANSMISSION SIGNAL.

EFFECT ON DATA:
FOR PHASE-MODULATED DATA TRANSMISSION, RESULTS CAN BE DISASTROUS ERROR RATES, ESPECIALLY WHEN SPURTS OF JITTER NEAR OR EXCEED 10 DEGREES. EFFECT ON FREQUENCY-MODULATED DATA IS NOMINAL.

EFFECT ON VOICE:
NEGLIGIBLE, IMPERCEPTIBLE.

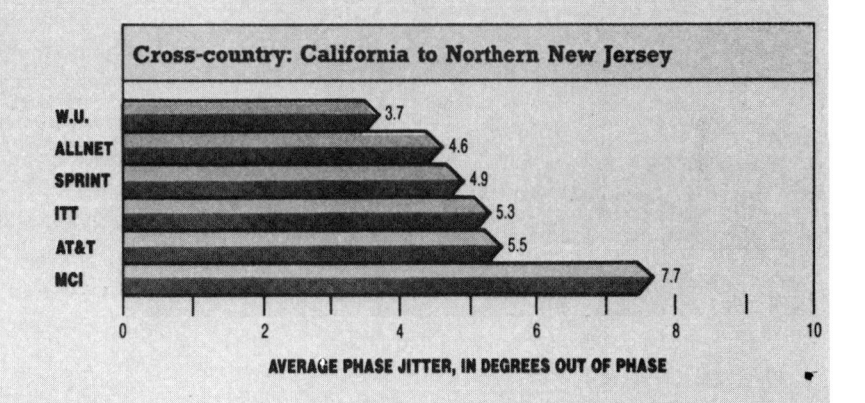

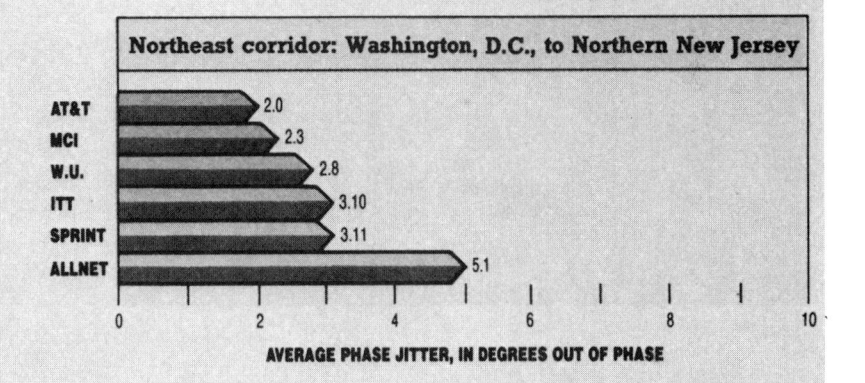

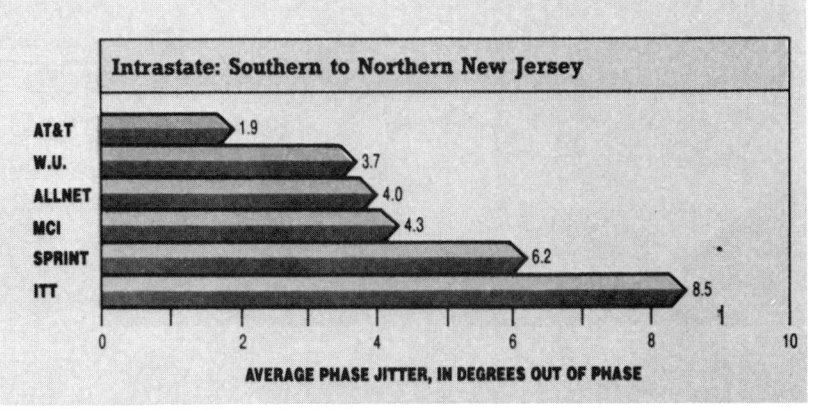

Cross-country: California to Northern New Jersey

Carrier	Average phase jitter
W.U.	3.7
ALLNET	4.6
SPRINT	4.9
ITT	5.3
AT&T	5.5
MCI	7.7

Northeast corridor: Washington, D.C., to Northern New Jersey

Carrier	Average phase jitter
AT&T	2.0
MCI	2.3
W.U.	2.8
ITT	3.10
SPRINT	3.11
ALLNET	5.1

Intrastate: Southern to Northern New Jersey

Carrier	Average phase jitter
AT&T	1.9
W.U.	3.7
ALLNET	4.0
MCI	4.3
SPRINT	6.2
ITT	8.5

Table 14: Peak-to-average ratio (P/AR)

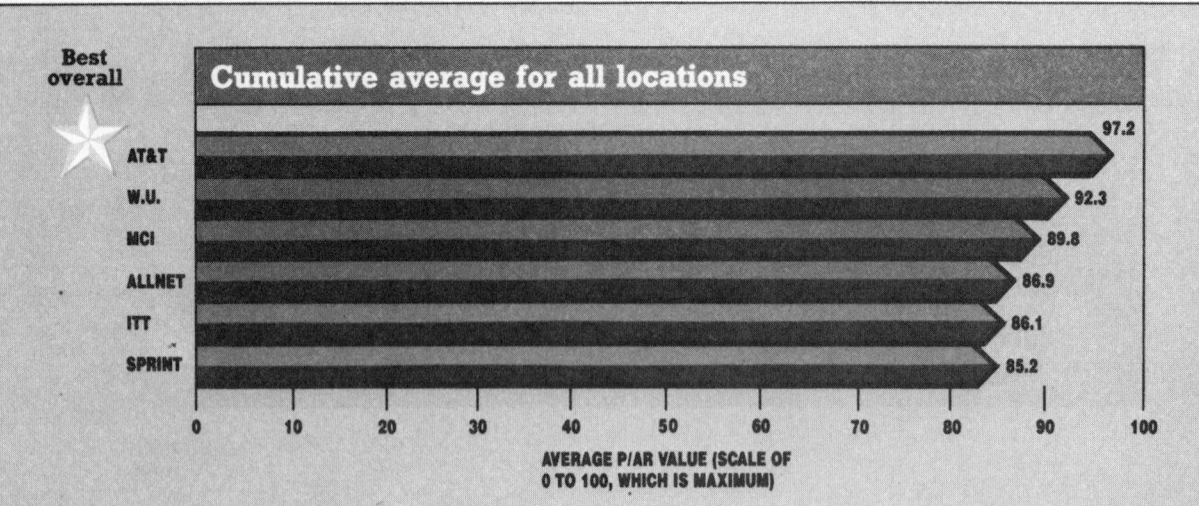

Cumulative average for all locations

Best overall ★

Carrier	Average P/AR value
AT&T	97.2
W.U.	92.3
MCI	89.8
ALLNET	86.9
ITT	86.1
SPRINT	85.2

AVERAGE P/AR VALUE (SCALE OF 0 TO 100, WHICH IS MAXIMUM)

WHAT MEASUREMENT SHOWS:
A COMPOSITE TEST FOR GAUGING OVERALL DATA-CARRYING CAPABILITY, P/AR VALUE INCLUDES ASSESSMENT OF DATA-SIGNAL OVERLAP (INTERMODULATION, OR HARMONIC, DISTORTION), TIME DELAY VARIATION BY FREQUENCY (ENVELOPE DELAY), AND BANDWIDTH LIMITATIONS OF THE CHANNEL.

EFFECT ON DATA:
A BAD P/AR MEASUREMENT INDICATES THERE ARE SOME PROBLEMS THAT WILL AFFECT OPTIMUM DATA TRANSMISSION; MEASUREMENT DOES NOT PINPOINT WHICH CHARACTERISTIC IS THE PROBLEM, HOWEVER.

EFFECT ON VOICE:
COMPONENTS OF P/AR THAT AFFECT VOICE INCLUDE NOISE LEVEL (BACKGROUND 'STATIC') AND HARMONIC DISTORTION, THOUGH THESE ARE NOT NECESSARILY PRESENT IN A BAD P/AR VALUE.

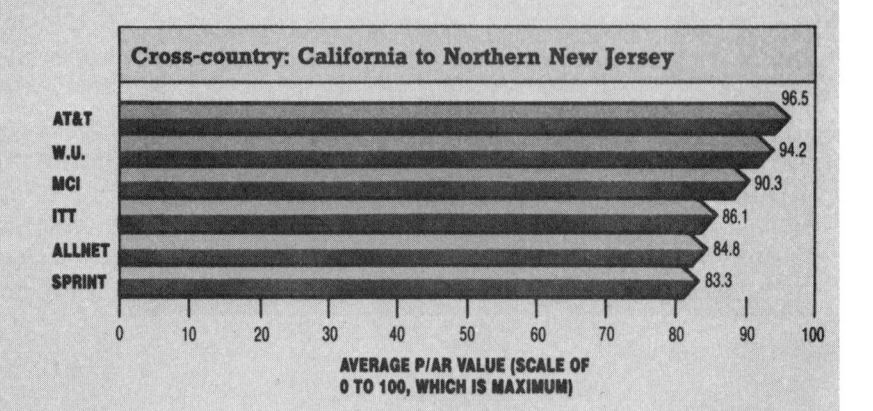

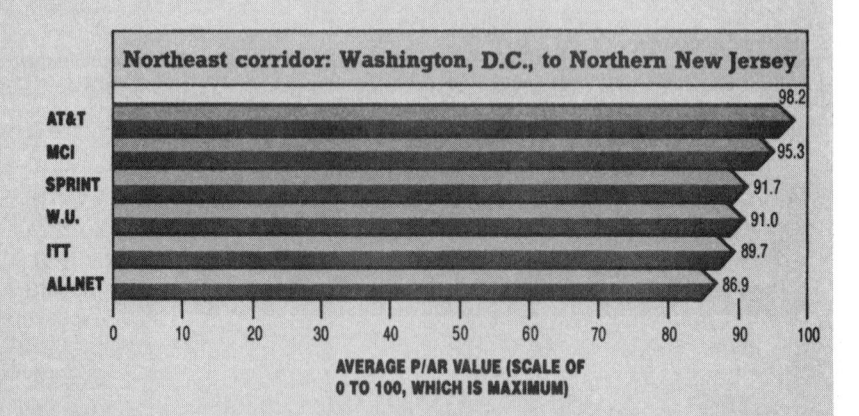

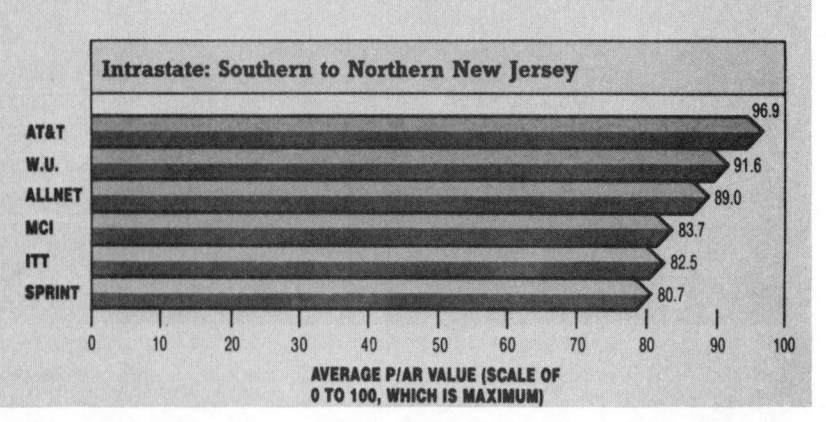

Cross-country: California to Northern New Jersey

Carrier	Average P/AR value
AT&T	96.5
W.U.	94.2
MCI	90.3
ITT	86.1
ALLNET	84.8
SPRINT	83.3

AVERAGE P/AR VALUE (SCALE OF 0 TO 100, WHICH IS MAXIMUM)

Northeast corridor: Washington, D.C., to Northern New Jersey

Carrier	Average P/AR value
AT&T	98.2
MCI	95.3
SPRINT	91.7
W.U.	91.0
ITT	89.7
ALLNET	86.9

AVERAGE P/AR VALUE (SCALE OF 0 TO 100, WHICH IS MAXIMUM)

Intrastate: Southern to Northern New Jersey

Carrier	Average P/AR value
AT&T	96.9
W.U.	91.6
ALLNET	89.0
MCI	83.7
ITT	82.5
SPRINT	80.7

AVERAGE P/AR VALUE (SCALE OF 0 TO 100, WHICH IS MAXIMUM)

Table 15: Summary comparison—analog signal quality

	OVERALL, ALL LOCATIONS	CROSS-COUNTRY	NORTHEAST CORRIDOR	INTRASTATE, NEW JERSEY
RECEIVED SIGNAL STRENGTH (dBm: DECIBELS REFERENCED TO 1 MILLIWATT)	SPRINT (−22.5) AT&T (−23.0) W.U. (−23.6) MCI (−24.57) ALLNET (−24.58) ITT (−25.9)	SPRINT (−20.8) AT&T (−21.6) ALLNET (−22.66) MCI (−22.69) W.U. (−22.7) ITT (−25.8)	ALLNET (−22.1) SPRINT (−22.7) W.U. (−23.40) AT&T (−23.41) ITT (−25.6) MCI (−25.9)	AT&T (−23.7) SPRINT (−24.0) MCI (−25.13) W.U. (−25.16) ITT (−26.7) ALLNET (−27.6)
SIGNAL LOSS "DISTORTION" (dB DIFFERENCE BETWEEN MID AND HIGH OR LOW FREQUENCY)	AT&T (1.62) SPRINT (1.91) W.U. (2.07) ALLNET (2.22) MCI (2.65) ITT (3.89)	AT&T (0.91) W.U. (1.19) ALLNET (1.58) SPRINT (1.59) MCI (1.66) ITT (4.22)	SPRINT (1.92) AT&T (2.41) W.U. (2.58) ALLNET (2.98) MCI (3.60) ITT (3.74)	AT&T (1.61) SPRINT (2.17) ALLNET (2.46) W.U. (2.63) MCI (2.70) ITT (3.57)
SIGNAL-TO-NOISE RATIO (dB OF SIGNAL STRENGTH OVER NOISE LEVEL)	AT&T (33.8) W.U. (33.0) MCI (32.1) ITT (31.8) SPRINT (31.1) ALLNET (31.0)	W.U. (32.4) ITT (29.96) ALLNET (29.92) AT&T (29.86) SPRINT (29.56) MCI (27.8)	AT&T (36.0) MCI (35.3) ITT (33.8) W.U. (33.5) SPRINT (33.1) ALLNET (31.0)	AT&T (35.2) MCI (33.3) W.U. (33.2) ALLNET (32.1) ITT (31.0) SPRINT (30.5)
PHASE JITTER (AVERAGE DEGREES OUT OF PHASE)	AT&T (3.0) W.U. (3.4) ALLNET (4.5) MCI (4.6) SPRINT (4.7) ITT (5.6)	W.U. (3.7) ALLNET (4.6) SPRINT (4.9) ITT (5.3) AT&T (5.5) MCI (7.7)	AT&T (2.0) MCI (2.3) W.U. (2.8) ITT (3.10) SPRINT (3.11) ALLNET (5.1)	AT&T (1.9) W.U. (3.7) ALLNET (4.0) MCI (4.3) SPRINT (6.2) ITT (8.5)
PEAK-TO-AVERAGE RATIO (SCALE OF 1 TO 100)	AT&T (97.2) W.U. (92.3) MCI (89.8) ALLNET (86.9) ITT (86.1) SPRINT (85.2)	AT&T (96.5) W.U. (94.2) MCI (90.3) ITT (86.1) ALLNET (84.8) SPRINT (83.3)	AT&T (98.2) MCI (95.3) SPRINT (91.7) W.U. (91.0) ITT (89.7) ALLNET (86.9)	AT&T (96.9) W.U. (91.6) ALLNET (89.0) MCI (83.7) ITT (82.5) SPRINT (80.7)

rankings from this year, then the greatest change is noted for Allnet, MCI, Sprint, and Western Union.

Allnet showed a considerable drop in several areas: first-try call completion and bit and block error rate. Similarly, the results for Sprint show that it dropped significantly in the categories of connect time, signal-to-noise ratio, and peak-to-average ratio. Western Union showed perhaps the most significant improvement, rising significantly in the ratings for connect time, first-try call completion, and bit and block error rate. MCI's most remarkable change was in a significantly improved bit error rate.

Billing

In the 1984 study, we found many small anomalies in the billing accuracy of the different carriers. This year, there were very few, although some other billing problems became evident.

Because of the way our test calls were made this year—10XXX dialing via equal-access central offices—the local telephone company was the agency that was supposed to bill for these gypsy calls. This was true for Allnet, AT&T, ITT, and Western Union, but not for MCI or Sprint (which billed us directly after presumably retrieving the call records from the local telephone companies).

All calls billed by MCI and Sprint were accurate: They were all rounded up to the next full minute but otherwise were accurate. Sprint did not bill us for calls of less than one minute (it began charging us for a two-minute call when actual duration was a minute and a half). MCI did charge us for calls that were less than a minute in duration, even where the modems couldn't connect and aborted the call after about five seconds.

What was especially noteworthy was the long delay in billing by the BOCs on behalf of the long-distance carriers. For each BOC—C&P Telephone, N. J. Bell, and Pacific Bell—the billing for calls placed over AT&T (also via 10XXX dialing) was always included in the first bill received after the calls were placed (within about two to three weeks). But charges for the other BOC-billed carriers did not appear until much later.

Two months after the calls were made, bills from the BOCs started to include the calls placed via ITT and Western Union. And as this article went into production, fully three months after the calls were made, none of the BOCs had yet billed us for calls we made via Allnet. The complaints by AT&T's competitors concerning BOC billing of 10XXX calls made via equal-access central offices, therefore, would appear to be warranted. ■

We wish to thank all those at CCMI, without whose help this study would not have been possible. And thanks also to Hewlett-Packard and Tektronix for the loan of their equipment, personnel, expertise, and help in interpreting the test results.

Table 16: Effect of phase hits on bit-error performance

	CARRIER	PHASE HITS	BER	BLOCK ERROR RATE	PERCENT CONNECTIONS WITH PHASE HITS
OVERALL ALL LOCATIONS	MCI				
	W.U.				
	AT&T				
	SPRINT				
	ALLNET				
	ITT				
CROSS-COUNTRY	W.U.				
	SPRINT				
	MCI				
	ITT				
	AT&T				
	ALLNET				
NORTHEAST CORRIDOR	MCI				
	ITT				
	W.U.				
	AT&T				
	SPRINT				
	ALLNET		—	—	
INTRASTATE NEW JERSEY	AT&T				
	MCI				
	W.U.				
	SPRINT				
	ALLNET				
	ITT				

AVERAGE NUMBER OF PHASE HITS, PER 15 MINUTES OF MEASURED TIME
- LESS THAN 1
- FROM 1 TO 5
- FROM 5 TO 15
- MORE THAN 15

BLOCK ERROR RATE; AVERAGE NUMBER OF BLOCKS WITH ERRORS, PER 1,000 BLOCKS

BIT ERROR RATE (BER)
- EXCELLENT = BETTER THAN 1 IN 10^5
- GOOD = FROM 1 IN 10^4 TO 1 IN 10^5
- FAIR = FROM 1 IN 10^3 TO 1 IN 10^4
- POOR = WORSE THAN 1 IN 10^3

PERCENT OF CONNECTIONS WHERE PHASE HITS WERE MEASURED
- LESS THAN 1 PERCENT
- FROM 1 TO 5 PERCENT
- FROM 5 TO 25 PERCENT
- MORE THAN 25 PERCENT

Table 17: Impulse noise

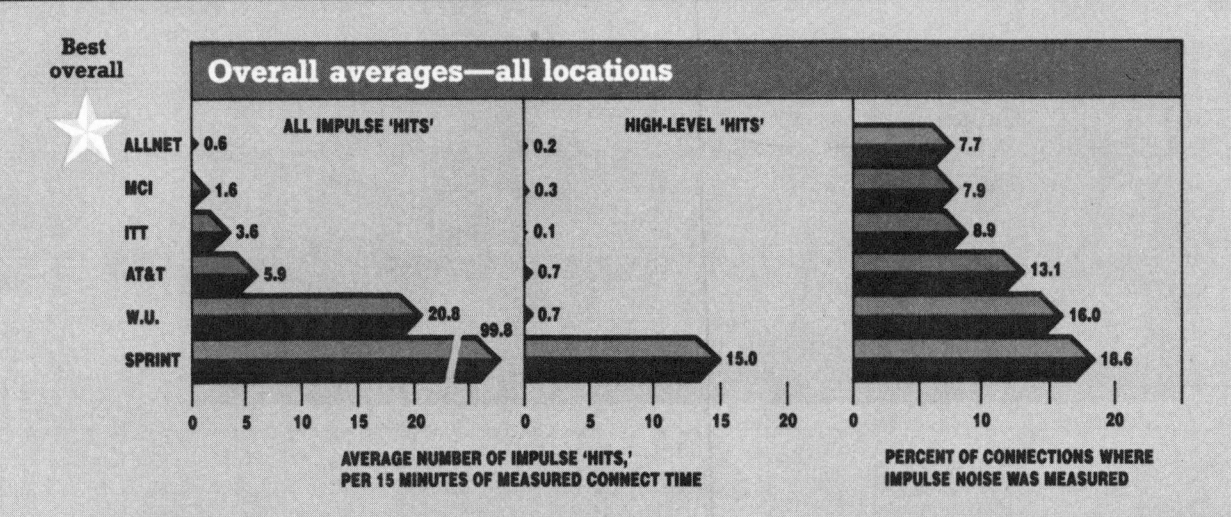

Best overall ★ ALLNET

Overall averages—all locations

ALL IMPULSE 'HITS' (Average number of impulse 'hits,' per 15 minutes of measured connect time)
- ALLNET: 0.6
- MCI: 1.6
- ITT: 3.6
- AT&T: 5.9
- W.U.: 20.8
- SPRINT: 99.8

HIGH-LEVEL 'HITS'
- ALLNET: 0.2
- MCI: 0.3
- ITT: 0.1
- AT&T: 0.7
- W.U.: 0.7
- SPRINT: 15.0

PERCENT OF CONNECTIONS WHERE IMPULSE NOISE WAS MEASURED
- ALLNET: 7.7
- MCI: 7.9
- ITT: 8.9
- AT&T: 13.1
- W.U.: 16.0
- SPRINT: 18.6

WHAT MEASUREMENT SHOWS:
BRIEF SURGES OF VOLTAGE, WHICH ARE SHARP 'SPIKES' IN THE BACKGROUND NOISE OF THE CHANNEL; 'HITS' LAST FOR ONLY A FEW MILLISECONDS AND CAN BE AS STRONG (LOUD) AS THE MAIN SIGNAL.

EFFECT ON DATA:
HIGH-LEVEL IMPULSE NOISE CAUSES ERROR BY CHANGING THE BINARY VALUE OF DATA TO THE RECEIVER.

EFFECT ON VOICE:
DISTINCT, AUDIBLE 'STATIC' AND CRACKLING ON THE LINE.

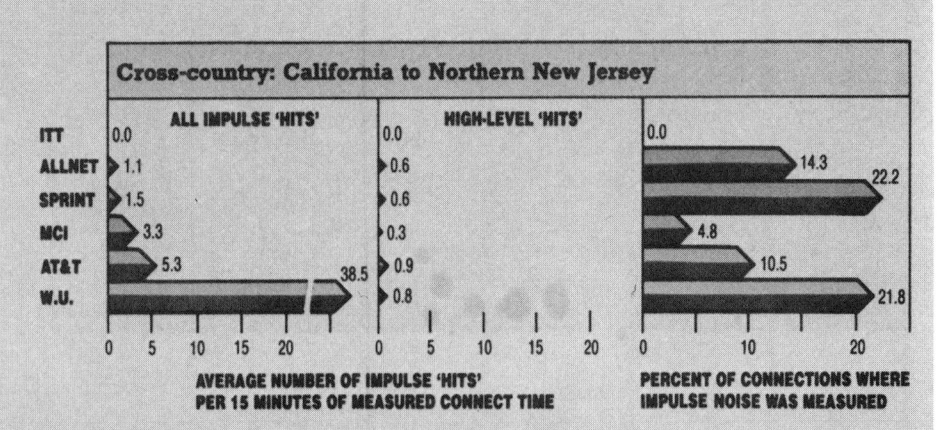

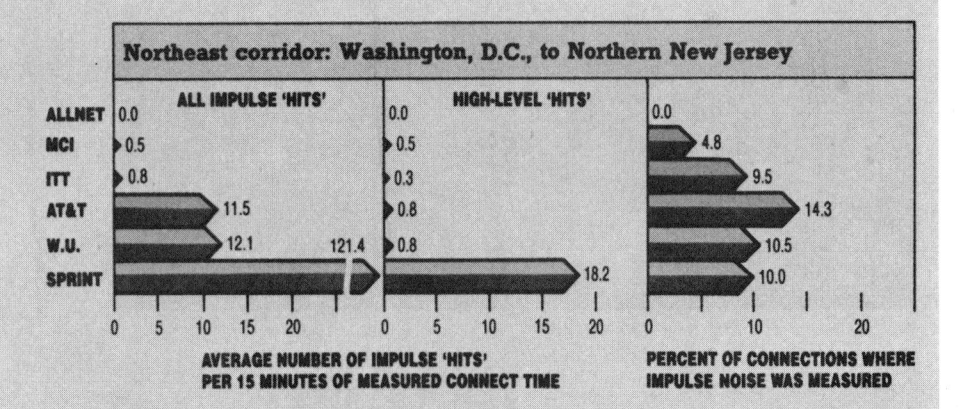

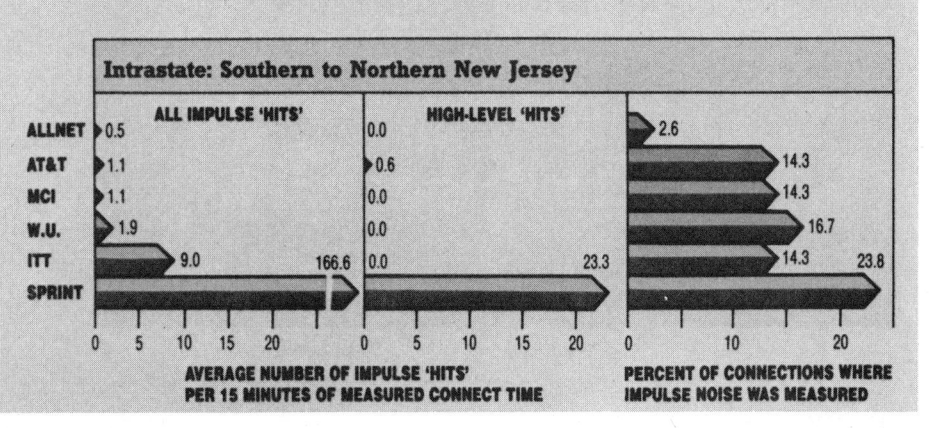

Cross-country: California to Northern New Jersey

ALL IMPULSE 'HITS'
- ITT: 0.0
- ALLNET: 1.1
- SPRINT: 1.5
- MCI: 3.3
- AT&T: 5.3
- W.U.: 38.5

HIGH-LEVEL 'HITS'
- ITT: 0.0
- ALLNET: 0.6
- SPRINT: 0.6
- MCI: 0.3
- AT&T: 0.9
- W.U.: 0.8

PERCENT OF CONNECTIONS WHERE IMPULSE NOISE WAS MEASURED
- ITT: 0.0
- ALLNET: 14.3
- SPRINT: 22.2
- MCI: 4.8
- AT&T: 10.5
- W.U.: 21.8

Northeast corridor: Washington, D.C., to Northern New Jersey

ALL IMPULSE 'HITS'
- ALLNET: 0.0
- MCI: 0.5
- ITT: 0.8
- AT&T: 11.5
- W.U.: 12.1
- SPRINT: 121.4

HIGH-LEVEL 'HITS'
- ALLNET: 0.0
- MCI: 0.5
- ITT: 0.3
- AT&T: 0.8
- W.U.: 0.8
- SPRINT: 18.2

PERCENT OF CONNECTIONS WHERE IMPULSE NOISE WAS MEASURED
- ALLNET: 0.0
- MCI: 4.8
- ITT: 9.5
- AT&T: 14.3
- W.U.: 10.5
- SPRINT: 10.0

Intrastate: Southern to Northern New Jersey

ALL IMPULSE 'HITS'
- ALLNET: 0.5
- AT&T: 1.1
- MCI: 1.1
- W.U.: 1.9
- ITT: 9.0
- SPRINT: 166.6

HIGH-LEVEL 'HITS'
- ALLNET: 0.0
- AT&T: 0.6
- MCI: 0.0
- W.U.: 0.0
- ITT: 0.0
- SPRINT: 23.3

PERCENT OF CONNECTIONS WHERE IMPULSE NOISE WAS MEASURED
- ALLNET: 2.6
- AT&T: 14.3
- MCI: 14.3
- W.U.: 16.7
- ITT: 14.3
- SPRINT: 23.8

Table 18: Change in ratings—1984 to 1986[1]

CRITERIA		ALLNET	AT&T	ITT	MCI	SPRINT	W.U.
QUICKEST CONNECT TIME	1984: CROSS-COUNTRY	4	1	5	2	3	6
	1986: CROSS-COUNTRY	2	1	6	4	5	3
	OVERALL	4	1	5	3	6	2
FIRST-TRY CALL COMPLETION	1984: CROSS-COUNTRY	1	2	6	3	4	5
	1986: CROSS-COUNTRY	3	1	6	4	5	2
	OVERALL	6	1	5	3	4	2
BEST BIT ERROR RATE	1984: CROSS-COUNTRY	1	3	6	5	2	4
	1986: CROSS-COUNTRY	5	6	4	3	2	1
	OVERALL	5	3	6	1	4	2
BEST BLOCK ERROR RATE	1984: CROSS-COUNTRY	1	2	6	5	3	4
	1986: CROSS-COUNTRY	5	6	3	4	2	1
	OVERALL	5	4	6	1	3	2
STRONGEST RECEIVED SIGNAL	1984: CROSS-COUNTRY	4	2	3	5	1	6
	1986: CROSS-COUNTRY	3	2	6	4	1	5
	OVERALL	5	2	6	4	1	3
BEST SIGNAL-TO-NOISE RATIO	1984: CROSS-COUNTRY	5	1	6	4	3	2
	1986: CROSS-COUNTRY	3	4	2	6	5	1
	OVERALL	6	1	4	3	5	2
BEST PEAK-TO-AVERAGE RATIO VALUE	1984: CROSS-COUNTRY	5	1	6	4	2	3
	1986: CROSS-COUNTRY	5	1	4	3	6	2
	OVERALL	4	1	5	3	6	2
LEAST PHASE JITTER	1984: CROSS-COUNTRY	3	1	6	5	4	2
	1986: CROSS-COUNTRY	2	5	4	6	3	1
	OVERALL	3	1	6	4	5	2

☐ DROPPED AT LEAST TWO PLACES IN RANKING, BOTH CROSS-COUNTRY AND OVERALL, FROM 1984 RANKING.

☐ IMPROVED AT LEAST TWO PLACES IN RANKING, BOTH CROSS-COUNTRY AND OVERALL, FROM 1984 RANKING.

(1) TESTING IN 1984 WAS DONE ONLY CROSS-COUNTRY, FROM SAN FRANCISCO TO NEW YORK. ALSO, TESTING IN 1984 INCLUDED TWO CARRIERS—SBS AND TELESAVER—WHICH WERE NOT TESTED THIS YEAR. THEIR RANKING IN THE 1984 TEST RESULTS HAS BEEN DELETED, AND THOSE OF THE REMAINING SIX CARRIERS ADJUSTED ACCORDINGLY.

David M. Rappaport, Arthur Andersen & Co., Chicago, Ill.

Voice mail: Key tool or costly toy?

Computer-based telephone message devices, priced at upwards of a million dollars, can amply earn their keep, but prospective buyers have a lot of homework to do.

Network managers, whose ultimate goal is to support improved white-collar performance, have a relatively new candidate for their repertoire of useful office technologies. Voice mail, which is essentially a computer-based, multi-user, intelligent answering machine, aims at enhancing the usefulness of that perennial professional appendage—the telephone.

Users of voice mail can exchange verbal information without being on the telephone at the same time. Voice-mail products are related to simple tape-based answering machines but are far more functional. They work by accepting verbal input, digitizing it (perhaps using a compression algorithm), and then storing it on a disk. Voice mail replaces telephone slips with digitized spoken messages that can be ordered, scanned, forwarded, annotated, and otherwise managed.

The voice-digitization equipment used for processing messages is fairly expensive. For example, low-end microcomputer boards that support voice mail on a single line run as much as a thousand dollars or more, while high-end products may cost a million dollars or more. And because the technology is new, its potential is largely unknown. Therefore, network managers should evaluate voice-mail products currently on the market in terms of their features, prices, limitations, and benefits. They can then use such techniques as surveys and pilot tests (trial runs) to assess how valuable voice mail might be in increasing their organizations' productivity.

Voice mail promises to reduce a major time-waster of busy professionals: telephone tag. Long episodes of telephone tag waste time, cause frustration, and cost money. Studies show that up to three-quarters of all business telephone calls are not completed on the first try. Since callers are more likely to leave long, detailed messages with a machine than with a person who may be busy, voice mail can reduce the overall number of calls made.

Another key advantage of voice mail is that calling and called parties need not be present at the same time. For instance, a voice-mail user may leave for a trip at three in the afternoon, arriving at a hotel well after five o'clock. Even though the home office is closed, the user can call in and retrieve important telephone messages that may have been left in the interim. Messages from other users on the same voice-mail processor can be answered or instructions can be left for a coworker to return specific calls and convey certain information. That way, the traveler can attend an early morning meeting the next day, knowing that yesterday's crises are being handled.

Similarly, important information required early in the morning in Europe need no longer mean a 2 A.M. telephone call from the United States. The message can be left on voice mail at the end of the business day in the United States (late night in Europe) and picked up when the European office opens the next morning.

Besides overcoming time-zone differences, voice mail enables people to make better use of their telephones. The majority of intracompany telephone calls do not require an immediate response. Voice mail allows people to leave information for busy colleagues without interrupting them.

User view

The best candidates for voice mail, those most likely to use it to its fullest, are those people who are away from their desks frequently, traveling or attending meetings, or those who need to be in regular communication with others in their department or work group. The least likely candidates for voice mail are managers who either rarely venture from their desks or have personal secretaries.

The immediate effect that voice mail will have on users is more subtle than that of other office technologies. It does not require additional terminal equipment, since users can access voice mail directly through their standard touch-

1. Command tree. *Pressing digits on a telephone keypad, a user can review and manage incoming messages (answering them, sending them on, saving them for later use, or purging them), create and send outgoing messages to one or more recipients (using standard delivery or special services), or ask for help along the way.*

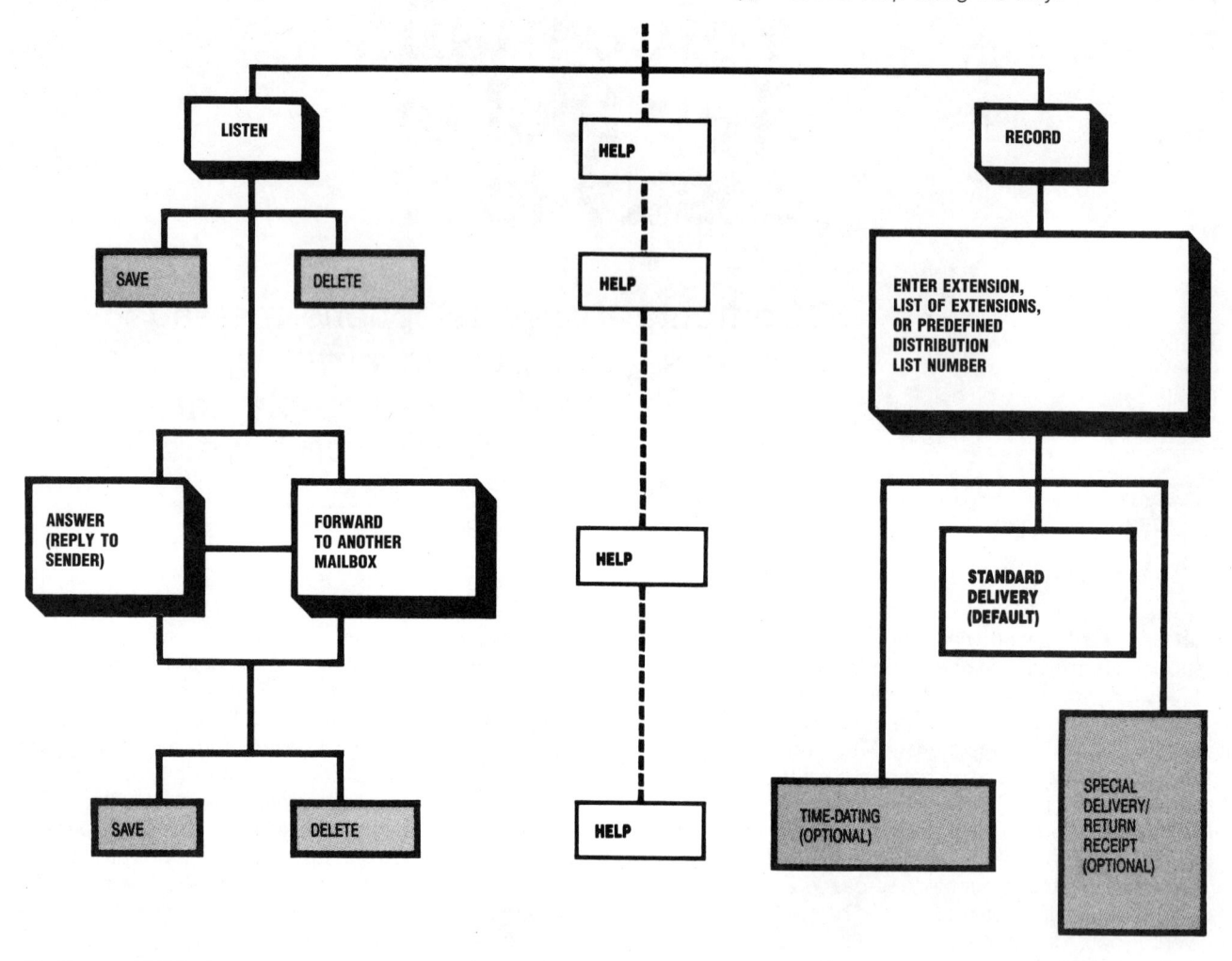

tone telephones. New users need only register with the company's voice-mail administrator and receive a unique mailbox number and secret password to get started.

Commands are entered via the telephone keypad in a logical order (Fig. 1). Through single-digit commands, users can listen to incoming messages or create new ones. Typically, they first access their mailboxes, listen to each new message (played back exactly as spoken by the sender), and then act on it. This may entail deleting the message, saving it for later reference, answering it, or forwarding it to other users.

To record a message, users enter the mailbox number(s) of the intended recipient(s), choose whether the message is to be delivered immediately or at some future time, and then speak the message into the telephone. At each level of the command hierarchy, a HELP function may also be available, pointing the way to documentation or providing directions on how to use specific functions.

In addition to user mailboxes and the HELP function, a typical voice-mail configuration might include a supplemental mailbox where users can leave comments for the voice-mail administrator about problems they have had or suggestions about changes to the voice-mail operation.

In addition to the actual telephones, a voice-mail network consists of a central computer, disk storage, and voice-processing equipment, with lines running to the telephone company central office (CO) or private branch exchange (PBX), as shown in Figure 2. The CO attachments permit access to voice mailboxes from outside the PBX network, while the internal lines can channel PBX calls to an electronic message center.

Configuration

Messages that enter through the input/output (I/O) ports are digitized by the voice coder/decoders (codecs) and tagged with pertinent details (source and destination identifiers, date and time information, and so on). The message processors control the codecs, handle time-outs (releasing the line after a predefined period of inactivity), and, in some implementations, differentiate between tone and voice commands. They may also omit the pauses that occur during the digitized message and use sophisticated compression algorithms to make the best use of available disk space for storing recorded messages (see "Inside an internetworking voice-mail processor").

Many of the functions performed by a voice-mail unit duplicate those of a modern digital PBX, so there are economic reasons to combine the hardware. For example, the central processing unit in the PBX could run additional software to handle voice mail. Since I/O logic is already in place to control the ports on the PBX, additional I/O ports could be installed for voice mail. Some PBXs even have codecs already in place for converting the analog voice signals to a digital format for internal switching. For these reasons, voice mail is appearing as an optional feature of digital PBXs.

Integrated, cooperating, standalone

The desired degree of integration between the voice mail-and PBX equipment is a major decision facing any prospective buyer. As stated, voice-mail products can be integrated into PBX equipment. They can also be purchased as standalone units. Between these two extremes are voice-mail processors that cooperate with one or more major PBXs. Though the cooperating units work more intimately with the PBXs than do the standalone devices, they rely on separate hardware and may come from separate vendors.

In the case of standalones, only voice and ring signals are passed between the PBX and the voice-mail unit, which the PBX views as a series of telephones. Cooperating units exchange commands with the PBX, either via a separate set of control links or via inband tones sent over the voice channels joining them. Totally integrated voice-mail units reside in the PBX, passing information via internal memory (changing status bits at certain memory locations).

Each approach represents a trade-off. Standalone units can benefit smaller companies that do not require PBX equipment. Like the standalones, cooperating devices may be attractive to a company that already owns or has a long-term lease on its PBX. However, integrated voice-mail products offer the advantage of lower incremental cost, since the units can share hardware components and such functionality as message-waiting notification. (Since a PBX controls its station sets, the integrated units may let the user's telephone emit a flashing light or a special tone to indicate that messages are waiting in the voice mailbox.)

The principle trade-off between the integrated type of voice-mail units and the other two types is vendor flexibility. Integration restricts the buyer to the capabilities offered by the PBX supplier. By providing devices that cooperate with certain PBXs, the makers of standalone units can achieve comparable functionality while offering the buyer greater PBX-shopping flexibility.

The standalone unit, as shown in Figure 2, allows the user to choose how many and which telephone lines will be routed through the PBX as opposed to going directly to voice mail. At first, it might seem that the more lines through the PBX, the better since such PBX functions as call queuing and hunt groups can then be applied to managing voice-mail traffic as well. However, it means that the PBX must have adequate line-handling capacity for present and future applications, which may place unnecessary cost demands on PBX hardware. Also, inbound Wide Area Telecommunications Service (WATS) lines may be misused if all lines are routed through the PBX. People could dial the toll-free number devoted to voice mail and then reroute their calls out of the PBX.

Finally, certain products can automatically reroute unanswered calls to voice mail. For example, a call to an unanswered or busy telephone can be passed automatically to the voice-mail unit, which answers the call with an individualized announcement and allows the caller to record a message.

In this mode, the only indication that the caller is talking to voice mail rather than to an answering machine might be the request to hit a key at the end of the message if transfer back to an operator is desired. The ability to be routed back to a human operator, known as an escape mechanism, is a major benefit of integrated units. While cooperating voice-mail products can also do this, they typically require the caller to rekey the extension that had

2. Inside view. *The computer that runs a voice-mail network stores digitized audio on disk and exchanges signals with the lines via specialized components.*

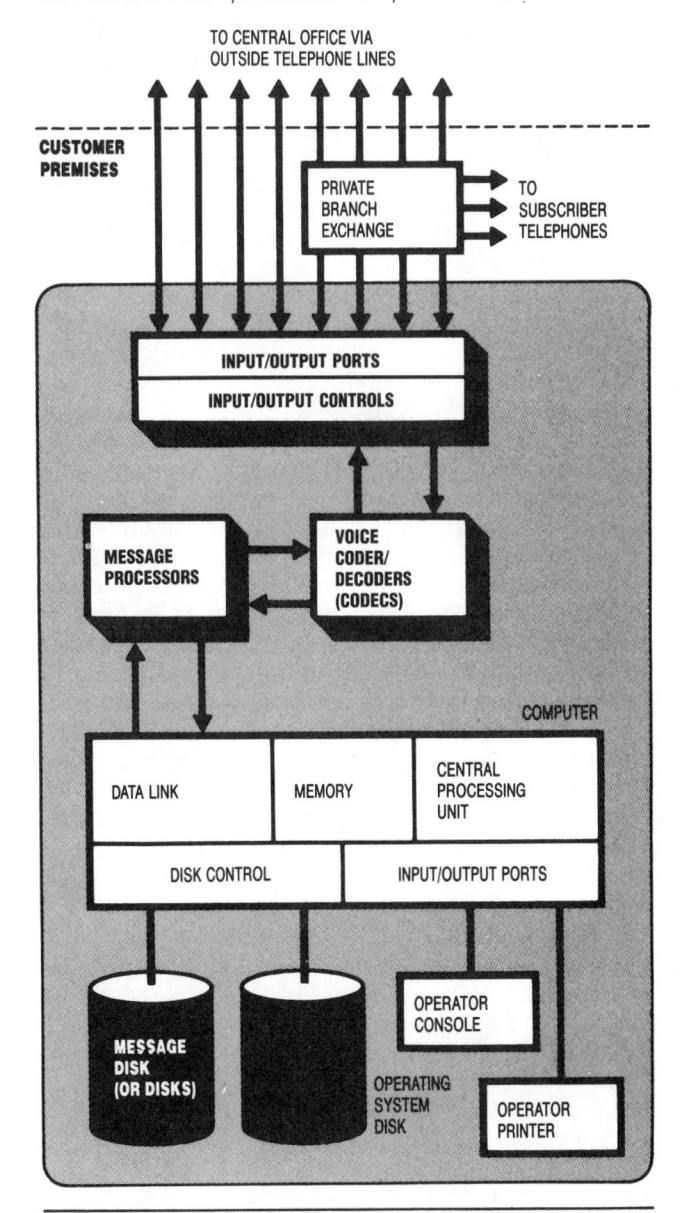

Inside an internetworking voice-mail processor

In small voice-mail configurations of 50 to 100 users, message processing is a fairly simple matter of digitizing analog inputs for storage, manipulation, and eventual playback. But small networks tend not to remain small. Demand for voice mail is growing both vertically (users seek larger voice-mail networks) and horizontally (as usage increases, vendors and their customers find the need to tie networks together). Because of this, vendors of voice-mail equipment typically try to offer a range of compatible products that can be used for large network applications and provide internetworking capability.

The pioneer voice-mail vendor and one of the market's major players, Dallas-based VMX Inc., has until recently been the only vendor talking about internetwork voice-mail links. VMX produces the largest voice-mail devices available today, based on a product architecture that uses multiple, distributed microcomputers, and proprietary encoding and compression techniques. These devices serve medium to extra-large user networks—from about 1,000 to more than 5,000 users—with computing power provided at the level of a Digital Equipment Corp. VAX 11/780 minicomputer.

Components

Claiming to be the only vendor currently shipping products with more than 24 voice ports, VMX offers configurations that range from 8- to 64-port models. In VMX units, each port employs an eight-bit microprocessor for port control. The microprocessor answers calls, converts touch-tone signals to instructions for the voice-mail central processing unit, monitors sessions in progress, and interprets user commands. It also handles voice synthesis and dialing for voice-mail processor-initiated outgoing calls (which are obviously useful in telemarketing applications).

Each port also contains a microprocessor custom-designed by VMX for digitizing analog signals. This chip employs a proprietary compression technique, a form of adaptive delta modulation (ADM), that samples and compares analog waveform modulations in order to make the best use of memory capacity—an important factor in large configurations of several thousand users, each of whom may have a dozen recorded telephone messages stored.

Adaptive delta modulation works similarly to the more widely known pulse-code modulation (PCM) used by the public telephone network. But where PCM requires eight bits for each sample (at a sample rate of 8,000 samples per second), ADM is able to record its information in just one bit per sample.

This is because ADM notes not the waveform amplitude but the change in amplitude, whether increased or decreased, over the previous samples—information that can be shown with a single binary switch denoting a change up or down.

To work properly, however, ADM requires a higher sampling rate, at least 16 kbit/s, necessitating powerful input/output (I/O) processing hardware. In a 64-port VMX unit, for instance, the ADM technique requires I/O throughput of at least 1 Mbit/s (64 ports at 16 kbit/s each), which is why each port must employ a dedicated microprocessor, as is the case in the VMX voice-mail units.

Three additional 16-bit microprocessors coordinate the port processors by performing administrative, database-access, and control functions, all of which are needed in dynamic, multiple-user environments. To keep track of call activities, each 16-bit chip uses from one-half to one Mbyte of memory, depending on the extent of functionality (which features out of the entire feature set the user has loaded, how the voice-mail unit has been configured, and so on).

VMX uses a standard storage-module-device interface that allows connection to a variety of popular hard-disk units, which are used primarily to store digitized messages. Disk capacity ranges up to about two gigabytes for larger configurations. With the ADM compression technique, two gigabytes can hold about 170 hours of messages or, at the average rate of one to one-and-a-half minutes per message after the compression of pauses, as many as 10,200 messages.

The linking of multiple voice-mail networks is accomplished by adding location-dependent digits to user mailbox addresses, much like the telephone area codes that precede local telephone numbers. In this fashion, directory databases need not be replicated throughout all networks; each local network maintains its own local database, with the extra digits pointing to the appropriate network.

For example, a typical mailbox address may contain four digits for intranetwork routing, with an additional one or two digits used to link to another network. Upward capacity for VMX products is currently 15 digits (including the prefix), more than enough for even the largest of VMX's current customers. Large multi-unit users typically combine three or four 64-port units with two or three 32-port products and average betwen 5,000 and 10,000 messages per day.

My machine will talk to yours

Although linking geographically dispersed networks together does not present any directory/addressing problems, it does tend to drive up basic product hardware requirements. VMX claims that the greater the extent of multinetwork linking, the higher the percentage of messages that go outside local networks. Thus additional channels and ports are needed to accommodate internetwork traffic as the number of users on each network grows.

To keep link costs down, VMX uses analog transmission between voice networks. This permits one VMX unit to literally talk to another over any voice-grade dial-up line. According to VMX, at the 16-kbit/s digitization rate, digital transmission would require modems and high-speed 56-kbit/s lines. With analog communications, the only equipment cost for multinetwork applications is that of two to four ports by means of which each network converses with the others.

originally been called, which can be an annoyance. (For example, a caller may dial a company's main number and get a central operator, who puts the call through without mentioning the extension. When the voice-mail unit asks for the desired extension, the caller does not know it and has to call back.) Standalone units cannot offer an escape mechanism, since they are regarded by the PBX as a bank of answering machines. Once a call is routed to them, they have no way to tell the PBX to take back the call.

Gauging feature richness

As voice-mail products grow to maturity, the mix of features necessary for their efficient use is becoming better understood. The following are key items for comparing vendors, as well as some of the more important features currently offered:

■ User capacity and price range. As the number of users increases, so do the requirements for message storage, communications ports, and, hence, the cost. Conversely, limiting message length and increasing the efficiency of the voice digitization algorithm reduces the amount of storage needed for a given number of users.

■ Distribution lists. This feature is measured by how many lists a user can access, as well as how many names each list can carry. Capabilities needed typically relate directly to company size; that is, the larger the company, the greater the distribution-list capacity required.

■ Deferred message delivery. This feature allows users to request specific delivery times for outgoing messages. Deferred delivery is helpful if the caller is sending the same announcement to several people and wants to ensure the message is available at a particular time of day.

For example, a supervisor might call in a message just before leaving for a trip on, say, Wednesday, and defer its delivery until Friday morning. One reason for this would be if the message announces a promotion as of 8:30 Friday morning and contains instructions on resulting departmental changes that affect a number of employees.

■ Message confirmation, annotation, and forwarding. These features, akin to those offered by electronic mail, point to the fundamental power of the voice-mail processor. Confirmation can be achieved manually or automatically. Users can periodically ask via a keystroke if all sent messages have been received, or they can request automatic confirmation at the time of sending, much like attaching a return card to a letter.

Annotation is a form of editing that lets a message recipient add information to a stored message and then forward it. (Forwarding is another common voice-mail feature.) A typical example is where the user prefaces a message with instructions to "please take care of this call" and forwards it to an associate.

■ Verbalized directory service. This feature offers a means of obtaining a mailbox number by simply entering a user name via the keypad. Since, with most voice-mail units, the sender must know the recipient's mailbox number, this capability is quite useful.

■ Message length. Maximum message length can range up to about six minutes or more, depending on storage and processing capacity and on the compression algorithms used. Obviously, the greater the possible message length, the easier the service is to use.

■ Networking. Allowing multiple voice-mail units to be interconnected into a network is particularly advantageous for larger corporations where the total voice-mail user population may exceed the capacity of a single configuration. A key networking issue is in routing messages between local and remote locations. Ideally, recipient locations should be transparent to the caller, with the network handling any internal forwarding of messages.

Users should weigh the relative merits of all voice-mail features according to their actual requirements (Table 1). For instance, in typical applications, a robust implementation of the distribution-list feature is far less valuable than the type-of-message-preview function. The table shows distribution-list capacities ranging from eight to 999 lists per user. This reflects the internal addressing method used (a simple function to alter) and should be evaluated in the context of real applications. Most people do not need more than eight or 10 different broadcast lists at one time.

On the other hand, the type-of-message-preview feature determines how much time and effort it will take the user to plow through a voice-mail in-box. This can be a tedious, oft-repeated chore, so the preview feature is worth examining closely when comparing voice-mail products. The quality of information provided when previewing a caller list can be critical in helping the user get through messages quickly and easily (perhaps finding and acting on a lurking, critical "hot potato" message).

Most products announce how many messages are in a mailbox. Some offer a second level of detail, enumerating messages by user name, saying, for example, "Pat Jones: four messages; Jean Doe: one message," and so on. This can be very handy for users dealing with a particularly pressing project. The user may want to go directly to the vice president's messages and play those back in the correct chronological order, since the last message may rebut something that was said earlier.

Slightly less helpful are those preview functions that do not summarize the waiting messages by name but simply list all messages in chronological order. This leaves it up to the user to scroll through them sequentially.

A feature usually considered of much less significance, and therefore not shown on the table, is archival capability. There are many means of storing information, from desk drawers and file cabinets to word processing and electronic-mail machines, so a facility for filing voice-mail messages may not be necessary. By and large, users find that the kind of simple note taking they do during routine telephone conversations is the most efficient means of storing voice messages.

In general, the features shown in the table represent brief comparisons of relatively well-known, deliverable products. Users should base their choices on real application needs and examine vendors closely when evaluating voice-mail offerings.

Drawbacks

Prospective buyers should also be aware of the disadvantages of voice mail. Because the operation of voice mail requires using dual-tone multi-frequency (touch-tone) telephone signals, it can be difficult for some outside callers and Centrex users to access. People with rotary-dial telephones (and those calling from Europe, where touch-

Table 1: A sampling of voice-mail products and their features

	AT&T INFORMATION SYSTEMS, AUDIX[1]	CENTIGRAM CORP., VOICEMEMO	IBM, AUDIO DISTRIBUTION SYSTEM[4]	INTECOM INC., INTEMAIL	ROLM, PHONEMAIL
MINIMUM CONFIGURATION PORTS / STORAGE (HOURS) / COST	8 / 25 / $140,000	4 / 4 / $25,000	4 / 30 / $90,000	8 / 16 / $130,000	8 / 7.5 / $80,000
MAXIMUM CONFIGURATION PORTS / STORAGE (HOURS) / COST	32 / 196 / $430,000	14 / 14 / $60,000	16 / 101 / $280,000	64 / 170 / $650,000	16 / 51.5 / $150,000
MAXIMUM POTENTIAL USERS[2]	4,000	500	3,000	8,000	1,000
MAXIMUM MESSAGE LENGTH (IN MINUTES)	20	5 FOR NONUSERS, 15 FOR USERS	5.75	3 (CAN BE INCREASED IF NEEDED)	VARIES WITH CONFIGURATION
INTERFACE CAPABILITIES	INTEGRATED INTO AT&T PBXs[3]	STANDALONE. ALSO PBX INTERFACE (COOPERATION) AVAILABLE WITH NTI, SIEMENS, AND LEXAR PBXs	STANDALONE OR SERIES/1-BASED WITH PBX INTERFACE (COOPERATION)	COMPLETELY INTEGRATED INTO INTECOM IBX	INTEGRATED INTO ROLM CBX OR CAN BE PURCHASED AS STANDALONE
NETWORKING	NOT AT THIS TIME	NOT AT THIS TIME	NO	SUPPORTS NETWORKING CAPABILITIES OF IBX	FUTURE RELEASE
DISTRIBUTION LISTS	999 LISTS PER USER, 250 USERS PER LIST	9 LISTS PER USER, 125 USERS PER LIST	25 LISTS PER USER, 10 USERS PER LIST	10 LISTS PER USER, UNLIMITED NUMBER OF USERS PER LIST	100 LISTS PER USER, UNLIMITED NUMBER OF USERS PER LIST
DEFERRED MESSAGE DELIVERY	YES	NO	YES	YES	NO
MESSAGE DELIVERY CONFIRMATION	YES	YES, BUT ONLY IF SENT VIA A DISTRIBUTION LIST	YES	YES	YES, BUT ONLY IF SPECIFIED THAT MESSAGE REQUIRES RETURN NOTIFICATION
MESSAGE CENTER FOR NONSUBSCRIBERS	CAN CALL FORWARD TO MAILBOX, NONUSERS RECEIVE USER'S GREETING	DIRECT DIAL NUMBER WITH USER NUMBER KEYED IN FROM ANOTHER STATION OR MESSAGES MANUALLY REDIRECTED	NONUSERS MAY BE ROUTED TO RECEPTIONIST OR ANOTHER VOICE-MAIL EXTENSION	CAN CALL FORWARD TO MAILBOX, NONUSERS RECEIVE USER'S GREETING	DIRECT INTEGRATION WITH ROLM CBX, CALLS ROUTED DIRECTLY INTO MAILBOX WITH NO REENTRY OF EXTENSION
OUTSIDE MESSAGE DELIVERY	NO	NO, BUT CAN ACTIVATE PAGER	TO AUTHORIZED PHONE NUMBERS ONLY	YES, SUPPORTS INTERNATIONAL DIRECT DIALING	NOT DIRECTLY OVER OUTSIDE LINES, ONLY VIA PBX
ANNOTATE AND FORWARD MESSAGE	YES	YES	YES, BUT CAN SPECIFY MESSAGE AS NONFORWARDABLE	YES	YES, BUT CAN SPECIFY MESSAGE AS NONFORWARDABLE
VOICE-MAIL SUBSCRIBER DIRECTORY	CAN LOOK UP NAMES AND NUMBERS, CAN ADDRESS MESSAGES BY USER NAME	NO	YES, SPELL OUT USER'S LAST NAME	NO	NO
TYPE OF MESSAGE PREVIEW	SCAN MESSAGE HEADERS AND ACCESS DESIRED MESSAGES	CAN SKIP TO NEXT MESSAGE UPON HEARING HEADER (DATE, AUTHOR) OR PARTIAL MESSAGE	MESSAGES LISTED LAST IN/FIRST OUT. CAN SELECT LISTENING ORDER	SCAN MESSAGE HEADERS AND SKIP TO NEXT MESSAGE IF DESIRED	CAN SKIP TO NEXT MESSAGE UPON HEARING HEADER (DATE, AUTHOR)
MESSAGE WAITING INDICATOR	MESSAGE-WAITING LAMP	PHONE LAMP IF PBX IS INTEGRATED. ATTACHED LAMP BOX IS AVAILABLE.	NO	AUDIO OR VISUAL INDICATIONS DEPENDING ON INTECOM PHONE TYPE	LAMP, DIAL-TONE INTERRUPT ON CBX DEPENDING ON PHONE SET

1. AVAILABLE FOR SYSTEM/85 AND DIMENSION.
2. NUMBER OF USERS DEPENDENT ON GRADE OF SERVICE DESIRED AND CHARACTERISTICS OF USERS.
3. FOUR-PORT CONFIGURATION AVAILABLE THROUGH INDEPENDENT DISTRIBUTORS.
4. NO LONGER MARKETED IN THE UNITED STATES; AVAILABLE OVERSEAS ONLY. ROLM PHONEMAIL NOW MARKETED BY IBM.
5. MAXIMUM OF EIGHT 16-PORT UNITS IN A SINGLE CLUSTER.
6. EIGHTY CLUSTERS MAY BE IN A NETWORK.

SPERRY, VOICE INFORMATION PROCESSING (VIPS)	VMX INC., VOICE MESSAGE EXCHANGE	WANG, DIGITAL VOICE EXCHANGE (DVX)
2 7.5 $125,000	8 16 $92,500	4 20 $90,000
12 30 $225,000	64 170 $600,000	128[5] 640 $1.5 MILLION
1,600	8,000	2,000
6.5	5	6
STANDALONE, PBX INTERFACE, OR COMPUTER INTERFACE FOR SPERRYLINK	STANDALONE, PBX INTERFACE, OR FULL INTEGRATION WITH SEVERAL PBXs	STANDALONE
FUTURE RELEASE	YES	YES[6]
10 LISTS PER USER 100 USERS PER LIST	10 LISTS PER USER UNLIMITED NUMBER OF USERS PER LIST	8 LISTS PER USER 32 USERS PER LIST
YES	YES	YES
YES, VIPS LISTS MESSAGES IN CHRONOLOGICAL ORDER ON SPERRYLINK DISPLAY	YES	YES
SUPPORTS CALL FORWARDING DIRECTLY TO SUBSCRIBER'S BOX WITH NO REENTRY OF EXTENSION	MESSAGES REDIRECTED MANUALLY OR BY CALLER KEYING IN NUMBER OF USER EXTENSION	CAN CALL FORWARD TO DVX EXTENSION, OUTSIDERS MUST REENTER MAILBOX NUMBER
YES	YES	YES
YES, BUT CAN SPECIFY MESSAGE AS NONFORWARDABLE	YES	YES
YES, SPELL OUT USER'S LAST NAME	NO	YES, SPELL OUT USER'S LAST NAME
LISTS MESSAGES IN CHRONOLOGICAL ORDER. SPERRYLINK DISPLAYS MESSAGES FOR SELECTION OF REVIEW ORDER	CAN REVIEW AUTHORS AND SELECT MESSAGES AS DESIRED	TELLS NUMBER OF MESSAGES PER CALLER
SPERRYLINK DISPLAY CAN INDICATE MESSAGE WAITING. NO TELEPHONE NOTICE AT PRESENT	AVAILABLE WITH SEVERAL PBXS AND CENTREX	SEPARATE UNIT AVAILABLE FOR MESSAGE LAMP

PBX = PRIVATE BRANCH EXCHANGE
CBX = COMPUTERIZED BRANCH EXCHANGE

tone signals may be pitched differently than in the United States) need to use a tone generator, available from many voice-mail vendors for around $10.

For that matter, the control of voice mail through tone generation is not perfect. Some female voices can actually trigger voice-mail commands, typically those initiated with the asterisk and pound control keys, because their vocal pitch is so close to the tones used by telephones.

Also, voice mail can intimidate outside callers. Many people simply refuse to respond to household answering machines. Such people may not want to deal with a voice-mail service that they perceive as artificial or nonhuman.

Other disadvantages include the very real threat of voice junk mail (the kinds of messages that would not get past a human secretary or receptionist may make it into the voice mailbox). Users may also regret the inability to edit voice-mail messages. Real editing is not currently feasible with voice mail. If dissatisfied with the message as recorded, the caller must erase it and record again.

The final issue, perhaps the most important one, concerns the image that the company wants to project to outsiders. Some experts believe that voice mail is best kept inside the company. Callers from the outside typically prefer to speak to someone than to leave a message. They also usually find it easier to accept the news that the person they called is not available if a human being, rather than a machine, breaks it to them. Besides, a voice mailbox rarely answers questions concerning the called party's whereabouts.

For some companies, the benefits of improved communications might outweigh the negatives. Voice mail can project an aura of high technology. It can also impress outside callers who have their own voice-mail equipment or who are intrigued by the use of automation. Still, the opportunity to escape computer messaging and return to a human receptionist should always be present.

Trial run
It is critical that installation of the voice-mail network reflect an accurate understanding of user needs. This understanding can be achieved by means of a pilot study, in which a sample set of users is given access to voice-mail services for a limited period of time.

Prospective buyers should consider using a voice-mail service bureau for the pilot project. Most vendors of voice-mail equipment should be willing to supply a list of such companies (presumably those that use their gear). Service-bureau pilots can let users access a full range of features while avoiding the investment in equipment and site preparation until the benefits can be tested.

The only hurdle is that, once the project is finished and an inhouse voice-mail network configured, users may have to learn slightly different commands that go with the new equipment. For example, the command to create a message may require hitting first the pound key and then the number two (#-2) on one vendor's model, while the same command for another vendor is activated by hitting pound-one (#-1). Since key layouts for the various voice-mail processors on the market are not standard, changing vendors can cause short-term confusion.

There is also the danger of misapplying voice mail. Although it need not be made available to every employee,

Table 2: Post-installation survey extract

FEATURES LIST	OPINION		USAGE		REASONS		
	LIKE	DISLIKE	USE	DON'T USE	EASY TO USE	HARD TO USE	OTHER REASONS
A. Recording messages	☐	☐	☐	☐	☐	☐	
B. Receiving messages	☐	☐	☐	☐	☐	☐	
C. Saving, then later playing back messages	☐	☐	☐	☐	☐	☐	
D. Forwarding messages with an introduction	☐	☐	☐	☐	☐	☐	
E. Replying to messages	☐	☐	☐	☐	☐	☐	
F. Distribution lists	☐	☐	☐	☐	☐	☐	
G. Sending messages to nonusers' telephones (dial out)	☐	☐	☐	☐	☐	☐	
H. Confirmation of message receipt	☐	☐	☐	☐	☐	☐	
I. Automated voice-mail directory assistance	☐	☐	☐	☐	☐	☐	
J. Review message before sending	☐	☐	☐	☐	☐	☐	
K. Scanning messages & selectively listening to them in desired order	☐	☐	☐	☐	☐	☐	
L. Delayed delivery	☐	☐	☐	☐	☐	☐	
M. Disable touch-tones during message playback[1]	☐	☐	☐	☐	☐	☐	
N. Extended absence notification[2]	☐	☐	☐	☐	☐	☐	
O. Changing your password	☐	☐	☐	☐	☐	☐	
P. Restart session[3]	☐	☐	☐	☐	☐	☐	
Q. Name/date/time of sender given before each message	☐	☐	☐	☐	☐	☐	

1. TO OVERCOME PROBLEM OF NOISY LINE THAT CAN SEEM TO GENERATE COMMANDS.
2. INFORMS SENDERS OF RECIPIENT'S EXTENDED ABSENCE BEFORE THEY SEND.
3. SWITCH TO A SECOND MAILBOX FROM REMOTE CALL-IN WITHOUT REDIALING.

manager, or executive, voice mail can be perceived as a sign of prestige and can cause trouble as a result. Therefore, all company personnel should be notified about the meaning and use of the new service.

A question of use

Unlike other current office technologies, voice mail is so new that users typically cannot determine in advance what features they need or how useful the service might be. Also, the cost of voice mail can be difficult to justify. While its chief benefits, better communications and fewer interruptions, are intangible, hardware costs that can easily exceed $100,000 are highly visible. To document the benefits of voice-mail operation, companies should undertake surveys before and after installation of the pilot setup and of the service eventually selected. They should also issue formal reports summarizing voice mail's impact and making suggestions for further use of voice mail in the organization.

A pre-installation survey, for instance, should focus on each user's normal telephone activities. It should include questions concerning the estimated number of user telephone calls from inside and outside the company, the estimated number of messages that could be handled by voice mail, and the percentage of calls requiring concurrent two-way discussions.

Post-installation questionnaires are more detailed and comprehensive. Some questions call for numerical answers, such as the frequency of voice-mail use, estimates of misapplications (how many messages would have been more effectively sent in writing or via two-way telephone discussions), and estimates of erroneous or incomplete

messages received. Other questions, such as the following, should give users opportunities for more reflective, textual responses:
- Do you feel the product is easy to use?
- How comfortable are you using voice mail?
- Do you feel it saves you time? If so, how much per day?
- Do you find the message-center feature useful?
- Do any of your clients use it? If so, how do they feel about it?
- Did the voice-mail training session provide adequate preparation?
- Do you feel voice mail has created a more- or less-personal work environment?
- Has voice mail helped or hindered your communications with peers? With superiors? Subordinates? Clients?
- Has voice mail helped you respond to a critical situation?
- Has it had any effect on how you do your job?

Still other postinstallation questions refer to specific features, with solicitations of user opinions on ease and frequency of use. Table 2 shows a sample portion of a larger survey form that allows for numerical responses as well as commentary.

Obviously, the larger the user base, the more extensive the analysis and the greater the pre-installation effort and cost required for successful voice-mail implementation. However, more users also means more likelihood that voice mail will have a positive effect on day-to-day communications. This is because voice mail is only truly effective when enough users of the service have to communicate with other users on a regular basis. They can then get the full benefit of voice mail by making it an integral part of their work, the reason for installing voice mail.

The future of voice

Eventually, voice mail will become just another feature in an all-encompassing business communications architecture, one that will also include electronic mail, general telephone service, data access, and the usual computing capabilities. All of these will be accessible through a single multifunction workstation, with electronic communications finally able to accurately mirror human patterns of text, voice, and image interchange.

Voice mail will be a common feature in all business telephone services, from PBX products to public telephone exchanges. And because voice and electronic mail are natural partners for office communications, they will frequently be combined into standalone equipment. Ten years from now a business user will be able to speak a message into a voice recognition port, play it back for editing, and then select the mode of distribution, whether by voice or electronic mail. ■

David M. Rappaport is a partner at Arthur Andersen & Co., an international accounting and consulting firm. He heads the firm's telecommunications consulting practice worldwide, dealing with clients in all industries and covering voice, data, and image communications. He has been working in the data processing and communications industry for nearly 20 years. Prior to joining Arthur Andersen in 1975, he was manager of data processing at the University of Illinois Medical Center. He holds a B. S. in electrical engineering and an M. S. in computer science from the University of Illinois.

William Stallings, Comp/Comm Consulting, Great Falls, Va.

Here is one way to get a close estimate of a data link's efficiency

The term "a" is defined: It means a packet's propagation time per unit of total transmission time. Several link types are analyzed.

Network planners and designers need a simple, accurate technique for estimating the utilization efficiency of a data communications link. Because of the generally high cost of communications facilities, and because the designer or planner has some control over how that facility is used, it is important to be able to quickly and easily compare the efficiencies of various approaches over various media.

Utilization efficiency may be defined as the link utilization in aggregate bits per second divided by the link's capacity—the maximum bits per second that the link can carry. The key to link-utilization efficiency lies in the mechanics of the data link control protocol between two devices that exchange data. This protocol provides mechanisms for:

Flow control. The sending device must not transmit at a rate faster than the receiving device can absorb.

Error control. Bit errors introduced by the transmission mechanisms must be corrected.

Access control. In the case of a local area network (LAN), access by multiple devices to a link must be controlled if the desired throughput is to be maintained.

A single parameter, usually labeled a (derived and defined mathematically later) in the literature, is the basis of the utilization-efficiency technique.

In analyzing data link performance, the two most useful parameters are the data rate, R, of the medium and the average signal-propagation delay, D (proportional to distance), between stations on the link. In fact, it is the product of these two terms, $R \times D$, that is the single most important parameter for determining the performance of a data link. Other factors being equal, a network's utilization efficiency will be the same, for example, for both a 50-Mbit/s, 1-km bus and a 10-Mbit/s, 5-km bus.

Note that the resulting product of the data rate times the delay equals the length of the transmission medium expressed in bits—that is, the number of bits that may be in transit between two nodes at any "snapshot" in time. For example, assuming a propagation velocity, V, of 2×10^8 meters/second (about 6.6×10^8 feet/second), a 500-meter (about 0.3 mile) Ethernet network (10 Mbit/s) has a bit length of 25; that is, 500 meters $\div$ (2×10^8 meters/second) $\times$ 10 Mbit/s = 25 bits. Both a 1-km (about 0.6 mile) Hyperchannel (50 Mbit/s) and a typical 5-km (in each direction—totaling 10 km or about 6 miles) broadband LAN (5 Mbit/s) run about 250 bits. (Hyperchannel is the name of a common LAN from Network Systems of Minneapolis.)

A useful way of viewing the $R \times D$ parameter is to consider the length of the medium compared with the typical packet transmitted. This comparison makes it possible to distinguish between various protocols: those geared for a LAN (typically one packet at a time in transit), those designed for a multiprocessor-backplane bus (which accommodate a maximum of a few bits in transit), and those designed for a satellite link (which accommodate several packets in transit). Intuitively, it can be seen that each protocol is different.

Compare LANs to multiprocessor computers. Events occur almost simultaneously in the latter: When one component begins to transmit, the others know it almost immediately. For LANs, the relative time gap in awareness requires a complex medium-access control protocol.

Now compare LANs to satellite links. To have any hope of efficiency, the satellite link must allow multiple packets to be in transit simultaneously. This places specific requirements on the link layer protocol, which must deal with a sequence of outstanding packets waiting to be acknowledged. LAN protocols generally allow only one packet at a time to be in transit or, at the most, a few for some ring protocols. Again, the transmission scheme affects the access protocol.

The length of the medium, or data path, expressed in

1. Utilization and a. *If frame-transmission time is 1, the propagation time is a. With a less than 1, the link's "bit length" is less than that of the frame. At $t_0 + a$, the transmitted frame's leading edge reaches the receiving station. At $t_0 + 1$, the transmission is completed. In the case of a greater than 1, the two steps are reversed.*

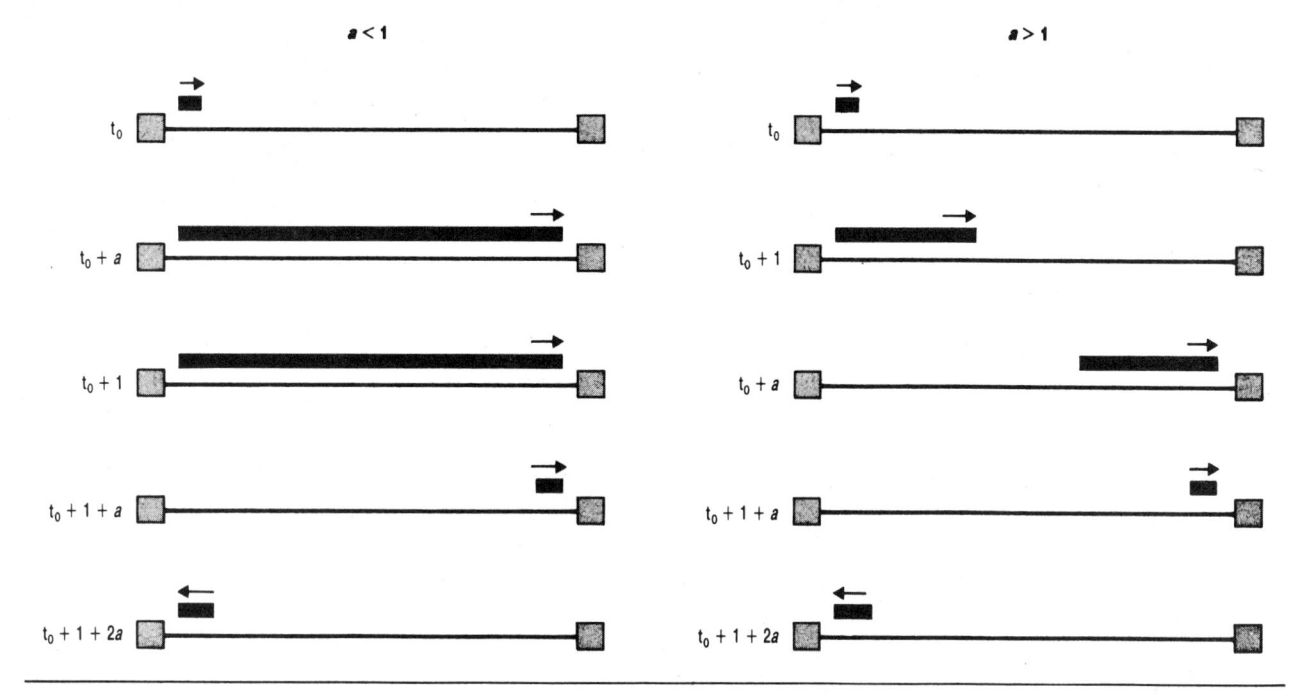

bits, divided by the bit length, L, of the typical packet is usually denoted by a. That is, $a = (R \times D) \div L$. But D is the propagation time on the medium (worst case), and L/R is the time it takes a transmitter to get an entire packet out onto the medium (transmission time). So, note that $a = D \div L/R$, or $a =$ (propagation time) $\div$ (transmission time).

Consider the case of two stations connected by a point-to-point link, with one station transmitting data frames to the other. We wish to determine the maximum potential efficiency if there are no errors. In that case, the efficiency depends solely on the flow-control policy.

One at a time

The simplest form of flow control is stop-and-wait. The sender transmits frames one at a time and then waits for an acknowledgment to each frame before sending the next one. Suppose that a long message is to be sent as a sequence of frames $f_1, f_2, \ldots, f_n$. In a polling procedure, the following sequence occurs:
- Station S_1 polls station S_2.
- S_2 responds with f_1.
- S_1 sends an acknowledgment.
- S_2 sends f_2.
- S_1 acknowledges ...
  .
  .
  .
- S_2 sends f_n.
- S_1 acknowledges.

The total time to send the data is:
$T_D = T_I + nT_F$
where
T_I = time to initiate sequence = $t_{prop} + t_{poll} + t_{proc}$
n = total number of frames transmitted

T_F = time to send one frame and receive an acknowledgment = $t_{prop} + t_{frame} + t_{proc} + t_{prop} + t_{ack} + t_{proc}$
t_{prop} = propagation delay
t_{poll} = time to transmit a poll signal
t_{proc} = processing time at each station to react to an incoming event
t_{frame} = time to transmit a frame
t_{ack} = time to transmit an acknowledgment

To simplify matters, we ignore a few terms. For a long sequence of frames, T_I is relatively small and can be dropped. Let us assume that the processing time between transmission and reception is relatively negligible, and that the acknowledgment frame is very small—typically no more than one-tenth of the message frame. Then we can express the total time to send the data as:

$T_D = n(2t_{prop} + t_{frame})$

Of that time, only $n \times t_{frame}$ is actually spent transmitting data—the rest is overhead. The utilization efficiency, U, of the line equals $(n \times t_{frame}) \div n(2t_{prop} + t_{frame})$. Noting that $a = t_{prop} \div t_{frame}$, the utilization efficiency expression reduces to $U = 1 \div (1 + 2a)$. This is the maximum possible utilization efficiency of the link.

In Figure 1, which illustrates this last expression for U, frame-transmission time is normalized to 1. Hence, the propagation time is a. First consider the case of $a < 1$; in this instance the link's "bit length" is less than that of the frame. A station begins transmitting a frame at time t_0. At $t_0 + a$, the leading edge of the frame reaches the receiving station, while the sending station is still in the process of transmitting the frame. At $t_0 + 1$, the sending station has completed transmission. At $t_0 + 1 + a$, the receiving station has received the entire frame and immediately transmits a small acknowledgment frame. This acknowledgment arrives back at the sending station at $t_0 + 1 +$

2a. Total elapsed time: 1 + 2a. Total transmission time: 1. Hence, the utilization efficiency is 1 ÷ (1 + 2a). The case of $a > 1$ achieves the same result.

Let us consider a few examples. At one extreme is a satellite link. The round-trip propagation time is about 270 milliseconds (ms). A typical digital transmission service is 56 kbit/s, and a 4,000-bit frame is within the typical range. Thus frame-transmission time equals: 4,000 ÷ 56,000 = 71 ms, and a = 270 ÷ 71 = 3.8. Thus the maximum utilization efficiency, 1 ÷ (1 + 2a), is 0.12. This utilization efficiency is about the smallest value of a that one might expect for a satellite link.

Recent and proposed satellite services use much shorter transmission times: from 6 ms down to 125 microseconds. For these times, a equals 45 and 2,160, respectively. Thus for the simple stop-and-wait acknowledgment protocol described above, efficiency could be as low as 0.0002!

At the other distance extreme is the LAN. Distances range from 0.1 to 10 km (0.06 to about 6 miles), with data rates of 100 kbit/s to 10 Mbit/s. Using the same propagation velocity, V, used earlier of 2×10^8 meters/second and a frame size of 500 bits, the value of a is in the range 10^{-4} to 1. Typical values are 0.01 to 0.1. For the latter range, utilization efficiency is in the range 0.83 to 0.98.

We can see that LANs are inherently quite efficient, whereas satellite links are not. As a final example, let us consider digital data transmission via a modem over a voice-grade line. A practical upper bound on data rate is 9.6 kbit/s. We can again use $V = 2 + 10^8$ meters/second. Again, let us consider a 500-bit frame. Such transmission is used for distances anywhere from a few tens of meters to thousands of kilometers (one meter = about 3.3 feet). If we pick, say, as a short distance d = 100 meters, then a = (100 meters ÷ 2×10^8 meters/sec) ÷ (500 bits ÷ 9.6 kbit/s) [or (9.6 kbit/s × 100 meters) ÷ (2×10^8 meters/sec × 500 bits)] = 9.6×10^{-6}, and utilization efficiency is effectively unity. Even in a long-distance case, such as d = 5,000 km (about 3,000 miles), we have a = ($9,600 \times 5 \times 10^6$) ÷ ($2 \times 10^8 \times 500$) = 0.48 and the efficiency is then equal to 0.5.

We can conclude that, in some cases, the simple stop-and-wait acknowledgment procedure provides adequate line utilization efficiency. For those cases in which this does not hold, a more elaborate procedure is desirable.

Many at a time

To improve efficiency, a "sliding-window" technique can be used. This allows multiple frames to be in transit at one time.

Let us examine how this might work for two stations, A and B, connected via a full-duplex link. Station B allocates seven buffers for reception instead of the one in the example of one frame at a time discussed above. Thus B can accept seven frames, and A is allowed to send seven frames without waiting for an acknowledgment. To keep track of which frames have been acknowledged, each is labeled with a sequence number in the range 0 to 7 (modulo 8). B acknowledges a frame by sending an acknowledgment that includes the sequence number of the next frame expected. Thus, if B returns the sequence number 5, this acknowledges receipt of frame number 4 and says that B is now expecting frame number 5.

This scheme can be used to acknowledge multiple frames. For example, B could receive frames 2, 3, and 4, but withhold acknowledgment until frame 4 arrives. By then returning sequence number 5, B acknowledges frames 2, 3, and 4 at one time. Station A maintains a list of sequence numbers that it is allowed to send, and B maintains a list of sequence numbers that it is prepared to receive. Each of these lists can be thought of as a "window" of frames.

As a further example of sliding-window operation, consider stations A and B having, initially, seven-frame windows. After transmitting three frames with no acknowledgment, A has shrunk its window to four frames. When frame 2 is acknowledged, A is back up to seven frames.

Later, B decides to temporarily restrict data flow to three frames (with the maximum window size capability remaining at seven). This is readily accomplished by withholding additional acknowledgments.

The efficiency of the line now depends on both N, the window size, and a. For convenience, let us again normalize frame transmission time to a value of 1; thus the propagation time is a.

Figure 2 illustrates the efficiency of a full-duplex, point-to-point line. Station A begins to emit a sequence of frames at time t_0. The leading edge of the first frame reaches station B at $t_0 + a$. The first frame is entirely absorbed by $t_0 + a + 1$. Assuming negligible processing time, station B can immediately acknowledge the first frame (ACK1). Let us also assume that the acknowledgment frame is so small that its transmission time is negligible. Then the ACK1 reaches station A at $t_0 + 2a + 1$.

There are two cases:

Case 1: $N > (2a + 1)$. The acknowledgment for frame 1 reaches station A before it has exhausted its window. Thus A can transmit continuously with no pause.

Case 2: $N < (2a + 1)$. Station A exhausts its window at $t_0 + N$ and cannot send additional frames until $t_0 + 2a + 1$. Thus the line utilization efficiency is N time units out of a period of (2a + 1) time units.

From the above, we can state that:

U = 1 when window size $N > (2a + 1)$

U = N ÷ (2a + 1) when $N < (2a + 1)$

Typically, an *n*-bit field provides for the sequence number, and the maximum window size is $N = 2^n - 1$. Figure 3 shows the maximum efficiency achievable for window sizes of 1, 7, and 127 as a function of a. A window size of 1, of course, corresponds to the simple protocol (one frame at a time) discussed earlier. A window size of 7 (three binary bits) should be adequate for most applications. A window size of 127 (seven binary bits) is adequate for some satellite links. If a larger window size is needed, the value of a should be determined for it, from which the utilization efficiency may then be calculated.

Make reception errorless

The imperfect world of data transmission requires that the flow-control techniques described above must be refined to include error control. Two refinements are added:

Error detection: An error-detecting code, such as a cyclic redundancy check (CRC), is added to each transmitted frame.

Automatic repeat request (ARQ): Whenever an error is detected, the receiving device automatically requests that

2. Full-duplex efficiency. Station A begins a sequence of frames. Station B acknowledges the first frame (ACK1), which reaches station A at $t_0 + 2a + 1$. With a window size of $N > (2a + 1)$, A can transmit continuously; if $N < (2a + 1)$, A exhausts its window at $t_0 + N$ and cannot send additional frames until $t_0 + 2a + 1$.

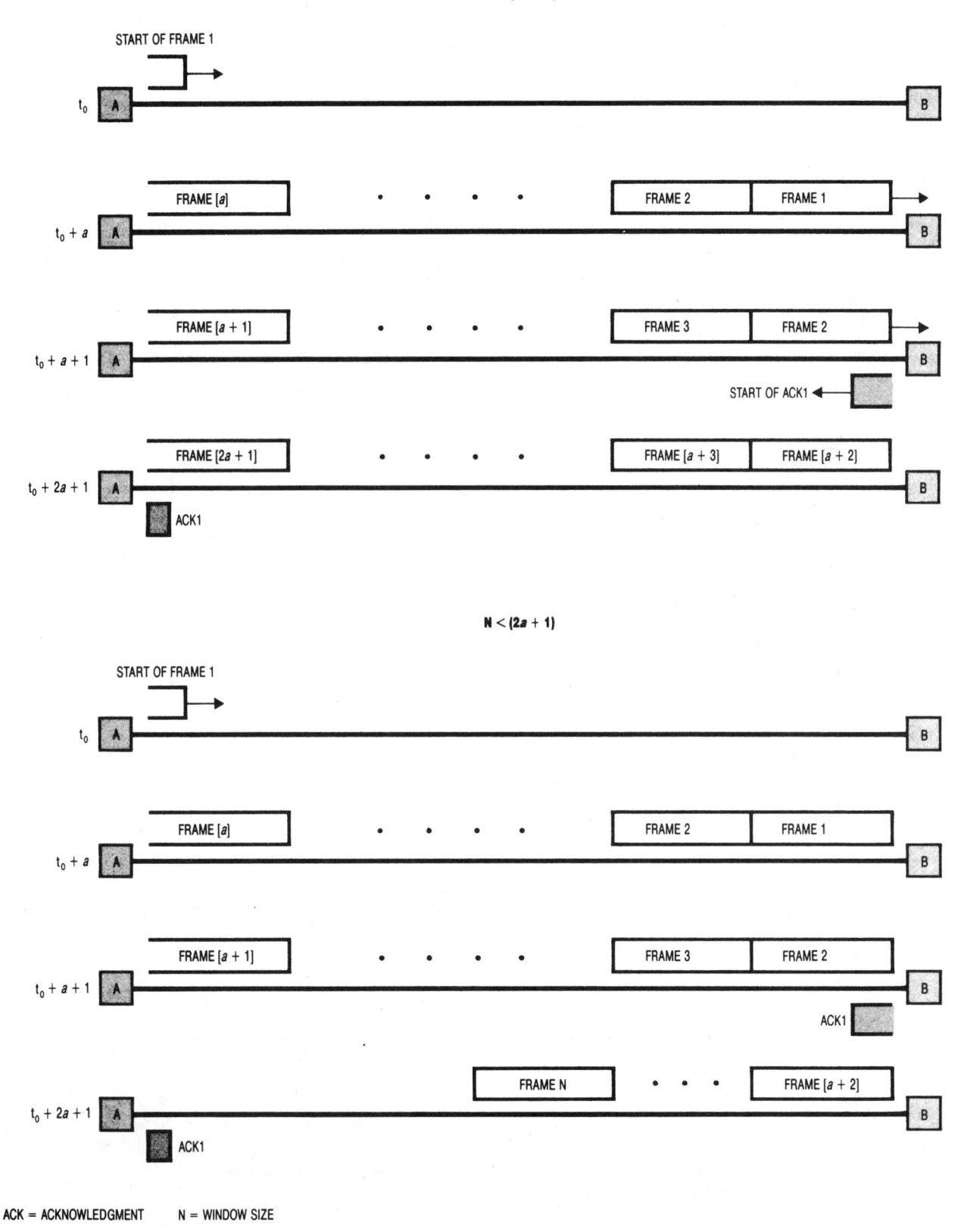

$N > (2a + 1)$

$N < (2a + 1)$

ACK = ACKNOWLEDGMENT   N = WINDOW SIZE

the frame be transmitted another time.

ARQ is a straightforward approach to error correction that results in the conversion of an unreliable data link into a reliable one. Three versions of ARQ are in popular use: stop-and-wait, go-back-N continuous, and selective-repeat continuous.

Stop-and-wait ARQ uses the simple acknowledgment scheme described above: The sending station transmits a single frame and then must await an acknowledgment. No other data frames can be sent until the receiving station's reply arrives at the transmitting station. The receiver sends a positive acknowledgment (ACK) if the frame is correct and a negative acknowledgment (NAK) if otherwise.

One modification is needed: The transmitted frame could be so corrupted by noise as not to be received at all, in which case the receiver will not acknowledge. To account for this possibility, the sender is equipped with a timer. After a frame is transmitted, the sender waits for an ACK or NAK. If no recognizable response is received within a preset time, then the same frame is sent again. Note that this scheme requires that the transmitter retain a copy of a transmitted frame until an ACK is received for that frame.

The principal advantage of stop-and-wait ARQ is its simplicity. Its principal disadvantage, as discussed earlier, is that this is inefficient for large a. The sliding-window technique can be adapted to provide more efficient line utilization. In this context, it is referred to as continuous ARQ.

One variant of continuous ARQ is known as go-back-N ARQ. In this technique, a station may send a series of frames determined by window size. If the receiving station detects an errored frame, it sends a NAK for that frame. The receiving station will discard all future incoming frames until the frame in error is correctly received. Thus the transmitting station, when it receives a NAK, must retransmit the frame in error plus all succeeding frames.

With go-back-N ARQ, each individual frame need not be acknowledged. For example, station A sends frames 0, 1, 2, and 3. Station B responds with ACK0 after frame 0 but does not respond to frames 1 and 2. After frame 3 is received, B issues ACK3, indicating that frame 3 and all previous frames are accepted.

Selective-repeat ARQ provides a more refined approach than go-back-N. The only frames retransmitted are those that receive a NAK. This would appear to be more efficient than the go-back-N approach. But the receiver must have (1) sufficient storage to save post-NAK frames until the errored frame is retransmitted and (2) the logic for reinserting the frame in the proper sequence. The transmitter, too, will require more complex logic to be able to send frames out of sequence. Because of such complications, the go-back-N algorithm is more commonly used.

Performance

Both go-back-N and selective-repeat appear to be more efficient than stop-and-wait. Let us develop some approximations to determine the degree of improvement to be expected.

First, consider stop-and-wait ARQ. With no errors, the maximum utilization efficiency, as developed above, is $1 \div (1 + 2a)$. We must take into account the fact that some frames are repeated because of errors. To do this, note that the utilization efficiency U can be defined as
$U = T_f \div T_t$
where,
T_f = the time for the transmitter to emit a single frame.
T_t = the total time that the line is engaged in the transmission of a single frame.

For error-free operation using stop-and-wait ARQ, we have the utilization equation:
$U = T_f \div (T_f + 2T_p)$
where,
T_p is the propagation time.
Dividing by T_f and remembering that $a = T_p \div T_f$, we again have $U = 1 \div (1 + 2a)$.

If errors occur, we must modify the utilization equation to:
$U = T_f \div (N_r \times T_t)$. N_r is the expected number of transmissions of a frame. Thus for stop-and-wait ARQ we have:
$U = 1 \div [N_r(1 + 2a)]$.

A simple expression for N_r can be derived by considering the probability, p, that a single frame is in error. If we assume that ACKs and NAKs are never in error, the probability that it will take exactly i attempts to transmit a frame successfully is $p^{i-1}(1 - p)$. After some additional mathematical manipulations, $N_r = 1 \div (1 - p)$. Therefore, for stop-and-wait ARQ, $U = (1 - p) \div (1 + 2a)$.

For selective-repeat ARQ, we can use the same reasoning as applied to stop-and-wait ARQ (above). That is, the error-free equations must be divided by N_r, which equals $1 \div (1 - p)$. So, for selective-repeat ARQ:
$U = 1 - p$ when $N > 2a + 1$
$U = 1[N(1 - p)] \div (2a + 1)$ when $N < 2a + 1$.

The same reasoning will still apply for go-back-N ARQ, but we must be more careful in approximating N_r. Each error causes retransmission of K frames rather than one frame. Thus N_r now equals the number of transmitted frames to successfully transmit one frame. After more algebraic manipulations, $N_r = (1 - p + Kp) \div (1 - p)$.

By studying Figure 2, the reader should conclude that K is approximately equal to $2a + 1$ for $N > (2a + 1)$, and $K = N$ for $N < (2a + 1)$. Thus, for go-back-N ARQ:

3. Window size, efficiency. *A window size of 1 means one frame at a time; 7 is adequate for most applications; 127 is adequate for some satellite links.*

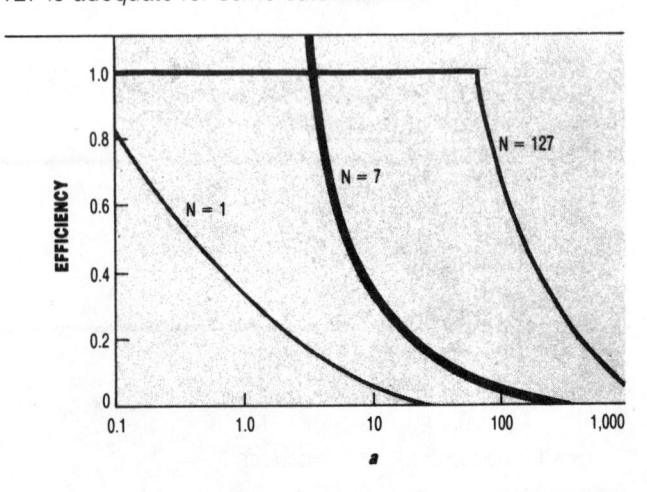

4. Comparing error-control techniques. *Note that for a window size of N = 1, both selective-repeat ARQ and go-back-N ARQ reduce to stop-and-wait ARQ.*

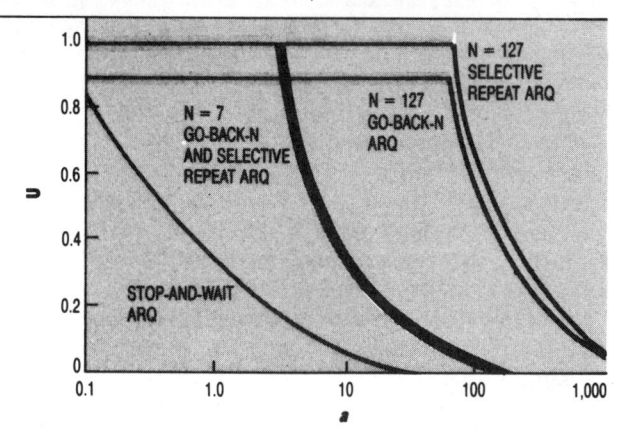

ARQ = AUTOMATIC REPEAT REQUEST
N = WINDOW SIZE

$U = (1 - p) \div (1 + 2ap)$ when $N > 2a + 1$
$U = [N(1 - p)] \div (2a + 1)(1 - p + Np)]$ when $N < 2a + 1$

Note that for N = 1, both selective-repeat and go-back-N reduce to stop-and-wait. Figure 4 compares these three error control techniques for a value of $p = 10^{-3}$ (the probability of a single errored frame is one in 1,000). This figure and the equations are only approximations. For example, we have ignored errors in acknowledgment frames and, in the case of go-back-N, errors in retransmitted frames other than the frame initially in error. Nevertheless, the results have been found to be very close to those produced by a more careful analysis.

Ethernet et al.

With LANs—especially carrier-sense multiple access (CSMA) LANs—the efficiency calculation is somewhat altered: We are dealing with multiple peer stations, any one of which can transmit at any time. Thus the concern is in the efficiency with which multiple users can share the link. A new consideration comes into play: the access control technique.

Consider a perfectly efficient access mechanism that allows only one transmission at a time. As soon as one transmission is over, another node begins transmitting. Furthermore, the transmission is pure data—no overhead bits. What is the maximum possible utilization efficiency of the network? It can be expressed as the ratio of total throughput of the network to the data rate capacity, R (remember, L = bit length of packet; D = propagation delay):

U = throughput $\div R$
   $= [L \div (\text{propagation} + \text{transmission time})] \div R$
   $= [L \div (D + L/R)] \div R$
   $= 1 \div (1 + a)$.

So, utilization efficiency varies inversely with a. This can be grasped intuitively by restudying Figure 1. Consider the case of a baseband bus with two stations as far apart as possible (worst case) that take turns sending packets. Consider first $a < 1$. If we normalize the packet transmission time to equal one, then the sequence of events can be expressed as follows [remember, a = (propagation time) $\div$ (transmission time)]:

1. A station begins transmission at t_0.
2. Reception begins at $t_0 + a$.
3. Transmission is completed at $t_0 + 1$.
4. Reception ends at $t_0 + 1 + a$.
5. The other station begins transmitting.

If $a > 1.0$, event 2 occurs *after* event 3. In any case, the total time for one "turn" is $1 + a$ (transmission time + propagation time). But the transmission time is only 1, resulting in a utilization efficiency of $1 \div (1 + a)$.

The same effect can be seen to apply to a ring network. Here we assume that one station transmits and then waits to receive its own transmission before any other station transmits. The identical sequence of events outlined above applies.

The LAN-performance implication of the expression for utilization efficiency, $1 \div (1 + a)$, is shown in Figure 5. The axes of the plot are:

S, the throughput or total rate of data being transmitted on the medium, normalized to a maximum of one by dividing by the medium's data rate capacity, and

G, the offered load to the LAN—that is, the total rate of data presented for transmission—similarly normalized.

The ideal (theoretical) case is $a = 0$, which allows 100 percent utilization efficiency. It can be seen that, as offered load increases, throughput remains equal to offered load up to the full capacity of the network (when $S = G = 1$) and then remains at $S = 1$. For any value of a, the throughput saturates at $S = 1 \div (1 + a)$.

So we can say that an upper bound on utilization efficiency is $1 \div (1 + a)$, regardless of the medium-access protocol used. Two caveats: First, this assumes that the maximum propagation time is incurred on each transmission. Second, it assumes that only one transmission may occur at a time. These assumptions are not always true. Nevertheless, the formula $1 \div (1 + a)$ is almost always a valid upper bound. This is true because overhead-incurred delays due to the medium-access protocol that is used more than make up for any possible inaccuracies of these assumptions.

5. Throughput vs. traffic load. *The throughput saturates at $1 \div (1 + a)$. The theoretical ideal case has $a = 0$ (allowing 100 percent utilization efficiency).*

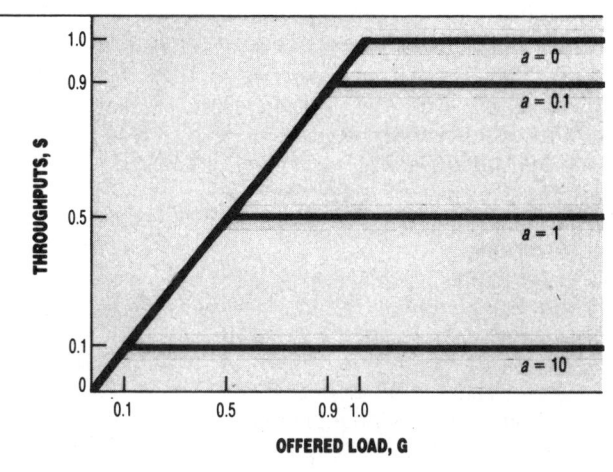

6. Throughput vs. a. *For both protocols, carrier-sense multiple access with collision detection and token passing, throughput declines as a increases.*

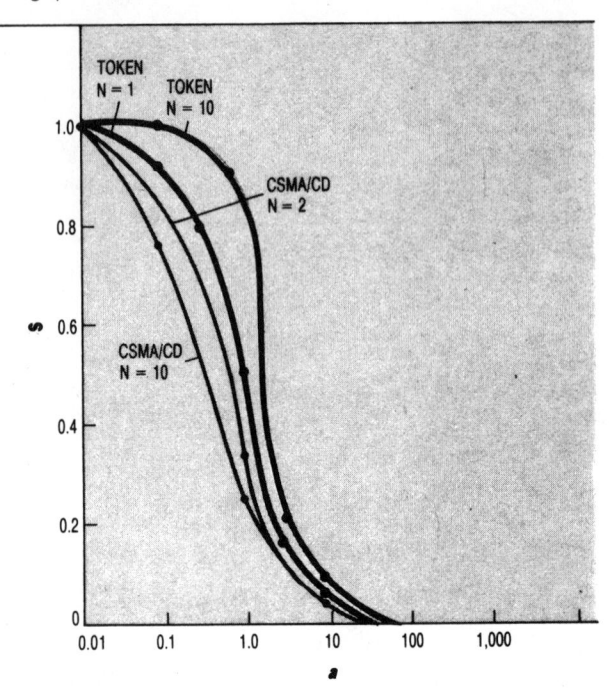

CSMA/CD = CARRIER SENSE MULTIPLE ACCESS WITH COLLISION DETECTION
N = NUMBER OF ACTIVE STATIONS
S = MAXIMUM ACHIEVABLE THROUGHPUT

The overhead is unavoidable. Packets must include address and synchronization bits, and there is administrative overhead for controlling the protocol.

By the development of two relatively simple performance models, some insight should be gained into the relative performances of three leading LAN protocols: CSMA/CD, token bus, and token ring. For the performance models, we shall assume a LAN with N active stations and a maximum normalized propagation delay of a (transmission time = 1). To simplify the analysis, assume that each station is always prepared to transmit a packet. This allows the development of an expression for maximum achievable throughput (S). This is not the sole figure of merit for a LAN, but it is the single most analyzed one. And it does permit useful performance comparisons.

Comparing techniques

First, consider a token ring. Time on the ring will alternate between data packet transmission and token passing. A cycle is a data packet followed by a token.

Definitions:
C = average time for one cycle.
T_1 = average time to transmit a data packet.
T_2 = average time to pass a token.

The average cycle rate is $1/C = 1 \div (T_1 + T_2)$. Intuitively, maximum throughput $S = T_1 \div (T_1 + T_2)$. That is, the throughput, normalized to network capacity (throughput ÷ data rate capacity), is just the fraction of time that is spent transmitting data.

Packet transmission time is normalized to equal one, and propagation time equals a. For $a < 1$, a station transmits a packet at time t_0, receives the leading edge of its own packet at $t_0 + a$, and completes transmission at $t_0 + 1$.

The station then emits a token, which takes an average time of a/N to reach the next station. Thus, one cycle takes $1 + a/N$, and the transmission time is 1. So, maximum throughput $S = 1 \div (1 + a/N)$.

For $a > 1$, the reasoning is slightly different. A station transmits at t_0, completes transmission at $t_0 + 1$, and receives the leading edge of its frame at $t_0 + a$. At that point, it is free to emit a token, which takes an average time of a/N to reach the next station. The cycle time C is therefore $a + a/N$, and $S = 1 \div [a(1 + 1/N)]$.

The above reasoning applies equally well to a token bus: We assume that the logical order of stations may be treated the same as the physical order of the token ring stations. Token-passing time is the same expression: a/N.

For CSMA/CD, consider time on the medium to be organized into slots the length of which is twice the end-to-end propagation delay ($2 \times a$). This slot time is the maximum time—from the start of transmission—required to detect a collision. With N active stations, if each station always has a packet to transmit, and does so, clearly there will be nothing but collisions on the line. Therefore assume that each station restrains itself to transmit only during an available slot with a probability p.

Time on the medium consists of two types of intervals. First is a transmission interval, which lasts $1/2a$ slots. (Each

7. Throughput vs. N. *Token-passing performance actually improves as N (the number of active stations) increases, because less time is spent in token passing.*

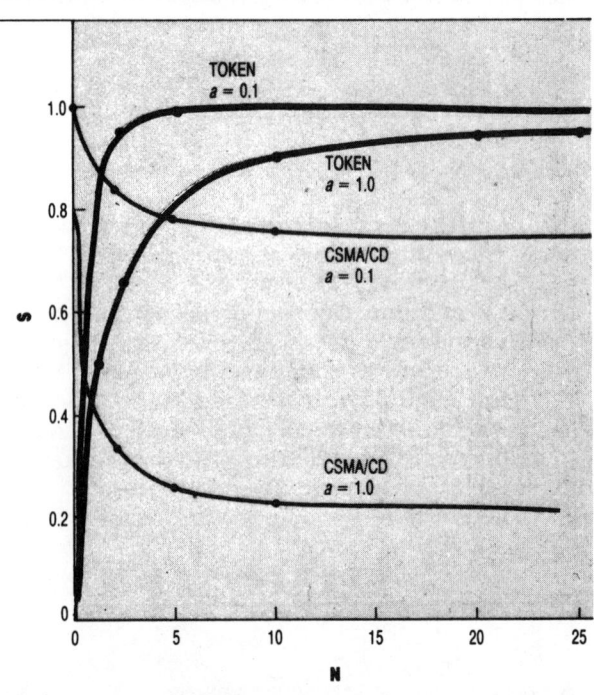

CSMA/CD = CARRIER SENSE MULTIPLE ACCESS WITH COLLISION DETECTION
N = NUMBER OF ACTIVE STATIONS
S = MAXIMUM ACHIEVABLE THROUGHPUT

slot = 2a time units. One time unit = 1/2a slots.) Second is a contention interval, which is a sequence of slots with either a collision or no transmission in each slot. The throughput is just the proportion of the total time that is spent in transmission intervals.

Solo try

To determine the average length of a contention interval, we begin by computing A, the probability that exactly one station attempts a transmission in a slot, and therefore acquires the medium. This is just the binomial probability that any one station attempts to transmit and the others do not. The probability expression A evolves to:

$A = Np(1 - p)^{N-1}$

This function takes on a maximum when $p = 1/N$:

$A = (1 - 1/N)^{N-1}$

Why the maximum? Because we want to calculate the maximum throughput of the medium. This will be achieved if we maximize the probability of successful seizure of the medium. This says that the following rule should be enforced: During periods of heavy usage, a station should restrain its offered load to 1/N. (This assumes that each station knows the value of N.)

Now we can estimate the mean length of a contention interval, w, in slots. The expression for the mean length, E, evolves to the following expression:

$E[w] = (1 - A)/A$.

We can now determine the maximum utilization efficiency, which is just the length of a transmission interval as a proportion of a cycle consisting of a transmission and a contention interval.

The maximum throughput for CSMA/CD: $S = 1/2\,a \div [1/2\,a + (1 - A)/A] = 1 \div [1 + 2a(1 + A)/A]$.

It is interesting to note the asymptotic value of S as N increases. For token passing, the limit of S (the maximum throughput), as N approaches infinity, equals one for $a < 1$; equals $1/a$ for $a > 1$. For the CSMA/CD case, the expression for the limit of S, as N approaches infinity, works out to $1 \div (1 + 3.44a)$.

Figure 6 shows normalized throughput (throughput ÷ data rate) as a function of a for various values of N, for both token passing and CSMA/CD. For both protocols, throughput declines as a increases, as expected. But the dramatic difference between the two protocols is seen in throughput versus N (Fig. 7). Token-passing performance actually improves as a function of N, because less time is spent in token passing. Conversely, the performance of CSMA/CD decreases because of the increased likelihood of collision or of no transmission. ∎

William Stallings, president of Comp/Comm Consulting, received a B. S. in electrical engineering from Notre Dame and a Ph. D. in computer science from M. I. T. He is the author of seven books, including Data and Computer Communications *(Macmillan, 1985).*

Peter Hansen, Doelz Networks Inc., Irvine, Calif.

Making the most of post-divestiture tariffs for data network design

Just as AT&T's Tariff 260 is now obsolete, so are network designs and equipment based on the old tariff structure

Communications managers whose existing networks were designed and built around AT&T's pre-divestiture tariffs are probably paying higher telephone bills today than they need to. A design based on the old tariffs might have been the optimum economic solution at the time, but much has changed since divestiture. To minimize the total ownership costs of data networks now, users are compelled under the "new rules" of the post-divestiture tariffs to use strategies and types of equipment different from what has been used in the past.

A tariff is a schedule of rates or charges. For communications-equipment vendors and data-network users, the impact of AT&T's post-divestiture private-line tariffs on data circuits has been significant. And key to successfully navigating the new tariff environment is an understanding of the rules governing the carrier routing of interstate private-line channels, as well as the equipment underlying the whole process.

Saving money, of course, is the goal. When a tariffed service becomes popular and widely used, equipment vendors invariably begin to make products that help network operators reduce monthly operating costs based on the price structure of that particular tariff.

AT&T's pre-divestiture Tariff 260 was in effect for years. Subsequently, the shape of many networks was determined by the way equipment vendors designed their products to conform to the tariff. Quite simply, equipment was designed to take advantage of savings opportunities that existed within AT&T's Tariff 260 structure.

New tariffs and costs

All this changed on January 1, 1984, when Judge Harold H. Greene's AT&T-divestiture order, compelling AT&T to separate its manufacturing, research, and local operating units into separate business units, went into effect. Tariff 260 was rendered obsolete by April 1985, when AT&T was allowed to put into place its new interstate privateline Tariffs 9, 10, and 11.

The divestiture order restricted the types of services each of AT&T's business units could offer. It also created entirely new service types. In addition, many of the Tariff 260 features, which had been viewed as opportunities for saving money, were eliminated in the replacement Tariffs 9, 10, and 11. Under the new tariffs, many private data-network telephone bills shot up from 20 percent to 50 percent.

The major impact of Judge Greene's restructuring of AT&T was the creation of distinct communications "transport" and "local distribution" service segments. The divested Bell operating companies, or BOCs, are generally restricted to providing local distribution service within the United States' 164 Local Access and Transport Areas (LATAs). They are prohibited from providing service between LATAs.

Such service is provided by the inter-LATA carriers— AT&T Communications, MCI, Western Union, U S Sprint, and others—which became transport carriers authorized to handle traffic between LATAs. These carriers are also now known as interoffice carriers, or IOCs.

Local exchange

The now-divested BOCs do not have the exclusive franchise to service all of the land area within a LATA. Besides the BOCs, independent telephone companies are also allowed to service certain portions within a given LATA. Under the generic title of local exchange carrier, BOCs can offer service in more than one LATA. But to do so they must use an IOC to carry traffic between LATAs.

An IOC, such as MCI or U S Sprint, then, provides service between inter-LATA-carrier terminating facilities, called central offices (COs), which are situated in the different LATAs. If a carrier has several COs within the same LATA, then it may offer intra-LATA service between its COs, as well. The IOC's terminating facilities, typically installed in or near a BOC central office, are generically called points of presence (POPs) or points of interface.

Figure 1 illustrates the difference between the relatively uncomplicated leased-line arrangement under AT&T's Tariff 260 (A), which was in effect until last year, and the new structure under AT&T's Tariffs 9, 10, and 11 (B).

When a customer wants to establish a leased data circuit, he may find himself dealing with:

1. A single local exchange carrier, or LEC, if his circuit stays within a LATA.
2. One LEC and one IOC, if his circuit crosses one or more LATAs but stays within the LEC's operating boundaries (these are usually intrastate circuits).
3. Two LECs and one IOC (most common on interstate circuits).
4. Two LECs and multiple IOCs (which could be the case with long-distance interstate circuits).

Demarcation

Each carrier is inherently responsible only for its portion of the total leased data circuit. If the data circuit breaks, the customer is responsible for dealing with each carrier (one at a time) to find the source of the impairment.

Alternately, the customer can contract with one of the involved carriers to provide end-to-end coordination services. The fee for this coordination, however, can range from as little as a flat $30 per month in the case of a point-to-point circuit, to as much as 5 percent to 10 percent of a multipoint circuit's total monthly cost.

In addition, a customer can order an end-to-end circuit from any of the LECs or IOCs involved with any portion of the circuit. However, the price he pays for the circuit will vary depending on which carrier is asked for the price quotation. Experience has shown that, as a general rule, one of the LECs at either end of the circuit will typically offer the lowest total price.

Figure 2 compares the configuration and pricing of a 50-mile digital channel. In the first case, the circuit is installed and furnished entirely by a local exchange carrier (pricing here is based on Pacific Bell tariffs in effect in California). In the second case, a 50-mile Dataphone digital service circuit furnished by an interoffice carrier is shown. The costs (AT&T's are used) are much higher.

Hierarchy

Central offices form a hierarchy (Fig. 3). Starting at the bottom, Class 5 COs—known as end offices, or wire centers—serve customers directly within a portion of a city. If a circuit passes through a Class 5 end office without going through switching equipment, the CO serves only as a wire center. However, if the user's private data circuit needs access to features or facilities not available in his CO, then the local loop is bridged over an interoffice channel to a CO that does offer the needed service.

Class 4 offices will typically serve an entire large city or else several smaller cities. In the central-office hierarchy, all the Class 5 offices in a city are connected to a common Class 4 office. Class 4 offices—also known as toll centers—contain the equipment necessary to generate customer billing data.

1. Before and after. *Divestiture separated AT&T from the Bell operating companies (BOCs). The figure shows the structure of a leased line under the pre-divestiture Tariff 260 (A), and post-divestiture Tariffs 9, 10, and 11 (B).*

(A) INTEREXCHANGE LINE (TARIFF 260)

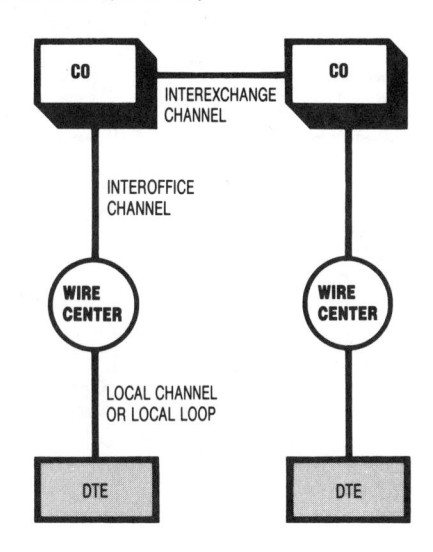

(B) INTER-LATA LINE (TARIFF 9, 10, 11)

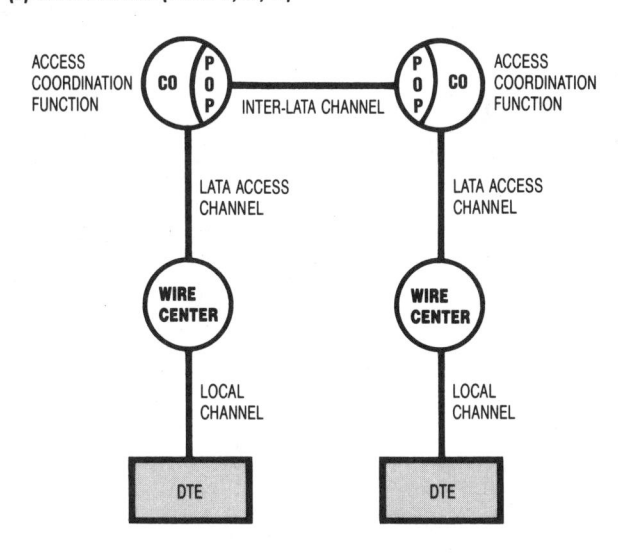

CO = CENTRAL OFFICE (LOCAL EXCHANGE CARRIER)
DTE = DATA TERMINAL EQUIPMENT (CUSTOMER PREMISES)
POP = POINT OF PRESENCE (INTEROFFICE CARRIER)

Continuing up the hierarchy, Class 3 central offices—also known as tandem offices—typically serve an entire metropolitan area. A Class 2 office can serve an entire state, and a Class 1 office can serve a large region comprised of several states.

Class 1 and 2 offices now are operated by interoffice carriers. A Class 3 tandem office, however, may be operated either by a local exchange carrier or an IOC. Class 4 and 5 COs are owned and operated by LECs.

Some Class 5 offices are big enough to contain an entire exchange. The first six digits of a standard North American 10-digit telephone number define an exchange. Every state

2. Depends on who you call. *Given the new leased-line rules, customers may order dedicated channels from either their local telephone company or an interexchange carrier. The figure shows the difference in a 50-mile DDS circuit if within a LATA and arranged by a local carrier (A), and one that goes between LATAs (B).*

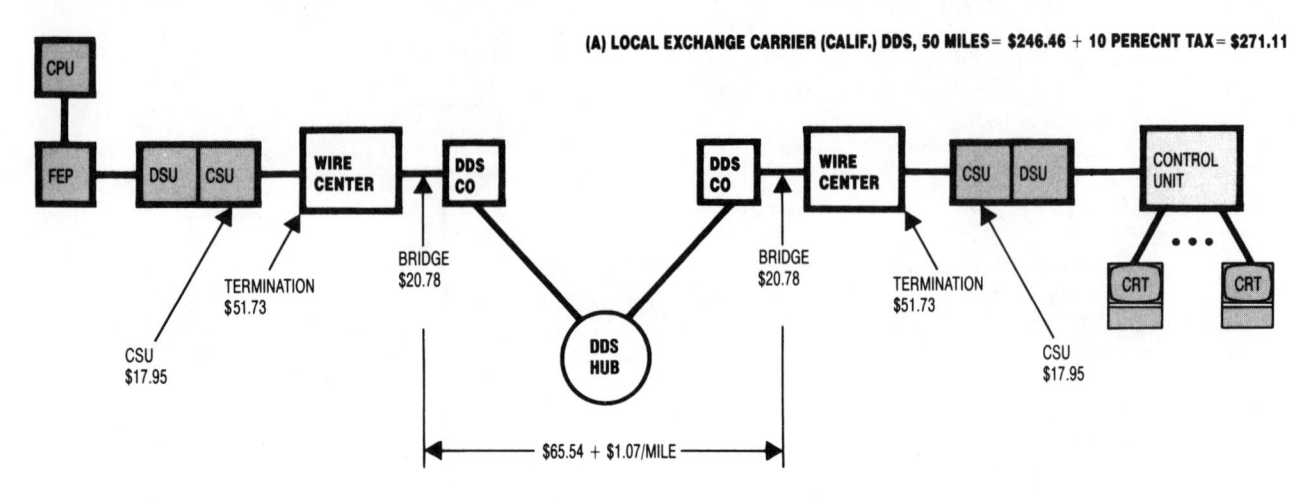

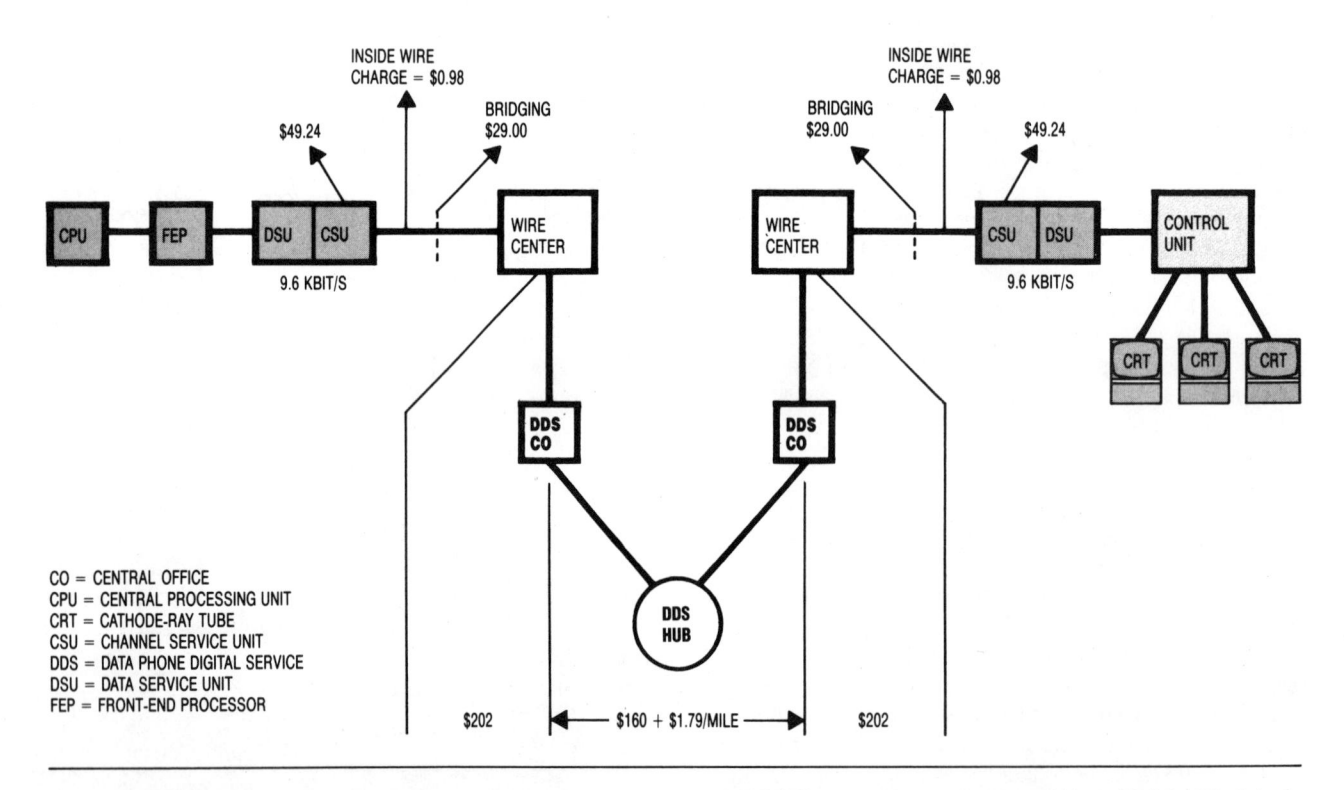

(A) LOCAL EXCHANGE CARRIER (CALIF.) DDS, 50 MILES = $246.46 + 10 PERECNT TAX = $271.11

(B) INTRASTATE, INTER-LATA DDS, 50-MILE CIRCUIT = $722.44 + 10 PERCENT TAX = $794.68

CO = CENTRAL OFFICE
CPU = CENTRAL PROCESSING UNIT
CRT = CATHODE-RAY TUBE
CSU = CHANNEL SERVICE UNIT
DDS = DATA PHONE DIGITAL SERVICE
DSU = DATA SERVICE UNIT
FEP = FRONT-END PROCESSOR

is divided into one or more numbering plan areas, or area codes, which are always the first three digits of a long-distance phone number.

In a given telephone number, the first six digits, designated by the nomenclature NPA-NXX, describe a given exchange—either a Class 4 or 5 CO. The last four digits identify the unique customer's line.

Intrastate, inter-LATA
Since January 1, 1984, only IOCs may carry traffic across LATA boundaries (invariably between Class 3 or Class 4 COs). When an independent phone company (not a BOC or RBOC) owns an exchange within a Bell LATA, it is free to choose whether it will be part of that LATA. And this determines whether its CO becomes an intra-LATA or inter-LATA exchange. If enough of these independent telephone companies are contiguous to one another, they could choose to form their own LATA.

Before divestiture, IOCs could provide only interstate services. But since the breakup, IOCs are allowed to enter intrastate, inter-LATA markets. Also, prior to the breakup, a BOC could provide intrastate, interexchange circuits. Now, however, a BOC can provide service only between exchanges within the same LATA.

3. Central office hierarchy. *Class 5, or end, offices are connected to a common Class 4, both part of the local exchange carrier's network. A Class 3 tandem office serves an entire metropolitan, often LATA-wide, area.*

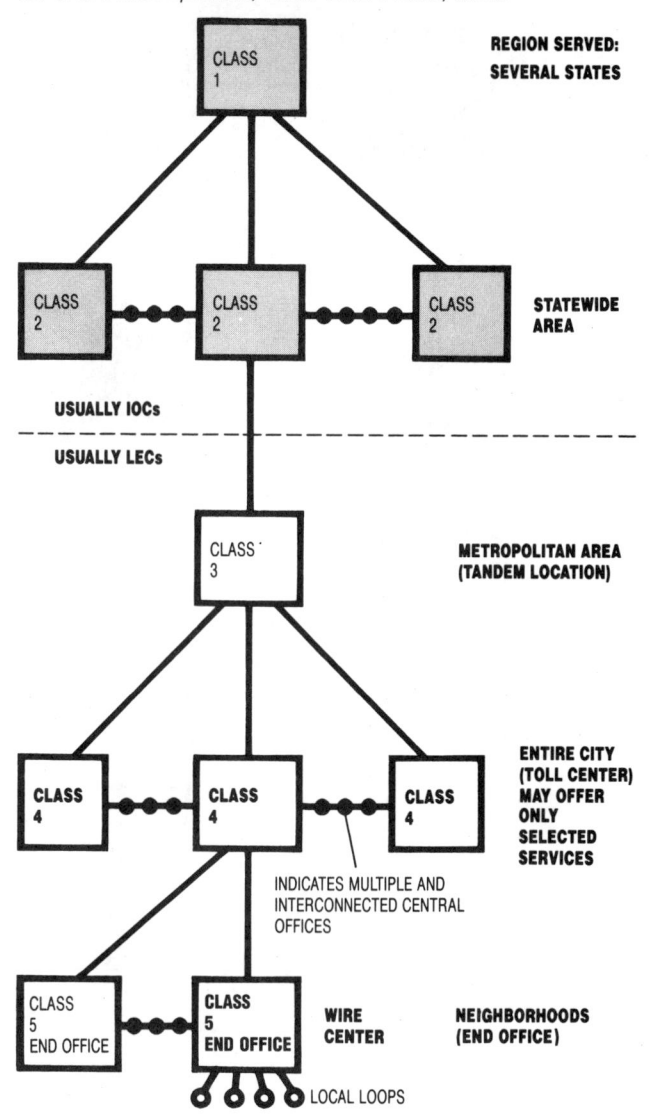

- CLASS 1 — REGION SERVED: SEVERAL STATES
- CLASS 2 — STATEWIDE AREA
- USUALLY IOCs / USUALLY LECs
- CLASS 3 — METROPOLITAN AREA (TANDEM LOCATION)
- CLASS 4 — ENTIRE CITY (TOLL CENTER) MAY OFFER ONLY SELECTED SERVICES
- INDICATES MULTIPLE AND INTERCONNECTED CENTRAL OFFICES
- CLASS 5 END OFFICE — WIRE CENTER — NEIGHBORHOODS (END OFFICE)
- LOCAL LOOPS

Special access tariff, or SAT, is the new name given to the pricing schedule for local, intra-LATA circuits—those short-haul, dedicated channels provided invariably by the local exchange carrier where both ends are within the same LATA. A network designer today needs access to more than 800 SATs when designing a nationwide network. By comparison, the same designer needed access to only about 200 such tariffs prior to divestiture.

Customer
POPs' network design now requires adoption of different strategies in different LATAs, based upon the special access tariffs of the LATA and the types of services offered by a given central office. What's more, in installing an end-to-end circuit, a customer may now place an order with a given LEC or IOC but specify a different carrier for each segment.

A few LATAs in the United States cross state lines, and the LECs in these LATAs are subsequently permitted to provide interexchange circuits that cross the state lines. An example is LATA number 228, which incorporates metropolitan Philadelphia (Pennsylvania) with most of the state of Delaware.

As mentioned, IOCs set up their point of presence in or near a CO. In some COs, since the LEC provides POP floor space to IOCs under the same roof, special tariffs have been developed to allow customer-provided equipment to also be located within the telephone company CO—creating, in effect, a customer POP.

This co-location of customer gear in a CO is one of the really big new opportunities made possible in Tariffs 9, 10, and 11. A corporate customer with heavy metropolitan networking requirements can really benefit—both from a cost and a reliability point of view—by establishing its own CO POP.

The local exchange carrier may also realize some facilities savings through a customer CO POP arrangement. When a concentrator is installed in a CO POP, the LEC does not then have to run multiple local loops from the CO to the customer's premises—where the concentrator would otherwise have to be located. In addition, most telephone-line failures are in the local loops. So, with a concentrator at a CO POP, a customer's local-network reliability can be enhanced even as costs are reduced. This is especially true if the customer equipment supports a 48-volt d.c. power supply, since some COs offer a connection service to their main uninterruptible battery-bank power supply.

Some LECs, however, simply do not have the floor space needed in their central offices to offer in-building POPs to the IOCs, let alone to all the customers who may want one. It is then up to the individual IOC to locate its own POP as near to the CO as possible and then run inter-office channels from POP to CO. AT&T, naturally, has at least one POP in every LATA. The other IOCs are present in selected LATAs.

Multipoint
When a customer orders a multipoint leased telephone line, he can have either the IOC or the LEC provide the bridging services. (A multipoint circuit is a series of point-to-point circuits that have been bridged together.) Again, the IOC and the LEC will frequently charge different rates for the same type of bridging. A price range of $6 to $30 per bridge is typical.

A typical long-distance multipoint line, with all the assorted interconnection and bridging components, as established under existing leased-line tariffs, is shown in Figure 4.

Since most multiplexers, concentrators, and data switches can support and perform the type of bridging needed for multipoint circuits, this is another incentive for customers to explore the possibility of strategically positioning this gear in their own CO POP, if space is available.

AT&T's new Tariff 9 contains rate information for the setup of inter-LATA circuit segments that span from an originating (AT&T) POP to a terminating POP. AT&T invariably uses its POP closest to each customer premises for computing the IOC channel mileage. This is called AT&T default routing.

In some cases the closest AT&T POP lies in the opposite

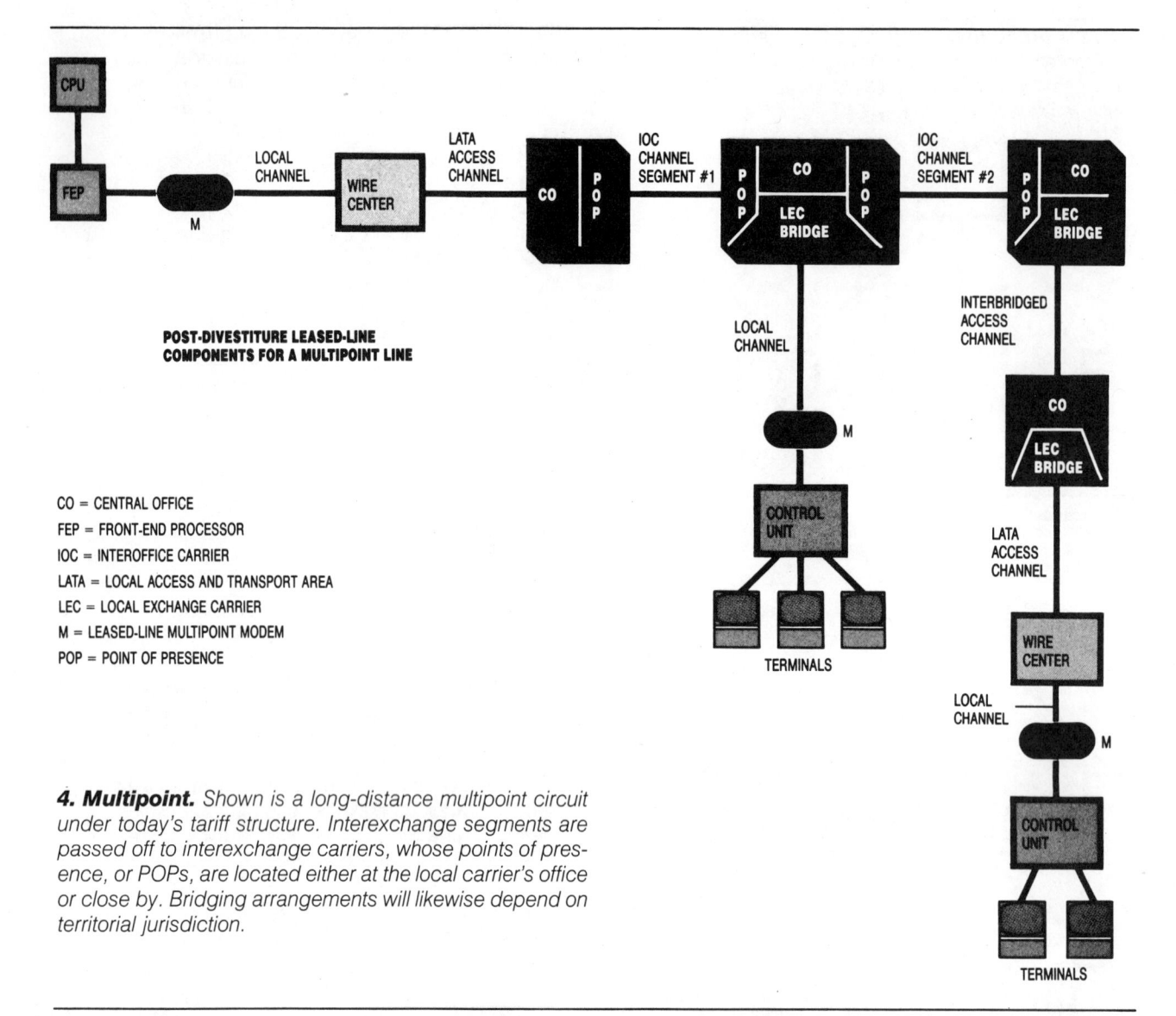

POST-DIVESTITURE LEASED-LINE COMPONENTS FOR A MULTIPOINT LINE

CO = CENTRAL OFFICE
FEP = FRONT-END PROCESSOR
IOC = INTEROFFICE CARRIER
LATA = LOCAL ACCESS AND TRANSPORT AREA
LEC = LOCAL EXCHANGE CARRIER
M = LEASED-LINE MULTIPOINT MODEM
POP = POINT OF PRESENCE

4. Multipoint. *Shown is a long-distance multipoint circuit under today's tariff structure. Interexchange segments are passed off to interexchange carriers, whose points of presence, or POPs, are located either at the local carrier's office or close by. Bridging arrangements will likewise depend on territorial jurisdiction.*

direction from the circuit route that would minimize the customer's costs. In these cases the customer pays twice: once to have his local loop extended in the wrong direction, and then again as the circuit heads back in the correct direction.

The customer has the option, however, of telling AT&T to use a particular route that he specifies. This process is called "customer-defined routing." There have been numerous cases and reports since divestiture, usually involving customers in rural areas, where leased-line telephone bills have been reduced by 15 percent to 20 percent merely through the application of customer-specified routing.

Tariff 10 lists the location of AT&T POPs, rate centers, and service availability. Tariff 11 contains rate information for AT&T-arranged, local-channel portions of inter-LATA circuits, in addition to access coordination functions. A bridged local channel is used to connect an LEC bridging hub to an AT&T POP, while an interbridged local channel is used to connect two LEC bridging hubs. Different tariffs apply for bridged and interbridged channels.

The telephone companies long ago created a vertical and horizontal (V&H) coordinate system that lays a grid, with line spacings of one-tenth of a mile, over the entire United States. Every LEC central office and IOC point of presence today is represented through a set of V&H coordinates. The following formula is universally used to calculate the channel mileage distance between two COs for the purposes of pricing a leased line:

$$\text{Distance (in miles)} = \sqrt{\frac{(V1-V2)^2 + (H1-H2)^2}{10}}$$

V1 and H1 are the coordinates for one central office, V2 and H2 the coordinates for the other. Companies including McGraw-Hill's Center for Communications Management Information and Economics and Technology Inc. offer information services that include a comprehensive listing of the V&H coordinates and services offered for every CO in the United States (including the non-Bell, independent telephone companies).

Both companies also offer computer services that calculate the optimum "customer-specified routing" between two user sites, and so does another firm, Contel Information Services, based in Great Neck, N.Y.

Access Interoffice carriers, in addition to charging customers based on the raw mileage of the circuits they lease, generally also include monthly charges for such basics as interconnection and access. IOCs are allowed to charge an End User Access Charge (EUAC), which averages about $25 per month per local "access" channel. Some may also seek a Message Station Equipment Charge (MSEC), which currently averages about $11 per end for both ends of an IOC channel.

Unless a customer can prove that there is no potential for his local channel to connect to the local exchange carrier's switched network (as might be done through, say, a PBX), then the customer's local channel is subject to the EUAC and MSEC charges, which are included in the IOC's monthly bill.

A leased data line terminating on a customer premises data PBX could, in all likelihood, access the LEC's network and would, therefore, be subject to the extra charges. If an end user's circuit goes into equipment incapable of switching the data line traffic, however, the customer should apply for a waiver of the EUAC and MSEC charges (on both local channels connecting to the IOC segment, if applicable). It takes perseverance and follow-up to check telephone bills and then fight for the removal of these charges.

Miscellaneous charges

Other options covered under AT&T's new tariffs (and separately priced) include: line conditioning; signaling (for PBX tie lines); inside wiring; special access surcharge; echo control; diverse routing; and something called "avoidance."

C-type and D-type conditioning charges apply to both local and interoffice line segments. C-2 conditioning, common on most data circuits, is typically $6 per local channel and $30 per IOC segment.

Signaling charges are typically $13 per local channel and $23 per IOC segment. Even if a PBX tie line is sometimes used for data, use of the circuit for switched-voice traffic still requires signaling.

Inside wiring fees are applicable whenever a customer has engaged the local exchange carrier to furnish internal wiring within the user's premises. Every telephone customer now has a "point of demarcation," usually inside a wiring closet or in the basement, which jurisdictionally separates the customer's inside wiring from the telephone company's. But if the customer asks the LEC to run wiring from the point of demarcation to some other location within the premises, then the inside wiring fee applies. Most states have permitted tariffs specifying charges of about $3 to $4 per month per local channel for this "inside wiring" service.

The special access surcharge, like the end-user access charge, applies if the data circuit has access to the public switched network. It typically costs $25 per month per local channel. If the SAS is charged, then the Message Station Recovery charge, typically $8 to $13, also applies.

Echo control costs about $26 per interoffice channel segment of a leased line. The charge pays for echo suppression equipment employed by the IOC on the interoffice transmission facility. If a customer runs two telephone circuits between the same two sites (primary for backup redundancy) and wants assurance that the two lines will not fail simultaneously, then he can ask for diverse routing. This applies generally only to the interoffice line segment, and typically costs $31 per IOC segment. With diverse routing, customer interoffice channels will not be routed through the same COs or over common transmission facilities.

The weakest link

The weak point with diverse routing, however, lies typically in the local channels. Whether it is even possible to have local channels routed through separate wire centers is something that varies from one LEC to another. Diverse routing's main purpose is to protect the subscriber from such problems as CO power failures, CO fires, microwave antenna outages, and icing of antennas in the winter.

For about $62 per interoffice segment, the IOC will promise to avoid a specified geographic location (set of COs) when it routes a subscriber circuit. Users should not confuse this "avoidance" option with route diversity, however. While diversity keeps parallel segments separate from each other, avoidance assures that a given channel will simply not pass through certain COs.

A user can request both diversity and avoidance from an IOC. The COs that many customers are fond of avoiding include, for example, those in areas adjacent to Los Angeles International Airport. Between rainy season floods, constant digging for major new construction (which often results in cut cables) and unusually heavy microwave interference from aerospace companies, data circuits routed near Los Angeles Airport have tended to have a disproportionately high error and outage record.

Equipment and technology

The transmission equipment typically employed with analog leased lines has advanced technologically, especially in the area of line diagnostics, to the point that certain savings in transmission-facility costs is also now possible. Still, problems of compatibility arising from the use of proprietary diagnostic methods can offset savings that can result from using unconditioned, instead of conditioned, lines and paying for alternate-routing or access-coordination services.

For years, modem vendors have been using secondary channels to perform "out-of-band" network diagnostics, performance measurement, and network control. But link management via secondary channel has required that all internetworked modems use the same primary and secondary channels. It was nearly impossible to mix modems from different vendors and still achieve cohesive network management and diagnostic control.

The secondary-channel approach to modem network management (an Open Systems Interconnection Layer 1 technical-control method) has also had the effect of excluding multiplexers and concentrators from participating in the network management schemes employed by modem vendors. As a result, several multiplexer and concentrator vendors (including Codex, Infotron, Doelz, and Digital Communications Associates) have developed their own network technical control and management schemes and products, most of which employ link-layer (OSI Layer 2) communications techniques.

Modern network multiplexer and concentration equipment is software-controlled, which also makes it possible to achieve new efficiencies over otherwise static and fixed-bandwidth leased-line networks. One way is through the use of virtual point-to-point and multipoint circuits.

A port on one network concentrator can be "software-connected" to a port on a another node, thus enabling many virtual circuits to be transmitted over a single trunk line between two communications nodes. Individual virtual circuits can even be assigned different priorities, so that when a trunk is saturated, lower-priority transmission ports can be flow-controlled. A concentrator-based network featuring virtual-circuit line sharing is shown in Figure 5.

Other new features supported by several vendors' network-concentration gear today include alternate path routing; trunk-load leveling; and automatic alarms that respond to such events as carrier loss, modem streaming, data loss, and too many link-level transmit/retransmit frame errors. In addition, some concentration processors permit different protocols to be specified/supported on a port-by-port basis, thus enabling multiple protocols (and different logical sessions) to be intermixed on the same trunk circuits between network nodes.

Case in multipoint

In one particular networking environment, a large bank had employed separate, discrete data networks for a variety of different data-traffic types. There were an asynchronous burglar and fire alarm network, which used a polled protocol; separate transmission facilities for automated teller machines (ATMs), employing synchronous IBM binary synchronous communications (BSC) or synchronous data link control (SDLC) protocols; and another network for asynchronous CRTs and printers in branches that connected to a different computer from the one handling the automated teller machines.

In this case, separate data networks radiated from the same bank computer center to the same set of branch banks—but each using its own protocol, each connected to a different computer, and each using a different set of phone lines.

Through the introduction of strategically placed network concentrators, the customer, Seattle First National Bank (Seafirst), was able to cut leased telephone-line costs nearly in half. Using virtual circuits between network concentrators (made by Doelz), computers from Tandem, IBM, Honeywell, and Mosler all were able to share a common physical network, though each still communicated—via virtual circuits—with separate terminals and controllers in each bank branch.

Poll-passing

Many large data networks still spend most of their time (and bandwidth) carrying polls from computers to device cluster controllers. Increasingly, however, host-end communications processors are being used to eat much of this polling traffic, while the multiplexer or concentrator at the cluster-controller end of the network regenerates these polls.

The net effect of this technique is the elimination of the polling traffic from the network trunk circuits. And the absence of polling traffic over the network backbone can,

5. Virtual circuits. *With sophisticated software-controlled multiplexer switches available today, a single leased channel can carry a multitude of different traffic types.*

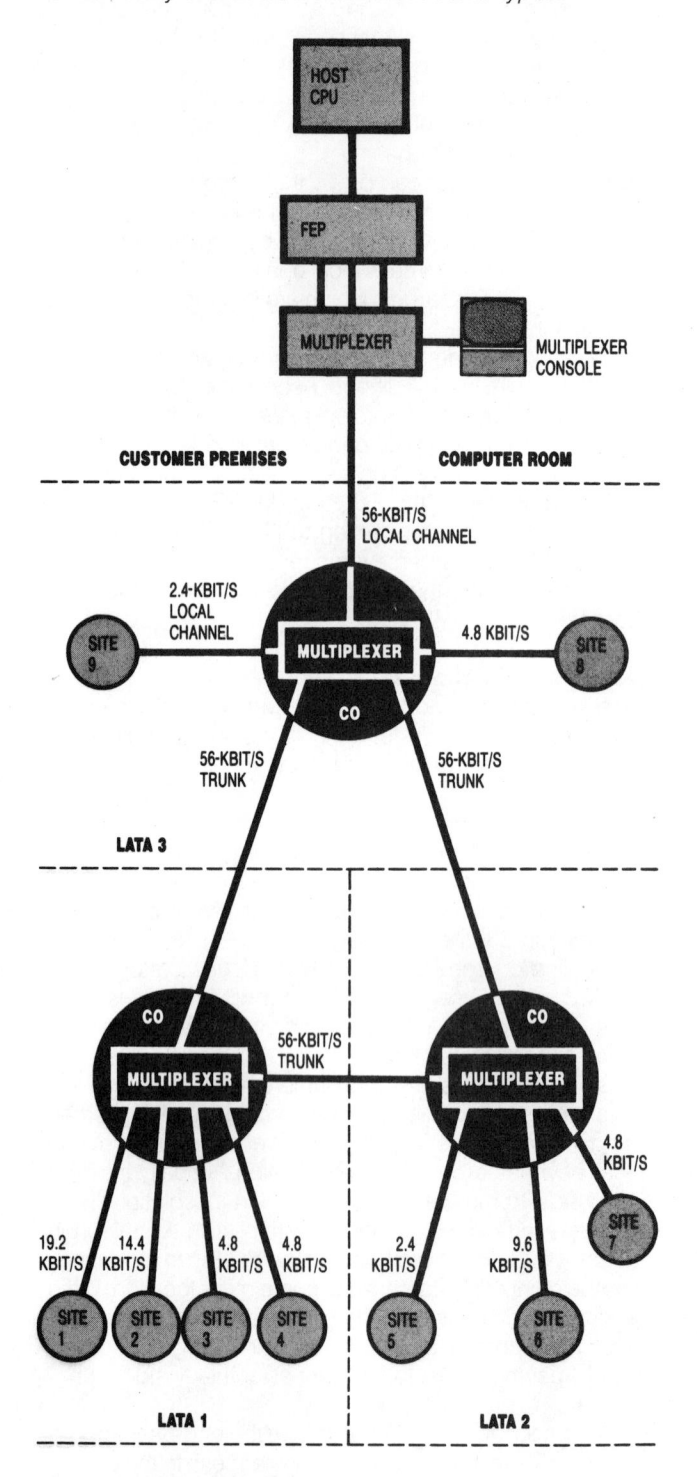

VIRTUAL CIRCUIT 1 = SITES 8 AND 9
                   2 = SITES 1 AND 2
                   3 = SITES 3 AND 4
                   4 = SITES 5, 6, AND 7

CO = CENTRAL OFFICE
CPU = CENTRAL PROCESSING UNIT
FEP = FRONT-END PROCESSOR

in some vendor implementations, improve network response time and bandwidth utilization.

Most of the techniques and efficiency-enhancing measures delineated here can be used in a wide variety of network environments. The optimization of leased-line trunks through intelligent nodal concentrators requires the acquisition of special hardware, of course, but several vendors now provide such sophisticated gear and prices are generally dropping.

New opportunities
Networkers should consider the benefits of placing multiplexing or concentration gear at a central-office point of presence, a new opportunity made possible under current local and Federal Communications Commission (FCC) tariffs. This can eliminate, for example, the requirement for much of the LEC or IOC circuit bridging. Only multipoint tail circuits would continue to require bridging. The customer sites within an exchange or wire-center serving area could then be connected to the POP-based equipment with simpler, and less expensive, local channels not requiring physical bridging.

In such a network, trunk circuits connecting customer gear in various POPs can run at higher speeds than the local channels, the data rate of which would be consistent with their aggregate data load. It generally is no longer necessary to run a whole network at one single speed.

Another strategy, available with newer concentration gear, is to place a mux within a LATA and then concentrate intra-LATA circuits into that mux. Inter-LATA trunk circuits are then used only to connect muxes. The tariffs for intra-LATA circuits are usually lower in price than for inter-LATA circuits and tend to be more reliable. Also, by minimizing the number of inter-LATA circuits, the user simplifies the troubleshooting of broken links. (A local carrier can usually fix an intra-LATA circuit much faster than when it has to coordinate with an IOC and another LEC on inter-LATA-circuit troubleshooting.)

When a customer adopts the architecture of a mux within a LATA to concentrate intra-LATA service sites, he gains the freedom to change any given phone line's speed to meet changing bandwidth demands of that one line. He can fine tune his network, making small incremental changes where they have the most impact. Most muxes also now support multiple trunks between the same two muxes. So, when a given trunk is being taxed at the highest available speed, the user can simply add another trunk between muxes. ■

Peter Hansen, until recently director of product marketing for Doelz Network Inc. and previously a systems engineer for Hewlett-Packard, is now an independent communications consultant living in Walnut, Calif., holds a degree in computer science from West Coast University in Los Angeles.

Darlane Hoffman, Metropolitan Life, New York, N.Y.

Squeezing line costs via data compression

A technology that promised to cut the number of leased lines by 50 percent sounded too good to be true. Here's how one insurance company evaluated it.

The data communications team at Metropolitan Life Insurance had heard that, in theory, data compression could work miracles with throughput rates and line costs. But the company already had a well-running network—and making it more economical would have to be done without compromising the existing performance, reliability, and quality.

The team proceeded to put data compression to the test. As with any other new technology or advanced equipment, they applied a sequence of evaluation techniques: goal setting, testing, analysis, internal company communications, and, in this case, implementation.

Data compression follows the principle that seven- and eight-bit characters can be represented, on average, by four bits. Thus, the throughput potential of character compression is 2 to 1, since twice as many characters can be transmitted in a given amount of time if they contain four bits rather than eight.

The compression factor theoretically doubles the effective bandwidth of a communications link, creating excess line capacity. For example, a 9.6-kbit/s link with the capacity of carrying 1,200 eight-bit characters per second should, after the addition of data compression technology, be able to handle a data flow of 2,400 four-bit characters per second. If the 2-to-1 compression ratio is achieved, one 9.6-kbit/s line would be able to do the work of two.

Met Life first learned of data compression technology in 1984. The cost and operational benefits that it promised were readily apparent—a 50 percent reduction in the number of leased lines would offer tremendous savings. And having fewer lines to manage would considerably simplify network management.

In early 1984, Met Life began preliminary research into data compression technology. The data communications team soon realized that going from 9.6- to 14.4-kbit/s modems in conjunction with data compression could yield up to a 66 percent reduction in transmission lines.

These benefits were too tempting to ignore. The team then decided to find out whether data compression could perform as promised.

When the investigation into data compression technology began, the standard Met Life circuit configuration included Codex LSI 9.6-kbit/s modems (see "The shaping of Met Life's network took shape"). Use of the SDLC protocol allowed modem-sharing devices to be introduced, so that two IBM 3X74 control units (up to 32 IBM 327X terminals) could be supported via one analog "3002" leased-line circuit and with just one port on the IBM 3725 front-end processor (FEP).

Monitoring network performance at the FEP port level was accomplished with Avant-Garde's Net/Alert, and FEP port switching and modem sharing were done with that company's Net/Switch (Fig. 1).

The Met Life data communications team widened its investigation beyond data compression technology to include statistical multiplexing, a technology that also promised line reductions. In all, the team evaluated equipment from three different vendors: two offering data compression technology; one with a statistical multiplexing application.

In the first phases of testing and evaluation, the team aimed to answer the following questions:
- What performance levels can be expected from data compression and statistical multiplexing in terms of increased data throughput? And with what effect on response time and error rate?
- Would the equipment affect Met Life's network in ways that could not be predicted before testing?
- Could real 19.2-kbit/s throughput be achieved?
- Could Met Life's standard requirement for a five-second response time be met with no increase in error rates?

The first limited field test occurred on a 3002 analog

The shaping of Met Life's network

The data communications network at Metropolitan Life Insurance Co. includes 425 point-to-point analog circuits, 180 multipoint analog circuits (having 1,200 nodes), two 230-kbit/s analog circuits, eight 56-kbit/s digital circuits, and eleven 56-kbit/s satellite channels used for data file transmission. In 1982, the company implemented SDLC, which resulted in a 50 percent reduction in point-to-point network costs. More recently, the use of IBM's Multisystem Networking Facility (MSNF) has enabled access to remote databases without the need to install additional circuits.

Taking charge

Soon after the AT&T divestiture in 1984, Met Life realized that it would have to take greater responsibility for its own network management, and cost containment became one of its highest priorities. By marrying a wide range of products and services and by putting a good team of data communications personnel in charge, Met Life has been able to achieve advances in network management while stemming the escalating rise of network costs.

Met Life is in the process of installing a T1 network in a ring configuration made up of 12 T1 circuits (see figure). The use of intelligent T1 multiplexers from Cohesive/Digital Communications Associates Inc. allows dynamic bandwidth allocation, alternate routing, and improved diagnostics. Tests using 19.2-kbit/s modems will be conducted and remote concentration of tail circuits will be implemented with the intention of further reducing network costs.

At present, the company's network management is addressed through IBM's Network Communications Control Facility, Network Problem Determination Application, and Network Performance Monitor; Avant-Garde's Net/Alert and Net/Switch; and Codex's Distributed Network Control System. And for network management enhancement, IBM's NetView is still anxiously awaited.

Managing these network activities requires true team effort. The team at Met Life is comprised of teleprocessing software engineers, data center network analysts, and network operators along with corporate data communications systems engineers.

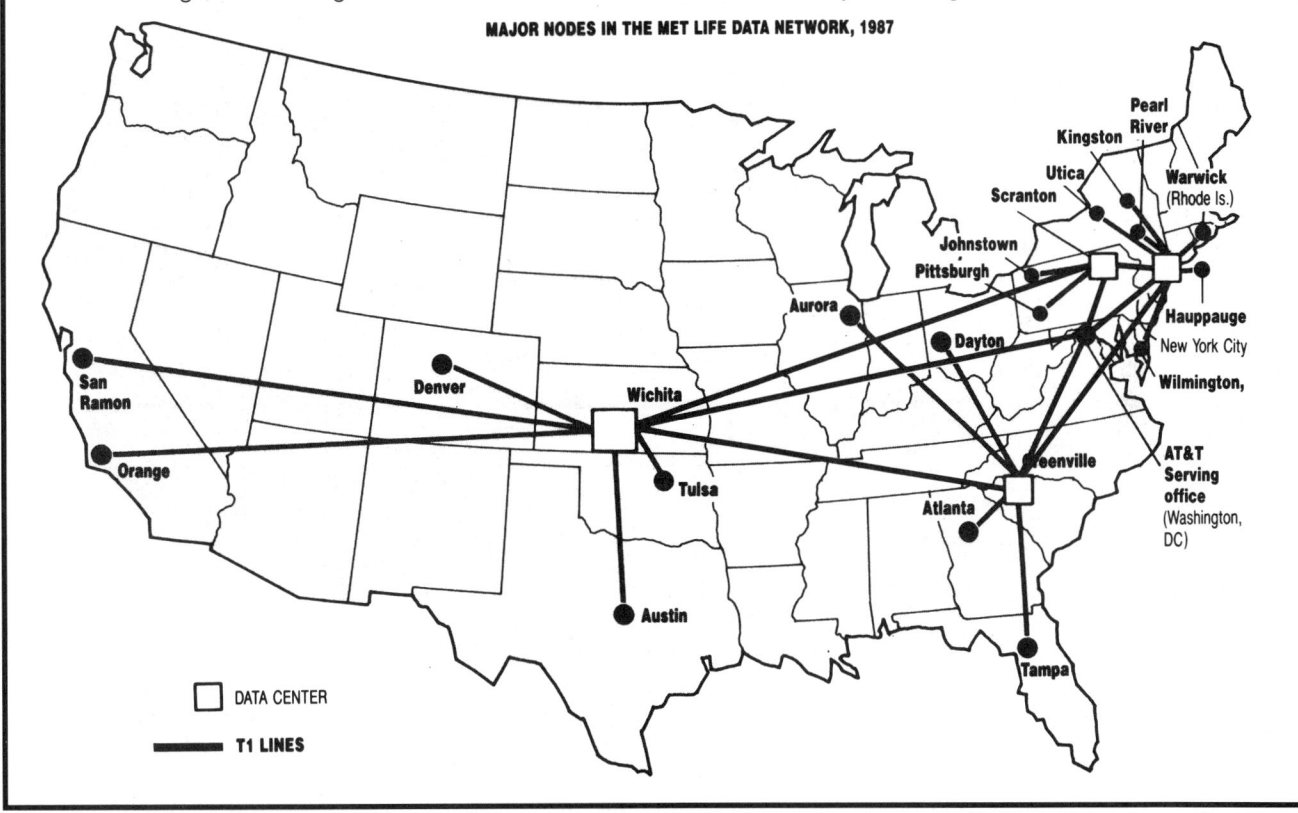

MAJOR NODES IN THE MET LIFE DATA NETWORK, 1987

circuit spanning a four-mile distance between Met Life's Greenville, S. C., data processing center and a Met Life health insurance claims data entry business office, also in Greenville, S. C. The testing was run at 9.6 kbit/s.

For the next test, the data communications team simulated field conditions in its data communications laboratory at the New York home office. They measured performance levels with the Avant-Garde network management equipment. The Net/Alert served as the primary data collection tool during the two-year study portion of the project.

The records showing how each vendor's equipment performed, and how it was subsequently rated, are illustrated in Table 1.

Racal-Vadic (Milpitas, Calif.) supplied a data compressor, the Scotsman II, with integral statistical multiplexing. The equipment did not, however, meet the objective performance requirements, and there were some operational difficulties with the firmware. Ironically, this equipment was the most expensive that was studied in this project.

Codex (Mansfield, Mass.) supplied a statistical multi-

1. Met Life circuit, before. *The company's standard circuit configuration included Codex LSI 9.6-kbit/s modems. Monitoring network performance at the port level of the front-end processor was accomplished with Avant-Garde's Net/Alert; while FEP port switching and modem sparing were achieved with that company's Net/Switch.*

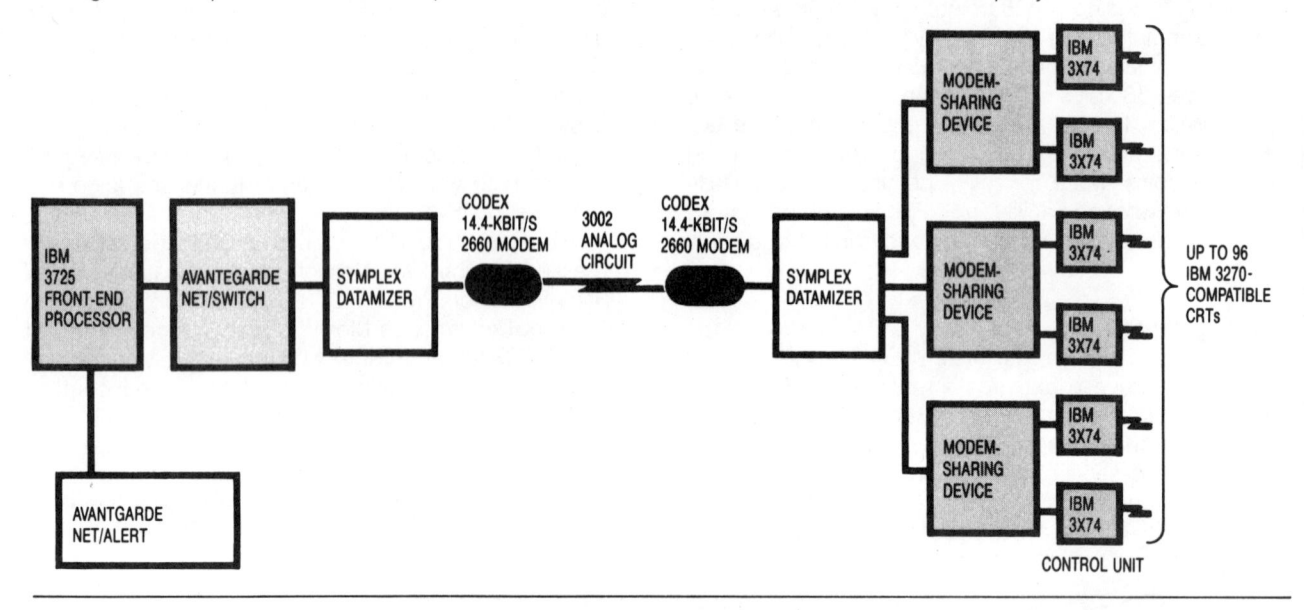

plexer, Model 6005. The company claimed its equipment would deliver a two-to-one throughput reduction. On lines operating at 5 percent to 10 percent of capacity, the claim proved true. However, on heavily used lines (greater than 40 percent of capacity) throughput degraded rapidly and fell short of the two-to-one claim. Codex's product was very reliable, and the Met Life team judged it a solid product but one that did not quite meet all the requirements. It was the least expensive equipment tested.

Symplex (Ann Arbor, Mich.) supplied a data compressor, the Datamizer, which has an integral statistical multiplexer. The company claimed a two-to-one increase in throughput capacity. The Met Life team found this to be true. In fact, throughputs of two-and-a-half-to-one were achieved on a line running at 9.6 kbit/s and three-to-one on a line running at 14.4 kbit/s. In each test the lines were identical. Speed variation was achieved through the use of modems. For the most part, the Symplex product was found reliable and easy to use. Its cost fell midway between that of Racal-Vadic and Codex devices.

Figure 2 shows the cable configuration that Met Life used between the matrix switch, Net/Alert, Symplex's data

Table 1: Performance of test equipment

VENDOR AND NAME OF PRODUCT	METHOD OF OPERATION	MANUFACTURER'S CLAIM	TEST RESULTS	RESPONSE TIME	ERROR RATE	COST	COMMENTS
RACAL-VADIC (SCOTSMAN II)	DATA COMPRESSION WITH STATISTICAL MULTIPLEXING	INCREASE THROUGHPUT BY 2:1 ON 9.6-KBIT/S LINES	<2:1	>5 SECONDS	INCREASED	GREATER THAN CODEX OR SYMPLEX PRODUCT	FIRMWARE (PROMs) PRESENTED DIFFICULTIES DURING OPERATION HARDWARE DID NOT PERFORM TO LEVELS OF CODEX OR SYMPLEX PRODUCTS
CODEX (MODEL 6005)	STATISTICAL MULTIPLEXING	INCREASE THROUGHPUT BY 2:1 ON 9.6-KBIT/S LINES	2:1 ON LIGHTLY UTILIZED LINES <2:1 ON HEAVILY UTILIZED LINES	5 SECONDS ON LIGHTLY UTILIZED LINES >5 SECONDS ON HEAVILY UTILIZED LINES	NO INCREASE	LESS THAN RACAL-VADIC OR SYMPLEX PRODUCT	PERFORMANCE DEGRADED RAPIDLY ON HEAVILY UTILIZED LINES—<2:1 THROUGHPUT AND >5-SECOND RESPONSE TIME HARDWARE VERY RELIABLE A GOOD, SOLID PIECE OF EQUIPMENT
SYMPLEX (DATAMIZER)	DATA COMPRESSION WITH STATISTICAL MULTIPLEXING	INCREASE THROUGHPUT BY 2:1 ON 9.6-KBIT/S LINES	2:1 TO 2.5:1 ON 9.6-KBIT/S LINES 3:1 ON 14.4-KBIT/S LINES	<5 SECONDS	NO INCREASE	MIDDLE RANGE	HIGHLY RELIABLE EASY TO OPERATE PERFORMANCE EXCEEDED MANUFACTURER'S CLAIMS

2. Cable configuration. *Net/Alert ribbon cables attach to the host ports of the data compressor. Cables from the matrix switch mate to the Net/Alert ribbon cables.*

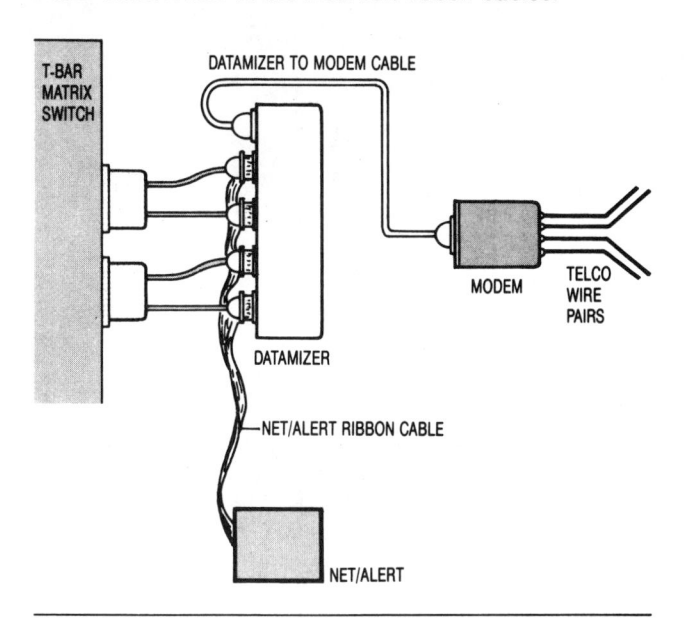

compressor, and the modem. The Net/Alert ribbon cables are first attached to the host ports of the data compressor. The cables from the matrix switch are then mated to the Net/Alert ribbon cables. Matrix switch ports feed directly to the front-end processor ports. Since this portion of the configuration did not change with the installation of the data compressors, response time and line monitoring functions remained unchanged.

Squeeze play
Symplex uses what the company says is an auto-adaptive multilevel data-compression algorithm called SCC tabling. SCC tabling includes character-level compression, string-level compression, header and trailer stripping, poll compression, run-length encoding, and several other forms of data compression—all apparently able to work together to achieve optimal throughput performance. The company says its equipment is compatible with all protocols.

The Met Life data communications team decided to find out if Symplex Datamizers would work with more than just the SDLC protocol. Experimenting with several of Met Life's bisynchronous circuits (the company has 60 that support a VM environment used in applications program development), the team found that the Symplex devices would, in fact, work in a bisynchronous environment.

The data compressor is a standalone device operating on the digital side of the line. The Met Life team considered this a plus, even though most company data communications currently operate in an analog circuit/modem environment. Should the company decide to migrate to digital circuits, only the modems would need to be replaced—with DSU/CSU gear. The data compressors would remain in place. In further testing, the team placed data compressors, one on each end of a line running from its Greenville data processing center to the New York home office.

Figure 3 shows the circuit configuration with 14.4-kbit/s modems and the data compressors. All the initial tests showed that the compression gear, coupled with 14.4-kbit/s transmission, tripled throughput. This remained "invisible" to the data being transmitted, and response time remained within Met Life's five-second service level requirements. Error rates did not noticeably increase.

The Met Life team designed a test situation involving 25 production lines running between its three data processing centers—Greenville, S. C., Scranton, Pa., and Wichita, Kan.—and the New York home office. Their performance goals for this test were:
■ The 50 units (all Datamizers installed in pairs on 25 lines) would be tested to see if they could yield at least a 95-percent hardware reliability level during the three-month test.
■ As part of this testing, the equipment could not cause response time to increase from present levels. This was measured via Net/Alert network management equipment by compiling data on individual circuits, both before and after the installation of the data compressors. Error rates could not significantly increase either.

At Met Life's claims-processing business offices, each IBM control unit handles up to 16 terminals. Installing compressors allowed the company to double the number of control units (and therefore the number of terminals) connected through a single circuit. Since the compressors achieved at least a two-to-one compression ratio, the simultaneous introduction of Codex 2660 modems operating at 14.4 kbit/s allowed them to add two additional control units to each circuit. The net result was a three-to-one circuit reduction allowing support of six control units (up to 96 terminals) on one analog 3002 circuit.

During the three-month testing period Met Life collected statistical data on circuit performance. One of Met Life's systems engineers wrote a special program for statistical analysis of the data. A summer intern was hired for the sole job of entering the mountains of data into the database.

Can it break?
The team then wanted to see if they could make the links supported by the compression equipment fail. Personnel at the data processing center were asked to choose those circuits in the network that had the highest stress. Those lines showing more than 80 percent utilization during peak periods were defined as high-stress lines; the company's average network utilization ranges between 40 and 50 percent of capacity.

For the first time during the testing process the team placed data compression equipment in working Met Life business offices. There were no technical support people at the test locations after initial installation, since the team wanted to see how business people worked with the equipment.

The results of the stress tests were invaluable to the project. This phase of the testing process showed that there was some degradation in response time. The Datamizers were no longer transparent to the data, and response time fell below the acceptable level of five seconds.

Fine-tuning the software solved this problem. First, adjustments were made to the size of the message packets being sent from the data compressors. Before the alteration, the 3X74 control units were receiving broken message packets that caused an unacceptable level of requests for

3. Circuit configuration, after. Addition of 14.4-kbit/s modems and Symplex Datamizer data compressors tripled throughput and was "invisible" to the data being transmitted. Response time remained within Met Life's five-second service level requirements, and error rates for transmittal data did not increase.

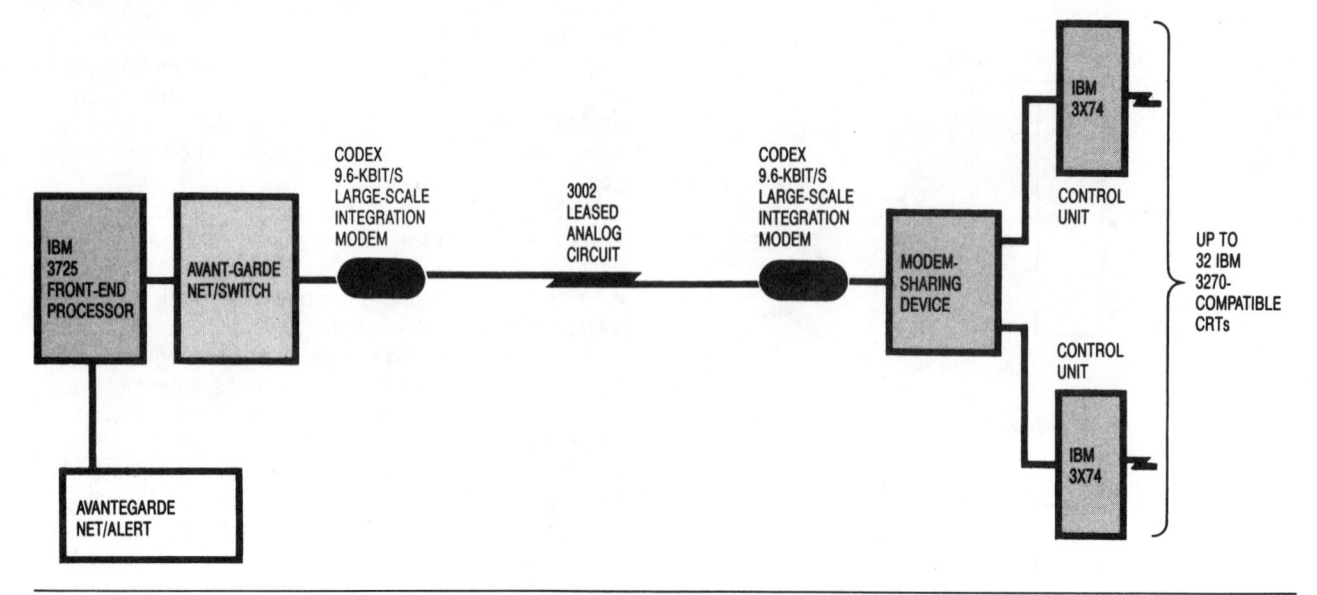

retransmission. The "new" packet size gave the control units a continuous stream of data, and this virtually eliminated requests for retransmission.

Also adjusted was the receive clock metering function on the data compressors to enable speed metering between 9.6 and 12 kbit/s. Slow metering, which allows data speeds between 1.2 and 9.6 kbit/s, was disabled. The team eventually found speed metering to be optimum on Met Life's heavily utilized circuits as it maximized throughput. Fifty-three circuits were involved in this phase of the project. Before the project began, each was carrying more than 10,000 transactions per day. Using Codex 2660 14.4-kbit/s modems and Datamizer data compressors, Met Life increased the capacity of these lines to more than 30,000 transactions each day.

The company's mandatory quality and reliability levels were slightly eroded at this level, however. The data communications team refined its network definition in response, stipulating that only three circuits—whose additive transaction level remained under 30,000 transactions per day—would be combined in a three-to-one compression arrangement.

Point to point
Integrating the new data compression technology into Met Life's nationwide point-to-point network was the next task of the data communications team.

Using guidelines set forth in Met Life's Quality Improvement Program the data communications team began holding a series of planning meetings. The majority of the strategies used to implement data compression in the company's network came out of these meetings.

One main strategic imperative was to include the active participation of data communications network personnel from the three data processing centers, since these people are key ingredients to the success of the project. They actively participated in the planning sessions and the hardware installations. This process generated a tremendous amount of enthusiasm. Everyone became involved in what was to be an exciting new project, and their subsequent high-quality performance reflected this initial enthusiasm.

The group decided that, to minimize disruption of office operations, the new equipment should be installed in the business office on weekends and after working hours. They also decided that each office should retain all circuits for one month after the Datamizers were installed to allow for rapid backout should problems develop. The data communications team did not want to impose the new technology; rather, they wanted business personnel to have the opportunity of either accepting or rejecting it. The subsequent 100-percent companywide acceptance of the equipment proved the wisdom of this choice.

At key points during the testing and evaluation of the project, the data communications team documented progress to top Met Life management personnel. This course also proved advantageous, for when the team sought approval for full implementation of the data compression equipment they had merely to write a two-page report reflecting their goals and plans. The details of the study had already been conveyed.

One of the concerns raised by top management during final review was that the new approach would be tying two to three times more users to a single circuit. Equipment failures on such heavily used lines could cause similarly intense problems. However, the team was able to ascertain the performance of the equipment with dual-dial backup arrangements.

In the dual-dial scheme, two long-distance calls are manually dialed over voice-grade lines. The lines are then patched into the network with one line sending and the other receiving. SDLC protocol requires that the two lines be used. The Met Life team found that they could maintain dual-dial connection at 14.4 kbit/s in almost all cases.

Table 2: Datamizer installation, 1986 report

INSTALL DATE	OFFICE	NUMBER OF LINES		SAVINGS	
		BEFORE	REMOVED	MONTHLY	YEARLY
PILOT	VARIOUS	34	34	$24,350	$292,200
3/86	SAN RAMON, CALIF.	7	4	4,464	53,568
5/86	NEW YORK, N.Y.	31	20	15,236	182,832
	*UTICA, N.Y.	11	6	3,624	43,488
	WARWICK, ILL.	34	17	14,814	177,768
6/86	GREENVILLE, S.C.	9	5	755	9,060
	TULSA, OKLA.	15	9	10,392	124,704
7/86	DENVER, COLO.	10	3	3,387	40,644
	KINGSTON, N.Y.	21	11	10,516	126,192
	AURORA, ILL.	28	16	14,155	169,860
	PEARL RIVER, N.Y.	14	7	6,342	76,104
8/86	**UTICA, N.Y.	25	14	13,762	165,144
	TAMPA, FLA.	34	17	16,961	203,532
9/86	DAYTON, OHIO	13	6	5,498	65,976
	HAUPPAUGE, N.Y.	17	7	6,426	77,112
	JOHNSTOWN, PA.	17	10	7,542	90,504
10/86	PITTSBURGH, PA.	13	6	4,854	58,248
	ORANGE, CALIF.	10	6	7,608	91,296
11/86	AUSTIN, TEX.	5	3	2,670	32,040
	WILMINGTON, DEL.	6	3	2,427	29,124
TOTAL		354	204	$175,783	$2,109,396

*METROPOLITAN PROPERTY AND LIABILITY DIVISION
**GROUP DIVISION

Hence, there was virtually no degradation of performance in this backup mode.

In addition, well-documented test results indicated that Symplex equipment could be expected to perform at a reliability level of 95 percent. Furthermore, each office would be stocking spare components. These would fill the gap between the time of a failure and the arrival within 24 hours of a new unit, since Symplex guarantees overnight delivery of replacement units.

Only two partial failures have occurred with the equipment since the testing period. Both times a power source failed. The Datamizers automatically switched to backup power source and communications were not interrupted. Automatic warning functions alerted Met Life staff to the problem, and both units were replaced within 24 hours.

The Met Life data communications team completed the implementation phase of the data compression project in October 1986. Twenty business offices throughout the United States were targeted as having sufficient network concentration to warrant installation of the data compressors.

The team's original goal of achieving a three-to-one reduction in the number of leased lines was not completely realized, since some offices had only one or two lines to a data processing center (Table 2). A 60 percent reduction, however, was achieved. In addition, at most offices capacity has increased because of the spare ports available on the Datamizer. (Each has four ports, but a full configuration, as defined by Met Life, uses three. Hence, the spare.)

Along with implementing the data compression technology, the project enabled the data communications team to reconfigure Met Life's network. Many of the company's offices began their initial operations with two or three lines and then rapidly exanded to 20 or more lines. Here are some of the additional benefits that the project yielded:
- Identification and labeling of all components of each business office circuit;
- Removal of old unused cable;
- Production of up-to-date maps locating all equipment;
- Increase in the hands-on operational skills of local support staff.

Overall the team enhanced the ability of business office management to administer their local data communications environment, and they improved the ability of the data center and business office personnel to work together to solve problems.

Another important benefit of this project has been the new working relationship established between the data communications team and Met Life's field personnel. The relationship that grew during implementation of the data compression equipment helps members work together to solve other problems as well. Furthermore, the team now knows the individual configurations and capabilities of each office. This knowledge continues contributing to improved data communications services within Met Life. ■

Darlane Hoffman is Technical Manager of Data Communications Services at Metropolitan Life Insurance Co., where she has worked for 14 years. She earned her B. A. in mathematics from South Dakota State College and an M. B. A. from Wichita State College. She holds the Fellow of Life Management Institute designation and is a member of the Association of Data Communications Users.

Joseph Braue, DATA COMMUNICATIONS

1987: The year when networking became part of the bottom line

SNA made prime time, deals consummated the marriage of computers and communications, and the fiber glut begot bargains

The stock market crash on Black Monday will taint most memories of 1987. But the crash also showed why 1987 was the year that networking came out of the wiring closet and into the mainstream of world commerce.

Controversial strategies blamed for the crash, such as program trading and portfolio insurance, were made possible by networks that tied together in one global exchange the financial centers of the United States and the rest of the world.

1987 was the year that networking was introduced on a grand scale to the masses. IBM's marketers shelved the Tramp, a standalone micro user, and brought the M*A*S*H crew out of retirement to demonstrate the joys of personal computer connectivity. And it was the year that SNA (Systems Network Architecture) made prime-time TV in a series of Wang Laboratories Inc. commercials where yuppie salespeople mentioned the IBM networking acronym as if it were as recognizable as flouride or Nutrasweet.

In addition to 60-second sales pitches, corporate America was being told by consultants and academics—preaching the gospel of using information technology as a strategic weapon—that most critical policy decisions to be made reside in building the company network.

Marriages of necessity

Networks became critical because of the fusing of computers and communications. Many pages have been soiled with the announcement of the marriage, but last year, in policy statements, product announcements, and business mergers, evidence of the fusion reached a critical mass.

No longer could a vendor specialize in one type of networking technology. To position themselves for future growth, local area network vendor 3Com merged with large net vendor Bridge Communications. Digital Communications Associates, a formidable micro-to-mainframe link and T1 multiplexer power, bought LAN vendor Fox Research—and nearly snagged Ungermann-Bass as well. LAN vendor Novell Inc. bought micro-to-mainframe link power CXI Inc. (see Table 1).

The three largest computer companies all made significant moves to extend their reach into wide area network communications. The biggest of them, IBM, entered a joint marketing agreement with T1 multiplexer maker Network Equipment Technologies. And the second-largest computer maker (in total sales), Unisys, bought T1 multiplexer vendor leader Timeplex. IBM and number three Digital Equipment Corp. began preaching to the industry about how they would connect their machines into future Integrated Services Digital Networks.

AT&T outdistances foes

As the computer vendors sought to better tie computers and multiplexers to offer private networks more functionality, public network giant AT&T did just that by weaving new hybrid public-private networks for customers such as General Electric and E. I. Du Pont de Nemours & Co.

For GE, AT&T is running System 85 PBXs from its own points of presence and allowing the customer unprecedented control over the customer's own network. In the Du Pont deal, AT&T is tying together the user's NET T1 multiplexers. In both cases, AT&T faced regulatory opposition from computer manufacturers and from its long-distance competitors.

The third post-divestiture year still finds AT&T earning most of its living off its long-distance arm. Vittorio Cassoni, president of AT&T's Data Systems Group, spent 1987 trying to convince industry analysts that AT&T had come up with the right plan to succeed in computers.

While AT&T's long-distance operations have never been in better health, especially with the news that the Federal Communications Commission wants to lift the rate-of-return cap on AT&T's long-distance profits, the same could not

1987 in review

Table 1: Important deals of 1987

Month	Deal
JANUARY	NOVELL INC. (LANs) AGREES TO ACQUIRE CXI INC. (MICRO-TO-MAINFRAME LINKS)
FEBRUARY	
MARCH	BOLT BERANEK AND NEWMAN INC. (PACKET SWITCHES) AGREES TO BUY NETWORK SWITCHING SYSTEMS (T1 SWITCH STARTUP)
APRIL	FRENCH GOVERNMENT CHOOSES L.M. ERICCSON (SWITCHES) TO BUY STATE-OWNED CGCT (SWITCHES)
MAY	
JUNE	IBM (COMPUTERS) AND NETWORK EQUIPMENT TECHNOLOGIES (T1 MULTIPLEXERS) ANNOUNCE JOINT MARKETING AND DEVELOPMENT AGREEMENT
JULY	DIGITAL COMMUNICATIONS ASSOCIATES INC. (MICRO-TO-MAINFRAME LINKS AND T1 MUXES) AGREES TO BUY FOX RESEARCH INC. (LANs); MICOM SYSTEMS INC. (DATA PBXs) AGREES TO ACQUIRE SPECTRUM DIGITAL CORP. (T1 MULTIPLEXERS); CONTEL CORP. (NETWORKS) AGREES TO BUY EQUATORIAL COMMUNICATIONS INC. (SATELLITE EARTH STATIONS) AND COMSAT INTERNATIONAL COMMUNICATIONS INC. (EARTH STATIONS) A PART OF COMSAT CORP.; 3COM CORP. (MICRO LANs) AND BRIDGE COMMUNICATIONS (LARGE LANs) AGREE TO MERGE.
AUGUST	
SEPTEMBER	MCI COMMUNICATIONS CORP. (LONG-DISTANCE CARRIER) AGREES TO BUY GENERAL ELECTRIC'S RCA GLOBAL COMMUNICATIONS CORP. (TELEX); PLESSEY CO. PLC AND THE GENERAL ELECTRIC CO. PLC (NO RELATION TO US GE) AGREE TO MERGE TELECOMMUNICATIONS BUSINESSES.
OCTOBER	
NOVEMBER	UNISYS CORP. (COMPUTERS) AGREES TO ACQUIRE TIMEPLEX INC. (T1 MULTIPLEXERS)
DECEMBER	

be said of its two closest long-distance competitors, MCI and U S Sprint, both of which posted mediocre financial returns for most of the year.

MCI founder William McGowan was out of action for the better part of 1987, recuperating from a heart transplant operation, while the corporate parents of U S Sprint—GTE Corp. and United Telecommunications—replaced Charles Skibo in Sprint's executive suite with United Telecom executive Robert Snedaker.

With networking serving as both the medium and the message in 1987, the nascent science of network management was the subject of many announcements, but end-to-end network management of multivendor devices remained elusive. A worldwide Open Systems Interconnection set of standards for network management was still in the definition stage in 1987.

While more vendors announced linkages into IBM's Netview management scheme, the first real pieces of which were only just being delivered by year's end, AT&T revealed an outline of its network-management scheme, called Unified Network Management Architecture. As public network proprietor, AT&T uniquely united public network customer-controlled reconfiguration capabilities with network management of its Dataphone II products, Starlan LANs, as well as providing access to IBM's Netview. Of course, this was just an announcement, and the product won't be here until sometime this year.

The vapor index rises

Not that AT&T's strategy was unique. Indeed, 1987 was the year of the preannouncement. In what formerly would have been branded vaporware, vendors in 1987 couched preannouncements as necessary aids to let customers in on future plans. IBM went so far as to make a preannouncement in April of a preannouncement in November, concerning when in 1988 its OS/2 operating system would be made available.

1987 was also the year in which IBM introduced conceivably the largest vaporware package in the history of the industry: Systems Application Architecture (SAA), a blueprint for the creation—by IBM, third-party software developers, and users—of common applications that would run across IBM's incompatible micro-, mini-, and mainframe computer architectures. Although its SNA foundation is already firmly in place, experts say it will take at least 10 years to reach SAA's promised land, the same amount of time SNA took to attain de facto standard prominence.

IBM shuffles the micro deck

Closer to reality is IBM's software foundation for microcomputer communications, OS/2. In April, IBM changed the rules for the microcomputer industry by introducing the PS/2, a new microcomputer line that runs the same MS-DOS software as existing machines but with a new "microchannel" architecture that is incompatible with its previous IBM PC, PC/XT, and PC/AT, all three of which have been discontinued in the past year.

Micro communications vendors quickly changed their PC adapter cards to fit the microchannel. Nevertheless, there is no evidence that it will be needed until OS/2, the next-generation operating system for the PS/2, and OS/2 applications become available. That is because the main

1987 in review

benefit of the microchannel is the tighter coupling between the PS/2's 80286 or 80386 microprocessors and the adapters for external devices and network links that will fit in the PS/2's slots.

In December, IBM began shipping the kernel of the operating system, jointly developed with Microsoft. Later this year will come OS/2 Extended Edition, which consists of a database management package and a smorgasbord of communications options. The hope is that applications will be developed using the OS/2 kernel, the Extended Edition, and the Presentation Manager, a Macintosh-like user interface that will truly enable micros to seamlessly handle multiple tasks and multiple sessions with other computers across the network.

IBM has indicated it will not use a micro network operating system developed by Microsoft and 3Com, called LAN Manager, as it did in reselling Microsoft's MS-Net operating system for the first generation of its PCs. It did announce the interim LAN Server program, however, and much of 1987 saw endless speculation about how network operating system powers Novell Inc., Microsoft, and IBM would coexist with OS/2.

For its part, Apple Computer finally made good on promises it made under the more pragmatic leadership of John Scully to allow the Macintosh to network with corporate America, most of which is using MS-DOS computers linked by Ethernet LANs. Two new Apple models, the Macintosh SE and Macintosh II, can run MS-DOS programs and can now handle expansion cards that make them more direct participants in Ethernet LANs.

ISDN several steps closer

Another spotlight contributing to the overall prominence of networking was Integrated Services Digital Network. There was more progress on the long road to ISDN, the proposed universal digital service of the 1990s. With somewhat closer cooperation between American and foreign standards makers, significant gains were made in areas such as the critical "U" interface (between the ISDN network and the user premises) and the Layer 2 link protocols, otherwise known as the Link Access Procedure-D, or LAPD. Stan-

In 1987, Federal District Court Judge Harold H. Greene (A) again played Scrooge to the Baby Bells; FCC Chairman Dennis Patrick (B) pushed deregulation; MCI Chairman William McGowan (C) recouperated from a heart transplant; IBM PC czar William Lowe (D) changed the rules of the micro game; a new data encryption policy from former National Security Adviser John Poindexter (E) failed; Unisys CEO Michael Blumenthal (F) bought mux maker Timeplex; and the stock exchange on Black Monday (G) reflected the new importance of networking.

dards makers, in fact, were already moving on to the next great standards objective, broadband ISDN.

1987 saw, though prematurely, a plethora of products claiming to be ISDN compatible. AT&T, Bellcore, the regional Bell operating companies (RBOCs), and many vendors were testing the devices individually to see if they could work together. The various ISDN trials proceeded apace in 1987, and plans were made to turn projects into commercial ISDN service in 1988.

But ISDN's limits were also evident. The trials, to date, all basically involve enhanced Centrex services. Nynex and other RBOCs began studying the job of renovating the millions of local loops to handle ISDN's demands. AT&T announced it was delaying its plan to deliver a primary-rate ISDN service in 1987 for another 18 to 24 months. And a chance for an early agreement on a worldwide broadband ISDN standard interface of 150 Mbit/s was lost.

Floating the intelligent network
While plans for the universal digital interface between users and the network progressed, the BOCs and the common carriers continued their study of future public network enhancements. The RBOCs readied their plans for Open Network Architecture (due next month to the FCC), the scheme to open up the local loop so that third parties will have a chance to offer enhanced services to users on an equal footing with the telephone companies.

Bellcore, central office switch makers, and the RBOCs debated Bellcore's Intelligent Network/2, a plan to add intelligence to the public network with the aid of Signaling System 7, the advanced signaling architecture envisioned with ISDN. IBM, which entered into agreements with Siemens and Ericsson, and DEC made plans to sell telephone companies the decentralized computers and databases needed to implement such a system.

More BOCs decided to start local packet-switching networks. Several forged links with larger players, such as BellSouth Corp.'s link with nationwide packet player Telenet Communications Corp., while a new company, Globenet, announced links with all the RBOCs. In addition, more of the BOCs announced local switched 56-kbit/s service, extending the reach for the underutilized AT&T nationwide version of the service. Finally, AT&T announced that it would link its switched 56-kbit/s service internationally, hooking up with France Transcom's equivalent high-speed dialup digital service.

1987 will also be known as the year that the fiber glut finally paid off for users. All that extra fiber bandwidth from the massive fiber building projects of the last few years resulted in huge discounts on long-distance services by carriers.

And where discounts weren't, or couldn't be, reflected in tariffed prices, service vendors came up with new and unique ways to offer users breaks, including cheaper financing, offering new features and special support staff.

Technological advances
As in every past year, in 1987, advancements were made in making communications products faster, cheaper, and better. The biggest scientific achievement of the year, the quantum leap in the temperatures of materials that would support superconductance properties, promised major communications benefits in future undersea cables and satellites.

More in the present tense, vendor demonstrations at the Hannover Fair in Germany and at Telecom '87 in Geneva proved that X.400, the International Telegraph and Telephone Consultative Committee (CCITT) electronic-mail standard, works. Major public electronic-mail providers announced support for the computer vendors' electronic-mail applications.

In 1987, Silicon Valley was the scene of a race between LAN vendors to develop reliable ways of carrying CSMA/CD (carrier-sense multiple access with collision detection), IEEE 802.3-sanctified, 10-Mbits Ethernet transmissions over conventional 24-gauge unshielded twisted-pair wiring. Vendors such as Synoptics Communications Inc. and 3Com led the way in developing the technique, which uses conventional telephone wiring rather than Ethernet cable-bus topology. The big loser could be AT&T's Starlan twisted-pair LAN.

Many of the technical advances of the year involved fiber optics. IBM researchers announced the "chiplet," a tiny gallium-arsenide hybrid optical receiver that is capable of accepting data at up to 3 Gbit/s, twice the current speed of similar devices.

For longer distance transmission, British Telecom announced a laboratory version of a purely optical lightwave regenerator, a device that eliminates the need to periodically convert light signals to electric pulses for reshaping and retiming and then convert them back again. AT&T announced it had brought on line the 1.7-Gbit/s FT Series G fiber optic transmission system between two central offices in Illinois. AT&T claims it is the world's fastest lightwave digital transmission system. And taking the fiber jockeys down a notch was the report that a prehistoric nemesis, sharks, had chewed up segments of AT&T's prototype fiber cables laying on the Atlantic Ocean floor.

Land-mobile satellite services, which could be used for low-level data communications into remote areas, was given a boost by the World Administrative Radio Council on Mobile Services, which reallocated parts of the low-frequency L band from aeronautical and maritime use to LMSS. Under the watchful eye of the FCC, eight vendors have agreed to submit a joint operating agreement for LMSS in the United States.

Competition spurred price decreases in many product areas in 1987, as it does every year. The most salient, perhaps, occurred in high-speed leased-line V.32 modems, where Universal Data Systems led competitors to drop prices from about $2,500 to about $1,500. The modems, which were priced at more than $3,000 when they first were introduced about two years ago, are expected to drop to about $1,000 this year.

The judge and the commissioner
All the technological changes made the jobs of regulatory bodies even harder. In the United States, the top regulation issue (see Table 2) involved Federal District Court Judge Harold H. Greene putting the RBOCs' development of new information services and products on hold, saying he was protecting the public from their monopolistic grip on the local loop. In Europe, despite advancements in England, an opening in France, and slight movement in Germany,

1987 in review

Table 2: Regulatory highlights of 1987

FCC CHAIRMAN MARK FOWLER ANNOUNCES HE WILL RESIGN	JANUARY
JUSTICE DEPT. ISSUES REVIEW OF AT&T DIVESTITURE ADVOCATING THAT JUDGE HAROLD H. GREENE ALLOW REGIONAL BELL OPERATING COMPANIES TO EXPAND BUSINESSES BEYOND LOCAL LOOP. "HUBER REPORT" RELEASED	FEBRUARY
PRESIDENT REAGAN ANNOUNCES HE WILL NAME FCC COMMISSIONER DENNIS PATRICK AS FCC CHAIRMAN	
IN HISTORIC SWITCH, MCI COMMUNICATIONS CORP ADVOCATES DEREGULATION OF AT&T	MARCH
WHITE HOUSE INFORMS CONGRESSIONAL SUBCOMMITTEE THAT DIRECTIVE TO PLACE MILITARY CONTROL OVER COMMERCIAL NETWORK DATA ENCRYPTION STANDARD HAS BEEN DROPPED	
	APRIL
	MAY
EUROPEAN COMMISSION ISSUES "GREEN PAPER" URGING COST-BASED PRICING FOR COMMUNICATIONS SERVICES	JUNE
FCC PROPOSES THAT THE VALUE-ADDED NETWORK VENDORS' EXEMPTION FROM ACCESS CHARGES BE LIFTED	
U.S. SENATE VOTES TO SUSPEND DOMESTIC SALES OF ALL TOSHIBA PRODUCTS IN RETALIATION FOR ILLEGAL SALE OF MACHINERY TO SOVIETS	
	JULY
FCC FLOATS PLAN TO LIFT RATE-OF-RETURN RESTRAINT ON AT&T LONG-DISTANCE PROFITS AND REPLACE IT WITH PRICE CEILINGS	AUGUST
JUDGE GREENE REJECTS BOCs BID TO OFFER INFORMATION SERVICES, ONLY ALLOWS THEM LIMITED GATEWAYS AND TRANSMISSION LINKS	SEPTEMBER
FRENCH PTT OPENS UP VALUE-ADDED NETWORK SERVICES MARKET TO COMPETITION WHILE GERMAN BUNDESPOST OPENS UP CUSTOMER PREMISES EQUIPMENT (CPE) MARKET BUT KEEPS ITS NETWORK MONOPOLY	
	OCTOBER
FCC SEEKS COMMENT ON PLAN TO IMPOSE ACCESS CHARGES ON PRIVATE NETWORKS THAT SWITCH LONG-DISTANCE CALLS INTO THE LOCAL LOOP	NOVEMBER
JUDGE GREENE UPHOLDS BOC MANUFACTURING BAN AND SAYS FURTHER THAT THE BABY BELLS CAN'T DESIGN PRODUCTS	DECEMBER

governments mostly sought to protect national monopolies on public networks.

In the first triennial review of the AT&T divestiture agreement, Greene rejected the position of the Justice Department and its consultant, Peter Huber, who argued that enough time had passed for the BOCs to be given more freedom to expand their businesses beyond their local-loop franchises. Greene prohibited anything but a low-level entry into information services and ruled that the original decree prohibited the Baby Bells from designing, as well as manufacturing, equipment.

These rulings spawned legal appeals from the BOCs and raised the ire of their Congressional supporters in Washington, D. C., who are signaling a second campaign to turn Greene's communications oversight responsibility to the more laissez-faire-minded FCC.

In Fowler's footsteps

The driving force behind federal communications deregulation, FCC Chairman Mark Fowler, resigned in early 1987 and was replaced by Dennis Patrick, who continued his predecessor's policies. Patrick wasted no time in showing his mettle by advancing three controversial policies that are currently undergoing comment from interested parties. The Patrick-led FCC floated rules to end the free ride that value-added networks (VANs) and some private networks had been given concerning paying access charges for connection to the local loop.

Thinking that the rate-of-return cap on AT&T's long-distance profits discouraged innovation and lower prices, Patrick sought to replace it with a procedure that would instead regulate prices for AT&T services. The theory is that with the prospect of making more money, AT&T would have more incentive to streamline long-distance operations and save consumers money and offer better services. This time it was Patrick who ran afoul of Congressional power brokers, who feared the rule would give AT&T a license to print money.

European monopolies hold firm

European fans of deregulation were cheered by a European Commission report that recommended that the 12-member nations open up competition in the customer premises equipment markets and base the charges for telecommunications services strictly on the cost of providing them. Currently, there are wide discrepancies for various communications services from one country to another. In addition, the French Postal, Telegraph, and Telephone (PTT) agency announced plans to open up the VAN market for competition. In Germany, a long-awaited Bundespost report on deregulation also advocated CPE deregulation.

But on the whole, all three 1987 forays into deregulation left intact the state-run public network monopolies. More European users reacted by secretly ignoring the more onerous PTT restrictions. And a few users, including two American banks, chose to either relocate or not to build networking centers in West Germany because of the country's restrictive regulatory atmosphere. That users would go to such lengths demonstrated once again how important networking had become to corporate bottom lines in 1987. ■

Donald F. Blumberg, D. F. Blumberg & Associates Inc., Fort Washington, Pa.

Looking ahead: Network planning for the 1990s

Forewarned is forearmed. Wanted: Integrated suppliers that can maintain and support all requirements on both public and private networks.

Where are the network services and technology markets heading? And how can you benefit from forecasting with reasonable accuracy the likely trends and turns of these markets? The possible scenarios are not that many, and hedging your bets can yield a workable network strategy to carry you into the next century. Consider the following:

If the voice and data communications networks and services market hasgrown dramatically since the last decade, it is in orbit today. The U. S. telecommunications market, including voice and data services over public networks and expenditures for private network technology and support services, topped $150 billion in 1986, a figure estimated to hit $300 billion by 1990 (Fig. 1). The fastest-growing area of the total telecommunications market has been the technology and services for private networks, which have expanded since the start of the 1980s. Behind that has been the purchase of data services from the public network.

Now, as U. S. industry approaches the century's last decade, fully integrated networks that link voice and data communications with data processing, office automation, building and process control, and other types of networks will be combined through a mixture of private bypass and public switch (virtual) networks. The result? Complete capability for network communications (Fig. 2).

Essentially, three issues have caused a shift toward full commitment to network technology and networks in the U. S. business world:
- Deregulation of the public telecommunications network.
- Blurring of technological distinctions between data processing, office automation, telecommunications, and control networks technology through the common use of network-oriented microprocessors.
- More and more technology has been tied onto the network's backbone as a framework for doing business.

The latter is certainly the most important, particularly in the segments of banking, finance, manufacturing, transportation and distribution, retail, and hospitals and healthcare facilities, where voice and data network technology is an essential ingredient of day-to-day operations.

Providers of equipment and services for public switched and private networks, in their response to this situation, are developing a vast array of new technology: T1; subsets of ISDN (Integrated Services Digital Network) such as 23 B+D—the primary rate (B standing for basic, or 64 kbit/s, while D is 64 kbit/s again); Signaling System No. 7 (see "Inside SS No. 7: A detailed look at ISDN's signaling system plan," DATA COMMUNICATIONS, October 1985, p. 120); and advanced value-added network and packet-switched network technology. These are the innovations that are leading the way to fully integrated voice and data networks. Can there be any question that the use of integrated networks, both public and private, will continue to burgeon in the next century?

The longer term

In terms of the overall time-line trends, at least three scenarios could emerge by the start of the 1990s:
Scenario A: growing dependence on private network technology for overall management and control of business operations.
Scenario B: growing dependence on the advanced technology offering of the public switched network.
Scenario C: the integrated use of both public and private networks technology to achieve innovative and productive improvements within the business operation.

Each scenario will involve different levels of commitment and revenue allocations, depending on the final choices of the market by the end of the 1980s. Since some user segments will opt for one of the three alternatives, general trends and options in vendor and supplier delivery must be

1. Burgeoning. *The growing market for telecommunications hardware, software, services, and support represents a rich opportunity. Demand may hit $300 billion by 1990.*

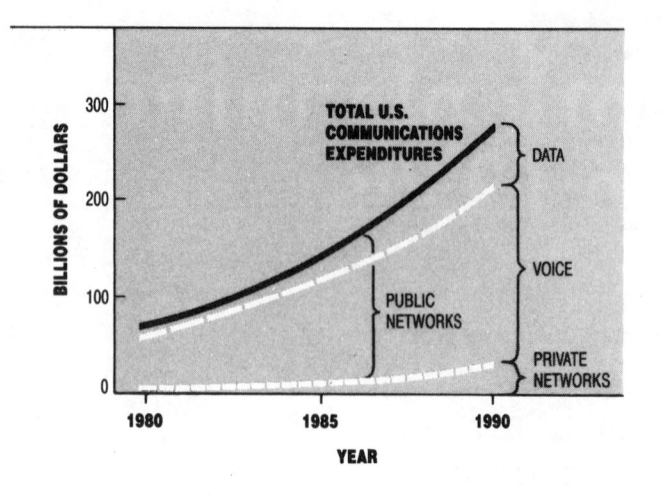

explored in order to establish an optimum market operating strategy to move into future conditions.

The future direction of network growth and implementation will be highly influenced by the strategic choices taken by major competitors, especially Digital Equipment Corp., IBM, Honeywell, and, of course, AT&T. The window of opportunity created by customer choices associated with Scenarios A, B, and C, particularly from 1987 through 1990, will be the framework within which such strategies will be tested.

The present competitive posture appears to suggest that the new network service strategy can be adapted and efficiently developed by the major regional Bell operating companies and by the leading hardware-oriented network and data processing developers named above.

Competition in networks and service market

The rapid evolution in both the application and use of network technology in industry, as Figure 2 illustrates, has led the way to a marked change in the application, use, and creation of a broad new range of alternatives for

2. Change. *A significant change over the past two decades relating to the implementation and use of networks was caused by the Carterfone decision in the 1970s, deregulation of the telephone networks in the early 1980s, and the current commitment to ISDN. As a result, the 1990s will flaunt fully integrated public and private networks.*

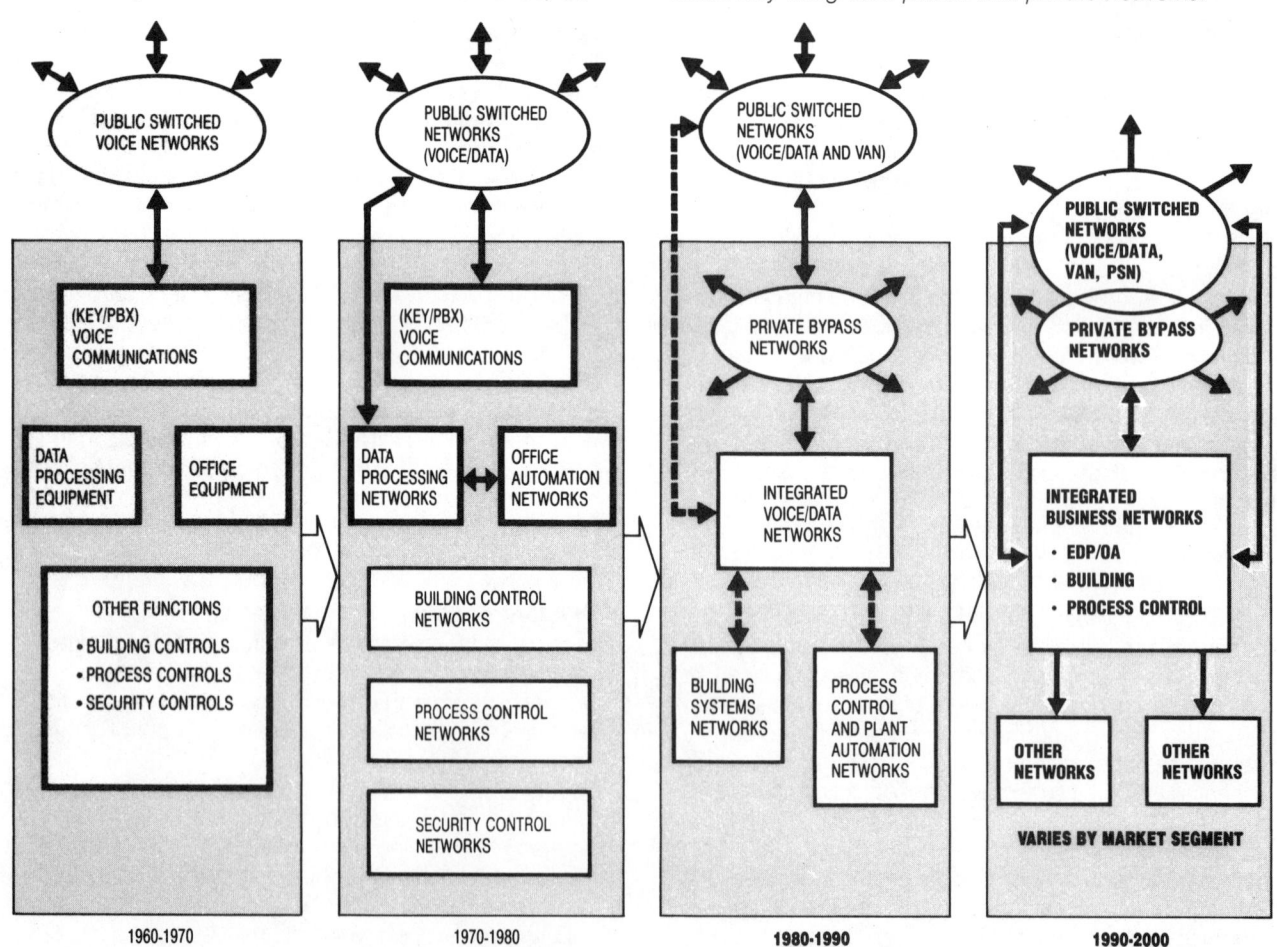

EDP = ELECTRONIC DATA PROCESSING PSN = PACKET-SWITCHED NETWORK
PBX = PRIVATE BRANCH EXCHANGE VAN = VALUE-ADDED NETWORK

Technology's revolution

Regulatory, technical, and economic factors have created a revolution in the use of network technology since the 1970s. Included in influences that have changed the course of the telecommunications industry have been:
- The Carterfone decision, which provided for the user's ability to connect privately owned equipment to the public switched network.
- Very large-scale integration and integrated electronic chips reduced system costs, and the microprocessor increased power.
- The introduction and increasing sophistication of data processing networking technology using broadband capacity.
- The introduction of on-line real-time systems technology based on military development and use of command control and intelligence network systems applications in the 1960s and early 1970s.
- The growing fear of economic and performance changes created as a result of the anticipation of deregulation of the public telephone network.
- The actual deregulation of the U. S. telephone network in 1983.
- The post-divestiture decisions of Judge Harold H. Greene, who has continued to approve the expansion of the regional Bell operating companies' role in everything but product.

These factors have combined to provoke an enormous change in the application and use of network systems and technology, as Figure 2 illustrates. Privately owned interconnect equipment, including key sets and automated PBXs, has emerged over the past several years to control and link internal voice communications to the public switched network.

New public switched network technology for data communications (Telpak, for one) created the ability to communicate from internal data processing networks through the public switched networks to their locations. Complex distributed data processing networks and office automation and word processing networks have also seen the light of day during this period, as have independent building control, process control, security, and industrial automation networks that use electromechanical or electronics-based Programmed Logic Controllers.

Strong economic incentives created by public carriers in general and AT&T in particular (including AT&T's attempts to integrate key and PBX users to new types of switches through progressive pricing and continued transmission line-rate increases) led to the emergence, in the late 1970s and in this decade's early years, of private bypass networks that connected to and made use of public switched networks where economically reasonable.

In essence, private bypass networks were made financially and operationally feasible through the combination of the Carterfone decision, the availability of low-cost interconnect switch technology, and the concern about potential effects of deregulation on businesses whose reliance on telecommunications had become crucial. AT&T's rate increases were no small matter either.

By the end of the last decade, economic projections of private network applications and use by industrial corporations showed a potential payback of less than two years. This incentive, coupled with the fear of reduced quality under future deregulation, created a strong commitment to private network development in the early 1980s.

This decade has also seen the emergence of internal integrated voice/data networks, independent distributed data networks, building control systems networks, and process control/plant automation networks linked to the voice/data networks through new local area network technology.

---

network delivery and implementation to the customer base.

The array of public network technology and service vendors has been increased through the separation of AT&T from the regulated Bell operating companies, now the seven regional Bell operating companies (RBOCs). Judge Harold H. Greene's recent decisions, prohibiting RBOCs' move into manufacturing, have, in effect, given them but one route into the market, and that is through services and distribution.

The breaking apart further educated mature, sophisticated users—those having large networks, including Tymshare, MCI, and U S Sprint. In addition, this new range of value-added and special network suppliers has widened its service options. In the area of private network equipment technology and services, for example, there are more suppliers of data processing, telecommunications, and office automation products so that customers can get assistance in constructing their own network technology and networks.

Indeed, a new class of telecommunications network integrators and turnkey/facility managers, along with independent service organizations, has come into being. Some customer organizations have even made such large investments in private networks that they are beginning to offer their excess capacity to other users, thereby adding to the competitive number of value-added special network suppliers.

A new breed

The result? A third class of competitor is in the making, due to the availability of public and private-switched network technologies and their alternatives, as well as the user's increasing dependence on the network. Now is an age that needs a Sears Roebuck of services to meet all of that user's needs. Such new network service providers will focus the breadth of services required to design and engineer, install, maintain, and repair networks using a mixture of public and private network technology.

The new category consists of the "Enterprise" elements of some of the more aggressive RBOCs—Bell Atlantic, BellSouth, and Pactel, among them—subject only to Judge Greene's constraints. It will also embrace the more asser-

Impact of future scenarios and customer focus on vendor positioning

FUTURE SCENARIOS	VENDOR POSITIONING		
	PRIVATE NETWORK TECHNOLOGY ■ PRODUCT ■ SALES	PUBLIC NETWORK TECHNOLOGY ■ VALUE-ADDED NETWORKS ■ INTEGRATED SERVICES DIGITAL NETWORK ■ OTHER	INTEGRATED SYSTEMS/SERVICE SUPPLIER ■ SERVICE ORIENTED ■ PUBLIC/PRIVATE MCX
A. FOCUS AND DEPENDENCE ON *PRIVATELY OWNED* NETWORK CONTROL	HIGH POTENTIAL	LOW POTENTIAL	INTERMEDIATE POTENTIAL
B. FOCUS AND DEPENDENCE ON FUNCTIONAL SERVICE COST OR TRANSMISSION SERVICE QUALITY CONTROL VIA PUBLIC NETWORK	LOW POTENTIAL	HIGH POTENTIAL	HIGH POTENTIAL
C. FOCUS AND DEPENDENCE OF NETWORK AS KEY STRATEGY OF BUSINESS OR PRODUCTIVITY IMPROVEMENT THROUGH GENERAL NETWORK TECHNOLOGY	INTERMEDIATE POTENTIAL	INTERMEDIATE POTENTIAL	HIGH POTENTIAL

tive and innovative hardware and technology vendors, primarily those of Digital Equipment Corp.'s ilk.

The network service supplier positioning represents an attempt to take advantage of the possibilities in Scenario C, described above: network service becoming a key factor in the user's vendor-selection process.

In essence, the application and use of the public network has continued to grow, creating new vendors of public network technology out of old-line organizations (AT&T, BOCs, RBOCs, and other telephone companies), as well as value-added special-network suppliers (Tymshare, MCI, and U S Sprint among them).

Just behind, just ahead
However, even greater expansion has taken place in private network technology, allowing IBM, Northern Telecom, Ungermann-Bass, Scientific Atlanta, Honeywell, and Wang into the ring of product suppliers. All this was provoked by the Carterfone decision, deregulation, increasing availability of high-capacity, low-cost microprocessor technology, and increasing dependence on networks as the key mechanism for delivering business networks and services (see "Technology's revolution"). Market segments whose needs are being filled by these new vendor organizations are, typically, banks, hospitals, and manufacturers of all sizes.

As this decade comes to a close, the potential for a full adaptation of a wide spectrum of new networking technology—including ISDN, 23 B+D, SS No. 7, T1, value-added networks, and packet-switched technology—could lead to one of three situations:
■ A lowering of the growth rate in the purchase of public network services and a big hike in growth of private networks and technology.
■ A marked increase in the use of the newer public networks technology and services and a limitation on any further expansion in private network technology.
■ Ongoing growth in the application, merging, and use of public and private network technology.

In essence, a new vendor capability has arisen that involves the delivery and management of many integrated network-based services for design, engineering, installation, and maintenance. This service-related strategy represents the optimum positioning in the marketplace (see table).

In short, users are demanding integrated networks. Some have already constructed their own, while others are more likely to use ISDN services through the public switched network vendors. A company will either acquire its own telecommunications network hardware or make use of the virtual capabilities of the public switched network, or, most likely, a combination of both.

Given the trends described above, the service that glues them all together has become more important than the components themselves. Users that want both public and private-switching access need to turn to organizations offering a strategic array of integrated network capabilities—Honeywell, DEC, BellSouth, or Bell Atlantic—that can not only provide design, engineering, and installation, but also maintenance and repair services. In other words, users that are building or leasing their own networks would have to worry about getting things fixed should something go wrong. Those users wanting public and private networks must turn to a service organization equipped with integrated services.

Private networks structure and direction
The typical private network structure involves a network backbone generally controlled by either an advanced integrated voice/data PBX or an on-line digital communications control processor unit. Either one involves linkages to long-haul bypass, local area networks, or local PBX/key networks. With few exceptions, these networks have become quite large in every vertical market segment in the U.S. industrial base with a large number of voice, office automation, data processing, and related equipment tied to the network.

In specific vertical market segments, the network topology, structure, size, and voice-versus-data orientation differ. These differences are driven by heating, ventilation, and air-conditioning networks, process control, and building automation. Vertical market segments with the highest degree of private network service—banking/finance, manufacturing, hospitals, and education—are adding sales terminals, electronic cash registers, automatic teller machines, scanners (vertical petrochemical manufacturers), and CAD/CAMs, all of which are being tied to the basic network backbone. As a result, service requirements will vary.

There are also significant differences in the levels of networks in operation, with hospitals, banking/finance, manufacturing, and education representing the vertical market segments with the highest degree of private network service.

The data also clearly shows that U. S. private networks are in a stage of major transition. Approximately 40 percent of the nation's industry has some level of networking in operation. The other 60 percent, however, are not now planning to use private networks and apparently are continuing to depend on the public network for voice and data communications services. In addition, a large number of these organizations operate more than one network and thus are in critical need of network integration and expansion services to make effective use of the technology that has already been installed.

Perhaps equally important is the fact that the applications running on the network are crucial to day-to-day operations, and there is an intense reliance on the network technology. Thus, the levels of interest in planning and design service, implementation services, and maintenance/repair services for networks are very high. The most intense interest seems to be in the ability to not only maintain but also repair a network in the event of partial or total network failure.

Future trends for technology and service

Unquestionably, the integrated service vendor that can maintain and support the network in operation is sorely needed. While ISDN and T1 technologies will influence the application and use of the public networks, the rise in service revenue from the public switched technologies still do not outstrip the demand for general network equipment service. It is estimated that T1 demand will grow to about $4 billion to $5 billion by 1989. However, network support services, including design and engineering, installation, and maintenance and repair service support of private networks, will be close to $15 billion. The network service strategy appears prominent.

In conclusion, the changing and emerging structure of the application and use of networks in major industrial market segments throughout the United States strongly suggests the growing commitment to, and use of, networks and network technology involving both public and private-switched services. ∎

Donald F. Blumberg is the president of D. F. Blumberg Associates, an international consulting firm that specializes in strategies and market research for telecommunications manufacturers, networks operators, and service organizations. He holds an M. B. A. in management from the Wharton School at the University of Pennsylvania.

Industry Watch

Turning your network into a profit center: Users as vendors

Can Sears, Westinghouse, Texas Air succeed where others have failed?

Twenty-five years ago, Newhouse Broadcasting, a publishing house that also operates a string of cable television networks in the Northeast, bought Eastern Microwave, a small, Syracuse, N.Y.-based transmission company, to supply video signals to those cable networks.

Over the years, Eastern Microwave has grown into one of the largest microwave companies in the Northeast. It has expanded its cable television business to include networks other than those owned by its Newhouse parent and has become a supplier of voice and data lines to interexchange carriers and private-line resellers.

Now in a major expansion, Eastern Microwave has agreed to buy U.S. Sprint's northeastern microwave network—between Washington, D. C., and New York City. The acquisition not only broadens Eastern Microwave's network, it also gives the company the incentive and means to actively pursue the commercial accounts it has largely ignored until now (see figure).

Eastern Microwave's history is illustrative because the company has done many of the things analysts say that large corporations planning to resell capacity on their private networks will have to do if they want to succeed in the highly competitive long-distance communications business. (Newhouse is a private company and does not release its or Eastern Microwave's financial results.)

Among the major user corporations that are moving into the long-distance business, as well as other information networking services, are Sears Roebuck & Co., Westinghouse Electric Corp., Texas Air Corp., AMR Inc., and Covia Inc., operator of the Apollo airline-reservation network for United Airlines.

■ **Grand scheme.** These corporations are much larger than Eastern Microwave or even Newhouse. And their plans are much more ambitious. They are looking to be national, even international, suppliers of communications services.

A number of electric utilities are also reselling capacity on their fiber optic and microwave networks to common carriers and commercial accounts (see "Power companies double as telecommunications suppliers").

Size alone, however, is no guarantee that these companies will be dazzling successes with their communications ventures. There have been other corporate giants—such as Eastman Kodak Co., Navistar (formerly International Harvester), and Security Pacific National Bank—that have tried and failed to become communications suppliers. Others, including Citicorp and Atlantic Richfield Co., have looked at it and backed away.

■ **Refrigerators.** Indeed, the record of corporations trying to act as telephone companies has been so checkered that Ken Bosomworth, head of International Resource Development (New Canaan, Conn.) doubts that Sears, Westinghouse, or any of the big companies now contemplating the move will succeed.

"Someday the top people at Sears

IndustryWatch

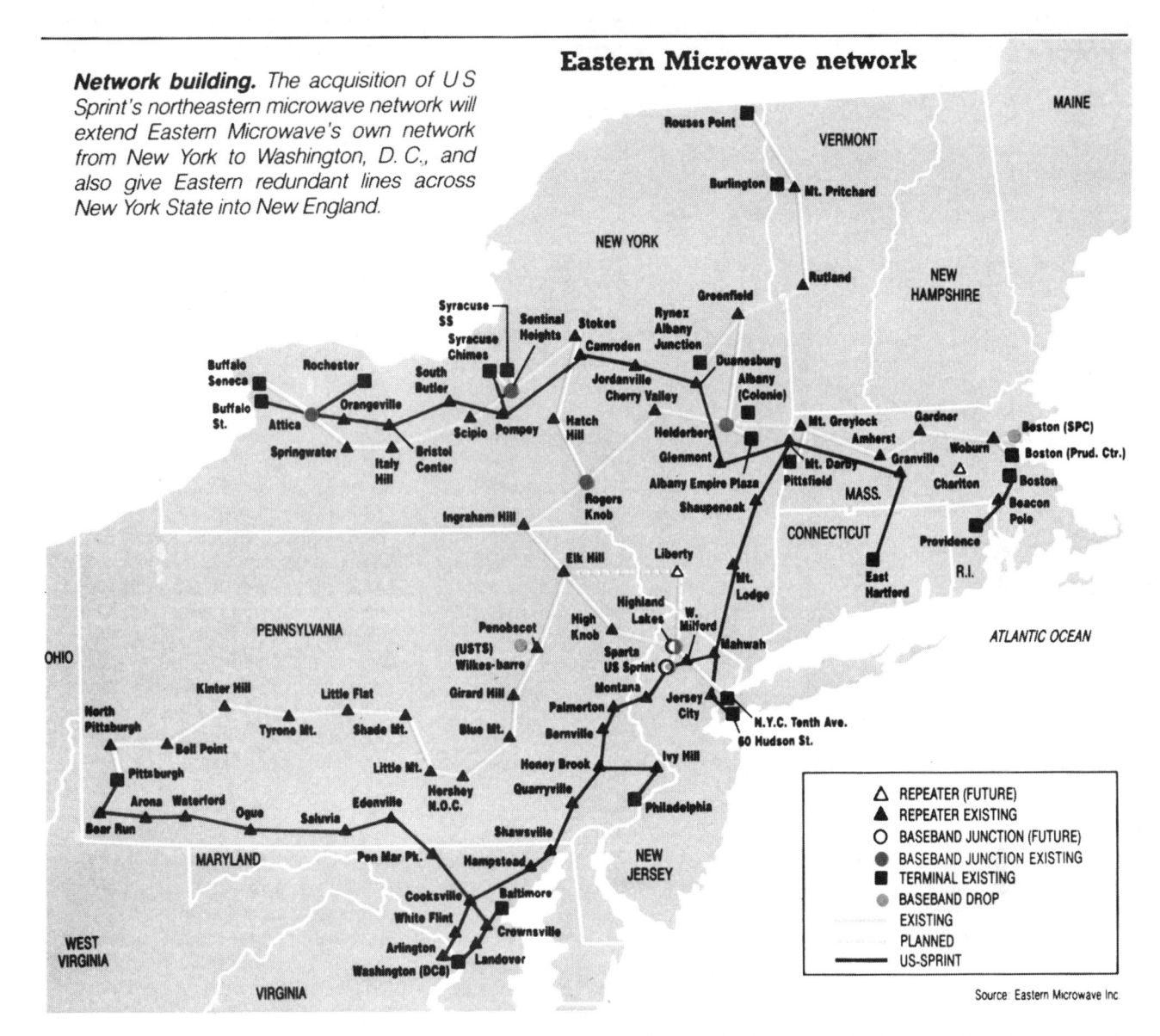

Eastern Microwave network

Network building. The acquisition of US Sprint's northeastern microwave network will extend Eastern Microwave's own network from New York to Washington, D.C., and also give Eastern redundant lines across New York State into New England.

Legend:
- △ REPEATER (FUTURE)
- ▲ REPEATER EXISTING
- ○ BASEBAND JUNCTION (FUTURE)
- ● BASEBAND JUNCTION EXISTING
- ■ TERMINAL EXISTING
- ● BASEBAND DROP
- EXISTING
- PLANNED
- US-SPRINT

Source: Eastern Microwave Inc.

will say: 'Should we put another $100 million into telecommunications or should we put it into building a new refrigerator?'" Bosomworth says. "I think they will put it into the refrigerator."

Potential customers also have to worry that if a corporation runs into problems in its main business, it may look at the communications service as an expense to be cut, says Robert Ellis, president of the Aries Group, a Rockville, Md., consultancy.

That happened with Xerox Corp. almost a decade ago when it tried to sell Xten, a pioneer bypass communications service. When Japanese competition shriveled Xerox's copier sales, the company dropped Xten to concentrate on its primary line of business, Ellis points out.

Telecommunications managers at Los Angeles-based Atlantic Richfield (Arco) and Atlanta, Ga.-based Delta Airlines also are suspicious of spending their companies' resources outside of their bread-and-butter businesses.

Arco trades capacity on its communications network—especially when offshore oil platforms are involved—with other oil and pipeline companies, but Joseph Coleman, Arco's telecommunications manager, says that's as far as the company will go into the communications business.

"Our business is finding and developing oil, not selling telecommunications," he argues. If an external communications venture began to earn money, then external customers could become more important to the communications department than internal customers, whose dollar contributions cannot be seen so easily, he contends.

H. R. Woodward, telecommunications manager at Delta Airlines, feels it is one thing for an airline to provide customer-reservation services to travel agents but quite a different matter to become a full-fledged communications company.

■ **Big chance.** Howard Anderson, managing director of the Yankee Group consultancy in Boston, agrees that becoming a supplier—and not just a buyer—of services is a treacherous business that has been marked

IndustryWatch

Power companies double as telecommunications suppliers

Retailers and airlines are not the only businesses that want to become communications carriers and suppliers. A number of electric utilities are looking to cash in on excess communications capacity on their private fiber optic and microwave networks by selling it to private corporations and institutions, as well as to telephone carriers.

Most of the utilities seem to be trying to recapture some of the costs of building their own internal communications networks. Instead of offering switched services, they are concentrating on reselling bulk private transmission facilities or offering only private line service. Some utilities say they will only take on customers who are near their network nodes, in which case they will extend their networks to pick up these accounts.

At least two holding companies, however, are treating their telecommunications operations as separate profit centers. One is Scana Corp. (Columbia, S.C.); the other is Iowa Resources Inc. (Des Moines).

Also in the Midwest, five utilities collectively own Norlight, a general partnership that operates and markets a 670-mile fiber optic network between Chicago, Milwaukee, Minneapolis, and other cities. Its customers include common carriers, other utilities, corporations, universities, and public agencies.

■ **Private carriers.** Public Service of Oklahoma (PSO, Tulsa) does not have a separate subsidiary marketing its network, but it is leasing out T1 date links on its integrated fiber and microwave network. To keep its status as a private carrier, however, PSO sells only "dark" fiber to common carriers who then put their own lightwave and electronics gear on the lines and manage the traffic themselves.

"Lit" fiber is leased to utilities, energy companies, and manufacturers—basically "noncarriers". PSO's 100 miles of fiber are centered in Tulsa, but its microwave towers traverse 800 miles of Oklahoma.

The Federal Communications Commission recently rejected an attempt by Oklahoma state regulators to control PSO. The case, according to other utilities, will determine how they offer their networks, although some say state regulators are less stringent than the FCC.

Baltimore Gas & Electric Co. wants to resell some of the capacity on its new fiber optic network in Baltimore to either commercial customers or common carriers. And Dominion Resources Inc. (Richmond, Va.) has talked to AT&T and other long-distance carriers about leasing them bandwidth on its network. Part of its fiber network in Richmond is still analog, but the newer sections being installed now are digital. The analog section was installed to be compatible with an existing analog microwave setup.

■ **Deal making.** Scana, the parent of South Carolina Electric & Gas Co., has another subsidiary, called MPX Systems Inc., that sells T1 through DS-3 capacity facilities to common carriers.

Timothy Jones, MPX's marketing manager, says that the company has signed up all the major carriers, including AT&T, MCI Communications Inc., Lightnet, and SouthernNet Inc.

MPX has also worked with SouthernNet, Alabama Gas & Electric Co. and Georgia Power & Electric Co. to install fiber networks in those two states for SouthernNet.

Mark Bergman, an analyst at Volpe & Covington, a San Francisco-based investment house, says that Scana owns about 15 percent of the stock of SouthernNet and a part of the carrier's network that Scana leases back to SouthernNet. In return, it will receive a percentage of revenues derived from the facilities for 40 years. MCI and other carriers are negotiating with several other utilities for similar deals, he adds.

Iowa Resources, parent of Iowa Power & Electric, created IOR Telecom Inc., to run its fiber optic network in the Des Moines area both for the utility and as a distinct business.

Ron Johnson, vice president at IOR Telecom, says that IOR is planning to put fiber optic networks into two other Midwestern cities. He would not name them, however.

Carriers such as AT&T, MCI, and Williams Telecommunications Co. (Tulsa, Okla.) buy capacity from IOR, but the company is also selling private links on its network to commercial concerns such as Meredith Corp. and the Principal Financial Group. It also installs private fiber optic links between buildings in other parts of Iowa, using either its own rights of way or those of other utilities. Customers then own the facilities or else lease them from IOR, Johnson says.

Johnson says Iowa Power put in fiber lines in 1979 because it could not "clean up" existing data lines. IOR Telecom was set up in 1984 to take advantage of opportunities in the newly deregulated telecommunications industry. "We are not simply reselling excess capacity but are developing capacity for resale," he adds.

In downtown Des Moines, IOR installs only dark fiber for its customers, this because many find it more advantagous to put their own electronics on a fiber than to buy an already digitized line, says Johnson. On lines running outside the downtown area, IOR also provides T1 and DS-3 service.

Johnson says IOR is growing rapidly and is "comfortable" with its financial results, but he would not disclose them. The network has been in operation since early 1987.

Norlight has been in operation since March 1987 and, according to President Gary Henshue, will reach the "cash break-even" point this year. About 4 percent of its revenue comes from data services, but it is a rapidly growing part of its business, he says. Norlight provides DS-0 through DS-3 private-line facilities only. —*J.T.M.*

IndustryWatch

by many mistakes. But he believes that by providing service and reliability at a lower price than smaller companies can achieve on their own, Sears and Westinghouse can make money.

"They are building a major business at Westinghouse that I think can do $10 million to $20 million," he says. "They are not just reselling capacity. They are offering higher reliability at the same cost."

Corporate resellers have to overcome internal suspicions from their customers' communications departments, Anderson adds, and they must demonstrate that they can provide a higher level of service than their in-house users were satisfied with.

Anderson says he is working with at least four other companies that are planning to enter the communications reselling business. The goal of these resellers is to use the same human and physical communications resources they have purchased to provide their own corporate networks, he notes.

Eastern Microwave became a communications company by building a network to serve its parent. When it branched out, it continued to operate in the business it already knew—providing head-end signal retransmission for cable television.

■ **Resellers.** And when it moved into the broader communications business, Eastern Microwave first went after large buyers, such as interexchange carriers and resellers, where sales are not as difficult or as time-consuming as they are in the corporate market (see "Resellers: A threat and a chance for user/vendors").

(Art Dunham, vice president of System One, the Texas Air communications subsidiary, says that regional re-sellers will be among his primary customer targets to buy extra space on the 10,000-mile backbone fiber optic network that System One obtained from Lightnet and Williams Telecommunications. Dunham sees System One acting as packager of customers for these resellers.)

Eastern Microwave has only gradually moved toward the tougher commercial accounts business. And all the while, the company has confined its business operations to markets in the Northeast that it knows.

It was a strategy of pursuing niche markets, both operationally and geographically.

The company is very conservative, says Richard Whalen, Eastern Microwave's director of marketing, and has no plans to expand beyond the Northeast. It has been offered and has

Bits flying. Art Dunham, System One's vice president, says it handles 780 million data transactions monthly.

rejected microwave networks in California and elsewhere by Wang Communications Inc. (Lowell, Mass.), the unit of Wang Laboratories Inc. that has been providing bypass services.

■ **New markets.** Whalen says that he is out to hire four or five former AT&T salespeople to handle user sales. Eastern Microwave has installed private networks for Chemical Bank and Manufacturers Hanover Trust.

"They did not exactly come over the transom, but this is a very conservative company, and we have devoted most of our time to the OCC [other common carriers] resellers," says Whalen. "We have not knocked on doors to sell our service. We have just maintained our network and provided service to existing customers. The Sprint acquisition will let us expand."

The time is right to go after commercial business because the trend is for corporations to have hybrid networks made up of private and public facilities Whalen says.

He adds that the fiasco following the May 8 fire in the Hinsdale, Ill., central office that knocked out telephone service for large parts of the Midwest for an extended time may now convince skeptics of the need for alternate routes.

■ **Local boys.** Whalen won't say how much Eastern Microwave is paying. Sprint for its microwave network, but he does confide that Eastern Microwave will provide services to Sprint. He will not discuss what those services are, but he does say about Sprint's boast of having a fiber optic-only network: "That's their plan."

"Eastern Microwave recognized that it could not be a full-purpose long-distance marketeer," says Ellis of the Aries Group. "You cannot compete across the board against AT&T. That's the mistake that Sprint has made. It is a little like juggling six axes. One of them may fall and cut your head off.

"Eastern also had one big captive account that covered its expenses and gave it the economies of scale to break into the communications business," says Ellis. "Everything else it sold [after servicing its parent company] was sold at the margin [at low pricing but with good profits]."

Another example of the same niche strategy, Ellis adds, is Schneider Freightways, of Green Bay, Wis. The trucking company developed special software and a nationwide network to coordinate routing of its trucks.

The communications division grew so much that it ate more than 2 percent of the parent's revenues. So Schneider decided to spin it off as Schneider Communications. "It was a reseller originally but is now a facilities-based carrier serving the Green Bay-to-Chicago area," says Ellis.

■ **Bigger fish.** Obviously, Sears, Westinghouse, Texas Air, and other huge businesses have different game plans than an Eastern Microwave or a Schneider Freightways, but they are following many of the same principles used by their smaller predecessors.

For instance, Sears Communications Co. has confined its business to

IndustryWatch

Resellers: A threat and a chance for user/vendors

All but written off several years ago as a doomed industry, long-distance resellers and regional communications carriers have shown a surprising resilience that makes them both competitive threats to, and potential customers of, corporations trying to resell excess capacity on their own networks.

By specializing in niche and especially regional markets, several leading resellers have become highly profitable companies that derive most of their revenues from the same accounts that corporate resellers also target.

These traditional resellers have the advantage of established relations with their customers. And many of them also are no longer just resellers—having replaced large segments of their leased lines with their own facilities, or having merged with regional carriers who have their own, often fiber-based, networks.

■ **Partners/competitors.** Mark Bergman, an analyst at the San Francisco investment house of Volpe & Covington, sees the success of these resellers and regional carriers due in part to users' demand for services rather than just cheap transmission. He says that corporate resellers, such as Sears Communications Inc., would compete for only a small part of the regional carriers' business and might even be large purchasers of the regional's services.

Wall Street, which seemed to lose interest in the reseller industry shortly after divestiture, is giving it more attention now. For instance, Teleconnect Co. (Cedar Rapids, Iowa) in May sold some $66 million worth of stock in an oversubscribed initial stock offering.

One of the most successful resellers, Teleconnect increased its sales from $20 million in 1983 to $168 million in 1987. It also largely replaced leased analog lines with its own digital fiber and microwave network. Most of its fiber lines are leased from Williams Telecommunications Co., although it also is a "huge user" of Norlight, the Midwest fiber consortium, and in addition uses lines from MCI Communications Inc.

■ **Switched only.** Teleconnect provides only switched services and does not plan to offer private lines in the near future, says Al Beach, senior vice president.

Switching voice traffic was the route that Advanced Telecommunications Corp. (Atlanta) took as it grew from a $20 million company in 1984 to one with $82 million by the end of 1987. Advanced Telecommunications is now merging with Microtel Corp. (Boca Raton, Fla.) in a stock swap valued at about $144 million. Microtel owns its own fiber optic network in the South. Combined revenues for the new corporation at the end of March 1988 were $167 million.

■ **Reseller/buyer.** The merger is a reverse image of the path that SouthernNet (Atlanta) has taken in the last two years to expand its operations. SouthernNet, which built its own fiber network along the east coast in 1986, bought two large resellers in South Carolina in 1986 and 1987. Then, in January 1988, it acquired Southland Communications, which owns a reseller operation in Florida, its own fiber optic network, and a telephone company.

Bergman says that Southern Net's acquisition strategy has been so successful that it has ignited a trend in the industry. Including the Advanced Telecommunications-Microtel merger. He predicts that other local carriers will merge with resellers in their regions.

Bergman says that companies reselling capacity on their own netwroks would be competing for only a small portion of the business of local carriers and resellers. No more than 10 percent to 15 percent of the regional carriers' revenues come from the large corporations that Sears and other corporate resellers target, he says.

Data services, in fact, have not typically been the strong suit of resellers or regional carriers. Resellers have concentrated on marketing switched voice-grade connections or bulk transport to other long-distance carriers.

Stephen Raville, chairman of Advanced Telecommunications, says that data has only accounted for about 5 percent of the company's business. He adds that about 30 percent of Microtel's revenue comes from private line sales and that some of the traffic on those lines has to be data.

Microtel's private-line offerings range from voice-grade and low-speed data links up to 56 kbit/s, T1 and DS-3 digital channels. The company has marketed to businesses in Florida and Georgia, and about 85 percent of its revenues came from commercial accounts.

Advanced Telecommunications serves residential and commercial accounts in nine southern states from Georgia to Texas, plus Kansas. Commercial business provides about three quarters of its revenues.

The merged companies will more actively pursue the data market, Raville says, but one reason for the merger is to drive down transmission costs and to have the digital circuitry that will be needed to provide ISDN services.

Bergman estimates that voice traffic accounts for 70 percent of SouthernNet's traffic but that data traffic is growing by 50 percent annually, versus only 10 percent growth for voice. SouthernNet's private-line operations have very high-gross margins—about 70 percent—he estimates.

Beach, at Teleconnect, says the company doesn't know how much of its business is data-related. Digital channels can be used to carry voice or data running at voice-grade levels, he says. Teleconnect's salespeople have not been trained to sell data; there is enough business on the voice side to keep them busy, he adds. —J.T.M.

IndustryWatch

offering dial-up and dedicated connections for data services over its SNA (IBM-based Systems Network Architecture) network. It does not provide voice services or design networks that will not run over its lines.

AMR Information Services, a sister corporation of American Airlines, however, has set up a network for Florafax International (Tulsa) that uses AMR's computers but runs outside its network. It also has designed a point-of-sale network for a manufacturer with access to an AMR mainframe for information processing. That manufacturer will resell the product to retailers.

■ **Tap resources.** The services are being offered through AMR General Computing and Network Services, a new group that was established last January to market both the data processing and communications resources of AMR.

Then there's Covia Inc., the United Airlines subsidiary that is developing a product for marketing to external customers. It has a very applications-oriented strategy. Mark Teflian, manager of the operation, says that applications will be part of transport and that transport itself is simply one set of a network's services.

He says that voice could be included as an application that might be integrated with point-of-sale service, as it is with airline ticket sales. Covia would design networks, he says. "We are not going to just sell bandwidth," he insists.

United's parent plans to sell Covia to a group of five airlines, including United. And until that sale is approved by federal regulators, Covia cannot actually move ahead with its plans. Teflian says he expects to announce products and customers later this year.

■ **National network.** Larry Jellen, vice president of marketing at Sears, says he is targeting customers with remote sites who "have a need for some backup and control, and who feel more comfortable that we can do it at a lower price than they can." The number of remote sites can be as low as five or as high as 1,000, he adds.

"The advantage to our network is that we have connections pretty much anywhere in the United States, including Alaska and Hawaii—even into Puerto Rico and Canada," Jellen says. Pricing is the same regardless of where a customer is located and is based on connections and kilo-characters of data passed over the network.

Sears does not provide X.25 services because there has not been much demand from its customers—and, Jellen admits, because Sears is not very good at it.

At one time, Sears did resell service on point-to-point, 4.8-kbit/s and 9.6-kbit/s lines. But it dropped that business because it could not add a lot of value to basic connections. Jellen points out that Sears could not "load up" the low-speed lines the way it can its T1 backbone network. Sears ended voice service three years ago because profits were too thin.

■ **No bumping.** Sears's backbone network has more than 80 T1 links that now carry voice and data traffic for the parent corporation and its other subsidiaries, such as Dean Witter Reynolds and Coldwell Banker. Jellen says that if it has to guarantee outside customers' service, Sears Communications will bump voice traffic off the private network and onto the public network.

Most of the Sears T1 lines come from US Sprint. Sears may even put in super-wideband DS-3 links for its major routes. The network has over 12,000 analog circuits for local access. Jellen says they are less expensive but as reliable as digital lines.

Multiplexers from Network Equipment Technologies Inc. (Redwood City, Calif.) handle traffic on the T1 lines, but 100 IBM 3725's are also used to switch traffic. Billing records are compiled from Netview and stored as VTAM records on IBM mainframes in the Arlington Heights, Ill., command center. (A back-up center runs simultaneously in Dallas.)

The guarantee Sears gives to its customers, Jellen says, is transit time across the network and availability. An entry-level, lower-priced service, which is popular with many customers, does not include a guaranteed response time. Sears guarantees only network availability. It does not restore an application if there is an outage.

The network serves 6,000 locations and 100,000 terminals for both internal and external customers. There are about 20 customers using the data transport, including Intel Corp. (Santa Clara, Calif.), Advanced Micro Devices (Sunnyvale, Calif.), and Kellwood Manufacturing Corp. (St. Louis, Mo.).

■ **Stupendous.** Ed Bowdin, telecommunications manager at Advanced Micro Devices, admits he sounds like an advertisement for Sears when he calls the service "stupendous." And he does not apologize for it. He says 30 AMD sales offices are connected over Sears's lines and that Sears goes out of its way to service a customer.

Bowdin and Phil Jones, who heads communications for Kellwood, another *Fortune* 500 company, praise Sears for helping them restore service to remote offices after the Hinsdale central office fire. Sears shipped out modems that allowed the offices to dial into Sears's mainframes over different lines, Bowdin and Jones say.

The two men add that Sears is able to deal more effectively with local telephone companies than they themselves are because Sears is established in areas where AMD or Kellwood have little influence. Jellen says that in Hinsdale, for example, Sears worked with telephone companies to obtain local telephone connections for its customers to the nearest Sears T1 switch.

■ **Good service.** Sears's customers are spared the capital expense of putting in their own equipment and the cost of hiring the staff to run their network, Jellen maintains. Bowdin at AMD admits that while he is paying Sears a premium for service, overall he is saving money on network management and getting reliable network service.

"The differential resource in communications is not technology—it is network management," agrees Ellis. Some customers will pay a premium to a company such as Sears to avoid the risks involved in trying to keep the staff it needs or to maintain its network.

Sears no longer does consulting, nor does it design networks that a customer would implement itself. Consultants have incentives to find clients and to do billable work, which salaried employees do not have, Jellen says, and consulting firms can more easily lay off workers.

Sears has been offering its data net-

IndustryWatch

working service for the last two years. Jellen would not disclose revenues, but he says Sears Communications would be profitable if it were not now adding sales and support personnel in an attempt to expand operations. Technical support personnel are more important than salespeople, Jellen adds. It is useless to bring on customers if they cannot receive adequate service.

■ **Profits only.** David Edison, executive vice president of the Westinghouse Electric organization that is reselling transport on that company's T1 network, says he will not take on a customer unless it is profitable. The unit was announced in March with Harlan Rosenzweig as its president to offer domestic and international voice and data service, including network management and value-added information products.

At a Probe Research Inc. conference in New York City last month, James Sever, manager of quality and reliability for Westinghouse, said that 20 percent of the corporation's private network capacity was excess.

Westinghouse does not hire salespeople for this operation, preferring to use network engineers to handle marketing, Edison maintains. Still, he says his top priority is to hire a vice president for marketing and sales.

There have been non-Westinghouse customers on the corporation's network for almost five years, says Edison, who adds he is looking for 100 new accounts. His target market is medium-size and larger businesses, although he says he is willing to price services specially for smaller customers.

"We don't want to take over anybody's network," Edison points out. "The value added is by pooling or sharing," he insists. "We are looking for opportunities in value-added sharing of facilities in specific areas or with specific services."

Westinghouse offers X.25 capability and plans compatibility with the CCITT's X.400 messaging standard. Basically, the company is selling a transport service, not applications.

Westinghouse's discounts on carrier prices can be as low as 5 percent or as high as 30 percent, according to Edison. "Generally, data service presents carriers with bigger profit margins. So on data, you can give good discounts, but on voice, the margins are tighter," he adds.

Westinghouse has 120 T1 links in its network. Data is handled by multiplexers from Timeplex Inc. (Woodcliff Lake, N.J.), while voice is routed through DSC Communications Inc. (Richardson, Tex.) T1 boxes.

Edison says external customers are treated equally with Westinghouse's own users. As an example, he cites an experience in Puerto Rico, where Westinghouse had the only available communications lines not affected by an island-wide outage. Service was apportioned equally among all users, he says.

Dunham at System One estimates that even without sales to outside customers, he will cut $200 million out of System One's expenses over the next five years by moving existing company traffic onto the new private network (and by consolidating operations). System One started marketing its excess communications capacity externally, to non-airline customers, after it set up its fiber optic network, and then installed an X.25 packet-switching network.

John T. Mulqueen

Timothy G. Zerbiec and Rosemary M. Cochran, Vertical Systems Group, Dedham, Mass.

Is a private T1 network the right *business* decision?

The bottom line: Cost is an issue, but only after you've screened the multiplexer vendors who can best satisfy your T1 network needs.

You've heard about the cost economies and benefits of private T1 networks. But before you go any further, you must ensure that such an investment would yield sound business benefits for your particular networking environment. First, evaluate your business requirements. Second, thoroughly understand the T1 (1.544-Mbit/s) multiplexer technology that will support these requirements.

Evaluate your business requirements by defining and measuring your risks. Decide if the network is a strategic necessity or a cost-saving tactic. For example, if you lost your voice and data networks right now, how long could your company operate? If your business is processing stock transactions or airline reservations, the answer is obvious.

Competitive business demands have created a trend toward T1 networks that enable companies to strategically leverage communications for their businesses. These networks enable companies to respond quickly to competitive pressures and customer needs.

As an example, let us look at two competing retail banks, First National Bank and Last National Bank, each with a branch office at a busy shopping mall. Each bank has two automatic teller machines (ATMs) that are constantly busy. Customers must often wait up to five minutes (it only *seems* longer) to use each machine. To alleviate customer complaints, Last National Bank decides to install a third ATM. The telecommunications manager places an order with the local telephone company for a new line and is quoted 45 days for installation.

Meanwhile, First National Bank has also decided to install another ATM. When it is ready to be connected to the ATM network, the bank's network operations manager simply provides a 4.8-kbit/s channel from First National's existing T1 network. This is accomplished from the bank's central operations center in minutes using interactive commands from a network management console. The result is that First National Bank has improved service to its existing customers and may even attract new customers from among those standing across the mall in the ATM lines at Last National Bank.

Increased user control has resulted in a trend toward the strategic use of T1 networks. At the beginning of this year, 871 of the largest 1,750 user organizations in the United States had installed T1 networks, a 22 percent increase over 1987. Of these 871 organizations, 362 have implemented backbone T1 networks, while the rest have point-to-point T1s.

'End' versus 'networking' muxes
Your business requirements determine the type of network capability you need. For relatively static configurations that provide the transport of applications between two or three locations, point-to-point networks can be implemented using less-featured, lower-cost "end" multiplexers. In contrast, "networking" multiplexers are used to build backbone networks that interconnect multiple locations and support multiple voice and data applications. The backbones have capabilities such as intelligent switching, automatic rerouting, and operations control.

More than 40 vendors offer end multiplexers. The leading vendors, based on worldwide market share—directly sold and through distributors—are Amdahl, Coastcom, Datatel, and Granger. Others are Aydin Monitor, Bayly, Dynatech, Fujitsu, Gandalf, Integrated Telecom, Megaring, NEC, Newbridge, Pulsecom, Scitec, Tau-tron, and Telco Systems. (Note: Addresses, telephone numbers, and contacts can be found in the DATA COMMUNICATIONS *Buyers' Guide*.)

Table 1 outlines a decision profile for each T1 multiplexer category. Based on an average cost of $26,390 per end multiplexer, the payback period for a point-to-point network

Table 1: T1 multiplexer decision profile

PARAMETER	END MULTIPLEXERS	NETWORKING MULTIPLEXERS
NETWORK TYPE	POINT-TO-POINT	BACKBONE
DECISION CRITERIA	COST, PRODUCT FEATURES	RELIABILITY, PRODUCT FEATURES, VENDOR SUPPORT, COST
DECISION MAKER	DEPARTMENT MANAGER, PURCHASING	CORPORATE COMMITTEE
DECISION CYCLE	LESS THAN 6 MONTHS	6 MONTHS +
AVERAGE PAYBACK PERIOD	LESS THAN 6 MONTHS	LESS THAN 1 YEAR
AVERAGE $/UNIT	$26,390	$80,420
CONFIGURATION OPTIONS	LIMITED	MANY, DESIGN REQUIRED
MODULARITY	LIMITED	SUBSTANTIAL
VENDOR EXPERTISE	EQUIPMENT	APPLICATIONS, EQUIPMENT
SUPPLIER	MANUFACTURER, MANUFACTURERS' REP., OEM, DISTRIBUTOR, CATALOG	MANUFACTURER, OEM

OEM = ORIGINAL EQUIPMENT MANUFACTURER
REP = REPRESENTATIVE

is typically less than six months. These networks are usually procured at a departmental level to support discrete voice and data applications. Also, many companies operate multiple point-to-point networks.

With the device viewed as a commodity purchase, your primary purchase-decision criteria for point-to-point networks are price and product features. End-multiplexer manufacturers such as Amdahl, Coastcom, Granger, and Tau-tron sell through direct sales organizations and manufacturers' representatives. End multiplexers are also available through distributors such as AT&T and the Bell operating companies and through catalogs such as that of Glasgal Communications of Northvale, N.J.

With the average cost of a networking multiplexer at about three times the cost of an end multiplexer—$80,420—backbone networks represent a comparatively major capital expenditure. Based on the latest tariffs, the average payback period for these networks has been cut to less than one year, half the period required as recently as two years ago. Because backbone networks support applications that affect multiple departments, a buy decision will most likely involve a management-level committee.

Backbone networks are typically deployed in three phases and planned for an average "locked-in" life of five to seven years. (In point-to-point networks, with the end multiplexer viewed as a commodity, it is readily replaced—typically when requirements change.) The pilot phase occurs during the first year, when networking multiplexers are deployed in four to six major locations and the functionality of the network is tested. The following year to 18 months is a "rollout" phase, when multiplexers are added to extend the network's capabilities to other company locations. At this point, the network is mature and enters a growth phase. New multiplexers are added more gradually, and existing equipment is upgraded with new features and additional capacities to support new applications.

Evaluate! Evaluate!

Because implementing a backbone network is a long-term commitment, consider your selection of a vendor as carefully as your selection of the multiplexer. In addition to equipment costs, carefully evaluate the vendor's financial stability, long-term product strategy, and the depth and experience of the support and service organizations. Before making your final decision, insist on customer references. Visit prospective vendors' headquarters to meet the executive management and tour the manufacturing and development facilities.

Based on market share, Timeplex, Network Equipment Technologies, and Digital Communications Associates are the leading networking multiplexer manufacturers. Avanti, General DataComm, Infotron, Micom Digital, Stratacom, and Tellabs have also developed networking multiplexers to target the backbone network market. The Vertical Systems Group projects a 54 percent increase in shipments of networking multiplexers for 1988, while shipments of end multiplexers will remain flat.

Computer and PBX manufacturers are also beginning to recognize the strategic importance of being the private-transport network vendor for major companies. These companies view T1 backbone networks as a platform on which corporate information movement resides, using network management as the cornerstone.

This private-transport strategy, which may be called vertically integrated rollout, has resulted in a number of recent acquisitions and strategic alliances. These include Unisys's acquisition of Timeplex, IBM's OEM (original equipment manufacturer) agreement with Network Equipment Technologies, and Digital Communications Associates' OEM agreement with Northern Telecom. There will be more of these relationships within the next year as other vendors plan their strategies.

Increased competition among the networking multiplexer vendors has also enlarged the task of analyzing product functionality and feature availability. No single feature will determine which product is best. And, unfortunately, you cannot rely on vendor brochures to make product comparisons or evaluate network capabilities. There are actually more than 100 elements that can be used to analyze the functionality of networking multiplexers. These elements can be categorized under five headings: Capacity, Capability, Configuration, Control, and Cost. Table 2 shows how the five headings can be used to map your business requirements to T1 multiplexer functionality.

■ *Capacity.* Ensure that the multiplexer has enough capacity to support the number of locations and applications,

Table 2: Matching requirements to functionality

	BUSINESS REQUIREMENTS	MULTIPLEXER FUNCTIONALITY
CAPACITY	• NUMBER OF LOCATIONS (CURRENT & PLANNED) • NUMBER OF USERS (CURRENT & PLANNED) • RELIABILITY • RESPONSE TIME OBJECTIVES	• NUMBER OF NODES PER NETWORK • NUMBER OF T1 AGGREGATES • NUMBER OF CIRCUITS PER AGGREGATE • THROUGHPUT • NODAL DELAY • BUS DESIGN • BUS SPEED
CAPABILITY	• APPLICATIONS AVAILABILITY • COMPETITIVE RESPONSE • USER CONNECTIVITY • DISASTER PLANNING	• NODAL INTELLIGENCE • AUTOMATED BANDWIDTH PROVISIONING • NETWORK SYNCHRONIZATION • CLASS OF SERVICE • INTEGRAL CHANNEL SERVICE UNIT (CSU) • EXTENDED SUPERFRAME FORMAT (ESF) SUPPORT • BINARY EIGHT ZERO SUBSTITUTION (B8ZS) SUPPORT • T3 INTERFACE

CONFIGURATION	• APPLICATIONS AVAILABILITY
	E.G.-ORDER PROCESSING
	   -OFFICE AUTOMATION
	   -RESERVATION SYSTEMS
	   -TELEMARKETING
	   -CAD/CAM
	   -VIDEOCONFERENCING
		• APPLICATIONS INTERFACES
		   -VOICE (E.G. ANALOG, DS-1)
		   -DATA (E.G. V.35, RS-232-C)
		• BANDWIDTH COMPRESSION
		• CHANNEL SYNCHRONIZATION
		• CONTROL-LEAD SUPPORT

CONTROL	• MANAGEMENT REPORTS • SECURITY • EQUIPMENT INVENTORY • VENDOR-SERVICE TRACKING	• NETWORK ADMINISTRATION • TECHNICAL-CONTROL FEATURES • ANALYSIS TOOLS

COST	• BUDGET
	   -EQUIPMENT
	   -OPERATIONS
	   -TRANSMISSION FACILITIES
		• NETWORK EQUIPMENT
		• NETWORK MANAGEMENT
		• INSTALLATION
		• TRAINING
		• MAINTENANCE
		• OTHER EQUIPMENT (E.G. CSUs, CHANNEL BANKS, ECHO CANCELLERS)

CAD = COMPUTER-AIDED DESIGN
CAM = COMPUTER-AIDED MANUFACTURING

---

both current and planned. Bigger is not always better. Variables such as the number of T1-aggregate ports, number of channel interfaces, and number of nodes (multiplexers) can all be used as measures of capacity. A multiplexer may be bigger in one category and lacking in another.

The number of nodes that can be configured as a single logical network is important to operations issues and will become even more important as your network grows. The numbers that appear in product literature can be misleading. Statements such as "There is no limit to the number of nodes in a network" suggest the vendor's assumption that the architecture it has implemented for initial networks will work for growing ones.

Investigate those assumptions; ask for an explanation of how the architecture works; get references of other customers who have already implemented large networks. Calculate the actual time required to perform tasks such as multiplexer reconfigurations (which include port re-assignments) and the time to reroute traffic from a failed transmission facility for both your current and planned topologies. These times may be acceptable for your five-node network today but unacceptable (rerouting time may be greater) when you expand the network to 20 nodes in the future.

Another variable for evaluating multiplexer capacity is the number of slots available for application channels (synchronous or asynchronous ports), common logic (logic common to other slots), and aggregate modules. The maximum number of input channels typically overstates real-world multiplexer-configuration capacities. Evaluate the node-site configuration, including its aggregates, channel interfaces, and redundancy. Too often, users find themselves buying another node because they forgot to consider the slot capacity required for logic redundancy or internodal links.

Real-world configurations for T1 multiplexers usually include redundancy for common logic and some—if not all—of the aggregates. Redundancy implementations that use one spare for N active modules will use fewer slots than implementations that use one-for-one protection. And do not forget future environmental requirements, such as power, air conditioning, and floor space.

If the number of module slots is at a premium, is it necessary to back up aggregate modules? Many network designers think multiple internodal links on a node will provide protection from individual link failure. The theory is that the transmission link is the more likely element to fail. Moreover, the bandwidth available on other aggregates will permit a redundant path to provide substitute connectivity. However, use this shortcut with care; weigh the use of a slot against the effects of the loss of internodal bandwidth caused by an aggregate-module failure.

The maximum number of circuits a multiplexer can carry on an aggregate link affects decisions on topological design. The actual number of circuits carried is a function of the application-circuit bandwidth, channel-framing technique, and overhead used by that vendor. Verify that the mix of circuits you need can be supported on a single aggregate. Errors in calculation may result in requiring additional T1 lines.

Circuit propagation times of 20 milliseconds or more introduce a noticeable echo on voice circuits or cause throughput delays for data. Delay is imposed by a multiplexer as it switches a circuit from a channel interface to an internodal link or when bypassing a circuit from one internodal link to another. Even geographically small networks—such as in campus environments or those with as few as five nodes—may route applications through many nodes to complete a transmission path, resulting in unacceptable circuit delay. To compensate for such a delay requires the installation of echo cancellers or other corrective equipment.

Variables such as throughput capacity (in Mbit/s) and internal bus design and speed influence the aggregate and channel-interface capacity and availability. A good bus design—recognized as such by being redundant and by how many functions it can handle—provides protection from bus failure and provides for the higher bandwidth necessary for future applications. Bus speed measures the capacity of a single switching shelf. (A shelf contains mux cards. A switching shelf is dedicated to T1 aggregate terminations and switching.) It also gives an indication of the maximum bandwidth that can be switched as a single channel or aggregate. Most network managers expect a five-to-seven-year life from their T1 networks. Therefore, a multiplexer design that provides no growth path could be disastrous.

■ *Capability*. The operations of a T1 backbone network show why users have embraced "do-it-yourself" common-carrier transport services. The single aspect that has made networking T1 multiplexers so revolutionary is the addition of nodal intelligence to a formerly manually controlled environment. The combination of software-controllable multiplexers and vendors' network-operations experience results in the automation of tedious network operations tasks. The benefits to the user include transmission-facilities usage optimization, automatic recovery of failed facilities, real-time arbitration for access to resources, and audit trails of anomalous network events.

Multiplexer intelligence is not a single feature but an integration of multiplexer capabilities. If there is a single area where multiplexer intelligence can be observed, it is in networking support. While you may be interested in the specific vendor implementation of networking support, focus instead on the vendor's understanding of user-operations issues. When evaluating functionality, you must measure the benefit to your organization of the substitution of nodal capability for personnel. Experience has shown that users of T1 networking-multiplexer backbone networks have the same or smaller staffing levels while increasing network services.

A large number (typically 10 to 15) of locations and complex intersite connectivity (such as a "mesh") are common with T1 backbone networks. Network-operations personnel usually require assistance in planning the utilization of transmission facilities. A circuit "router" function (part of the T1 mux), which examines potential paths for a circuit, must take into account many variables. These include permissible path length, call priority, status of network transmission facilities, and such special considerations as permissible delay, encryption, and transmission security. The goal is to let the network automatically arbitrate bandwidth management as much as possible. It is not practical to have a network operator manually interconnecting circuits.

Decide how much automation you're willing to buy. Use this checklist as an analysis aid to measure nodal intelligence:

■ *How much manual intervention is required to define a circuit? Do you specify only circuit endpoints, or all intermediary nodes as well?*
■ *Does the router automatically adapt to changes in the network topology?*
■ *Is there vulnerability because router intelligence is concentrated in only one node?*
■ *Try "what-if" scenarios to examine what it will take to run your network on either a normal or an exceptional (failure-mode) basis for currently planned functions.*
■ *How long does it take to reconnect all of the circuits on a failed internodal link? (Note: Use specific examples when comparing this parameter for networking multiplexers. The truly high-end [multifeatured] products jump off the page at you; their milliseconds reroute times are one to two orders of magnitude lower than less powerful competition.)*
■ *How is contention for internodal link resources (aggregates) handled? for channel resources (ports)?*

As networks expand in size, precautions must be taken to ensure that all transmission facilities work together. This is especially true if a number of different common carriers or a combination of public and private carriers (such as bypass microwave and fiber) are used.

The method the multiplexer employs to maintain synchronization with its neighbors is called network synchronization. This parameter must be addressed during the network's design phase, prior to transmission-facilities engineering, or the result may prove inadequate.

T1 multiplexers provide an economic division of T1 transmission bandwidth into multiple application services. Before the advent of networking multiplexers, the predominant method for multiplexing applications on T1 facilities constrained the minimum controllable circuit-bandwidth size to 24 DS-0 channels (64 kbit/s each) per T1.

This generation of networking T1 multiplexers permits as many as 573 individually controllable circuits of 2.4 kbit/s each per aggregate. While this has improved T1 line utilization, it has some disadvantages. To gain this level of efficiency, vendors have developed proprietary link communications formats—especially for an aggregate of fewer than 64 circuits—that are not compatible with those of other vendors at the internodal link level. This situation is similar to that of statistical multiplexers.

"Open" network T1 architectures (such as what Avanti is proposing) are on the horizon, but they address only the potential synergy between T1 multiplexers and carrier

> ## Don't settle for 'Yes, we support T3' or you'll be left trying to plug it in.

services. An "open-network gateway" could be used to connect two different vendors' networks. But this may incur a penalty of decreased T1 line efficiency—gateway overhead may reduce the usable bandwidth—and decreased network management functionality (different vendors usually have different, incompatible network management approaches). Therefore, carefully assess your long-term relationship with your vendor, because you will probably be relying exclusively on this vendor for the life of your network.

There has been considerable interest in a T1 line-framing technique, called extended superframe format (ESF), which increases the level of technical control capability on a T1 line (see "The hidden treasures of ESF," DATA COMMUNICATIONS, September 1986, p. 204). Line-quality statistics as well as a communications channel for messaging are its most promising features. While ESF is not yet universally available on carrier-provided T1 lines, ensure that your multiplexer will be compatible with ESF. Specifically, find out what it will be able to do with the ESF technical-control features.

Mentioned almost synonymously with ESF is binary eight zeros substitution (B8ZS). B8ZS is a T1 coding method that ensures sufficient signal transitions on a T1 line to maintain line-repeater synchronization, even when there are more than 15 consecutive zeros in the T1 data stream. Like ESF, it is not yet supported on all common carrier circuits. Ensure that your T1 multiplexer vendor has implemented this capability or a similar one.

Now that T1 networks have gained acceptance, users and vendors already want to up the ante with T3 (44.736 Mbit/s—commonly, 45 Mbit/s) internodal links. Today, T3 common carrier services are at the point of development that T1 service was in 1983, with users trying to decide how to apply all that bandwidth. Most T1 users cannot cost-justify T3 links yet. For those who can, the issue is how to compartmentalize its use.

T3 is available as M28 tariffed service—where the T3 channel is formatted as 28 T1s—or a single, large bit pipe running at 44.736 Mbit/s. For those who want it as M28, an external separate M13 (industry terminology for T3) multiplexer can take up to 28 T1 multiplexer interfaces and combine them into a single T3.

Some T1 mux vendors integrate M13 multiplexers into their T1 nodes. If a multiplexer's switching bus can handle T3, then the multiplexer has the potential of operating at T3.

Another way of incorporating T3 operation is by using multiple T1 interfaces. For users seeking to support individual application channels running at greater than T1 rates (such as bridging local area networks), they should determine if the multiplexer can accommodate T3 as a single 45-Mbit/s pipe. Vendors with a definitive T3 strategy will be able to explain their specific T3-support plans. Do not settle for a simple "Yes, we support T3" response, or you'll be left trying to figure out how to plug it in.

■ *Configuration.* Business applications for T1 networks include order processing, factory automation, office automation, reservation networks, telemarketing, and videoconferencing. The T1 network represents a bandwidth utility for implementing these applications.

Vocal support

T1 networks heralded the first practical method for toll-quality voice channels to be multiplexed with other voice and data channels on a single transmission facility. The applications include the support of inter-PBX trunks, tie lines, and foreign exchanges. The number of ports at a site may vary considerably. The interface may be analog—one circuit per line—or digital, per AT&T Technical Publication 62411, with a format of 24 circuits per line.

Look at product descriptions to see how voice is interfaced to the multiplexer. On a networking T1 multiplexer, the use of a digital-PBX interface reduces hardware clutter and supports a large number of channels—24 per T1, typically. These multiplexed interfaces allow direct connection to PBXs with a T1 interface or to channel banks that provide adaptation to analog facilities. Multiplexed interfaces have inconsistent technical-control functions. Features like channel loopbacks or level-check and -adjust may be lacking. T1 voice interfaces may not be cost effective at sites where the support of only a few voice channels is required. This cost may not help determine vendor selection, but it will, of course, affect the overall network cost.

Voice channels use large amounts of bandwidth: 64 kbit/s in a standard toll-quality channel using pulse code modulation (PCM). The advent of integrated-circuit digital signal processors has enabled voice compression to reduce this bandwidth requirement to 32 kbit/s—in some cases to 16 kbit/s. Adaptive differential pulse code modulation (ADPCM) is a widely accepted compression method, with vendors moving toward a common ANSI standard: T1.301. When evaluating a product, find out if it has ADPCM, the data rates it supports, and the level of quality it provides. A good measure of ADPCM quality is the group type of facsimile machine that can be supported (at least

Group 2—analog 3-minute transmission over the public telephone network) and the modem data rate supported (in the range of 2.4 to 4.8 kbit/s).

Other low-bit-rate voice digitization methods offer variable results and/or introduce other operational considerations, such as additional circuit delay. However, do not believe that ADPCM is a panacea. The issue is not the compatibility of your private network's ADPCM coding with that of other networks. (Chances are, your ADPCM codes will never leave the network.) ADPCM standards are important to ensure that a channel meets minimum service objectives.

Digital speech interpolation and packetized voice are growing in popularity. Bandwidth-compression methods offer from 2:1 to 10:1 gain in the number of channels on a T1 line. They provide adequate-quality speech but do not provide bandwidth compression for facsimile transmission and for modem traffic. Applications like voice order-entry are well suited to this type of transmission.

The term "integral transcoder" appears on many product descriptions. Transcoding refers to the ability to change PCM coding to ADPCM and vice versa. This is particularly important if T1 voice interfaces are used to access the network. These interfaces use PCM at 64 kbit/s, which would be costly for backbone transport at that low a data rate. Determine if the transcoder performs its function on a "per port" or "bundle" (a group of 11 channels) basis. Find out what the options are for converting PCM to some lower-bit-rate method. This capability will affect the flexibility and economy with which you will be able to route voice calls through the network.

Compatible interfaces
Besides voice, the network configuration is, of course, affected by data support. Care must be taken to ensure that the interfaces between the application and the network are compatible.

On a T1 network, a data channel looks like a private-line data link. The same channel-service objectives—such as availability, signal integrity, and acceptable delay—must be met by the multiplexer network for an application to work as before. When evaluating the adequacy of the data support on a networking mux, consider these items: bandwidth efficiency, channel transmission control, physical interface type, and circuit availability.

Data-interface specifications for multiplexers are generic. They all contain descriptions that address the rates they will support, types of electrical interfaces, and number of ports per [printed-circuit] card. Evaluate these items to ensure that your applications can be connected to the network at the physical level—Open Systems Interconnection (OSI) Layer 1. Although this issue may seem obvious, vendors often use adapters external to the multiplexer to accommodate different logical, physical, or mechanical configurations. This issue may interest your operations staff, which must inventory, install, and maintain the connecting cables.

Channel-control leads are important to data link level (OSI Layer 2) channel support. Determine how the multiplexer supports control-lead signaling. Some applications use control leads for ascertaining channel status or for flow control. Know how many leads you need per channel and how quickly a change in status must be recognized relative to the data. Vendors differ widely on the number of control leads supported and the amount of overhead required to support them.

Perhaps the most important aspect of data-channel support concerns how the multiplexer supports channels that cannot be synchronized to the network reference clock—which normally results in transmission errors. While many data applications incorporate protocols to ensure end-to-end data integrity, poor channel performance results in frequent retransmissions and, therefore, reduced throughput. Methods like "positive justification" use a small amount of additional bandwidth per channel to pass the data independent of network-reference timing. Other methods include buffers that provide protection for short-duration, transaction-oriented transmissions—typically less than 15 minutes. Ask vendors what their methods are, and ask them to explain what performance level (parameters such as recovery time and throughput) their methods will give your application. Failure to do so could result in the unavailability of a critical application.

T1 time-division multiplexers are not well suited to asynchronous-circuit support. Asynchronous traffic is better supported by a secondary network of statistical muxes or packet switches, with the aggregates of these network processors then passed as synchronous data channels onto the T1 backbone. For those applications that justify direct connection to the backbone, investigate how the multiplexer combines direct connection with other traffic.

Most T1 multiplexers that do support async convert the async channel to a synchronous one for transit across the backbone. Thus, even when the port is idle, it takes up backbone bandwidth. Some T1 multiplexers incorporate statistical muxes, which reduce this bandwidth penalty. Consider carefully the requirement for asynchronous support on your T1 backbone; there are many other economical and feature-rich methods—such as packet switching and statistical multiplexing—for supporting async.

■ *Control.* Beyond the cost economies of buying bandwidth at "wholesale" prices, T1 networks opened the doors to user control of the bandwidth resource. Communications-wise corporations use T1 networks to reduce the connectivity time between sites, thereby improving their responsiveness to business conditions. Similarly, they use the technical control capabilities of these networks to reduce the MTTR (mean time to repair). This reduction is accomplished by pinpointing network problems and involving the appropriate parties sooner to affect restoral of service. Automation in network management is blossoming as the cost of processor MIPS (million instructions per second) comes down and artificial intelligence and graphic presentations become prevalent.

It is incumbent on a corporate network designer to work with operations staff to develop an operations philosophy for ensuring the availability of services across the network. Each vendor has different experience, wisdom, and network control methods.

Table 3: Features of T1 networking multiplexers

	AT&T 740/745	AVANTI ONC/ONX	DCA SYSTEM 9000	GDC MEGA-SWITCH	INFOTRON NX4600	MICOM DX-500	NET IDNX/70	STRATACOM IPX	TIMEPLEX LINK/2
NUMBER OF AGGREGATES	1/16	1/16	36	16	21	8	96	16	6
AGGREGATE REDUNDANCY	1:1/1:N	1:1/1:N	1:1	1:1	1:1	1:1	1:1	M:N	1:1
THROUGHPUT CAPACITY	1.5 M/ 25 MBIT/S	2 M/ 25 MBIT/S	55.3 MBIT/S	24.6 MBIT/S	20.2 MBIT/S	12 MBIT/S	1,966 MBIT/S	160,000 PPS	12.3 MBIT/S
MAXIMUM NUMBER OF NODES IN NETWORK	250	100	80	128	64	125	250	63	160
SOURCE CHANNEL CAPACITY (PORTS)	128	128	136	512	4000	508	384	384	208
BYPASS DELAY (MICROSECONDS)	N.A./250	N.A./250	375V, 3,000D	250	2,000	500	5,000	2,750	2,000
MULTIPLEXING ORIENTATION	BIT/BYTE	BIT/BYTE	BYTE	BIT	BYTE	BIT	BYTE	PACKET	BYTE
ROUTER INTELLIGENCE	DISTRIBUTED	DISTRIBUTED	CENTRAL	DISTRIBUTED	DISTRIBUTED	DISTRIBUTED	DISTRIBUTED	DISTRIBUTED	DISTRIBUTED
ROUTER TYPE	TABLE	ALGORITHM	ALGORITHM	TABLE	TABLE	ALGORITHM	ALGORITHM	ALGORITHM	TABLE
ROUTE GENERATION	AUTOMATIC	AUTOMATIC	AUTOMATIC	MANUAL	MANUAL	AUTOMATIC	AUTOMATIC	AUTOMATIC	MANUAL
TIME TO REROUTE (SECONDS)	#	10	20	#	86	#	10	2	#
PARAMETERIZED ROUTING	NO	YES	YES	NO	NO	NO	YES	YES	NO
MAXIMUM NUMBER OF HOPS	N.A.	16	10	N.A.	N.A.	N.A.	12	10	7
PRIORITY LEVELS	3	16	16	NONE	NONE	64	4	NONE	16
PRIORITY BUMPING	YES (3)	NO	NO	NO	NO	NO	YES (4)	NO	NO
TRAFFIC BALANCING	NO	NO	NO	NO	NO	NO	NO	YES	NO
BANDWIDTH CONTENTION	NO	NO	NO	NO	YES	NO	YES	YES	YES
LOWEST BYPASS CHANNEL	N.A./ 64 KBIT/S	N.A./ 64 KBIT/S	2,667 BIT/S	75 BIT/S	300 BIT/S	400 BIT/S	1.2 KBIT/S	N.A.	50 BIT/S

D = DATA
N.A. = NOT APPLICABLE
PPS = PACKETS PER SECOND
V = VOICE
# = NOT DETERMINED

Source: VERTICAL SYSTEMS GROUP

The product descriptions and options have only enough depth to suggest which issues the vendors address. The nature of the capabilities offered by state-of-the-art network management requires an evaluation in itself.

Consider these questions when evaluating network management:

- How robust (feature-rich) is the telemetry for both node-to-node as well as node-to-network-management communications? The bandwidth must be adequate to propagate control messages with acceptable delay and be sufficient for current and future requirements.
- Do technical control features allow for timely and accurate problem diagnosis?
- Can you inventory network components as well as circuits currently in use?
- Is there a software administration program to help maintain down-loadable multiplexer code?
- How do you control network management access?
- Can you track trouble tickets, dispatch of repair personnel, and other maintenance activity?
- What is the capability to produce such management reports as on-line utilization and line-failure analysis?

- *Can the vendor produce working evidence of product claims (with user references)?*
- *Cost.* T1 multiplexer costs are as difficult to compare as T1 multiplexer features. Since the ultimate test of a multiplexer's functionality is how well it supports user applications, costs really can only be compared on a network-implementation level.

When considering costs, obtain itemized prices for each network node, all network management options, and any other equipment required to support your applications (such as channel service units, echo cancellers, channel banks, and satellite-transmission buffers). In addition to equipment costs, closely evaluate the costs of installing and maintaining the network and training your network operations staff.

To analyze and compare total network costs, nine networking multiplexer manufacturers were asked to propose implementations of a sample four-node network (see the figure). The nine were Avanti, Digital Communications Associates, General DataComm, Infotron, Micom, Network Equipment Technologies, Stratacom, Tellabs (represented by AT&T), and Timeplex. They also answered technical questions concerning applications support (see "Product analysis"). Network costs varied considerably from vendor to vendor. The total network list price, including basic network management, ranged from under $300,000 to over $500,000. Most of the vendors also offer sophisticated network management, which ranges from PC-based software packages at $1,000 to workstations with color graphics priced up to $80,000.

Check the extras

There was also a wide range in the cost proposed for installation, maintenance, and training. Installation costs for the sample network ranged to over $30,000. Monthly maintenance costs ranged from $2,200 to $5,800, based on standard business-day coverage (eight hours per day, Monday through Friday). These maintenance costs also assume that the network locations are all in major cities. If you have remote locations to support, check for vendor surcharges for nonmetropolitan areas. On a per-trainee basis, training costs ranged up to $1,300, not including travel, living, and other incidental expenses.

With this much variation in the costs for a small network configuration, it is no surprise that all of these vendors offer discounts. Expect a 20 to 30 percent discount on the total network price, plus other special offers on services such as training and project management. Also, do not forget to evaluate the costs of upgrades (such as additional port cards) and announced new product features. The bottom line is that cost is only an issue after you have qualified the vendors that can satisfy your T1 network requirements.

Once you've mapped your business requirements to T1 multiplexer functionality, the next step is to determine which product is best suited to your company's needs.

Products and product features shown in Table 3 (also see "Product analysis") are currently operable in user network environments. The units represent the top offerings available from these nine vendors as of March 15, 1988.

Many vendors have aggressive development plans, but planned products and features should not be compared to deliverable capabilities. The information for the matrix was compiled based on discussions with users and vendors, coupled with independent technical evaluations. A number of the products listed are also sold by other vendors under OEM agreements.

Although all of these multiplexers support backbone-network configurations, their networking architectures vary. Two of the products (described below) have an architecture that combines an end multiplexer and a cross-connect under common network management software. This is done to achieve networking-multiplexer functionality. A cross-connect assigns and redistributes 64-kbit/s channels among the internodal links in a T1 multiplexer network (see "The digital cross-connect: Cornerstone of future networks?" DATA COMMUNICATIONS, August 1987, p. 165).

AT&T's Acculink 740/745—consisting of the 740 end multiplexer and the 745 cross-connect—is manufactured by Tellabs and is an enhanced version of Tellabs's own Crossnet product.

The Avanti ONC/ONX is a combination of the Open Network Concentrator (formerly called Ultramux) and the Open Network Exchange cross-connect. These products are being delivered in this configuration until the ONX has channel interface support. Although other companies market separate end multiplexer and cross-connect combinations as a single solution, AT&T and Avanti are the market leaders.

The Timeplex Link/2 is included in the Table 3 comparison rather than the Link/100 because the latter was not deliverable by the March 15 cutoff date.

Routing intelligence

Networking multiplexers have automated capabilities that assist in the operation of T1 networks. One of the most important capabilities is that of having the network build circuits to support applications.

Business requirements mandate maximum network availability. The loss of a circuit—even for as little as 30 seconds—could result in a critical loss of revenue. Loss of a circuit for less critical networks may only present an inconvenience, and therefore recovery delays can be more readily tolerated.

A company's business requirements also reflect how much automation will be necessary in order to satisfy circuit requirements. The complexity of the network and the expertise of the operations staff necessitate a particular level of routing automation.

Network complexity and staff expertise lead to specific requirements for the multiplexer's routing intelligence. The network should be totally self-healing within a specified time frame, or it may be flexible enough to allow time for operator intervention. Circuit requirements that change on a daily basis mandate automatic route generation.

The networking multiplexers available today utilize various techniques to satisfy circuit-routing objectives. All of the multiplexers profiled in this article can automatically reroute around a failed transmission facility. The key issues

The alternatives. *To analyze and compare total network costs, nine networking multiplexer manufacturers were asked to propose implementations of this four-node network. They also answered application-support questions. Network costs, including basic network management, ranged from under $300,000 to over $500,000.*

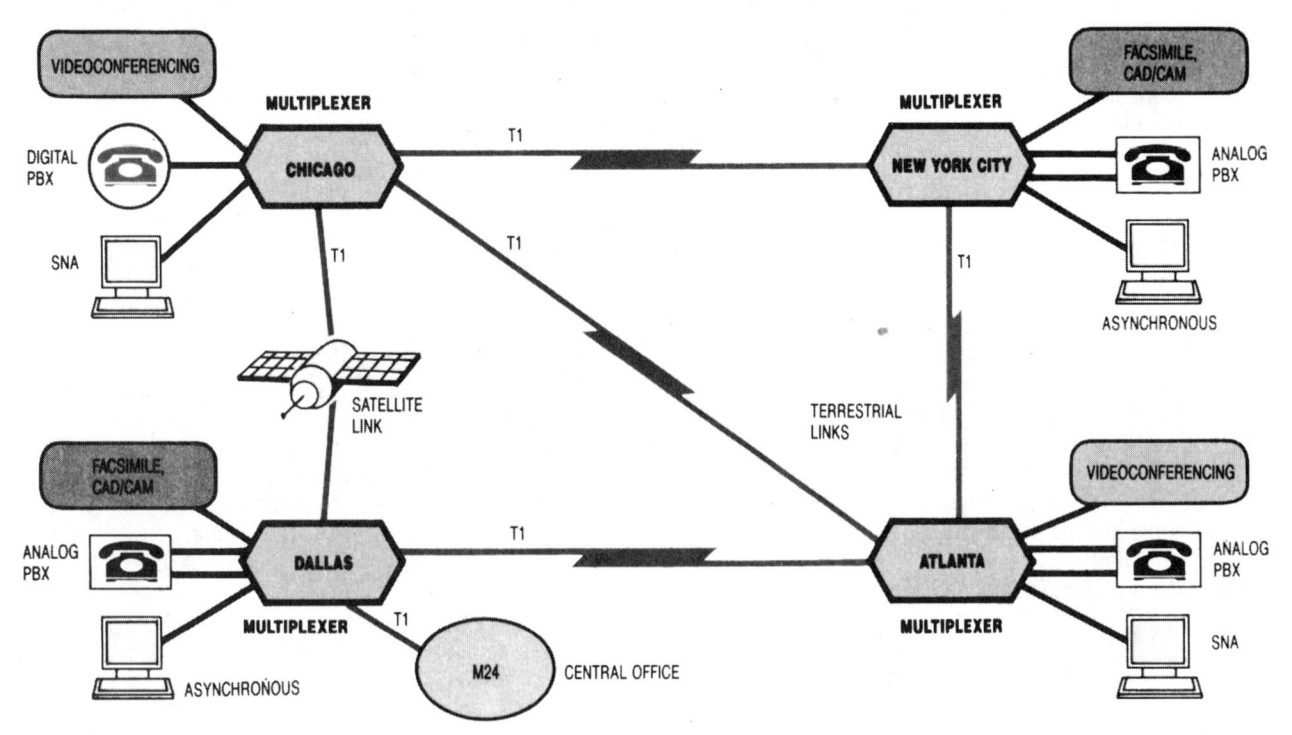

CAD/CAM = COMPUTER-AIDED DESIGN/COMPUTER-AIDED MANUFACTURING
M24 = AT&T SERVICE THAT SUPPLIES 24 VOICE CHANNELS FROM A T1 LINE
SNA = SYSTEMS NETWORK ARCHITECTURE

are the ease of implementing changes to the network and the speed at which the multiplexer restores circuits affected by an outage. The goal for circuit restoral in a data environment is typically 22 to 30 seconds, which is the time after which an SNA session will time-out. (SNA is IBM's Systems Network Architecture.) In a voice environment, the circuit-restoral requirements are dependent on service objectives. Some users may tolerate circuit drops and call again. Others may wish to keep calls in progress if connectivity can be reestablished in 10 seconds or less.

Circuit reroute times are determined by multiplexer factors such as the number of routes that can be calculated per second, the time to send circuit setup messages, and the time needed to reframe (time-align) an internodal trunk when adding a new circuit. For example, NET's IDNX can reroute four circuits per second; DCA's System 9000 can reroute two circuits per second. Reroute times will vary depending on the number of circuits affected and how many intermediate nodes must be traversed to restore the connection. In Table 3 (under "Time to reroute"), the amount of time to restore 40 affected circuits rerouted over an average of three intermediate links each is shown.

These numbers do not reflect the time necessary to determine that a facility is actually out of service. Many of these products allow the user to program this wait time before declaring transmission-facility failures; this time would be added to the times shown. Table 3 shows that only Avanti's ONX, DCA's System 9000, NET's IDNX, and Stratacom's IPX could satisfy a reconnect time of under 30 seconds for this example.

There are two philosophies for routing intelligence: distributed and centralized. Each can satisfy sophisticated network requirements but with varying levels of functionality. Distributed router intelligence specifies that each node in the network has routing responsibilities. A benefit of this implementation is a decrease in the network's vulnerability to node or trunk failures by not depending on a single site for circuit-routing functions.

Pluses and minuses

Typically, the origin end node is responsible for getting a circuit to its destination. Each node has either a table or algorithm that determines the route path for the circuit. An additional benefit of distributed intelligence is the ability to have multiple call-request queues, which decreases the time to reconnect circuits after a transmission facility or node failure. A disadvantage is that each node requires some additional processing capability, which drives up the cost per node by about 15 to 30 percent.

Centralized router intelligence can be based in a node or in host-based network management. In a centralized scheme, a single control point maintains current informa-

Product analysis

A comparison was made of the extent of networking support inherent in networking multiplexers supplied by nine different vendors. In four areas, the nine products have the same properties:
- *Redundant switching.* There is no traffic disruption with automatic cutover to backup logic. This is called "hot" redundant switching.
- *Nondisruptive capability.* The node's logical and/or physical configuration can be changed without disruption to circuits terminating or passing through that node.
- *Automatic rerouting.* No manual operator intervention is required to reconnect a circuit that has been disrupted by a node or line failure; connection is reestablished automatically.
- *Time-of-day switching.* Sometimes called bandwidth reservation, this network management feature allows the user to predefine circuit connectivity in order to ensure bandwidth availability when an application is to be run. The feature is most useful in networks where connectivity requirements are different for regular business hours than for nonprime time.

Where they differ. Table 3 compares the nine networking multiplexers in 18 other areas:
- *Number of aggregates* is the maximum number of non-redundant internodal physical interfaces supported by a single addressable node. The product's largest possible configuration is assumed. (For the AT&T and Avanti units with separate access and cross-connect elements, the number for each element is given: access/cross-connect.)
- *Aggregate redundancy* indicates the implementation of the redundant internodal aggregates. 1:1 means one backup card for each card protected. 1:N means one backup card can protect multiple cards. M:N means "M" cards can protect "N" cards on an as-needed basis. For products that employ separate access and cross-connect elements, two numbers are given: access/cross-connect.
- *Throughput capacity* is the total (input plus output) full-duplex switching capacity for a maximum configuration. The figures include source-channel-to-aggregate connections and aggregate-channel-to-aggregate-channel connections.
- *Maximum number of nodes in network* is the maximum number of individually addressable nodes that can be managed as a single network.
- *Source-channel capacity* is the maximum number of nonmultiplexed channels using a single physical interface per port that can be housed in a single addressable node.
- *Bypass delay* is the additional circuit propagation delay that occurs when a circuit is passed through a node from one internodal link to another. For those units where this delay varies depending on circuit transmission rate, a value for a 32-kbit/s circuit was calculated.
- *Multiplexing orientation* indicates the type of internodal aggregate multiplexing method used: bit-, byte-, or packet-interleaving.
- *Router intelligence* indicates the processing-information design required to route/build circuits for applications carried by the network. DISTRIBUTED means that responsibility for circuit routing may be handled by one of several network nodes. CENTRAL (or centralized) is where a single node has the routing responsibility.
- *Router type* indicates the method for selecting the internodal links that provide an end-to-end network connection. TABLE means a predefined list at the origin-end node or at the network control point. ALGORITHM means that a formula in a program running in the origin node or at the network control point determines the best network path at the time the circuit is requested.
- *Route generation* indicates whether path routes are op-

tion about network-resource utilization. One advantage of this is the ability to easily arbitrate bandwidth utilization on a network-wide basis. Updates to the central router affect the entire network, which guarantees consistency of routing decisions throughout the network. An additional benefit of centralized routing is a lower-cost implementation. (Distributed routing costs 15 to 30 percent more, as cited above.)

The disadvantages of centralized routing are its single-site vulnerability and the delay that can occur when many circuit paths are requested at the same time. In the case of the System 9000, a sophisticated centralized routing scheme allows a single node to select circuit paths for all circuits in the network. This central node can be any of the nodes in the network. If that node should be lost, a new node is automatically chosen to take over the routing responsibilities. Centralized router intelligence located in the network management host is being developed by GDC for the Megaswitch and by Newbridge for their 3600 series.

(This is a low-end multiplexer—with fewer features—not included in Table 3.)

Two methods are used for the selection of circuit paths: either tables or algorithms. An algorithm calculates the circuit path based on current topology and user-defined parameters. Algorithmic implementations simplify route selection. For example, the operator of an algorithm-based multiplexer, such as the IDNX, System 9000, or IPX, need only specify the end points of a circuit and the path will automatically be determined. Topology and resource changes are automatically taken into consideration for subsequent path selections.

Global and neighbor

Algorithmic routing implementations vary. Those in the IDNX, IPX, System 9000, and ONX decide the best route for a circuit based on global knowledge of the network topology. Alternatively, Micom's DX-500 uses "nearest-neighbor" algorithms, where each node knows the inter-

erator initiated or are automatically generated and maintained by the network.

■ *Time to reroute* is the time required to reconnect 40 circuits (representative of traffic on an internodal link) that have just been disrupted by a T1 outage. This measurement takes into account the time to select the new route and to send the appropriate circuit-setup messages. It does *not* include the time that the nodes wait to ensure that the outage is "hard." Since the time calculation may depend on the length of the new path selected, an average value of three links was chosen. The time values listed were calculated with vendor-supplied reroute factors. The latter included the time to establish a frame on an internodal link and the time to "look up" a new path. (# denotes that no calculable or credible reroute time could be ascertained for those units so indicated. Vertical Systems estimates these reroute times to be 30 seconds to minutes for the 40-circuit-reconnection example.)

■ *Parameterized routing* indicates the ability to define and route circuits using path-selection variables. Typical variables are: media (such as copper wire, microwave, optical fiber, and satellite), security requirements (encrypted or unencrypted), and delay tolerance (such as a circuit not to exceed a certain delay threshold). Units with algorithmic routers take these parameters into account in the selection of circuit paths. Parameterized routing is most important in complex-topology networks, to gain efficient use of internodal transmission facilities while ensuring that basic applications support objectives are met.

■ *Maximum number of hops* indicates the maximum number of internodal links that a circuit can traverse to complete an end-to-end connection. This number becomes an issue when the total of internodal links becomes large (exceeds 10), since end-to-end connectivity is achieved by completing connections as a continuous series of internodal line segments between endpoint nodes. Even geographically small networks (such as in campus environments or those with as few as five nodes) may involve a large number of internodal links. (Where "not applicable" is listed, vendor-supplied information states that the multiplexer imposes no restrictions on the maximum number of hops. Units with N. A. have not had sufficient field exposure to provide verifiable numbers.)

■ *Priority levels* indicates the ability to specify, as part of the circuit definition, a priority to arbitrate contention for multiple-circuit router requests. One example: When an internodal link fails and the multiple circuits that had been traversing that link must be rerouted. This feature determines who will get reconnected first when a node or link fails. (Four priority levels is a practical maximum.)

■ *Priority bumping* indicates the ability to override existing circuits on an internodal link when a new circuit needs that internodal bandwidth to complete its connection. (The number next to the Yes listings is the number of priority-bumping levels.)

■ *Traffic balancing* is a traffic-distribution ability for load-leveling in internodal trunk utilization. For time-division multiplexed internodal trunks, traffic balancing does not affect throughput. Its benefit: It reduces the number of circuits that may be affected by an internodal link failure. *Bandwidth contention* is the ability to oversubscribe a transmission facility by configuring multiple ports with a total bandwidth exceeding that of the transmission facility. With all available bandwidth in use, any subsequent active ports would receive a busy signal until the session of one of the busy ports is completed.

■ *Lowest bypass channel* is the lowest data rate (smallest bandwidth) that can be transferred from one internodal link to another. The smaller the number, the greater the potential for internodal link and nodal multiplexing-bus efficiency.

connections of only those nodes directly connected to it. Routing algorithms that are based on global network topology result in lower reroute times than neighbor-node algorithms, since the end-to-end route is known immediately. Neighbor-node algorithms seek the best possible route from neighbors on a link-by-link basis. This may result in multiple connection attempts. The disadvantage of global algorithm methods is the large amount of messaging necessary to keep all nodes current on network conditions.

The table-driven multiplexer is less flexible in a frequently changing network. Table-driven multiplexers use predefined paths for routing circuits to other nodes. Tables that determine a circuit's path are stored either in each node or in central network management. These tables must be updated when nodes or lines are added or deleted. Route selection must be defined manually by the network operator or by an automated process under operator control. Complex networks that require many alternate-route scenarios and daily table updates become difficult to manage.

The number of alternate routes that can be stored is important. Timeplex's Link/2 can store up to eight alternate routes—this is typical. For example, a network connecting three sites in a triangle would require a maximum of four routes in each node—one directly to each adjacent node, and one alternate for each. As the complexity of the network increases, the number of possible alternate routes also increases. A limited number of alternate routes may require the network operator to manually configure new table entries during a node or line outage. Table-driven multiplexers, such as the Link/2, GDC's Megaswitch, and Infotron's NX4600 are best suited to networks with infrequent topology or circuit changes.

Networking boons
All of the products represented in Table 3 have basic features such as time-of-day switching. This feature allows the network to automatically reallocate bandwidth and circuit connectivity based on the time of day.

The more advanced features that have emerged include: parameterized routing, priority levels, priority bumping, traffic balancing, and bandwidth contention (all defined in "Product analysis"). These features allow users to gain greater control over a circuit's initial and fallback routing.

For example, the IDNX allows the user to specify class-of-service parameters for each circuit. These parameters can be used to specify that a particular circuit travel only over optical-fiber, encrypted links. If the network topology cannot support the circuit request, the IDNX automatically responds with an alert.

In another example, an IPX network uses each fast packet to carry information that specifies the maximum permissible circuit delay. (In fast packet switching—analogous to statistical multiplexing—there is no time alignment among transmissions [while providing internodal transport]. This constrasts with TDM [time-division multiplexing], where the transmissions are all time-aligned.) The intermediate nodes will adjust fast packet queuing to control this delay.

A priority-level example: The Link/2 allows the user to specify one of 16 priority levels for each circuit. The user may place important data applications higher in the queue than voice to avoid time-out restrictions and maintain data circuit availability. Without this feature, the user has no control over the order in which circuits are placed back in service.

Priority bumping enhances the user's control during facility outages. This feature allows a higher-priority circuit to override or "bump" a lower-priority circuit when its bandwidth is required. For example, with the 740/745, three levels of bumping are available. When a facility failure decreases the available bandwidth, the multiplexer automatically arbitrates which circuits stay up and which get bumped. Four levels is the practicable, usable number for priority bumping.

Priority levels are often confused with priority bumping levels. While a circuit may have a priority level of one and get reconnected first, it may eventually be bumped by a circuit from another node with a higher bumping priority.

As for the traffic-balancing feature, it is supported only by the IPX as part of its fast-packet architecture.

Nodal architectures

"Multiplexing orientation" is how the product interleaves information on its internodal aggregate links. Almost all the multiplexers in Table 3 utilize TDM; the one exception: Stratacom's IPX, which uses a fast-packet architecture. The TDM units are either bit- or byte-oriented. This means that the multiplexer puts individual channels in either one-bit-oriented or eight-bit (byte)-oriented time slots.

Bit multiplexing is more efficient, but it is implemented using vendor-proprietary formats. This limits flexibility for trunk interconnection to carrier services (such as CCR, AT&T's Customer Controlled Reconfiguration, and M24, AT&T's multiplexed voice interface) or to other vendors' equipment (PBXs, channel banks, other T1 multiplexers). It is possible to be bit-oriented and still be DACS (Digital Access and Cross-connect System)-compatible at the DS-0 level, as is the case with the 740 and the ONC. These products bit-interleave within DS-0 boundaries, but they do not interface to channelized services such as M24 and subrate (below DS-0) digital multiplexing.

Compared to the TDM approach of the other vendors, Stratacom's IPX fast-packet architecture is a unique multiplexing technique. It assembles a T1 transmission stream into fast packets, using a standard T1 frame (193 bits), with one bit used for framing. Each 192-bit fast packet contains a destination address as well as the T1 channel's data or voice information. The fast-packet architecture permits faster reroute times than are found with circuit-switched products (see Table 3, "Time to reroute"). It also provides bandwidth efficiency (more channels per given bandwidth, accomplished more efficiently than with TDM) for low-speed channels (9.6 kbit/s and below). A disadvantage of this architecture is its relatively long (about 2.7 milliseconds) intermediate-node circuit delays and lengthy (also about 2.7 milliseconds) processing delays for low-speed channels at the origin node.

Network delays can impose restrictions on applications support. Circuit-delay time is based on the combination of propagation time across transmission facilities and the bypass delay imposed by intermediate multiplexers. Some data applications, such as CAD/CAM (computer-aided design/computer-aided manufacturing), are delay-sensitive. A round-trip delay of 20 milliseconds on a voice circuit can produce echo that is annoying during a phone conversation. This delay corresponds to that of a terrestrial circuit of about 1,800 miles and results in a need for echo cancellers as additional equipment on voice channels.

For some multiplexers, this problem manifests itself at shorter (less than 1,800 miles) distances. On a 32-kbit/s voice circuit, an IDNX imposes a bypass delay of 5 milliseconds per intermediate node, and the IPX imposes a delay of 2.75 milliseconds. These intermediate-node delays are high enough to warrant the use of echo cancellation equipment at shorter internodal distances (the equivalent of 450 miles shorter for each IDNX and 250 miles shorter for each IPX).

A different way

The System 9000 employs an approach that does not impose, on voice-signal information, the delays that are incurred when multiplexing low-speed data (see Table 3, "Bypass delay"). Voice samples are placed in the multiplexer's internodal transmission frame so that an intermediate node can recognize them and pass them immediately. In order to achieve the granularity (the ability to divide the T1 "pipe" into lower-data-rate channels) necessary for low-speed data, the data is buffered and placed in a subframe. The result is an intermediate delay of 0.375 millisecond for voice circuits, compared with 3 milliseconds for data circuits.

The lowest allowable bypass channel reveals another aspect of a multiplexer's flexibility. The lowest bypass channel is the bandwidth of the envelope carrying a single circuit. A large envelope (such as 64 kbit/s in the ONX and the 740/745) could create bandwidth inefficiency when a single low-speed circuit needs its own path. A small envelope, like 50 bit/s in the Link/2, creates a flexible and efficient bandwidth utilization.

Most network applications mandate redundancy of internodal link modules. The total number of aggregates supported is affected by the redundancy scheme employed (see *Aggregate redundancy* in "Product analysis"). The 1:N scheme is not only more flexible but more cost-effective. For example, the IDNX can support 96 non-redundant internodal links and employs a 1:1 redundancy scheme. Therefore, a fully redundant IDNX could support 48 aggregates (with 48 backups). The 1:N redundancy scheme utilized in the 745 and the ONX allows the user to create a backup arrangement where one module can back up as many as 15 aggregates (15 backed up by one).

The maximum number of aggregates and the maximum number of channels a multiplexer can hold does not necessarily reflect the multiplexer's true capacity (see *Throughput capacity* in "Product analysis"). The throughput capacity values in Table 3 give a relative measure of the real "work" potential of each multiplexer. These values can be used in network planning to determine the number of nodes required to support a given location. Other elements that can affect actual capacity are physical slot restrictions and the logical number of circuits that can be managed by the node.

After determining individual nodal requirements, you should determine the maximum number of nodes that will be required in the network. Restrictions may be imposed by the network management scheme or by the multiplexer's addressing capabilities. Circuits will be required to traverse a number of intermediate links to get to the destination. Under normal conditions, the number of hops should be kept to a minimum, but alternative reroute scenarios could cause the circuit to be routed over many more links than intended. Most corporate backbone networks require support of at least five intermediate hops. ■

Tim Zerbiec is a principal and vice president for technology at the Vertical Systems Group, a consulting and market research firm. He is responsible for managing technology-related offerings. Zerbiec has over 17 years of telecommunications engineering, research, and management experience.

Rosemary Cochran, also a principal of Vertical Systems, is responsible for managing the firm's marketing-related offerings. She has 14 years of experience in data services and telecommunications.

This article is based on material abstracted from the report, "T1 Multiplexer Industry Analysis: 1988," available from Vertical Systems Group, One Dedham Place, Dedham, Mass. 02026; telephone: (617) 329-0900.

User survey exclusive

Ratings reveal carriers are worlds apart

In the opinion of more than 650 users recently surveyed, AT&T's dial-up circuit quality and leased-line reliability are unsurpassed. But while AT&T's technical ducks are clearly in order, it seems it needs work in customer relations: Users said the planning assistance they got from carriers including MCI and ALC/Allnet was better than from AT&T.

These and many other findings, from Datapro Research Corp.'s soon-to-be-published annual users' ratings of communications carriers and value-added networks (VANs), show that all carriers are not equal in the eyes of users—this despite the post-divestiture regulatory policy in the United States, which places all long-distance (interexchange) carriers on equal footing.

Based on the final 2,149 communications-carrier user ratings received (many users rated more than one carrier), the U.S. long-distance market is apparently evolving into what is effectively a three-horse race. For switched and dial-up communications (see Pie chart 1), AT&T's response share indicates it holds slightly more than 60 percent, down from nearly 70 percent in last year's survey. Second-place MCI's response share has climbed, however, from roughly 12 percent last year to nearly 19 percent, and US Sprint's has also edged up, from about 9 percent to roughly 12 percent.

For the leased analog line and digital facilities market, including T1 (see Pie chart 2), AT&T garnered two-thirds of the response share; MCI users comprised 10 percent of the total.

Users were also asked to provide information on and rate any value-added networks (VANs) that they used. After screening, there were 564 VAN responses and ratings, which were then tabulated. The breakdown of these responses (see Pie chart 3), shows Telenet and Tymnet each with about a quarter of the total and the datanetwork offerings of Compuserve and IBM (its Information Network) accounting for another quarter.

The communications carriers' ratings are summarized in Table 1; those for the VANs are shown in Table 2. The rated criteria are nearly the same for both: quality of customer support/planning assistance; the reliability of facilities; the carrier's ability to find and repair problems; and overall performance. VAN users were additionally asked to rate ease of use, while the carrier questionnaire asked users to assess the quality of the carrier's facilities. All of these ratings were tabulated using a weighted

Switched and dial-up pie:
Response share, by carrier*

- AT&T 61.0%
- MCI 18.7%
- US SPRINT 12.0%
- ALL OTHER 6.5%
- ALC (ALLNET) 1.8%

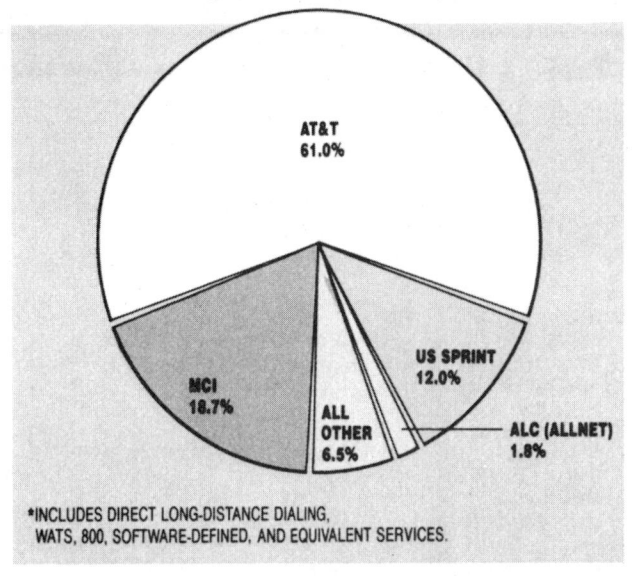

*INCLUDES DIRECT LONG-DISTANCE DIALING, WATS, 800, SOFTWARE-DEFINED, AND EQUIVALENT SERVICES.

Digital and leased facilities:
Response share, by carrier

- AT&T 68.1%
- MCI 10.3%
- ALL OTHER 12.1%
- US SPRINT 4.0%
- ITT (USTS) 3.3%
- WESTERN UNION 2.2%

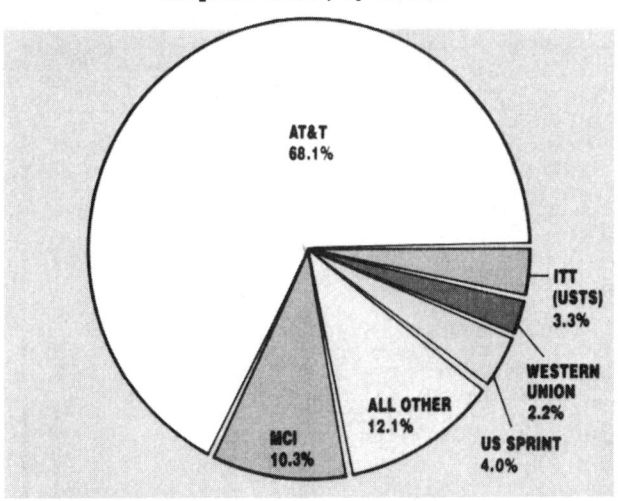

Value-added data networks:
Response share, by VAN

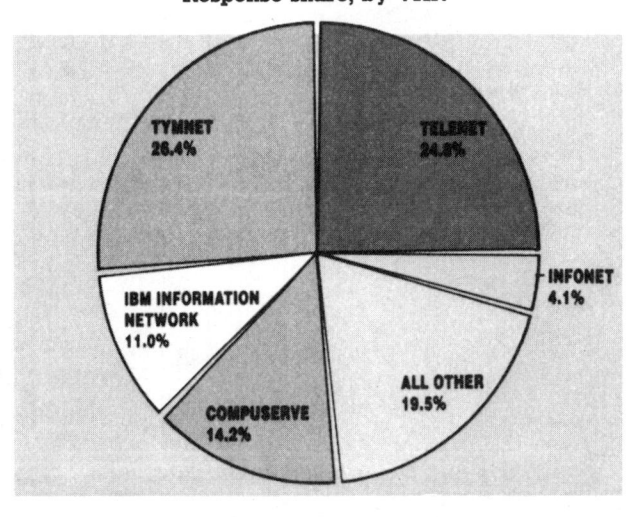

- TYMNET 26.4%
- TELENET 24.8%
- IBM INFORMATION NETWORK 11.0%
- COMPUSERVE 14.2%
- ALL OTHER 19.5%
- INFONET 4.1%

User survey exclusive

Table 1. User ratings of transmission facilities

CARRIER AND SERVICE	NUMBER OF USER RESPONSES	PLANNING ASSISTANCE					CIRCUIT QUALITY					CIRCUIT RELIABILITY					PROBLEM DIAGNOSIS					OVERALL PERFORMANCE					WOULD YOU RECOMMEND THIS CARRIER?		
		WA	E	G	F	P	WA	E	G	F	P	WA	E	G	F	P	WA	E	G	F	P	WA	E	G	F	P	YES	NO	UNDECIDED
AT&T—																													
DIAL-UP LONG DISTANCE SERVICE	346	2.9	74	171	68	20	3.2	118	184	31	3	3.2	116	176	38	5	2.9	67	169	76	14	3.1	77	222	34	3	294	11	41
WATS-TYPE SERVICE	176	2.9	39	92	32	11	3.3	64	96	12	0	3.2	57	94	20	0	2.9	26	104	30	6	3.1	33	118	19	0	151	8	17
800/IN WATS SERVICE	187	2.9	37	100	36	12	3.3	65	101	14	2	3.3	65	99	17	1	2.9	30	102	38	6	3.1	32	133	14	2	163	7	17
SOFTWARE-DEFINED/ VIRTUAL NETWORK SERVICE	22	3.0	5	9	6	0	3.3	7	13	1	0	3.2	8	10	2	1	3.0	6	8	6	0	3.3	6	13	1	0	18	1	3
LEASED TELEGRAPH GRADE LINE	29	2.7	2	16	9	1	3.0	3	23	2	0	3.0	5	20	2	1	2.6	2	15	10	1	2.9	1	24	2	1	21	3	5
LEASED VOICE GRADE LINE	223	2.8	33	117	50	18	3.1	54	137	24	1	3.1	53	130	32	1	2.8	25	127	49	14	3.0	27	157	31	1	178	10	35
LEASED WIDEBAND LINE	67	2.9	13	37	12	4	3.2	19	40	6	0	3.2	19	41	5	0	3.0	12	38	14	0	3.1	11	48	5	0	56	3	8
DIGITAL DATA SERVICE (UP TO 56K BPS)	139	2.9	33	70	23	11	3.3	53	74	8	0	3.2	48	70	16	2	2.9	29	69	28	8	3.1	33	92	9	2	120	7	12
T1 SERVICE	120	2.9	22	69	21	8	3.1	32	67	16	2	3.1	32	66	14	5	2.8	18	59	33	7	3.0	21	78	17	1	97	4	19
SATELLITE SERVICE	15	3.2	5	8	2	0	3.4	8	5	2	0	3.3	8	5	1	1	3.1	5	8	1	1	3.3	7	7	0	1	13	1	1
OTHER	12	2.5	1	5	4	1	3.1	3	7	2	0	2.7	1	7	3	1	2.7	1	6	5	0	2.8	1	8	3	0	9	0	3
SUBTOTAL	1,336	2.9	264	694	263	86	3.2	426	747	118	8	3.2	412	718	150	18	2.9	221	705	290	57	3.1	249	900	135	11	1,120	55	161
MCI—																													
DIAL-UP LONG DISTANCE SERVICE	115	2.9	22	55	28	1	2.9	25	54	24	5	3.0	24	56	21	4	2.9	17	56	22	5	3.0	22	62	22	2	76	8	31
WATS-TYPE SERVICE	73	2.9	15	37	17	1	3.0	21	29	15	3	3.0	16	38	10	2	2.9	15	28	19	2	3.0	15	41	10	1	60	4	9
800/IN WATS SERVICE	32	3.2	12	13	5	0	3.3	14	13	2	2	3.2	11	16	2	1	3.0	10	12	3	3	3.2	13	11	4	2	27	3	2
SOFTWARE-DEFINED/ VIRTUAL NETWORK SERVICE	5	3.2	2	2	1	0	3.0	2	2	0	1	3.2	2	2	1	0	3.0	2	1	2	0	3.2	2	2	1	0	4	1	0
LEASED TELEGRAPH GRADE LINE	4	3.0	1	2	1	0	3.3	2	1	1	0	2.8	0	3	1	0	2.8	0	3	1	0	2.8	0	3	1	0	2	1	1
LEASED VOICE GRADE LINE	34	2.8	7	16	9	2	2.9	9	13	8	2	2.8	5	18	7	2	2.9	6	17	8	1	2.9	5	19	7	1	26	3	5
LEASED WIDEBAND LINE	11	2.9	3	4	4	0	2.8	3	4	3	1	2.6	2	4	4	1	2.5	2	3	4	2	2.6	2	5	2	2	7	2	2
DIGITAL DATA SERVICE (UP TO 56K BPS)	14	2.7	2	7	4	1	3.1	7	4	0	3	2.9	4	7	1	2	3.0	3	8	3	0	3.0	3	8	3	0	10	3	1
T1 SERVICE	24	3.2	10	8	5	0	3.2	11	9	2	2	3.1	8	12	2	2	3.1	9	9	5	1	3.2	9	11	3	1	20	2	2
SATELLITE SERVICE	6	2.2	0	1	5	0	2.5	0	3	3	0	2.3	0	2	4	0	2.2	0	2	3	1	2.3	0	2	4	0	3	0	3
OTHER	3	3.5	1	1	0	0	3.7	2	1	0	0	3.7	2	1	0	0	3.5	1	1	0	0	3.7	2	1	0	0	2	0	1
SUBTOTAL	321	3.0	75	146	79	5	3.0	96	133	58	19	3.0	74	159	53	14	2.9	65	140	70	15	3.0	73	165	57	9	237	27	57
US SPRINT—																													
DIAL-UP LONG DISTANCE SERVICE	77	2.7	6	38	24	3	2.9	20	25	20	4	2.7	12	32	19	6	2.5	5	31	25	7	2.7	9	34	22	4	44	16	17
WATS-TYPE SERVICE	52	2.5	3	23	20	4	3.1	17	20	12	1	2.9	10	28	10	2	2.5	3	25	15	6	2.7	6	27	13	3	28	12	12
800/IN WATS SERVICE	15	2.8	1	10	4	0	2.9	5	4	3	2	2.9	4	6	3	1	2.9	1	10	3	0	2.9	4	6	3	1	8	4	3
LEASED VOICE GRADE LINE	12	2.4	1	3	8	0	2.8	2	5	4	0	2.8	2	5	4	0	2.5	1	4	6	0	2.5	1	5	4	1	5	5	2
DIGITAL DATA SERVICE (UP TO 56K BPS)	6	2.3	0	2	4	0	2.7	0	4	2	0	2.8	0	5	1	0	2.2	0	2	3	1	2.8	0	5	1	0	3	2	1
T1 SERVICE	16	2.6	1	8	7	0	2.9	4	6	6	0	2.6	2	7	6	1	2.2	0	6	7	3	2.7	1	9	6	0	11	2	3
OTHER	6	2.4	0	2	3	0	2.6	0	4	0	1	2.6	0	4	0	1	2.4	0	3	1	1	2.4	0	3	1	1	3	1	2
SUBTOTAL	184	2.6	12	86	70	7	2.9	48	68	47	8	2.8	30	87	43	11	2.5	10	81	60	18	2.7	21	89	50	10	102	42	40
ALC (ALLNET/LEXITEL)—																													
DIAL-UP LONG DISTANCE SERVICE	11	3.3	3	7	0	0	3.0	3	5	3	0	2.9	2	5	3	0	2.9	3	5	2	1	3.0	3	6	1	1	7	1	3
WATS-TYPE SERVICE	6	3.0	2	3	0	1	2.8	2	1	3	0	3.4	3	1	1	0	3.0	1	3	1	0	3.3	3	2	1	0	4	1	1
800/IN WATS SERVICE	4	3.5	2	2	0	0	3.3	2	1	1	0	3.3	2	1	1	0	2.7	0	2	1	0	3.3	2	1	1	0	3	0	1
OTHER	2	3.0	1	0	1	0	3.5	1	1	0	0	3.5	1	1	0	0	3.5	1	1	0	0	3.5	1	1	0	0	2	0	0
SUBTOTAL	23	3.2	8	12	1	1	3.0	8	8	7	0	3.1	8	8	5	0	3.0	5	11	4	1	3.2	9	10	3	1	16	2	5

WEIGHTED AVERAGE (WA) IS BASED ON ASSIGNING A WEIGHT OF 4 TO EACH USER RATING OF EXCELLENT (E), 3 TO GOOD (G), 2 TO FAIR (F), AND 1 TO POOR (P).

CARRIER AND SERVICE	NUMBER OF USER RESPONSES	PLANNING ASSISTANCE					CIRCUIT QUALITY					CIRCUIT RELIABILITY					PROBLEM DIAGNOSIS					OVERALL PERFORMANCE					WOULD YOU RECOMMEND THIS CARRIER?		
		WA	E	G	F	P	WA	E	G	F	P	WA	E	G	F	P	WA	E	G	F	P	WA	E	G	F	P	YES	NO	UNDECIDED
ITT (USTS)—																													
LEASED TELEGRAPH GRADE LINE	6	2.5	0	4	1	1	2.7	1	3	1	1	2.7	1	3	1	1	2.2	0	3	1	2	2.2	0	3	1	2	1	2	3
LEASED VOICE GRADE LINE	10	2.2	0	4	4	2	2.5	0	5	5	0	2.4	1	3	5	1	2.4	1	3	5	1	2.4	0	5	4	1	4	1	5
LEASED WIDEBAND LINE	4	3.0	1	2	1	0	3.0	1	2	1	0	2.5	0	2	2	0	3.0	1	2	1	0	2.5	0	2	2	0	2	0	2
DIGITAL DATA SERVICE (UP TO 56K BPS)	3	2.7	0	2	1	0	2.7	0	2	1	0	2.7	0	2	1	0	3.0	0	3	0	0	3.0	0	3	0	0	1	0	2
T1 SERVICE	5	2.6	0	3	2	0	2.6	0	3	2	0	2.4	0	2	3	0	2.4	0	2	3	0	2.4	0	2	3	0	1	1	3
SATELLITE SERVICE	4	2.8	0	3	1	0	3.0	1	2	1	0	3.0	1	2	1	0	2.3	0	2	1	1	2.8	0	3	1	0	1	0	3
OTHER	6	2.4	0	2	3	0	2.0	0	1	3	1	2.2	0	1	4	0	2.2	0	2	2	1	2.0	0	1	3	1	2	1	3
SUBTOTAL	38	2.5	1	20	13	3	2.6	3	18	14	2	2.5	3	15	17	2	2.4	2	17	13	5	2.4	0	19	14	4	12	5	21
WESTERN UNION—																													
DIAL-UP LONG DISTANCE SERVICE	4	3.0	2	0	2	0	3.0	2	0	2	0	3.0	2	0	2	0	2.8	2	0	1	1	3.0	2	0	2	0	3	1	0
LEASED TELEGRAPH GRADE LINE	10	2.5	0	6	31	1	2.9	1	6	2	0	3.0	2	5	2	0	2.3	0	4	4	1	2.8	0	7	2	0	7	1	2
LEASED VOICE GRADE LINE	5	2.6	0	3	2	0	2.8	0	3	1	0	2.8	0	3	1	0	2.3	0	1	3	0	2.8	0	3	1	0	3	1	1
DIGITAL DATA SERVICE (UP TO 56K BPS)	4	2.8	1	1	2	0	2.5	0	2	2	0	2.5	0	2	2	0	2.5	1	1	1	1	2.5	0	2	2	0	2	2	0
SATELLITE SERVICE	4	2.3	0	1	2	0	2.7	1	0	2	0	2.7	1	0	2	0	2.3	1	0	1	1	2.7	1	0	2	0	1	2	1
OTHER	7	2.7	2	0	4	0	2.3	1	1	3	1	2.2	0	2	3	1	2.3	1	2	1	2	2.3	0	2	4	0	1	4	2
SUBTOTAL	34	2.6	5	11	15	1	2.7	5	12	12	1	2.7	5	12	12	1	2.4	5	8	11	6	2.7	3	14	13	0	17	11	6
CONTEL ASC—																													
SATELLITE SERVICE	8	2.9	3	2	2	1	2.8	3	2	1	2	3.1	2	5	1	0	2.6	3	1	2	2	2.6	2	2	3	1	4	2	2
OTHER	5	2.2	0	1	4	0	2.2	0	2	2	1	2.4	1	1	2	1	2.2	1	0	3	1	2.4	1	1	2	1	2	1	2
SUBTOTAL	13	2.6	3	3	6	1	2.5	3	4	3	3	2.8	3	6	3	1	2.5	4	1	5	3	2.5	3	3	5	2	6	3	4
OTHER—																													
DIAL-UP LONG DISTANCE SERVICE	47	2.7	5	25	7	6	3.0	14	20	7	3	3.0	12	21	9	1	2.8	7	21	12	2	2.9	8	23	9	2	32	9	6
WATS-TYPE SERVICE	18	2.9	2	11	2	1	3.3	7	6	3	0	3.2	6	6	3	0	2.9	2	11	1	1	3.0	2	11	2	0	15	0	3
800/IN WATS SERVICE	5	2.3	0	1	3	0	3.0	1	2	1	0	3.0	1	2	1	0	2.5	0	2	2	0	2.5	0	2	2	0	3	1	1
SOFTWARE-DEFINED/VIRTUAL NETWORK SERVICE	4	2.8	1	1	2	0	3.3	2	1	1	0	3.3	2	1	1	0	2.8	1	1	2	0	3.3	2	1	1	0	2	1	1
LEASED TELEGRAPH GRADE LINE	4	3.3	1	3	0	0	3.3	2	1	1	0	3.0	1	2	1	0	3.3	1	3	0	0	3.3	1	3	0	0	4	0	0
LEASED VOICE GRADE LINE	37	2.6	7	13	9	6	3.1	13	13	9	0	3.0	11	15	7	2	2.7	5	17	10	3	2.8	6	16	13	0	21	5	11
LEASED WIDEBAND LINE	6	3.0	0	6	0	0	3.3	3	2	1	0	3.0	1	4	1	0	2.8	1	4	0	1	3.0	1	4	1	0	4	1	1
DIGITAL DATA SERVICE (UP TO 56K BPS)	25	2.8	5	12	7	1	3.4	12	10	2	0	3.4	10	13	1	0	3.0	5	15	3	1	3.2	7	15	2	0	22	1	2
T1 SERVICE	30	2.8	5	15	8	2	3.1	8	15	3	1	2.9	5	17	3	2	2.5	3	12	10	3	2.9	4	17	6	1	20	2	8
SATELLITE SERVICE	11	3.2	5	3	1	1	2.9	2	5	3	0	2.8	1	6	3	0	2.8	3	3	3	1	2.8	1	6	3	0	6	2	3
OTHER	13	3.0	3	4	3	0	3.2	3	6	1	0	3.0	2	6	2	0	2.8	2	5	2	1	3.0	2	6	2	0	4	3	6
SUBTOTAL	200	2.8	34	94	42	17	3.1	67	81	32	4	3.1	52	93	32	5	2.8	30	94	45	13	2.9	34	104	41	3	133	25	42
GRAND TOTAL	2,149	2.8	402	1,066	489	121	3.1	656	1,071	291	45	3.1	587	1,098	315	52	2.8	342	1,057	498	118	3.0	392	1,304	318	40	1,643	170	336

User survey exclusive

Table 2: User ratings of value-added networks (VANs)

CARRIER AND SERVICE	TOTAL NUMBER OF USER RESPONSES	CUSTOMER SUPPORT					EASE OF USE					RELIABILITY					PROBLEM DIAGNOSIS/REPAIR RESPONSE TIME					OVERALL PERFORMANCE					WOULD YOU RECOMMEND THIS CARRIER?		
		WA	E	G	F	P	WA	E	G	F	P	WA	E	G	F	P	WA	E	G	F	P	WA	E	G	F	P	YES	NO	UNDECIDED
TELENET—																													
PUBLIC DIAL ACCESS	97	2.7	13	46	30	4	2.9	14	57	19	2	2.8	14	53	22	3	2.6	8	42	31	8	2.8	12	52	25	3	73	7	17
PRIVATE DIAL ACCESS	16	3.1	5	8	2	1	3.0	4	8	4	0	3.1	4	9	3	0	2.7	4	5	5	2	2.9	3	9	4	0	12	1	3
DEDICATED ACCESS	27	2.8	6	10	9	1	3.1	8	13	5	0	3.2	9	12	5	0	2.8	4	12	10	0	3.0	5	15	6	0	18	0	9
SUBTOTAL	140	2.8	24	64	41	6	3.0	26	78	28	2	2.9	27	74	30	3	2.6	16	59	46	10	2.8	20	76	35	3	103	8	29
TYMNET—																													
PUBLIC DIAL ACCESS	101	2.9	19	49	25	2	3.0	19	53	21	1	2.9	19	49	23	3	2.7	13	43	29	7	2.9	16	56	20	2	77	6	18
PRIVATE DIAL ACCESS	25	3.1	5	17	3	0	3.1	6	15	4	0	3.1	8	11	6	0	2.7	3	12	7	1	2.9	3	18	3	1	20	4	1
DEDICATED ACCESS	23	3.0	7	10	6	0	3.0	6	12	5	0	3.1	7	12	4	0	2.9	5	11	7	0	2.9	4	13	6	0	16	1	6
SUBTOTAL	149	3.0	31	76	34	2	3.0	31	80	30	1	3.0	34	72	33	3	2.7	21	66	43	8	2.9	23	89	29	3	115	11	25
INFONET—																													
PUBLIC DIAL ACCESS	14	3.0	3	9	1	1	3.0	4	6	2	1	2.8	2	8	2	1	2.7	2	7	2	2	2.7	2	7	2	2	11	2	1
DEDICATED ACCESS	9	2.9	2	5	1	1	3.0	3	3	3	0	2.8	1	5	3	0	2.9	2	4	3	0	2.8	1	5	3	0	7	1	1
SUBTOTAL	23	3.0	5	14	2	2	3.0	7	9	5	1	2.8	3	13	5	1	2.8	4	11	5	2	2.7	3	12	5	2	18	3	2
COMPUSERVE—																													
PUBLIC DIAL ACCESS	69	2.9	13	40	12	3	2.9	12	41	11	3	3.1	15	43	8	1	2.9	10	39	12	2	3.0	11	44	9	1	53	1	15
PRIVATE DIAL ACCESS	8	3.6	5	3	0	0	3.6	5	3	0	0	3.4	3	5	0	0	3.4	4	3	1	0	3.3	3	4	1	0	6	1	1
DEDICATED ACCESS	3	3.0	1	1	1	0	3.3	1	2	0	0	2.7	0	2	1	0	2.7	1	1	0	1	2.7	0	2	1	0	2	1	0
SUBTOTAL	80	3.0	19	44	13	3	3.0	18	46	11	3	3.1	18	50	9	1	2.9	15	43	13	3	3.0	14	50	11	1	61	3	16
IBM INFORMATION NETWORK—																													
PUBLIC DIAL ACCESS	24	3.1	8	9	4	1	3.0	7	11	3	2	3.2	8	11	3	0	3.1	8	11	2	2	3.2	8	11	4	0	16	1	7
PRIVATE DIAL ACCESS	7	3.4	3	4	0	0	3.3	3	3	1	0	3.3	2	5	0	0	3.2	3	1	2	0	3.1	2	4	1	0	5	0	2
DEDICATED ACCESS	31	3.2	9	19	2	1	3.1	11	12	6	1	3.3	9	20	1	0	3.2	11	14	4	1	3.2	9	19	2	0	25	2	4
SUBTOTAL	62	3.2	20	32	6	2	3.1	21	26	10	3	3.3	19	36	4	0	3.1	22	26	8	3	3.2	19	34	7	0	46	3	13
OTHER—																													
PUBLIC DIAL ACCESS	53	3.1	15	28	8	1	3.1	14	27	9	1	3.0	9	33	9	0	2.8	7	31	12	2	3.1	12	31	7	1	42	2	9
PRIVATE DIAL ACCESS	24	2.9	5	11	5	1	2.9	6	9	5	2	3.0	6	11	5	0	2.9	6	8	6	1	2.9	6	9	6	1	14	4	6
DEDICATED ACCESS	33	3.0	6	21	5	1	3.1	8	19	4	0	3.2	10	16	5	0	3.0	7	18	5	1	3.1	9	17	5	0	30	0	3
SUBTOTAL	110	3.0	26	60	18	3	3.0	28	56	17	3	3.1	26	59	19	0	2.9	20	58	21	4	3.1	27	58	17	2	86	6	18
GRAND TOTAL	564	3.0	125	290	114	18	3.0	131	294	102	13	3.0	126	305	100	8	2.8	98	262	137	30	3.0	106	316	105	11	427	34	103

WEIGHTED AVERAGE (WA) IS BASED ON ASSIGNING A WEIGHT OF 4 TO EACH USER RATING OF EXCELLENT (E), 3 TO GOOD (G), 2 TO FAIR (F), AND 1 TO POOR (P).

Methodology

Questionnaires were mailed in April 1988 to a selected group of DATA COMMUNICATIONS' subscribers. By the cutoff date, May 30, approximately 800 completed forms had been received (many with multiple carrier and/or VAN ratings). After screening and validation, a total of 674 valid forms remained, yielding a total of 564 responses on VANs and 2,149 common carrier responses.

These were tabulated by DataVision Research of Princeton, N.J.

Datapro strongly suggests that this information be used with discretion. Acquisition decisions should be made ony after further investigation. For more information, contact Datapro Research Corp., 1805 Underwood Blvd., Delran, N.J. 08705 (609) 764-0100.

User survey exclusive

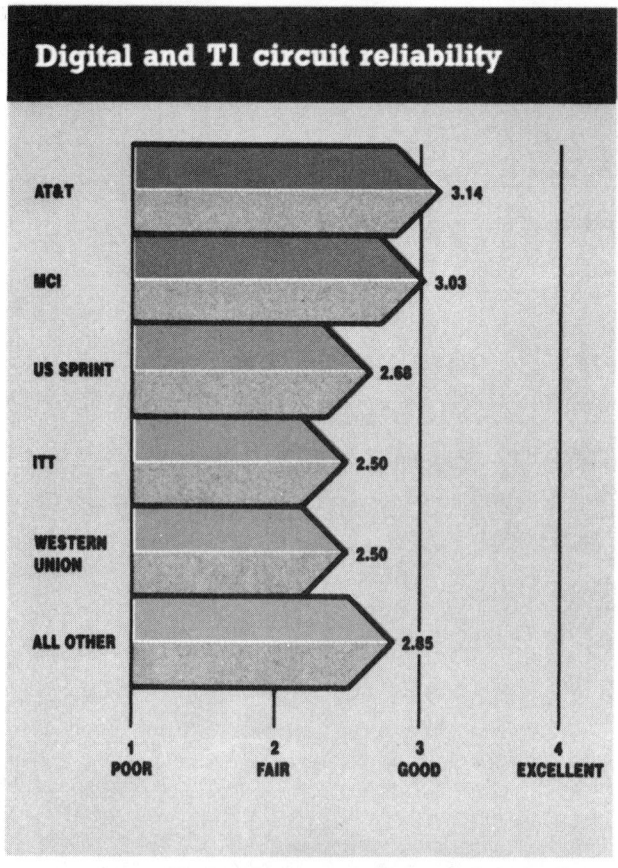

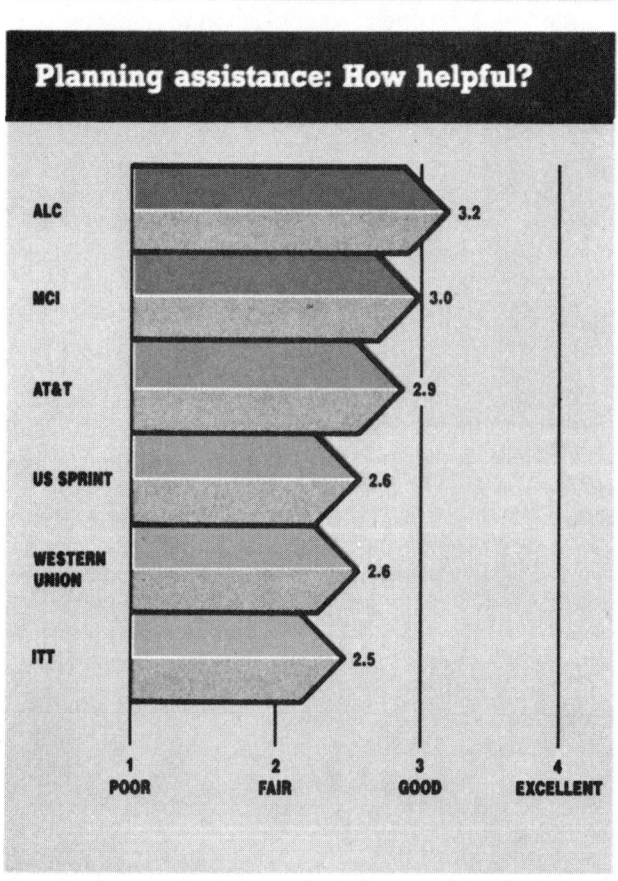

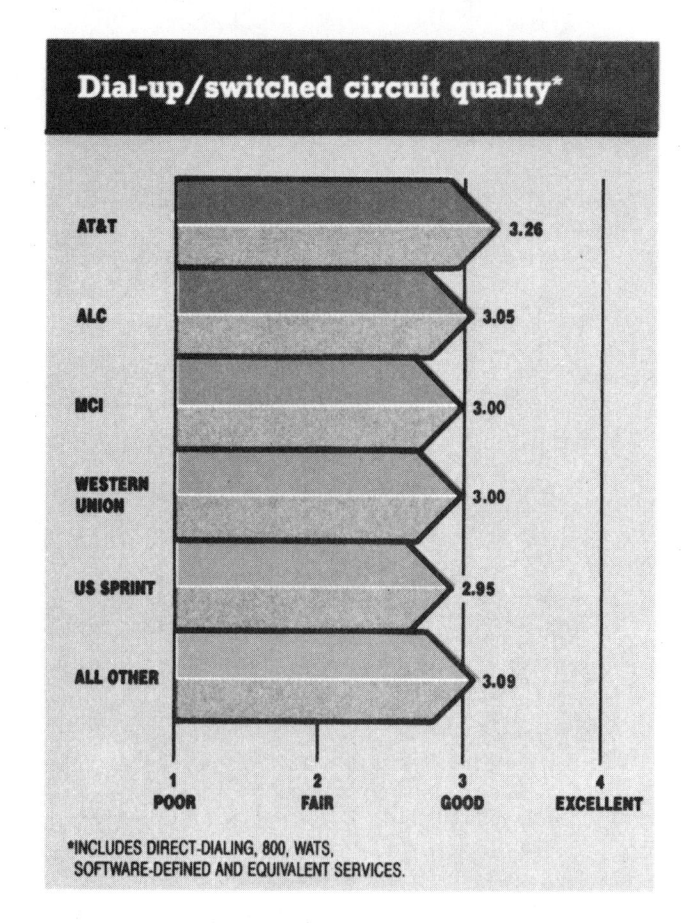

Digital and T1 circuit reliability

Carrier	Rating
AT&T	3.14
MCI	3.03
US SPRINT	2.68
ITT	2.50
WESTERN UNION	2.50
ALL OTHER	2.85

Scale: 1 POOR — 2 FAIR — 3 GOOD — 4 EXCELLENT

Dial-up/switched circuit quality*

Carrier	Rating
AT&T	3.26
ALC	3.05
MCI	3.00
WESTERN UNION	3.00
US SPRINT	2.95
ALL OTHER	3.09

Scale: 1 POOR — 2 FAIR — 3 GOOD — 4 EXCELLENT

*INCLUDES DIRECT-DIALING, 800, WATS, SOFTWARE-DEFINED AND EQUIVALENT SERVICES.

Planning assistance: How helpful?

Carrier	Rating
ALC	3.2
MCI	3.0
AT&T	2.9
US SPRINT	2.6
WESTERN UNION	2.6
ITT	2.5

Scale: 1 POOR — 2 FAIR — 3 GOOD — 4 EXCELLENT

average system, where an "excellent" rating was assigned a value of 4; 3 for "good;" 2 for "fair;" and 1 for "poor" (see "Methodology").

Users were also asked whether they would recommend the rated carrier to others. The responses were much more diverse for common carriers than for VANs. For example, Telenet actually received the lowest "yes" percentage of the VANs that were individually identified: 73.8 percent. But the best rated VAN, Infonet, didn't do much better—receiving a 78.3 percent "yes" recommendation share.

The common carrier ratings, on the other hand, show that users are much more discriminating in rating their carriers. Not even a third of the ITT customers surveyed would recommend that carrier (31 percent would, versus 55 percent that said they would not). At the other end of the scale, AT&T's customers seemed the happiest: Nearly 84 percent would recommend AT&T, while 12 percent said they wouldn't. MCI came in second behind AT&T, with a 74 percent positive recommendation. The next-best regarded was ALC (Allnet), with a 70 percent "yes" vote.

AT&T also enjoyed the top user ratings for both the reliability of its digital and T1 circuits (see Fig. 4, "Digital and T1 circuit reliability") and its dial-up and switched-network quality (Fig. 5, "Dial-up/switched circuit quality"). One of the few areas where AT&T users were less than completely pleased was in the quality of planning assistance they receive from AT&T (see Fig. 5, "Planning assistance: How helpful?"). In this case, both MCI and ALC received higher user marks than did AT&T.

James Randall, Independent Consultant, Jamesburg, N. J.

Making your network make money

Business-savvy communications managers can produce revenue for their company by turning a corporate overhead utility into a product-offering profit center.

In the next decade and into the next century, the technical data communications manager seeking personal success will be measured by more than hard-won technical competence. Bottom-line performance will be of growing importance and will determine advancement up the corporate ladder, as well as the bonuses and prestige earned en route.

It is communications professionals who devise methods for interconnecting all the dispersed computing power in the postdivestiture world. Unfortunately, data and telecommunications operations have grown so fast, often with a damn-the-torpedoes mandate, that apart from panicked intervention to keep the information flowing, there is precious little time to plan with care, manage with an eye on the bottom line, and operate the way Ma Bell taught us: recover all direct costs, plus a guaranteed return on investment (ROI).

This new corporate profit center can be called the Private Multi-Vendor Network/Computer Utility and Asset (PMVN/CU&A). And it can be exploited for maximum bottom-line contribution, career advancement, and growth within the corporate milieu.

Predivestiture

Remember the characteristics of predivestiture telecommunications services provided by "the phone company"? Immediate dial tone, the envy of the rest of the world; transparent support services from equipment installation to new product introductions; a monopoly on all parts of the telecommunications service utility from the handset on your desk to the private branch exchange at the receptionist's station in the lobby to Centrex and "long lines" services connecting to the rest of the world.

These are some of the qualities that made Ma Bell unequaled as a communications service provider. And a very effective quasi-political lobbying effort on behalf of all the Bell operating companies ensured that the fundamental revenue concept upon which Ma Bell built her success—recovery of all direct costs plus a guaranteed ROI—was the way tariffs were determined.

The telephone operating companies wielded considerable power before their state public utility commissions (PUCs). Consider, for example, the following anecdote.

About a dozen years ago, Robert Abrams, the current Attorney General of the State of New York, was Borough President of Bronx County, New York. He was seeking the Democratic nomination for the state office of Attorney General. Abrams was "running against the phone company" and the Public Service Commission (New York's PUC) as a consumer advocate. He made some radio commercials in which he asked: "In the last four years, do you know how many tariff increases the New York Telephone Company requested from your Public Service Commission?" After a pause, Abrams answered his own question: "91," he said. Then he asked: "Do you know how many were granted?" Then, after an even longer pause, Abrams responded: "91."

It is memorable after all these years because it stated simply how powerful the phone company was in getting its way before a public utility commission, whose job ostensibly was to protect the interest of the subscriber while simultaneously ensuring that New York Telephone received a fair return on its investment.

Postdivestiture

The character of the postdivestiture telecommunications and computer world is a very different one for the technical data communications professional. Management Information Systems (MIS), the data processing and data distributing corporate department, is being challenged for control of the larger portion of the overall MIS-PMVN/CU&A budget. The challengers are the communications activities, which are growing apace with distributed computer power—the direct result of the microcomputer revolution. These two contenders now rival MIS in laying claim to the larger share of growing budgets.

The immediate challenge for technical data communications managers is to seize the opportunities presented by the growth of distributed computing and the spread of network capabilities to gain control of the asset. How? By controlling the budget for that asset.

Divestiture has permitted many new players to enter the communications/computer business. This multivendor supermarket can be exploited by savvy technical data communications managers for lower product-acquisition costs, better performance and service support, and growing profits for their profit centers, provided any one vendor can be prevented from becoming too dominant.

Keeping qualified vendors at arm's length is essential to the independent crafting of your own performance specifications. Tables can be turned by exploiting the desirable features of informal proposals to identify the best products and services offering those features. This process will also enable you to discover the needs unique to your organization, which may require costly features only available on special order.

Then, with clear product or service requirements stipulated in the request for quote (RFQ) format, communications managers are in control because they have defined what they want to buy. With a number of proposals in hand from qualified suppliers, outright purchase, try-and-buy, beta-site testing, leasing, and other ways of obtaining equipment and services can be considered.

Defining requirements, determining which of your users would pay to connect to your network, and exploiting multivendor offerings for the best deal is a far cry from the overhead allocations of data and telecommunications costs, which is the most persistent and conspicuous hangover from the predivestiture days (see table). Data communications managers are faced with four major challenges: marketing the network, designing the network, setting performance standards, and building a team. This will determine the success of the PMVN/CU&A.

Marketing the network
To the technical professionals who manage today's networks, the idea of marketing the network may appear a violation of the inherent service mentality. But marketing the network is central to making the PMVN/CU&A produce revenue (see Fig. 1). This is the revenue that affects the bottom line and ultimately, when properly accounted for and managed, leads to profits for the PMVN/CU&A profit center, profits for the corporation as a whole, and bonuses and promotions for the leader of the PMVN/CU&A department and those who help produce the revenue. These profits may not be trivial either in the amount or the proof they provide of the value of the PMVN/CU&A.

Such an approach is almost universal in companies that develop products, and it is almost totally absent from organizations whose tradition is providing a service. What worked well for the phone company—recovery of costs plus a guaranteed ROI—must be updated and streamlined: There is no PUC to lobby for tariff increases to cover costs and ROI. The PMVN/CU&A that is managed like a high-tech product company has the best chance of producing the growing revenue (see Fig. 2).

The first and most important step toward producing that growing revenue is to determine what your users want and what they will pay for it. Hence, the marketing approach.

Marketing is the activity of determining what potential users want and are likely to pay for, then marshalling the resources to satisfy those needs at a profit.

Selling, by contrast, is the activity of developing leads, qualifying them as prospects, determining if they can both make a buy decision and meet the payments, and getting a favorable decision—closing.

The selling skills that will be needed by technical data communications managers in the 1990s, will be almost equally divided between getting plans, proposals, and budgets approved by upper management and persuading associates to actively innovate in the planning and implementation of the vision for PMVN/CU&A.

Selling the plan
Selling a marketing plan and a budget (which, combined, are the PMVN/CU&A business plan) is a far cry from product-flogging. The kind of selling required is the kind done every day to investment bankers by entrepreneurs seeking financing; to foundations that are importuned regularly to support one scholar's efforts over another's; or to senior management in high-tech companies by those seeking support of a new product.

Anyone who has an agenda and needs the approval and collaboration of those in more powerful and influential

Comparison between innovative and traditional leadership

INNOVATIVE	TRADITIONAL
VISIONARY	SOMEONE ELSE'S VISION
DEFINES OBJECTIVES	REACT TO UNQUALIFIED DEMANDS
INITIATES TASK-ORIENTED PROBLEM-SOLVING	AD HOC, CHAOTIC ACTIVITY
SETS SPECIFIC STANDARDS FOR PERFORMANCE OF THE NETWORK AS WELL AS REVENUE PRODUCTION	OBSCURE, ELASTIC, OR NONEXISTENT STANDARDS
ESTABLISHES LINE OF AUTHORITY REWARD ARRANGEMENTS	STAFF/ADVISORY FUNCTION: HIERARCHICAL CONTROL, OFTEN PUNITIVE
BUILDS ROUNDTABLE, TEAMWORKING ATMOSPHERE IN WHICH INDEPENDENT SOLUTIONS ARE ENCOURAGED	JOB DESCRIPTION IS PRIMARY PREOCCUPATION OF CONTENDING GROUP MEMBERS
IMAGINATIVE, CREATIVE	RESPONSIVE, REFLEXIVE

1. Marketing cycle. *Marketing a service that has always been available may seem unnecessary, so the transition from telecommunications provider to profit-and-loss department may be formidable. The flow chart, freely adapted from product development use, describes the steps that will meet demand and command user fees.*

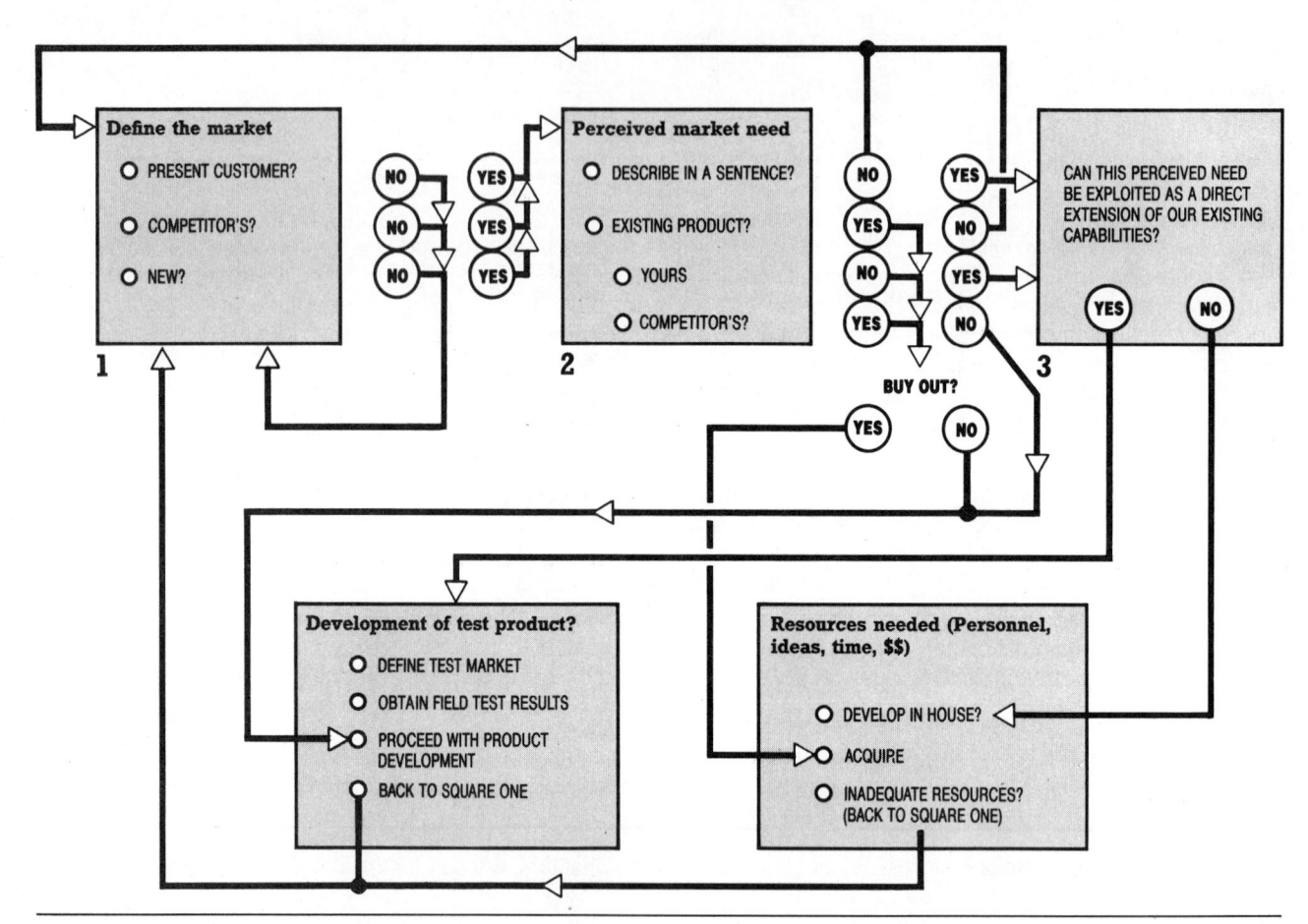

positions must know how to sell and how to sell with facts, plans, and quantified data — not hopes.

Managers also have to know how to sell their associates. Associates are not easily motivated, but they can be led. The essence of leadership is having formed an inner vision of what needs to be accomplished and then articulating that vision for all those involved with the PMVN/CU&A enterprise. In such a collegial, peer-to-peer endeavor, there are no bosses, only a collection of variously gifted associates, painstakingly selected to implement the shared vision.

Of course, teams are not built from scratch. Within existing telecommunications and distributed computing operations are established ways of doing things and people who are doing them. Any heavy-handed approach to force change implies dissatisfaction or disapproval with current staff and methods.

The conversion of department members to your way of thinking involves consulting with them, factoring in their ideas, fully disclosing your program, and continuously seeking their agreement as plans evolve. This constitutes selling your associates at the most diplomatic level. Commitment to seeking their counsel in order to make the PMVN/CU&A a success is the underlying principle. This way, rewards and challenges are shared.

This type of leadership was the way Dr. J. Robert Oppenheimer ran the Manhattan Project. He had his opponents within the project, but he always heard them out. An amazing level of loyalty emerged. It is also the way the famous Skunk Works of Silicon Valley and other high-tech venues are run.

Once all the managerial essentials are in place (see "The essential PMVN/CU&A"), attention can be turned to the design of the network, the group's bread and butter.

Designing the network

The essentials are in place. The data communications manager is the leader, reporting to upper management. The manager is responsible for the nuts and bolts of the PMVN/CU&A, as well as the growing revenue it produces. The manager is rewarded based on how well the PMVN/CU&A works and how much revenue it produces.

No doubt, an extensive and active telecommunications infrastructure is already in place, passing voice, data, and images in both directions and to all kinds of users throughout the organization. This existing network must be audited to discover its extent, where it goes, who's connected to it, and what kind of information is passed and with what quality.

This auditing function is time-consuming and fraught with ambiguity. Many pieces of terminal equipment will be orphans; no one will admit to owning them—so budget-center assignment is impractical. Depreciation allowances thus become nebulous. Cost or salvage as the determinant in assigning value to the asset becomes a guessing game. Never mind. The network in place *must* be rendered in block diagram form, with as much detail and accuracy as possible.

A standard for current valuation of equipment should be established, so that replacement cost, salvage cost, and title assignments can be determined. Time spent on determining an asset's worth should be measured against the asset's reasonable present value. For example, a laser printer priced at $5,000 a few years ago may now have a replacement cost that is a fraction of that.

The laboriously gathered information, the end product of systematic site surveys, will be the first data entries for a computer-based inventory control system, which will track all network and terminal equipment by manufacturer's name, model number, serial number, and to whom it is assigned (itself tracked by name, telephone extension, physical location, and profit center to which the device is charged) plus the purchase price and date of purchase. It is essential that this database be kept clean and up to date.

All requests for terminal moves, circuit additions, and cutovers, as well as the purchase of all data terminal equipment (DTE) and data circuit-terminating equipment (DCE) assets, should be entered into the inventory system *before* any new microcomputers, modems, or local area network servers are purchased.

A tall order? Indeed. Finding out just what's on a network and who has it is the most difficult part of designing the network. Some network managers may feel that this is beneath them. But consider that it was always Ma Bell's way of tracking products, which consisted of terminal equipment and network services. She knew who had what and what it was earning in revenue.

Once managers have a reasonably good idea of what makes up their networks and what's connected to them, the marketing plan dictates the next step: establishing schedules and priorities for implementing that plan.

All work to be done, whether it is as simple as moving a microcomputer from one location to another or as intricate and potentially disastrous as a major hot cutover, should be described in writing as a task (Fig. 3).

Once the task is described and the degree of difficulty or the level of commitment of in-house resources is estimated, a decision can be made regarding the use of in-house employees. Alternatively, bids can be solicited from qualified consultants.

The bottom-line mentality encourages the use of outside specialists. Only a cost-passthrough organization with no bottom-line concerns can have on staff every conceivable kind of specialist. In fact, service organizations, such as the predivestiture telephone companies, could afford to do just that since they operated on the recovery of all direct costs, plus a guaranteed ROI. Your products, PMVN/CU&A services, however, operate most effectively when all costs are accurately identified and built into user fees.

2. Engineering cycle. *While regarded as an arcane offshoot of science, replete with complex formulas, incomprehensible handbooks, and practitioners who are out of touch with reality, engineering is simply codified trial and error. The flow chart depicts how a design is refined until an optimum result is obtained.*

The engineering cycle in product development

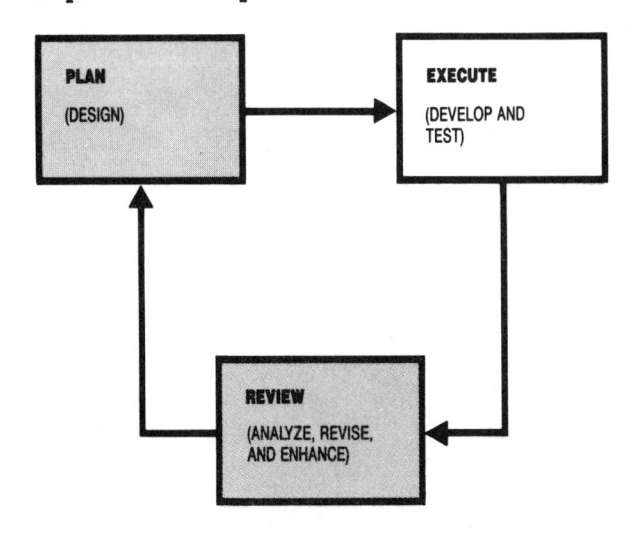

PLAN (DESIGN) → EXECUTE (DEVELOP AND TEST) → REVIEW (ANALYZE, REVISE, AND ENHANCE) → (back to PLAN)

The PMVN/CU&A as a revenue producer, developed using the engineering cycle

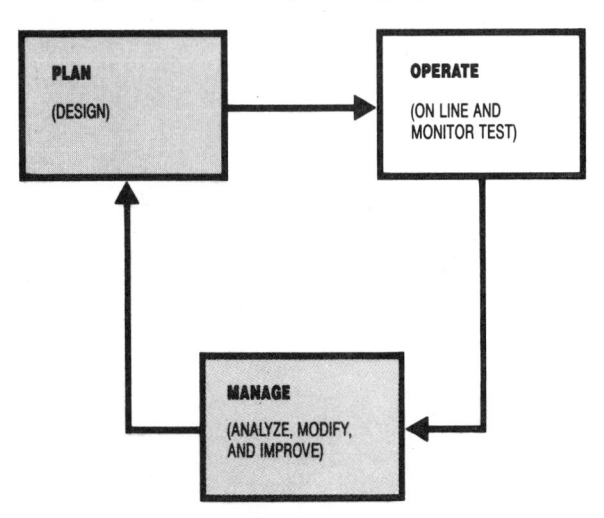

PLAN (DESIGN) → OPERATE (ON LINE AND MONITOR TEST) → MANAGE (ANALYZE, MODIFY, AND IMPROVE) → (back to PLAN)

PMVN/CU&A = PRIVATE MULTIVENDOR NETWORK/COMPUTER UTILITY AND ASSET

> **The essential PMVN/CU&A**
> **The management**
> - A data communications manager reporting to upper management.
> - A business plan blessed by upper management.
> - A team that is committed to implementing a shared vision.
> - A marketing group that has carefully solicited and recorded what the users want and what they will pay for.
>
> **The network**
> - A topographical schematic of the PMVN/CU&A, using industry-accepted symbols, showing all nodes, switches, multiplexers, and circuits, including data rate capacity and supplier.
> - An inventory of both data circuit-terminating equipment and data terminal equipment connection arrangements determining which is which, listing manufacturer, model and serial numbers, and physical location, plus who and what budget center is assigned to it.
> - Gateways to other networks and common carrier services.
> - LANs: location, type, and manufacturer.
> - Network control and monitoring functions. Location, types of information delivered, and remote network control capability.

Averaged costs, plus general and administrative costs, all have a negative impact on profit. Assigned costs for defined tasks can be marked up to carry your profit burden and contribute to positive bottom-line growth. Properly selected consultants can do jobs that you need to have done and add to the bottom line. The design of networks and the pre-installation modeling of networks, however, should be kept in-house. After all, who knows your business better than you do?

If anything goes wrong, data communications managers want to be able to identify the problem quickly, determining how it happened without having to deal with alibis and explanations. All network control and monitoring design and testing should also be kept in-house.

Any network design that is undertaken, either as an enhancement of existing capability or an expansion of the existing network topography or carrying capacity, should always be viewed as temporary. This is where the brains of the PMVN/CU&A team should be applied so that the improvements in a given technology can be factored in without too much rework and new users can be accommodated with a minimum of surprises.

A horror story

Broadband metropolitan two-way transmission appeared to network designers to be an added benefit when made available through the installation of CATV networks a decade ago. Had designers modeled part of the proposed network, however, its pitfalls would have become apparent.

LAN designers in New York City, for example, saw this broadband CATV backbone snaking its way through tunnels, subways, and sewers as a boon to the development of broadband LAN technology. It seemed much more promising than the limited data-carrying capacity of a 10-Mbit/s Ethernet.

Entertainment programming, however, is one-way, send-only distribution, whereas data transmission requires both transmit and receive over the same cable infrastructure. Thus was born the split-band division of the CATV radio frequency spectrum; one portion for transmission and another for reception.

Unfortunately, this scheme drastically reduced the number of subscribers that could get point-to-point or multidrop service because the useful, effective bandwidth had been reduced to accommodate both transmit and receive capabilities on a single CATV infrastructure.

Using the existing CATV scheme as a point of departure, the engineers at Wang Laboratories thought they had a better way: a two-cable network where one cable carried all outbound traffic (transmit) and a second, identical cable carrying all inbound traffic (receive).

The advantage to broadband LAN users was that they could use full CATV bandwidth on each leg. This benefit seemed so overwhelmingly favorable that the slight drawback of requiring a matched pair of circuit interface devices (CIDs) for every LAN access seemed a mere quibble. It was more than a quibble, however, and the additional cost and complexity of this approach remains a problem.

Making sure that CIDs track together and are installed together and in fact, like misshapen twins, go through life together for optimum performance, was a feat of technical wizardry that simply could not be delivered because of the inherent instabilities of commercially available CIDs at the time. Technical improvements in drift control have markedly improved CIDs from a variety of sources. But unfortunately, the stigma associated with unstable CIDs affected the acceptance of Wangnet (and, therefore, the size of the Wangnet user base). The limited installed base, in fact, is insufficient to justify retrofitting it with vastly improved CIDs.

Success story

Twisted-pair cable can be substituted for more expensive coaxial cable through the addition of a simple, passive, impedance-transforming device (which effectively connects a balanced circuit to an unbalanced circuit, hence the acronym, balun). IBM 3270-type terminals and terminal controllers, usually interconnected via coax, can now, over distances of several hundred feet, be interconnected using baluns with existing twisted-pair installed in the wall or floor. Multiple twisted pairs can be freshly installed, with designated pairs set aside for specific services such as voice, while other pairs are used as an effective low-cost, baseband LAN.

Savings can be significant, particularly if the existing twisted pair is disconnected or additional capacity is lying unused in the walls and floors of a site. However, maximum home-run distances—the total distance the cable must run,

3. Request form. *To earn revenue, you must know where terminal equipment is, who is using it, and the nature of the work being performed. A record such as this one will make it easier to quote prices and assign personnel to tasks.*

```
A TYPICAL PMVN/CU&A WORK ORDER REQUEST FORM

 ┌─────────────────────────────────────────────────────────────────┐
 │  NAME_____ LOCATION_____ EXTENSION_____│
 │                                                                 │
 │  BUDGET CENTER_____ EMPLOYEE NUMBER_____ NETWORK ACCESS NUMBER_____│
 │  BRIEF DESCRIPTION OF WORK TO BE PERFORMED:                     │
 │                                                                 │
 │                                                                 │
 │  DESIRED COMPLETION DATE_____  WILL YOUR PROFIT CENTER PAY PREMIUM RATES │
 │                                 FOR OVERTIME?  YES____ NO____   │
 │  PRESENT USER EQUIPMENT CONNECTED TO THE PMVN/CU&A              │
 │           ITEM             MODEL #          SERIAL#             │
 │  Terminal Equipment                                             │
 │                                                                 │
 │                                                                 │
 │  Video Display Terminals                                        │
 │                                                                 │
 │                                                                 │
 │  Personal Computers                                             │
 │                                                                 │
 │                                                                 │
 │  Word Processors                                                │
 │                                                                 │
 │                                                                 │
 │  Printers                                                       │
 │                                                                 │
 │                                                                 │
 │  Modems, Circuit Interface Devices                              │
 │                                                                 │
 │                                                                 │
 │  Budget Center Manager_____         │
 │  Do you want a written quote?  Yes ____  No ____                │
 │  For PMVN/CU&A Operating Group Only:                            │
 │  Site Survey required?  Yes ____  No ____                       │
 │  Inventory Control verifies equipment complement? Yes ____ NO ___│
 │  Technician Assigned_____                │
 └─────────────────────────────────────────────────────────────────┘
```

turns and all, from the farthest point to, say, a terminal controller—must be tested in situ for the maximum data rate that can be transmitted before unacceptable waveform degradation takes place.

In this success story, the installing contractor, an outside consultant to a major bank, set up a test bench in unoccupied space at the site and proceeded to drive various twisted-pair/balun combinations at a variety of lengths, some in excess of the longest to be used in the application with a variable-rate data source. The consultant monitored the outputs for attenuation versus length, rise-time degradation, leading and trailing edge overshoot, and group delay for a standard pulse.

A clear narrative describing the tests was written by the outside consultant's project engineer. Simple block diagrams showing the test configurations were part of the test procedure. Oscilloscope pictures were taken and annotated with calibration data for amplitude and time. And all of this was done *before* any cable was pulled into the walls and floors.

Several valued results accrue to both the client and the contractor when such a sensible engineering approach is employed. There is a recorded baseline of data showing actual, live performance for actual cable and balun configurations. Testing is done off-line, before installation. As the actual installation proceeds, before-and-after comparisons can be made to benchmark the live network for future performance reference. The client and the contractor are then more comfortable with each other because they are confident of results, which are predictable.

Setting standards

The PMVN/CU&A should be viewed as a product that offers certain features to the range of users established by the marketing plan. This product should appeal to current users and be capable of being enhanced and expanded for future users and needs.

Largely because of the hangover from Ma Bell's regime, networks are viewed as vast, complex, and impenetrable conglomerations of technology incapable of being understood in the whole, save by the geniuses who designed them.

This is nonsense. If the product approach is used, then the PMVN/CU&A can be described in written form (using both specifications and block diagrams). It becomes a utility performing certain information-carrying functions and describes in detail what is required of a potential user's equipment to become attached to the PMVN/CU&A.

The written specifications should be widely dispersed, updated on a regular basis, and reduced to a computer model with graphics. The topography of the current network should include a well-labeled block diagram complete with place names, node locations, backbone and tail-circuit data-rate capacity, equipment names, and model numbers. An inventory summary should be published and could even be made available through a microcomputer-accessible electronic mailbox. A user directory should also be made widely available to all users and potential users in the same way.

When users and potential users know what's available to them, word of mouth takes over, and given the choice, instead of going their own way, most users will become rate payers.

Of course, good old word of mouth works both ways. Lousy service gets talked about, too. The leaders of the PMVN/CU&A are Ma Bell's progeny, so why not make every effort to emulate her most significant achievement: telecommunications services of unrivaled quality that also produce abundant and growing profits.

Multiple vendors of varying size are available to supply most PMVN/CU&A products. A well-crafted RFQ or request for proposal (RFP) solicits their interest in your project on your terms. The recent Pentagon scandals center around a single, powerful customer being manipulated by its contractors and unscrupulous consultants and being told by suppliers what it wanted.

One way to avoid being overly dominated by a single vendor is to be sure that all vendors get the same information on your firm and its organization by publishing information on key people, including their relationships to each other. Horror stories abound of the favored vendor dominating the customer, providing excessive support that supplants an independent in-house capability and eventually leads to sloppy procurement practices and lazy, dependent buyers.

Until there is an independent way of establishing specifications without the collaboration and collusion of suppliers and consultants, bribery and chicanery will continue. There is a lesson here for those who aspire to make the PMVN/CU&A pay. That lesson is: If you don't understand the existing and emerging technologies that can be applied to your PMVN/CU&A well enough to write your own specifications and RFQs and RFPs, go back to school or into another line of work.

Qualified consultants can be a boon to the leaders of the PMVN/CU&A. They have several qualities to recommend them. It is the temporary nature of the assignment and the consultant's own specialized skills that helps yield an independent judgment of hardware and software proposals.

The drawback to retaining consultants is that they can become family members, bringing some perceived indispensable skill to the enterprise. If the talents retained are so critical or in such constant demand, develop a job description, discuss it with PMVN/CU&A staff, and see if one of them won't take it on. If the consensus is that someone is really needed, go into the market and find one.

Finally, network monitoring and control in the PMVN/CU&A is similar to documentation where high-tech products are concerned: It is often the last thing considered. Documentation of a product from its inception is de rigueur

---

Words to live by

A few pearls of wisdom for technical data communications professionals who want to control their destiny and grow in their careers:

- He or she who fashions the budget controls the asset—and vice versa.
- He or she who helps increase revenue is entitled to a portion of the increase.
- Staff jobs seldom lead to the top.
- Line jobs are risky and exciting and lead to the top.
- Forget not your technical education, so that bits, bytes, baluns, and modems cease to have a meaning grounded in reality and become buzzwords to impress the uninitiated.
- Heed the advice of technicians, even unto the third helper, for theirs is a wisdom unknown to technocrats.
- Remember, networks improve and grow by cutting over live circuits.
- The self-made man or woman relieves the Almighty of an enormous responsibility. No one's accomplishments are solely his or her own.

---

in any well-run engineering and product-development operation. Like inventory control, network monitoring and control is a top-priority project and should begin right up there with the inventory site survey.

Revenue is the other half of the PMVN/CU&A performance standard. Revenue measures the productivity of an asset. In the case of the PMVN/CU&A, the criteria for measuring performance have traditionally been technical standards: data rates available, downtime, bit error rate, and so on.

If the PMVN/CU&A is to turn a profit, become market-driven, and provide growth opportunities for the professionals who run it, attention has to be paid to revenue and other bottom-line considerations, such as depreciation, book value, title, ROI, and cost/benefit analysis. As much time needs to be spent on these financial matters as on developing the network.

Remember the marketing plan, site survey, and inventory system? This is where they pay off. If managers know who has what equipment where, then they can begin to develop an internal tariff structure (that's right, just like Ma Bell). They can charge users "connect" charges—such as for connect time or for the total number of packets accepted and delivered—and equipment charges (if the asset belongs to PMVN/CU&A). In addition, communications managers can begin to factor into those charges some portion of the total cost for underwriting capital expenditures.

Managers can begin to write down the capital advanced or borrowed to build or enhance the PMVN/CU&A. This will be based on the marketing plan. The improvements should reflect user wants, and their capital budget will show where the income will come from to pay both long-term (capital) and current expenses.

When senior managers are presented with a plan crafted around technology plus a marketing plan and a predicted source of revenue, they will be hearing language they understand.

Just as the PMVN/CU&A should be monitored for performance from its inception, so should the revenue production side be subject to regular review. In any income-producing activity, managers should be looking for growth (both in absolute amounts and in increases over the previous accounting period), return on invested capital, break-even point, overhead underwriting, and delinquent accounts.

Sometimes, however, a write-off and write-down is appropriate. A major New York bank installed a broadband LAN that is now only being used to 12 percent of its capacity. The bank is in a Catch-22 situation. It cannot attract enough users because of the network's poor performance reputation, so the bank cannot demonstrate to upper management that it can bring in revenue to pay for the necessary improvements. The only alternative is to take full depreciation against the asset and, when it reaches zero, close it down.

Building a team
Bureaucracies are defined by structure, lines of authority, and superior/subordinate reporting arrangements. When growth is minuscule and the tasks repetitive and mundane, this structure can get the job done. Bureaucracies are resistant to change, innovation, or expansion of function, and this is why they ultimately require subsidies to keep them operating.

The communications team that is envisioned in this article will be entrepreneurial, performance-driven, and dollar- and promotion-rewarded (see "Words to live by"). Problems are seen as opportunities, a challenge wherein skills can be tested, something new learned, or both.

Several preconditions must exist to get the PMVN/CU&A functioning as needed. Autonomy must be granted by upper management with one person reporting to the chief executive officer or the chief financial officer. MIS and the PMVN/CU&A have to be combined on some equitable basis—for the division of function as well as rewards. Naming a chief information officer is a start, provided someone is qualified who also has a direct line to the CEO.

In the future, there will be a growing pool of technical professionals trained in both engineering and computer science who are eager to take a leadership role. The new MIS-PMVN/CU&A can and should be on a par with other profit/loss departments in status, visibility, and budget. ■

James Randall has a BSEE from Brooklyn Polytechnic Institute. He has worked in sales and engineering in the electronics industry for more than 30 years.

Nicholas John Lippis III, Digital Equipment Corp., Littleton, Mass.

Coping with the cost realities of multipoint networks

Changing tariffs have eroded multipoint's traditional cost advantage. Here's an analysis of digital vs. analog tariffs from an avid point-to-point proponent.

Budgets for multipoint data networks are being squeezed hard these days by reversals in tariff economics—effects of the introduction of new technologies and systemic changes in corporate data communications requirements. One question facing managers of multipoint networks is: How high can multipoint network costs climb before the plethora of point-to-point alternatives becomes too tempting?

Multipoint data networks have been employed in the United States for the better part of three decades and typically comprise a series of polled devices that share the same communications channel. The devices on a multipoint network are most often remote slave terminals tied together to a master unit by the physical bridging of their dedicated analog lines at the exchange company's facilities, or in some cases, on the customer's premises.

Data transmissions on a multipoint network are available to all locations simultaneously, with line access controlled by the master device. Multipoint is sometimes referred to as "multidrop," and its applications include remote 3270 terminals, automatic teller machines, airline reservation systems, and alarm systems.

In the past, the appeal of multipoint stemmed from reduced cost arising from bridging located in either long-haul carrier points-of-presence (POP) or local exchange carrier central offices. With multipoint topologies, for example, a host's communications processor can exchange data with geographically dispersed equipment through a synchronous modem and a bridged voice-grade line at low cost.

During the 1980s, corporations began to understand the benefits of distributed computing based on a peer-to-peer relationship between CPUs. At the same time, microprocessor technology made computing resources directly available to end users. These trends have helped to accelerate the current shift from batch-oriented computing and timesharing to the distributed computing of clients and servers, aided byLANs using multiple protocols such as Local Area Transport, DECnet, TCP/IP, Xerox Network Systems, and SNA.

The two styles of computing, peer-to-peer and hierarchical, can coexist and, to a limited degree, interoperate with each other in the same network. But traditionally, peer-to-peer and hierarchical computing have been implemented on distinctly different topologies—multipoint versus point-to-point—for economic reasons (see Fig. 1A).

Point-to-point connections can be analog, but are increasingly digital, in which case methods such as T1 and Dataphone Digital Service (DDS) are employed to connect only two devices on a dedicated channel. Multiple devices are supported with point-to-point when the two points have multiplexing capability.

As these digital services come to dominate the industry, vendors are developing ways to superimpose *logical* multipoint networks on *physical* point-to-point topologies. With the wide availability of T1 services from the regional BOCs, AT&T,and the other common carriers, the technology and infrastructure is in place to integrate both styles of computing in a single wide area network (WAN).

Figure 1B illustrates how separate multipoint, point-to-point, and voice networks can be integrated on a T1 backbone with tail circuits serving the low-traffic offices. Smaller branch-office networks have remote cluster controllers for wide-area terminal-to-host connections through logical multipoint sessions. Distributed computers coexist in their peer-to-peer fashion over point-to-point links in these same branch offices.

In general, the efficiencies of point-to-point technology are reflected in the attractiveness of tariffs for T1 service versus analog private lines. Tariffs of analog private lines

Tariff model

Figure A details the various connections and mileage-sensitive charges associated with establishing cost for an intraLATA multipoint topology using local exchange bridging. Figure B provides the same information for an interLATA topology.

The physical parameters associated with the circuits discussed here are: 3002 type for multipoint voice-grade private line; C-2 and D-5 conditioning where applicable; and type I local channels for Dataphone Digital Service (DDS).

■ **Descriptions and assumptions.** The interLATA model was based on the following assumptions: The distance between LEC COs and AT&T POP locations were selected to be zero miles at one end and 50 miles at the other end. This assumption was chosen because zero-mile and 50-mile distances between a LEC CO and POP are extreme cases, so that the range of various LEC CO–POP distance combinations has been built into the model. The distances from the LEC CO to the customers' premises were alternated from zero to eight miles. The zero- and eight-mile distances correspond to a calculated average access charge.

The intraLATA model is based on these assumptions:
■ Access to interLATA service was derived by averaging the intraLATA cost of service in Massachusetts, New York, and Georgia through tariff 11.
■ The distance between the AT&T POP and LEC CO was fixed at eight miles.
■ The distance between one customer premises and AT&T POP was fixed at zero miles.

In comparing intraLATA multipoint to DDS and voice-grade point-to-point service, the following assumptions were made:
■ The distance between the AT&T POP and LEC CO was fixed at zero miles.
■ The topology was four-drop multipoint.

This last was important only because the circuit cost between the POP and the LEC CO is shared among the various customers that the LEC bridging hub serves. So a number of customers' premises had to be assumed and extra circuit costs applied when comparing multipoint circuit charges to point-to-point. Again, multiplexers, concentrators, or other transmission equipment were not included in the analysis.

In comparing interLATA multipoint-to-T1 services, the following assumptions were made:
■ The same distances between POP and LEC CO chosen in the multipoint interLATA model were used.
■ The cost of T1 multiplexing equipment or other transmission equipment was not included.

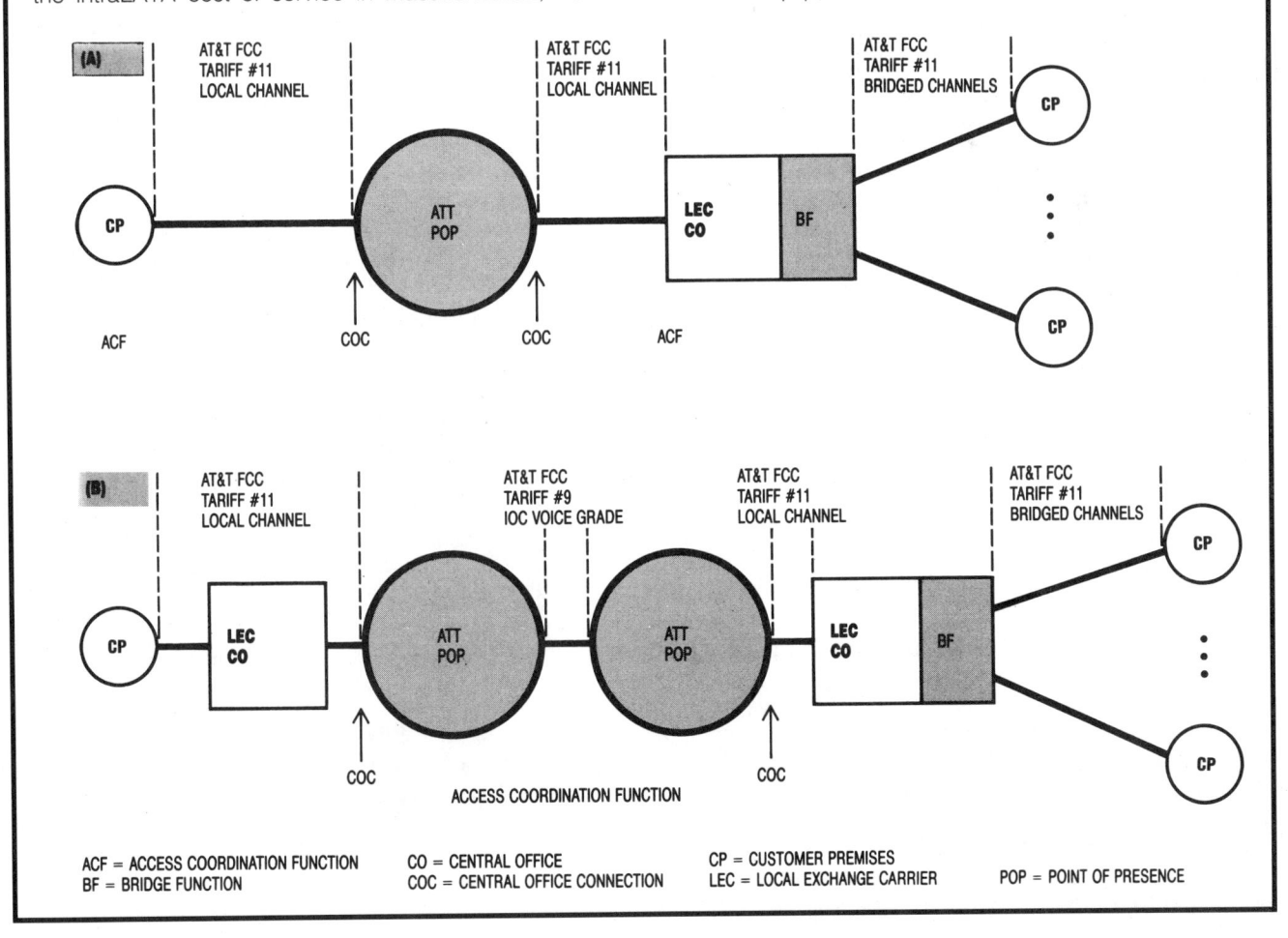

ACF = ACCESS COORDINATION FUNCTION
BF = BRIDGE FUNCTION
CO = CENTRAL OFFICE
COC = CENTRAL OFFICE CONNECTION
CP = CUSTOMER PREMISES
LEC = LOCAL EXCHANGE CARRIER
POP = POINT OF PRESENCE

1. Separate and together. *Predigital networks (A) support peer-to-peer and master-slave data on separate topologies. In B, multiple transmission networks are consolidated.*

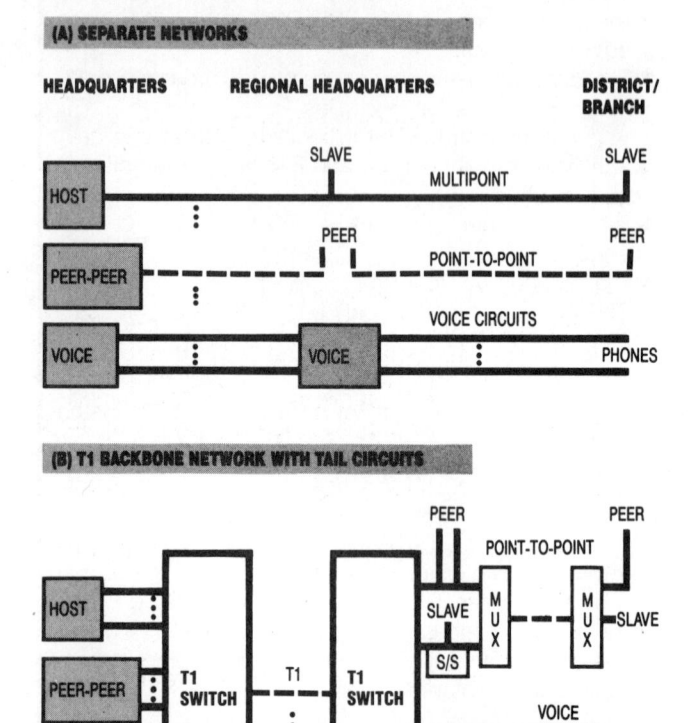

(A) SEPARATE NETWORKS

(B) T1 BACKBONE NETWORK WITH TAIL CIRCUITS

S/S = SIGNAL SPLITTER

(multipoint communications is a subset of analog private line) have been steadily increasing after the divestiture of AT&T with the introduction of tariffs 9, 10, and 11.

Figure 2 illustrates that anywhere from 12 to 24 analog circuits can be economically displaced by a single AT&T Accunet T1.5 circuit within a 25-to-1,700-mile range. The box enclosing the linear curve in this figure serves to highlight the cost-justification region, that is, where T1 proves financially viable compared with analog voice-grade service.

Breaking even

Corporate T1 backbone networks have emerged since 1984 to take advantage of evolving pricing structures. In this scheme, the cost of a T1 circuit is less expensive than a dozen voice-grade analog circuits plus some combination of data circuits (such as 56 kbit/s and 9.6 kbit/s).

Since multipoint communications can support only one protocol (such as Synchronous Data Link Control or bisynchronous) at any particular time, bandwidth splitting of the line through time-division or statistical multiplexers is not an option as it is in point-to-point transmission topologies. Multipoint networks are limited in their support and integration of various protocol stacks (such as DECnet, SNA, TCP/IP, and X.25), requiring additional network capacity. Point-to-point transmission supports multiplexing and packet-switching technology. Finally, the cost incentives that once made multipoint networks attractive are eroding.

Figure 3 illustrates the break-even curves between the price of 9.6-kbit/s and 56-kbit/s DDS compared with multipoint bridged circuits. These curves show where point-to-point physical topologies become attractive when juxtaposed with multipoint topologies.

In the intraLATA (Local Access Transport Area) domain, one or two multipoint drops into a customer's premises could be displaced by one 9.6-kbit/s DDS circuit within a range of 0 to 100 miles. For example, in the retail banking industry, multiple multipoint networks into a customer's premises are not uncommon, since separate networks are needed for automatic teller machines, teller terminals, and security.

Further, two or three multipoint drops into a customer's premises could be displaced by one 56-kbit/s DDS circuit within the 0-to-100-mile range. Finally, one multipoint drop into a customer's premises could be displaced by one voice-grade circuit—a local channel—within the 0-to-100-mile range. Point-to-point tail circuits employing concentrating, multiplexing, or packet-switching technology from a smaller office accessing the corporate T1 backbone would provide cost savings compared to multiple separate networks.

2. T1 vs. analog, circa 1988. *For many organizations, T1 service can be cost-justified by replacing 12 to 24 analog private lines over a 25-to-1,700-mile range.*

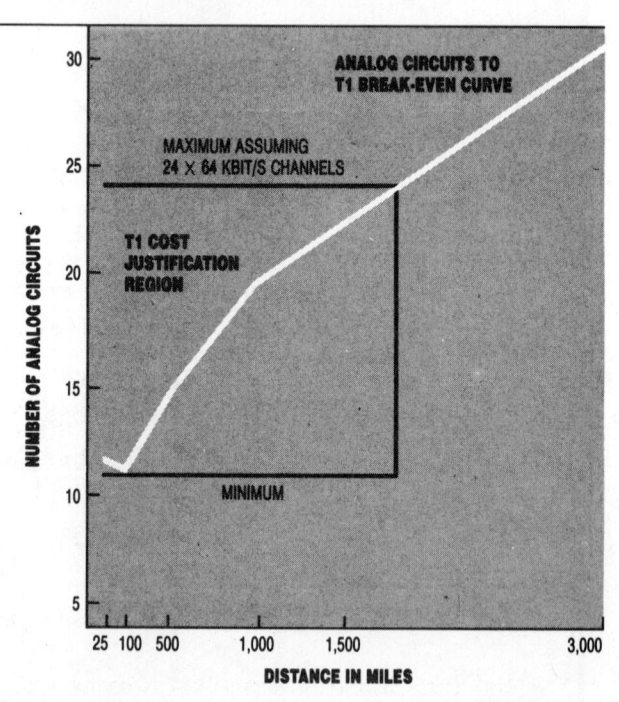

3. IntraLATA DDS vs. analog, circa 1988. *A 56-kbit/s DDS circuit can be cost-justifed with two or three multipoint lines. For a 9.6-kbit/s circuit, only one or two are needed.*

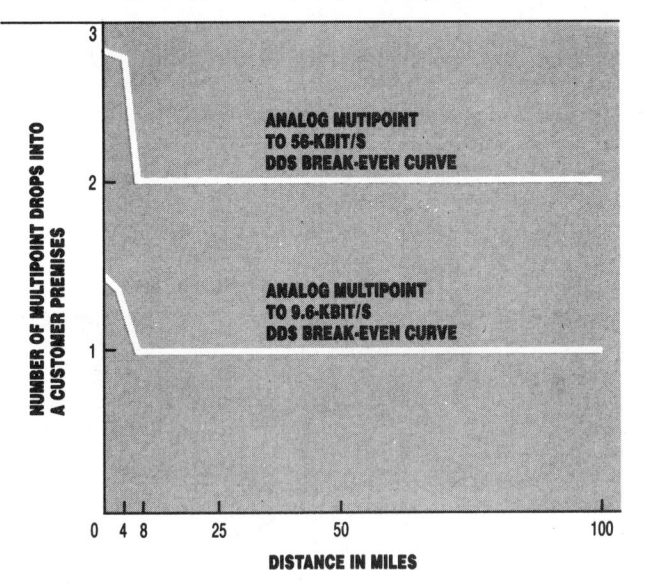

DDS = DATAPHONE DIGITAL SERVICE

In contrast to the increasing cost of multipoint, AT&T Accunet T1.5 service has decreased for intraLATA service in a 0-to-100-mile range by an average 10 percent between 1987 and 1988 and 16 percent between 1988 and 1989. Further, the cost for interLATA Accunet T1.5 service over a range of 25 to 3,000 miles has decreased an average 4 percent between 1987 and 1988 and 21 percent between 1988 and 1989.

The cost of 56-kbit/s DDS intraLATA service has decreased by 2 percent from 1987 to 1988 and by 29 percent from 1988 to 1989. InterLATA 56-kbit/s costs dropped 2 percent from 1987 to 1988 and 25 percent from 1988 to 1989, respectively. Changes in the costs of circuit installations have favored T1 over 56-kbit/s or 9.6-kbit/s DDS: a decrease of 9 percent as opposed to increases of 1 percent in 1987 and 13 percent in 1988.

AT&T recently filed with the FCC to overhaul its DDS rates, reducing by 4.5 percent the cost of interLATA 2.4-kbit/s, 4.8-kbit/s, and 9.6-kbit/s offerings, and reducing by 4 percent the monthly mileage cost for 56-kbit/s service while also decreasing central office connection charges. Discounts based on volume and yearly fixed-pricing schedules are also requested. If this filing is enacted, the curves of Figure 3 will shift downward, making digital connections to the smaller branch offices even more economically attractive.

Multipoint cost trends

Thus far only the comparative costs of various private-line services have been discussed. In order to provide insight into what network managers might expect of communications costs for multipoint networks in the coming years, one must also look at the prevailing trends in tariff structures.

This analysis was performed on a typical six-drop multipoint network topology modeled to provide communications cost trends from 1984 to 1988. This hypothetical topology spans New York, Massachusetts, and Georgia, supporting six of a network manager's sites.

This network could support a mainframe and front-end processor in New York with cluster controllers in two Massachusetts and three Georgia sites providing distributed wide-area terminal-to-host access.

Circuit costs increased 18 percent between 1985 and 1987 and 14 percent from 1987 to 1988. If conditioning had been employed, a compound annual growth rate of 20 percent would have resulted during 1985 and 1988. The cost to install this multipoint network has increased 123 percent between 1985 and 1987 and 36 percent between 1987 and 1988.

Figure 4 shows the breakdown of circuit charges for the six-drop multipoint network topology. The long-distance carriers' share of the total circuit cost of multipoint topologies (and other analog private-line services) is getting smaller all the time and is now lower than the charges levied to connect a customer's premises to the long-haul carrier.

Before divestiture it was economical for a network manager to add links and grow multipoint networks, but today's increasingly large installation and local access costs make alternative transmission topologies and technologies attractive. The local access from a customer's premises to a local exchange carrier central office (LEC CO) destined for long-haul is now becoming a resource that wants multiplexed bandwidth—which is an impossibility with multipoint, polled networks.

Increases from 1984 to 1988

Figures 5A and 5B illustrate the increasing communications costs of intra- and interLATA multipoint networks. These graphs show the rising cost of an eight-drop multipoint network from 1984 to 1988 as a function of distance. See "Tariff model" for details of the method used to generate these graphs.

Figure 5A shows an average increase of 65 percent from 1985 to 1987 and 23 percent from 1987 to 1988 in intraLATA circuit charges for this eight-drop multipoint network in the 0-to-100-mile range. The 1984 cost line provides a reference point since, before divestiture, most central offices belonged to AT&T. The cost of providing local access was then subsidized by the long-haul network to realize universal service.

Figure 5B shows average increases of 27 percent from 1985 to 1987 and 17 percent from 1987 to 1988 in interLATA multipoint circuit charges for distances from 25 to 3,000 miles. Notice that the 1984 interLATA circuit cost was higher than 1988 cost in the high-distance mileage. This is because the long-haul part of the network subsidized the local access.

After divestiture, long-haul private-line prices were reduced, only to be increased beyond the original 1984 price in 1988, except for the high-mileage limit. The initial postdivestiture price drop occurred because long-distance

4. Circuit-cost breakdown. *Between 1984 and 1988 the cost of local access has increased to the point where now it is the single largest component of total link price. Considering the expense of the resource, local access should be as flexible as possible to support multiple data protocols on point-to-point circuits.*

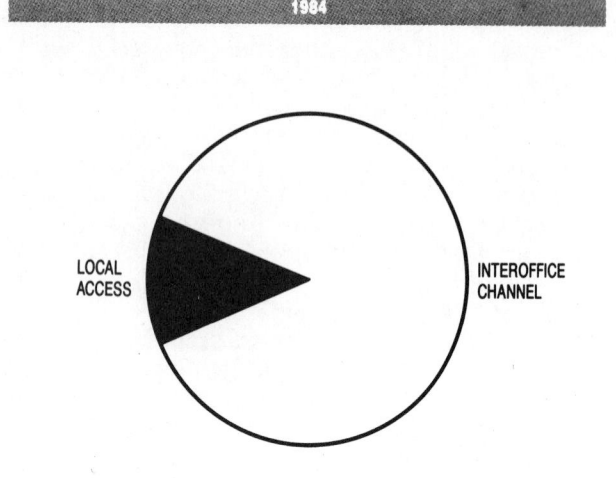

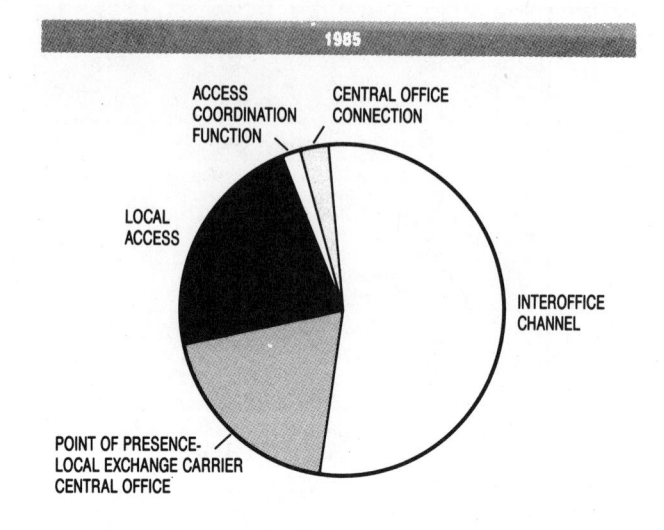

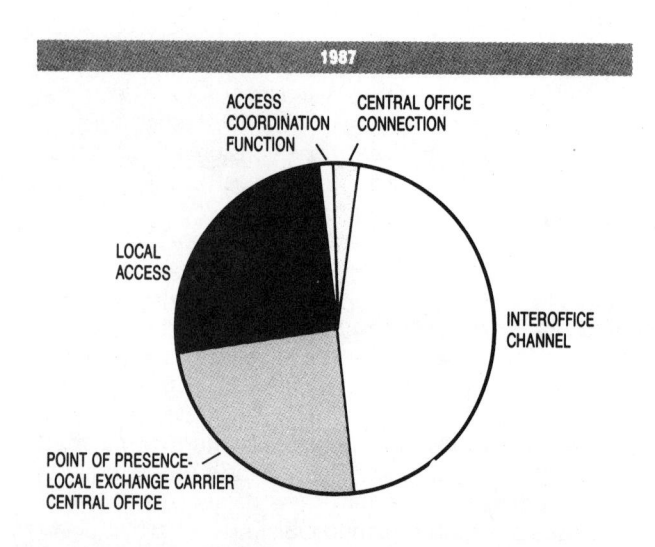

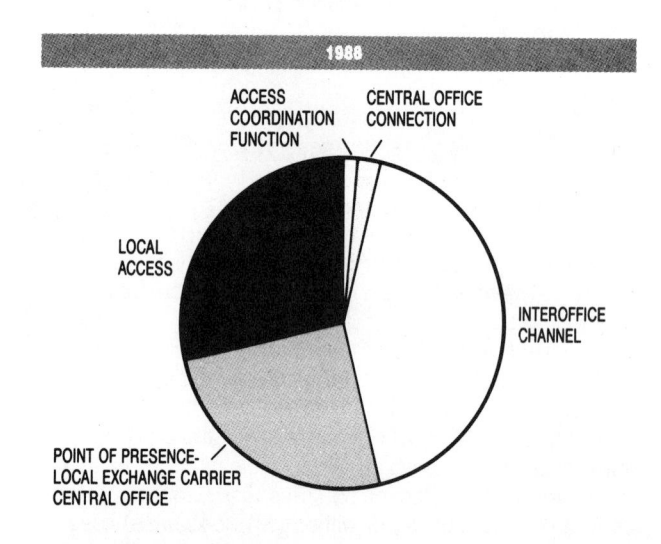

lines no longer subsidized local ones; but the cost of maintaining the circuits has gone up, naturally causing prices to rise. The carriers' long-term strategy appears to be to move away from private lines in preparation for ISDN.

Conditioning costs rise

As applications demanded more bandwidth from the network, network managers turned to higher-speed modems and conditioned lines (see "Breaking the throughput barrier: Which high-speed modems test best," DATA COMMUNICATIONS, May 1988). However, the cost of a conditioned line has skyrocketed from 1985 to 1988 to the point where it is now comparable to that of an additional analog line.

The cost of conditioning affects both point-to-point and multipoint transmission. But point-to-point networks, because of their ability to support multiple protocol stacks through multiplexing, are better suited to taking advantage of increased bandwidth.

A compound annual growth rate of 37 percent for the eight-drop intraLATA multipoint circuit (Fig. 5) from 1985 to 1988 resulted when C-2 and D-5 conditioning was applied. A compound annual growth rate of 31 percent for this eight-drop interLATA multipoint topology, with line conditioning applied, occurred from 1985 to 1988. There is an increasingly severe cost penalty for employing conditioning, thus limiting the bandwidth of the line in most multiple multipoint networks.

As MIS departments deploy more distributed computing, traffic volume grows. Greater traffic volumes degrade the performance of multipoint networks, creating a need for network redesigns. Those redesigns result in fewer drops on a multipoint line, so that networks evolve to point-to-point topologies. This change in computing and distributed applications drives network managers to ana-

5. Intra- and interLATA costs. IntraLATA access costs (A) continue to increase year after year. InterLATA access costs (B) show that long-haul private-line prices fell after the divestiture of AT&T in 1984. In spite of this reduction, access costs for circuits less than 2,000 miles long have incrementally grown past their original levels.

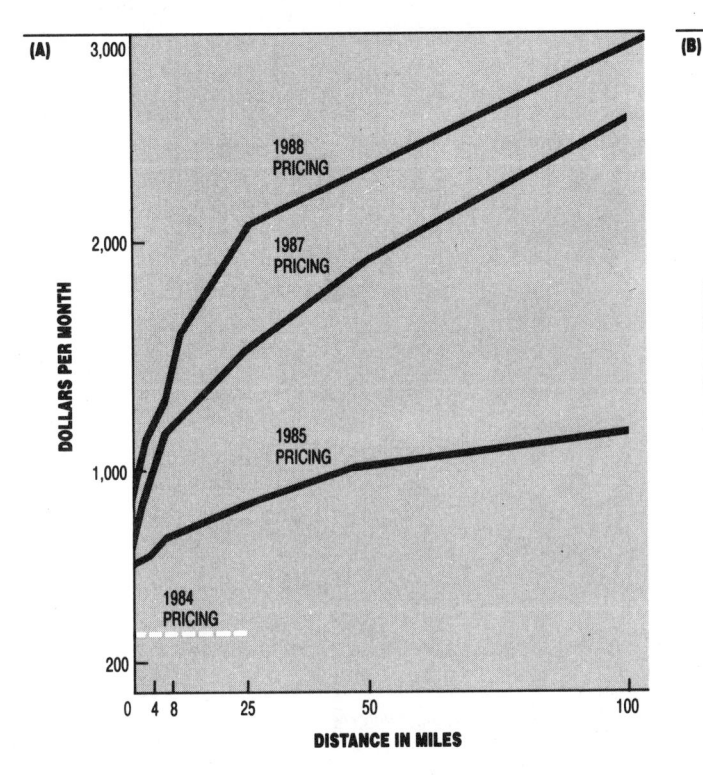

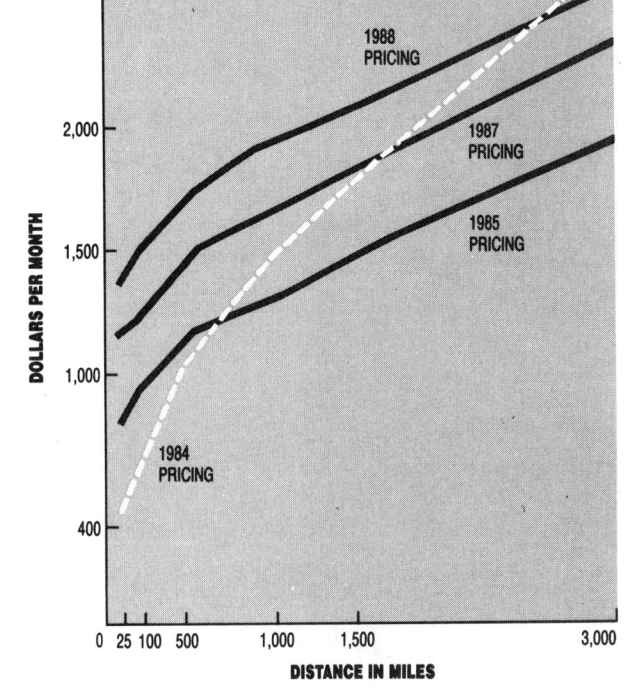

lyze alternative transmission topologies that provide either a migration path or coexistence of multipoint-to-point-to-point communication.

Based on the break-even analysis of Figures 2 and 3, users who have more than three multipoint drops are ripe for a change in transmission topology and technology.

Alternative topologies

Figure 6A depicts an alternative transmission topology that is finding wide acceptance within the financial services industry, but the topology would work in any industry where the reporting chain moves from regional offices to regional headquarters to a central headquarters.

The WAN in Figure 6A illustrates an interLATA T1 backbone network connecting a headquarters to its regional branch office. The smaller branch offices are brought into the T1 backbone via a variety of private-line, point-to-point services such as DDS, hubless DDS, or analog lines.

Displacing multiple multipoint branch networks with one point-to-point line through the use of concentrators, multiplexers, or packet switching provides the network manager with lower communications cost (depending on the number of lines being displaced), better control, and more flexibility. Concentrator and packet-switching technology can provide a logical multipoint network over a physical point-to-point line so that the communications can be shared with master-slave devices and distributed computing devices.

The acceptance of this form of topology likely stems from the fact that the higher charges for analog circuits make it cost-effective to displace a minimum of two multipoint lines entering a customer's premises at either 4.8 or 9.6 kbit/s with one line operating at 14.4 kbit/s with concentrators or packet switching.

Packet switching, a more robust architecture than multipoint, offers improved redundancy and diversity, can work with both peer-to-peer and master-slave networks, and provides a migration path to OSI.

Hubless DDS services are available from the RBOCs that support the configuration illustrated in Figure 6A. The difference between hubless and hubbed DDS is that the latter was introduced as a high-availability service with routing performed at hub locations to allow for 24-hour-a-day monitoring and control. Since there are approximately 125 hub cities in the United States, a user would almost invariably have to pay extra mileage costs, since the service was tariffed from premises to hub to hub to premises.

The hubless services are routed from premises to premises, eliminating the hub and extra-circuit charges. Low bit-error rates in the 10-6 range are being cited for hubless DDS. Ameritech, Bell Atlantic, BellSouth, Nynex, Southwestern Bell, and US West have offered hubless DDS services for more than two years, pricing them 38 percent to 57 percent below the cost of hubbed DDS. Pacific Telesis has recently introduced its hubless DDS service—

6. Topologies. *An alternative transmission topology (A) displaces multipoint networks. This architecture uses T1 between headquarters and regional headquarters, employing T1 multiplexing. Another topology (B) displaces multipoint with POP multiplexing through AT&T M24 multiplexing at lower speeds to smaller offices.*

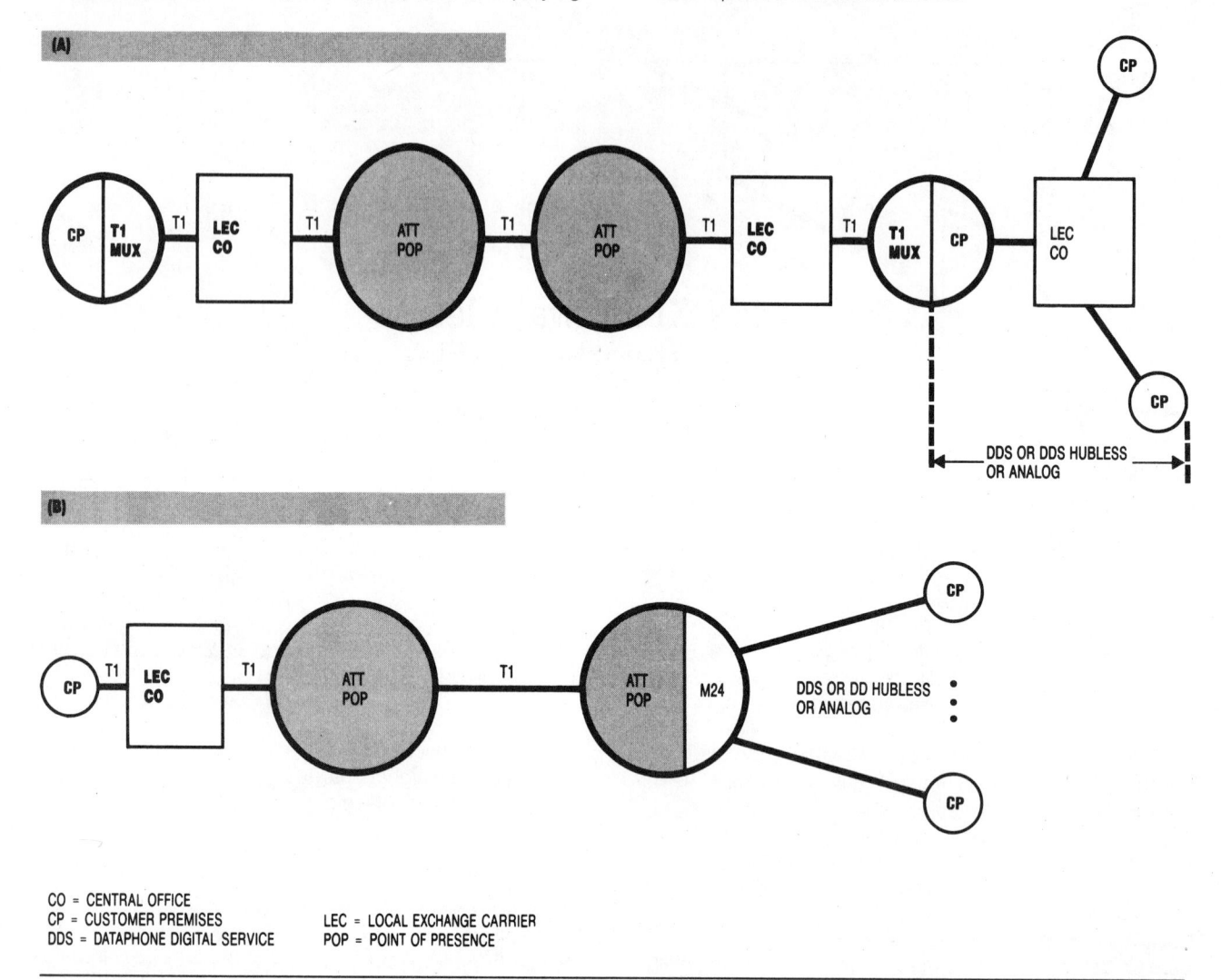

CO = CENTRAL OFFICE
CP = CUSTOMER PREMISES
DDS = DATAPHONE DIGITAL SERVICE
LEC = LOCAL EXCHANGE CARRIER
POP = POINT OF PRESENCE

Advanced Digital Network—and in June, AT&T introduced the Accunet Spectrum of Digital Services with hubless DDS at 64 kbit/s.

Figure 6B shows yet another alternative to traditional multipoint topologies. This one is somewhat limited by the fact that AT&T currently provides 56-kbit/s DDS service out of its POP through the M24 office function option. M24 demultiplexes a T1 circuit down into twenty-four 64-kbit/s channels.

Note that AT&T has started to offer lower speeds (2.4 kbit/s, 4.8 kbit/s, and 9.6 kbit/s) within the subrate digital multiplex office function at substantially reduced rates. The service has not yet been phased in at all POP locations. Some of the RBOCs are offering multiplexing options within their LEC CO, moving the T1 termination and multiplexing functions out of the customer's premises and into the LEC CO. There is not yet enough data to project how successful these topologies will be.

Under current regulation in the telephony industry, strategic pricing will probably continue, making the job of migrating traditional multipoint networks to digital platforms a priority for many organizations. ■

Nicholas Lippis is a strategic analyst with DEC's Networks and Communications Distributed Systems Strategic Planning Group. Previously, he was the telecommunications consultant responsible for DEC's T1 and fiber private network. He received his BSEE from Boston University and his MSEE from Boston University and MIT.

Acknowledgments:
The author thanks Rolf McClellan, Joe Yanushpolsky, Fred Goldstein, Ken Pappas, Steve Sullivan, and Rich Rosenboum from DEC (Littleton, Mass.); Warren Cohen of DEC (Concord, Mass.); and Leo Bourdelais and Don Weatherbee of AT&T (Waltham, Mass.) for their time and effort contributing to the models developed in this article.

Cheryl L. Sommer, Computer Task Group Inc., Schaumburg, Ill.

The hidden costs of using an SNA backbone for X.25 traffic

IBM's XI software lets SNA networks act like packet-switched data networks. But buyer beware...

An increasingly popular data communications scenario for multinational companies includes a combination of data traffic: packets that conform to the CCITT X.25 recommendation plus messages that follow the de facto standard of IBM's Systems Network Architecture. One way to integrate these dissimilar traffic types is via an IBM software program called XI, which stands for X.25 SNA Interconnection. Although in theory XI provides monetary savings to companies that already have an SNA backbone and two or more microcomputers or minicomputers that generate X.25 messages and must communicate with each other, using XI may not be worth the effort. This article discusses the relative advantages and disadvantages of using XI to support X.25 traffic over an SNA network.

The momentum behind the growth of X.25 networks began in 1976 when the CCITT and ISO adopted X.25 as a standard specifying the interface between data terminal equipment (DTE) and packet-switching data circuit-terminating equipment (DCE). Although the market for X.25 continues to grow, the demand for XI appears to be small.

IBM created XI in May 1986 as a custom PRPQ (program request for price quotation) for one user who needed to bridge the gap between X.25 and SNA. The product was released publicly in January 1988.

As of August 1988, the network strategies consulting practice of Ernst & Young located in Fairfax, Va., estimated that there were "12 IBM customers worldwide that had XI in operational mode, rather than merely in test." Other companies had licensed one copy and were evaluating it, according to industry analyst David Passmore, but they had not chosen to put it into production use. Passmore expects that the number of XI users may have doubled in 1989. He also notes that XI customers are typically large multinational companies that have SNA backbone networks.

Indeed, XI is of primary interest to users having a certain set of network conditions. If a user has non-IBM equipment that supports X.25 and can connect this gear more economically via an existing SNA backbone than by building a separate X.25 backbone or using a public X.25 network, then XI is a likely networking alternative.

In standard X.25 networks the DCE serves as the access point to a packet-switched data network (PSDN). When XI is used, the access point is not needed since the XI program—which resides in the front-end processor (FEP)—itself acts as the PSDN. Data packets travel from an X.25 DTE (for example, a VAX minicomputer from Digital Equipment Corp.) via, for example, a 56-kbit/s leased line, to the FEP in the SNA network. From there they travel via, say, a T1 line to another FEP and on to another DTE that "talks" X.25.

It is at the point when the packet reaches the FEP that the XI program alters the packet's format. Here, XI fools the X.25 host into thinking that the SNA network is really an X.25 network. XI lets the packet retain its X.25 format, but it also adds the information necessary to have it look like an SNA packet. Thus the X.25 packet is encapsulated into an SNA packet (see Fig. 1). When the packet reaches the distant site, it again runs through the Network Control Program (NCP) and the XI software before reaching its final destination: the remote X.25 DTE.

When a user chooses the XI approach to integrating X.25 and SNA networks, the software must be installed in all 372X and 3745 FEPs that are directly attached to DTEs in the network. To install XI, system programmers at the user's site incorporate it into the NCP generation in IBM front-end processors.

The onetime charge for XI is approximately $22,000, with volume discounts. The monthly license fee is $615. One copy is needed for each front-end processor that is attached to a DTE, but it does not have to be installed in any

SNA

1. X.25 packets posing as SNA. *XI treats X.25 packets as part of the path information units flowing through SNA. Here, the XI-Packet Level Protocol header plus the X.25 packet are encapsulated within the basic link unit that SNA normally transfers. Before going to its destination DTE, the packet passes through the NCP and the XI software again.*

LINK HEADER (3 BYTES)	TRANSMISSION HEADER (26 BYTES)	REQUEST/ RESPONSE HEADER (3 BYTES)	XI-PLP HEADER (3 OR 4 BYTES*)	X.25 PACKET	LINK TRAILER (3 BYTES)

X.25 PORTION (XI-PLP HEADER + X.25 PACKET)

PLP = PACKET-LEVEL PROTOCOL
XI = X.25 SNA INTERCONNECTION
*MODULO 8 TAKES 3 BYTES. MODULO 128 TAKES 4 BYTES

2. X.25 without the network cloud. *Here, a DTE in proximity to an NCP overseas sends X.25 messages to a DTE in the United States—without using a packet-switched network. XI is installed in all 327X and 3745 front-end processors directly attached to DTEs. The XI data travels from one XI node to another over a DXE-DXE session.*

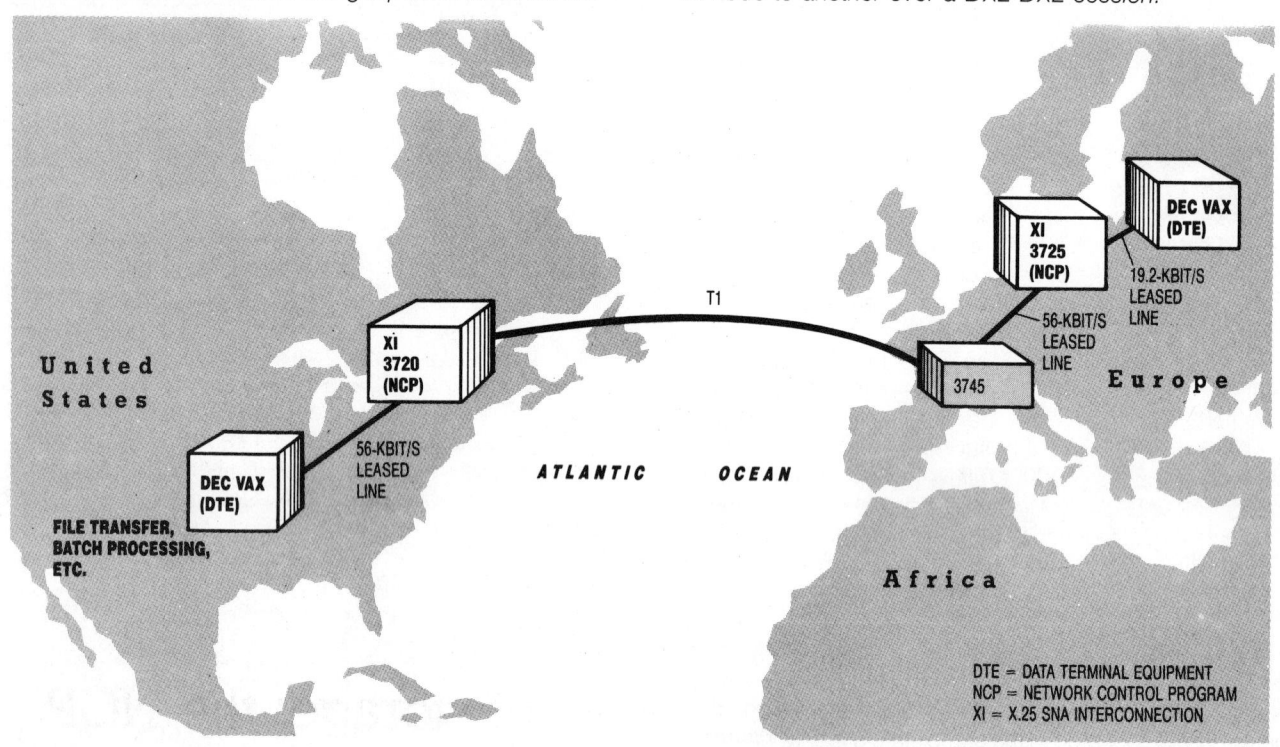

DTE = DATA TERMINAL EQUIPMENT
NCP = NETWORK CONTROL PROGRAM
XI = X.25 SNA INTERCONNECTION

intermediate FEPs (those that may pass on X.25 data but are not directly attached to the DTE).

The 372X and/or 3745 FEPs can be connected to each other by one or more communications lines (see Fig. 2). These transport the packet to the portion of XI that handles packet-level communications so that it may communicate with the similar portion of XI on another XI node. (An XI node is defined as an FEP that has XI installed on it.) The part of XI that receives packets is known as the Data Exchange Equipment (DXE). Thus, X.25 data travels from one XI node to another over the DXE-DXE session. A logical grouping of the communications links that connect two FEPs in the context of an XI installation is defined by IBM as a "transmission group."

The difficulty associated with installing, configuring, maintaining, and tuning XI may outweigh the monetary savings that it ostensibly provides. Tuning is the term used for getting response times and general network performance to be as efficient as possible. On an XI network, tuning involves repeatedly modifying appropriate network parameters until the user is satisfied, as much as is possible, with the performance. If the XI network is not tuned properly, unsatisfactory response time may irritate interactive users. A response should be received within two seconds after a user strikes the Enter key.

Dual specialty

In some cases, the particular DTE (such as a micro- or minicomputer) targeted for use with XI has not been tested for compatibility with the product. Specifications of the DTE and XI may differ on, for example, the handling of the virtual call reset procedure—even though both technically comply

SNA

with the CCITT X.25 recommendation. In this case, the user would have to start troubleshooting the problem both with the vendor of the DTE and with IBM. One of them must concede to change the way their device handles the procedure.

Hidden costs also lie in the fact that specialized personnel may have to be employed to maintain an XI network. In addition to an administrator for the DTE network, there should be an individual knowledgeable in both X.25 and SNA. There is also the cost of coordination between the XI support team and the DTE support team for configuration, implementation, maintenance, and testing. This cost increases if the two support teams are part of—or are owned by—different organizations.

In contrast, these personnel considerations need not concern the user when the DTEs communicate via either X.25 or a proprietary protocol. For example, when VAX computers are the DTEs, the administrator overseeing the VAX is usually the network administrator as well. This individual need be familiar only with the packet network being used: for example, X.25 or DECnet.

In considering whether or not to install XI, the network administrator should perform a cost analysis. For example, how much would be spent on X.25 service if XI were not installed? The figure would include costs associated with subscription to a PSDN (such as cost of a switched virtual circuit or a permanent virtual circuit plus the cost of any special services currently used by the installation, such as subscription to closed user groups).

The cost analysis should also reflect the amount of X.25 traffic that flows over the existing network and what the projected volumes are, what amount of X.25 traffic the current SNA backbone can support, and if any additional SNA resources must be acquired.

When considering XI as an alternative to X.25 service from a network provider, it is also important to weigh the XI approach against two versions of X.25 service: subscription to an X.25 PSDN as well as X.25 service via a leased line. However, a leased-line solution may not be plausible if DTEs are located at international locations, since the computer site may have to wait an unacceptable amount of time for the lines to be installed. Moreover, if file transfers between DTEs located in the same state are only intermittent, XI would not be a wise choice, since the DTEs might spend only about $800 per month between them for a switched virtual circuit from Telenet.

Coping with XI

Assuming that XI does suit a user's network requirements—and after the costly process of planning and design has ended—the system programmer must tune the software. One caveat is that XI consumes resources (for example, memory for storage space) on the IBM host and the FEPs.

Although transporting X.25 packets via XI does not inherently consume significantly more resources than does passing an equivalent amount of SNA traffic, adding X.25 to the SNA backbone necessarily increases the total volume of traffic. The SNA network planner must accommodate the increased traffic by adding processing power and storage space regardless of the source of the data.

An increase in traffic requires more processing power in the FEP. There is no specific formula that can be used to determine the amount of additional power required. It varies widely depending on the number of FEPs in the network and what devices are already attached to them.

XI traffic also places an additional load on the intermediate network node links and requires additional storage space in IBM front-end processors. The XI program itself requires extra storage in the FEP. Depending on the version, XI code takes up at least 175 Kbytes of storage (the NCP takes about a half a megabyte). Additional storage is also required for configured XI resources, such as logical channels, and for buffer space used by data traffic. Configuring spare resources also consumes storage capacity.

For example, suppose there is a network requirement for connecting a an X.25 DTE to a 3720 processor as access to an SNA backbone. If the links already attached to this device carry a substantial amount of XI traffic, then attaching another link may not be a workable solution. If files will be transferred to and from the newly attached DTE on a continual basis, and if file transfers typically use 53 percent of a 56-kbit/s link that is already connected from a DTE to the 3720, then perhaps another FEP should be added to the network to handle the load of extra XI traffic. As the utilization of the links connecting the DTE to the FEP approaches 60 percent, the NCP becomes overloaded—since buffer storage requirements tend to increase exponentially after that point.

To manage XI nodes, the user must also have Network Supervisory Function (NSF) software. This is a set of SNA host-resident instructions. Management of XI nodes includes activating and deactivating, modifying, and displaying and logging information about XI resources defined in the NCP.

> **XI code takes up at least 175 Kbytes of storage; the NCP takes about 500.**

Network Supervisory Function runs as a subtask on an MVS/SP or MVS/XA host under either NetView or the Network Communications Control Facility (NCCF). The XI program communicates with the FEP via a logical connection that lies between Node Management Functions (NMF) and NSF. (For an overview of XI network components, see Fig. 3.)

Node Management Functions will typically perform the tasks that are requested by NSF, such as a display of a DCE Physical Unit, which is a resource in the Network Control Program representing the connection to a DTE. But the NMF will also send error information to host applications. For example, it will send "record formatted maintenance statistics," which are records containing alert messages that are typically displayed when a network

SNA

component fails. These messages will be sent to the Network Problem Determination Application, which is the hardware monitor in NetView.

The SNA host that runs NSF must also have extra storage and processing power available to handle increased NetView functionality. Network Supervisory Function consumes a minimum of 43 Kbytes, which includes the NSF programs and the configuration data.

A little out of tune . . .

How much effort a system programmer invests in tuning depends on how important it is for the application to have adequate response time. The programmer must consider the needs of the user. For example, which application is used most? What tuning parameters are available on this application? If the X.25 users are mainly doing batch operations, then tuning may not be as crucial. However, if interactive sessions are common, performance tuning is mandatory and can be quite involved.

3. The logical makeup of XI nets. *The XI program communicates with the front-end processor via a logical connection, shown here as the NSF-NMF session. NSF runs as a subtask on a host under NetView or the NCCF.*

DCE = DATA CIRCUIT-TERMINATING EQUIPMENT
DTE = DATA TERMINAL EQUIPMENT
DXE = DATA EXCHANGE EQUIPMENT
NCCF = NETWORK COMMUNICATIONS CONTROL FACILITY
NCP = NETWORK CONTROL PROGRAM
NMF = NODE MANAGEMENT FUNCTIONS
NSF = NETWORK SUPERVISORY FUNCTION
MVS/SP = MULTIPLE VIRTUAL STORAGE/SYSTEM PRODUCT
MVS/XA = MULTIPLE VIRTUAL STORAGE FOR EXTENDED ARCHITECTURE
VTAM = VIRTUAL TELECOMMUNICATIONS ACCESS UNIT
XI = X.25 SNA INTERCONNECTION

Little documentation exists on how to tune the XI product. Whoever does the tuning must depend on a personal store of knowledge about SNA and X.25. Moreover, the resident IBM system engineers at the user's site sometimes do not have sufficient expertise in the field of X.25 via SNA to assist in tuning an XI network.

This lack of expertise may be because IBM support personnel are not expected to be proficient in both X.25 and SNA. Some larger IBM customers have worked directly with the XI developers at IBM for assistance in tuning the software.

Parsing performance factors

The network parameters that affect performance can be divided into two categories: X.25 and SNA. The former are configured in the DTE as well as in the XI software itself. The X.25 parameters that most affect throughput are packet size, frame-window size, and packet-window size.

Modulo 8 or 128 (which refers to a method of sequencing frames and packets) also affects performance and is related to window size.

The modulo is configurable in both X.25 and SNA networks. In the modulo method, the virtual circuit of a DTE may have no more than a window's worth of frames and packets outstanding at any particular time for each direction, where the window can be a maximum of modulo minus one. The XI product supports both Modulo 8 and 128 at the packet level, but it supports only Modulo 8 at the frame level.

Various other values, such as timers and retry counts, are less important in affecting throughput than are the previously mentioned values. Some of these values default and can no longer be modified. For example, before XI became a licensed product it had a tuning macro, which was called EXIGTUNE and was not included in the public release. It allowed the user to choose various timer values and to select parameters that affect operation of the XI code. These values can affect performance.

The SNA parameters that affect response time are mainly those that relate to the DXE-DXE session. This, of course, is the only way that the XI components of two NCPs can communicate. Not only do XI parameters affect this session, but so do some of the values that would normally affect any data transmission between one NCP and another.

Be aware that what may be optimal

SNA

for XI traffic is not necessarily optimal for standard SNA traffic, and vice versa. One drawback is that XI does not differentiate between batch or interactive traffic: All traffic has equal priority in terms of the program operation because XI data flows over a single session.

To discover why certain network parameters may not be optimal for XI traffic even though they work quite well for SNA traffic, start by determining how X.25 packets flow through the SNA network:
- Determine how well the NCP buffer size accommodates X.25 packets.
- Determine how the XI Packet Level Protocol header and X.25 packets are grouped into path information units flowing through the SNA network.
- Take into consideration NCP buffer sizes and the X.25 packet size.

With these figures in mind, a system programmer may find that XI results in a significant waste of buffer space because it is not always possible to change the NCP buffer size to accommodate X.25 data. A recommended NCP buffer size for XI using a 128-byte packet is 192 bytes, which includes user data plus SNA headers, X.25 headers, and the NCP buffer prefix. This differs from a recommended buffer size for standard SNA traffic, which is usually 128 bytes.

The programmer may also find that small packets (128 bytes or less) minimize the amount of buffer space that the intermediate network node uses but that small packets increase the load on the NCP. Processing time is increased because with small packets, more blocks are created that need to be taken off one intermediate network node link and put on another link.

Delayed packets

A similar effect is usually apparent in the DTE itself. XI usually sends one or more packets in a single block. Although the load on the NCP is heavier with small packets, large packets can create problems in that they generate large blocks traveling through the network. This can delay the passage of shorter blocks, possibly creating unpredictable response times.

The delay occurs when shorter blocks have to wait for longer blocks when they are resequenced at the other end of the transmission group. This could also happen if the blocks are the same size but the line speeds within the transmission group are considerably different.

A flow control mechanism called Virtual Route Pacing affects flow control throughout the network and applies to XI as well as to standard SNA traffic. Tuning this mechanism can be quite complex, and numerous publications provide guidance for the process.

Session-level pacing (that is, regulating the amount of requests that the NCP sends to the logical unit), however, is not used on the DXE-DXE session. Instead, XI uses a parameter that also affects flow control. It is called the Transit parameter. XI uses this parameter to calculate the DXE-DXE window size, which is determined for each virtual circuit.

In brief, the Transit parameter is equal to the sum of the transmission delays of the transmission groups that are supporting the virtual route, the average waiting times in the outbound queues of the intermediate network nodes in the transmission groups, and the switching time in the FEPs. Switching time is the amount of time it takes to remove a packet from the intermediate network node input queue and deposit it in the output queue, plus the time to take the packet off the virtual route and put it in the XI queue for the DTE.

To track XI data

Once the parameters affecting performance for a particular application are identified, how does the programmer figure out what effect they have? A plan for tuning should be developed that varies and monitors each parameter while

> **The NCP buffer for XI should be 192 bytes; SNA traffic needs a 128-byte buffer.**

keeping the other values constant. Performance tools are provided, for example, in IBM's NetView product line. Although Node Management Function is a component of NetView, it is of little use in performance monitoring. It does not provide any data as a function of time increments small enough to provide a performance measure.

In other words, the programmer can use NMF to count the packets being transmitted—but not to determine when a user gets a response. An IBM product called NetView Performance Monitor can give line utilizations, but it does not explicitly track XI data. The most useful tools are either protocol analyzers, which can perform remote response-time monitoring or monitoring of packets from the DTE. Such analyzers are best because the user is the ultimate judge, and many DTEs provide utilities that monitor response times.

Each installation will have to determine whether or not it wants to use an XI network as an alternative to subscribing to an X.25 service or using a leased-line X.25 network. The network planners will have to decide whether the perceived monetary savings is worth the complexity that XI introduces to the network. And keep in mind that even at the end of elaborate performance monitoring, a user still may not see adequate response times. Yes, XI lets you share bandwidth with standard SNA traffic (which can be very cost-effective), but sometimes it might be easier to have guaranteed response time, pay for the extra bandwidth, and not have to deal with the complications. ■

Cheryl Sommer is a telecommunications consultant for Computer Task Group Inc. (headquartered in Buffalo, N.Y.). Her XI experience is derived from work at Motorola Inc. Schaumburg, Ill.), where she designed, implemented, and tested an XI network, using DEC and Tandem computers as DTEs. She has a BS in computer engineering from the University of Michigan in Ann Arbor.

Index

Page references in this index are to the first pages of the articles in which the subjects appear, or to special sections devoted to those subjects.

Acronyms, 106
Artificial intelligence, 162, 255, 264
Availability, 118, 494
Boeing Computer Services, 75
Business architectures, 6, 23, 62, 75, 95, 125, 361, 633
Cabling, 386, 409
Carrier comparisons, 561, 647
CATV (Community-antenna television), 224, 396
Cellular service, 502
Compression, 610
Computer-aided design and engineering (CAD/CAM), 146, 162, 428
Connectivity guidelines, 254
Contention management, 26
Corporation for Open Systems, 298
Design programs, 154, 339
Digital Equipment Corp. (DEC), 37, 361
Direct connections, 250
Dual-bus network, 90
Electronic mail, 2, 30, 95
Facsimile, 418
Fiber optics, 82, 195, 361, 480
Forecasting, 150, 154
Gateways, 396
Glossaries, 32, 40, 163, 440
Human factors, 255, 354
IBM, 6, 30, 48
IEEE (Institute of Electrical and Electronic Engineers), 82
IEEE committee 802.6, 82, 90
IEEE project 802, 82, 379
Imaging, 418
Integrated Services Digital Network (ISDN), 130, 214, 284, 324, 456, 488, 529
International Organization for Standardization, 105
Internetworking, 177, 233, 379, 468, 510, 588
Lasers, 195
Light-emitting diodes, 195
Load splitting, 188
Long-distance, 561

Manufacturing Automation Protocol, 296
Marketing, 652
Market assessments, 554, 616, 621, 626
Models, 275
Multivendor networks, 62, 233, 243, 475
Object-oriented design, 545
Open networks, 62, 68
Open Systems Interconnection, 105, 142, 177, 295, 437, 534, 545
Packet switching (see also X.25 networks), 286
Parallel processing, 37, 44
Port selectors, 243
Power transients, 267
Productivity, 2, 354
Protocol analysis, 291
Remote control, 448
Routing, 303, 510, 516
Satellites, 315, 375
Security, 167
Segmentation, 516
Signaling, 134, 142, 403
Slotted ring architecture, 23, 82
Switched 56-kbit/s service, 16
System Network Architecture, 6, 30, 48, 70, 295, 468, 534, 667
Tariffs, 602, 660
Technical and Office Protocol (TOP), 75, 105, 296
Time-staged delivery, 2
Token rings, 478
T1 networks, 125, 190, 209, 229, 303, 324, 391, 494, 633
T3 networks, 494
User ratings, network management, 475
Video, 418
Vendor comparisons, 16, 191, 209, 339, 450, 508, 516, 585, 633
Voice and data, 214
Voice mail, 585
Wiring, 310, 479, 484
Xerox Network Systems (XNS), 522
X.25 networks, 70, 95, 150, 315, 329, 366, 488, 534, 667